DIMENSION	SI	SI/ENGLISH
Power	1 W = 1 J/s 1 kW = 1000 W = 1.341 hp	1 kW = 3412.2 Btu/h = 0.73756 lbf·ft/s 1 hp = 550 lbf·ft/s = 0.7068 Btu/s = 42.41 Btu/min = 2544.5 Btu/h = 0.74570 kW
Pressure	1 Pa = 1 N/m^2 1 kPa = 10^3 Pa = 10^{-3} MPa 1 atm = 101.325 kPa = 1.01325 bars = 760 mmHg at 0°C	1 Pa = 1.4504 × 10^{-4} psia = 0.020886 lbf/ft^2 1 psia = 144 lbf/ft^2 1 atm = 14.696 psia = 29.92 inHg at 32°F
Specific Heat	1 kJ/(kg·°C) = 1 kJ/(kg·K) = 1 J/(g·°C)	1 Btu/(lbm·°F) = 4.1868 kJ/(kg·°C) 1 kJ/(kg·°C) = 0.23885 Btu/(lbm·°F) = 0.23885 Btu/(lbm·R)
Specific Volume	1 m^3/kg = 1000 L/kg = 1000 cm^3/g	1 m^3/kg = 16.02 ft^3/lbm
Temperature	T (K) = T (°C) + 273.15 ΔT (K) = ΔT (°C)	T (R) = T (°F) + 459.67 T (°F) = 1.8 T (°C) + 32 ΔT (°F) = ΔT (R) = 1.8 ΔT (K)
Velocity	1 m/s = 3.60 km/h	1 m/s = 3.2808 ft/s = 2.237 mi/h 1 mi/h = 1.609 km/h
Volume	1 m^3 = 1000 L = 10^6 cm^3 (cc)	1 m^3 = 6.1022 × 10^4 in^3 = 35.313 ft^3 = 264.17 gal (U.S.) 1 U.S. gallon = 231 in^3 = 3.7853 L

Thermodynamics: An Engineering Approach

THERMODYNAMICS
An Engineering Approach

Dr. Yunus A. Çengel
Department of Mechanical Engineering
University of Nevada–Reno

Dr. Michael A. Boles
Department of Mechanical and Aerospace Engineering
North Carolina State University

McGRAW-HILL BOOK COMPANY
New York | St. Louis | San Francisco | Auckland | Bogotá | Caracas
Colorado Springs | Hamburg | Lisbon | London | Madrid | Mexico
Milan | Montreal | New Delhi | Oklahoma City | Panama | Paris
San Juan | São Paulo | Singapore | Sydney | Tokyo | Toronto

Thermodynamics:
An Engineering Approach

1 2 3 4 5 6 7 8 9 0 HALHAL 8 9 3 2 1 0 9 8

ISBN 0-07-010356-9

This book was set in Times Roman by General Graphic Services, Inc.
The editors were John Corrigan and James W. Bradley;
the designer was Nicholas Krenitsky;
the production supervisor was Denise L. Puryear.
Arcata Graphics/Halliday was printer and binder.

Library of Congress Cataloging-in-Publication Data

Çengel, Yunus A.
Thermodynamics: an engineering approach.

Includes bibliographies and index.
1. Thermodynamics. I. Boles, Michael A.
II. Title.
TJ265.C43 1989 621.402′1 88-13282
ISBN 0-07-010356-9

About the Authors

Yunus A. Çengel received his Ph.D. in mechanical engineering from North Carolina State University in 1984. He then joined the faculty of mechanical engineering at the University of Nevada–Reno, where he has been teaching undergraduate and graduate courses in thermodynamics and heat transfer while conducting research. He published several technical papers in the areas of radiation heat transfer, natural convection, and solar energy. Dr. Çengel has been voted the outstanding teacher by the ASME student sections in both North Carolina State University and the University of Nevada–Reno. He is a member of the American Society of Mechanical Engineers.

Michael A. Boles is Associate Professor of Mechanical and Aerospace Engineering at North Carolina State University. He earned his Ph.D. in mechanical engineering at North Carolina State University. Dr. Boles has received numerous awards and citations for excellence as an engineering educator. He is a past recipient of the SAE Ralph R. Teetor Educational Award and a member of the Academy of Outstanding Teachers at North Carolina State University. In 1987 Dr. Boles was selected as an Alumni Distinguished Professor at North Carolina State University. Dr. Boles specializes in heat transfer and has been involved in the analytical and numerical solution of phase change and drying of porous media heat transfer. He is a member of the American Society of Mechanical Engineers, the American Society for Engineering Education, and Sigma Xi.

Contents

10 ■ REFRIGERATION CYCLES 493

11 ■ THERMODYNAMIC PROPERTY RELATIONS 535

12 ■ GAS MIXTURES 571

Preface

Thermodynamics is a basic science that deals with energy and has long been an essential part of engineering curricula all over the world. This introductory text is intended for use in undergraduate engineering courses and contains sufficient material for two sequential courses in thermodynamics.

The traditional classical, or macroscopic, approach is used throughout this text, with microscopic arguments serving in the supporting role where appropriate. This approach is more in line with the students' intuition and makes learning the subject matter much easier.

Thermodynamics is often perceived as a difficult subject, and the majority of students dread the experience. The authors believe, on the contrary, that thermodynamics is a simple subject, and an observant mind should have no difficulty understanding it. After all, the principles of thermodynamics are based on our everyday experiences and experimental observations. Thermodynamics is a mature basic science, and the topics covered in introductory texts are well established. Primarily, the texts differ only in the approach used. In this text, a more physical, intuitive approach is used throughout. Frequently, parallels are drawn between the subject matter and students' everyday experiences so that they can relate the subject matter to what they already know.

Yesterday's engineer spent a major portion of his or her time substituting values into formulas and obtaining numerical results. But it will not be long before all the formula manipulations and number crunching are left to the computers. Tomorrow's engineer will have to have a clear understanding and firm grasp of the basic principles in order to understand, formulate, and interpret the results of even the most complex problems. A conscious effort is made to lead students in this direction.

The material in the text is introduced at a level that average students can follow comfortably. It speaks *to* the students, not over them. In fact, the material is self-instructive, thus freeing the instructor to use class time more productively.

The order of coverage is from simple to general. That is, it starts with the simplest case and adds complexities one at a time. In this way, the basic principles are repeatedly applied to different systems, and students master how to apply the principles instead of how to simplify a general formula. Since thermodynamic principles are based on experimental observations, all derivations in this text are based on physical grounds; thus they are easy to follow and understand.

Figures are important learning tools that help the students to get the picture. They attract attention and stimulate curiosity and interest. The text makes effective use of graphics and probably contains more figures and illustrations than any other thermodynamics book. Some of the figures do not function in the traditional sense. Rather, they serve as a means of emphasizing some key statements that would otherwise go unnoticed or as paragraph summaries. The popular cartoon feature ''Blondie'' is used to make some important points in a humorous way and also to break the ice and ease the nerves. Who says studying thermodynamics cannot be fun?

Each chapter contains numerous worked-out examples that clarify the material and illustrate the use of the basic principles. A consistent and systematic approach is used in the solution of the example problems, with particular attention to the proper use of units. A sketch and a process diagram are included for most examples to clearly illustrate the geometry and the type of process involved.

The subject material is covered in a logical order. First, various concepts are reviewed and some new ones are defined in order to establish a firm basis for the development of thermodynamic principles. Then the properties of pure substances are discussed and the use of property tables is illustrated. At this point the ideal-gas approximation is introduced, together with other equations of state, and the deviation from ideal-gas behavior is examined through the use of compressibility charts. After the introduction of heat and work, the conservation of energy principle is developed for a closed system. Following a discussion of flow energy, the conservation of energy principle for control volumes is developed, first for steady-flow systems and then for general unsteady-flow systems. The development of the second-law relations follow the same order, with special emphasis given to entropy generation. The concepts of availability, reversible work, and irreversibility are developed using familiar examples before they are applied to more complex engineering systems. The principles of thermodynamics are then applied to various areas of engineering.

A short summary is included at the end of each chapter for a quick overview of basic concepts and important relations; this is followed by a list of references that are appropriate for the level of students studying thermodynamics for the first time. Summaries can also be used as formula sheets during exams by instructors who prefer closed-book tests and yet want to make the equations available to the students.

The end-of-chapter problems are grouped under specific topics in the order they are covered to make problem selection easier for both instructors and students. The problems within each group start with concept questions, indicated by "C," to check the students' level of understanding of basic concepts. The problems involving numerical calculations are arranged in increasing complexity, with the later ones requiring more comprehensive analysis. A number of problems that require the use of a computer are also included throughout the text so that instructors who wish to incorporate computers into the course may do so. Answers to selected problems are listed immediately following the problem for convenience to the students.

A unique problem and responsibility faces today's engineers in English-speaking countries: they must think, understand, and promote metric, yet be comfortable with and be able to communicate in the old English system. The phrase "old English system" is quite appropriate since even England has fully converted to the SI (Système International) or the metric system. In recognition of the fact that English units are still widely used in some industries, both SI and English units are used in this text, with an emphasis on SI. The material in this text can be covered using combined SI/English units or SI units alone, depending on the preference of the instructor. The property tables and charts in the appendix are presented in both units, except the ones that involve dimensionless quantities. Problems, tables, and charts in English units are designated by "E" after the number for easy recognition. Frequently used conversion factors are listed on the inside cover pages of the text for easy reference.

We would like to acknowledge with appreciation the numerous and valuable comments, suggestions, criticism, and praise of the following academic reviewers: John Hester, California State University, Sacramento; Wesley M. Rohrer, University of Pittsburgh; Larry Simmons, University of Portland; David A. Walker, Lehigh University; and William Worek, University of Illinois at Chicago. Their suggestions have greatly helped us improve the finished product. We also would like to acknowledge the comments and suggestions made by the numerous students who have used various versions of this text at the University of Nevada–Reno. Special thanks are due to Ann Tomiyasu for carefully and patiently reviewing the entire manuscript, Celaleddin Gökçek for his contributions to the chapter on air conditioning, and to Judy Wirtz for her expert typing. Finally, we would like to express our appreciation to our wives, Zehra and Sylvia, for their continued patience, understanding, encouragement, and support throughout the preparation of this text.

Yunus A. Çengel
Michael A. Boles

Nomenclature

a	Acceleration, m/s^2
a	Specific Helmholtz function, $u - Ts$, kJ/kg
A	Area, m^2
A	Helmholtz function, $U - TS$, kJ
AF	Air-fuel ratio
c	Speed of sound, m/s
C	Specific heat, kJ/(kg·K)
C_p	Constant pressure specific heat, kJ/(kg·K)
C_v	Constant volume specific heat, kJ/(kg·K)
C_{p0}	Ideal-gas constant pressure specific heat, kJ/(kg·K)
C_{v0}	Ideal-gas constant volume specific heat, kJ/(kg·K)
COP	Coefficient of performance
COP_R	Coefficient of performance of a refrigerator
COP_{HP}	Coefficient of performance of a heat pump
d	Exact differential
d, D	Diameter, m
e	Specific total energy, kJ/kg
E	Total energy, kJ
f	Function
F	Force, N
FA	Fuel-air ratio

g	Gravitational acceleration, m/s^2
g	Specific Gibbs function, $h - Ts$, kJ/kg
G	Total Gibbs function, $H - TS$, kJ
h	Height, m
h	Specific enthalpy, $u + Pv$, kJ/kg
H	Total enthalpy, $U + PV$, kJ
HHV	Higher heating value, kJ/kmol fuel
$\bar{h}_c$	Enthalpy of combustion, kJ/kmol fuel
$\bar{h}_f$	Enthalpy of formation, kJ/kmol
$\bar{h}_R$	Enthalpy of reaction, kJ/kmol
i	Specific irreversibility, kJ/kg
I	Electric current, A
I	Total irreversibility, kJ
k	Specific heat ratio, C_p/C_v
K_p	Equilibrium constant
ke	Specific kinetic energy, $V^2/2$, kJ/kg
KE	Total kinetic energy, $mV^2/2$, kJ
L	Length
LHV	Lower heating value, kJ/kmol fuel
m	Mass, kg
$\dot{m}$	Mass flow rate, kg/s
M	Molar mass, kg/kmol
MEP	Mean effective pressure, kPa
mf	Mass fraction
n	Polytropic exponent
N	Number of moles, kmol
P	Pressure, kPa
P_{cr}	Critical pressure, kPa
P_i	Partial pressure, kPa
P_m	Mixture pressure, kPa
P_r	Relative pressure
P_R	Reduced pressure
P_v	Vapor pressure, kPa
P_0	Surroundings pressure, kPa
pe	Specific potential energy, gz, kJ/kg
PE	Total potential energy, mgz, kJ
q	Specific heat transfer, kJ/kg
Q	Total heat transfer, kJ
Q_H	Heat transfer with high-temperature body, kJ
Q_L	Heat transfer with low-temperature body, kJ
$\dot{Q}$	Heat transfer rate, kW

r	Compression ratio
r_p	Pressure ratio
r_c	Cutoff ratio
R	Gas constant, kJ/(kg·K)
R_u	Universal gas constant, kJ/(kmol·K)
s	Displacement, m
s	Specific entropy, kJ/(kg·K)
s_{gen}	Specific entropy generation, kJ/(kg·K)
S	Total entropy, kJ/K
S_{gen}	Total entropy generation, kJ/K
t	Time, s
T	Temperature, °C or K
T_{cr}	Critical temperature, K
T_{db}	Dry-bulb temperature, °C
T_{dp}	Dew-point temperature, °C
T_R	Reduced temperature
T_{wb}	Wet-bulb temperature, °C
T_0	Surroundings temperature, °C or K
T_H	Temperature of high-temperature body, K
T_L	Temperature of low-temperature body, K
u	Specific internal energy, kJ/kg
U	Total internal energy, kJ
v	Specific volume, m^3/kg
v_{cr}	Critical specific volume, m^3/kg
v_r	Relative specific volume
v_R	Pseudo-reduced specific volume
V	Total volume, m^3
$\mathbf{V}$	Velocity, m/s
w	Specific work, kJ/kg
W	Total work, kJ
$\dot{W}$	Power, kW
W_{in}	Work input, kJ
W_{out}	Work output, kJ
W_{rev}	Reversible work, kJ
x	Quality
y	Mole fraction
z	Elevation, m
Z	Compressibility factor
Z_h	Enthalpy departure factor
Z_s	Entropy departure factor

Greek Letters

β	Volume expansivity, 1/K
Δ	Finite change in a quantity
δ	Differential of a path function
ϵ	Degree of completion of a chemical reaction
η_{th}	Thermal efficiency
η_{II}	Second-law efficiency
θ	Total energy of a flowing fluid, kJ/kg
α	Isothermal compressibility, 1/kPa
μ	Joule-Thomson coefficient, K/kPa
μ	Chemical potential, kJ/kg
ν	Stoichiometric coefficient
ρ	Density, kg/m^3
ρ_s	Specific weight or relative density
τ	Torque, Nm
ϕ	Relative humidity
ϕ	Specific closed system availability, kJ/kg
Φ	Total closed system availability, kJ
ψ	Stream availability, kJ/kg
ω	Specific or absolute humidity, kg H_2O/kg dry air

Subscripts

a	Air
abs	Absolute
act	Actual
atm	Atmospheric
av	Average
c	Combustion
cr	Critical point property
cv	Control volume
e	Exit conditions
f	Saturated liquid
fg	Difference in property between saturated liquid and saturated vapor
g	Saturated vapor
gen	Generation
H	High temperature (as in T_H and Q_H)
i	Inlet conditions
i	ith component
int	Internally

irrev	Irreversible
L	Low temperature (as in T_L and Q_L)
m	Mixture
r	Relative
R	Reduced
rev	Reversible
s	Isentropic
sat	Saturated
surr	Surroundings
sys	System
v	Water vapor
0	Dead state
1	Initial or inlet state
2	Final or exit state

Superscripts

. (dot)	Quantity per unit time
- (bar)	Quantity per unit mole
°	Standard reference state
′	Pseudo quantity (as in P'_{cr})

Thermodynamics: An Engineering Approach

Basic Concepts of Thermodynamics

CHAPTER 1

Every science has a unique vocabulary associated with it, and thermodynamics is no exception. Precise definition of basic concepts forms a sound foundation for the development of a science and prevents possible misunderstandings. In this chapter the unit systems that will be used are reviewed, and the basic concepts of thermodynamics such as system, energy, property, state, process, cycle, pressure, and temperature are explained. Careful study of these concepts is essential for a good understanding of the topics in the following chapters.

1-1 ■ THERMODYNAMICS AND ENERGY

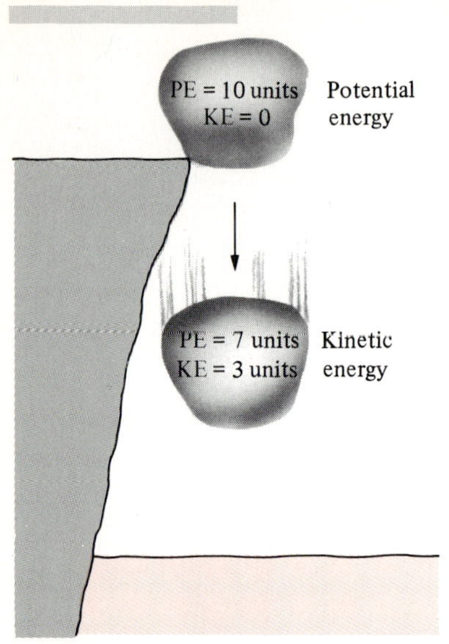

FIGURE 1-1

Energy cannot be created or destroyed; it can only change forms (the first law).

FIGURE 1-2*

Conservation of energy principle for the human body.

FIGURE 1-3

Heat can flow only from hot to cold bodies (the second law).

Thermodynamics can be defined as the science of energy. Although everybody has a feeling of what energy is, it is difficult to give a precise definition for it. Energy can be viewed as the capacity to do work or as the ability to cause changes.

One of the most fundamental laws of nature is the **conservation of energy** principle. It simply states that during an interaction, energy can change from one form to another but that the total amount of energy remains constant. That is, energy cannot be created or destroyed. A rock falling off a cliff, for example, picks up speed as a result of its potential energy being converted to kinetic energy (Fig. 1-1). The conservation of energy principle also forms the backbone of the diet industry: a person who has a greater energy input (food) than energy output (exercise) will gain weight (store energy in the form of fat), and a person who has a smaller energy input than output will lose weight (Fig. 1-2).

Thermodynamics deals with the conversion of energy from one form to another. It also deals with various properties of substances and the changes in these properties as a result of energy transformations. Like all sciences, thermodynamics is based on experimental observations. The findings from these observations have been expressed as some basic laws. The **first law of thermodynamics**, for example, is simply an expression of the conservation of energy principle. The **second law of thermodynamics** asserts that processes occur in a certain direction but not in the reverse direction. A cup of hot coffee left on a table in an office, for example, eventually cools, but a cup of cool coffee on the same table never gets hot by itself (Fig. 1-3).

Although the principles of thermodynamics have been in existence since the creation of the universe, thermodynamics did not emerge as a science until about 1700 when the first attempts to build a steam engine were made in England by T. Savery and T. Newcomen. These engines were very slow and inefficient, but they opened the way for the development of a new science. The term *thermodynamics* was first used in a publication by Lord Kelvin in 1849. The first thermodynamics textbook was written in 1859 by W. Rankine, a professor at the University of Glasgow. The greatest progress in thermodynamics was made in the early 1900s when it was stripped of all the erroneous theories and developed into a mature science.

It is well known that a substance consists of a large number of particles called *molecules*. The properties of the substance naturally depend on the behavior of these particles. For example, the pressure of a gas in a container is the result of momentum transfer between the molecules and the walls of the container. But one does not need to know the behavior of the gas particles to determine the pressure in the container. It would be sufficient to attach a pressure gage to the container. This macroscopic approach to the study of thermodynamics which does not require a knowledge of the behavior of individual particles is called **classical thermody-**

*BLONDIE cartoons are reprinted with special permission of King Features Syndicate, Inc.

namics It provides a direct and easy way to the solution of engineering problems. A more elaborate approach, based on the average behavior of large groups of individual particles, is called **statistical thermodynamics**. This microscopic approach is rather involved and is used in this text only in the supporting role.

Application Areas of Thermodynamics

Every engineering activity involves an interaction between energy and matter, thus it is hard to imagine an area which does not relate to thermodynamics in some respect. Therefore, developing a good understanding of thermodynamic principles has long been an essential part of engineering education.

One does not need to go very far to see some application areas of thermodynamics. In fact, one does not need to go anywhere. These areas are right where one lives. An ordinary house is, in some respects, an exhibition hall filled with thermodynamic wonders. Many ordinary household utensils and appliances are designed, in whole or in part, by using the principles of thermodynamics. Some examples include the electric or gas range, the heating and air-conditioning systems, the refrigerator, the humidifier, the pressure cooker, the water heater, the shower, the iron, and even the computer, the TV, and the VCR set. On a larger scale, thermodynamics plays a major part in the design and analysis of automotive engines, rockets, jet engines, and conventional or nuclear power plants (Fig. 1-4). We should also mention the human body as an interesting application area of thermodynamics.

1-2 ■ A NOTE ON DIMENSIONS AND UNITS

Any physical quantity can be characterized by **dimensions**. The arbitrary magnitudes assigned to the dimensions are called **units**. Some basic di-

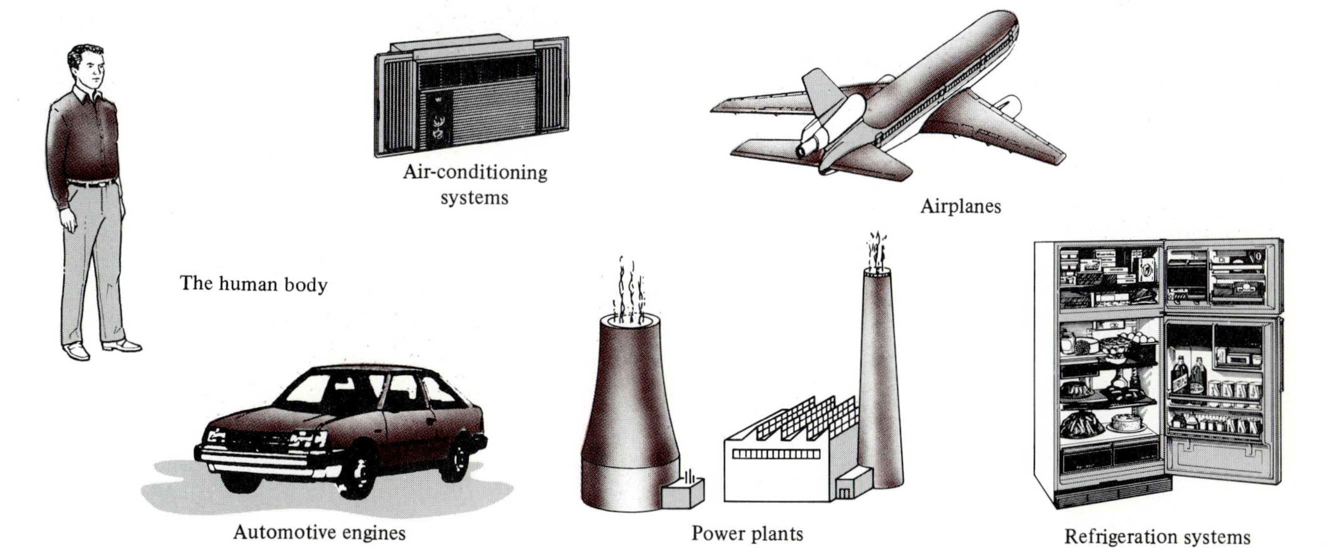

FIGURE 1-4
Some application areas of thermodynamics.

mensions such as mass m, length L, time t, and temperature T are selected as **primary dimensions** (Fig. 1-5), while others such as velocity $\mathbf{V}$, energy E, and volume V are expressed in terms of the primary dimensions and are called **secondary dimensions**, or **derived dimensions**.

m = mass (kg)
L = length (m)
t = time (s)
T = temperature (°C)

FIGURE 1-5
Some primary dimensions and their units.

A number of unit systems have been developed over the years. Despite strong efforts in the scientific and engineering community to unify the world with a single unit system, two sets of units are still in common use today: the **English system** which is also known as the *United States Customary System* (USCS) and the metric **SI** (from *Le Système International d'Unités*) which is also known as the *International System*. The SI is a simple and logical system based on a decimal relationship between the various units, and it is being used for scientific and engineering work in most of the industrialized nations, including England. The English system, however, has no numerical base, and various units in this system are related to each other rather arbitrarily (12 in in 1 ft, 16 oz in 1 lb, 4 qt in 1 gal, etc.) which makes it confusing and difficult to learn. The United States is one of the three countries (the other two are Burma and Brunei) that have not yet fully converted to the metric system (Fig. 1-6).

FIGURE 1-6
The United States is in a transition period to the full use of SI.

The recent trend toward the metric system in the United States seems to have started in 1968 when Congress, in response to what was happening in the rest of the world, passed a Metric Study Act. Congress continued to promote a voluntary switch to the metric system by passing the Metric Conversion Act in 1975. The industries that are heavily involved in international trade (such as the automotive, soft drink, and liquor industries) have been quick in converting to the metric system for economic reasons (having a single worldwide design, fewer sizes, smaller inventories, etc.). Today, nearly all the cars manufactured in the United States are metric. Most car owners probably do not realize this until they try an inch socket wrench on a metric bolt. Most industries, however, resisted the change, thus slowing down the conversion process.

Presently the United States is a dual-system society, and it will stay that way until the transition to the metric system is completed. This puts an extra burden on today's engineering students since they are expected to retain their understanding of the English system while learning, thinking, and working in terms of the SI. Given the position of the engineers in the transition period, both unit systems are used in this text with particular emphasis on SI units.

As pointed out earlier, the SI is based on a decimal relationship between units. The prefixes used to express the multiples of the various units are listed in Table 1-1. They are standard for all units, and the student is encouraged to memorize them because of their widespread use (Fig. 1-7).

TABLE 1-1
Standard prefixes in SI units

Multiple	Prefix
10^{12}	tera, T
10^{9}	giga, G
10^{6}	mega, M
10^{3}	kilo, k
10^{-2}	centi, c
10^{-3}	milli, m
10^{-6}	micro, μ
10^{-9}	nano, n
10^{-12}	pico, p

Some SI and English Units

In SI, the units of mass, length, and time are the kilogram (kg), meter (m), and second (s), respectively. The respective units in the English

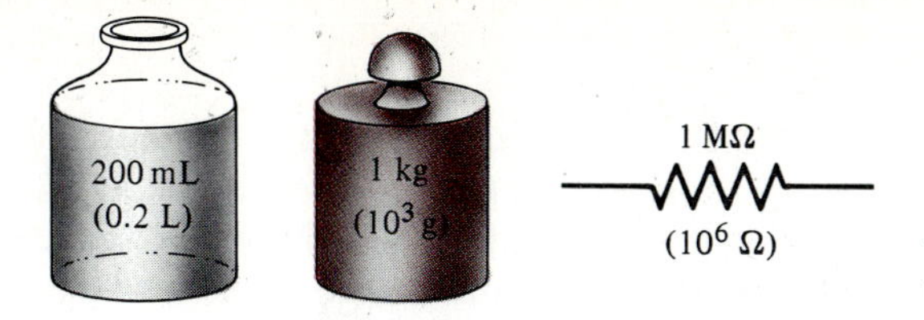

FIGURE 1-7

The SI unit prefixes are used in all branches of engineering.

system are the pound-mass (lbm), foot (ft), and second (s or sec). The mass and length units in the two systems are related to each other by

$$1 \text{ lbm} = 0.45359 \text{ kg}$$
$$1 \text{ ft} = 0.3048 \text{ m}$$

In the English system, force is usually considered to be one of the primary dimensions and is assigned a nonderived unit. This is a source of confusion and error that necessitates the use of a conversion factor (g_c) in many formulas. To avoid this nuisance, we consider force to be a secondary dimension whose unit is derived from Newton's second law, i.e.,

$$\text{Force} = (\text{mass})(\text{acceleration})$$

or

$$F = ma \tag{1-1}$$

In SI, the force unit is the **newton** (N), and it is defined as *the force required to accelerate a mass of 1 kg at a rate of 1 m/s²*. In the English system, the force unit is the **pound-force** (lbf) and is defined as *the force required to accelerate a mass of 32.174 lbm (1 slug) at a rate of 1 ft/s²*. That is,

$$1 \text{ N} = 1 \text{ kg·m/s}^2$$
$$1 \text{ lbf} = 32.174 \text{ lbm·ft/s}^2$$

This is illustrated in Fig. 1-8.

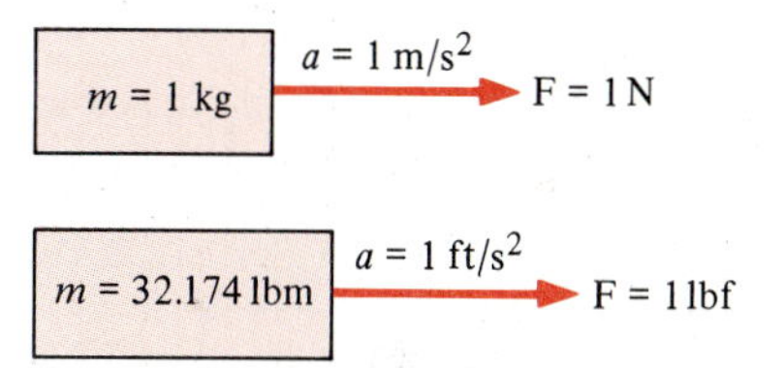

FIGURE 1-8

The definition of the force units.

The term **weight** is often incorrectly used to express mass, particularly by the "weight watchers." Unlike mass, weight W is a *force*. It is the gravitational force applied to a body (Fig. 1-9), and its magnitude is determined from Newton's second law,

$$W = mg \qquad (\text{N}) \tag{1-2}$$

where m is the mass of the body and g is the local gravitational acceleration (g is 9.807 m/s² or 32.174 ft/s² at sea level and 45° latitude). The weight of a unit volume of a substance is called the **specific weight** w and is determined from $w = \rho g$, where ρ is density.

The mass of a body will remain the same regardless of its location in the universe. Its weight, however, will change with a change in gravitational acceleration. A body will weigh less on top of a mountain since g decreases with altitude. On the surface of the moon, an astronaut will

FIGURE 1-9

The ordinary bathroom scale measures the gravitational force applied on a body.

weigh about one-sixth of what she or he normally weighs on earth (Fig. 1-10).

FIGURE 1-10
A body weighing 150 pounds on earth will weigh only 25 pounds on the moon.

At sea level a mass of 1 kg will weigh 9.807 N, as illustrated in Fig. 1-11. A mass of 1 lbm, however, will weigh 1 lbf which misleads people to believe that pound-mass and pound-force can be used interchangeably as pound (lb), which is a major source of error in the English system.

Work, which is a form of energy, can simply be defined as force times distance; therefore, it has the unit "newton-meter (N·m)," which is called a *joule* (J). That is,

$$1\ \text{J} = 1\ \text{N}\cdot\text{m}$$

A more common unit for energy in SI is the kilojoule (1 kJ = 10^3 J). In the English system, the energy unit is the Btu (British thermal unit), which is defined as the energy required to raise the temperature of 1 lbm of water at 68°F by 1°F. The magnitudes of the kilojoule and Btu are almost identical (1 Btu = 1.055 kJ).

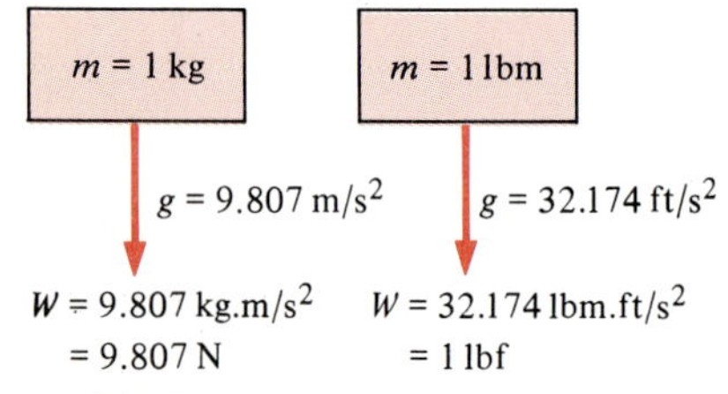

FIGURE 1-11
The weight of a unit mass at sea level.

Dimensional Homogeneity

We all know from grade school that apples and oranges do not add. But we somehow manage to do it (by mistake, of course). In engineering, all equations must be *dimensionally homogeneous*. That is, every term in an equation must have the same units (Fig. 1-12). If, at some stage of an analysis, we find ourselves in a position to add two quantities that have different units, it is a clear indication that we have made an error at an earlier stage. So checking units can serve as a valuable tool to spot errors.

EXAMPLE 1-1

While solving a problem, a person ended up with the following equation at some stage:

$$E = 25\ \text{kJ} + 7\ \text{kJ/kg}$$

where E is the total energy and has the unit of kilojoules. Determine the error that may have caused it.

Solution The two terms on the right-hand side do not have the same units, and therefore they cannot be added to obtain the total energy. Multiplying the last term by mass will eliminate the kilograms in the denominator, and the whole equation will become dimensionally homogeneous, i.e., every term in the equation will have the same unit. Obviously this error was caused by forgetting to multiply the last term by mass at an earlier stage.

FIGURE 1-12
To be dimensionally homogeneous, all the terms in an equation must have the same units.

We all know from experience that units can give terrible headaches if they are not used carefully in solving a problem. But with some attention and skill, units can be used to our advantage. They can be used to check formulas; they can even be used to derive formulas, as explained in the following example.

EXAMPLE 1-2

A tank is filled with oil whose density is $\rho = 850 \text{ kg/m}^3$. If the volume of the tank is $V = 2 \text{ m}^3$, determine the amount of mass m in the tank.

Solution A sketch of the system described above is given in Fig. 1-13. Suppose we forgot the formula that relates mass to density and volume. But we know that mass has the unit of kilograms. That is, whatever calculations we do, we should end up with the unit of kilograms. Putting the given information into perspective, we have

$$\rho = 850 \text{ kg/m}^3 \qquad V = 2 \text{ m}^3$$

It is obvious that we can eliminate m^3 and end up with kg by multiplying these two quantities. Therefore, the formula we are looking for is

$$m = \rho V$$

Thus,

$$m = (850 \text{ kg/m}^3)(2 \text{ m}^3) = 1700 \text{ kg}$$

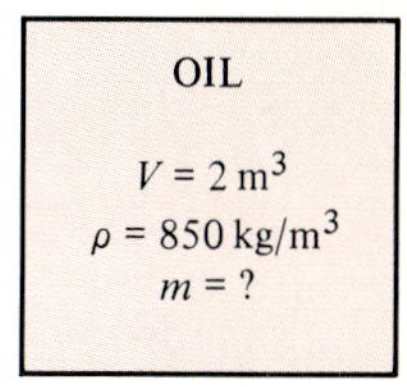

FIGURE 1-13
Sketch for Example 1-2.

The student should keep in mind that a formula which is not dimensionally homogeneous is definitely wrong, but a dimensionally homogeneous formula is not necesssarily right.

1-3 ■ CLOSED AND OPEN SYSTEMS

thermodynamic system, or simply a **system**, is defined as *a quantity of matter or a region in space chosen for study*. The region outside the system is called the **surroundings**. The real or imaginary surface that separates the system from its surroundings is called the **boundary**. These terms are illustrated in Fig. 1-14. The boundary of a system can be *fixed* or *movable*.

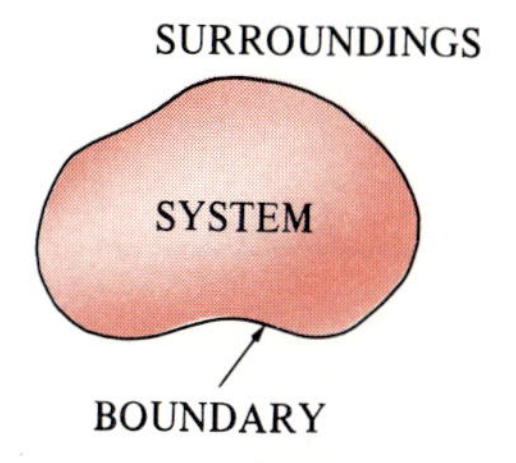

FIGURE 1-14
System, surroundings, and boundary.

Systems may be considered to be *closed* or *open*, depending on whether a fixed mass or a fixed volume in space is chosen for study. A **closed system** (also known as a **control mass**) consists of a fixed amount of mass, and no mass can cross its boundary. That is, no mass can enter or leave a closed system, as shown in Fig. 1-15. But energy, in the form of heat or work, can cross the boundary, and the volume of a closed system does not have to be fixed. If, as a special case, even energy is not allowed to cross the boundary, that system is called an **isolated system**.

Consider the piston-cylinder device shown in Fig. 1-16. Let us say that we would like to find out what happens to the enclosed gas when it is heated. Since we are focusing our attention on the gas, it is our system. The inner surfaces of the piston and the cylinder form the boundary, and since no mass is crossing this boundary, it is a closed system. Notice that energy may cross the boundary, and part of the boundary (the inner surface of the piston, in this case) may move. Everything outside the gas, including the piston and the cylinder, is the surroundings.

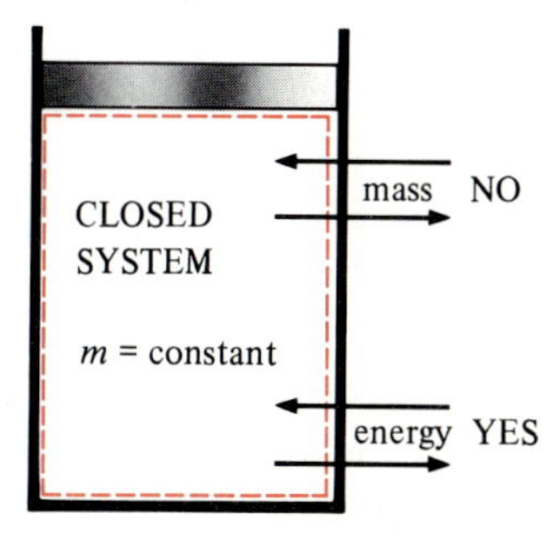

FIGURE 1-15
Mass cannot cross the boundaries of a closed system, but energy can.

An **open system**, or a **control volume**, as it is often called, is a properly selected region in space. It usually encloses a device which involves mass

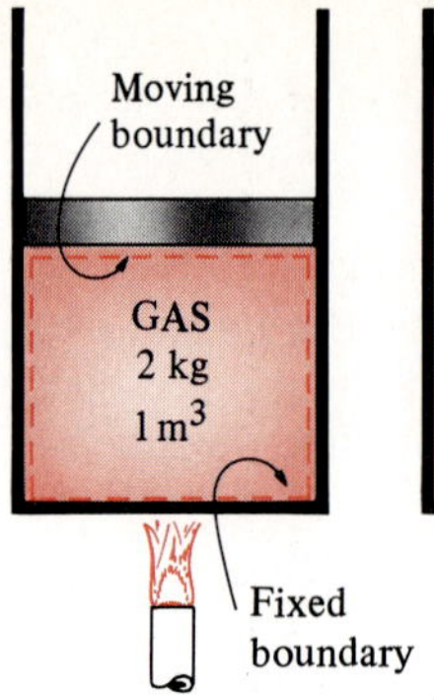

Moving boundary

GAS
2 kg
1 m^3

GAS
2 kg
3 m^3

Fixed boundary

FIGURE 1-16
A closed system with a moving boundary.

flow such as a compressor, turbine, or nozzle. Flow through these devices is best studied by selecting the region within the device as the control volume. Both mass and energy can cross the boundary of a control volume, which is called a **control surface**. This is illustrated in Fig. 1-17.

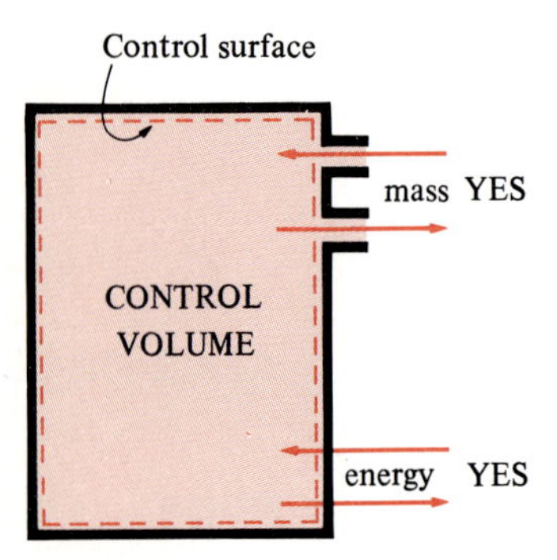

FIGURE 1-17
Both mass and energy can cross the boundaries of a control volume.

As an example of an open system, consider the water heater shown in Fig. 1-18. Let us say that we would like to determine how much heat we must add to the water in the tank in order to supply a steady stream of hot water. Since hot water will leave the tank and be replaced by cold water, it is not convenient to choose a fixed mass as our system for the analysis. Instead, we can concentrate our attention on the volume formed by the interior surfaces of the tank and consider the hot and cold water streams as mass leaving and entering the control volume. The interior surfaces of the tank form the control surface for this case, and mass is crossing the control surface at two locations.

The thermodynamic relations that are applicable to closed and open systems are different. Therefore, it is extremely important that we recognize the type of system we have before we start analyzing it.

In all thermodynamic analyses, the systems under study *must* be defined carefully. In most cases, the system investigated is quite simple and obvious, and defining the system may seem like a tedious and unnecessary task. In other cases, however, the system under study may be rather involved, and a proper choice of the system may greatly simplify the analysis.

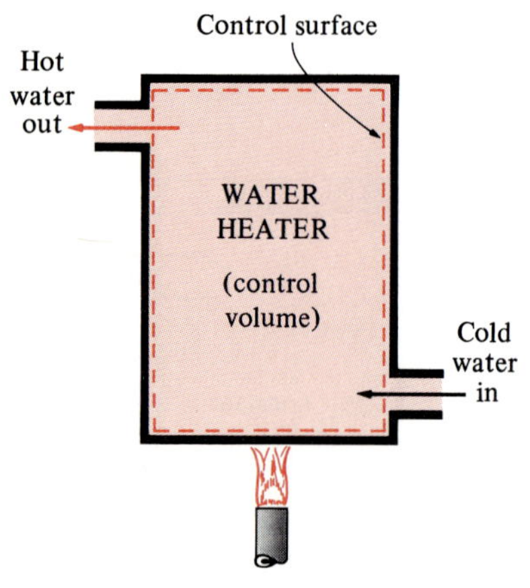

FIGURE 1-18
An open system (a control volume) with one inlet and one exit.

1-4 ■ FORMS OF ENERGY

Energy can exist in numerous forms such as thermal, mechanical, kinetic, potential, electric, magnetic, chemical, and nuclear, and their sum constitutes the **total energy** E of a system. The total energy of a system on a *unit mass* basis is denoted by e and is defined as

$$e = \frac{E}{m} \qquad \text{(kJ/kg)} \tag{1-3}$$

Thermodynamics provides no information about the absolute value of the total energy of a system. It only deals with the *change* of the total energy, which is what matters in engineering problems. Thus the total energy of a system can be assigned a value of zero ($E = 0$) at some convenient reference point. The change in total energy of a system is independent of the reference point selected. The decrease in the potential energy of a falling rock, for example, depends on only the elevation difference and not the reference level chosen.

In thermodynamic analysis, it is often helpful to consider the various forms of energy that make up the total energy of a system in two groups: *macroscopic* and *microscopic*. The **macroscopic** forms of energy, on one hand, are those a system possesses as a whole with respect to some outside reference frame, such as kinetic and potential energies (Fig. 1-19). The **microscopic** forms of energy, on the other hand, are those related to the molecular structure of a system and the degree of the molecular activity, and they are independent of outside reference frames. The sum of all the microscopic forms of energy is called the **internal energy** of a system and is denoted by U.

FIGURE 1-19
The macroscopic energy of an object changes with velocity and elevation.

The macroscopic energy of a system is related to motion and the influence of some external effects such as gravity, magnetism, electricity, and surface tension. The energy that a system possesses as a result of its motion relative to some reference frame is called **kinetic energy** KE. When all parts of a system move with the same velocity, the kinetic energy is expressed as

$$\mathrm{KE} = \frac{m\mathbf{V}^2}{2} \qquad \text{(kJ)} \tag{1-4}$$

or, on a unit mass basis,

$$\mathrm{ke} = \frac{\mathbf{V}^2}{2} \qquad \text{(kJ/kg)} \tag{1-5}$$

where the upright bold **V** denotes the velocity of the system relative to some fixed reference frame.

The energy that a system possesses as a result of its elevation in a gravitational field is called **potential energy** PE and is expressed as

$$\mathrm{PE} = mgz \qquad \text{(kJ)} \tag{1-6}$$

or, on a unit mass basis,

$$\mathrm{pe} = gz \qquad \text{(kJ/kg)} \tag{1-7}$$

where g is the gravitational acceleration and z is the elevation of the center of gravity of a system relative to some arbitrarily selected reference plane.

The magnetic, electric, and surface tension effects are significant in some specialized cases only and are not considered in this text. In the absence of these effects, the total energy of a system consists of the kinetic, potential, and internal energies and is expressed as

$$E = U + \text{KE} + \text{PE} = U + \frac{m\mathbf{V}^2}{2} + mgz \qquad \text{(kJ)} \qquad (1\text{-}8)$$

or, on a unit mass basis,

$$e = u + \text{ke} + \text{pe} = u + \frac{\mathbf{V}^2}{2} + gz \qquad \text{(kJ/kg)} \qquad (1\text{-}9)$$

Most closed systems remain stationary during a process and thus experience no change in their kinetic and potential energies. Closed systems whose velocity and elevation of the center of gravity remain constant during a process are frequently referred to as **stationary systems**. The change in the total energy ΔE of a stationary system is identical to the change in its internal energy ΔU. In this text a closed system is assumed to be stationary unless it is specifically stated otherwise.

Some Physical Insight to Internal Energy

Internal energy is defined above as the sum of all the microscopic forms of energy of a system. It is related to the molecular structure and the degree of molecular activity, and it may be viewed as the sum of the kinetic and potential energies of the molecules.

To have a better understanding of internal energy, let us examine a system at the molecular level. The individual molecules of a system, in general, will move around with some velocity, vibrate about each other, and rotate about an axis during their random motion. Associated with these motions are translational, vibrational, and rotational kinetic energies, the sum of which constitutes the kinetic energy of a molecule. The portion of the internal energy of a system associated with the kinetic energies of the molecules is called the **sensible energy** (Fig. 1-20). The average velocity and the degree of activity of the molecules are proportional to the temperature of the gas. Thus, at higher temperatures the molecules will possess higher kinetic energies, and as a result the system will have a higher internal energy.

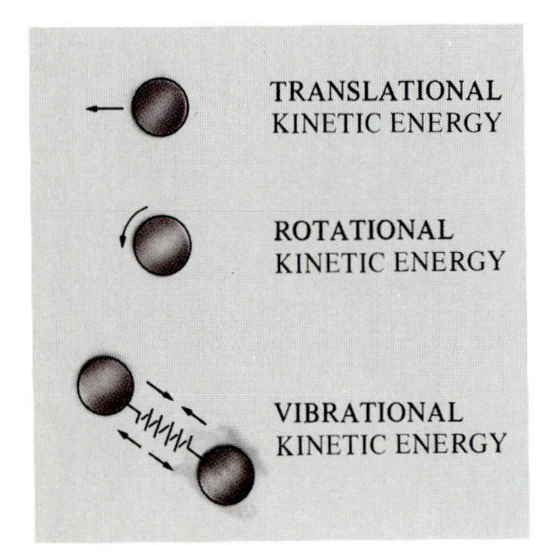

FIGURE 1-20
The various forms of molecular energy that make up sensible internal energy.

The internal energy is also associated with the intermolecular forces between the molecules of a system. These are the forces that bind the molecules to each other, and, as one would expect, they are strongest in solids and weakest in gases. If sufficient energy is added to the molecules of a solid or liquid, they will overcome these molecular forces and break away, turning the system to a gas. This is a phase-change process. Because of this added energy, a system in the gas phase is at a higher internal energy level than it is in the solid or the liquid phase. The internal energy associated with the phase of a system is called **latent energy**.

The changes mentioned above can occur without a change in the chemical composition of a system. Most thermodynamic problems fall into this category, and one does not need to pay any attention to the forces binding the atoms in a molecule. The internal energy associated with the atomic bonds in a molecule is called **chemical** (or **bond**) **energy**. During a chemical reaction, such as a combustion process, some chemical bonds are destroyed while others are formed. As a result, the internal energy changes.

We should also mention the tremendous amount of internal energy associated with the bonds within the nucleus of the atom itself (Fig. 1-21). This energy is called **nuclear energy** and is released during nuclear reactions. Obviously, we need not be concerned with nuclear energy in thermodynamics unless, of course, we have a fusion or fission reaction on our hands.

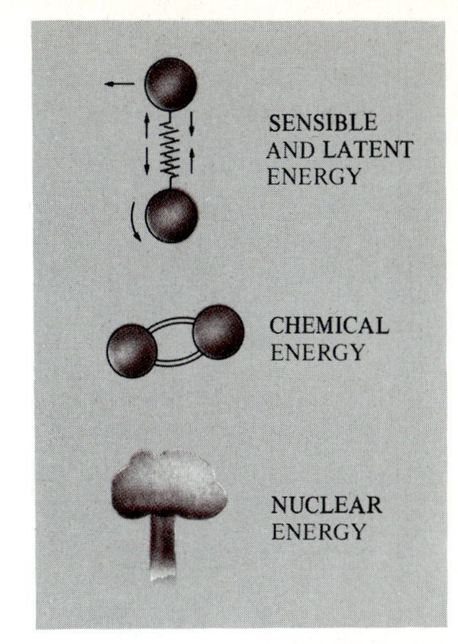

FIGURE 1-21

The internal energy of a system is the sum of all forms of the microscopic energies.

1-5 ■ PROPERTIES OF A SYSTEM

Any characteristic of a system is called a **property**. Some familiar examples are pressure P, temperature T, volume V, and mass m. The list can be extended to include less familiar ones such as viscosity, thermal conductivity, modulus of elasticity, thermal expansion coefficient, electric resistivity, and even velocity and elevation.

Not all properties are independent, however. Some are defined in terms of other ones. For example, **density** is defined as *mass per unit volume:*

$$\rho = \frac{m}{V} \qquad (\text{kg/m}^3) \tag{1-10}$$

Sometimes the density of a substance is given relative to the density of a better known substance. Then it is called **specific gravity**, or **relative density**, and is defined as *the ratio of the density of a substance to the density of some standard substance at a specified temperature* (usually water at 4°C for which $\rho_{H_2O} = 1000$ kg/m³). That is,

$$\rho_s = \frac{\rho}{\rho_{H_2O}} \tag{1-11}$$

Note that specific gravity is a dimensionless quantity.

A more frequently used property in thermodynamics is the **specific volume**. It is the reciprocal of density (Fig. 1-22) and is defined as *the volume per unit mass:*

$$v = \frac{V}{m} = \frac{1}{\rho} \qquad (\text{m}^3/\text{kg}) \tag{1-12}$$

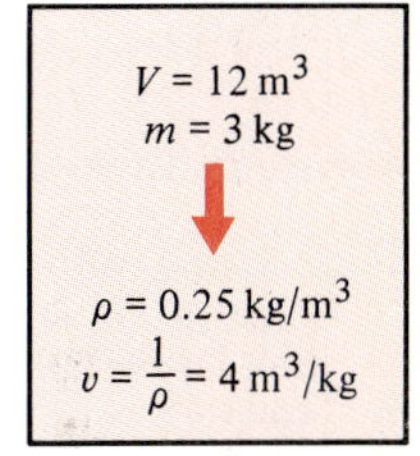

FIGURE 1-22

Density is mass per unit volume; specific volume is volume per unit mass.

Note that in classical thermodynamics, the atomic structure of a substance (thus the spaces between and within the molecules) is disregarded, and the substance is viewed to be a continuous, homogeneous matter with no microscopic holes, i.e., a **continuum**. This idealization is valid as long as we work with volumes, areas, and lengths which are large relative to the intermolecular spacings.

Properties are considered to be either *intensive* or *extensive*. **Intensive properties** are those which are independent of the size of a system such as temperature, pressure, and density (Fig. 1-23). **Extensive properties** vary directly with the size—or extent—of the system. Mass m, volume V, and total energy E are some examples of extensive properties

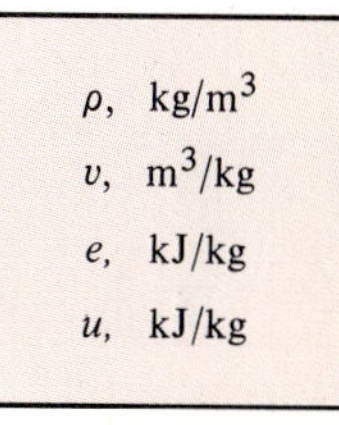

FIGURE 1-23

Intensive properties are independent of the size of the system.

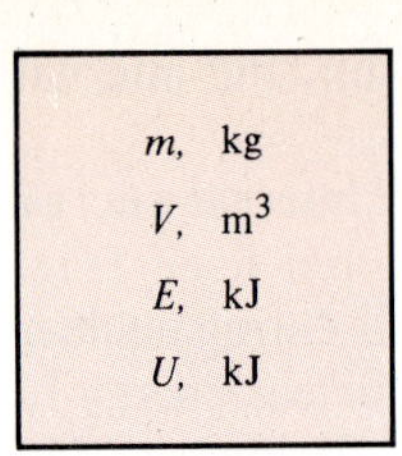

FIGURE 1-24
The larger the system, the larger the extensive properties.

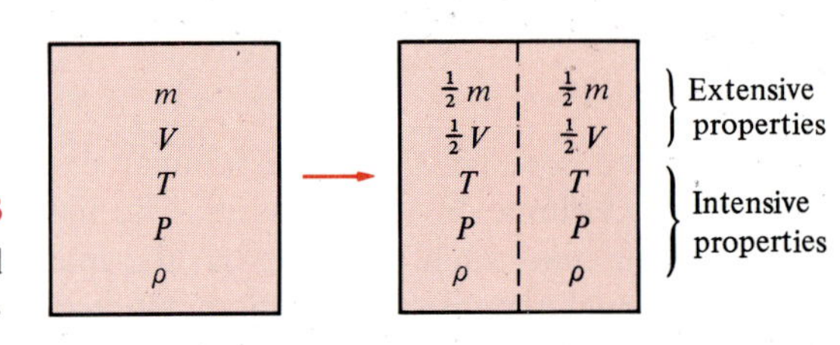

FIGURE 1-25
Criteria to differentiate intensive and extensive properties.

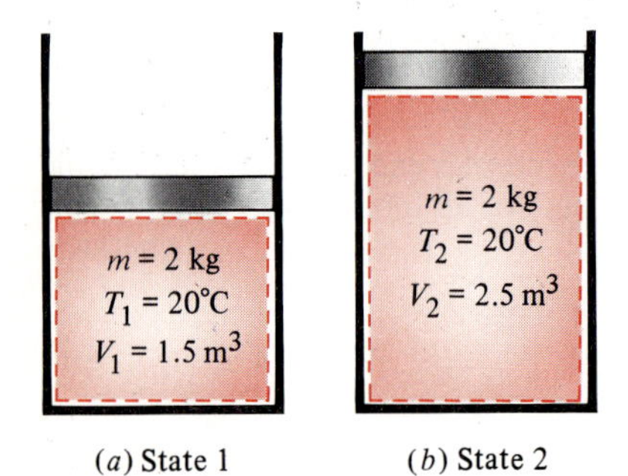

FIGURE 1-26
A system at two different states.

(Fig. 1-24). An easy way to determine whether a property is intensive or extensive is to divide the system into two equal parts with a partition, as shown in Fig. 1-25. Each part will have the same value of intensive properties as the original system, but half the value of the extensive properties.

Generally, uppercase letters are used to denote extensive properties (with mass m being a major exception), and lowercase letters are used for intensive properties (with pressure P and temperature T being the obvious exceptions).

Extensive properties per unit mass are called **specific properties**. Some examples of specific properties are specific volume ($v = V/m$), specific total energy ($e = E/m$), and specific internal energy ($u = U/m$).

1-6 ■ STATE AND EQUILIBRIUM

Consider a system which is not undergoing any change. At this point, all the properties can be measured or calculated throughout the entire system, which gives us a set of properties that completely describe the condition, or the **state**, of the system. At a given state, all the properties of a system have fixed values. If the value of even one property changes, the state will change to a different one. In Fig. 1-26 a system is shown in two different states.

Thermodynamics deals with **equilibrium** states. The word *equilibrium* implies a state of balance. In an equilibrium state there are no unbalanced potentials (or driving forces) within the system. A system which is in equilibrium experiences no changes when it is isolated from its surroundings.

There are many types of equilibrium, and a system is not in thermodynamic equilibrium unless the conditions of all the relevant types of equilibrium are satisfied (Fig. 1-27). For example, a system is in **thermal equilibrium** if the temperature is the same throughout the entire system,

I WONDER IF MY BODY IS IN THERMODYNAMIC EQUILIBRIUM...
I GUESS NOT!

FIGURE 1-27
A system which involves changes with time is not in equilibrium.

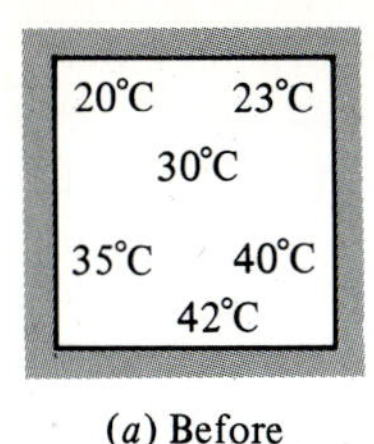

(*a*) Before (*b*) After

FIGURE 1-28
A closed system reaching thermal equilibrium.

as shown in Fig. 1-28. That is, the system involves no temperature differentials, which is the driving force for heat flow. **Mechanical equilibrium** is related to pressure, and a system is in mechanical equilibrium if there is no change in pressure at any point of the system with time. However, the pressure may vary within the system with elevation as a result of gravitational effects. But the higher pressure at a bottom layer is balanced by the extra weight it must carry, and, therefore, there is no imbalance of forces. The variation of pressure as a result of gravity in most thermodynamic systems is relatively small and usually disregarded. If a system involves two phases, it is **phase equilibrium** when the mass of each phase reaches an equilibrium level and stays there. Finally, a system is in **chemical equilibrium** if its chemical composition does not change with time, i.e., no chemical reactions occur. A system will not be in equilibrium unless all the relevant equilibrium criteria are satisfied (Fig. 1-29).

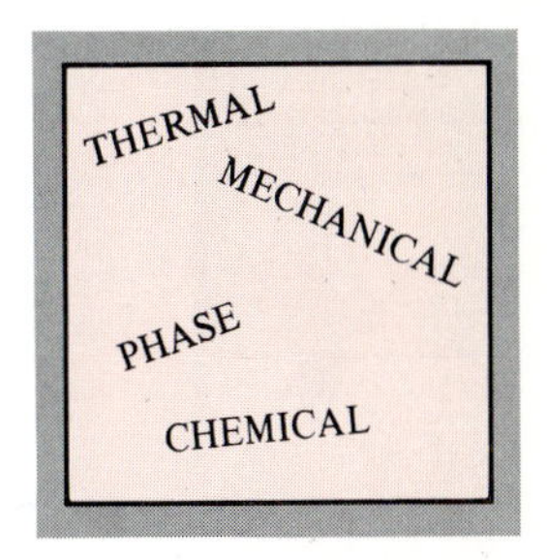

FIGURE 1-29
A system is not in thermodynamic equilibrium unless it satisfies the conditions for all modes of equilibrium.

1-7 ■ PROCESSES AND CYCLES

Any change that a system undergoes from one equilibrium state to another is called a **process**, and the series of states through which a system passes during a process is called the **path** of the process (Fig. 1-30). To describe a process completely, one should specify the initial and final states of the process, as well as the path it follows, and the interactions with the surroundings.

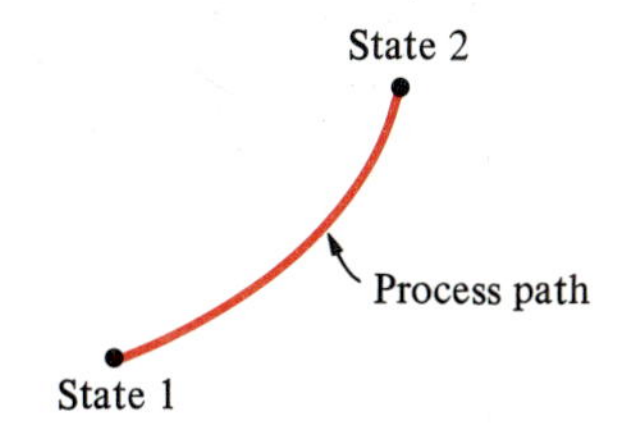

FIGURE 1-30
A process between states 1 and 2 and the process path.

When a process proceeds in such a manner that the system remains infinitesimally close to an equilibrium state at all times, it is called a **quasi-static**, or **quasi-equilibrium, process**. A quasi-equilibrium process can be viewed as a sufficiently slow process which allows the system to adjust itself internally so that properties in one part of the system do not change any faster than those at other parts.

This is illustrated in Fig. 1-31. When a gas in a piston-cylinder device is compressed suddenly, the molecules near the face of the piston will not have enough time to escape and they will have to pile up in a small region in front of the piston, thus creating a high-pressure region there. Because of this pressure difference, the system can no longer be said to be in equilibrium, and this makes the entire process non-quasi-equilibrium. However, if the piston is moved slowly, the molecules will have sufficient time to redistribute and there will not be a molecule pileup in front of the

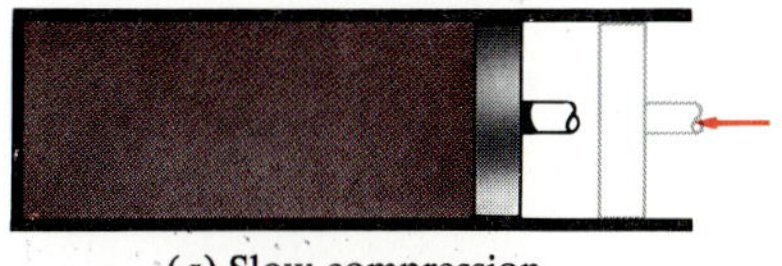

(*a*) Slow compression (quasi-equilibrium)

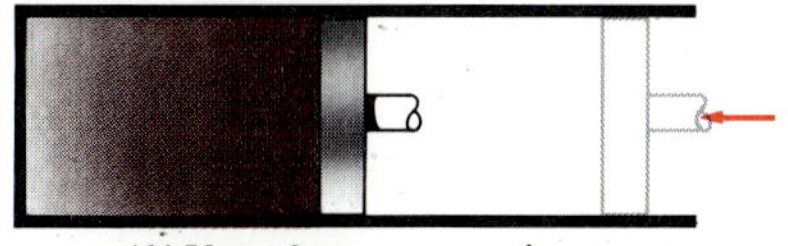

(*b*) Very fast compression (non-quasi-equilibrium)

FIGURE 1-31
Quasi-equilibrium and non-quasi-equilibrium compression processes.

piston. As a result, the pressure inside the cylinder will always be uniform and will rise at the same rate at all locations. Since equilibrium is maintained at all times, this is a quasi-equilibrium process.

It should be pointed out that a quasi-equilibrium process is an idealized process and is not a true representation of an actual process. But many actual processes closely approximate it, and they can be modeled as quasi-equilibrium with negligible error. Engineers are interested in quasi-equilibrium processes for two reasons. First, they are easy to analyze; second, work-producing devices deliver the most work when they operate on quasi-equilibrium processes (Fig. 1-32). Therefore, quasi-equilibrium processes serve as standards to which actual processes can be compared.

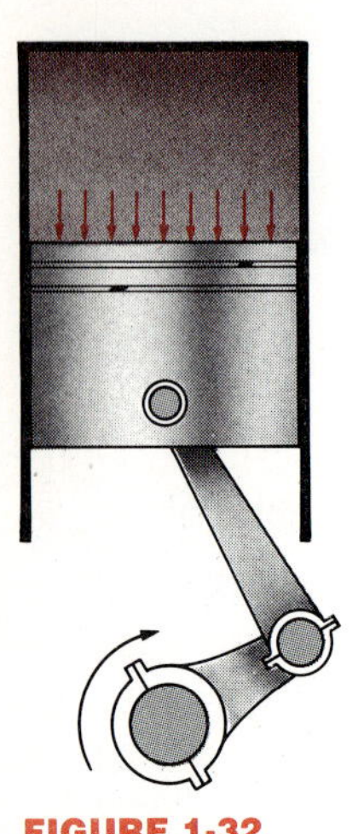

FIGURE 1-32
Work-producing devices operating in a quasi-equilibrium manner deliver the most work.

Process diagrams which are plotted by employing thermodynamic properties as coordinates are very useful in visualizing the processes. Some common properties that are used as coordinates are temperature T, pressure P, and volume V (or specific volume v). Figure 1-33 shows the P-V diagram of a compression process of a gas.

Note that the process path indicates a series of equilibrium states which the system passes through during a process and has significance for quasi-equilibrium processes only. For non-quasi-equilibrium processes, we are not able to specify the states through which the system passes during the process and so we cannot speak of a process path. A non-quasi-equilibrium process is denoted by a dashed line between the initial and final states, as illustrated in Fig. 1-34, instead of a solid line.

The prefix *iso-* is often used to designate a process for which a particular property remains constant. An **isothermal process**, for example, is a process during which the temperature T remains constant, an **isobaric process** is a process during which the pressure P remains constant, and an **isochoric** (or **isometric**) **process** is a process during which the specific volume v remains constant.

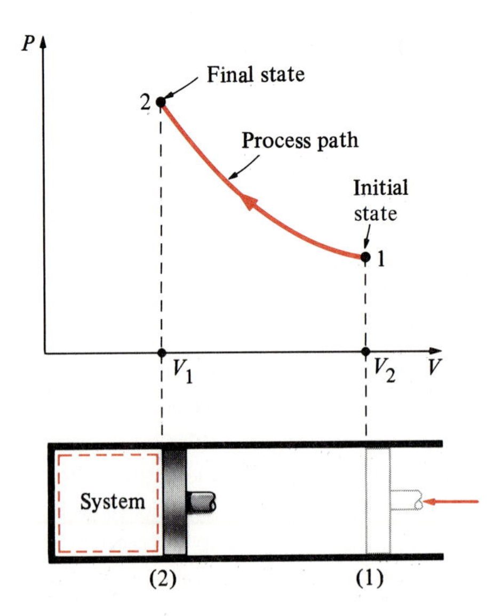

FIGURE 1-33
The P-V diagram of a compression process.

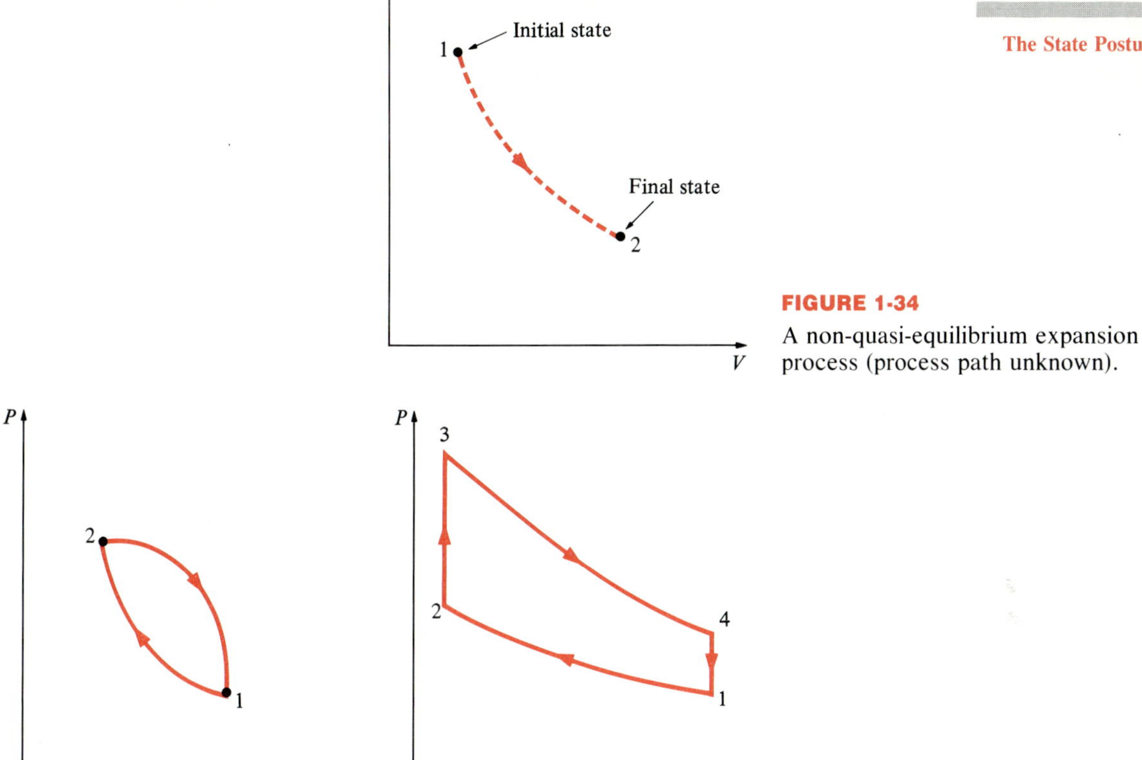

FIGURE 1-34
A non-quasi-equilibrium expansion process (process path unknown).

FIGURE 1-35
Two thermodynamic cycles.

A system is said to have undergone a **cycle** if it returns to its initial state at the end of the process. That is, for a cycle the initial and final states are identical. The cycle in Fig. 1-35*a* consists of two processes, and the one in Fig. 1-35*b* consists of four processes (this is the ideal cycle for gasoline engines and is analyzed in Chap. 8).

1-8 ■ THE STATE POSTULATE

As noted earlier, the state of a system is described by its properties. But we know from experience that we do not need to specify all the properties in order to fix a state. Once a sufficient number of properties are specified, the rest of the properties assume certain values automatically. That is, specifying a certain number of properties is sufficient to fix a state. The number of properties required to fix the state of a system is given by the **state postulate**:

The state of a simple compressible system is completely specified by two independent, intensive properties.

A system is called a **simple compressible system** in the absence of electrical, magnetic, gravitational, motion, and surface tension effects. These effects are due to external force fields and are negligible for most engineering problems. Otherwise, an additional property needs to be specified for each effect which is significant. If the gravitational effects are to be considered, for example, the elevation z needs to be specified in addition to the two properties necessary to fix the state.

The state postulate requires that the two properties specified be **independent** to fix the state. Two properties are independent if one property can be varied while the other one is held constant. Temperature and specific volume, for example, are always independent properties, and together they can fix the state of a simple compressible system (Fig. 1-36). Temperature and pressure, however, are independent properties for single-phase systems, but are dependent properties for multiphase systems. At sea level (P = 1 atm), water boils at 100°C, but on a mountaintop where the pressure is lower, water boils at a lower temperature. That is, $T = f(P)$ during a phase-change process, thus temperature and pressure are not sufficient to fix the state of a two-phase system. Phase-change processes are discussed in detail in the next chapter.

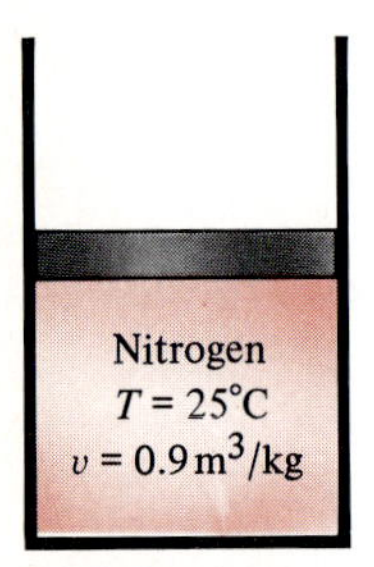

FIGURE 1-36
The state of nitrogen is fixed by two independent, intensive properties.

1-9 ■ PRESSURE

Pressure is *the force exerted by a fluid per unit area*. We speak of pressure only when we deal with a gas or a liquid. The counterpart of pressure in solids is *stress*. For a fluid at rest, the pressure at a given point is the same in all directions. The pressure in a fluid increases with depth as a result of the weight of the fluid, as shown in Fig. 1-37. This is due to the fluid at lower levels carrying more weight than the fluid at upper levels. The pressure varies in the vertical direction as a result of gravitational effects, but there is no variation in the horizontal direction. The pressure in a tank containing a gas may be considered to be uniform since the weight of the gas is too small to make a significant difference (Fig. 1-38).

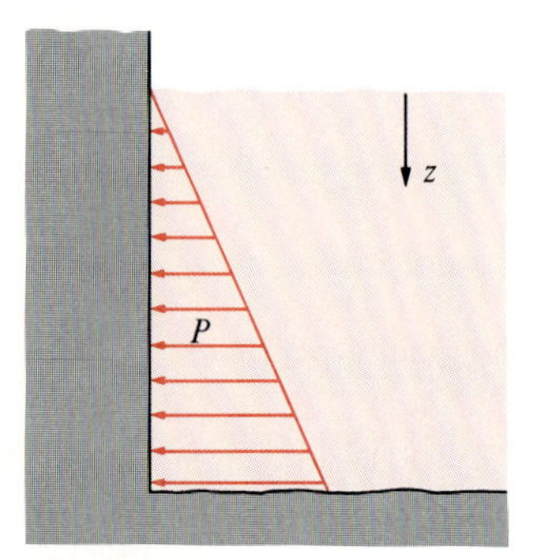

FIGURE 1-37
The pressure of a fluid at rest increases with depth (as a result of added weight).

Since pressure is defined as force per unit area, it has the unit of newtons per square meter (N/m^2), which is called a *pascal* (Pa). That is,

$$1 \text{ Pa} = 1 \text{ N/m}^2$$

The pressure unit pascal is too small for pressures encountered in practice; therefore, its multiples *kilopascal* (1 kPa = 10^3 Pa) and *megapascal* (1 MPa = 10^6 Pa) are commonly used. Two other common pressure units are the bar and standard atmosphere:

$$1 \text{ bar} = 10^5 \text{ Pa} = 0.1 \text{ MPa} = 100 \text{ kPa}$$

$$1 \text{ atm} = 101{,}325 \text{ Pa} = 101.325 \text{ kPa} = 1.01325 \text{ bars}$$

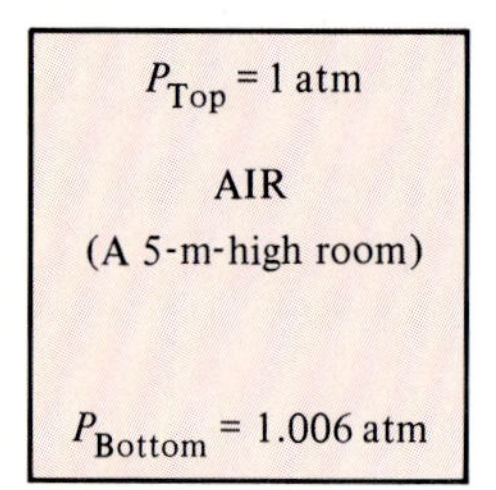

FIGURE 1-38
In a container filled with a gas the variation of pressure with height is negligible.

In the English system, the pressure unit is *pound-force per square inch* (lbf/in^2, or psi), and 1 atm = 14.696 psi.

The actual pressure at a given position is called the **absolute pressure**, and it is measured relative to absolute vacuum, i.e., absolute zero pressure. Most pressure-measuring devices, however, are calibrated to read zero in the atmosphere (Fig. 1-39), and so they indicate the difference between the absolute pressure and the local atmospheric pressure. This difference is called the **gage pressure**. Pressures below atmospheric pressure are called **vacuum pressures** and are measured by vacuum gages which indicate the difference between the atmospheric pressure and the absolute pressure. Absolute, gage, and vacuum pressures are all positive quantities and are related to each other by

$$P_{\text{gage}} = P_{\text{abs}} - P_{\text{atm}} \quad \text{(for pressures above } P_{\text{atm}}\text{)} \tag{1-13}$$

$$P_{\text{vac}} = P_{\text{atm}} - P_{\text{abs}} \quad \text{(for pressures below } P_{\text{atm}}\text{)} \tag{1-14}$$

This is illustrated in Fig. 1-40.

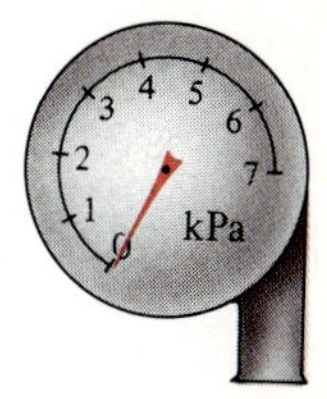

FIGURE 1-39

A pressure gage which is open to the atmosphere reads zero.

EXAMPLE 1-3

A vacuum gage connected to a chamber reads 5.8 psi at a location where the atmospheric pressure is 14.5 psi. Determine the absolute pressure in the chamber.

Solution The absolute pressure is easily determined from Eq. 1-14:

$$P_{\text{abs}} = P_{\text{atm}} - P_{\text{vac}} = (14.5 - 5.8) \text{ psi} = 8.7 \text{ psi}$$

In thermodynamic relations and tables, absolute pressure is almost always used. Throughout this text, the pressure P will denote absolute pressure unless it is otherwise specified. Often the letters "a" (for absolute pressure) and "g" (for gage pressure) are added to pressure units (such as psia and psig) in order to clarify what is meant.

Manometer

Small and moderate pressure differences are often measured by using a device known as a **manometer**, which mainly consists of a glass or plastic U-tube containing a fluid such as mercury, water, alcohol, or oil. To keep

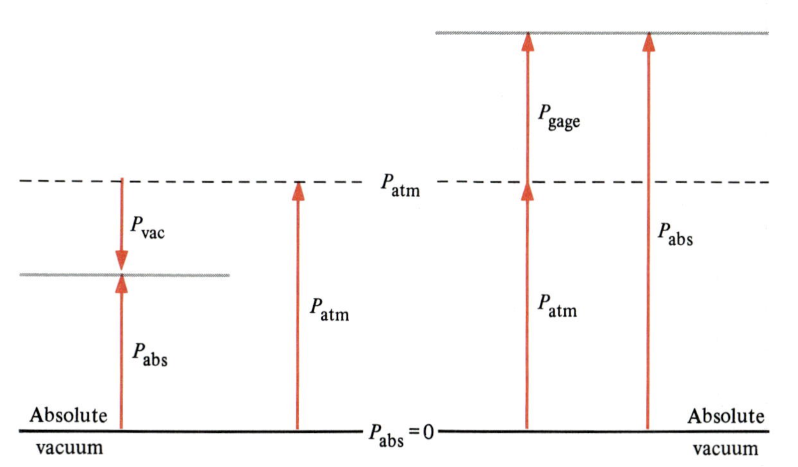

FIGURE 1-40

Absolute, gage, and vacuum pressures.

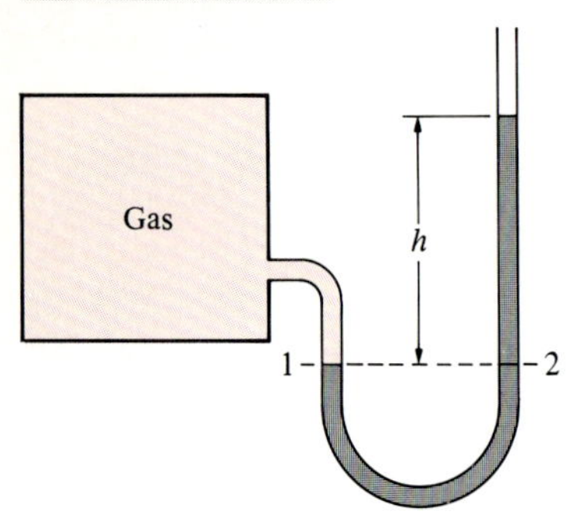

FIGURE 1-41
The basic manometer.

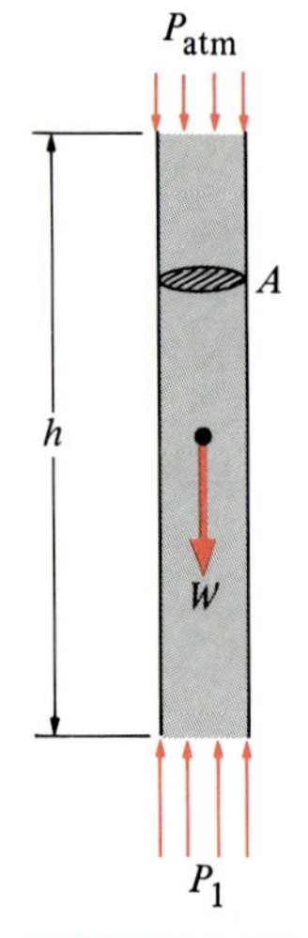

FIGURE 1-42
The free-body diagram of a fluid column of height h.

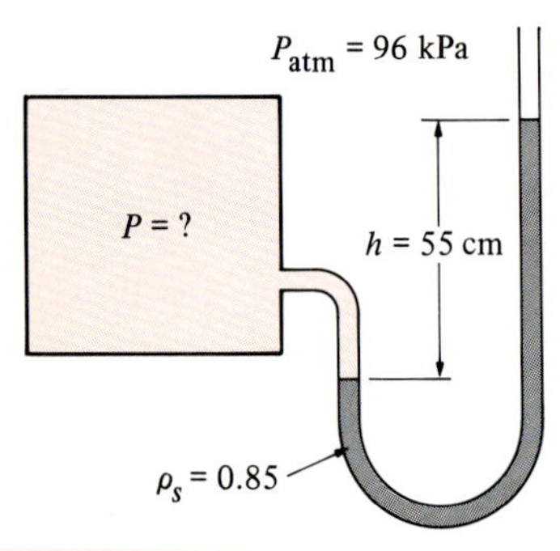

FIGURE 1-43
Sketch for Example 1-4.

the size of the manometer at a manageable level, heavy fluids such as mercury are used if large pressure differences are anticipated.

Consider the manometer shown in Fig. 1-41 which is used to measure the pressure in the tank. Since the gravitational effects of gases are negligible, the pressure anywhere in the tank and at position 1 has the same value. Furthermore, since pressure in a fluid does not vary in the horizontal direction within a fluid, the pressure at 2 is the same as the pressure at 1, or $P_2 = P_1$.

The differential fluid column of height h is in static equilibrium, and its free-body diagram is shown in Fig. 1-42. A force balance in the vertical direction gives

$$AP_1 = AP_{atm} + W$$

where $$W = mg = \rho V g = \rho A h g$$

Thus, $$P_1 = P_{atm} + \rho g h \qquad \text{(kPa)} \tag{1-15}$$

In the above relations, W is the weight of the fluid column, ρ is the density of the fluid and is assumed to be constant, g is the local gravitational acceleration, A is the cross-sectional area of the tube, and P_{atm} is the atmospheric pressure. The pressure difference can be expressed as

$$\Delta P = P_1 - P_{atm} = \rho g h \qquad \text{(kPa)} \tag{1-16}$$

Note that the cross-sectional area of the tube has no effect on the height differential h, and thus the pressure exerted by the fluid.

EXAMPLE 1-4

A manometer is used to measure the pressure in a tank. The fluid used has a specific gravity of 0.85, and the manometer column height is 55 cm, as shown in Fig. 1-43. If the local atmospheric pressure is 96 kPa, determine the absolute pressure within the tank.

Solution The gravitational acceleration is not specified, so we assume the standard value of 9.807 m/s². The density of the fluid is obtained by multiplying its specific gravity by the density of water, which is taken to be 1000 kg/m³:

$$\rho = (\rho_s)(\rho_{H_2O}) = (0.85)(1000 \text{ kg/m}^3) = 850 \text{ kg/m}^3$$

From Eq. 1-15,

$$P = P_{atm} + \rho g h$$
$$= 96 \text{ kPa} + (850 \text{ kg/m}^3)(9.807 \text{ m/s}^2)(0.55 \text{ m})\left(\frac{1 \text{ kPa}}{1000 \text{ N/m}^2}\right)$$

100.6 kPa

Barometer

The atmospheric pressure is measured by a device called a **barometer**; thus the atmospheric pressure is often called the *barometric pressure*.

As Torricelli (1608–1647) discovered some centuries ago, the atmospheric pressure can be measured by inverting a mercury-filled tube into a mercury container which is open to the atmosphere, as shown in Fig. 1-44. The pressure at point B is equal to the atmospheric pressure, and the pressure at C can be taken to be zero since there is only mercury vapor above point C and the pressure it exerts is negligible. Writing a force balance in the vertical direction gives

$$P_{atm} = \rho g h \qquad \text{(kPa)} \tag{1-17}$$

where ρ is the density of mercury, g is the local gravitational acceleration, and h is the height of the mercury column above the free surface. Note that the length and the cross-sectional area of the tube have no effect on the height of the fluid column of a barometer (Fig. 1-45).

A frequently used pressure unit is the *standard atmosphere,* which is defined as the pressure produced by a column of mercury 760 mm in height at 0°C (ρ_{Hg} = 13,595 kg/m³) under standard gravitational acceleration (g = 9.807 m/s²). If water were used to measure the standard atmospheric pressure instead of mercury, a water column of about 10.3 m would be needed. Pressure is sometimes expressed (especially by weather forecasters) in terms of the height of the mercury column. The standard atmospheric pressure, for example, is 760 mmHg (29.92 inHg) at 0°C.

Remember that the atmospheric pressure at a location is simply the weight of the air above that location per unit surface area. Therefore, it changes not only with elevation but also with weather conditions.

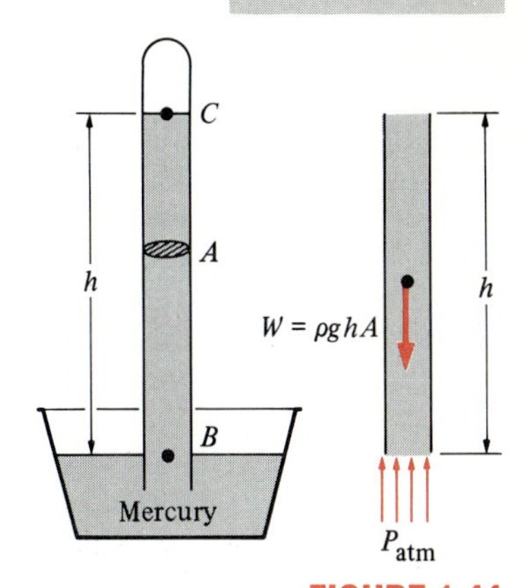

FIGURE 1-44
The basic barometer.

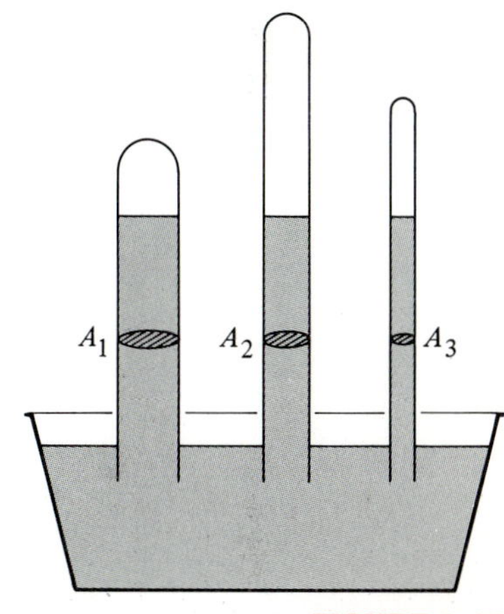

FIGURE 1-45
The length or the cross-sectional area of the tube has no effect on the height of the fluid column of a barometer.

EXAMPLE 1-5

Determine the atmospheric pressure at a location where the barometric reading is 740 mmHg and the gravitational acceleration is g = 9.7 m/s². Assume the temperature of mercury to be 10°C at which its density is 13,570 kg/m³.

Solution From Eq. 1-17, the atmospheric pressure is determined to be

$$\begin{aligned} P_{atm} &= \rho g h \\ &= (13{,}570\ \text{kg/m}^3)(9.7\ \text{m/s}^2)(0.74\ \text{m})\left(\frac{1\ \text{N}}{1\ \text{kg}\cdot\text{m/s}^2}\right)\left(\frac{1\ \text{kPa}}{1000\ \text{N/m}^2}\right) \\ &= 97.41\ \text{kPa} \end{aligned}$$

EXAMPLE 1-6

The piston of a piston-cylinder device containing a gas has a mass of 60 kg and a cross-sectional area of 0.04 m², as shown in Fig. 1-46. The local atmospheric pressure is 0.97 bar, and the gravitational acceleration is 9.8 m/s².

(*a*) Determine the pressure inside the cylinder.

(*b*) If some heat is transferred to the gas and its volume doubles, do you expect the pressure inside the cylinder to change?

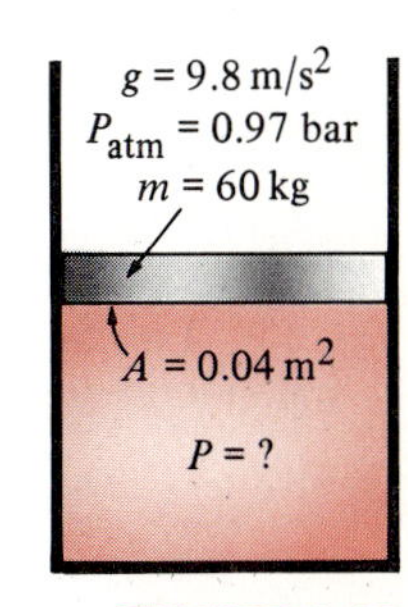

FIGURE 1-46
Sketch for Example 1-6.

Solution (*a*) The gas pressure in the piston-cylinder device depends on the atmospheric pressure and the weight of the piston. Drawing the free-body diagram of the piston (Fig. 1-47) and balancing the vertical forces yield

$$PA = P_{atm}A + W$$

$$P = P_{atm} + \frac{mg}{A}$$

$$= 0.97 \text{ bar} + \frac{(60 \text{ kg})(9.8 \text{ m/s}^2)}{0.04 \text{ m}^2}\left(\frac{1 \text{ N}}{1 \text{ kg}\cdot\text{m/s}^2}\right)\left(\frac{1 \text{ bar}}{10^5 \text{ N/m}^2}\right)$$

1.117 bars

(*b*) The volume change will have no effect on the free-body diagram drawn in part (*a*), and therefore the pressure inside the cylinder will remain the same.

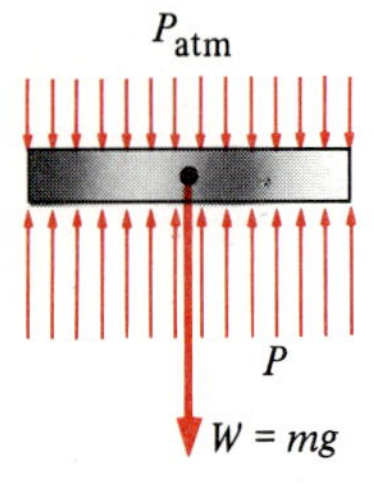

FIGURE 1-47
Free-body diagram of the piston.

1-10 ■ TEMPERATURE AND THE ZEROTH LAW OF THERMODYNAMICS

Although we are familiar with **temperature** as a measure of "hotness" or "coldness," it is not easy to give an exact definition for it. Based on our physiological sensations, we express the level of temperature qualitatively with words like *freezing cold, cold, warm, hot,* and *red-hot*. However, we cannot assign numerical values to temperatures based on our sensations alone. Furthermore, our senses may be misleading. A metal chair, for example, will feel much colder than a wooden one even when both are at the same temperature.

Fortunately, several properties of materials change with temperature in a repeatable and predictable way, and this forms the basis for accurate temperature measurement. The commonly used mercury-in-glass thermometer, for example, is based on the expansion of mercury with temperature. Temperature is also measured by using several other temperature-dependent properties.

It is common experience that a cup of hot coffee left on the table eventually cools off and a cold drink eventually warms up. That is, when a body is brought into contact with another body which is at a different temperature, heat is transferred from the body at higher temperature to the one at lower temperature until both bodies attain the same temperature (Fig. 1-48). At that point, the heat transfer stops, and the two bodies are said to have reached **thermal equilibrium**. The equality of temperature is the only requirement for thermal equilibrium.

The **zeroth law of thermodynamics** states that if two bodies are in thermal equilibrium with a third body, they are also in thermal equilibrium

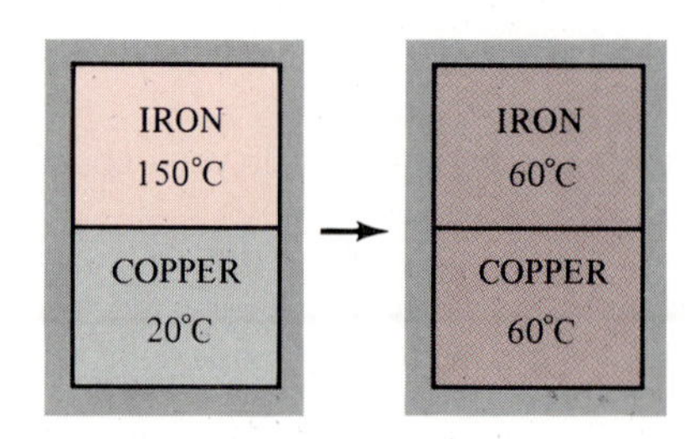

FIGURE 1-48
Two bodies reaching thermal equilibrium after being brought into contact.

with each other (Fig. 1-49). It may seem silly that such an obvious fact is called one of the basic laws of thermodynamics. However, it cannot be concluded from the other laws of thermodynamics, and it serves as a basis for the validity of temperature measurement. By replacing the third body with a thermometer, the zeroth law can be restated as *two bodies are in thermal equilibrium if both have the same temperature reading even if they are not in contact.*

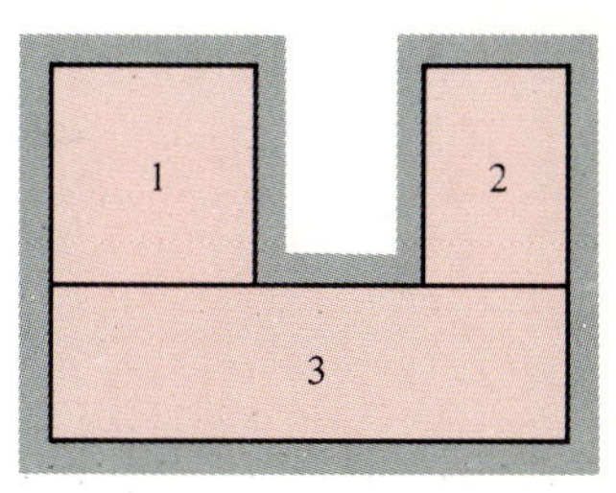

FIGURE 1-49
The zeroth law of thermodynamics.

Temperature Scales

Temperature scales enable scientists to use a common basis for temperature measurements, and several have been introduced throughout history. All temperature scales are based on some easily reproducible states such as the freezing and boiling points of water, which are also called the *ice point* and the *steam point,* respectively. A mixture of ice and water which is in equilibrium with air saturated with vapor at 1-atm pressure is said to be at the ice point, and a mixture of liquid water and water vapor (with no air) in equilibrium at 1-atm pressure is said to be at the steam point.

The temperature scales used in the SI and in the English system today are the **Celsius scale** (formerly called the *centigrade scale;* named after the Swedish astronomer A. Celsius who devised it) and the **Fahrenheit scale** (named after the German instrument maker G. Fahrenheit, 1686–1736), respectively. On the Celsius scale, the ice and steam points are assigned the values of 0 and 100°C, respectively. The corresponding values on the Fahrenheit scale are 32 and 212°F. These are often referred to as *two-point scales* since temperature values are assigned at two different points.

A more useful temperature scale in thermodynamics is the **absolute temperature scale**. As the name implies, there are no negative temperatures on an absolute temperature scale, and the lowest attainable temperature is absolute zero. The absolute temperature scale in the SI is the **Kelvin scale** (named after W. Thomson, 1824–1907, also known as Lord Kelvin). The temperature unit on this scale is the *kelvin* which is designated by K without the degree symbol.

A sealed, rigid tank containing a gas at a low pressure can be used as a thermometer to measure absolute temperature. This device is called the *constant-volume gas thermometer,* and it is based on the principle that the temperature of a gas at low pressures is proportional to its pressure. Thus when the pressure reading is halved, so is the absolute temperature. It can be determined by extrapolation that as the absolute pressure in a constant-volume gas thermometer approaches zero, the absolute temperature reading will also approach zero (Fig. 1-50). The reading of a thermometer with a Celsius scale, however, will approach −273.15°C. Therefore, a temperature of −273.15°C corresponds to an absolute temperature of 0 K. The Kelvin scale is related to the Celsius scale by

$$T(\text{K}) = T(^\circ\text{C}) + 273.15 \tag{1-18}$$

In the English system, the absolute temperature scale is the **Rankine**

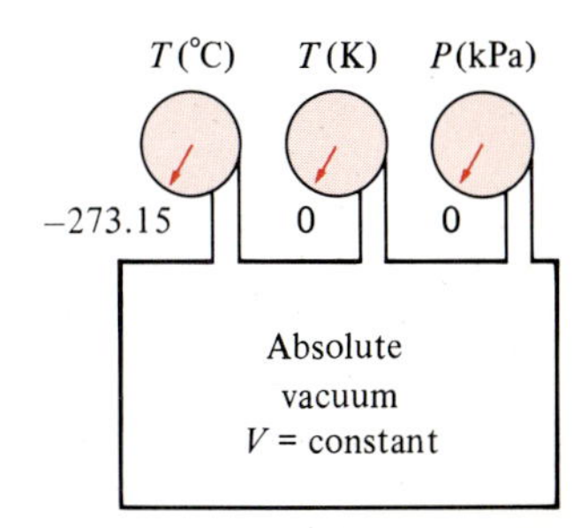

FIGURE 1-50
A constant-volume gas thermometer would read −273.15°C at absolute zero pressure.

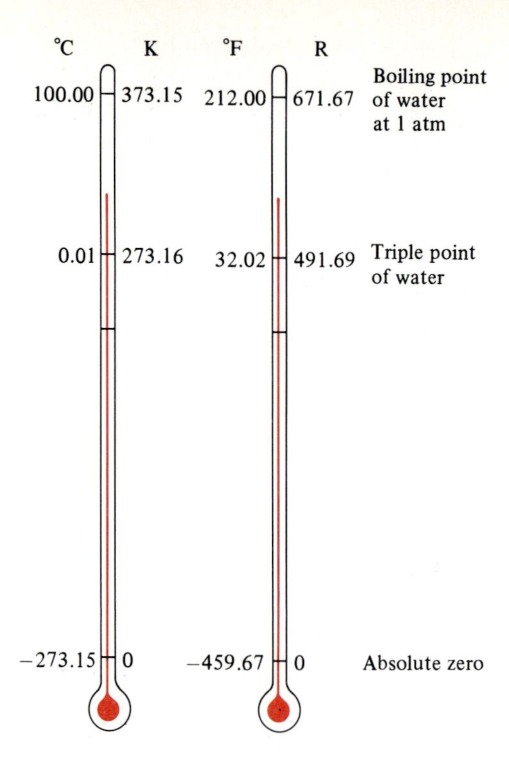

FIGURE 1-51
Comparison of temperature scales.

scale (named after W. J. M. Rankine, 1820–1872). It is related to the Fahrenheit scale by

$$T(\mathrm{R}) = T(°\mathrm{F}) + 459.67 \qquad (1\text{-}19)$$

It is common practice to round the constant in Eq. 1-18 to 273 and the one in Eq. 1-19 to 460.

The temperature scales in two unit systems are related by

$$T(\mathrm{R}) = 1.8\,T(\mathrm{K}) \qquad (1\text{-}20)$$

and

$$T(°\mathrm{F}) = 1.8T(°\mathrm{C}) + 32 \qquad (1\text{-}21)$$

A comparison of various temperature scales is given in Fig. 1-51. A more elaborate development of the absolute temperature scale is undertaken in Chap. 5 in conjunction with the second law of thermodynamics.

At the Tenth Conference on Weights and Measures in 1954, the Celsius scale was redefined in terms of a single fixed point and the absolute temperature scale. The selected single point is the *triple point* of water (the state at which all three phases of water coexist in equilibrium), which is assigned the value 0.01°C. The magnitude of the degree is defined from the absolute temperature scale. As before, the boiling point of water at 1-atm pressure is 100.00°C. Thus the new Celsius scale is essentially the same as the old one.

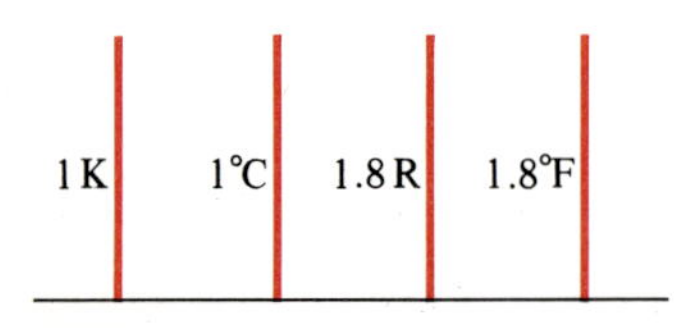

FIGURE 1-52
Comparison of magnitudes of various temperature units.

Note that the magnitudes of each division of 1 K and 1°C are identical (Fig. 1-52). Therefore, when we are dealing with temperature differences ΔT, the temperature interval on both scales is the same. Raising the temperature of a substance by 10°C is the same as raising it by 10 K. That is,

$$\Delta T(\mathrm{K}) = \Delta T(^\circ\mathrm{C}) \tag{1-22}$$

Similarly,

$$\Delta T(\mathrm{R}) = \Delta T(^\circ\mathrm{F}) \tag{1-23}$$

Some thermodynamic relations involve the temperature T and often the question arises of whether it is in K or °C. If the relation involves temperature differences (such as $a = b\,\Delta T$), it makes no difference and either can be used. But if the relation involves temperatures only instead of temperature differences (such as $a = bT$), then K must be used. When in doubt, it is always safe to use K because there are virtually no situations in which the use of K is incorrect, but there are many thermodynamic relations which will yield erroneous result if °C is used.

EXAMPLE 1-7

During a heating process, the temperature of a system rises by 10°C. Express this rise in temperature in K, °F, and R.

Solution This problem deals with temperature changes, which are identical in Kelvin and Celsius scales. Then from Eq. 1-22,

$$\Delta T(\mathrm{K}) = \Delta T(^\circ\mathrm{C}) = 10\ \mathrm{K}$$

The temperature changes in Fahrenheit and Rankine scales are also identical and are related to the changes in Celsius and Kelvin scales through Eqs. 1-20 and 1-23:

$$\Delta T(\mathrm{R}) = 1.8\,\Delta T(\mathrm{K}) = (1.8)(10) = 18\ \mathrm{R}$$

and

$$\Delta T(^\circ\mathrm{F}) = \Delta T(\mathrm{R}) = 18^\circ\mathrm{F}$$

1-11 ■ SUMMARY

In this chapter, the basic concepts of thermodynamics are introduced and discussed. *Thermodynamics* is the science that primarily deals with energy. The *first law of thermodynamics* is an expression of the conservation of energy principle, and the *second law* asserts that a process occurs in a certain direction.

A system of fixed mass is called a *closed system,* or *control mass,* and a system that involves mass transfer across its boundaries is called an *open system,* or *control volume*. The mass-dependent properties of a system are called *extensive properties* and the others, *intensive properties*. *Density* is mass per unit volume, and *specific volume* is volume per unit mass.

The sum of all forms of energy of a system is called *total energy* which is considered to consist of internal, kinetic, and potential energies. *Internal energy* represents the molecular energy of a system and may exist in sensible, latent, chemical, and nuclear forms.

A system is said to be in *thermodynamic equilibrium* if it maintains thermal, mechanical, phase, and chemical equilibrium. Any change from one state to another is called a *process*. A process with identical end states is called a *cycle*. During a *quasi-static* or *quasi-equilibrium process,* the system remains practically in equilibrium at all times. The state of a simple, compressible system is completely specified by two independent, intensive properties.

Force per unit area is called *pressure,* and its unit is the pascal. The absolute, gage, and vacuum pressures are related by

$$P_{gage} = P_{abs} - P_{atm} \qquad \text{(kPa)} \tag{1-13}$$

$$P_{vac} = P_{atm} - P_{abs} \qquad \text{(kPa)} \tag{1-14}$$

Small to moderate pressure differences are measured by a *manometer,* and a differential fluid column of height h corresponds to a pressure difference of

$$\Delta P = \rho g h \qquad \text{(kPa)} \tag{1-16}$$

where ρ is the fluid density and g is the local gravitational acceleration. The atmospheric pressure is measured by a *barometer* and is determined from

$$P_{atm} = \rho g h \qquad \text{(kPa)} \tag{1-17}$$

where h is the height of the liquid column above the free surface.

The *zeroth law of thermodynamics* states that two bodies are in thermal equilibrium if both have the same temperature reading even if they are not in contact.

The temperature scales used in the SI and the English system today are the *Celsius scale* and the *Fahrenheit scale,* respectively. The absolute temperature scale in the SI is the *Kelvin scale* which is related to the Celsius scale by

$$T(\text{K}) = T(^\circ\text{C}) + 273.15 \tag{1-18}$$

In the English system, the absolute temperature scale is the *Rankine scale,* which is related to the Fahrenheit scale by

$$T(\text{R}) = T(^\circ\text{F}) + 459.67 \tag{1-19}$$

The magnitudes of each division of 1 K and 1°C are identical, and so are the magnitudes of each division of 1 R and 1°F. That is,

$$\Delta T(\text{K}) = \Delta T(^\circ\text{C}) \tag{1-22}$$

and

$$\Delta T(\text{R}) = \Delta T(^\circ\text{F}) \tag{1-23}$$

REFERENCES AND SUGGESTED READING

1 W. Z. Black and J. G. Hartley, *Thermodynamics,* Harper and Row, New York, 1985.

2 J. B. Jones and G. A. Hawkins, *Engineering Thermodynamics,* 2d ed., Wiley, New York, 1986.

3 D. C. Look, Jr., and H. J. Sauer, Jr., *Engineering Thermodynamics*, PWS Engineering, Boston, 1986.

4 G. D. Meixel, "The State of Metric Transition," *Compressed Air*, vol. 92, no. 2, pp. 25–28, 1987.

5 G. J. Van Wylen and R. E. Sonntag, *Fundamentals of Classical Thermodynamics*, 3d ed., Wiley, New York, 1985.

6 K. Wark, *Thermodynamics*, 5th ed., McGraw-Hill, New York, 1988.

PROBLEMS*

Thermodynamics and Energy

1-1C Why does a bicyclist pick up speed on a downhill road even when he is not pedaling? Does this violate the conservation of energy principle?

1-2C An office worker claims that a cup of cold coffee on his table warmed up to 80°C by picking up energy from the surrounding air which is at 25°C. Is there any truth to his claim? Does this process violate any thermodynamic laws?

1-3C A bicyclist claims to have picked up speed on a level road on a calm day even if she was not pedaling. Is her claim valid? Does it violate any thermodynamic laws?

1-4C What is the difference between the classical and the statistical approaches to thermodynamics?

Mass, Force, and Acceleration

1-5C What is the difference between pound-mass and pound-force?

1-6C What is the net force acting on a car cruising at a constant velocity of 70 km/h (*a*) on a level road and (*b*) on an uphill road?

1-7 What is the force required to accelerate a mass of 15 kg at a rate of 7 m/s^2? *Answer:* 105 N

1-7E What is the force required to accelerate a mass of 30 lbm at a rate of 20 ft/s^2? *Answer:* 18.65 lbf

1-8 A 5-kg plastic tank that has a volume of 0.2 m^3 is filled with liquid water. Assuming the density of water is 1000 kg/m^3, determine the weight of the combined system.

1-9 Determine the mass and the weight of the air contained in a room whose dimensions are 6 m by 6 m by 8 m. Assume the density of the air is 1.16 kg/m^3. *Answer:* 334.1 kg, 3277 N

*Students are encouraged to answer *all* the concept "C" questions.

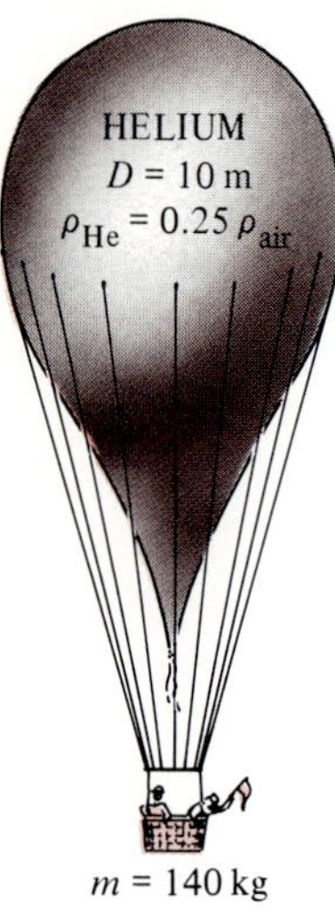

FIGURE P1-11

1-9E Determine the mass and the weight of the air contained in a room whose dimensions are 15 ft by 20 ft by 20 ft. Assume the density of the air is 0.0724 lbm/ft^3.

1-10 At 45° latitude, the gravitational acceleration as a function of elevation z above sea level is given by $g = a - bz$, where $a = 9.807$ m/s^2 and $b = 3.32 \times 10^{-6}$ s^{-2}. Determine the height above sea level where the weight of a subject will decrease by 1 percent. *Answer:* 29,539 m

1-11 Balloons are often filled with helium gas because it weighs only about one-fourth of what air weighs under identical conditions. The buoyancy force which can be expressed as $F_b = \rho_{air} g V_{balloon}$ will push the balloon upward. If the balloon has a diameter of 10 m and carries two people, 70 kg each, determine the acceleration of the balloon when it is first released. Assume the density of air is $\rho = 1.16$ kg/m^3, and neglect the weight of the ropes and the cage. *Answer:* 10.6 m/s^2

1-11E Balloons are often filled with helium gas because it weighs only about one-fourth of what air weighs under identical conditions. The buoyancy force which can be expressed as $F_B = \rho_{air} g V_{balloon}$ will push the balloon upwards. If the balloon has a diameter of 30 ft and carries two people, 140 lbm each, determine the acceleration of the balloon when it is first released. Assume the density of air is $\rho = 0.0724$ lbm/ft^3, and neglect the weight of the ropes and the cage. *Answer:* 29.3 ft/s^2

1-12 Determine the maximum amount of load, in kg, the balloon described in Prob. 1-11 can carry. *Answer:* 455.5 kg

1-13 A 75-kg astronaut who weighs 735 N on earth took his bathroom scale (a spring scale) and a beam scale (compares masses) to the moon where the local gravity is $g = 1.67$ m/s^2. Determine how much he will weigh (*a*) on the spring scale, and (*b*) on the beam scale.
Answer: (*a*) 125.25 N; (*b*) 735.5 N

1-13E A 150-lbm astronaut who weighs 145 lbf on earth took his bathroom scale (a spring scale) and a beam scale (compares masses) to the moon where the local gravity is $g = 5.48$ ft/s^2. Determine how much he will weigh (*a*) on the spring scale, and (*b*) on the beam scale.
Answer: (*a*) 25.5 lbf; (*b*) 150 lbf

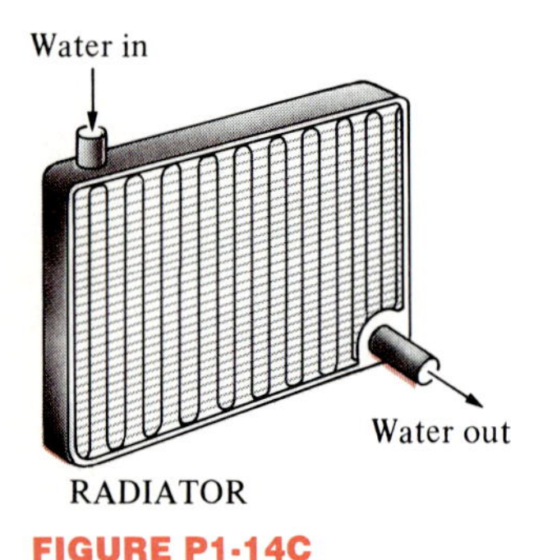

FIGURE P1-14C

Forms of Energy, Systems, State, Properties

1-14C The energy generated in the engine of a car is rejected to the air by the radiator through the circulating water. Should the radiator be analyzed as a closed system or as an open system? Explain.

1-15C A can of soft drink at room temperature is put into the refrigerator so that it will cool. Would you model the can of soft drink as a closed system or as an open system? Explain.

1-16C Can mass cross the boundaries of a closed system? How about energy?

1-17C Portable electric heaters are commonly used to heat small rooms. Explain the energy transformation involved during this heating process.

1-18C Consider the process of heating water on top of an electric range. What are the forms of energy involved during this process? What are the energy transformations that take place?

1-19C A burning candle can be viewed as an energy transformer. What are the energy transformations involved during this process?

1-20C What is the difference between the macroscopic and microscopic forms of energy?

1-21C What is total energy? Identify the different forms of energy that constitute the total energy.

1-22C List the effects that contribute to the internal energy of a system.

1-23C What is the difference between intensive and extensive properties?

1-24C For a system to be in thermodynamic equilibrium, do the temperature and the pressure have to be the same everywhere?

1-25C What is a quasi-equilibrium process? What is its importance in engineering?

1-26C Define the isothermal, isobaric, and isochoric processes.

1-27C What is the state postulate?

1-28C Is the state of the air in an insulated room completely specified by the temperature and the pressure? Explain.

Pressure

1-29C What is the difference between gage pressure and absolute pressure?

1-30 A pressure gage connected to a tank reads 3.5 bars at a location where the barometric reading is 75 cmHg. Determine the absolute pressure in the tank. Take $\rho_{Hg} = 13{,}590\ kg/m^3$. *Answer:* 4.5 bars

1-30E A pressure gage connected to a tank reads 50 psi at a location where the barometric reading is 29.1 inHg. Determine the absolute pressure in the tank. Take $\rho_{Hg} = 848.4\ lbm/ft^3$. *Answer:* 64.29 psia

1-31 A pressure gage connected to a tank reads 600 kPa at a location where the atmospheric pressure is 98 kPa. Determine the absolute pressure in the tank.

1-32 A vacuum gage connected to a tank reads 30 kPa at a location where the barometric reading is 755 mmHg. Determine the absolute pressure in the tank. Take $\rho_{Hg} = 13{,}590\ kg/m^3$. *Answer:* 70.6 kPa

1-32E A vacuum gage connected to a tank reads 5.4 psia at a location where the barometric reading is 28.5 inHg. Determine the absolute pressure in the tank. Take ρ_{Hg} = 848.4 lbm/ft³. *Answer:* 8.6 psia

1-33 The basic barometer can be used to measure the height of a building. If the barometric readings at the top and at the bottom of a building are 730 and 755 mmHg, respectively, determine the height of the building. Assume an average air density of 1.18 kg/m³.

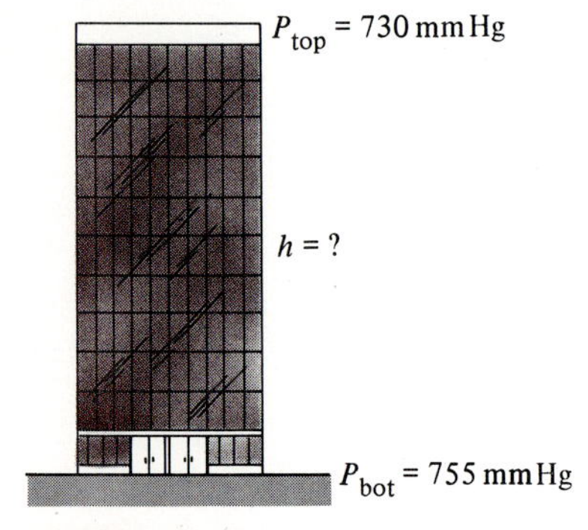

FIGURE P1-33

1-34 The basic barometer can be used as an altitude-measuring device in airplanes. The ground control reports a barometric reading of 753 mmHg while the pilot's reading is 690 mmHg. Estimate the altitude of the plane from ground level if the average air density is 1.20 kg/m³ and g = 9.8 m/s². *Answer:* 714 m

1-34E The basic barometer can be used as an altitude-measuring device in airplanes. The ground control reports a barometric reading of 28.9 inHg while the pilot's reading is 26.0 inHg. Estimate the altitude of the plane from ground level if the average air density is 0.075 lbm/ft³ and g = 31.6 ft/s². *Answer:* 2737 ft

1-35 The barometer of a mountain hiker reads 955 mbars at the beginning of a hiking trip and 820 mbars at the end. Neglecting the effect of altitude on local gravitational acceleration, determine the vertical distance climbed. Assume an average air density of 1.20 kg/m³ and take g = 9.7 m/s². *Answer:* 1160 m

1-35E The barometer of a mountain hiker reads 14.6 psia at the beginning of a hiking trip and 13.9 psia at the end. Neglecting the effect of altitude on local gravitational acceleration, determine the vertical distance she climbed. Assume an average air density of 0.074 lbm/ft³ and take g = 31.8 ft/s². *Answer:* 1378 ft

1-36 Determine the pressure exerted on a diver at 30 m below the free surface of the sea. Assume a barometric pressure of 101 kPa and a specific gravity of 1.03 for the seawater. *Answer:* 404.0 kPa

1-37 Determine the pressure exerted on the surface of a submarine cruising 100 m below the free surface of the sea. Assume that the barometric pressure is 101 kPa and the specific gravity of the seawater is 1.03.

1-37E Determine the pressure exerted on the surface of a submarine cruising 300 ft below the free surface of the sea. Assume that the barometric pressure is 14.7 psia and the specific gravity of the seawater is 1.03.

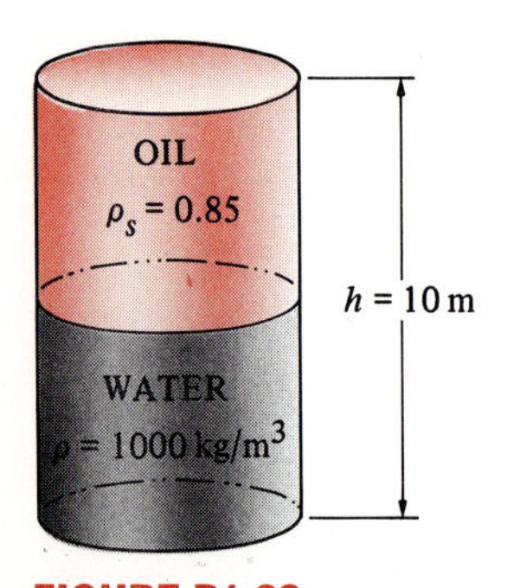

FIGURE P1-38

1-38 The lower half of a 10-m-high cylindrical container is filled with water (ρ = 1000 kg/m³) and the upper half with oil that has a specific gravity of 0.85. Determine the pressure difference between the top and bottom of the cylinder. *Answer:* 90.7 kPa

1-39 A vertical, frictionless piston-cylinder device contains a gas at 350 kPa. The atmospheric pressure outside is 1 bar, and the piston area is 20 cm^2. Determine the mass of the piston. Assume standard gravitational acceleration.

1-40 A gas is contained in a vertical, frictionless piston-cylinder device. The piston has a mass of 5 kg and cross-sectional area of 25 cm^2. A compressed spring above the piston exerts a force of 75 N on the piston. If the atmospheric pressure is 98 kPa, determine the pressure inside the cylinder. *Answer:* 147.6 kPa

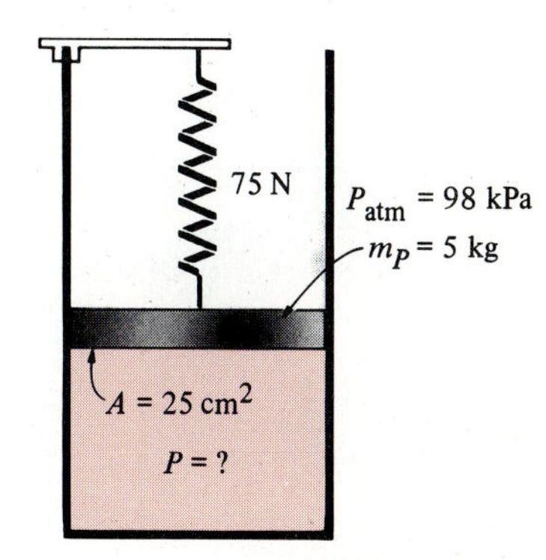

FIGURE P1-40

1-40E A gas is contained in a vertical, frictionless piston-cylinder device. The piston has a mass of 10 lbm and a cross-sectional area of 4 in^2. A compressed spring above the piston exerts a force of 20 lbf on the piston. If the atmospheric pressure is 14.5 psia, determine the pressure inside the cylinder. *Answer:* 22.0 psia

1-41 Both a gage and a manometer are attached to a gas tank to measure its pressure. If the reading on the pressure gage is 80 kPa, determine the distance between the two fluid levels of the manometer if the fluid is (*a*) mercury (ρ = 13,600 kg/m^3) and (*b*) water (ρ = 1000 kg/m^3).

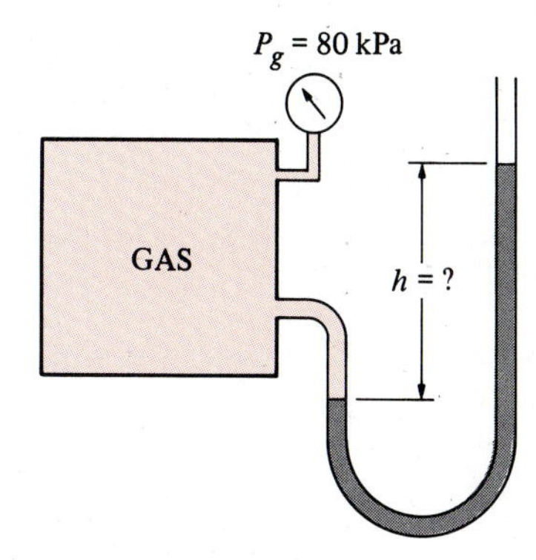

FIGURE P1-41

1-42 A manometer containing oil (ρ = 850 kg/m^3) is attached to a tank filled with air. If the oil-level difference between the two columns is 45 cm and the atmospheric pressure is 98 kPa, determine the absolute pressure of the air in the tank. *Answer:* 101.75 kPa

1-42E A manometer containing oil (ρ = 53 lbm/ft^3) is attached to a tank filled with air. If the oil-level difference between the two columns is 20 in and the atmospheric pressure is 14.6 psia, determine the absolute pressure of the air in the tank. *Answer:* 15.21 psia

1-43 A mercury manometer (ρ = 13,600 kg/m^3) is connected to an air duct to measure the pressure inside. The difference in the manometer levels is 15 mm, and the atmospheric pressure is 100 kPa.

(*a*) Judging from Fig. P1-43, determine if the pressure in the duct is above or below the atmospheric pressure.

(*b*) Determine the absolute pressure in the duct.

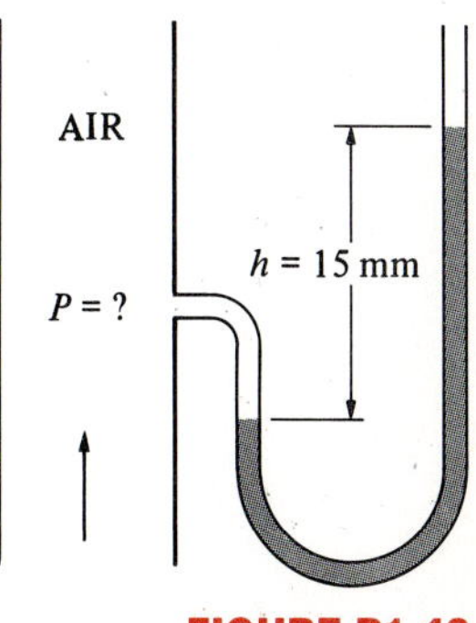

FIGURE P1-43

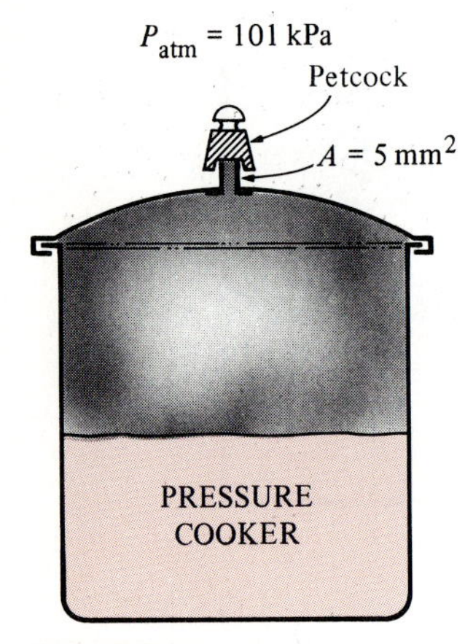

FIGURE P1-44

1-44 A pressure cooker cooks a lot faster than an ordinary pan by maintaining a higher pressure and temperature inside. The lid of a pressure cooker is well sealed, and steam can escape only through an opening in the middle of the lid. A separate piece of certain mass, the petcock, sits on top of this opening and prevents steam from escaping until the pressure force overcomes the weight of the rocker. The periodic escape of the steam in this manner prevents any potentially dangerous pressure buildup and keeps the pressure inside at a constant value.

Determine the mass of the petcock of a pressure cooker whose operation pressure is 100 kPa gage and has an opening cross-sectional area of 5 mm². Assume an atmospheric pressure of 101 kPa, and draw the free-body diagram of the petcock. *Answer:* 51.0 g

Temperature

1-45C What is the zeroth law of thermodynamics?

1-46C What are the ordinary and absolute temperature scales in the SI and the English system?

1-47C Consider an alcohol and a mercury thermometer that read exactly 0°C at the ice point and 100°C at the steam point. The distance between the two points is divided into 100 equal parts in both thermometers. Do you think these thermometers will give exactly the same reading at a temperature of, say, 60°C? Explain.

1-48 The deep body temperature of a healthy person is 37°C. What is it in kelvins? *Answer:* 310 K

1-48E The deep body temperature of a healthy person is 98.6°F. What is it in Rankine? *Answer:* 558.3 R

1-49 Consider a system whose temperature is 22°C. Express this temperature in kelvins.

1-49E Consider a system whose temperature is 22°C. Express this temperature in R, K, and °F.

1-50 The temperature of a system rises by 20°C during a heating process. Express this rise in temperature in kelvins. *Answer:* 20 K

1-50E The temperature of a system rises by 45°F during a heating process. Express this rise in temperature in R, K, and °C.
Answers: 45 R, 25 K, 25°C

1-51 The temperature of a system drops by 10°C during a cooling process. Express this drop in temperature in kelvins.

1-51E The temperature of a system drops by 20°F during a cooling process. Express this drop in temperature in K, R, and °C.

Properties of Pure Substances

CHAPTER

2

In this chapter the concept of pure substances is introduced and the various phases as well as the physics of phase-change processes are discussed. Various property diagrams and P-v-T surfaces of pure substances are illustrated. The use of property tables is explained, and the hypothetical substance "ideal gas" and the ideal-gas equation of state are discussed. The compressibility factor, which accounts for the deviation of real gases from ideal-gas behavior is introduced, and its use is illustrated. Finally, some of the best known equations of state are presented.

2-1 ■ PURE SUBSTANCE

A substance that has a fixed chemical composition throughout is called a **pure substance** Water, nitrogen, helium, and carbon dioxide, for example, are all pure substances.

A pure substance does not have to be of a single chemical element or compound, however. A mixture of various chemical elements or compounds also qualifies as a pure substance as long as the mixture is homogeneous. Air, for example, is a mixture of several gases, but it is often considered to be a pure substance because it has a uniform chemical composition (Fig. 2-1). However, the mixture of oil and water is not a pure substance. Since oil is not soluble in water, it will collect on top of the water, forming two chemically dissimilar regions.

FIGURE 2-1
Nitrogen and gaseous air are pure substances.

A mixture of two or more phases of a pure substance is still a pure substance as long as the chemical composition of all phases is the same (Fig. 2-2). A mixture of ice and liquid water, for example, is a pure substance because both phases have the same chemical composition. A mixture of liquid air and gaseous air, however, is not a pure substance since the composition of liquid air is different from the composition of gaseous air, and thus the mixture is no longer chemically homogeneous. This is due to different components in air having different condensation temperatures at a specified pressure.

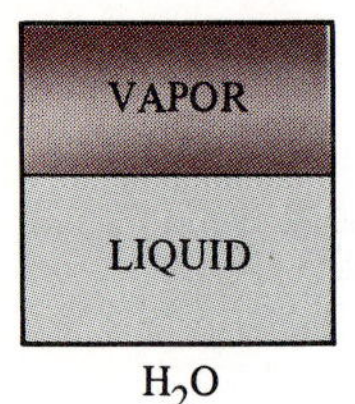

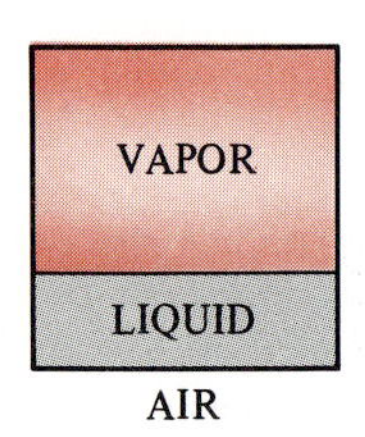

FIGURE 2-2
A mixture of liquid and gaseous water is a pure substance, but a mixture of liquid and gaseous air is not.

2-2 ■ PHASES OF A PURE SUBSTANCE

We all know from experience that substances exist in different phases. At room temperature and pressure, copper is a solid, mercury is a liquid, and nitrogen is a gas. Under different conditions, each may appear in a different phase. Even though there are three principal phases—solid, liquid, and gas—a substance may have several phases within a principal phase, each with a different molecular structure. Carbon, for example, may exist as graphite or diamond in the solid phase. Helium has two liquid phases; iron has three solid phases. Ice may exist at seven different phases at high pressures. A phase is identified as having a distinct molecular arrangement which is homogeneous throughout and separated from the others by easily identifiable boundary surfaces. The two phases of H_2O in iced water represent a good example of this.

When studying phases or phase changes in thermodynamics, one does not need to be concerned with the molecular structure and behavior of different phases. However, it is very helpful to have some understanding of the molecular phenomena involved in each phase, and a brief discussion of phase transformations is given below.

It is often stated that molecular bonds are the strongest in solids and the weakest in gases. One reason is that molecules in solids are closely packed together whereas they are separated by great distances in gases.

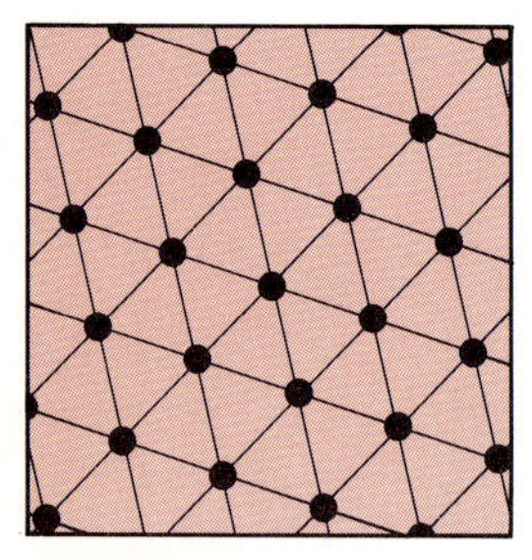

FIGURE 2-3
The molecules in a solid are kept at fixed positions by large intermolecular forces.

The molecules in a **solid** are arranged in a three-dimensional pattern (lattice) which is repeated throughout the solid (Fig. 2-3). Because of the small distances between molecules in a solid, the attractive forces of

FIGURE 2-4

In a solid, the attractive and the repulsive forces between the molecules tend to maintain them at relatively constant distances from each other.

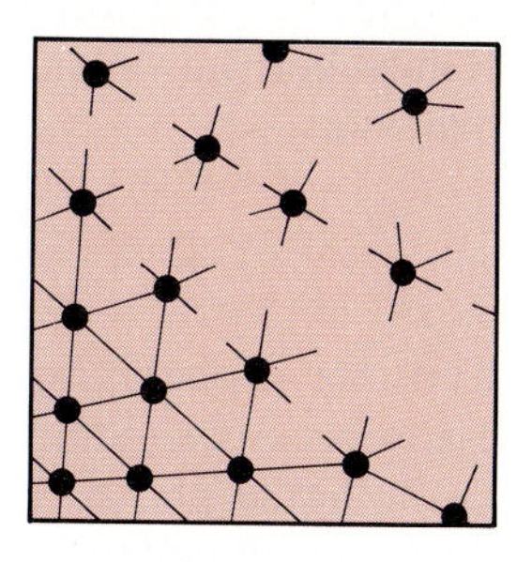

FIGURE 2-5

At high temperatures, molecules overcome the intermolecular forces and break away.

molecules on each other are large and keep the molecules at fixed positions within the solid (Fig. 2-4). Note that the attractive forces between molecules turn to repulsive forces as the distance between the molecules approaches zero, thus preventing the molecules from piling up on top of each other. Even though the molecules in a solid cannot move relative to each other, they continually oscillate about their equilibrium position. The velocity of the molecules during these oscillations depends on the temperature. At sufficiently high temperatures, the velocity (and thus the momentum) of the molecules may reach a point where the intermolecular forces are partially overcome and groups of molecules break away (Fig. 2-5). This is the beginning of the melting process.

The molecular spacing in the **liquid** phase is not much different from that of the solid phase, except the molecules are no longer at fixed positions relative to each other. In a liquid, chunks of molecules float about each other (Fig. 2-6); however, the molecules maintain an orderly structure within each chunk and retain their original positions with respect to one another. The distances between molecules generally experience a slight increase as a solid turns liquid, with water being a rare exception.

In the **gas** phase, the molecules are far apart from each other, and a molecular order is nonexistent. Gas molecules move about at random, continually colliding with each other and the walls of the container they are in (Fig. 2-7). Particularly at low densities, the intermolecular forces are very small, and collisions are the only mode of interaction between the molecules. Molecules in the gas phase are at a considerably higher energy level than they are in the liquid or solid phases. Therefore, the gas must release a large amount of its energy before it can condense or freeze.

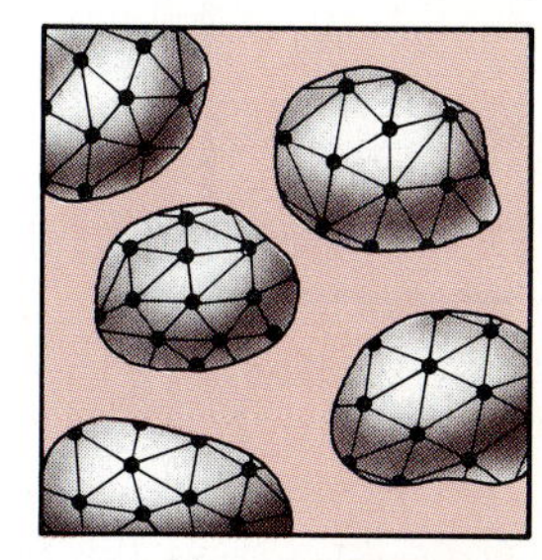

FIGURE 2-6

In the liquid phase, chunks of molecules float about each other.

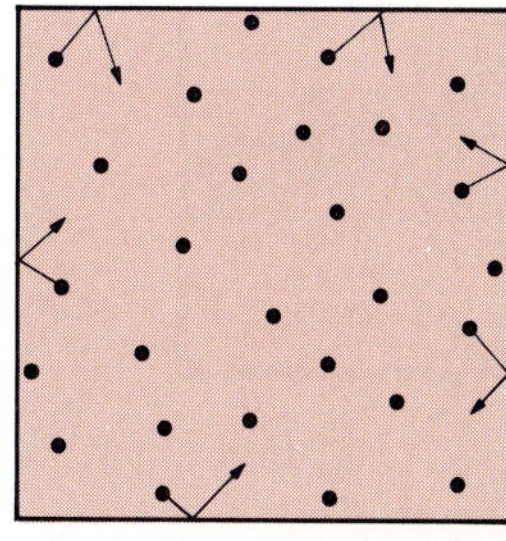

FIGURE 2-7

In the gas phase, the molecules are at the highest energy level and move about at random.

2-3 ■ PHASE-CHANGE PROCESSES OF PURE SUBSTANCES

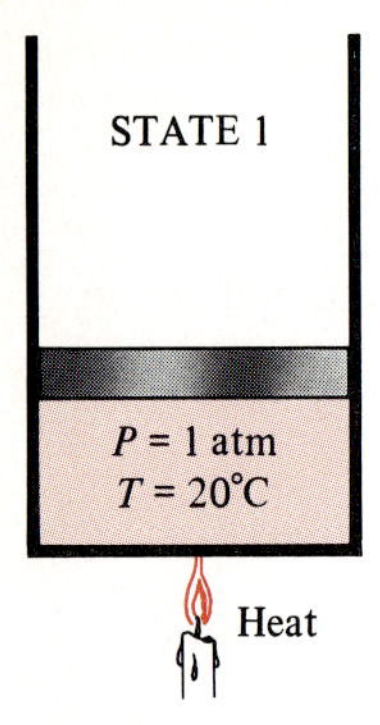

FIGURE 2-8

At 1 atm and 20°C water exists in the liquid phase (*compressed liquid*).

There are many practical situations where two phases of a pure substance coexist in equilibrium. Water exists as a mixture of liquid and vapor in the boiler and the condenser of a steam power plant. The refrigerant turns from liquid to vapor in the freezer of a refrigerator. Even though many homeowners consider the freezing of water in underground pipes as the most important phase-change process, attention in this section is focused on the liquid and vapor phases and the mixture of these two. As a familiar substance, water will be used to demonstrate the basic principles involved. Remember, however, that all pure substances exhibit the same general behavior.

Compressed Liquid and Saturated Liquid

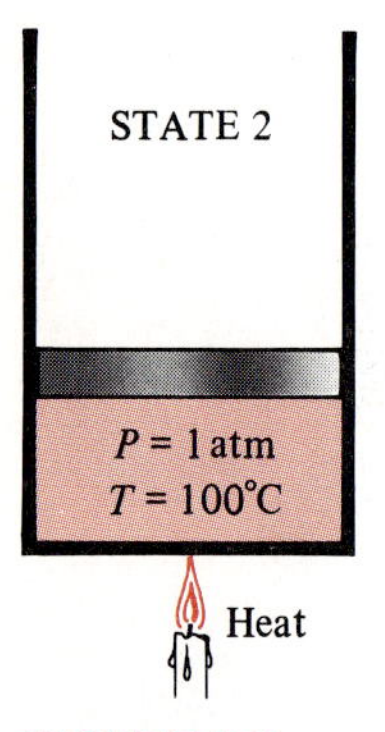

FIGURE 2-9

At 1-atm pressure and 100°C, water exists as a liquid which is ready to vaporize (*saturated liquid*).

Consider a piston-cylinder device containing liquid water at 20°C and 1-atm pressure (state 1, Fig. 2-8). Under these conditions, water exists in the liquid phase, and it is called a **compressed liquid**, or a **subcooled liquid**, meaning that it is *not about to vaporize*. Heat is now added to the water until its temperature rises to, say 40°C. As the temperature rises, the liquid water will expand slightly, and so its specific volume will increase. To accommodate this expansion, the piston will move up slightly. The pressure in the cylinder remains constant at 1 atm during this process since it depends on the outside barometric pressure and the weight of the piston, both of which are constant. Water is still a compressed liquid at this state since it has not started to vaporize.

As more heat is added, the temperature will keep rising until it reaches 100°C (state 2, Fig. 2-9). At this point water is still a liquid, but any heat addition, no matter how small, will cause some of the liquid to vaporize. That is, a phase-change process from liquid to vapor is about to take place. A liquid which is *about to vaporize* is called a **saturated liquid**. Therefore, state 2 is a saturated liquid state.

Saturated Vapor and Superheated Vapor

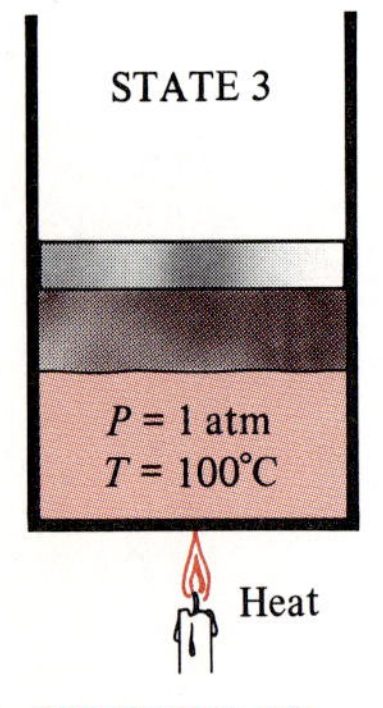

FIGURE 2-10

As more heat is added, part of saturated liquid vaporizes (*saturated liquid–vapor mixture*).

Once boiling starts, the temperature will stop rising until the liquid is completely vaporized. That is, the temperature will remain constant during the entire phase-change process if the pressure is held constant. This can easily be verified by placing a thermometer into boiling water on top of a stove. At sea level (P = 1 atm), the thermometer will always read 100°C if the pan is uncovered or covered with a light lid. During a vaporization (boiling) process, the only change we will observe is a large increase in the volume and a steady decline in the liquid level as a result of more liquid turning to vapor.

Midway about the vaporization line (state 3, Fig. 2-10), the cylinder contains equal amounts of liquid and vapor. As we continue adding heat, the vaporization process will continue until the last drop of liquid is vaporized (state 4, Fig. 2-11). At this point, the entire cylinder is filled with vapor which is on the borderline of the liquid phase. Any heat loss from this vapor, no matter how small, will cause some of the vapor to condense (phase change from vapor to liquid). A vapor which is *about to condense* is called a **saturated vapor**. Therefore, state 4 is a saturated vapor state. A substance at states between 2 and 4 is often referred to as a **saturated liquid–vapor mixture** since the *liquid and vapor phases coexist in equilibrium* at these states.

Once the phase-change process is completed, we are back to a single-phase region again (this time vapor), and further transfer of heat will result in an increase in both the temperature and the specific volume (Fig. 2-12). At state 5, the temperature of the vapor is, let us say, 300°C; and if we transfer some heat from the vapor, the temperature may drop some but no condensation will take place as long as the temperature remains above 100°C (for $P = 1$ atm). A vapor which is *not about to condense* (i.e., not a saturated vapor) is called a **superheated vapor**. Therefore, water at state 5 is a superheated vapor. The constant-pressure phase-change process described above is illustrated on a *T-v* diagram in Fig. 2-13.

If the entire process described above is reversed by cooling the water while maintaining the pressure at the same value, the water will go back to state 1, retracing the same path, and in so doing, the amount of heat released will exactly match the amount of heat added during the heating process.

In our daily life, water implies liquid water and steam implies water vapor. In thermodynamics, however, both water and steam usually mean only one thing: H_2O.

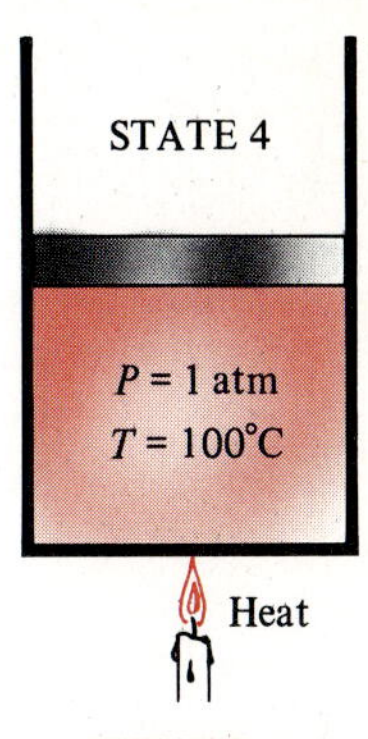

FIGURE 2-11

At 1-atm pressure, the temperature remains constant at 100°C until the last drop of liquid is vaporizied (*saturated vapor*).

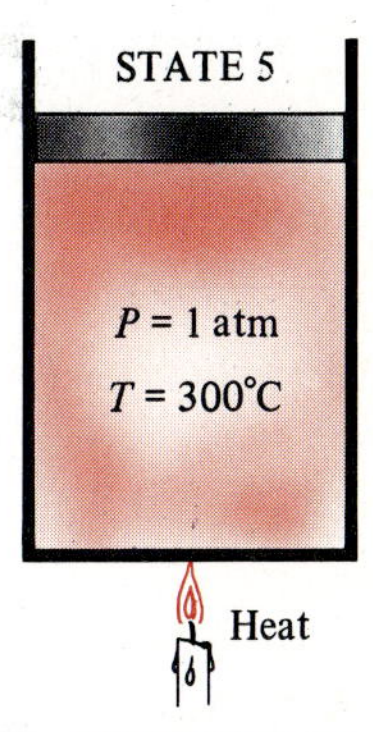

FIGURE 2-12

As more heat is added, the temperature of the vapor starts to rise (*superheated vapor*).

Saturation Temperature and Saturation Pressure

It probably came as no surprise to you that the water started "boiling" at 100°C. Strictly speaking, the statement "water boils at 100°C" is incorrect. The correct statement is "water boils at 100°C at 1-atm pressure."

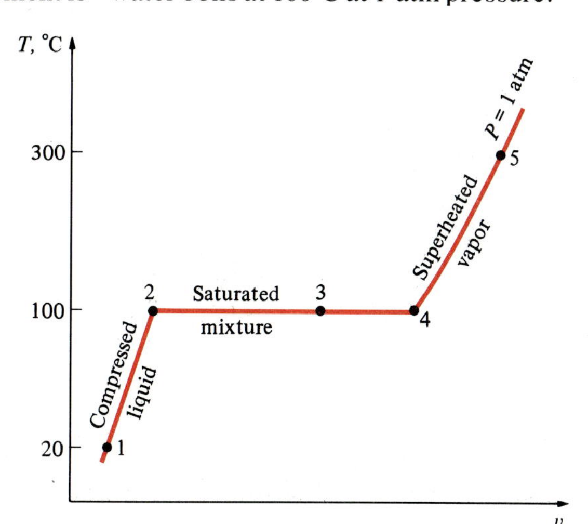

FIGURE 2-13

T-v diagram for the heating process of water at constant pressure.

FIGURE 2-14

At a fixed pressure, a pure substance boils at a fixed temperature (*saturation temperature*).

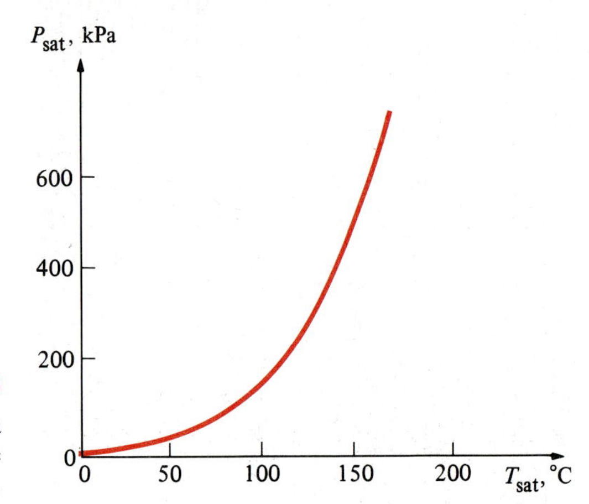

FIGURE 2-15

The liquid-vapor saturation curve of a pure substance (numerical values are for water).

The only reason the water started boiling at 100°C was because we held the pressure constant at 1 atm (101.35 kPa). If the pressure inside the cylinder were raised to 500 kPa by adding weights on top of the piston, the water would start boiling at 151.9°C. That is, *the temperature at which water starts boiling depends on the pressure; therefore, if the pressure is fixed, so is the boiling temperature* (Fig. 2-14).

At a given pressure, the temperature at which a pure substance starts boiling is called the **saturation temperature** T_{sat}. Likewise, at a given temperature, the pressure at which a pure substance starts boiling is called the **saturation pressure** P_{sat}. At a pressure of 101.35 kPa, T_{sat} is 100°C. Conversely, at a temperature of 100°C, P_{sat} is 101.35 kPa.

During a phase-change process, pressure and temperature are obviously dependent properties, and there is a definite relation between them, that is, $T_{sat} = f(P_{sat})$. A plot of T_{sat} vs. P_{sat}, such as the one given for water in Fig. 2-15, is called a **liquid-vapor saturation curve**. A curve of this kind is characteristic of all pure substances.

It is clear from Fig. 2-15 that T_{sat} increases with P_{sat}. Thus a substance at higher pressures will boil at higher temperatures. In the kitchen, higher boiling temperatures mean shorter cooking times and energy savings (Fig. 2-16). A beef stew, for example, may take 1 to 2 h to cook in a regular pan which operates at 1-atm pressure, but only 20 to 30 min in a pressure cooker operating at 2-atm absolute pressure (corresponding boiling temperature: 120°C).

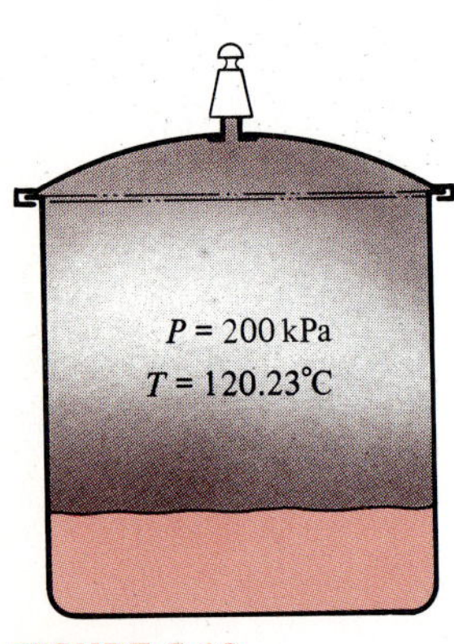

FIGURE 2-16

A pressure cooker maintains a higher pressure, thus a higher temperature.

2-4 ■ PROPERTY DIAGRAMS FOR PHASE-CHANGE PROCESSES

The variations of properties during phase-change processes are best studied and understood with the help of property diagrams. Below we develop and discuss the *T-v*, *P-v*, and *P-T* diagrams for pure substances.

1 The *T-v* Diagram

The phase-change process of water at 1-atm pressure was described in detail in the last section and plotted on a *T-v* diagram in Fig. 2-13. Now we repeat this process at different pressures to develop the *T-v* diagram for water.

Let us add weights on top of the piston until the pressure inside the cylinder reaches 1 MPa. At this high pressure, water will have a somewhat smaller specific volume than it did at 1-atm pressure. As heat is added to the water at this new pressure, the process will follow a path which looks very much like the process path at 1-atm pressure, as shown in Fig. 2-17, but there are some noticeable differences. First, water will start boiling at a much higher temperature (179.9°C) at this pressure. Second, the specific volume of the saturated liquid is larger, and the specific volume of the saturated vapor is smaller than the corresponding values at 1-atm pressure. That is, the horizontal line that connects the saturated liquid and saturated vapor states is much shorter.

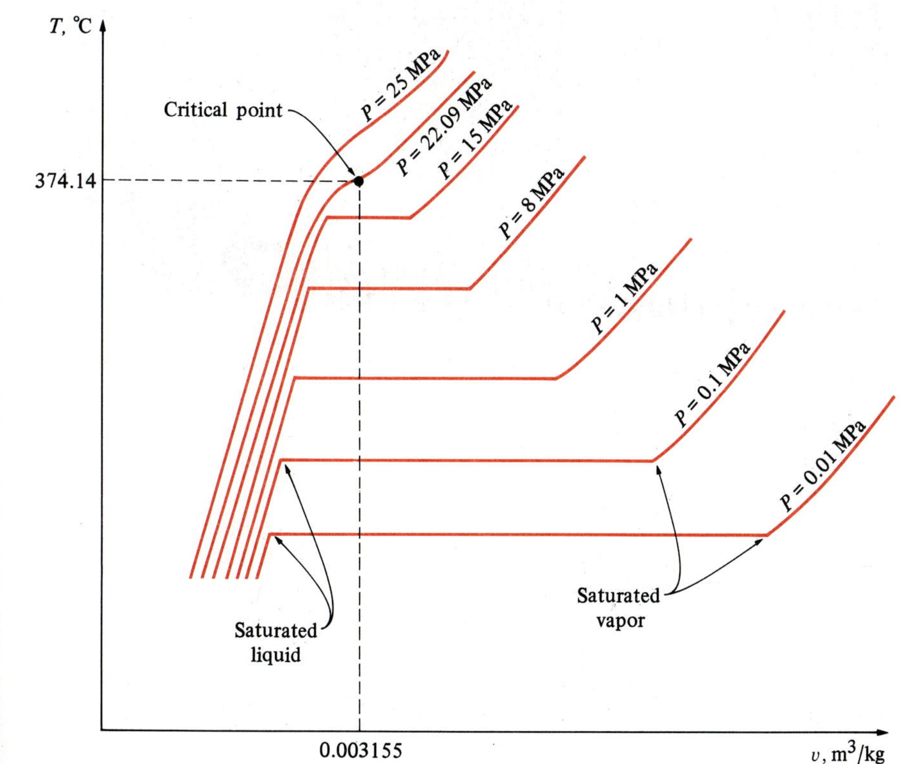

FIGURE 2-17

T-v diagram of constant-pressure phase-change processes of a pure substance at various pressures (numerical values are for water).

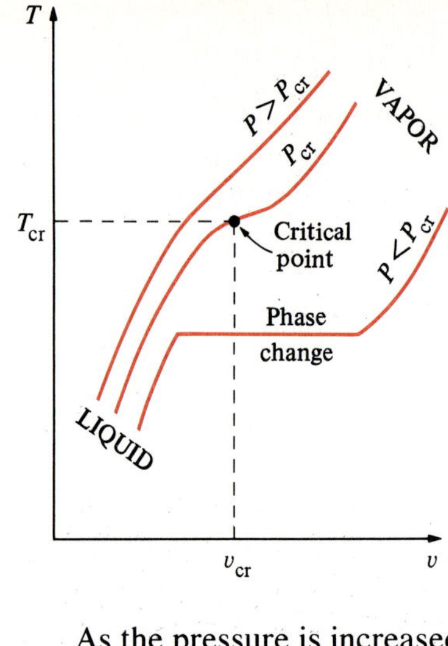

FIGURE 2-18
At supercritical pressures ($P > P_{cr}$), there is no distinct phase-change (boiling) process.

As the pressure is increased further, this saturation line will continue to get shorter, as shown in Fig. 2-17, and it will become a point when the pressure reaches 22.09 MPa for the case of water. This point is called the **critical point**, and it may be defined as *the point at which the saturated liquid and saturated vapor states are identical.*

The temperature, pressure, and specific volume of a substance at the critical point are called, respectively, the *critical temperature* T_{cr}, *critical pressure* P_{cr}, and *critical specific volume* v_{cr}. The critical-point properties of water are P_{cr} = 22.09 MPa, T_{cr} = 374.14°C, and v_{cr} = 0.003155 m^3/kg. For helium they are 0.23 MPa, −267.85°C, and 0.01444 m^3/kg. The critical properties for various substances are given in Table A-1 in the Appendix.

At pressures above the critical pressure, there will not be a distinct phase-change process (Fig. 2-18). Instead, the specific volume of the substance will continually increase, and at all times there will be only one phase present. Eventually, it will resemble a vapor, but we can never tell when the change has occurred. Above the critical state there is no line that separates the compressed liquid region and the superheated vapor region. However, it is customary to refer to the substance as superheated vapor at temperatures above the critical temperature and as compressed liquid at temperatures below the critical temperature (Fig. 2-19).

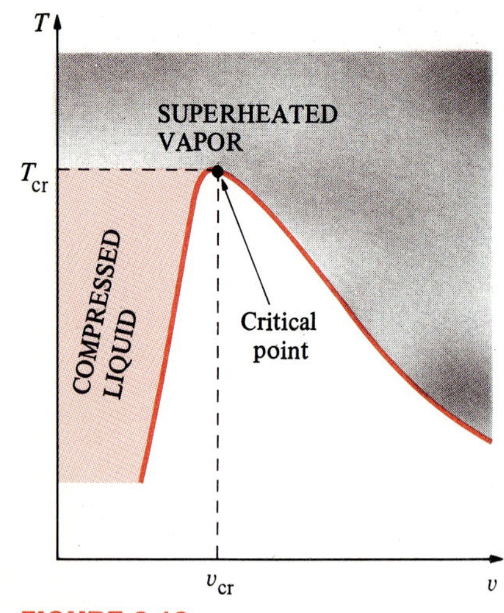

FIGURE 2-19
Above the critical temperature, a substance is considered to be a superheated vapor.

The saturated liquid states in Fig. 2-17 can be connected by a line which is called the **saturated liquid line**, and saturated vapor states in the same figure can be connected by another line which is called the **saturated vapor line**. These two lines meet each other at the critical point, forming a dome as shown in Fig. 2-20. All the compressed liquid states are located in the region to the left of the saturated liquid line, and it is called the **compressed liquid region**. All the superheated vapor states are located to the right of the saturated vapor line which is called the **superheated vapor region**. In these two regions, the substance exists in a single phase, a liquid or a vapor. All the states that involve both phases in equilibrium are located under the dome which is called the **saturated liquid–vapor mixture region**, or the **wet region**.

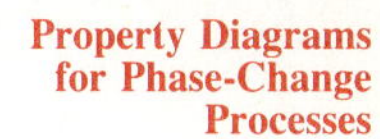

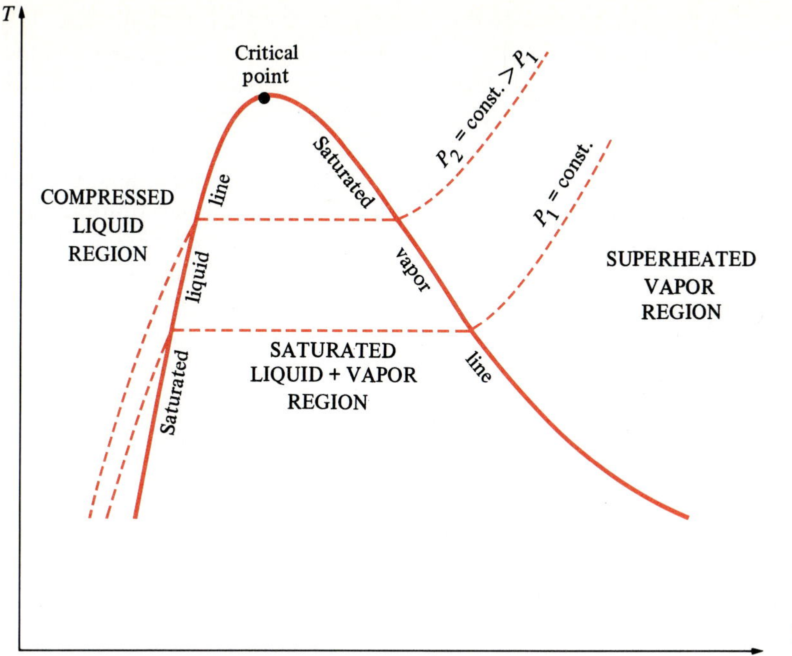

FIGURE 2-20

T-v diagram of a pure substance.

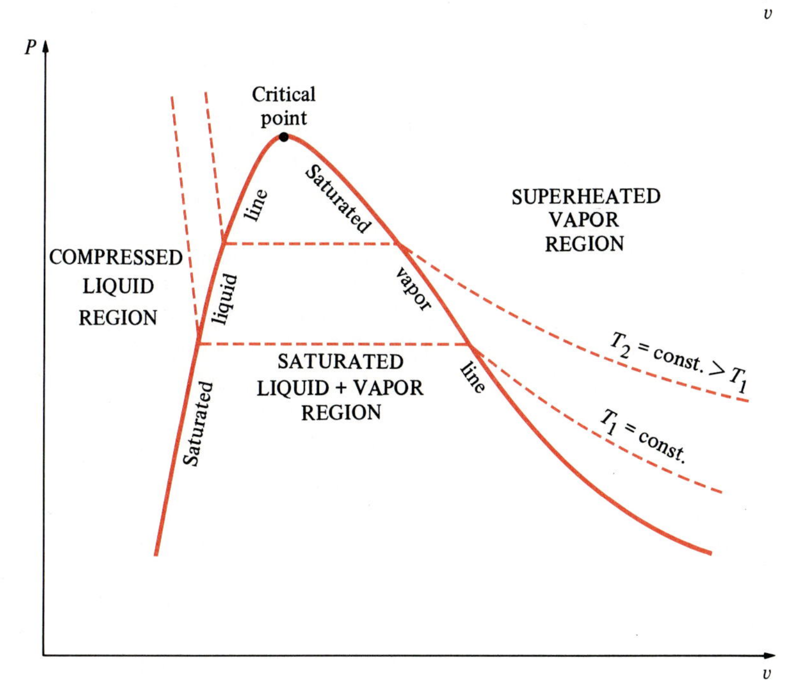

FIGURE 2-21

P-v diagram of a pure substance.

2 The *P-v* Diagram

The general shape of the *P-v* diagram of a pure substance is very much like the *T-v* diagram, but the T = constant lines on this diagram have a downward trend, as shown in Fig. 2-21.

Consider again a piston-cylinder device that contains liquid water at 1 MPa and 150°C. Water at this state exists as a compressed liquid. Now the weights on top of the piston are removed one by one so that the pressure inside the cylinder decreases gradually (Fig. 2-22). The water is allowed to exchange heat with the surroundings, so that its temperature remains constant. As the pressure decreases, the volume of the water will increase slightly. When the pressure reaches the saturation-pressure value at the specified temperature (0.4758 MPa), the water will start to boil. During this vaporization process, both the temperature and the pressure remain constant, but the specific volume increases. Once the last drop of liquid is vaporized, further reduction in pressure results in a further increase in specific volume. Notice that during the phase-change process, we did not remove any weights. Doing so would cause the pressure and therefore the temperature to drop [since $T_{sat} = f(P_{sat})$], and the process would no longer be isothermal.

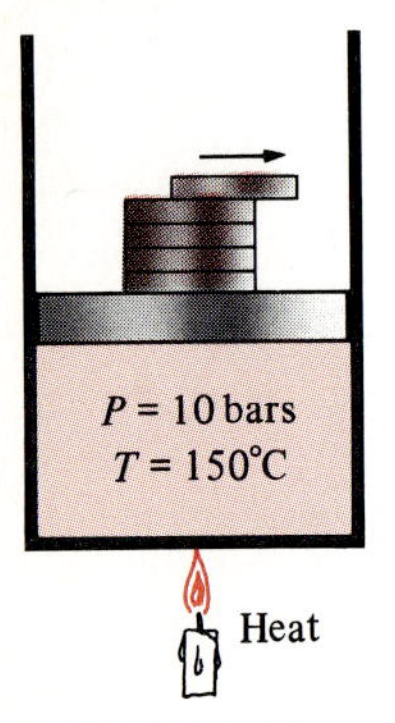

FIGURE 2-22
The pressure in a piston-cylinder device can be reduced by reducing the weight of the piston.

If the process is repeated for other temperatures, similar paths will be obtained for the phase-change processes. Connecting the saturated liquid and the saturated vapor states by a curve, we obtain the *P-v* diagram of a pure substance, as shown in Fig. 2-21.

Extending the Diagrams to Include the Solid Phase

The two equilibrium diagrams developed so far represent the equilibrium states involving the liquid and the vapor phases only. But these diagrams can easily be extended to include the solid phase as well as the solid-liquid and the solid-vapor saturation regions. The basic principles discussed in conjunction with the liquid-vapor phase-change process equally apply to the solid-liquid and solid-vapor phase-change processes (Fig. 2-23). Most substances contract during a solidification (i.e., freezing) process. Others, like water, expand as they freeze. The *P-v* diagrams for both groups of substances are given in Figs. 2-24 and 2-25. These two diagrams differ only in the solid-liquid saturation region. The *T-v* diagrams look very much like the *P-v* diagrams, especially for substances that contract on freezing.

FIGURE 2-23
Iced water is a saturated solid-liquid mixture.

The fact that water expands upon freezing has vital consequences in nature. If water contracted on freezing as most other substances do, the ice formed would be heavier than the liquid water and it would settle to the bottom of rivers, lakes, or oceans instead of floating at the top. The sun's rays would never reach these ice layers, and the bottoms of many rivers, lakes, or oceans would be covered with ice year round, seriously disrupting marine life.

We are all familiar with two phases being in equilibrium, but under some conditions all three phases of a pure substance coexist in equilibrium (Fig. 2-26). On *P-v* or *T-v* diagrams, these triple-phase states form a line called the **triple line**. The states on the triple line of a substance have the same pressure and temperature but different specific volumes. The triple line appears as a point on the *P-T* diagrams and, therefore, is often called the **triple point**. For water, the triple-point temperature and pressure val-

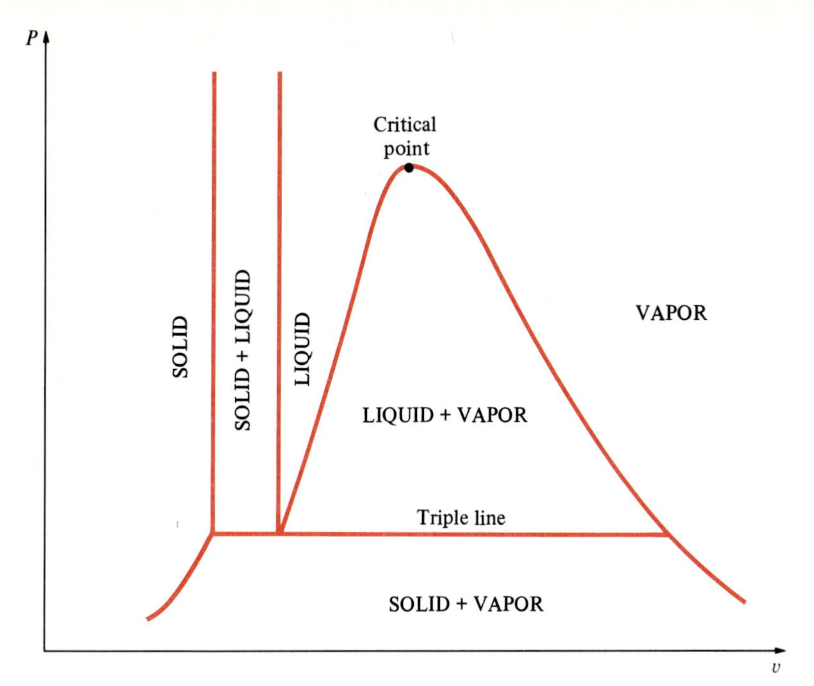

FIGURE 2-24

P-v diagram of a substance that contracts on freezing.

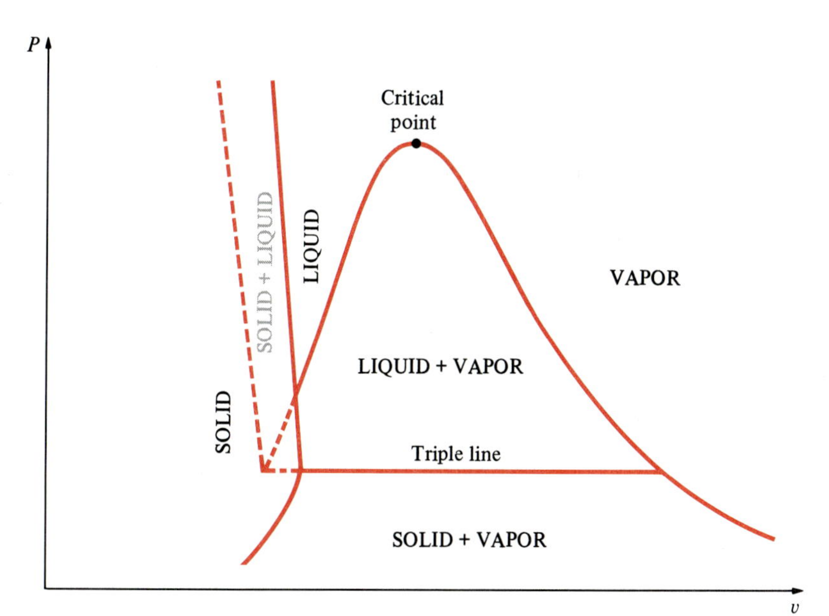

FIGURE 2-25

P-v diagram of a substance that expands on freezing (such as water).

ues are 0.01°C and 0.6113 kPa, respectively. That is, all three phases of water will exist in equilibrium only if the temperature and pressure have precisely these values. No substance can exist in the liquid phase at pressures below the triple-point pressure. The same can be said for temperature for substances that contract on freezing.

There are two ways a substance can pass from the solid to vapor

FIGURE 2-26

At triple pressure and temperature, a substance exists in three phases in equilibrium.

phase: either it melts first into a liquid and subsequently evaporates, or it evaporates directly without melting first. The latter occurs at pressures below the triple-point pressure, since a pure substance cannot exist in the liquid phase at those pressures (Fig. 2-27). Passing from the solid phase directly into the vapor phase is called **sublimation**. For substances that have a triple-point pressure above the atmospheric pressure such as solid CO_2 (dry ice), sublimation is the only way to change from the solid to vapor phase at atmospheric conditions.

FIGURE 2-27
At low pressures (below triple-point value), solids evaporate without melting first (*sublimation*).

3 The *P-T* Diagram

Figure 2-28 shows the *P-T* diagram of a pure substance. This diagram is often called the **phase diagram** since all three phases are separated from each other by three lines. The sublimation line separates the solid and the vapor regions, the vaporization line separates the liquid and vapor regions, and the melting (or fusion) line separates the solid and liquid regions. These three lines meet at the triple point, where all three phases coexist in equilibrium. The vaporization line ends at the critical point because no distinction can be made between liquid and vapor phases above the critical point. Substances that expand and contract on freezing differ only in the melting line on the *P-T* diagram.

2-5 ■ THE *P-v-T* SURFACE

In Chap. 1 we indicated that the state of a simple compressible substance is fixed by any two independent, intensive properties. Once the two appropriate properties are fixed, all the other properties become dependent

FIGURE 2-28
P-T diagram of pure substances.

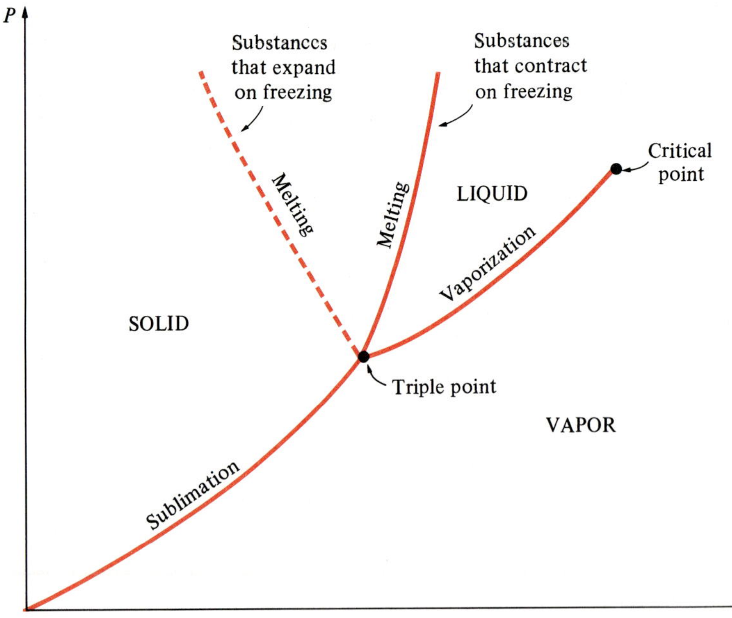

properties. Remembering that any equation with two independent variables in the form $z = z(x, y)$ represents a surface in space, we can represent the P-v-T behavior of a substance as a surface in space, as shown in Figs. 2-29 and 2-30. Here T and v may be viewed as the independent variables (the base) and P as the dependent variable (the height).

All the points on the surface represent equilibrium states. All states along the path of a quasi-equilibrium process lie on the P-v-T surface since such a process must pass through equilibrium states. The single-phase

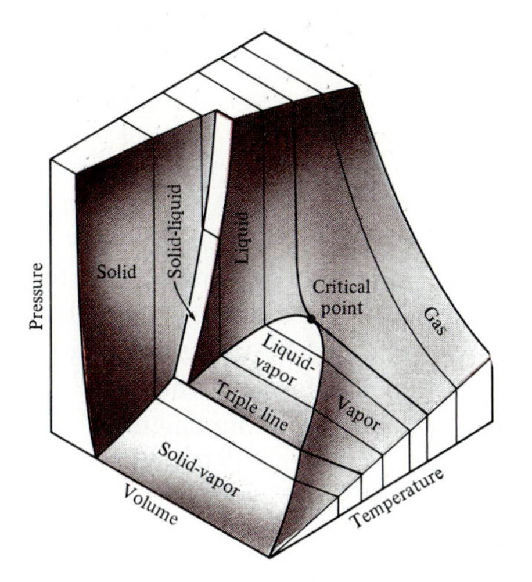

FIGURE 2-29
P-v-T surface of a substance that *contracts* on freezing.

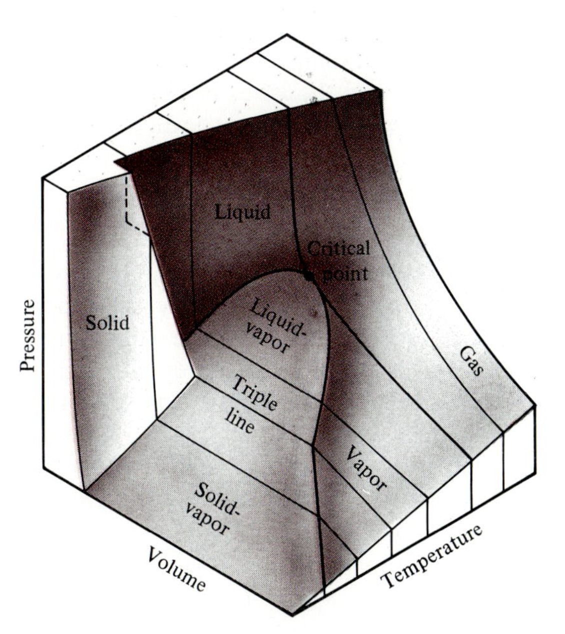

FIGURE 2-30
P-v-T surface of a substance that *expands* on freezing (like water).

regions appear as curved surfaces on the P-v-T surface, and the two-phase regions as surfaces perpendicular to the P-T plane. This is expected since the projections of two-phase regions on the P-T plane are lines.

All the two-dimensional diagrams we have discussed so far are merely projections of this three-dimensional surface onto the appropriate planes. A P-v diagram is just a projection of the P-v-T surface on the P-v plane, and a T-v diagram is nothing more than the bird's-eye view of this surface. The P-v-T surfaces present a great deal of information at once, but in a thermodynamic analysis it is more convenient to work with two-dimensional diagrams, such as the P-v and T-v diagrams.

2-6 ■ PROPERTY TABLES

For most substances, the relationships among thermodynamic properties are too complex to be expressed by simple equations. Therefore, properties are frequently presented in the form of tables. Some thermodynamic properties can be measured easily, but others cannot be measured directly and are calculated by using the relations that relate them to measurable properties. The results of these measurements and calculations are presented in tables in a convenient format. In the following discussion, the steam tables will be used to demonstrate the use of thermodynamic property tables. Property tables of other substances are used in the same manner.

For each substance, the thermodynamic properties are listed in more than one table. In fact, a separate table is prepared for each region of interest such as the superheated vapor, compressed liquid, and saturated (mixture) regions. Property tables are given in the Appendix in both SI and English units. The tables in English units carry the same number as the corresponding tables in SI, followed by an identifier E. Tables A-6 and A-6E, for example, list properties of superheated water vapor, the former in SI and the latter in English units. Before we get into the discussion of property tables, we will define a new property called *enthalpy*.

Enthalpy—A Combination Property

A person looking at the tables carefully will notice two new properties: enthalpy h and entropy s. Entropy is a property associated with the second law of thermodynamics, and we will not use it until it is properly defined in Chap. 6. However, it is appropriate to introduce enthalpy at this point.

In the analysis of certain types of processes, particularly in power generation and refrigeration (Fig. 2-31), we frequently encounter the combination of properties $U + PV$. For the sake of simplicity and convenience, this combination is defined as a new property, **enthalpy**, and given the symbol H:

$$H = U + PV \qquad \text{(kJ)} \tag{2-1}$$

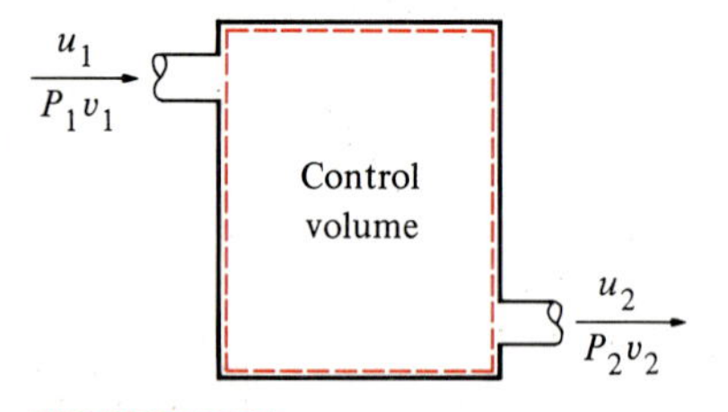

FIGURE 2-31
The combination $u + Pv$ is frequently encountered in the analysis of control volumes.

or, per unit mass,

$$h = u + Pv \qquad \text{(kJ/kg)} \tag{2-2}$$

Both the total enthalpy H and specific enthalpy h are simply referred to as enthalpy since the context will clarify which one is meant. Notice that the equations given above are dimensionally homogeneous. That is, the unit of the pressure-volume product may differ from the unit of the internal energy by only a factor (Fig. 2-32). For example, it can be easily shown that 1 kPa·m³ = 1 kJ. In some tables encountered in practice, the internal energy u is frequently not listed, but it can always be determined from $u = h - Pv$.

It should be mentioned that some properties are assigned a value of zero at some arbitrarily chosen reference point. Therefore, it is possible that different tables list different values for these properties at the same state because of a different reference state chosen. However, in thermodynamics we are concerned with only the *changes* in properties, and not the property values themselves. Therefore, the reference point chosen is of no consequence in calculations as long as we use values from a single set of tables. Some properties may even have negative values as a result of the reference point chosen.

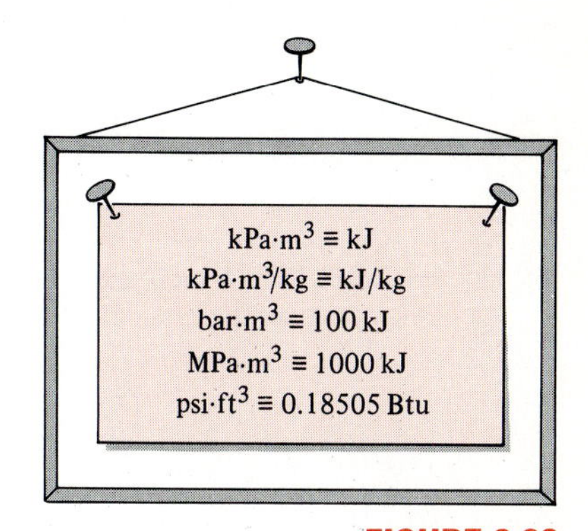

FIGURE 2-32
The product *pressure* × *volume* has energy units.

1a Saturated Liquid and Saturated Vapor States

The properties of saturated liquid and saturated vapor for water are listed in Tables A-4 and A-5. Both tables give the same information. The only difference is that in Table A-4 properties are listed under temperature and in Table A-5 under pressure. Therefore, it is more convenient to use Table A-4 when temperature is given and Table A-5 when pressure is given. The use of Table A-4 is illustrated in Fig. 2-33.

The subscript f is used to denote properties of a saturated liquid, and

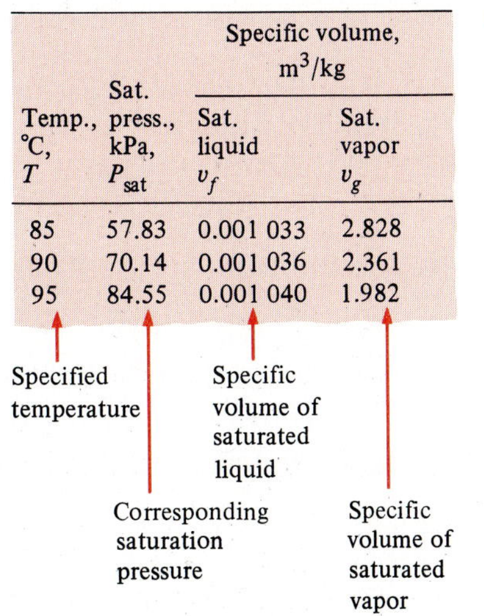

		Specific volume, m³/kg	
Temp., °C, T	Sat. press., kPa, P_{sat}	Sat. liquid v_f	Sat. vapor v_g
85	57.83	0.001 033	2.828
90	70.14	0.001 036	2.361
95	84.55	0.001 040	1.982

FIGURE 2-33
A partial list of Table A-4.

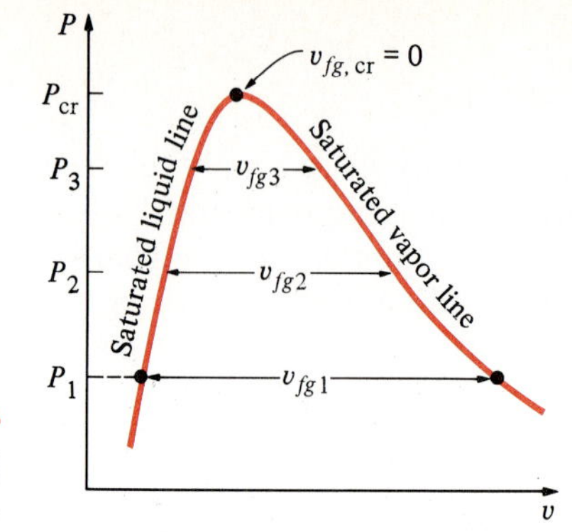

FIGURE 2-34

v_{fg} decreases as P or T increases, and becomes zero at the critical point.

the subscript g to denote the properties of saturated vapor. These symbols are commonly used in thermodynamics and originated from German. Another subscript commonly used is fg which denotes the difference between the saturated vapor and saturated liquid values of the same property (Fig. 2-34). For example,

$$v_f = \text{specific volume of saturated liquid}$$
$$v_g = \text{specific volume of saturated vapor}$$
$$v_{fg} = \text{difference between } v_g \text{ and } v_f \text{ (that is, } v_{fg} = v_g - v_f\text{)}$$

The quantity h_{fg} is called the **enthalpy of vaporization** (or latent heat of vaporization). It represents the amount of energy needed to vaporize a unit mass of saturated liquid at a given temperature or pressure. It decreases as the temperature or pressure increases, and it becomes zero at the critical point.

EXAMPLE 2-1

A rigid tank contains 50 kg of saturated liquid water at 90°C. Determine the pressure in the tank and the volume of the tank.

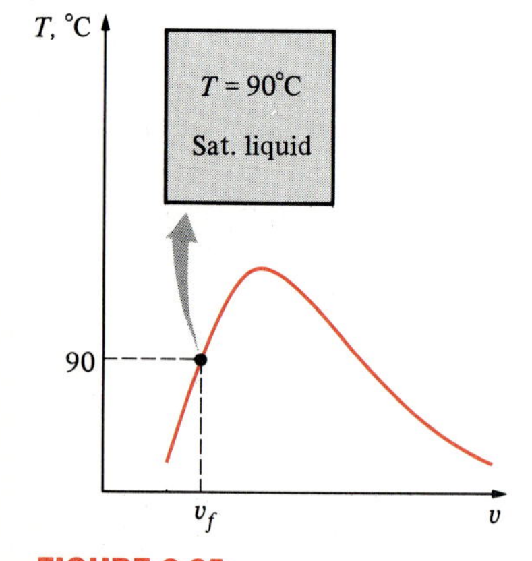

FIGURE 2-35

Schematic and T-v diagram for Example 2-1.

Solution The state of the saturated liquid water is shown on a T-v diagram in Fig. 2-35. Since saturation conditions exist in the tank, the pressure must be the saturation pressure at 90°C:

$$P = P_{\text{sat @ 90°C}} = \mathbf{70.14\ kPa} \qquad \text{(Table A-4)}$$

The specific volume of the saturated liquid at 90°C is

$$v = v_{f\text{ @ 90°C}} = 0.001036\ \text{m}^3/\text{kg} \qquad \text{(Table A-4)}$$

Then the total volume of the tank is determined from

$$V = mv = (50\ \text{kg})(0.001036\ \text{m}^3/\text{kg}) = \mathbf{0.0518\ m^3}$$

EXAMPLE 2-2

A piston-cylinder device contains 2 ft³ of saturated water vapor at 50-psia pressure. Determine the temperature of the vapor and the mass of the vapor inside the cylinder.

Solution The state of the saturated water vapor is shown on a P-v diagram in Fig. 2-36. Since the cylinder contains saturated vapor at 50 psia, the temperature inside must be the saturation temperature at this pressure:

$$T = T_{\text{sat @ 50 psia}} = 281.03°\text{F} \qquad \text{(Table A-5E)}$$

The specific volume of the saturated vapor at 50 psia is

$$v = v_{g\text{ @ 50 psia}} = 8.518 \text{ ft}^3/\text{lbm} \qquad \text{(Table A-5E)}$$

Then the mass of water vapor inside the cylinder becomes

$$m = \frac{V}{v} = \frac{2 \text{ ft}^3}{8.518 \text{ ft}^3/\text{lbm}} = 0.235 \text{ lbm}$$

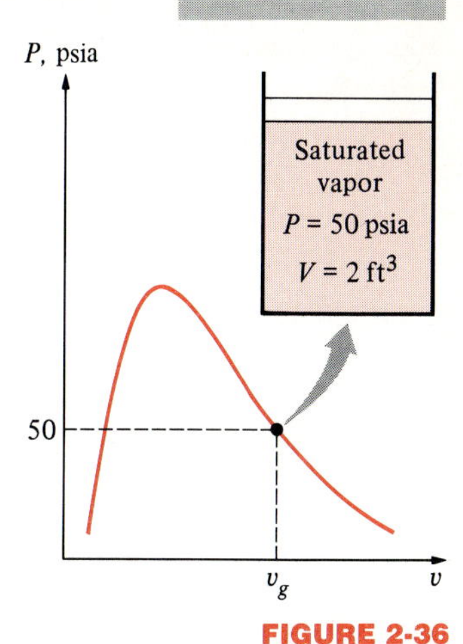

FIGURE 2-36

Schematic and P-v diagram for Example 2-2.

EXAMPLE 2-3

A mass of 200 g of saturated liquid water is completely vaporized at a constant pressure of 100 kPa. Determine (*a*) the volume change and (*b*) the amount of energy added to the water.

Solution (*a*) The process described is illustrated on a P-v diagram in Fig. 2-37. The volume change per unit mass during a vaporization process is v_{fg}, which is the difference between v_g and v_f. Reading these values from Table A-5 at 100 kPa and substituting yield

$$v_{fg} = v_g - v_f = (1.6940 - 0.001043) \text{ m}^3/\text{kg} = 1.6930 \text{ m}^3/\text{kg}$$

Thus, $\Delta V = mv_{fg} = (0.2 \text{ kg})(1.6930 \text{ m}^3/\text{kg}) = 0.3386 \text{ m}^3$

Note that we have considered the first four decimal digits of v_f and disregarded the rest. This is because v_g has significant numbers to the first four decimal places only, and we do not know the numbers in the other decimal places. Taking v_f as it is would mean that we are assuming $v_g = 1.694000$, which is not necessarily the case. It could very well be that $v_g = 1.694038$ since this number, too, would truncate to 1.6940. All the digits in our result (1.6930) are significant. But if we used v_f as it is, we would obtain $v_{fg} = 1.692957$, which falsely implies that our result is accurate to the sixth decimal place.

(*b*) The amount of energy needed to vaporize the unit mass of a substance at a given pressure is the enthalpy of vaporization at that pressure which, at 100 kPa, is $h_{fg} = 2258.0$ kJ/kg. Thus the amount of energy added is

$$mh_{fg} = (0.2 \text{ kg})(2258 \text{ kJ/kg}) = 451.6 \text{ kJ}$$

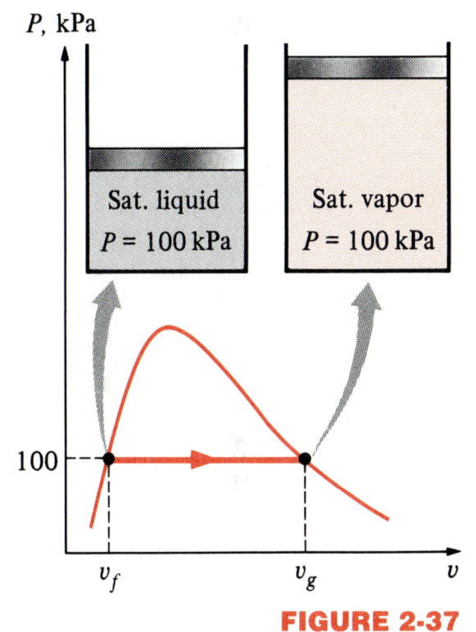

FIGURE 2-37

Schematic and P-v diagram for Example 2-3.

1b Saturated Liquid–Vapor Mixture

During a vaporization process, a substance exists as part liquid and part vapor. That is, it is a mixture of saturated liquid and saturated vapor (Fig. 2-38). To analyze this mixture properly, we need to know the proportions of the liquid and vapor phases in the mixture. This is done by defining a new property called the **quality** x as the ratio of the mass of vapor to the total mass of the mixture:

$$x = \frac{m_{\text{vapor}}}{m_{\text{total}}} \qquad (2\text{-}3)$$

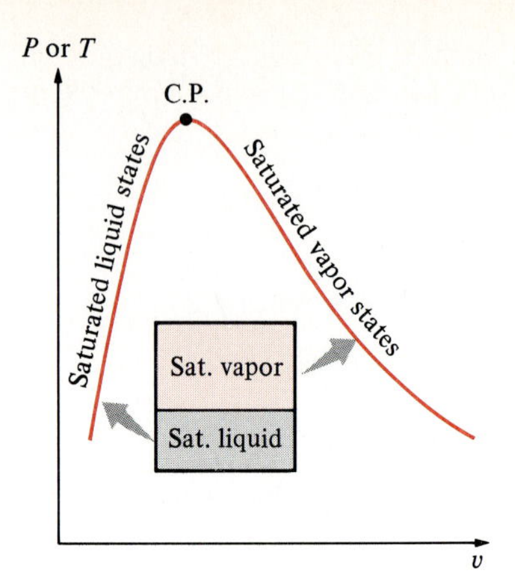

FIGURE 2-38

The relative amounts of liquid and vapor phases in a saturated mixture are specified by *quality x*.

where $$m_{\text{total}} = m_{\text{liquid}} + m_{\text{vapor}} = m_f + m_g$$

Quality has significance for saturated mixtures only. It has no meaning in the compressed liquid or superheated vapor regions. Its value is always between 0 and 1 (Fig. 2-39). The quality of a system that consists of saturated liquid is 0 (or 0 percent), and the quality of a system consisting of saturated vapor is 1 (or 100 percent). In saturated mixtures, quality can serve as one of the two independent intensive properties needed to describe a state. Note that *the properties of the saturated liquid are the same whether it exists alone or in a mixture with saturated vapor*. During the vaporization process, only the amount of saturated liquid changes, not its properties. The same can be said about a saturated vapor.

A saturated mixture can be treated as a combination of two subsystems: the saturated liquid and the saturated vapor. However, the amount of mass for each phase is usually not known. Therefore, it is often more

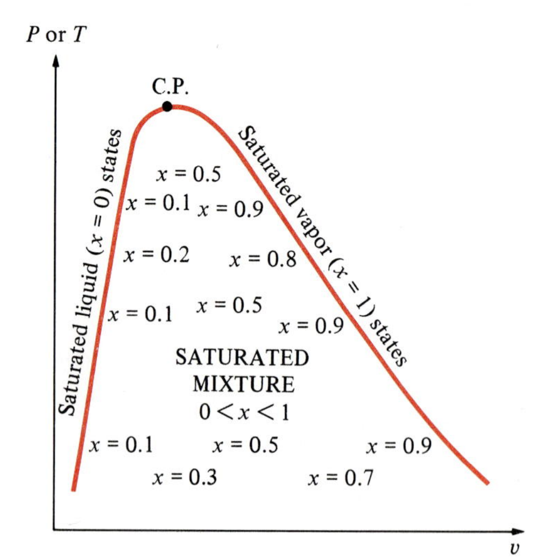

FIGURE 2-39

Quality can have values between 0 and 1, and it has significance only at states under the saturation curve.

convenient to imagine that the two phases are mixed very well, forming a homogeneous appearance (Fig. 2-40). Then the properties of this "mixture" will simply be the average properties of the saturated liquid–vapor mixture under consideration. Here is how it is done:

Consider a tank that contains a saturated liquid–vapor mixture. The volume occupied by saturated liquid is V_f, and the volume occupied by saturated vapor is V_g. The total volume V is the sum of these two:

$$V = V_f + V_g$$

$$V = mv \longrightarrow m_t v_{av} = m_f v_f + m_g v_g$$

$$m_f = m_t - m_g \longrightarrow m_t v_{av} = (m_t - m_g)v_f + m_g v_g$$

Dividing by m_t yields

$$v_{av} = (1 - x)v_f + xv_g$$

since $x = m_g/m_t$. This relation can also be expressed as

$$v_{av} = v_f + xv_{fg} \qquad (\text{m}^3/\text{kg}) \tag{2-4}$$

where $v_{fg} = v_g - v_f$. Solving for quality, we obtain

$$x = \frac{v_{av} - v_f}{v_{fg}} \tag{2-5}$$

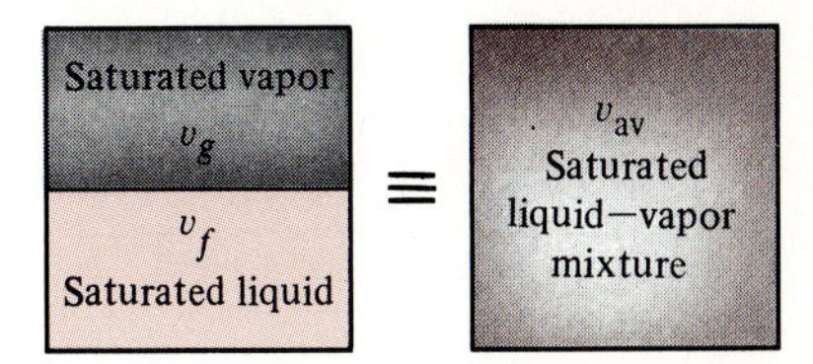

FIGURE 2-40

A two-phase system can be treated as a homogeneous mixture for computational purposes.

Based on this equation, quality can be related to the horizontal distances on a *P-v* or *T-v* diagram (Fig. 2-41). At a given temperature or pressure, the numerator of Eq. 2-5 is the distance between the actual state and the saturated liquid state, and the denominator is the length of the entire horizontal line that connects the saturated liquid and saturated vapor states. A state of 50 percent quality will lie in the middle of this horizontal line.

The analysis given above can be repeated for internal energy and enthalpy with the following results:

$$u_{av} = u_f + xu_{fg} \qquad (\text{kJ/kg}) \tag{2-6}$$

$$h_{av} = h_f + xh_{fg} \qquad (\text{kJ/kg}) \tag{2-7}$$

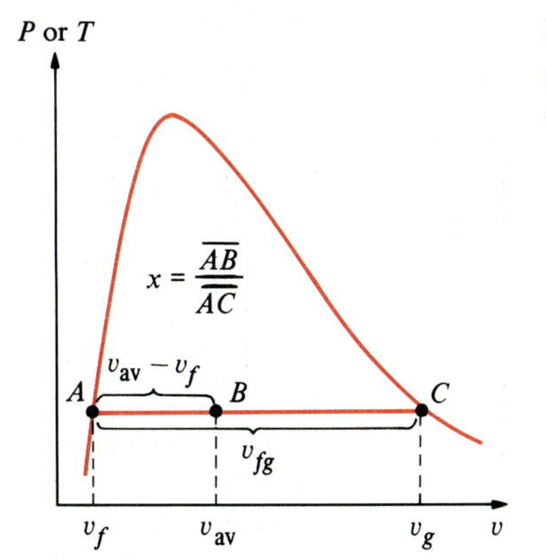

FIGURE 2-41

Quality is related to the horizontal distances on *P-v* and *T-v* diagrams.

All the results are of the same format, and they can be summarized in a single equation as

$$y_{\text{av}} = y_f + xy_{fg} \tag{2-8}$$

where y is v, u, or h. The subscript "av" (for "average") is usually dropped for simplicity. The values of the average properties of the mixtures are always between the values of the saturated liquid and the saturated vapor properties (Fig. 2-42). That is,

$$y_f \leqslant y_{\text{av}} \leqslant y_g$$

Finally, all the saturated-mixture states are located under the saturation curve, and to analyze saturated mixtures, all we need are saturated liquid and saturated vapor data (Tables A-4 and A-5 in the case of water).

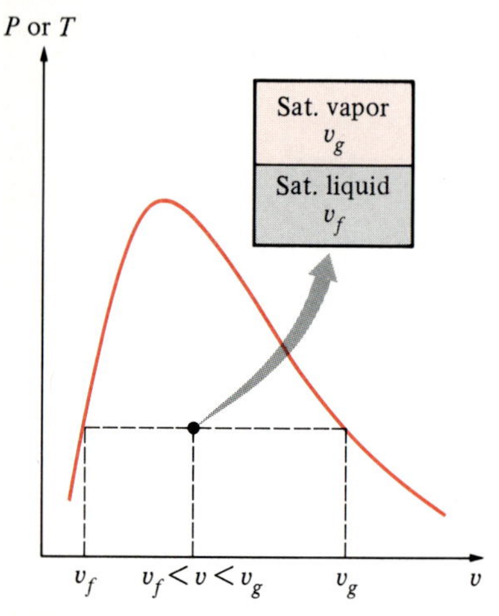

FIGURE 2-42
The v value of a saturated liquid–vapor mixture lies between the v_f and v_g values at the specified T or P.

EXAMPLE 2-4

A rigid tank contains 10 kg of water at 90°C. If 8 kg of the water is in the liquid form and the rest is in the vapor form, determine (*a*) the pressure in the tank and (*b*) the volume of the tank.

Solution (*a*) The state of the saturated liquid–vapor mixture is shown in Fig. 2-43. Since the two phases coexist in equilibrium, we have a saturated mixture and the pressure must be the saturation pressure at the given temperature:

$$P = P_{\text{sat @ 90°C}} = 70.14 \text{ kPa} \qquad \text{(Table A-4)}$$

(*b*) At 90°C, v_f and v_g values are $v_f = 0.001036 \text{ m}^3/\text{kg}$ and $v_g = 2.361 \text{ m}^3/\text{kg}$ (Table A-4).

One way of finding the volume of the tank is to determine the volume occupied by each phase and then add them:

$$\begin{aligned} V &= V_f + V_g = m_f v_f + m_g v_g \\ &= (8 \text{ kg})(0.001 \text{ m}^3/\text{kg}) + (2 \text{ kg})(2.361 \text{ m}^3/\text{kg}) \\ &= 4.73 \text{ m}^3 \end{aligned}$$

Another way is to first determine the quality x, then the average specific volume v, and finally the total volume:

$$x = \frac{m_g}{m_t} = \frac{2 \text{ kg}}{10 \text{ kg}} = 0.2$$

$$\begin{aligned} v &= v_f + xv_{fg} \\ &= 0.001 \text{ m}^3/\text{kg} + (0.2)[(2.361 - 0.001) \text{ m}^3/\text{kg}] \\ &= 0.473 \text{ m}^3/\text{kg} \end{aligned}$$

and $$V = mv = (10 \text{ kg})(0.473 \text{ m}^3/\text{kg}) = 4.73 \text{ m}^3$$

The first method appears to be easier for this case since the masses of each phase are given. But in most cases, the masses of each phase are not available, and the second method becomes more convenient.

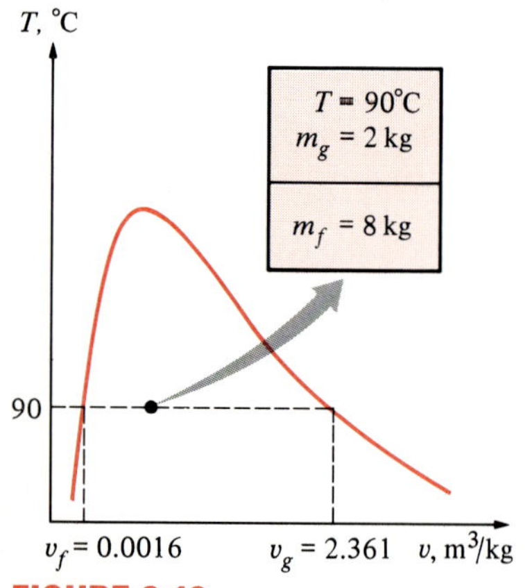

FIGURE 2-43
Schematic and T-v diagram for Example 2-4.

EXAMPLE 2-5

An 80-L vessel contains 4 kg of refrigerant-12 at a pressure of 160 kPa. Determine (*a*) the temperature of the refrigerant, (*b*) the quality, (*c*) the enthalpy of the refrigerant, and (*d*) the volume occupied by the vapor phase.

Solution (*a*) The state of the saturated liquid–vapor mixture is shown in Fig. 2-44. At this point we do not know whether the refrigerant is in the compressed liquid, superheated vapor, or saturated mixture region. This can be determined by comparing a suitable property to the saturated liquid and saturated vapor data. From the information given, we can determine the specific volume:

$$v = \frac{V}{m} = \frac{0.080 \text{ m}^3}{4 \text{ kg}} = 0.02 \text{ m}^3/\text{kg}$$

At 160 kPa, we read

$$v_f = 0.0006876 \text{ m}^3/\text{kg} \qquad \text{(Table A-12)}$$
$$v_g = 0.1031 \text{ m}^3/\text{kg}$$

Obviously $v_f < v < v_g$, and, therefore, the refrigerant is in the saturated mixture region. Thus the temperature must be the saturation temperature at the specified pressure:

$$T = T_{\text{sat @ 160 kPa}} = -18.49°\text{C}$$

(*b*) Quality can be determined from Eq. 2-4:

$$x = \frac{v - v_f}{v_{fg}} = \frac{0.02 - 0.0007}{0.1031 - 0.0007} = 0.188$$

(*c*) At 160 kPa, we also read from Table A-12 that $h_f = 19.18$ kJ/kg and $h_{fg} = 160.23$ kJ/kg. Then from Eq. 2-7, the enthalpy is

$$\begin{aligned} h &= h_f + xh_{fg} \\ &= 19.18 \text{ kJ/kg} + (0.188)(160.23 \text{ kJ/kg}) \\ &= 49.3 \text{ kJ/kg} \end{aligned}$$

(*d*) The mass of the vapor can be determined from Eq. 2-3:

$$m_g = xm_t = (0.188)(4 \text{ kg}) = 0.752 \text{ kg}$$

and the volume occupied by the vapor phase is

$$V_g = m_g v_g = (0.752 \text{ kg})(0.1031 \text{ m}^3/\text{kg}) = 0.0775 \text{ m}^3 \quad \text{(or 77.5 L)}$$

The rest of the volume (2.5 L) is occupied by the liquid.

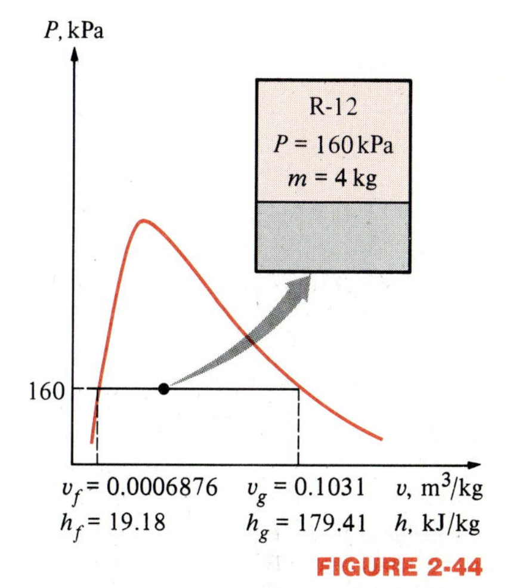

FIGURE 2-44

Schematic and *P-v* diagram for Example 2-5.

Property tables are also available for saturated solid–vapor mixtures. Properties of saturated ice–water vapor mixtures, for example, are listed in Table A-8. Saturated solid–vapor mixtures can be handled just as saturated liquid–vapor mixtures.

2 Superheated Vapor

In the region to the right of the saturated vapor line, a substance exists as superheated vapor. Since the superheat region is a single-phase region (vapor phase only), temperature and pressure are no longer dependent properties and they can conveniently be used as the two independent properties in the tables. The format of the superheated vapor tables is illustrated in Fig. 2-45.

In these tables, the properties are listed versus temperature for selected pressures starting with the saturated vapor data. The saturation temperature is given in parentheses following the pressure value.

Superheated vapor is characterized by

Lower pressures ($P < P_{sat}$ at a given T)

Higher temperatures ($T > T_{sat}$ at a given P)

Higher specific volumes ($v > v_g$ at a given P or T)

Higher internal energies ($u > u_g$ at a given P or T)

Higher enthalpies ($h > h_g$ at a given P or T)

T, °C	v, m^3/kg	u, kJ/kg	h, kJ/kg
	$P = 0.1$ MPa (99.63°C)		
Sat.	1.6940	2506.1	2675.5
100	1.6958	2506.7	2676.2
150	1.9364	2582.8	2776.4
⋮	⋮	⋮	⋮
1300	7.260	4683.5	5409.5
	$P = 0.5$ MPa (151.86°C)		
Sat.	0.3749	2561.2	2748.7
200	0.4249	2642.9	2855.4
250	0.4744	2723.5	2960.7

FIGURE 2-45

A partial listing of Table A-6.

EXAMPLE 2-6

Determine the internal energy of water at 20 psia and 400°F.

Solution At 20 psia, the saturation temperature is 227.96°F. Since $T > T_{sat}$, the water is in the superheated vapor region. Then the internal energy is determined from the superheated vapor table (Table A-6E) to be

$$u = 1145.1 \text{ Btu/lbm}$$

at the given temperature and pressure.

EXAMPLE 2-7

Determine the temperature of water at a state of P = 0.5 MPa and h = 2890 kJ/kg.

Solution At 0.5 MPa, the enthalpy of saturated water vapor is h_g = 2748.7 kJ/kg. Since $h > h_g$, as shown in Fig. 2-46, we again have superheated vapor. Under 0.5 MPa in Table A-6 we read

T, °C	h, kJ/kg
200	2855.4
250	2960.7

Obviously, the temperature is between 200 and 250°C. By linear interpolation it is determined to be

$$T = 216.4°\text{C}$$

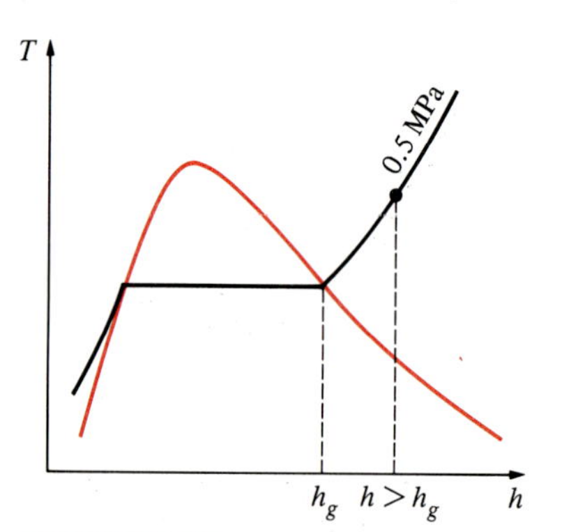

FIGURE 2-46

At a specified P, superheated vapor exists at a higher h than the saturated vapor (Example 2-7).

3 Compressed Liquid

There are not many data for compressed liquid in the literature, and Table A-7 is the only compressed liquid table in this text. The format of Table A-7 is very much like the format of the superheated vapor tables. One reason for the lack of compressed liquid data is the relative independence of compressed liquid properties from pressure. Variation of properties of compressed liquid with pressure is very mild. Increasing the pressure 100 times often causes properties to change less than 1 percent. The property most affected by pressure is enthalpy.

In the absence of compressed liquid data, a general approximation is *to treat compressed liquid as saturated liquid at the given temperature* (Fig. 2-47). This is because the compressed liquid properties depend on temperature more strongly than they do on pressure. Thus,

$$y \cong y_{f\,@\,T}$$

for compressed liquids where y is v, u, or h. In general, a compressed liquid is characterized by

Higher pressures ($P > P_{sat}$ at a given T)

Lower temperatures ($T < T_{sat}$ at a given P)

Lower specific volumes ($v < v_f$ at a given P or T)

Lower internal energies ($u < u_f$ at a given P or T)

Lower enthalpies ($h < h_f$ at a given P or T)

But these effects are not as pronounced as they are for the superheated vapor.

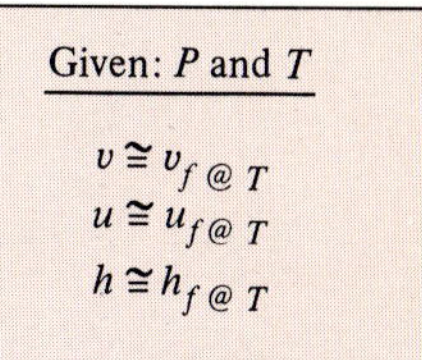

FIGURE 2-47

A compressed liquid may be approximated as a saturated liquid at the same temperature.

EXAMPLE 2-8

Determine the internal energy of compressed liquid water at 80°C and 5 MPa, using (*a*) data from the compressed liquid table and (*b*) saturated liquid data. What is the error involved in the second case?

Solution At 80°C, the saturation pressure of water is 47.39 kPa, and since 5 MPa $> P_{sat}$, we obviously have compressed liquid as shown in Fig. 2-48.

(*a*) From the compressed liquid table (Table A-7)

$$\left.\begin{array}{l} P = 5\text{ MPa} \\ T = 80^\circ\text{C} \end{array}\right\} \quad u = 333.72\text{ kJ/kg}$$

(*b*) From the saturation table (Table A-4), we read

$$u \cong u_{f\,@\,80^\circ\text{C}} = 334.86\text{ kJ/kg}$$

The error involved is

$$\frac{334.86 - 333.72}{333.72} \times 100 = 0.34\%$$

which is less than 1 percent.

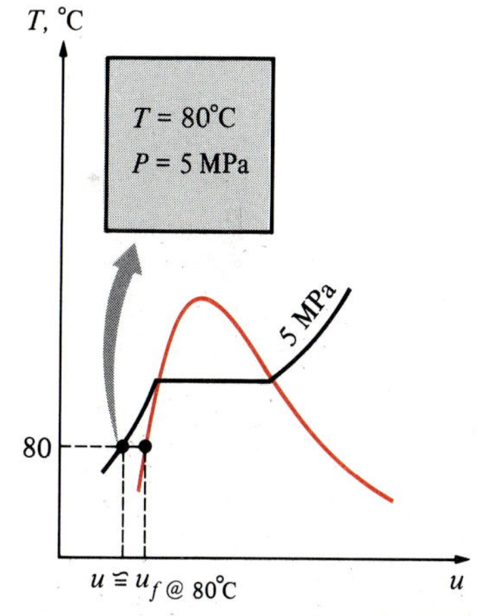

FIGURE 2-48

Schematic and T-v diagram for Example 2-8.

2-7 ■ THE IDEAL-GAS EQUATION OF STATE

One way of reporting property data for pure substances is to list values of properties at various states. The property tables provide very accurate information about the properties, but they are very bulky and vulnerable to typographical errors. A more practical and desirable approach would be to have some simple relations among the properties that are sufficiently general and accurate.

Any equation that relates the pressure, temperature, and specific volume of a substance is called an **equation of state** (Fig. 2-49). There are several equations of state, some simple and others very complex. The simplest and best known equation of state for substances in the gas phase is the ideal-gas equation of state. This equation predicts the *P-v-T* behavior of a gas quite accurately within some properly selected region.

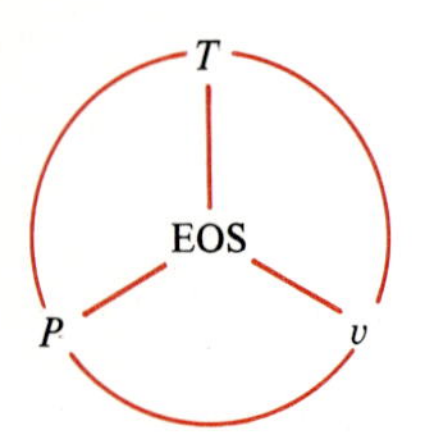

FIGURE 2-49
An equation of state is a relation among temperature, pressure, and specific volume.

Gas and *vapor* are often used as synonymous words. The vapor phase of a substance is customarily called a *gas* when it is above the critical temperature. *Vapor* usually implies a gas which is not far from a state of condensation.

In 1662 Robert Boyle, an Englishman, observed during his experiments with a vacuum chamber that the pressure of gases is inversely proportional to their volume. About 150 years later, J. Charles and J. Gay-Lussac, Frenchmen, experimentally determined that at low pressures the volume of a gas is proportional to its temperature. That is,

$$P = R\left(\frac{T}{v}\right)$$

or

$$Pv = RT \tag{2-9}$$

where the constant of proportionality R is called the **gas constant**. Equation 2-9 is called the **ideal-gas equation of state**, or simply the **ideal-gas relation**, and a gas which obeys this relation is called an **ideal gas**. In this equation, P is the absolute pressure, T is the absolute temperature, and v is the specific volume.

The gas constant R is different for each gas (Fig. 2-50) and is determined from

$$R = \frac{R_u}{M} \qquad [\text{kJ/(kg·K) or kPa·m}^3\text{/(kg·K)}] \tag{2-10}$$

where R_u is the **universal gas constant** and M is the molar mass (also called *molecular weight*) of the gas. Constant R_u is the same for all substances, and its value is

$$R_u = \begin{cases} 8.314\ \text{kJ/(kmol·K)} \\ 8.314\ \text{kPa·m}^3\text{/(kmol·K)} \\ 0.08314\ \text{bar·m}^3\text{/(kmol·K)} \\ 1.986\ \text{Btu/(lbmol·R)} \\ 10.73\ \text{psia·ft}^3\text{/(lbmol·R)} \\ 1545\ \text{ft·lbf/(lbmol·R)} \end{cases} \tag{2-11}$$

Substance	R, kJ/(kg·K)
Air	0.2870
Helium	2.0770
Argon	0.2081
Nitrogen	0.2968

FIGURE 2-50
Different substances have different gas constants.

The **molar mass** M can simply be defined as *the mass of one mole* (also called a *gram-mole*, abbreviated gmol) *of a substance in grams*, or, *the mass of one kmol* (also called a *kilogram-mole,* abbreviated kgmol) *in kilograms*. In English units, it is the mass of 1 lbmol (1 pound-mole = 0.4536 kmol) in lbm (1 pound-mass = 0.4536 kg). Notice that the molar mass of a substance has the same numerical value in both unit systems because of the way it is defined. When we say the molar mass of nitrogen is 28, it simply means the mass of 1 kmol of nitrogen is 28 kg, or the mass of 1 lbmol of nitrogen is 28 lbm. That is, M = 28 kg/kmol = 28 lbm/lbmol. The mass of a system is equal to the product of its molar mass M and the mole number N:

$$m = MN \qquad \text{(kg)} \tag{2-12}$$

The values of R and M for several substances are given in Table A-1.

The ideal-gas equation of state can be written in several different forms:

$$V = mv \longrightarrow PV = mRT \tag{2-13}$$

$$mR = (MN)R = NR_u \longrightarrow PV = NR_uT \tag{2-14}$$

$$V = N\bar{v} \longrightarrow P\bar{v} = R_uT \tag{2-15}$$

where $\bar{v}$ is the molar specific volume, i.e., the volume per unit mole (in m^3/kmol or ft^3/lbmol). A bar above a property will denote values on a unit-mole basis throughout this text (Fig. 2-51).

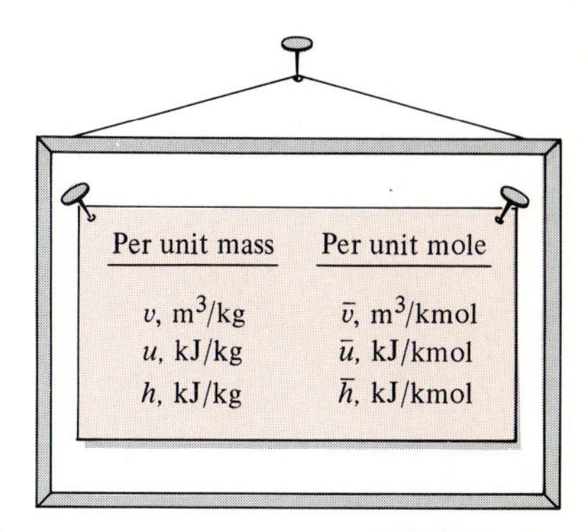

FIGURE 2-51
Properties per unit mole are denoted with a bar on the top.

By writing Eq. 2-13 twice for a fixed mass and simplifying, the properties of an ideal gas at two different states are related to each other by

$$\frac{P_1V_1}{T_1} = \frac{P_2V_2}{T_2} \tag{2-16}$$

An ideal gas is an *imaginary* substance that obeys the relation $Pv = RT$ (Fig. 2-52). It has been experimentally observed that the ideal-gas relation given above closely approximates the P-v-T behavior of real gases at low densities. At low pressures and high temperatures, the density of a gas decreases, and the gas behaves as an ideal gas under these conditions. What constitutes low pressure and high temperature is explained in the next section.

FIGURE 2-52
The ideal-gas relation often is not applicable to real gases; thus, exercise caution when you are using it.

In the range of practical interest, many familiar gases such as air, nitrogen, oxygen, hydrogen, helium, argon, neon, krypton, and even heavier gases such as carbon dioxide can be treated as ideal gases with negligible error (often less than 1 percent). Dense gases such as water vapor in steam power plants and refrigerant vapor in refrigerators, however, should not be treated as ideal gases. Instead, the property tables should be used for these substances.

EXAMPLE 2-9

Determine the mass of the air in a room whose dimensions are 4 m × 5 m × 6 m at 100 kPa and 25°C.

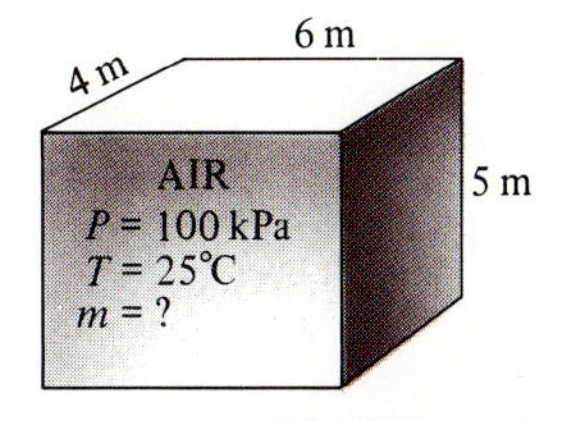

FIGURE 2-53
Schematic for Example 2-9.

Solution A sketch of the room is given in Fig. 2-53. Air at specified conditions can be treated as an ideal gas. From Table A-1, the gas constant of air is R = 0.287 kPa·m^3/(kg·K), and the absolute temperature is T = 25°C + 273 =

298 K. The volume of the room is

$$V = (4 \text{ m})(5 \text{ m})(6 \text{ m}) = 120 \text{ m}^3$$

By substituting these values into Eq. 2-13, the mass of air in the room is determined to be

$$m = \frac{PV}{RT} = \frac{(100 \text{ kPa})(120 \text{ m}^3)}{[0.287 \text{ kPa}\cdot\text{m}^3/(\text{kg}\cdot\text{K})](298 \text{ K})} = 140.3 \text{ kg}$$

Is Water Vapor an Ideal Gas?

This question cannot be answered with a simple yes or no. The error involved in treating water vapor as an ideal gas is calculated and plotted in Fig. 2-54. It is clear from this figure that at pressures below 10 kPa,

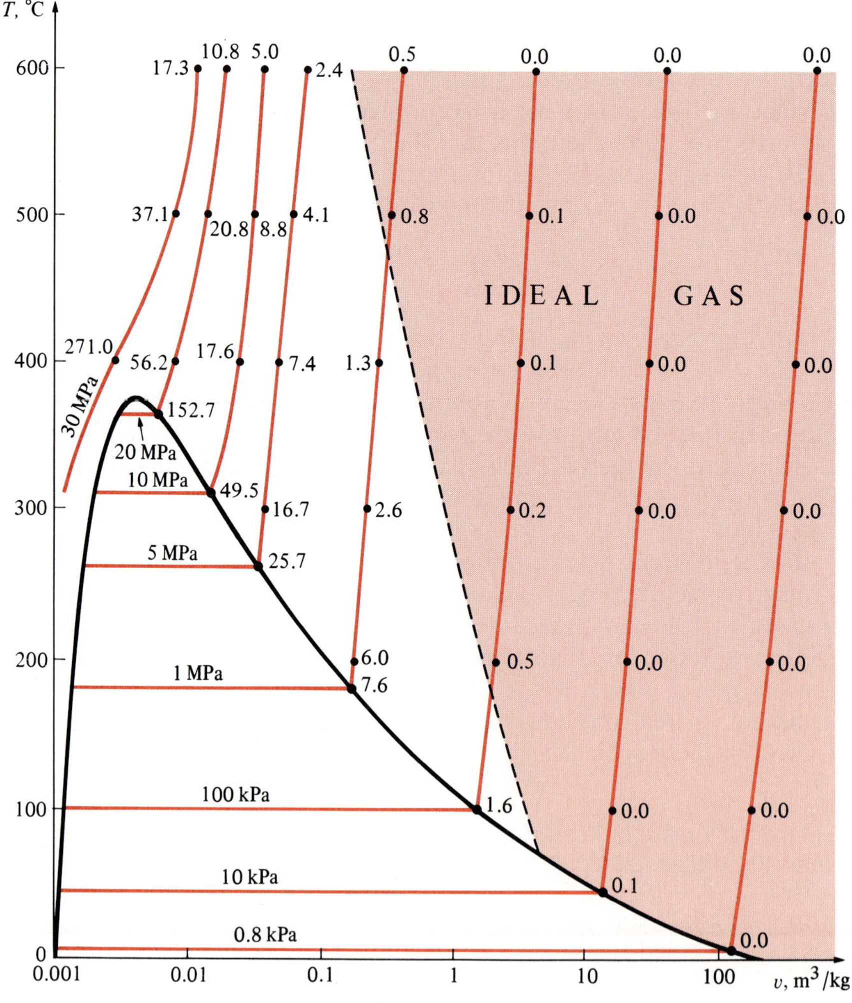

FIGURE 2-54 Percentage of error involved in assuming steam to be an ideal gas, and the region where steam can be treated as an ideal gas with less than 1 percent error (percentage of error = $[(|v_{\text{table}} - v_{\text{ideal}}|)/v_{\text{table}}] \times 100$).

water vapor can be treated as an ideal gas, regardless of its temperature, with negligible error (less than 0.1 percent). But at higher pressures, the ideal-gas assumption yields unacceptable errors, particularly in the vicinity of the critical point and the saturated vapor line (over 100 percent). Therefore, in air-conditioning applications, the water vapor in the air can be treated as an ideal gas with essentially no error since the pressure of the water vapor is very low. In steam power plant applications, however, the pressures involved are usually very high; therefore, ideal-gas relations should not be used.

2-8 ■ COMPRESSIBILITY FACTOR— A Measure of Deviation from Ideal-Gas Behavior

he ideal-gas equation is very simple and thus very convenient to use. ut as illustrated in Fig. 2-54, gases deviate from ideal-gas behavior sigficantly at states near the saturation region and the critical point. This deviation from ideal-gas behavior at a given temperature and pressure can accurately be accounted for by the introduction of a correction factor called the **compressibility factor** Z. It is defined as

$$Z = \frac{Pv}{RT} \tag{2-17}$$

or

$$Pv = ZRT \tag{2-18}$$

It can also be expressed as

$$Z = \frac{v_{\text{actual}}}{v_{\text{ideal}}} \tag{2-19}$$

where $v_{\text{ideal}} = RT/P$. Obviously, $Z = 1$ for ideal gases. For real gases Z can be greater than or less than unity (Fig. 2-55). The farther away Z is from unity, the more the gas deviates from ideal-gas behavior.

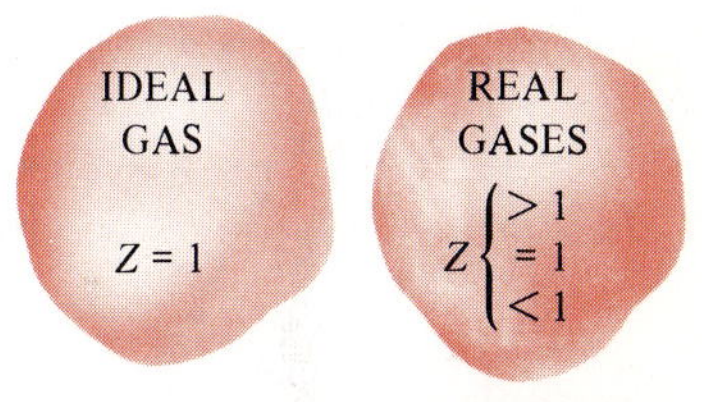

FIGURE 2-55
The compressibility factor is unity for ideal gases.

We have repeatedly said that gases follow the ideal-gas equation closely at low pressures and high temperatures. But what exactly constitutes low pressure or high temperature? Is $-100°C$ a low temperature? It definitely is for most substances, but not for air. Air (or nitrogen) can be treated as an ideal gas at this temperature and atmospheric pressure with an error under 1 percent. This is because nitrogen is well over its critical temperature ($-147°C$) and away from the saturation region. But at this temperature and pressure, most substances would exist in the solid phase. Therefore, the pressure or temperature of a substance is high or low relative to its critical temperature or pressure.

Gases behave differently at a given temperature and pressure, but they behave very much the same at temperatures and pressures normalized with respect to their critical temperatures and pressures. The normalization is done as

$$P_R = \frac{P}{P_{\text{cr}}} \quad \text{and} \quad T_R = \frac{T}{T_{\text{cr}}} \tag{2-20}$$

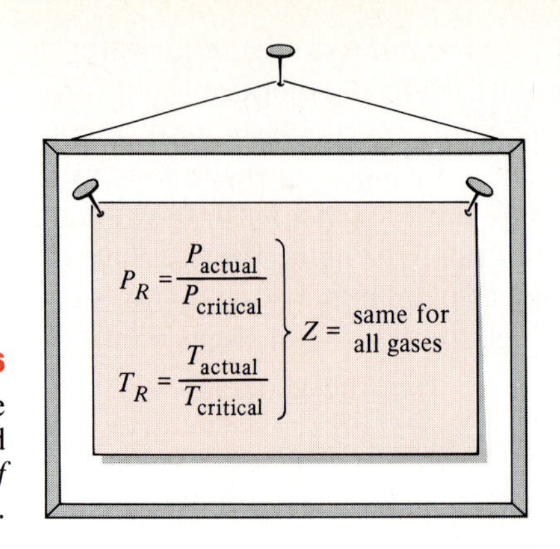

FIGURE 2-56

The compressibility factor is the same for all gases at the same reduced pressure and temperature (*principle of corresponding states*).

Here P_R is called the **reduced pressure** and T_R the **reduced temperature**. The Z factor for all gases is approximately the same at the same reduced pressure and temperature (Fig. 2-56). This is called the **principle of corresponding states**. In Fig. 2-57, the experimentally determined Z values

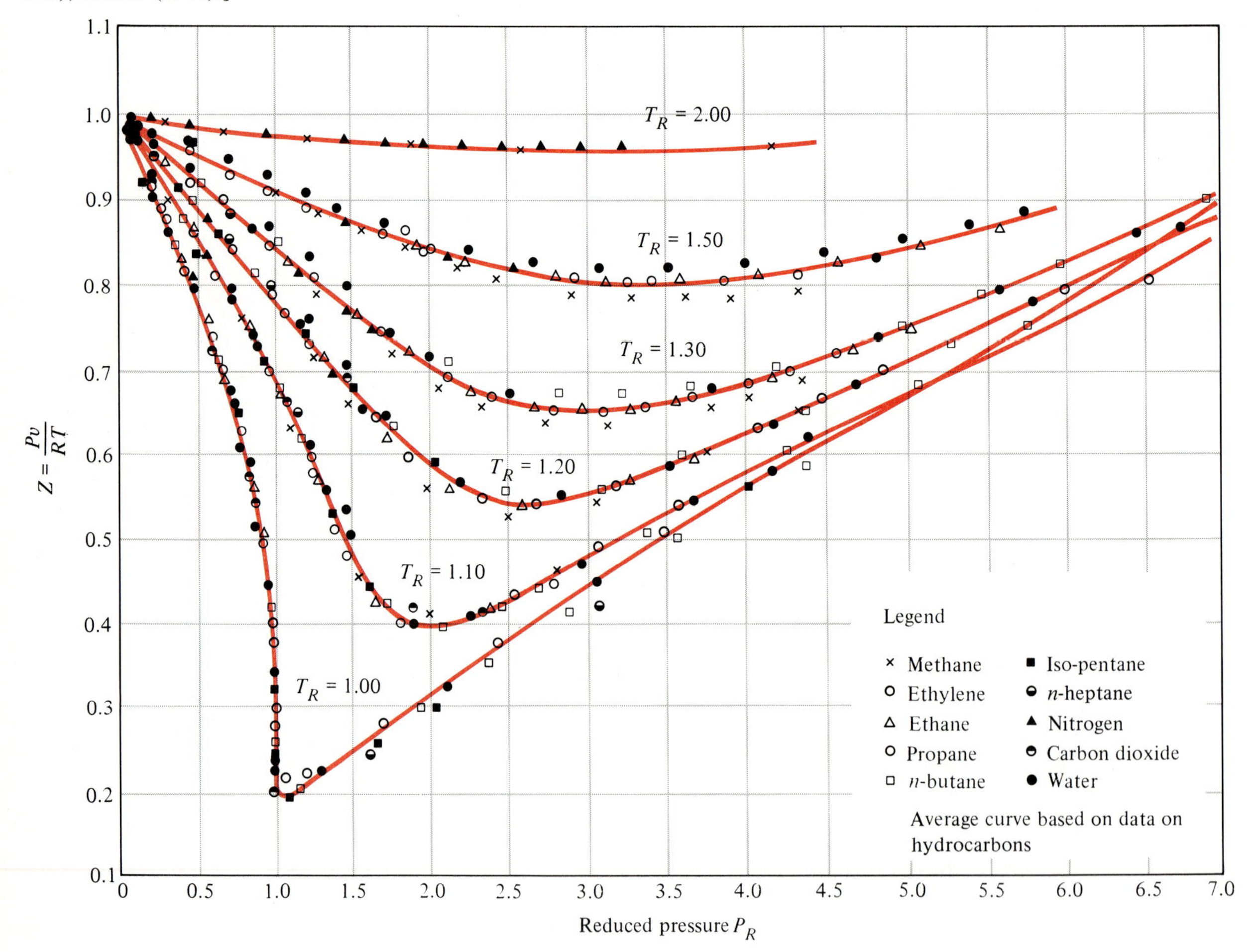

FIGURE 2-57

Comparison of Z factors for various gases. [*Source:* Gour-Jen Su, "Modified Law of Corresponding States," *Ind. Eng. Chem.* (Intern. Ed.), 38:803 (1946).]

are plotted against P_R and T_R for several gases. The gases seem to obey the principle of corresponding states reasonably well. By curve-fitting all the data, we obtain the **generalized compressibility chart** which can be used for all gases. This chart is given in the Appendix in three separate parts (Fig. A-30*a*, *b*, and *c*), each for a different range of reduced pressures for more accurate reading. The use of a compressibility chart requires a knowledge of critical-point data, and the results obtained are accurate to within a few percent.

The following observations can be made from the generalized compressibility chart:

1 At very low pressures ($P_R << 1$), the gases behave as an ideal gas regardless of temperature (Fig. 2-58).

2 At high temperatures ($T_R > 2$), ideal-gas behavior can be assumed with good accuracy regardless of pressure (except when $P_R >> 1$).

3 The deviation of a gas from ideal-gas behavior is greatest in the vicinity of the critical point (Fig. 2-59).

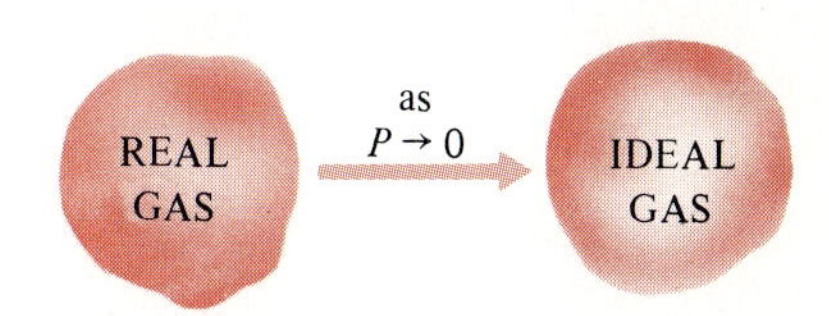

FIGURE 2-58

At very low pressures, all gases approach ideal-gas behavior (regardless of their temperature).

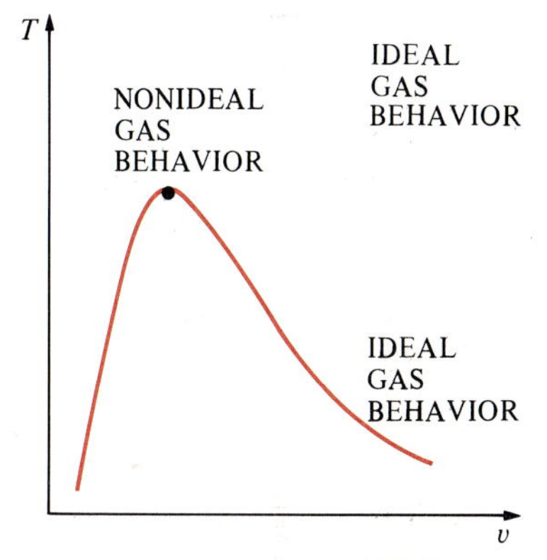

FIGURE 2-59

Gases deviate from the ideal-gas behavior most in the neighborhood of the critical point.

EXAMPLE 2-10

Determine the specific volume of refrigerant-12 at 1 MPa and 50°C, using (*a*) the refrigerant-12 tables, (*b*) the ideal-gas equation of state, and (*c*) the generalized compressibility chart. Also determine the error involved in parts (*b*) and (*c*).

Solution A sketch of the system is given in Fig. 2-60. The gas constant, the critical pressure, and the critical temperature of refrigerant-12 are determined from Table A-1 to be

$$R = 0.0688 \text{ kPa·m}^3/(\text{kg·K})$$
$$P_{cr} = 4.01 \text{ mPa}$$
$$T_{cr} = 384.7 \text{ K}$$

(*a*) The specific volume of refrigerant-12 at the specified state is determined from Table A-13 to be

$$\left.\begin{array}{l} P = 1 \text{ MPa} \\ T = 50°\text{C} \end{array}\right\} \quad v = 0.01837 \text{ m}^3/\text{kg}$$

This is the experimentally determined value, and thus it is the most accurate.

(*b*) The specific volume of the refrigerant-12 under the ideal-gas assumption is determined from the ideal-gas relation (Eq. 2-9) to be

$$v = \frac{RT}{P} = \frac{[0.0688 \text{ kPa·m}^3/(\text{kg·K})](323 \text{ K})}{1000 \text{ kPa}} = 0.02222 \text{ m}^3/\text{kg}$$

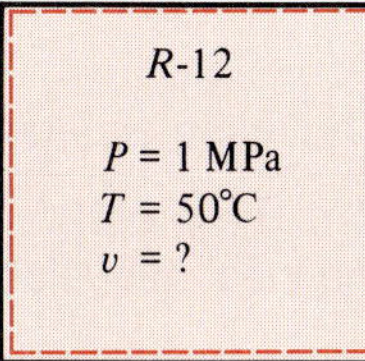

FIGURE 2-60

Schematic for Example 2-10.

Therefore, treating the refrigerant-12 vapor as an ideal gas would result in an error of (0.02222 − 0.01837)/0.01837 = 0.208, or 20.8 percent in this case.

(*c*) To determine the correction factor Z from the compressibility chart, we first need to calculate the reduced pressure and temperature:

$$\left.\begin{aligned} P_R &= \frac{P}{P_{cr}} = \frac{1 \text{ MPa}}{4.01 \text{ MPa}} = 0.249 \\ T_R &= \frac{T}{T_{cr}} = \frac{323 \text{ K}}{384.7 \text{ K}} = 0.840 \end{aligned}\right\} \quad Z = 0.83 \qquad \text{(Fig. A-30)}$$

Thus, $\quad v = Zv_{\text{ideal}} = (0.83)(0.02222 \text{ m}^3/\text{kg}) = 0.01844 \text{ m}^3/\text{kg}$

The error in this result is less than 1 percent. Therefore, in the absence of exact tabulated data, the generalized compressibility chart can be used with confidence.

When P and v, or T and v, are given instead of P and T, the generalized compressibility chart can still be used to determine the third property, but it would involve tedious trial and error. Therefore, it is very convenient to define one more reduced property called the **pseudo-reduced specific volume** v_R as

$$v_R = \frac{v_{\text{actual}}}{RT_{cr}/P_{cr}} \tag{2-21}$$

$$\left.\begin{aligned} P_R &= \frac{P}{P_{cr}} \\ v_R &= \frac{v}{RT_{cr}/P_{cr}} \end{aligned}\right\} \begin{aligned} &Z = \ldots \\ &\text{(Fig. A-30)} \end{aligned}$$

FIGURE 2-61
The compressibility factor can also be determined from a knowledge of P_R and v_R.

Note that v_R is defined differently from P_R and T_R. It is related to T_{cr} and P_{cr} instead of v_{cr}. Lines of constant v_R are also added to the compressibility charts, and this enables one to determine T or P without having to resort to time-consuming iterations (Fig. 2-61).

EXAMPLE 2-11

Determine the pressure of water vapor at 600°F and 0.514 ft³/lbm, using (*a*) the steam tables, (*b*) the ideal-gas equation, and (*c*) the generalized compressibility chart.

Solution A sketch of the system is given in Fig. 2-62. The gas constant, the critical pressure, and the critical temperature of steam are determined from Table A-1E to be

$$\begin{aligned} R &= 0.5956 \text{ psia·ft}^3/(\text{lbm·R}) \\ P_{cr} &= 3204 \text{ psia} \\ T_{cr} &= 1165.3 \text{ R} \end{aligned}$$

H_2O

$T = 600°\text{F}$
$v = 0.514 \text{ ft}^3/\text{lbm}$
$P = ?$

FIGURE 2-62
Schematic for Example 2-11.

(*a*) The pressure of steam at the specified state is determined from Table A-6E to be

$$\left.\begin{aligned} v &= 0.514 \text{ ft}^3/\text{lbm} \\ T &= 600°\text{F} \end{aligned}\right\} \quad P = 1000 \text{ psia}$$

This is the experimentally determined value, and thus it is the most accurate.

(*b*) The pressure of steam under the ideal-gas assumption is determined from the ideal-gas relation (Eq. 2-9) to be

$$P = \frac{RT}{v} = \frac{[0.5956 \text{ psia}\cdot\text{ft}^3/(\text{lbm}\cdot\text{R})](1060 \text{ R})}{0.514 \text{ ft}^3/\text{lbm}} = 1228.3 \text{ psia}$$

Therefore, treating the steam as an ideal gas would result in an error of (1228.3 − 1000)/1000 = 0.228, or 22.8 percent in this case.

(c) To determine the correction factor Z from the compressibility chart (Fig. A-30), we first need to calculate the pseudo-reduced specific volume and the reduced temperature:

$$v_R = \frac{v_{\text{actual}}}{RT_{\text{cr}}/P_{\text{cr}}} = \frac{(0.514 \text{ ft}^3/\text{lbm})(3204 \text{ psia})}{[0.5956 \text{ psia}\cdot\text{ft}^3/(\text{lbm}\cdot\text{R})](1165.3 \text{ R})} = 2.373$$

$$T_R = \frac{T}{T_{\text{cr}}} = \frac{1060 \text{ R}}{1165.3 \text{ R}} = 0.91$$

$$\Bigg\} \quad P_R = 0.33$$

Thus, $$P = P_R P\text{cr} = (0.33)(3204 \text{ psia}) = 1057.3 \text{ psia}$$

Using the compressibility chart reduced the error from 22.8 to 5.7 percent, which is acceptable for most engineering purposes (Fig. 2-63). A bigger chart, of course, would give better resolution and reduce the reading errors. Notice that we did not have to determine Z in this problem since we could read P_R directly from the chart.

	P, psia
Exact	1000.0
Z chart	1057.3
Ideal gas	1228.3

(from Example 2-11)

FIGURE 2-63

Results obtained by using the compressibility chart are usually within a few percent of the experimentally determined values.

2-9 ■ OTHER EQUATIONS OF STATE

The ideal-gas equation of state is very simple, but its range of applicability is limited. It is desirable to have equations of state that represent the P-v-T behavior of substances accurately over a larger region with no limitations. Such equations are naturally more complicated. Several equations have been proposed for this purpose (Fig. 2-64), but we shall discuss only three: the *van der Waals* equation because it is one of the earliest, the *Beattie-Bridgeman* equation of state because it is one of the best known and is reasonably accurate, and the *Benedict-Webb-Rubin* equation because it is one of the more recent and is very accurate.

Van der Waals Equation of State

The van der Waals equation of state was proposed in 1873, and it has two constants which are determined from the behavior of a substance at the critical point. The van der Waals equation of state is given by

$$\left(P + \frac{a}{v^2}\right)(v - b) = RT \tag{2-22}$$

Van der Waals intended to improve the ideal-gas equation of state by including two of the effects not considered in the ideal-gas model: the

van der Waals
Berthelet
Redlich-Kwang
Beattie-Bridgeman
Benedict-Webb-Rubin
Strobridge
Virial

FIGURE 2-64

Several equations of state are proposed throughout the history.

intermolecular attraction forces and the volume occupied by the molecules themselves. The term a/v^2 accounts for the intermolecular attraction forces, and b accounts for the volume occupied by the gas molecules. In a room at atmospheric pressure and temperature, the volume actually occupied by molecules is only about one-thousandth of the volume of the room. As the pressure increases, the volume occupied by the molecules becomes an increasingly significant part of the total volume. Van der Waals proposed to correct this by replacing v in the ideal-gas relation with the quantity $v - b$, where b represents the volume occupied by the gas molecules per unit mass.

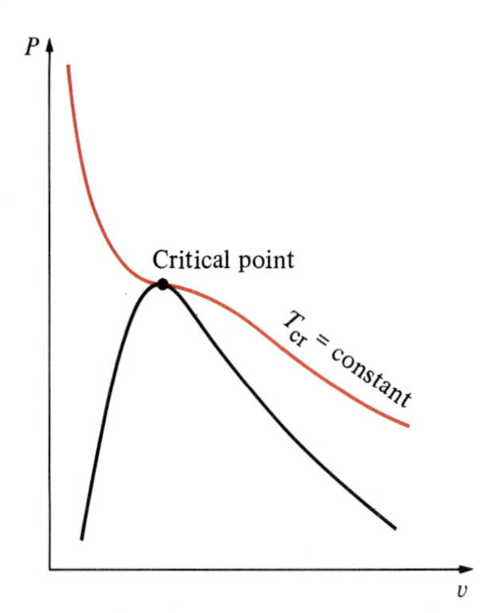

FIGURE 2-65
Critical isotherm of a pure substance has an inflection point at the critical state.

The determination of the two constants appearing in this equation is based on the observation that the critical isotherm on a P-v diagram has a horizontal inflection point at the critical point (Fig. 2-65). Thus the first and the second derivatives of P with respect to v at the critical point must be zero. That is,

$$\left(\frac{\partial P}{\partial v}\right)_{T=T_{cr}=\text{const}} = 0 \quad \text{and} \quad \left(\frac{\partial^2 P}{\partial v^2}\right)_{T=T_{cr}=\text{const}} = 0 \tag{2-23}$$

By performing the differentiations and eliminating v_{cr}, constants a and b are determined to be

$$a = \frac{27R^2T_{cr}^2}{64P_{cr}} \quad \text{and} \quad b = \frac{RT_{cr}}{8P_{cr}} \tag{2-24}$$

The constants a and b can be determined for any substance from the critical-point data alone (Table A-1).

The accuracy of van der Waals equation of state is often inadequate, but it can be improved by using values of a and b that are based on the actual behavior of the gas over a wider range instead of a single point. Despite its limitations, the van der Waals equation of state has a historical value in that it was one of the first attempts to model the behavior of real gases. Van der Waals equation of state can also be expressed on a unit-mole basis by replacing the v in Eq. 2-22 by $\bar{v}$, and the R in Eqs. 2-22 and 2-24 by R_u.

Beattie-Bridgeman Equation of State

The Beattie-Bridgeman equation, proposed in 1928, is an equation of state based on five experimentally determined constants. It was proposed in the form of

$$P = \frac{R_uT}{\bar{v}^2}\left(1 - \frac{c}{\bar{v}T^3}\right)(\bar{v} + B) - \frac{A}{\bar{v}^2} \tag{2-25}$$

where

$$A = A_0\left(1 - \frac{a}{\bar{v}}\right) \quad \text{and} \quad B = B_0\left(1 - \frac{b}{\bar{v}}\right) \tag{2-26}$$

The constants appearing in the above equation are given in Table A-29*a* for various substances. The Beattie-Bridgeman equation is known to be reasonably accurate for densities up to about $0.8\rho_{cr}$, where ρ_{cr} is the density of the substance at the critical point.

Benedict-Webb-Rubin Equation of State

Benedict, Webb, and Rubin extended the Beattie-Bridgeman equation in 1940 by raising the number of constants to eight. It is expressed as

$$P = \frac{R_u T}{\bar{v}} + \left(B_0 R_u T - A_0 - \frac{C_0}{T^2}\right)\frac{1}{\bar{v}^2} + \frac{bR_u T - a}{\bar{v}^3} + \frac{a\alpha}{\bar{v}^6} + \frac{c}{\bar{v}^3 T^2}\left(1 + \frac{\gamma}{\bar{v}^2}\right)e^{-\gamma/\bar{v}^2} \quad (2\text{-}27)$$

The values of the constants appearing in this equation are given in Table A-29*b*. This equation can handle substances at densities up to about $2.5\rho_{cr}$. In 1962, *Strobridge* further extended this equation by raising the number of constants to 16 (Fig. 2-66).

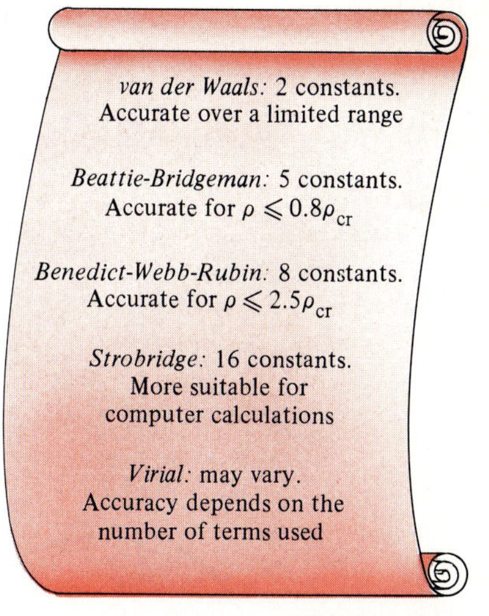

FIGURE 2-66
Complex equations of state represent the *P-v-T* behavior of gases more accurately over a wider range.

Virial Equation of State

The equation of state of a substance can also be expressed in a series form as

$$P = \frac{RT}{v} + \frac{a(T)}{v^2} + \frac{b(T)}{v^3} + \frac{c(T)}{v^4} + \frac{d(T)}{v^5} + \cdots \quad (2\text{-}28)$$

This and similar equations are called the *virial equations of state,* and the coefficients $a(T)$, $b(T)$, $c(T)$, etc., which are functions of temperature alone are called *virial coefficients*. These coefficients can be determined experimentally or theoretically from statistical mechanics. Obviously as the pressure approaches zero, all the virial coefficients will vanish and the equation will reduce to the ideal-gas equation of state. The *P-v-T* behavior of a substance can be represented accurately with the virial equation of state over a wider range by including a sufficient number of terms. All equations of state discussed above are applicable to the gas phase of the substances only, and thus should not be used for liquids or liquid-vapor mixtures.

Complex equations represent the *P-v-T* behavior of substances reasonably well and are very suitable for digital computer applications. For hand calculations, however, it is suggested that the reader use the property tables or the simpler equations of state for convenience. This is particularly true for specific-volume calculations since all the equations above are implicit in v and will require a trial-and-error approach. The accuracy of the van der Waals, Beattie-Bridgeman, and Benedict-Webb-Rubin equations of state is illustrated in Fig. 2-67. It is obvious from this figure that the Benedict-Webb-Rubin equation of state is the most accurate.

EXAMPLE 2-12

Predict the pressure of nitrogen gas at T = 175 K and v = 0.00375 m³/kg on the basis of (*a*) the nitrogen table, (*b*) the ideal-gas equation of state, (*c*) the van der Waals equation of state, (*d*) the Beattie-Bridgeman equation of state, and (*e*) the Benedict-Webb-Rubin equation of state.

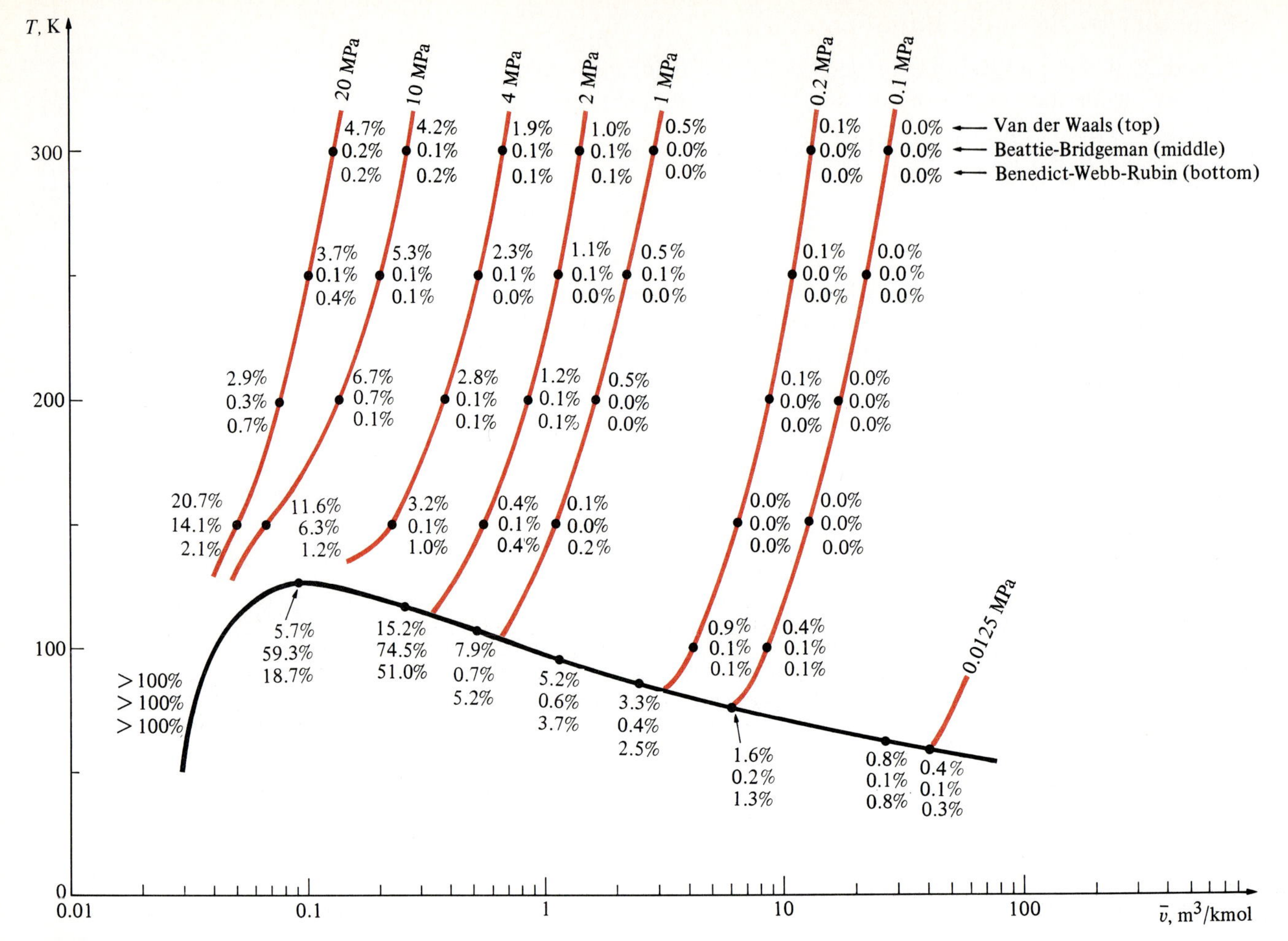

FIGURE 2-67

Percentage of error involved in various equations of state for nitrogen (% error $= [(|v_{\text{table}} - v_{\text{equation}}|)/v_{\text{table}}] \times 100$).

Solution (*a*) From the superheated nitrogen table (Table A-16), the pressure of nitrogen is determined to be

$$\left.\begin{array}{l} T = 175\text{ K} \\ v = 0.00375\text{ m}^3/\text{kg} \end{array}\right\} \quad P = 10\text{ MPa} = 10{,}000\text{ kPa}$$

This is the experimentally determined value, and thus it is the most accurate.

(*b*) By using the ideal-gas equation of state (Eq. 2-9), the pressure is found to be

$$P = \frac{RT}{v} = \frac{[0.297\text{ kPa}\cdot\text{m}^3/(\text{kg}\cdot\text{K})](175\text{ K})}{0.00375\text{ m}^3/\text{kg}} = 13{,}860\text{ kPa}$$

which is in error by 38.6 percent.

(*c*) The van der Waals constants for nitrogen are determined from Eq. 2-24 to be

$$a = 0.175\text{ m}^6\cdot\text{kPa/kg}^2$$
$$b = 0.00138\text{ m}^3/\text{kg}$$

From Eq. 2-22,

$$P = \frac{RT}{v - b} - \frac{a}{v^2} = 9465 \text{ kPa}$$

which is in error by 5.4 percent.

(*d*) The constants in the Beattie-Bridgeman equation are determined from Table A-29*a* to be

$$A = 102.29$$
$$B = 0.05378$$
$$c = 4.2 \times 10^4$$

Also, $\bar{v} = Mv = (28.013 \text{ kg/kmol})(0.00375 \text{ m}^3/\text{kg}) = 0.10505 \text{ m}^3/\text{kmol}$. Substituting these values into Eq. 2-25, we obtain

$$P = \frac{R_uT}{\bar{v}^2}\left(1 - \frac{c}{\bar{v}T^3}\right)(\bar{v} + B) - \frac{A}{\bar{v}^2} = 10{,}110 \text{ kPa}$$

which is in error by 1.1 percent.

(*e*) The constants in the Benedict-Webb-Rubin equation are determined from Table A-29*b* to be

$$a = 2.54 \qquad A_0 = 106.73$$
$$b = 0.002328 \qquad B_0 = 0.04074$$
$$c = 7.379 \times 10^4 \qquad C_0 = 8.164 \times 10^5$$
$$\alpha = 1.272 \times 10^{-4} \qquad \gamma = 0.0053$$

Substituting these values into Eq. 2-27, we obtain

$$P = \frac{R_uT}{\bar{v}} + \left(B_0R_uT - A_0 - \frac{C_0}{T^2}\right)\frac{1}{\bar{v}^2} + \frac{bR_uT - a}{\bar{v}^3} + \frac{a\alpha}{\bar{v}^6} + \frac{c}{\bar{v}^3T^2}\left(1 + \frac{\gamma}{\bar{v}^2}\right)e^{-\gamma/\bar{v}^2}$$
$$= 10{,}009 \text{ kPa}$$

which is in error by only 0.09 percent. Thus the accuracy of the Benedict-Webb-Rubin equation of state is rather impressive in this case.

2-10 ■ SUMMARY

A substance that has a fixed chemical composition throughout is called a *pure substance*. A pure substance exists in different phases depending on its energy level. In the liquid phase, a substance which is not about to vaporize is called a *compressed* or *subcooled liquid*. In the gas phase, a substance which is not about to condense is called a *superheated vapor*. During a phase-change process, the temperature and pressure of a pure substance are dependent properties. At a given pressure, a substance boils at a fixed temperature which is called the *saturation temperature*. Likewise, at a given temperature, the pressure at which a substance starts boiling is called the *saturation pressure*. During a phase-change process, both the liquid and the vapor phases coexist in equilibrium, and under this condition the liquid is called *saturated liquid* and the vapor *saturated vapor*.

In a saturated liquid–vapor mixture, the mass fraction of the vapor phase is called the *quality* and is defined as

$$x = \frac{m_{\text{vapor}}}{m_{\text{total}}} \qquad (2\text{-}3)$$

The quality may have values between 0 (saturated liquid) and 1 (saturated vapor). It has no meaning in the compressed liquid or superheated vapor regions. In the saturated mixture region, the average value of any intensive property y is determined from

$$y = y_f + xy_{fg} \qquad (2\text{-}8)$$

where f stands for saturated liquid and g for saturated vapor.

In the absence of compressed liquid data, a general approximation is to treat a compressed liquid as a saturated liquid at the given *temperature*, i.e.,

$$y \cong y_{f\,@\,T}$$

where y stands for v, u, or h.

The state beyond which there is no distinct vaporization process is called the *critical point*. At supercritical pressures, a substance gradually and uniformly expands from the liquid to vapor phase. All three phases of a substance coexist in equilibrium at states along the *triple line* characterized by triple-line temperature and pressure. Various properties of some pure substances are listed in the Appendix. As can be noticed from these tables, the compressed liquid has lower v, u, and h values than the saturated liquid at the same T or P. Likewise, superheated vapor has higher v, u, and h values than the saturated vapor at the same T or P.

Any relation among the pressure, temperature, and specific volume of a substance is called an *equation of state*. The simplest and best known equation of state is the *ideal-gas equation of state,* given as

$$Pv = RT \qquad (2\text{-}9)$$

where R is the gas constant. Caution should be exercised in using this relation since an ideal gas is a fictitious substance. Real gases exhibit ideal-gas behavior at relatively low pressures and high temperatures.

The deviation from ideal-gas behavior can be properly accounted for by using the *compressibility factor* Z, defined as

$$Z = \frac{Pv}{RT} \quad \text{or} \quad Z = \frac{v_{\text{actual}}}{v_{\text{ideal}}} \qquad (2\text{-}17, 2\text{-}19)$$

The Z factor is approximately the same for all gases at the same *reduced temperature* and *reduced pressure*, which are defined as

$$T_R = \frac{T}{T_{\text{cr}}} \quad \text{and} \quad P_R = \frac{P}{P_{\text{cr}}} \qquad (2\text{-}20)$$

where P_{cr} and T_{cr} are the critical pressure and temperature, respectively. This is known as the *principle of corresponding states*. When either P or

T is unknown, it can be determined from the compressibility chart with the help of the *pseudo-reduced specific volume,* defined as

$$v_R = \frac{v_{\text{actual}}}{RT_{\text{cr}}/P_{\text{cr}}} \tag{2-21}$$

The P-v-T behavior of substances can be represented more accurately by the more complex equations of state. Three of the best known are

van der Waals:

$$\left(P + \frac{a}{v^2}\right)(v - b) = RT \tag{2-22}$$

where

$$a = \frac{27R^2T_{\text{cr}}^2}{64P_{\text{cr}}} \quad \text{and} \quad b = \frac{RT_{\text{cr}}}{8P_{\text{cr}}} \tag{2-24}$$

Beattie-Bridgeman:

$$P = \frac{R_uT}{\overline{v}^2}\left(1 - \frac{c}{\overline{v}T^3}\right)(\overline{v} + B) - \frac{A}{\overline{v}^2} \tag{2-25}$$

where

$$A = A_0\left(1 - \frac{a}{\overline{v}}\right) \quad \text{and} \quad B = B_0\left(1 - \frac{b}{\overline{v}}\right) \tag{2-26}$$

Benedict-Webb-Rubin:

$$P = \frac{R_uT}{\overline{v}} + \left(B_0R_uT - A_0 - \frac{C_0}{T^2}\right)\frac{1}{\overline{v}^2} + \frac{bR_uT - a}{\overline{v}^3} + \frac{a\alpha}{\overline{v}^6} + \frac{c}{\overline{v}^3T^2}\left(1 + \frac{\gamma}{\overline{v}^2}\right)e^{-\gamma/\overline{v}^2} \tag{2-27}$$

The constants appearing in the Beattie-Bridgeman and Benedict-Webb-Rubin equations are given in Table A-29 for various substances.

REFERENCES AND SUGGESTED READING

1 W. Z. Black and J. G. Hartley, *Thermodynamics,* Harper and Row, New York, 1985.

2 M. D. Burghardt, *Engineering Thermodynamics with Applications,* Harper and Row, New York, 1986.

3 J. B. Jones and G. A. Hawkins, *Engineering Thermodynamics,* 2d ed., Wiley, New York, 1986.

4 J. R. Howell and R. O. Buckius, *Fundamentals of Engineering Thermodynamics,* McGraw-Hill, New York, 1987.

5 G. J. Van Wylen and R. E. Sonntag, *Fundamentals of Classical Thermodynamics,* 3d ed., Wiley, New York, 1985.

6 K. Wark, *Thermodynamics,* 5th ed., McGraw-Hill, New York, 1988.

PROBLEMS*

Pure Substances, Phase-Change Processes, Phase Diagrams

2-1C Is iced water a pure substance? Why?

2-2C Is sweetened tea a pure substance? Explain.

2-3C At what phase is a substance at the highest energy level?

2-4C What is the difference between saturated liquid and compressed liquid?

2-5C What is the difference between saturated vapor and superheated vapor?

2-6C Is there any difference between the properties of saturated vapor at a given temperature and the vapor of a saturated mixture at the same temperature?

2-7C Is there any difference between the properties of saturated liquid at a given temperature and the liquid of a saturated mixture at the same temperature?

2-8C Is it true that water boils at higher temperatures at higher pressures? Explain.

2-9C If the pressure of a substance is increased during a boiling process, will the temperature also increase or will it remain constant? Why?

2-10C Why are the temperature- and pressure-dependent properties in the saturated mixture region?

2-11C What is the difference between the critical point and the triple point?

2-12C Is it possible to have liquid water at 0.5-kPa pressure?

2-13C Is it possible to have water vapor at $-10°C$?

2-14C A househusband is cooking beef stew for his family in a pan which is (*a*) uncovered, (*b*) covered with a light lid, and (*c*) covered with a heavy lid. For which case will the cooking time be the shortest? Why?

2-15C A rigid tank contains some liquid water at 20°C. The rest of the tank is filled with atmospheric air. Is it possible to boil the water in the tank without raising its temperature? How?

2-16C A well-sealed rigid tank contains some water and air at atmospheric pressure. The tank is now heated, and the water starts boiling. Will the temperature in the tank remain constant during this boiling process? Why?

2-17C How does the boiling process at supercritical pressures differ from the boiling process at subcritical pressures?

---

*Students are encouraged to answer *all* the concept "C" questions.

2-18C Can the enthalpy of a substance at a given state be determined from a knowledge of u, P, and v data? How?

2-19C Does the reference point selected for the properties of a substance have any effect in thermodynamic analysis? Why?

2-20C What is the physical significance of h_{fg}? Can it be obtained from a knowledge of h_f and h_g?

2-21C Is it true that it takes more energy to vaporize 1 kg of saturated liquid at 100°C than it would to vaporize it at 120°C?

2-22C What is quality? Does it have any meaning in the superheated vapor region?

2-23C Which process requires more energy: completely vaporizing 1 kg of saturated liquid at 1-atm pressure or completely vaporizing 1 kg of saturated liquid at 8-atm pressure?

2-24C Does h_{fg} change with pressure? How?

2-25C Can quality be expressed as the ratio of the volume occupied by the vapor phase to the total volume?

2-26C In the absence of compressed liquid tables, how is the specific volume of a compressed liquid at a given P and T determined?

2-27 Complete the following table for H_2O:

T, °C	P, kPa	v, m³/kg	Phase description
60		3.25	
	150		Saturated vapor
200	300		
110	500		

2-27E Complete the following table for H_2O:

T, °F	P, psia	v, ft³/lbm	Phase description
150		15.2	
	50		Saturated vapor
400	60		
250	100		

2-28 Complete the following table for H_2O:

T, °C	P, kPa	u, kJ/kg	Phase description
110		2300	
	300		Saturated liquid
250	200		
90	800		

2-28E Complete the following table for H_2O:

T, °F	P, psia	u, Btu/lbm	Phase description
250		983	
	50		Saturated liquid
400	60		
300	250		

2-29 Complete the following table for H_2O:

T, °C	P, kPa	h, kJ/kg	x	Phase description
	200		0.7	
140		1800		
	1000		0.0	
80	400			
	600	3165.7		

2-30 Complete the following table for refrigerant-12:

T, °C	P, kPa	v, m³/kg	Phase description
−20	500		
4		0.012	
	200		Saturated vapor
80	700		

2-30E Complete the following table for refrigerant-12:

T, °F	P, psia	v, ft³/lbm	Phase description
0	75		
10		0.8	
	30		Saturated vapor
80	15		

2-31 Complete the following table for refrigerant-12:

T, °C	P, kPa	u, kJ/kg	Phase description
40		130	
−15			Saturated liquid
	200	188.4	
20	3000		

2-32 Complete the following table for refrigerant-12:

T, °C	P, kPa	h, kJ/kg	x	Phase description
	180	75		
−10			0.4	
15	800			
	800	213.29		
5			1.0	

2-32E Complete the following table for refrigerant-12:

T, °F	P, psia	h, Btu/lbm	x	Phase description
	30	75		
10			0.4	
40	60			
	120	98.2		
40			1.0	

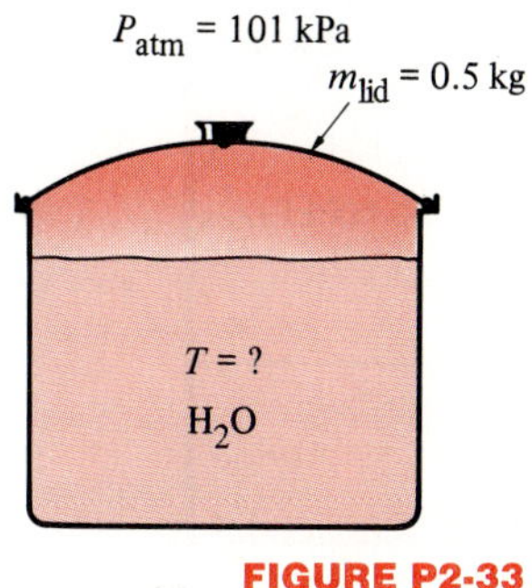

FIGURE P2-33

2-33 A cooking pan whose inner diameter is 20 cm is filled with water and covered with a 0.5-kg lid. If the local atmospheric pressure is 101 kPa, determine the temperature at which the water will start boiling when it is heated. *Answer:* 99.9°C

2-34 Water is being heated in a vertical piston-cylinder device. The piston has a mass of 20 kg and a cross-sectional area of 100 cm^2. If the local atmospheric pressure is 100 kPa, determine the temperature at which the water will start boiling.

2-35 A rigid tank with a volume of 2.5 m^3 contains 5 kg of saturated liquid–vapor mixture of water at 75°C. Now the water is slowly heated. Determine the temperature at which the liquid in the tank is completely vaporized. Also show the process on a T-v diagram with respect to saturation lines. *Answer:* 140.7°C

2-35E A rigid tank with a volume of 30 ft^3 contains 1.5 lbm of saturated liquid–vapor mixture of water at 190°F. Now the water is slowly heated. Determine the temperature at which the liquid in the tank is completely vaporized. Also show the process on a T-v diagram with respect to saturation lines. *Answer:* 228.4°F.

2-36 A rigid tank with a volume of 0.07 m^3 contains 1 kg of refrigerant-12 vapor at 150 kPa. The refrigerant-12 is now allowed to cool. Determine the pressure when the refrigerant first starts condensing. Also show the process on a P-v diagram with respect to saturation lines.

2-36E A rigid tank with a volume of 3 ft^3 contains 1 lbm of refrigerant-12 vapor at 20 psia. The refrigerant-12 is now allowed to cool. Determine the pressure when the refrigerant first starts condensing. Also show the process on a P-v diagram with respect to saturation lines.

2-37 A rigid vessel contains 2 kg of refrigerant-12 at 800 kPa and 50°C. Determine the volume of the vessel and the total internal energy. *Answers:* 0.0481 m^3, 388.38 kJ

2-38 A 200-L rigid tank contains 5 kg of water at 150 kPa. Determine (*a*) the temperature, (*b*) the total enthalpy, and (*c*) the mass of each phase.

2-38E A 5-ft^3 rigid tank contains 5 lbm of water at 20 psia. Determine (*a*) the temperature, (*b*) the total enthalpy, and (*c*) the mass of each phase.

2-39 A 0.5-m^3 vessel contains 10 kg of refrigerant-12 at −20°C. Determine (*a*) the pressure, (*b*) the total internal energy, and (*c*) the volume occupied by the liquid phase.
Answers: (*a*) 150.93 kPa, (*b*) 836.5 kJ, (*c*) 0.00373 m^3

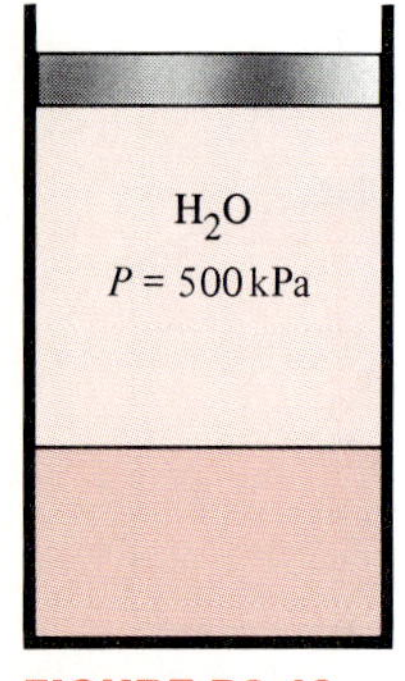

FIGURE P2-40

2-40 A piston-cylinder device contains 0.1 m^3 of liquid water and 0.9 m^3 of water vapor in equilibrium at 500 kPa. Heat is added at constant pressure until the temperature reaches 200°C.

(*a*) What is the initial temperature of the water?
(*b*) Determine the total mass of the water.
(*c*) Calculate the final volume.
(*d*) Show the process on a P-v diagram with respect to saturation lines.

2-41 Superheated water vapor at 1 MPa and 300°C is allowed to cool at constant volume until the temperature drops to 150°C. At the final state, determine (*a*) the pressure, (*b*) the quality, and (*c*) the enthalpy. Also show the process on a T-v diagram with respect to saturation lines. *Answers:* (*a*) 475.8 kPa, (*b*) 0.656, (*c*) 2030.5 kJ/kg

2-41E Superheated water vapor at 180 psia and 500°F is allowed to cool at constant volume until the temperature drops to 250°F. At the final state, determine (*a*) the pressure, (*b*) the quality, and (*c*) the enthalpy. Also show the process on a T-v diagram with respect to saturation lines. *Answers:* (*a*) 29.82 psia, (*b*) 0.219, (*c*) 425.7 Btu/lbm

2-42 A piston-cylinder device initially contains 50 L of liquid water at 25°C and 300 kPa. Heat is added to the water at constant pressure until the entire liquid is vaporized.

(*a*) What is the mass of the water?
(*b*) What is the final temperature?
(*c*) Determine the total enthalpy change.
(*d*) Show the process on a T-v diagram with respect to saturation lines.

Answers: (*a*) 49.85 kg, (*b*) 133.55°C, (*c*) 130,627 kJ

2-42E A piston-cylinder device initially contains 2 ft^3 of liquid water at 70°F and 50 psia. Heat is added to the water at constant pressure until the entire liquid is vaporized.

(*a*) What is the mass of the water?
(*b*) What is the final temperature?
(*c*) Determine the total enthalpy change.

(*d*) Show the process on a *T-v* diagram with respect to saturation lines.

2-43 A 2-L rigid tank contains 1 kg of saturated liquid–vapor mixture of water at 50°C. The water is now slowly heated until it exists in a single phase. At the final state, will the water be in the liquid phase or the vapor phase? What would your answer be if the volume of the tank were 200 L instead of 2 L?

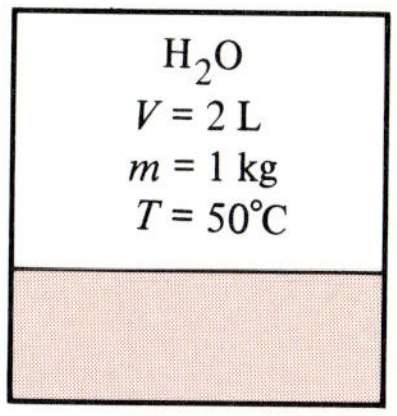

H_2O
$V = 2$ L
$m = 1$ kg
$T = 50$°C

FIGURE P2-43

2-44 A 0.5-m^3 rigid vessel initially contains saturated liquid–vapor mixture of water at 100°C. The water is now heated until it reaches the critical state. Determine the mass of the liquid water and the volume occupied by the liquid at the initial state. *Answers:* 158.28 kg, 0.165 m^3

2-44E A 5-ft^3 rigid vessel initially contains saturated liquid–vapor mixture of water at 212°F. The water is now heated until it reaches the critical state. Determine the mass of the liquid water and the volume occupied by the liquid at the initial state. *Answers:* 98.8 lbm, 1.65 ft^3

2-45 Determine the specific volume, internal energy, and enthalpy of compressed liquid water at 100°C and 15 MPa using the saturated liquid approximation. Compare these values to the ones obtained from the compressed liquid tables.

2-46 Determine the specific volume, internal energy, and enthalpy of compressed liquid water at 60°C and 10 MPa using the saturated liquid approximation. Compare these values to those obtained from the compressed liquid tables.

2-46E Determine the specific volume, internal energy, and enthalpy of compressed liquid water at 100°F and 2000 psia using the saturated liquid approximation. Compare these values to those obtained from the compressed liquid tables.

2-47 A 10-kg mass of superheated refrigerant-12 at 0.8 MPa and 40°C is cooled at constant pressure until it exists as a compressed liquid at 20°C.

(*a*) Show the process on a *T-v* diagram with respect to saturation lines.

(*b*) Determine the change in volume.

(*c*) Find the change in total internal energy.

Answers: (*b*) -0.221 m^3, (*c*) -1333.7 kJ

2-48 A 0.5-m^3 rigid tank contains saturated mixture of refrigerant-12 at 200 kPa. If the saturated liquid occupies 10 percent of the volume, determine the quality and the total mass of the refrigerant in the tank.

2-48E A 15-ft^3 rigid tank contains saturated mixture of refrigerant-12 at 30 psia. If the saturated liquid occupies 10 percent of the volume, determine the quality and the total mass of the refrigerant in the tank.

2-49 A piston-cylinder device contains 0.8 kg of steam at 300°C and 1 MPa. Steam is cooled at constant pressure until one-half of the mass condenses.

(*a*) Show the process on a *T*-*v* diagram.
(*b*) Find the final temperature.
(*c*) Determine the volume change.

2-50 A rigid tank contains water vapor at 200°C and an unknown pressure. When the tank is cooled to 150°C, the vapor starts condensing. Estimate the initial pressure in the tank. *Answer:* 0.544 MPa

2-50E A rigid tank contains water vapor at 400°F and an unknown pressure. When the tank is cooled to 300°F, the vapor starts condensing. Estimate the initial pressure in the tank. *Answer:* 77.6 psia

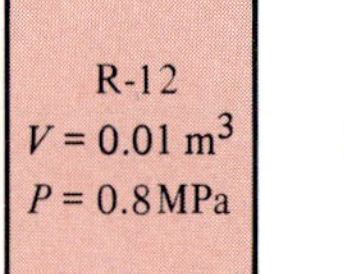

FIGURE P2-51

2-51 A tank whose volume is unknown is divided into two parts by a partition. One side of the tank contains 0.01 m³ of refrigerant-12 which is a saturated liquid at 0.8 MPa, while the other side is evacuated. The partition is now removed, and the refrigerant fills the entire tank. If the final state of the refrigerant is 25°C and 200 kPa, determine the volume of the tank. *Answer:* 1.26 m³

2-52 Write a computer program to express $T_{sat} = f(P_{sat})$ for steam as a fifth-degree polynomial where the pressure is in kPa and the temperature is in °C. Use tabulated data from Table A-4. What is the accuracy of this equation?

Ideal Gas

2-53C Under what conditions is the ideal-gas assumption suitable for real gases?

2-54C What is the difference between R and R_u? How are these two related?

2-55C What is the differencc bctwccn mass and molar mass? How are these two related?

2-56 A spherical balloon with a diameter of 6 m is filled with helium at 20°C and 200 kPa. Determine the mole number and the mass of the helium in the balloon. *Answers:* 9.28 kmol, 37.15 kg

2-56E A spherical balloon with a diameter of 25 ft is filled with helium at 70°F and 30 psia. Determine the mole number and the mass of the helium in the balloon. *Answers:* 43.16 lbmol, 172.6 lbm

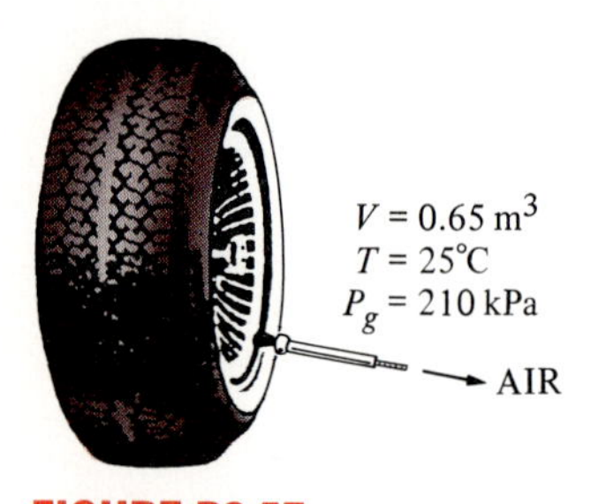

FIGURE P2-57

2-57 The pressure in an automobile tire depends on the temperature of the air in the tire. When the air temperature is 25°C, the pressure gage reads 210 kPa. If the volume of the tire is 0.65 m³, determine the pressure rise in the tire when the air temperature in the tire rises to 50°C. Also determine the amount of air that must be bled off to restore pressure to its original value at this temperature. Assume the atmospheric pressure to be 100 kPa.

2-58 The air in an automobile tire with a volume of 0.70 m³ is at 30°C and 150 kPa (gage). Determine the amount of air that must be added to raise the pressure to the recommended value of 200 kPa (gage). Assume the atmospheric pressure to be 98 kPa and the temperature and the volume to remain constant. *Answer:* 0.40 kg

2-58E The air in an automobile tire with a volume of 20 ft^3 is at 90°F and 20 psig. Determine the amount of air that must be added to raise the pressure to the recommended value of 30 psig. Assume the atmospheric pressure to be 14.6 psia and the temperature and the volume to remain constant. *Answer:* 0.981 lbm

2-59 A 0.5-m^3 rigid tank containing hydrogen at 20°C and 600 kPa is connected by a valve to another 0.5-m^3 rigid tank that holds hydrogen at 30°C and 150 kPa. Now the valve is opened, and the system is allowed to reach thermal equilibrium with the surroundings which are at 15°C. Determine the final pressure in the tank.

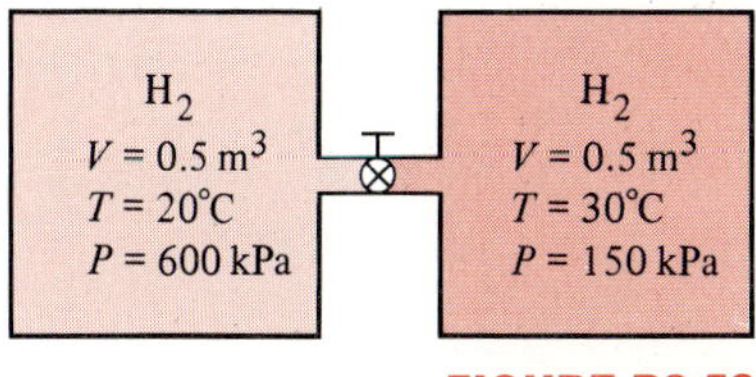

FIGURE P2-59

2-60 A 20-m^3 tank contains nitrogen at 25°C and 800 kPa. Some nitrogen is allowed to escape until the pressure in the tank drops to 600 kPa. If the temperature at this point is 20°C, determine the amount of nitrogen that has escaped. *Answer:* 42.9 kg

2-61 The pressure gage on a 0.8-m^3 oxygen tank reads 200 kPa. Determine the amount of oxygen in the tank if the temperature is 18°C and the atmospheric pressure is 95 kPa.

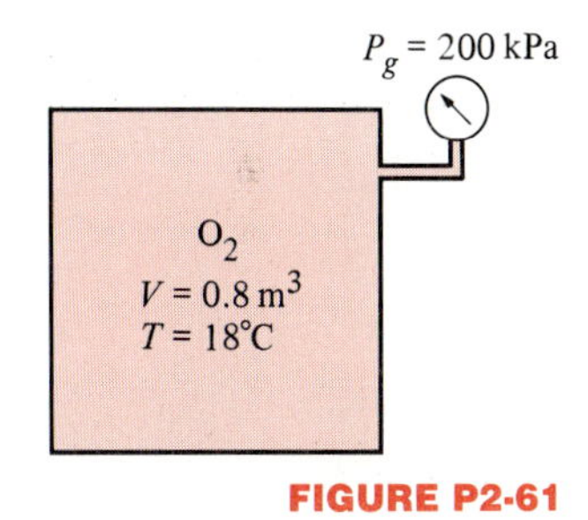

FIGURE P2-61

2-61E The pressure gage on a 20-ft^3 oxygen tank reads 30 psig. Determine the amount of oxygen in the tank if the temperature is 65°F and the atmospheric pressure is 14.5 psia.

2-62 A rigid tank contains 10 kg of air at 150 kPa and 20°C. More air is added to the tank until the pressure and temperature rise to 250 kPa and 30°C, respectively. Determine the amount of air added to the tank. *Answer:* 6.12 kg

2-62E A rigid tank contains 20 lbm of air at 20 psia and 70°F. More air is added to the tank until the pressure and temperature rise to 35 psia and 90°F, respectively. Determine the amount of air added to the tank. *Answer:* 13.73 lbm

2-63 A 500-L rigid tank contains 8 kg of air at 35°C. Determine the reading on the pressure gage if the atmospheric pressure is 96 kPa.

2-64 A 1-m^3 tank containing air at 25°C and 500 kPa is connected through a valve to another tank containing 5 kg of air at 35°C and 200 kPa. Now the valve is opened, and the entire system is allowed to reach thermal equilibrium with the surroundings which are at 20°C. Determine the volume of the second tank and the final equilibrium pressure of air. *Answer:* 284.1 kPa

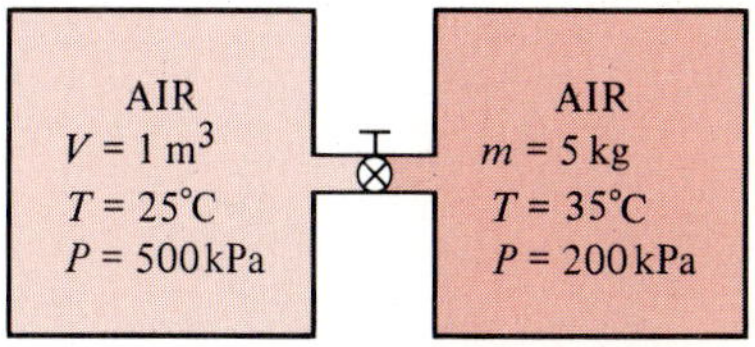

AIR
$V = 1\ m^3$
$T = 25°C$
$P = 500$ kPa

AIR
$m = 5$ kg
$T = 35°C$
$P = 200$ kPa

FIGURE P2-64

Compressibility Factor

2-65C What is the physical significance of the compressibility factor Z?

2-66C What is the principle of corresponding states?

2-67C What are the reduced pressure and reduced temperature?

2-68 Determine the specific volume of superheated water vapor at 10 MPa and 400°C, using (*a*) the ideal-gas equation, (*b*) the generalized compressibility chart, and (*c*) the steam tables. Also determine the error involved in the first two cases. *Answers:* (*a*) 0.03106 m^3/kg, 17.6 percent; (*b*) 0.02609 m^3/kg, 1.2 percent; (*c*) 0.02641 m^3/kg

2-68E Determine the specific volume of superheated water vapor at 1500 psia and 600°F, using (*a*) the ideal-gas equation, (*b*) the generalized compressibility chart, and (*c*) the steam tables. Also determine the error involved in the first two cases. *Answers:* (*a*) 0.4209 ft^3/lbm, 49.5 percent; (*b*) 0.3073 ft^3/lbm, 9.1 percent; (*c*) 0.2816 ft^3/lbm

2-69 Determine the specific volume of refrigerant-12 vapor at 1.6 MPa and 120°C based on (*a*) the ideal-gas equation, (*b*) the generalized compressibility chart, and (*c*) the experimental data from tables. Also determine the error involved in the first two cases. *Answers:* (*a*) 0.01690 m^3/kg, 15.7 percent; (*b*) 0.01453 m^3/kg, 0.5 percent; (*c*) 0.01461 m^3/kg

2-70 Determine the specific volume of nitrogen gas at 10 MPa and 150 K based on (*a*) the ideal-gas equation, (*b*) the generalized compressibility chart, and (*c*) the experimental data from tables. Also determine the error involved in the first two cases. *Answers:* (*a*) 0.004452 m^3/kg, 86.4 percent; (*b*) 0.002404 m^3/kg, 0.7 percent; (*c*) 0.002388 m^3/kg

2-71 Determine the specific volume of superheated water vapor at 1.6 MPa and 225°C based on (*a*) the ideal-gas equation, (*b*) the generalized compressibility chart, and (*c*) the steam tables. Determine the error involved in the first two cases.

2-72 Steam at 400°C has a specific volume of 0.02 m^3/kg. Determine the pressure of the steam based on (*a*) the ideal-gas equation, (*b*) the generalized compressibility chart, and (*c*) the steam tables.
Answers: (*a*) 15,529 kPa, (*b*) 12,591 kPa, (*c*) 12,500 kPa

2-73 Refrigerant-12 at 0.7 MPa has a specific volume of 0.0281 m^3/kg. Determine the temperature of the refrigerant based on (*a*) the ideal-gas equation, (*b*) the generalized compressibility chart, and (*c*) the refrigerant tables.

2-73E Refrigerant-12 at 300 psia has a specific volume of 0.194 ft^3/lbm. Determine the temperature of the refrigerant based on (*a*) the ideal-gas equation, (*b*) the generalized compressibility chart, and (*c*) the refrigerant tables.

2-74 A 0.014-m^3 tank contains 1 kg of refrigerant-12 at 110°C. Determine the pressure of the refrigerant, using (*a*) the ideal-gas equation, (*b*) the generalized compressibility chart, and (*c*) the refrigerant tables. *Answers:* (*a*) 1.882 MPa, (*b*) 1.604 MPa, (*c*) 1.6 MPa

2-75 Somebody claims that oxygen gas at 140 K and 2 MPa can be treated as an ideal gas with an error of less than 10 percent. Is this claim valid?

2-75E Somebody claims that oxygen gas at 280 R and 300 psia can be treated as an ideal gas with an error of less than 10 percent. Is this claim valid?

2-76 What is the percentage of error involved in treating carbon dioxide at 3 MPa and 10°C as an ideal gas? *Answer:* 25 percent

Other Equations of State

2-77C What is the physical significance of the two constants that appear in the van der Waals equation of state? On what basis are they determined?

2-78 A 3.27-m^3 tank contains 100 kg of nitrogen at 225 K. Determine the pressure in the tank, using (*a*) the ideal-gas equation, (*b*) the van der Waals equation of state, (*c*) the Beattie-Bridgeman equation, and (*d*) the nitrogen table.

2-79 A 1-m^3 tank contains 2.841 kg of steam at 0.6 MPa. Determine the temperature of the steam, using (*a*) the ideal-gas equation, (*b*) the van der Waals equation, and (*c*) the steam tables.
Answers: (*a*) 457.6 K, (*b*) 465.9 K, (*c*) 473 K

2-80 Refrigerant-12 at 0.7 MPa has a specific volume of 0.0281 m^3/kg. Determine the temperature of the refrigerant based on (*a*) the ideal-gas equation, (*b*) the van der Waals equation, and (*c*) the refrigerant tables.

2-80E Refrigerant-12 at 100 psia has a specific volume of 0.4556 ft^3/lbm. Determine the temperature of the refrigerant based on (*a*) the ideal-gas equation, (*b*) the van der Waals equation, and (*c*) the refrigerant tables.

2-81 Nitrogen at 150 K has a specific volume of 0.041884 m^3/kg. Determine the pressure of the nitrogen, using (*a*) the ideal-gas equation, (*b*) the Beattie-Bridgeman equation, and (*c*) the nitrogen table.
Answers: (*a*) 1063 kPa, (*b*) 1000.4 kPa, (*c*) 1000 kPa

2-81E Nitrogen at 350 R has a specific volume of 0.115 ft^3/lbm. Determine the pressure of nitrogen, using (*a*) the ideal-gas equation, (*b*) the Beattie-Bridgeman equation, and (*c*) the nitrogen table.

2-82 Write a computer program to determine the pressure of a substance at a given temperature and specific volume, using the Benedict-Webb-Rubin equation. Check your program by evaluating the pressure of nitrogen at 10 different states and comparing them to the tabulated values.

2-83 Write a computer program to determine the specific volume of a substance at a given temperature and pressure, using the Beattie-Bridgeman equation. Check your program by evaluating the specific volume of nitrogen at 10 different states and comparing them to the tabulated values.

The First Law of Thermodynamics-Closed Systems

CHAPTER 3

The first law of thermodynamics is a statement of the conservation of energy principle. In this chapter the concepts of heat and work are introduced, and various work modes, including moving boundary work, are studied. The first-law relation for closed systems is developed in a step-by-step manner. Specific heats are defined, and relations are obtained for the internal energy and enthalpy of ideal gases in terms of specific heats and temperature. This approach is also applied to solids and liquids which are approximated as incompressible substances.

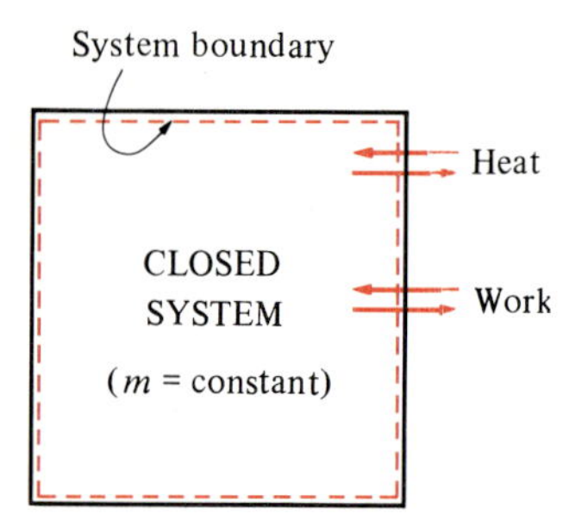

FIGURE 3-1

Energy can cross the boundaries of a closed system in the form of heat and work.

3-1 ■ INTRODUCTION TO THE FIRST LAW OF THERMODYNAMICS

In Chap. 1 it was pointed out that energy can be neither created nor destroyed; it can only change forms. This principle is based on experimental observations and is called the *first law of thermodynamics,* or the *conservation of energy principle.* The first law can simply be stated as follows: During an interaction between a system and its surroundings, the amount of energy gained by the system must be exactly equal to the amount of energy lost by the surroundings.

Energy can cross the boundary of a closed system in two distinct forms: *heat* and *work* (Fig. 3-1). It is important to distinguish between these two forms of energy. Therefore, they will be discussed first, to form a sound basis for the development of the first law of thermodynamics.

3-2 ■ HEAT

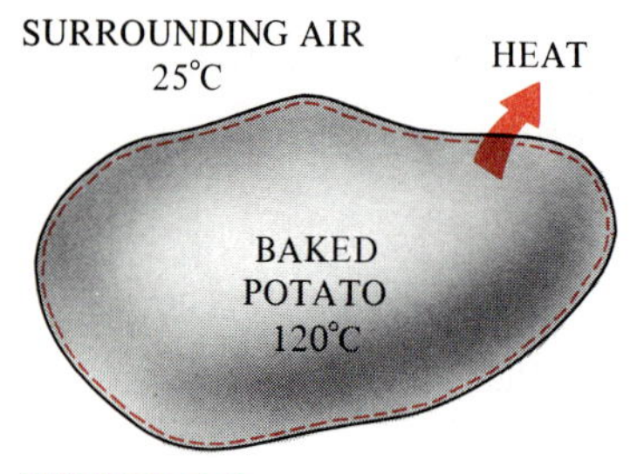

FIGURE 3-2

Heat is transferred from hot bodies to colder ones by virtue of a temperature difference.

We know from experience that a can of cold soda left on a table eventually warms up and that a hot baked potato on the same table cools down (Fig. 3-2). That is, when a body is left in a medium which is at a different temperature, energy transfer takes place between the body and the surrounding medium until thermal equilibrium is established, i.e., the body and the medium reach the same temperature. The direction of energy transfer is always from the higher temperature body to the lower temperature one. For the case of the baked potato, energy will be leaving the potato until it cools to room temperature. Once the temperature equality is established, the energy transfer stops. In the processes described above, energy is said to be transferred in the form of heat.

Heat is defined as *the form of energy that is transferred between two systems (or a system and its surroundings) by virtue of a temperature difference.* That is, an energy interaction is heat only if it takes place because of a temperature difference. Then it follows that there cannot be any heat transfer between two systems that are at the same temperature.

The meaning of the term *heat* in thermodynamics is quite different from the everyday usage. In our daily life, heat is often used to mean internal energy (heat content of a fuel, heat rises, birds preserve their body heat, etc.). In thermodynamics, however, heat and internal energy are two different things. Energy is a property, but heat is not. A body contains energy, but not heat. Energy is associated with a state; heat is associated with a process.

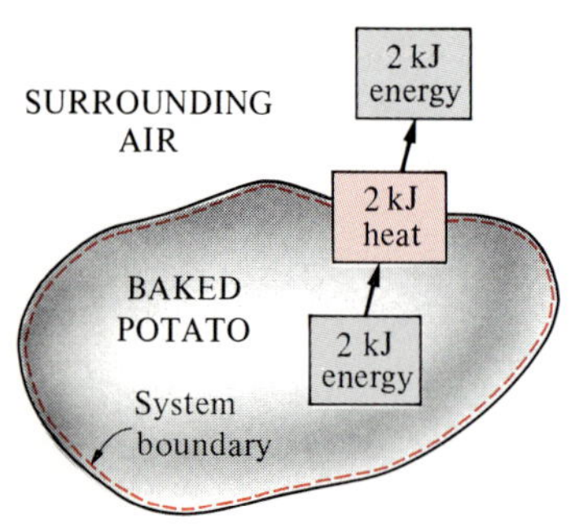

FIGURE 3-3

Energy is recognized as heat only as it crosses the system boundary.

Heat is energy in transition. It is recognized only as it crosses the boundaries of a system. Consider the hot baked potato one more time. The potato contains energy, but this energy is called heat only as it passes through the skin of the potato (the system boundary) to reach the air, as shown in Fig. 3-3. Once in the surroundings, the heat becomes part of the internal energy of the surroundings. Thus in thermodynamics the term *heat* simply means *heat transfer.*

The strict usage of the word *heat* applies to its *noun* form only. The verb *to heat* means *to raise the temperature or to add energy,* which is

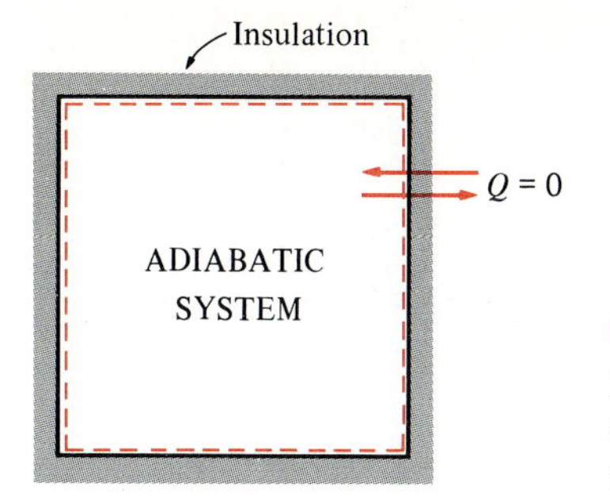

FIGURE 3-4

During an adiabatic process, a system exchanges no heat with its surroundings.

consistent with its use in everyday life and other branches of science. Also the transfer of heat into a system is called *heat addition,* and the transfer of heat out of a system is called *heat rejection.*

A process during which there is no heat transfer is called an **adiabatic process** (Fig. 3-4). There are two ways a process can be adiabatic: Either the system is well insulated so that only a negligible amount of heat can pass through the boundary, or both the system and the surroundings are at the same temperature and therefore there is no driving force (temperature difference) for heat transfer. An adiabatic process should not be confused with an isothermal process. Even though there is no heat transfer during an adiabatic process, the energy content and thus the temperature of a system can still be changed by other means such as work.

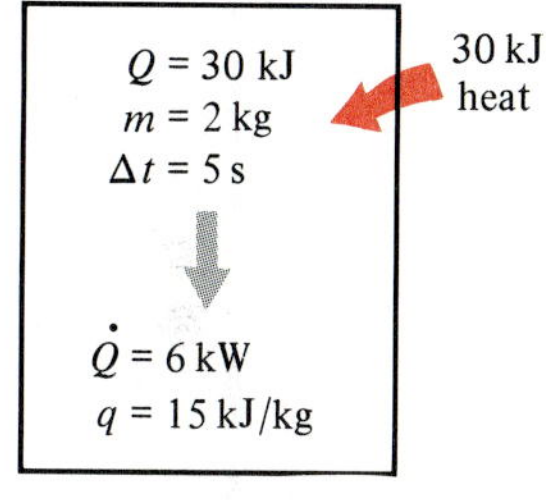

FIGURE 3-5

The relationships among q, Q, and $\dot{Q}$.

As a form of energy, heat has energy units, kJ (or Btu) being the most common one. The amount of heat transferred during the process between two states (states 1 and 2) is denoted Q_{12}, or just Q. Heat transfer *per unit mass* of a system is denoted q and is determined from

$$q = \frac{Q}{m} \qquad \text{(kJ/kg)} \tag{3-1}$$

Sometimes it is desirable to know the *rate of heat transfer* (the amount of heat transferred per unit time) instead of the total heat transferred over some time interval (Fig. 3-5). The heat transfer rate is denoted $\dot{Q}$, where the raised dot stands for the time derivative, or "per unit time." The heat transfer rate $\dot{Q}$ has the unit kJ/s, which is equivalent to kW.

Heat is a directional quantity, thus the equation $Q = 5$ kJ tells us nothing about the direction of heat flow unless we adopt a **sign convention**. The universally accepted sign convention for heat is as follows: *Heat transfer to a system is positive, and heat transfer from a system is negative* (Fig. 3-6). That is, any heat transfer which increases the energy of a system is positive, and any heat transfer which decreases the energy of a system is negative.

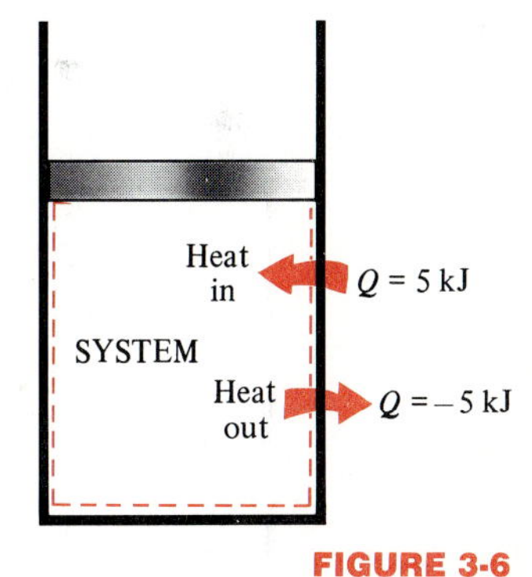

FIGURE 3-6

Sign convention for heat: positive if to the system, negative if from the system.

3-3 ■ WORK

Work, like heat, is an energy interaction between a system and its surroundings. As mentioned earlier, energy can cross the boundary of a closed system in the form of heat or work. Therefore, *if the energy cross-*

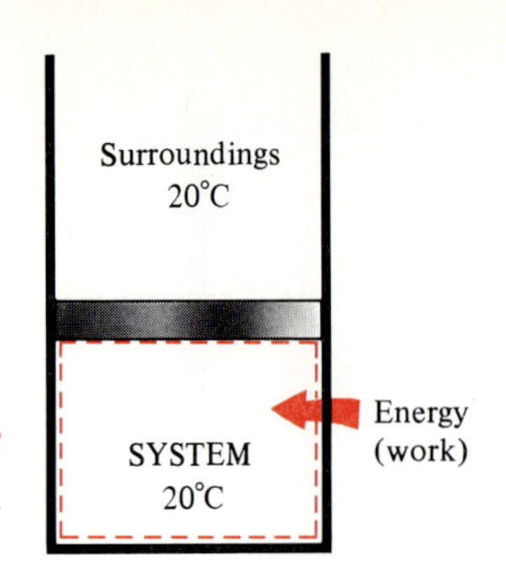

FIGURE 3-7
An energy interaction which is not heat is work.

ing the boundary is not heat, it must be work (Fig. 3-7). Heat is easy to recognize: Its driving force is a temperature difference between the system and its surroundings. Then we can simply say that an energy interaction which is not caused by a temperature difference between a system and its surroundings is work. A rising piston, a rotating shaft, and an electric wire crossing the system boundaries are all associated with work interactions.

$W = 30$ kJ
$m = 2$ kg
$\Delta t = 5$ s
30 kJ work
$\dot{W} = 6$ kW
$w = 15$ kJ/kg

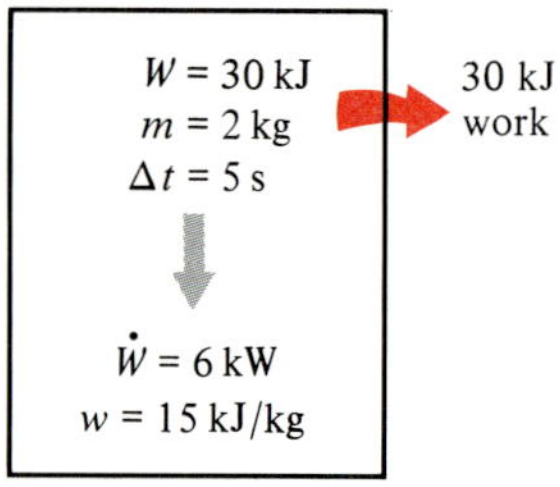

FIGURE 3-8
The relationships among w, W, and $\dot{W}$.

Work is also a form of energy like heat and, therefore, has energy units such as kJ. The work done during a process between states 1 and 2 is denoted W_{12}, or simply W. The work done *per unit mass* of a system is denoted w and is defined as

$$w = \frac{W}{m} \qquad \text{(kJ/kg)} \tag{3-2}$$

The work done *per unit time* is called **power** and denoted $\dot{W}$ (Fig. 3-8). The unit of power is kJ/s, or kW.

The production of work by a system is viewed as a desirable, positive effect and the consumption of work as an undesirable, negative effect. The sign convention for work adapted in this text reflects this philosophy: *Work done by a system is positive, and work done on a system is negative* (Fig. 3-9). By this convention, the work produced by car engines, hydraulic, steam, or gas turbines is positive, and the work consumed by compressors, pumps, and mixers is negative. In other words, work produced by a system during a process is positive, and work consumed is negative. Notice that the energy of a system decreases as it does work and increases as work is done on the system. The reader is reminded that some texts use the opposite sign convention for work.

Sometimes the identifiers *in* and *out* are used to indicate the direction of any heat or work interaction in place of the negative and positive signs. Heat transfer to a system is denoted Q_{in}, and heat transfer from a system Q_{out}. A heat loss of 5 kJ can be expressed as either $Q = -5$ kJ or $Q_{out} = 5$ kJ. Likewise, a work output of 5 kJ can be expressed as $W = 5$ kJ or $W_{out} = 5$ kJ. When we are dealing with work-consuming devices such as compressors and pumps, the negative sign associated with the work term can conveniently be avoided by speaking of work (or

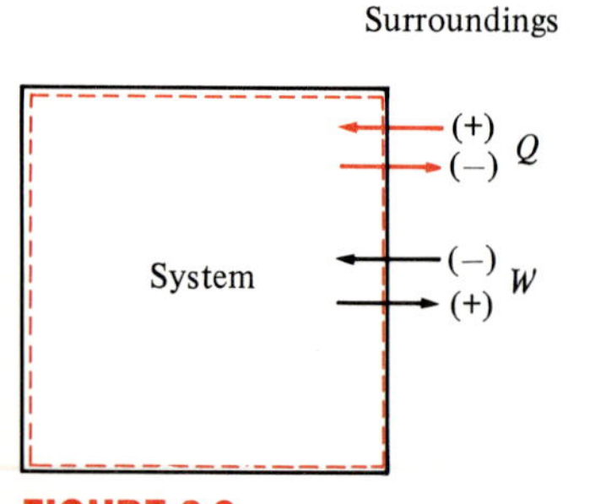

FIGURE 3-9
Sign convention for heat and work.

power) input instead of work done (for example, $\dot{W}_{in} = 2$ kW instead of $\dot{W} = -2$ kW).

Heat and work are *interactions* between a system and its surroundings, and there are many similarities between the two:

1 Both are recognized at the boundaries of the system as they cross them. That is, both heat and work are *boundary* phenomena.

2 Systems possess energy, but not heat or work. That is, heat and work are *transient* phenomena.

3 Both are associated with a *process,* not a state. Unlike properties, heat or work has no meaning at a state.

4 Both are *path functions* (i.e., their magnitudes depend on the path followed during a process as well as the end states).

Path functions have **inexact differentials** designated by the symbol δ. Therefore, a differential amount of heat or work is represented by δQ or δW, respectively, instead of dQ or dW. Properties, however, are **point functions** (i.e., they depend on the state only, and not on how a system reaches that state), and they have **exact differentials** designated by the symbol d. A small change in volume, for example, is represented by dV and the total volume change during a process between states 1 and 2 is

$$\int_1^2 dV = V_2 - V_1 = \Delta V$$

That is, the volume change during process 1-2 is always the volume at state 2 minus the volume at state 1, regardless of the path followed (Fig. 3-10). The total work done during process 1-2, however, is

$$\int_1^2 \delta W = W_{12} \qquad (not\ \Delta W)$$

That is, the total work is obtained by following the process path and adding the differential amounts of work (δW) done along the way. The integral of δW is *not* $W_2 - W_1$ (i.e., the work at state 2 minus work at state 1), which is meaningless since work is not a property and systems do not possess work at a state.

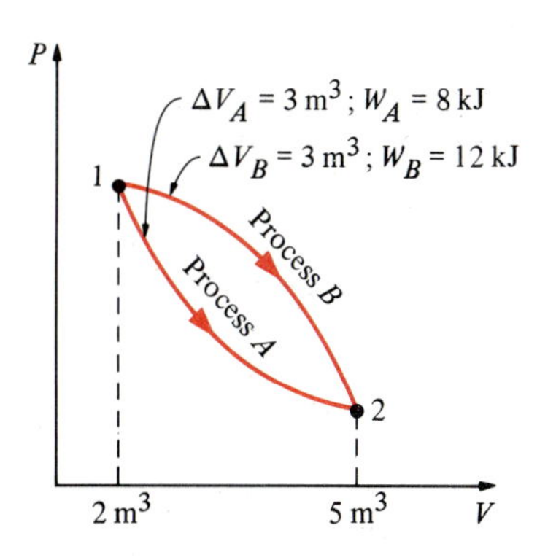

FIGURE 3-10
Properties are point functions; but heat and work are path functions (their magnitudes depend on the path followed).

EXAMPLE 3-1

A candle is burning in a well-insulated room. Taking the room (the air plus the candle) as the system, determine (*a*) if there is any heat transfer during this burning process and (*b*) if there is any change in the internal energy of the system.

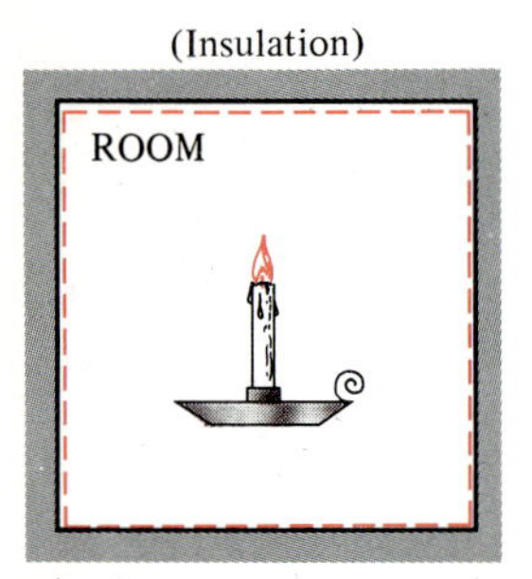

FIGURE 3-11
Schematic for Example 3-1.

Solution (*a*) The interior surfaces of the room form the system boundary, as indicated by the dashed lines in Fig. 3-11. As pointed out earlier, heat is recognized as it crosses the boundaries. Since the room is well insulated, we have an adiabatic system and no heat will pass through the boundaries. Therefore, $Q = 0$ for this process.

(*b*) As discussed in Chap. 1, the internal energy involves energies that exist in various forms (sensible, latent, chemical, nuclear). During the process described above, part of the chemical energy is converted to sensible energy. That is, part of the internal energy of the system is changed from one form to another. Since there is no increase or decrease in the total internal energy of the system, $\Delta U = 0$ for this process.

EXAMPLE 3-2

A potato which is initially at room temperature (25°C) is being baked in an oven which is maintained at 200°C, as shown in Fig. 3-12. Is there any heat transfer during this baking process?

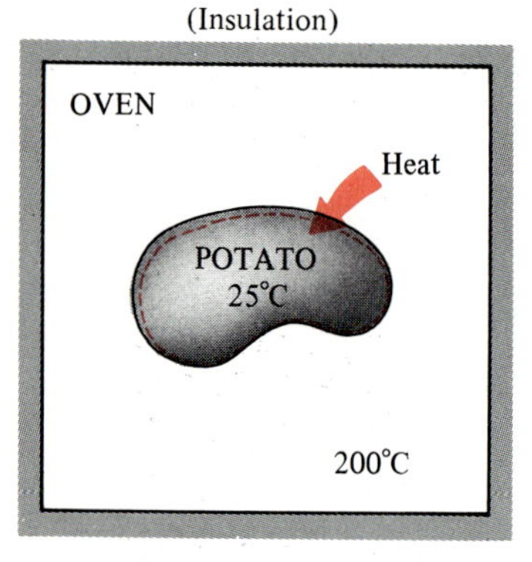

FIGURE 3-12
Schematic for Example 3-2.

Solution This is not a well-defined problem since the system is not specified. Let us assume that we are observing the potato, which will be our system. Then the skin of the potato will be the system boundary. Part of the energy in the oven will pass through the skin to the potato. Since the driving force for this energy transfer is a temperature difference, this is a heat transfer process.

EXAMPLE 3-3

A well-insulated electric oven is being heated through its heating element. If the entire oven, including the heating element, is taken to be the system, determine whether this is a heat or work interaction.

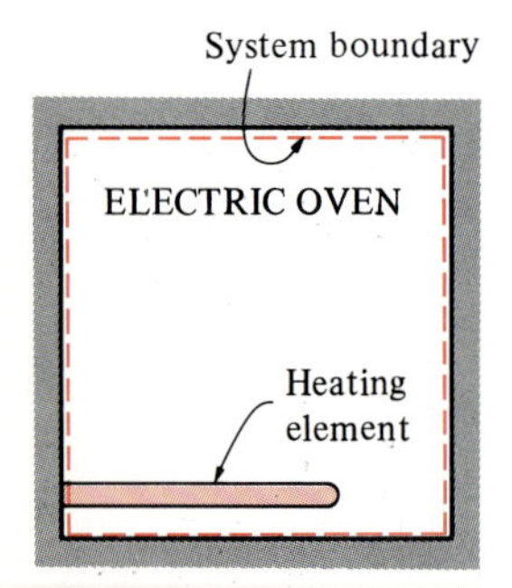

FIGURE 3-13
Schematic for Example 3-3.

Solution For this problem the interior surfaces of the oven form the system boundary, as shown in Fig. 3-13. The energy content of the oven obviously increases during this process, as evidenced by a rise in temperature. This energy transfer to the oven is not caused by a temperature difference between the oven and the surrounding air. Instead, it is caused by negatively charged particles called *electrons* crossing the system boundary and thus doing work. Therefore, this is a work interaction.

EXAMPLE 3-4

Answer the question in Example 3-3 if the system is taken as only the air in the oven without the heating element.

Solution This time the system boundary will include the outer surface of the heating element and will not cut through it, as shown in Fig. 3-14. Therefore, no electrons will be crossing the system boundary at any point. Instead, the energy generated in the interior of the heating element will be transferred to the air around it as a result of the temperature difference between the heating element and the air in the oven. Therefore, this is a heat transfer process.

For both cases, the amount of energy transfer to the air is the same. These two examples show that the same interaction can be heat or work depending on how the system is selected.

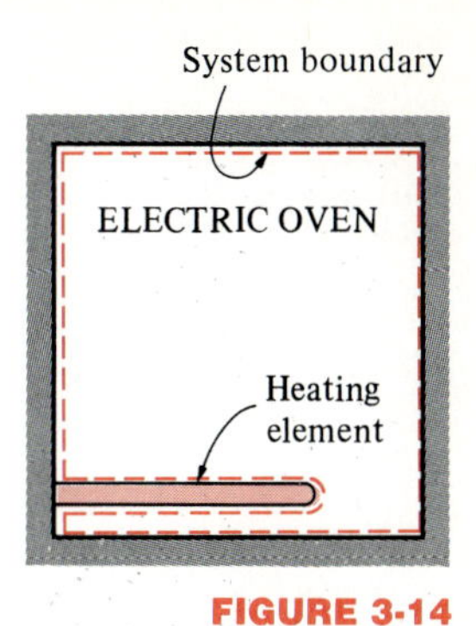

FIGURE 3-14
Schematic for Example 3-4.

Electrical Work

It was pointed out in Example 3-3 that electrons crossing the system boundary do electrical work on the system. In an electric field, electrons in a wire move under the effect of electromotive forces, doing work. When N electrons move through a potential difference V, the electrical work done is

$$W_e = VN \qquad \text{(kJ)}$$

which can also be expressed in the rate form as

$$\dot{W}_e = VI \qquad \text{(kW)} \tag{3-3}$$

where $\dot{W}_e$ is the electrical power and I is the number of electrons flowing per unit time, i.e., the current (Fig. 3-15). In general, both V and I vary with time, and the electrical work done during a time interval Δt is expressed as

$$W_e = \int_1^2 VI\,dt \qquad \text{(kJ)} \tag{3-4}$$

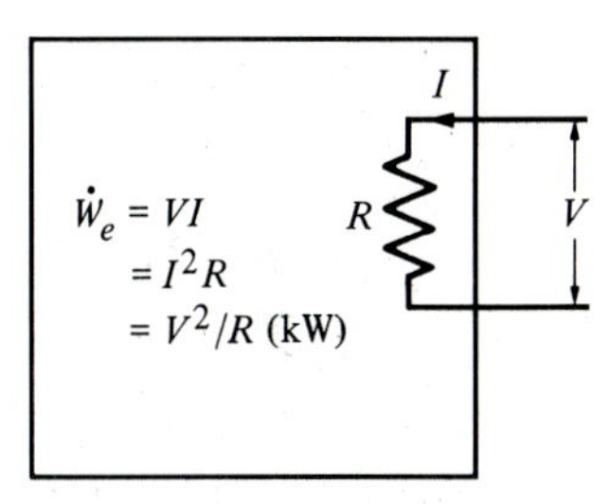

FIGURE 3-15
Electrical power in terms of resistance R, current I, and potential difference V.

If both V and I remain constant during the time interval Δt, this equation will reduce to

$$W_e = VI\,\Delta t \qquad \text{(kJ)} \tag{3-5}$$

EXAMPLE 3-5

A small tank containing iced water at 0°C is placed in the middle of a large, well-insulated tank filled with oil, as shown in Fig. 3-16. The entire system is initially

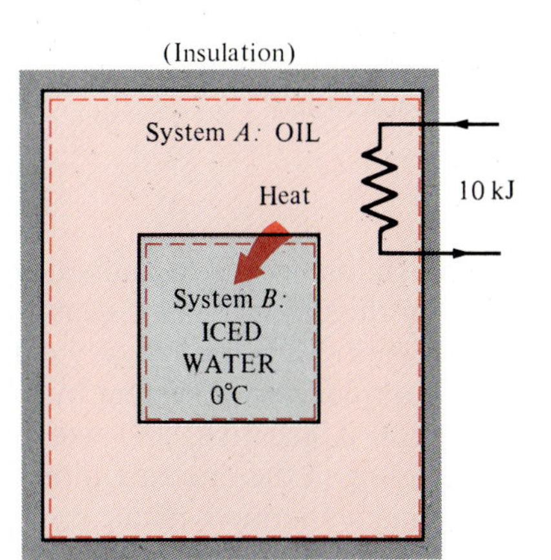

FIGURE 3-16

in thermal equilibrium at 0°C. The electric heater in the oil is now turned on, and 10 kJ of electrical work is done on the oil. After a while, it is noticed that the entire system is again at 0°C, but some ice in the small tank has melted. Considering the oil to be system *A* and the iced water to be system *B*, discuss the heat and work interactions for system *A*, system *B*, and the combined system (oil and iced water).

Solution The boundaries of each system are indicated by dashed lines in the figure. Notice that the boundary of system *B* also forms the inner part of the boundary of system *A*.

System A: When the heater is turned on, electrons cross the outer boundary of system *A*, doing electrical work. This work is done on the system, and therefore $W_A = -10$ kJ. Because of this added energy, the temperature of the oil will rise, creating a temperature gradient which results in a heat flow process from the oil to the iced water through their common boundary. Since the oil is restored to its initial temperature of 0°C, the energy lost as heat must equal the energy gained as work. Therefore, $Q_A = -10$ kJ (or $Q_{A,\text{out}} = 10$ kJ).

System B: The only energy interaction at the boundaries of system *B* is the heat flow from system *A*. All the heat lost by the oil is gained by the iced water. Thus, $W_B = 0$ and $Q_B = +10$ kJ.

Combined system: The outer boundary of system *A* forms the entire boundary of the combined system. The only energy interaction at this boundary is the electrical work. Since the tank is well insulated, no heat will cross this boundary. Therefore, $W_{\text{comb}} = -10$ kJ and $Q_{\text{comb}} = 0$. Notice that the heat flow from the oil to the iced water is an internal process for the combined system and, therefore, is not recognized as heat. It is simply the redistribution of the internal energy.

3-4 ■ MECHANICAL FORMS OF WORK

There are several different ways of doing work, each in some way related to a force acting through a distance (Fig. 3-17). In elementary mechanics, the work done by a constant force F on a body which is displaced a distance s in the direction of the force is given by

$$W = Fs \qquad \text{(kJ)} \tag{3-6}$$

If the force F is not constant, the work done is obtained by adding (i.e., integrating) the differential amounts of work (force times the differential displacement ds):

$$W = \int_1^2 F\, ds \qquad \text{(kJ)} \tag{3-7}$$

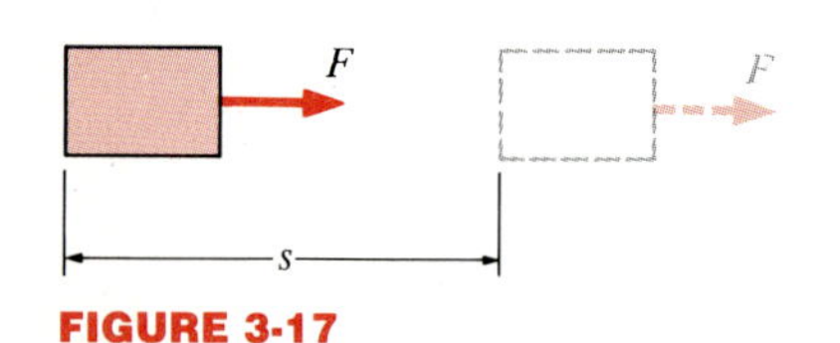

FIGURE 3-17
The work done is proportional to the force applied (F) and the distance traveled (s).

Obviously one needs to know how the force varies with displacement to perform this integration. Equations 3-6 and 3-7 give only the magnitude of the work. The sign is easily determined from physical considerations: The work done on a system by an external force acting in the direction of motion is negative, and work done by a system against an external force acting in the opposite direction to motion is positive.

In many thermodynamic problems, mechanical work is the only form

FIGURE 3-18
If there is no movement, no work is done.

of work involved. It is associated with the movement of the boundary of a system or with the movement of the entire system as a whole (Fig. 3-18). Some common forms of mechanical work are discussed below.

1 Moving Boundary Work

One form of mechanical work frequently encountered in practice is associated with the expansion or compression of a gas in a piston-cylinder device. During this process, part of the boundary (the inner face of the piston) moves back and forth. Therefore, the expansion and compression work is often called **moving boundary work**, or simply **boundary work** (Fig. 3-19). Some prefer to call it the $P\,dV$ work for reasons explained below. The moving boundary work is the primary form of work involved in automobile engines. During their expansion, the combustion gases force the piston to move, which in turn forces the crank shaft to rotate.

The moving boundary work associated with real engines or compressors cannot be determined exactly from a thermodynamic analysis alone because the piston usually moves at very high speeds, making it difficult for the gas inside to maintain equilibrium, and there is friction between the piston and the cylinder. Then the states that the system passes through during the process cannot be specified, and no process path can be drawn. Work, being a path function, cannot be determined analytically without a knowledge of the path. Therefore, the boundary work in real engines or compressors is determined by direct measurements.

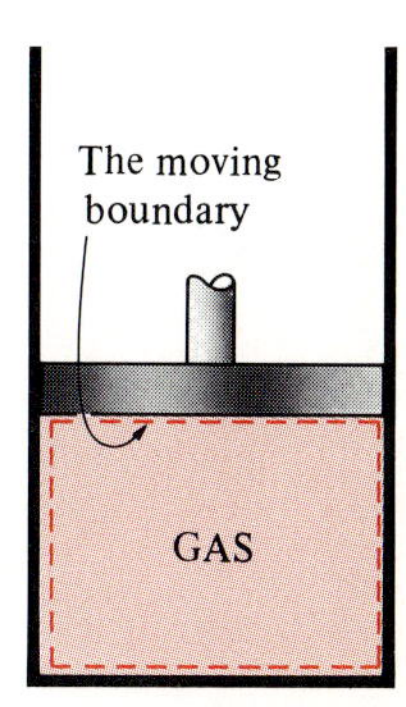

FIGURE 3-19
The work associated with a moving boundary is called *boundary work.*

In this section we analyze the moving boundary work for a *quasi-equilibrium process,* a process during which the system remains in equilibrium at all times. The quasi-equilibrium process is closely approximated by real engines, especially when the piston moves at low velocities. Under identical conditions, the work output of the engines is found to be a maximum, and the work input to the compressors to be a minimum, when quasi-equilibrium processes are used in place of non-quasi-equilibrium processes. Below, the work associated with a moving boundary is evaluated for a quasi-equilibrium process.

Consider the gas enclosed in the piston-cylinder arrangement shown in Fig. 3-20. The initial pressure of the gas is P, the total volume is V, and the cross-sectional area of the piston is A. If the piston is allowed to move a distance ds in a quasi-equilibrium manner, the differential work done during this process is

$$\delta W_b = F\,ds = PA\,ds = P\,dV \qquad (3\text{-}8)$$

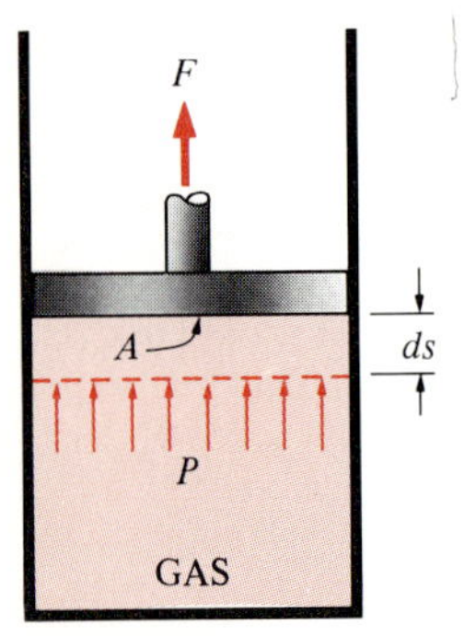

FIGURE 3-20

A gas does a differential amount of work δW_b as it forces the piston to move by a differential amount ds.

That is, the boundary work in the differential form is equal to the product of the absolute pressure P and the differential change in the volume dV of the system. This expression also explains why the moving boundary work is sometimes called the $P\,dV$ work.

Note in Eq. 3-8 that P is the absolute pressure which is always positive. However, the volume change dV is positive during an expansion process (volume increasing) and negative during a compression process (volume decreasing). Thus the boundary work is positive during an expansion process and negative during a compression process, which is consistent with the sign convention adopted for work.

The total boundary work done during the entire process as the piston moves is obtained by adding all the differential works from the initial state to the final state:

$$W_b = \int_1^2 P\,dV \qquad \text{(kJ)} \qquad (3\text{-}9)$$

This integral can be evaluated only if we know the functional relationship between P and V during the process. That is, $P = f(V)$ should be available. Note that $P = f(V)$ is simply the equation of the process path on a P-V diagram.

The quasi-equilibrium expansion process described above is shown on a P-V diagram in Fig. 3-21. On this diagram, the differential area dA is equal to $P\,dV$, which is the differential work. The total area A under the process curve 1-2 is obtained by adding these differential areas:

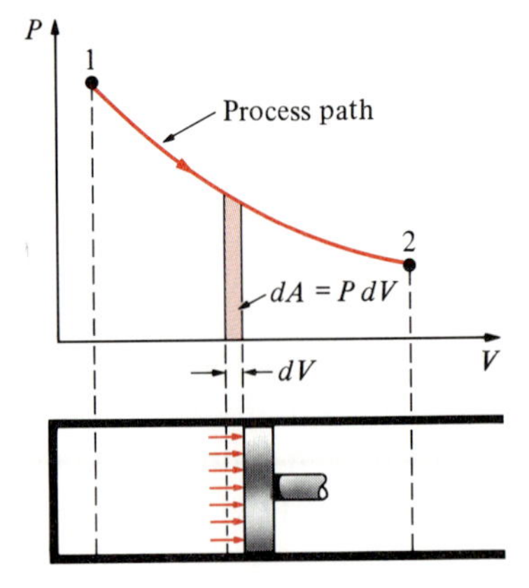

FIGURE 3-21

The area under the process curve on a P-V diagram represents the boundary work.

$$\text{Area} = A = \int_1^2 dA = \int_1^2 P\,dV$$

A comparison of this equation with Eq. 3-9 reveals the *the area under the process curve on a P-V diagram is equal, in magnitude, to the work done during a quasi-equilibrium expansion or compression process of a closed system.* (On the *P-v* diagram, it represents the boundary work done per unit mass.)

A gas can follow several different paths as it expands from state 1 to state 2. In general, each path will have a different area underneath it, and since this area represents the magnitude of the work, the work done will be different for each process (Fig. 3-22). This is expected since work is a path function (i.e., it depends on the path followed as well as the end states). If work were not a path function, no cyclic devices (car engines, power plants) could operate as work-producing devices. The work produced by these devices during one part of the cycle would have to be consumed during another part, and there would be no net work output. The cycle shown in Fig. 3-23 produces a net work output because the work done by the system during the expansion process (area under path *A*) is greater than the work done on the system during the compression part of the cycle (area under path *B*), and the difference between these two is the net work done during the cycle (the color area).

If the relationship between *P* and *V* during an expansion or a compression process is given in terms of experimental data instead of in a functional form, obviously we cannot perform the integration analytically. But we can always plot the *P-V* diagram of the process, using these data points, and calculate the area underneath graphically to determine the work done.

The use of the boundary work relation (Eq. 3-9) is not limited to the quasi-equilibrium processes of gases only. It can also be used for solids and liquids.

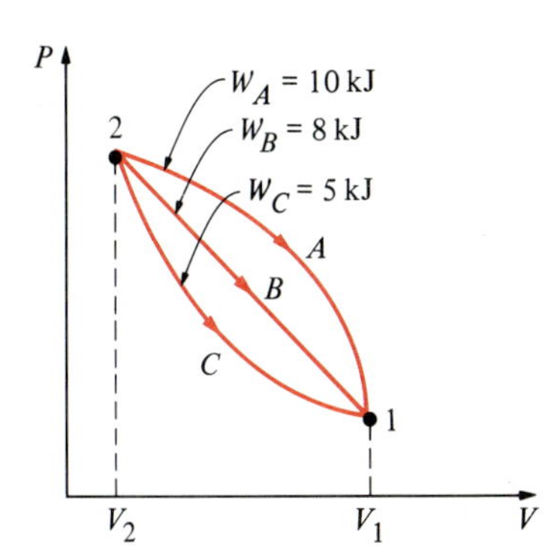

FIGURE 3-22

The boundary work done during a process depends on the path followed as well as the end states.

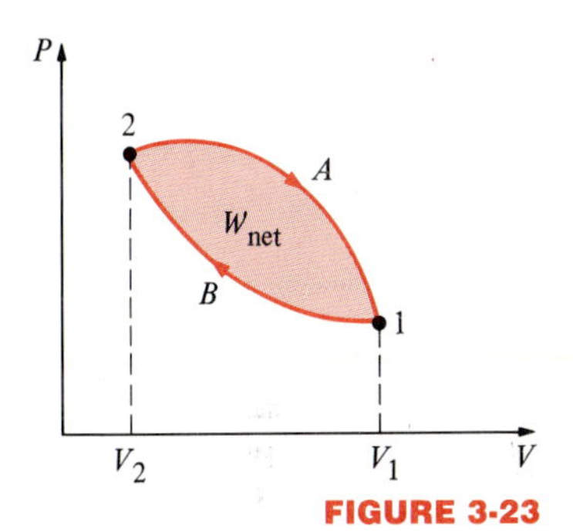

FIGURE 3-23

The net work done during a cycle is the difference between the work done by the system and the work done on the system.

EXAMPLE 3-6

A rigid tank contains air at 500 kPa and 150°C. As a result of heat transfer to the surroundings, the temperature and pressure inside the tank drop to 65°C and 400 kPa, respectively. Determine the boundary work done during this process.

Solution A sketch of the system and the *P-V* diagram of the process are shown

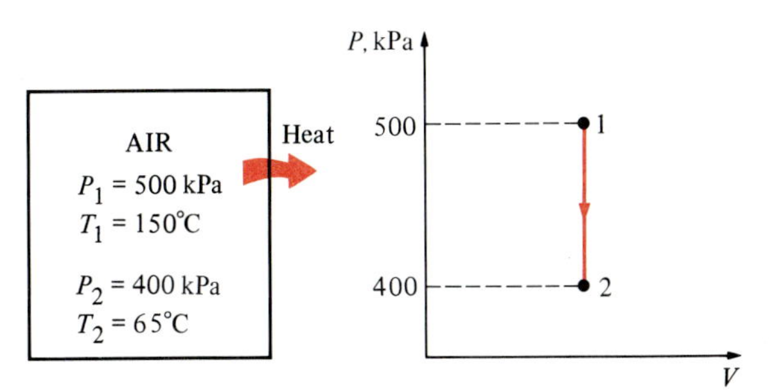

FIGURE 3-24

Schematic and *P-V* diagram for Example 3-6.

in Fig. 3-24. Assuming the process to be quasi-equilibrium, the boundary work can be determined from Eq. 3-9:

$$W_b = \int_1^2 P \cancel{dV}^{\,0} = 0$$

This is expected since a rigid tank has a constant volume and $dV = 0$ in the above equation. Therefore, there is no boundary work done during this process. That is, the boundary work done during a constant-volume process is always zero. This is also evident from the *P-V* diagram of the process (the area under the process curve is zero).

EXAMPLE 3-7

A frictionless piston-cylinder device contains 10 lbm of water vapor at 60 psia and 320°F. Heat is now added to the steam until the temperature reaches 400°F. If the piston is not attached to a shaft and its mass is constant, determine the work done by the steam during this process.

Solution A sketch of the system and the *P-V* diagram of the process are shown in Fig. 3-25. Even though it is not explicitly stated, the pressure of the steam within the cylinder remains constant during this process since both the atmospheric pressure and the weight of the piston remain constant. Therefore, this is a constant-pressure process, and from Eq. 3-9

$$W_b = \int_1^2 P\,dV = P_0 \int_1^2 dV = P_0(V_2 - V_1)$$

or

$$W_b = mP_0(v_2 - v_1) \tag{3-10}$$

since $V = mv$. From the superheated vapor table (Table A-6E), the specific volumes are determined to be $v_1 = 7.485$ ft³/lbm at state 1 (60 psia, 320°F) and $v_2 = 8.353$ ft³/lbm at state 2 (60 psia, 400°F). Substituting these values yields

$$W_b = (10 \text{ lbm})(60 \text{ psia}) \left[(8.353 - 7.485) \text{ ft}^3\text{/lbm}\right] \left(\frac{1 \text{ Btu}}{5.404 \text{ psia}\cdot\text{ft}^3}\right)$$

$$= 96.4 \text{ Btu}$$

The positive sign indicates that the work is done by the system. That is, the steam used 96.4 Btu of its energy to do this work. The magnitude of this work could also be determined by calculating the area under the process curve on the *P-V* diagram, which is $P_0\,\Delta V$ for this case.

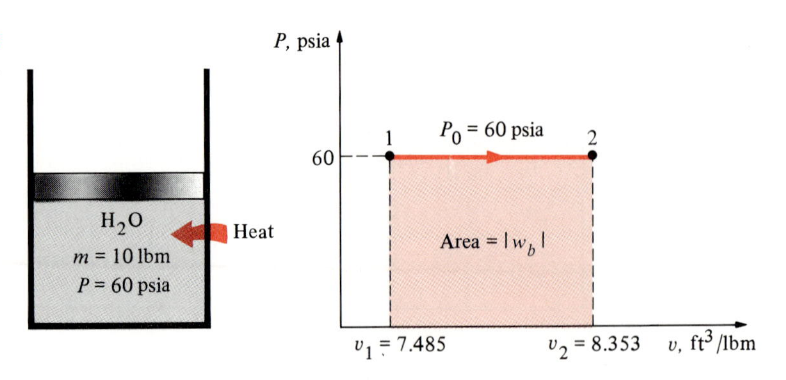

FIGURE 3-25
Schematic and *P-v* diagram for Example 3-7.

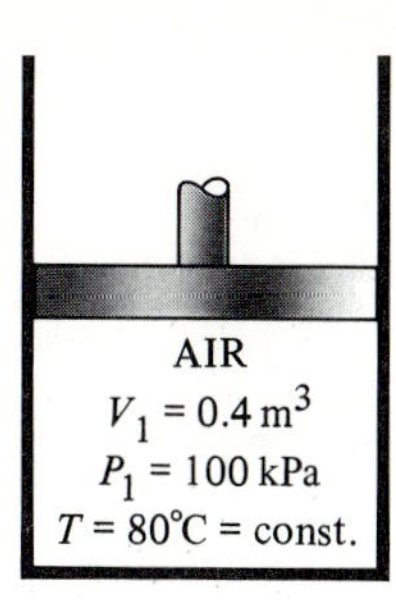

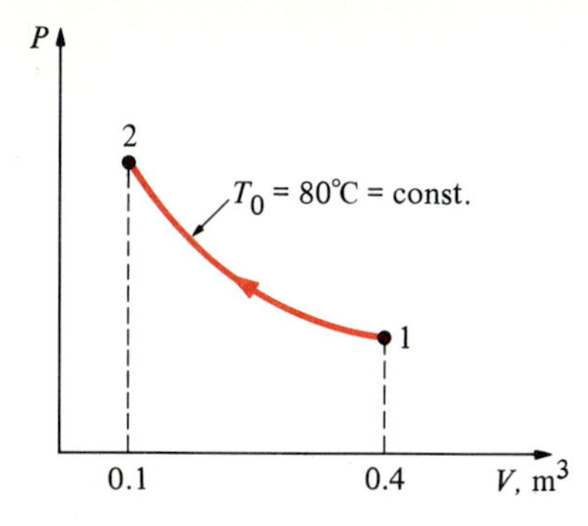

FIGURE 3-26

Schematic and *P-V* diagram for Example 3-8.

EXAMPLE 3-8

A piston-cylinder device initially contains 0.4 m^3 of air at 100 kPa and 80°C. The air is now compressed to 0.1 m^3 in such a way that the temperature inside the cylinder remains constant. Determine the work done during this process.

Solution A sketch of the system and the *P-V* diagram of the process are shown in Fig. 3-26. At the specified conditions, air can be considered to be an ideal gas since it is at a high temperature and low pressure relative to its critical-point values ($T_{cr} = -147°C$, $P_{cr} = 3390$ kPa for nitrogen, the main constituent of air). For an ideal gas at constant temperature T_0,

$$PV = mRT_0 = C \qquad \text{or} \qquad P = \frac{C}{V}$$

where C is a constant. Substituting this into Eq. 3-9, we have

$$W_b = \int_1^2 P\,dV = \int_1^2 \frac{C}{V}\,dV = C\int_1^2 \frac{dV}{V} = C\ln\frac{V_2}{V_1} = P_1V_1\ln\frac{V_2}{V_1} \qquad (3\text{-}11)$$

In the above equation, P_1V_1 can be replaced by P_2V_2 or mRT_0. Also, V_2/V_1 can be replaced by P_1/P_2 for this case since $P_1V_1 = P_2V_2$.

Substituting the numerical values into the above equation yields

$$W_b = (100\text{ kPa})(0.4\text{ m}^3)\left(\ln\frac{0.1}{0.4}\right)\left(\frac{1\text{ kJ}}{1\text{ kPa}\cdot\text{m}^3}\right)$$

$$-55.45\text{ kJ}$$

The negative sign indicates that this work is done on the system, which is always the case for compression processes.

EXAMPLE 3-9

During expansion and compression processes of real gases, pressure and volume are often related by $PV^n = C$, where n and C are constants. A process of this kind is called a polytropic process. Develop a general expression for the work done during a polytropic process.

Solution A sketch of the system and the *P-V* diagram of the process are shown in Fig. 3-27. The pressure for a polytropic process can be expressed as

$$P = CV^{-n}$$

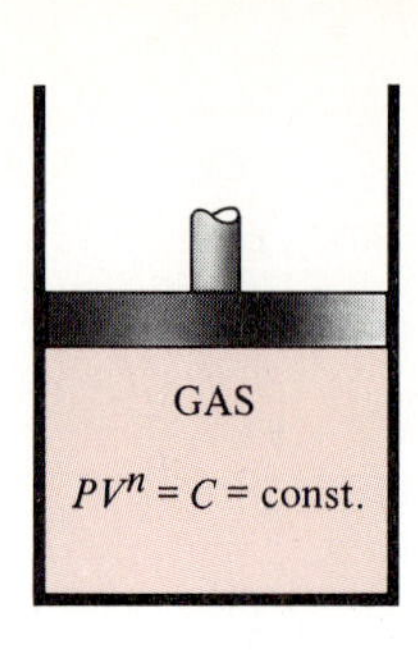

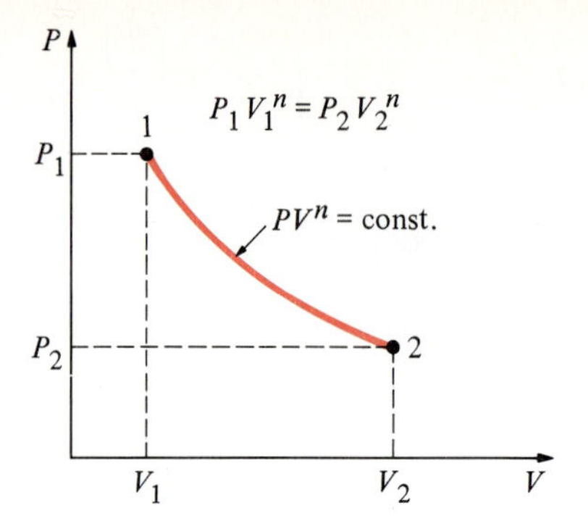

FIGURE 3-27

Schematic and *P-V* diagram for Example 3-9.

Substituting this relation into Eq. 3-9, we obtain

$$W_b = \int_1^2 P\,dV = \int_1^2 CV^{-n}\,dV = C\frac{V_2^{-n+1} - V_1^{-n+1}}{-n+1} = \frac{P_2V_2 - P_1V_1}{1-n} \qquad (3\text{-}12)$$

since $C = P_1V_1^n = P_2V_2^n$. For an ideal gas ($PV = mRT$), this equation can also be written as

$$W_b = \frac{mR(T_2 - T_1)}{1-n} \qquad \text{(kJ)} \qquad (3\text{-}13)$$

The special case of $n = 1$ is equivalent to the isothermal process discussed in the previous example.

2 Gravitational Work

The **gravitational work** can be defined as the work done by or against a gravitational force field. In a gravitational field, the force acting on a body is

$$F = mg$$

where m is the mass of the body and g is the acceleration of gravity, which is assumed to be constant. Then the work required to raise this body from level z_1 to level z_2 is

$$W_g = \int_1^2 F\,dz = mg\int_1^2 dz = mg(z_2 - z_1) \qquad \text{(kJ)} \qquad (3\text{-}14)$$

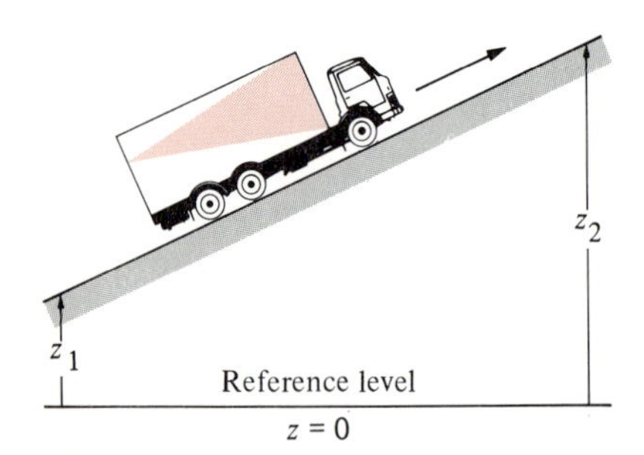

FIGURE 3-28

Vehicles require more power (gravitational work per unit time) as they climb a hill.

where $z_2 - z_1$ is the vertical distance traveled (Fig. 3-28). This expression is easily recognized as the *change in potential energy*. We conclude from Eq. 3-14 that the gravitational work depends on the end states only and is independent of the path followed. Also, the work done is equal, in magnitude, to the change in the potential energy of the system.

The sign of the gravitational work is determined by inspection: *positive* if done by the system (as the system falls) and *negative* if done on the system (as the system is raised). The potential energy of a system increases when gravitational work is done on it.

EXAMPLE 3-10

Determine the work done by a person to lift a 50-lbm suitcase shown in Fig. 3-29 by 1 ft.

Solution By assuming a standard gravitational acceleration and using Eq. 3-14, the work done is

$$W_g = mg(z_2 - z_1)$$

$$= (50 \text{ lbm})(32.174 \text{ ft/s}^2)(1 \text{ ft}) \left(\frac{1 \text{ Btu}}{25{,}037 \text{ ft}^2/\text{s}^2}\right)$$

$$0.064 \text{ Btu}$$

That is, 0.064 Btu of work is needed to perform this task. The potential energy of the system (the suitcase) increases by 0.064 Btu during this process.

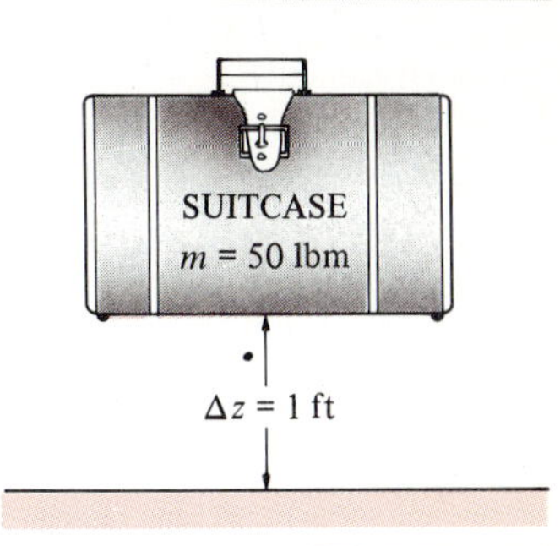

FIGURE 3-29
Schematic for Example 3-10.

3 Accelerational Work

The work associated with a change in velocity of a system is defined as the **accelerational work**. The accelerational work required to accelerate a body of mass m from an initial velocity of $\mathbf{V}_1$ to a final velocity of $\mathbf{V}_2$ (Fig. 3-30) is determined from the definition of acceleration and Newton's second law:

$$\left.\begin{aligned} F &= ma \\ a &= \frac{d\mathbf{V}}{dt} \end{aligned}\right\} \quad F = m\frac{d\mathbf{V}}{dt}$$

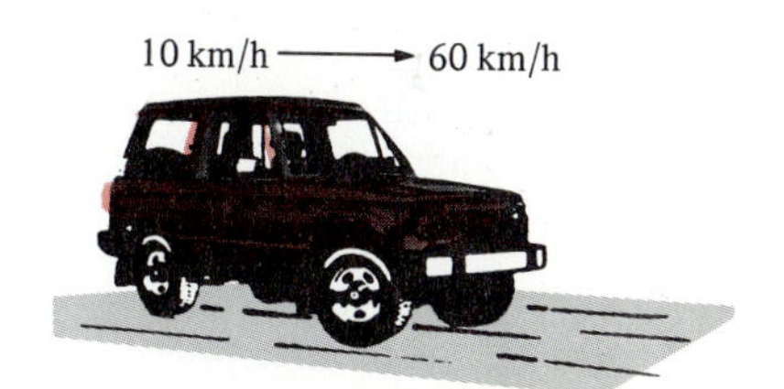

FIGURE 3-30
Vehicles require more power (accelerational work per unit time) as they accelerate.

The differential displacement ds is related to velocity $\mathbf{V}$ by

$$\mathbf{V} = \frac{ds}{dt} \longrightarrow ds = \mathbf{V}\,dt$$

Substituting the F and ds relations into the work expression (Eq. 3-7), we obtain

$$W_a = \int_1^2 F\,ds = \int_1^2 \left(m\frac{d\mathbf{V}}{dt}\right)(\mathbf{V}\,dt)$$

$$= m\int_1^2 \mathbf{V}\,d\mathbf{V} = \tfrac{1}{2}m(\mathbf{V}_2^2 - \mathbf{V}_1^2) \qquad \text{(kJ)} \qquad (3\text{-}15)$$

The work done to accelerate a body is independent of path and is equivalent to the *change in the kinetic energy* of the body.

The sign of accelerational work is determined by inspection: *positive* if done by the system and *negative* if done on the system.

EXAMPLE 3-11

Determine the power required to accelerate a 900-kg car shown in Fig. 3-31 from rest to a velocity of 80 km/h in 20 s on a level road.

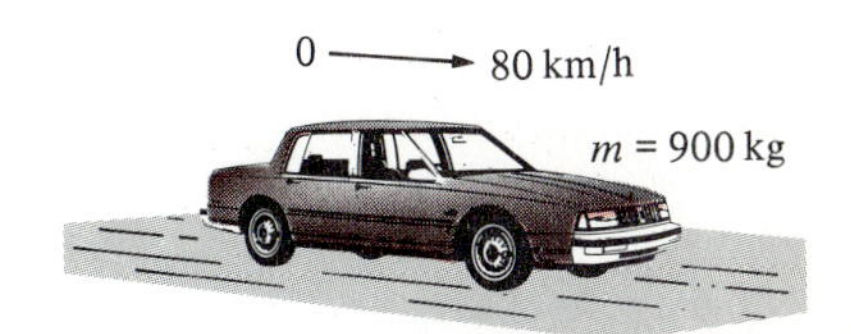

FIGURE 3-31
Schematic for Example 3-11.

Solution The accelerational work is determined from Eq. 3-15 to be

$$W_a = \tfrac{1}{2}m(V_2^2 - V_1^2) = \tfrac{1}{2}(900\text{ kg})\left[\left(\frac{80{,}000\text{ m}}{3600\text{ s}}\right)^2 - 0^2\right]\left(\frac{1\text{ kJ}}{1000\text{ kg}\cdot\text{m}^2/\text{s}^2}\right)$$
$$= 222.2\text{ kJ}$$

The average power is determined from

$$\dot{W}_a = \frac{W_a}{\Delta t} = \frac{222.2\text{ kJ}}{20\text{ s}} = 11.1\text{ kW} \quad (\text{or } 14.9\text{ hp})$$

This is in addition to the power required to combat friction, rolling resistance, and other imperfections.

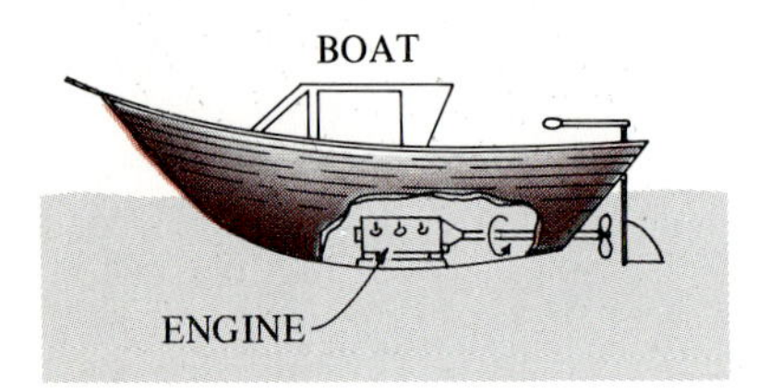

FIGURE 3-32

Energy transmission through rotating shafts is commonly encountered in practice.

4 Shaft Work

Energy transmission with a rotating shaft is very common in engineering practice (Fig. 3-32). Often the torque τ applied to the shaft is constant, which means that the force F applied is also constant. For a specified constant torque, the work done during n revolutions is determined as follows: A force F acting through a moment arm r generates a torque τ (Fig. 3-33) which is determined from

$$\tau = Fr \longrightarrow F = \frac{\tau}{r}$$

This force acts through a distance s which is related to the radius r by

$$s = (2\pi r)n$$

Then the shaft work is determined from Eq. 3-6:

$$W_{sh} = Fs = \left(\frac{\tau}{r}\right)(2\pi rn) = 2\pi n\tau \qquad \text{(kJ)} \tag{3-16}$$

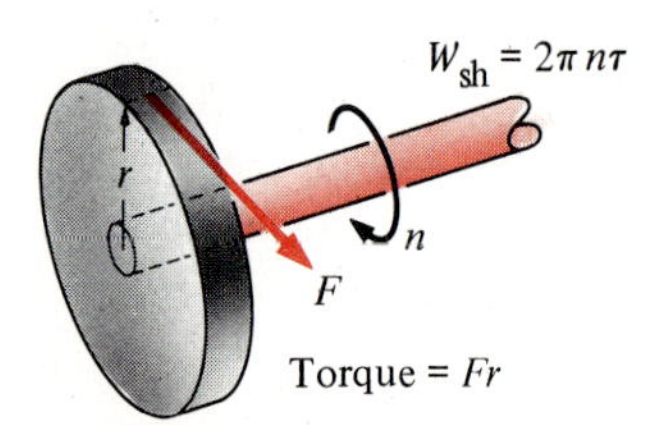

FIGURE 3-33

Shaft work is proportional to the torque applied and the number of revolutions of the shaft.

The power transmitted through the shaft is the shaft work done per unit time, which can be expressed as

$$\dot{W}_{sh} = 2\pi\dot{n}\tau \qquad \text{(kW)} \tag{3-17}$$

where $\dot{n}$ is the number of revolutions per unit time.

The sign of the shaft work is also determined by inspection: *positive* if done by the system and *negative* if done on the system.

EXAMPLE 3-12

Determine the power transmitted through the shaft of a car when the torque applied is 200 N·m and the shaft rotates at a rate of 4000 revolutions per minute (rpm).

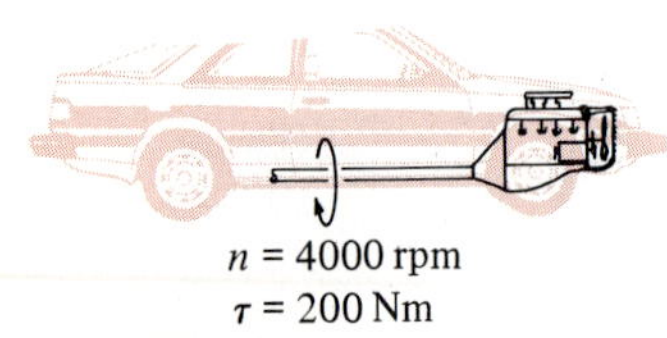

FIGURE 3-34

Schematic for Example 3-12.

Solution A sketch of the car is given in Fig. 3-34. The shaft power is determined from Eq. 3-17:

$$\dot{W}_{sh} = 2\pi\dot{n}\tau = (2\pi)\left(4000\frac{1}{\text{min}}\right)(200\text{ N}\cdot\text{m})\left(\frac{1\text{ min}}{60\text{ s}}\right)\left(\frac{1\text{ kJ}}{1000\text{ N}\cdot\text{m/s}}\right)$$

$$83.7\text{ kW} \quad (\text{or } 112.2\text{ hp})$$

This is the magnitude of the power transmitted through the shaft of the car. The sign of the shaft work depends on the choice of the system.

5 Spring Work

It is common knowledge that when a force is applied on a spring, the length of the spring changes (Fig. 3-35). When the length of the spring changes by a differential amount dx under the influence of a force F, the work done is

$$\delta W_{\text{spring}} = F\,dx \tag{3-18}$$

To determine the total spring work, we need to know a functional relationship between F and x. For linear elastic springs, the displacement x is proportional to the force applied (Fig. 3-36). That is,

$$F = kx \qquad \text{(N)} \tag{3-19}$$

where k is the spring constant and has the unit kN/m. The displacement x is measured from the undisturbed position of the spring (that is, $x = 0$ when $F = 0$). Substituting Eq. 3-19 into Eq. 3-18 and integrating yield

$$W_{\text{spring}} = \tfrac{1}{2}k(x_2^2 - x_1^2) \qquad \text{(kJ)} \tag{3-20}$$

where x_1 and x_2 are the initial and the final displacements of the spring, respectively. Both x_1 and x_2 are measured from the undisturbed position of the spring.

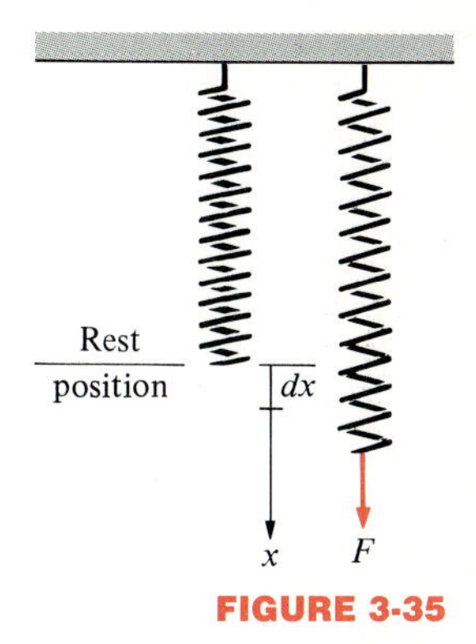

FIGURE 3-35
Elongation of a spring under the influence of a force.

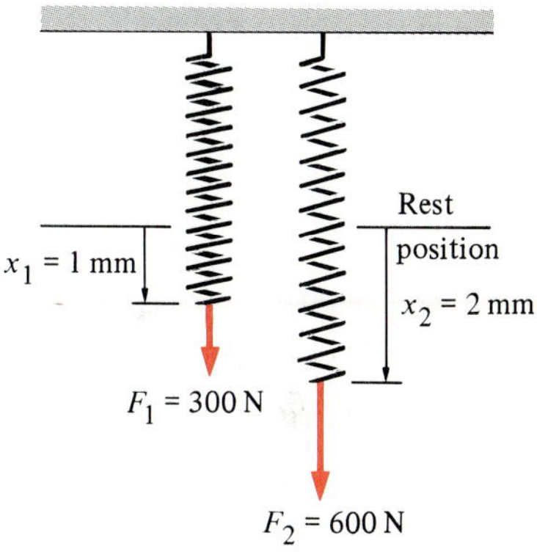

FIGURE 3-36
The displacement of a linear spring doubles when the force is doubled.

EXAMPLE 3-13

A piston-cylinder device contains 0.05 m^3 of a gas initially at 200 kPa. At this state a linear spring which has a spring constant of 150 kN/m is touching the piston but exerting no force on it. Now heat is transferred to the gas, causing the piston to rise and to compress the spring until the volume inside the cylinder doubles. If the cross-sectional area of the piston is 0.25 m^2, determine (*a*) the final pressure inside the cylinder, (*b*) the total work done by the gas, and (*c*) the fraction of this work done against the spring to compress it.

Solution (*a*) A sketch of the system and the *P-V* diagram of the process are shown in Fig. 3-37. The enclosed volume at the final state is

$$V_2 = 2V_1 = (2)(0.05\ \text{m}^3) = 0.1\ \text{m}^3$$

Then the displacement of the piston (and the spring) becomes

$$x = \frac{\Delta V}{A} = \frac{(0.1 - 0.05)\ \text{m}^3}{0.25\ \text{m}^2} = 0.2\ \text{m}$$

The force applied by the linear spring at the final state is determined from Eq. 3-19 to be

$$F = kx = (150\ \text{kN/m})(0.2\ \text{m}) = 30\ \text{kN}$$

The additional pressure applied by the spring on the gas at this state is

$$P = \frac{F}{A} = \frac{30\ \text{kN}}{0.25\ \text{m}^2} = 120\ \text{kPa}$$

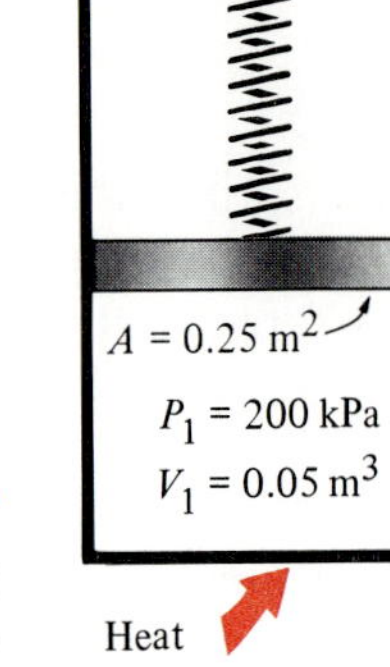

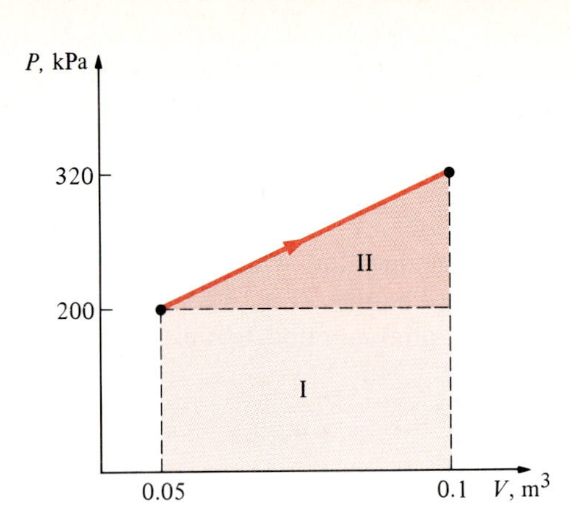

FIGURE 3-37
Schematic and *P-V* diagram for Example 3-13.

Without the spring, the pressure of the gas would remain constant at 200 kPa while the piston is rising. But under the effect of the spring, the pressure rises linearly from 200 kPa to

$$200 + 120 = 320 \text{ kPa}$$

at the final state.

(*b*) An easy way of finding the work done is to plot the process on a *P-V* diagram and find the area under the process curve. From Fig. 3-37, the area under the process curve (a trapezoid) is determined to be

$$|W| = \text{area} = \frac{(200 + 320) \text{ kPa}}{2}[(0.1 - 0.05) \text{ m}^3]\left(\frac{1 \text{ kJ}}{1 \text{ kPa·m}^3}\right) = 13 \text{ kJ}$$

The sign of the work is determined, by inspection, to be positive since it is done by the system.

(*c*) The work represented by the rectangular area (region I) is done against the piston and the atmosphere, and the work represented by the triangular area (region II) is done against the spring. Thus

$$W_{\text{spring}} = \tfrac{1}{2}[(320 - 200) \text{ kPa}](0.05 \text{ m}^3)\left(\frac{1 \text{ kJ}}{1 \text{ kPa·m}^3}\right) = 3 \text{ kJ}$$

This result could also be obtained from Eq. 3-20:

$$W_{\text{spring}} = \tfrac{1}{2}k(x_2^2 - x_1^2)$$

$$= \tfrac{1}{2}(150 \text{ kN/m})[(0.2 \text{ m})^2 - 0^2]\left(\frac{1 \text{ kJ}}{1 \text{ kN·m}}\right) = 3 \text{ kJ}$$

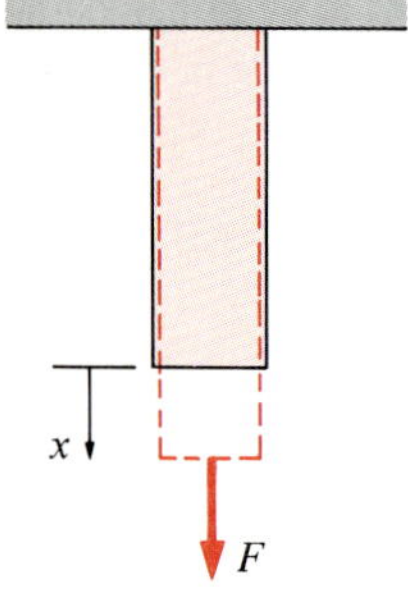

FIGURE 3-38
Solid bars behave as springs under the influence of a force.

Work Done on Elastic Solid Bars

Solids are often modeled as linear springs because under the action of the force they contract or elongate, as shown in Fig. 3-38, and when the force is lifted, they return to their original lengths, like a spring. This is true as long as the force is in the elastic range, i.e., not large enough to cause permanent (plastic) deformations. Therefore, the equations given for a linear spring can also be used for elastic solid bars. Often the work associated with an elastic solid bar is given in terms of stress (force per unit area) and strain (the ratio of elongation to the original length) for convenience.

3-5 ■ THE FIRST LAW OF THERMODYNAMICS

So far we have considered various forms of energy such as heat Q, work W, and total energy E individually, and no attempt has been made to relate them to each other during a process. The *first law of thermodynamics,* also known as *the conservation of energy principle,* provides a sound basis for studying the relationships among the various forms of energy and energy transformations. Based on experimental observations, the first law of thermodynamics states that *energy can be neither created nor destroyed; it can only change forms.* Therefore, every bit of energy should be accounted for during a process. The first law cannot be proved mathematically, but no process in nature is known to have violated the first law, and this should be considered as sufficient proof.

We all know that a rock at some elevation possesses some potential energy, and part of this potential energy is converted to kinetic energy as the rock falls (Fig. 3-39). Experimental data show that the decrease in potential energy ($mg\,\Delta z$) exactly equals the increase in kinetic energy $[m(V_2^2 - V_1^2)/2]$ when the air resistance is negligible, thus confirming the conservation of energy principle. The first-law relation for closed systems is developed below with the help of some familiar examples.

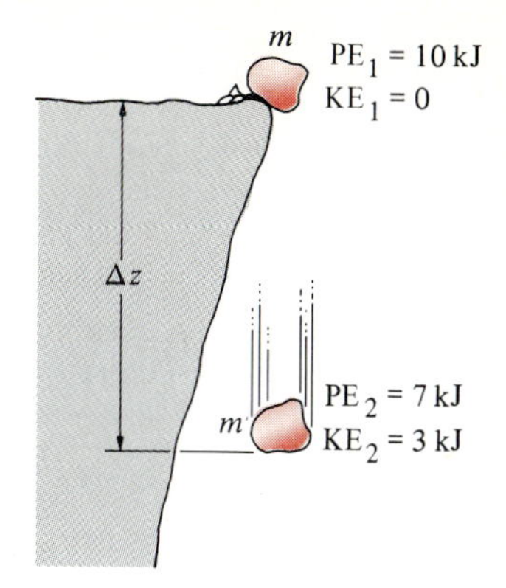

FIGURE 3-39

Energy cannot be created or destroyed; it can only change forms.

Let us consider first some processes that involve heat transfer but no work interactions. The potato in the oven that we have discussed previously is a good example for this case (Fig. 3-40). As a result of heat transfer to the potato, the energy of the potato will increase. If we disregard any mass transfer (moisture loss from the potato), the increase in the total energy of the potato becomes equal to the amount of heat transfer. That is, if 5 kJ of heat is transferred to the potato, the energy increase of the potato will also be 5 kJ. Therefore, the conservation of energy principle for this case can be expressed as $Q = \Delta E$.

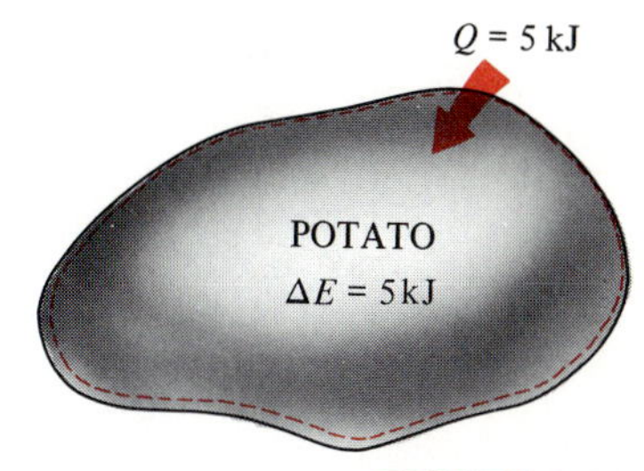

FIGURE 3-40

The increase in the energy of a potato in an oven is equal to the amount of heat transferred to it.

As another example, consider the heating of water in a pan on top of a range (Fig. 3-41). If 15 kJ of heat is transferred to the water from the heating element and 3 kJ of it is lost from the water to the surrounding air, the increase in energy of the water will be equal to the net heat transfer to the water, which is 12 kJ. That is, $Q = Q_{net} = \Delta E$.

The above conclusions can be summarized as follows: *In the absence of any work interactions between a system and its surroundings, the amount of net heat transfer is equal to the change in energy of a closed system.* That is,

$$Q = \Delta E \qquad \text{when } W = 0 \tag{3-21}$$

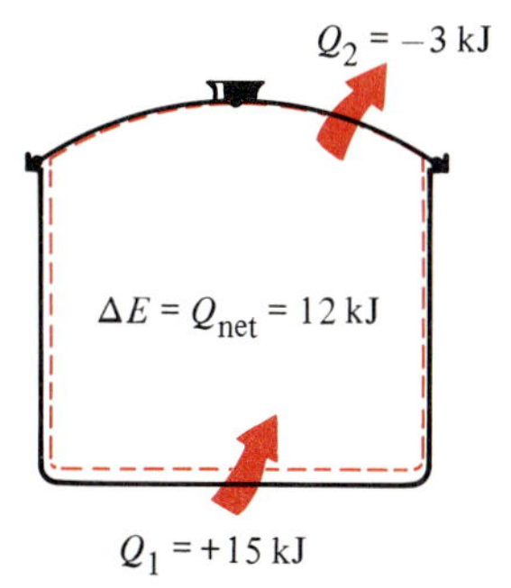

FIGURE 3-41

In the absence of any work interactions, energy change of a system is equal to the net heat transfer.

Now consider a well-insulated (i.e., adiabatic) room heated by an

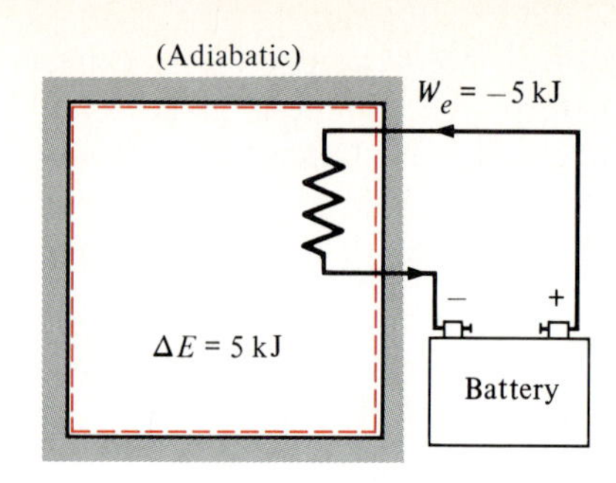

FIGURE 3-42

The work (electrical) done on an adiabatic system is equal to the increase in the energy of the system.

electric heater as our system (Fig. 3-42). As a result of electrical work done, the energy of the system will increase. Since the system is adiabatic and cannot have any heat interactions with the surroundings ($Q = 0$), the conservation of energy principle dictates that the electrical work done on the system must equal the increase in energy of the system. That is, $-W_e = \Delta E$.

The negative sign is due to the sign convention that work done on a system is negative. This ensures that work done on a system increases the energy of the system and work done by a system decreases it.

(Adiabatic)

$\Delta E = 8$ kJ

$W_{pw} = -8$ kJ

FIGURE 3-43

The work (shaft) done on an adiabatic system is equal to the increase in the energy of the system.

Now let us replace the electric heater with a paddle wheel (Fig. 3-43). As a result of the stirring process, the energy of the system will increase. Again since there is no heat interaction between the system and its surroundings ($Q = 0$), the paddle-wheel work done on the system must show up as an increase in the energy of the system. That is, $-W_{pw} = \Delta E$.

Many of you have probably noticed that the temperature of air rises when it is compressed (Fig. 3-44). This is because energy is added to the air in the form of boundary work. In the absence of any heat transfer ($Q = 0$), the entire boundary work will be stored in the air as part of its total energy. The conservation of energy principle again requires that $-W_b = \Delta E$.

It is clear from the foregoing examples that *for adiabatic processes, the amount of work done is equal to the change in the energy of a closed system*. That is,

$$-W = \Delta E \qquad \text{when } Q = 0 \tag{3-22}$$

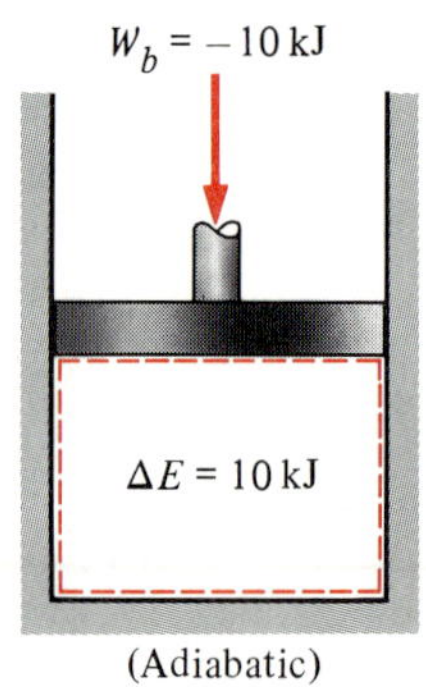

FIGURE 3-44

The work (boundary) done on an adiabatic system is equal to the increase in the energy of the system.

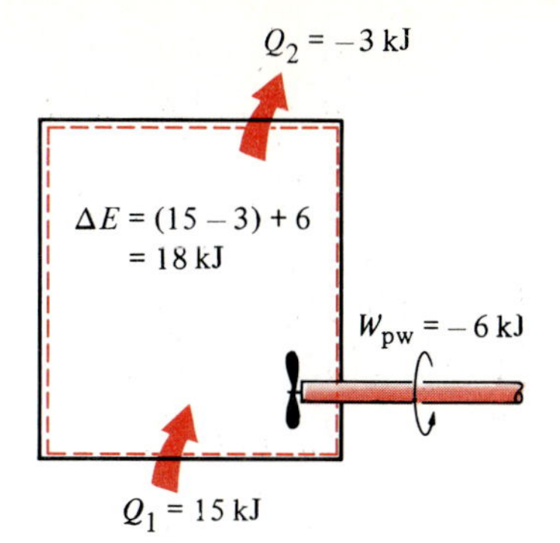

FIGURE 3-45
The energy change of a system during a process is equal to the *net* work and heat transfer between the system and its surroundings.

Now we are in a position to consider simultaneous heat and work interactions. As you may have already guessed, when a system involves both heat and work interactions during a process, their contributions are simply added. That is, if a system receives 12 kJ of heat while a paddle wheel does 6 kJ of work on the system, the net increase in energy of the system for this process will be 18 kJ (Fig. 3-45).

To generalize our conclusions, the **first law of thermodynamics**, or the **conservation of energy principle** for a closed system or a fixed mass, may be expressed as follows:

$$\begin{bmatrix} \text{Net energy transfer} \\ \text{to (or from) the system} \\ \text{as heat and work} \end{bmatrix} = \begin{bmatrix} \text{Net increase (or decrease)} \\ \text{in the total energy} \\ \text{of the system} \end{bmatrix}$$

or

$$Q - W = \Delta E \quad \text{(kJ)} \tag{3-23}$$

where

Q = net heat transfer across system boundaries ($= \Sigma Q_{in} - \Sigma Q_{out}$)
W = net work done in all forms ($= \Sigma W_{out} - \Sigma W_{in}$)
ΔE = net change in total energy of system, $E_2 - E_1$

As discussed in Chap. 1, the total energy E of a system is considered to consist of three parts: internal energy U, kinetic energy KE, and potential energy PE. Then the change in total energy of a system during a process can be expressed as the sum of the changes in its internal, kinetic, and potential energies:

$$\Delta E = \Delta U + \Delta \text{KE} + \Delta \text{PE} \quad \text{(kJ)} \tag{3-24}$$

Substituting this relation into Eq. 3-23, we obtain

$$Q - W = \Delta U + \Delta \text{KE} + \Delta \text{PE} \quad \text{(kJ)} \tag{3-25}$$

where $\Delta U = m(u_2 - u_1)$
$\Delta \text{KE} = \frac{1}{2}m(\mathbf{V}_2^2 - \mathbf{V}_1^2)$
$\Delta \text{PE} = mg(z_2 - z_1)$

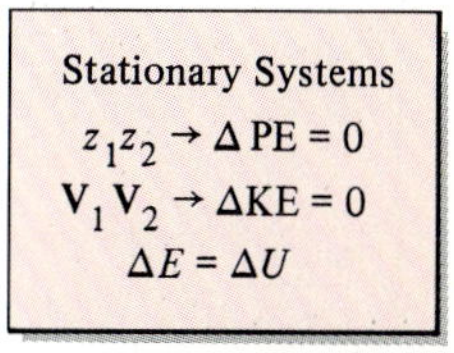

Most closed systems encountered in practice are stationary, i.e., they do not involve any changes in their velocity or the elevation of their center of gravity during a process (Fig. 3-46). Thus for **stationary closed systems**,

Stationary Systems
$z_1 z_2 \rightarrow \Delta \text{PE} = 0$
$\mathbf{V}_1 \mathbf{V}_2 \rightarrow \Delta \text{KE} = 0$
$\Delta E = \Delta U$

FIGURE 3-46
For stationary systems, $\Delta \text{KE} = \Delta \text{PE} = 0$; thus $\Delta E = \Delta U$.

the changes in kinetic and potential energies are negligible (that is, $\Delta KE = \Delta PE = 0$), and the first-law relation reduces to

$$Q - W = \Delta U \qquad \text{(kJ)} \tag{3-26}$$

If the initial and final states are specified, the internal energies u_1 and u_2 can easily be determined from property tables or some thermodynamic relations.

Sometimes it is convenient to consider the work term in two parts: W_{other} and W_b, where W_{other} represents all forms of work except the boundary work. (This distinction has important bearings with regard to the second law of thermodynamics, as is discussed in later chapters.) Then the first law takes the following form:

$$Q - W_{other} - W_b = \Delta E \qquad \text{(kJ)} \tag{3-27}$$

It is extremely important that the *sign convention* be observed for heat and work interactions. Heat flow to a system and work done by a system are positive, and heat flow from a system and work done on a system are negative. A system may involve more than one form of work during a process. The only form of work whose sign we do not need to be concerned with is the boundary work W_b as defined by Eq. 3-9. Boundary work calculated by using Eq. 3-9 will always have the correct sign. The signs of other forms of work are determined by inspection.

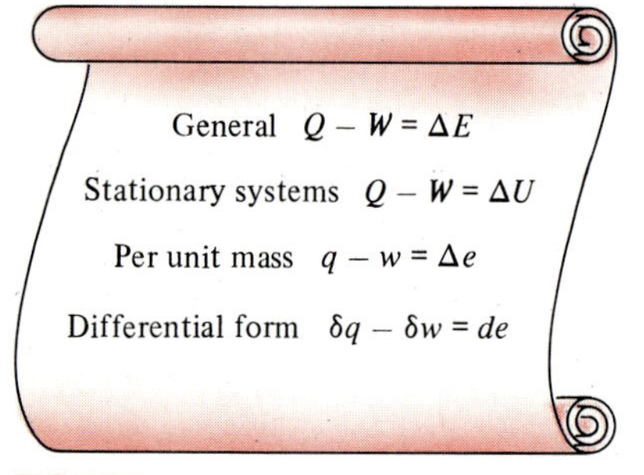

FIGURE 3-47

Various forms of the first-law relation for closed systems.

Other Forms of the First-Law Relation

The first-law relation for closed systems can be written in various forms (Fig. 3-47). Dividing Eq. 3-23 by the mass of the system, for example, gives the first-law relation on a **unit-mass** basis as

$$q - w = \Delta e \qquad \text{(kJ/kg)} \tag{3-28}$$

The **rate form** of the first law is obtained by dividing Eq. 3-23 by the time interval Δt and taking the limit as $\Delta t \to 0$. It yields

$$\dot{Q} - \dot{W} = \frac{dE}{dt} \qquad \text{(kW)} \tag{3-29}$$

where $\dot{Q}$ is the rate of net heat transfer, $\dot{W}$ is the power, and dE/dt is the rate of change of total energy.

Equation 3-23 can be expressed in the **differential form** as

$$\delta Q - \delta W = dE \qquad \text{(kJ)} \tag{3-30}$$

or

$$\delta q - \delta w = de \qquad \text{(kJ/kg)} \tag{3-31}$$

For a **cyclic process**, the initial and final states are identical, and therefore $\Delta E = E_2 - E_1 = 0$. Then the first-law relation for a cycle simplifies to

$$Q - W = 0 \qquad \text{(kJ)} \tag{3-32}$$

That is, the net heat transfer and the net work done during a cycle must be equal (Fig. 3-48).

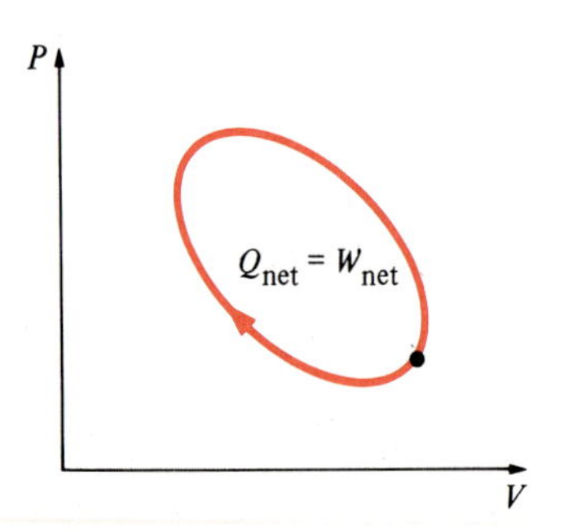

FIGURE 3-48

For a cycle $\Delta E = 0$, thus $Q = W$.

As energy quantities, heat and work are not that different, and you are probably wondering why we keep differentiating these two. After all, the change in the energy content of a system is equal to the amount of energy that crosses the system boundaries, and it makes no difference whether the energy crosses the boundary as heat or work. It seems as if the first-law relations would be much simpler if we had just one quantity which we could call *energy interaction* to represent both heat and work. Well, from the first-law point of view, heat and work are not different at all. But from the second-law point of view, heat and work are very different, as is discussed in later chapters.

EXAMPLE 3-14

A rigid tank contains a hot fluid which is cooled while being stirred by a paddle wheel. Initially, the internal energy of the fluid is 800 kJ. During the cooling process, the fluid loses 500 kJ of heat, and the paddle wheel does 100 kJ of work on the fluid. Determine the final internal energy of the fluid. Neglect the energy stored in the paddle wheel.

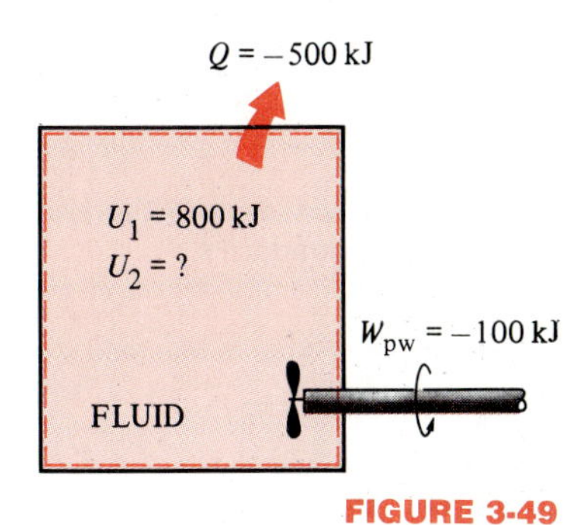

FIGURE 3-49
Schematic for Example 3-14.

Solution We choose the fluid in the tank as our system. The system boundaries are indicated in Fig. 3-49.

Since no mass crosses the boundary, this is a closed system (a fixed mass). There is no mention of motion. Therefore, we assume that the closed system is stationary and that the potential and kinetic energy changes are zero. By applying the conservation of energy principle as given by Eq. 3-25 to this process, U_2 is determined to be

$$\begin{aligned} Q - W &= \Delta U + \cancelto{0}{\Delta KE} + \cancelto{0}{\Delta PE} \\ &= U_2 - U_1 \\ -500 \text{ kJ} - (-100 \text{ kJ}) &= U_2 - 800 \text{ kJ} \\ U_2 &= 400 \text{ kJ} \end{aligned}$$

Note that the heat transfer is negative since it is from the system; so is the work, since it is done on the system.

3-6 ■ A SYSTEMATIC APPROACH TO PROBLEM SOLVING

To this point, we have concentrated our efforts on understanding the basics of thermodynamics. Armed with this knowledge, we are now in a position to tackle significant engineering problems. Knowledge is certainly an essential part of problem solving. But thermodynamic problems, particularly the complicated ones, also require a systematic approach. By using a step-by-step approach, an engineer can solve a series of simple problems instead of one large, formidable problem (Fig. 3-50).

The proper approach to solving thermodynamic problems is illustrated below with the help of a sample problem. Readers are urged to master this approach and use it zealously, since this will help them avoid some of the common pitfalls of problem solving.

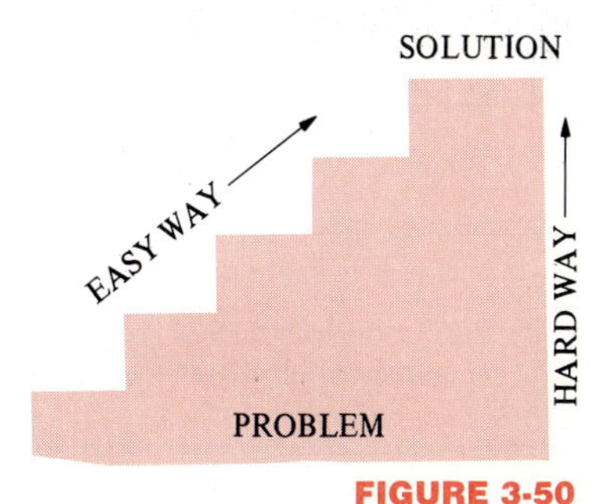

FIGURE 3-50
A step-by-step approach can greatly simplify problem solving.

SAMPLE PROBLEM

A 0.1-m^3 rigid tank contains steam initially at 500 kPa and 200°C. The steam is now allowed to cool until its temperature drops to 50°C. Determine the amount of heat transfer during this process and the final pressure in the tank.

Step 1: Draw a Sketch and Identify the System

It is always a good practice to start solving a problem by drawing a sketch of the physical system. It does not have to be something elaborate, but it should resemble the physical system on hand. The system that is about to be analyzed should be identified on the sketch by drawing its boundaries by dashed lines (Fig. 3-51). In this way the region or system to which the conservation equations are applied is clearly specified.

For simple problems, the choice of the system is quite obvious (the steam in the tank, in this case), but large problems may involve several devices and even several different substances, and may require separate analyses for various parts of the system. For such cases, it is essential that the system be identified before each analysis.

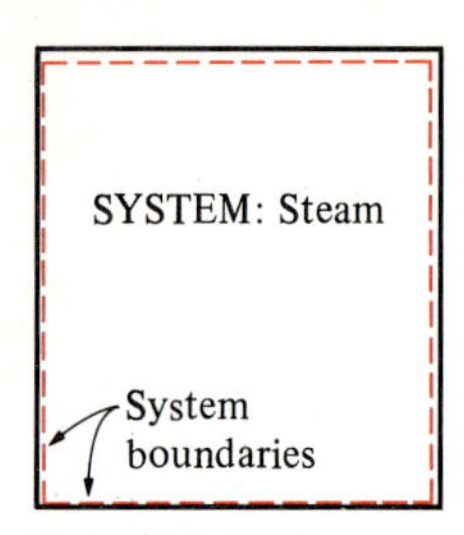

FIGURE 3-51
Step 1: Draw a sketch of the system and system boundaries.

Step 2: List the Given Information on the Sketch

In a typical problem, the hard information is scattered, and listing the given data with the proper symbols on the sketch enables one to see the entire problem at once (Fig. 3-52). Heat and work interactions, if any, should also be indicated on the sketch with the proper signs.

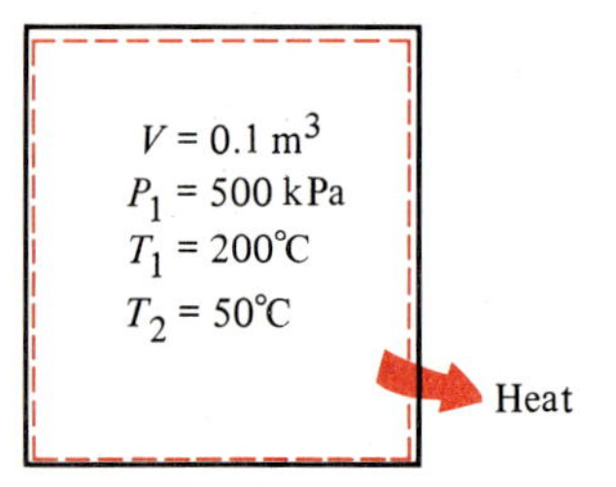

FIGURE 3-52
Step 2: List the given information on the sketch.

Step 3: Check for Special Processes

During a process not all the properties change. Also, not all processes involve heat transfer and various work interactions simultaneously. Often a key property, such as temperature or pressure, remains constant during the process, and this greatly simplifies the analysis. In our case, both the temperature and the pressure vary but the specific volume remains constant since rigid tanks have a fixed volume (V = constant), and the mass of our system is fixed (m = constant) (Fig. 3-53). Then it follows that

$$v = \frac{V}{m} = \text{constant} \longrightarrow v_2 = v_1$$

If the process were isothermal, we would have $T_2 = T_1$, and if it were adiabatic, we would have $Q = 0$. Our system involves no moving boundaries or any other forms of work, thus the work term is zero, $W = 0$.

$v_1 = v_2$
$W = 0$

FIGURE 3-53
Step 3: Look for simplifications.

Step 4: State any Assumptions

The simplifying assumptions that are made to solve a problem should be stated and fully justified. Assumptions whose validity is questionable should be avoided, if possible. Some commonly made assumptions in thermodynamics are assuming the process to be quasi-equilibrium, neglecting the changes in kinetic and potential energies of a system, treating a gas as an ideal gas, and neglecting the heat transfer to or from insulated systems.

The system in our case can be assumed to be stationary since there is no indication to the contrary. Thus the changes in kinetic and potential energies can be neglected (Fig. 3-54).

Assume:

(1) $\Delta PE = 0$ (since there is no mention of elevation change).

(2) $\Delta KE = 0$ (since there is no mention of velocity change).

FIGURE 3-54
Step 4: Make realistic assumptions, if necessary.

Step 5: Apply the Conservation Equations

Now we are ready to apply the conservation equations such as the conservation of mass and conservation of energy. We should start with the most general forms of the equations and simplify them, using the applicable assumptions (Fig. 3-55). The numerical values should not be introduced into the equations before they are reduced to their simplest forms.

Mass: $m_2 = m_1 = m$

Energy: $Q - \overset{0}{\cancel{W}} = \Delta U + \overset{0}{\Delta \cancel{KE}} + \overset{0}{\Delta \cancel{PE}}$

$$Q = m(u_2 - u_1)$$

FIGURE 3-55

Step 5: Apply relevant conservation equations and simplify them.

Step 6: Draw a Process Diagram

Process diagrams, such as the *P-v* or *T-v* diagrams, are extremely helpful in visualizing the initial and final states of a system and the path of the process. If a property remains constant during a process, it should be apparent on the diagram. For our problem, the volume remains constant during the process, and so it will appear as a vertical-line segment on a *T-v* diagram (Fig. 3-56). For pure substances, the process diagrams should be plotted relative to the saturation lines. That way, the region where a substance is found at each state will be apparent. From this diagram it is clear that steam is a superheated vapor at the initial state and a saturated mixture at the final state.

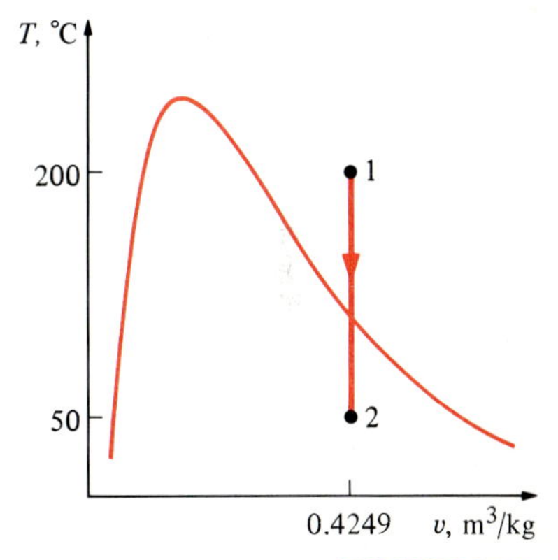

FIGURE 3-56

Step 6: Show the process on a property diagram.

Step 7: Determine the Required Properties and Unknowns

The unknown properties at any state can be determined with the help of thermodynamic relations or tables. Thermodynamic relations are usually valid over some limited range, and therefore their validity should be checked before they are used, to prevent any errors. The thermodynamic relation that is used incorrectly most often is probably the ideal-gas relation. Even though its use is limited to gases at low pressures (relative to the critical-point value), some use it carelessly for substances that are not even in the gas phase.

When reading properties from the steam or refrigerant tables, we first need to know whether the substance exists as a superheated vapor, a compressed liquid, or a saturated mixture. This can easily be done by comparing the given properties to the corresponding saturation properties.

In the sample problem, the temperature and pressure at the initial state are given (Fig. 3-57). Steam exists as a superheated vapor at this state since the

State 1: $P_1 = 500$ kPa, $T_1 = 200°C$ $\Big\}$ $v_1 = 0.4249$ m^3/kg, $u_1 = 2642.9$ kJ/kg

State 2: $v_2 = v_1 = 0.4269$ m^3/kg

$T_2 = 50°C \rightarrow v_f = 0.001$ m^3/kg

$v_g = 12.03$ m^3/kg

$u_f = 209.32$ kJ/kg

$u_g = 2443.5$ kJ/kg

(Table A-4)

$P_2 = P_{sat}$ @ 50°C= **12.349 kPa**

$v_2 = v_f + x_2 v_{fg}$

$0.4249 = 0.001 + x_2(12.03 - 0.001)$

$x_2 = 0.0352$

$u_2 = u_f + x_2 u_{fg}$

$= 209.32 + (0.0352)(2443.5 - 209.32)$

$= 288.0$ kJ/kg

$$m = \frac{V}{v} = \frac{0.1 \text{ m}^3}{0.4249 \text{ m}^3/\text{kg}} = 0.235 \text{ kg}$$

$Q = m(u_2 - u_1)$

$= (0.235 \text{ kg})(288 - 264.9 \text{ kJ/kg}$

$=$ **-553.4 kJ**

FIGURE 3-57

Step 7: Determine the required properties, and solve the problem.

temperature is greater than the saturation temperature at the given pressure (that is, 200°C > $T_{sat\ @\ 500\ kPa}$ = 151.9°C). Then the initial values of the specific volume and internal energy are readily determined from the superheated vapor tables. At the final state, steam exists as a saturated mixture since the specific volume v_2 is greater than v_f but less than v_g at the final temperature, that is, $v_f < v_2 < v_g$.

Particular attention should be paid to the units of various quantities when the numerical values are substituted into the equations. The majority of the errors at this stage are due to using inconsistent units. Finally, any unreasonable results should be interpreted as an indication of possible errors, and the analysis should be checked.

The approach described above is consistently used in the example problems that follow without explicitly stating each step. For some problems, some of the steps may not be applicable or necessary, and they may be skipped. However, we cannot overemphasize the importance of a logical and orderly approach to problem solving. The majority of the difficulties encountered while solving a problem are not due to a lack of knowledge; they are due to a lack of coordination. The individual studying thermodynamics is strongly encouraged to follow these steps in problem solving until he or she develops a personal systematic approach that works best.

EXAMPLE 3-15

A piston-cylinder device contains 25 g of saturated water vapor which is maintained at a constant pressure of 300 kPa. A resistance heater within the cylinder is turned on and passes a current of 0.2 A for 5 min from a 120-V source. At the same time, a heat loss of 3.7 kJ occurs. (*a*) Show that for a closed system the boundary work W_b and the change in internal energy ΔU in the first-law relation can be combined into one term, ΔH, for a constant-pressure process. (*b*) Determine the final temperature of the steam.

Solution The water and the resistance wires contained within the piston-cylinder device are selected as the system, and the system boundaries are indicated by dashed lines in Fig. 3-58. Also shown in this figure is the *P-v* diagram

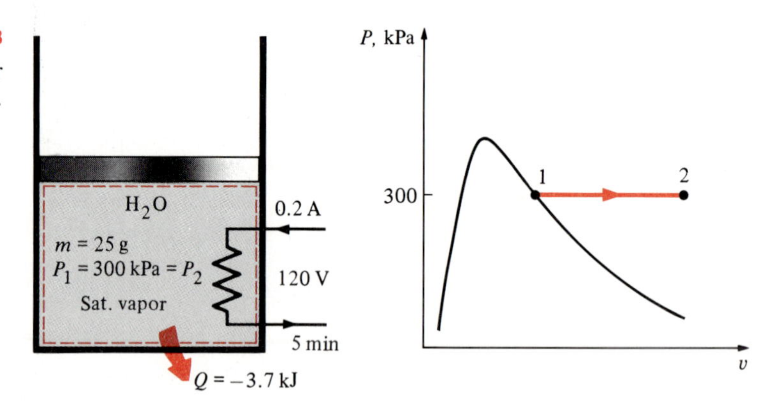

FIGURE 3-58 Schematic and *P-v* diagram for Example 3-15.

for the process. The electric wires make up a very small fraction of the system; thus the internal energy change of the system is the internal energy change of water only.

(*a*) Neglecting the changes in kinetic and potential energies and expressing the work as the sum of boundary and other forms of work, Eq. 3-25 simplifies to

$$Q - W = \Delta U + \cancel{\Delta KE}^{\,0} + \cancel{\Delta PE}^{\,0}$$
$$Q - W_{\text{other}} - W_b = U_2 - U_1$$

For a constant-pressure process, the boundary work is given by Eq. 3-10 as $W_b = P_0(V_2 - V_1)$. Substituting this into the above relation gives

$$Q - W_{\text{other}} - P_0(V_2 - V_1) = U_2 - U_1$$

But

$$P_0 = P_2 = P_1 \longrightarrow Q - W_{\text{other}} = (U_2 + P_2V_2) - (U_1 + P_1V_1)$$

Also $H = U + PV$, and thus

$$Q - W_{\text{other}} = H_2 - H_1 \qquad \text{(kJ)} \tag{3-33}$$

which is the desired relation (Fig. 3-59). *This equation is very convenient to use in the analysis of closed systems undergoing a constant-pressure quasi-equilibrium process since the boundary work is automatically taken care of by the enthalpy terms, and one no longer needs to determine it separately.*

(*b*) For our case, the only other form of work is the electrical work which can be determined from Eq. 3-5:

$$W_e = VI\,\Delta t = (120\text{ V})(0.2\text{ A})(300\text{ s})\left(\frac{1\text{ kJ}}{1000\text{ VA·s}}\right) = 7.2\text{ kJ}$$

State 1: $\left.\begin{array}{l} P_1 = 300\text{ kPa} \\ \text{sat. vapor} \end{array}\right\} \quad h_1 = h_{g\,@\,300\text{ kPa}} = 2725.3\text{ kJ/kg} \qquad \text{(Table A-5)}$

The enthalpy at the final state can be determined from the conservation of energy relation for closed systems undergoing a constant-pressure process (Eq. 3-33):

$$Q - W_e = m(h_2 - h_1)$$
$$-3.7\text{ kJ} - (-7.2\text{ kJ}) = (0.025\text{ kg})(h_2 - 2725.3\text{ kJ/kg})$$
$$h_2 = 2865.3\text{ kJ/kg}$$

Now the final state is completely specified since we know both the pressure and the enthalpy. The temperature at this state is

State 2: $\left.\begin{array}{l} P_2 = 300\text{ kPa} \\ h_2 = 2865.3\text{ kJ/kg} \end{array}\right\} \quad T_2 = 200°\text{C} \qquad \text{(Table A-6)}$

Therefore, the steam will be at 200°C at the end of this process.

Strictly speaking, the potential energy change of the steam is not zero for this process since the center of gravity of the steam rose somewhat. Assuming an elevation change of 1 m (which is rather unlikely), the change in the potential energy of the steam would be (from Eq. 3-14) 0.0002 kJ, which is very small compared to the other terms in the first-law relation. Therefore, in problems of this kind, the potential energy term is always neglected.

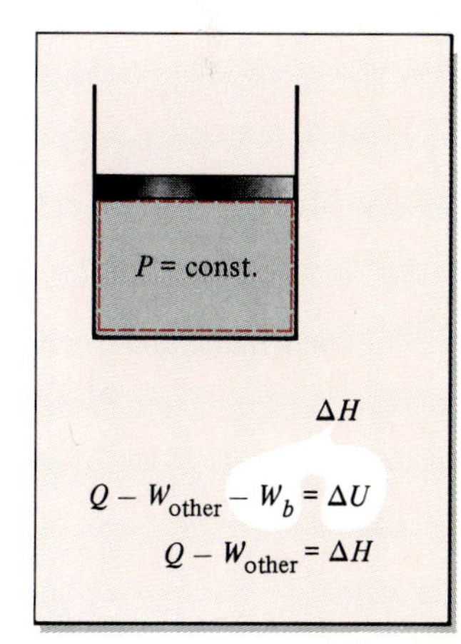

FIGURE 3-59

For a closed system undergoing a quasi-equilibrium P = constant process, $\Delta U + W_b = \Delta H$.

EXAMPLE 3-16

A rigid tank with a volume of 3 ft^3 is initially filled with refrigerant-12 at 120 psia and 140°F. The refrigerant is now cooled to 20°F. Determine (*a*) the mass of the refrigerant, (*b*) the final pressure in the tank, and (*c*) the heat transferred from the refrigerant.

Solution We take the refrigerant in the tank as our system. A sketch of the system and the *T*-*v* diagram of the process are given in Fig. 3-60. Since no mass is entering or leaving the tank during the process, this is a closed system.

(*a*) To find the mass, we need to know the specific volume of the refrigerant at the initial state, which is determined from Table A-13E:

State 1:

$$\left.\begin{array}{l} P_1 = 120 \text{ psia} \\ T_1 = 140°\text{F} \end{array}\right\} \quad \begin{array}{l} v_1 = 0.389 \text{ ft}^3/\text{lbm} \\ u_1 = 86.098 \text{ Btu/lbm} \end{array}$$

Thus,

$$m = \frac{V}{v_1} = \frac{3 \text{ ft}^3}{0.389 \text{ ft}^3/\text{lbm}} = 7.71 \text{ lbm}$$

(*b*) This is a constant-volume process, and therefore $v_2 = v_1 = 0.389$ ft^3/lbm.

At 20°F:

$$\begin{array}{ll} v_f = 0.01130 \text{ ft}^3/\text{lbm} & u_f = 12.79 \text{ Btu/lbm} \\ v_g = 1.0988 \text{ ft}^3/\text{lbm} & u_g = 72.12 \text{ Btu/lbm} \end{array} \qquad \text{(Table A-11E)}$$

The refrigerant is a saturated liquid–vapor mixture at the final state since $v_f < v_2 < v_g$. Therefore, the pressure must be the saturation pressure at 20°F:

$$P_2 = P_{\text{sat @ 20°F}} = 35.736 \text{ psia} \qquad \text{(Table A-11E)}$$

(*c*) Assuming $\Delta\text{KE} = \Delta\text{PE} = 0$ and realizing that there are no boundary or other forms of work interactions, we see that the first-law relation (Eq. 3-25) simplifies to

$$Q - W^{\nearrow 0} = \Delta U + \Delta\text{KE}^{\nearrow 0} + \Delta\text{PE}^{\nearrow 0}$$

$$Q = m(u_2 - u_1)$$

To determine u_2, we first need to know the quality x_2 at the final state. It is determined from

$$x_2 = \frac{v_2 - v_f}{v_{fg}} = \frac{0.389 - 0.0113}{1.0988 - 0.0113} = 0.348$$

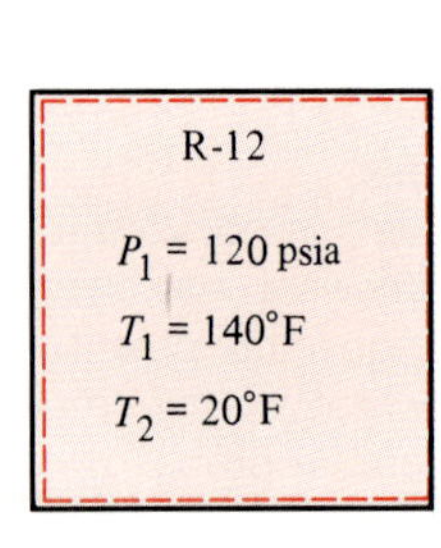

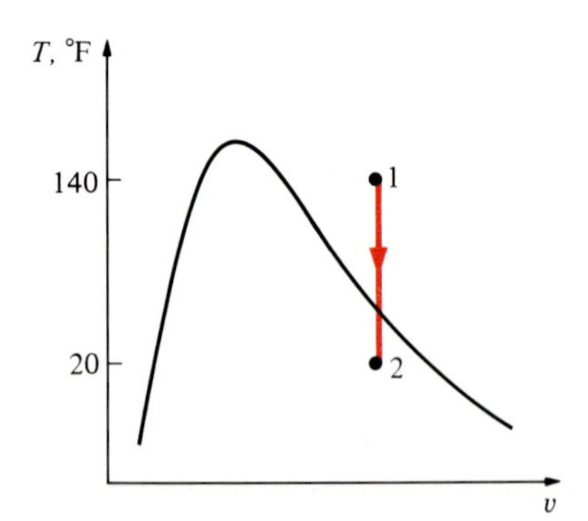

FIGURE 3-60
Schematic and *T*-*v* diagram for Example 3-16.

That is, 34.8 percent of the refrigerant is in the vapor form at the final state. Then

$$u_2 = u_f + x_2 u_{fg}$$
$$= 12.79 \text{ Btu/lbm} + (0.348)[(72.12 - 12.79) \text{ Btu/lbm}]$$
$$= 33.44 \text{ Btu/lbm}$$

Finally, substituting these values into the first-law relation will give us the heat transfer:

$$Q = (7.71 \text{ lbm})[(33.44 - 86.098) \text{ Btu/lbm}]$$
$$= -406.0 \text{ Btu}$$

The negative sign indicates that heat is leaving the system.

EXAMPLE 3-17

A rigid tank is divided into two equal parts by a partition. Initially, one side of the tank contains 5 kg of water at 200 kPa and 25°C, and the other side is evacuated. The partition is then removed, and the water expands into the entire tank. The water is allowed to exchange heat with its surroundings until the temperature in the tank returns to the initial value of 25°C. Determine (*a*) the volume of the tank, (*b*) the final pressure, and (*c*) the heat transfer for this process.

Solution We take the water in the tank as our system. A sketch of the system as well as a *P*-*v* diagram of the process is given in Fig. 3-61.

(*a*) Initially the water in the tank exists as a compressed liquid since its pressure (200 kPa) is greater than the saturation pressure at 25°C (3.169 kPa). Approximating the compressed liquid as a saturated liquid at the given temperature, we find

$$v_1 \cong v_{f\,@\,25^\circ C} = 0.001003 \text{ m}^3/\text{kg} \cong 0.001 \text{ m}^3/\text{kg} \qquad \text{(Table A-4)}$$

Then the initial volume of the water is

$$V_1 = mv_1 = (5 \text{ kg})(0.001 \text{ m}^3/\text{kg}) = 0.005 \text{ m}^3$$

The total volume of the tank is twice this amount:

$$V_{tank} = (2)(0.005 \text{ m}^3) = 0.01 \text{ m}^3$$

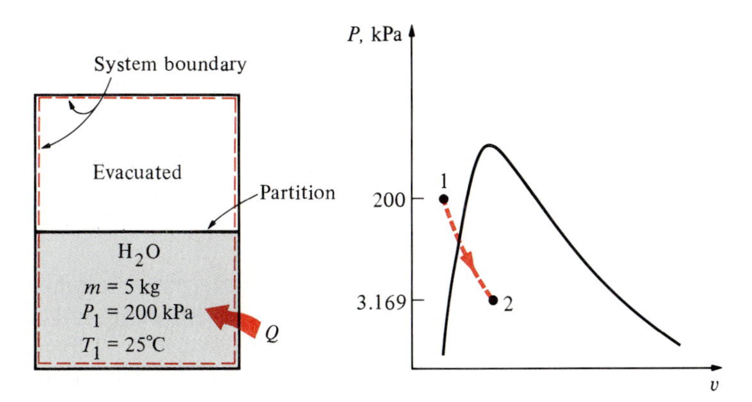

FIGURE 3-61

Schematic and *P*-*v* diagram for Example 3-17.

(*b*) At the final state, the specific volume of the water is

$$v_2 = \frac{V_2}{m} = \frac{0.01\ \text{m}^3}{5\ \text{kg}} = 0.002\ \text{m}^3/\text{kg}$$

which is twice the initial value of the specific volume. This result is expected since the volume doubles while the amount of mass remains constant.

At 25°C: $v_f = 0.001003\ \text{m}^3/\text{kg}$ and $v_g = 43.36\ \text{m}^3/\text{kg}$ (Table A-4)

Since $v_f < v_2 < v_g$, the water is a saturated liquid–vapor mixture at the final state, and thus the pressure is the saturation pressure at 25°C:

$$P_2 = P_{\text{sat @ 25°C}} = 3.169\ \text{kPa} \qquad \text{(Table A-4)}$$

(*c*) At this point we need to specify the system to which the conservation of energy equation will be applied. This time we choose the water and the evacuated space to be our system. The boundaries of this system coincide with the inner surfaces of the tank walls. By neglecting the changes in kinetic and potential energies, the first-law relation for this closed system reduces to

$$Q - \cancel{W}^{\,0} = \Delta U + \Delta \cancel{KE}^{\,0} + \Delta \cancel{PE}^{\,0}$$

$$Q = m(u_2 - u_1)$$

Notice that even though the water is expanding during this process, the system chosen involves fixed boundaries only (the dashed lines) and therefore the moving boundary work is zero. Then $W = 0$ since the system does not involve any other forms of work. (Can you reach the same conclusion by choosing the water as our system?)

Initially,

$$u_1 \cong u_{f\,@\,25°\text{C}} = 104.88\ \text{kJ/kg}$$

The quality at the final state is determined from the specific-volume information:

$$x_2 = \frac{v_2 - v_f}{v_{fg}} = \frac{0.002 - 0.001}{43.36 - 0.001} = 2.3 \times 10^{-5}$$

Then

$$\begin{aligned} u_2 &= u_f + x_2 u_{fg} \\ &= 104.88\ \text{kJ/kg} + (2.3 \times 10^{-5})(2304.9\ \text{kJ/kg}) \\ &= 104.93\ \text{kJ/kg} \end{aligned}$$

Substituting yields

$$Q = (5\ \text{kg})[(104.93 - 104.88)\ \text{kJ/kg}] = 0.25\ \text{kJ}$$

The positive sign indicates that heat is transferred to the water.

3-7 ■ SPECIFIC HEATS

We know from experience that it takes different amounts of energy to raise the temperature of identical masses of different substances by one degree. For example, we need about 4.5 kJ of energy to raise the temperature of 1 kg iron from 20 to 30°C, whereas it takes about 9 times this energy (41.8 kJ to be exact) to raise the temperature of 1 kg of liquid water by the same amount (Fig. 3-62). Therefore, it is desirable to have a property that will enable us to compare the energy storage capabilities of various substances. This property is the specific heat.

The **specific heat** is defined as *the energy required to raise the temperature of a unit mass of a substance by one degree* (Fig. 3-63). In general, this energy will depend on how the process is executed. In thermodynamics, we are interested in two kinds of specific heats: **specific heat at constant volume** C_v and **specific heat at constant pressure** C_p.

Physically, the specific heat at constant volume C_v can be viewed as *the energy required to raise the temperature of the unit mass of a substance by one degree as the volume is maintained constant*. The energy required to do the same as the pressure is maintained constant is the specific heat at constant pressure C_p. This is illustrated in Fig. 3-64. The specific heat at constant pressure C_p is always greater than C_v because at constant pressure the system is allowed to expand and the energy for this expansion work must also be supplied to the system.

Now we will attempt to express the specific heats in terms of other thermodynamic properties. First, consider a stationary closed system undergoing a constant-volume process ($w_b = 0$). The first-law relation for this process can be expressed in the differential form as

$$\delta q - \delta w_{\text{other}} = du$$

The left-hand side of this equation ($\delta q - \delta w_{\text{other}}$) represents the amount of energy transferred to the system in the form of heat and/or work. From the definition of C_v, this energy must be equal to $C_v\, dT$, where dT is the differential change in temperature. Thus,

$$C_v\, dT = du \qquad \text{at constant volume}$$

or

$$C_v = \left(\frac{\partial u}{\partial T}\right)_v \tag{3-34}$$

Similarly, an expression for the specific heat at constant pressure C_p can be obtained by considering a constant-pressure process ($w_b + \Delta u = \Delta h$).

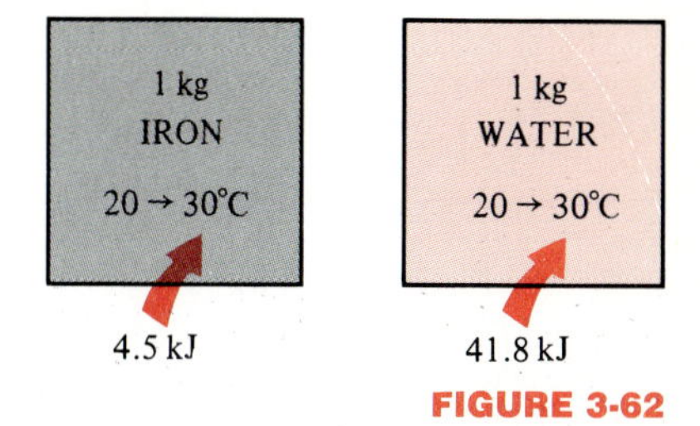

FIGURE 3-62
It takes different amounts of energy to raise the temperature of different substances by the same amount.

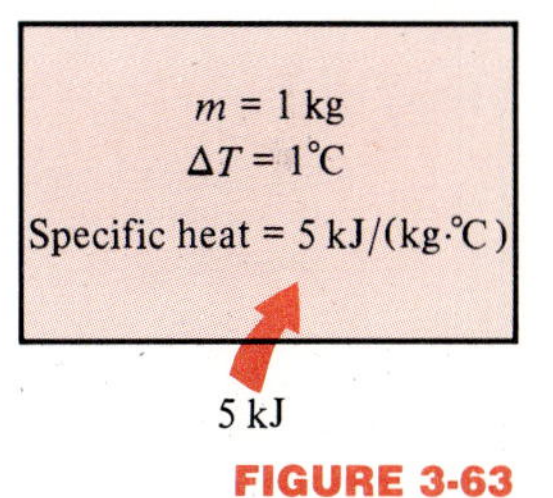

FIGURE 3-63
Specific heat is the energy required to raise the temperature of a unit mass of a substance by one degree in a specified way.

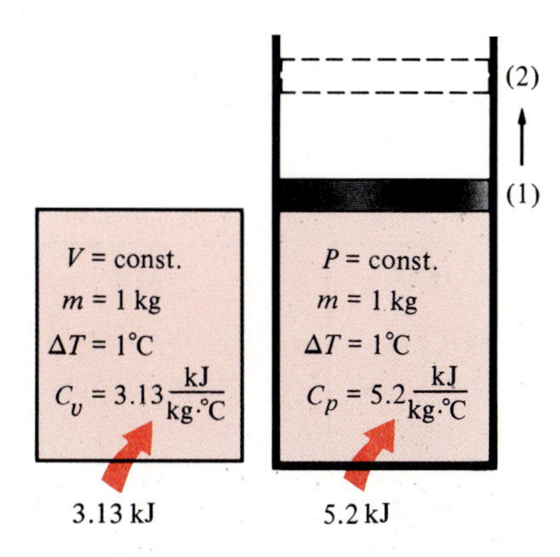

FIGURE 3-64
Constant-volume and constant-pressure specific heats C_v and C_p (values given are for helium gas).

It yields

$$C_p = \left(\frac{\partial h}{\partial T}\right)_p \tag{3-35}$$

Equations 3-34 and 3-35 are the defining equations for C_v and C_p, and their interpretation is given in Fig. 3-65.

Note that C_v and C_p are expressed in terms of other properties, thus they must be properties themselves. Like any other property, the specific heats of a substance depend on the state which, in general, is specified by two independent, intensive properties. That is, the energy required to raise the temperature of a substance by one degree will be different at different temperatures and pressures (Fig. 3-66). But this difference is usually not very large.

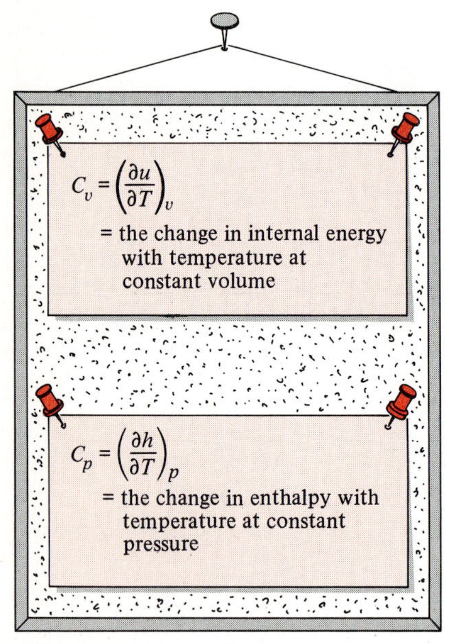

FIGURE 3-65
Mathematical definitions of C_v and C_p.

A few observations can be made from Eqs. 3-34 and 3-35. First, these equations are *property relations* and as such *are independent of the type of processes*. They are valid for *any* substance undergoing *any* process. The only relevance C_v has to a constant-volume process is that C_v happens to be the energy transferred to a system during a constant-volume process per unit mass per unit degree rise in temperature. This is how the values of C_v are determined. This is also how the name *specific heat at constant volume* originated. Likewise, the energy transferred to a system per unit mass per unit degree temperature rise during a constant-pressure process happens to be equal to C_p. This is how the values of C_p can be determined and also explains the origin of the name *specific heat at constant pressure*.

Another observation that can be made from Eqs. 3-34 and 3-35 is that C_v is related to the changes in internal energy and C_p to the changes in enthalpy. In fact, it would be more proper to define C_v as *the change in the specific internal energy of a substance per unit change in temperature at constant volume*. Likewise, C_p can be defined as *the change in the specific enthalpy of a substance per unit change in temperature at constant pressure*. In other words, C_v is a measure of the variation of internal energy of a substance with temperature, and C_p is a measure of the variation of enthalpy of a substance with temperature.

Both the internal energy and enthalpy of a substance can be changed by the transfer of *energy* in any form, with heat being only one of them. Therefore, the term *specific energy* is probably more appropriate than the term *specific heat,* which implies that energy is transferred (and even stored) in the form of heat.

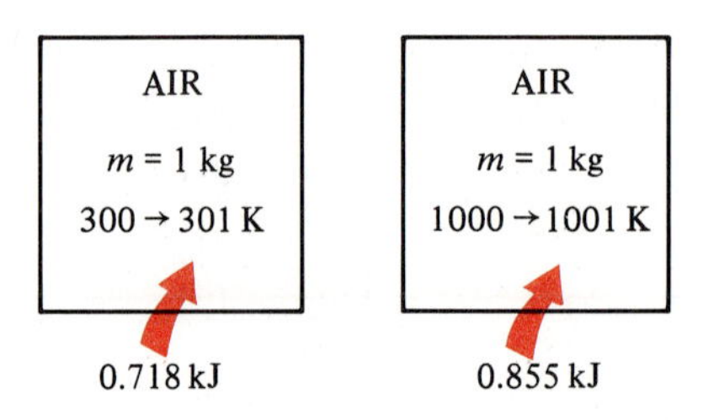

FIGURE 3-66
The specific heat of a substance changes with temperature.

A common unit for specific heats is kJ/(kg·°C) or kJ/(kg·K). Notice that these two units are *identical* since $\Delta T(°C) = \Delta T(K)$, and 1°C change in temperature is equivalent to a change of 1 K (see Sec. 1-10). The specific heats are sometimes given on a *molar basis*. They are denoted by $\overline{C}_v$ and $\overline{C}_p$ and have the unit kJ/(kmol·°C) or kJ/(kmol·K).

3-8 ■ INTERNAL ENERGY, ENTHALPY, AND SPECIFIC HEATS OF IDEAL GASES

In Chap. 2, we defined an ideal gas as a substance whose temperature, pressure, and specific volume are related by

$$Pv = RT$$

It has been demonstrated mathematically (Chap. 11) and experimentally (Joule, 1843) that for an ideal gas the internal energy is a function of the temperature only. That is,

$$u = u(T) \tag{3-36}$$

In his classical experiment, Joule submerged two tanks connected with a pipe and a valve in a water bath, as shown in Fig. 3-67. Initially, one tank contained air at a high pressure, and the other tank was evacuated. When thermal equilibrium was attained, he opened the valve to let air pass from one tank to the other until the pressures equalized. Joule observed no change in the temperature of the water bath and assumed that no heat was transferred to or from the air. Since there was also no work done, he concluded that the internal energy of the air did not change even though the volume and the pressure changed. Therefore, he reasoned, the internal energy is a function of temperature only and not a function of pressure or specific volume. (Joule later showed that for gases which deviate from ideal-gas behavior significantly, the internal energy is not a function of temperature alone.)

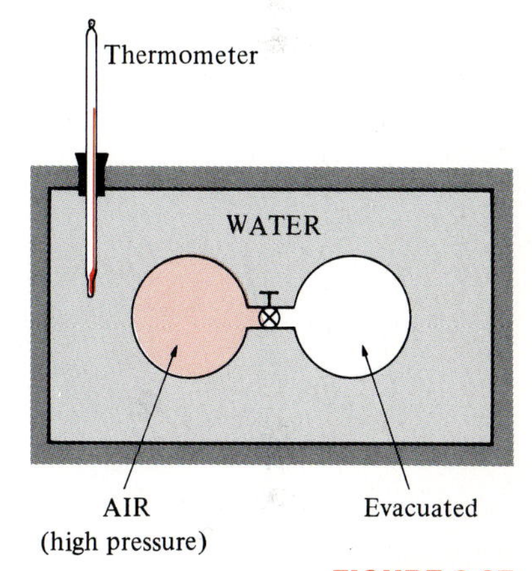

FIGURE 3-67

Schematic of the experimental apparatus used by Joule.

Using the definition of enthalpy and the equation of state of an ideal gas, we have

$$\left.\begin{aligned} h &= u + Pv \\ Pv &= RT \end{aligned}\right\} \quad h = u + RT$$

Since R is constant and $u = u(T)$, it follows that the enthalpy of an ideal gas is also a function of temperature only:

$$h = h(T) \tag{3-37}$$

$u = u(T)$
$h = h(T)$
$C_v = C_v(T)$
$C_p = C_p(T)$

FIGURE 3-68

For ideal gases, u, h, C_v, and C_p vary with temperature only.

Since u and h depend only on temperature for an ideal gas, the specific heats C_v and C_p also depend, at most, on temperature only. Therefore, *at a given temperature u, h, C_v, and C_p of an ideal gas will have fixed values regardless of the specific volume or pressure* (Fig. 3-68). Thus for ideal gases, the partial derivatives in Eqs. 3-34 and 3-35 can be replaced by ordinary derivatives. Then the differential changes in the internal energy and enthalpy of an ideal gas can be expressed as

$$du = C_v(T)\,dT \tag{3-38}$$

and

$$dh = C_p(T)\,dT \tag{3-39}$$

The change in internal energy or enthalpy for an ideal gas during a process from state 1 to state 2 is determined by integrating these equations:

$$\Delta u = u_2 - u_1 = \int_1^2 C_v(T)\,dT \qquad \text{(kJ/kg)} \tag{3-40}$$

and

$$\Delta h = h_2 - h_1 = \int_1^2 C_p(T)\,dT \qquad \text{(kJ/kg)} \tag{3-41}$$

To carry out these integrations, we need to have relations for C_v and C_p as functions of temperature.

At low pressures all real gases approach ideal-gas behavior, and therefore their specific heats depend on temperature only. The specific heats of real gases at low pressures are called *ideal-gas specific heats,* or *zero-pressure specific heats,* and are often denoted C_{p0} and C_{v0}. Accurate analytical expressions for ideal-gas specific heats, based on direct measurements or calculations from statistical behavior of molecules, are available and are given as a third-degree polynomial in the Appendix (Table A-2*c*) for several gases. A plot of $\overline{C}_{p0}(T)$ data for some common gases is given in Fig. 3-69.

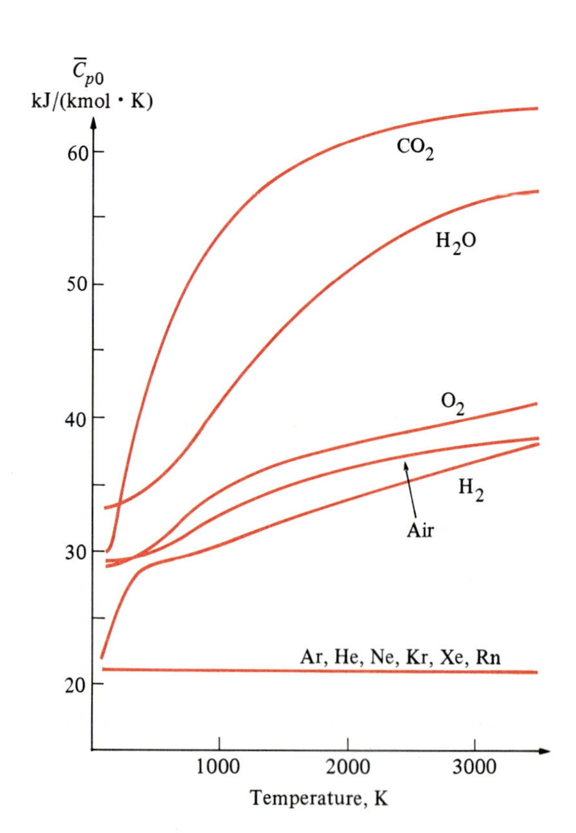

FIGURE 3-69

Ideal-gas constant-pressure specific heats for some gases (see Table A-2*c* for $\overline{C}_{p0}$ equations).

The use of ideal-gas specific heat data is limited to low pressures, but these data can also be used at moderately high pressures with reasonable accuracy as long as the gas does not deviate from ideal-gas behavior significantly.

The integrations in Eqs. 3-40 and 3-41 are straightforward but rather time-consuming and thus impractical. To avoid these laborious calculations, u and h data for a number of gases have been tabulated over small temperature intervals. These tables are obtained by choosing an arbitrary reference point and performing the integrations in Eqs. 3-40 and 3-41 by treating state 1 as the reference state. In the ideal-gas tables given in the Appendix, zero kelvin is chosen as the reference state, and both the enthalpy and the internal energy are assigned zero values at that state (Fig. 3-70). The choice of the reference state has no effect on Δu or Δh calculations. The u and h data are given in kJ/kg for air (Table A-17) and in kJ/kmol for other gases (N_2, O_2, CO_2, CO, H_2O, and H_2). The unit kJ/kmol is very convenient in the thermodynamic analysis of chemical reactions.

AIR

T, K	u, kJ/kg	h, kJ/kg
0	0	0
.	.	.
.	.	.
300	214.17	300.19
310	221.25	310.24
.	.	.
.	.	.

FIGURE 3-70

In the preparation of ideal-gas tables, 0 K is chosen as the reference temperature.

Some observations can be made from Fig. 3-69. First, the specific heats of gases with complex molecules (molecules with two or more atoms) are higher and increase with temperature. Also the variation of specific heats with temperature is smooth and may be approximated as linear over small temperature intervals (a few hundred degrees or less). Then the specific heat functions in Eqs. 3-40 and 3-41 can be replaced by the constant average specific heat values. Now the integrations in these equations can be performed, yielding

$$u_2 - u_1 = C_{v,\text{av}}(T_2 - T_1) \quad \text{(kJ/kg)} \tag{3-42}$$

and

$$h_2 - h_1 = C_{p,\text{av}}(T_2 - T_1) \quad \text{(kJ/kg)} \tag{3-43}$$

The specific heat values for some common gases are listed as a function of temperature in Table A-2*b*. The average specific heats $C_{p,\text{av}}$ and $C_{v,\text{av}}$ are evaluated from this table at the average temperature $(T_1 + T_2)/2$, as shown in Fig. 3-71. If the final temperature T_2 is not known, the specific heats may be evaluated at T_1 or at an anticipated average temperature. Then T_2 can be determined by using these specific heat values. The value of T_2 can be refined, if necessary, by evaluating the specific heats at the new average temperature.

Another way of determining the average specific heats is to evaluate them at T_1 and T_2 and then take their average. Usually both methods give reasonably good results, and one is not necessarily better than the other.

Another observation that can be made from Fig. 3-69 is that the ideal-gas specific heats of *monatomic gases* such as argon, neon, and helium remain constant over the entire temperature range. Thus Δu and Δh of monatomic gases can easily be evaluated from Eqs. 3-42 and 3-43.

Note that the Δu and Δh relations given above are not restricted to any kind of process. They are valid for all processes. The presence of the constant-volume specific heat C_v in an equation should not lead one to believe that this equation is valid for a constant-volume process only. On the contrary, the relation $\Delta u = C_{v,\text{av}}\,\Delta T$ is valid for *any* ideal gas

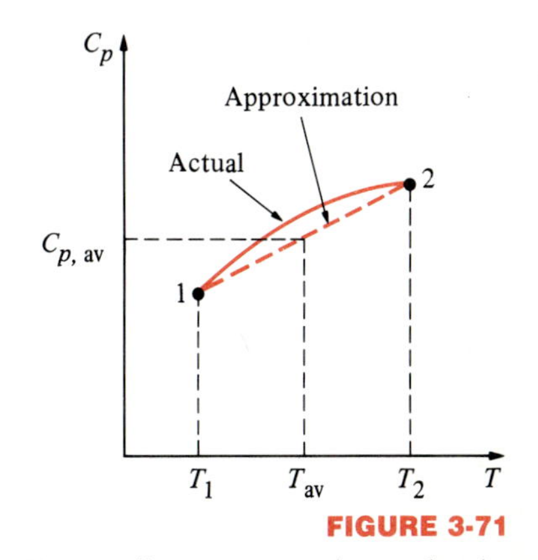

FIGURE 3-71

For small temperature intervals, the specific heats may be assumed to vary linearly with temperature.

undergoing *any* process (Fig. 3-72). A similar argument can be given for C_p and Δh.

To summarize, there are three ways to determine the internal energy and enthalpy changes of ideal gases (Fig. 3-73):

1 By using the tabulated u and h data. This is the easiest and most accurate way when tables are readily available.

2 By using the C_v or C_p relations as a function of temperature and performing the integrations. This is very inconvenient for hand calculations, but quite desirable for computerized calculations. The results obtained are very accurate.

3 By using average specific heats. This is very simple and certainly very convenient when property tables are not available. The results obtained are reasonably accurate if the temperature interval is not very large.

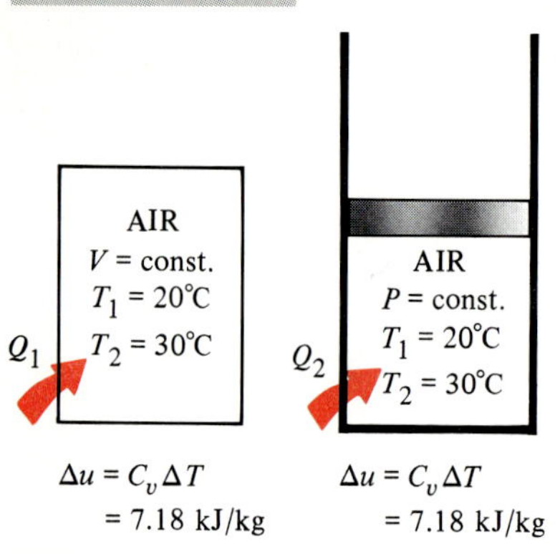

FIGURE 3-72

The relation $\Delta u = C_v\,\Delta T$ is valid for *any* kind of process, constant-volume or not.

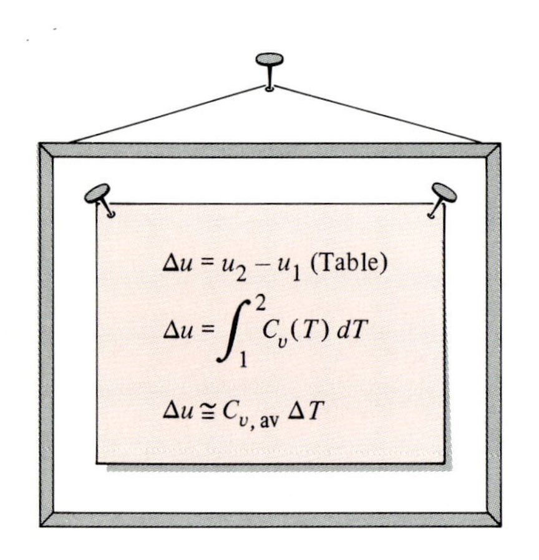

Three ways of calculating Δu.

Specific-Heat Relations of Ideal Gases

A special relationship between C_p and C_v for ideal gases can be obtained by differentiating the relation $h = u + RT$, which yields

$$dh = du + R\,dT$$

Replacing dh by $C_p\,dT$ and du by $C_v\,dT$ and dividing the resultant expression by dT, we obtain

$$C_p = C_v + R \qquad [\text{kJ/(kg·K)}] \tag{3-44}$$

This is an important relationship for ideal gases since it enables us to determine C_v from a knowledge of C_p and the gas constant R.

When the specific heats are given on a molar basis, R in the above equation should be replaced by the universal gas constant R_u (Fig. 3-74). That is,

$$\overline{C}_p = \overline{C}_v + R_u \qquad [\text{kJ/(kmol·K)}] \tag{3-45}$$

At this point, we introduce another ideal-gas property called the **specific heat ratio** k, defined as

$$k = \frac{C_p}{C_v} \tag{3-46}$$

AIR at 300K

$C_v = 0.718$ kJ/(kg · K), $R = 0.287$ kJ/(kg · K) } $C_p = 1.005$ kJ/(kg · K)

or,

$\overline{C}_v = 20.80$ kJ/(kmol · K), $R_u = 8.314$ kJ/(kmol · K) } $\overline{C}_p = 29.114$ kJ/(kmol · K)

FIGURE 3-74

The C_p of an ideal gas can be determined from a knowledge of C_v and R.

The specific heat ratio also varies with temperature, but this variation is very small. For monatomic gases, its value is essentially constant at 1.667. Many diatomic gases, including air, have a specific heat ratio of about 1.4 at room temperature.

EXAMPLE 3-18

Air at 300 K and 200 kPa is heated at constant pressure to 600 K. Determine the change in internal energy of air per unit mass, using (*a*) data from the air table (Table A-17), (*b*) the functional form of the specific heat (Table A-2*c*), and (*c*) the average specific heat value (Table A-2*b*).

Solution At specified conditions, air can be considered to be an ideal gas since it is at a high temperature and low pressure relative to its critical-point values ($T_{cr} = -147°C$, $P_{cr} = 3390$ kPa for nitrogen, the main constituent of air). The internal energy change Δu of ideal gases depends on the initial and final temperatures only, and not on the type of process. Thus the solution given below is valid for any kind of process.

(*a*) One way of determining the change in internal energy of air is to read the u values at T_1 and T_2 from Table A-17 and take the difference:

$$u_1 = u_{@\,300\text{ K}} = 214.07\text{ kJ/kg}$$
$$u_2 = u_{@\,600\text{ K}} = 434.78\text{ kJ/kg}$$

Thus, $\Delta u = u_2 - u_1 = (434.78 - 214.07)\text{ kJ/kg} = 220.71\text{ kJ/kg}$

(*b*) The change in internal energy of air, using the functional form of the specific heat, is determined as follows: The $\overline{C}_p(T)$ of air is given in Table A-2*c* in the form of a third-degree polynomial expressed as

$$\overline{C}_p(T) = a + bT + cT^2 + dT^3$$

where $a = 28.11$, $b = 0.1967 \times 10^{-2}$, $c = 0.4802 \times 10^{-5}$, and $d = -1.966 \times 10^{-9}$. From Eq. 3-45,

$$\overline{C}_v(T) = \overline{C}_p - R_u = (a - R_u) + bT + cT^2 + dT^3$$

From Eq. 3-40,

$$\Delta\overline{u} = \int_1^2 \overline{C}_v(T)\,dT$$
$$= \int_{T_1}^{T_2} [(a - R_u) + bT + cT^2 + dT^3]\,dT$$

Performing the integration and substituting the values, we obtain

$$\Delta\overline{u} = 6447.15\text{ kJ/kmol}$$

The change in the internal energy on a unit-mass basis is determined by dividing this value by the molar mass of air (Table A-1):

$$\Delta u = \frac{\Delta\overline{u}}{M} = \frac{6447.15\text{ kJ/kmol}}{28.97\text{ kg/kmol}} = 222.55\text{ kJ/kg}$$

which differs from the exact result by 0.8 percent.

(*c*) The average value of the constant-volume specific heat $C_{v,\text{av}}$ is determined from Table A-2*b* at the average temperature $(T_1 + T_2)/2 = 450$ K to be

$$C_{v,\text{av}} = C_{v\,@\,450\text{ K}} = 0.733\text{ kJ/(kg·K)}$$

Thus, $\Delta u = C_{v,\text{av}}(T_2 - T_1) = [0.733 \text{ kJ/(kg·K)}][(600 - 300) \text{ K}]$
$= 219.9 \text{ kJ/kg}$

This answer differs from the exact result (220.71 kJ/kg) by only 0.4 percent. This close agreement is not surprising since the assumption that C_v varies linearly with temperature is a reasonable one at temperature intervals of only a few hundred degrees. If we had used the C_v value at $T_1 = 300$ K instead of at T_{av}, the result would be 215.4 kJ/kg, which is in error by about 2 percent. Errors of this magnitude are acceptable for most engineering purposes.

EXAMPLE 3-19

An insulated rigid tank initially contains 1.5 lbm of helium at 80°F and 50 psia. A paddle wheel with a power rating of 0.02 hp is operated within the tank for 30 min. Determine (*a*) the final temperature and (*b*) the final pressure of the helium gas.

Solution We take the helium gas within the tank as our system, which is a stationary closed system. A sketch of the system and the *P*-*v* diagram of the process are shown in Fig. 3-75. The helium gas at the specified conditions can be considered to be an ideal gas since it is at a very high temperature relative to its critical-point temperature ($T_{cr} = -451$°F for helium).

(*a*) The amount of paddle-wheel work done on the system is

$$W_{pw} = \dot{W}_{pw}\,\Delta t = (-0.02 \text{ hp})(0.5 \text{ h})\left(\frac{2545 \text{ Btu/h}}{1 \text{ hp}}\right) = -25.45 \text{ Btu}$$

Since we have no moving boundaries, the boundary work is zero ($W_b = 0$), and the heat losses can be neglected since the system is well insulated ($Q = 0$). The system is assumed to be stationary, thus the changes in kinetic and potential energies are also zero ($\Delta KE = \Delta PE = 0$). Then the conservation of energy equation for this closed system reduces to

$$\cancelto{0}{Q} - W_{pw} - \cancelto{0}{W_b} = \Delta U + \cancelto{0}{\Delta KE} + \cancelto{0}{\Delta PE}$$
$$-W_{pw} = m(u_2 - u_1) \cong mC_{v,\text{av}}(T_2 - T_1)$$

As we pointed out earlier, the ideal-gas specific heats of monatomic gases (helium being one of them) are constant. The C_v value of helium is determined from Table A-2E*a* to be $C_v = 0.753$ Btu/(lbm·°F). Substituting this and other known quantities

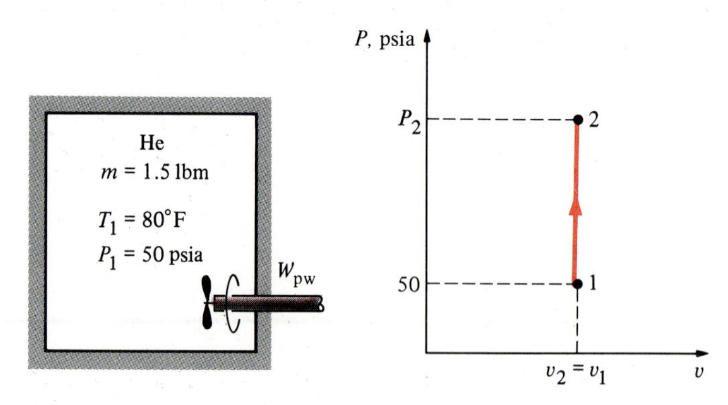

FIGURE 3-75
Schematic and *P*-*v* diagram for Example 3-19.

into the above energy equation, we obtain

$$-(-25.45 \text{ Btu}) = (1.5 \text{ lbm})[0.753 \text{ Btu/(lbm·°F)}](T_2 - 80°\text{F})$$

$$T_2 = 102.5°\text{F}$$

(*b*) The final pressure is determined from the ideal-gas relation

$$\frac{P_1V_1}{T_1} = \frac{P_2V_2}{T_2}$$

where V_1 and V_2 are identical and cancel. Then the final pressure becomes

$$\frac{50 \text{ psia}}{(80 + 460) \text{ R}} = \frac{P_2}{(102.5 + 460) \text{ R}}$$

$$P_2 = 53.1 \text{ psia}$$

EXAMPLE 3-20

A piston-cylinder device initially contains 0.5 m^3 of nitrogen gas at 400 kPa and 27°C. An electric heater within the device is turned on and is allowed to pass a current of 2 A for 5 min from a 120-V source. Nitrogen expands at constant pressure, and a heat loss of 2800 J occurs during the process. Determine the final temperature of the nitrogen, using data from the nitrogen table (Table A-18).

Solution This time we take the nitrogen in the piston-cylinder device as our system. A sketch of the system and the *P-V* diagram of the process are given in Fig. 3-76. At the specified conditions, the nitrogen gas can be considered to be an ideal gas since it is at a high temperature and low pressure relative to its critical-point values ($T_{cr} = -147°\text{C}$, $P_{cr} = 3390$ kPa).

First, let us determine the electrical work done on the nitrogen:

$$W_e = VI\,\Delta t = (120 \text{ V})(2 \text{ A})(5 \times 60 \text{ s})\left(\frac{1 \text{ kJ}}{1000 \text{ VA·s}}\right) = -72 \text{ kJ}$$

The negative sign is added because the work is done on the system.

The number of moles of nitrogen is determined from the ideal-gas relation:

$$N = \frac{P_1V_1}{R_uT_1} = \frac{(400 \text{ kPa})(0.5 \text{ m}^3)}{[8.314 \text{ kJ/(kmol·K)}](300 \text{ K})} = 0.080 \text{ kmol}$$

When gases other than air are involved, it is more convenient to work with mole numbers instead of masses since all the *u* and *h* data are given on a mole basis

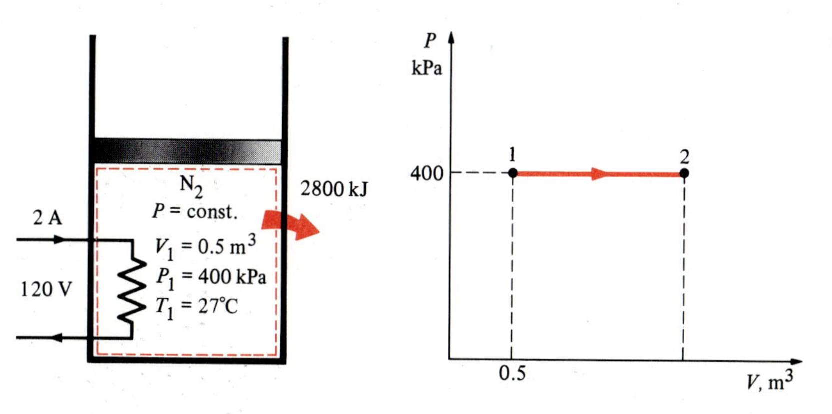

FIGURE 3-76

Schematic and *P-V* diagram for Example 3-20.

(Fig. 3-77). Assuming no changes in kinetic and potential energies ($\Delta KE = \Delta PE = 0$), the conservation of energy equation for this closed system can be written as

$$Q - W_e - W_b = \Delta U$$

For the constant-pressure process of a closed system, $\Delta U + W_b$ is equivalent to ΔH. Thus,

$$Q - W_e = \Delta H = m(h_2 - h_1) = N(\bar{h}_2 - \bar{h}_1)$$

From the nitrogen table, $\bar{h}_1 = \bar{h}_{@\,300\,K} = 8723$ kJ/kmol. The only unknown quantity in the above equation is $\bar{h}_2$, and it is found to be

$$-2.8 \text{ kJ} - (-72 \text{ kJ}) = (0.08 \text{ kmol})(\bar{h}_2 - 8723 \text{ kJ/kmol})$$
$$\bar{h}_2 = 9588 \text{ kJ/kmol}$$

The temperature corresponding to this enthalpy value is

$$T_2 = 329.7 \text{ K} = 56.7°\text{C}$$

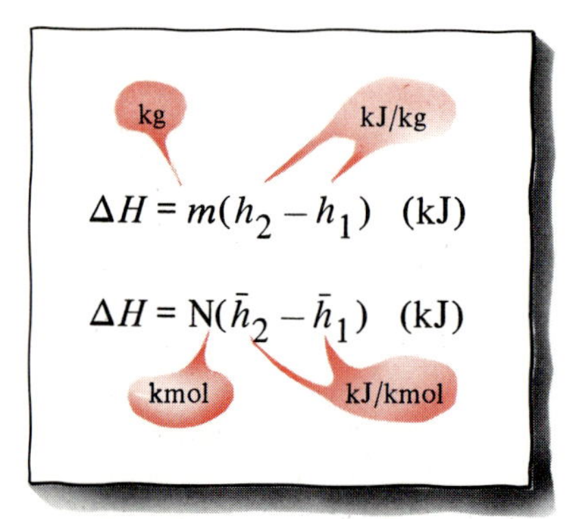

FIGURE 3-77

Two equivalent ways of determining the total enthalpy change ΔH.

EXAMPLE 3-21

A piston-cylinder arrangement initially contains air at 150 kPa and 27°C. At this state, the piston is resting on a pair of stops, as shown in Fig. 3-78, and the enclosed volume is 400 L. The mass of the piston is such that a 350-kPa pressure is required to move it. The air is now heated until its volume has doubled. Determine (*a*) the final temperature, (*b*) the work done by the air, and (*c*) the total heat added.

Solution The air contained within the piston-cylinder device is the obvious choice for the system, and since no mass is crossing the system boundaries, it is a closed system. Under the given conditions, the air may be assumed to behave as an ideal gas since it is at a high temperature and low pressure relative to its critical-point values ($T_{cr} = -147°C$, $P_{cr} = 3390$ kPa for nitrogen, the main constituent of air). This process can be considered in two parts: a constant-volume process during which the pressure rises to 350 kPa and a constant-pressure process during which the volume doubles.

(*a*) The final temperature can be determined easily by using the ideal-gas relation

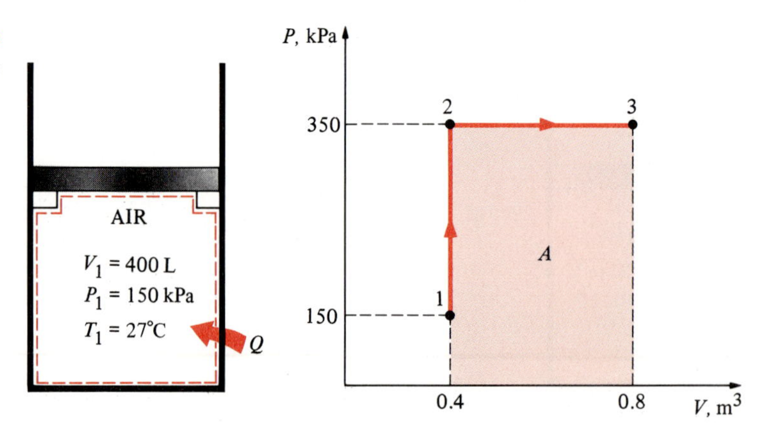

FIGURE 3-78

Schematic and P-V diagram for Example 3-21.

between states 1 and 3 in the following form:

$$\frac{P_1V_1}{T_1} = \frac{P_3V_3}{T_3} \longrightarrow \frac{(150 \text{ kPa})(V_1)}{300 \text{ K}} = \frac{(350 \text{ kPa})(2V_1)}{T_3}$$

$$T_3 = 1400 \text{ K}$$

(*b*) The work done could be determined by integration of Eq. 3-9, but for this case it is much easier to find it from the area under the process curve on a *P*-*V* diagram, shown in Fig. 3-78:

$$A = (V_2 - V_1)(P_2) = (0.4 \text{ m}^3)(350 \text{ kPa}) =$$

Therefore,

$$W_{13} = 140 \text{ kJ}$$

The work is done by the system (to raise the piston and to push the atmospheric air out of the way), thus it is positive.

(*c*) The total heat transfer can be determined from the first-law relation written between the initial and the final states. Assuming $\Delta\text{KE} = \Delta\text{PE} = 0$,

$$Q_{13} - W_{13} = U_3 - U_1 = m(u_3 - u_1)$$

The mass of the system can be determined from the ideal-gas equation of state:

$$m = \frac{P_1V_1}{RT_1} = \frac{(150 \text{ kPa})(0.4 \text{ m}^3)}{[0.287 \text{ kPa·m}^3/(\text{kg·K})](300 \text{ K})} = 0.697 \text{ kg}$$

The internal energies are determined from the air table (Table A-17) to be

$$u_1 = u_{@\,300\text{ K}} = 214.07 \text{ kJ/kg}$$

$$u_3 = u_{@\,1400\text{ K}} = 1113.52 \text{ kJ/kg}$$

Thus,

$$Q_{13} - 140 \text{ kJ} = (0.697 \text{ kg})[(1113.52 - 214.07) \text{ kJ/kg}]$$

$$Q_{13} = 766.9 \text{ kJ}$$

The positive sign indicates that heat is transferred to the system.

3-9 INTERNAL ENERGY, ENTHALPY, AND SPECIFIC HEATS OF SOLIDS AND LIQUIDS

A substance whose specific volume (or density) is constant is called an **incompressible substance**. The specific volumes of solids and liquids essentially remain constant during a process (Fig. 3-79). Therefore, liquids and solids can be approximated as incompressible substances without sacrificing much in accuracy. The constant-volume assumption should be taken to imply that the energy associated with the volume change, such as the boundary work, is negligible compared with other forms of energy. Otherwise, this assumption would be ridiculous for studying the thermal stresses in solids (caused by volume change with temperature) or analyzing liquid-in-glass thermometers.

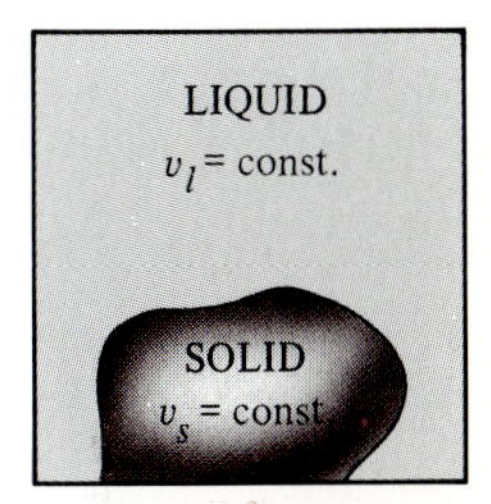

FIGURE 3-79
The specific volumes of incompressible substances remain constant during a process.

It can be mathematically shown (Chap. 11) that the constant-volume and constant-pressure specific heats are identical for incompressible substances (Fig. 3-80). Therefore, for solids and liquids the subscripts on C_p

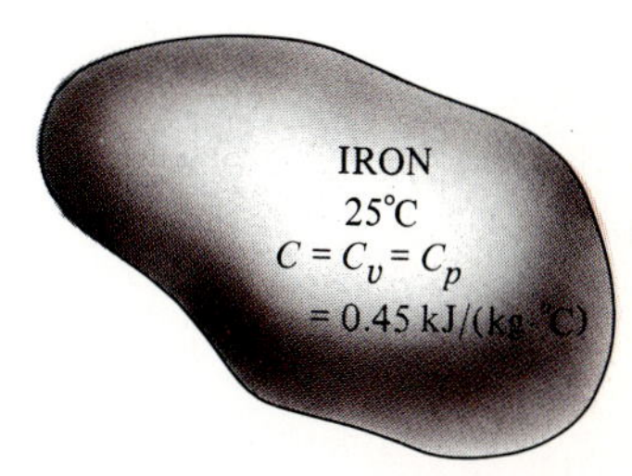

FIGURE 3-80
The C_v and C_p values of incompressible substances are identical and are denoted by C.

and C_v can be dropped, and both specific heats can be represented by a single symbol C. That is,

$$C_p = C_v = C \tag{3-47}$$

This result could also be deduced from the physical definitions of constant-volume and constant-pressure specific heats. Specific heat values for several common liquids and solids are given in Table A-3.

Like those of ideal gases, the specific heats of incompressible substances depend on temperature only. Thus the partial differentials in the defining equation of C_v (Eq. 3-34) can be replaced by ordinary differentials, which yields

$$du = C_v\, dT = C(T)\, dT \tag{3-48}$$

The change in internal energy between states 1 and 2 is then obtained by integration:

$$\Delta u = u_2 - u_1 = \int_1^2 C(T)\, dT \qquad \text{(kJ/kg)} \tag{3-49}$$

The variation of specific heat C with temperature should be known before this integration can be carried out. For small temperature intervals, a C value at the average temperature can be used and treated as a constant, yielding

$$\Delta u \cong C_{\text{av}}(T_2 - T_1) \qquad \text{(kJ/kg)} \tag{3-50}$$

The enthalpy change of incompressible substances (solids or liquids) during process 1-2 can be determined from the definition of enthalpy ($h = u + Pv$) to be

$$h_2 - h_1 = (u_2 - u_1) + v(P_2 - P_1) \tag{3-51}$$

since $v_1 = v_2 = v$. It can also be expressed in a compact form as

$$\Delta h = \Delta u + v\, \Delta P \qquad \text{(kJ/kg)} \tag{3-52}$$

The second term ($v\,\Delta P$) in Eq. 3-52 is often small compared with the first term (Δu) and can be neglected without significant loss in accuracy.

For a constant-temperature process ($\Delta T = 0$), the internal energy change of an incompressible substance is zero. Then from Eq. 3-51 the change in enthalpy will be $h_2 - h_1 = v(P_2 - P_1)$. By taking state 2 as the compressed liquid state and state 1 as the saturated liquid state at the same temperature, the enthalpy of the compressed liquid at a given P and T can be determined from

$$h_{@\,P,T} \cong h_{f\,@\,T} + v_{f\,@\,T}(P - P_{\text{sat}}) \tag{3-53}$$

where P_{sat} is the saturation pressure at the given temperature. This is an improvement over the assumption that the enthalpy of the compressed liquid could be taken as h_f at the given temperature (that is, $h_{@\,P,T} \cong h_{f\,@\,T}$). However, the contribution of the last term is often very small, so it is usually neglected.

EXAMPLE 3-22

A 50-kg iron block at 80°C is dropped into an insulated tank which contains 0.5 m³ of liquid water at 25°C. Determine the temperature when thermal equilibrium is reached.

Solution We take the iron block and the water as our system. The inner surfaces of the tank walls form the system boundary, as shown in Fig. 3-81. Since the tank is insulated, no heat will cross these boundaries ($Q = 0$). Also, since there is no movement of the boundary ($W_b = 0$) and no indication of other forms of work ($W_{other} = 0$), the work term for this process is zero ($W = 0$). Then the conservation of energy equation for this process will reduce to

$$\cancel{Q}^{\,0} - \cancel{W}^{\,0} = \Delta U \qquad \text{or} \qquad \Delta U = 0$$

The total internal energy U is an extensive property, and therefore it can be expressed as the sum of the internal energies of the parts of the system. Then the total internal energy change of the system is

$$\Delta U_{sys} = \Delta U_{\text{iron}} + \Delta U_{\text{water}} = 0$$
$$[mC(T_2 - T_1)]_{\text{iron}} + [mC(T_2 - T_1)]_{\text{water}} = 0$$

The specific volume of liquid water at or about room temperature can be taken to be 0.001 m³/kg. Then the mass of the water is

$$m_{\text{water}} = \frac{V}{v} = \frac{0.5\ \text{m}^3}{0.001\ \text{m}^3/\text{kg}} = 500\ \text{kg}$$

The specific heats of iron and liquid water are determined from Table A-3 to be $C_{\text{iron}} = 0.45$ kJ/(kg·°C) and $C_{\text{water}} = 4.184$ kJ/(kg·°C). Substituting these values into the energy equation, we obtain

$$(50\ \text{kg})[0.45\ \text{kJ/(kg·°C)}](T_2 - 80°\text{C}) + (500\ \text{kg})[4.184\ \text{kJ/(kg·°C)}](T_2 - 25°\text{C}) = 0$$

$$T_2 = 25.6°\text{C}$$

Therefore, when thermal equilibrium is established, both the water and iron will be at 25.6°C. The small rise in water temperature is due to its large mass and large specific heat.

FIGURE 3-81

Schematic for Example 3-22.

EXAMPLE 3-23

Determine the enthalpy of liquid water at 100°C and 15 MPa (*a*) from the compressed liquid tables, (*b*) by approximating it as a saturated liquid, and (*c*) by using the correction given by Eq. 3-53.

Solution At 100°C, the saturation pressure of water is 101.35 kPa, and since $P > P_{\text{sat}}$, the water exists as a compressed liquid at the specified state.

(*a*) From the compressed liquid tables, we read

$$\left.\begin{array}{l} P = 15\ \text{MPa} \\ T = 100°\text{C} \end{array}\right\} \quad h = 430.28\ \text{kJ/kg} \qquad \text{(Table A-7)}$$

This is the exact value.

(*b*) Approximating the compressed liquid as a saturated liquid at 100°C, as is commonly done, we obtain

$$h \cong h_{f\,@\,100°C} = 419.04 \text{ kJ/kg}$$

This value is in error by about 2.6 percent.

(*c*) From Eq. 3-53,

$$\begin{aligned} h_{@\,P,T} &= h_{f\,@\,T} + v_f(P - P_{sat}) \\ &= (419.04 \text{ kJ/kg}) + (0.001 \text{ m}^3\text{/kg})[(15{,}000 - 101.35) \text{ kPa}]\left(\frac{1 \text{ kJ}}{1 \text{ kPa}\cdot\text{m}^3}\right) \\ &= 434.60 \text{ kJ/kg} \end{aligned}$$

The correction term reduced the error from 2.6 to about 1 percent. But this improvement in accuracy is often not worth the extra effort involved.

3-10 ■ SUMMARY

The first law of thermodynamics is an expression of the conservation of energy principle. Energy can cross the boundaries of a closed system in the form of heat or work. If the energy transfer is due to a temperature difference between a system and its surroundings, it is *heat;* otherwise, it is *work*. Heat transfer to a system and work done by a system are positive; heat transfer from a system and work done on a system are negative.

Various forms of work are expressed as follows:

Electrical work	$W_e = VI\,\Delta t$	(kJ)	(3-5)
Boundary work	$W_b = \int_1^2 P\,dV$	(kJ)	(3-9)
Gravitational work (= ΔPE)	$W_g = mg(z_2 - z_1)$	(kJ)	(3-14)
Accelerational work (= ΔKE)	$W_a = \frac{1}{2}m(\mathbf{V}_2^2 - \mathbf{V}_1^2)$	(kJ)	(3-15)
Shaft work	$W_{sh} = 2\pi n\tau$	(kJ)	(3-16)
Spring work	$W_{spring} = \frac{1}{2}k(x_2^2 - x_1^2)$	(kJ)	(3-20)

For the *polytropic process* (Pv^n = constant) of real gases, the boundary work can be expressed as

$$W_b = \frac{P_2V_2 - P_1V_1}{1 - n} \quad \text{(kJ)} \tag{3-12}$$

The *first law of thermodynamics* for a closed system is given by

$$Q - W = \Delta U + \Delta KE + \Delta PE \quad \text{(kJ)} \tag{3-25}$$

where $W = W_{other} + W_b$
$\Delta U = m(u_2 - u_1)$
$\Delta PE = mg(z_2 - z_1)$
$\Delta KE = \frac{1}{2}m(\mathbf{V}_2^2 - \mathbf{V}_1^2)$

For a *constant-pressure process,* $W_b + \Delta U = \Delta H$. Thus,

$$Q - W_{other} = \Delta H + \Delta KE + \Delta PE \qquad \text{(kJ)} \qquad (3\text{-}33)$$

The amount of energy needed to raise the temperature of a unit mass of a substance by one degree is called the *specific heat at constant volume* C_v for a constant-volume process and the *specific heat at constant pressure* C_p for a constant-pressure process. They are defined as

$$C_v = \left(\frac{\partial u}{\partial T}\right)_v \quad \text{and} \quad C_p = \left(\frac{\partial h}{\partial T}\right)_P \qquad (3\text{-}34, 3\text{-}35)$$

For ideal gases u, h, C_v, and C_p are functions of temperature alone. The Δu and Δh of ideal gases can be expressed as

$$\Delta u = u_2 - u_1 = \int_1^2 C_v(T)\,dT \cong C_{v,av}(T_2 - T_1) \qquad \text{(kJ/kg)} \qquad (3\text{-}40, 3\text{-}42)$$

$$\Delta h = h_2 - h_1 = \int_1^2 C_p(T)\,dT \cong C_{p,av}(T_2 - T_1) \qquad \text{(kJ/kg)} \qquad (3\text{-}41, 3\text{-}43)$$

For ideal gases, C_v and C_p are related by

$$C_p = C_v + R \qquad [\text{kJ/(kg}\cdot\text{K)}] \qquad (3\text{-}44)$$

where R is the gas constant. The *specific heat ratio* k is defined as

$$k = \frac{C_p}{C_v} \qquad (3\text{-}46)$$

For *incompressible substances* (liquids and solids), both the constant-pressure and constant-volume specific heats are identical and denoted by C:

$$C_p = C_v = C \qquad [\text{kJ/(kg}\cdot\text{K)}] \qquad (3\text{-}47)$$

The Δu and Δh of incompressible substances are given by

$$\Delta u = \int_1^2 C(T)\,dT \cong C_{av}(T_2 - T_1) \qquad \text{(kJ/kg)} \qquad (3\text{-}49, 3\text{-}50)$$

$$\Delta h = \Delta u + v\,\Delta P \qquad \text{(kJ/kg)} \qquad (3\text{-}52)$$

REFERENCES AND SUGGESTED READING

1 W. Z. Black and J. G. Hartley, *Thermodynamics,* Harper and Row, New York, 1985.

2 J. B. Jones and G. A. Hawkins, *Engineering Thermodynamics,* 2d ed., Wiley, New York, 1986.

3 J. R. Howell and R. O. Buckius, *Fundamentals of Engineering Thermodynamics,* McGraw-Hill, New York, 1987.

4 G. J. Van Wylen and R. E. Sonntag, *Fundamentals of Classical Thermodynamics,* 3d ed., Wiley, New York, 1985.

5 K. Wark, *Thermodynamics,* 5th ed., McGraw-Hill, New York, 1988.

PROBLEMS*

Heat and Work

3-1C Is the process of burning wood in a stove an energy creation or an energy transformation process?

3-2C In what forms can energy cross the boundaries of a system?

3-3C When is the energy crossing the boundaries of a system heat and when is it work?

3-4C What is an adiabatic process?

3-5C What are the sign conventions for heat and work?

3-6C A gas in a piston-cylinder device is compressed, and as a result its temperature rises. Is this a heat or work interaction?

3-7C A room is heated by an iron which is left plugged in. Is this a heat or work interaction? Take the entire room, including the iron, as the system.

3-8C A room is heated as a result of solar radiation coming in through the windows. Is this a heat or work interaction?

3-9C An insulated room is heated by burning candles. Is this a heat or work interaction? Take the entire room, including the candles, as the system.

3-10C What are point and path functions? Give some examples.

Boundary Work

3-11C On a *P*-*v* diagram, what does the area under the process curve represent?

3-12C Explain why the work output is greater when a process takes place in a quasi-equilibrium manner.

3-13C Is the boundary work associated with constant-volume systems always zero?

---

*Students are encouraged to answer *all* the concept "C" questions.

3-14C An ideal gas at a given state expands to a fixed final volume first at constant pressure and then at constant temperature. For which case is the work done greater?

3-15C Show that 1 kPa·m³ = 1 kJ.

3-16 A mass of 3 kg of saturated water vapor at 200 kPa is heated at constant pressure until the temperature reaches 400°C. Calculate the work done by the steam during this process. *Answer:* 398.2 kJ

3-16E A mass of 5 lbm of saturated water vapor at 40 psia is heated at constant pressure until the temperature reaches 600°F. Calculate the work done by the steam during this process. *Answer:* 191.9 Btu

3-17 A frictionless piston-cylinder device initially contains 200 L of saturated liquid refrigerant-12. The piston is free to move, and its mass is such that it maintains a pressure of 800 kPa on the refrigerant. The refrigerant is now heated until its temperature rises to 50°C. Calculate the work done during this process. *Answer:* 4775 kJ

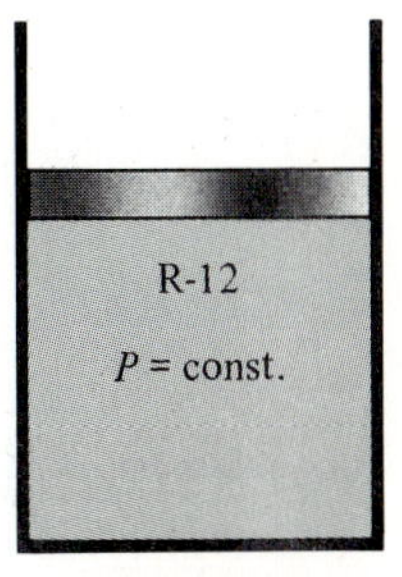

FIGURE P3-17

3-18 A frictionless piston-cylinder device contains 8 kg of superheated water vapor at 500 kPa and 300°C. Steam is now cooled at constant pressure until 70 percent of it, by mass, condenses. Determine the work done during this process.

3-18E A frictionless piston-cylinder device contains 12 lbm of superheated water vapor at 60 psia and 500°F. Steam is now cooled at constant pressure until 70 percent of it, by mass, condenses. Determine the work done during this process.

3-19 A mass of 0.8 kg of air at 100 kPa and 25°C is contained in a gas-tight, frictionless piston-cylinder device. The air is now compressed to a final pressure of 500 kPa. During the process heat is transferred from the air such that the temperature inside the cylinder remains constant. Calculate the work done during this process. *Answer:* −110.1 kJ

3-20 Nitrogen at an initial state of 300 K, 150 kPa, and 0.2 m³ is compressed slowly in an isothermal process to a final pressure of 800 kPa. Determine the work done during this process.

3-20E Nitrogen at an initial state of 70°F, 20 psia, and 5 ft³ is compressed slowly in an isothermal process to a final pressure of 100 psia. Determine the work done during this process.

3-21 A gas is compressed from an initial volume of 0.38 m³ to a final volume of 0.1 m³. During the quasi-equilibrium process, the pressure changes with volume according to the relation $P = aV + b$, where $a = -1200$ kPa/m³ and $b = 600$ kPa. Calculate the work done during this process (*a*) by plotting the process on a P-v diagram and finding the area under the process curve and (*b*) by performing the necessary integrations.

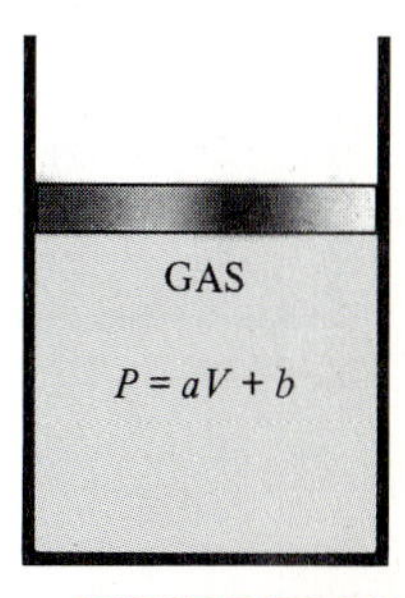

FIGURE P3-21

3-22 During an expansion process, the pressure of a gas changes from 100 to 900 kPa according to the relation $P = aV + b$, where $a =$

1 MPa/m³ and b is a constant. If the initial volume of the gas is 0.2 m³, calculate the work done during the process. *Answer:* 400 kJ

3-22E During an expansion process, the pressure of a gas changes from 15 to 100 psia according to the relation $P = aV + b$, where $a = 5$ psia/ft³ and b is a constant. If the initial volume of the gas is 7 ft³, calculate the work done during the process. *Answer:* 188.8 Btu

3-23 During some actual expansion and compression processes in piston-cylinder devices, the gases have been observed to satisfy the relationship $PV^n = C$, where n and C are constants. Calculate the work done when a gas expands from a state of 100 kPa and 0.05 m³ to a final volume of 0.3 m³ for the case of $n = 1.2$.

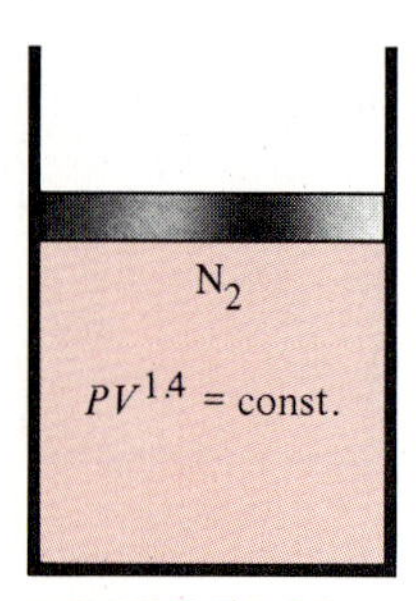

FIGURE P3-24

3-24 A frictionless piston-cylinder device contains 2 kg of nitrogen at 100 kPa and 300 K. Nitrogen is now compressed slowly according to the relation $PV^{1.4}$ = constant until it reaches a final temperature of 360 K. Calculate the work done during this process. *Answer:* −89.0 kJ

3-24E A frictionless piston-cylinder device contains 5 lbm of nitrogen at 14.7 psia and 550 R. Nitrogen is now compressed slowly according to the relation $PV^{1.4}$ = constant until it reaches a final temperature of 700 R. Calculate the work done during this process.
Answer: −132.9 Btu

3-25 The equation of state of a gas is given as $\bar{v}(P + 10/\bar{v}^2) = R_uT$, where the units of $\bar{v}$ and P are m³/kmol and bars, respectively. Now 0.5 kmol of this gas is expanded in a quasi-equilibrium manner from 2 to 4 m³ at a constant temperature of 300 K. Determine (*a*) the unit of the quantity 10 in the equation and (*b*) the work done during this isothermal expansion process.

3-26 Carbon dioxide contained in a piston-cylinder device is compressed from 0.3 to 0.1 m³. During the process, the pressure and volume are related by $P = aV^{-2}$, where $a = 8$ kPa·m⁶. Calculate the work done on the carbon dioxide during this process. *Answer:* −53.3 kJ

3-26E Carbon dioxide contained in a piston-cylinder device is compressed from 10 to 3 ft³. During the process, the pressure and volume are related by $P = aV^{-2}$, where $a = 175$ psia·ft⁶. Calculate the work done on the carbon dioxide during this process.

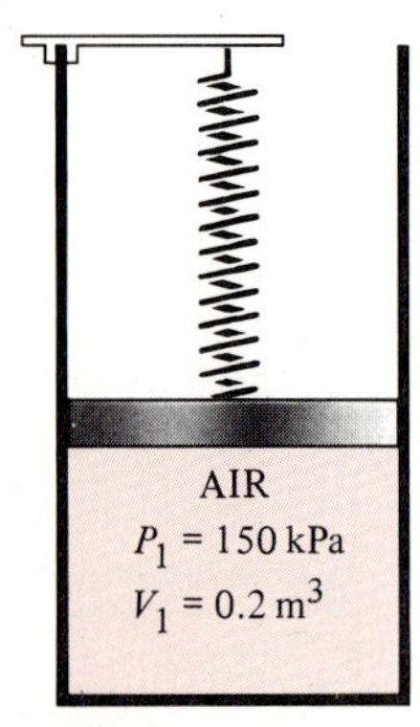

FIGURE P3-27

3-27 A frictionless piston-cylinder device initially contains air at 150 kPa and 0.2 m³. At this state, a linear spring ($F \propto x$) is touching the piston but exerts no force on it. The air is now heated to a final state of 0.5 m³ and 600 kPa. Determine (*a*) the total work done by the air and (*b*) the work done against the spring. Also show the process on a P-v diagram. *Answers:* (*a*) 112.5 kJ, (*b*) 67.5 kJ

3-28 Hydrogen is contained in a piston-cylinder device at 100 kPa and 1 m³. At this state, a linear spring ($F \propto x$) with a spring constant of 200 kN/m is touching the piston but exerts no force on it. The cross-sectional area of the piston is 0.8 m². Heat is transferred to the hydrogen, causing

it to expand until its volume doubles. Determine (*a*) the final pressure, (*b*) the total work done by the hydrogen, and (*c*) the fraction of this work done against the spring. Also show the process on a *P-v* diagram.

3-28E Hydrogen is contained in a piston-cylinder device at 14.7 psia and 15 ft^3. At this state, a linear spring ($F \propto x$) with a spring constant of 15,000 lbf/ft is touching the piston but exerts no force on it. The cross-sectional area of the piston is 3 ft^2. Heat is transferred to the hydrogen, causing it to expand until its volume doubles. Determine (*a*) the final pressure, (*b*) the total work done by the hydrogen, and (*c*) the fraction of this work done against the spring. Also show the process on a *P-v* diagram.

3-29 A piston-cylinder device contains 50 kg of water at 150 kPa and 25°C. The cross-sectional area of the piston is 0.1 m^2. Heat is now transferred to the water, causing part of it to evaporate and expand. When the volume reaches 0.2 m^3, the piston reaches a linear spring whose spring constant is 100 kN/m. More heat is transferred to the water until the piston rises 20 cm more. Determine (*a*) the final pressure and temperature and (*b*) the work done during this process. Also show the process on a *P-V* diagram. *Answers:* (*a*) 350 kPa, 138.88°C; (*b*) 27.5 kJ

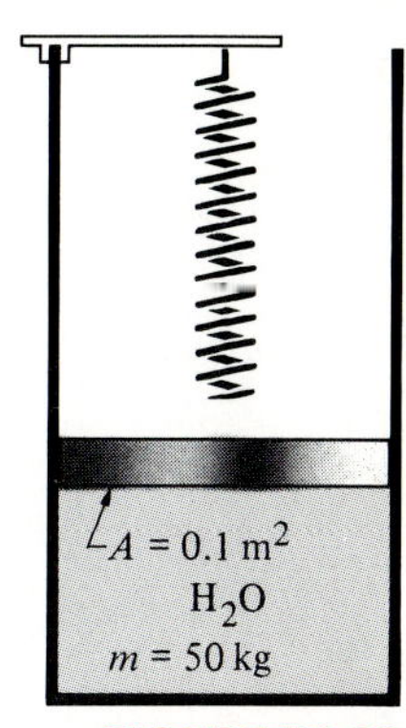

FIGURE P3-29

3-30 A piston-cylinder device with a set of stops contains 5 kg of refrigerant-12. Initially, 4 kg of the refrigerant is in the liquid form, and the temperature is −10°C. Now heat is transferred slowly to the refrigerant until the piston hits the stops, at which point the volume is 200 L. Determine (*a*) the temperature when the piston first hits the stops and (*b*) the work done during this process. Also show the process on a *P-v* diagram. *Answers:* (*a*) −10°C, (*b*) 26.3 kJ

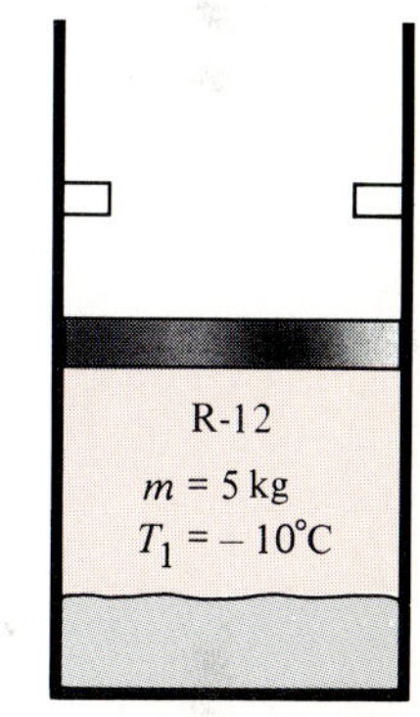

FIGURE P3-30

3-30E A piston-cylinder device with a set of stops contains 10 lbm of refrigerant-12. Initially, 8 lbm of the refrigerant is in the liquid form, and the temperature is 10°F. Now heat is transferred slowly to the refrigerant until the piston hits the stops, at which point the volume is 7 ft^3. Determine (*a*) the temperature when the piston first hits the stops and (*b*) the work done during this process. Also show the process on a *P-v* diagram.

3-31 A mass of 10 kg of saturated liquid–vapor mixture of water is contained in a piston-cylinder device at 100 kPa. Initially, 4 kg of the water is in the liquid phase, and the rest is in the vapor phase. Heat is now transferred to the water, and the piston, which is resting on a set of stops, starts moving when the pressure inside reaches 200 kPa. Heat transfer continues until the total volume increases by 20 percent. Determine (*a*) the initial and final temperatures, (*b*) the mass of liquid water when the piston first starts moving, and (*c*) the work done during this process. Also show the process on a *P-v* diagram.

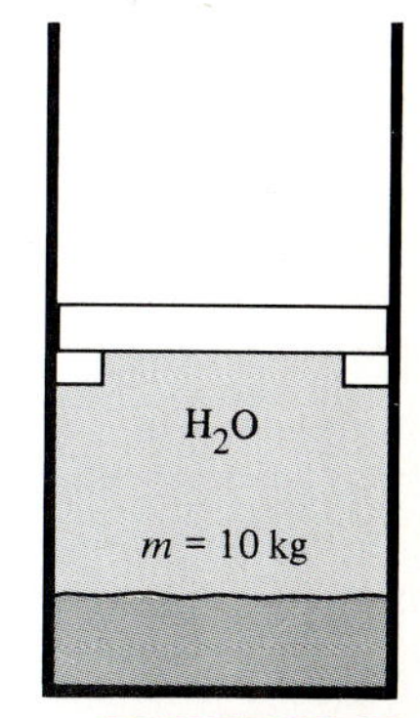

FIGURE P3-31

3-32 A spherical balloon contains 5 kg of air at 200 kPa and 500 K. The balloon material is such that the pressure inside is always proportional to the square of the diameter. Determine the work done when the volume of the balloon doubles as a result of heat transfer. *Answer:* 936 kJ

3-32E A spherical balloon contains 10 lbm of air at 30 psia and 800 R. The balloon material is such that the pressure inside is always proportional to the square of the diameter. Determine the work done when the volume of the balloon doubles as a result of heat transfer.
Answer: 715.3 Btu

3-33 A frictionless piston-cylinder device contains 10 kg of saturated refrigerant-12 vapor at 50°C. The refrigerant is then allowed to expand isothermally by gradually decreasing the pressure in a quasi-equilibrium manner to a final value of 500 kPa. Determine the work done during this process (*a*) by using the experimental specific volume data from the tables and (*b*) by treating the refrigerant vapor as an ideal gas. Also determine the error involved in the latter case.

3-34 Write a computer program to determine the boundary work done by a gas during an expansion process. The measured pressure and volume values at various states are as follows: 300 kPa, 1 L; 290 kPa, 1.1 L; 270 kPa, 1.2 L; 250 kPa, 1.4 L; 220 kPa, 1.7 L; and 200 kPa, 2 L.

Other Forms of Work

3-35C A car is accelerated from rest to 85 km/h in 10 s. Would the work done on the car be different if it were accelerated to the same speed in 5 s?

3-36C Lifting a weight to a height of 20 m takes 20 s for one crane and 10 s for another. Is there any difference in the amount of work done on the weight by each crane?

3-37 Determine the work required to accelerate a 1000-kg car from rest to 85 km/h on a level road. *Answer:* 278.7 kJ

3-37E Determine the work required to accelerate a 2000-lbm car at rest to 60 mi/h on a level road.

3-38 Determine the work required to accelerate a 2000-kg car from 20 to 70 km/h on an uphill road with a vertical rise of 40 m.

3-39 Determine the torque applied to the shaft of a car that transmits 400 kW and rotates at a rate of 3000 rpm. *Answer:* 1273 N·m

3-39E Determine the torque applied to the shaft of a car that transmits 450 hp and rotates at a rate of 3000 rpm.

3-40 Determine the work required to deflect a linear spring with a spring constant of 50 kN/m by 50 cm from its rest position.

3-41 The engine of a 1500-kg automobile has a power rating of 75 kW. Determine the time required to accelerate this car from rest to a speed of 85 km/h at full power on a level road. Is your answer realistic?

3-41E The engine of a 2000-lbm automobile has a power rating of 100 hp. Determine the time required to accelerate this car from rest to a speed of 60 mi/h at full power on a level road. Is your answer realistic?
Answer: 4.37 s

3-42 A ski lift has a one-way length of 1 km and a vertical rise of 200 m. The chairs are spaced 20 m apart, and each chair can seat three people. The lift is operating at a steady speed of 10 km/h. Neglecting friction and air drag and assuming that the average mass of each loaded chair is 250 kg, determine the power required to operate this ski lift. Also estimate the time required to accelerate this ski lift to its operating speed when it is first turned on. Assume $g = 9.6 \text{ m/s}^2$.

3-43 Determine the power required for a 2000-kg car to climb a 100-m-long uphill road with a slope of 30° (from horizontal) in 10 s (*a*) at a constant velocity, (*b*) from rest to a final velocity of 30 m/s, and (*c*) from 35 m/s to a final velocity of 5 m/s. Disregard friction, air drag, and rolling resistance. *Answers:* (*a*) 98.07 kW, (*b*) 188.07 kW, (*c*) −21.93 kW

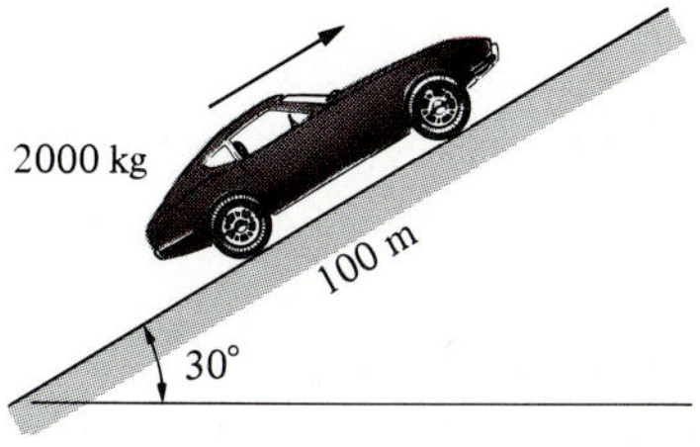

2000 kg
100 m
30°

FIGURE P3-43

3-43E Determine the power required for a 3000-lbm car to climb a 300-ft-long uphill road with a slope of 30° (from horizontal) in 10 s (*a*) at a constant velocity, (*b*) from rest to a final velocity of 30 mi/h, and (*c*) from 35 mi/h to a final velocity of 5 mi/h. Disregard friction, air drag, and rolling resistance.

3-44 A damaged 1200-kg car is being towed by a truck. Neglecting the friction, air drag, and rolling resistance, determine the extra power required (*a*) for constant velocity on a level road, (*b*) for constant velocity of 50 km/h on a 30° (from horizontal) uphill road, and (*c*) to accelerate on a level road from stop to 80 km/h in 15 s.
Answers: (*a*) 0, (*b*) 81.7 kW, (*c*) 19.75 kW

Closed-System Energy Analysis: General Systems

3-45C For a cycle, is the net work necessarily zero? For what kind of systems will this be the case?

3-46C Under what conditions is the relation $Q - W_{\text{other}} = H_2 - H_1$ valid for a closed system?

3-47C On a hot summer day, a student turns his fan on when he leaves his room in the morning. When he returns in the evening, will the room be warmer or cooler than the neighboring rooms? Why? Assume all the doors and windows are kept closed.

3-48C Consider two identical rooms, one with a refrigerator in it and the other without one. If all the doors and windows are closed, will the room that contains the refrigerator be cooler or warmer than the other room? Why?

3-49C Consider a can of soft drink that is dropped from the top of a tall building. Will the temperature of the soft drink increase as it falls, as a result of decreasing potential energy?

3-50 Water is being heated in a closed pan on top of a range while being stirred by a paddle wheel. During the process, 30 kJ of heat is added to the water, and 5 kJ of heat is lost to the surrounding air. The paddle-

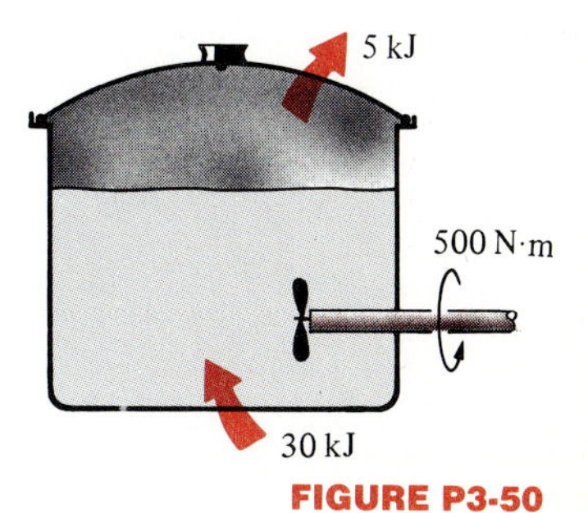

FIGURE P3-50

wheel work amounts to 500 N·m. Determine the final energy of the system if its initial energy is 10 kJ. *Answer:* 35.5 kJ

3-51 A vertical piston-cylinder device contains water and is being heated on top of a range. During the process, 50 kJ of heat is transferred to the water, and heat losses from the side walls amount to 8 kJ. The piston rises as a result of evaporation, and 5 kJ of boundary work is done. Determine the change in the energy of the water for this process.

3-51E A vertical piston-cylinder device contains water and is being heated on top of a range. During the process, 50 Btu of heat is transferred to the water, and heat losses from the side walls amount to 8 Btu. The piston rises as a result of evaporation, and 5 Btu of boundary work is done. Determine the change in the energy of the water for this process. *Answer:* 37 Btu

3-52 Fill in the missing data for each of the following processes of a closed system between states 1 and 2. (Everything is in kJ.)

	Q	W	E_1	E_2	ΔE
(*a*)	18	−6		35	
(*b*)	−10			4	−15
(*c*)		12	3		32
(*d*)	25		14		10

3-53 Fill in the missing data for each of the following processes of a closed system between states 1 and 2. (Everything is in kJ.)

	Q	W	E_1	E_2	ΔE
(*a*)		18	6		20
(*b*)	5		20		35
(*c*)	25	−10		40	
(*d*)	−9			12	−15

3-54 A closed system undergoes a cycle consisting of two processes. During the first process, 30 kJ of heat is transferred to the system while the system does 50 kJ of work. During the second process, 35 kJ of work is done on the system.

(*a*) Determine the heat transfer during the second process.
(*b*) Calculate the net work and net heat transfer for the cycle.

Answers: (*a*) −15 kJ; (*b*) 15 kJ, 15 kJ

3-54E A closed system undergoes a cycle consisting of two processes. During the first process, 30 Btu of heat is transferred to the system while the system does 50 Btu of work. During the second process, 35 Btu of work is done on the system.

(*a*) Determine the heat transfer during the second process.
(*b*) Calculate the net work and net heat transfer for the cycle.

3-55 A closed system undergoes a cycle consisting of three processes.

During the first process which is adiabatic, 50 kJ of work is done on the system. During the second process, 200 kJ of heat is transferred to the system while no work interaction takes place. And during the third process, the system does 90 kJ of work as it returns to its initial state.

(*a*) Determine the heat transfer during the last process.
(*b*) Determine the net work done during this cycle.

3-56 A classroom that normally contains 50 people is to be air conditioned with window air-conditioning units of 4-kW rating. A person at rest may be assumed to dissipate heat at a rate of about 360 kJ/h. There are 20 light bulbs in the room, each with a rating of 100 W. The rate of heat transfer to the classroom through the walls and the windows is estimated to be 18,000 kJ/h. If the room air is to be maintained at a constant temperature of 21°C, determine the number of window air-conditioning units required. *Answer:* 3 units

Closed-System Energy Analysis: Saturation, Compressed Liquid, and Superheat Data

3-57 The radiator of a steam heating system has a volume of 15 L and is filled with superheated vapor at 200 kPa and 200°C. At this moment both the inlet and exit valves to the radiator are closed. Determine the amount of heat that will be transferred to the room when the steam pressure drops to 100 kPa. Also show the process on a *P-v* diagram with respect to saturation lines. *Answer:* −12.6 kJ

FIGURE P3-57

3-57E The radiator of a steam heating system has a volume of 0.5 ft^3 and is filled with superheated vapor at 40 psia and 400°F. At this moment both the inlet and exit valves to the radiator are closed. Determine the amount of heat that will be transferred to the room when the steam pressure drops to 14.7 psia. Also show the process on a *P-v* diagram with respect to saturation lines.

3-58 A 0.5-m^3 rigid tank contains refrigerant-12 initially at 200 kPa and 40 percent quality. Heat is now transferred to the refrigerant until the pressure reaches 800 kPa. Determine (*a*) the mass of the refrigerant in the tank and (*b*) the amount of heat transferred. Also show the process on a *P-v* diagram with respect to saturation lines.

3-59 A 0.6-m^3 rigid tank initially contains refrigerant-12 in the saturated vapor form at 0.8 MPa. As a result of heat transfer from the refrigerant, the pressure drops to 200 kPa. Show the process on a *P-v* diagram with respect to saturation lines, and determine (*a*) the final temperature, (*b*) the amount of refrigerant that has condensed, and (*c*) the heat transfer. *Answers:* (*a*) −12.53°C, (*b*) 20.4 kg, (*c*) −3360 kJ

3-59E A 20-ft^3 rigid tank initially contains refrigerant-12 in the saturated vapor form at 120 psia. As a result of heat transfer from the refrigerant, the pressure drops to 30 psia. Show the process on a *P-v* diagram with respect to saturation lines, and determine (*a*) the final temperature, (*b*) the amount of refrigerant that has condensed, and (*c*) the heat transfer.

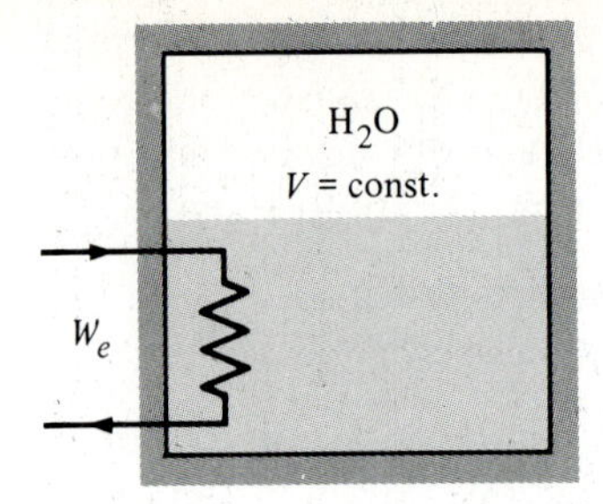

FIGURE P3-60

3-60 A well-insulated rigid tank contains 2 kg of saturated liquid–vapor mixture of water at 100 kPa. Initially, three-quarters of the mass is in the liquid phase. An electric resistor placed in the tank is connected to a 120-V source, and a current of 5 A flows through the resistor when the switch is turned on. Determine how long it will take to vaporize all the liquid in the tank. Also, show the process on a *T-v* diagram with respect to saturation lines.

3-60E A well-insulated rigid tank contains 5 lbm of saturated liquid–vapor mixture of water at 14.7 psia. Initially, three-quarters of the mass is in the liquid phase. An electric resistor placed in the tank is connected to a 120-V source, and a current of 5 A flows through the resistor when the switch is turned on. Determine how long it will take to vaporize all the liquid in the tank. Also, show the process on a *T-v* diagram with respect to saturation lines. *Answer:* 101.8 min

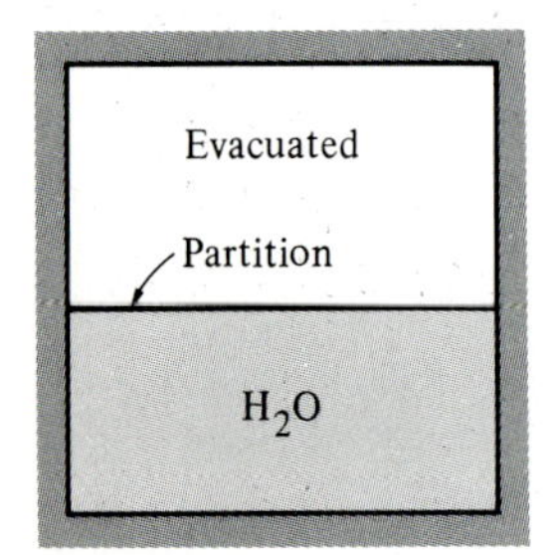

FIGURE P3-61

3-61 An insulated tank is divided into two parts by a partition. One part of the tank contains 0.5 kg of compressed liquid water at 60°C and 600 kPa while the other part is evacuated. The partition is now removed, and the water expands to fill the entire tank. Determine the final temperature of the water and the volume of the tank for a final pressure of 10 kPa.

3-62 A piston-cylinder device contains 5 kg of refrigerant-12 at 800 kPa and 60°C. The refrigerant is now cooled at constant pressure until it exists as a liquid at 20°C. Determine the amount of heat transfer, and show the process on a *T-v* diagram with respect to saturation lines.
Answer: −829.25 kJ

3-63 A piston-cylinder device contains 0.2 kg of water intially at 800 kPa and 0.08 m^3. Now 180 kJ of heat is transferred to the water while its pressure is kept constant. Determine the final temperature of the water. Also show the process on a *T-v* diagram with respect to saturation lines.

3-63E A piston-cylinder device contains 0.5 lbm of water initially at 120 psia and 2 ft^3. Now 200 Btu of heat is transferred to the water while its pressure is held constant. Determine the final temperature of the water. Also show the process on a *T-v* diagram with respect to saturation lines.

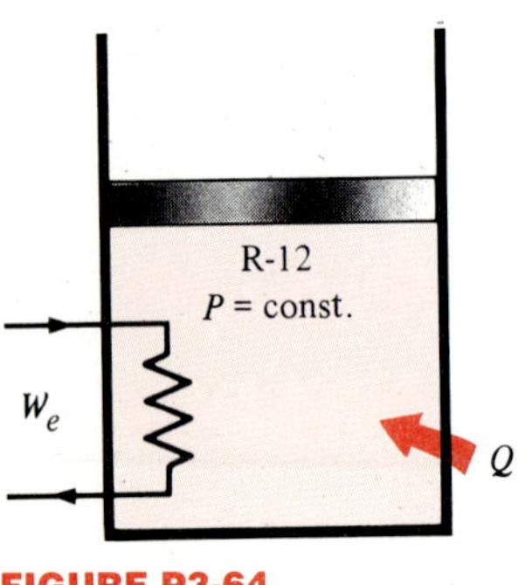

FIGURE P3-64

3-64 A mass of 10 kg of saturated refrigerant-12 vapor is contained in a piston-cylinder device at 200 kPa. Now 150 kJ of heat is transferred to the refrigerant at constant pressure while a 110-V source supplies current to a resistor within the cylinder for 5 min. Determine the current supplied if the final temperature is 70°C. Also, show the process on a *T-v* diagram with respect to saturation lines. *Answer:* 11.34 A

3-65 An insulated piston-cylinder device contains 2 L of saturated liquid water at a constant pressure of 150 kPa. Water is stirred by a paddle wheel while a current of 8 A flows for 20 min through a resistor placed in the water. If one-half of the liquid is evaporated during this constant-pressure process and the paddle-wheel work amounts to 25 kJ, determine the voltage of the source. Also, show the process on a *P-v* diagram with respect to saturation lines. *Answer:* 217.7 V

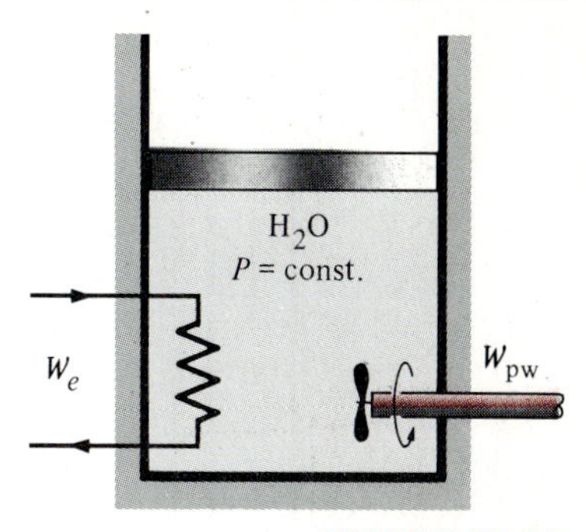

FIGURE P3-65

3-65E An insulated piston-cylinder device contains 0.07 ft³ of saturated liquid water at a constant pressure of 20 psia. Water is stirred by a paddle wheel while a current of 8 A flows for 20 min through a resistor placed in the water. If one-half of the liquid is evaporated during this constant-pressure process and the paddle-wheel work amounts to 25 Btu, determine the voltage of the source. Also, show the process on a *P-v* diagram with respect to saturation lines.

3-66 A piston-cylinder device contains steam initially at 1 MPa, 350°C, and 1.5 m³. Steam is allowed to cool at constant pressure until it first starts condensing. Show the process on a *T-v* diagram with respect to saturation lines, and determine (*a*) the mass of the steam, (*b*) the final temperature, and (*c*) the amount of heat transfer.

3-66E A piston-cylinder device contains steam initially at 180 psia, 500°F, and 50 ft³. Steam is allowed to cool at constant pressure until it first starts condensing. Show the process on a *T-v* diagram with respect to saturation lines. Determine (*a*) the mass of the steam, (*b*) the final temperature, and (*c*) the amount of heat transfer.
Answers: (*a*) 16.44 kg, (*b*) 373.13°F, (*c*) − 1201.7 Btu

3-67 A mass of 0.2 kg of saturated refrigerant-12 is contained in a piston-cylinder device at 200 kPa. Initially, 75 percent of the mass is in the liquid phase. Now heat is transferred to the refrigerant at constant pressure until the cylinder contains vapor only. Show the process on a *P-v* diagram with respect to saturation lines. Determine (*a*) the volume occupied by the refrigerant initially, (*b*) the work done, and (*c*) the total heat transfer.

3-68 A piston-cylinder device initially contains steam at 200 kPa, 150°C, and 0.1 m³. At this state, a linear spring ($F \propto x$) is touching the piston but exerts no force on it. Heat is now slowly transferred to the steam, causing the pressure and the volume to rise to 500 kPa and 0.12 m³, respectively. Show the process on a *P-v* diagram with respect to saturation lines, and determine (*a*) the final temperature, (*b*) the work done by the steam, and (*c*) the total heat transferred.
Answers: (*a*) 977.6°C, (*b*) 7 kJ, (*c*) 155.8 kJ

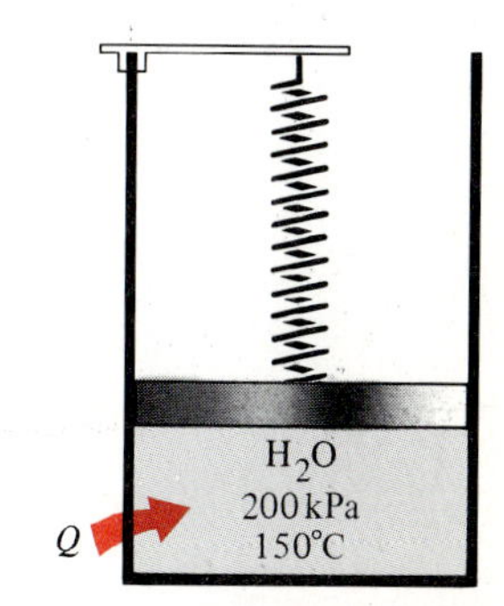

FIGURE P3-68

3-69 A piston-cylinder device initially contains 0.5 m³ of saturated water vapor at 200 kPa. At this state, the piston is resting on a set of stops, and the mass of the piston is such that a pressure of 300 kPa is required to move it. Heat is now slowly transferred to the steam until the volume

doubles. Show the process on a *P-v* diagram with respect to saturation lines and determine (*a*) the final temperature, (*b*) the work done during this process, and (*c*) the total heat transfer.
Answers: (*a*) 878.90°C, (*b*) 150 kJ, (*c*) 875 kJ

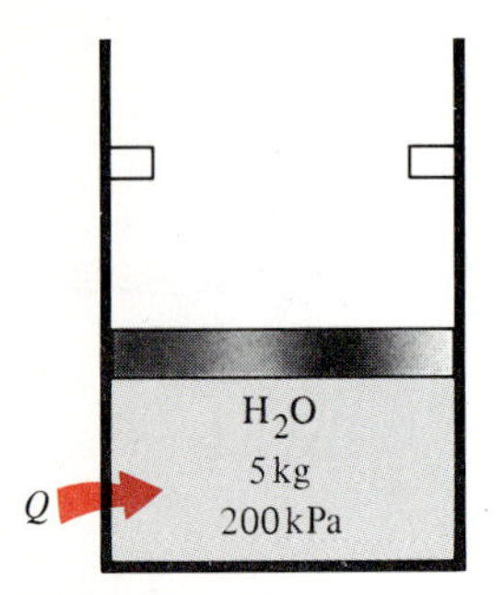

FIGURE P3-70

3-70 A piston-cylinder device with a set of stops on the top contains 5 kg of saturated liquid water at 200 kPa. Heat is now transferred to the water, causing some of the liquid to evaporate and move the piston up. When the piston reaches the stops, the enclosed volume is 100 L. More heat is added until the pressure is doubled. Show the process on a *P-v* diagram with respect to saturation lines, and determine (*a*) the amount of liquid at the final state, if any, (*b*) the final temperature, and (*c*) the total work and heat transfer.

3-70E A piston-cylinder device with a set of stops on the top contains 10 lbm of saturated liquid water at 20 psia. Heat is now transferred to the water, causing some of the liquid to evaporate and move the piston up. When the piston reaches the stops, the enclosed volume is 3 ft^3. More heat is added until the pressure is doubled. Show the process on a *P-v* diagram with respect to saturation lines. Determine (*a*) the amount of liquid at the final state, if any, (*b*) the final temperature, and (*c*) the total work and heat transfer.

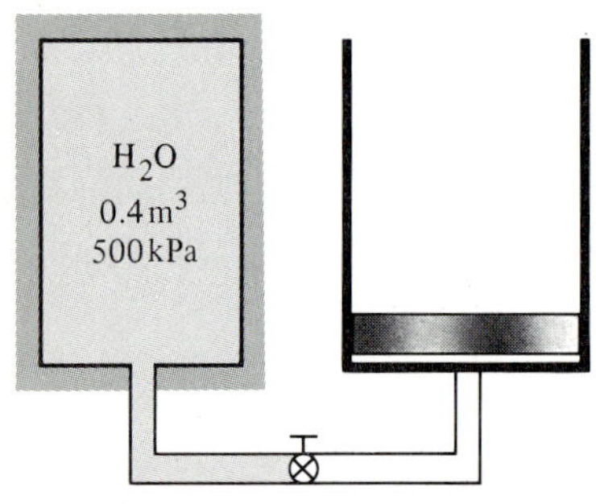

FIGURE P3-71

3-71 An insulated tank containing 0.4 m^3 of saturated water vapor at 500 kPa is connected to an initially evacuated insulated piston-cylinder device. The mass of the piston is such that a pressure of 200 kPa is required to raise the piston. Now the valve is opened slightly, and part of the steam flows to the cylinder, pushing the piston up. This process continues until the pressure in the tank falls to 200 kPa. Determine the temperature of the steam after thermal equilibrium is established.

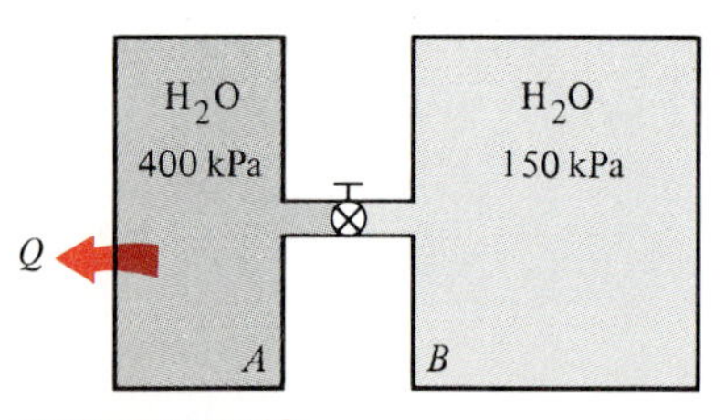

FIGURE P3-72

3-72 Two rigid tanks are connected by a valve. Tank *A* contains 0.2 m^3 of water at 400 kPa and 80 percent quality. Tank *B* contains 0.5 m^3 of water at 150 kPa and 250°C. The valve is now opened, and the two tanks eventually come to the same state. Determine the pressure and the amount of heat transfer when the system reaches thermal equilibrium with the surroundings at 25°C. *Answers:* 3.169 kPa, −2170 kJ

Specific Heats, Δu, and Δh of Ideal Gases

3-73C Is the relation $\Delta U = mC_{v,\text{av}}\,\Delta T$ restricted to constant-volume processes only, or can it be used for any kind of process of an ideal gas?

3-74C Is the relation $\Delta H = mC_{p,\text{av}}\,\Delta T$ restricted to constant-pressure processes only, or can it be used for any kind of process of an ideal gas?

3-75C Show that for an ideal gas $\overline{C}_p = \overline{C}_v + R_u$.

3-76C Is the energy required to heat air from 295 to 305 K the same as the energy required to heat it from 345 to 355 K? Assume the pressure remains constant in both cases.

3-77C What is the correct unit of C_v in the relation $\Delta U = mC_v \Delta T$—kJ/(kg·°C) or kJ/(kg·K)?

3-78C A fixed mass of an ideal gas is heated from 50 to 80°C at a constant pressure of (*a*) 1 atm and (*b*) 3 atm. For which case do you think the energy required will be greater? Why?

3-79C A fixed mass of an ideal gas is heated from 50 to 80°C at a constant volume of (*a*) 1 m^3 and (*b*) 3 m^3. For which case do you think the energy required will be greater? Why?

3-80C A fixed mass of an ideal gas is heated from 50 to 80°C (*a*) at constant volume and (*b*) at constant pressure. For which case do you think the energy required will be greater? Why?

3-81 Determine the enthalpy change Δh of nitrogen, in kJ/kg, as it is heated from 400 to 600 K, using (*a*) the empirical data for h from the nitrogen table (Table A-18), (*b*) the empirical specific heat equation as a function of temperature (Table A-2*c*), (*c*) the C_p value at the average temperature (Table A-2*b*), and (*d*) the C_p value at room temperature (Table A-2*a*). Also determine the percentage error involved in each case. *Answers:* (*a*) 211.4 kJ/kg; (*b*) 212.7 kJ/kg, 0.6 percent; (*c*) 211.2 kJ/kg, 0.1 percent; (*d*) 207.8 kJ/kg, 1.7 percent

3-82 Determine the enthalpy change Δh of oxygen, in kJ/kg, as it is heated from 500 to 800 K, using (*a*) the empirical data for h from the oxygen table (Table A-19), (*b*) the empirical specific heat equation as a function of temperature (Table A-2*c*), (*c*) the C_p value at the average temperature (Table A-2*b*), and (*d*) the C_p value at room temperature (Table A-2*a*).

3-82E Determine the enthalpy change Δh of oxygen, in Btu/lbm, as it is heated from 800 to 1500 R, using (*a*) the empirical data for h from the oxygen table (Table A-19E), (*b*) the empirical specific heat equation as a function of temperature (Table A-2E*c*), (*c*) the C_p value at the average temperature (Table A-2E*b*), and (*d*) the C_p value at room temperature (Table A-2E*a*). *Answers:* (*a*) 169.2 Btu/lbm, (*b*) 170.1 Btu/lbm, (*c*) 178.5 Btu/lbm, (*d*) 153.3 Btu/lbm

3-83 Determine the internal energy change Δu of hydrogen, in kJ/kg, as it is heated from 400 to 1000 K, using (*a*) the empirical data for u from the hydrogen table (Table A-22), (*b*) the empirical specific heat equation as a function of temperature (Table A-2*c*), (*c*) the C_v value at average temperature (Table A-2*b*), and (*d*) the C_v value at room temperature (Table A-2*a*). Also determine the percentage error involved in each case.

3-83E Determine the internal energy change Δu of hydrogen, in Btu/lbm, as it is heated from 700 to 1500 R, using (*a*) the empirical data for u from the hydrogen table (Table A-22E), (*b*) the empirical specific heat equation as a function of temperature (Table A-2E*c*), (*c*) the C_v value at average temperature (Table A-2E*b*), and (*d*) the C_v value at room temperature (Table A-2E*a*). Also determine the percentage error involved in each case.

3-84 Write a computer program to express the variation of specific heat $\overline{C}_p$ of air with temperature as a third-degree polynomial, using the data in Table A-2*b*. Compare your result to that given in Table A-2*c*.

Closed-System Energy Analysis: Ideal Gases

3-85 A rigid tank contains 5 kg of air at 100 kPa and 27°C. The air is now heated until its pressure doubles. Determine (*a*) the volume of the tank and (*b*) the amount of heat transfer.

3-85E A rigid tank contains 10 lbm of air at 14.7 psia and 80°F. The air is now heated until its pressure doubles. Determine (*a*) the volume of the tank and (*b*) the amount of heat transfer.
Answers: (*a*) 136.1 ft^3, (*b*) 948.9 Btu

3-86 A 0.5-m^3 rigid tank contains hydrogen at 500 kPa and 500 K. The gas is now cooled until its temperature drops to 300 K. Determine (*a*) the final pressure in the tank and (*b*) the amount of heat transfer.

3-87 A 4 m × 5 m × 6 m room is to be heated by a baseboard resistance heater. It is desired that the resistance heater be able to raise the air temperature in the room from 7 to 25°C within 20 min. Assuming no heat losses from the room and an atmospheric pressure of 100 kPa, determine the required power of the resistance heater. Assume constant specific heats at room temperature. *Answer:* 1.61 kW

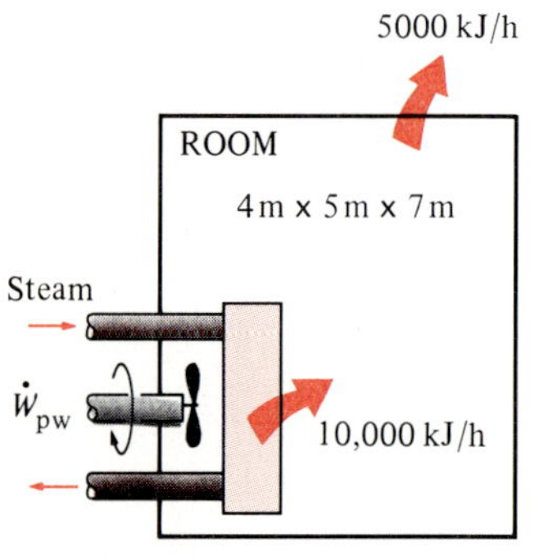

FIGURE P3-88

3-88 A 4 m × 5 m × 7 m room is heated by the radiator of a steam-heating system. The steam radiator transfers heat at a rate of 10,000 kJ/h, and a 100-W fan is used to distribute the warm air in the room. The rate of heat loss from the room is estimated to be about 5000 kJ/h. If the initial temperature of the room air is 10°C, determine how long it will take for the air temperature to rise to 20°C. For air, use the specific heat data at room temperature.

3-88E A 12 ft × 15 ft × 20 ft room is heated by the radiator of a steam-heating system. The steam radiator transfers heat at a rate of 10,000 Btu/h, and a 100-W fan is used to distribute the warm air in the room. The heat losses from the room are estimated to be at a rate of about 5000 Btu/h. If the initial temperature of the room air is 45°F, determine how long it will take for the air temperature to rise to 75°F. For air, use the specific heat data at room temperature. *Answer:* 978 s

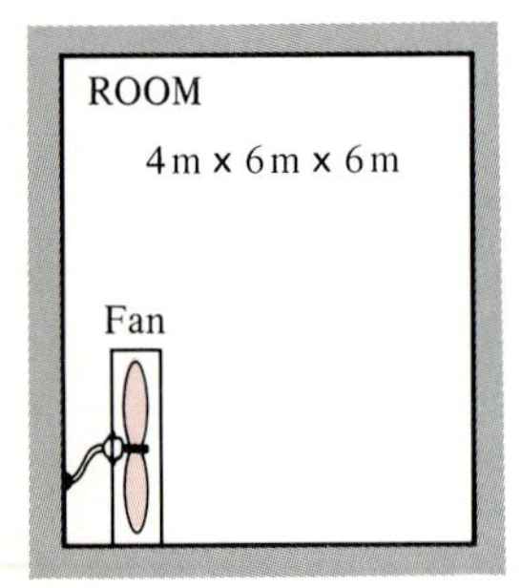

FIGURE P3-89

3-89 A student living in a 4 m × 6 m × 6 m dormitory room turns on her 200-W fan before she leaves the room on a summer day, hoping that the room will be cooler when she comes back in the evening. Assuming all the doors and windows are tightly closed and disregarding any heat transfer through the walls and the windows, determine the temperature in the room when she comes back 8 h later. Use specific heat values at room temperature, and assume the room to be at 100 kPa and 15°C in the morning when she leaves. *Answer:* 61°C

3-90 A 0.3-m^3 tank contains oxygen initially at 100 kPa and 27°C. A paddle wheel within the tank is rotated until the pressure inside rises to 150 kPa. During the process 2 kJ of heat is lost to the surroundings. Using the oxygen table, determine the paddle-wheel work done. Neglect the energy stored in the paddle wheel. *Answer:* −40.94 kJ

3-90E A 10-ft^3 tank contains oxygen initially at 14.7 psia and 80°F. A paddle wheel within the tank is rotated until the pressure inside rises to 20 psia. During the process 20 Btu of heat is lost to the surroundings. Using the oxygen table, determine the paddle-wheel work done. Neglect the energy stored in the paddle wheel.

3-91 An insulated rigid tank is divided into two equal parts by a partition. Initially, one part contains 2 kg of an ideal gas at 500 kPa and 70°C, and the other part is evacuated. The partition is now pulled out, and the gas expands into the entire tank. Determine the final temperature and pressure in the tank.

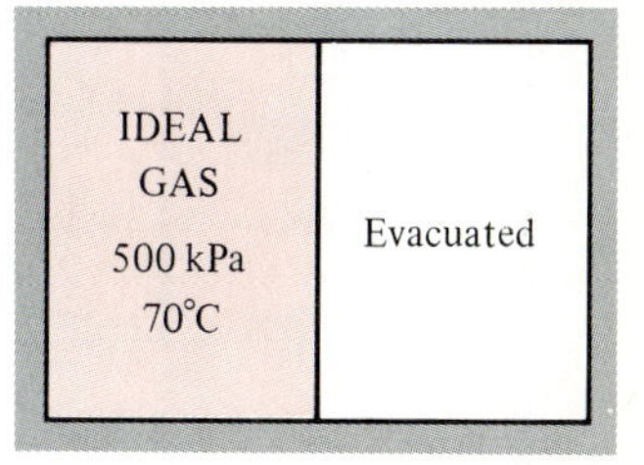

FIGURE P3-91

3-91E An insulated rigid tank is divided into two equal parts by a partition. Initially, one part contains 5 lbm of an ideal gas at 75 psia and 150°F, and the other part is evacuated. The partition is now pulled out, and the gas expands into the entire tank. Determine the final temperature and pressure in the tank.

3-92 A piston-cylinder device, whose piston is resting on top of a set of stops, initially contains 0.5 kg of helium gas at 100 kPa and 25°C. The mass of the piston is such that 500 kPa of pressure is required to raise it. How much heat must be transferred to the helium before the piston starts rising? *Answer:* 1857 kJ

3-93 An insulated piston-cylinder device contains 200 L of air at 200 kPa and 25°C. A paddle wheel within the cylinder is rotated until 15 kJ of work is done on the air while the pressure is held constant. Determine the final temperature of the air using data from the air table. Neglect the energy stored in the paddle wheel.

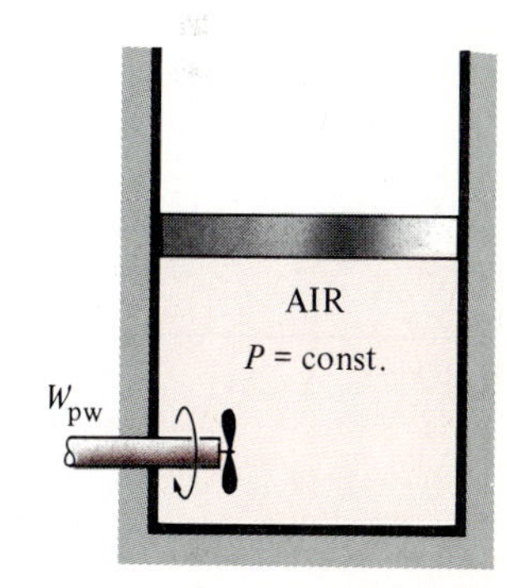

FIGURE P3-93

3-94 A piston-cylinder device contains 0.8 m^3 of nitrogen at 300 kPa and 327°C. The nitrogen is now allowed to cool at constant pressure until the temperature drops to 77°C. Using data from the nitrogen table, determine the heat transfer. *Answer:* −355.2 kJ

3-94E A piston-cylinder device contains 25 ft^3 of nitrogen at 50 psia and 700°F. Nitrogen is now allowed to cool at constant pressure until the temperature drops to 140°F. Using data from the nitrogen table, determine the heat transfer.

3-95 A mass of 10 kg of air in a piston-cylinder device is heated from 25 to 77°C by passing current through a resistance heater inside the cylinder. The pressure inside the cylinder is held constant at 200 kPa during the process, and a heat loss of 5 kJ occurs. Determine the electric energy supplied, in kWh. *Answer:* 0.147 kWh

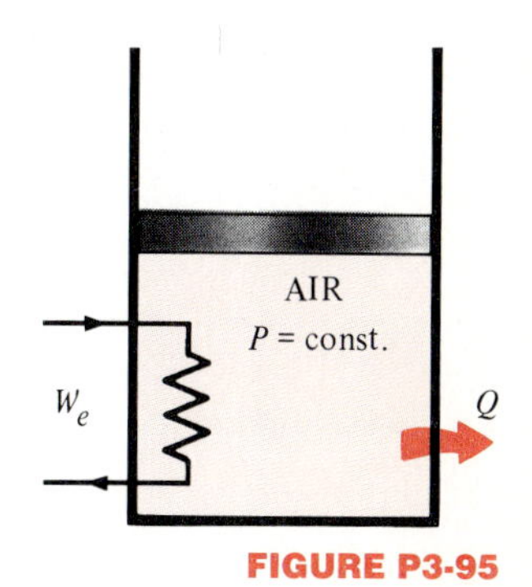

FIGURE P3-95

3-96 An insulated piston-cylinder device initially contains 0.3 m^3 of carbon dioxide at 200 kPa and 27°C. An electric switch is turned on, and a 110-V source supplies current to a resistance heater inside the cylinder for a period of 10 min. The pressure is held constant during the process, while the volume is doubled. Determine the current that passes through the resistance heater.

3-96E An insulated piston-cylinder device initially contains 10 ft^3 of carbon dioxide at 30 psia and 80°F. An electric switch is turned on, and a 110-V source supplies current to a resistance heater inside the cylinder for 10 min. The pressure is held constant during the process, while the volume is doubled. Determine the current that passes through the resistance heater. *Answer:* 4.6 A

3-97 A piston-cylinder device contains 0.8 kg of nitrogen initially at 100 kPa and 27°C. The nitrogen is now compressed slowly in a polytropic process during which $PV^{1.3}$ = constant until the volume is reduced by one-half. Determine the work done and the heat transfer for this process.

FIGURE P3-98

3-98 A piston-cylinder device contains helium gas initially at 150 kPa, 20°C, and 0.5 m^3. The helium is now compressed in a polytropic process (PV^n = constant) to 400 kPa and 140°C. Determine the heat transfer for this process. *Answer:* −11.2 kJ

3-98E A piston-cylinder device contains helium gas initially at 25 psia, 70°F, and 15 ft^3. The helium is now compressed in a polytropic process (PV^n = constant) to 60 psia and 300°F. Determine the heat transfer for this process.

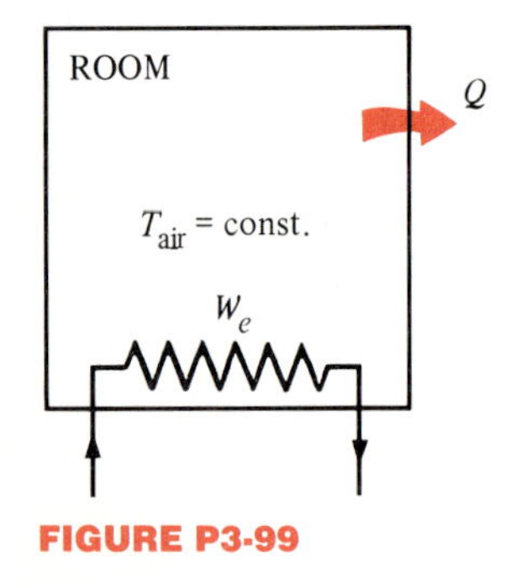

FIGURE P3-99

3-99 A room is heated by a baseboard resistance heater. When the heat losses from the room on a winter day amount to 10,000 kJ/h, the air temperature in the room remains constant even though the heater operates continuously. Determine the power rating of the heater, in kW.

3-100 A piston-cylinder device contains 0.2 m^3 of air at 200 kPa and 50°C. Heat is transferred to the air in the amount of 30 kJ as the air expands isothermally. Determine the amount of boundary work done during this process. *Answer:* 30 kJ

3-100E A piston-cylinder device contains 6 ft^3 of air at 30 psia and 150°F. Heat is transferred to the air in the amount of 30 Btu as the air expands isothermally. Determine the amount of boundary work done during this process.

3-101 A piston-cylinder device contains 4 kg of argon at 500 kPa and 25°C. During a quasi-equilibrium, isothermal expansion process, 10 kJ of boundary work is done by the system, and 2 kJ of paddle-wheel work is done on the system. Determine the heat transfer for this process. *Answer:* 8 kJ

3-101E A piston-cylinder device contains 10 lbm of argon at 75 psia and 80°F. During a quasi-equilibrium, isothermal expansion process, 10 Btu of boundary work is done by the system, and 2 Btu of paddle-wheel work is done on the system. Determine the heat transfer for this process.

3-102 A frictionless piston-cylinder device and a rigid tank initially contain 10 kg of an ideal gas each at the same temperature, pressure, and volume. It is desired to raise the temperatures of both systems by 10°C. Determine the amount of extra heat that must be supplied to the air in the cylinder which is maintained at constant pressure to achieve this. Assume the molar mass of the gas is 25.

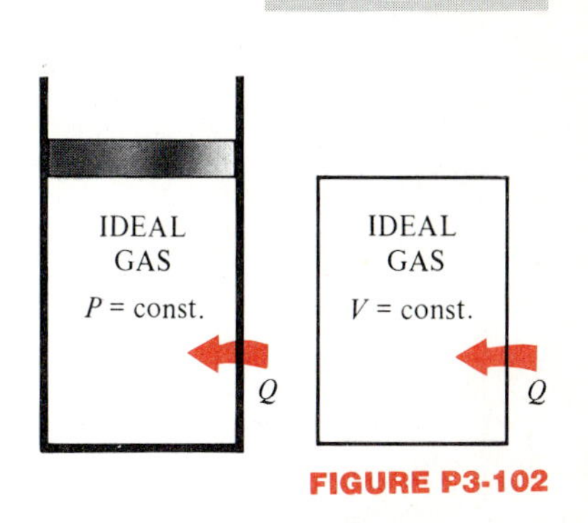

FIGURE P3-102

3-103 A piston-cylinder device whose piston is resting on a set of stops, initially contains 3 kg of air at 200 kPa and 27°C. The mass of the piston is such that a pressure of 400 kPa is required to move it. Heat is now transferred to the air until its volume doubles. Determine the work done by the air and the total heat transferred to the air during this process. Also show the process on a P-v diagram. *Answers:* 516 kJ, 2674 kJ

3-104 A piston-cylinder device, with a set of stops on the top, initially contains 3 kg of air at 200 kPa and 27°C. Heat is now transferred to the air, and the piston rises until it hits the stops, at which point the volume is twice the initial volume. More heat is transferred until the pressure inside the cylinder also doubles. Determine the work done and the amount of heat transfer for this process. Also show the process on a P-v diagram.

3-104E A piston-cylinder device, with a set of stops on the top, initially contains 5 lbm of air at 30 psia and 80°F. Heat is now transferred to the air, and the piston rises until it hits the stops, at which point the volume is twice the initial volume. More heat is transferred until the pressure inside the cylinder also doubles. Determine the work done and the amount of heat transfer for this process. Also show the process on a P-v diagram. *Answers:* 184.9 Btu, 1731 Btu

3-105 A piston-cylinder device contains 0.5 m³ of helium gas initially at 100 kPa and 25°C. At this position, a linear spring is touching the piston but exerts no force on it. Heat is now transferred to helium until both the pressure and the volume triple. Determine (*a*) the work done and (*b*) the amount of heat transfer for this process. Also show the process on a P-v diagram. *Answers:* (*a*) 200 kJ, (*b*) 801.4 kJ

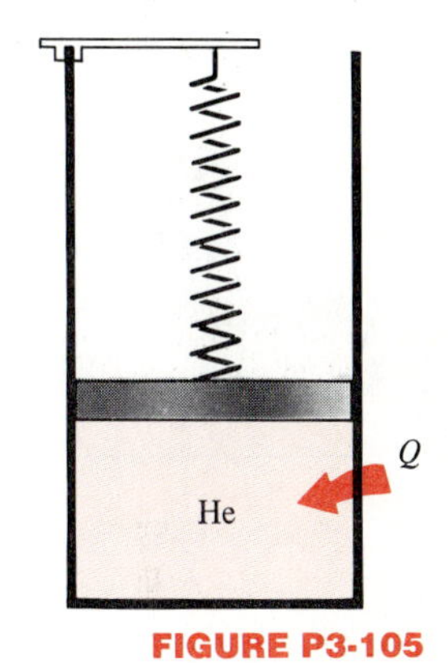

FIGURE P3-105

3-106 A rigid tank containing 0.5 m³ of air at 300 kPa and 20°C is connected by a valve to a piston-cylinder device with zero clearance. The mass of the piston is such that a pressure of 200 kPa is required to raise the piston. The valve is now opened slightly, and air is allowed to flow into the cylinder until the pressure in the tank drops to 200 kPa. During this process, heat is exchanged with the surroundings such that the entire air remains at 20°C at all times. Determine the heat transfer for this process.

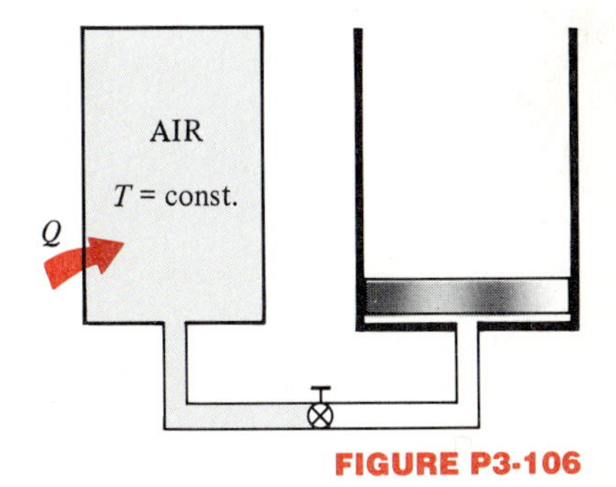

FIGURE P3-106

Closed-System Energy Analysis: Solids and Liquids

3-107 An unknown mass of aluminum at 75°C is dropped into an insulated tank which contains 100 L of water at 25°C and atmospheric pressure. If the final equilibrium temperature is 30°C, determine the mass of the aluminum. Assume the density of liquid water to be 1000 kg/m³.
Answer: 51.7 kg

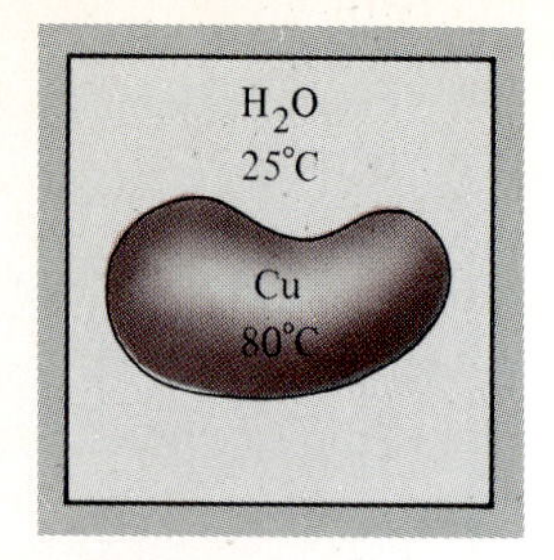

FIGURE P3-108

3-108 A 30-kg mass of copper at 80°C is dropped into an insulated tank which contains 100 kg of water at 25°C. Determine the final equilibrium temperature in the tank.

3-108E A 60-lbm mass of copper at 200°F is dropped into an insulated tank which contains 150 lbm of water at 70°F. Determine the final equilibrium temperature in the tank. *Answer:* 74.6°F

3-109 A 20-kg mass of iron at 100°C is brought into contact with 20 kg of aluminum at 200°C in an insulated enclosure. Determine the final equilibrium temperature of the combined system.

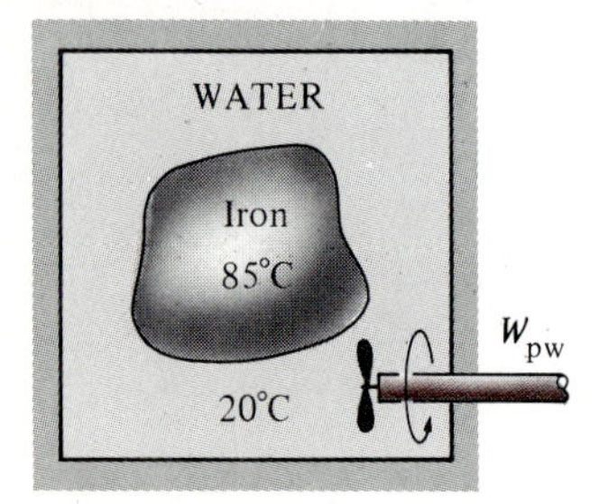

FIGURE P3-110

3-110 An unknown mass of iron at 85°C is dropped into an insulated tank that contains 100 L of water at 20°C. At the same time, a paddle wheel driven by a 200-W motor is activated to stir the water. Thermal equilibrium is established after 20 min with a final temperature of 24°C. Determine the mass of the iron. Neglect the energy stored in the paddle wheel, and take the water density to be 1000 kg/m^3. *Answer:* 52.2 kg

3-111 A 30-kg mass of copper at 70°C and a 20-kg mass of iron at 80°C are dropped into a tank containing 300 L of water at 20°C. If 10 kJ of heat is lost to the surroundings during the process, determine the final equilibrium temperature. *Answer:* 20.90°C

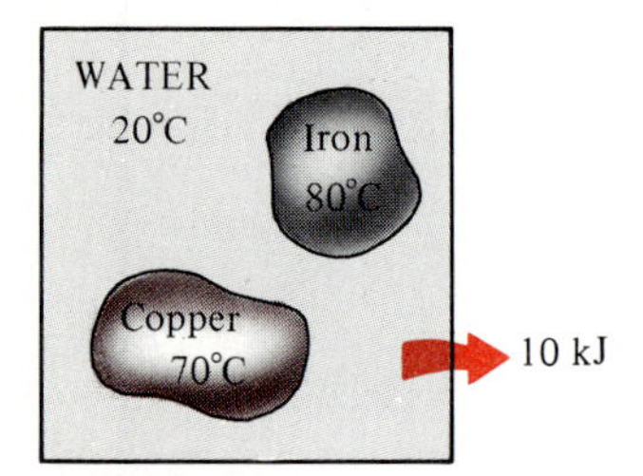

FIGURE P3-111

3-111E A 70-lbm mass of copper at 160°F and a 50-lbm mass of iron at 200°F are dropped into a tank containing 5 ft^3 of water at 70°F. If 10 Btu of heat is lost to the surroundings during the process, determine the final equilibrium temperature.

The First Law of Thermodynamics-Control Volumes

CHAPTER 4

In Chap. 3 we discussed the energy interactions between a system and its surroundings, and the conservation of energy principle for closed (nonflow) systems. In this chapter, we extend the analysis to systems that involve mass flow across their boundaries, i.e., control volumes. The conservation of energy equation for a general control volume can be rather involved and intimidating. Therefore, we treat the control volume energy analysis in two stages. First, we consider the steady-flow process which is the model process for many engineering devices such as turbines, compressors, and heat exchangers. Second, we discuss the general unsteady-flow processes with particular emphasis on the uniform-flow process, which is the model process for commonly encountered charging and discharging processes.

4-1 ■ THERMODYNAMIC ANALYSIS OF CONTROL VOLUMES

A large number of engineering problems involve mass flow in and out of a system and, therefore, are modeled as *control volumes* (Fig. 4-1). A water heater, a car radiator, a turbine, and a compressor all involve mass flow and should be analyzed as control volumes (open systems) instead of as control masses (closed systems). In general, *any arbitrary region in space* can be selected as a control volume. There are no concrete rules for the selection of control volumes, but the proper choice certainly makes the analysis much easier. If we were to analyze the flow of air through a nozzle, for example, a good choice for the control volume would be the region within the nozzle.

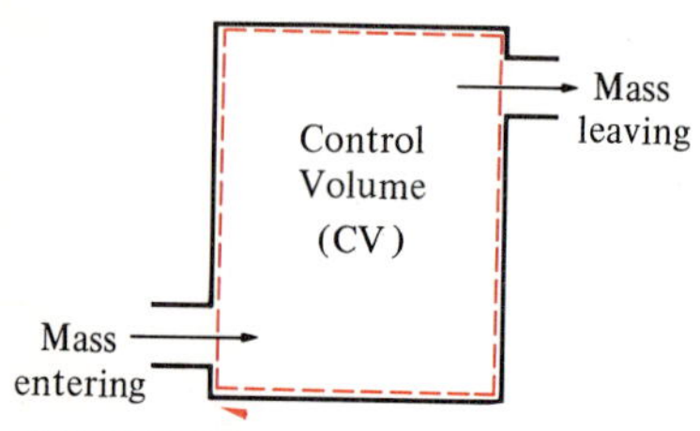

FIGURE 4-1
Mass may flow into and out of a control volume.

The boundaries of a control volume are called a *control surface,* and they can be real or imaginary. In the case of a nozzle, the inner surface of the nozzle forms the real part of the boundary, and the entrance and exit areas form the imaginary part, since there are no physical surfaces there (Fig. 4-2).

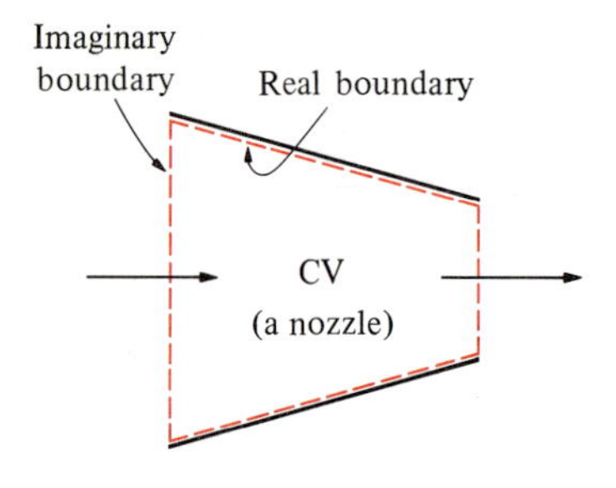

FIGURE 4-2
Real and imaginary boundaries of a control volume.

A control volume can be fixed in size and shape, as in the case of a nozzle, or it may involve a moving boundary, as shown in Fig. 4-3. Most control volumes, however, have fixed boundaries and thus do not involve any moving boundary work. A control volume may also involve heat and work interactions just as a closed system, in addition to mass interaction.

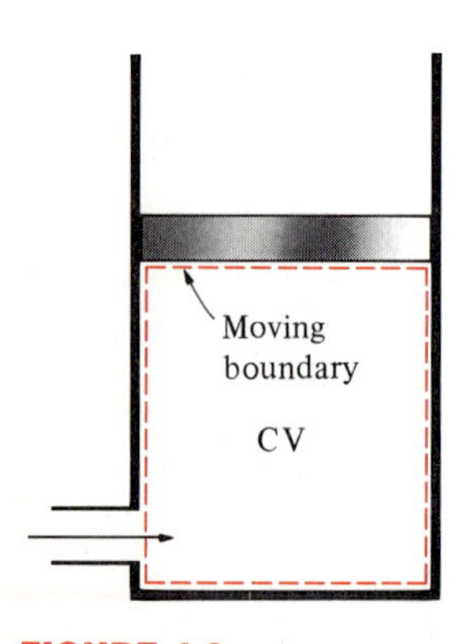

FIGURE 4-3
Some control volumes involve moving boundaries.

A large variety of thermodynamic problems may be solved by the control volume analysis. Even though it is possible to derive the relevant equations for the most general case and simplify them for special cases, many students are intimidated early in the analysis by the complexities involved and find them difficult to digest. The method that we use is a step-by-step approach, starting with the simplest case and adding complexities one at a time.

The terms *steady* and *uniform* are used extensively in this chapter, and thus it is important to have a clear understanding of their meanings. The term *steady* implies *no change with time*. The opposite of steady is *unsteady,* or *transient*. The term *uniform,* however, implies *no change with location* over a specified region. These meanings are consistent with their everyday use (steady girlfriend, uniform distribution, etc.).

An overview of the conservation of mass and the conservation of energy principles for control volumes is given below.

Conservation of Mass Principle

The conservation of mass is one of the most fundamental principles in nature. We are all familiar with this principle, and it is not difficult to understand. As the saying goes, you cannot have your cake and eat it, too! A person does not have to be an engineer to figure out how much vinegar-and-oil dressing she is going to have if she mixes 100 g of oil with 25 g of vinegar. Even chemical equations are balanced on the basis of the conservation of mass principle (Fig. 4-4). When 16 kg of oxygen reacts with 2 kg of hydrogen, 18 kg of water is formed. In an electrolysis process, the water will separate back to 2 kg of hydrogen and 16 kg of oxygen.

Mass, like energy, is a conserved property, and it cannot be created or destroyed. However, mass m and energy E can be converted to each other according to the famous formula proposed by Einstein:

$$E = mc^2 \tag{4-1}$$

where c is the speed of light. This equation suggests that the mass of a system will change when its energy changes. However, for all energy interactions encountered in practice, with the exception of nuclear reactions, the change in mass is extremely small and cannot be detected by even the most sensitive devices. For example, when 1 kg of water is formed from oxygen and hydrogen, the amount of energy released is 15,879 kJ, which corresponds to a mass of 1.76×10^{-10} kg. A mass of this magnitude is beyond the accuracy required by practically all engineering calculations and thus can be disregarded.

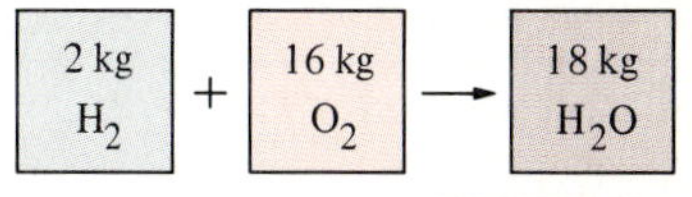

FIGURE 4-4

Mass is conserved even during chemical reactions.

For closed systems, the conservation of mass principle is implicitly used by requiring that the mass of the system remain constant during a process. For control volumes, however, mass can cross the boundaries, and so we must keep track of the amount of the mass entering and leaving the control volume (Fig. 4-5). The **conservation of mass principle** for a control volume (CV) undergoing a process can be expressed as

$$\begin{pmatrix} \text{Total} \\ \text{mass entering} \\ \text{CV} \end{pmatrix} - \begin{pmatrix} \text{Total} \\ \text{mass leaving} \\ \text{CV} \end{pmatrix} = \begin{pmatrix} \text{Net change} \\ \text{in mass within} \\ \text{CV} \end{pmatrix}$$

or

$$\Sigma m_i - \Sigma m_e = \Delta m_{\text{CV}} \tag{4-2}$$

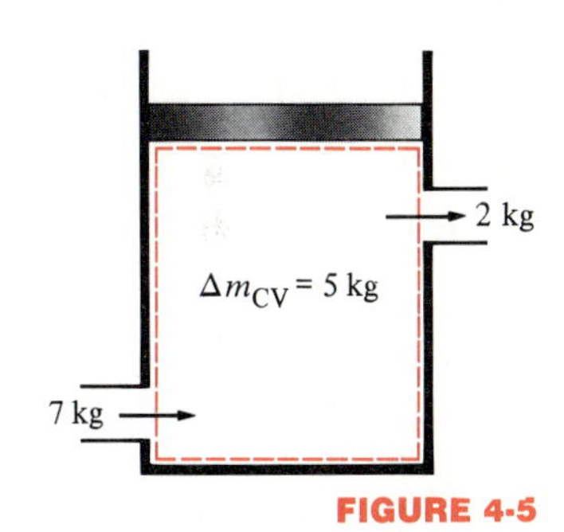

FIGURE 4-5

Conservation of mass principle for a control volume.

where the subscripts i, e, and CV stand for *inlet, exit,* and *control volume,* respectively. The conservation of mass equation could also be expressed in the rate form by expressing the quantities per unit time. Equation 4-2 is a verbal statement of the conservation of mass principle for a general control volume undergoing *any* process.

A person who can balance a checkbook (by keeping track of deposits and withdrawals, or simply by observing the "conservation of money" principle) should have no difficulty in applying the conservation of mass principle to thermodynamic systems (Fig. 4-6). The conservation of mass equation is often referred to as the *continuity equation* in fluid mechanics.

FIGURE 4-6

We are surrounded by numerous conservation principles in life.

Mass and Volume Flow Rates

The amount of mass flowing through a cross section per unit time is called the **mass flow rate** and is denoted $\dot{m}$. As before, the dot over a symbol is used to indicate a quantity per unit time.

A liquid or a gas flows in and out of a control volume through pipes or ducts. The mass flow rate of a fluid flowing in a pipe or duct is proportional to the cross-sectional area A of the pipe or duct, the density ρ, and the velocity $\mathbf{V}$ of the fluid. The mass flow rate through a differential area dA can be expressed as

$$d\dot{m} = \rho \mathbf{V}_n \, dA \tag{4-3}$$

where $\mathbf{V}_n$ is the velocity component normal to dA. The mass flow rate through the entire cross-sectional area of the pipe or duct is obtained by integration:

$$\dot{m} = \int_A \rho \mathbf{V}_n \, dA \qquad \text{(kg/s)} \tag{4-4}$$

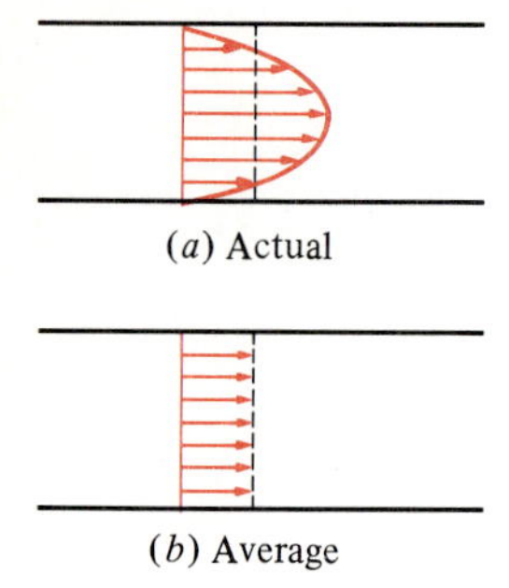

FIGURE 4-7

Actual and average velocity profiles for flow in a pipe (the mass flow rate is the same for both cases).

The density of a fluid flowing in a pipe usually remains constant over the cross section, but this is not the case for velocity. Due to viscous effects (friction between fluid layers), the velocity will vary from zero at the wall to a maximum at the center. For most practical purposes, however, an equivalent average value of velocity can be used for the entire cross section (Fig. 4-7). Then the integration in Eq. 4-4 can be performed, to yield

$$\dot{m} = \rho \mathbf{V}_{av} A \qquad \text{(kg/s)} \tag{4-5}$$

where ρ = density, kg/m^3 ($= 1/v$)
$\mathbf{V}_{av}$ = average fluid velocity normal to A, m/s
A = cross-sectional area, m^2

The volume of the fluid flowing through a cross section per unit time is called the **volume flow rate** $\dot{V}$ (Fig. 4-8) and is given by

$$\dot{V} = \int_A \mathbf{V}_n \, dA = \mathbf{V}_{av} A \qquad \text{(m}^3\text{/s)} \tag{4-6}$$

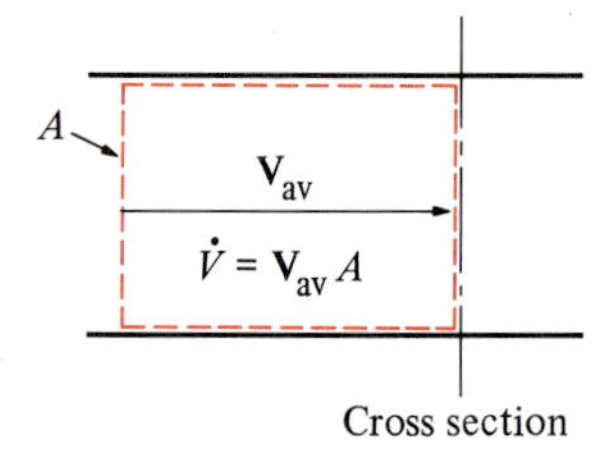

FIGURE 4-8

The volume flow rate is the volume of fluid flowing through a cross section per unit time.

The mass and volume flow rates are related by

$$\dot{m} = \rho \dot{V} = \frac{\dot{V}}{v} \tag{4-7}$$

This relation is analogous to $m = V/v$, which is the relation between the mass and the volume of a fluid.

For simplicity, we drop the subscript on the average velocity. Unless otherwise stated, $\mathbf{V}$ denotes the average velocity in the flow direction. Also A denotes the cross-sectional area normal to the flow direction.

Conservation of Energy Principle

We have already discussed the conservation of energy principle, or the first law of thermodynamics, in relation to a closed system. As we pointed out in Chap. 3, the energy of a closed system can be changed by heat or work interactions only, and the change in the energy of a closed system during a process is equal to the net heat and work transfer across the system boundary. This was expressed as

$$Q - W = \Delta E$$

For control volumes, however, an additional mechanism can change the energy of a system: *mass flow in and out of the control volume* (Fig. 4-9). When mass enters a control volume, the energy of the control volume increases because the entering mass carries some energy with it. Likewise,

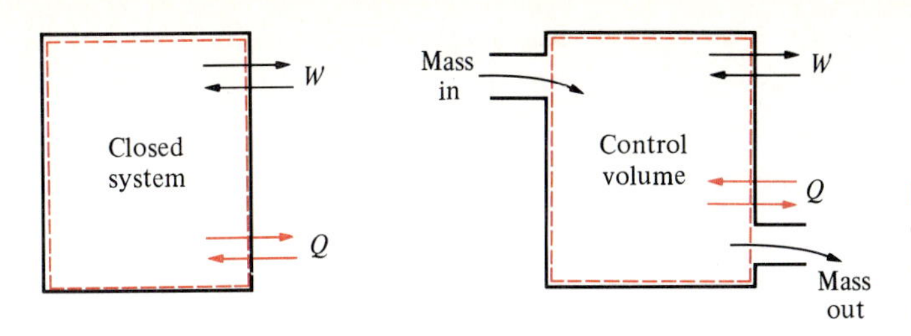

FIGURE 4-9
The energy content of a control volume can be changed by mass flow as well as heat and work interactions.

when some mass leaves the control volume, the energy contained within the control volume decreases because the leaving mass takes *out* some energy with it. For example, when some hot water is taken out of a water heater and is replaced by the same amount of cold water, the energy content of the hot-water tank (the control volume) decreases as a result of this mass interaction.

Then the conservation of energy equation for a control volume undergoing a process can be expressed as

$$\begin{pmatrix}\text{Total energy}\\ \text{crossing boundary}\\ \text{as heat and work}\end{pmatrix} + \begin{pmatrix}\text{Total energy}\\ \text{of mass}\\ \text{entering CV}\end{pmatrix} - \begin{pmatrix}\text{Total energy}\\ \text{of mass}\\ \text{leaving CV}\end{pmatrix} = \begin{pmatrix}\text{Net change}\\ \text{in energy}\\ \text{of CV}\end{pmatrix}$$

or

$$Q - W + \Sigma E_{\text{in}} - \Sigma E_{\text{out}} = \Delta E_{\text{CV}} \qquad \text{(4-8)}$$

Obviously if no mass is entering or leaving the control volume, the second and third terms drop out and the above equation reduces to the one given for a closed system. Despite its simple appearance, Eq. 4-8 is applicable to *any* control volume undergoing *any* process. This equation can also be expressed in the rate form by expressing the quantities above per unit time.

Heat transfer to or from a control volume should not be confused with the energy transported with mass into and out of a control volume. Remember that heat is the form of energy transferred as a result of a temperature difference between the control volume and the surroundings.

A control volume, like a closed system, may involve one or more forms of work at the same time (Fig. 4-10). If the boundary of the control volume is stationary, as is often the case, the moving boundary work is zero. Then the work term will involve, at most, shaft work and electrical work for simple compressible systems. As before, when the control volume is insulated, the heat transfer term becomes zero.

The energy required to push fluid into or out of a control volume is called the *flow work,* or *flow energy*. It is considered to be part of the energy transported with the fluid and is discussed below.

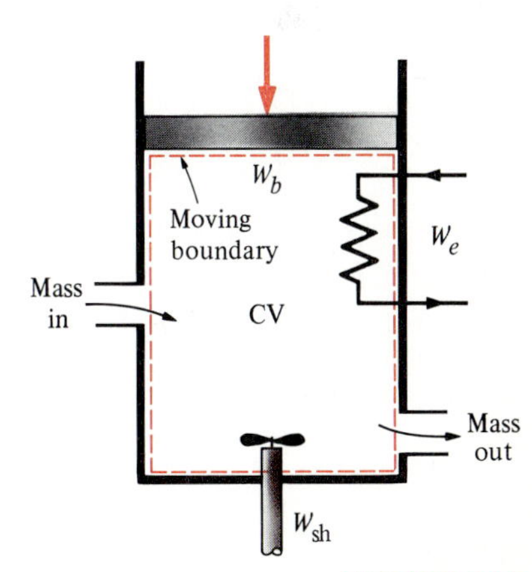

FIGURE 4-10
A control volume may involve boundary work in addition to electrical and shaft work.

Flow Work

Unlike closed systems, control volumes involve mass flow across their boundaries, and some work is required to push the mass into or out of the control volume. This work is known as the **flow work**, or **flow energy**, and is necessary for maintaining a continuous flow through a control volume.

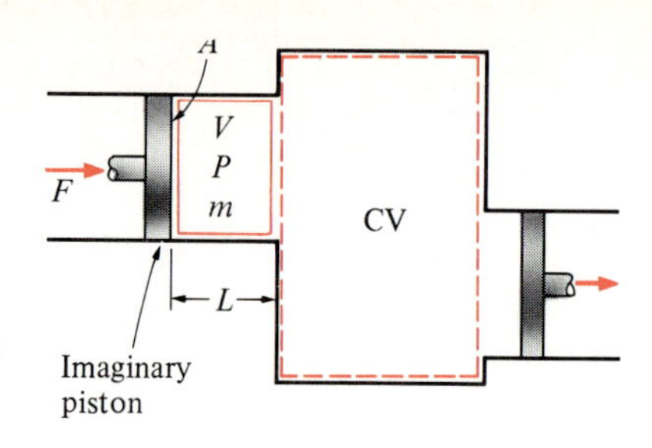

FIGURE 4-11

Schematic for flow work.

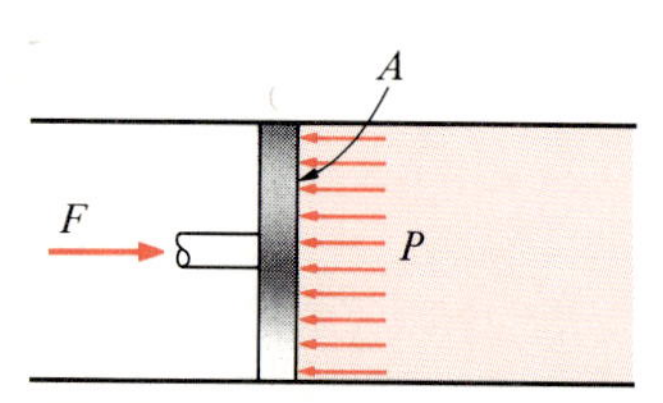

FIGURE 4-12

In the absence of acceleration, the force applied on a fluid by a piston is equal to the force applied on the piston by the fluid.

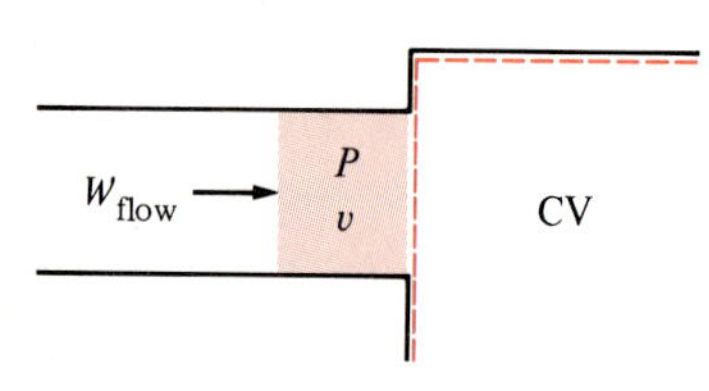

(*a*) Before entry

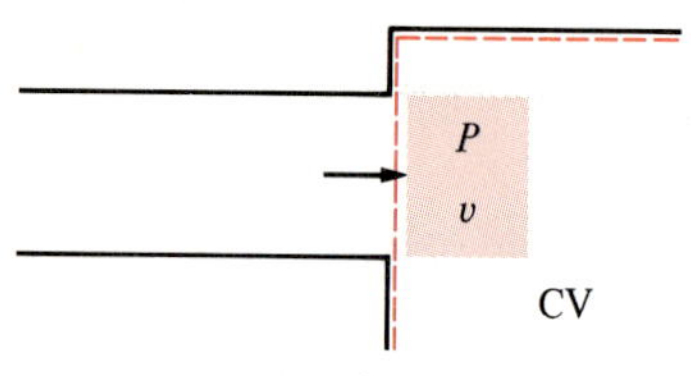

(*b*) After entry

FIGURE 4-13

Flow work is the energy needed to push a fluid into or out of a control volume, and it is equal to Pv.

To obtain a relation for flow work, consider a fluid element of volume V as shown in Fig. 4-11. The fluid immediately upstream will force this fluid element to enter the control volume; thus it can be regarded as an imaginary piston. The fluid element can be chosen to be sufficiently small so that it has uniform properties throughout.

If the fluid pressure is P and the cross-sectional area of the fluid element is A (Fig. 4-12), the force applied on the fluid element by the imaginary piston is

$$F = PA \tag{4-9}$$

To push the entire fluid element into the control volume, this force must act through a distance L. Thus the work done in pushing the fluid element across the boundary (i.e., the flow work) is

$$W_{\text{flow}} = FL = PAL = PV \qquad \text{(kJ)} \tag{4-10}$$

The flow work per unit mass is obtained by dividing both sides of this equation by the mass of the fluid element:

$$w_{\text{flow}} = Pv \qquad \text{(kJ/kg)} \tag{4-11}$$

The flow work relation is the same whether the fluid is pushed into or out of the control volume (Fig. 4-13).

It is interesting that unlike other work quantities, flow work is expressed in terms of the properties. In fact, it is the product of two properties of the fluid. For that reason some people view it as a *combination property* (like enthalpy) and refer to it as *flow energy, convected energy,* or *transport energy* instead of flow work. Others, however, argue rightfully that the product Pv represents energy for flowing fluids only and does not represent any form of energy for nonflow (closed) systems. Therefore, it should be treated as work. This controversy is not likely to end, but it is comforting to know that both arguments yield the same result for the energy equation. In the discussions that follow, we consider the flow energy to be part of the energy of a flowing fluid, since this greatly simplifies the derivation of the energy equation for control volumes.

Total Energy of a Flowing Fluid

As we discussed in Chap. 1, the total energy of a simple compressible system consists of three parts: internal, kinetic, and potential energies (Fig. 4-14). On a unit-mass basis, it is expressed as

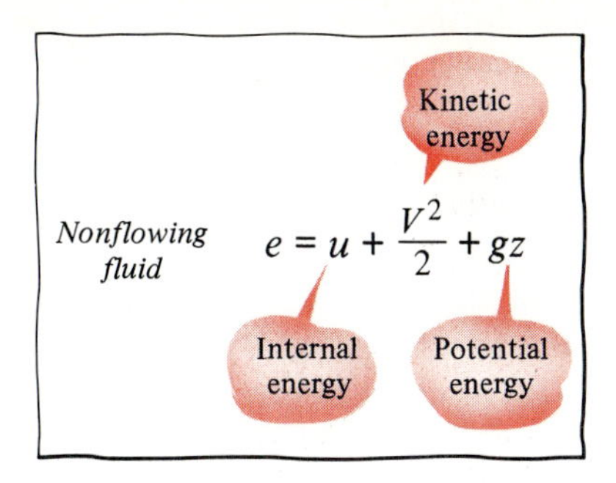

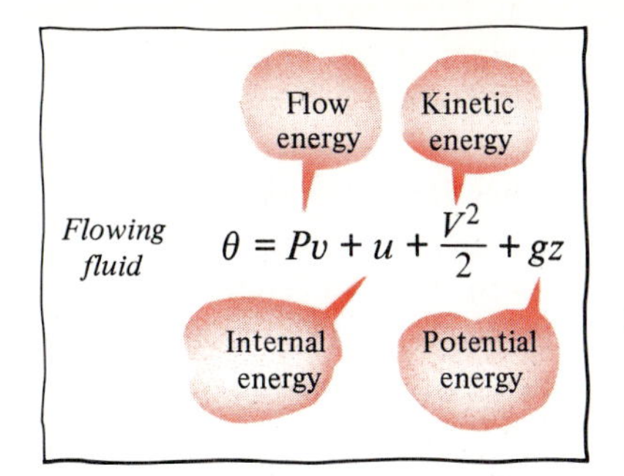

FIGURE 4-14

The total energy consists of three parts for a nonflowing fluid and four parts for a flowing fluid.

$$e = u + \text{ke} + \text{pe} = u + \frac{\mathbf{V}^2}{2} + gz \qquad \text{(kJ/kg)} \qquad (4\text{-}12)$$

where **V** is the velocity and z is the elevation of the system relative to some external reference point.

The fluid entering or leaving a control volume possesses an additional form of energy—the *flow energy Pv,* as discussed above. Then the total energy of a **flowing fluid** on a unit-mass basis (denoted θ) becomes

$$\theta = Pv + e = Pv + (u + \text{ke} + \text{pe})$$

But the combination $Pv + u$ has been previously defined as the enthalpy h. So the above relation reduces to

$$\theta = h + \text{ke} + \text{pe} = h + \frac{\mathbf{V}^2}{2} + gz \qquad \text{(kJ/kg)} \qquad (4\text{-}13)$$

By using the enthalpy instead of the internal energy to represent the energy of a flowing fluid, one does not need to be concerned about the flow work. The energy associated with pushing the fluid into or out of the control volume is automatically taken care of by enthalpy. In fact, this is the main reason for defining the property enthalpy. From now on, the energy of a fluid stream flowing into or out of a control volume is represented by Eq. 4-13, and no reference will be made to flow work or flow energy. Thus the work term W in the control-volume energy equations will represent all forms of work (boundary, shaft, electrical, etc.) except flow work.

4-2 ■ THE STEADY-FLOW PROCESS

A large number of engineering devices such as turbines, compressors, and nozzles operate for long periods of time under the same conditions, and they are classified as *steady-flow devices*.

Processes involving steady-flow devices can be represented reasonably well by a somewhat idealized process, called the **steady-flow process**. A steady-flow process can be defined as *a process during which a fluid flows through a control volume steadily* (Fig. 4-15). That is, the fluid properties can change from point to point within the control volume, but

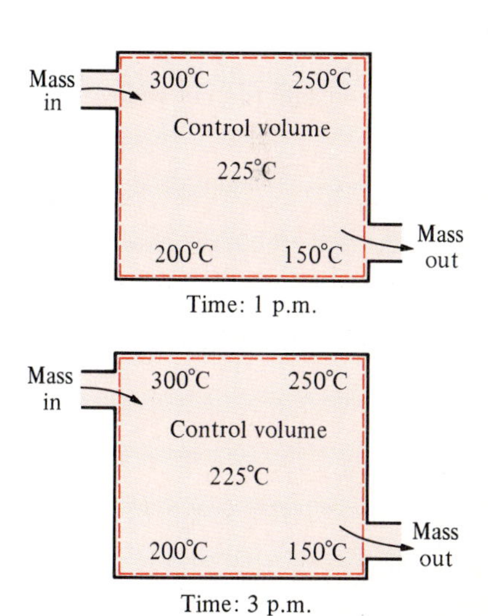

FIGURE 4-15

During a steady-flow process, fluid properties within the control volume may change with position, but not with time.

at any fixed point they remain the same during the entire process. (Remember, *steady* means *no change with time.*) A steady-flow process is characterized by the following:

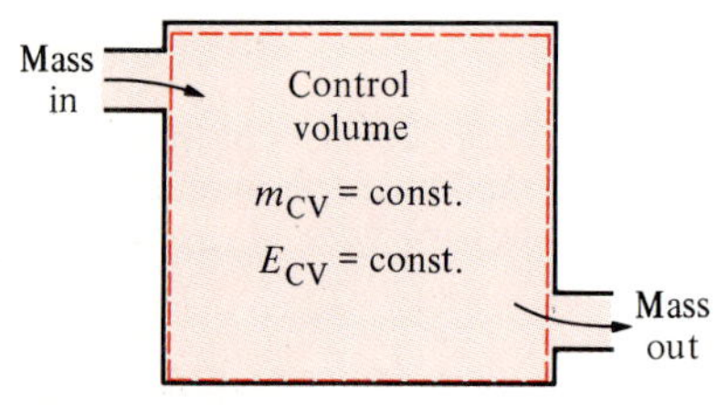

FIGURE 4-16

Under steady-flow conditions, the mass and energy contents of a control volume remain constant.

1 No properties (intensive or extensive) *within the control volume* change with time. Thus the volume V, the mass m, and the total energy content E of the control volume remain constant during a steady-flow process (Fig. 4-16). As a result, the boundary work is zero for steady-flow systems (since V_{CV} = constant), and the total mass or energy entering the control volume must be equal to the total mass or energy leaving it (since m_{CV} = constant and E_{CV} = constant). These observations greatly simplify the analysis.

2 No properties change at the *boundaries* of the control volume with time. Thus the fluid properties at an inlet or an exit will remain the same during the entire process. The properties may, however, be different at different openings (inlets and exits). They may even vary over the cross section of an inlet or an exit. But all properties, including the velocity and elevation, must remain constant with time at a fixed position. It follows that the mass flow rate of the fluid at an opening must remain constant during a steady-flow process (Fig. 4-17). As an added simplification, the fluid properties at an opening are usually considered to be uniform (at some average value) over the cross section. Thus the fluid properties at an inlet or exit may be specified by the average single values.

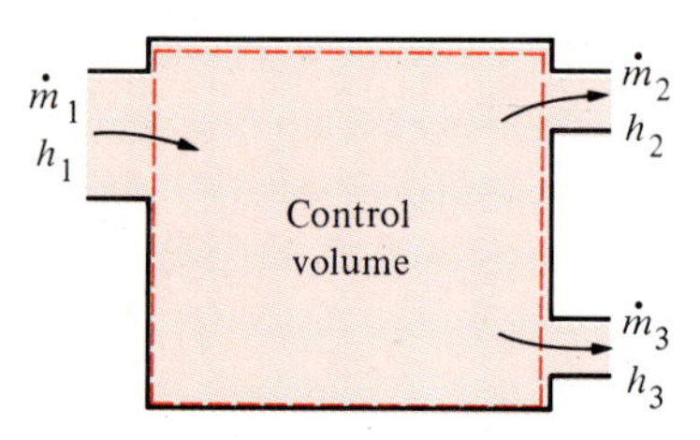

FIGURE 4-17

Under steady-flow conditions, the fluid properties at an inlet or exit remain constant (do not change with time).

3 The heat and work interactions between a steady-flow system and its surroundings do not change with time. Thus the power delivered by a system and the rate of heat transfer to or from a system remain constant during a steady-flow process.

Some cyclic devices, such as reciprocating engines or compressors, do not satisfy any of the conditions stated above since the flow at the inlets and the exits will be pulsating and not steady. However, the fluid properties vary with time in a periodic manner, and the flow through these devices can still be analyzed as a steady-flow process by using time-averaged values for the properties and the heat flow rates through the boundaries (Fig. 4-18).

Steady-flow conditions can be closely approximated by devices that

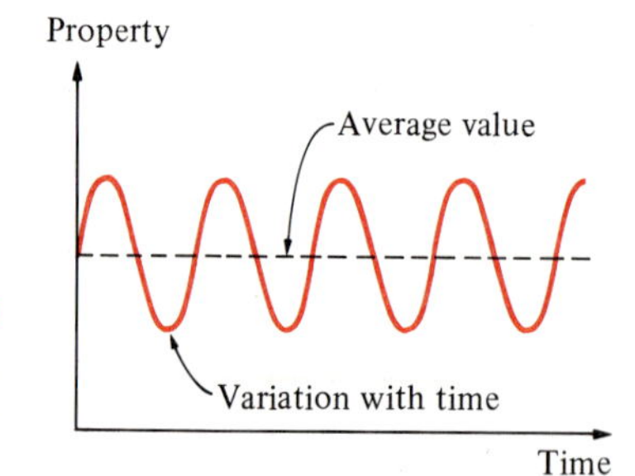

FIGURE 4-18

A process that involves periodic changes in fluid properties can also be analyzed as a steady flow process.

are intended for continuous operation such as turbines, pumps, boilers, condensers, and heat exchangers of steam power plants. The equations that are developed later in this section can be used for these and similar devices once the transient start-up period is completed and a steady operation is established.

Conservation of Mass

During a steady-flow process, the total amount of mass contained within a control volume does not change with time (m_{CV} = constant). Then the conservation of mass principle requires that the total amount of mass entering a control volume equal the total amount of mass leaving it (Fig. 4-19). For a garden hose nozzle, for example, the amount of water entering the nozzle is equal to the amount of water leaving it under steady operation (Fig. 4-20).

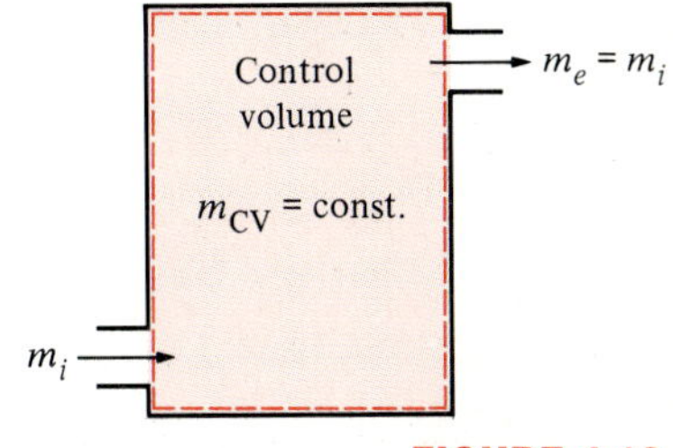

FIGURE 4-19

During a steady-flow process, the amount of mass entering the control volume equals the amount of mass leaving.

When dealing with steady-flow processes, we are not interested in the amount of mass that flows in and out of a device over a time; instead, we are interested in the amount of mass flowing per unit time, i.e., *the mass flow rate* $\dot{m}$. The **conservation of mass principle** for a general steady-flow system with multiple inlets and exits (Fig. 4-21) can be expressed in the rate form as

$$\begin{pmatrix}\text{Total mass}\\ \text{entering CV}\\ \text{per unit time}\end{pmatrix} = \begin{pmatrix}\text{Total mass}\\ \text{leaving CV}\\ \text{per unit time}\end{pmatrix}$$

or

$$\Sigma \dot{m}_i = \Sigma \dot{m}_e \tag{4-14}$$

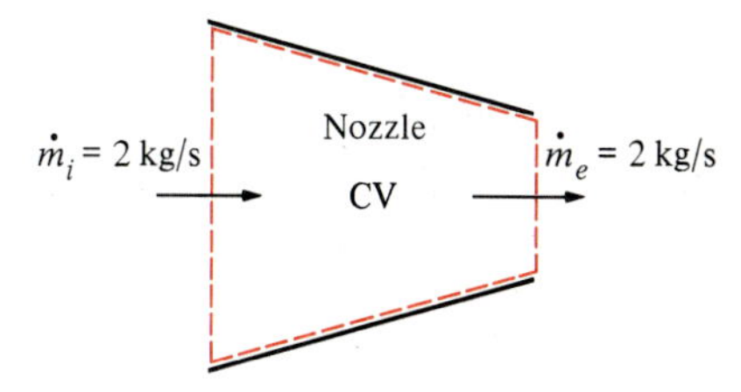

FIGURE 4-20

Conservation of mass principle for the nozzle of a garden hose under steady operation.

where the subscript i stands for *inlet* and e for *exit*. Most engineering devices such as nozzles, diffusers, turbines, compressors, and pumps involve a single stream (only one inlet and one exit). For these cases, we denote the inlet state by the subscript 1 and the exit state by the subscript 2. We also drop the summation signs. Then Eq. 4-14 reduces, for single-stream steady-flow systems, to

$$\dot{m}_1 = \dot{m}_2 \qquad \text{(kg/s)} \tag{4-15}$$

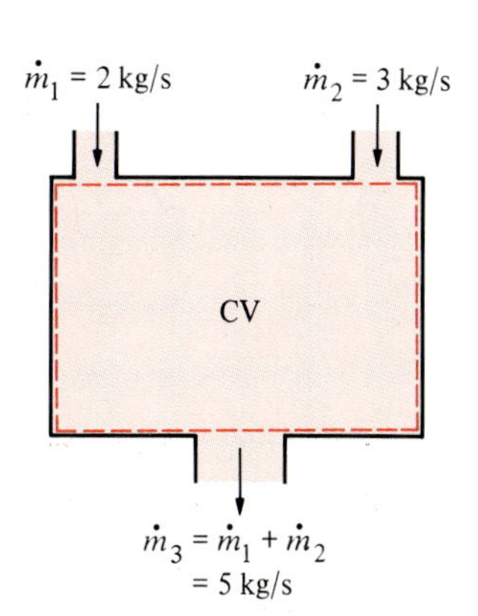

FIGURE 4-21

Conservation of mass principle for a two-inlet one-exit steady-flow system.

or
$$\rho_1 \mathbf{V}_1 A_1 = \rho_2 \mathbf{V}_2 A_2 \tag{4-16}$$

or
$$\frac{1}{v_1} \mathbf{V}_1 A_1 = \frac{1}{v_2} \mathbf{V}_2 A_2 \tag{4-17}$$

where ρ = density, kg/m^3
v = specific volume, m^3/kg (= $1/\rho$)
$\mathbf{V}$ = average flow velocity in flow direction, m/s
A = cross-sectional area normal to flow direction, m^2

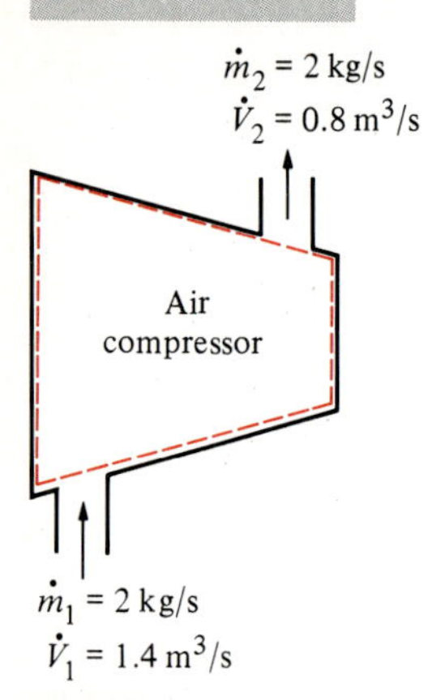

FIGURE 4-22

During a steady-flow process, volume flow rates are not necessarily conserved.

The reader is reminded that there is no such thing as a "conservation of volume" principle. Therefore, the volume flow rates ($\dot{V} = \mathbf{V}A$, m^3/s) into and out of a steady-flow device may be different. The volume flow rate at the exit of an air compressor will be much less than that at the inlet even though the mass flow rate of air through the compressor is constant (Fig. 4-22). This is due to the higher density of air at the compressor exit. For liquid flow, however, the volume flow rates, as well as the mass flow rates, remain constant since liquids are essentially incompressible (constant-density) substances. Water flow through the nozzle of a garden hose is a good example for the latter case.

Conservation of Energy

It was pointed out earlier that during a steady-flow process the total energy content of a control volume remains constant (E_{CV} = constant, as shown in Fig. 4-23). That is, the change in the total energy of the control volume during such a process is zero ($\Delta E_{CV} = 0$). Thus the amount of energy entering a control volume in all forms (heat, work, mass transfer) must be equal to the amount of energy leaving it for a steady-flow process.

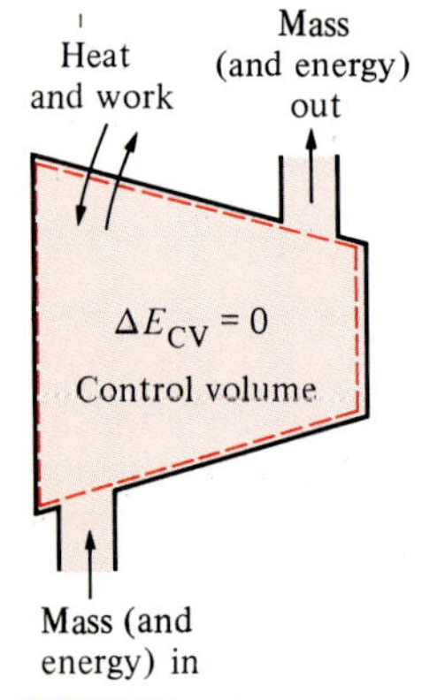

FIGURE 4-23

During a steady-flow process, the rate of energy flow into the control volume equals the rate of energy flow out of it.

Consider, for example, an ordinary electric hot-water heater under steady operation, as shown in Fig. 4-24. A cold-water stream with a mass flow rate $\dot{m}$ is continuously flowing into the water heater, and a hot-water stream of the same mass flow rate is continuously flowing out of it. The water heater (the control volume) is losing heat to the surrounding air at a rate of $\dot{Q}$, and the electric heating element is doing electrical work (heating) on the water at a rate of $\dot{W}$. On the basis of the conservation of

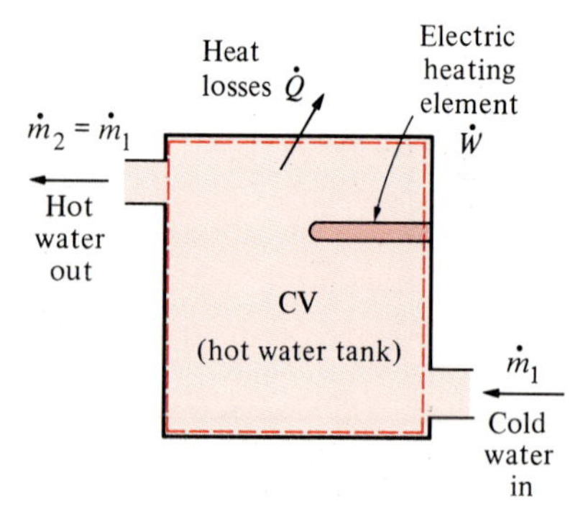

FIGURE 4-24

A water heater under steady operation.

energy principle, we can say that *the water stream will experience an increase in its total energy as it flows through the water heater, which is equal to the electric energy supplied to the water minus the heat losses.*

By this line of reasoning, the first law of thermodynamics or the **conservation of energy principle** for a general steady-flow sytem with multiple inlets and exits can verbally be expressed as

$$\begin{pmatrix}\text{Total energy}\\ \text{crossing boundary}\\ \text{as heat and work}\\ \text{per unit time}\end{pmatrix} = \begin{pmatrix}\text{Total energy}\\ \text{transported out of}\\ \text{CV with mass}\\ \text{per unit time}\end{pmatrix} - \begin{pmatrix}\text{Total energy}\\ \text{transported into}\\ \text{CV with mass}\\ \text{per unit time}\end{pmatrix}$$

or

$$\dot{Q} - \dot{W} = \Sigma \dot{m}_e \theta_e - \Sigma \dot{m}_i \theta_i \qquad (4\text{-}18)$$

where θ is the total energy of the flowing fluid, including the flow work, per unit mass (Fig. 4-25). It can also be expressed as

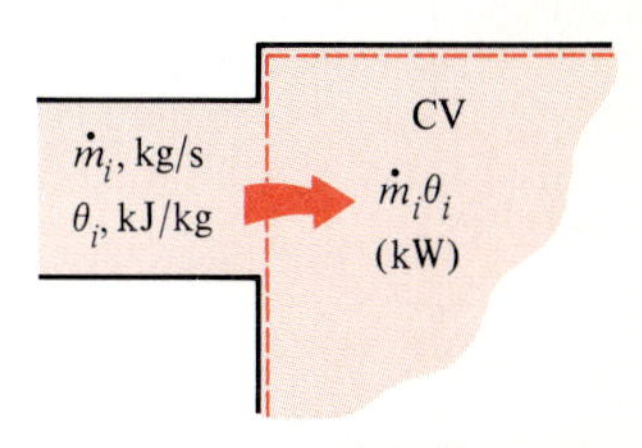

FIGURE 4-25
The product $\dot{m}_i\theta_i$ is the energy transported into the control volume by mass per unit time.

$$\dot{Q} - \dot{W} = \underbrace{\sum \dot{m}_e \left(h_e + \frac{\mathbf{V}_e^2}{2} + gz_e\right)}_{\text{for each exit}} - \underbrace{\sum \dot{m}_i \left(h_i + \frac{\mathbf{V}_i^2}{2} + gz_i\right)}_{\text{for each inlet}} \quad \text{(kW)} \qquad (4\text{-}19)$$

since $\theta = h + \text{ke} + \text{pe}$ (Eq. 4-13). Equation 4-19 is the general form of the first-law relation for steady-flow processes.

For single-stream (one-inlet, one-exit) systems the summations over the inlets and the exits drop out, and the inlet and exit states in this case are denoted by subscripts 1 and 2, respectively, for simplicity. The mass flow rate through the entire control volume remains constant ($\dot{m}_1 = \dot{m}_2$) and is denoted $\dot{m}$. Then the conservation of energy equation for *single-stream steady-flow systems* becomes

$$\dot{Q} - \dot{W} = \dot{m}\left[h_2 - h_1 + \frac{\mathbf{V}_2^2 - \mathbf{V}_1^2}{2} + g(z_2 - z_1)\right] \quad \text{(kW)} \qquad (4\text{-}20)$$

or

$$\dot{Q} - \dot{W} = \dot{m}(\Delta h + \Delta\text{ke} + \Delta\text{pe}) \quad \text{(kW)} \qquad (4\text{-}21)$$

Dividing these equations by $\dot{m}$, we obtain the first-law relation on a unit-mass basis as

$$q - w = h_2 - h_1 + \frac{\mathbf{V}_2^2 - \mathbf{V}_1^2}{2} + g(z_2 - z_1) \quad \text{(kJ/kg)} \qquad (4\text{-}22)$$

or

$$q - w = \Delta h + \Delta\text{ke} + \Delta\text{pe} \quad \text{(kJ/kg)} \qquad (4\text{-}23)$$

where

$$q = \frac{\dot{Q}}{\dot{m}} \quad \text{(heat transfer per unit mass, kJ/kg)} \qquad (4\text{-}24)$$

and

$$w = \frac{\dot{W}}{\dot{m}} \quad \text{(work done per unit mass, kJ/kg)} \qquad (4\text{-}25)$$

If the fluid experiences a negligible change in its kinetic and potential energies as it flows through the control volume (that is, $\Delta\text{ke} \cong 0$, $\Delta\text{pe} \cong 0$), then the energy equation for a single-stream steady-flow system reduces further to

$$q - w = \Delta h \quad \text{(kJ/kg)} \qquad (4\text{-}26)$$

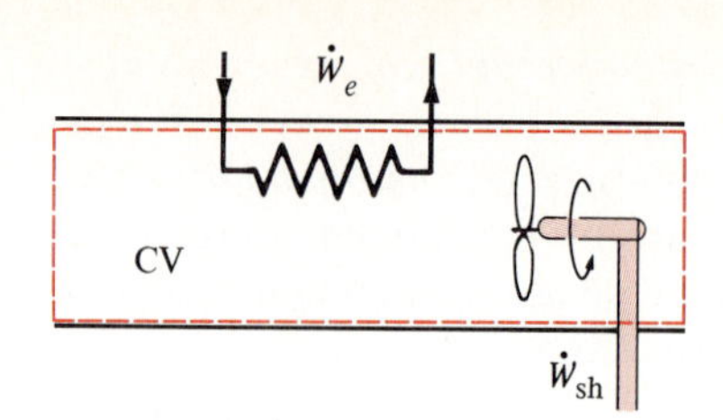

FIGURE 4-26

Under steady operation, shaft work and electrical work are the only forms of work a simple compressible system may involve.

This is the simplest form of the first-law relation for control volumes. Its form resembles the first-law relation for closed systems except that Δu is replaced by Δh in this case.

The various terms appearing in the above equations are as follows:

$\dot{Q}$ = **rate of heat transfer between the control volume and its surroundings.** When the control volume is losing heat (as in the case of the water heater), $\dot{Q}$ is negative. If the control volume is well insulated (i.e, adiabatic), then $\dot{Q} = 0$.

$\dot{W}$ = **power.** For steady-flow devices, the volume of the control volume is constant, thus there is no boundary work involved. The work required to push mass into and out of the control volume is also taken care of by using enthalpies for the energy of fluid streams instead of internal energies. Then $\dot{W}$ represents the remaining forms of work done per unit time (Fig. 4-26). Many steady-flow devices, such as turbines, compressors, and pumps, transmit power through a shaft, and $\dot{W}$ simply becomes the shaft power for those devices. If the control surface is crossed by electric wires (as in the case of an electric water heater), $\dot{W}$ will represent the electrical work done per unit time. If neither is present, then $\dot{W} = 0$.

$\Delta h = h_{exit} - h_{inlet}$. The enthalpy change of a fluid can easily be determined by reading the enthalpy values at the exit and inlet states from the tables. For ideal gases, it may be approximated by $\Delta h = C_{p,av}(T_2 - T_1)$. Note that (kg/s)(kJ/kg) ≡ kW.

$\Delta ke = (\mathbf{V}_2^2 - \mathbf{V}_1^2)/2$. The unit of kinetic energy is m²/s², which is equivalent to J/kg (Fig. 4-27). The enthalpy is usually given in kJ/kg. To add these two quantities, the kinetic energy should be expressed in kJ/kg. This is easily accomplished by dividing it by 1000.

A velocity of 45 m/s corresponds to a kinetic energy of only 1 kJ/kg, which is a very small value compared with the enthalpy values encountered in practice. Thus, the kinetic energy term at low velocities can be neglected. When a fluid stream enters and leaves a steady-flow device at about the same velocity ($\mathbf{V}_1 \cong \mathbf{V}_2$), the change in the kinetic energy is close to zero regardless of the velocity. Caution should be exercised at high velocities, however, since small changes in velocities may cause significant changes in kinetic energy (Fig. 4-28).

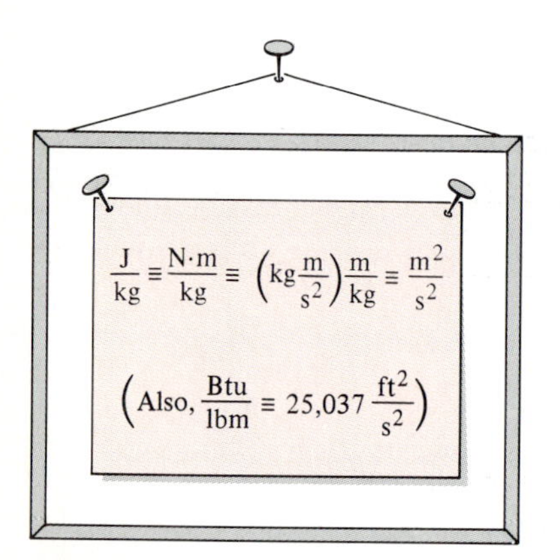

FIGURE 4-27

The units m^2/s^2 and J/kg are equivalent.

$\mathbf{V}_1$ m/s	$\mathbf{V}_2$ m/s	Δke kJ/kg
0	45	1
50	67	1
100	110	1
200	205	1
500	502	1

FIGURE 4-28

At very high velocities, even small changes in velocities may cause significant changes in the kinetic energy of the fluid.

$\Delta pe = g(z_2 - z_1)$. A similar argument can be given for the potential energy term. A potential energy change of 1 kJ/kg corresponds to an elevation difference of 102 m. The elevation difference between the inlet and exit of most industrial devices such as turbines and compressors is well below this value, and the potential energy term is always neglected for these devices. The only time the potential energy term is significant is when a process involves pumping a fluid to high elevations. This is particularly true for systems involving negligible heat transfer.

4-3 ■ SOME STEADY-FLOW ENGINEERING DEVICES

Many engineering devices operate essentially under the same conditions for long periods of time. The components of a steam power plant (turbines, compressors, heat exchangers, and pumps), for example, operate nonstop for months before the system is shut down for maintenance (Fig. 4-29). Therefore, these devices can be conveniently analyzed as steady-flow devices.

In this section, some common steady-flow devices are described, and the thermodynamic aspects of the flow through them are analyzed. The conservation of mass and the conservation of energy principles for these devices are illustrated with examples.

1 Nozzles and Diffusers

Nozzles and diffusers are commonly utilized in jet engines, rockets, spacecraft, and even garden hoses. A **nozzle** is a device that *increases the velocity of a fluid* at the expense of pressure. **diffuser** is a device that *increases the pressure of a fluid* by slowing it down. That is, nozzles and diffusers perform opposite tasks. The cross-sectional area of a nozzle

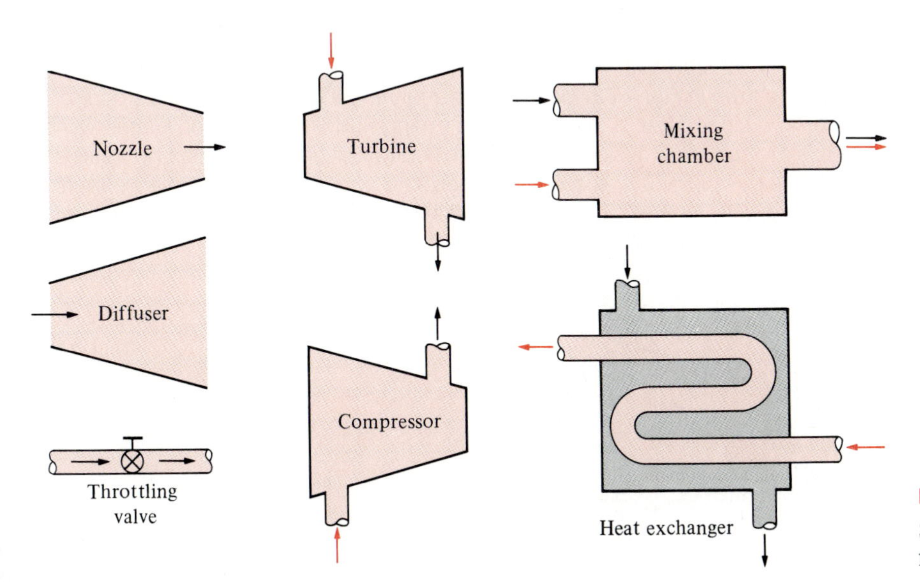

FIGURE 4-29
Steady-flow devices operate steadily for long periods.

decreases in the flow direction for subsonic flows and increases for supersonic flows. The reverse is true for diffusers. The different behavior of the fluid for supersonic flows is explained in Chap. 16. Figure 4-30 shows a nozzle and a diffuser schematically.

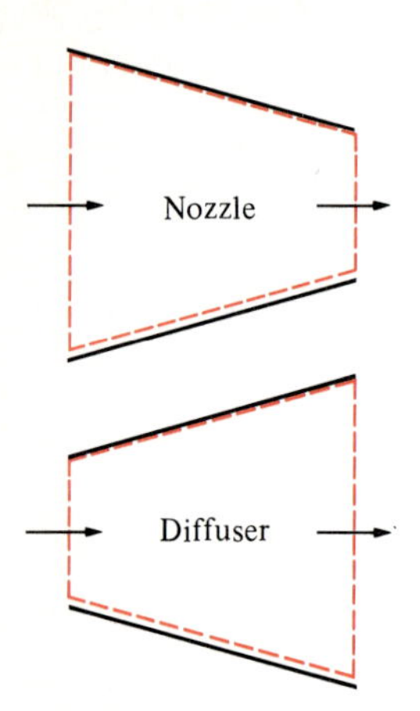

FIGURE 4-30

Schematic of a nozzle and diffuser for subsonic flows (velocities under the speed of sound).

The relative importance of the terms appearing in the energy equation for nozzles and diffusers is as follows:

$\dot{Q} \cong 0$. The rate of heat transfer between the fluid flowing through a nozzle or a diffuser and the surroundings is usually very small, even when these devices are not insulated. This is mainly due to the fluid's having high velocities and thus not spending enough time in the device for any significant heat transfer to take place. Therefore, in the absence of any heat transfer data, the flow through nozzles and diffusers may be assumed to be adiabatic.

$\dot{W} = 0$. The work term for nozzles and diffusers is zero since these devices basically are properly shaped ducts and they involve no shaft or electric resistance wires.

$\Delta ke \neq 0$. Nozzles and diffusers usually involve very high velocities, and as a fluid passes through a nozzle or diffuser, it experiences large changes in its velocity (Fig. 4-31). Therefore, the kinetic energy changes must be accounted for in analyzing the flow through these devices.

$\Delta pe \cong 0$. The fluid usually experiences little or no change in its elevation as it flows through a nozzle or a diffuser, and therefore the potential energy term can be neglected.

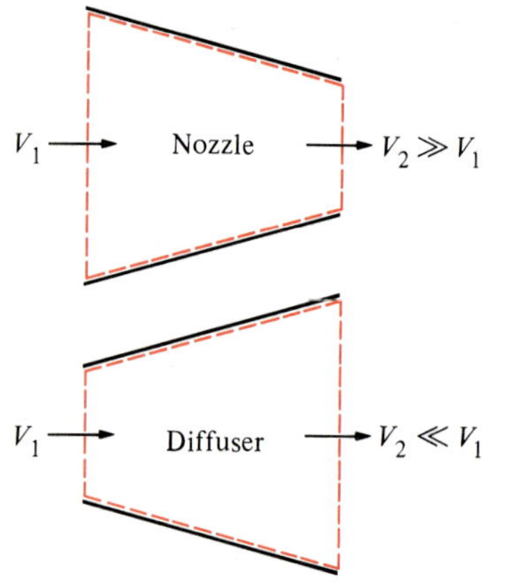

FIGURE 4-31

Nozzles and diffusers are shaped so that they cause large changes in fluid velocities and thus kinetic energies.

EXAMPLE 4-1

Air at 10°C and 80 kPa enters the diffuser of a jet engine steadily with a velocity of 200 m/s. The inlet area of the diffuser is 0.4 m^2. The air leaves the diffuser with a velocity that is very small compared with the inlet velocity. Determine (*a*) the mass flow rate of the air and (*b*) the temperature of the air leaving the diffuser.

Solution The region within the diffuser is selected as the system, and its boundaries are shown in Fig. 4-32. Mass is crossing the boundaries, thus it is a *control volume*. And since there is no observable change within the control volume with time, it is a *steady-flow system*. At the specified conditions, the air can be treated as an *ideal gas* since it is at a high temperature and low pressure relative to its critical values ($T_{cr} = -147$°C and $P_{cr} = 3390$ kPa for nitrogen, the main constituent of air).

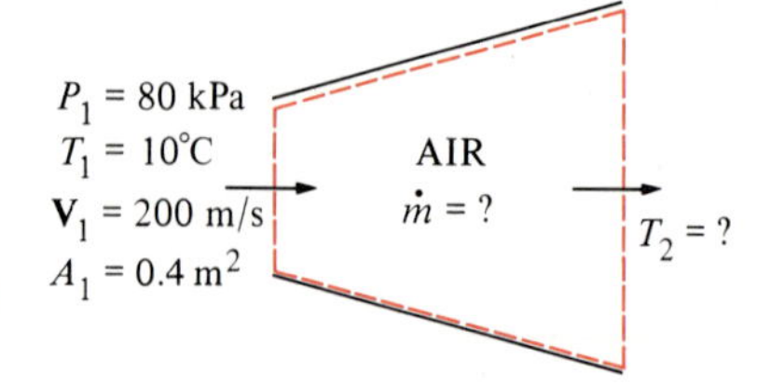

FIGURE 4-32

Schematic for Example 4-1.

(*a*) To determine the mass flow rate, we need to find the specific volume of the air first. This is determined from the ideal-gas relation at the inlet conditions:

$$v_1 = \frac{RT_1}{P_1} = \frac{[0.287 \text{ kPa}\cdot\text{m}^3/(\text{kg}\cdot\text{K})](283 \text{ K})}{80 \text{ kPa}} = 1.015 \text{ m}^3/\text{kg}$$

Then from Eq. 4-5,

$$\dot{m} = \frac{1}{v_1}\mathbf{V}_1A_1 = \frac{1}{1.015 \text{ m}^3/\text{kg}}(200 \text{ m/s})(0.4 \text{ m}^2) = 78.8 \text{ kg/s}$$

Since the flow is steady, the mass flow rate through the entire diffuser will remain constant at this value.

(*b*) A diffuser normally involves no shaft or electrical work ($w = 0$), negligible heat transfer ($q \cong 0$), and a small (if any) elevation change between the inlet and the exit ($\Delta\text{pe} \cong 0$). Then the conservation of energy relation on a unit-mass basis for this single-stream steady-flow system (Eq. 4-23) reduces to

$$\cancelto{0}{q} - \cancelto{0}{w} = \Delta h + \Delta\text{ke} + \cancelto{0}{\Delta\text{pe}}$$

$$0 = h_2 - h_1 + \frac{\mathbf{V}_2^2 - \mathbf{V}_1^2}{2}$$

The exit velocity of a diffuser is usually small compared to the inlet velocity ($\mathbf{V}_2 << \mathbf{V}_1$); thus the kinetic energy at the exit can be neglected. The enthalpy of air at the diffuser inlet is determined from the air table (Table A-17) to be

$$h_1 = h_{@\,283\text{ K}} = 283.14 \text{ kJ/kg}$$

Substituting, we get

$$h_2 = 283.14 \text{ kJ/kg} - \frac{0 - (200 \text{ m/s})^2}{2}\left(\frac{1 \text{ kJ/kg}}{1000 \text{ m}^2/\text{s}^2}\right)$$

$$= 303.14 \text{ kJ/kg}$$

From Table A-17, the temperature corresponding to this enthalpy value is

$$T_2 = 303.1 \text{ K}$$

which shows that the temperature of the air increased by about 20°C as it was slowed down in the diffuser. The temperature rise of the air is mainly due to the conversion of kinetic energy to internal energy.

EXAMPLE 4-2

Steam at 250 psia and 700°F steadily enters a nozzle whose inlet area is 0.2 ft². The mass flow rate of the steam through the nozzle is 10 lbm/s. Steam leaves the nozzle at 200 psia with a velocity of 900 ft/s. The heat losses from the nozzle per unit mass of the steam are estimated to be 1.2 Btu/lbm. Determine (*a*) the inlet velocity and (*b*) the exit temperature of the steam.

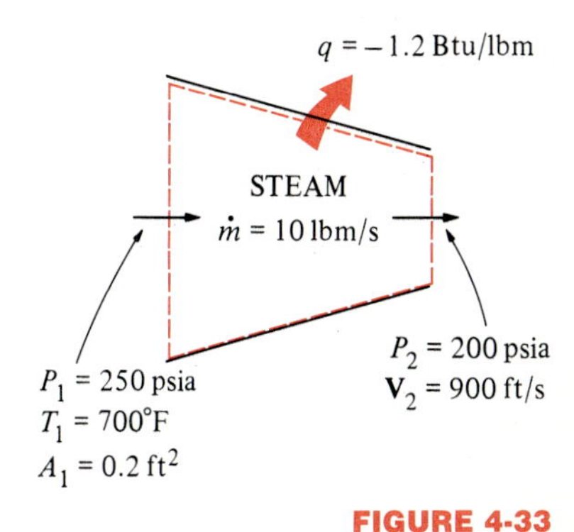

FIGURE 4-33
Schematic for Example 4-2.

Solution This time we take the region within the nozzle as our system with boundaries as indicated in Fig. 4-33. Mass is crossing the boundaries, and thus it is a *control volume*; and there is no observable change within the control volume with time, so it is a *steady-flow system*.

(*a*) The inlet velocity is determined from Eq. 4-5. But first we need to determine the specific volume of the steam at the nozzle inlet:

$$\left.\begin{array}{l} P_1 = 250 \text{ psia} \\ T_1 = 700°\text{F} \end{array}\right\} \quad \begin{array}{l} v_1 = 2.688 \text{ ft}^3/\text{lbm} \\ h_1 = 1371.1 \text{ Btu/lbm} \end{array} \qquad \text{(Table A-6E)}$$

Then from Eq. 4-5,

$$\dot{m} = \frac{1}{v_1}\mathbf{V}_1 A_1$$

$$10 \text{ lbm/s} = \frac{1}{2.688 \text{ ft}^3/\text{lbm}}(\mathbf{V}_1)(0.2 \text{ ft}^2)$$

$$\mathbf{V}_1 = 134.4 \text{ ft/s}$$

(*b*) A nozzle typically involves no shaft or electrical work ($w = 0$) and a small (if any) elevation change between the inlet and the exit ($\Delta\text{pe} \cong 0$). Then the conservation of energy relation (Eq. 4-23) for this single-stream steady-flow system reduces to

$$q - \cancelto{0}{w} = \Delta h + \Delta\text{ke} + \cancelto{0}{\Delta\text{pe}}$$

$$q = h_2 - h_1 + \frac{\mathbf{V}_2^2 - \mathbf{V}_1^2}{2}$$

Thus,

$$-1.2 \text{ Btu/lbm} = h_2 - 1371.1 \text{ Btu/lbm} + \frac{(900 \text{ ft/s})^2 - (134.4 \text{ ft/s})^2}{2}\left(\frac{1 \text{ Btu/lbm}}{25{,}037 \text{ ft}^2/\text{s}^2}\right)$$

which yields

$$h_2 = 1354.1 \text{ Btu/lbm}$$

Then,

$$\left.\begin{array}{l} P_2 = 200 \text{ psia} \\ h_2 = 1354.1 \text{ Btu/lbm} \end{array}\right\} \quad T_2 = 661.9°\text{F} \qquad \text{(Table A-6E)}$$

Therefore, the temperature of steam will drop by 38.1°F as it flows through the nozzle. This drop in temperature is mainly due to the conversion of internal energy to kinetic energy. (The heat loss is too small to cause any significant effect in this case.)

2 Turbines and Compressors

In steam, gas, or hydroelectric power plants, the device that drives the electric generator is the turbine. As the fluid passes through the turbine, work is done against the blades which are attached to the shaft. As a result, the shaft rotates, and the turbine produces work. The work done in a turbine is positive since it is done by the fluid.

Compressors, as well as pumps and fans, are devices used to increase the pressure of a fluid. Work is supplied to these devices from an external source through a rotating shaft. Therefore, the work term for compressors is negative since work is done on the fluid. Even though these three devices function similarly, they do differ in the tasks they perform. A *fan* increases the pressure of a gas slightly and is mainly used to move a gas around. A *compressor* is capable of compressing the gas to very high pressures. *Pumps* work very much like compressors except that they handle liquids instead of gases.

For turbines and compressors, the relative magnitudes of the various terms appearing in the energy equation are as follows:

$\dot{Q} \cong 0$. The heat transfer for these devices is generally small relative to the shaft work unless there is intentional cooling (as for the case of a compressor). An estimated value based on the experimental studies can be used in the analysis, or the heat transfer may be neglected if there is no intentional cooling.

$\dot{W} \neq 0$. All these devices involve rotating shafts crossing their boundaries, therefore the work term is important. For turbines, $\dot{W}$ represents the power output; for pumps and compressors, it represents the power input.

$\Delta \mathrm{pe} \cong 0$. The potential energy change that a fluid experiences as it flows through turbines, compressors, fans, and pumps is usually very small and is normally neglected.

$\Delta \mathrm{ke} \cong 0$. The velocities involved with these devices, with the exception of turbines, are usually too low to cause any significant change in the kinetic energy. The fluid velocities encountered in most turbines are very large, and the fluid experiences a significant change in its kinetic energy. However, this change is usually very small relative to the change in enthalpy, and thus it is often disregarded.

EXAMPLE 4-3

Air at 100 kPa and 280 K is compressed steadily to 600 kPa and 400 K. The mass flow rate of the air is 0.02 kg/s, and a heat loss of 16 kJ/kg occurs during the process. Assuming the changes in kinetic and potential energies are negligible, determine the necessary power input to the compressor.

Solution We choose the region within the compressor as our system, and its boundaries are indicated by dashed lines in Fig. 4-34. Mass is crossing the boundaries, thus it is a *control volume;* and there is no observable change within the control volume with time, so it is a *steady-flow system.* At the specified conditions, the air can be treated as an *ideal gas* since it is at a high temperature and low pressure relative to its critical values ($T_{cr} = -147°C$ and P_{cr} = 3390 kPa for nitrogen, the main constituent of air).

It is stated that $\Delta \mathrm{ke} \cong 0$ and $\Delta \mathrm{pe} \cong 0$. Then the conservation of energy equation (Eq. 4-23) for this single-stream steady-flow system reduces to

$$q - w = \Delta h + \cancelto{0}{\Delta \mathrm{ke}} + \cancelto{0}{\Delta \mathrm{pe}}$$
$$= h_2 - h_1$$

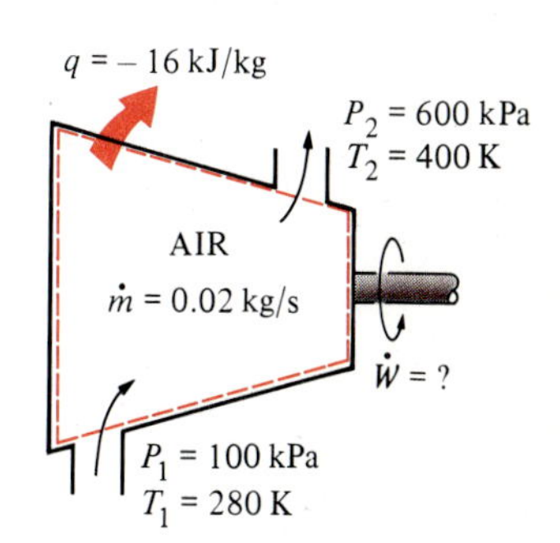

FIGURE 4-34
Schematic for Example 4-3.

The enthalpy of an ideal gas depends on temperature only, and the enthalpies of the air at the specified temperatures are determined from the air table (Table A-17) to be

$$h_1 = h_{@\,280\,\mathrm{K}} = 280.13 \text{ kJ/kg}$$
$$h_2 = h_{@\,400\,\mathrm{K}} = 400.98 \text{ kJ/kg}$$

Substituting yields

$$-16 \text{ kJ/kg} - w = (400.98 - 280.13) \text{ kJ/kg}$$
$$w = -136.85 \text{ kJ/kg}$$

This is the work done on the air per unit mass. The power input to the compressor is determined by multiplying this value by the mass flow rate:

$$\dot{W} = \dot{m}w = (0.02 \text{ kg/s})(-136.85 \text{ kJ/kg}) = -2.74 \text{ kW}$$

EXAMPLE 4-4

The power output of an adiabatic steam turbine is 5 MW, and the inlet and the exit conditions of the steam are as indicated in Fig. 4-35.

(*a*) Compare the magnitudes of Δh, Δke, and Δpe.

(*b*) Determine the work done per unit mass of the steam flowing through the turbine.

(*c*) Calculate the mass flow rate of the steam.

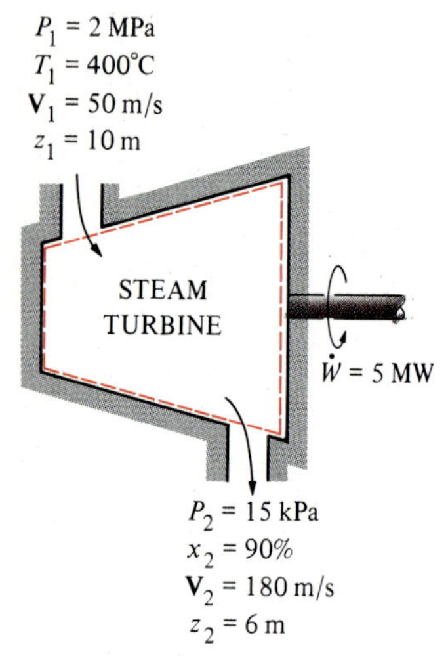

FIGURE 4-35
Schematic for Example 4-4.

Solution This time we select the region within the turbine as our system. Its boundaries are indicated by dashed lines in Fig. 4-35. Mass is crossing the boundaries, thus it is a *control volume.* There is no indication of any change within the control volume with time, thus it is a *steady-flow system.*

(*a*) At the inlet, steam is in a superheated vapor state, and its enthalpy is

$$\left.\begin{array}{l} P_1 = 2 \text{ MPa} \\ T_1 = 400°\text{C} \end{array}\right\} \quad h_1 = 3247.6 \text{ kJ/kg} \qquad \text{(Table A-6)}$$

At the turbine exit, we obviously have a saturated liquid–vapor mixture at 15-kPa pressure. The enthalpy at this state is

$$h_2 = h_f + x_2 h_{fg} = [225.94 + (0.9)(2373.1)] \text{ kJ/kg} = 2361.73 \text{ kJ/kg}$$

Then

$$\Delta h = h_2 - h_1 = (2361.73 - 3247.6) \text{ kJ/kg} = -885.87 \text{ kJ/kg}$$

$$\Delta \text{ke} = \frac{\mathbf{V}_2^2 - \mathbf{V}_1^2}{2} = \frac{(180 \text{ m/s})^2 - (50 \text{ m/s})^2}{2}\left(\frac{1 \text{ kJ/kg}}{1000 \text{ m}^2/\text{s}^2}\right) = 14.95 \text{ kJ/kg}$$

$$\Delta \text{pe} = g(z_2 - z_1) = (9.807 \text{ m/s}^2)[(6 - 10)\text{m}]\left(\frac{1 \text{ kJ/kg}}{1000 \text{ m}^2/\text{s}^2}\right) = -0.04 \text{ kJ/kg}$$

Two observations can be made from the above results. First, the change in potential energy is insignificant in comparison to the changes in enthalpy and kinetic energy. This is typical for most engineering devices. Second, as a result of low pressure and thus high specific volume, the steam velocity at the turbine exit can be very high. Yet the change in kinetic energy is a small fraction of the change in enthalpy (less than 2 percent in our case) and is therefore often neglected.

(*b*) The work done per unit mass for this one-inlet, one-exit device can be determined from the first-law relation on a unit-mass basis (Eq. 4-23):

$$\cancelto{0}{q} - w = \Delta h + \Delta \text{ke} + \Delta \text{pe}$$
$$w = -(-885.87 + 14.95 - 0.04) \text{ kJ/kg} = 870.96 \text{ kJ/kg}$$

(c) The required mass flow rate for a 5-MW power output is determined from

$$\dot{m} = \frac{\dot{W}}{w} = \frac{5000\ \text{kJ/s}}{870.96\ \text{kJ/kg}} = 5.74\ \text{kg/s}$$

3 Throttling Valves

Throttling valves are *any kind of flow-restricting devices* that cause a significant pressure drop in the fluid. Some familiar examples are ordinary adjustable valves, capillary tubes, and porous plugs (Fig. 4-36). Unlike turbines, they produce a pressure drop without involving any work. The pressure drop in the fluid is often accompanied by a *large drop in temperature,* and for that reason throttling devices are commonly used in refrigeration and air-conditioning applications. The magnitude of the temperature drop (or, sometimes, the temperature rise) during a throttling process is governed by a property called the *Joule-Thomson coefficient,* which is discussed in Chap. 11.

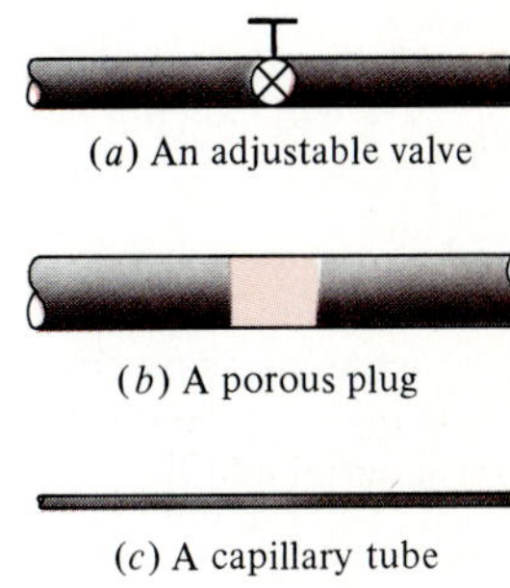

(*a*) An adjustable valve

(*b*) A porous plug

(*c*) A capillary tube

FIGURE 4-36

Throttling valves are devices that cause large pressure drops in the fluid.

Throttling valves are usually small devices, and the flow through them may be assumed to be adiabatic ($q \cong 0$) since there is neither sufficient time nor large enough area for any effective heat transfer to take place. Also, there is no work done ($w = 0$), and the change in potential energy, if any, is very small ($\Delta\text{pe} \cong 0$). Even though the exit velocity is often considerably higher than the inlet velocity, in many cases, the increase in kinetic energy is insignificant ($\Delta\text{ke} \cong 0$). Then the conservation of energy equation for this single-stream steady-flow device reduces to

$$h_2 \cong h_1 \qquad \text{(kJ/kg)} \tag{4-27}$$

That is, enthalpy values at the inlet and exit of a throttling valve are the same. For this reason, a throttling valve is sometimes called an *isenthalpic device*.

To gain some insight into how throttling affects fluid properties, let us express Eq. 4-27 as follows:

$$u_1 + P_1 v_1 = u_2 + P_2 v_2$$

or $\qquad$ Internal energy + flow energy = constant

Thus the final outcome of a throttling process depends on which of the two quantities increases during the process. If the flow energy increases during the process ($P_2 v_2 > P_1 v_1$), it can do so at the expense of the internal energy. As a result, internal energy decreases, which is usually accompanied by a drop in temperature. If the product Pv decreases, the internal energy and the temperature of a fluid will increase during a throttling process. In the case of an ideal gas $h = h(T)$, and thus the temperature has to remain constant during a throttling process (Fig. 4-37).

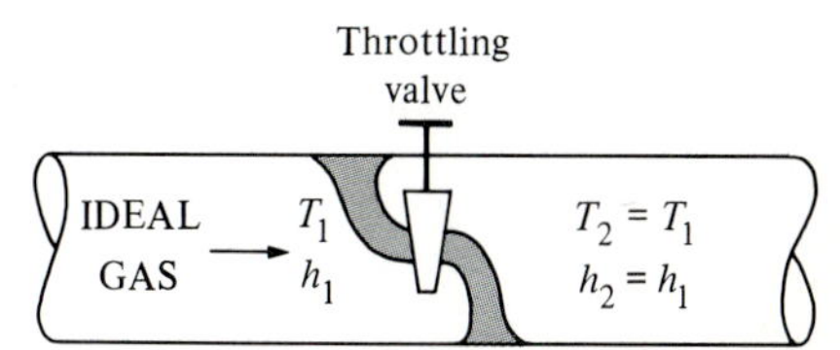

FIGURE 4-37

The temperature of an ideal gas does not change during a throttling (h = constant) process since $h = h(T)$.

EXAMPLE 4-5

Refrigerant-12 enters the capillary tube of a refrigerator as saturated liquid at 0.8 MPa and is throttled to a pressure of 0.12 MPa. Determine the quality of the refrigerant at the final state and the temperature drop during this process.

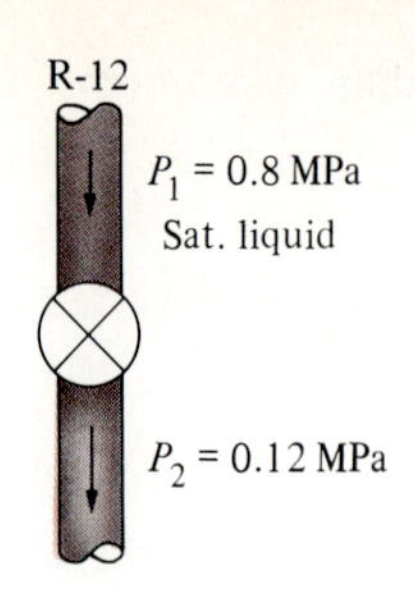

FIGURE 4-38

Schematic for Example 4-5.

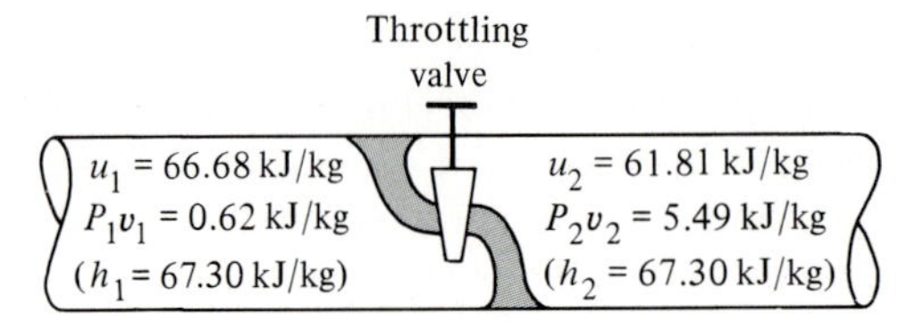

FIGURE 4-39

During a throttling process, the enthalpy (flow energy + internal energy) of a fluid remains constant. But internal and flow energies may be converted to each other.

Solution A sketch of the system is given in Fig. 4-38. A capillary tube is a simple flow-restricting device which is commonly used in refrigeration applications to cause a large pressure drop in the refrigerant. Flow through a capillary tube is a throttling process; thus the enthalpy of the refrigerant remains constant (Fig. 4-39).

At inlet: $\left.\begin{array}{l} P_1 = 0.8 \text{ MPa} \\ \text{sat. liquid} \end{array}\right\} \quad \begin{array}{l} T_1 = T_{\text{sat @ 0.8 MPa}} = 32.74°\text{C} \\ h_1 = h_{f\text{ @ 0.8 MPa}} = 67.30 \text{ kJ/kg} \end{array}$ (Table A-12)

At exit: $P_2 = 0.12 \text{ MPa} \longrightarrow h_f = 12.66 \text{ kJ/kg} \qquad T_{\text{sat}} = -25.74°\text{C}$
$(h_2 = h_1) \qquad h_g = 176.14 \text{ kJ/kg}$

Obviously $h_f < h_2 < h_g$; thus the refrigerant exists as a saturated mixture at the exit state. The quality at this state is determined from

$$x_2 = \frac{h_2 - h_f}{h_{fg}} = \frac{67.30 - 12.66}{176.14 - 12.66} = 0.334$$

Since the exit state is a saturated mixture at 0.12 MPa, the exit temperature must be the saturation temperature at this pressure, which is −25.74°C. Then the temperature change for this process becomes

$$\Delta T = T_2 - T_1 = (-25.74 - 32.74)°\text{C} = -58.48°\text{C}$$

That is, the temperature of the refrigerant drops by 58.48°C during this throttling process. Notice that 33.4 percent of the refrigerant vaporizes during this throttling process, and the energy needed to vaporize this refrigerant is absorbed from the refrigerant itself.

4a Mixing Chambers

In engineering applications, mixing two streams of fluids is not a rare occurrence. The section where the mixing process takes place is commonly referred to as a **mixing chamber**. The mixing chamber does not have to be a distinct "chamber." An ordinary T-elbow or a Y-elbow in a shower, for example, serves as the mixing chamber for the cold- and hot-water streams (Fig. 4-40).

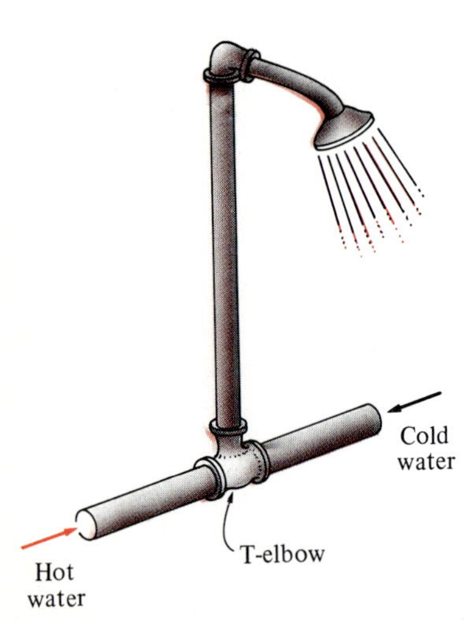

FIGURE 4-40

The T-elbow of an ordinary shower serves as the mixing chamber for the hot- and the cold-water systems.

The conservation of mass principle for a mixing chamber requires that the sum of the incoming mass flow rates equal the mass flow rate of the outgoing mixture.

Mixing chambers are usually well insulated ($q \cong 0$) and do not involve any kind of work ($w = 0$). Also, the kinetic and potential energies of the fluid streams are usually negligible (ke $\cong 0$, pe $\cong 0$). Then all there is left in the energy equation (Eq. 4-19) is the total energies of the incoming streams and the outgoing mixture. The conservation of energy principle requires that these two equal each other. Therefore, the conservation of energy equation becomes analogous to the conservation of mass equation for this case.

EXAMPLE 4-6

Consider an ordinary shower where hot water at 140°F is tempered with cold water at 50°F. If it is desired that a steady stream of warm water at 110°F be supplied, determine the ratio of the mass flow rates of the hot to cold water. Assume the heat losses from the mixing chamber to be negligible and the mixing to take place at a pressure of 20 psia.

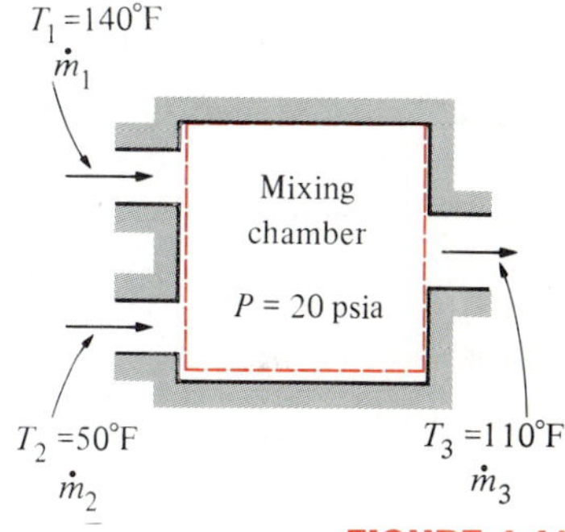

FIGURE 4-41
Schematic for Example 4-6.

Solution We take the mixing chamber as our system whose boundaries are indicated by dashed lines in Fig. 4-41. Mass is crossing the boundaries, thus it is a *control volume;* and there is no indication of any change within the control volume with time, thus it is a *steady-flow system.*

The conservation of mass equation (Eq. 4-14) for this multiple-stream steady-flow system is

$$\Sigma \dot{m}_i = \Sigma \dot{m}_e$$

or

$$\dot{m}_1 + \dot{m}_2 = \dot{m}_3$$

No heat or work is crossing the boundaries ($\dot{Q} \cong 0$, $\dot{W} = 0$), and the kinetic and potential energies are considered to be negligible (ke $\cong 0$, pe $\cong 0$). Then the conservation of energy equation for this steady-flow system reduces to

$$\cancel{\dot{Q}}^{0} - \cancel{\dot{W}}^{0} = \sum \dot{m}_e \left(h_e + \cancel{\frac{\mathbf{V}_e^2}{2}}^{0} + \cancel{gz_e}^{0}\right) - \sum \dot{m}_i \left(h_i + \cancel{\frac{\mathbf{V}_i^2}{2}}^{0} + \cancel{gz_i}^{0}\right)$$

$$\Sigma \dot{m}_i h_i = \Sigma \dot{m}_e h_e$$

$$\dot{m}_1 h_1 + \dot{m}_2 h_2 = \dot{m}_3 h_3$$

or

$$\dot{m}_1 h_1 + \dot{m}_2 h_2 = (\dot{m}_1 + \dot{m}_2) h_3$$

Dividing this equation by $\dot{m}_2$ yields

$$y h_1 + h_2 = (y + 1) h_3$$

where $y = \dot{m}_1/\dot{m}_2$ is the desired mass flow rate ratio.

The saturation temperature of water at 20 psia is 227.96°F. Since the temperatures of all three streams are below this value ($T < T_{sat}$), the water in all three streams exists as a compressed liquid (Fig. 4-42). A compressed liquid can be approximated as a saturated liquid at the given temperature. Thus,

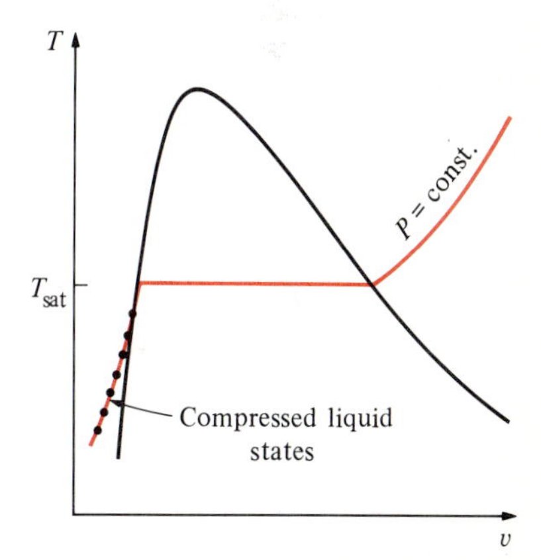

FIGURE 4-42
A substance exists as a compressed liquid at temperatures below the saturation temperatures at the given pressure.

$$h_1 \cong h_{f\,@\,140°F} = 107.96 \text{ Btu/lbm}$$
$$h_2 \cong h_{f\,@\,50°F} = 18.06 \text{ Btu/lbm}$$
$$h_3 \cong h_{f\,@\,110°F} = 78.02 \text{ Btu/lbm}$$

Solving for y and substituting yields

$$y = \frac{h_3 - h_2}{h_1 - h_3} = \frac{78.02 - 18.06}{107.96 - 78.02} = 2.0$$

Thus the mass flow rate of the hot water must be twice the mass flow rate of the cold water for the mixture to leave at 110°F.

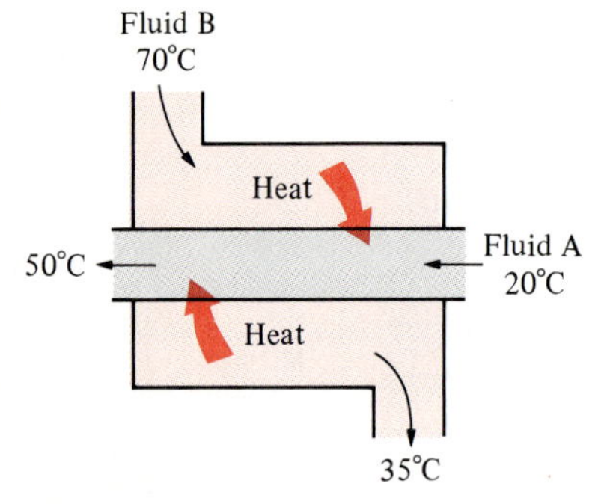

FIGURE 4-43

A heat exchanger can be as simple as two concentric pipes of different diameters.

4b Heat Exchangers

As the name implies, **heat exchangers** are devices where two moving fluid streams exchange heat without mixing. Heat exchangers are widely used in various industries, and they come in various designs.

The simplest form of a heat exchanger is a *double-tube* (also called *tube-and-shell*) *heat exchanger,* shown in Fig. 4-43. It is composed of two concentric pipes of different diameters. One fluid flows in the inner pipe, and the other in the annular space between the two pipes. Heat is transferred from the hot fluid to the cold one through the wall separating them. Sometimes the inner tube makes a couple of turns inside the shell to increase the heat transfer area, thus the rate of heat transfer. The mixing chambers discussed earlier are sometimes classified as *direct-contact* heat exchangers.

The conservation of mass principle for a heat exchanger in steady operation requires that the sum of the inbound mass flow rates equal the sum of the outbound mass flow rates. This principle can also be expressed as follows: *Under steady operation, the mass flow rate of each fluid stream flowing through a heat exchanger remains constant.*

Heat exchangers typically involve no work interactions ($w = 0$) and negligible kinetic and potential energy changes ($\Delta ke \cong 0$, $\Delta pe \cong 0$) for each fluid stream. The heat transfer rate associated with heat exchangers depends on how the control volume is selected. Heat exchangers are intended for heat transfer between two fluids *within* the device, and the outer shell is usually well insulated to prevent any heat loss to the surrounding medium.

When the entire heat exchanger is selected as the control volume,

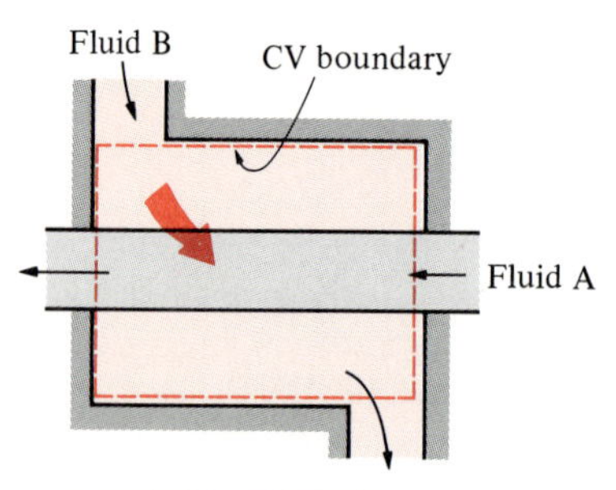

(*a*) System: Entire heat exchanger ($Q_{CV} = 0$)

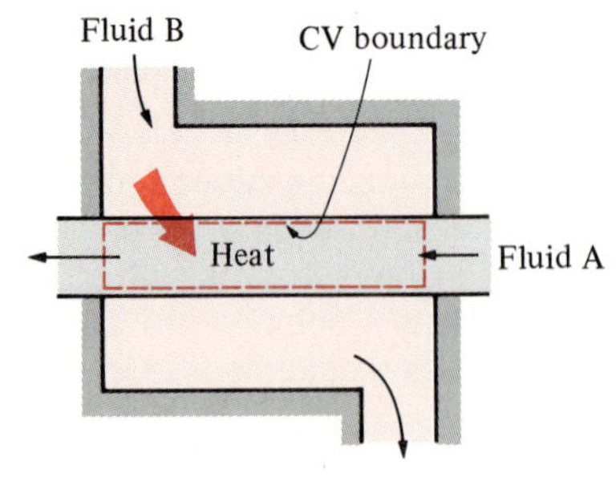

(*b*) System: Fluid A ($Q_{CV} \neq 0$)

FIGURE 4-44

The heat transfer associated with a heat exchanger may be zero or nonzero depending on how the system is selected.

$\dot{Q}$ becomes zero, since the boundary for this case lies just beneath the insulation and little or no heat crosses the boundary (Fig. 4-44). If, however, only one of the fluids is selected as the control volume, then heat will cross this boundary as it flows from one fluid to the other and $\dot{Q}$ will not be zero. In fact, $\dot{Q}$ in this case will be the amount of heat transfer between the two fluids.

EXAMPLE 4-7

Refrigerant-12 is to be cooled by water in a condenser. The refrigerant enters the condenser with a mass flow rate of 6 kg/min at 1 MPa and 70°C and leaves at 35°C. The cooling water enters at 300 kPa and 15°C and leaves at 25°C. Neglecting any pressure drops, determine (*a*) the mass flow rate of the cooling water required and (*b*) the heat transfer rate from the refrigerant to water.

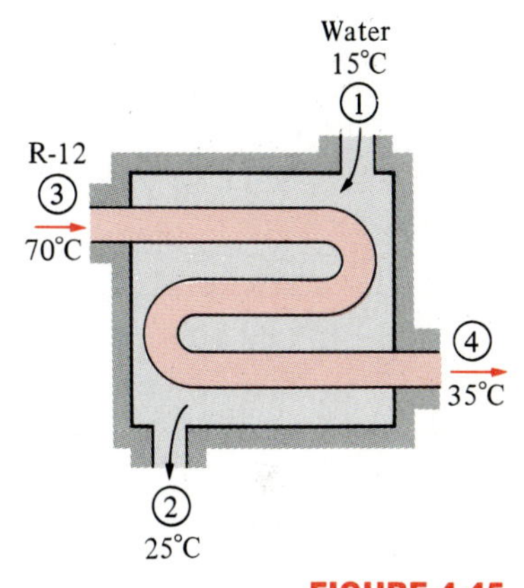

FIGURE 4-45
Schematic for Example 4-7.

Solution There are several possibilities for selecting the control volume for this multiple-stream steady-flow device. The proper choice for the control volume will depend on the situation at hand. Again no work crosses the boundary, and the kinetic and potential energies are considered negligible.

(*a*) To determine the mass flow rate of water, we choose the entire heat exchanger as our control volume, as shown in Fig. 4-45. This is a good choice since the equations will involve only one unknown—the mass flow rate. The conservation of mass principle requires that the mass flow rate of each fluid stream remain constant:

$$\dot{m}_1 = \dot{m}_2 = \dot{m}_w$$
$$\dot{m}_3 = \dot{m}_4 = \dot{m}_R$$

Heat exchangers are ordinarily well insulated. The boundaries of the control volume selected lie just beneath the insulation, so no heat will cross the boundaries. Then the conservation of energy equation reduces to

$$\cancelto{0}{\dot{Q}} - \cancelto{0}{\dot{W}} = \sum \dot{m}_e \left(h_e + \cancelto{0}{\frac{\mathbf{V}_e^2}{2}} + \cancelto{0}{gz_e} \right) - \sum \dot{m}_i \left(h_i + \cancelto{0}{\frac{\mathbf{V}_i^2}{2}} + \cancelto{0}{gz_i} \right)$$

$$\Sigma \dot{m}_i h_i = \Sigma \dot{m}_e h_e$$

or

$$\dot{m}_w h_1 + \dot{m}_R h_3 = \dot{m}_w h_2 + \dot{m}_R h_4$$

Rearranging, we have

$$\dot{m}_w (h_1 - h_2) = \dot{m}_R (h_4 - h_3)$$

Now we need to determine the enthalpies at all four states. Water exists as a compressed liquid at both the inlet and the exit since the temperatures at both locations are below the saturation temperature of water at 300 kPa (133.55°C). Approximating the compressed liquid as a saturated liquid at the given temperature, we have

$$\begin{aligned} h_1 &\cong h_{f\,@\,15^\circ\mathrm{C}} = 62.99 \text{ kJ/kg} \\ h_2 &\cong h_{f\,@\,25^\circ\mathrm{C}} = 104.89 \text{ kJ/kg} \end{aligned} \qquad \text{(Table A-4)}$$

The refrigerant enters the condenser as a superheated vapor and leaves as a compressed liquid at 35°C. From refrigerant-12 tables,

$$\left.\begin{array}{l} P_3 = 1\text{ MPa} \\ T_3 = 70°\text{C} \end{array}\right\} \quad h_3 = 225.32\text{ kJ/kg} \qquad \text{(Table A-13)}$$

$$\left.\begin{array}{l} P_4 = 1\text{ MPa} \\ T_4 = 35°\text{C} \end{array}\right\} \quad h_4 \cong h_{f\,@\,35°\text{C}} = 69.49\text{ kJ/kg} \qquad \text{(Table A-11)}$$

Substituting, we find

$$\dot{m}_w(62.99 - 104.89)\text{ kJ/kg} = (6\text{ kg/min})[(69.49 - 225.32)\text{ kJ/kg}]$$
$$\dot{m}_w = 22.3\text{ kg/min}$$

(*b*) To determine the heat transfer from the refrigerant to the water, we have to choose a control volume whose boundary lies on the path of the heat flow, since heat is recognized as it crosses the boundaries. We can choose the volume occupied by either fluid as our control volume. For no particular reason, we choose the volume occupied by the water. All the assumptions stated earlier apply, except that the heat flow is no longer zero. Then the conservation of energy equation (Eq. 4-21) for this single-stream steady-flow system reduces to

$$\dot{Q} - \cancel{\dot{W}}^{0} = \dot{m}_w(\Delta h + \cancel{\Delta ke}^{0} + \cancel{\Delta pe}^{0})$$
$$\dot{Q} = \dot{m}_w(h_2 - h_1) = (22.3\text{ kg/min})[(104.89 - 62.99)\text{ kJ/kg}]$$
$$= 934.4\text{ kJ/min}$$

Had we chosen the volume occupied by the refrigerant as the control volume (Fig. 4-46), we would have obtained the same result with the opposite sign since the heat gained by the water is equal to the heat lost by the refrigerant.

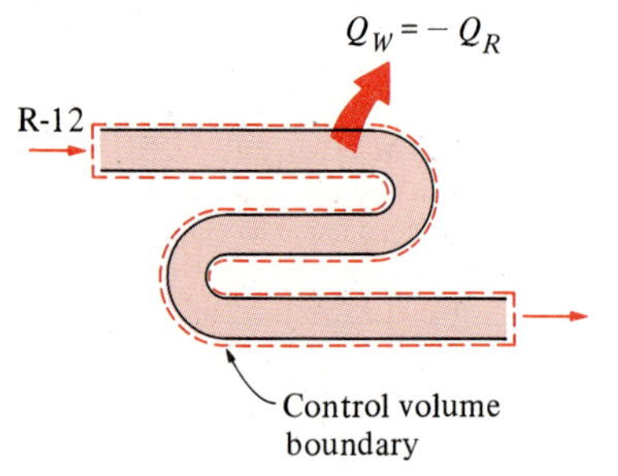

FIGURE 4-46
In a heat exchanger, the heat transfer depends on the choice of the control volume.

5 Pipe and Duct Flow

The transport of liquids or gases in pipes and ducts is of great importance in many engineering applications. Flow through a pipe or a duct usually satisfies the steady-flow conditions and thus can be analyzed as a steady-flow process. This, of course, excludes the transient start-up and shut-down periods. The control volume can be selected to coincide with the interior surfaces of the portion of the pipe or the duct that we are interested in analyzing.

When flow through pipes or ducts is analyzed, the following points should be considered:

$\dot{Q} \neq 0$. Under normal operating conditions, the amount of heat gained or lost by the fluid may be very significant, particularly if the pipe or duct is long (Fig. 4-47). Sometimes heat transfer is desirable and is the sole purpose of the flow. Water flow through the pipes in the furnace of a power plant, the flow of refrigerant in a freezer, and the flow in heat exchangers are some examples of this case. At other times, heat transfer is undesirable, and the pipes or ducts are insulated to prevent any heat loss or gain, particularly when the temperature difference between the flowing fluid and the surroundings is large. Heat transfer in this case is negligible.

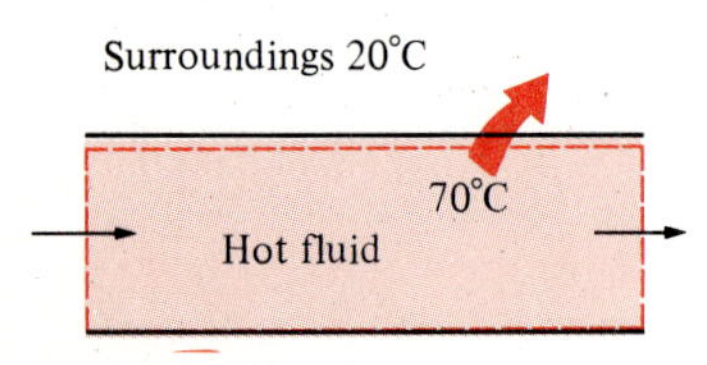

FIGURE 4-47
Heat losses from a hot fluid flowing through an uninsulated pipe or duct to the cooler environment may be very significant.

$\dot{W} \neq 0$. If the control volume involves a heating section (electric wires), a fan, or a pump (shaft), the work term interactions should be considered (Fig. 4-48). Of these, fan work is usually small and often neglected. If the control volume involves none of these work devices, the work term is zero.

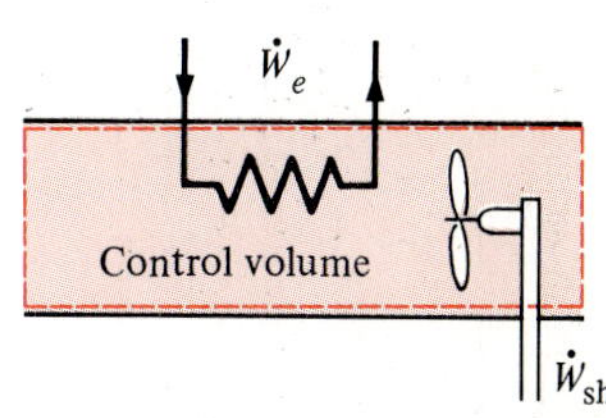

FIGURE 4-48

Pipe or duct flow may involve more than one form of work at the same time.

$\Delta ke \cong 0$. The velocities involved in pipe and duct flow are relatively low, and the kinetic energy changes are usually insignificant. This is particularly true when the pipe or duct diameter is constant and the heating effects are negligible. Kinetic energy changes may be significant, however, for gas flow in ducts with variable cross-sectional areas.

$\Delta pe \neq 0$. In pipes and ducts, the fluid may undergo a considerable elevation change. Thus the potential energy term may be significant (Fig. 4-49). This is particularly true for flow through insulated pipes and ducts where the heat transfer does not overshadow other effects.

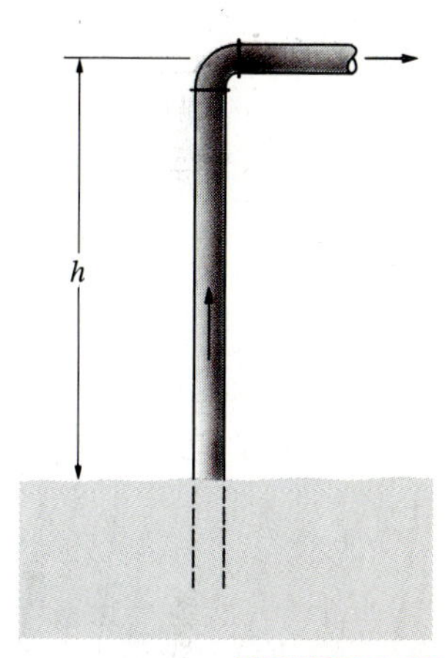

FIGURE 4-49

A fluid may experience a large elevation (thus potential energy) change as it flows through a pipe or duct.

EXAMPLE 4-8

The electric heating systems used in many houses consist of a simple duct with resistance wires. Air is heated as it flows over resistance wires. Consider a 15-kW electric heating system. Air enters the heating section at 100 kPa and 17°C with a volume flow rate of 150 m³/min. If heat is lost from the air in the duct to the surroundings at a rate of 200 W, determine the exit temperature of air.

Solution We select the heating section portion of the duct, shown in Fig. 4-50, as our control volume. By neglecting the kinetic and potential energy changes, the conservation of energy equation for this single-stream steady-flow system simplifies to

$$\dot{Q} - \dot{W} = \dot{m}(\Delta h + \cancel{\Delta ke}^{\,0} + \cancel{\Delta pe}^{\,0}) = \dot{m}(h_2 - h_1)$$

At the specified conditions, the air can be treated as an *ideal gas* since it is at a high temperature and low pressure relative to its critical values ($T_{cr} = -147°C$ and $P_{cr} = 3390$ kPa for nitrogen, the main constituent of air). Then the specific volume of the air at the inlet becomes

$$v_1 = \frac{RT_1}{P_1} = \frac{[0.287\ \text{kPa·m}^3/(\text{kg·K})](290\ \text{K})}{100\ \text{kPa}} = 0.832\ \text{m}^3/\text{kg}$$

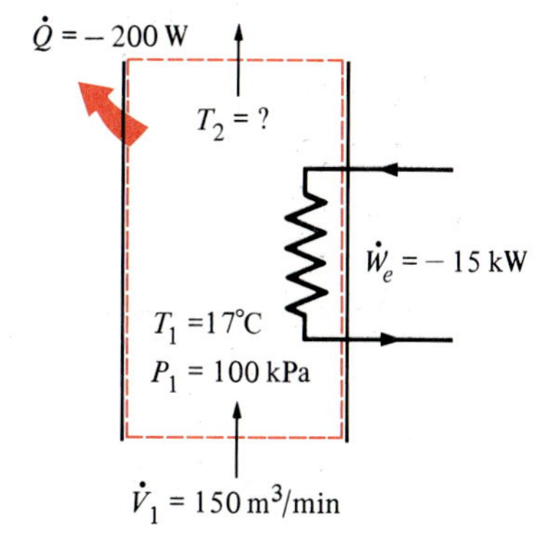

FIGURE 4-50

Schematic for Example 4-8.

The mass flow rate of the air through the duct is determined from

$$\dot{m} = \frac{\dot{V}_1}{v_1} = \frac{150 \text{ m}^3/\text{min}}{0.832 \text{ m}^3/\text{kg}}\left(\frac{1 \text{ min}}{60 \text{ s}}\right) = 3.0 \text{ kg/s}$$

At the temperatures encountered in heating and air-conditioning systems, Δh can be replaced by $C_p \Delta T$, where $C_p = 1.005$ kJ/(kg·°C) (the value at room temperature) with negligible error (Fig. 4-51). Then the energy equation takes the following form:

$$\dot{Q} - \dot{W} = \dot{m}C_p(T_2 - T_1)$$

Substituting the known quantities, we see that the exit temperature of the air is

$$-0.2 \text{ kJ/s} - (-15 \text{ kJ/s}) = (3 \text{ kg/s})[1.005 \text{ kJ/(kg·°C)}](T_2 - 17°\text{C})$$

$$T_2 = 21.9°\text{C}$$

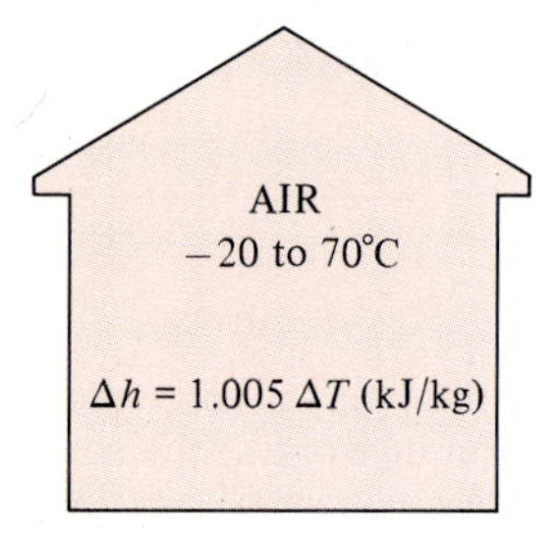

FIGURE 4-51

The error involved in $\Delta h = C_p \Delta T$, where $C_p = 1.005$ kJ/(kg·°C) is less than 0.5 percent for air in the temperature range −20 to 70°C.

EXAMPLE 4-9

In rural areas, water is often extracted from underground by pumps. Consider an underground water source whose free surface is 60 m below ground level. The water is to be raised 5 m above the ground by a pump. The diameter of the pipe is 10 cm at the inlet and 15 cm at the exit. Neglecting any heat interaction with the surroundings and frictional heating effects, determine the power input to the pump required for a steady flow of water at a rate of 15 L/s (= 0.015 m³/s).

Solution We take the region within the pipes and the pump as our system. The system boundaries are indicated by dashed lines in Fig. 4-52. Mass is crossing the boundaries of the system, so it is a *control volume*. There is no indication of any change within the control volume with time, thus it is a *steady-flow system*.

The density of liquid water at or about room temperature can be taken to be constant at 1000 kg/m³ with negligible error. Then the mass flow rate and the flow velocities become

$$\dot{m} - \rho\dot{\mathbf{V}} = (1000 \text{ kg/m}^3)(0.015 \text{ m}^3/\text{s}) - 15 \text{ kg/s}$$

$$\mathbf{V}_1 = \frac{\dot{m}}{\rho_1 A_1} = \frac{15 \text{ kg/s}}{(1000 \text{ kg/m}^3)[\pi(0.1 \text{ m})^2/4]} = 1.9 \text{ m/s}$$

$$\mathbf{V}_2 = \frac{\dot{m}}{\rho_2 A_2} = \frac{15 \text{ kg/s}}{(1000 \text{ kg/m}^3)[\pi(0.15 \text{ m})^2/4]} = 0.85 \text{ m/s}$$

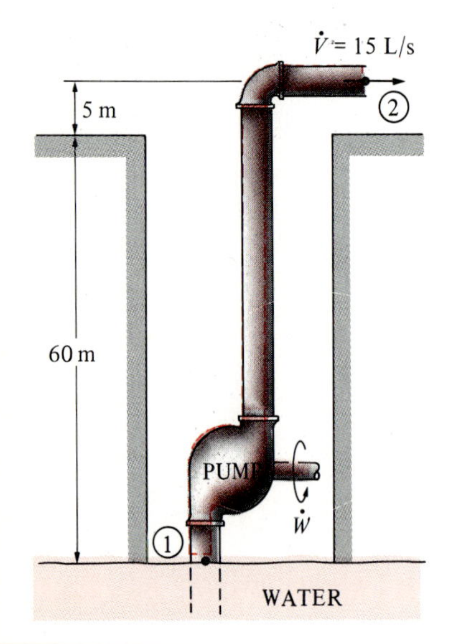

FIGURE 4-52

Schematic for Example 4-9.

As pointed out in Chap. 3, liquids can be treated as incompressible substances (v = constant). Thus their enthalpy change can be expressed as

$$\begin{aligned} h_2 - h_1 &= (u_2 + P_2 v_2) - (u_1 + P_1 v_1) \\ &= (u_2 - u_1) + v(P_2 - P_1) \\ &= C(T_2 - T_1) + v(P_2 - P_1) \end{aligned}$$

since $\Delta u = C\,\Delta T$. Then the conservation of energy equation for liquids for a steady-flow process becomes

$$\dot{Q} - \dot{W} = \dot{m}\left[C(T_2 - T_1) + v(P_2 - P_1) + \frac{\mathbf{V}_2^2 - \mathbf{V}_1^2}{2} + g(z_2 - z_1)\right]$$

For our case, there is no heat transfer ($\dot{Q} = 0$) and no change in temperature ($T_2 = T_1$). Furthermore, we have atmospheric pressure at both the inlet and the exit ($P_1 = P_2 = P_{\text{atm}}$). Then the above equation simplifies to

$$\dot{W} = -\dot{m}\left[\frac{\mathbf{V}_2^2 - \mathbf{V}_1^2}{2} + g(z_2 - z_1)\right]$$

Substituting gives

$$\begin{aligned}\dot{W} &= -(15 \text{ kg/s})\left[\frac{(0.85 \text{ m/s})^2 - (1.9 \text{ m/s})^2}{2} + (9.807 \text{ m/s}^2)(65 \text{ m})\right] \\ &= -(15 \text{ kg/s})(-1.44 \text{ m}^2/\text{s}^2 + 637.46 \text{ m}^2/\text{s}^2)\left(\frac{1 \text{ kJ/kg}}{1000 \text{ m}^2/\text{s}^2}\right) \\ &= -9.54 \text{ kW}\end{aligned}$$

This is the required pump work. It is interesting to note how small the kinetic energy term can be relative to the potential energy term when the process involves a liquid undergoing a considerable elevation change. This is typical for many actual processes. We should also note that the frictional losses for flow problems through pipes and ducts can be very significant. Therefore, in reality, we would need a larger pump to overcome this extra resistance to flow (Fig. 4-53). Frictional losses are treated in detail in fluid mechanics courses.

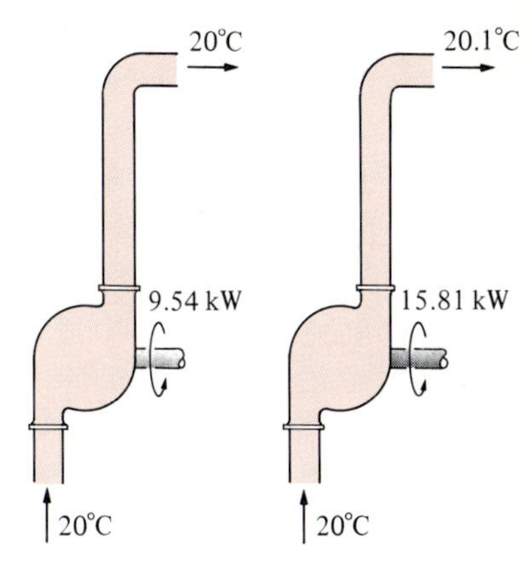

FIGURE 4-53

An increase of 0.1°C in water temperature as a result of frictional effects would increase the pumping power requirements of the system in Example 4-9 by 66 percent.

4-4 ■ UNSTEADY-FLOW PROCESSES

During a steady-flow process, no changes occur within the control volume; thus one does not need to be concerned about what is going on within the boundaries. Not having to worry about any changes within the control volume with time greatly simplifies the analysis.

Many processes of interest, however, involve *changes* within the control volume with time. Such processes are called **unsteady-flow**, or **transient-flow processes**. The steady-flow relations developed in Sec. 4-2 are obviously not applicable to these processes. When an unsteady-flow process is analyzed, it is important to keep track of the mass and energy contents of the control volume as well as the energy interactions across the boundary.

Some familiar unsteady-flow processes are the charging of rigid vessels from supply lines (Fig. 4-54), discharging a fluid from a pressurized vessel, driving a gas turbine with pressurized air stored in a large container, inflating tires or balloons, and even cooking with an ordinary or pressure cooker.

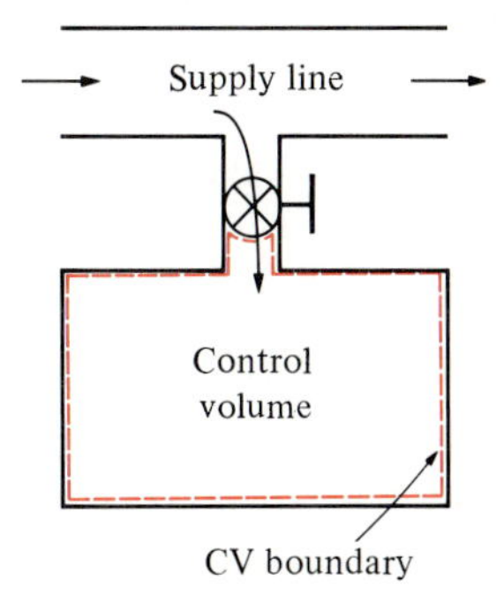

FIGURE 4-54

Charging of a rigid tank from a supply line is an unsteady-flow process since it involves changes within the control volume.

Unlike steady-flow processes, unsteady-flow processes start and end over some finite time instead of continuing indefinitely. Therefore in this section, we deal with changes that occur over some time interval Δt instead of with the rate of changes (changes per unit time). An unsteady-flow system, in some respects, is similar to a closed system, except that the mass within the system boundaries does not remain constant during a process.

Another difference between steady- and unsteady-flow systems is that steady-flow systems are fixed in space, in size, and in shape. Unsteady-flow systems, however, are not (Fig. 4-55). They are usually stationary; i.e., they are fixed in space, but they may involve moving boundaries and thus boundary work. Next we develop the conservation of mass and the conservation of energy equations for a general unsteady-flow process.

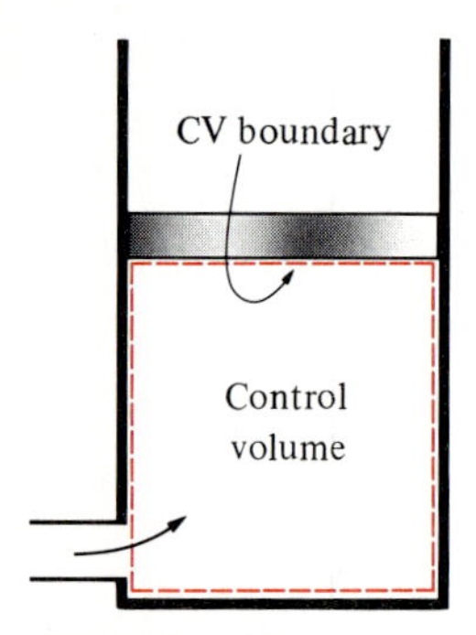

FIGURE 4-55

The shape and size of a control volume may change during an unsteady-flow process.

Conservation of Mass

Unlike the case of steady-flow processes, the amount of mass within the control volume *does* change with time during an unsteady-flow process. The degree of change depends on the amount of mass that enters and leaves the control volume during the process.

Consider an ordinary bathtub, for example, that is half full of water, as shown in Fig. 4-56. The total mass of water in the tub at this point is given as $m_1 = 150$ kg. Let us take the volume occupied by the water as our control volume. Obviously the upper boundary of the control volume (the free surface) will move up or down depending on the amount of water in the bathtub. Now both the tap and the drain plug are opened, allowing some mass transfer to and from the bathtub. The water level in the bathtub will rise or drop depending on which stream (inlet or exit) dominates. After some time Δt, both the tap and the drain plug are closed. The amount of water that entered the bathtub is measured to be $m_i = 50$ kg, and the amount that drained out is $m_e = 30$ kg.

Disregarding any water loss by evaporation, the amount of water m_2 in the bathtub at the end of the process is determined on the basis of the conservation of mass principle, which can be expresssed as follows: *The increase in the water mass in the bathtub is equal to the net mass flow into the bathtub:*

$$m_i - m_e = (m_2 - m_1)_{\text{bathtub}}$$
$$50 \text{ kg} - 30 \text{ kg} = m_2 - 150 \text{ kg}$$
$$m_2 = 170 \text{ kg}$$

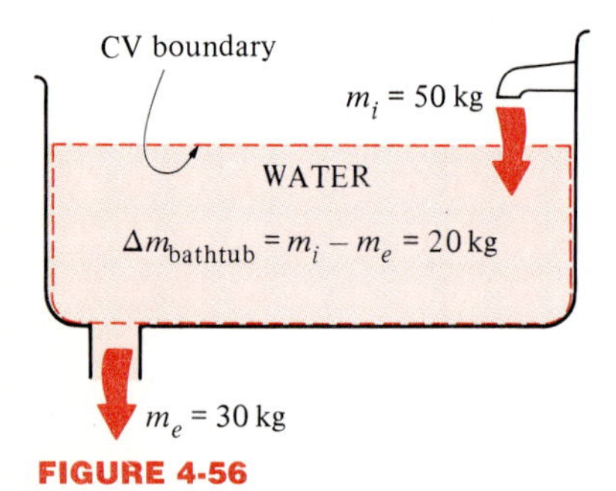

FIGURE 4-56

Conservation of mass principle for an ordinary bathtub.

Thus the **conservation of mass principle** for a control volume (CV) undergoing *any* unsteady-flow process for a time interval Δt can be expressed as

$$\begin{pmatrix}\text{Total mass}\\ \text{entering CV}\\ \text{during } \Delta t\end{pmatrix} - \begin{pmatrix}\text{Total mass}\\ \text{leaving CV}\\ \text{during } \Delta t\end{pmatrix} = \begin{pmatrix}\text{Net change in}\\ \text{mass within CV}\\ \text{during } \Delta t\end{pmatrix}$$

or

$$\Sigma m_i - \Sigma m_e = \Delta m_{\text{CV}} \tag{4-28}$$

or $$\Sigma m_i - \Sigma m_e = (m_2 - m_1)_{CV} \quad \text{(kg)} \qquad (4\text{-}29)$$

where the subscripts are

i = inlet
e = exit
1 = initial state of CV
2 = final state of CV

Often one or more terms in Eq. 4-29 are zero. For example, $m_i = 0$ if no mass enters the CV during the process, $m_e = 0$ if no mass leaves the CV during the process, and $m_1 = 0$ if the CV is initially evacuated.

The conservation of mass principle for a general unsteady-flow process can also be expressed in the rate form by dividing each term in Eq. 4-28 by the time interval Δt and taking the limit as $\Delta t \to 0$:

$$\sum \dot{m}_i - \sum \dot{m}_e = \frac{dm_{CV}}{dt} \quad \text{(kg/s)} \qquad (4\text{-}30)$$

For the special case of a steady-flow process ($dm_{CV}/dt = 0$), this equation will reduce to Eq. 4-14, the conservation of mass equation for steady-flow processes.

Conservation of Energy

Unlike the case of steady-flow processes, the energy content of a control volume changes with time during an unsteady-flow process. The degree of change depends on the amount of energy transfer across the system boundaries as heat and work as well as on the amount of energy transported into and out of the control volume by mass during the process. When analyzing an unsteady-flow process, we must keep track of the energy content of the control volume as well as the energies of the incoming and outgoing streams.

To clarify the various forms of energy interactions involved, let us reconsider the bathtub discussed earlier and redrawn in Fig. 4-57. The energy content of the water initially present in the bathtub is the initial energy of the control volume (say, $E_1 = 500$ kJ). Some heat will probably be lost from the water in the bathtub to the ground, the surrounding air, etc. (say, $Q = -50$ kJ). If the water level is rising, some boundary work will be done against the atmospheric air to push it out of the way (say, $W_b = 10$ kJ). This is the only form of work involved in this case since no shafts or electric wires are crossing the boundaries. The energy content of the control volume will increase as mass flows in and will decrease as mass flows out (say, $\Theta_i = 300$ kJ and $\Theta_e = 100$ kJ, where Θ represents the energy transported with mass).

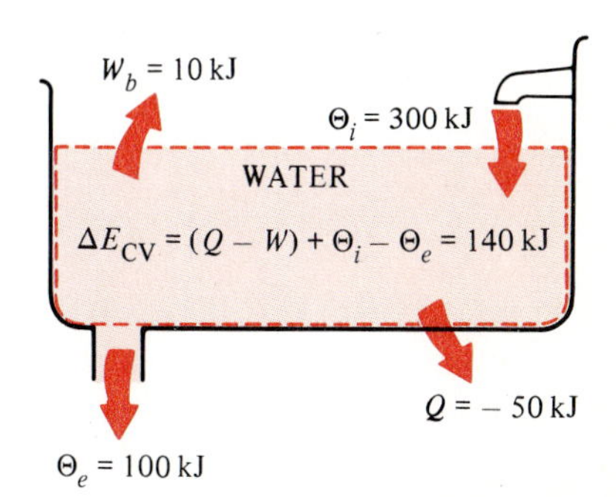

FIGURE 4-57
Conservation of energy principle for an ordinary bathtub.

Disregarding any water loss by evaporation, the energy content of the water in the bathtub at the end of the process (E_2) is determined on the basis of the conservation of energy principle, which can be expressed as follows: *The increase in the energy content of the bathtub is equal to the net energy flow into the bathtub:*

$$
\begin{aligned}
Q - W + \Theta_i - \Theta_e &= (E_2 - E_1)_{\text{bathtub}} \\
-50 \text{ kJ} - 10 \text{ kJ} + 300 \text{ kJ} - 100 \text{ kJ} &= E_2 - 500 \text{ kJ} \\
E_2 &= 640 \text{ kJ}
\end{aligned}
$$

Thus the **conservation of energy principle** for a control volume undergoing any unsteady-flow process for a time interval Δt can be expressed as

$$
\begin{pmatrix}\text{Total energy}\\ \text{crossing boundary}\\ \text{as heat and work}\\ \text{during } \Delta t\end{pmatrix} + \begin{pmatrix}\text{Total energy}\\ \text{transported by}\\ \text{mass into CV}\\ \text{during } \Delta t\end{pmatrix} - \begin{pmatrix}\text{Total energy}\\ \text{transported by}\\ \text{mass out of CV}\\ \text{during } \Delta t\end{pmatrix} = \begin{pmatrix}\text{Net change}\\ \text{in energy}\\ \text{of CV}\\ \text{during } \Delta t\end{pmatrix}
$$

or

$$
Q - W + \Sigma\Theta_i - \Sigma\Theta_e = \Delta E_{\text{CV}} \qquad \text{(kJ)} \tag{4-31}
$$

where Θ represents the *total* energy transported by mass into or out of a control volume through an inlet or exit during the process.

In Eq. 4-31 the heat and work terms (Q and W) can be determined by external measurements. The total energy of the control volume at the beginning and at the end of the process (E_1 and E_2) can also be determined easily by measuring the relevant properties of the substance at these two states. The total energy transported by mass into or out of the control volume (Θ_i, Θ_e), however, is not as easy to determine since the properties of the mass at each inlet or exit may be changing with time as well as over the cross section. Thus the only way to determine the energy transport through an opening as a result of mass flow is to consider sufficiently small differential masses δm that have uniform properties and to add their total energies.

The total energy of a flowing fluid of mass δm is $\theta\ \delta m$, where $\theta = h + \text{ke} + \text{pe}$ is the total energy of the fluid per unit mass. Then the total energy transported by mass through an inlet or exit (Θ_i or Θ_e) is obtained by integration. At an inlet, for example, it becomes

$$
\Theta_i = \int_{m_i} \theta_i\, \delta m_i = \int_{m_i} \left(h_i + \frac{\mathbf{V}_i^2}{2} + gz_i\right) \delta m_i
$$

or

$$
\Theta_i = \int_0^{\Delta t} \left(h_i + \frac{\mathbf{V}_i^2}{2} + gz_i\right) \dot{m}_i\, dt \tag{4-32}
$$

Repeating this at each inlet and exit and substituting the results into Eq. 4-31, we obtain

$$
\begin{aligned}
Q - W = &\sum \int_{m_e} \left(h_e + \frac{\mathbf{V}_e^2}{2} + gz_e\right) \delta m_e \\
&- \sum \int_{m_i} \left(h_i + \frac{\mathbf{V}_i^2}{2} + gz_i\right) \delta m_i + \Delta E_{\text{CV}} \qquad \text{(kJ)}
\end{aligned} \tag{4-33}
$$

One needs to know how the properties of the mass at the inlets and the exits change during the process in order to perform these integrations.

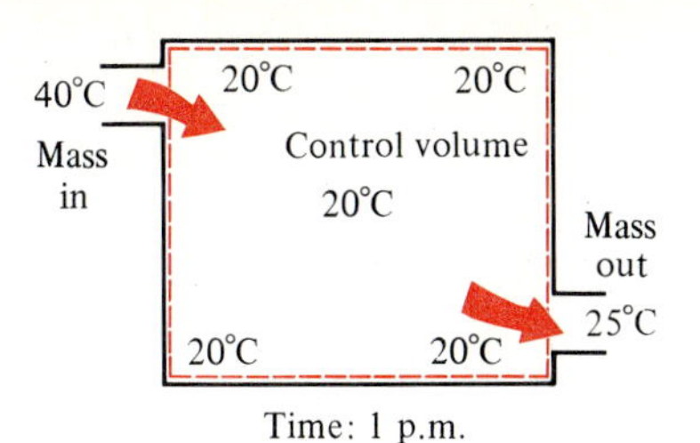

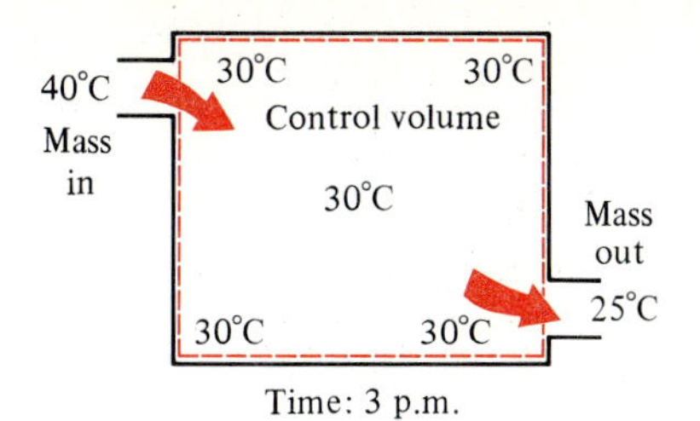

FIGURE 4-58
A control volume undergoing a uniform-flow process.

A Special Case: Uniform-Flow Processes

The unsteady-flow processes are, in general, difficult to analyze because the integrations in Eq. 4-33 are difficult to perform. Some unsteady-flow processes, however, can be represented reasonably well by another simplified model—the **uniform-flow process**. A uniform-flow process involves the following idealizations, which greatly simplify the analysis:

1 At any instant during the process, the state of the control volume is uniform (i.e., it is the same throughout). The state of the control volume may change with time, but it will do so uniformly (Fig. 4-58). Consequently, the state of the mass leaving the control volume at any instant is the same as the state of the mass in the control volume at that instant. (This assumption is in contrast to the steady-flow assumption which requires that the state of the control volume change with location but not with time.)

2 The fluid properties may differ from one inlet or exit to another, but the fluid flow at an inlet or exit is *uniform and steady*. That is, the properties do not change with time or position over the cross section of an inlet or exit. If they do, they are averaged and treated as constants for the entire process.

Under these idealizations, the integrations in Eq. 4-33 can easily be performed, and the conservation of energy equation for a uniform-flow process becomes

$$Q - W = \sum m_e \left(h_e + \frac{\mathbf{V}_e^2}{2} + gz_e \right) - \sum m_i \left(h_i + \frac{\mathbf{V}_i^2}{2} + gz_i \right) + (m_2 e_2 - m_1 e_1)_{\mathrm{CV}} \qquad \text{(kJ)} \quad (4\text{-}34)$$

This equation can also be expressed in the rate form by dividing each term by the time interval Δt and taking the limit as $\Delta t \to 0$.

When the kinetic and potential energy changes associated with the control volume and the fluid streams are negligible, Eq. 4-34 reduces to

$$Q - W = \Sigma m_e h_e - \Sigma m_i h_i + (m_2 u_2 - m_1 u_1)_{\mathrm{CV}} \qquad \text{(kJ)} \quad (4\text{-}35)$$

Again, the various subscripts appearing in the above relations are i =

inlet, e = exit, 1 = initial state, and 2 = final state of the control volume.

Notice that if no mass is entering or leaving the control volume ($m_i = m_e = 0$), the first two terms on the right-hand side of the above relation drop out and this equation reduces to the first-law relation for closed systems (Fig. 4-59).

Brief descriptions of the various terms appearing in the above equations are as follows:

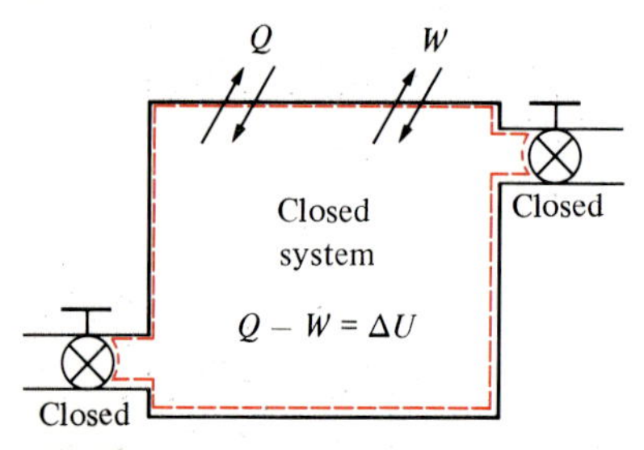

FIGURE 4-59

The energy equation of a uniform-flow system reduces to that of a closed system when all the inlets and exits are closed.

Q = **total heat transfer between the control volume and the surroundings during the process.** It is negative if heat is leaving the control volume and zero if the control volume is well insulated.

W = **total work associated with the control volume.** It may involve electrical work, shaft work, and even boundary work if the boundaries of the control volume move during the process (Fig. 4-60). It is zero for a control volume that involves no moving boundaries, shafts, or electric resistors.

m_e = **mass leaving the control volume.** It is zero if no mass leaves the control volume during the process.

m_i = **mass entering the control volume.** It is zero if no mass enters the control volume during a process.

$U_1 = m_1u_1$ = **total initial internal energy of the control volume.** It is zero for a control volume which is initially evacuated.

$U_2 = m_2u_2$ = **total final internal energy of the control volume.**

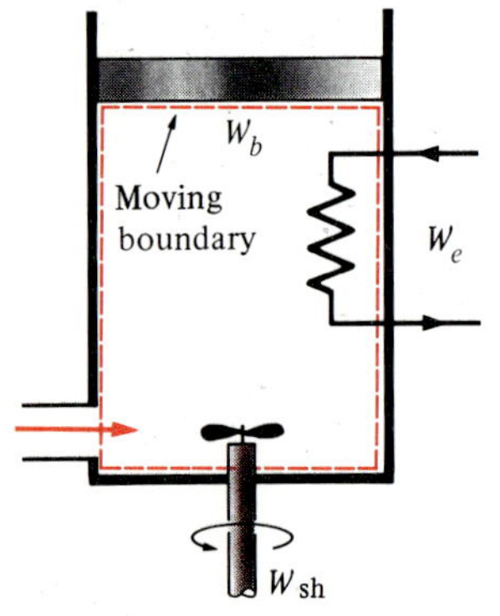

FIGURE 4-60

A uniform-flow system may involve electrical, shaft, and boundary work all at once.

Even though both the steady-flow and uniform-flow processes are somewhat idealized, many actual processes can be approximated reasonably well by one of these with satisfactory results. The degree of satisfaction depends upon the desired accuracy and the degree of validity of the assumptions made.

Engineers often find themselves in a position to choose between a quick, simple analysis with simplifying assumptions, at the expense of some accuracy, and an accurate, in-depth analysis with minimal assumptions, at the expense of time and extra effort. The right choice will depend on the situation at hand.

EXAMPLE 4-10

A rigid, insulated tank which is initially evacuated is connected through a valve to a supply line that carries steam at 1 MPa and 300°C. Now the valve is opened, and steam is allowed to flow slowly into the tank until the pressure reaches 1 MPa, at which point the valve is closed. Determine the final temperature of the steam in the tank.

Solution We choose the region within the tank as the control volume. The boundaries of the control volume are indicated by the dashed lines in Fig. 4-61.

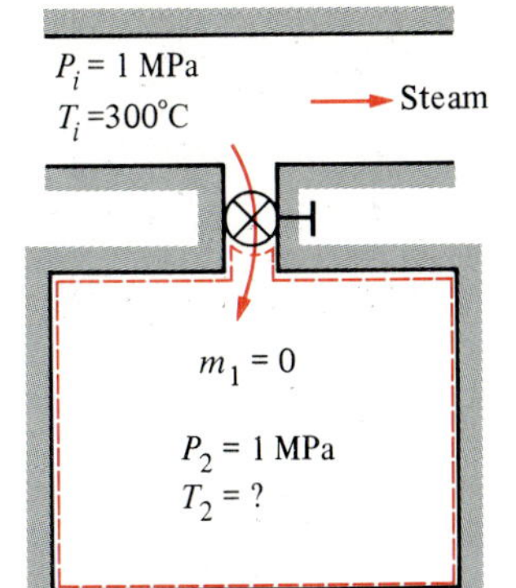

(*a*) Flow of steam into an evacuated tank

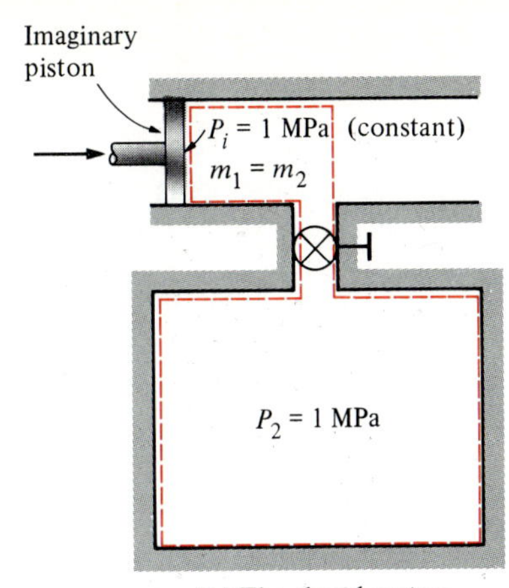

(*b*) The closed-system equivalence

FIGURE 4-61
Schematic for Example 4-10.

This is not a steady-flow process since the state of the control volume changes during this process.

The inlet conditions of the steam during this process remain constant, and the state within the control volume may be assumed to be changing uniformly since the flow process is taking place slowly. Therefore, this process can be analyzed as a *uniform-flow process*.

The tank is initially evacuated ($m_1 = 0$), and no mass is leaving the system ($m_e = 0$). Then the conservation of mass equation (Eq. 4-29) for this one-inlet, no-exit uniform-flow system reduces to

$$\Sigma m_i - \cancelto{0}{\Sigma m_e} = (m_2 - \cancelto{0}{m_1})_{CV}$$
$$m_i = m_2$$

That is, the final mass within the tank is equal to the amount of mass that entered the tank during the process.

The tank is insulated ($Q = 0$), no work is involved ($W = 0$), the initial internal energy is zero ($U_1 = m_1 u_1 = 0$ since $m_1 = 0$), and the kinetic and potential energies can be neglected since there is no indication that they are significant. Then the conservation of energy equation (Eq. 4-35) for this case reduces to

$$\cancelto{0}{Q} - \cancelto{0}{W} = \cancelto{0}{\Sigma m_e h_e} - \Sigma m_i h_i + (m_2 u_2 - \cancelto{0}{m_1 u_1})_{CV}$$
$$0 = -m_i h_i + m_2 u_2$$

But $m_i = m_2$. Thus,

$$u_2 = h_i$$

That is, the final internal energy of the steam in the tank is equal to the enthalpy of the steam entering the tank.

The enthalpy of the steam at the inlet state is

$$\left.\begin{aligned} P_i &= 1 \text{ MPa} \\ T_i &= 300°\text{C} \end{aligned}\right\} \quad h_i = 3051.2 \text{ kJ/kg} \qquad \text{(Table A-6)}$$

This is equal to u_2. Since we now know two properites at the final state, it is fixed and the temperature at this state is determined from the same table:

$$\left.\begin{aligned} P_2 &= 1 \text{ MPa} \\ u_2 &= 3051.2 \text{ kJ/kg} \end{aligned}\right\} \quad T_2 = 456.2°\text{C}$$

That is, the temperature of the steam in the tank has increased by 156.2°C. This result may be surprising at first, and you may be wondering where the energy to raise the temperature of the steam came from. The answer lies in the enthalpy term $h = u + Pv$. Part of the energy represented by enthalpy is the flow energy Pv, and this flow energy is converted to sensible internal energy once the flow ceases to exist in the control volume, and it shows up as an increase in temperature (Fig. 4-62).

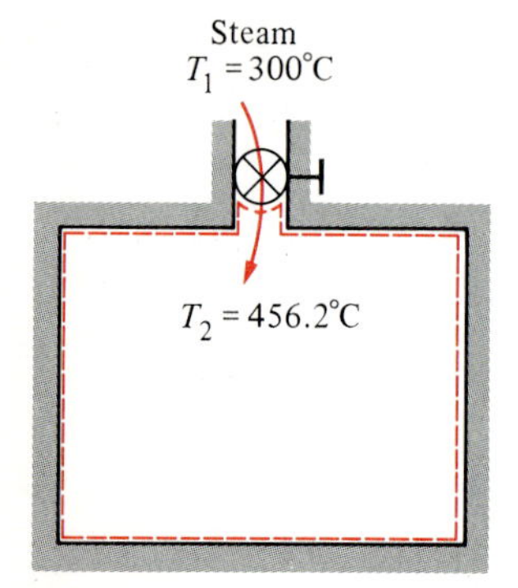

FIGURE 4-62

The temperature of steam rises from 300 to 456.2°C as it enters a tank as a result of flow energy being converted to internal energy.

Alternative solution This problem can also be solved by considering the region within the tank and the mass that is destined to enter the tank as a closed system, as shown in Fig. 4-61*b*. Since no mass crosses the boundaries, viewing this as a closed system is appropriate.

During the process, the steam upstream (the imaginary piston) will push the enclosed steam in the line into the tank at a constant pressure of 1 MPa. Then the boundary work done during this process is

$$W_b = \int_1^2 P_i \, dV = P_i(V_2 - V_1) = P_i[V_{\text{tank}} - (V_{\text{tank}} + V_i)] = -P_i V_i$$

where V_i is the volume occupied by the steam before it enters the tank and P_i is the pressure at the moving boundary (the imaginary piston face). Dividing this equation by the mass m_i, we obtain the boundary work per unit mass of the steam:

$$w_b = -P_i v_i$$

Writing the conservation of energy equation for a closed system on a unit-mass basis and simplifying yield

$$\cancelto{0}{q} - \cancelto{0}{w_{\text{other}}} - w_b = u_2 - u_1$$

or

$$u_2 = u_1 - w_b = u_i + P_i v_i - h_i$$

since the initial state of the system is simply the line conditions of the steam. This result is identical to the one obtained with the uniform-flow analysis. Once again, the temperature rise is caused by the so-called flow energy or flow work which is the energy required to push the substance into the tank.

EXAMPLE 4-11

A pressure cooker is a pan that cooks food much faster than ordinary pans by maintaining a higher pressure and temperature during cooking. The pressure inside the pan is controlled by a pressure regulator (the petcock) which keeps the pressure at a constant level by periodically allowing some steam to escape, thus preventing any excess pressure buildup.

A certain pressure cooker has a volume of 6 L and an operating pressure of 75 kPa gage. Initially, it contains 1 kg of water. Heat is supplied to the pressure cooker at a rate of 500 W for 30 min after the operating pressure is reached. Assuming an atmospheric pressure of 100 kPa, determine (*a*) the temperature at which cooking takes place and (*b*) the amount of water left in the pressure cooker at the end of the process.

Solution An obvious choice for the control volume is the region within the pressure cooker, as shown in Fig. 4-63. Some changes occur within the control volume during the process; thus this is not a steady-flow process. However, the properties of the steam leaving the control volume (saturated vapor at the cooker pressure) remain constant during the entire cooking process, and the properties within the control volume change uniformly. Therefore, this process can be analyzed as a *uniform-flow process*.

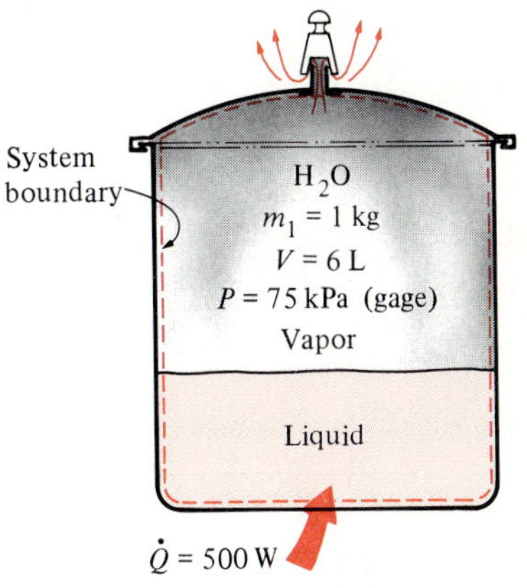

FIGURE 4-63
Schematic for Example 4-11.

(*a*) The absolute pressure within the cooker is

$$P_{abs} = P_{gage} + P_{atm} = 75 \text{ kPa} + 100 \text{ kPa} = 175 \text{ kPa}$$

Since saturation conditions exist in the cooker at all times (Fig. 4-64), the cooking temperature must be the saturation temperature corresponding to this pressure. From Table A-5, it is

$$T = T_{sat\ @\ 175\ kPa} = 116.06°C$$

which is about 16°C higher than the ordinary cooking temperature.

(*b*) The conservation of mass equation for this one-exit, no-inlet uniform-flow system simplifies to

$$\cancelto{0}{\Sigma m_i} - \Sigma m_e = (m_2 - m_1)_{CV}$$
$$m_e = (m_1 - m_2)_{CV}$$

That is, the amount of mass leaving the cooker is equal to the difference between the initial and final masses of the steam within the cooker.

There is no shaft, electrical, or boundary work for this process ($W = 0$), and the kinetic and potential energies are negligible. Then the conservation of energy equation (Eq. 4-35) for this case simplifies to

$$Q - \cancelto{0}{W} = \Sigma m_e h_e - \cancelto{0}{\Sigma m_i h_i} + (m_2 u_2 - m_1 u_1)_{CV}$$
$$Q = m_e h_e + (m_2 u_2 - m_1 u_1)_{CV}$$
$$= (m_1 - m_2) h_e + (m_2 u_2 - m_1 u_1)_{CV}$$

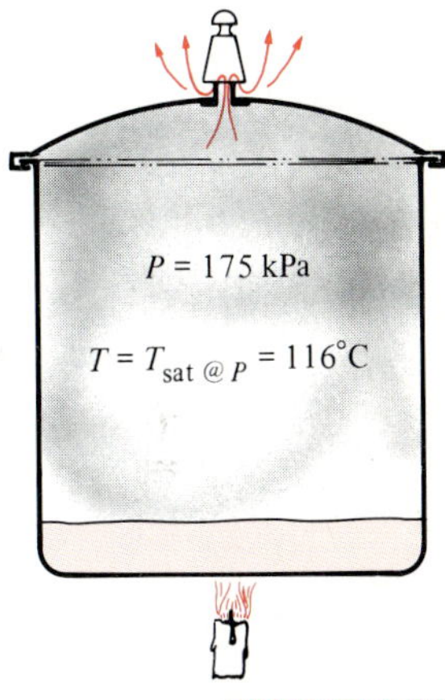

FIGURE 4-64
As long as there is liquid in a pressure cooker, the saturation conditions exist and the temperature remains constant at the saturation temperature.

The amount of heat transfer during this process is found from

$$Q = \dot{Q}\,\Delta t = (0.5 \text{ kJ/s})(30 \times 60 \text{ s}) = 900 \text{ kJ}$$

Steam leaves the pressure cooker as saturated vapor at 175 kPa at all times (Fig. 4-65). Thus,

$$h_e = h_{g\ @\ 175\ kPa} = 2700.6 \text{ kJ/kg}$$

The initial internal energy is found after the quality is determined:

$$v_1 = \frac{V}{m_1} = \frac{0.006 \text{ m}^3}{1 \text{ kg}} = 0.006 \text{ m}^3/\text{kg}$$

$$x_1 = \frac{v_1 - v_f}{v_{fg}} = \frac{0.006 - 0.001}{1.004 - 0.001} = 0.005$$

Thus, $u_1 = u_f + x_1 u_{fg} = 486.8 + (0.005)(2038.1) \text{ kJ/kg} = 497.0 \text{ kJ/kg}$

and $U_1 = m_1 u_1 = (1 \text{ kg})(497 \text{ kJ/kg}) = 497 \text{ kJ}$

The mass of the system at the final state is $m_2 = V/v_2$. Substituting this into the energy equation yields

$$Q = \left(m_1 - \frac{V}{v_2}\right) h_e + \left(\frac{V}{v_2} u_2 - m_1 u_1\right)$$

There are two unknowns in this equation, u_2 and v_2. Thus we need to relate them to a single unknown before we can determine these unknowns. Assuming there is still some liquid water left in the cooker at the final state (i.e., saturation conditions exist), v_2 and u_2 can be expressed as

$$v_2 = v_f + x_2 v_{fg} = 0.001 + x_2(1.004 - 0.001)\ \text{m}^3/\text{kg}$$
$$u_2 = u_f + x_2 u_{fg} = 486.8 + x_2(2038.1)\ \text{kJ/kg}$$

Notice that during a boiling process at constant pressure, the properties of each phase remain constant (only the amounts change). When these expressions are substituted into the above energy equation, x_2 becomes the only unknown, and it is determined to be

$$x_2 = 0.009$$

Thus $\quad v_2 = 0.001 + (0.009)(1.004 - 0.0001)\ \text{m}^3/\text{kg} = 0.010\ \text{m}^3/\text{kg}$

and $\quad m_2 = \dfrac{V}{v_2} = \dfrac{0.006\ \text{m}^3}{0.01\ \text{m}^3/\text{kg}} = \mathbf{0.6\ kg}$

Therefore, after 30 minutes there is 0.6 kg water (liquid + vapor) left in the pressure cooker.

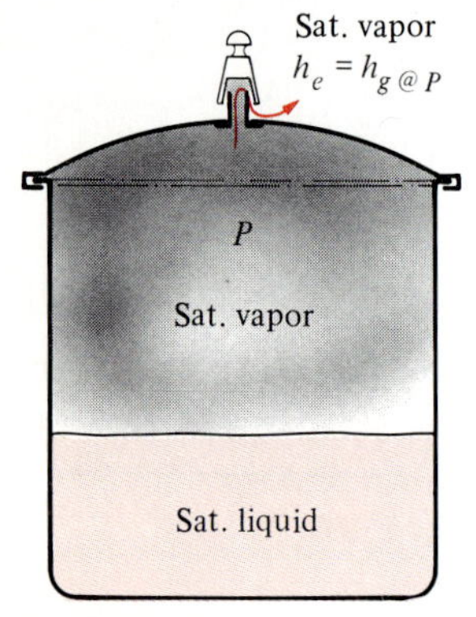

FIGURE 4-65

In a pressure cooker, the enthalpy of the exiting steam is $h_{g\,@\,P}$ (enthalpy of the saturated vapor at the given pressure).

EXAMPLE 4-12

A 1-m³ rigid tank contains air initially at 300 kPa and 300 K. A valve is opened, and air is allowed to escape slowly until the pressure in the tank drops to the atmospheric pressure of 100 kPa. The air in the tank is observed to have undergone a polytropic process (Pv^n = constant) with $n = 1.2$. Determine the heat transfer for this process (*a*) by utilizing the uniform-flow assumptions and (*b*) without utilizing the uniform-flow assumptions.

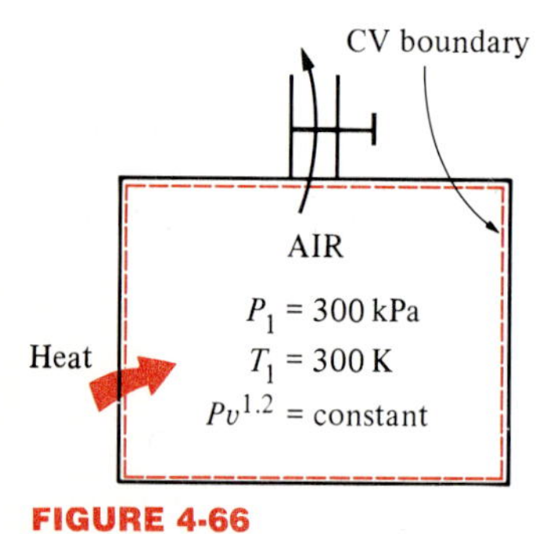

FIGURE 4-66

Schematic for Example 4-12.

Solution The obvious choice for the control volume in this case is the region within the tank. The control volume boundaries are indicated in Fig. 4-66. This is an unsteady-flow process since the properties within the control volume change during the process. At the specified conditions, the air can be treated as an *ideal gas* since it is at a high temperature and low pressure relative to its critical values ($T_{cr} = -147$°C and $P_{cr} = 3390$ kPa for nitrogen, the main constituent of air).

No mass is entering the control volume ($m_i = 0$). Thus the conservation of mass equation for this one-exit, no-inlet unsteady-flow system simplifies to

$$\cancelto{0}{\Sigma m_i} - \Sigma m_e = (m_2 - m_1)_{CV}$$
$$m_e = (m_1 - m_2)_{CV}$$

By using the ideal-gas equation ($Pv = RT$), the polytropic relation Pv^n = const. can also be expressed as

$$TP^{(1-n)/n} = C = \text{const.} \qquad \text{or} \qquad \frac{T_2}{T_1} = \left(\frac{P_2}{P_1}\right)^{(n-1)/n}$$

Then the final temperature of the air in the tank is

$$T_2 = T_1 \left(\frac{P_2}{P_1}\right)^{(n-1)/n} = (300 \text{ K}) \left(\frac{100 \text{ kPa}}{300 \text{ kPa}}\right)^{(1.2-1)/1.2} = 249.8 \text{ K}$$

The initial and final masses of the air in the tank are, respectively,

$$m_1 = \frac{P_1V_1}{RT_1} = \frac{(300 \text{ kPa})(1 \text{ m}^3)}{[0.287 \text{ kPa·m}^3/(\text{kg·K})](300 \text{ K})} = 3.484 \text{ kg}$$

$$m_2 = \frac{P_2V_2}{RT_2} = \frac{(100 \text{ kPa})(1 \text{ m}^3)}{[0.287 \text{ kPa·m}^3/(\text{kg·K})](249.8 \text{ K})} = 1.395 \text{ kg}$$

and $$m_e = (m_1 - m_2)_{CV} = (3.484 - 1.395) \text{ kg} = 2.089 \text{ kg}$$

(*a*) Under uniform-flow assumptions, the state of the control volume changes uniformly, and the properties of the mass leaving the control volume remain constant during the process. The second condition is obviously not satisfied for this process since the conditions of air leaving the control volume change from 300 kPa and 300 K at the beginning to 100 kPa and 249.8 K at the end. But, to take advantage of the simplicity of the uniform-flow analysis, we assume the air to leave the tank at the average temperature of (300 + 249.8)/2 = 274.9 K during the entire process. (The error involved in this approximation is determined in the next part.)

The control volume involves no work interactions ($W = 0$), and the kinetic and potential energies can be disregarded (ke ≅ 0, pe ≅ 0). Then the conservation of energy equation for this one-exit, no-inlet uniform-flow system simplifies to

$$Q - \cancelto{0}{W} = \Sigma m_e h_e - \cancelto{0}{\Sigma m_i h_i} + (m_2u_2 - m_1u_1)_{CV}$$

$$Q = m_e h_e + (m_2u_2 - m_1u_1)_{CV}$$

From Table A-17,

$$u_1 = u_{@\,300\text{ K}} = 214.07 \text{ kJ/kg}$$
$$u_2 = u_{@\,249.8\text{ K}} = 178.14 \text{ kJ/kg}$$
$$h_e \cong h_{@\,T_{av}} = h_{@\,274.9\text{ K}} = 275.02 \text{ kJ/kg}$$

Substituting gives

$$Q = (2.089 \text{ kg})(275.02 \text{ kJ/kg}) + (1.395 \text{ kg})(178.14 \text{ kJ/kg}) - (3.484 \text{ kg})(214.07 \text{ kJ/kg})$$

$$= (574.5 + 248.5 - 745.8) \text{ kJ} = 77.2 \text{ kJ}$$

That is, 77.2 kJ of heat is transferred to the air in the tank during this process. Notice that the air temperature drops to −23°C during this process, which suggests that compressed air at room temperature can be used for refrigeration purposes.

(*b*) Now we repeat the solution without assuming that the properties of air leaving the tank are constant at some average value. The solution will differ only in the evaluation of the energy transported out of the control volume, that is, Θ_e. Now it will be determined by integration from Eq. 4-32:

$$\Theta_e = \int_{m_e} \left(h_e + \cancelto{0}{\frac{\mathbf{V}_e^2}{2}} + \cancelto{0}{gz_e}\right) \delta m_e = \int_{m_e} h_e \,\delta m_e = \int_{m_e} C_p T_e \,\delta m_e$$

Notice that this equation reduces to $\Theta_e = h_e m_e$ when h_e is constant. To perform the integration in the above relation, we need to find a relation between T_e and δm_e.

The differential mass leaving the control volume δm_e is equal to the negative of the differential change in the mass of the control volume:

$$m_{CV} = \frac{PV}{RT} \longrightarrow dm_{CV} = \frac{V}{R}\left(\frac{dP}{T} - \frac{P\,dT}{T^2}\right) = -\delta m_e \qquad (\text{since } V = \text{const.})$$

Thus, $$\Theta_e = \int_1^2 C_p T \frac{V}{R}\left(\frac{P\,dT}{T^2} - \frac{dP}{T}\right) = \int_1^2 \frac{kV}{k-1}\left(\frac{P\,dT}{T} - dP\right)$$

since $C_p/R = k/(k-1)$, where k is the specific heat ratio and T_e at any time during the process is equal to the temperature of air within the control volume T.

The specific heat ratio k of air practically remains constant during this process (it changes from 1.400 at 300 K to 1.401 at 250 K); thus it can be taken out of the integral together with the volume:

$$\Theta_e = \frac{kV}{k-1}\left(\int_1^2 \frac{P\,dT}{T} - \int_1^2 dP\right)$$

The second integration is straightforward and yields $P_2 - P_1$. The first integration can also be performed easily after P is expressed in terms of T, by using the polytropic relation $TP^{(1-n)/n}$ = constant. It yields $(P_2 - P_1)(n - 1)/n$. By substituting these results, Θ_e is determined to be

$$\Theta_e = \frac{kV(P_1 - P_2)}{(k-1)n}$$

Thus, $$\Theta_e = \frac{1.4(1\text{ ms}^3)[(300 - 100)\text{ kPa}]}{(1.4-1)(1.2)}\left(\frac{1\text{ kJ}}{1\text{ kPa}\cdot\text{m}^3}\right) = 583.3\text{ kJ}$$

This result differs from the one obtained by assuming constant enthalpy at the average temperature (574.5 kJ) by only about 1.5 percent. A difference of this magnitude is typical for many unsteady-flow processes. Therefore, the uniform-flow approximations can be used with confidence in engineering analysis, particularly when the properties do not change very much between the initial and the final states. The heat transfer in this case becomes

$$Q = (583.3 + 248.5 - 745.8)\text{ kJ} = 86.0\text{ kJ}$$

4-5 ■ SUMMARY

In this chapter, we have discussed the conservation of mass and the conservation of energy principles for *control volumes*. Mass carries energy with it, thus the energy content changes when mass enters or leaves the control volume.

Mass flow through a cross section per unit time is called the *mass flow rate* and is denoted $\dot{m}$. It is expressed as

$$\dot{m} = \rho \mathbf{V}_{av} A \qquad (\text{kg/s}) \tag{4-5}$$

where ρ = density, kg/m^3 ($= 1/v$)
$\mathbf{V}_{av}$ = average fluid velocity normal to A, m/s
A = cross-sectional area, m^2

The fluid volume flowing through a cross section per unit time is called the *volume flow rate* $\dot{V}$. It is given by

$$\dot{V} = \int_A \mathbf{V}_n \, dA = \mathbf{V}_{av} A \qquad (\text{m}^3/\text{s}) \tag{4-6}$$

The mass and volume flow rates are related by

$$\dot{m} = \rho \dot{V} = \frac{\dot{V}}{v}$$

Thermodynamic processes involving control volumes can be considered in two groups: steady-flow processes and unsteady-flow processes. During a *steady-flow process,* the fluid flows through the control volume steadily, experiencing no change with time at a fixed position. The mass and energy content of the control volume remains constant during a steady-flow process. The conservation of mass and energy equations for steady-flow processes are expressed as

$$\Sigma \dot{m}_i = \Sigma \dot{m}_e \qquad (\text{kg/s}) \tag{4-14}$$

$$\dot{Q} - \dot{W} = \underbrace{\sum \dot{m}_e \left(h_e + \frac{\mathbf{V}_e^2}{2} + gz_e \right)}_{\text{for each exit}} - \underbrace{\sum \dot{m}_i \left(h_i + \frac{\mathbf{V}_i^2}{2} + gz_i \right)}_{\text{for each inlet}} \qquad (\text{kW}) \tag{4-19}$$

where the subscript i stands for *inlet* and e for *exit*. These are the most general forms of the equations for steady-flow processes. For single-stream (one-inlet, one-exit) systems such as nozzles, diffusers, turbines, compressors, and pumps, they simplify to

$$\dot{m}_1 = \dot{m}_2 \qquad (\text{kg/s}) \tag{4-15}$$

or

$$\frac{1}{v_1} \mathbf{V}_1 A_1 = \frac{1}{v_2} \mathbf{V}_2 A_2 \tag{4-17}$$

and

$$\dot{Q} - \dot{W} = \dot{m} \left[h_2 - h_1 + \frac{\mathbf{V}_2^2 - \mathbf{V}_1^2}{2} + g(z_2 - z_1) \right] \qquad (\text{kW}) \tag{4-20}$$

$$q - w = h_2 - h_1 + \frac{\mathbf{V}_2^2 - \mathbf{V}_1^2}{2} + g(z_2 - z_1) \qquad (\text{kJ/kg}) \tag{4-22}$$

or

$$q - w = \Delta h + \Delta \text{ke} + \Delta \text{pe} \qquad (\text{kJ/kg}) \tag{4-23}$$

where

$$q = \frac{\dot{Q}}{\dot{m}} \qquad \text{heat transfer per unit mass, kJ/kg} \tag{4-25}$$

and

$$w = \frac{\dot{W}}{\dot{m}} \qquad \text{work done per unit mass, kJ/kg} \tag{4-26}$$

In the above relations, subscripts 1 and 2 denote the inlet and exit states, respectively.

The steady-flow process is the model process for flow through nozzles, diffusers, turbines, compressors, fans, pumps, pipes, throttling valves, mixing chambers, and heat exchangers.

For *unsteady-flow processes,* the conservation of mass and energy equations are

$$\Sigma m_i - \Sigma m_e = (m_2 - m_1)_{\text{CV}} \qquad \text{(kg)} \tag{4-29}$$

$$Q - W = \sum \int_{m_e} \left(h_e + \frac{\mathbf{V}_e^2}{2} + gz_e \right) \delta m_e - \sum \int_{m_i} \left(h_i + \frac{\mathbf{V}_i^2}{2} + gz_i \right) \delta m_i + \Delta E_{\text{CV}} \qquad \text{(kJ)} \tag{4-33}$$

The various subscripts appearing in the above equations are i = inlet, e = exit, 1 = initial state, and 2 = final state of the control volume.

Often one or more terms in Eq. 4-29 are zero. For example, $m_i = 0$ if no mass enters the CV during the process, $m_e = 0$ if no mass leaves the CV during the process, and $m_1 = 0$ if the CV is initially evacuated.

The unsteady-flow processes are, in general, difficult to analyze because the integrations in Eq. 4-33 are difficult to perform. Some unsteady-flow processes, however, can be represented by another simplified model called the *uniform-flow process*. During a uniform-flow process, the state of the control volume may change with time, but it may do so uniformly. Also the fluid properties at the inlets and the exits are assumed to remain constant during the entire process. The conservation of energy equation for a uniform-flow process reduces to

$$Q - W = \sum m_e \left(h_e + \frac{\mathbf{V}_e^2}{2} + gz_e \right) - \sum m_i \left(h_i + \frac{\mathbf{V}_i^2}{2} + gz_i \right) + (m_2 e_2 - m_1 e_1)_{\text{CV}} \qquad \text{(kJ)} \tag{4-34}$$

When the kinetic and potential energy changes associated with the control volume and the fluid streams are negligible. Equation 4-34 simplifies to

$$Q - W = \Sigma m_e h_e - \Sigma m_i h_i + (m_2 u_2 - m_1 u_1)_{\text{CV}} \qquad \text{(kJ)} \tag{4-35}$$

REFERENCES AND SUGGESTED READING

1 W. Z. Black and J. G. Hartley, *Thermodynamics,* Harper & Row, New York, 1985.

2 J. R. Howell and R. O. Buckius, *Fundamentals of Engineering Thermodynamics,* McGraw-Hill, New York, 1987.

3 J. B. Jones and G. A. Hawkins, *Engineering Thermodynamics,* 2d ed., Wiley, New York, 1986.

4 D. C. Look, Jr., and H. J. Sauer, Jr., *Engineering Thermodynamics,* PWS Engineering, Boston, 1986.

5 W. C. Reynolds and H. C. Perkins, *Engineering Thermodynamics,* 2d ed., McGraw-Hill, New York, 1977.

6 G. J. Van Wylen and R. E. Sonntag, *Fundamentals of Classical Thermodynamics,* 3d ed., Wiley, New York, 1985.

7 K. Wark, *Thermodynamics,* 5th ed., McGraw-Hill, New York, 1988.

PROBLEMS*

General Control Volume Analysis

4-1C How does a control volume differ from a closed system?

4-2C Express the conservation of mass principle for a control volume verbally.

4-3C Define mass and volume flow rates. How do they differ?

4-4C The velocity of a liquid flowing in a circular pipe of radius R varies from zero at the wall to a maximum at the pipe center. The velocity distribution in the pipe can be represented as $V(r)$, where r is the radial distance from the pipe center. Based on the definition of mass flow rate $\dot{m}$, obtain a relation for the average velocity in terms of $V(r)$, R, and r.

4-5C Express the conservation of energy principle for a control volume verbally.

4-6C What are the different mechanisms for transferring energy to or from a control volume?

4-7C What is flow energy? Do fluids at rest possess any flow energy?

4-8C How do the energies of flowing fluids and a fluid at rest compare? Name the specific forms of energy associated with each case.

4-9C Consider a room filled with warm air. Some cold air leaks into the room now, and the air temperature drops somewhat. No air leaks out during the process. Does the room contain more or less energy now? Explain.

Steady-Flow Processes

4-10C When is the flow through a control volume steady?

4-11C How is a steady-flow system characterized?

4-12C Can a steady-flow system involve boundary work?

*Students are encouraged to answer *all* the concept "C" questions.

Nozzles and Diffusers

4-13C A diffuser is an adiabatic device that decreases the kinetic energy of the fluid by slowing it down. What happens to this *lost* kinetic energy?

4-14C The kinetic energy of a fluid increases as it is accelerated in an adiabatic nozzle. Where does this energy come from?

4-15C Is heat transfer to or from the fluid desirable as it flows through a nozzle? How will heat transfer affect the fluid velocity at the nozzle exit?

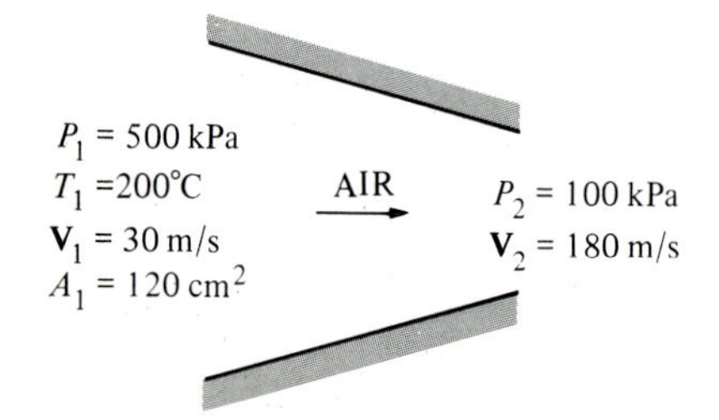

FIGURE P4-16

4-16 Air enters an adiabatic nozzle steadily at 500 kPa, 200°C, and 30 m/s and leaves at 100 kPa and 180 m/s. The inlet area of the nozzle is 120 cm^2. Determine (*a*) the mass flow rate through the nozzle, (*b*) the exit temperature of the air, and (*c*) the exit area of the nozzle.
Answers: (*a*) 1.326 kg/s, (*b*) 184.60°C, (*c*) 96.7 cm^2

4-16E Air enters an adiabatic nozzle steadily at 75 psia, 400°F, and 100 ft/s and leaves at 15 psia and 500 ft/s. The inlet area of the nozzle is 20 in^2. Determine (*a*) the mass flow rate through the nozzle, (*b*) the exit temperature of the air, and (*c*) the exit area of the nozzle.

4-17 Steam at 5 MPa and 500°C enters a nozzle steadily with a velocity of 80 m/s, and it leaves at 2 MPa and 400°C. The inlet area of the nozzle is 38 cm^2, and heat is being lost at a rate of 8 kJ/s. Determine (*a*) the mass flow rate of the steam, (*b*) the exit velocity of the steam, and (*c*) the exit area of the nozzle.

4-18 Carbon dioxide enters an adiabatic nozzle steadily at 1 MPa and 500°C with a mass flow rate of 6000 kg/h and leaves at 100 kPa and 450 m/s. The inlet area of the nozzle is 40 cm^2. Determine (*a*) the inlet velocity and (*b*) the exit temperature.
Answers: (*a*) 60.8 m/s, (*b*) 685.8 K

4-19 Air enters a nozzle steadily at 300 kPa, 77°C, and 50 m/s and leaves at 100 kPa and 320 m/s. The heat loss from the nozzle is estimated to be 3.2 kJ/kg of air flowing. The inlet area of the nozzle is 100 cm^2. Determine (*a*) the exit temperature of air and (*b*) the exit area of the nozzle.
Answers: (*a*) 24.2°C, (*b*) 39.7 cm^2

4-19E Air enters a nozzle steadily at 50 psia, 140°F, and 150 ft/s and leaves at 14.7 psia and 900 ft/s. The heat loss from the nozzle is estimated to be 6.5 Btu/lbm of air flowing. The inlet area of the nozzle is 0.1 ft^2. Determine (*a*) the exit temperature of air and (*b*) the exit area of the nozzle. *Answers:* (*a*) 441.7 R, (*b*) 0.0417 ft^2

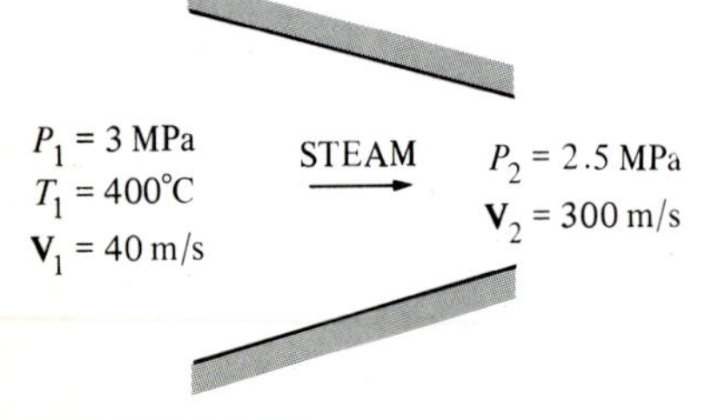

FIGURE P4-21

4-20 Refrigerant-12 at 800 kPa and 80°C enters an adiabatic nozzle steadily with a velocity of 15 m/s and leaves at 300 kPa and 30°C. Determine (*a*) the exit velocity and (*b*) the ratio of the inlet to exit area A_1/A_2.

4-21 Steam at 3 MPa and 400°C enters an adiabatic nozzle steadily with a velocity of 40 m/s and leaves at 2.5 MPa and 300 m/s. Determine (*a*) the exit temperature and (*b*) the ratio of the inlet to exit area A_1/A_2.

4-21E Steam at 500 psia and 900°F enters an adiabatic nozzle steadily with a velocity of 120 ft/s and leaves at 400 psia and 900 ft/s. Determine (*a*) the exit temperature and (*b*) the ratio of the inlet to exit area A_1/A_2.

4-22 Air at 600 kPa and 500 K enters an adiabatic nozzle that has an inlet-to-exit area ratio of 2:1 with a velocity of 120 m/s and leaves with a velocity of 380 m/s. Determine (*a*) the exit temperature and (*b*) the exit pressure of the air. *Answers:* (*a*) 436.5 K, (*b*) 330.8 kPa

4-23 Air at 100 kPa and 127°C enters an adiabatic diffuser steadily at a rate of 8000 kg/h and leaves at 120 kPa. The velocity of the airstream is decreased from 230 to 30 m/s as it passes through the diffuser. Find (*a*) the exit temperature of the air and (*b*) the exit area of the diffuser.

4-24 Air at 80 kPa and −8°C enters an adiabatic diffuser steadily with a velocity of 200 m/s and leaves with a low velocity at a pressure of 95 kPa. The exit area of the diffuser is 5 times the inlet area. Determine (*a*) the exit temperature and (*b*) the exit velocity of the air.

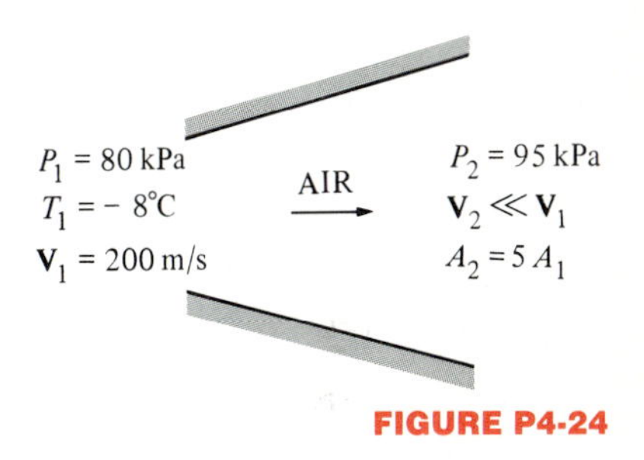

FIGURE P4-24

4-24E Air at 13 psia and 20°F enters an adiabatic diffuser steadily with a velocity of 600 ft/s and leaves with a low velocity at a pressure of 14.5 psia. The exit area of the diffuser is 5 times the inlet area. Determine (*a*) the exit temperature and (*b*) the exit velocity of the air.

4-25 Air at 80 kPa, 27°C, and 220 m/s enters a diffuser at a rate of 2.5 kg/s and leaves at 42°C. The exit area of the diffuser is 400 cm^2. The air is estimated to lose heat at a rate of 18 kJ/s during this process. Determine (*a*) the exit velocity and (*b*) the exit pressure of the air.
Answers: (*a*) 62.0 m/s, (*b*) 91.1 kPa

4-26 Nitrogen gas at 80 kPa and 7°C enters an adiabatic diffuser steadily with a velocity of 200 m/s and leaves at 105 kPa and 22°C. Determine (*a*) the exit velocity of the nitrogen and (*b*) the ratio of the inlet to exit area A_1/A_2.

4-27 Steam at 0.1 MPa and 150°C enters a diffuser with a velocity of 180 m/s and leaves as saturated vapor at 120°C with a velocity of 50 m/s. The exit area of the diffuser is 0.08 m^2. Determine (*a*) the mass flow rate of the steam, (*b*) the rate of heat transfer, and (*c*) the inlet area of the diffuser.

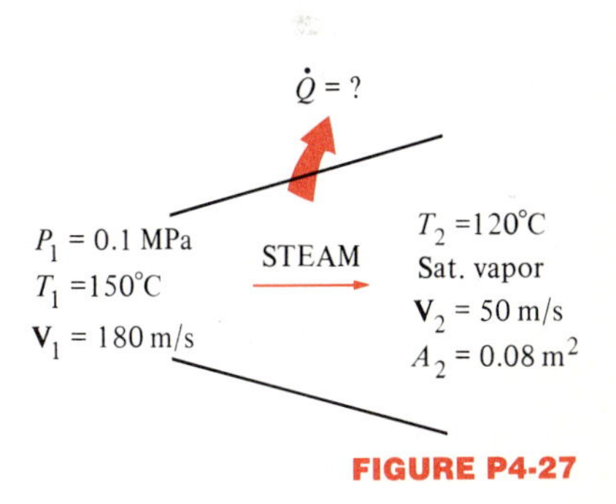

FIGURE P4-27

4-27E Steam at 14.7 psia and 320°F enters a diffuser with a velocity of 500 ft/s and leaves as saturated vapor at 240°F with a velocity of 100 ft/s. The exit area of the diffuser is 120 in^2. Determine (*a*) the mass flow rate of the steam, (*b*) the rate of heat transfer, and (*c*) the inlet area of the diffuser. *Answers:* (*a*) 5.1 lbm/s, (*b*) −235.8 Btu/s, (*c*) 46.1 in^2

4-28 Refrigerant-12 enters a diffuser steadily as saturated vapor at 700 kPa with a velocity of 120 m/s, and it leaves at 800 kPa and 40°C. The refrigerant is gaining heat at a rate of 12 kW as it passes through the diffuser. If the exit area is 30 percent greater than the inlet area, determine (*a*) the exit velocity and (*b*) the mass flow rate of the refrigerant.
Answers: (*a*) 84.3 m/s, (*b*) 3.26 kg/s

Turbines and Compressors

4-29C Consider an adiabatic turbine operating steadily. Does the work output of the turbine have to be equal to the decrease in the energy of the steam?

4-30C Consider a steam turbine operating on a steady-flow process. Would you expect the temperatures at the turbine inlet and exit to be the same?

4-31C Consider an air compressor operating on a steady-flow process Would you expect the air density to be the same at the compressor inlet and exit?

4-32C Consider an air compressor operating on a steady-flow process. How would you compare the volume flow rates of the air at the compressor inlet and exit?

4-33C Will the temperature of air rise as it is compressed by an adiabatic compressor? Why?

4-34C Somebody proposes the following system to cool a house in the summer: Compress the regular outdoor air, let it cool back to the outdoor temperature, pass it through a turbine, and discharge the cold air leaving the turbine into the house. From a thermodynamic point of view, is there anything wrong with this proposed system?

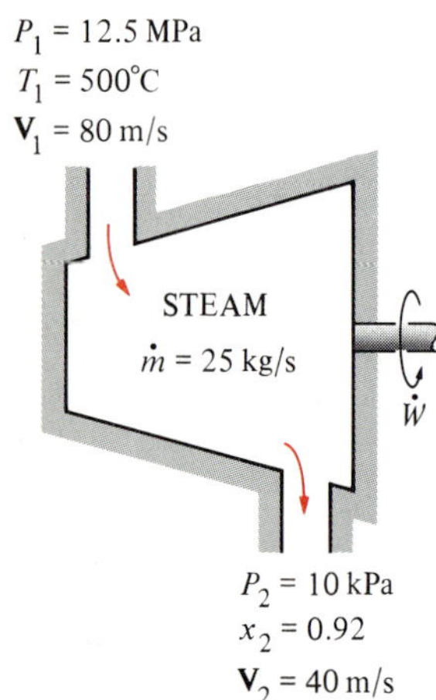

FIGURE P4-35

4-35 Steam flows steadily through an adiabatic turbine. The inlet conditions of the steam are 12.5 MPa, 500°C, and 80 m/s, and the exit conditions are 10 kPa, 92 percent quality, and 40 m/s. The mass flow rate of the steam is 25 kg/s. Determine (*a*) the change in kinetic energy, (*b*) the power output, and (*c*) the turbine inlet area.
Answers: (*a*) −2.4 kJ/kg, (*b*) 23.8 MW, (*c*) 0.0080 m^2

4-36 Steam enters an adiabatic turbine at 10 MPa and 400°C and leaves at 20 kPa with a quality of 90 percent. Neglecting the changes in kinetic and potential energies, determine the mass flow rate required for a power output of 1 MW. *Answer:* 1.384 kg/s

4-36E Steam enters an adiabatic turbine at 1250 psia and 800°F and leaves at 5 psia with a quality of 90 percent. Neglecting the changes in kinetic and potential energies, determine the mass flow rate required for a power output of 1 MW. *Answer:* 2.75 lbm/s

4-37 Steam enters a turbine steadily at 10 MPa and 550°C with a velocity of 60 m/s and leaves at 25 kPa with a quality of 95 percent. A heat loss of 30 kJ/kg occurs during the process. The inlet area of the turbine is 150 cm^2, and the exit area is 1400 cm^2. Determine (*a*) the mass flow rate of the steam, (*b*) the exit velocity, and (*c*) the power output.

4-38 Steam flows steadily through a turbine at a rate of 25,000 kg/h, entering at 8 MPa and 450°C and leaving at 30 kPa as saturated vapor. If the power generated by the turbine is 4 MW, determine the rate of heat loss from the steam. *Answer:* 491 kW

4-38E Steam flows steadily through a turbine at a rate of 45,000 lbm/h, entering at 1000 psia and 900°F and leaving at 5 psia as saturated vapor. If the power generated by the turbine is 4 MW, determine the rate of heat loss from the steam.

4-39 Steam enters an adiabatic turbine at 10 MPa and 500°C at a rate of 3 kg/s and leaves at 20 kPa. If the power output of the turbine is 2 MW, determine the temperature of the steam at the exit state. Neglect kinetic energy changes. *Answer:* 110.8°C

4-39E Steam enters an adiabatic turbine at 1000 psia and 900°F at a rate of 10 lbm/s and leaves at 5-psia pressure. If the power output of the turbine is 2 MW, determine the temperature (or quality, if saturated), of the steam at the exit state. Neglect kinetic energy changes.

4-40 Argon gas enters steadily an adiabatic turbine at 800 kPa and 400°C with a velocity of 100 m/s and leaves at 150 kPa with a velocity of 200 m/s. The inlet area of the turbine is 60 cm². If the power output of the turbine is 250 kW, determine the exit temperature of the argon.
Answer: 231°C

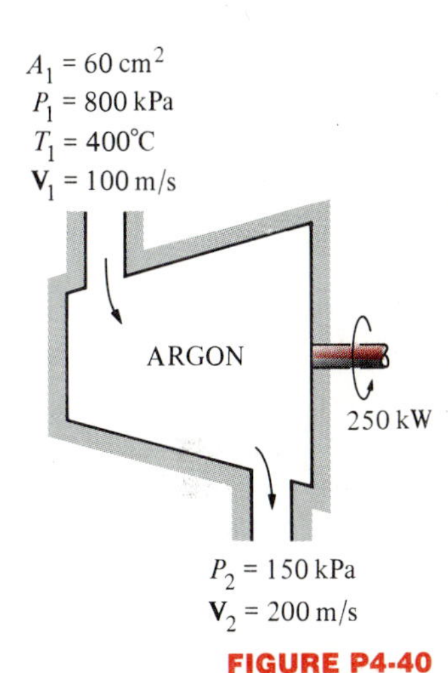

FIGURE P4-40

4-41 Air flows steadily through an adiabatic turbine, entering at 1 MPa, 500°C, and 120 m/s and leaving at 150 kPa, 150°C, and 250 m/s. The inlet area of the turbine is 80 cm². Determine (*a*) the mass flow rate of the air and (*b*) the power output of the turbine.

4-41E Air flows steadily through an adiabatic turbine, entering at 150 psia, 900°F, and 350 ft/s and leaving at 20 psia, 300°F, and 700 ft/s. The inlet area of the turbine is 0.1 ft². Determine (*a*) the mass flow rate of the air and (*b*) the power output of the turbine.

4-42 Refrigerant-12 enters an adiabatic compressor as saturated vapor at −20°C and leaves at 0.7 MPa and 70°C. The mass flow rate of the refrigerant is 0.25 kg/s. Determine (*a*) the power input to the compressor and (*b*) the volume flow rate of the refrigerant at the compressor inlet.

4-43 Refrigerant-12 enters an adiabatic compressor at 100 kPa and −20°C with a volume flow rate of 3 m³/min and leaves at a pressure of 800 kPa. The power input to the compressor is 15 kW. Determine (*a*) the mass flow rate of the refrigerant and (*b*) the exit temperature.

4-43E Refrigerant-12 enters an adiabatic compressor at 15 psia and 20°F with a volume flow rate of 10 ft³/s and leaves at a pressure of 120 psia. The power input to the compressor is 60 hp. Find (*a*) the mass flow rate of the refrigerant and (*b*) the exit temperature.

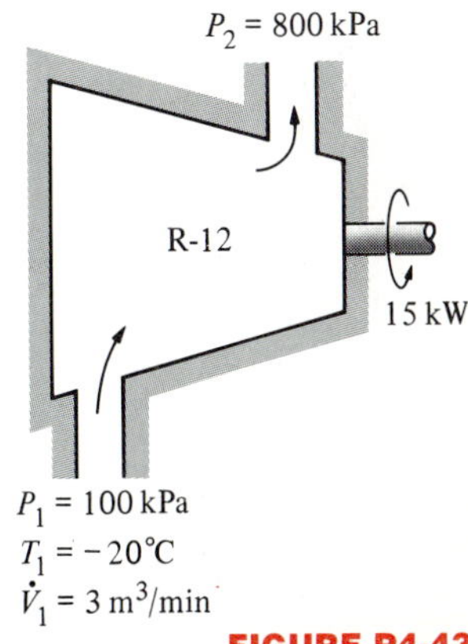

FIGURE P4-43

4-44 Air enters the compressor of a gas-turbine power plant at ambient conditions of 96 kPa and 25°C with a low velocity and exits at 1 MPa and 327°C with a velocity of 120 m/s. The compressor is cooled at a rate of 1500 kJ/min, and the power input to the compressor is 300 kW. Determine the mass flow rate of air through the compressor. *Answer:* 0.870 kg/s

4-45 Air is compressed from 100 kPa and 22°C to a pressure of 1 MPa while being cooled at a rate of 16 kJ/kg by circulating water through the compressor casing. The volume flow rate of the air at the inlet conditions is 150 m^3/min, and the power input to the compressor is 500 kW. Determine (*a*) the mass flow rate of the air and (*b*) the temperature at the compressor exit. *Answers:* (*a*) 2.95 kg/s, (*b*) 174°C

4-45E Air is compressed from 14.7 psia and 60°F to a pressure of 150 psia while being cooled at a rate of 10 Btu/lbm by circulating water through the compressor casing. The volume flow rate of the air at the inlet conditions is 5000 ft^3/min, and the power input to the compressor is 700 hp. Determine (*a*) the mass flow rate of the air and (*b*) the temperature at the compressor exit. *Answers:* (*a*) 6.36 lbm/s, (*b*) 883 R

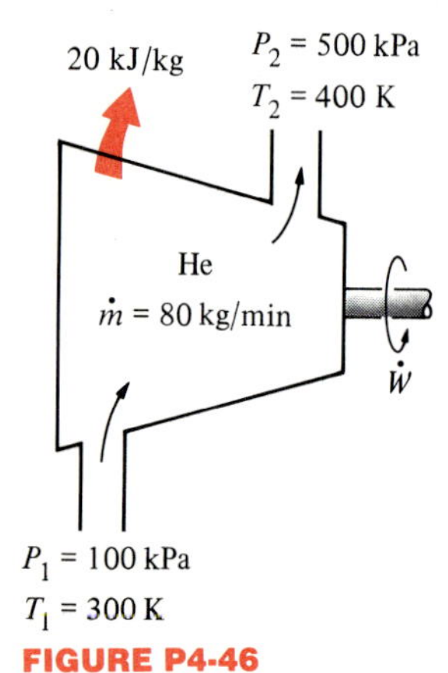

FIGURE P4-46

4-46 Helium is to be compressed from 100 kPa and 300 K to 500 kPa and 400 K. A heat loss of 20 kJ/kg occurs during the compression process. Neglecting kinetic energy changes, determine the power input required for a mass flow rate of 80 kg/min.

4-47 Carbon dioxide enters an adiabatic compressor at 100 kPa and 300 K at a rate of 0.5 kg/s and leaves at 600 kPa and 450 K. Neglecting kinetic energy changes, determine (*a*) the volume flow rate of the carbon dioxide at the compressor inlet and (*b*) the power input to the compressor. *Answers:* (*a*) 0.28 m^3/s, (*b*) 68.8 kW

Throttling Valves

4-48C Why are throttling devices commonly used in refrigeration and air-conditioning applications?

4-49C During a throttling process, the temprerature of a fluid drops from 30 to −20°C. Can this process occur adiabatically?

4-50C Would you expect the temperature of air to drop as it undergoes a steady-flow throttling process?

4-51C Would you expect the temperature of a liquid to change as it is throttled? How?

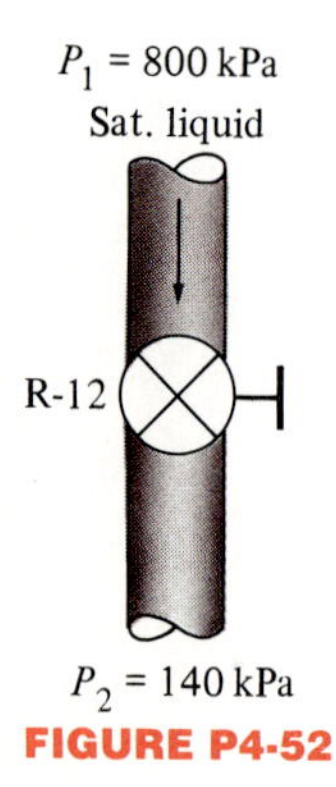

FIGURE P4-52

4-52 Refrigerant-12 is throttled from the saturated liquid state at 800 kPa to a pressure of 140 kPa. Determine the temperature drop during this process and the final specific volume of the refrigerant.
Answers: 54.65°C, 0.0375 m^3/kg

4-53 Refrigerant-12 at 800 kPa and 25°C is throttled to a temperature of −20°C. Determine the pressure and the internal energy of the refrigerant at the final state. *Answers:* 151 kPa, 55.3 kJ/kg

4-53E Refrigerant-12 at 120 psia and 90°F is throttled to a temperature of 0°F. Determine the pressure and the internal energy of the refrigerant at the final state. *Answers:* 23.85 psia, 26.6 Btu/lbm

4-54 A well-insulated valve is used to throttle steam from 10 MPa and 550°C to 8 MPa. Determine the final temperature of the steam.
Answer: 541.8°C

4-55 Air at 1 MPa and 25°C is throttled to the atmospheric pressure of 100 kPa. Determine the final temperature of the air.

4-55E Air at 150 psia and 80°F is throttled to the atmospheric pressure of 14.7 psia. Determine the final temperature of the air.

Mixing Chambers and Heat Exchangers

4-56C When two fluid streams are mixed in a mixing chamber, can the mixture temperature be lower than the temperature of both streams? How?

4-57C Consider a steady-flow mixing process. Under what conditions will the energy transported into the control volume by the incoming streams be equal to the energy transported out of it by the outgoing stream?

4-58C Consider a steady-flow heat exchanger involving two different fluid streams. Under what conditions will the amount of heat lost by one fluid be equal to the amount of heat gained by the other?

4-59 A hot-water stream at 70°C enters a mixing chamber with a mass flow rate of 0.6 kg/s where it is mixed with a stream of cold water at 20°C. If it is desired that the mixture leave the chamber at 42°C, determine the mass flow rate of the cold-water stream. Assume all the streams are at a pressure of 300 kPa. *Answer:* 0.76 kg/s

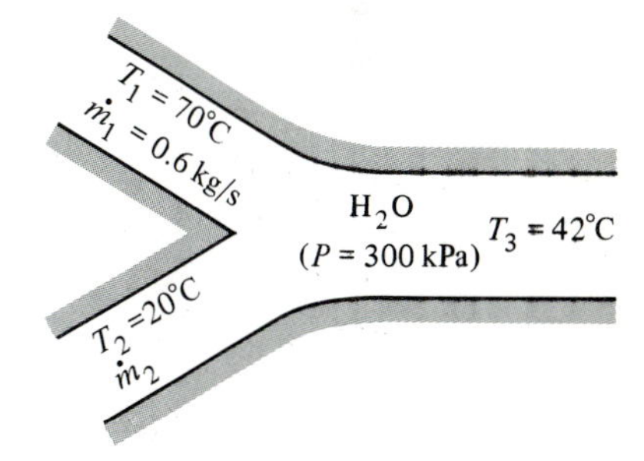

FIGURE P4-59

4-59E A hot-water stream at 180°F enters a mixing chamber with a mass flow rate of 1.2 lbm/s, where it is mixed with a stream of cold water at 60°F. If it is desired that the mixture leave the chamber at 110°F, determine the mass flow rate of the cold-water stream. Assume all the streams are at a pressure of 50 psia. *Answer:* 1.68 lbm/s

4-60 Liquid water at 200 kPa and 20°C is heated in a chamber by mixing it with superheated steam at 200 kPa and 300°C. Cold water enters the chamber at a rate of 2.5 kg/s. If the mixture leaves the mixing chamber at 60°C, determine the mass flow rate of the superheated steam required. *Answer:* 0.15 kg/s

4-61 In steam power plants, open feedwater heaters are frequently utilized to heat the feedwater by mixing it with steam bled off the turbine at some intermediate stage. Consider an open feedwater heater that operates at a pressure of 600 kPa. Feedwater at 50°C and 600 kPa is to be heated with superheated steam at 200°C and 600 kPa. In an ideal feedwater heater, the mixture leaves the heater as saturated liquid at the feedwater pressure. Determine the ratio of the mass flow rates of the feedwater and the superheated vapor for this case. *Answer:* 4.73

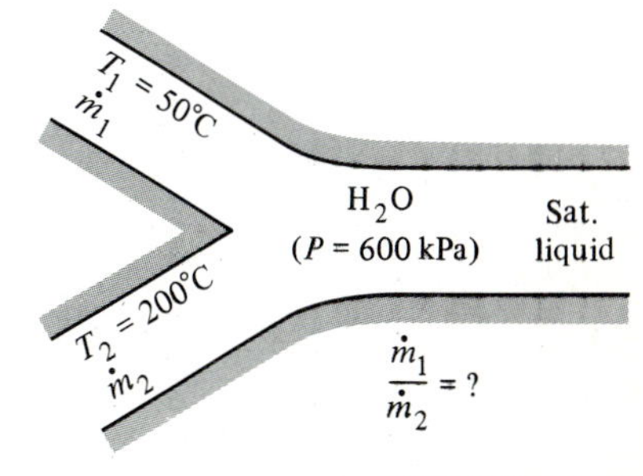

FIGURE P4-61

4-62 Water at 25°C and 300 kPa is heated in a chamber by mixing it with saturated water vapor at 300 kPa. If both streams enter the mixing chamber at the same mass flow rate, determine the temperature and the quality of the exiting stream.

4-62E Water at 50°F and 50 psia is heated in a chamber by mixing it with saturated water vapor at 50 psia. If both streams enter the mixing chamber at the same mass flow rate, determine the temperature and the quality of the exiting stream. *Answers:* 281°F, 0.374

4-63 A stream of refrigerant-12 at 1 MPa and 8°C is mixed with another stream at 1 MPa and 60°C. If the mass flow rate of the cold stream is twice that of the hot one, determine the temperature and the quality of the exit stream.

4-64 Refrigerant-12 at 1 MPa and 80°C is to be cooled to 1 MPa and 30°C in a condenser by air. The air enters at 100 kPa and 27°C with a volume flow rate of 800 m³/min and leaves at 95 kPa and 60°C. Determine the mass flow rate of the refrigerant. *Answer:* 183.1 kg/min

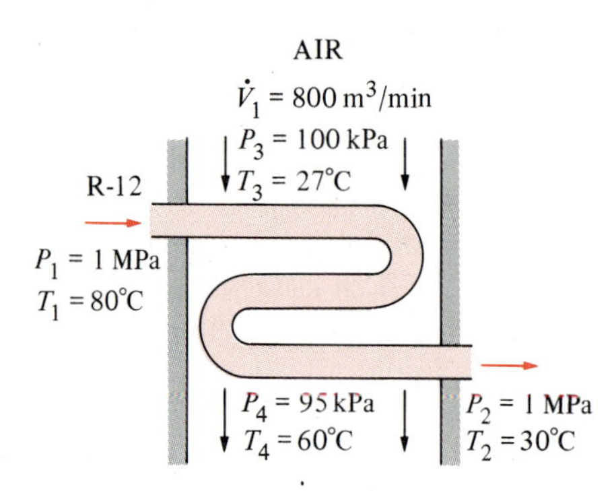

FIGURE P4-64

4-65 Air enters the evaporator section of a window air conditioner at 100 kPa and 27°C with a volume flow rate of 12 m³/min. Refrigerant-12 at 140 kPa with a quality of 30 percent enters the evaporator at a rate of 2 kg/min and leaves as saturated vapor at the same pressure. Determine (*a*) the exit temperature of the air and (*b*) the rate of heat transfer from the air. *Answers:* (*a*) 15.5°C, (*b*) 518.2 kJ/min

4-65E Air enters the evaporator section of a window air conditioner at 14.7 psia and 90°F with a volume flow rate of 200 ft³/min. Refrigerant-12 at 20 psia with a quality of 30 percent enters the evaporator at a rate of 4 lbm/min and leaves as saturated vapor at the same pressure. Determine (*a*) the exit temperature of the air and (*b*) the rate of heat transfer from the air. *Answers:* (*a*) 33.7°F, (*b*) 195 Btu/min

4-66 Refrigerant-12 at 800 kPa, 70°C, and 6 kg/min is cooled by water in a condenser until it exits as a saturated liquid at the same pressure. The cooling water enters the condenser at 300 kPa and 15°C and leaves at 25°C at the same pressure. Determine the mass flow rate of the cooling water required to cool the refrigerant. *Answer:* 23 kg/min

4-67 In a steam heating system, air is heated by being passed over some tubes through which steam flows steadily. Steam enters the heat exchanger at 200 kPa and 200°C at a rate of 8 kg/min, and it leaves at 180 kPa and 100°C. Air enters at 100 kPa and 25°C and leaves at 47°C. Determine the volume flow rate of air at the inlet. *Answer:* 758 m^3/min

4-67E In a steam heating system, air is heated by being passed over some tubes through which steam flows steadily. Steam enters the heat exchanger at 30 psia and 400°F at a rate of 15 lbm/min and leaves at 25 psia and 212°F. Air enters at 14.7 psia and 80°F and leaves at 130°F. Determine the volume flow rate of air at the inlet.

4-68 Steam enters the condenser of a steam power plant at 20 kPa and a quality of 95 percent with a mass flow rate of 20,000 kg/h. It is to be cooled by water from a nearby river by circulating the water through the tubes within the condenser. To prevent thermal pollution, the river water is not allowed to experience a temperature rise above 10°C. If the steam is to leave the condenser as saturated liquid at 20 kPa, determine the mass flow rate of cooling water required. *Answer:* 17,866 kg/min

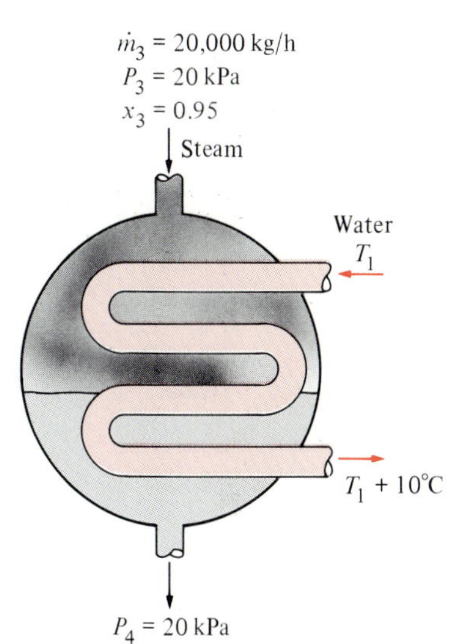

FIGURE P4-68

4-68E Steam enters the condenser of a steam power plant at 3 psia and a quality of 95 percent with a mass flow rate of 40,000 lbm/h. It is to be cooled by water from a nearby river by circulating the water through the tubes within the condenser. To prevent thermal pollution, the river water is not allowed to experience a temperature rise above 18°F. If the steam is to leave the condenser as saturated liquid at 3 psia, determine the mass flow rate of the cooling water required.

4-69 In large gas-turbine power plants, air is preheated by the exhaust gases in a heat exchanger called the *regenerator* before it enters the combustion chamber. Air enters the regenerator at 1 MPa and 550 K at a mass flow rate of 900 kg/min. Heat is transferred to the air at a rate of 3500 kJ/s. Exhaust gases enter the regenerator at 150 kPa and 800 K and leave at 130 kPa and 600 K. Treating the exhaust gases as air, determine (*a*) the exit temperature of the air and (*b*) the mass flow rate of the exhaust gases. *Answers:* (*a*) 769 K, (*b*) 16.3 kg/s

4-70 In large steam power plants, the feedwater is frequently heated in a closed feedwater heater by using steam extracted from the turbine at some stage. Steam enters the feedwater heater at 1 MPa and 200°C and leaves as saturated liquid at the same pressure. Feedwater enters the heater at 2.5 MPa and 50°C and leaves at 10°C below the exit temperature of the steam. Determine the ratio of the mass flow rates of the extracted steam and the feedwater.

Pipe and Duct Flow

4-71 A 5 m × 6 m × 8 m room is to be heated by an electric resistance heater placed in a short duct in the room. Initially, the room is at 15°C, and the local atmospheric pressure is 98 kPa. The room is losing heat steadily to the outside at a rate of 300 kJ/min. A 200-W fan circulates the air steadily through the duct and the electric heater at an average mass flow rate of 50 kg/min. The duct can be assumed to be adiabatic, and there is no air leaking in or out of the room. If it takes 15 min for the room air to reach an average temperature of 25°C, find (*a*) the power rating of the electric heater and (*b*) the temperature rise that the air experiences each time it passes through the heater.

4-72 A building with an internal volume of 400 m³ is to be heated by a 30-kW electric resistance heater placed in a duct inside the building. Initially, the air in the building is at 14°C, and the local atmospheric pressure is 95 kPa. The building is losing heat to the surroundings at a steady rate of 450 kJ/min. Air is forced to flow through the duct and the heater steadily by a 250-W fan, and it experiences a temperature rise of 5°C each time it passes through the duct, which may be assumed to be adiabatic.

(*a*) How long will it take for the air inside the building to reach an average temperature of 24°C?

(*b*) Determine the average mass flow rate of air through the duct.

Answers: (*a*) 146 s, (*b*) 6.02 kg/s

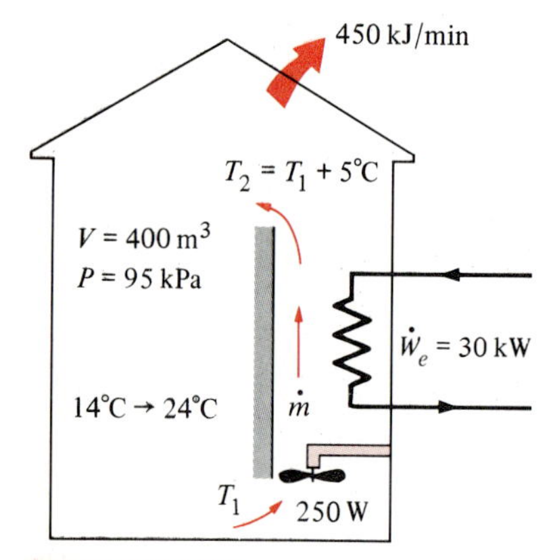

FIGURE P4-72

4-73 A house has an electric heating system that consists of a 500-W fan and an electric resistance heating element placed in a duct. Air flows steadily through the duct at a rate of 1 kg/s and experiences a temperature rise of 7°C. The rate of heat loss from the air in the duct is estimated to be 300 W. Determine the power rating of the electric resistance heating element. *Answer:* 6.835 kW

4-73E A house has an electric heating system that consists of a 500-W fan and an electric resistance heating element placed in a duct. Air flows through the duct at a rate of 2 lbm/s and experiences a temperature rise of 12°F. The rate of heat loss from the air in the duct is estimated to be 0.2 Btu/s. Determine the power rating of the electric heating element.

4-74 A hair dryer is basically a duct in which a few layers of electric resistors are placed. A small fan pulls the air in and forces it through the resistors where it is heated. Air enters a 1200-W hair dryer at 100 kPa and 22°C and leaves at 47°C. The cross-sectional area of the hair dryer at the exit is 25 cm². Neglecting the power consumed by the fan and the heat losses through the walls of the hair dryer, determine (*a*) the volume flow rate of air at the inlet and (*b*) the velocity of the air at the exit.
Answers: (*a*) 0.0404 m³/kg, (*b*) 17.5 m/s

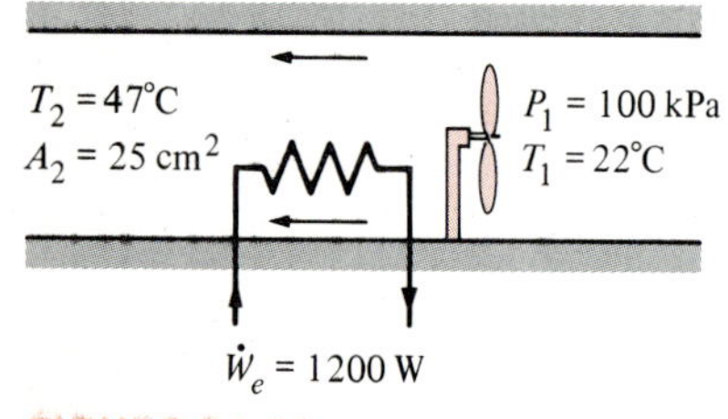

FIGURE P4-74

4-75 The ducts of an air heating system pass through an unheated area. As a result of heat losses, the temperature of the air in the duct drops by 3°C. If the mass flow rate of air is 150 kg/min, determine the rate of heat loss from the air to the cold environment.

4-76 Air enters the duct of an air-conditioning system at 105 kPa and 12°C at a volume flow rate of 12 m³/min. The diameter of the duct is 20 cm, and heat is transferred to the air in the duct from the surroundings at a rate of 2 kJ/s. Determine (*a*) the velocity of the air at the duct inlet and (*b*) the temperature of the air at the exit.
Answers: (*a*) 6.37 m/s, (*b*) 19.74°C

4-76E Air enters the duct of an air-conditioning system at 15 psia and 50°F at a volume flow rate of 450 ft^3/min. The diameter of the duct is 10 in, and heat is transferred to the air in the duct from the surroundings at a rate of 2 Btu/s. Determine (*a*) the velocity of the air at the duct inlet and (*b*) the temperature of the air at the exit.

4-77 It is proposed to have a water heater that consists of an insulated pipe of 3-cm diameter and an electric resistor inside. Cold water at 15°C enters the heating section steadily at a rate of 15 L/min. If water is to be heated to 45°C, determine (*a*) the power rating of the resistance heater and (*b*) the average velocity of the water in the pipe.

4-78 Water is heated in an insulated, constant-diameter tube by a 15-kW electric resistance heater. If the water enters the heater steadily at 12°C and leaves at 85°C, determine the mass flow rate of water.

4-78E Water is heated in an insulated, constant-diameter tube by a 15-kW electric resistance heater. If the water enters the heater steadily at 50°F and leaves at 150°F, determine the mass flow rate of water.

4-79 Water is to be pumped from a well to the top of a 200-m-tall building. There is a 15-kW pump available in the basement, and the water surface level in the well is 40 m below ground level. Neglecting any heat transfer and frictional effects, determine the maximum flow rate of water than can be maintained by this pump.

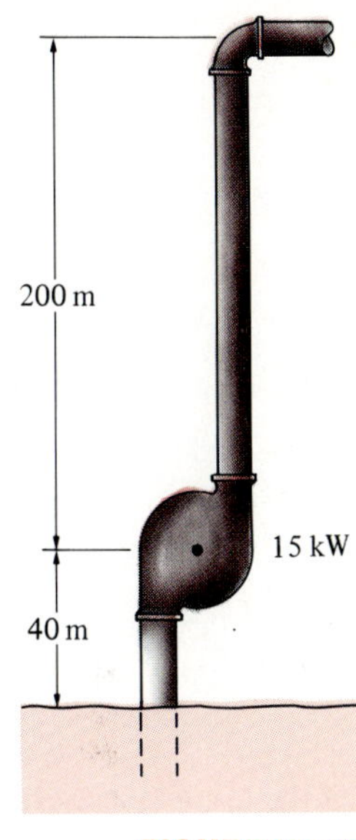

FIGURE P4-79

4-80 Steam enters a long, horizontal pipe with an inlet diameter of D_1 = 10 cm at 1 MPa and 300°C with a velocity of 6 m/s. Further downstream, the conditions are 800 kPa and 250°C, and the diameter is D_2 = 8 cm. Determine (*a*) the mass flow rate of the steam and (*b*) the rate of heat transfer. *Answers:* (*a*) 0.183 kg/s, (*b*) −18.5 kJ/s

4-81 A 5-kW pump is used to raise the elevation of a lake's water by 25 m from the free surface of the lake. The pipe inlet is 2 m below the free surface. The temperature of water increases by 0.05°C during this process as a result of the frictional effects. Neglecting any heat transfer and kinetic energy changes, determine the mass flow rate of the water. *Answer:* 11.0 kg/s

4-81E A 5-hp pump is used to raise the elevation of a lake's water by 75 ft from the free surface of the lake. The pipe inlet is 6 ft below the free surface. The temperature of water increases by 0.1°F during this process as a result of the frictional effects. Neglecting any heat transfer and kinetic energy changes, determine the mass flow rate of the water.

4-82 The free surface of the water in a well is 20 m below the ground level. This water is to be pumped steadily to an elevation of 30 m above the ground level. Neglecting any heat transfer, kinetic energy changes, and frictional effects, determine the power input to the pump required for a steady flow of water at a rate of 1.5 m^3/min. *Answer:* 12.3 kW

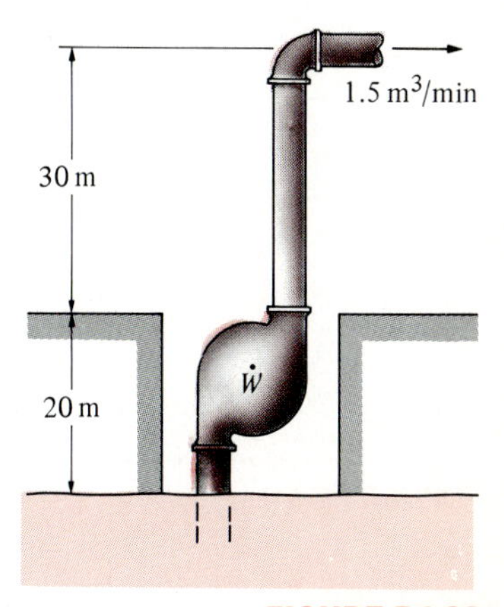

FIGURE P4-82

4-82E The free surface of water in a well is 60 ft below the ground level.

This water is to be pumped steadily to an elevation of 100 ft above the ground level. Neglecting any heat transfer, kinetic energy changes, and frictional effects, determine the power input to the pump required for a steady flow of water at a rate of 50 ft^3/min. *Answer:* 15.1 hp

Unsteady-Flow Processes

4-83C Does the amount of mass entering a control volume have to be equal to the amount of mass leaving during an unsteady-flow process? How about the amounts of energy entering and leaving?

4-84C Under what conditions can an unsteady-flow process be approximated as a uniform-flow process?

4-85C Can a uniform-flow system involve boundary work?

4-86C The valve of an initially evacuated, adiabatic rigid tank is opened, and air at 30°C flows in. When the pressure inside the tank reaches atmospheric pressure, the air temperature in the tank increases to 150°C. Explain what caused the temperature of the air to increase.

4-87C When a can which contains a refrigerant at 500 kPa and 25°C is slightly opened and refrigerant is allowed to escape, a layer of ice forms outside the can. Explain how that happens.

4-88C The valve of an insulated rigid vessel containing air at a high pressure is slightly opened, allowing some air to escape. Will the temperature of air in the tank change during this process? How?

Charging Processes

4-89 An insulated rigid tank is initially evacuated. A valve is opened, and atmospheric air at 100 kPa and 25°C enters the tank until the pressure in the tank reaches 100 kPa, at which point the valve is closed. Determine the final temperature of the air in the tank. Assume constant specific heats. *Answer:* 417 K

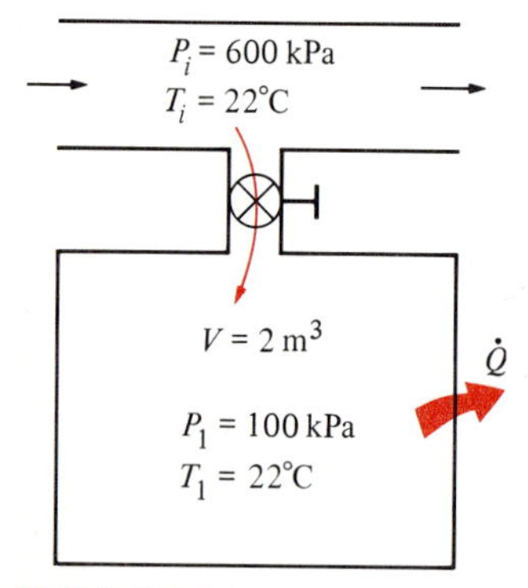

FIGURE P4-90

4-90 A 2-m^3 rigid tank initially contains air at 100 kPa and 22°C. The tank is connected to a supply line through a valve. Air is flowing in the supply line at 600 kPa and 22°C. The valve is opened, and air is allowed to enter the tank until the pressure in the tank reaches the line pressure, at which point the valve is closed. A thermometer placed in the tank indicates that the air temperature at the final state is 77°C. Determine (*a*) the mass of air that has entered the tank and (*b*) the amount of heat transfer. *Answers:* (*a*) 9.58 kg, (*b*) −339.4 kJ

4-90E A 70-ft^3 rigid tank initially contains air at 14.7 psia and 80°F. The tank is connected to a supply line through a valve. Air is flowing in the supply line at 75 psia and 80°F. The valve is opened, and air is allowed to enter the tank until the pressure in the tank reaches the line pressure, at which point the valve is closed. A thermometer placed in the tank indicates that the air temperature at the final state is 140°F. Determine (*a*) the mass of air that has entered the tank and (*b*) the amount of heat transfer. *Answers:* (*a*) 18.48 lbm, (*b*) −441 Btu

4-91 A 0.2-m^3 rigid tank initially contains refrigerant-12 at 8°C. At this state, 60 percent of the mass is in the vapor phase, and the rest is in the liquid phase. The tank is connected by a valve to a supply line where refrigerant at 1.2 MPa and 100°C flows steadily. Now the valve is opened slightly, and the refrigerant is allowed to enter the tank. When the pressure in the tank reaches 800 kPa, the entire refrigerant in the tank exists in the vapor phase only. At this point the valve is closed. Determine (*a*) the final temperature in the tank, (*b*) the mass of refrigerant that has entered the tank, and (*c*) the heat transfer between the system and the surroundings.

4-92 A 0.1-m^3 rigid tank initially contains saturated water vapor at 120°C. The tank is connected by a valve to a supply line that carries steam at 1 MPa and 300°C. Now the valve is opened, and steam is allowed to enter the tank. Heat transfer takes place with the surroundings such that the temperature in the tank remains constant at 120°C at all times. The valve is closed when it is observed that one-half of the volume of the tank is occupied by liquid water. Determine (*a*) the final pressure in the tank, (*b*) the amount of steam that has entered the tank, and (*c*) the amount of heat transfer.

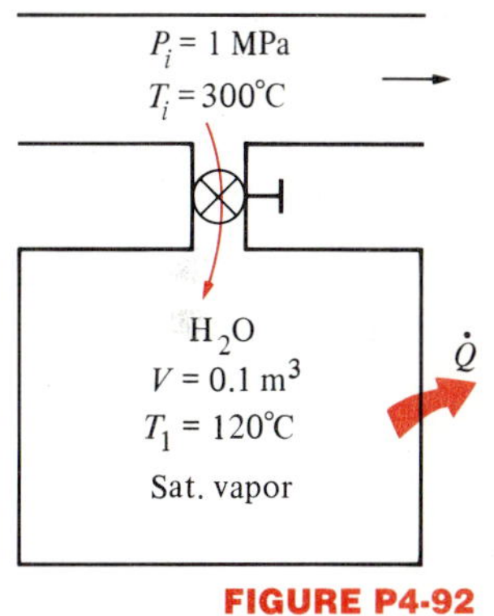

FIGURE P4-92

4-92E A 4-ft^3 rigid tank initially contains saturated water vapor at 250°F. The tank is connected by a valve to a supply line that carries steam at 160 psia and 400°F. Now the valve is opened, and steam is allowed to enter the tank. Heat transfer takes place with the surroundings such that the temperature in the tank remains constant at 250°F at all times. The valve is closed when it is observed that one-half of the volume of the tank is occupied by liquid water. Find (*a*) the final pressure in the tank, (*b*) the amount of steam that has entered the tank, and (*c*) the amount of heat transfer. *Answers:* (*a*) 29.82 psia, (*b*) 117.5 lbm, (*c*) −117,539 Btu

4-93 A balloon initially contains 65 m^3 helium gas at atmospheric conditions of 100 kPa and 22°C. The balloon is connected by a valve to a large reservoir that supplies helium gas at 150 kPa and 25°C. Now the valve is opened, and helium is allowed to enter the balloon until the pressure equilibrium with the helium at the supply line is reached. The material of the balloon is such that its volume increases linearly with pressure. If no heat transfer takes place during this process, determine the final temperature in the balloon. *Answer:* 318.6 K

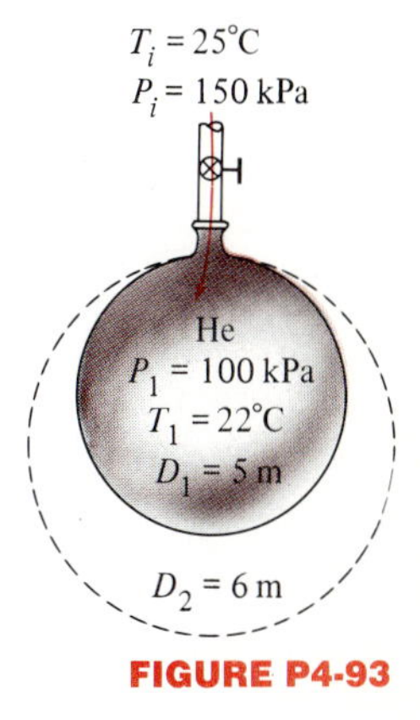

FIGURE P4-93

4-94 A vertical piston-cylinder device initially contains 0.01 m^3 of steam at 200°C. The mass of the frictionless piston is such that it maintains a constant pressure of 500 kPa inside. Now steam at 1 MPa and 350°C is allowed to enter the cylinder from a supply line until the volume inside doubles. Neglecting any heat transfer that may have taken place during

the process, determine (*a*) the final temperature of the steam in the cylinder and (*b*) the amount of mass that has entered.
Answers: (*a*) 262.6°C, (*b*) 0.0176 kg

4-95 An insulated, vertical piston-cylinder device initially contains 5 kg of water, 4 kg of which is in the vapor phase. The mass of the piston is such that it maintains a constant pressure of 300 kPa inside the cylinder. Now steam at 1 MPa and 400°C is allowed to enter the cylinder from a supply line until all the liquid in the cylinder has vaporized. Determine (*a*) the final temperature in the cylinder and (*b*) the mass of the steam that has entered. *Answers:* (*a*) 133.6°C, (*b*) 4.0 kg

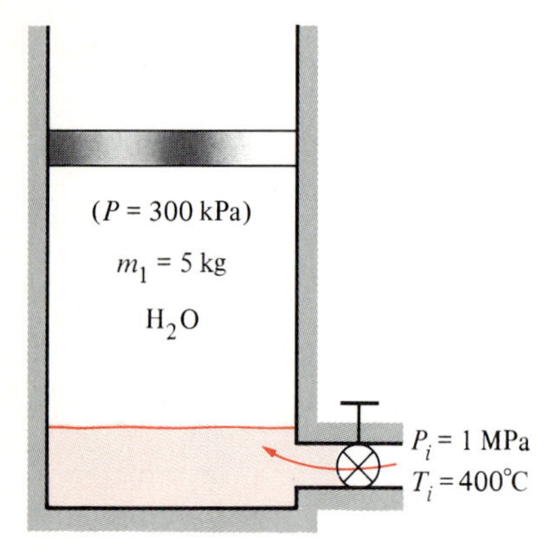

4-95E An insulated, vertical piston-cylinder device initially contains 10 lbm of water, 8 lbm of which is in the vapor phase. The mass of the piston is such that it maintains a constant pressure of 50 psia inside the cylinder. Now steam at 120 psia and 600°F is allowed to enter the cylinder from a supply line until all the liquid in the cylinder has vaporized. Determine (*a*) the final temperature in the cylinder and (*b*) the mass of the steam that has entered.

4-96 A 0.1-m^3 rigid tank initially contains refrigerant-12 at 1 MPa and 100 percent quality. The tank is connected by a valve to a supply line that carries refrigerant at 1.4 MPa and 30°C. Now the valve is opened, and the refrigerant is allowed to enter the tank. The valve is closed when it is observed that the tank contains saturated liquid at 1.2 MPa. Determine (*a*) the mass of the refrigerant that has entered the tank and (*b*) the amount of heat transfer. *Answers:* (*a*) 115.7 kg, (*b*) −15,424 kJ

4-97 An insulated, vertical piston-cylinder device initially contains 0.2 m^3 of air at 200 kPa and 22°C. At this state, a linear spring touches the piston but exerts no force on it. The cylinder is connected by a valve to a line that supplies air at 800 kPa and 22°C. The valve is opened, and air from the high-pressure line is allowed to enter the cylinder. The valve is turned off when the pressure inside the cylinder reaches 600 kPa. If the enclosed volume inside the cylinder doubles during this process, determine (*a*) the mass of air that entered the cylinder and (*b*) the final temperature of the air inside the cylinder.

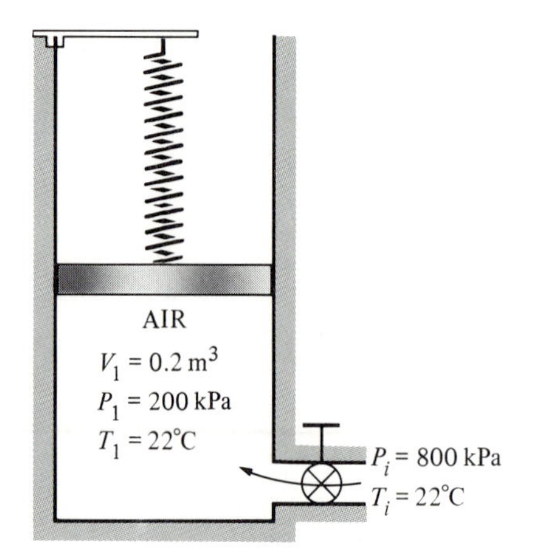

FIGURE P4-97

4-98 A 0.4-m³ rigid tank is filled with saturated liquid water at 200°C. A valve at the bottom of the tank is opened, and liquid is withdrawn from the tank. Heat is transferred to the water such that the temperature in the tank remains constant. Determine the amount of heat that must be transferred by the time one-half of the total mass has been withdrawn.

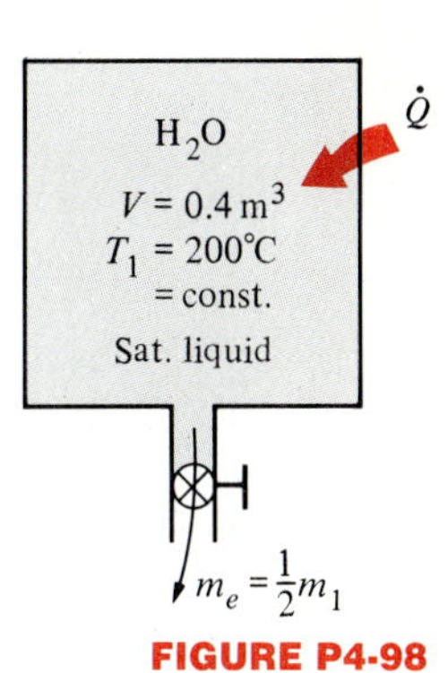

FIGURE P4-98

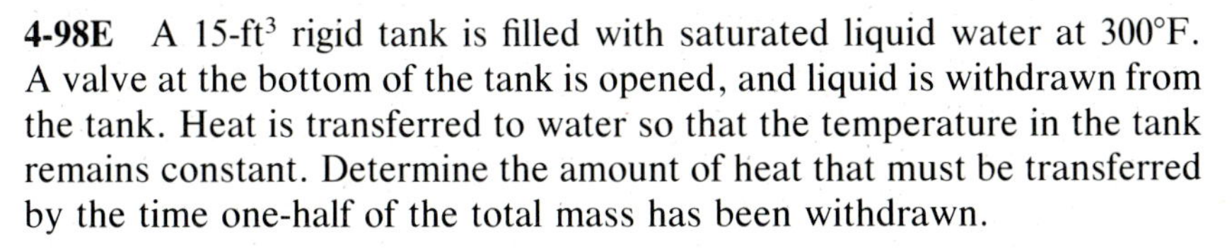

4-98E A 15-ft³ rigid tank is filled with saturated liquid water at 300°F. A valve at the bottom of the tank is opened, and liquid is withdrawn from the tank. Heat is transferred to water so that the temperature in the tank remains constant. Determine the amount of heat that must be transferred by the time one-half of the total mass has been withdrawn.

4-99 A 0.1-m³ rigid tank contains saturated refrigerant-12 at 800 kPa. Initially, 20 percent of the volume is occupied by liquid and the rest by vapor. A valve at the bottom of the tank is now opened, and liquid is withdrawn from the tank. Heat is transferred to the refrigerant such that the pressure inside the tank remains constant. The valve is closed when no liquid is left in the tank and vapor starts to come out. Determine the total heat transfer for this process. *Answer:* 121.8 kJ

4-100 A 0.1-m³ rigid tank contains saturated refrigerant-12 at 800 kPa. Initially, 20 percent of the volume is occupied by liquid and the rest by vapor. A valve at the top of the tank is now opened, and vapor is allowed to escape slowly from the tank. Heat is transferred to the refrigerant such that the pressure inside the tank remains constant. The valve is closed when the last drop of liquid in the tank is vaporized. Determine the total heat transfer for this process. *Answer:* 3418 kJ

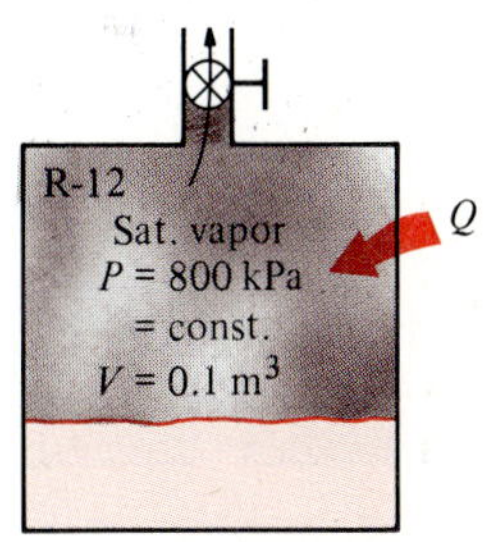

FIGURE P4-100

4-100E A 4-ft³ rigid tank contains saturated refrigerant-12 at 100 psia. Initially, 20 percent of the volume is occupied by liquid and the rest by vapor. A valve at the top of the tank is now opened, and vapor is allowed to escape slowly from the tank. Heat is transferred to the refrigerant such that the pressure inside the tank remains constant. The valve is closed when the last drop of liquid in the tank is vaporized. Determine the total heat transfer for this process. *Answer:* 40,108 Btu

4-101 A 0.2-m³ rigid tank equipped with a pressure regulator contains steam at 2 MPa and 300°C. The steam in the tank is now heated. The regulator keeps the steam pressure constant by letting out some steam, but the temperature inside rises. Determine the amount of heat transferred when the steam temperature reaches 500°C.

4-102 A 5-L pressure cooker has an operating pressure of 200 kPa. Initially, 20 percent of the volume is occupied by liquid and the rest by vapor. The cooker is placed on a heating unit which supplies heat to the water inside at a rate of 400 W. Determine how long it will take for the liquid in the pressure cooker to be depleted (i.e., the cooker contains only saturated vapor at the final state). *Answer:* 1.44 h

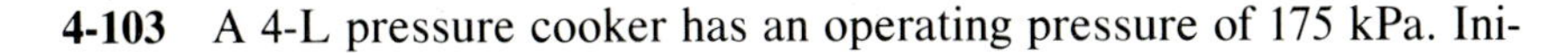

4-103 A 4-L pressure cooker has an operating pressure of 175 kPa. Initially, one-half of the volume is filled with liquid and the other half with vapor. If it is desired that the pressure cooker not run out of liquid water for 1 h, determine the highest rate of heat transfer allowed.

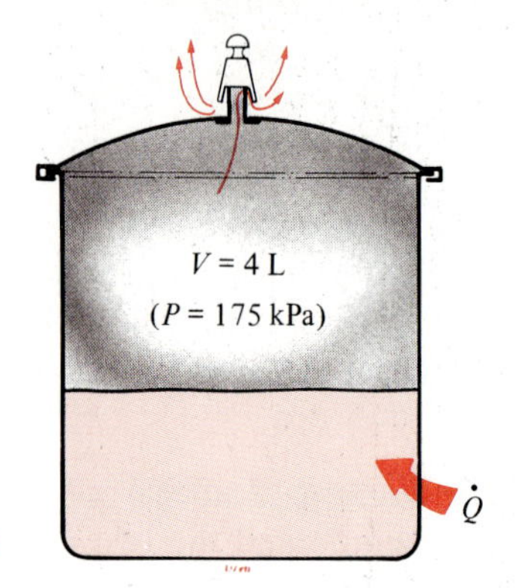

FIGURE P4-103

4-103E A 0.15-ft³ pressure cooker has an operating pressure of 30 psia. Initially, one-half of the volume is filled with liquid and the other half with vapor. If it is desired that the pressure cooker not run out of liquid water for 1 h, determine the highest rate of heat transfer allowed.

4-104 An insulated 0.2-m³ tank contains helium at 1200 kPa and 50°C. A valve is now opened, allowing some helium to escape. The valve is closed when one-half of the initial mass has escaped. Determine the final temperature and pressure in the tank. *Answers:* 205 K, 383 kPa

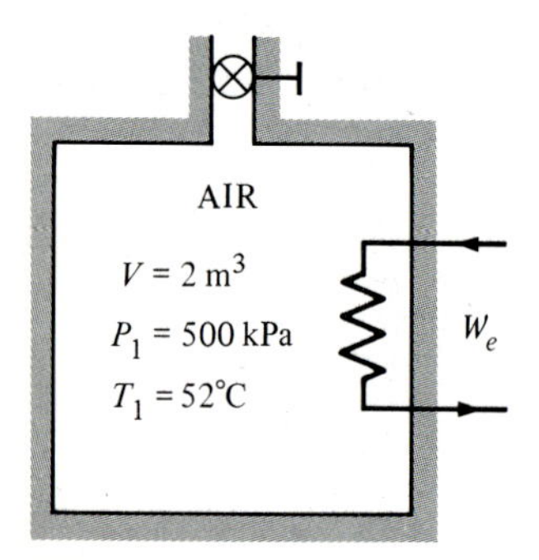

FIGURE P4-105

4-105 An insulated 2-m³ rigid tank contains air at 500 kPa and 52°C. A valve connected to the tank is now opened, and air is allowed to escape until the pressure inside drops to 200 kPa. The air temperature during this process is maintained constant by an electric resistance heater placed in the tank. Determine the electrical work done during this process. *Answer:* 600 kJ

4-105E An insulated 60-ft³ rigid tank contains air at 75 psia and 120°F. A valve connected to the tank is now opened, and air is allowed to escape until the pressure inside drops to 30 psia. The air temperature during this process is maintained constant by an electric resistance heater placed in the tank. Determine the electrical work done during this process.

4-106 A vertical piston-cylinder device initially contains 0.1 m³ of air at 27°C. The mass of the piston is such that it maintains a constant pressure of 400 kPa inside. Now a valve connected to the cylinder is opened, and air is allowed to escape until the volume inside the cylinder is decreased by one-half. Heat transfer takes place during the process, so that the temperature of the air in the cylinder remains constant. Determine (*a*) the amount of air that has left the cylinder and (*b*) the amount of heat transfer. *Answers:* (*a*) 0.232 kg, (*b*) 0

4-107 A balloon initially contains 10 m³ of helium gas at 150 kPa and 27°C. Now a valve is opened, and helium is allowed to escape slowly until the pressure inside drops to 100 kPa, at which point the valve is closed. During this process the volume of the balloon decreases by 15 percent. The balloon material is such that the volume of the balloon changes linearly with pressure in this range. If the heat transfer during this process is negligible, find (*a*) the final temperature of the helium in the balloon and (*b*) the amount of helium that has escaped.

4-108 A vertical piston-cylinder device initially contains 0.5 m³ of steam at 1 MPa and 200°C. A linear spring at this point applies full force to the piston. A valve connected to the cylinder is now opened, and steam is allowed to escape. As the piston moves down, the spring unwinds, and at the final state the pressure drops to 800 kPa and the volume to 0.3 m³. If at the final state the cylinder contains saturated vapor only, determine (*a*) the initial and final masses in the cylinder and (*b*) the amount and direction of any heat transfer.

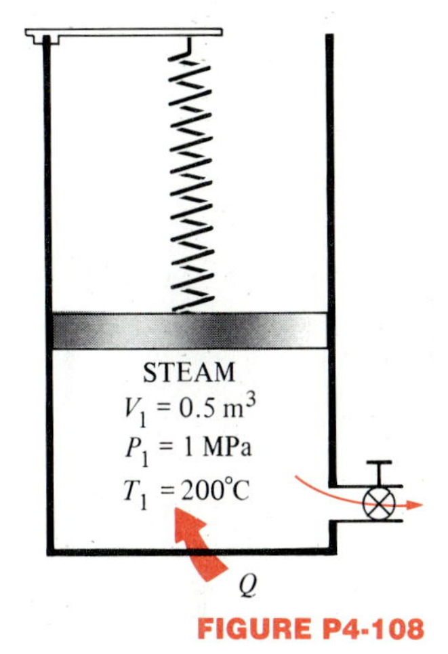

FIGURE P4-108

4-108E A vertical piston-cylinder device initially contains 15 ft³ of steam at 120 psia and 500°F. A linear spring at this point applies full force to the piston. A valve connected to the cylinder is now opened, and steam is allowed to escape. As the piston moves down, the spring unwinds, and at the final state the pressure drops to 100 psia and the volume to 10 ft³. If at the final state the cylinder contains saturated vapor only, determine (*a*) the initial and final masses in the cylinder and (*b*) the amount and direction of any heat transfer.

4-109 Pressurized air stored in a 10,000-m³ cave at 500 kPa and 400 K is to be used to drive a turbine at times of high demand for electric power. If the turbine exit conditions are 100 kPa and 300 K, determine the amount of work delivered by the turbine when the air pressure in the cave drops to 300 kPa. Assume both the cave and the turbine to be adiabatic. *Answer:* 980.8 kJ

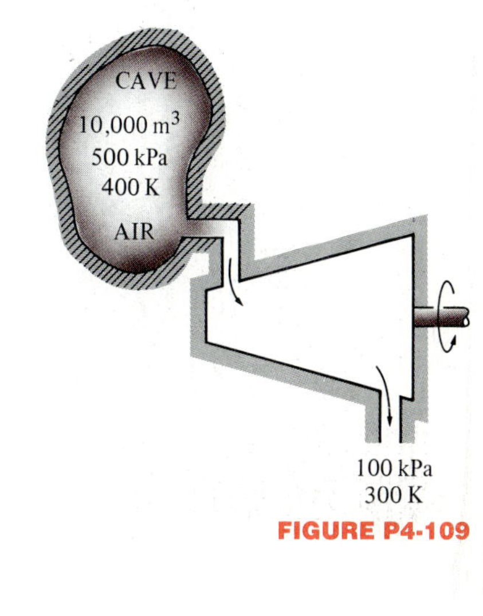

FIGURE P4-109

4-110 A vertical piston-cylinder device initially contains 0.4 m³ of steam at 200°C. The mass of the piston is such that it maintains a constant pressure of 300 kPa. Now a valve is opened, and steam is allowed to escape. Heat transfer takes place during the process so that the temperature inside remains constant. If the final volume is 0.1 m³, determine (*a*) the amount of steam that has escaped and (*b*) the amount of heat transfer. *Answers:* (*a*) 0.419 kg, (*b*) 0

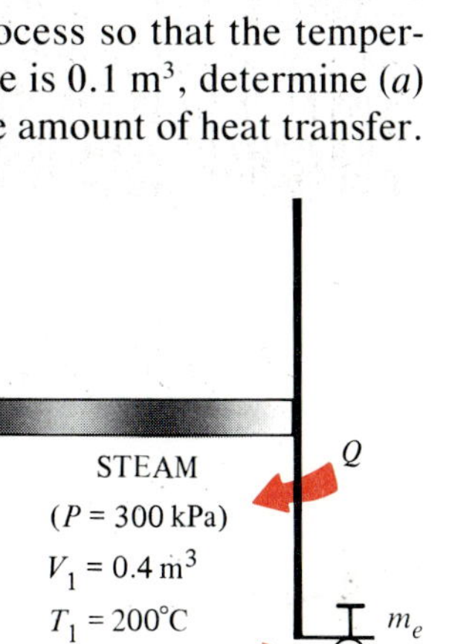

FIGURE P4-110

4-110E A vertical piston-cylinder device initially contains 1.2 ft^3 of steam at 400°F. The mass of the piston is such that it maintains a constant pressure of 60 psia. Now a valve is opened, and steam is allowed to escape. Heat transfer takes place during the process so that the temperature inside remains constant. If the final volume is 0.3 ft^3, calculate (*a*) the amount of steam that has escaped and (*b*) the amount of heat transfer. *Answers:* (*a*) 0.108 lbm, (*b*) 0

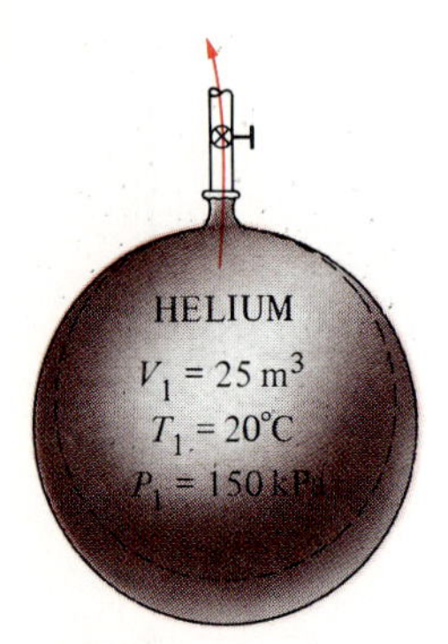

FIGURE P4-111

4-111 A spherical balloon initially contains 25 m^3 of helium gas at 20°C and 150 kPa. A valve is now opened, and the helium is allowed to escape slowly. The valve is closed when the pressure inside the balloon drops to the atmospheric pressure of 100 kPa. The elasticity of the balloon material is such that the pressure inside the balloon during the process varies with volume according to the relation $P = a + bV$, where $a = -100$ kPa and b is a constant. Disregarding any heat transfer, determine (*a*) the final temperature in the balloon and (*b*) the mass of helium that has escaped. *Answers:* (*a*) 183.4 K, (*b*) 0.911 kg

4-112 Write a computer program to solve Prob. 4-111, using a stepwise approach. Use (*a*) 5, (*b*) 20, and (*c*) 50 increments for pressure between the initial value of 150 kPa and the final value of 100 kPa. Take the starting point of the first step as the initial state of the helium (150 kPa, 20°C, and 25 m^3). The starting point of the second step is the state of the helium at the end of the first step, and so on. Compare your results with those obtained by using the uniform-flow approximation (i.e., a one-step solution).

The Second Law of Thermodynamics

CHAPTER

5

To this point we have focused our attention on the first law of thermodynamics, which requires that energy be conserved during a process. In this chapter we introduce the second law of thermodynamics, which asserts that processes occur in a certain direction and that energy has quality as well as quantity. A process cannot take place unless it satisfies both the first and second laws of thermodynamics. In this chapter, the thermal energy reservoirs, reversible and irreversible processes, heat engines, refrigerators, and heat pumps are introduced first. Various statements of the second law are followed by a discussion of perpetual-motion machines and the absolute thermodynamic temperature scale. The Carnot cycle is introduced next, and the Carnot principles are examined. Finally, idealized Carnot heat engines, refrigerators, and heat pumps are discussed.

5-1 ■ INTRODUCTION TO THE SECOND LAW OF THERMODYNAMICS

In the preceding two chapters, we applied the *first law of thermodynamics*, or the *conservation of energy principle*, to processes involving closed and open systems. As pointed out repeatedly in those chapters, energy is a conserved property, and no process is known to have taken place in violation of the first law of thermodynamics. Therefore, it is reasonable to conclude that a process must satisfy the first law to occur. However, as explained below, satisfying the first law alone does not ensure that the process will actually take place.

FIGURE 5-1
A cup of hot coffee does not get hotter in a cooler room.

It is a common experience that a cup of hot coffee left in a cooler room eventually cools off (Fig. 5-1). This process satisfies the first law of thermodynamics since the amount of energy lost by the coffee is equal to the amount gained by the surrounding air. Now let us consider the reverse process—the hot coffee getting even hotter in a cooler room as a result of heat transfer from the room air. We all know that this process never takes place. Yet, doing so would not violate the first law as long as the amount of energy lost by the air is equal to the amount gained by the coffee.

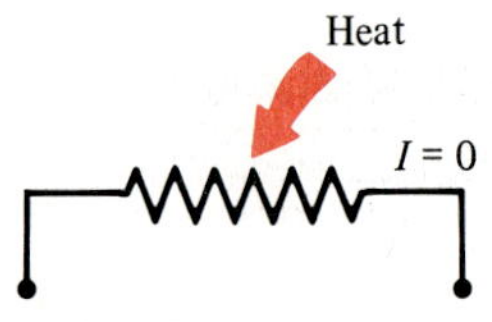

FIGURE 5-2
Transferring heat to a wire will not generate electricity.

As another familiar example, consider the heating of a room by the passage of current through an electric resistor (Fig. 5-2). Again, the first law dictates that the amount of electric energy supplied to the resistance wires be equal to the amount of energy transferred to the room air as heat. Now let us attempt to reverse this process. It will come as no surprise that transferring some heat to the wires will not cause an equivalent amount of electric energy to be generated in the wires, even though doing so would not violate the first law.

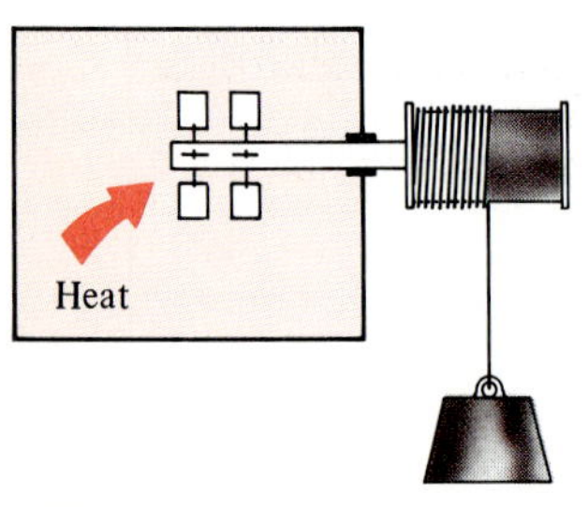

FIGURE 5-3
Transferring heat to a paddle wheel will not cause it to rotate.

Finally, consider a paddle-wheel mechanism that is operated by the fall of a mass (Fig. 5-3). The paddle wheel rotates as the mass falls and stirs a fluid within an insulated container. As a result, the potential energy of the mass decreases, and the internal energy of the fluid increases in accordance with the conservation of energy principle. However, the reverse process, raising the mass by transferring heat from the fluid to the paddle wheel, does not occur in nature, although doing so would not violate the first law of thermodynamics.

FIGURE 5-4
Processes occur in a certain direction, and not in the reverse direction.

It is clear from the above that processes proceed in a *certain direction* and not in the reverse direction (Fig. 5-4). The first law places no restriction on the direction of a process, but satisfying the first law does not ensure that that process will actually occur. This inadequacy of the first law to identify whether a process can take place is remedied by introducing another general principle, the *second law of thermodynamics*. We show later in this chapter that the reverse processes discussed above violate the second law of thermodynamics. This violation is easily detected with the help of a property, called *entropy*, defined in the next chapter. *A process will not occur unless it satisfies both the first and the second laws of thermodynamics* (Fig. 5-5).

There are numerous valid statements of the second law of thermodynamics. Two such statements are presented and discussed later in this chapter in relation to some engineering devices that operate on cycles.

The use of the second law of thermodynamics is not limited to identifying the direction of processes, however. The second law also asserts that energy has *quality* as well as quantity. The first law is concerned with the quantity of energy and the transformations of energy from one form to another with no regard to its quality. Preserving the quality of energy is a major concern to engineers, and the second law provides the necessary means to determine the quality as well as the degree of degradation of energy during a process. As discussed later in this chapter, more of high-temperature energy can be converted to work, and thus it has a higher quality than the same amount of energy at a lower temperature.

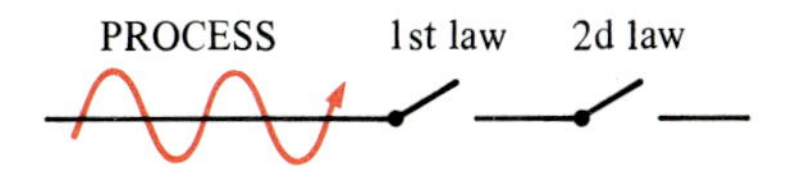

FIGURE 5-5
A process must satisfy both the first and second laws of thermodynamics to continue.

The second law of thermodynamics is also used in determining the *theoretical limits* for the performance of commonly used engineering systems, such as heat engines and refrigerators, as well as predicting the *degree of completion* of chemical reactions.

5-2 ■ THERMAL ENERGY RESERVOIRS

In the development of the second law of thermodynamics, it is very convenient to have a hypothetical body with a relatively large *thermal energy capacity* (mass × specific heat) that can supply or absorb finite amounts of energy as heat without undergoing any change in temperature. Such a body is called a **thermal energy reservoir**, or just a **reservoir**. In practice, large bodies of water such as oceans, lakes, and rivers as well as the atmospheric air can be modeled accurately as thermal energy reservoirs because of their large thermal energy storage capabilities or thermal masses (Fig. 5-6). The *atmosphere,* for example, does not warm up as a result of heat losses from residential buildings in winter. Likewise, megajoules of waste energy dumped in large rivers by power plants do not cause any significant change in water temperature.

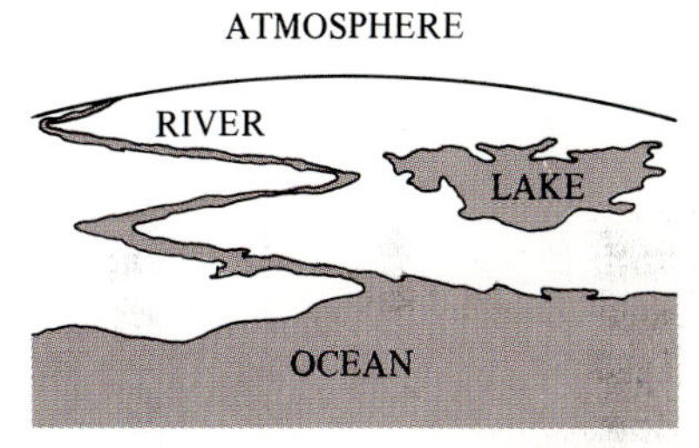

FIGURE 5-6
Bodies with relatively large thermal masses can be modeled as thermal energy reservoirs.

A *two-phase system* can be modeled as a reservoir also since it can absorb and release large quantities of heat while remaining at constant temperature. Another familiar example of a thermal energy reservoir is the *industrial furnace*. The temperatures of most furnaces are carefully controlled, and they are capable of supplying large quantities of thermal energy as heat in an essentially isothermal manner. Therefore, they can be modeled as reservoirs.

A body does not actually have to be very large to be considered a reservoir. Any physical body whose thermal energy capacity is large relative to the amount of energy it supplies or absorbs can be modeled as one. The air in a room, for example, can be treated as a reservoir in the analysis of the heat dissipation from a TV set in the room, since the amount of heat transfer from the TV set to the room air is not large enough to have a noticeable effect on the room air temperature (Fig. 5-7).

FIGURE 5-7
A body does not have to be large to be modeled as a thermal energy reservoir.

A reservoir that supplies energy in the form of heat is called a **source**, and one that absorbs energy in the form of heat is called a **sink** (Fig.

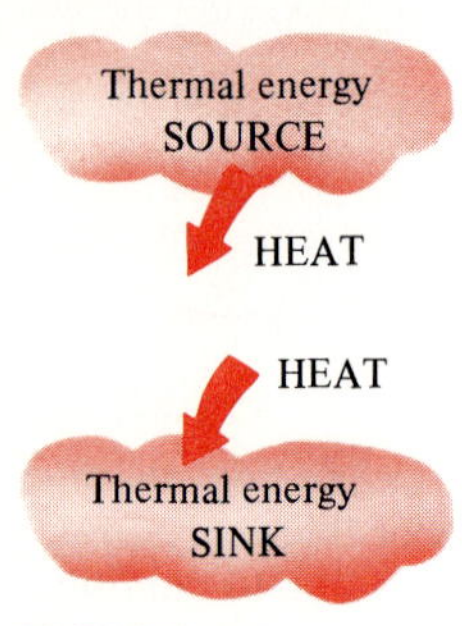

FIGURE 5-8

A source supplies energy in the form of heat, and a sink absorbs it.

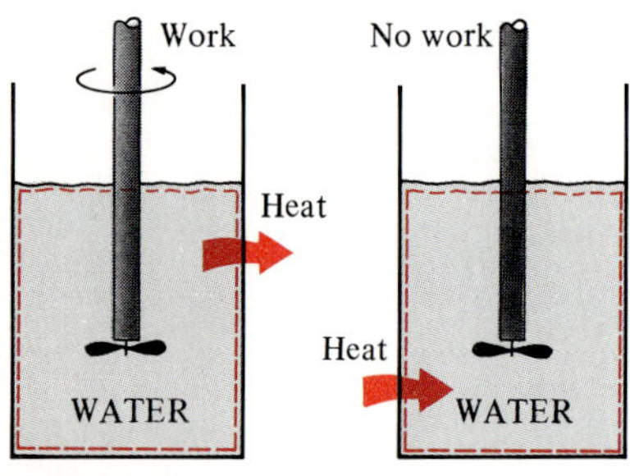

FIGURE 5-9

Work can be converted to heat directly and completely, but the reverse is not true.

5-8). Thermal energy reservoirs are often referred to as **heat reservoirs** since they supply or absorb energy in the form of heat.

Heat transfer from industrial sources to the environment is of major concern to environmentalists as well as to engineers. Irresponsible management of waste energy can significantly increase the temperature of portions of the environment, causing what is called *thermal pollution*. If it is not carefully controlled, thermal pollution can seriously disrupt marine life in lakes and rivers. However, by careful design and management, the waste energy dumped into large bodies of water can be used to significantly improve the quality of marine life by keeping the local temperature increases within safe and desirable levels.

5-3 ■ HEAT ENGINES

As pointed out in Sec. 5-1, work can easily be converted to other forms of energy, but converting other forms of energy to work is not that easy. The mechanical work done by the shaft shown in Fig. 5-9, for example, is first converted to the internal energy of the water. This energy may then leave the water as heat. We know from experience that any attempt to reverse this process will fail. That is, transferring heat to the water will not cause the shaft to rotate. From this and other observations we conclude that work can be converted to heat directly and completely, but converting heat to work requires the use of some special devices. These devices are called **heat engines**.

Heat engines differ considerably from one another, but all can be characterized by the following (Fig. 5-10):

1 They receive heat from a high-temperature source (solar energy, oil furnace, nuclear reactor, etc.).

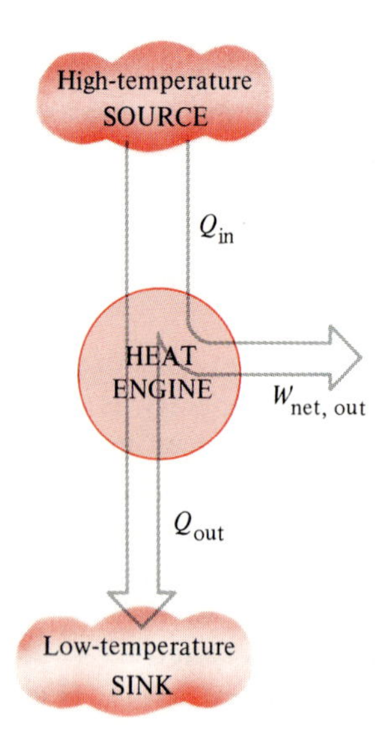

FIGURE 5-10

Part of the heat received by a heat engine is converted to work, while the rest is rejected to a sink.

2 They convert part of this heat to work (usually in the form of a rotating shaft).

3 They reject the remaining waste heat to a low-temperature sink (the atmosphere, rivers, etc.).

4 They operate on a cycle.

Heat engines and other cyclic devices usually involve a fluid to and from which heat is transferred while undergoing a cycle. This fluid is called the **working fluid**.

The term *heat engine* is often used in a broader sense to include work-producing devices that do not operate in a thermodynamic cycle. Engines that involve internal combustion such as gas turbines and car engines fall into this category. These devices operate in a mechanical cycle but not in a thermodynamic cycle since the working fluid (the combustion gases) does not undergo a complete cycle. Instead of being cooled to the initial temperature, the exhaust gases are purged and replaced by fresh air-and-fuel mixture at the end of the cycle.

The work-producing device that best fits into the definition of a heat engine is the *steam power plant* which is an external-combustion engine. That is, the combustion process takes place outside the engine, and the thermal energy released during this process is transferred to the steam as heat. The schematic of a basic steam power plant is shown in Fig. 5-11. This is a rather simplified diagram, and the discussion of actual steam power plants with all their complexities is left to Chap. 9. The various quantities shown on this figure are as follows:

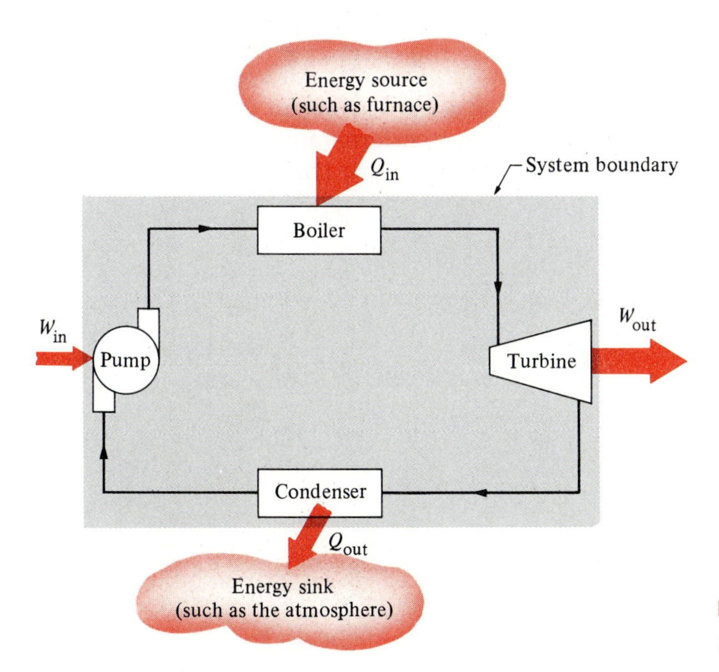

FIGURE 5-11
Schematic of a steam power plant.

Q_{in} = amount of heat supplied to steam in boiler from a high-temperature source (furnace)
Q_{out} = amount of heat rejected from steam in condenser to a low-temperature sink (the atmosphere, a river, etc.)
W_{out} = amount of work delivered by steam as it expands in turbine
W_{in} = amount of work required to compress water to boiler pressure

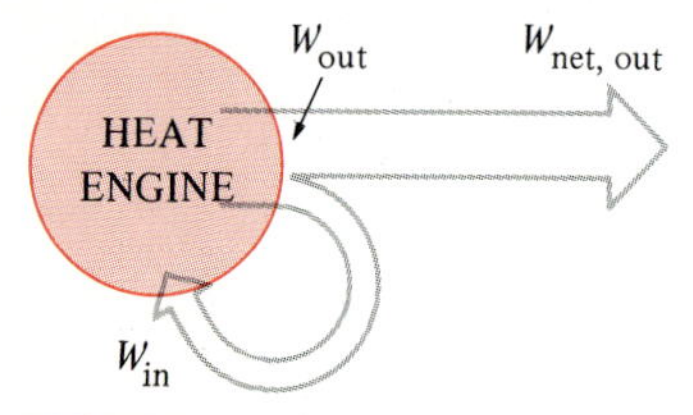

FIGURE 5-12

A portion of the work output of a heat engine is consumed internally to maintain continuous operation.

Notice that the directions of the heat and work interactions are indicated by the subscripts *in* and *out*. Therefore, all four quantities described above are always *positive*.

The net work output of this power plant is simply the difference between the total work output of the plant and the total work input (Fig. 5-12):

$$W_{net,out} = W_{out} - W_{in} \quad (kJ) \tag{5-1}$$

The net work can also be determined from the heat transfer data alone. The four components of the steam power plant involve mass flow in and out, and therefore they should be treated as open systems. These components, together with the connecting pipes, however, always contain the same fluid (not counting the steam that may leak out, of course). No mass enters or leaves this combination system, which is indicated by the shaded area on Fig. 5-11; thus it can be analyzed as a closed system. Recall that for a closed system undergoing a cycle, the change in internal energy ΔU is zero, and therefore the net work output of the system is also equal to the net heat transfer to the system:

$$W_{net,out} = Q_{in} - Q_{out} \quad (kJ) \tag{5-2}$$

Thermal Efficiency

In Eq. 5-2, Q_{out} represents the magnitude of the energy wasted in order to complete the cycle. But Q_{out} is never zero, thus the net work output of a heat engine is always less than the amount of heat input. That is, only part of the heat transferred to the heat engine is converted to work. *The fraction of the heat input that is converted to net work output is a measure of the performance of a heat engine and is called the* **thermal efficiency** η_{th} (Fig. 5-13).

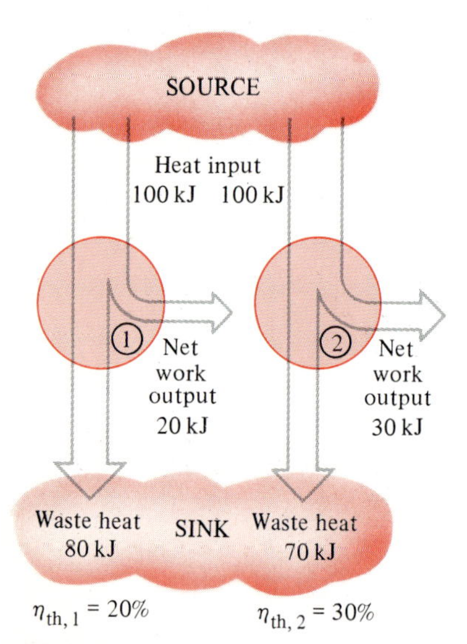

FIGURE 5-13

Some heat engines perform better than others (convert more of the heat they receive to work).

Performance or efficiency, in general, can be expressed in terms of the desired output and the required input as (Fig. 5-14)

$$\text{Performance} = \frac{\text{desired output}}{\text{required input}} \tag{5-3}$$

For heat engines, the desired output is the net work output, and the required input is the amount of heat supplied to the working fluid. Then the thermal efficiency of a heat engine can be expressed as

$$\text{Thermal efficiency} = \frac{\text{net work output}}{\text{total heat input}}$$

or

$$\eta_{th} = \frac{W_{net,out}}{Q_{in}} \tag{5-4}$$

It can also be expressed as

$$\eta_{th} = 1 - \frac{Q_{out}}{Q_{in}} \tag{5-5}$$

since $W_{net,out} = Q_{in} - Q_{out}$.

Cyclic devices of practical interest such as heat engines, refrigerators, and heat pumps operate between a high-temperature medium (or reservoir) at temperature T_H and a low-temperature medium (or reservoir) at temperature T_L. To bring uniformity to the treatment of heat engines, refrigerators, and heat pumps, we define the following two quantities:

Q_H = *magnitude* of heat transfer between cyclic device and high-temperature medium at temperature T_H

Q_L = *magnitude* of heat transfer between cyclic device and low-temperature medium at temperature T_L

Notice that both Q_L and Q_H are defined as *magnitudes* and therefore are *positive quantities*. The direction of Q_H and Q_L is easily determined by inspection, and we do not need to be concerned about their signs. Then the net work output and thermal efficiency relations for any heat engine (shown in Fig. 5-15) can also be expressed as

$$W_{net,out} = Q_H - Q_L \tag{5-6}$$

and

$$\eta_{th} = \frac{W_{net,out}}{Q_H} \tag{5-7}$$

or

$$\eta_{th} = 1 - \frac{Q_L}{Q_H} \tag{5-8}$$

The thermal efficiency of a heat engine is always less than unity since both Q_L and Q_H are defined as positive quantities.

Thermal efficiency is a measure of how efficiently a heat engine converts the heat that it receives to work. Heat engines are built for the purpose of converting heat to work, and engineers are constantly trying to improve the efficiencies of these devices since increased efficiency means less fuel consumption and thus lower fuel bills.

The thermal efficiencies of work-producing devices are amazingly low. Ordinary spark-ignition automobile engines have a thermal efficiency of about 20 percent. That is, an automobile engine converts, at an average, about 20 percent of the chemical energy of the gasoline to mechanical work. This number is about 30 percent for diesel engines and large gas-turbine plants and 40 percent for large steam power plants. Thus, even with the most efficient heat engines available today, more than one-half of the energy supplied ends up in the rivers, lakes, or the atmosphere as waste or unusable energy (Fig. 5-16).

FIGURE 5-14
The definition of performance is not limited to thermodynamics only.

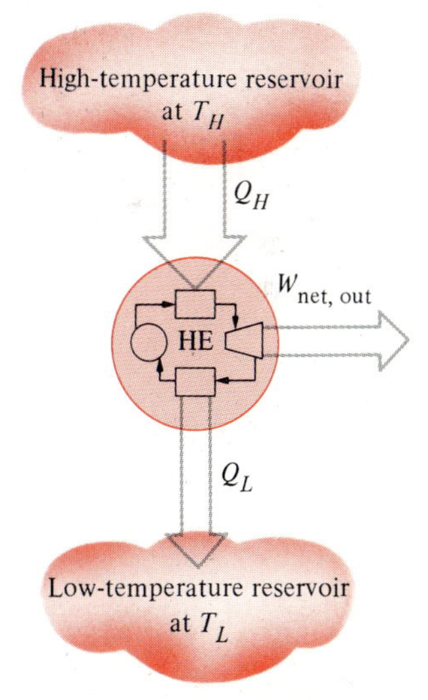

FIGURE 5-15
Schematic of a heat engine.

Can We Save Q_{out}?

In a steam power plant, the condenser is the device where large quantities of waste heat is rejected to rivers, lakes, or the atmosphere. Then one may ask, can we not just take the condenser out of the plant and save all that waste energy? The answer to this question is, unfortunately, a firm *no* for the simple reason that without the cooling process in a condenser the cycle cannot be completed. (Cyclic devices such as steam power plants cannot run continuously unless the cycle is completed.) This is demonstrated below with the help of a simple heat engine.

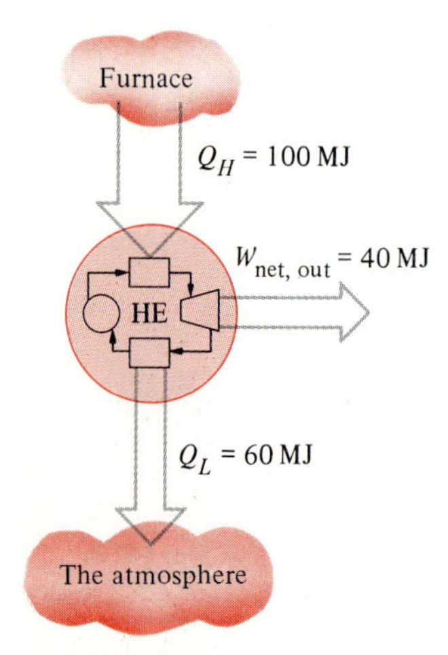

FIGURE 5-16
Even the most efficient heat engines reject most of the heat they receive as waste heat.

Consider the simple heat engine shown in Fig. 5-17 that is used to lift weights. It consists of a piston-cylinder device with two sets of stops. The working fluid is the gas contained within the cylinder. Initially, the gas temperature is 30°C. The piston, which is loaded with the weights, is resting on top of the lower stops. Now 100 kJ of heat is transferred to the gas in the cylinder from a source at 100°C, causing it to expand and to raise the loaded piston until the piston reaches the upper stops, as shown in the figure. At this point the load is removed, and the gas temperature is observed to be 90°C.

The work done on the load during this expansion process is equal to the increase in its potential energy, say 15 kJ. Even under ideal conditions (weightless piston, no friction, no heat losses, and quasi-equilibrium expansion), the amount of heat supplied to the gas is greater than the work done since part of the heat supplied is used to raise the temperature of the gas.

Now let us try to answer the following question: *Is it possible to transfer the 85 kJ of excess heat at 90°C back to the reservoir at 100°C for later use?* If it is, then we will have a heat engine that can have a thermal efficiency of 100 percent under ideal conditions. The answer to this question is again *no*, for the very simple reason that heat always flows from a high-temperature medium to a low-temperature one, and never the other way around. Therefore, we cannot cool this gas from 90 to 30°C by transferring heat to a reservoir at 100°C. Instead, we have to bring the system into contact with a low-temperature reservoir, say at 20°C, so that the gas can return to its initial state by rejecting its 85 kJ of excess energy as heat to this reservoir. This energy cannot be recycled, and it is properly called *waste energy*.

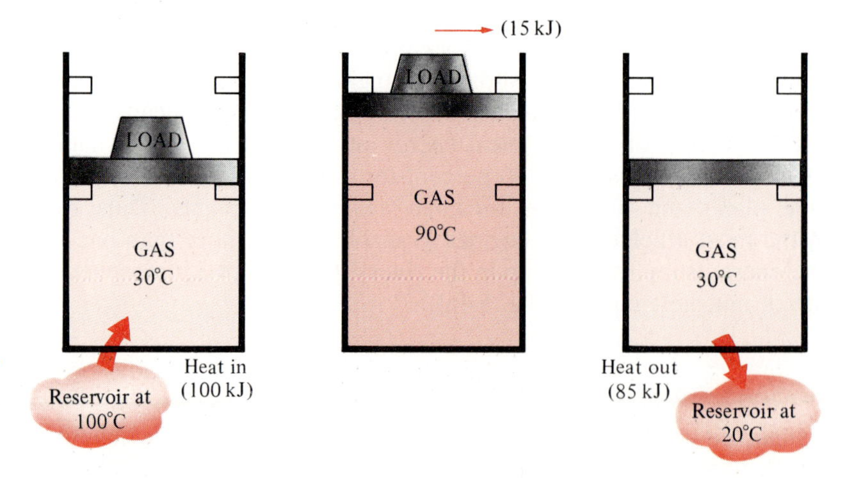

FIGURE 5-17
A heat-engine cycle cannot be completed without rejecting some heat to a low-temperature sink.

We conclude from the above discussion that every heat engine must *waste* some energy by transferring it to a low-temperature reservoir in order to complete the cycle, even under idealized conditions. The requirement that a heat engine exchange heat with at least two reservoirs for continuous operation forms the basis for the Kelvin-Planck expression of the second law of thermodynamics discussed later in this section.

EXAMPLE 5-1

Heat is transferred to a heat engine from a furnace at a rate of 80 MW. If the rate of waste heat rejection to a nearby river is 50 MW, determine the net power output and the thermal efficiency for this heat engine.

Solution A schematic of the heat engine is given in Fig. 5-18. The furnace serves as the high-temperature reservoir for this heat engine and the river as the low-temperature reservoir. Then the given quantities can be expressed in rate form as

$$\dot{Q}_H = 80 \text{ MW} \qquad \text{and} \qquad \dot{Q}_L = 50 \text{ MW}$$

Neglecting the heat losses that may occur from the working fluid as it passes through the pipes and other components, the net power output of this heat engine is determined from Eq. 5-6 to be

$$\dot{W}_{\text{net,out}} = \dot{Q}_H - \dot{Q}_L = (80 - 50) \text{ MW} = 30 \text{ MW}$$

Then the thermal efficiency is easily determined from Eq. 5-7:

$$\eta_{\text{th}} = \frac{\dot{W}_{\text{net,out}}}{\dot{Q}_H} = \frac{30 \text{ MW}}{80 \text{ MW}} = 0.375 \quad (\text{or } 37.5\%)$$

That is, this heat engine converts 37.5 percent of the heat it receives to work.

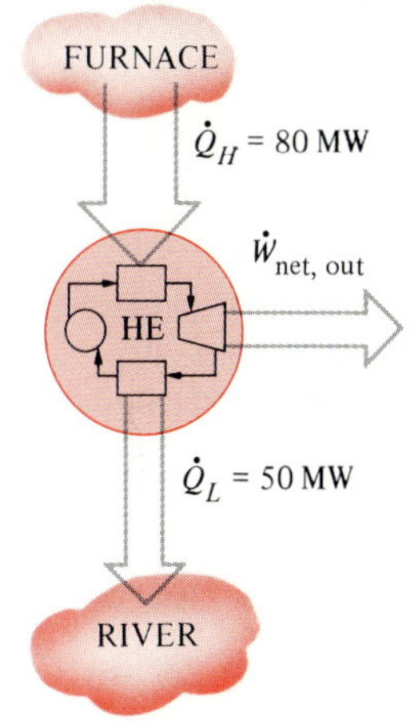

FIGURE 5-18
Schematic for Example 5-1.

EXAMPLE 5-2

A car engine with a power output of 65 hp has a thermal efficiency of 24 percent. Determine the fuel consumption rate of this car if the fuel has an energy content of 19,000 Btu/lbm (that is, 19,000 Btu of energy is released for each lbm of fuel burned).

Solution A schematic of the car engine is given in Fig. 5-19. The car engine is powered by converting 24 percent of the chemical energy released during the combustion process to work. The amount of energy input required to produce a power output of 65 hp is determined from the definition of thermal efficiency (Eq. 5-7):

$$\dot{Q}_H = \frac{\dot{W}_{\text{net,out}}}{\eta_{\text{th}}} = \frac{65 \text{ hp}}{0.24}\left(\frac{2545 \text{ Btu/h}}{1 \text{ hp}}\right) = 689{,}262 \text{ Btu/h}$$

To supply energy at this rate, the engine must burn fuel at a rate of

$$\dot{m} = \frac{689{,}262 \text{ Btu/h}}{19{,}000 \text{ Btu/lbm}} = 36.3 \text{ lbm/h}$$

since 19,000 Btu of thermal energy is released for each lbm of fuel burned.

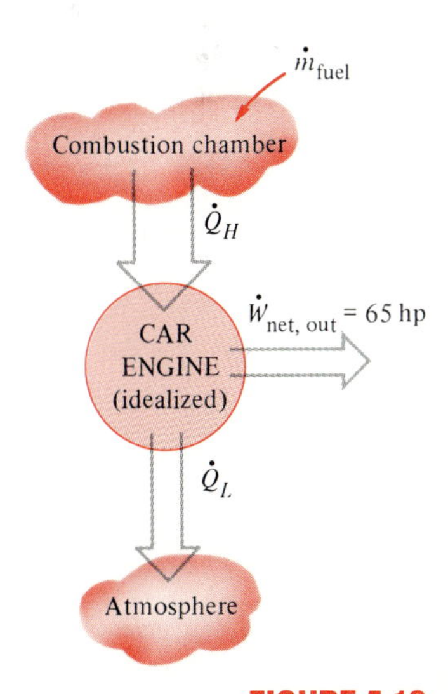

FIGURE 5-19
Schematic for Example 5-2.

The Second Law of Thermodynamics: Kelvin-Planck Statement

We have demonstrated earlier with reference to the heat engine shown in Fig. 5-17 that, even under ideal conditions, a heat engine must reject some heat to a low-temperature reservoir in order to complete the cycle. That is, no heat engine can convert all the heat it receives to useful work. This limitation on the thermal efficiency of heat engines forms the basis for the Kelvin-Planck statement of the second law of thermodynamics, which is expressed as follows:

> *It is impossible for any device that operates on a cycle to receive heat from a single reservoir and produce an equivalent amount of work.*

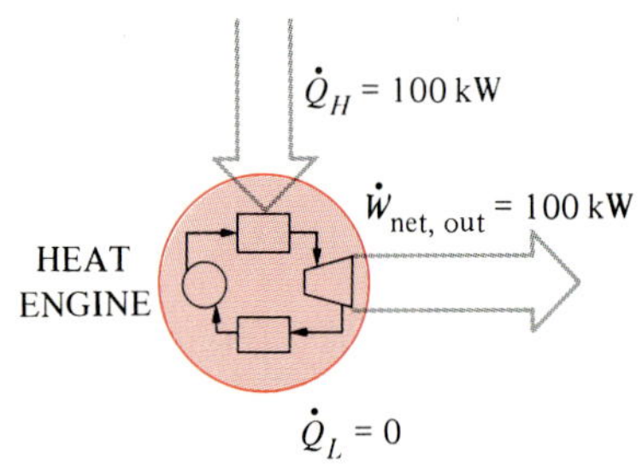

FIGURE 5-20

A heat engine that violates the Kelvin-Planck statement of the second law.

That is, a heat engine must exchange heat with a low-temperature sink as well as a high-temperature source to keep operating. The Kelvin-Planck statement can also be expressed as *no heat engine can have a thermal efficiency of 100 percent* (Fig. 5-20), or as *for a power plant to operate, the working fluid must exchange heat with the environment as well as the furnace*.

Note that the impossibility of having a 100 percent efficient heat engine is not due to friction or other dissipative effects. It is a limitation that applies to both the idealized and the actual heat engines. Later in this chapter we develop a relation for the maximum thermal efficiency of a heat engine. We also demonstrate that this maximum value depends on the reservoir temperatures only.

5-4 ■ REFRIGERATORS AND HEAT PUMPS

We all know from experience that heat flows in the direction of decreasing temperature, i.e., from high-temperature media to low-temperature ones. This heat transfer process occurs in nature without requiring any devices. The reverse process, however, cannot occur by itself. The transfer of heat from a low-temperature medium to a high-temperature one requires special devices called **refrigerators**.

Refrigerators, like heat engines, are cyclic devices. The working fluid used in the refrigeration cycle is called a **refrigerant**. The most frequently used refrigeration cycle is the *vapor-compression refrigeration cycle* which involves four main components: a compressor, a condenser, an expansion valve, and an evaporator, as shown in Fig. 5-21.

The refrigerant enters the compressor as a vapor and is compressed to the condenser pressure. It leaves the compressor at a relatively high temperature and cools down and condenses as it flows through the coils of the condenser by rejecting heat to the surrounding medium. It then enters a capillary tube where its pressure and temperature drop drastically due to the throttling effect. The low-temperature refrigerant then enters

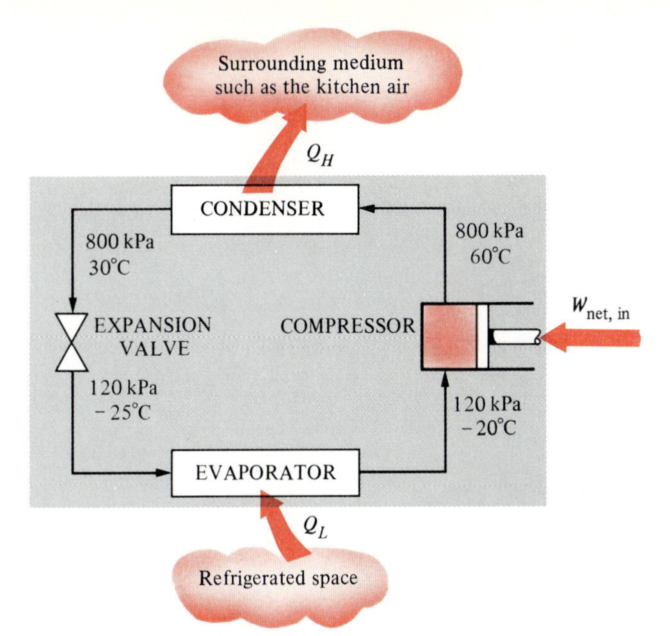

FIGURE 5-21
Basic components of a refrigeration system and typical operating conditions.

the evaporator, where it evaporates by absorbing heat from the refrigerated space. The cycle is completed as the refrigerant leaves the evaporator and reenters the compressor.

In a household refrigerator, the freezer compartment where heat is picked up by the refrigerant serves as the evaporator, and the coils behind the refrigerator where heat is dissipated to the kitchen air as the condenser.

A refrigerator is shown schematically in Fig. 5-22. Here Q_L is the magnitude of the heat removed from the refrigerated space at temperature T_L, Q_H is the magnitude of the heat rejected to the warm environment at temperature T_H, and $W_{net,in}$ is the net work input to the refrigerator. As discussed before, Q_L and Q_H represent magnitudes and so are positive quantities.

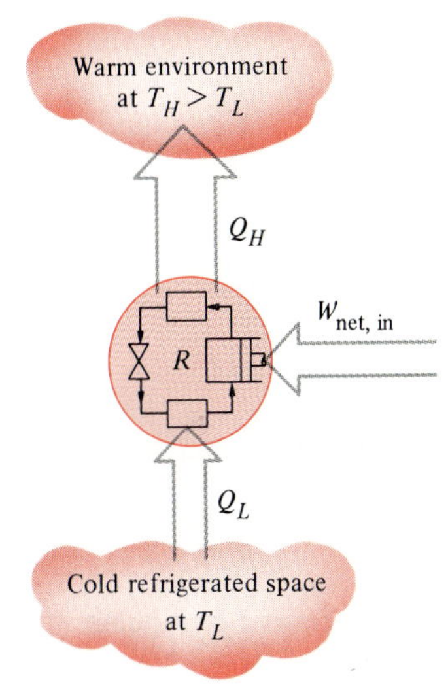

FIGURE 5-22
Schematic of a refrigerator.

Coefficient of Performance

The *efficiency* of a refrigerator is expressed in terms of the **coefficient of performance** (COP), denoted by COP_R. The objective of a refrigerator is to remove heat (Q_L) from the refrigerated space (Fig. 5-23). To accomplish this objective, it requires a work input of $W_{net,in}$. Then the COP of a refrigerator can be expressed as

$$COP_R = \frac{\text{desired output}}{\text{required input}} = \frac{Q_L}{W_{net,in}} \qquad (5\text{-}9)$$

This relation can also be expressed in rate form by replacing Q_L by $\dot{Q}_L$ and $W_{net,in}$ by $\dot{W}_{net,in}$.

The conservation of energy principle for a cyclic device requires that

$$W_{net,in} = Q_H - Q_L \qquad \text{(kJ)} \qquad (5\text{-}10)$$

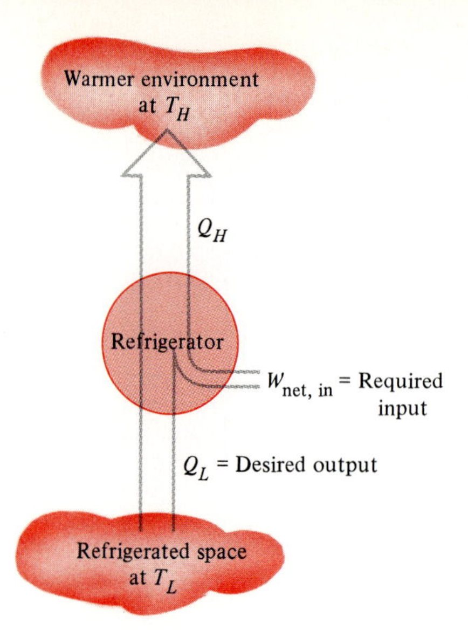

FIGURE 5-23
The objective of a refrigerator is to remove heat Q_L from the refrigerated space.

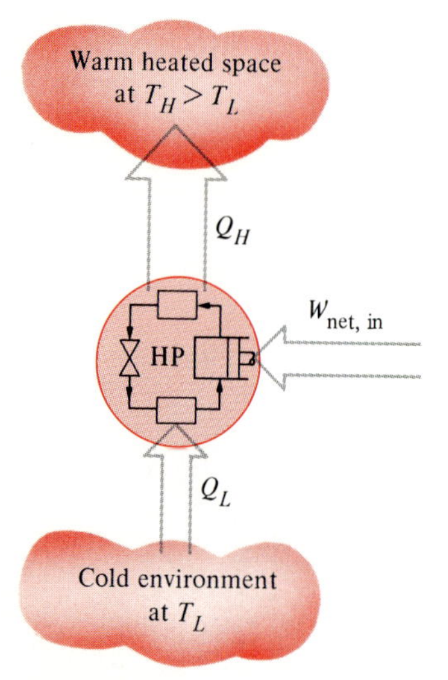

FIGURE 5-24
Schematic of a heat pump.

Then the COP relation can also be expressed as

$$\text{COP}_R = \frac{Q_L}{Q_H - Q_L} = \frac{1}{Q_H/Q_L - 1} \tag{5-11}$$

Notice that the value of COP_R can be *greater than unity*. That is, the amount of heat removed from the refrigerated space can be greater than the amount of work input. This is in contrast to the thermal efficiency which can never be greater than 1. In fact, one reason for expressing the efficiency of a refrigerator by another term—the coefficient of performance—is the desire to avoid the oddity of having efficiencies greater than unity.

Heat Pumps

Another device that transfers heat from a low-temperature medium to a high-temperature one is the **heat pump**, shown schematically in Fig. 5-24. Refrigerators and heat pumps operate on the same cycle but differ in their objectives. The objective of a refrigerator is to maintain the refrigerated space at a low temperature by removing heat from it. Discharging this heat to a higher temperature medium is merely a necessary part of the operation, not the purpose. The objective of a heat pump, however, is to maintain a heated space at a high temperature. This is accomplished by absorbing heat from a low-temperature source, such as well water or cold outside air in winter, and supplying this heat to the high-temperature medium such as a house (Fig. 5-25).

An ordinary refrigerator that is placed in the window of a house with its door open to the cold outside air in winter will function as a heat pump since it will try to cool the outside by absorbing heat from it and rejecting this heat into the house through the coils behind it (Fig. 5-26).

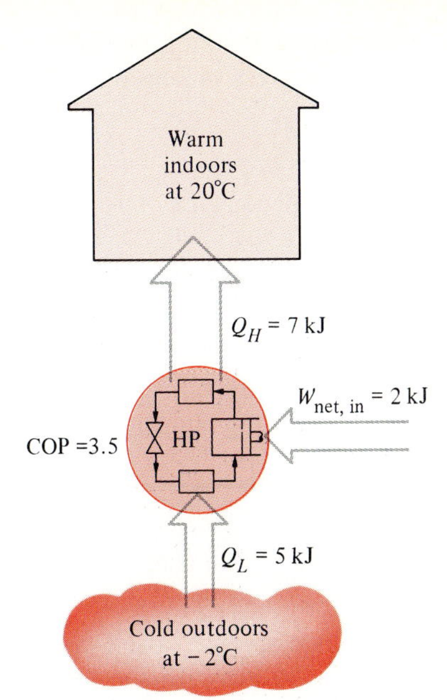

FIGURE 5-25
The work supplied to a heat pump is used to extract energy from the cold outdoors and carry it into the warm indoors.

The measure of performance of a heat pump is also expressed in terms of the **coefficient of performance** COP_{HP}, defined as

$$COP_{HP} = \frac{\text{desired output}}{\text{required input}} = \frac{Q_H}{W_{net,in}} \tag{5-12}$$

which can also be expressed as

$$COP_{HP} = \frac{Q_H}{Q_H - Q_L} = \frac{1}{1 - Q_L/Q_H} \tag{5-13}$$

A comparison of Eqs. 5-9 and 5-12 reveals that

$$COP_{HP} = COP_R + 1 \tag{5-14}$$

for fixed values of Q_L and Q_H. This relation implies that the coefficient of performance of a heat pump is always greater than unity since COP_R is a positive quantity. That is, a heat pump will function, at worst, as a resistance heater, supplying as much energy to the house as it consumes. In reality, however, part of Q_H is lost to the outside air through piping and other devices, and COP_{HP} may drop below unity when the outside air temperature is too low. When this happens, the system usually switches to a resistance heating mode. Most heat pumps in operation today have seasonally averaged COP of 2 to 3.

FIGURE 5-26
When installed backward, an air conditioner will function as a heat pump.

Air conditioners are basically refrigerators whose refrigerated space is a room or a building instead of the food compartment. A window air conditioning unit cools a room by absorbing heat from the room air and discharging it to the outside. The same air conditioning unit can be used as a heat pump in winter by installing it backward. In this mode, the unit will pick up heat from the cold outside and deliver it to the room. Air conditioning systems that are equipped with proper controls operate as air conditioners in summer and as heat pumps in winter.

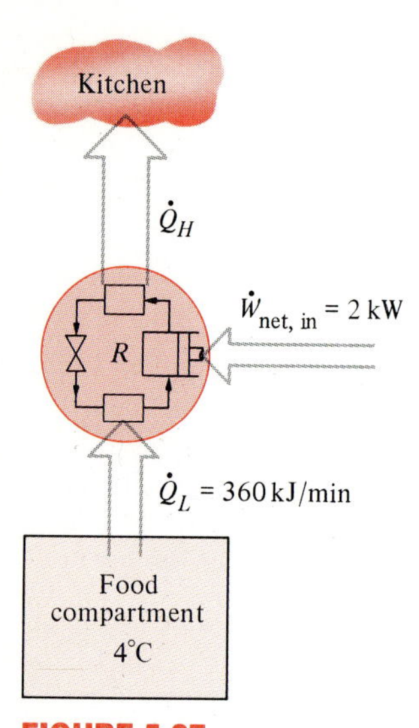

FIGURE 5-27
Schematic for Example 5-3.

EXAMPLE 5-3

The food compartment of a refrigerator, shown in Fig. 5-27, is maintained at 4°C by removing heat from it at a rate of 360 kJ/min. If the required power input to the refrigerator is 2 kW, determine (*a*) the coefficient of performance of the refrigerator and (*b*) the rate of heat discharge to the room that houses the refrigerator.

Solution (*a*) The coefficient of performance of a refrigerator is defined by Eq. 5-9, which can be expressed in rate form as

$$\text{COP}_R = \frac{\dot{Q}_L}{\dot{W}_{\text{net,in}}} = \frac{360 \text{ kJ/min}}{2 \text{ kW}}\left(\frac{1 \text{ kW}}{60 \text{ kJ/min}}\right) = 3$$

That is, 3 kJ of heat is removed from the refrigerated space for each kJ of work supplied.

(*b*) The rate at which heat is discharged to the room that houses the refrigerator is determined from the conservation of energy relation for cyclic devices (Eq. 5-10), expressed in rate form as

$$\dot{Q}_H = \dot{Q}_L + \dot{W}_{\text{net,in}} = 360 \text{ kJ/min} + (2 \text{ kW})\left(\frac{60 \text{ kJ/min}}{1 \text{ kW}}\right) = 480 \text{ kJ/min}$$

Notice that both the energy removed from the refrigerated space as heat and the energy supplied to the refrigerator as electrical work eventually show up in the room air and become part of the internal energy of the air. This demonstrates that energy can change from one form to another, can move from one place to another, but is never destroyed during a process.

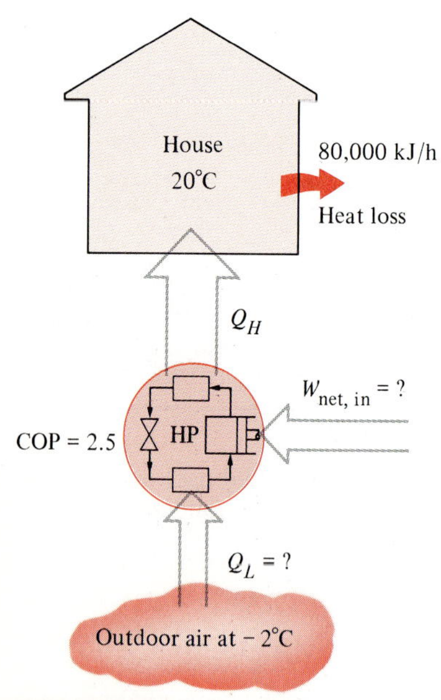

FIGURE 5-28
Schematic for Example 5-4.

EXAMPLE 5-4

A heat pump is used to meet the heating requirements of a house and maintain it at 20°C. On a day when the outdoor air temperature drops to −2°C, the house is estimated to lose heat at a rate of 80,000 kJ/h. If the heat pump under these conditions has a COP of 2.5, determine (*a*) the power consumed by the heat pump and (*b*) the rate at which heat is extracted from the cold outdoor air.

Solution (*a*) The power consumed by this heat pump, shown in Fig. 5-28, can be determined from the definition of the coefficient of performance of a heat pump (Eq. 5-12), expressed in rate form as

$$\dot{W}_{\text{net,in}} = \frac{\dot{Q}_H}{\text{COP}_{\text{HP}}} = \frac{80{,}000 \text{ kJ/h}}{2.5} = 32{,}000 \text{ kJ/h} \quad (\text{or } 8.9 \text{ kW})$$

(*b*) The house is losing heat at a rate of 80,000 kJ/h. If the house is to be maintained at a constant temperature of 20°C, the heat pump must deliver heat to the house at the same rate, i.e., at a rate of 80,000 kJ/h. Then the rate of heat transfer from the outdoor air is determined from the conservation of energy principle for a cyclic device (Eq. 5-10):

$$\dot{Q}_L = \dot{Q}_H - \dot{W}_{\text{net,in}} = (80{,}000 - 32{,}000) \text{ kJ/h} = 48{,}000 \text{ kJ/h}$$

That is, 48,000 of the 80,000 kJ/h heat delivered to the house is actually extracted from the cold outdoor air. Therefore, we are paying only for the 32,000-kJ/h energy, which is supplied as electrical work to the heat pump. If we were to use an electric resistance heater instead, we would have to supply the entire 80,000 kJ/h to the resistance heater as electric energy. This would mean a heating bill which is 2.5 times higher. This explains the popularity of heat pumps as heating systems and why they are preferred to simple electric resistance heaters despite their considerably higher initial cost.

The Second Law of Thermodynamics: Clausius Statement

There are two classical statements of the second law—the Kelvin-Planck statement, which is related to heat engines and discussed in the preceding section, and the Clausius statement, which is related to refrigerators or heat pumps. The Clausius statement is expressed as follows:

> *It is impossible to construct a device that operates in a cycle and produces no effect other than the transfer of heat from a lower temperature body to a higher temperature body.*

It is common knowledge that heat does not, of it own volition, flow from a cold medium to a warmer one. The Clausius statement does not imply that a cyclic device that transfers heat from a cold medium to a warmer one is impossible to construct. In fact, this is precisely what a common household refrigerator does. It simply states that a refrigerator will not operate unless its compressor is driven by an external power source, such as an electric motor (Fig. 5-29). This way, the net effect on the surroundings involves the consumption of some energy in the form of work, in addition to the transfer of heat from a colder body to a warmer one. That is, it leaves a trace in the surroundings. Therefore, a household refrigerator is in complete compliance with the Clausius statement of the second law.

Both the Kelvin-Planck and the Clausius statements of the second law are negative statements, and a negative statement cannot be proved. Like any other physical law, the second law of thermodynamics is based on experimental observations. To date, no experiment has been conducted that contradicts the second law, and this should be taken as sufficient evidence of its validity.

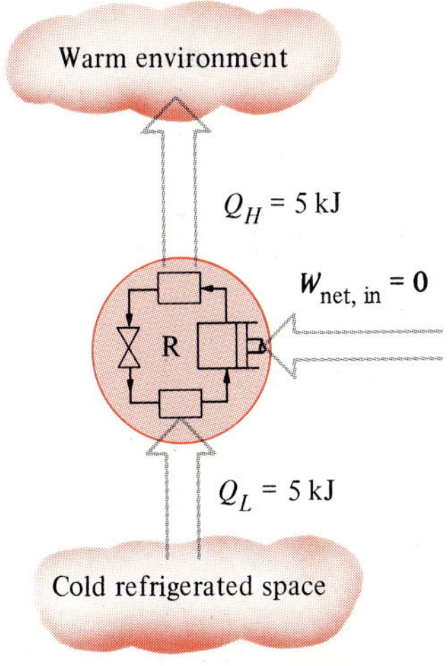

FIGURE 5-29
A refrigerator that violates the Clausius statement of the second law.

Equivalence of the Two Statements

The Kelvin-Planck and the Clausius statements are equivalent in their consequences, and either statement can be used as the expression of the second law of thermodynamics. Any device that violates the Kelvin-Planck statement also violates the Clausius statement, and vice versa. This can be demonstrated as follows:

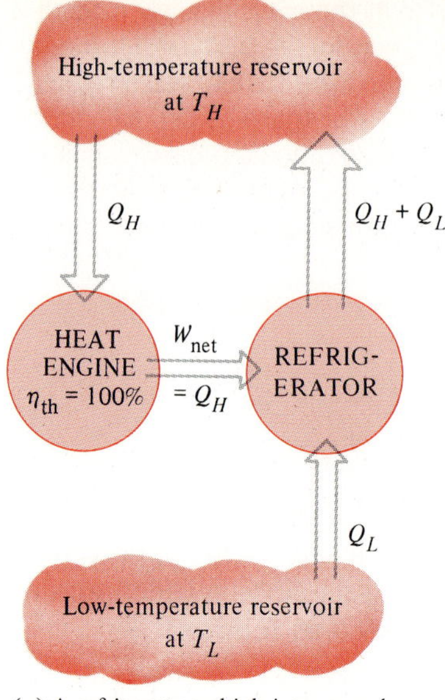

(*a*) A refrigerator which is powered by a 100% efficient heat engine

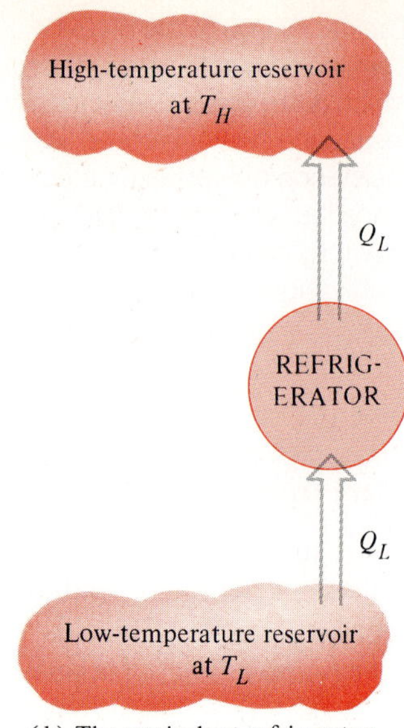

(*b*) The equivalent refrigerator

FIGURE 5-30
Proof that the violation of the Kelvin-Planck statement leads to the violation of the Clausius statement.

Consider the heat-engine–refrigerator combination shown in Fig. 5-30*a*, operating between the same two reservoirs. The heat engine is assumed to have, in violation of the Kelvin-Planck statement, a thermal efficiency of 100 percent, and therefore it converts all the heat Q_H it receives to work W. This work is now supplied to a refrigerator that removes heat in the amount of Q_L from the low-temperature reservoir and rejects heat in the amount of $Q_L + Q_H$ to the high-temperature reservoir. During this process, the high-temperature reservoir receives a net amount of heat Q_L (the difference between $Q_L + Q_H$ and Q_H). Thus the combination of these two devices can be viewed as a refrigerator, as shown in Fig. 5-30*b,* that transfers heat in an amount of Q_L from a cooler body to a warmer one without requiring any input from outside. This is clearly a violation of the Clausius statement. Therefore, a violation of the Kelvin-Planck statement results in the violation of the Clausius statement.

It can also be shown in a similar manner that a violation of the Clausius statement leads to the violation of the Kelvin-Planck statement. Therefore, the Clausius and the Kelvin-Planck statements are two equivalent expressions of the second law of thermodynamics.

5-5 ■ PERPETUAL-MOTION MACHINES

We have repeatedly stated that a process cannot take place unless it satisfies both the first and second laws of thermodynamics. Any device that violates either law is called a **perpetual-motion machine**, and despite numerous attempts, no perpetual-motion machine is known to have worked. But this has not stopped inventors from trying to create new ones.

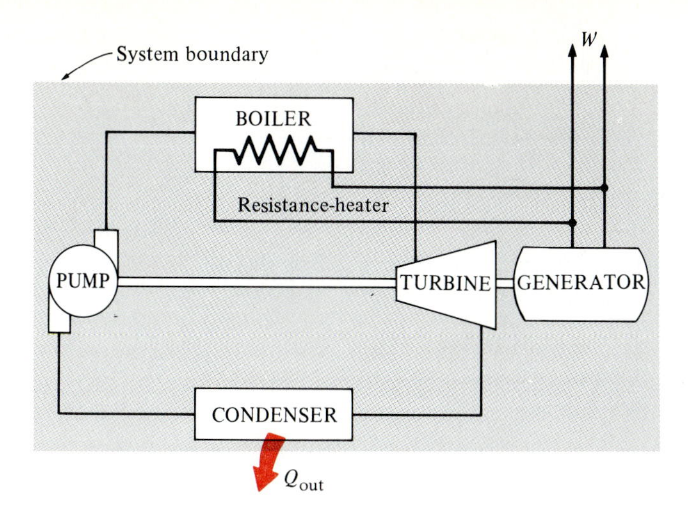

FIGURE 5-31
A perpetual-motion machine that violates the first law of thermodynamics (PMM1).

A device that violates the first law of thermodynamics (by *creating* energy) is called a **perpetual-motion machine of the first kind** (PMM1), and a device that violates the second law of thermodynamics is called a **perpetual-motion machine of the second kind** (PMM2).

Consider the steam power plant shown in Fig. 5-31. It is proposed to heat the steam by resistance heaters placed inside the boiler, instead of by the energy supplied from fossil or nuclear fuels. Part of the electricity generated by the plant is to be used to power the resistors as well as the pump. The rest of the electric energy is to be supplied to the electric network as the net work output. The inventor claims that once the system is started, this power plant will produce electricity indefinitely without requiring any energy input from the outside.

Well, here is an invention that could solve the world's energy problem—if it works, of course. A careful examination of this invention reveals that the system enclosed by the shaded area is continuously supplying energy to the outside at a rate of $\dot{Q}_{out} + \dot{W}_{net,out}$ without receiving any energy. That is, this system is creating energy at a rate of $\dot{Q}_{out} + \dot{W}_{net,out}$, which is clearly a violation of the first law. Therefore, this wonderful device is nothing more than a PMM1 and does not warrant any further consideration.

Now let us consider another novel idea by the same inventor. Convinced that energy cannot be created, the inventor suggests the following modification which will greatly improve the thermal efficiency of that power plant without violating the first law. Aware that more than one-half of the heat transferred to the steam in the furnace is discarded in the condenser to the environment, the inventor suggests getting rid of this wasteful component and sending the steam to the pump as soon as it leaves the turbine, as shown in Fig. 5-32. This way, all the heat transferred to the steam in the boiler will be converted to work, and thus the power plant will have a theoretical efficiency of 100 percent. The inventor realizes that some heat losses and friction between the moving components are

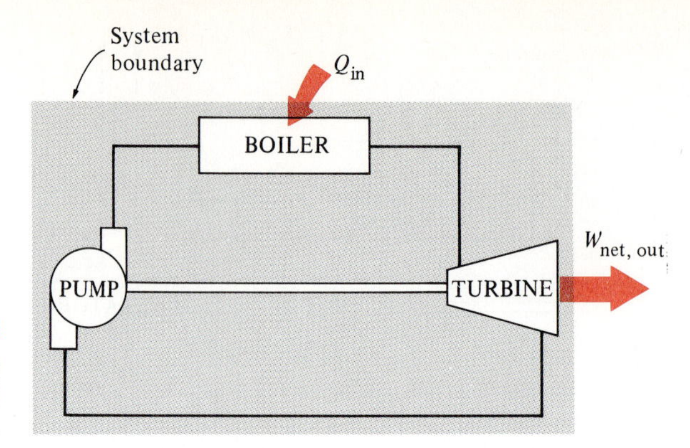

FIGURE 5-32

A perpetual-motion machine that violates the second law of thermodynamics (PMM2).

unavoidable and that these effects will hurt the efficiency somewhat, but still expects the efficiency to be no less than 80 percent (as opposed to 40 percent in actual power plants) for a carefully designed system.

Well, the possibility of doubling the efficiency would certainly be very tempting to plant managers and, if not properly trained, they would probably give this idea a chance, since intuitively they see nothing wrong with it. A student of thermodynamics, however, will immediately label this device as a PMM2, since it works on a cycle and does a net amount of work while exchanging heat with a single reservoir (the furnace) only. It satisfies the first law but violates the second law, and therefore it will not work.

Countless perpetual-motion machines have been proposed throughout history, and many more are being proposed. Some proposers have even gone so far as patenting their inventions, only to find out that what they actually have in their hands is a worthless piece of paper. These people have innovative minds, but they usually lack a formal engineering training, which is very unfortunate. No one is immune from being deceived by an innovative perpetual-motion mechanism. But, as the saying goes, if something sounds too good to be true, it probably is.

5-6 ■ REVERSIBLE AND IRREVERSIBLE PROCESSES

The second law of thermodynamics states that no heat engine can have an efficiency of 100 percent. Then one may ask, What is the highest efficiency that a heat engine can possibly have? Before we can answer this question, we need to define an idealized process first, which is called the *reversible process*.

The processes that were discussed in Sec. 5-1 occurred in a certain direction. Once having taken place, these processes cannot reverse themselves spontaneously and restore the system to its initial state. For this reason, they are classified as *irreversible processes*. Once a cup of hot coffee cools, it will not heat up retrieving the heat it lost from the surroundings. If it could, the surroundings, as well as the system (coffee), would be restored to their original condition, and this would be a reversible process.

A **reversible process** is defined as a *process which can be reversed without leaving any trace on the surroundings* (Fig. 5-33). That is, both the system *and* the surroundings are returned to their initial states at the end of the reverse process. This is possible only if the net heat *and* net work exchange between the system and the surroundings is zero for the combined (original and reverse) process. Processes that are not reversible are called **irreversible processes**.

It should be pointed out that a system can be restored to its initial state following a process, regardless of whether the process is reversible or irreversible. But for reversible processes, this restoration is made without leaving any net change on the surroundings, whereas for irreversible processes, the surroundings usually do some work on the system and therefore will not return to their original state.

Reversible processes actually do not occur in nature. They are merely *idealizations* of actual processes. Reversible processes can be approximated by actual devices, but they can never be achieved. That is, all the processes occurring in nature are irreversible. You may be wondering, then, *why* we are bothering with such fictitious processes. There are two reasons. First, they are easy to analyze, since a system passes through a series of equilibrium states during a reversible process; second, they serve as idealized models to which actual processes can be compared.

Engineers are interested in reversible processes because work-producing devices such as car engines and gas or steam turbines *deliver the most work,* and work-consuming devices such as compressors, fans, and pumps *require least work* when reversible processes are used instead of irreversible ones (Fig. 5-34).

Reversible processes can be viewed as *theoretical limits* for the corresponding irreversible ones. Some processes are more irreversible than others. We may never be able to have a reversible process, but we may certainly approach it. The more closely we approximate a reversible process, the more work delivered by a work-producing device or the less work required by a work-consuming device.

The concept of reversible processes leads to the definition of *second-law efficiency* for actual processes, which is the degree of approximation to the corresponding reversible processes. This enables us to compare the performance of different devices that are designed to do the same task on the basis of their efficiencies. The better the design, the lower the irreversibilities and the higher the second-law efficiency.

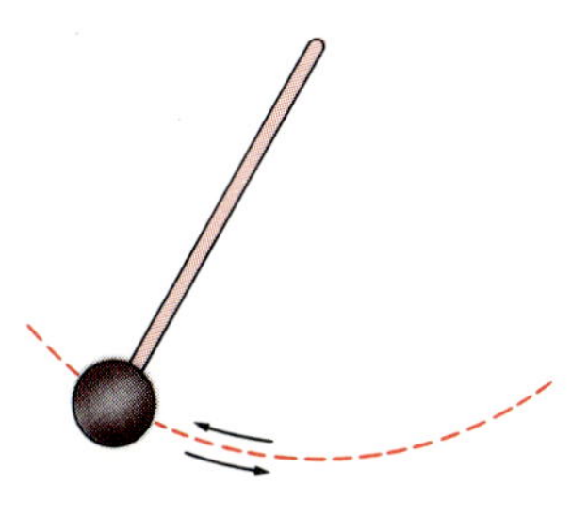

(*a*) Frictionless pendulum

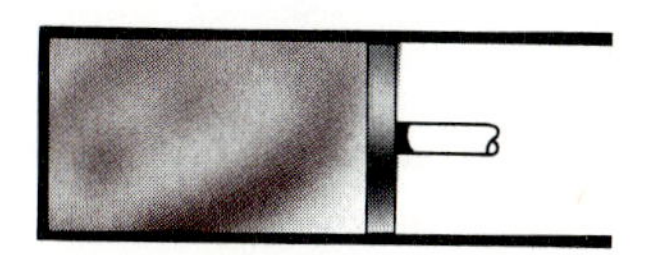

(*b*) Quasi-equilibrium expansion and compression of a gas

FIGURE 5-33

Two familiar reversible processes.

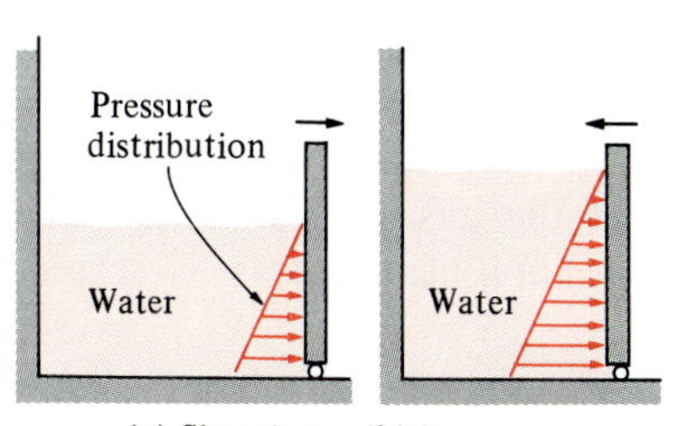

(*a*) Slow (reversible) process

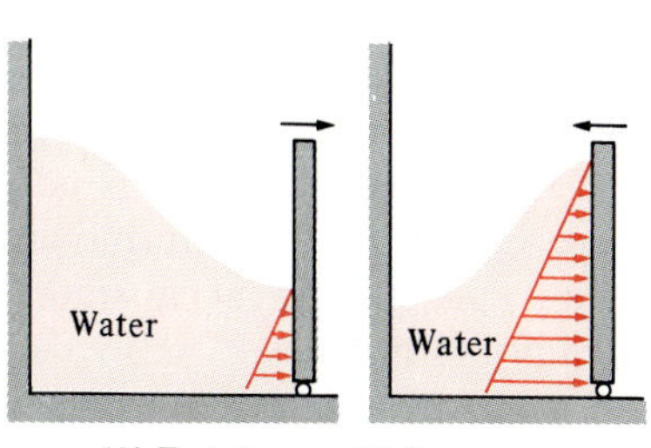

(*b*) Fast (irreversible) process

FIGURE 5-34

Reversible processes deliver the most work and consume the least.

Irreversibilities

The factors that cause a process to be irreversible are called **irreversibilities**. They include friction, unrestrained expansion, mixing of two gases, heat transfer across a finite temperature difference, electric resistance, inelastic deformation of solids, and chemical reactions. The presence of any of these effects renders a process irreversible. A reversible process involves none of these. Some of the frequently encountered irreversibilities are discussed briefly below.

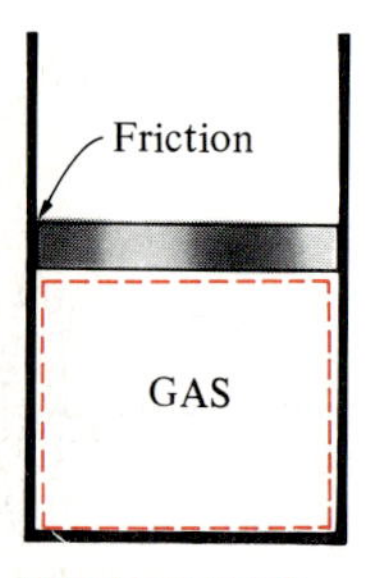

FIGURE 5-35
Friction renders a process irreversible.

Friction

Friction is a familiar form of irreversibility associated with bodies in motion. When two bodies in contact are forced to move relative to each other (a piston in a cylinder for example, as shown in Fig. 5-35), a friction force which opposes the motion develops at the interface of these two bodies, and some work is needed to overcome this friction force. The energy supplied as work is eventually converted to heat during the process and is transferred to the bodies in contact, as evidenced by a temperature rise at the interface. When the direction of the motion is reversed, the bodies will be restored to their original position, but the interface will not cool, and heat will not be converted back to work. Instead, more of the work will be converted to heat while overcoming the friction forces which also oppose the reverse motion. Since the system (the moving bodies) and the surroundings cannot be returned to their original states, this process is irreversible. Therefore, any process that involves friction is irreversible. The larger the friction forces involved, the more irreversible the process is.

Friction does not always involve two solid bodies in contact. It is also encountered between a fluid and solid and even between the layers of a fluid moving at different velocities. A considerable fraction of the power produced by a car engine is used to overcome the friction (the drag force) between the air and the external surfaces of the car, and it eventually becomes part of the internal energy of the air. It is not possible to reverse this process and recover that lost power, even though doing so would not violate the conservation of energy principle.

Non-quasi-Equilibrium Expansion and Compression

In Chap. 1, we defined a quasi-equilibrium process as one during which the system remains infinitesimally close to a state of equilibrium at all times. Consider a frictionless adiabatic piston-cylinder device that contains a gas. Now the piston is pushed into the cylinder, compressing the gas. If the piston velocity is not very high, the pressure and the temperature will increase uniformly throughout the gas. Since the system is always maintained at a state close to equilibrium, this is a quasi-equilibrium process.

Now the external force on the piston is slightly decreased, allowing the gas to expand. The expansion process will also be *quasi-equilibrium* if the gas is allowed to expand slowly. When the piston returns to its original position, all the boundary ($P\,dV$) work done on the gas during compression is returned to the surroundings during expansion. That is, the net work for the combined process is zero. Also, there has been no heat transfer involved during this process, and thus both the system and the surroundings will return to their initial states at the end of the reverse process. Therefore, the slow frictionless adiabatic expansion or compression of a gas is a reversible process.

Now let us repeat this adiabatic process in a *non-quasi-equilibrium* manner, as shown in Fig. 5-36. If the piston is pushed in very rapidly, the gas molecules near the piston face will not have sufficient time to escape, and they will pile up in front of the piston. This will raise the pressure near the piston face, and as a result, the pressure there will be higher than the pressure in other parts of the cylinder. The nonuniformity of pressure will render this process non-quasi-equilibrium. The actual boundary work is a function of pressure, as measured at the piston face. Because of this higher pressure value at the piston face, a non-quasi-equilibrium compression process will require a larger work input than the corresponding quasi-equilibrium one. When the process is reversed by letting the gas expand rapidly, the gas molecules in the cylinder will not be able to follow the piston as fast, thus creating a low-pressure region before the piston face. Because of this low-pressure value at the piston face, a non-quasi-equilibrium process will deliver less work than a corresponding reversible one. Consequently, the work done by the gas during expansion is less than the work done by the surroundings on the gas during compression, and thus the surroundings have a net work deficit. When the piston returns to its initial position, the gas will have excess internal energy, equal in magnitude to the work deficit of the surroundings.

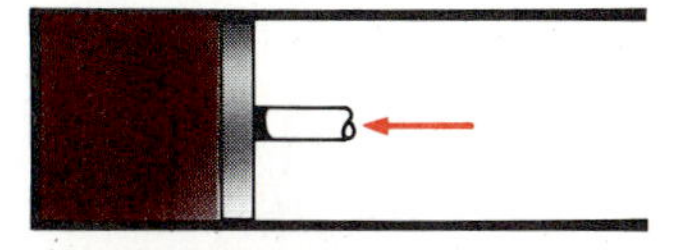

(*a*) Fast compression

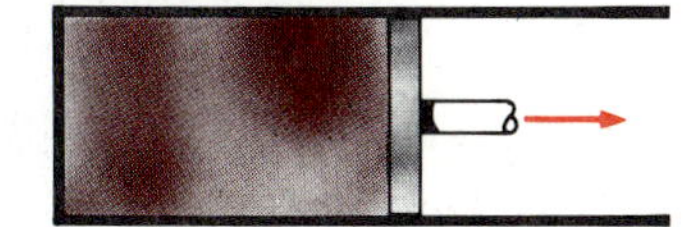

(*b*) Fast expansion

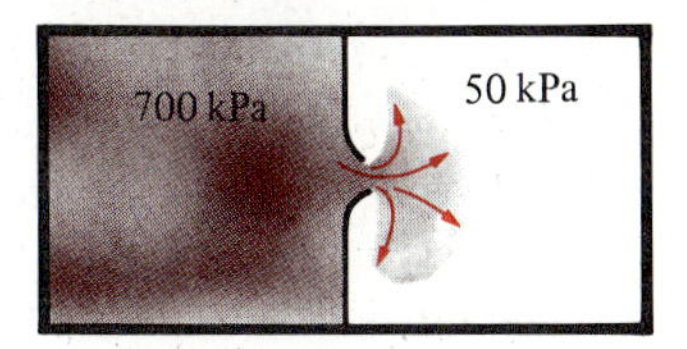

(*c*) Unrestrained expansion

FIGURE 5-36
Irreversible compression and expansion processes.

The system can easily be returned to its initial state by transferring this excess internal energy to the surroundings as heat. But the only way the surroundings can be returned to their initial condition is by completely converting this heat to work, which can only be done by a heat engine that has an efficiency of 100 percent. This, however, is impossible to do, even theoretically, since it would violate the second law of thermodynamics. Since only the system, not both the system and the surroundings, can be returned to its initial state, we conclude that the adiabatic non-quasi-equilibrium expansion or compression of a gas is irreversible.

Another example of non-quasi-equilibrium expansion processes is the unrestrained expansion of a gas separated from a vacuum by a membrane, as shown in Fig. 5-36*c*. When the membrane is ruptured, the gas fills the entire tank. The only way to restore the system to its original state is to compress it to its initial volume, while transferring heat from the gas until it reaches its initial temperature. From the conservation of energy considerations, it can easily be shown that the amount of heat transferred from the gas equals the amount of work done on the gas by the surroundings. The restoration of the surroundings involves conversion of this heat completely to work, which would violate the second law. Therefore, unrestrained expansion of a gas is an irreversible process.

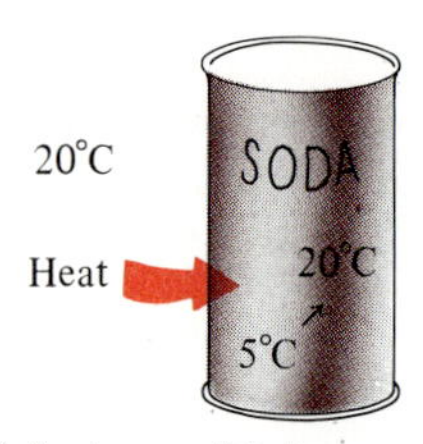

(*a*) An irreversible heat transfer process

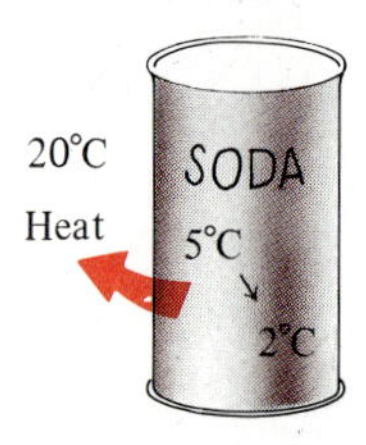

(*b*) An impossible heat transfer process

FIGURE 5-37

(*a*) Heat transfer through a temperature difference is irreversible, and (*b*) the reverse process is impossible.

Heat Transfer

Another form of irreversibility familiar to us all is heat transfer through a finite temperature difference. Consider a can of cold soda left in a warm room, for example, as shown in Fig. 5-37. Heat will flow from the warmer room air to the cooler soda. The only way this process can be reversed and the soda restored to its original temperature is to provide refrigeration, which requires some work input. At the end of the reverse process, the soda will be restored to its initial state, but the surroundings will not be. The internal energy of the surroundings will increase by an amount equal in magnitude to the work supplied to the refrigerator. The restoration of the surroundings to its initial state can be done only by converting this excess internal energy completely to work, which is impossible to do without violating the second law. Since only the system, not both the system and the surroundings, can be restored to its initial condition, heat transfer through a finite temperature difference is an irreversible process.

Heat transfer can occur only when there is a temperature difference between a system and its surroundings. Therefore, it is physically impossible to have a reversible heat transfer process. But a heat transfer process becomes less and less irreversible as the temperature difference between the two bodies approaches zero. Then heat transfer through a differential temperature difference dT can be considered to be reversible. As dT approaches zero, the process can be reversed in direction (at least theoretically) without requiring any refrigeration. Notice that reversible heat transfer is a conceptual process and cannot be duplicated in the laboratory.

The smaller the temperature difference between two bodies, the smaller the heat transfer rate will be. When the temperature difference is small, any significant heat transfer will require a very large surface area and a very long time. Therefore, even though approaching reversible heat transfer is desirable from a thermodynamic point of view, it is impractical and not economically feasible.

Internally and Externally Reversible Processes

A process is an interaction between a system and its surroundings, and a reversible process involves no irreversibilities associated with either of them.

A process is called **internally reversible** if no irreversibilities occur within the boundaries of the system during the process. During an internally reversible process, a system proceeds through a series of equilibrium states, and when the process is reversed, the system passes through exactly the same equilibrium states while returning to its initial state. That is, the paths of the forward and reverse processes coincide for an internally reversible process. The quasi-equilibrium process discussed earlier is an example of an internally reversible process.

A process is called **externally reversible** if no irreversibilities occur outside the system boundaries during the process. Heat transfer between a reservoir and a system is an externally reversible process, since the temperature of the reservoir remains constant during this process.

A process is called **totally reversible**, or simply **reversible**, if it involves no irreversibilities within the system or its surroundings (Fig. 5-38). A totally reversible process involves no heat transfer through a finite temperature difference, no non-quasi-equilibrium changes, and no friction or other dissipative effects.

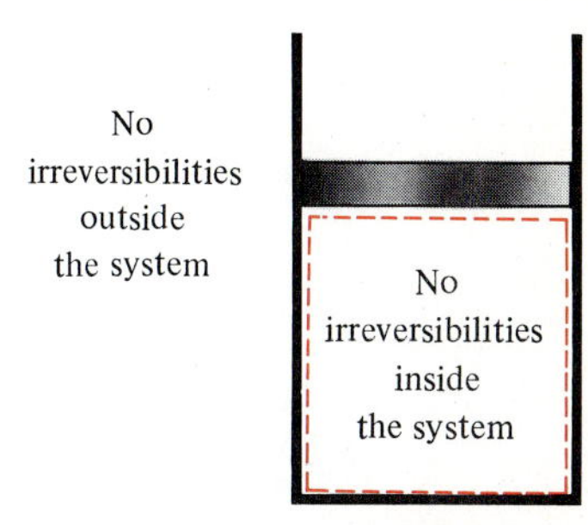

FIGURE 5-38
A reversible process involves no internal and external irreversibilities.

As an example, consider the transfer of heat to two identical systems that are undergoing a constant-pressure (thus constant-temperature) phase-change process, as shown in Fig. 5-39. Both processes are internally reversible, since both take place isothermally and both pass through exactly the same equilibrium states. The first process shown is externally reversible also, since heat transfer for this process takes place through an infinitesimal temperature difference dT. The second process, however, is externally irreversible, since it involves heat transfer through a finite temperature difference ΔT.

5-7 ■ THE CARNOT CYCLE

We mentioned earlier that heat engines are cyclic devices and that the working fluid of a heat engine returns to its initial state at the end of each cycle. Work is done by the working fluid during one part of the cycle and on the working fluid during another part. The difference between these two is the net work delivered by the heat engine. The efficiency of a heat-engine cycle greatly depends on how the individual processes that make up the cycle are executed. The net work, thus the cycle efficiency, can be maximized by using processes that require the least amount of work and deliver the most, that is, by using *reversible processes*. Therefore, it is no surprise that the most efficient cycles are reversible cycles, i.e., cycles that consist entirely of reversible processes.

Reversible cycles cannot be achieved in practice because the irre-

Thermal energy reservoir at 20.0 ... .01°C

(*a*) Totally reversible

Thermal energy reservoir at 30°C

(*b*) Internally reversible

FIGURE 5-39
Totally and internally reversible heat transfer processes.

versibilities associated with each process cannot be eliminated. However, reversible cycles provide upper limits on the performance of real cycles. Heat engines and refrigerators that work on reversible cycles serve as models to which actual heat engines and refrigerators can be compared. Reversible cycles also serve as starting points in the development of actual cycles and are modified as needed to meet certain requirements.

Probably the best known reversible cycle is the **Carnot cycle**, first proposed in 1824 by a French engineer Sadi Carnot. The theoretical heat engine that operates on the Carnot cycle is called the **Carnot heat engine**. The Carnot cycle is composed of four reversible processes—two isothermal and two adiabatic—and it can be executed either in a closed or a steady-flow system.

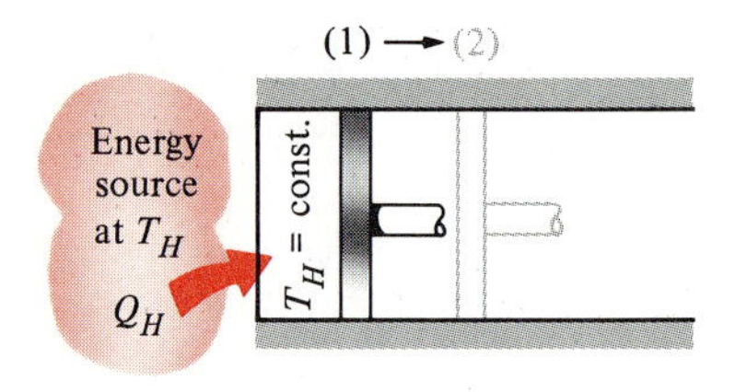

(a) Process 1-2

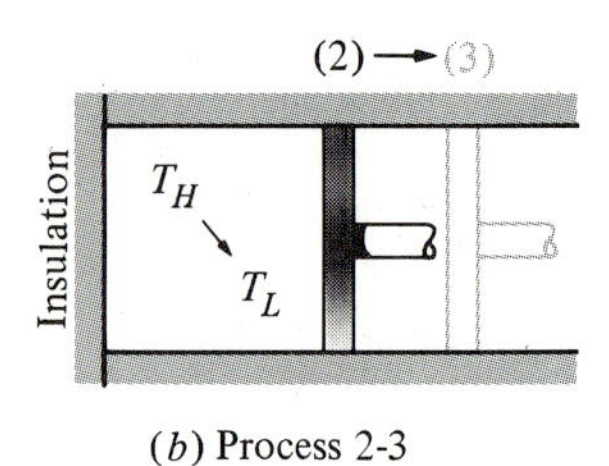

(b) Process 2-3

(c) Process 3-4

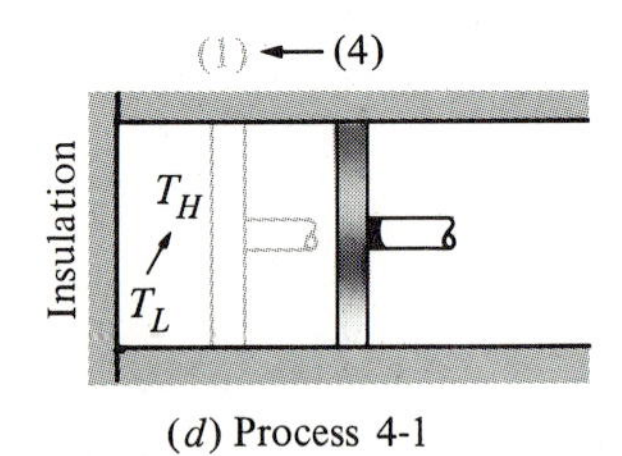

(d) Process 4-1

FIGURE 5-40

Execution of the Carnot cycle in a closed system.

Consider a closed system that consists of a gas contained in an adiabatic piston-cylinder device, as shown in Fig. 5-40. The insulation of the cylinder head is such that it may be removed to bring the cylinder into contact with reservoirs to provide heat transfer. The four reversible processes that make up the Carnot cycle are as follows:

Reversible isothermal expansion (process 1-2, T_H = constant). Initially (state 1) the temperature of the gas is T_H, and the cylinder head is in close contact with a source at temperature T_H. The gas is allowed to expand slowly, doing work on the surroundings. As the gas expands, the temperature of the gas tends to decrease. But as soon as the temperature drops by an infinitesimal amount dT, some heat flows from the reservoir into the gas, raising the gas temperature to T_H. Thus, the gas temperature is kept constant at T_H. Since the temperature difference between the gas and the reservoir never exceeds a differential amount dT, this is a reversible heat transfer process. It continues until the piston reaches position 2. The amount of total heat transferred to the gas during this process is Q_H.

Reversible adiabatic expansion (process 2-3, temperature drops from T_H to T_L). At state 2, the reservoir that was in contact with the cylinder head is removed and replaced by insulation so that the system becomes adiabatic. The gas continues to expand slowly, doing work on the surroundings until its temperature drops from T_H to T_L (state 3). The piston is assumed to be frictionless and the process to be quasi-equilibrium, so the process is reversible as well as adiabatic.

Reversible isothermal compression (process 3-4, T_L = constant). At state 3, the insulation at the cylinder head is removed, and the cylinder is brought into contact with a sink at temperature T_L. Now the piston is pushed inward by an external force, doing work on the gas. As the gas is compressed, its temperature tends to rise. But as soon as it rises by an infinitesimal amount dT, heat flows from the gas to the sink, causing the gas temperature to drop to T_L. Thus the gas temperature is maintained constant at T_L. Since the temperature difference between the gas and the sink never exceeds a differential amount

dT, this a reversible heat transfer process. It continues until the piston reaches position 4. The amount of heat rejected from the gas during this process is Q_L.

Reversible adiabatic compression (process 4-1, temperature rises from T_L to T_H). State 4 is such that when the low-temperature reservoir is removed and the insulation is put back on the cylinder head and the gas is compressed in a reversible manner, the gas returns to its initial state (state 1). The temperature rises from T_L to T_H during this reversible adiabatic compression process, which completes the cycle.

The P-v diagram of this cycle is shown in Fig. 5-41. Remembering that on a P-v diagram the area under the process curve represents the boundary work for quasi-equilibrium (internally reversible) processes, we see that the area under curve 1-2-3 is the work done by the gas during the expansion part of the cycle, and the area under curve 3-4-1 is the work done on the gas during the compression part of the cycle. The area enclosed by the path of the cycle (area 1-2-3-4-1) is the difference between these two and represents the net work done during the cycle.

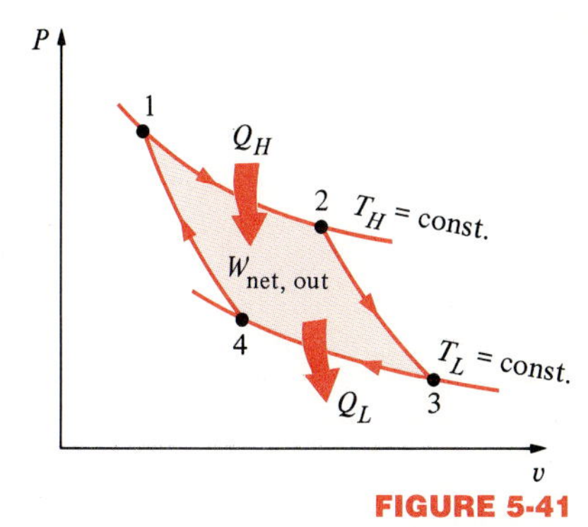

FIGURE 5-41
P-v diagram of the Carnot cycle.

Notice that if we acted stingily and compressed the gas at state 3 adiabatically instead of isothermally in an effort *to save* Q_L, we would end up back at state 2, retracing the process path 3-2. By doing so we would save Q_L, but we would not be able to obtain any net work output from this engine. This illustrates once more the necessity of a heat engine exchanging heat with at least two reservoirs at different temperatures to operate in a cycle and produce a net amount of work.

The Carnot cycle can also be executed in a steady-flow system. It is discussed in Chap. 8 in conjunction with other power cycles.

Being a reversible cycle, the Carnot cycle is the most efficient cycle operating between two specified temperature limits. Even though the Carnot cycle cannot be achieved in reality, the efficiency of actual cycles can be improved by attempting to approximate the Carnot cycle more closely.

The Reversed Carnot Cycle

The Carnot heat-engine cycle described above is a totally reversible cycle. Therefore, all the processes that comprise it can be *reversed*, in which case it becomes the **Carnot refrigeration cycle**. This time, the cycle remains exactly the same, except that the directions of any heat and work interactions are reversed: Heat in the amount of Q_L is absorbed from the low-temperature reservoir, heat in the amount of Q_H is rejected to a high-temperature reservoir, and a work input of $W_{\text{net,in}}$ is required to accomplish all this.

The P-v diagram of the reversed Carnot cycle is the same as the one given for the Carnot cycle, except that the directions of the processes are reversed, as shown in Fig. 5-42.

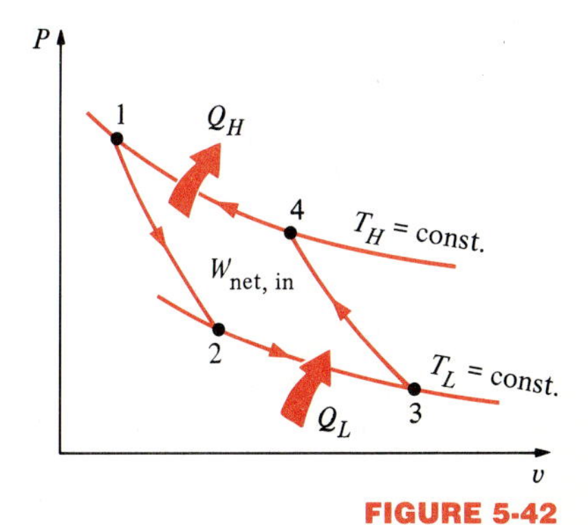

FIGURE 5-42
P-v diagram of the reverse Carnot cycle.

5-8 ■ THE CARNOT PRINCIPLES

The second law of thermodynamics places limitations on the operation of cyclic devices as expressed by the Kelvin-Planck and Clausius statements. A heat engine cannot operate by exchanging heat with a single reservoir, and a refrigerator cannot operate without a net work input from an external source.

We can draw valuable conclusions from these statements. Two conclusions pertain to the thermal efficiency of reversible and irreversible (i.e., actual) heat engines, and they are known as the **Carnot principles** (Fig. 5-43). They are expressed as follows:

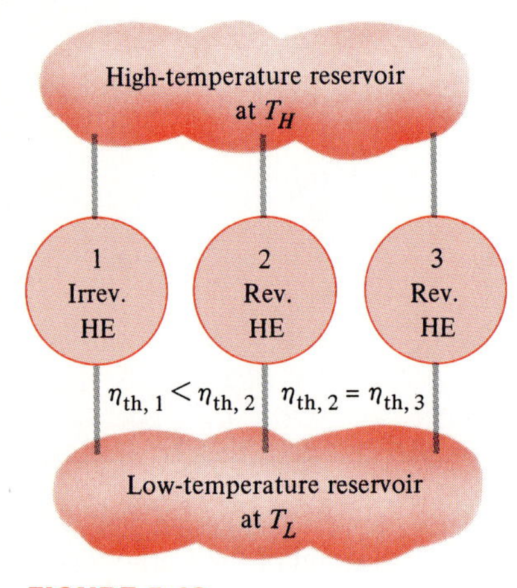

FIGURE 5-43
The Carnot principles.

1 The efficiency of an irreversible heat engine is always less than the efficiency of a reversible one operating between the same two reservoirs.

2 The efficiencies of all reversible heat engines operating between the same two reservoirs are the same.

These two statements can be proved by demonstrating that the violation of either statement results in the violation of the second law of thermodynamics.

To prove the first statement, consider two heat engines operating between the same reservoirs, as shown in Fig. 5-44. One engine is reversible, and the other is irreversible. Now each engine is supplied with the same amount of heat Q_H. The amount of work produced by the reversible heat engine is W_{rev}, and the amount produced by the irreversible one is W_{irrev}.

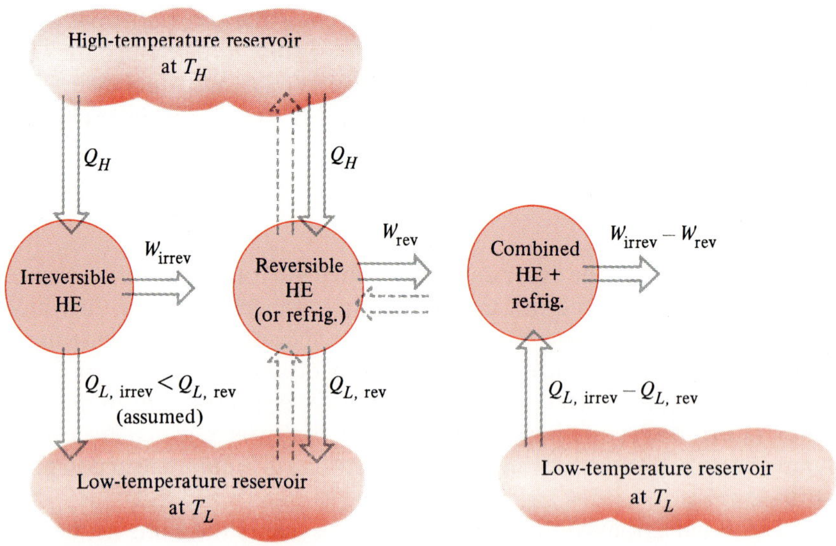

(*a*) A reversible and an irreversible heat engine operating between the same two reservoirs (the reversible heat engine is then reversed to run as a refrigerator)

(*b*) The equivalent combined system

FIGURE 5-44
Proof of the first Carnot principle.

In violation of the first Carnot principle, we assume that the irreversible heat engine is more efficient than the reversible one (that is, $\eta_{th,irrev} > \eta_{th,rev}$) and thus delivers more work than the reversible one. Now let the reversible heat engine be reversed and operate as a refrigerator. This refrigerator will receive a work input of W_{rev} and reject heat to the high-temperature reservoir. Since the refrigerator is rejecting heat in the amount of Q_H to the high-temperature reservoir and the irreversible heat engine is receiving the same amount of heat from this reservoir, the net heat exchange for this reservoir is zero. Thus it could be eliminated by having the refrigerator discharge Q_H directly into the irreversible heat engine.

Now considering the refrigerator and the irreversible engine together, we have an engine that produces a net work in the amount of $W_{irrev} - W_{rev}$ while exchanging heat with a single reservoir—a violation of the Kelvin-Planck statement of the second law. Therefore, our initial assumption that $\eta_{th,irrev} > \eta_{th,rev}$ is incorrect. Then we conclude that no heat engine can be more efficient than a reversible heat engine operating between the same reservoirs.

The second Carnot principle can also be proved in a similar manner. This time, let us replace the irreversible engine by another reversible engine that is more efficient and thus delivers more work than the first reversible engine. By following through the same reasoning as above, we will end up having an engine that produces a net amount of work while exchanging heat with a single reservoir, which is a violation of the second law. Therefore we conclude that no reversible heat engine can be more efficient than another reversible heat engine operating between the same two reservoirs, regardless of how the cycle is completed or the kind of working fluid used.

5-9 ■ THE ABSOLUTE THERMODYNAMIC TEMPERATURE SCALE

A temperature scale that is independent of the properties of the substances that are used to measure temperature is called an **absolute temperature scale**. Such a temperature scale offers great conveniences in thermodynamic calculations, and its derivation is given below.

The second Carnot principle discussed in Sec. 5-8 states that all reversible heat engines have the same thermal efficiency when operating between the same two reservoirs (Fig. 5-45). That is, the efficiency of a reversible engine is independent of the working fluid employed and its properties, the way the cycle is executed, or the type of reversible engine used. Since energy reservoirs are characterized by their temperatures, the thermal efficiency of reversible heat engines is a function of the reservoir temperatures only. That is,

$$\eta_{th,rev} = g(T_H, T_L)$$

or

$$\frac{Q_H}{Q_L} = f(T_H, T_L) \tag{5-15}$$

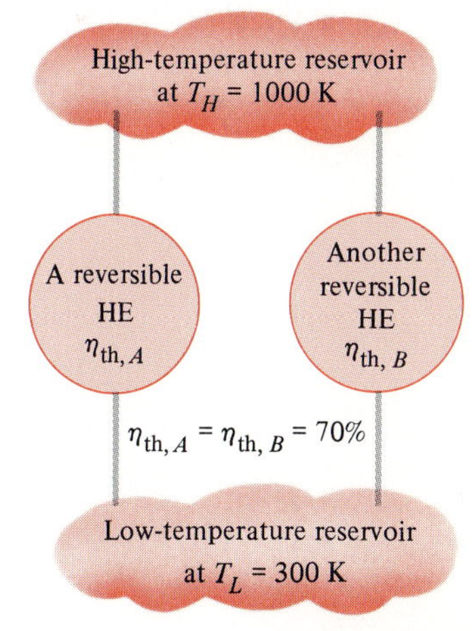

FIGURE 5-45
All reversible heat engines operating between the same two reservoirs have the same efficiency (the second Carnot principle).

since $\eta_{th} = 1 - Q_L/Q_H$. In these relations T_H and T_L are the temperatures of the high- and low-temperature reservoirs, respectively.

The functional form of $f(T_H,T_L)$ can be developed with the help of the three reversible heat engines shown in Fig. 5-46. Engines A and C are supplied with the same amount of heat Q_1 from the high-temperature reservoir at T_1. Engine C rejects Q_3 to the low-temperature reservoir at T_3. Engine B receives the heat Q_2 rejected by engine A at temperature T_2 and rejects heat in the amount of Q_3 to a reservoir at T_3.

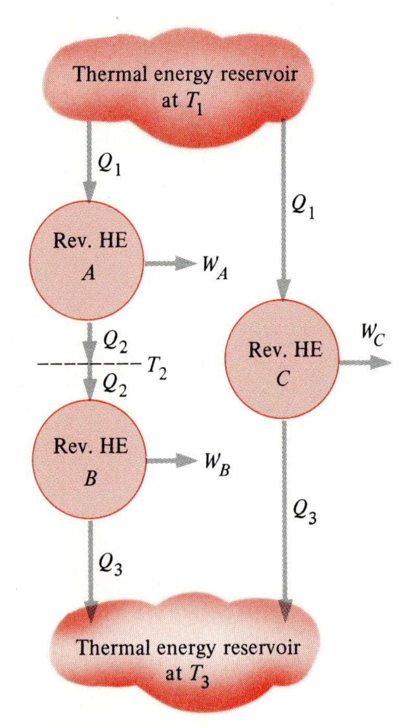

FIGURE 5-46

The arrangement of heat engines used to develop the absolute temperature scale.

The amounts of heat rejected by engines B and C must be the same since engines A and B can be combined into one reversible engine operating between the same reservoirs as engine C and thus the combined engine will have the same efficiency as engine C. Since the heat input to engine C is the same as the heat input to the combined engines A and B, both systems must reject the same amount of heat.

Applying Eq. 5-15 to all three engines separately, we obtain

$$\frac{Q_1}{Q_2} = f(T_1,T_2) \qquad \frac{Q_2}{Q_3} = f(T_2,T_3) \qquad \text{and} \qquad \frac{Q_1}{Q_3} = f(T_1,T_3)$$

Now consider the identity

$$\frac{Q_1}{Q_3} = \frac{Q_1}{Q_2}\frac{Q_2}{Q_3}$$

which corresponds to

$$f(T_1,T_3) = f(T_1,T_2)\cdot f(T_2,T_3)$$

A careful examination of this equation reveals that the left-hand side is a function of T_1 and T_3, and therefore the right-hand side must also be a function of T_1 and T_3 only, and not T_2. That is, the value of the product on the right-hand side of this equation is independent of the value of T_2. This condition will be satisfied only if the function f has the following form:

$$f(T_1,T_2) = \frac{\phi(T_1)}{\phi(T_2)} \qquad \text{and} \qquad f(T_2,T_3) = \frac{\phi(T_2)}{\phi(T_3)}$$

so that $\phi(T_2)$ will cancel from the products of $f(T_1,T_2)$ and $f(T_2,T_3)$, yielding

$$\frac{Q_1}{Q_3} = f(T_1,T_3) = \frac{\phi(T_1)}{\phi(T_3)} \tag{5-16}$$

This relation is much more specific than Eq. 5-15 for the functional form of Q_1/Q_3 in terms of T_1 and T_3.

For a reversible heat engine operating between two reservoirs at temperatures T_H and T_L, Eq. 5-16 can be written as

$$\frac{Q_H}{Q_L} = \frac{\phi(T_H)}{\phi(T_L)} \tag{5-17}$$

This is the only requirement that the second law places on the ratio of heat flows to and from the reversible heat engines. Several functions $\phi(T)$

will satisfy this eq' ation, and the choice is completely arbitrary. Lord Kelvin first proposed taking $\phi(T) = T$ to define a thermodynamic temperature scale as (F'g. 5-47)

$$\left(\frac{Q_H}{Q_L}\right)_{\text{rev}} = \frac{T_H}{T_L} \tag{5-18}$$

This temperature scale is called the **Kelvin scale**, and the temperatures on this scale are called **absolute temperatures**. On the Kelvin scale, the temperature ratios depend on the ratios of heat transfer between a reversible heat engine and the reservoirs and are independent of the physical properties of any substance. On this scale, temperatures vary between zero and infinity.

The absolute temperature scale is not completely defined by Eq. 5-18 since it gives us only a ratio of absolute temperatures. We also need to know the magnitude of a kelvin degree. At the International Conference on Weights and Measures held in 1954, the triple point of water (the state at which all three phases of water exist in equilibrium) was assigned the value 273.16 K. The *magnitude of a kelvin degree* is defined as 1/273.16 of the temperature interval between absolute zero and the triple-point temperature of water. The magnitudes of temperature units on the Kelvin and Celsius scales are identical (1 K ≡ 1°C). The temperatures on these two scales differ by a constant 273.15:

$$T(°\text{C}) = T(\text{K}) - 273.15 \tag{5-19}$$

Even though the absolute temperature scale is defined with the help of the reversible heat engines, it is not possible, nor is it practical, to actually operate such an engine to determine numerical values on the absolute temperature scale. Absolute temperatures can be measured accurately by other means, such as the constant-volume ideal-gas thermometer discussed in Chap. 1 together with extrapolation techniques. The validity of Eq. 5-18 can be demonstrated from physical considerations for a reversible cycle using an ideal gas as the working fluid.

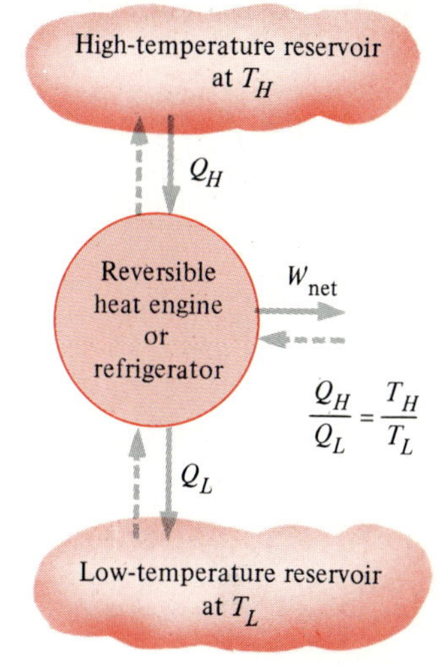

FIGURE 5-47
For reversible cycles, the heat transfer ratio Q_H/Q_L can be replaced by the absolute temperature ratio T_H/T_L.

5-10 ■ THE CARNOT HEAT ENGINE

The hypothetical heat engine that operates on the reversible Carnot cycle is called the **Carnot heat engine**. The thermal efficiency of any heat engine, reversible or irreversible, is given by Eq. 5-8 as

$$\eta_{\text{th}} = 1 - \frac{Q_L}{Q_H}$$

where Q_H is heat transferred to the heat engine from a high-temperature reservoir at T_H, and Q_L is heat rejected to a low-temperature reservoir at T_L. For reversible heat engines, the heat transfer ratio in the above relation can be replaced by the ratio of the absolute temperatures of the two reservoirs, as given by Eq. 5-18. Then the efficiency of a Carnot engine, or any other reversible heat engine, becomes

$$\eta_{\text{th,rev}} = 1 - \frac{T_L}{T_H} \tag{5-20}$$

This relation is often referred to as the **Carnot efficiency** since the Carnot heat engine is the best known reversible engine. *This is the highest efficiency a heat engine operating between the two thermal energy reservoirs at temperatures T_L and T_H can have* (Fig. 5-48). All irreversible (i.e., actual) heat engines operating between these temperature limits (T_L and T_H) will have lower efficiencies. An actual heat engine cannot reach this maximum theoretical efficiency value because it is impossible to completely eliminate all the irreversibilities associated with the actual cycle.

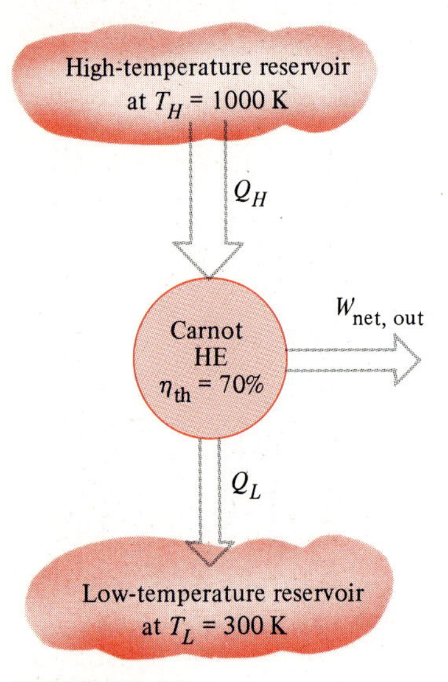

FIGURE 5-48

The Carnot heat engine is the most efficient of all heat engines operating between the same high- and low-temperature reservoirs.

Note that T_L and T_H in Eq. 5-20 are *absolute temperatures*. Using °C or °F for temperatures in this relation will give results grossly in error.

The thermal efficiencies of actual and reversible heat engines operating between the same temperature limits compare as follows (Fig. 5-49):

$$\eta_{\text{th}} \begin{cases} < \eta_{\text{th,rev}} & \text{irreversible heat engine} \\ = \eta_{\text{th,rev}} & \text{reversible heat engine} \\ > \eta_{\text{th,rev}} & \text{impossible heat engine} \end{cases} \tag{5-21}$$

Most work-producing devices (heat engines) in operation have efficiencies under 40 percent, which appear low relative to 100 percent. However, when the performance of actual heat engines is assessed, the efficiencies should not be compared to 100 percent; instead, they should be compared to the efficiency of a reversible heat engine operating between the same temperature limits—because this is the true theoretical upper limit for the efficiency, not the 100 percent.

The maximum efficiency of a steam power plant operating between T_H = 750 K and T_L = 300 K is 60 percent, as determined from Eq. 5-20. Compared to this value, an actual efficiency of 40 percent does not seem so bad, even though there is still plenty of room for improvement.

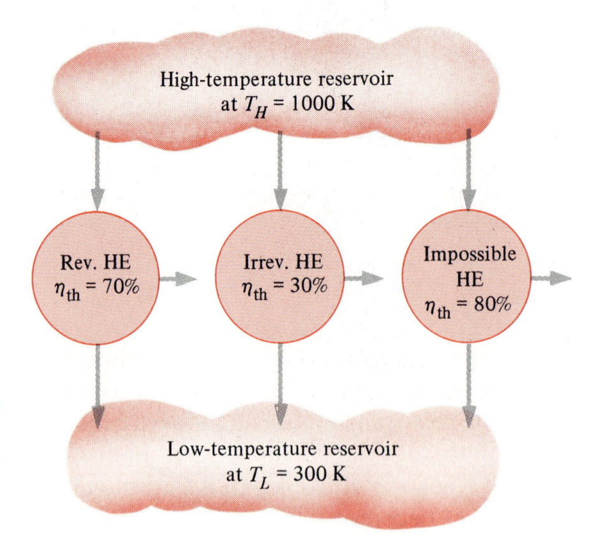

FIGURE 5-49

No heat engine can have a higher efficiency than a reversible heat engine operating between the same high- and low-temperature reservoirs.

It is obvious from Eq. 5-20 that the efficiency of a Carnot heat engine increases as T_H is increased, or as T_L is decreased. This is to be expected since as T_L decreases, so does the amount of heat rejected, and as T_L approaches zero, the Carnot efficiency approaches unity. This is also true for actual heat engines. *The thermal efficiency of actual heat engines can be maximized by supplying heat to the engine at the highest possible temperature* (limited by material strength) *and rejecting heat from the engine at the lowest possible temperature* (limited by the temperature of the cooling medium such as rivers, lakes, or the atmosphere).

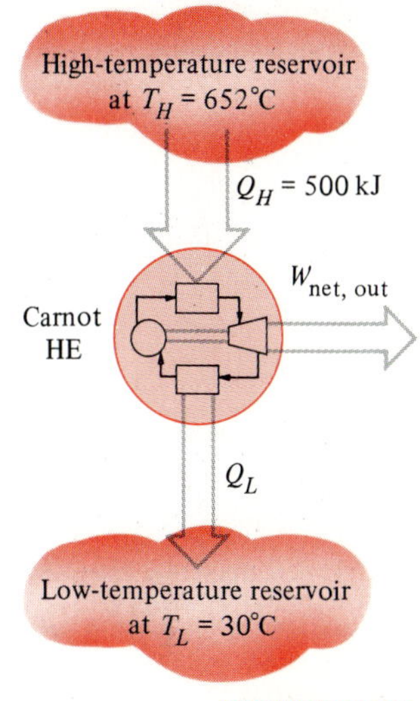

FIGURE 5-50
Schematic for Example 5-4.

EXAMPLE 5-5

A Carnot heat engine, shown in Fig. 5-50, receives 500 kJ of heat per cycle from a high-temperature source at 652°C and rejects heat to a low-temperature sink at 30°C. Determine (*a*) the thermal efficiency of this Carnot engine and (*b*) the amount of heat rejected to the sink per cycle.

Solution (*a*) The Carnot heat engine is a reversible heat engine, and so its efficiency can be determined from Eq. 5-20:

$$\eta_{\text{th},C} = \eta_{\text{th,rev}} = 1 - \frac{T_L}{T_H} = 1 - \frac{(30 + 273)\text{ K}}{(652 + 273)\text{ K}} = 0.672$$

That is, this Carnot heat engine converts 67.2 percent of the heat it receives to work.

(*b*) The amount of heat rejected Q_L by this reversible heat engine is easily determined from Eq. 5-28:

$$Q_{L,\text{rev}} = \frac{T_L}{T_H} Q_{H,\text{rev}} = \frac{(30 + 273)\text{ K}}{(652 + 273)\text{ K}} (500\text{ kJ}) = 163.8\text{ kJ}$$

Therefore, this Carnot heat engine discharges 163.8 kJ of the 500 kJ of heat it receives during each cycle to a sink.

The Quality of Energy

The Carnot heat engine in Example 5-5 receives heat from a source at 925 K and converts 67.2 percent of it to work while rejecting the rest (32.8 percent) to a sink at 303 K. Now let us examine how the thermal efficiency varies with the source temperature when the sink temperature is held constant.

The thermal efficiency of a Carnot heat engine that rejects heat to a sink at 303 K is evaluated at various source temperatures using Eq. 5-20 and is listed in Fig. 5-51. Clearly the thermal efficiency decreases as the source temperature is lowered. When heat is supplied to the heat engine at 500 instead of 925 K, for example, the thermal efficiency drops from 67.2 to 39.4 percent. That is, the fraction of heat that can be converted to work drops to 39.4 percent when the temperature of the source drops to 500 K. When the source temperature is 350 K, this fraction becomes a mere 13.4 percent.

These efficiency values show that energy has **quality** as well as quan-

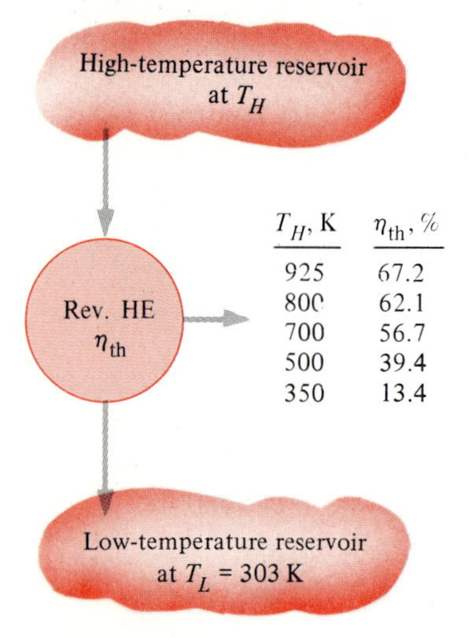

T_H, K	η_{th}, %
925	67.2
800	62.1
700	56.7
500	39.4
350	13.4

FIGURE 5-51

The fraction of heat that can be converted to work as a function of source temperature (for T_L = 303 K).

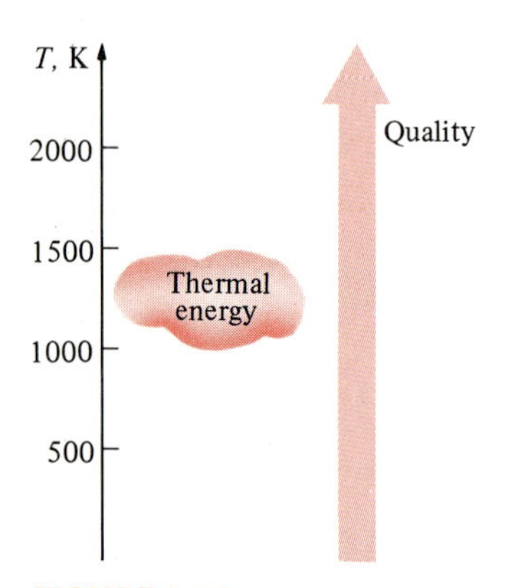

FIGURE 5-52

The higher the temperature of the thermal energy, the higher its quality.

tity. It is clear from the thermal efficiency values in Fig. 5-51 that *more of the high-temperature thermal energy can be converted to work. Therefore, the higher the temperature, the higher the quality of the energy* (Fig. 5-52).

Large quantities of solar energy, for example, can be stored in large bodies of water called *solar ponds* at about 350 K. This stored energy can then be supplied to a heat engine to produce work (electricity). However, the efficiency of solar pond power plants is very low (under 5 percent) because of the low quality of the energy stored in the source, and the construction and maintenance costs are relatively high. Therefore, they are not competitive even though the energy supply of such plants is free. The temperature (and thus the quality) of the solar energy stored could be raised by utilizing concentrating collectors, but the equipment cost in this case becomes prohibitive.

Work is a more valuable form of energy than heat since 100 percent of work can be converted to heat but only a fraction of heat can be converted to work. When heat is transferred from a high-temperature body to a lower temperature one, it is degraded since less of it now can be converted to work. For example, if 100 kJ of heat is transferred from a body at 1000 K to a body at 300 K, at the end we will have 100 kJ of thermal energy stored at 300 K, which has no practical value. But if this conversion is made through a heat engine, up to $1 - 300/1000 = 70$ percent of it could be converted to work, which is a more valuable form of energy. Thus 70 kJ of work potential is wasted as a result of this heat transfer, and energy is degraded. The degradation of energy during a process is discussed more fully in Chap. 7.

5-11 ■ THE CARNOT REFRIGERATOR AND HEAT PUMP

A refrigerator or a heat pump that operates on the reversed Carnot cycle is called a **Carnot refrigerator**, or a **Carnot heat pump**. The coefficient of performance of any refrigerator or heat pump, reversible or irreversible, is given by Eqs. 5-11 and 5-13 as

$$\text{COP}_\text{R} = \frac{1}{Q_H/Q_L - 1} \qquad \text{and} \qquad \text{COP}_\text{HP} = \frac{1}{1 - Q_L/Q_H}$$

where Q_L is the amount of heat absorbed from the low-temperature medium and Q_H is the amount of heat rejected to the high-temperature medium. The COPs of all reversible (such as Carnot) refrigerators or heat pumps can be determined by replacing the heat transfer ratios in the above relations by the ratios of the absolute temperatures of the high- and low-temperature media, as expressed by Eq. 5-18. Then the COP relations for reversible refrigerators and heat pumps become

$$\text{COP}_{\text{R,rev}} = \frac{1}{T_H/T_L - 1} \tag{5-22}$$

and

$$\text{COP}_{\text{HP,rev}} = \frac{1}{1 - T_L/T_H} \tag{5-23}$$

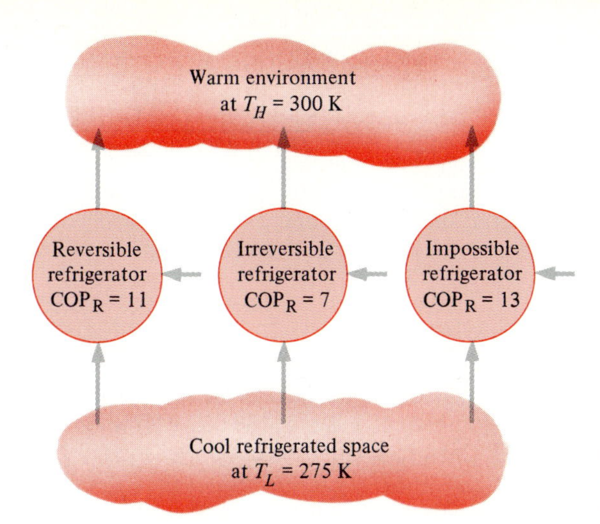

FIGURE 5-53
No refrigerator can have a higher COP than a reversible refrigerator operating between the same temperature limits.

These are the highest coefficients of performance that a refrigerator or a heat pump operating between the temperature limits of T_L and T_H can have. All actual refrigerators or heat pumps operating between these temperature limits (T_L and T_H) will have lower coefficients of performance (Fig. 5-53).

The coefficients of performance of actual and reversible (such as Carnot) refrigerators operating between the same temperature limits can be compared as follows:

$$\mathrm{COP_R}\begin{cases} < \mathrm{COP_{R,rev}} & \text{irreversible refrigerator} \\ = \mathrm{COP_{R,rev}} & \text{reversible refrigerator} \\ > \mathrm{COP_{R,rev}} & \text{impossible refrigerator} \end{cases} \tag{5-24}$$

A similar relation can be obtained for heat pumps by replacing all values of $\mathrm{COP_R}$ in Eq. 5-24 by $\mathrm{COP_{HP}}$.

The COP of a reversible refrigerator or heat pump is the maximum theoretical value for the specified temperature limits. Actual refrigerators or heat pumps may approach these values as their designs are improved, but they can never reach them.

As a final note, the COPs of both the refrigerators and the heat pumps decrease as T_L decreases. That is, it requires more work to absorb heat from lower temperature media. As the temperature of the refrigerated space approaches zero, the amount of work required to produce a finite amount of refrigeration approaches infinity and $\mathrm{COP_R}$ approaches zero.

EXAMPLE 5-6

An inventor claims to have developed a refrigerator that maintains the refrigerated space at 35°F while operating in a room where the temperature is 75°F and that has a COP of 13.5. Is there any truth to his claim?

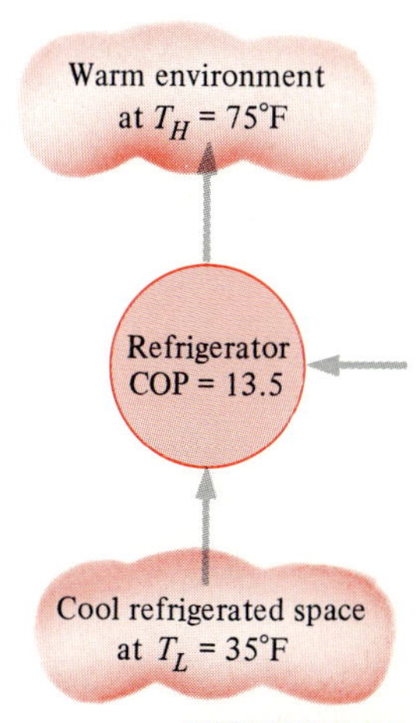

FIGURE 5-54
Schematic for Example 5-6.

Solution The performance of this refrigerator (shown in Fig. 5-54) can be evaluated by comparing it to a Carnot or any other reversible refrigerator operating between the same temperature limits:

$$\text{COP}_{\text{R,max}} = \text{COP}_{\text{R,rev}} = \frac{1}{T_H/T_L - 1}$$

$$= \frac{1}{(75 + 460 \text{ R})/(35 + 460 \text{ R}) - 1} = 12.4$$

This is the highest COP a refrigerator can have when removing heat from a cool medium at 35°F to a warmer medium at 75°F. Since the COP claimed by the inventor is above this maximum value, his claim is *false*.

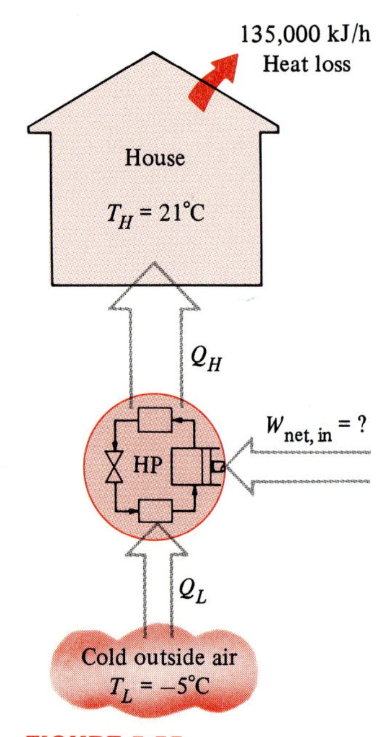

FIGURE 5-55
Schematic for Example 5-7.

EXAMPLE 5-7

A heat pump is to be used to heat a house during the winter, as shown in Fig. 5-55. The house is to be maintained at 21°C at all times. The house is estimated to be losing heat at a rate of 135,000 kJ/h when the outside temperature drops to −5°C. Determine the minimum power required to drive this heat pump unit.

Solution To maintain the house at a fixed temperature, the heat pump must supply the house with as much heat as it is losing. That is, the heat pump must reject heat to the house (the high-temperature reservoir) at a rate of Q_H = 135,000 kJ/h = 37.5 kW.

The power requirements will be minimum if a reversible heat pump is used to do the job. The COP of a reversible heat pump operating between the house (T_H = 21 + 273 = 294 K) and the outside air (T_L = −5 + 273 = 268 K) is, from Eq. 5-23,

$$\text{COP}_{\text{HP,rev}} = \frac{1}{1 - T_L/T_H} = \frac{1}{1 - (268 \text{ K}/294 \text{ K})} = 11.3$$

Then the required power input to this reversible heat pump is determined from the definition of the COP, Eq. 5-12:

$$\dot{W}_{\text{net,in}} = \frac{\dot{Q}_H}{\text{COP}_{\text{HP}}} = \frac{37.5 \text{ kW}}{11.3} = 3.32 \text{ kW}$$

That is, this heat pump can meet the heating requirements of this house by consuming electric power at a rate of 3.32 kW only. If this house were to be heated by electric resistance heaters instead, the power consumption rate would jump up 11.3 times to 37.5 kW. This is because in resistance heaters the electric energy is converted to heat at a one-to-one ratio. With a heat pump, however, energy is absorbed from the outside and carried to the inside using a refrigeration cycle that consumes only 3.32 kW. Notice that the heat pump does not create energy. It merely transports it from one medium (the cold outdoors) to another (the warm indoors).

5-12 ■ SUMMARY

The *second law of thermodynamics* states that processes occur in a certain direction, not in any direction. A process will not occur unless it satisfies both the first and the second laws of thermodynamics. Bodies that can absorb or reject finite amounts of heat isothermally are called *thermal energy reservoirs* or just *reservoirs*.

Work can be converted to heat directly, but heat can only be converted to work by some devices called heat engines. The *thermal efficiency* of a heat engine is defined as

$$\eta_{th} = \frac{W_{net,out}}{Q_H} = 1 - \frac{Q_L}{Q_H} \qquad (5\text{-}7, 5\text{-}8)$$

where $W_{net,out}$ is the net work output of the heat engine, Q_H is the amount of heat supplied to the engine, and Q_L is the amount of heat rejected by the engine.

Refrigerators and heat pumps are devices that absorb heat from low-temperature media and reject it to higher temperature ones. The efficiency of a refrigerator or a heat pump is expressed in terms of the *coefficient of performance,* which is defined as

$$\mathrm{COP_R} = \frac{Q_L}{W_{net,in}} = \frac{1}{Q_H/Q_L - 1} \qquad (5\text{-}9, 5\text{-}11)$$

$$\mathrm{COP_{HP}} = \frac{Q_H}{W_{net,in}} = \frac{1}{1 - Q_L/Q_H} \qquad (5\text{-}12, 5\text{-}13)$$

The *Kelvin-Planck statement* of the second law of thermodynamics states that no heat engine can produce a net amount of work while exchanging heat with a single reservoir only. The *Clausius statement* of the second law states that no device can transfer heat from a cooler body to a warmer one without leaving an effect on the surroundings.

Any device that violates the first or the second law of the thermodynamics is called a *perpetual-motion machine*.

A process is said to be *reversible* if both the system and the surroundings can be restored to their original conditions. Any other process is *irreversible*. The effects, such as friction, non-quasi-equilibrium expansion or compression, and heat transfer through a finite temperature difference render a process irreversible and are called *irreversibilities*.

The *Carnot cycle* is a reversible cycle that is composed of four reversible processes, two isothermal and two adiabatic. The *Carnot principles* state that the thermal efficiencies of all reversible heat engines operating between the same two reservoirs are the same, and that no heat engine is more efficient that a reversible one operating between the same two reservoirs. These statements form the basis for establishing an *absolute temperature scale,* also called the *Kelvin scale*, related to the heat transfers between a reversible device and the high- and low-temperature reservoirs by

$$\left(\frac{Q_H}{Q_L}\right)_{rev} = \frac{T_H}{T_L} \qquad (5\text{-}18)$$

Then the Q_H/Q_L ratio can be replaced by T_H/T_L for reversible devices, where T_H and T_L are the absolute temperatures of the high- and low-temperature reservoirs, respectively.

A heat engine that operates on the reversible Carnot cycle is called the *Carnot heat engine*. The thermal efficiency of a Carnot heat engine, as well as all other reversible heat engines, is given by

$$\eta_{\text{th,rev}} = 1 - \frac{T_L}{T_H} \tag{5-20}$$

This is the maximum efficiency a heat engine operating between two reservoirs at temperatures T_H and T_L can have.

The COPs of reversible refrigerators and heat pumps are given in a similar manner as

$$\text{COP}_{\text{R,rev}} = \frac{1}{T_H/T_L - 1} \tag{5-22}$$

and

$$\text{COP}_{\text{HP,rev}} = \frac{1}{1 - T_L/T_H} \tag{5-23}$$

Again, these are the highest COPs a refrigerator or a heat pump operating between the temperature limits of T_H and T_L can have.

REFERENCES AND SUGGESTED READING

1 W. Z. Black and J. G. Hartley, *Thermodynamics,* Harper & Row, New York, 1985.

2 J. R. Howell and R. O. Buckius, *Fundamentals of Engineering Thermodynamics,* McGraw-Hill, New York, 1987.

3 J. B. Jones and G. A. Hawkins, *Engineering Thermodynamics,* 2d ed., Wiley, New York, 1986.

4 G. J. Van Wylen and R. E. Sonntag, *Fundamentals of Classical Thermodynamics,* 3d ed., Wiley, New York, 1985.

5 K. Wark, *Thermodynamics*, 5th ed., McGraw-Hill, New York, 1988.

PROBLEMS*

Introduction to the Second Law of Thermodynamics

5-1C Does satisfying the first law of thermodynamics ensure that the process can actually take place? Explain.

5-2C Describe an imaginary process that satisfies the first law but violates the second law of thermodynamics.

5-3C Describe an imaginary process that satisfies the second law but violates the first law of thermodynamics.

*Students are encouraged to answer *all* concept "C" questions.

5-4C Describe an imaginary process that violates both the first and the second laws of thermodynamics.

5-5C An experimentalist claims to have raised the temperature of a small amount of water to 150°C, using high-pressure steam at 120°C. Is this a reasonable claim? Why? Assume no refrigerator or heat pump is used in the process.

5-6C A mechanic claims to have developed a car engine that runs on water instead of gasoline. What is your response to this claim?

Thermal Energy Reservoirs

5-7C What is a thermal energy reservoir? Give some examples.

5-8C In studying the process of boiling eggs, can the boiling water be treated as a thermal energy reservoir? Explain.

5-9C In studying the process of baking potatoes in a conventional oven, can the hot air in the oven be treated as a thermal energy reservoir? Explain.

5-10C In studying the energy generated by a TV set, what is a suitable choice for a thermal energy reservoir?

5-11C In studying the energy dissipated by a computer in a room, what is a suitable choice for a thermal energy reservoir?

Heat Engines and Thermal Efficiency

5-12C What are the characteristics of all heat engines?

5-13C What is the physical significance of the thermal efficiency of a heat engine, and how is it determined?

5-14C Describe two ways to determine the net work output of a heat engine.

5-15C Consider the process of baking potatoes in a conventional oven. How would you define the efficiency of the oven for this baking process?

5-16C Consider a pan of water being heated (*a*) by placing it on an electric range and (*b*) by placing a heating element in the water. Which method is a more efficient way of heating water? Explain.

5-17C Which one of these is a more efficient way of heating a house—burning wood in a fireplace or burning wood in a stove in the middle of the house?

5-18C Which is a more efficient way of converting electricity to light—using a light bulb or using a fluorescent tube?

5-19C Is it possible for a heat engine to operate without rejecting any waste heat to a low-temperature reservoir? Explain.

5-20C What is the Kelvin-Planck expression of the second law of thermodynamics?

5-21C Does a heat engine that has a thermal efficiency of 100 percent necessarily violate (*a*) the first law and (*b*) the second law of thermodynamics? Explain.

5-22C In the absence of any friction and other irreversibilities, can a heat engine have an efficiency of 100 percent? Explain.

5-23C Are the efficiencies of all the work-producing devices, including the hydroelectric power plants, limited by the Kelvin-Planck statement of the second law? Explain.

5-24C Baseboard heaters are basically electric resistance heaters and are frequently used in space heating. A homeowner claims that her 5-years old baseboard heaters have a conversion efficiency of 100 percent. Is this claim in violation of any thermodynamic laws? Explain.

5-25C Can a frictionless heat engine receive 100 kJ of energy from a thermal energy reservoir and convert it all to work? Would such an engine violate any thermodynamic laws? Explain.

5-26 A 600-MW steam power plant, which is cooled by a nearby river, has a thermal efficiency of 40 percent. Determine the rate of heat transfer to the river water. Will the actual heat transfer rate be higher or lower than this value? Why?

5-27 A steam power plant receives heat from a furnace at a rate of 280 GJ/h. Heat losses to the surrounding air from the steam as it passes through the pipes and other components are estimated to be about 8 GJ/h. If the waste heat is transferred to the cooling water at a rate of 145 GJ/h, determine (*a*) the net power output and (*b*) the thermal efficiency of this power plant. *Answers:* (*a*) 35.3 MW, (*b*) 45.4 percent

5-28 A car engine with a power output of 70 kW has a thermal efficiency of 22 percent. Determine the rate of fuel consumption if the energy content of the fuel is 44,000 kJ/kg.

5-28E A car engine with a power output of 90 hp has a thermal efficiency of 20 percent. Determine the rate of fuel consumption if the energy content of the fuel is 19,000 Btu/lbm.

5-29 A steam power plant with a power output of 100 MW consumes coal at a rate of 40 tons/h. If the energy content of the coal is 30,000 kJ/kg, determine the thermal efficiency of this plant (1 ton ≡ 1000 kg). *Answer:* 30.0 percent

5-30 A gas turbine has an efficiency of 17 percent and develops a power output of 6000 kW. Determine the fuel consumption rate of this gas turbine, in L/min, if the fuel has an energy content of 46,000 kJ/kg and a density of 0.8 g/cm^3.

5-31 An automobile engine consumes fuel at a rate of 20 L/h and delivers

60 kW of power to the wheels. If the fuel has an energy content of 44,000 kJ/kg and a density of 0.8 g/cm^3, determine the efficiency of this engine. *Answer:* 21.9 percent

5-31E An automobile engine consumes fuel at a rate of 5 gal/h and delivers 70 hp of power to the wheels. If the fuel has an energy content of 19,000 Btu/lbm and a density of 50 lbm/ft^3, determine the efficiency of this engine. *Answer:* 28.1 percent

5-32 Solar energy stored in large bodies of water, called solar ponds, is being used for generating electricity. If such a solar power plant has an efficiency of 3 percent and a net power output of 200 kW, determine the average value of the required solar energy collection rate, in kJ/h. *Answer:* 2.4×10^7 kJ/h

5-32E Solar energy stored in large bodies of water, called solar ponds, is being used for generating electricity. If such a solar power plant has an efficiency of 4 percent and a net power output of 300 kW, determine the average value of the required solar energy collection rate, in Btu/h.

Refrigerators and Heat Pumps

5-33C What is the difference between a refrigerator and a heat pump?

5-34C What is the difference between a refrigerator and an air conditioner?

5-35C In a refrigerator, heat is transferred from a lower temperature medium (the refrigerated space) to a higher temperature one (the kitchen air). Is this a violation of the second law of thermodynamics? Explain.

5-36C A heat pump is a device that absorbs energy from the cold outdoor air and transfers it to the warmer indoors. Is this a violation of the second law of thermodynamics? Explain.

5-37C Define the coefficient of performance of a refrigerator in words. Can it be greater than unity?

5-38C Define the coefficient of performance of a heat pump in words. Can it be greater than unity?

5-39C A heat pump that is used to heat a house has a COP of 2.5. That is, the heat pump delivers 2.5 kWh of energy to the house for each 1 kWh of electricity it consumes. Is this a violation of the first law of thermodynamics? Explain.

5-40C A refrigerator has a COP of 1.5. That is, the refrigerator removes 1.5 kWh of energy from the refrigerated space for each 1 kWh of electricity it consumes. Is this a violation of the first law of thermodynamics? Explain.

5-41C Show that $\mathrm{COP_{HP}} = \mathrm{COP_R} + 1$ when both the heat pump and the refrigerator have the same Q_L and Q_H values.

5-42C What is the Clausius expression of the second law of thermodynamics?

5-43C Show that the Kelvin-Planck and the Clausius expressions of the second law are equivalent.

5-44 A household refrigerator with a COP of 1.5 removes heat from the refrigerated space at a rate of 60 kJ/min. Determine (*a*) the electric power consumed by the refrigerator and (*b*) the rate of heat transfer to the kitchen air. *Answers:* (*a*) 0.67 kW, (*b*) 100 kJ/min

5-44E A household refrigerator with a COP of 1.8 removes heat from the refrigerated space at a rate of 55 Btu/min. Determine (*a*) the electric power consumed by the refrigerator and (*b*) the rate of heat transfer to the kitchen air. *Answers:* (*a*) 0.72 hp, (*b*) 85.56 Btu/min

5-45 An air conditioner removes heat steadily from a house at a rate of 750 kJ/min while drawing electric power at a rate of 6 kW. Determine (*a*) the COP of this air conditioner and (*b*) the rate of heat discharge to the outside air. *Answers:* (*a*) 2.08, (*b*) 1110 kJ/min

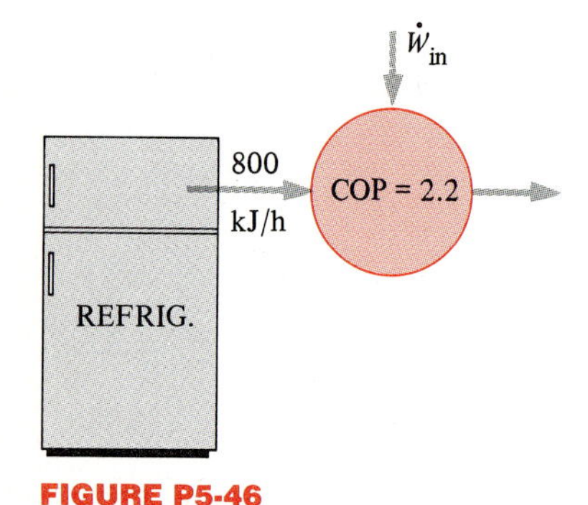

FIGURE P5-46

5-46 A household refrigerator runs one-fourth of the time and removes heat from the food compartment at an average rate of 800 kJ/h. If the COP of the refrigerator is 2.2, determine the power the refrigerator draws when running.

5-46E A household refrigerator runs one-fourth of the time and removes heat from the food compartment at an average rate of 800 Btu/h. If the COP of the refrigerator is 2.2, determine the power the refrigerator draws when running.

5-47 An air conditioning system is used to maintain a house at a constant temperature of 20°C. The house is gaining heat from outdoors at a rate of 20,000 kJ/h, and the heat generated in the house from the people, lights, and appliances amounts to 8000 kJ/h. For a COP of 2.5, determine the required power input to this air conditioning system.
Answer: 3.11 kW

5-48 Water enters an ice machine at 15°C and leaves as ice at −5°C. If the COP of the ice machine is 2.2 during this operation, determine the required power input for an ice production rate of 8 kg/h. (Note that 384 kJ of energy needs to be removed from each 1 kg of water at 15°C to turn it into ice at −5°C.)

5-48E Water enters an ice machine at 55°F and leaves as ice at 25°F. If the COP of the ice machine is 2.2 during this operation, determine the required power input for an ice production rate of 15 lbm/h. (Note that 169 Btu of energy needs to be removed from each 1 lbm of water at 55°F to turn it into ice at 25°F.)

5-49 A household refrigerator that has a power input of 450 W and a COP of 2.5 is to cool five large watermelons, 10 kg each, to 8°C. If the watermelons are initially at 20°C, determine how long it will take for the refrigerator to cool them. The watermelons can be treated as water whose specific heat is 4.2 kJ/(kg·°C). Is your answer realistic or optimistic? Explain. *Answer:* 2240 s

5-50 When a man returns to his well-sealed house on a summer day, he finds that the house is at 32°C. He turns on the air conditioner which cools the entire house to 20°C in 15 min. If the COP of the air conditioning system is 2.5, determine the power drawn by the air conditioner. Assume the entire mass within the house is equivalent to 800 kg of air for which C_v = 0.72 kJ/(kg·°C) and C_p = 1.0 kJ/(kg·°C).

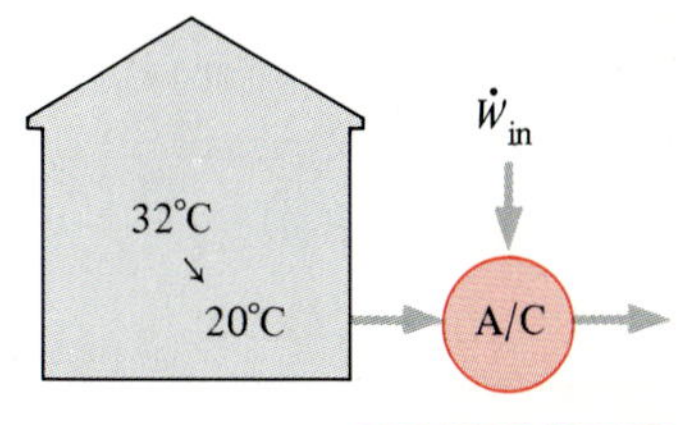

FIGURE P5-50

5-50E When a man returns to his well-sealed house on a summer day, he finds that the entire house is at 90°F. He turns on the air conditioner which cools the entire house to 70°F in 15 min. If the COP of the air conditioning system is 2.5, determine the power drawn by the air conditioner. Assume the entire mass within the house is equivalent to 1800 lbm of air for which C_v = 0.17 Btu/(lbm·°F) and C_p = 0.24 Btu/(lbm·°F).

5-51 Determine the COP of a refrigerator which removes heat from the food compartment at a rate of 5000 kJ/h for each 1 kW of power it consumes. Also determine the rate of heat rejection to the surrounding air.

5-52 Determine the COP of a heat pump which supplies energy to a house at a rate of 8000 kJ/h for each 1 kW of electric power it draws. Also determine the rate of energy absorption from the outdoor air.
Answers: 2.22, 4400 kJ/h

5-53 A house that was heated by electric resistance heaters consumed 1650 kWh of electric energy in a winter month. If this house were heated by a heat pump, instead, that has an average COP of 2.8, determine how much money the homeowner would have saved that month. Assume a price of 8.5¢/kWh for electricity.

5-54 A heat pump with a COP of 2.5 supplies energy to a house at a rate of 60,000 kJ/h. Determine (*a*) the electric power drawn by the heat pump and (*b*) the rate of heat removal from the outside air.
Answers: (*a*) 6.67 kW, (*b*) 36,000 kJ/h

5-54E A heat pump with a COP of 2.5 supplies energy to a house at a rate of 60,000 Btu/h. Determine (*a*) the electric power drawn by the heat pump and (*b*) the rate of heat removal from the outside air.
Answers: (*a*) 9.43 hp, (*b*) 36,000 Btu/h

5-55 A heat pump that is used to heat a house runs about one-third of the time. The house is losing heat at an average rate of 15,000 kJ/h. If the COP of the heat pump is 3.5, determine the power the heat pump draws when running.

5-55E A heat pump that is used to heat a house runs about one-third of the time. The house is losing heat at an average rate of 15,000 Btu/h. If the COP of the heat pump is 3.5, determine the power the heat pump draws when running.

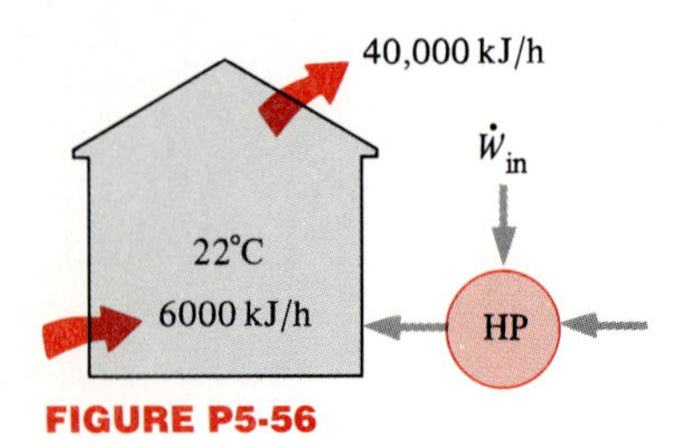

FIGURE P5-56

5-56 A heat pump is used to maintain a house at a constant temperature of 22°C. The house is losing heat to the outside air through the walls and the windows at a rate of 40,000 kJ/h while the energy generated within the house from people, lights, and appliances amounts to 6000 kJ/h. For a COP of 2.4, determine the required power input to the heat pump. *Answer:* 3.70 kW

5-56E A heat pump is used to maintain a house at a constant temperature of 70°F. The house is losing heat to the outside air through the walls and the windows at a rate of 40,000 Btu/h while the energy generated within the house from people, lights, and appliances amounts to 6000 Btu/h. For a COP of 2.4, determine the required power input to the heat pump.

5-57 A heat pump with a COP of 3.2 is used to heat a house. When running, the heat pump consumes power at a rate of 5 kW. If the temperature in the house is 7°C when the heat pump is turned on, how long will it take for the heat pump to raise the temperature of the house to 22°C? Is this answer realistic or optimistic? Explain. Assume the entire mass within the house (air, furniture, etc.) is equivalent to 1500 kg of air for which C_v = 0.72 kJ/(kg·°C) and C_p = 1.0 kJ/(kg·°C). *Answer:* 1012 s

Perpetual-Motion Machines

5-58C An inventor claims to have developed a resistance heater that supplies 1.2 kWh of energy to a room for each 1 kWh of electricity it consumes. Is this a reasonable claim, or has she developed a perpetual-motion machine? Explain.

5-59C It is common knowledge that the temperature of air rises as it is compressed. An inventor thought about using this high-temperature air to heat buildings. He used a compressor driven by an electric motor. The inventor claims that the compressed hot-air system is 12 percent more efficient than a resistance heating system that provides an equivalent amount of heating. Is this claim valid, or is this just another perpetual-motion machine? Explain.

Reversible and Irreversible Processes

5-60C A cold canned drink is left in a warmer room where its temperature rises as a result of heat transfer. Is this a reversible process? Explain.

5-61C A hot baked potato is left on a table where it cools to the room temperature. Is this a reversible or an irreversible process? Explain.

5-62C Why are engineers interested in reversible processes even though they can never be achieved?

5-63C Air is compressed from 100 to 800 kPa first in a reversible manner and then in an irreversible manner. Which case do you think will require more work input?

5-64C Why does a non-quasi-equilibrium compression process require a larger work input than the corresponding quasi-equilibrium one?

5-65C Why does a non-quasi-equilibrium expansion process deliver less work than the corresponding quasi-equilibrium one?

5-66C Describe internally, externally, and totally reversible processes.

The Carnot Cycle and Carnot Principles

5-67C What are the four processes that make up the Carnot cycle?

5-68C What are the four processes that make up the reversed Carnot cycle?

5-69C What are the two statements known as the Carnot principles?

5-70C Somebody claims to have developed a new reversible heat-engine cycle that has a higher theoretical efficiency than the Carnot cycle operating between the same temperature limits. How do you evaluate this claim?

5-71C Somebody claims to have developed a new reversible heat-engine cycle that has the same theoretical efficiency as the Carnot cycle operating between the same temperature limits. Is this a reasonable claim?

5-72C Is it possible to develop (*a*) an actual and (*b*) a reversible heat-engine cycle that is more efficient than a Carnot cycle operating between the same temperature limits? Explain.

Carnot Heat Engines

5-73C Is there any way to increase the efficiency of a Carnot heat engine other than by increasing T_H or decreasing T_L?

5-74C Consider two actual power plants operating with solar energy. Energy is supplied to one plant from a solar pond at 80°C and to the other from concentrating collectors that raise the water temperature to 600°C. Which of these power plants will have a higher efficiency and why?

5-75 A Carnot heat engine operates between a source at 1200 K and a sink at 300 K. If the heat engine is supplied with heat at a rate of 800 kJ/min, determine (*a*) the thermal efficiency and (*b*) the power output of this heat engine. *Answers:* (*a*) 75 percent, (*b*) 10 kW

5-75E A Carnot heat engine operates between a source at 2000 R and a sink at 440 R. If the heat engine is supplied with heat at a rate of 800 Btu/min, determine (*a*) the thermal efficiency and (*b*) the power output of this heat engine. *Answers:* (*a*) 78 percent, (*b*) 14.7 hp

5-76 A Carnot heat engine receives 500 kJ of heat from a source of unknown temperature and rejects 200 kJ of it to a sink at 17°C. Determine (*a*) the temperature of the source and (*b*) the thermal efficiency of the heat engine.

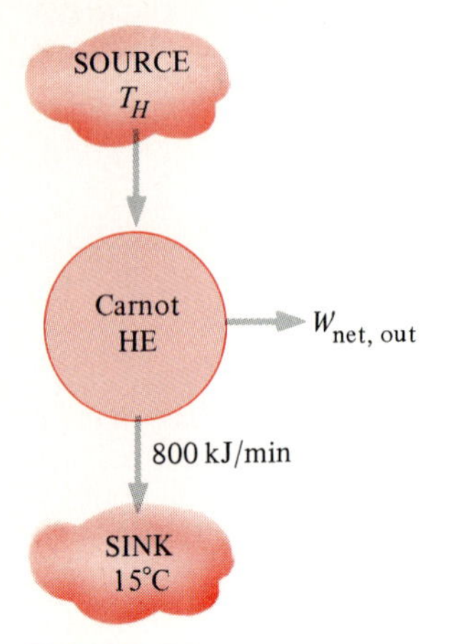

FIGURE P5-78

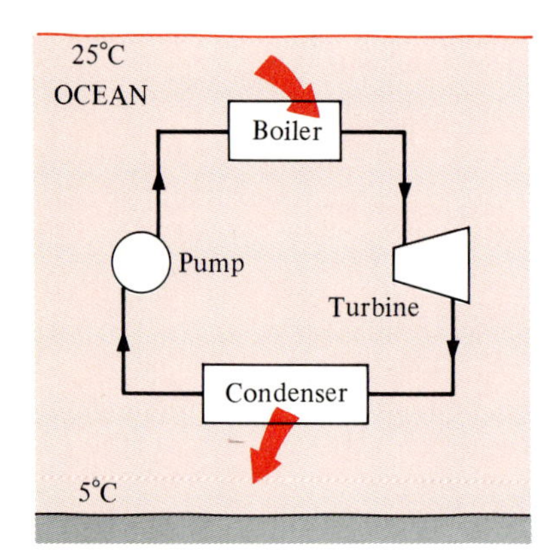

FIGURE P5-80

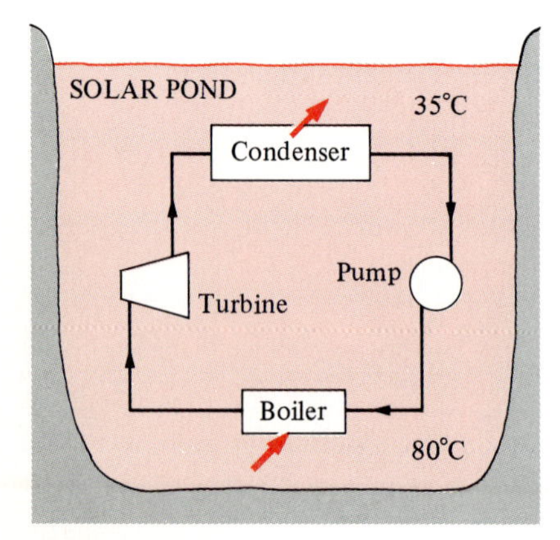

FIGURE P5-82

5-77 A heat engine operates between a source at 550°C and a sink at 25°C. If heat is supplied to the heat engine at a steady rate of 1200 kJ/min, determine the maximum power output of this heat engine.

5-78 A heat engine is operating on Carnot cycle and has a thermal efficiency of 55 percent. The waste heat from this engine is rejected to a nearby lake at 15°C at a rate of 800 kJ/min. Determine (*a*) the power output of the engine and (*b*) the temperature of the source.
Answers: (*a*) 16.3 kW, (*b*) 640 K

5-78E A heat engine is operating on Carnot cycle and has a thermal efficiency of 55 percent. The waste heat from this engine is rejected to a nearby lake at 60°F at a rate of 800 Btu/min. Determine (*a*) the power output of the engine and (*b*) the temperature of the source.
Answers: (*a*) 23.1 hp, (*b*) 1155.6 R

5-79 A Carnot heat engine is operating between a source at T_H and a sink at T_L. If it is desired to double the thermal efficiency of this engine, what should the new source temperature be? Assume the sink temperature is held constant.

5-80 In tropical climates, the water near the surface of the ocean remains warm throughout the year as a result of solar energy absorption. In the deeper parts of the ocean, however, the water remains at a relatively low temperature since the sun's rays cannot penetrate very far. It is proposed to take advantage of this temperature difference and construct a power plant that will absorb heat from the warm water near the surface and reject the waste heat to the cold water a few hundred meters below. Determine the maximum thermal efficiency of such a plant if the water temperatures at the two respective locations are 25 and 5°C.
Answer: 6.7 percent

5-81 An innovative way of power generation involves the utilization of geothermal energy—the energy of hot steam that exists naturally underground—as the heat source. If a supply of saturated steam at 125°C is discovered at a location where the environmental temperature is 25°C, determine the maximum thermal efficiency a geothermal power plant built at that location can have.

5-82 A promising method of power generation involves collecting and storing solar energy in large artificial lakes a few meters deep, called solar ponds. Solar energy is absorbed by all parts of the pond, and the water temperature rises everywhere. The top part of the pond, however, loses much of the heat it absorbs to the atmosphere, and as a result, its temperature drops. This cool water serves as insulation for the bottom part of the pond and helps trap the energy there. Usually, salt is planted at the bottom of the pond to prevent the rise of this hot water to the top. A power plant that uses an organic fluid, such as alcohol, as the working fluid can be operated between the top and the bottom portions of the pond. If the water temperature is 35°C near the surface and 80°C near the bottom of the pond, determine the maximum thermal efficiency that this power plant can have. Is it realistic to use 35 and 80°C for temperatures in the calculations? Explain. *Answer:* 12.7 percent

5-83 An inventor claims to have developed a heat engine that receives 600 kJ of heat from a source at 400 K and produces 200 kJ of net work while rejecting the waste heat to a sink at 300 K. Is this a reasonable claim? Why?

5-83E An inventor claims to have developed a heat engine that receives 600 Btu of heat from a source at 750 R and produces 200 Btu of net work while rejecting the waste heat to a sink at 550 R. Is this a reasonable claim? Why?

5-84 An experimentalist claims that, based on his measurements, a heat engine receives 300 kJ of heat from a source at 500 K, converts 160 kJ of it to work, and rejects the rest as waste heat to a sink at 300 K. Are these measurements reasonable? Why?

5-84E An experimentalist claims that, based on his measurements, a heat engine receives 300 Btu of heat from a source of 900 R, converts 160 Btu of it to work, and rejects the rest as waste heat to a sink at 540 R. Are these measurements reasonable? Why?

5-85 Consider two Carnot heat engines operating in series. The first engine receives heat from a reservoir at 1000 K and rejects the waste heat to another reservoir at temperature T. The second engine receives this energy rejected by the first one, converts some of it to work, and rejects the rest to a reservoir at 300 K. If the thermal efficiencies of both engines are the same, determine the temperature T. *Answer:* 547.7 K

Carnot Refrigerators and Heat Pumps

5-86C How can we increase the COP of a Carnot refrigerator?

5-87C What is the highest COP that a refrigerator operating between temperature levels T_L and T_H can have?

5-88C What is the highest COP that a heat pump operating between temperature levels T_L and T_H can have?

5-89C Does a household refrigerator operate more efficiently (i.e., remove more heat from the food compartment per unit of electric energy consumed) when the kitchen air is warm or cool?

5-90C Does a heat pump operate more efficiently (i.e., supply more heat to the house per unit of electric energy consumed) when the outdoor air temperature is moderately low or very low? Explain.

5-91C Discuss how lowering the thermostat setting of a refrigerator will affect the performance of the refrigerator (i.e., will the amount of heat removed from the food compartment per unit of electric energy consumed increase or decrease?).

5-92C Discuss how lowering the thermostat setting of a house heated by a heat pump will affect the performance of the heat pump (i.e, will the amount of heat supplied to the house by the heat pump per unit of electric energy consumed increase or decrease?).

5-93C In an effort to conserve energy in a heat-engine cycle, somebody suggests incorporating a refrigerator which will absorb some of the waste energy Q_L and transfer it to the energy source of the heat engine. Is this a smart idea? Explain.

5-94C It is well established that the thermal efficiency of a heat engine increases as the temperature at which heat is rejected from the heat engine T_L decreases. In an effort to increase the efficiency of a power plant, somebody suggests refrigerating the cooling water before it enters the condenser, where heat rejection takes place. Would you be in favor of this idea? Why?

5-95C It is well known that the thermal efficiency of heat engines increases as the temperature of the energy source increases. In an attempt to improve the efficiency of a power plant, somebody suggests transferring heat from the available energy source to a higher temperature medium by a heat pump before energy is supplied to the power plant. What do you think of this suggestion? Explain.

5-96 A Carnot refrigerator operates in a room in which the temperature is 22°C and consumes 2 kW of power when operating. If the food compartment of the refrigerator is to be maintained at 4°C, determine the rate of heat removal from the food compartment.

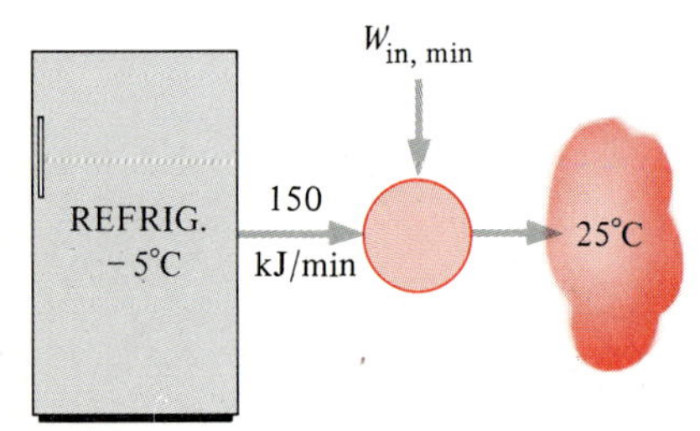

FIGURE P5-97

5-97 A refrigerator is required to remove heat from the cooled space at a rate of 150 kJ/min to maintain its temperature at −5°C. If the ambient air surrounding the refrigerator is at 25°C, determine the minimum power input required for this refrigerator. *Answer:* 0.28 kW

5-97E A refrigerator is required to remove heat from the cooled space at a rate of 150 Btu/min to keep its temperature at 25°F. If the ambient air surrounding the refrigerator is at 80°F, determine the minimum power input required for this refrigerator. *Answer:* 0.4 hp

5-98 An air conditioning system operating on the reversed Carnot cycle is required to transfer heat from a house at a rate of 750 kJ/min, to maintain its temperature at 20°C. If the outdoor air temperature is 35°C, determine the power required to operate this air conditioning system.
Answer: 0.64 kW

5-99 An air conditioning system is used to maintain a house at 22°C when the temperature outside is 32°C. If this air conditioning system draws 5 kW of power when operating, determine the maximum rate of heat removal from the house that it can provide.

5-99E An air conditioning system is used to maintain a house at 70°F when the temperature outside is 90°F. If this air conditioning system draws

5 hp of power when operating, determine the maximum rate of heat removal from the house that it can provide.

5-100 A Carnot refrigerator operates in a room in which the temperature is 25°C. The refrigerator consumes 500 W of power when operating and has a COP of 4.5. Determine (*a*) the rate of heat removal from the refrigerated space and (*b*) the temperature of the refrigerated space. *Answers:* (*a*) 135 kJ/min, (*b*) −29.2°C

5-101 An inventor claims to have developed a refrigeration system that removes heat from the cooled region at −5°C and transfers it to the surrounding air at 22°C while maintaining a COP of 7.5. Is this claim reasonable? Why? *Answer:* Yes

5-101E An inventor claims to have developed a refrigeration system that removes heat from the cooled region at 20°F and transfers it to the surrounding air at 75°F while maintaining a COP of 7.5. Is this claim reasonable? Why?

5-102 During an experiment conducted in a room at 25°C, a laboratory assistant measures that a refrigerator which draws 2 kW of power has removed 30,000 kJ of heat from the refrigerated space, which is maintained at −30°C. The running time of the refrigerator during the experiment was 20 min. Determine if these measurements are reasonable.

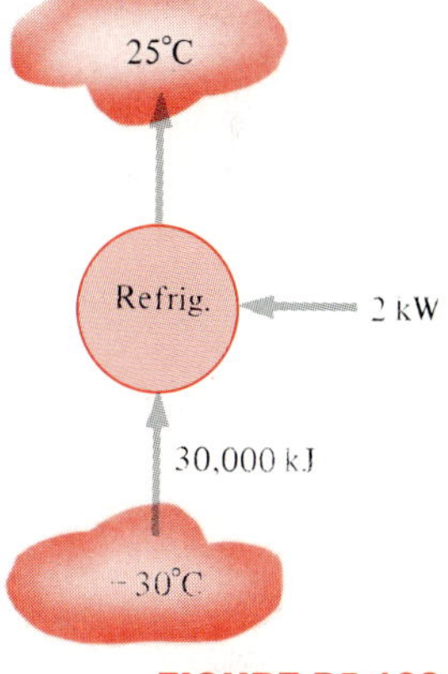

FIGURE P5-102

5-103 The COP of a refrigerator decreases as the temperature of the refrigerated space is decreased. That is, removing heat from a medium at a very low temperature will require a large work input. Determine the minimum work input required to remove 1 kJ of heat from liquid helium at 4 K when the outside temperature is 300 K. *Answer:* 74 kJ

5-104 An air conditioning system is used to maintain a house at 22°C when the temperature outside is 33°C. The house is gaining heat through the walls and the windows at a rate of 750 kJ/min, and the heat generation rate within the house from people, lights, and appliances amounts to 150 kJ/min. Determine the minimum power input required for this air conditioning system. *Answer:* 0.56 kW

5-104E An air conditioning sytem is used to maintain a house at 75°F when the temperature outside is 95°F. The house is gaining heat through the walls and the windows at a rate of 750 Btu/min, and the heat generation rate within the house from people, lights, and appliances amounts to 150 Btu/min. Determine the minimum power input required for this air conditioning system. *Answer:* 0.79 hp

5-105 A heat pump is used to heat a house and maintain it at 20°C. On a winter day when the outdoor air temperature is −5°C, the house is estimated to lose heat at a rate of 60,000 kJ/h. Determine the minimum power required to operate this heat pump.

5-105E A heat pump is used to heat a house and maintain it at 70°F. On

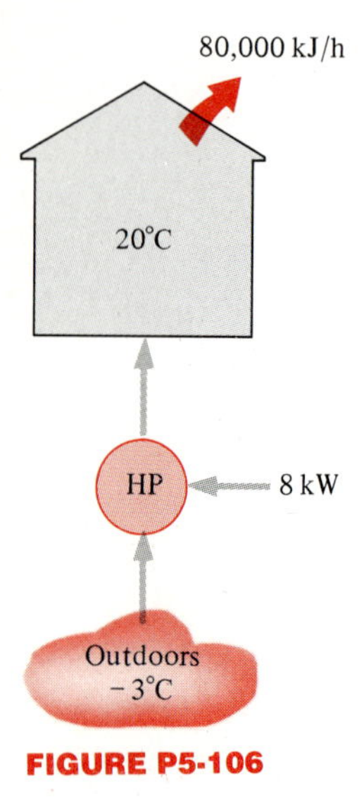

FIGURE P5-106

a winter day when the outdoor air temperature is 20°F, the house is estimated to lose heat at a rate of 60,000 Btu/h. Determine the minimum power required to operate this heat pump.

5-106 A heat pump is used to maintain a house at 20°C by extracting heat from the outside air on a day when the outside air temperature is −3°C. The house is estimated to lose heat at a rate of 80,000 kJ/h, and the heat pump consumes 8 kW of electric power when operating. Is this heat pump powerful enough to do the job?

5-107 A Carnot heat pump is used to heat and maintain a residential building at 22°C. An energy analysis of the house reveals that it loses heat at a rate of 2500 kJ/h per °C temperature difference between the indoors and the outdoors. For an outdoor temperature of 4°C, determine (*a*) the coefficient of performance and (*b*) the required power input to the heat pump. *Answers:* (*a*) 16.4, (*b*) 0.762 kW

5-107E A Carnot heat pump is used to heat and maintain a residential building at 75°F. An energy analysis of the house reveals that it loses heat at a rate of 2500 Btu/h per °F temperature difference between the indoors and the outdoors. For an outdoor temperature of 35°F, determine (*a*) the coefficient of performance and (*b*) the required power input to the heat pump. *Answers:* (*a*) 13.4, (*b*) 2.93 hp

5-108 The structure of a house is such that it loses heat at a rate of 3000 kJ/h per °C difference between the indoor and outdoor air temperatures. A heat pump that requires a power input of 8 kW is used to maintain this house at 22°C. Determine the lowest outdoors temperature for which the heat pump can meet the heating requirements of this house.
Answer: −31°C

5-109 The performance of a heat pump degrades (i.e., its COP decreases) as the temperature of the heat source decreases. This makes using heat pumps at locations with severe weather conditions unattractive. Consider a house that is heated and maintained at 20°C by a heat pump during the winter. What is the maximum COP for this heat pump if heat is extracted from the outdoor air at (*a*) 10°C, (*b*) −5°C, and (*c*) −30°C?

5-110 A heat pump is to be used for heating a house in winter. The house is to be maintained at 21°C at all times. When the temperature outdoors drops to −2°C, the heat losses from the house are estimated to be 80,000 kJ/h. Determine the minimum power required to run this heat pump if heat is extracted from (*a*) the outdoor air at −2°C and (*b*) the well water at 12°C.

5-110E A heat pump is to be used for heating a house in the winter. The house is to be maintained at 78°F at all times. When the temperature outdoors drops to 25°F, the heat losses from the house are estimated to be 80,000 Btu/h. Determine the minimum power required to run this heat pump if heat is extracted from (*a*) the outdoor air at 25°F and (*b*) the well water at 50°F.

5-111 A Carnot heat pump is to be used for heating a house and maintaining it at 20°C during the winter. On a day when the average outdoor temperature remains at about 2°C, the house is estimated to lose heat at a steady rate of 72,000 kJ/h. If the heat pump consumes 10 kW of power while operating, determine (*a*) how long the heat pump ran on that day; (*b*) the total heating costs, assuming an average price of 8.5¢/kWh for electricity; and (*c*) the heating cost for the same day if resistance heating is used instead of a heat pump.
Answers: (*a*) 2.94 h, (*b*) $2.50, (*c*) $40.80

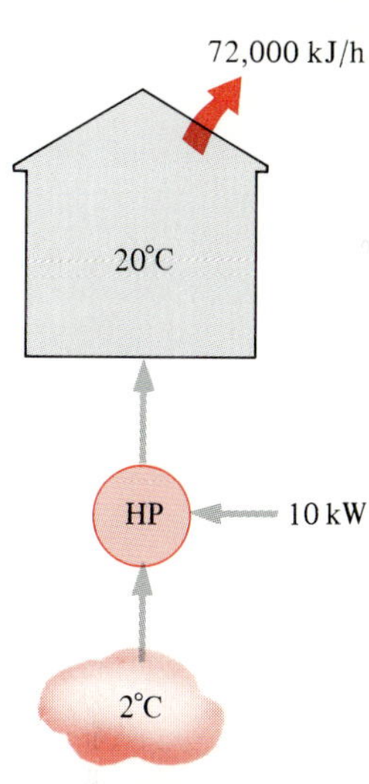

FIGURE P5-111

5-112 A Carnot heat engine receives heat at 750 K and rejects the waste heat to the environment at 300 K. The entire work output of the heat engine is used to drive a Carnot refrigerator which removes heat from the cooled space at −15°C at a rate of 400 kJ/min and rejects it to the same environment at 300 K. Determine (*a*) the rate of heat supplied to the heat engine and (*b*) the total rate of heat rejection to the environment.

5-113 A Carnot heat engine receives heat from a reservoir at 900°C at a rate of 700 kJ/min and rejects the waste heat to the ambient air at 27°C. The entire work output of the heat engine is used to drive a refrigerator that removes heat from the refrigerated space at −5°C and transfers it to the same ambient air at 27°C. Determine (*a*) the maximum rate of heat removal from the refrigerated space and (*b*) the total rate of heat rejection to the ambient air. *Answers:* (*a*) 4359 kJ/min, (*b*) 5059 kJ

5-113E A Carnot heat engine receives heat from a reservoir at 1700°F at a rate of 700 Btu/min and rejects the waste heat to the ambient air at 80°F. The entire work output of the heat engine is used to drive a refrigerator that removes heat from the refrigerated space at 20°F and transfers it to the same ambient air at 80°F. Determine (*a*) the maximum rate of heat removal from the refrigerated space and (*b*) the total rate of heat rejection to the ambient air.
Answers: (*a*) 4262 Btu/min, (*b*) 4962 Btu/min

5-114 A heat engine operates between two reservoirs at 727 and 17°C. One-half of the work output of the heat engine is used to drive a Carnot heat pump which removes heat from the cold surroundings at 2°C and transfers it to a house maintained at 22°C. If the house is losing heat at a rate of 80,000 kJ/h, determine the minimum rate of heat supply to the heat engine required to keep the house at 22°C.

5-115 Write a computer program to determine the maximum work that can be extracted from a pond containing 10^5 kg of water at 350 K when the temperature of the surroundings is 300 K. Notice that the temperature of water in the pond will be gradually decreasing as energy is extracted from it; therefore, the efficiency of the engine will be decreasing. Use temperature intervals of (*a*) 5 K, (*b*) 2 K, and (*c*) 1 K until the pond temperature drops to 300 K. Can you also solve this problem exactly by integration?

Entropy

CHAPTER

6

In Chap. 5 we introduced the second law of thermodynamics and applied it to cycles and cyclic devices. In this chapter, we apply the second law to processes. The first law of thermodynamics deals with the property *energy* and the conservation of it. The second law leads to the definition of a new property called *entropy*. Entropy is a somewhat abstract property, and it is difficult to give a physical description of it. Entropy is best understood and appreciated by studying its uses in commonly encountered engineering processes, and this is precisely what we intend to do.

This chapter starts with a discussion of the Clausius inequality, which forms the basis for the definition of entropy, and continues with the increase-in-entropy principle. Unlike energy, entropy is a nonconserved property, and there is no such thing as a *conservation of entropy principle*. Next, the entropy changes that take place during processes for pure substances, incompressible substances, and ideal gases are discussed, and a special class of idealized processes, called *isentropic processes,* are examined. Finally, the reversible steady-flow work and the isentropic efficiencies of various engineering devices such as turbines and compressors are discussed.

6-1 ■ THE CLAUSIUS INEQUALITY

The second law of thermodynamics often leads to expressions that involve inequalities. An irreversible (i.e., actual) heat engine, for example, is less efficient than a reversible one operating between the same two thermal energy reservoirs. Likewise, an irreversible refrigerator or a heat pump has a lower coefficient of performance (COP) than a reversible one operating between the same temperature limits. Another important inequality that has major consequences in thermodynamics is the **Clausius inequality**, which is expressed as

$$\oint \frac{\delta Q}{T} \leq 0$$

That is, *the cyclic integral of $\delta Q/T$ is always less than or equal to zero.* This inequality is valid for all cycles, reversible or irreversible. The symbol $\oint$ (integral symbol with a circle in the middle) is used to indicate that the integration is to be performed over the entire cycle. Any heat flow to or from a system can be considered to consist of differential amounts of heat. Then the cyclic integral of $\delta Q/T$ can be viewed as the sum of all these differential amounts of heat divided by the temperature at the boundary.

The validity of the Clausius inequality can easily be demonstrated with the help of two heat engines, one reversible and the other irreversible, both operating between a high-temperature reservoir at T_H and a low-temperature reservoir at T_L, as shown in Fig. 6-1. First, consider the reversible heat engine. The amount of heat received by this heat engine during a cycle is Q_H, the amount of heat rejected is Q_L, and the net work output is W_{rev}. Given the Q_H and Q_L are transferred at constant temperatures of T_H and T_L, respectively, the cyclic integral of $\delta Q/T$ for this reversible heat-engine cycle becomes

$$\oint \left(\frac{\delta Q}{T}\right)_{\text{rev}} = \int \frac{\delta Q_H}{T_H} - \int \frac{\delta Q_L}{T_L}$$

$$= \frac{1}{T_H}\int \delta Q_H - \frac{1}{T_L}\int \delta Q_L = \frac{Q_H}{T_H} - \frac{Q_L}{T_L} = 0$$

since $Q_H/T_H = Q_L/T_L$ for reversible cycles (Eq. 5-18). Thus, for a reversible heat-engine cycle

$$\oint \left(\frac{\delta Q}{T}\right)_{\text{rev}} = 0 \qquad (6\text{-}1)$$

Equation 6-1 is developed for a totally reversible heat engine, but it is equally valid for heat engines that are only internally reversible. In this case, T_H and T_L can be viewed as the temperature of the working fluid at locations where heat is received and rejected, respectively. Therefore, Eq. 6-1 can also be expressed as

$$\oint \left(\frac{\delta Q}{T}\right)_{\text{int rev}} = 0 \qquad (6\text{-}2)$$

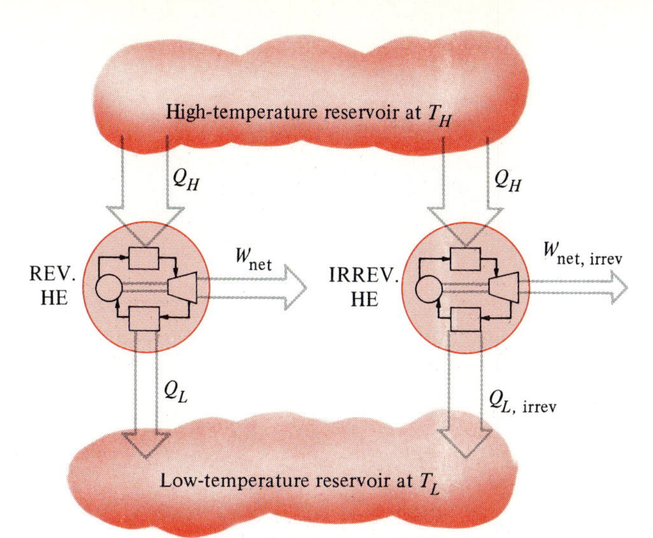

FIGURE 6-1
A reversible and an irreversible heat engine operating between the same two reservoirs.

Now consider the irreversible heat engine operating between the same thermal energy reservoirs as the reversible one and receiving the same amount of heat Q_H during a cycle. According to the first Carnot principle, the irreversible heat engine will deliver less net work and thus reject more waste heat. Therefore,

$$Q_{L,\text{irrev}} > Q_L$$

or

$$Q_{L,\text{irrev}} = Q_L + Q_{\text{diff}}$$

where Q_{diff} is a positive quantity. Performing the cyclic integral of $\delta Q/T$ for this irreversible heat engine yields

$$\oint \left(\frac{\delta Q}{T}\right)_{\text{irrev}} = \frac{Q_H}{T_H} - \frac{Q_{L,\text{irrev}}}{T_L} = \frac{Q_H}{T_H} - \frac{Q_L}{T_L} - \frac{Q_{\text{diff}}}{T_L} = -\frac{Q_{\text{diff}}}{T_L} < 0$$

since Q_{diff} is a positive quantity. Thus for an irreversible heat-engine cycle,

$$\oint \left(\frac{\delta Q}{T}\right)_{\text{irrev}} < 0$$

The Clausius inequality is obtained by combining this result with Eq. 6-1 as

$$\oint \frac{\delta Q}{T} \leqslant 0 \tag{6-3}$$

which is valid for all cycles. In the above relation the equality holds for totally or just internally reversible cycles and the inequality for the irreversible ones. In a similar manner, the validity of the Clausius inequality can also be shown by using a refrigeration cycle instead of a heat-engine cycle. This is left as an exercise for the student.

EXAMPLE 6-1

A heat engine receives 600 kJ of heat from a high-temperature source at

1000 K during a cycle. It converts 150 kJ of this heat to net work and rejects the remaining 450 kJ to a low-temperature sink at 300 K. Determine if this heat engine violates the second law of thermodynamics on the basis of (*a*) the Clausius inequality and (*b*) the Carnot principle.

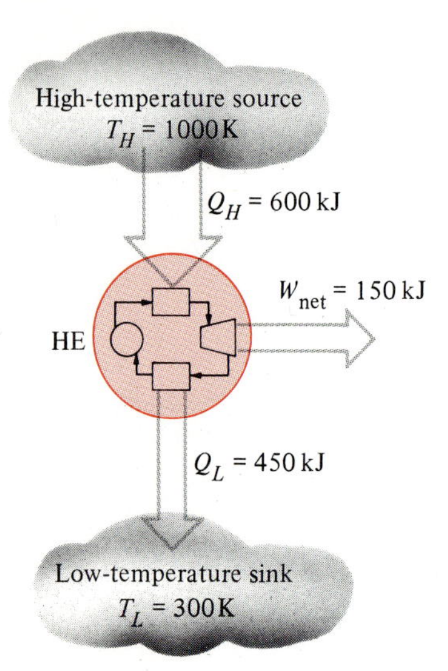

FIGURE 6-2
Schematic for Example 6-1.

Solution (*a*) A sketch of the heat engine is shown in Fig. 6-2. One way of determining whether this cycle violates the second law is to check if it violates the Clausius inequality. A cycle that violates the Clausius inequality also violates the second law.

By assuming that the temperature at locations where heat is crossing the boundaries of the heat engine is equal to the temperature of the reservoirs, the cyclic integral of $\delta Q/T$ for the heat-engine cycle under consideration is determined to be

$$\oint \frac{\delta Q}{T} = \frac{Q_H}{T_H} - \frac{Q_L}{T_L} = \frac{600 \text{ kJ}}{1000 \text{ K}} - \frac{450 \text{ kJ}}{300 \text{ K}} = -0.9 \text{ kJ/K}$$

Since the cyclic integral of $\delta Q/T$ is negative, this cycle satisfies the Clausius inequality and thus the second law of thermodynamics.

(*b*) Another way of determining whether this heat engine violates the second law is to check if the Carnot principle is satisfied. This is done by comparing the thermal efficiency of this heat engine to the thermal efficiency of a reversible heat engine, such as the Carnot engine, operating between the same temperature limits:

$$\eta_{th} = 1 - \frac{Q_L}{Q_H} = 1 - \frac{450 \text{ kJ}}{600 \text{ kJ}} = 0.25 \quad \text{or } 25\%$$

$$\eta_{th,rev} = 1 - \frac{T_L}{T_H} = 1 - \frac{300 \text{ K}}{1000 \text{ K}} = 0.70 \quad \text{or } 70\%$$

Again, this heat engine is in complete compliance with the second law of thermodynamics since $\eta_{th} < \eta_{th,rev}$. Note that a cycle that violates the Clausius inequality will also violate the Carnot principle.

6-2 ■ ENTROPY

The Clausius inequality discussed in Sec. 6-1 forms the basis for the definition of a new property called *entropy*.

To develop a relation for the definition of entropy, let us examine Eq. 6-2 more closely. Here we have a quantity whose cyclic integral is zero. Let us think for a moment what kind of quantities can have this characteristic. We know that the cyclic integral of *work* is not zero. (It is a good thing that it is not. Otherwise, heat engines that work on a cycle such as steam power plants would produce zero net work.) Neither is the cyclic integral of heat. As you may recall, these two quantities were defined in Chap. 3 as *path functions* since their magnitudes depend on the process path followed.

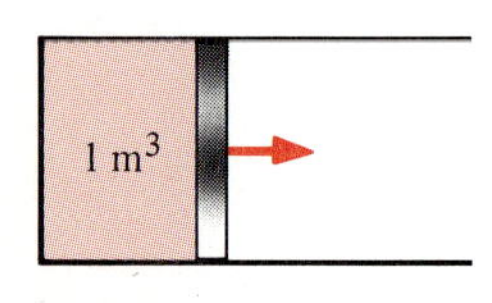

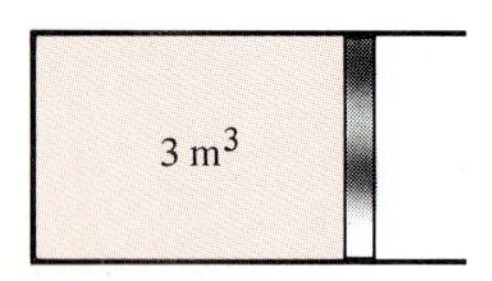

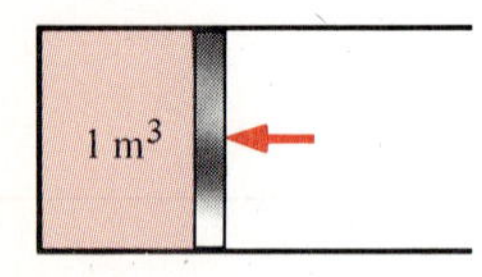

$\oint dV = \Delta V_{cycle} = 0$

FIGURE 6-3
The net change in volume (a property) during a cycle is always zero.

Now consider the volume occupied by a gas in a piston-cylinder device undergoing a cycle, as shown in Fig. 6-3. When the piston returns to its initial position at the end of a cycle, the volume of the gas also returns to its initial value. Thus the net change in volume during a cycle is zero. This is also expressed as

$$\oint dV = 0$$

That is, the cyclic integral of volume (or any other property) is zero. Conversely, a quantity whose cyclic integral is zero depends on the *state* only and not the process path, and thus it is a property. Therefore the quantity $(\delta Q/T)_{\text{int rev}}$ must represent a property in the differential form.

To demonstrate this point further, consider a cycle that consists of two internally reversible processes, A and B, as shown in Fig. 6-4. Applying Eq. 6-2 to this internally reversible cycle we obtain

$$\oint \left(\frac{\delta Q}{T}\right)_{\text{int rev}} = \int_1^2 \left(\frac{\delta Q}{T}\right)_A + \int_2^1 \left(\frac{\delta Q}{T}\right)_B = 0$$

Reversing the limits of the last integral and changing its sign to maintain equality yield

$$\int_1^2 \left(\frac{\delta Q}{T}\right)_A = \int_1^2 \left(\frac{\delta Q}{T}\right)_B$$

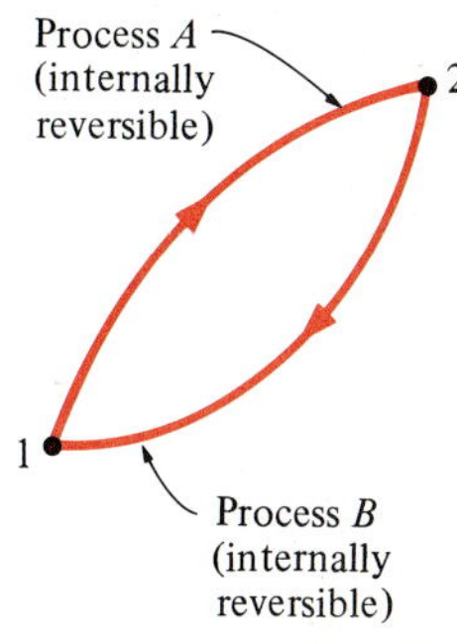

FIGURE 6-4

An internally reversible cycle composed of two internally reversible processes.

Since A and B are *any* two internally reversible process paths between states 1 and 2, the value of this integral depends on the end states only and not on the path followed. Therefore, it must represent the change of a property. This property is called **entropy**. It is designated S and is defined as

$$dS = \left(\frac{\delta Q}{T}\right)_{\text{int rev}} \qquad \text{(kJ/K)} \qquad \text{(6-4)}$$

Entropy is an extensive property of a system and sometimes is referred to as *total entropy*. Entropy per unit mass, designated s, is an intensive property and has the unit kJ/(kg·K). The term *entropy* is generally used to refer to both total entropy and entropy per unit mass since the context usually clarifies which one is meant.

The entropy change of a system during a process can be determined by integrating Eq. 6-4 between the initial and the final states:

$$\Delta S = S_2 - S_1 = \int_1^2 \left(\frac{\delta Q}{T}\right)_{\text{int rev}} \qquad \text{(kJ/K)} \qquad \text{(6-5)}$$

Notice that we have actually defined the *change* in entropy instead of entropy itself, just as we defined the change in energy instead of energy when we developed the first-law relation for closed systems in Chap. 3. Absolute values of entropy are determined on the basis of the third law of thermodynamics, which is discussed later in this chapter. Engineers are usually concerned with the *changes* in entropy. Therefore, the entropy of a substance can be assigned a zero value at some arbitrarily selected reference state, and the entropy values at other states can be determined

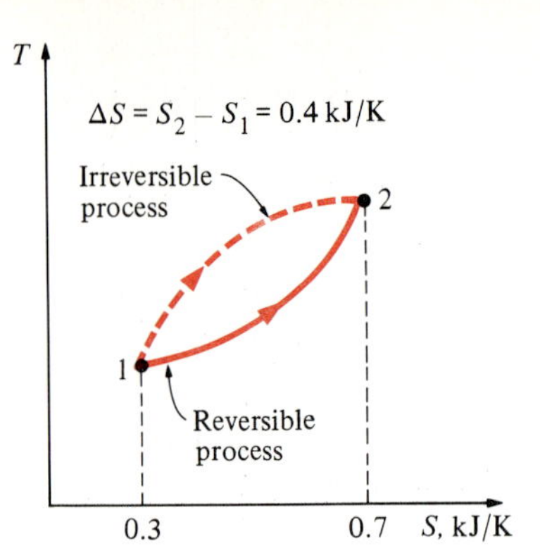

FIGURE 6-5

The entropy change between two specified states is the same whether the process is reversible or irreversible.

from Eq. 6-5 by choosing state 1 to be the reference state ($S = 0$) and state 2 to be the state at which entropy is to be determined.

To perform the integration in Eq. 6-5, one needs to know the relation between Q and T during a process. This relation is often not available, and the integral in Eq. 6-5 can be performed for a few cases only. For the majority of cases we have to rely on tabulated data for entropy.

Note that entropy is a property, and like all other properties, it has fixed values at fixed states. Therefore, the entropy change ΔS between two specified states is the same no matter what path, reversible or irreversible, is followed during a process (Fig. 6-5).

Also note that the integral of $\delta Q/T$ will give us the value of entropy change *only if* the integration is carried out along an *internally reversible* path between the two states. The integral of $\delta Q/T$ along an irreversible path is not a property, and in general, different values will be obtained when the integration is carried out along different irreversible paths. Therefore, even for irreversible processes, the entropy change should be determined by carrying out this integration along some convenient *imaginary* internally reversible path between the specified states.

EXAMPLE 6-2

An insulated rigid tank shown in Fig. 6-6 contains 5 kg of air at 15°C and 100 kPa. An electric resistance heater inside the tank is now turned on and kept on until the air temperature rises to 40°C. Determine the entropy change of air during this process.

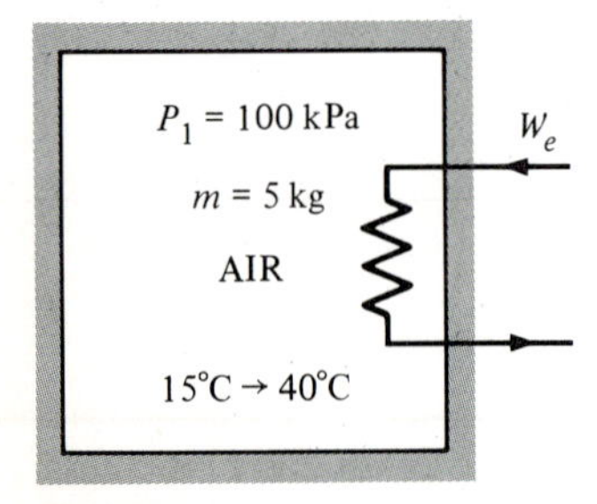

FIGURE 6-6

Schematic for Example 6-2.

Solution Probably the first thought that comes to mind for determining the entropy change of air is to use Eq. 6-5. Since the tank is insulated, $\delta Q = 0$ and the integral in Eq. 6-5 yields $\Delta S = 0$. This approach, of course, is not valid since the integral in Eq. 6-5 must be performed along an internally reversible path. The actual process is irreversible, as discussed in Chap. 5. The entropy change for this process should be evaluated by making up an internally reversible process between the specified end states and performing the integration along its path.

Here is a corresponding internally reversible process: The same effect could be achieved in an internally reversible manner by transferring heat to the air in the rigid tank from an external source until the air temperature rises to 40°C. No irreversibilities occur within the system boundaries during this process, thus it is internally reversible. The process also involves no work interactions. Thus the differential form of the first-law relation (Eq. 3-30) for this closed system reduces to

$$\delta Q - \cancelto{0}{\delta W} = dU$$
$$\delta Q = dU$$

Treating air as an ideal gas with constant specific heats, we have $dU = mC_v\,dT$. Thus,

$$\delta Q_{\text{int rev}} = mC_v\,dT$$

The entropy change is then determined from Eq. 6-5 to be

$$\Delta S = \int_1^2 \left(\frac{\delta Q}{T}\right)_{\text{int rev}} = \int_1^2 \frac{mC_v\,dT}{T} = mC_v \ln \frac{T_2}{T_1}$$
$$= (5\text{ kg})[0.717\text{ kJ/(kg·K)}] \ln \frac{313\text{ K}}{288\text{ K}} = \mathbf{0.30\text{ kJ/K}}$$

Therefore, the entropy change of air during this process is 0.30 kJ/K. The same result would be obtained if other internally reversible paths were used instead of the one we selected. This is quite a cumbersome way of determining the entropy change and is given here for demonstration purposes only. The entropy change for this and other processes can be determined very easily by using the relations developed later in this chapter.

A Special Case: Isothermal Heat Transfer Processes

We pointed out in Chap. 5 that isothermal heat transfer processes are internally reversible. Therefore, the entropy change of a system during an isothermal heat transfer process can be determined by performing the integration in Eq. 6-5:

$$\Delta S = \int_1^2 \left(\frac{\delta Q}{T}\right)_{\text{int rev}} = \int_1^2 \left(\frac{\delta Q}{T_0}\right)_{\text{int rev}} = \frac{1}{T_0}\int_1^2 (\delta Q)_{\text{int rev}}$$

which reduces to

$$\Delta S = \frac{Q}{T_0} \qquad \text{(kJ/K)} \tag{6-6}$$

where T_0 is the constant absolute temperature of the system and Q is the heat transfer for the internally reversible process. Equation 6-6 is particularly useful for determining the entropy changes of thermal energy reservoirs which can absorb or supply heat indefinitely at a constant temperature.

Notice that the entropy change of a system during an internally reversible process can be positive or negative depending on the direction of heat transfer. Heat transfer to a system (Q positive) will increase the entropy of that system whereas heat transfer from the system (Q negative) will decrease it (Fig. 6-7). In fact, losing heat is the only way the entropy of a system can be decreased.

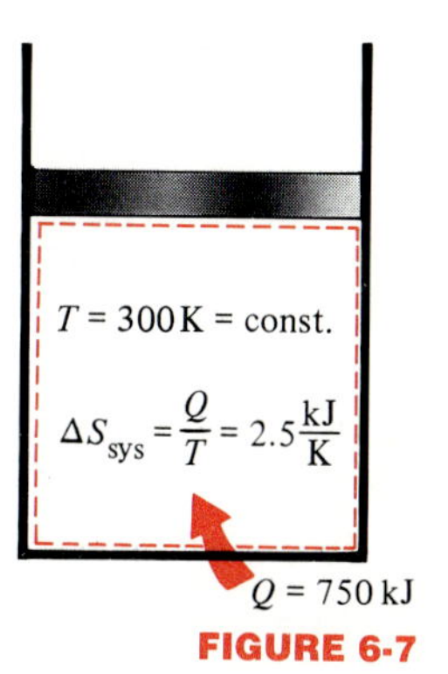

FIGURE 6-7
Determination of entropy change during an internally reversible, isothermal process.

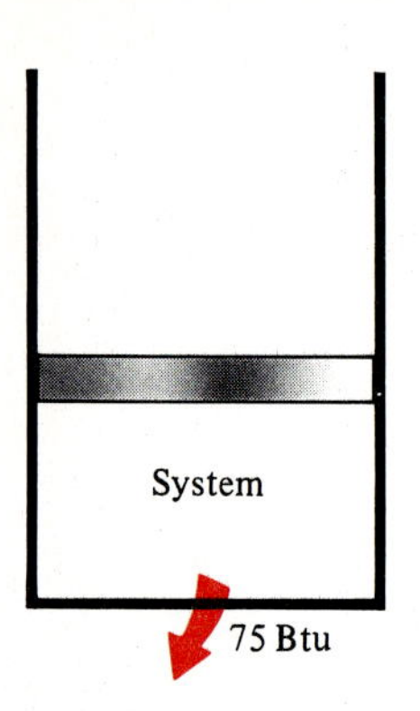

FIGURE 6-8

Schematic for Example 6-3.

EXAMPLE 6-3

During a process, a system rejects 75 Btu of heat to the surrounding air at 525 R, as shown in Fig. 6-8. Determine the entropy change of the surrounding air for this process.

Solution The surrounding air can absorb and reject large quantities of heat without experiencing any significant change in its temperature. Therefore, it can be treated as a thermal energy reservoir at 525 R, and its entropy change can be determined from Eq. 6-6 to be

$$\Delta S_{\text{surr}} = \frac{Q_{\text{surr}}}{T_{\text{surr}}} = \frac{+75\text{ Btu}}{525\text{ R}} = 0.143\text{ Btu/R}$$

Notice that Q_{surr}, and thus ΔS_{surr}, are positive quantities since the surrounding air is gaining heat.

6-3 ■ THE INCREASE-IN-ENTROPY PRINCIPLE

In Sec. 6-2 we used the equality portion of the Clausius inequality (Eq. 6-3) to define a new property called *entropy*. The inequality portion of Eq. 6-3 can now be used to develop one of the most fundamental principles of thermodynamics, called the *increase-in-entropy principle*. This is done by considering a closed system undergoing an irreversible cycle.

Consider a cycle that is made up of two processes, one internally reversible and the other irreversible, as shown in Fig. 6-9. This is an irreversible cycle since part of the cycle is irreversible. The Clausius inequality states that the cyclic integral of $\delta Q/T$ for this irreversible cycle is less than zero, i.e.,

$$\oint \left(\frac{\delta Q}{T}\right)_{\text{irrev}} < 0$$

or

$$\int_{1,A}^{2} \left(\frac{\delta Q}{T}\right)_{\text{irrev}} + \int_{2,B}^{1} \left(\frac{\delta Q}{T}\right)_{\text{int rev}} < 0$$

The second integral in the above relation is easily recognized as the entropy change $S_1 - S_2$. Thus,

$$\int_{1,A}^{2} \left(\frac{\delta Q}{T}\right)_{\text{irrev}} + S_1 - S_2 < 0$$

which can be rearranged as

$$\Delta S = S_2 - S_1 > \int_{1,A}^{2} \left(\frac{\delta Q}{T}\right)_{\text{irrev}} \tag{6-7}$$

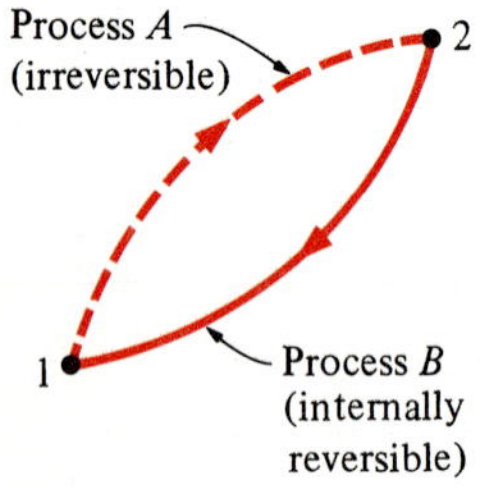

Process A (irreversible)
2
1
Process B (internally reversible)

FIGURE 6-9

A cycle composed of a reversible and an irreversible process.

We may conclude from Eq. 6-7 that *the entropy change of a closed system during an irreversible process is greater than the integral of $\delta Q/T$ evaluated for that process.*

The relation between the entropy change of a closed system and the

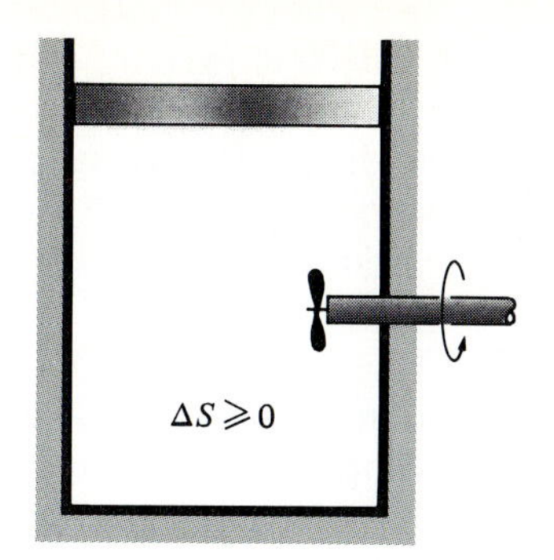

FIGURE 6-10

The entropy change of a system can never be negative during an adiabatic process.

integral of $\delta Q/T$ can be expressed for any process in the general form by combining Eqs. 6-5 and 6-7 as

$$\Delta S \geq \int_1^2 \frac{\delta Q}{T} \tag{6-8}$$

or, in differential form,

$$dS \geq \frac{\delta Q}{T} \tag{6-9}$$

where the equality holds for a totally or just internally reversible process and the inequality for an irreversible process. In the above relations, δQ represents a differential amount of actual heat transfer between a system and its surroundings, and T is the absolute temperature at the boundary.

Equation 6-8 has far-reaching implications in thermodynamics. For an adiabatic process, the heat transfer between a system and its surroundings is zero, and Eq. 6-8 reduces to

$$\Delta S_{\text{adiabatic}} \geq 0 \tag{6-10}$$

This equation can be expressed as *the entropy of an adiabatic closed system (a fixed mass) always increases or, in the limiting case, remains constant during a process*. In other words, it *never* decreases (Fig. 6-10). This is one of the expressions of the *increase-in-entropy principle*. Note that in the absence of any heat transfer, entropy change is due to irreversibilities only, and their effect is always to increase the entropy.

Equation 6-10 is extremely valuable in the second-law analysis of engineering processes, but it is limited to adiabatic processes of closed systems and therefore is not of general use. It would be very desirable to have a relation for the increase-in-entropy principle which is applicable to both open and closed systems as well as adiabatic and nonadiabatic systems. This is accomplished as follows:

An adiabatic system may consist of any number of subsystems (Fig. 6-11). A system plus its surroundings, for example, constitutes an adiabatic system since both can be enclosed by a sufficiently large arbitrary boundary across which there is no heat or mass transfer (Fig. 6-12).

A system and its surroundings can be viewed as the two subsystems

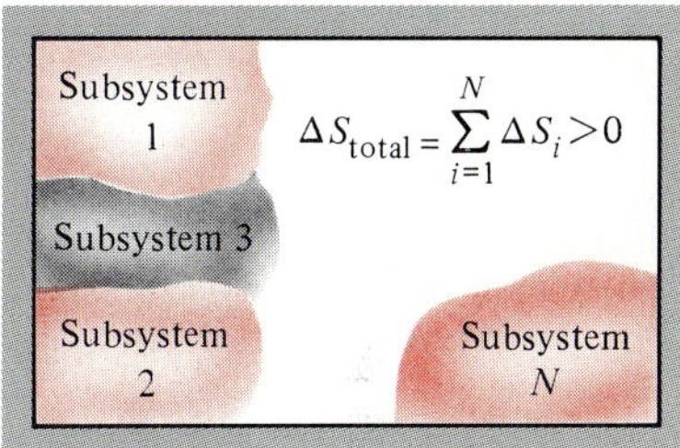

FIGURE 6-11

The entropy change of an adiabatic system is the sum of the entropy changes of its components, and is never less than zero.

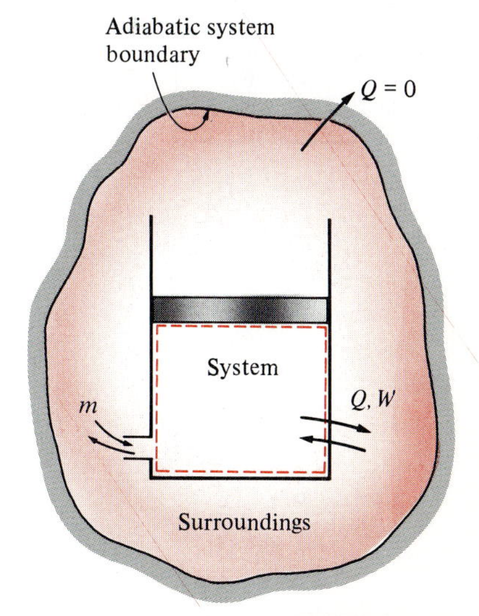

FIGURE 6-12

A system and its surroundings form an adiabatic system.

of an adiabatic system, and the entropy change of this adiabatic system during a process is the sum of the entropy changes of the system and its surroundings. This is called the **total entropy change** ΔS_{total} or **entropy generation** S_{gen}, and the increase-in-entropy principle for any process is expressed as

$$S_{\text{gen}} = \Delta S_{\text{total}} = \Delta S_{\text{sys}} + \Delta S_{\text{surr}} \geq 0 \qquad \text{(kJ/K)} \qquad (6\text{-}11)$$

Equation 6-11 is a general expression for the **increase-in-entropy principle** and is applicable to both closed (control mass) and open (control volume) systems since any system and its surroundings form an adiabatic system (Fig. 6-13). It states that *the total entropy change associated with a process must be positive or zero.* The equality holds for reversible processes and the inequality for irreversible ones.

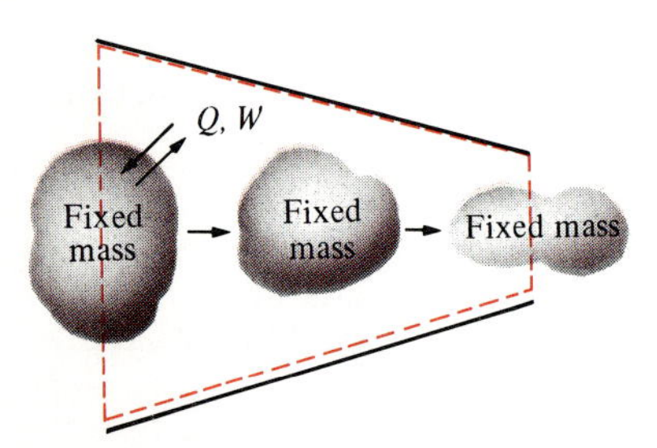

FIGURE 6-13

All relations developed for closed systems are also applicable to a fixed mass flowing through a control volume.

Since no actual process is truly reversible, we can conclude that the net entropy change for any process that takes place is positive, and therefore the entropy of the universe, which can be considered to be an adiabatic system, is continuously increasing. The more irreversible a process is, the larger the entropy generated during the process is. No entropy is generated during reversible processes ($S_{\text{gen}} = 0$).

Entropy increase of the universe is a major concern not only to engineers but also to philosophers and theologians since entropy is viewed as a measure of the disorder (or ''mixed-up-ness'') in the universe.

Equation 6-11 does not imply that the entropy of a system or the surroundings cannot decrease. The entropy change of a system or its surroundings *can* be negative during a process (Fig. 6-14); but their sum cannot. The increase-in-entropy principle can be summarized as follows:

$$S_{\text{gen}} = \Delta S_{\text{total}} \begin{cases} > 0 & \text{irreversible process} \\ = 0 & \text{reversible process} \\ < 0 & \text{impossible process} \end{cases}$$

This relation serves as a criterion in determining whether a process is reversible, irreversible, or impossible.

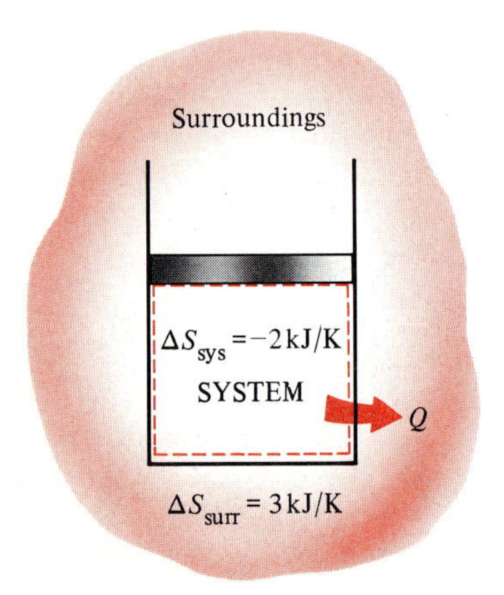

FIGURE 6-14

The entropy change of a system can be negative; but the sum $\Delta S_{\text{sys}} + \Delta S_{\text{surr}}$ cannot.

The Increase-in-Entropy Principle: Closed Systems

A closed system involves no mass flow across its boundaries, and its entropy change is simply the difference between the final and initial entropies of the system. The entropy of the surroundings of a closed system will be affected by a process only if the system exchanges heat with the surroundings during the process. The surroundings can usually be treated as a thermal energy reservoir at temperature T_{surr}, and its entropy change is related to heat transfer by Eq. 6-6.

The increase-in-entropy principle for a closed system is expressed as (Eq. 6-11)

$$S_{\text{gen}} = \Delta S_{\text{total}} = \Delta S_{\text{sys}} + \Delta S_{\text{surr}} \geq 0 \qquad \text{(kJ/K)}$$

where $\Delta S_{sys} = S_2 - S_1 = m(s_2 - s_1)$

$$\Delta S_{surr} = \frac{Q_{surr}}{T_{surr}}$$

If a closed system exchanges heat with other systems or reservoirs also, the total entropy change for the process is obtained by superposition, i.e., by adding the entropy changes of all the systems and reservoirs involved.

For an *adiabatic* process, $Q_{surr} = -Q_{sys} = 0$ and the increase-in-entropy principle reduces to

$$S_{gen} = \Delta S_{total} = m(s_2 - s_1) \geqslant 0$$

That is, the entropy of a closed system can never decrease during an adiabatic process.

EXAMPLE 6-4

A frictionless piston-cylinder device, shown in Fig. 6-15, contains a saturated mixture of water at 100°C. During a constant-pressure process, 600 kJ of heat is transferred to the surrounding air which is at 25°C. As a result, part of the water vapor contained in the piston-cylinder device condenses. Determine (*a*) the entropy change of the water, (*b*) the entropy change of the surrounding air during this process, and (*c*) whether this process is reversible, irreversible, or impossible.

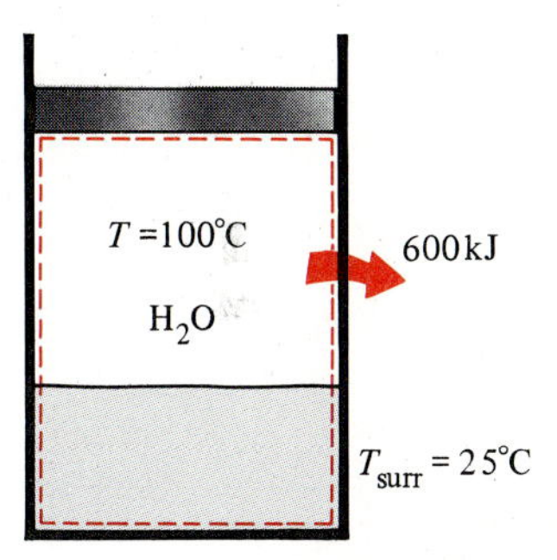

FIGURE 6-15

Schematic for Example 6-4.

Solution (*a*) The temperature of water remains constant at 100°C during this phase-change process since the pressure inside the piston-cylinder device is kept constant. This process is internally reversible since it involves no irreversibilities within the system boundaries. Therefore, the entropy change of water during this internally reversible, isothermal process can be determined from Eq. 6-6:

$$\Delta S_{water} = \frac{Q_{water}}{T_{water}} = \frac{-600 \text{ kJ}}{(100 + 273) \text{ K}} = -1.61 \text{ kJ/K}$$

Notice that Q_{water} is negative since heat is transferred from the water, and that the entropy of the system (water) is decreasing during this process.

(*b*) The entropy change of the surrounding air, which serves as the sink during this process, is determined in a similar manner. But this time the heat transfer term is positive since the heat lost by the system is gained by the surroundings:

$$Q_{surr} = -Q_{sys} = +600 \text{ kJ}$$

and
$$\Delta S_{surr} = \frac{Q_{surr}}{T_{surr}} = \frac{+600 \text{ kJ}}{(25 + 273) \text{ K}} = +2.01 \text{ kJ/K}$$

(*c*) To determine whether this process is reversible, irreversible, or impossible, we need to find the total entropy change for this process. From Eq. 6-11,

$$\Delta S_{total} = \Delta S_{sys} + \Delta S_{surr} = (-1.61 + 2.01) \text{ kJ/K}$$
$$= +0.4 \text{ kJ/K}$$

The total entropy change for this process is positive which indicates that this is an *irreversible* process. This result is expected since we know from experience that this process is not impossible, and that it is not reversible since it involves heat transfer through a finite temperature difference.

For the sake of discussion, let us examine the reverse process (i.e., the transfer of 600 kJ of heat from the surrounding air at 25°C to saturated water at 100°C) and see if the increase-in-entropy principle can detect the impossibility of this process. This time, the sign of the heat transfer terms will be reversed since the direction of heat flow is changed. This will make ΔS_{water} positive and ΔS_{surr} negative, and thus ΔS_{total} will become -0.4 kJ/K. The negative sign for the total entropy change indicates that the reverse process is *impossible* (Fig. 6-16).

To complete the discussion, let us consider the case where the surrounding air temperature is a differential amount below 100°C (say 99.999 . . . 9°C) instead of 25°C. This time, heat transfer from the saturated water to the surrounding air will take place through a differential temperature difference, rendering this process *reversible*. It can be easily shown that $\Delta S_{surr} = -\Delta S_{water}$ and thus $\Delta S_{total} = 0$ for this process. Therefore, this is a reversible process. Remember that reversible processes are idealized processes, and they can be approached but never reached in reality.

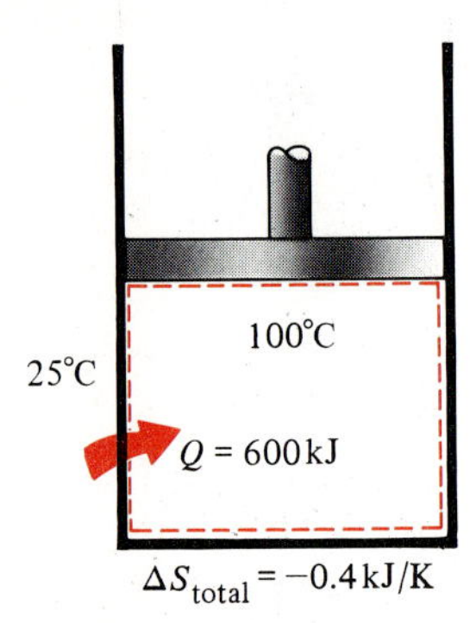

FIGURE 6-16

The total entropy change is negative for *all* impossible processes.

The Increase-in-Entropy Principle: Control Volumes

The second-law analysis of control volumes is similar to the one given above for closed systems, but this time we have to consider one more thing: *mass flow across the control volume boundaries*. Mass that enters or leaves a control volume carries some entropy (as well as energy) with it, and therefore mass transfer will affect the entropy content of both the control volume and the surroundings (Fig. 6-17). With this consideration, the entropy change of a control volume (CV) and its surroundings can be expressed as

$$\Delta S_{CV} = (S_2 - S_1)_{CV}$$

$$\Delta S_{surr} = \frac{Q_{surr}}{T_{surr}} + S_e - S_i$$

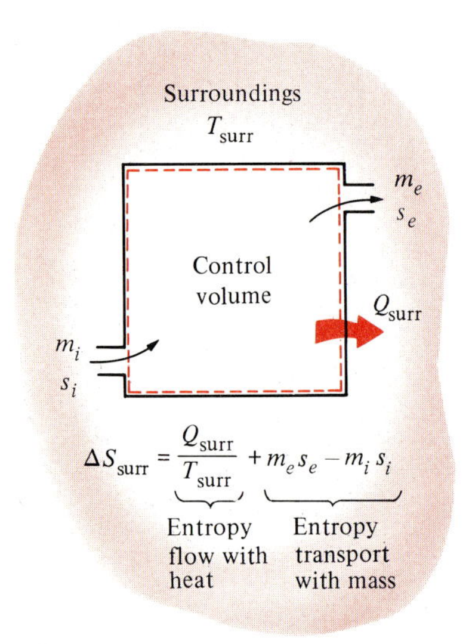

FIGURE 6-17

The entropy changes as a result of mass flow as well as heat flow.

where the subscripts are 1 = initial state, 2 = final state, i = inlet state, and e = exit state. Also,

S_i = total entropy transported into control volume (from surroundings) with mass entering control volume

S_e = total entropy transported out of control volume (into surroundings) with mass leaving control volume

Notice that the terms S_i and S_e do not appear in the entropy-change relation for the control volume since the value of the final entropy S_2 will reflect their influence. Substituting these two equations into Eq. 6-11, we obtain the *increase-in-entropy principle relation* for control volumes:

$$S_{gen} = \Delta S_{total} = (S_2 - S_1)_{CV} + S_e - S_i + \frac{Q_{surr}}{T_{surr}} \geq 0 \qquad \text{(kJ/K)}$$

In general, the fluid properties at an inlet or an exit may vary during a process. When this happens, the transported entropy terms should be determined by integration:

$$S_i = \sum \int_{m_i} s_i \, \delta m_i \qquad \text{and} \qquad S_e = \sum \int_{m_e} s_e \, \delta m_e$$

Then the general form of the *increase-in-entropy principle* for a control volume becomes

$$S_{gen} = (S_2 - S_1)_{CV} + \sum \int_{m_e} s_e \, \delta m_e - \sum \int_{m_i} s_i \, \delta m_i + \frac{Q_{surr}}{T_{surr}} \geqslant 0 \qquad \text{(kJ/K)} \qquad (6\text{-}12)$$

Uniform-Flow Processes

During a uniform-flow process, the state of the control volume changes uniformly, and fluid properties at any inlet or exit are assumed to remain constant. Then the general increase-in-entropy relation (Eq. 6-12) simplifies for a uniform-flow process to

$$S_{gen} = (m_2 s_2 - m_1 s_1)_{CV} + \sum m_e s_e - \sum m_i s_i + \frac{Q_{surr}}{T_{surr}} \geqslant 0 \qquad \text{(kJ/K)} \qquad (6\text{-}13)$$

Steady-Flow Processes

The fluid properties anywhere within the control volume, including the inlets and the exits, remain constant with respect to time during a steady-flow process. Therefore, the entropy change of a control volume is zero ($\Delta S_{CV} = 0$), and the total entropy change for a steady-flow process (ΔS_{total} or S_{gen}) becomes the entropy change of the surroundings only:

$$S_{gen} = \Delta S_{surr} = S_e - S_i + \frac{Q_{surr}}{T_{surr}} \geqslant 0$$

which can be expressed in the rate form as

$$\dot{S}_{gen} = \sum \dot{m}_e s_e - \sum \dot{m}_i s_i + \frac{\dot{Q}_{surr}}{T_{surr}} \geqslant 0 \qquad \text{(kW/K)} \qquad (6\text{-}14)$$

For a single-stream steady-flow device, this relation reduces to

$$\dot{S}_{gen} = \dot{m}(s_e - s_i) + \frac{\dot{Q}_{surr}}{T_{surr}} \geqslant 0 \qquad \text{(kW/K)} \qquad (6\text{-}15)$$

or

$$s_{gen} = s_e - s_i + \frac{q_{surr}}{T_{surr}} \geqslant 0 \qquad \text{[kJ/(kg·K)]} \qquad (6\text{-}16)$$

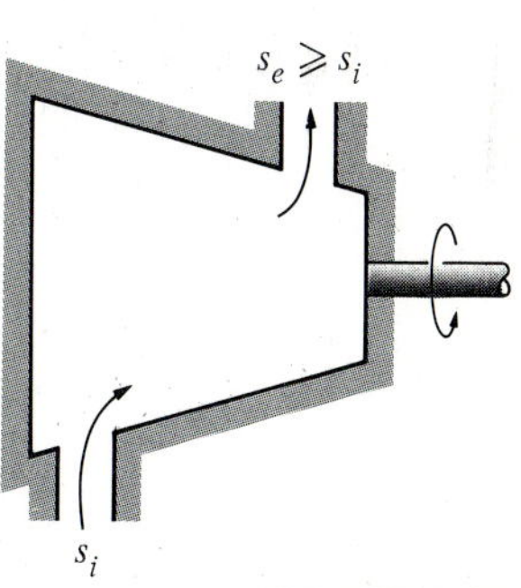

FIGURE 6-18
The entropy of a substance always increases (or remains constant in the case of reversible process) as it flows through a single-stream adiabatic steady-flow device.

Thus the entropy of a fluid will increase as it flows through an adiabatic steady-flow device as a result of irreversibilities (Fig. 6-18).

Notice that S_{gen} is not a property. Its value depends on how the process is executed, and thus it is a path function just as heat and work. In the above relations $\dot{S}_{gen}$ is the **rate of total entropy change**, or the **rate of entropy generation**. The equality of these relations applies to reversible processes, and the inequality to irreversible ones.

6-4 ■ CAUSES OF ENTROPY CHANGE

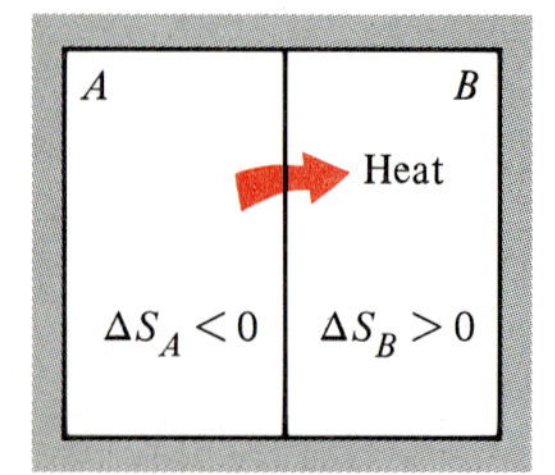

FIGURE 6-19
Heat transfer to a system increases its entropy, and heat transfer from a system decreases it.

It is clear from the previous discussions that two factors cause the entropy of a system to change: heat transfer and irreversibilities.

Heat transfer to a system increases the entropy of that system, and heat transfer from a system decreases it (Fig. 6-19). In fact, this is the only way the entropy of a system can be decreased. Entropy change as a result of reversible heat transfer is called *entropy flow* and does not involve any entropy generation. Therefore, the entropy of the universe does not increase as a result of entropy flow.

Irreversibilities such as friction, fast expansion or compression, and heat transfer through a finite temperature difference always cause the entropy to increase. Thus the entropy of a system cannot decrease during an adiabatic process (Fig. 6-20). Entropy generated during a process is due to irreversibilities, and $S_{gen} = 0$ for a reversible process.

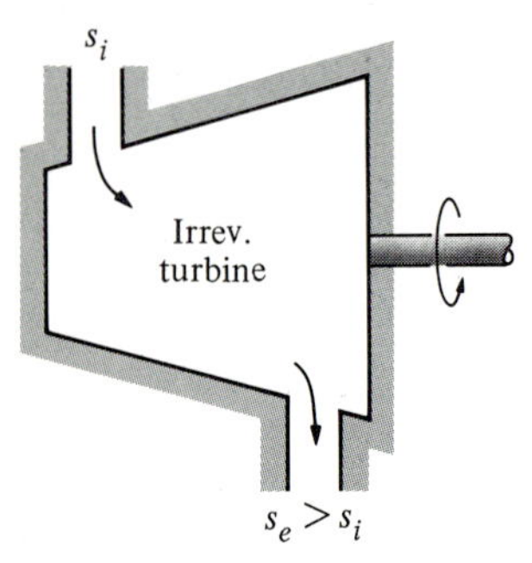

FIGURE 6-20
Irreversibilities always cause the entropy to increase.

If a process involves no heat transfer (*adiabatic*) and no irreversibilities within the system (*internally reversible*), the entropy of a system must remain constant during that process. Such a process is called an **internally reversible adiabatic** or **isentropic process**. An isentropic process is also an idealization, like the quasi-equilibrium process discussed in earlier chapters, and serves as a model for real processes.

Some Remarks about Entropy

In the light of the preceding discussions we can draw the following conclusions:

1 Processes can occur in a *certain* direction only, not in *any* direction. A process must proceed in the direction that complies with the increase-in-entropy principle, that is, $\Delta S_{total} \geqslant 0$. A process that violates this principle is impossible. This principle often forces chemical reactions to come to a halt before reaching completion, as discussed in Chap. 15.

2 Entropy is a *nonconserved property,* and there is *no* such thing as the *conservation of entropy principle*. Entropy is conserved during the idealized reversible processes only and increases during *all* actual processes. Therefore, the entropy of the universe is continuously increasing.

3 The performance of engineering systems is degraded by the presence of irreversibilities, and the *entropy generation* is a measure of the magnitudes of the irreversibilities present during that process. The greater the extent of irreversibilities, the greater the entropy generation. Therefore, entropy can be used as a quantitative measure of irreversibilities associated with a process. It is also used to establish criteria for the

performance of engineering devices. This point is further illustrated in the following example.

EXAMPLE 6-5

A thermal energy source at 800 K loses 2000 kJ of heat to a sink at (*a*) 500 K and (*b*) 750 K. Determine which heat transfer process is more irreversible.

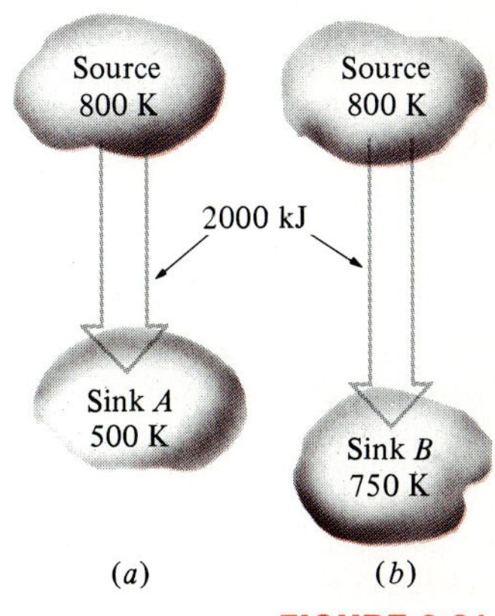

FIGURE 6-21
Schematic for Example 6-5.

Solution A sketch of the reservoirs is shown in Fig. 6-21. Both cases involve heat transfer through a finite temperature difference, and therefore both are irreversible. The magnitude of the irreversibility associated with each process can be determined by calculating the total entropy change for each case. The total entropy change for a heat transfer process involving two reservoirs (a source and a sink) is the sum of the entropy changes of each reservoir since the two reservoirs form an adiabatic system. Or, one reservoir can be considered as the system and the other as the surroundings.

The entropy change for each reservoir can be determined from Eq. 6-6 since each reservoir undergoes an internally reversible, isothermal process.

(*a*) For the heat transfer process to a sink at 500 K,

$$\Delta S_{source} = \frac{Q_{source}}{T_{source}} = \frac{-2000 \text{ kJ}}{800 \text{ K}} = -2.5 \text{ kJ/K}$$

$$\Delta S_{sink} = \frac{Q_{sink}}{T_{sink}} = \frac{+2000 \text{ kJ}}{500 \text{ K}} = +4.0 \text{ kJ/K}$$

and $$\Delta S_{total} = \Delta S_{source} + \Delta S_{sink} = (-2.5 + 4.0) \text{ kJ/K} = +1.5 \text{ kJ/K}$$

(*b*) Repeating the calculations in part (*a*) for a sink temperature of 750 K, we obtain

$$\Delta S_{source} = -2.5 \text{ kJ/K}$$

$$\Delta S_{sink} = +2.7 \text{ kJ/K}$$

and $$\Delta S_{total} = (-2.5 + 2.7) \text{ kJ/K} = +0.2 \text{ kJ/K}$$

The total entropy change for the process in part (*b*) is smaller, and so therefore it is less irreversible. This is expected since the process in (*b*) involves a smaller temperature difference and thus a smaller irreversibility.

The irreversibilities associated with both processes could be eliminated by operating a Carnot heat engine between the source and the sink. For this case it can be easily shown that $\Delta S_{total} = 0$.

6-5 ■ WHAT IS ENTROPY?

It is clear from the previous discussions that entropy is a useful property and serves as a valuable tool in the second-law analysis of engineering devices. But this does not mean that we know and understand entropy well. Because we do not. In fact, we cannot even give an adequate answer to the question, What is entropy? Not being able to describe entropy fully, however, does not take anything away from its usefulness. In Chap. 1, we could not define *energy* either, but it did not interfere with our understanding of energy transformations and the conservation of energy

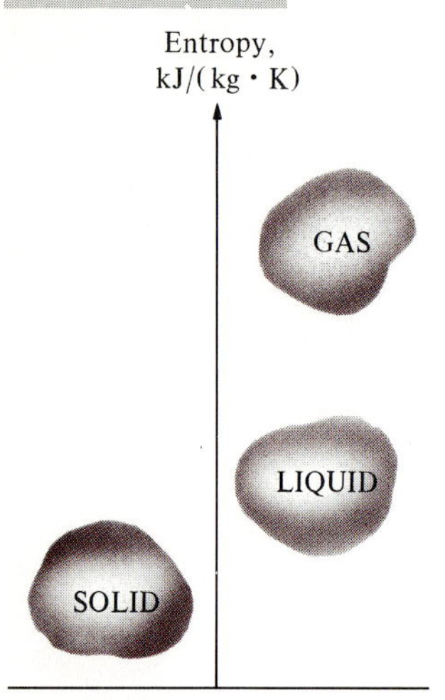

FIGURE 6-22
The level of molecular disorder (entropy) of a substance increases as it melts or evaporates.

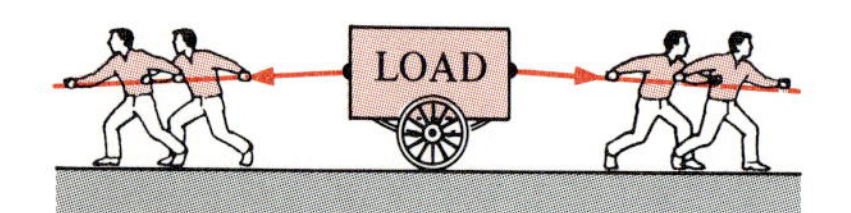

FIGURE 6-23
Disorganized energy does not create much useful effect, no matter how large it is.

principle. Granted, entropy is not a household word as energy is. But with continued use, our understanding of entropy will deepen, and our appreciation of it will grow. The discussion below will shed some light on the physical meaning of entropy by considering the microscopic nature of matter.

Entropy can be viewed as a measure of *molecular disorder,* or *molecular randomness.* As a system becomes more disordered, the positions of the molecules become less predictable and the entropy increases. Thus it is not surprising that the entropy of a substance is lowest in the solid phase and highest in the gas phase (Fig. 6-22). In the solid phase, the molecules of a substance continually oscillate about their equilibrium positions, but they cannot move relative to each other, and their position at any instant can be predicted with good certainty. In the gas phase, however, the molecules move about at random, collide with each other, and change direction, making it extremely difficult to predict accurately the microscopic state of a system at any instant. Associated with this molecular chaos is a high value of entropy.

Molecules in the gas phase possess a considerable amount of kinetic energy. But we know that no matter how large their kinetic energies are, the gas molecules will not rotate a paddle wheel inserted into the container and produce work. This is because the gas molecules, and the energy they carry with them, are disorganized. Probably the number of molecules trying to rotate the wheel in one direction at any instant is equal to the number of molecules that are trying to rotate it in the opposite direction, causing the wheel to remain motionless. Therefore, we cannot extract any useful work directly from disorganized energy (Fig. 6-23).

Now consider a rotating shaft shown in Fig. 6-24. This time the energy of the molecules is completely organized since the molecules of the shaft are rotating in the same direction together. This organized energy can readily be used to perform useful tasks such as raising a weight or generating electricity. Being an organized form of energy, work is free of disorder or randomness and thus free of entropy. *There is no entropy transfer associated with energy transfer as work.* Therefore, in the absence of any friction, the process of raising a weight by a rotating shaft

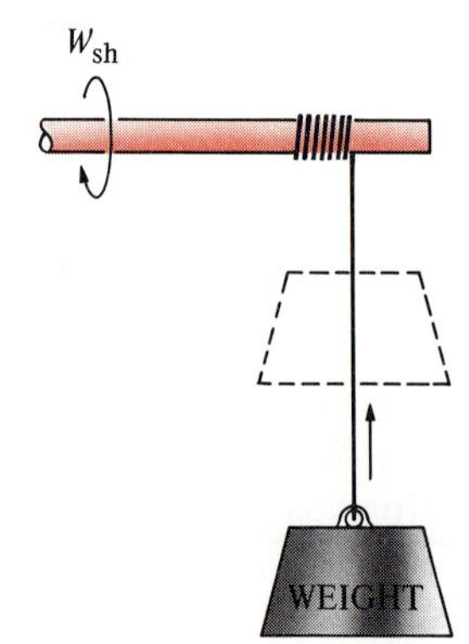

FIGURE 6-24
In the absence of friction, raising a weight by a rotating shaft does not create any disorder (entropy), and thus energy is not degraded during this process.

(or a flywheel) will not produce any entropy. Any process that does not produce a net entropy is reversible, and thus the process described above can be reversed by lowering the weight. Therefore, energy is not degraded during this process, and no potential to do work is lost.

Instead of raising a weight, let us operate the paddle wheel in a container filled with a gas, as shown in Fig. 6-25. The paddle-wheel work in this case will be converted to the internal energy of the gas, as evidenced by a rise in gas temperature, creating a higher level of molecular chaos and disorder in the container. This process is quite different from raising a weight since the organized paddle-wheel energy is now converted to a highly disorganized form of energy, which cannot be converted back to the paddle wheel as the rotational kinetic energy. Only a portion of this energy can be converted to work by partially reorganizing it through the use of a heat engine. Therefore, energy is degraded during this process, the ability to do work is reduced, molecular disorder is produced, and associated with all this is an increase in entropy.

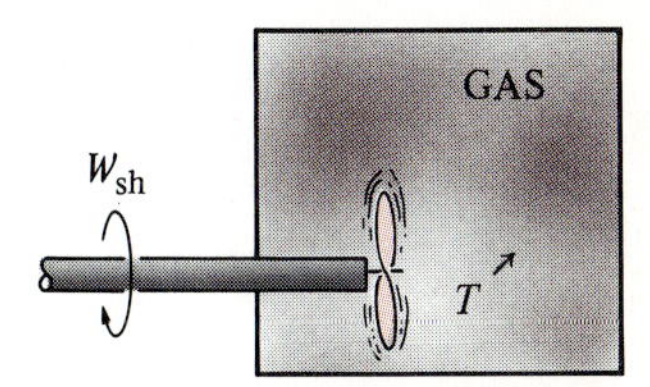

FIGURE 6-25
The paddle-wheel work done on a gas increases the level of disorder (entropy) of the gas, and thus energy is degraded during this process.

The *quantity* of energy is always preserved during an actual process (the first law), but the *quality* is bound to decrease (the second law). This decrease in quality is always accompanied by an increase in entropy. As an example, consider the transfer of 10 kJ of energy as heat from a hot medium to a cold one. At the end of the process, we will still have the 10 kJ of energy, but at a lower temperature and thus at a lower quality.

Heat is, in essence, a form of disorganized energy, and some disorganization (entropy) will flow with heat (Fig. 6-26). As a result, the entropy and the level of molecular disorder or randomness of the hot body will decrease while the entropy and the level of molecular disorder of the cold body increase. The second law requires that the increase in entropy of the cold body be greater than the decrease in entropy of the hot body, and thus the net entropy of the combined system (the cold body and the hot body) increases. That is, the combined system is at a state of greater disorder at the final state. Thus we can conclude that processes can occur only in the direction of increased overall entropy or molecular disorder. That is, the entire universe is getting more and more chaotic every day. This is a major concern not only to engineers but also to philosophers.

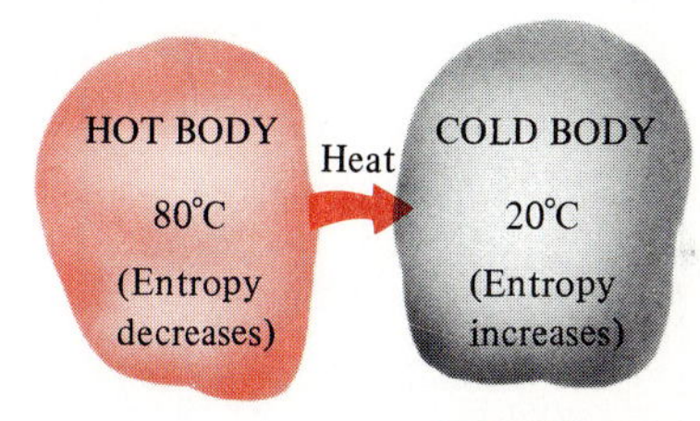

FIGURE 6-26
During a heat transfer process, the net disorder (entropy) increases. (The increase in the disorder of the cold body more than offsets the decrease in the disorder of the hot body.)

From a statistical point of view, entropy is a measure of molecular randomness, i.e., the uncertainty about the positions of molecules at any instant. Even in the solid phase, the molecules of a substance continually oscillate, creating an uncertainty about their position. These oscillations, however, fade as the temperature is decreased, and the molecules become completely motionless at absolute zero. This represents a state of ultimate molecular order (and minimum energy). Therefore, *the entropy of a pure crystalline substance at absolute zero temperature is zero* since there is no uncertainty about the state of the molecules at that instant (Fig. 6-27). This statement is known as the **third law of thermodynamics**. The third law of thermodynamics provides an absolute reference point for the determination of entropy. The entropy determined relative to this point is called **absolute entropy**, and it is extremely useful in the thermodynamic analysis of chemical reactions. Notice that the entropy of a substance

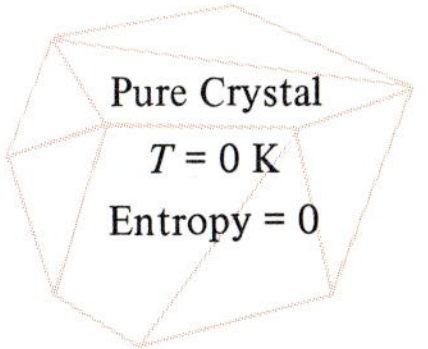

FIGURE 6-27
A pure substance at absolute zero temperature is in perfect order, and its entropy is zero (the third law of thermodynamics).

that is not pure crystalline (such as a solid solution) is not zero at absolute zero temperature. This is because more than one molecular configuration exists for such substances, which introduces some uncertainty about the microscopic state of the substance.

The concept of entropy as a measure of disorganized energy can also be applied to other areas. Iron molecules, for example, create a magnetic field around themselves. In ordinary iron, molecules are randomly aligned, and they cancel each other's magnetic effect. When iron is treated and the molecules are realigned, however, that piece of iron turns into a piece of magnet, creating a powerful magnetic field around it.

Entropy and Us

The concept of entropy may even be applied to human beings. Efficient people are those who lead low-entropy lives. They are neat and highly organized and have a place for everything. It takes minimum energy for them to find anything they need. Disorganized people, however, lead high-entropy lives and are highly inefficient. It takes them hours to find something they need in a hurry instead of seconds (sound familiar?). They are likely to create a bigger mess as they are searching, since they will probably conduct the search in a disorganized manner (Fig. 6-28). People leading high-entropy life styles are always on the run, never seem to catch up, and are more likely to die of a heart attack.

FIGURE 6-28
The use of entropy (disorganization, uncertainty) is not limited to thermodynamics.

You have probably noticed (with frustration) that some people seem to learn fast and remember what they learn well. We may call this low-entropy learning. To keep the entropy generation to a minimum, people must make a conscientious effort to file the new information properly by relating it to their existing knowledge base and create a solid information network in their minds. People who throw the information into their minds as they study, with no effort to secure it, may *think* they are learning. They are bound to discover otherwise when they need to locate the information during a test, for example. It is not an easy task to retrieve information from a knowledge base in the gas phase. A library with no filing system is like no library at all.

6-6 ■ PROPERTY DIAGRAMS INVOLVING ENTROPY

Property diagrams serve as great visual aids in the thermodynamic analysis of processes. We have used *P-v* and *T-v* diagrams extensively in previous chapters in conjunction with the first law of thermodynamics. In the second-law analysis, it is very helpful to plot the processes on diagrams for which one of the coordinates is entropy. The two diagrams used most extensively in the second-law analysis are the *temperature-entropy* and the *enthalpy-entropy* diagrams.

1 The *T-s* Diagram

Consider the defining equation of entropy (Eq. 6-4). It can be rearranged as

$$\delta Q_{\text{int rev}} = T\,dS \qquad \text{(kJ)} \tag{6-17}$$

As shown in Fig. 6-29, δQ_{rev} corresponds to a differential area on a *T-S* diagram. The total heat transfer during an internally reversible process is determined by integration to be

$$Q_{\text{int rev}} = \int_1^2 T\,dS \qquad \text{(kJ)} \tag{6-18}$$

which corresponds to the area under the process curve on a *T-S* diagram. Therefore we conclude that *the area under the process curve on a T-S diagram represents the internally reversible heat transfer*. This is somewhat analogous to reversible boundary work being represented by the area under the process curve on a *P-V* diagram. Note that the area under the process curve represents heat transfer for processes that are internally (or totally) reversible. It has no meaning for irreversible processes.

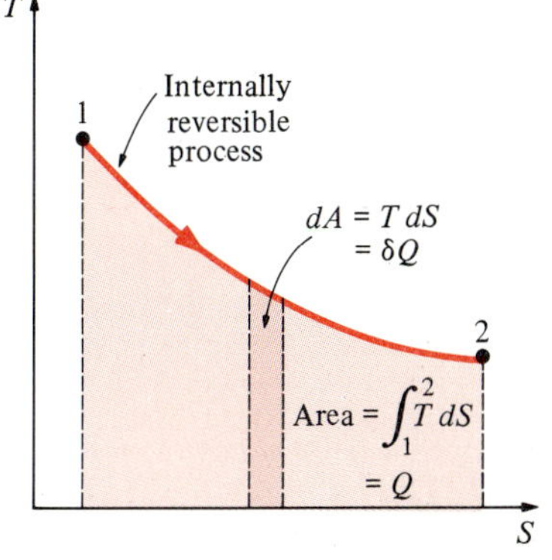

FIGURE 6-29
On a *T-S* diagram, the area under the process curve represents the heat transfer for internally reversible processes.

Equations 6-17 and 6-18 can also be expressed on a unit-mass basis as

$$\delta q_{\text{int rev}} = T\,ds \qquad \text{(kJ/kg)} \tag{6-19}$$

and

$$q_{\text{int rev}} = \int_1^2 T\,ds \qquad \text{(kJ/kg)} \tag{6-20}$$

To perform the integrations in Eqs. 6-18 and 6-20, one needs to know the relationship between T and s during a process. One special case for which these integrations can be performed easily is the *internally reversible isothermal process*. It yields

$$Q_{\text{int rev}} = T_0\,\Delta S \qquad \text{(kJ)} \tag{6-21}$$

or

$$q_{\text{int rev}} = T_0\,\Delta s \qquad \text{(kJ/kg)} \tag{6-22}$$

where T_0 is the constant temperature and ΔS is the entropy change of the system during the process.

In the relations above, T is the absolute temperature, which is always positive. Therefore, heat transfer during internally reversible processes is positive when entropy increases and negative when entropy decreases. An isentropic process on a *T-s* diagram is easily recognized as a vertical-line segment. This is expected since an isentropic process involves no heat transfer, and therefore the area under the process path must be zero (Fig. 6-30). The *T-s* diagrams serve as valuable tools for visualizing the second-law aspects of processes and cycles, and thus they are frequently used in thermodynamics. The *T-s* diagram of water is given in the Appendix in Fig. A-9.

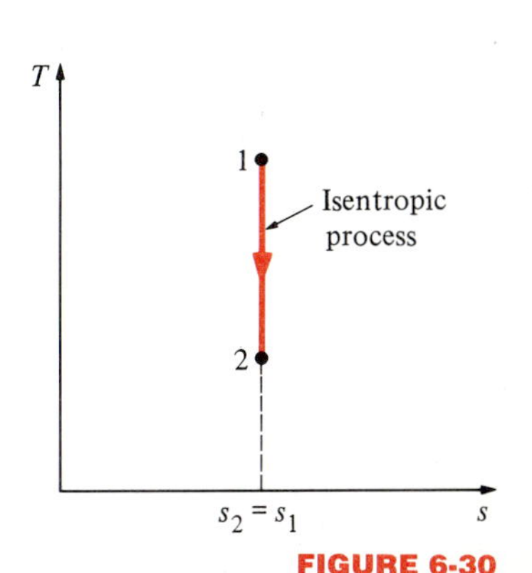

FIGURE 6-30
During an internally reversible, adiabatic (isentropic) process, the entropy of a system remains constant.

The general characteristics of a *T-s* diagram for the liquid and vapor regions of a pure substance are shown in Fig. 6-31. The following observations can be made:

1 At any point in a single phase region, the constant-volume lines are steeper than the constant-pressure lines.

2 In the saturated liquid–mixture region, the constant-pressure lines are parallel to the constant-temperature lines.

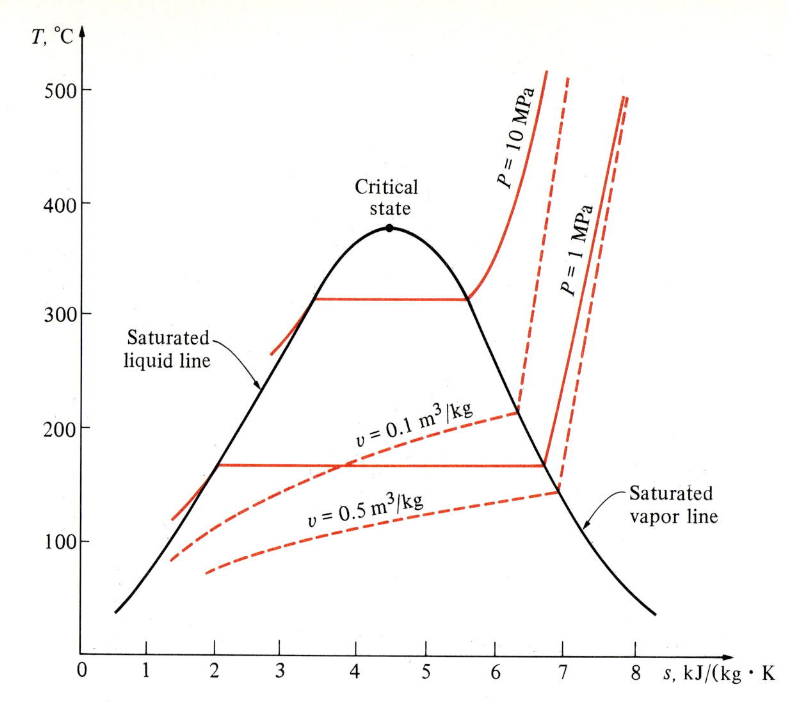

FIGURE 6-31
Schematic of a *T-s* diagram for water.

3 In the compressed liquid region, the constant-pressure lines almost coincide with the saturated liquid line.

The student is encouraged to remember the general appearance of constant-property lines on a *T-s* diagram since this diagram is used extensively in the forthcoming chapters.

EXAMPLE 6-6

Show the Carnot cycle on a *T-S* diagram and indicate the areas that represent the heat added Q_H, heat rejected Q_L, and the net work output $W_{net,out}$ on this diagram.

Solution You will recall from Chap. 5 that the Carnot cycle is made up of two reversible isothermal (T = constant) processes and two isentropic (s = constant) processes. These four processes form a rectangle on a *T-S* diagram, as shown in Fig. 6-32.

On a *T-S* diagram, the area under the process curve represents the heat transfer for that process. Thus the area *A*12*B* represents Q_H, the area *A*43*B* represents Q_L, and the difference between these two (the area in color) represents the net work since

$$W_{net,out} = Q_H - Q_L$$

Therefore, the area enclosed by the path of a cycle (area 1234) on a *T-S* diagram represents the net work. Recall from Chap. 3 that the area enclosed by the cycle also represents the net work on a *P-v* diagram.

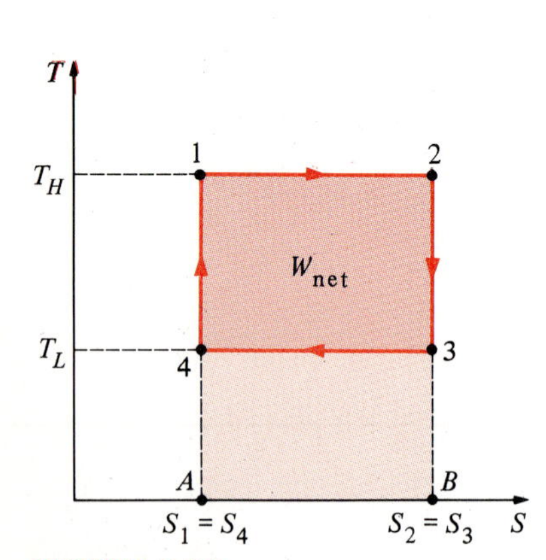

FIGURE 6-32
The *T-S* diagram of a Carnot cycle (Example 6-6).

2 The *h-s* Diagram

Another diagram commonly used in engineering is the enthalpy-entropy diagram, which is quite valuable in the analysis of steady-flow devices such as turbines, compressors, and nozzles. The coordinates of an *h-s* diagram represent two properties of major interest: enthalpy, which is a primary property in the first-law analysis of the steady-flow devices, and entropy, which is the property that accounts for irreversibilities during adiabatic processes. In analyzing the steady flow of steam through an adiabatic turbine, for example, the vertical distance between the inlet and the exit states (Δh) is a measure of the work output of the turbine, and the horizontal distance (Δs) is a measure of the irreversibilities associated with the process (Fig. 6-33).

The *h-s* diagram is also called a **Mollier diagram** after the German scientist R. Mollier (1863–1935). An *h-s* diagram is given in the Appendix for steam in Fig. A-10. The general features of an *h-s* diagram are illustrated in Fig. 6-34. On an *h-s* diagram, the constant-temperature lines are straight in the saturated liquid–vapor mixture region. They become almost horizontal in the superheated vapor region, particularly at low pressures. This is not surprising since steam approaches ideal-gas behavior as it moves away from the saturation region, and for ideal gases the enthalpy is a function of temperature only.

The *h-s* diagrams that are sufficiently large offer great convenience in determining properties with reasonable accuracy and are commonly used in practice.

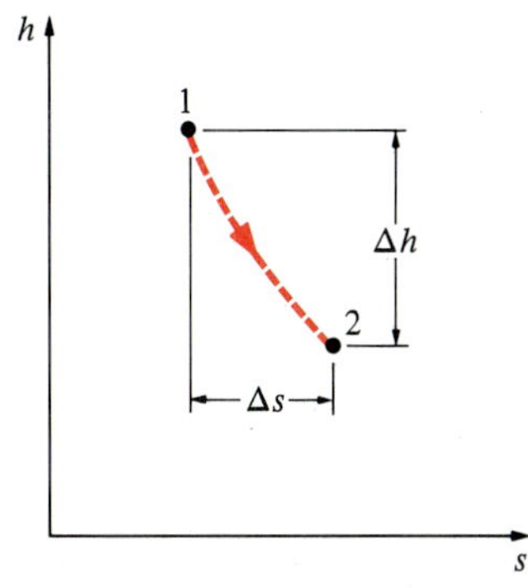

FIGURE 6-33

For adiabatic steady-flow devices, the vertical distance Δh on an *h-s* diagram is a measure of work, and the horizontal distance Δs is a measure of irreversibilities.

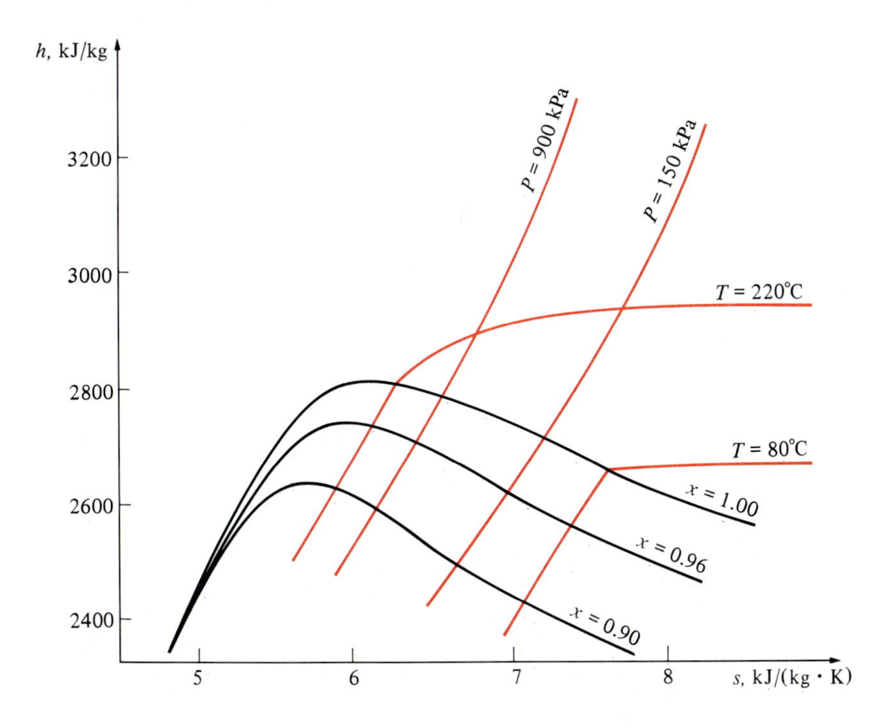

FIGURE 6-34

Schematic of an *h-s* diagram for water.

6-7 ■ THE $T\,ds$ RELATIONS

Earlier in this chapter it was shown that the quantity $(\delta Q/T)_{\text{int rev}}$ corresponds to a differential change in a property, called entropy. The entropy change for a process, then, was evaluated by integrating $\delta Q/T$ along some imaginary internally reversible path between the actual end states (Eq. 6-5). For isothermal internally reversible processes, this integration is straightforward. But when the temperature varies during the process, we have to have a relation between δQ and T to perform this integration. Finding such relations is what we intend to do in this section.

The differential form of the conservation of energy equation for a closed stationary system (a fixed mass) containing a simple compressible substance (Eq. 3-30) can be expressed for an internally reversible process as

$$\delta Q_{\text{int rev}} - \delta W_{\text{int rev}} = dU$$

But

$$\delta Q_{\text{int rev}} = T\,dS \qquad \text{(Eq. 6-17)}$$

$$\delta W_{\text{int rev}} = P\,dV \qquad \text{(Eq. 3-8)}$$

Thus,

$$T\,dS = dU + P\,dV$$

or

$$T\,ds = du + P\,dv \tag{6-23}$$

per unit mass. This equation is known as the first $T\,ds$, or *Gibbs, equation*. Notice that the only type of work interaction a simple compressible system may involve as it undergoes an internally reversible process is the quasi-equilibrium boundary work.

The second $T\,ds$ equation is obtained by eliminating du from Eq. 6-23 by using the definition of enthalpy ($h = u + Pv$):

$$\left.\begin{aligned} h = u + Pv &\longrightarrow dh = du + P\,dv + v\,dP \\ \text{Eq. 6-20} &\longrightarrow T\,ds - du + P\,dv \end{aligned}\right\} \quad T\,ds = dh - v\,dP \tag{6-24}$$

Equations 6-23 and 6-24 are extremely valuable since they relate entropy changes of a system to the changes in other properties. Unlike Eq. 6-4, they are property relations and therefore are independent of the type of the processes.

The $T\,ds$ relations above are developed with an internally reversible process in mind since the entropy change between two states must be evaluated along a reversible path. But the results obtained are valid for both reversible and irreversible processes since entropy is a property and the change in a property between two states is independent of the type of process the system undergoes. Equations 6-23 and 6-24 are relations between the properties of a unit mass of a simple compressible system as it undergoes a change of state, and they are applicable whether the change occurs in a closed or an open system (Fig. 6-35).

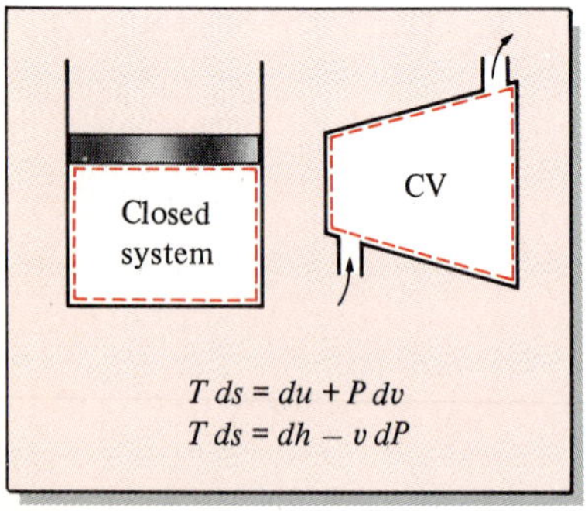

FIGURE 6-35
The $T\,ds$ relations are valid for both reversible and irreversible processes and for both closed and open systems.

Explicit relations for differential changes in entropy are obtained by solving for ds in Eqs. 6-23 and 6-24:

$$ds = \frac{du}{T} + \frac{P\,dv}{T} \tag{6-25}$$

and $$ds = \frac{dh}{T} - \frac{v\,dP}{T} \tag{6-26}$$

The entropy change during a process can be determined by integrating either of these equations between the initial and the final states. To perform these integrations, however, we must know the relationship between du or dh and the temperature (such as $du = C_v\,dT$ and $dh = C_p\,dT$ for ideal gases) as well as the equation of state for the substance (such as the ideal-gas equation of state $Pv = RT$). For substances for which such relations exist, such as ideal gases and incompressible substances, the integration of Eq. 6-25 or 6-26 is straightforward. This is done later in this chapter. For other substances, we have to rely on tabulated data.

The $T\,ds$ relations for nonsimple systems, i.e., systems that involve more than one mode of quasi-equilibrium work interactions, can be obtained in a similar fashion by including all the relevant quasi-equilibrium work modes.

6-8 ■ THE ENTROPY CHANGE OF PURE SUBSTANCES

The $T\,ds$ relations developed above are not limited to a particular substance in a particular phase. They are valid for all pure substances at any phase or combination of phases. The successful use of these relations, however, depends on the availability of the property relations between T and du or dh and the P-v-T behavior of the substance. For a pure substance, in general, these relations are too complicated, and this makes it impossible to obtain simple relations for entropy changes. The values of s, therefore, are determined from measurable property data following rather involved computations and are tabulated in exactly the same manner as the properties v, u, and h (Fig. 6-36).

The entropy values in the property tables are given relative to an arbitrary reference state. In steam tables the entropy of saturated liquid s_f at 0.01°C is assigned the value of zero. For refrigerant-12, the zero

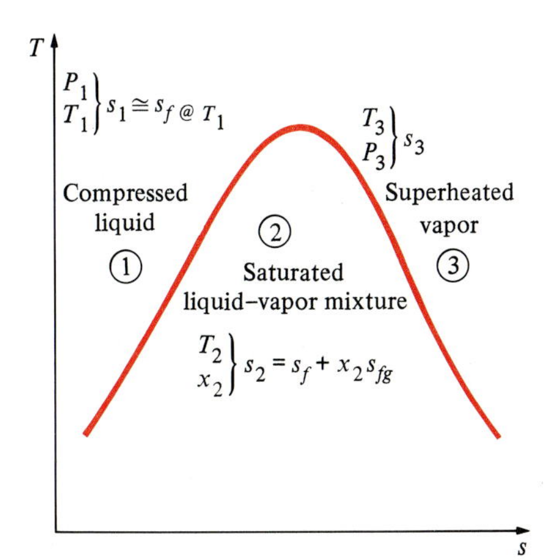

FIGURE 6-36

The entropy of a pure substance is determined from the tables, just as for any other property.

value is assigned to saturated liquid at $-40°C$. The entropy values become negative at temperatures below the reference value.

The value of entropy at a specified state is determined just as any other property. In the compressed liquid and superheated vapor regions, it can be obtained directly from the tables at the specified state. In the saturated mixture region, it is determined from

$$s = s_f + x s_{fg} \qquad [\text{kJ/(kg·K)}] \tag{6-27}$$

where x is the quality and s_f and s_{fg} values are listed in the saturation tables. In the absence of compressed liquid data, the entropy of the compressed liquid can be approximated by the entropy of the saturated liquid at the given temperature:

$$s_{@P,T} \cong s_{f@T} \tag{6-28}$$

The entropy change of a pure substance during a process is simply the difference between the entropy values at the final and initial states:

$$\Delta S = m(s_2 - s_1) \qquad (\text{kJ/K}) \tag{6-29}$$

or

$$\Delta s = s_2 - s_1 \qquad [\text{kJ/(kg·K)}] \tag{6-30}$$

Equation 6-30 is applicable to a closed system as well as to a unit mass passing through a control volume.

EXAMPLE 6-7

A rigid tank contains 5 kg of refrigerant-12 initially at 20°C and 140 kPa. The refrigerant is now cooled while being stirred until its pressure drops to 100 kPa. Determine the entropy change of the refrigerant during this process.

Solution A sketch of the system and the *T-s* diagram for the process are given in Fig. 6-37. The change in entropy of a substance during a process is simply the difference between the entropy values at the final and the initial states.

Recognizing that the specific volume remains constant during this process ($v_2 = v_1$), we see that the entropies of the refrigerant at both states are

State 1: $\left.\begin{array}{l} P_1 = 140 \text{ kPa} \\ T_1 = 20°\text{C} \end{array}\right\}$ $\begin{array}{l} s_1 = 0.8035 \text{ kJ/(kg·K)} \\ v_1 = 0.1397 \text{ m}^3\text{/kg} \end{array}$ (Table A-13)

State 2: $\begin{array}{l} P_2 = 100 \text{ kPa} \longrightarrow \\ (v_2 = v_1) \end{array}$ $\begin{array}{l} v_f = 0.0006719 \text{ m}^3\text{/kg} \\ v_g = 0.1600 \text{ m}^3\text{/kg} \end{array}$ (Table A-12)

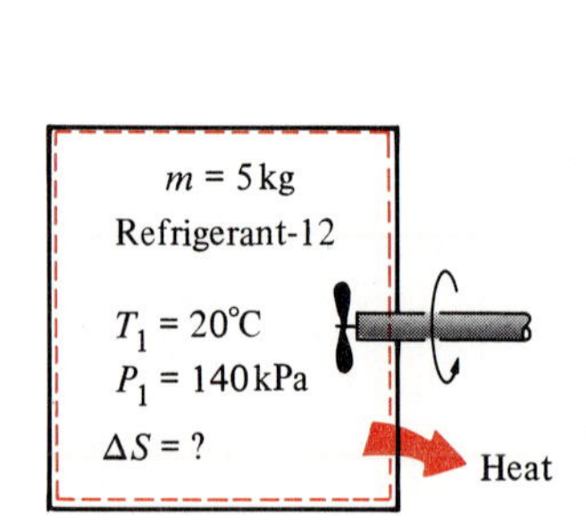

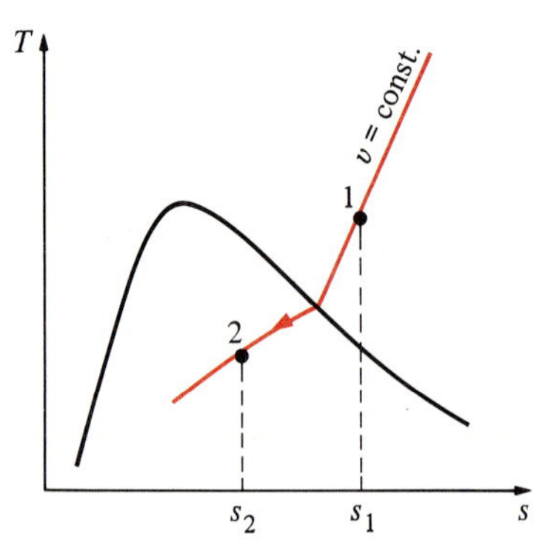

FIGURE 6-37

Schematic and *T-s* diagram for Example 6-7.

The refrigerant is a saturated liquid–vapor mixture at the final state since $v_f < v_2 < v_g$ at 100 kPa of pressure. Therefore, we need to determine the quality first:

$$x_2 = \frac{v_2 - v_f}{v_{fg}} = \frac{0.1397 - 0.0007}{0.1600 - 0.0007} = 0.873$$

Thus,

$$s_2 = s_f + x_2 s_{fg} = 0.0368 + (0.873)(0.6803) \text{ kJ/(kg·K)} = 0.6307 \text{ kJ/(kg·K)}$$

Then the entropy change of the refrigerant during this process is determined from Eq. 6-29 to be

$$\Delta S = m(s_2 - s_1) = (5 \text{ kg})[(0.6307 - 0.7977) \text{ kJ/(kg·K)}] = -0.835 \text{ kJ/K}$$

The negative sign indicates that the entropy of the system is decreasing during this process. This is not a violation of the second law, however, since it is the *total entropy change* ($\Delta S_{sys} + \Delta S_{surr}$) that cannot be negative.

EXAMPLE 6-8

A piston-cylinder device initially contains 3 lbm of liquid water at 20 psia and 70°F. The water is now heated at constant pressure by the addition of 3450 Btu of heat. Determine the entropy change of the water during this process.

Solution A sketch of the system and the *T-s* diagram for the process are given in Fig. 6-38. The water exists as a compressed liquid at the initial state since its pressure is greater than the saturation pressure (0.3632 psia) at 70°F. By approximating the compressed liquid as a saturated liquid at the given temperature, the properties at the initial state are

State 1: $\left.\begin{array}{l} P_1 = 20 \text{ psia} \\ T_1 = 70°\text{F} \end{array}\right\}$ $\begin{array}{l} s_1 \cong s_{f @ 70°\text{F}} = 0.07463 \text{ Btu/(lbm·R)} \\ h_1 \cong h_{f @ 70°\text{F}} = 38.09 \text{ Btu/lbm} \end{array}$ (Table A-4E)

At the final state, the pressure is still 20 psia, but we need one more property to fix the state. This property is determined from the first-law relation (Eq. 3-27) for closed systems:

$$Q - W_{other} - W_b = \Delta U + \Delta \text{KE} + \Delta \text{PE}$$

The closed system under consideration is stationary (ΔKE $=$ ΔPE $= 0$) and involves boundary work only ($W_{other} = 0$). Furthermore, for a closed system

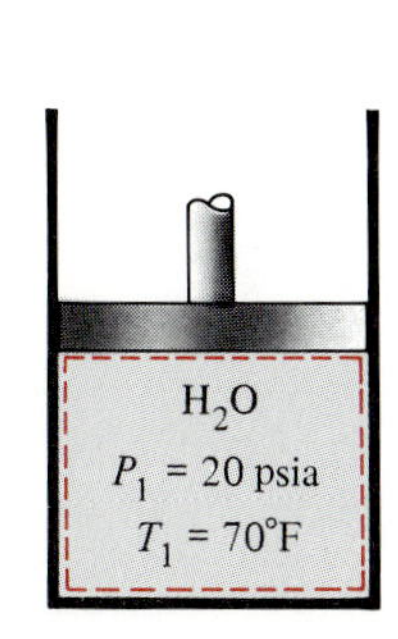

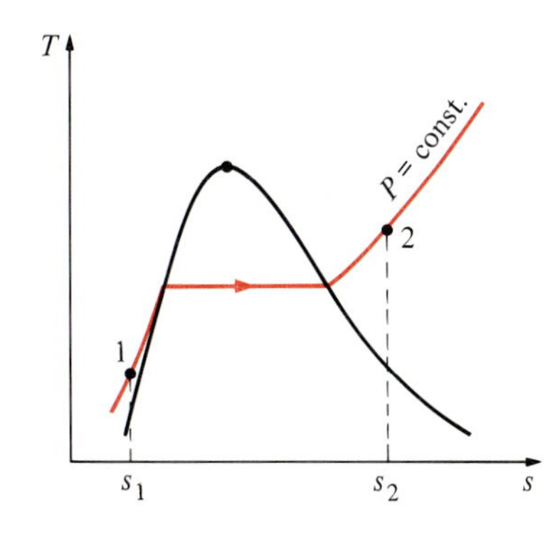

FIGURE 6-38
Schematic and *T-s* diagram for Example 6-8.

undergoing a constant-pressure process, $\Delta U + W_b = \Delta H$. Thus,

$$Q = m(h_2 - h_1)$$
$$3450 \text{ Btu} = (3 \text{ lbm})(h_2 - 38.09 \text{ Btu/lbm})$$
$$h_2 = 1188.1 \text{ Btu/lbm}$$

State 2: $\left.\begin{array}{l} P_2 = 20 \text{ psia} \\ h_2 = 1188.1 \text{ Btu/lbm} \end{array}\right\}$ $s_2 = 1.7759$ Btu/(lbm·R) (Table A-6E, interpolation)

Then the entropy change of the water during this process is determined from Eq. 6-29 to be

$$\Delta S = m(s_2 - s_1) = (3 \text{ lbm})[(1.7759 - 0.07463) \text{ Btu/(lbm·R)}]$$
$$= 5.1038 \text{ Btu/R}$$

EXAMPLE 6-9

Steam at 7 MPa and 450°C is throttled in a valve to a pressure of 3 MPa during a steady-flow process. Determine the entropy generation for this process, and check if the increase-in-entropy principle is satisfied.

Solution A sketch of the system and the *T-s* diagram for the process are given in Fig. 6-39. Remembering that throttling valves are essentially isenthalpic (h = constant) devices, we see that the entropy of steam at the inlet and the exit states are

State 1: $\left.\begin{array}{l} P_1 = 7 \text{ MPa} \\ T_1 = 450°\text{C} \end{array}\right\}$ $\begin{array}{l} h_1 = 3287.1 \text{ kJ/kg} \\ s_1 = 6.6327 \text{ kJ/(kg·K)} \end{array}$ (Table A-6)

State 2: $\left.\begin{array}{l} P_2 = 3 \text{ MPa} \\ h_2 = h_1 \end{array}\right\}$ $s_2 = 6.9919$ kJ/(kg·K) (Table A-6)

Heat transfer during a throttling process is usually negligible. Therefore, Eq. 6-16 reduces, for this case, to

$$s_{gen} = s_e - s_i + \frac{q_{surr}}{T_{surr}}\overset{\nearrow 0}{} = s_2 - s_1$$

Thus, $s_{gen} = (6.9919 - 6.6327)$ kJ/(kg·K) $=$ 0.3592 kJ/(kg·K)

This is the amount of entropy generated per unit mass of steam as it is throttled

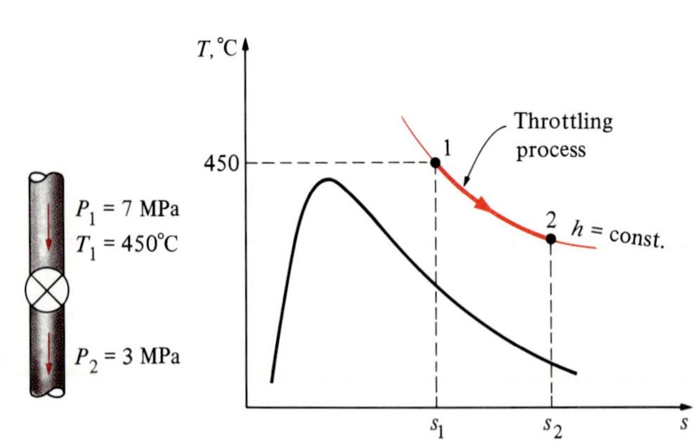

FIGURE 6-39
Schematic and *T-s* diagram for Example 6-9.

from the inlet state to the final pressure, and it is caused by unrestrained expansion.

The increase-in-entropy principle is obviously satisfied during this process since the total entropy change is not negative (or, there is no destruction of entropy).

Isentropic Processes of Pure Substances

We pointed out earlier that two factors can change the entropy of a system: heat transfer and irreversibilities. Then it follows that the entropy of a system will not change during an internally reversible, adiabatic process, which is called an *isentropic* (constant-entropy) *process*. An isentropic process appears as a vertical line on a T-s diagram.

Many engineering systems or devices such as pumps, turbines, nozzles, and diffusers are essentially adiabatic in their operation, and they perform best when the irreversibilities, such as the friction associated with the process, are minimized. Therefore, an isentropic process can serve as an appropriate model for actual processes. Also isentropic processes enable us to define efficiencies for processes to compare the actual performance of these devices to the performance under idealized conditions. This should be sufficient motivation for studying the isentropic processes.

No special relations exist for the isentropic processes of pure substances other than

$$s_2 = s_1 \qquad [\text{kJ/(kg·K)}] \tag{6-31}$$

except for some idealized cases that are discussed in the next two sections. A substance will have the same entropy value at the final state as it does at the initial state if the process is carried out in an isentropic manner.

It should be recognized that a reversible adiabatic process is necessarily isentropic ($s_2 = s_1$) but an isentropic process is not necessarily reversible adiabatic. (The entropy increase of a substance during a process as a result of irreversibilities may be offset by a decrease in entropy as a result of heat losses, for example.) However, the term *isentropic process* is customarily used in thermodynamics to imply an *internally reversible, adiabatic process*.

EXAMPLE 6-10

Steam enters an adiabatic turbine at 5 MPa and 450°C and leaves at a pressure of 1.4 MPa. Determine the work output of the turbine per unit mass of steam flowing through the turbine if the process is reversible and the changes in kinetic and potential energies are negligible.

Solution A sketch of the system and the T-s diagram for the process are given in Fig. 6-40. The work output of a turbine which is operating steadily can be determined from the conservation of energy relation for steady-flow devices expressed on a unit-mass basis (Eq. 4-23), which reduces to

$$\cancelto{0}{q} - w = \Delta h + \cancelto{0}{\Delta ke} + \cancelto{0}{\Delta pe}$$

$$w = h_1 - h_2$$

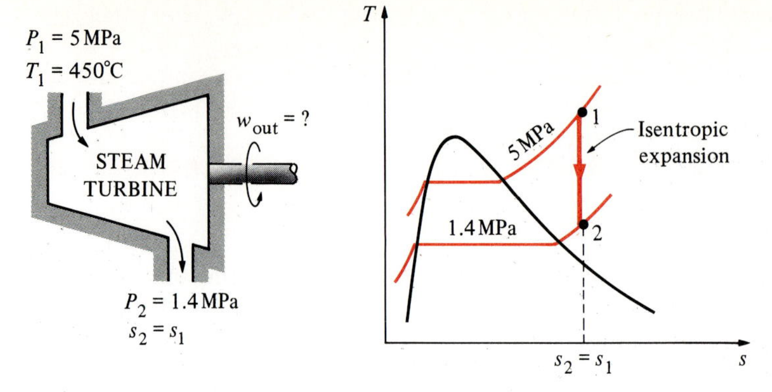

FIGURE 6-40
Schematic and *T-s* diagram for Example 6-10.

Thus to calculate the work, all we need to do is to determine the enthalpy values at the initial and the final states:

State 1: $\left.\begin{aligned} P_1 &= 5 \text{ MPa} \\ T_1 &= 450°\text{C} \end{aligned}\right\}$ $\begin{aligned} h_1 &= 3316.2 \text{ kJ/kg} \\ s_1 &= 6.8186 \text{ kJ/(kg·K)} \end{aligned}$ (Table A-6)

Only one property (pressure) is given at the final state, which is not sufficient to fix the state. However, this process is isentropic (constant entropy) since it is both reversible and adiabatic. Therefore, $s_2 = s_1 = 6.8186$ kJ/(kg·K). This provides the second property necessary to fix the final state:

State 2: $\left.\begin{aligned} P_2 &= 1.4 \text{ MPa} \\ s_2 &= 6.8186 \text{ kJ/(kg·K)} \end{aligned}\right\}$ $h_2 = 2966.6$ kJ/kg (Table A-6, interpolation)

Thus,
$$\begin{aligned} w &= (3316.2 - 2966.6) \text{ kJ/kg} \\ &= 349.6 \text{ kJ/kg} \end{aligned}$$

which is the work output of the turbine per unit mass of steam flowing through the turbine.

6-9 ■ THE ENTROPY CHANGE OF SOLIDS AND LIQUIDS

We mentioned in Sec. 3-9 that solids and liquids can be idealized as *incompressible substances* since their volumes remain essentially constant during a process. Thus $dv = 0$ for solids and liquids, and Eq. 6-26 for this case reduces to

$$ds = \frac{du}{T} = \frac{C\,dT}{T}$$

since $C_p = C_v = C$ for incompressible substances and $du = C\,dT$. The entropy change for a process is determined by integration:

$$s_2 - s_1 = \int_1^2 C(T)\frac{dT}{T} \qquad [\text{kJ/(kg·K)}] \tag{6-32}$$

The specific heat C of liquids and solids, in general, depends on temperature, and we need a relation for C as a function of temperature to perform this integration. In many cases, however, C may be treated as a constant at some average value over the given temperature range. Then the integration in Eq. 6-32 can be performed, to yield

$$s_2 - s_1 = C_{av} \ln \frac{T_2}{T_1} \qquad [\text{kJ/(kg·K)}] \qquad (6\text{-}33)$$

Note that the entropy change of a truly incompressible substance depends on temperature only. Equation 6-33 can be used to determine the entropy changes of solids and liquids with reasonable accuracy.

EXAMPLE 6-11

A 50-kg block of iron casting at 500 K is thrown into a large lake which is at a temperature of 285 K, as shown in Fig. 6-41. The iron block eventually reaches thermal equilibrium with the lake water. Assuming an average specific heat of 0.45 kJ/(kg·K) for the iron, determine (*a*) the entropy change of the iron block, (*b*) the entropy change of the lake water, and (*c*) the total entropy change for this process.

FIGURE 6-41
Schematic for Example 6-11.

Solution To determine the entropy change for the iron block and for the lake, first we need to know the final equilibrium temperature. Given that the thermal energy capacity of the lake is very large relative to that of the iron block, the lake will absorb all the heat rejected by the iron block without experiencing any change in its temperature. Therefore, the iron block will cool to 285 K during this process while the lake temperature remains constant at 285 K. Then the entropy changes are determined as follows:

(*a*) Like all solids, the iron block can be approximated as an incompressible substance, and thus its entropy change can be determined from Eq. 6-33:

$$\Delta S_{iron} = m(s_2 - s_1) = mC_{av} \ln \frac{T_2}{T_1}$$

$$= (50 \text{ kg})[0.45 \text{ kJ/(kg·K)}] \ln \frac{285 \text{ K}}{500 \text{ K}}$$

$$= -12.65 \text{ kJ/K}$$

(*b*) The lake water in this problem acts as a thermal energy reservoir, and its entropy change can be determined from Eq. 6-6. But first we need to determine the heat transfer to the lake. Taking the iron block as our system and disregarding the changes in the kinetic and potential energies, we see that the conservation of energy equation for this closed system reduces to

$$Q - \cancel{W}^{\,0} = \Delta U + \cancel{\Delta KE}^{\,0} + \cancel{\Delta PE}^{\,0}$$

$$Q_{iron} = mC_{av}(T_2 - T_1) = (50 \text{ kg})[0.45 \text{ kJ/(kg·K)}][(285 - 500) \text{ K}]$$

$$= -4837.5 \text{ kJ}$$

Then

$$Q_{lake} = -Q_{iron} = +4837.5 \text{ kJ}$$

and

$$\Delta S_{lake} = \frac{Q_{lake}}{T_{lake}} = \frac{+4837.5 \text{ kJ}}{285 \text{ K}} = 16.97 \text{ kJ/K}$$

(*c*) The total entropy change for this process is the sum of these two since the iron block and the lake together form an adiabatic system:

$$\Delta S_{total} = \Delta S_{iron} + \Delta S_{lake} = (-12.65 + 16.97) \text{ kJ/K}$$

$$= 4.32 \text{ kJ/K}$$

The positive sign for the total entropy change indicates that this is an irreversible process.

EXAMPLE 6-12

Water at 20 psia and 50°F enters a mixing chamber at a rate of 300 lbm/min, where it is mixed steadily with steam entering at 20 psia and 240°F. The mixture leaves the chamber at 20 psia and 130°F, and heat is lost to the surrounding air at 70°F at a rate of 180 Btu/min. Neglecting the changes in kinetic and potential energies, determine the rate of entropy generation for this process.

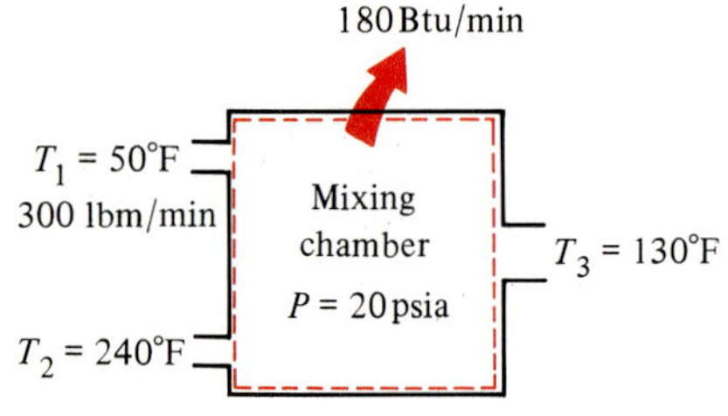

FIGURE 6-42
Schematic for Example 6-12.

Solution This is a steady-flow process, and a sketch of the mixing chamber and the system boundaries is shown in Fig. 6-42. The rate of total entropy change (or, the rate of entropy generation) for this process can be determined from Eq. 6-14. But first we need to determine the mass flow rate of the steam by applying the conservation of mass and conservation of energy equations (Eqs. 4-14 and 4-19) to this steady-flow system:

$$\Sigma \dot{m}_i = \Sigma \dot{m}_e \longrightarrow \dot{m}_1 + \dot{m}_2 = \dot{m}_3$$

and
$$\dot{Q} - \cancel{\dot{W}}^{\,0} = \Sigma \dot{m}_e h_e - \Sigma \dot{m}_i h_i \longrightarrow \dot{Q} = \dot{m}_3 h_3 - \dot{m}_2 h_2 - \dot{m}_1 h_1$$

The desired properties at the specified states are

State 1: $\left.\begin{array}{l} T_1 = 50°\text{F} \\ P_1 = 20\,\text{psia} \end{array}\right\}$ $\begin{array}{l} h_1 \cong h_{f\,@\,50°\text{F}} = 18.06\,\text{Btu/lbm} \\ s_1 \cong s_{f\,@\,50°\text{F}} = 0.03607\,\text{Btu/(lbm·R)} \end{array}$ (Table A-4E)

State 2: $\left.\begin{array}{l} T_2 = 240°\text{F} \\ P_2 = 20\,\text{psia} \end{array}\right\}$ $\begin{array}{l} h_2 = 1162.3\,\text{Btu/lbm} \\ s_2 = 1.7405\,\text{Btu/(lbm·R)} \end{array}$ (Table A-6E)

State 3: $\left.\begin{array}{l} T_3 = 130°\text{F} \\ P_3 = 20\,\text{psia} \end{array}\right\}$ $\begin{array}{l} h_3 \cong h_{f\,@\,130°\text{F}} = 97.98\,\text{Btu/lbm} \\ s_3 \cong s_{f\,@\,130°\text{F}} = 0.18172\,\text{Btu/(lbm·R)} \end{array}$ (Table A-4E)

Substituting, we get

$$\begin{aligned} -180\,\text{Btu/min} &= (\dot{m}_2 + 300\,\text{lbm/min})(97.98\,\text{Btu/lbm}) \\ &\quad - (300\,\text{lbm/min})(18.06\,\text{Btu/lbm}) \\ &\quad - \dot{m}_2(1162.3\,\text{Btu/lbm}) \end{aligned}$$

Thus
$$\dot{m}_2 = 22.7\,\text{lbm/min}$$

The total rate of entropy generation is obtained by substituting the values above into Eq. 6-14:

$$\dot{S}_{\text{gen}} = \sum \dot{m}_e s_e - \sum \dot{m}_i s_i + \frac{\dot{Q}_{\text{surr}}}{T_{\text{surr}}}$$

or
$$\begin{aligned} \dot{S}_{\text{gen}} &= \dot{m}_3 s_3 - \dot{m}_1 s_1 - \dot{m}_2 s_2 + \frac{\dot{Q}_{\text{surr}}}{T_{\text{surr}}} \\ &= (322.7\,\text{lbm/min})[0.18172\,\text{Btu/(lbm·R)}] \\ &\quad - (300\,\text{lbm/min})[0.03607\,\text{Btu/(lbm·R)}] \\ &\quad - (22.7\,\text{lbm/min})[1.7405\,\text{Btu/(lbm·R)}] + \frac{+180\,\text{Btu/min}}{530\,\text{R}} \\ &= 8.65\,\text{Btu/(min·R)} \end{aligned}$$

Therefore, entropy is generated during this process at a rate of 8.65 Btu/(min·R). This entropy generation is caused by the mixing of two fluid streams (an irreversible process) and the heat transfer between the mixing chamber and the surroundings through a finite temperature difference (another irreversible process). Notice that this process satisfies the increase-in-entropy principle (and thus the second law) since the entropy generation is positive.

Isentropic Processes of Solids and Liquids

A relation for isentropic (constant-entropy) processes of solids and liquids is obtained by setting the entropy-change relation (Eq. 6-33) equal to zero:

$$C_{av} \ln \frac{T_2}{T_1} = 0$$

or

$$T_2 = T_1 \tag{6-34}$$

That is, the temperature of a truly incompressible substance remains constant during an isentropic process. Therefore, the isentropic process of an incompressible substance is also isothermal. This behavior is closely approximated by solids and liquids.

6-10 ■ THE ENTROPY CHANGE OF IDEAL GASES

An expression for the entropy change of an ideal gas can be obtained from Eq. 6-25 or 6-26 by employing the property relations for ideal gases (Fig. 6-43). By substituting $du = C_v\, dT$ and $P = RT/v$ into Eq. 6-25, the differential entropy change of an ideal gas becomes

$$ds = C_v \frac{dT}{T} + R \frac{dv}{v}$$

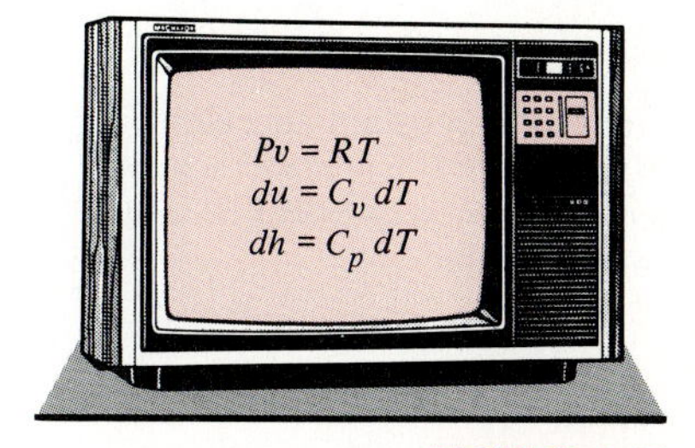

FIGURE 6-43
A broadcast from channel IG.

The entropy change for a process is obtained by integrating this relation between the end states:

$$s_2 - s_1 = \int_1^2 C_v(T) \frac{dT}{T} + R \ln \frac{v_2}{v_1} \tag{6-35}$$

A second relation for the entropy change of an ideal gas is obtained in a similar manner by substituting $dh = C_p\, dT$ and $v = RT/P$ into Eq. 6-26 and integrating. The result is

$$s_2 - s_1 = \int_1^2 C_p(T) \frac{dT}{T} - R \ln \frac{P_2}{P_1} \tag{6-36}$$

The specific heats of ideal gases, with the exception of monatomic gases, depend on temperature, and the integrals in Eqs. 6-35 and 6-36 cannot be performed unless the dependence of C_v and C_p on temperature is known. Even when the $C_v(T)$ and $C_p(T)$ functions are available, performing long integrations every time entropy change is calculated is not practical. Then two reasonable choices are left: either perform these integrations by simply assuming constant specific heats, or evaluate those integrals once and tabulate the results. Both approaches are presented below.

1 Constant Specific Heats: Approximate Treatment

Assuming constant specific heats for ideal gases is a common approximation, and we used this assumption before on several occasions. It usually simplifies the analysis greatly, and the price we pay for this convenience is some loss in accuracy. The magnitude of the error introduced by this assumption depends on the situation on hand. For example, for monatomic ideal gases such as helium, the specific heats are independent of temperature, and therefore the constant-specific-heat assumption introduces no error. For ideal gases whose specific heats vary almost linearly in the temperature range of interest, the possible error is minimized by using specific-heat values evaluated at the average temperature (Fig. 6-44). The results obtained in this way usually are sufficiently accurate for most ideal gases if the temperature range is not greater than a few hundred degrees.

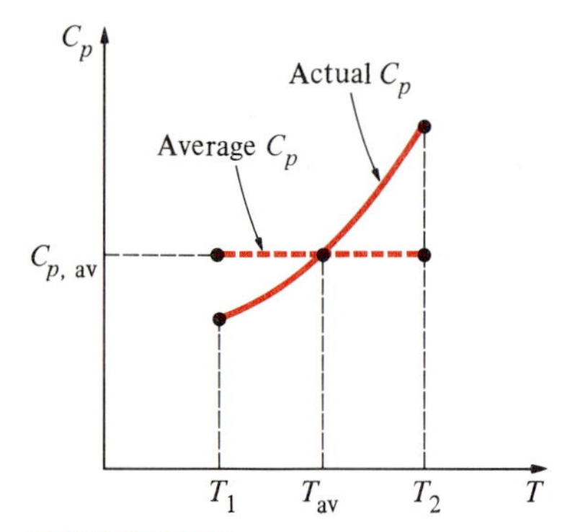

FIGURE 6-44

Under the constant-specific-heat assumption, the specific heat is assumed to be constant at some average value.

The entropy-change relations for ideal gases under the constant-specific-heat assumption are easily obtained by replacing $C_v(T)$ and $C_p(T)$ in Eqs. 6-35 and 6-36 by $C_{v,\text{av}}$ and $C_{p,\text{av}}$, respectively, and performing the integrations. We obtain

$$s_2 - s_1 = C_{v,\text{av}} \ln \frac{T_2}{T_1} + R \ln \frac{v_2}{v_1} \qquad [\text{kJ/(kg·K)}] \qquad (6\text{-}37)$$

and

$$s_2 - s_1 = C_{p,\text{av}} \ln \frac{T_2}{T_1} - R \ln \frac{P_2}{P_1} \qquad [\text{kJ/(kg·K)}] \qquad (6\text{-}38)$$

Entropy changes can also be expressed on a unit-mole basis by multiplying these relations by molar mass:

$$\bar{s}_2 - \bar{s}_1 = \overline{C}_{v,\text{av}} \ln \frac{T_2}{T_1} + R_u \ln \frac{v_2}{v_1} \qquad [\text{kJ/(kmol·K)}] \qquad (6\text{-}39)$$

and

$$\bar{s}_2 - \bar{s}_1 = \overline{C}_{p,\text{av}} \ln \frac{T_2}{T_1} - R_u \ln \frac{P_2}{P_1} \qquad [\text{kJ/(kmol·K)}] \qquad (6\text{-}40)$$

2 Variable Specific Heats: Exact Treatment

When the temperature change during a process is large and the specific heats of the ideal gas vary nonlinearly within the temperature range, the assumption of constant specific heats may lead to considerable errors in entropy-change calculations. For those cases, the variation of specific heats with temperature should be properly accounted for by utilizing accurate relations for the specific heats as a function of temperature. The entropy change during a process is then determined by substituting these $C_v(T)$ or $C_p(T)$ relations into Eq. 6-35 or 6-36 and performing the integrations.

Instead of performing these laborious integrals each time we have a new process, it would be very desirable and convenient to do these integrals once and tabulate the results. For this purpose, we choose a reference temperature of zero degrees absolute and define a function $s°$ as

$$s° = \int_0^T C_p(T) \frac{dT}{T} \qquad (6\text{-}41)$$

According to this definition, $s°$ is a function of temperature alone, and its value is zero at absolute zero temperature. The values of $s°$ are calculated at various temperatures from Eq. 6-41, and the results are tabulated in the Appendix as a function of temperature for several ideal gases. Given this definition, the integral Eq. 6-36 becomes

$$\int_1^2 C_p(T)\,\frac{dT}{T} = s_2^\circ - s_1^\circ$$

where s_2° is the value of $s°$ at T_2 and s_1° is the value at T_1. Thus,

$$s_2 - s_1 = s_2^\circ - s_1^\circ - R \ln \frac{P_2}{P_1} \qquad \text{[kJ/(kg·K)]} \qquad (6\text{-}42)$$

Note that unlike internal energy and enthalpy, the entropy of an ideal gas varies with specific volume or pressure as well as the temperature. Therefore, entropy cannot be tabulated as a function of temperature alone. The $s°$ values in the tables account for the temperature dependence of entropy (Fig. 6-45). The variation of entropy with pressure is accounted for by the last term in Eq. 6-42. Another relation for entropy change can be developed based on Eq. 6-35, but this would require the definition of another function and tabulation of its values, which is not practical.

Equation 6-42 can also be expressed on a unit-mole basis as

$$\bar{s}_2 - \bar{s}_1 = \bar{s}_2^\circ - \bar{s}_1^\circ - R_u \ln \frac{P_2}{P_1} \qquad \text{[kJ/(kmol·K)]} \qquad (6\text{-}43)$$

Equations 6-42 and 6-43 account for the variation of specific heat with temperature and therefore should be preferred in entropy-change calculations.

T, K	$s°(T)$, kJ/(kg · K)
⋮	⋮
300	1.70203
310	1.73498
320	1.76690
⋮	⋮

(Table A-17)

FIGURE 6-45
The entropy of an ideal gas depends on both T and P. The function $s°$ represents the temperature-dependent part of entropy only.

EXAMPLE 6-13

Nitrogen gas is compressed from an initial state of 100 kPa and 17°C to a final state of 600 kPa and 57°C. Determine the entropy change of the nitrogen during this compression process by using (*a*) exact property values from the nitrogen table and (*b*) average specific heats.

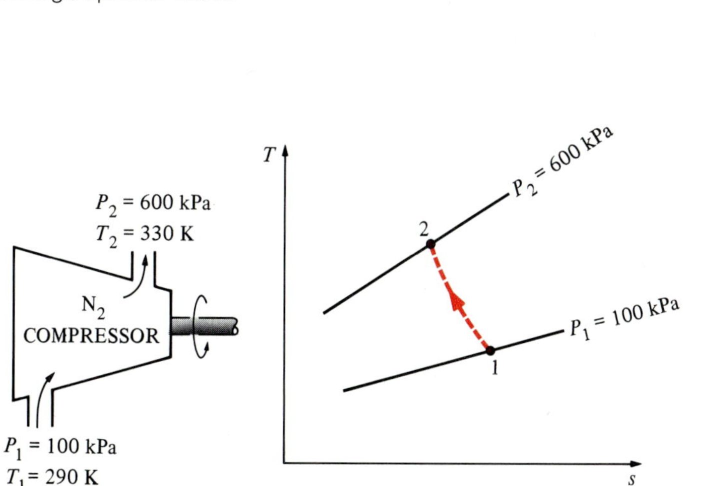

FIGURE 6-46
Schematic and T-s diagram for Example 6-13.

$\bar{s}^\circ = 194.459\,kJ/(kmol \cdot K)$
$M = 28.013\,kg/kmol$

⇓

$s^\circ = \frac{\bar{s}^\circ}{M} = 6.942\,kJ/(kg \cdot K)$

FIGURE 6-47
Properties per unit mole and properties per unit mass are related to each other through the molar mass of the substance.

Solution A sketch of the system and the *T-s* diagram for the process are given in Fig. 6-46. At specified conditions, nitrogen can be treated as an ideal gas since it is at a high temperature and low pressure relative to its critical values ($T_{cr} = -147°C$ and $P_{cr} = 3390$ kPa for nitrogen). Therefore, the entropy-change relations developed under the ideal-gas assumption are applicable.

(*a*) The properties of the nitrogen gas in the nitrogen table (Table A-18) are given on a unit-mole basis. Therefore, it is more convenient to determine the entropy change of the nitrogen by using Eq. 6-43 and then convert the results to the desired units (Fig. 6-47). Reading $\bar{s}^\circ$ values at given temperatures and substituting, we find

$$\bar{s}_2 - \bar{s}_1 = \bar{s}_2^\circ - \bar{s}_1^\circ - R_u \ln\frac{P_2}{P_1}$$

$$= [(194.459 - 190.695)\,kJ/(kmol{\cdot}K)] - [8.314\,kJ/(kmol{\cdot}K)]\ln\frac{600\,kPa}{100\,kPa}$$

$$= -11.133\,kJ/(kmol{\cdot}K)$$

Then $$s_2 - s_1 = \frac{\bar{s}_2 - \bar{s}_1}{M} = \frac{-11.133\,kJ/(kmol{\cdot}K)}{28.013\,kg/kmol} = -0.3974\,kJ/(kg{\cdot}K)$$

(*b*) The entropy change of the nitrogen during this process can also be determined approximately from Eq. 6-38 by using a C_p value at the average temperature of 37°C (Table A-2*b*) and treating it as a constant:

$$s_2 - s_1 = C_{p,av}\ln\frac{T_2}{T_1} - R\ln\frac{P_2}{P_1}$$

$$= [1.0394\,kJ/(kg{\cdot}K)]\ln\frac{330\,K}{290\,K} - [0.297\,kJ/(kg{\cdot}K)]\ln\frac{600\,kPa}{100\,kPa}$$

$$= -0.3978\,kJ/(kg{\cdot}K)$$

The two results above are almost identical since the change in temperature during this process is relatively small. When the temperature change is large, however, they may differ significantly. For those cases, Eq. 6-43 should be used instead of Eq. 6-38 since it accounts for the variation of specific heats with temperature.

Isentropic Processes of Ideal Gases

Several relations for the isentropic processes of ideal gases can be obtained by setting the entropy-change relations developed above equal to zero. Again, this is done first for the case of constant specific heats and then for the case of variable specific heats.

Constant Specific Heats: Approximate Treatment

When the constant-specific-heat assumption is valid, the isentropic relations for ideal gases are obtained by setting Eqs. 6-37 and 6-38 equal to zero. From Eq. 6-37,

$$\ln\frac{T_2}{T_1} = -\frac{R}{C_v}\ln\frac{v_2}{v_1}$$

which can be rearranged as

$$\ln \frac{T_2}{T_1} = \ln \left(\frac{v_1}{v_2}\right)^{R/C_v}$$

or

$$\left(\frac{T_2}{T_1}\right)_{s=\text{const.}} = \left(\frac{v_1}{v_2}\right)^{k-1} \tag{6-44}$$

since $R = C_p - C_v$, $k = C_p/C_v$, and thus $R/C_v = k - 1$.

Equation 6-44 is the *first isentropic relation* for ideal gases under the constant-specific-heat assumption. The *second isentropic relation* is obtained in a similar manner from Eq. 6-38 with the following result:

$$\left(\frac{T_2}{T_1}\right)_{s=\text{const.}} = \left(\frac{P_2}{P_1}\right)^{(k-1)/k} \tag{6-45}$$

The *third isentropic relation* is obtained by substituting Eq. 6-45 into Eq. 6-44 and simplifying:

$$\left(\frac{P_2}{P_1}\right)_{s=\text{const.}} = \left(\frac{v_1}{v_2}\right)^{k} \tag{6-46}$$

Equations 6-44 through 6-46 can also be expressed in a compact form as

$$Tv^{k-1} = \text{constant} \tag{6-47a}$$

$$TP^{(1-k)/k} = \text{constant} \tag{6-47b}$$

$$Pv^{k} = \text{constant} \tag{6-47c}$$

The specific heat ratio k, in general, varies with temperature, and so in the isentropic relations above an average k value for the given temperature range should be used.

Note that the isentropic relations above, as the name implies, are strictly valid for isentropic processes only when the constant-specific-heat assumption is appropriate (Fig. 6-48).

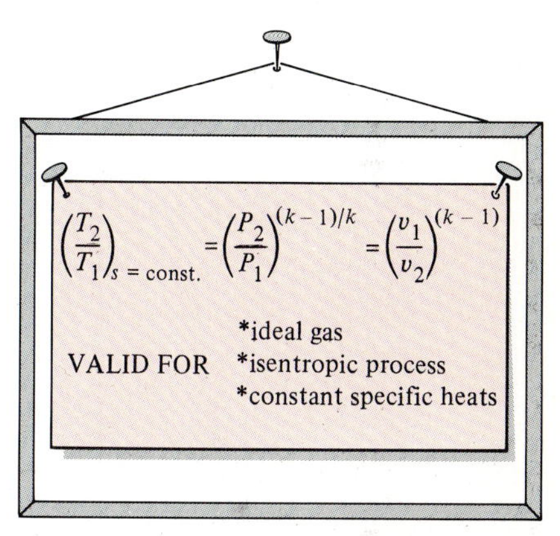

FIGURE 6-48
The isentropic relations of ideal gases are valid for the isentropic processes of ideal gases only.

Variable Specific Heats: Exact Treatment

When the constant-specific-heat assumption is not appropriate, the isentropic relations developed above will yield results that are not quite accurate. For such cases, we should use an isentropic relation obtained from an entropy-change equation that accounts for the variation of specific heats with temperature, namely, Eq. 6-42. Setting this equation equal to zero, we have

$$0 = s_2^\circ - s_1^\circ - R \ln \frac{P_2}{P_1}$$

or

$$s_2^\circ = s_1^\circ + R \ln \frac{P_2}{P_1} \tag{6-48}$$

where s_2° is the s° value at the end of the isentropic process. This is the relation for the isentropic processes of ideal gases under variable-specific-heat conditions.

Relative Pressure and Relative Specific Volume

Equation 6-48 provides an accurate way of evaluating property changes of ideal gases during isentropic processes as it accounts for the variation of specific heats with temperature. However, it involves tedious iterations when the volume ratio is given instead of the pressure ratio. This is quite an inconvenience in optimization studies, which usually require numerous repetitive calculations. To remedy this deficiency, we define two new dimensionless quantities associated with isentropic processes.

The definition of the first is based on Eq. 6-48, which can be rearranged as

$$\frac{P_2}{P_1} = \exp \frac{s_2^\circ - s_1^\circ}{R}$$

or

$$\frac{P_2}{P_1} = \frac{\exp(s_2^\circ/R)}{\exp(s_1^\circ/R)}$$

The quantity $\exp(s^\circ/R)$ is defined as the **relative pressure** P_r. With this definition the above relation becomes

$$\left(\frac{P_2}{P_1}\right)_{s=\text{const.}} = \frac{P_{r2}}{P_{r1}} \tag{6-49}$$

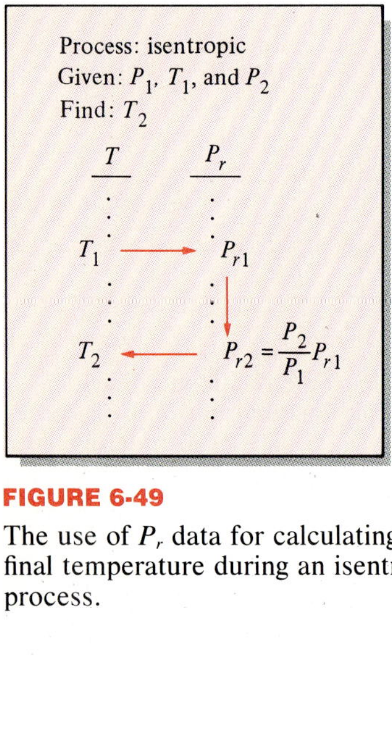

Note that the relative pressure P_r is a *dimensionless* quantity which is a function of temperature only since s° depends on temperature alone. Therefore, values of P_r can be tabulated against temperature. This is done for air in Table A-17. The use of P_r data is illustrated in Fig. 6-49.

Process: isentropic
Given: P_1, T_1, and P_2
Find: T_2

T | P_r
$T_1 \rightarrow P_{r1}$
$T_2 \leftarrow P_{r2} = \frac{P_2}{P_1} P_{r1}$

FIGURE 6-49
The use of P_r data for calculating the final temperature during an isentropic process.

Sometimes specific volume ratios are given instead of pressure ratios. This is particularly the case when automotive engines are analyzed. In such cases, one needs to work with volume ratios. Therefore, we define another quantity related to specific volume ratios for isentropic processes. This is done by utilizing the ideal-gas relation and Eq. 6-49:

$$\frac{v_2}{v_1} = \frac{T_2}{T_1}\frac{P_1}{P_2} = \frac{T_2}{T_1}\frac{P_{r1}}{P_{r2}} = \frac{T_2/P_{r2}}{T_1/P_{r1}}$$

The quantity T/P_r is a function of temperature only and is defined as **relative specific volume** v_r. Thus

$$\left(\frac{v_2}{v_1}\right)_{s=\text{const.}} = \frac{v_{r2}}{v_{r1}} \tag{6-50}$$

Equations 6-49 and 6-50 are strictly valid for isentropic processes of ideal gases only. They account for the variation of specific heats with temperature and therefore give more accurate results than Eqs. 6-44 through 6-47, developed under the constant-specific-heat assumption. The values of P_r and v_r are listed for air in Table A-17.

EXAMPLE 6-14

Air is compressed in an adiabatic piston-cylinder device from 22°C and 95 kPa in a reversible manner. If the compression ratio V_1/V_2 of this piston-cylinder device is 8, determine the final temperature of the air.

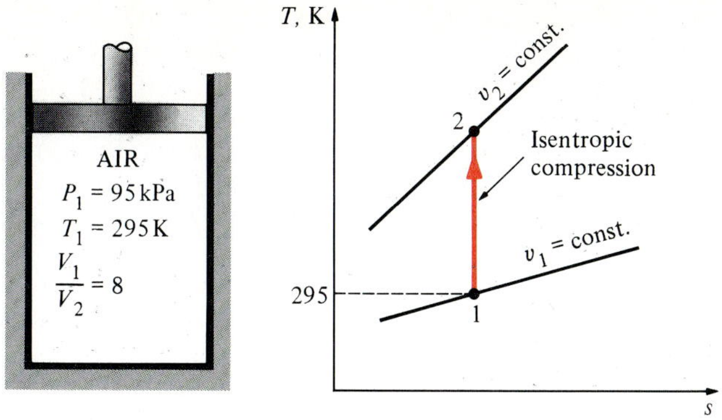

AIR
$P_1 = 95$ kPa
$T_1 = 295$ K
$\frac{V_1}{V_2} = 8$

T, K
$v_2 =$ const.
2
Isentropic compression
$v_1 =$ const.
295
1
s

FIGURE 6-50

Schematic and T-s diagram for Example 6-14.

Solution A sketch of the system and the T-s diagram for the process are given in Fig. 6-50. At specified conditions, air can be treated as an *ideal gas* since it is at a high temperature and low pressure relative to its critical values ($T_{cr} = -147°C$ and $P_{cr} = 3390$ kPa for nitrogen, the main constituent of air). Therefore the isentropic relations developed above for ideal gases are applicable.

This process is easily recognized as isentropic since it is both reversible and adiabatic. The final temperature for this isentropic process can be determined from Eq. 6-50 with the help of relative specific volume data (Table A-17), as illustrated in Fig. 6-51.

For closed systems $\qquad \frac{V_2}{V_1} = \frac{v_2}{v_1}$

At $T_1 = 295$ K, $\qquad v_{r1} = 647.9$

From Eq. 6-50,

$$v_{r2} = v_{r1}\left(\frac{v_2}{v_1}\right) = (647.9)\left(\frac{1}{8}\right) = 80.99 \longrightarrow T_2 = 662.7 \text{ K}$$

Therefore, the temperature of air will increase by 367.7°C during this process.

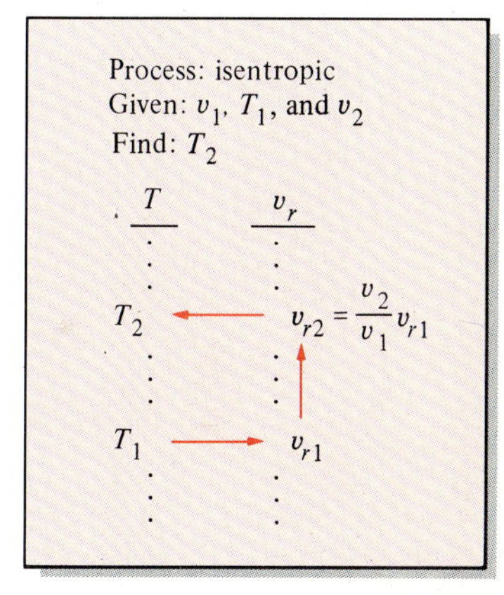

FIGURE 6-51

The use of v_r data for calculating the final temperature during an isentropic process (Example 6-14).

Alternative solution The final temperature could also be determined from Eq. 6-44 by assuming constant specific heats for the air:

$$\left(\frac{T_2}{T_1}\right)_{s=\text{const.}} = \left(\frac{v_1}{v_2}\right)^{k-1}$$

The specific heat ratio k also varies with temperature, and we need to use the value of k corresponding to the average temperature. However, the final temperature is not given, and so we cannot determine the average temperature in advance. For such cases, calculations can be started with a k value at the initial or the anticipated average temperature. This value could be refined later, if necessary, and the calculations can be repeated. We know that the temperature of the air will rise considerably during this adiabatic compression process, so we *guess* that the average temperature will be about 450 K. The k value at this anticipated average temperature is determined from Table A-2*b* to be 1.391. Then the final temperature of air becomes

$$T_2 = (295 \text{ K})(8)^{1.391-1} = 665.2 \text{ K}$$

This will give an average temperature value of 480.1 K, which is sufficiently close to the assumed value of 450 K. Therefore, it is not necessary to repeat the calculations by using the k value at this average temperature.

The result obtained by assuming constant specific heats for this case is in error by about 0.4 percent, which is rather small. This is not surprising since the temperature change of air is relatively small (only a few hundred degrees) and the specific heats of air vary almost linearly with temperature in this temperature range.

EXAMPLE 6-15

Helium gas is compressed in an adiabatic compressor from an initial state of 14 psia and 50°F to a final temperature of 320°F in a reversible manner. Determine the exit pressure of helium.

Solution A sketch of the system and the T-s diagram for the process are given in Fig. 6-52. At specified conditions, helium can be treated as an *ideal gas* since it is at a very high temperature relative to its critical temperature ($T_{cr} = -450°F$ for helium). Therefore the isentropic relations developed above for ideal gases are applicable. The specific heat ratio k of helium is 1.667 and is independent of temperature in the region where it behaves as an ideal gas. Thus the final pressure of helium can be determined from Eq. 6-45:

$$P_2 = P_1\left(\frac{T_2}{T_1}\right)^{k/(k-1)} = (14 \text{ psia})\left(\frac{780 \text{ R}}{510 \text{ R}}\right)^{1.667/0.667} = 40.5 \text{ psia}$$

6-11 ■ REVERSIBLE STEADY-FLOW WORK

Work and heat, in general, are path functions, and the heat transfer or the work done during a process depends on the path followed as well as on the properties at the end states. In Chap. 3, we discussed reversible (quasi-equilibrium) moving boundary work associated with closed systems and expressed it in terms of the fluid properties as (Eq. 3-9):

$$W_b = \int_1^2 P\, dV$$

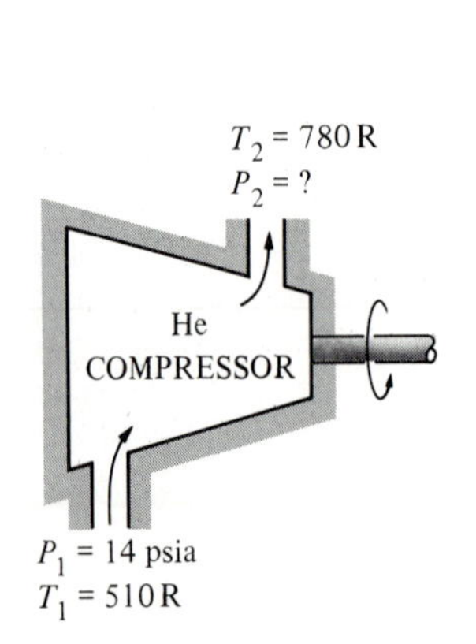

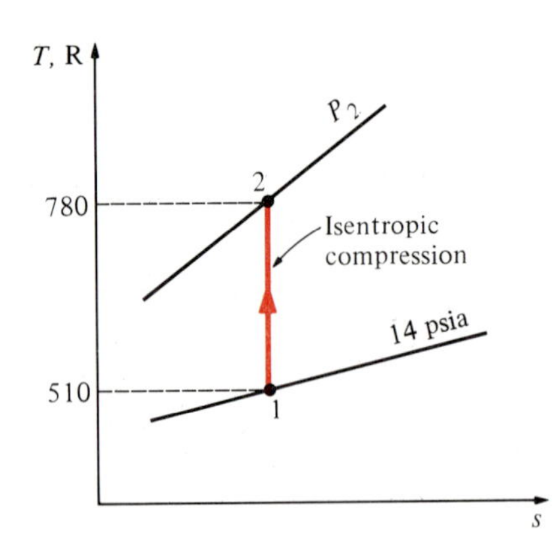

FIGURE 6-52
Schematic and T-s diagram for Example 6-15.

We mentioned that the quasi-equilibrium work interactions lead to the maximum work output for work-producing devices and the minimum work input for work-consuming devices.

It would also be very desirable and insightful to express the work associated with steady-flow devices in terms of fluid properties.

The conservation of energy equation for a steady-flow device undergoing an internally reversible process can be expressed in differential form as

$$\delta q_{rev} - \delta w_{rev} = dh + d\text{ke} + d\text{pe}$$

But $\left.\begin{aligned} \delta q_{rev} &= T\,ds && \text{(Eq. 6-15)} \\ T\,ds &= dh - v\,dP && \text{(Eq. 6-24)} \end{aligned}\right\} \quad \delta q_{rev} = dh - v\,dP$

Substituting this into the relation above and canceling dh yield

$$-\delta w_{rev} = v\,dP + d\text{ke} + d\text{pe}$$

Integrating, we find

$$w_{rev} = -\int_1^2 v\,dP - \Delta\text{ke} - \Delta\text{pe} \qquad \text{(kJ/kg)} \qquad (6\text{-}51)$$

When the changes in kinetic and potential energies are negligible, this equation reduces to

$$w_{rev} = -\int_1^2 v\,dP \qquad \text{(kJ/kg)} \qquad (6\text{-}52)$$

Equations 6-51 and 6-52 are relations for the reversible work associated with an internally reversible process in a steady-flow device. The resemblance between the $v\,dP$ in these relations and $P\,dv$ is striking. They should not be confused with each other, however, since $P\,dv$ is associated with reversible boundary work in closed systems (Fig. 6-53).

Obviously one needs to know v as a function of P for the given process to perform the integration in Eq. 6-51. When the working fluid is an *incompressible fluid,* the specific volume v remains constant during the process and can be taken out of the integration. Then Eq. 6-51 simplifies to

$$w_{rev} = v(P_1 - P_2) - \Delta\text{ke} - \Delta\text{pe} \qquad \text{(kJ/kg)} \qquad (6\text{-}53)$$

For the steady flow of a liquid through a device that involves no work interactions (such as a nozzle or a pipe section), the work term is zero, and the equation above can be expressed as

$$v(P_2 - P_1) + \frac{\mathbf{V}_2^2 - \mathbf{V}_1^2}{2} + g(z_2 - z_1) = 0 \qquad (6\text{-}54)$$

which is known as the **Bernoulli equation** in fluid mechanics. Equation 6-54 is developed for an internally reversible process and thus is applicable to incompressible fluids that involve no irreversibilities such as friction or shock waves. This equation can be modified, however, to incorporate these effects.

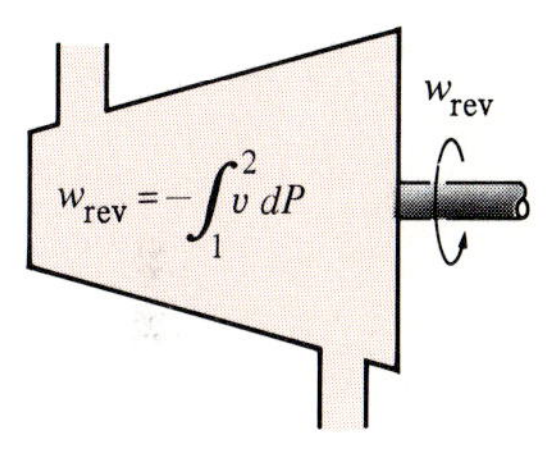

(*a*) Steady-flow system

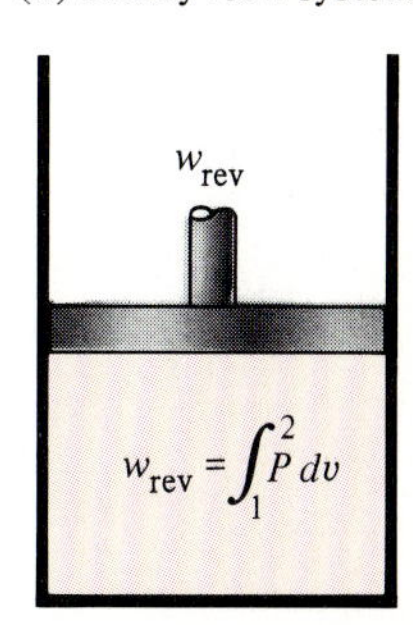

(*b*) Closed system

FIGURE 6-53
Reversible work relations for closed and steady-flow systems.

$$w = -\int_1^2 v\,dP$$
$$W = -\int_1^2 v\,dP$$
$$W = -\int_1^2 v\,dP$$

FIGURE 6-54

The larger the specific volume, the greater the work produced (or consumed) by a steady-flow device.

Equation 6-52 has far-reaching implications in engineering regarding devices that produce or consume work steadily such as turbines, compressors, and pumps. It is obvious from this equation that the reversible steady-flow work is closely associated with the specific volume of the fluid flowing through the device. *The larger the specific volume, the larger the reversible work produced or consumed by the steady-flow device* (Fig. 6-54). This conclusion is equally valid for actual steady-flow devices. Therefore, every effort should be made to keep the specific volume of a fluid as small as possible during a compression process to minimize the work input and as large as possible during an expansion process to maximize the work output.

In steam or gas power plants, the pressure rise in the pump or compressor is equal to the pressure drop in the turbine if we disregard the pressure losses in various other components. In steam power plants, the pump handles liquid, which has a very small specific volume, and the turbine handles vapor, whose specific volume is many times larger. Therefore, the work output of the turbine is much larger than the work input to the pump. This is one of the reasons for the overwhelming popularity of steam power plants in electric power generation.

If we were to compress the steam exiting the turbine back to the turbine inlet pressure before cooling it first in the condenser in order to "save" the heat rejected, we would have to supply all the work produced by the turbine back to the compressor. In reality, the required work input would be even greater than the work output of the turbine because of the irreversibilities present in both processes.

In gas power plants, the working fluid (typically air) is compressed in the gas phase, and a considerable portion of the work output of the turbine is consumed by the compressor. As a result, a gas power plant delivers less net work per unit mass of the working fluid.

EXAMPLE 6-16

Determine the work required to compress steam isentropically from 100 kPa to 1 MPa, assuming that the steam exists as (*a*) saturated liquid and (*b*) saturated vapor at the initial state. Neglect the changes in kinetic and potential energies.

Solution Sketches of the pump and the compressor are given in Fig. 6-55.

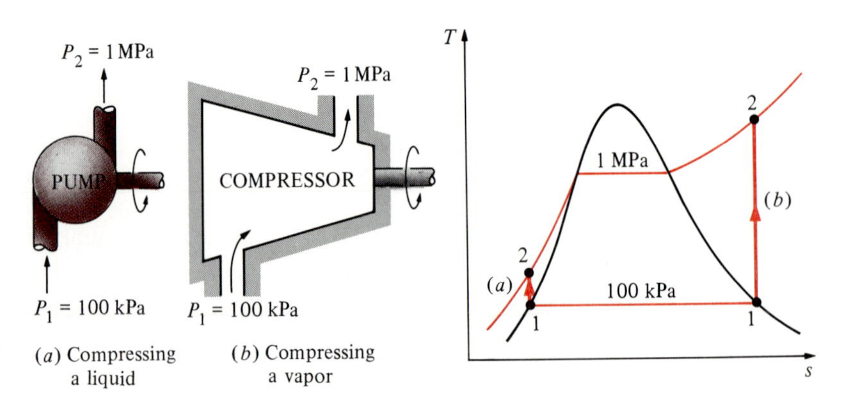

FIGURE 6-55

Schematic and *T-s* diagram for Example 6-16.

We expect the work input to the pump to be considerably smaller since it handles a liquid. The compression process is stated to be reversible, and the kinetic and potential energy changes are negligible. Thus Eq. 6-52 is applicable.

(*a*) In this case, the steam is a saturated liquid initially, and its specific volume is

$$v_1 = v_{f\,@\,100\text{ kPa}} = 0.001043 \text{ m}^3/\text{kg} \qquad \text{(Table A-5)}$$

which remains essentially constant during the process. Thus,

$$\begin{aligned} w_{rev} &= -\int_1^2 v\,dP \cong v_1(P_1 - P_2) \\ &= (0.001043 \text{ m}^3/\text{kg})[(100 - 1000)\text{ kPa}]\left(\frac{1\text{ kJ}}{1\text{ kPa·m}^3}\right) \\ &= -0.94 \text{ kJ/kg} \end{aligned}$$

(*b*) This time, the steam is a saturated vapor initially and remains a vapor during the entire compression process. Since the specific volume of a gas changes considerably during a compression process, we need to know how v varies with P to perform the integration in Eq. 6-52. This relation, in general, is not readily available. But for an isentropic process, it is easily obtained from the second $T\,ds$ relation by setting $ds = 0$:

$$\left.\begin{aligned} T\,ds &= dh - v\,dP \quad \text{(Eq. 6-23)} \\ ds &= 0 \quad \text{(isentropic process)} \end{aligned}\right\} \qquad v\,dP = dh$$

Thus, $$w_{rev} = -\int_1^2 v\,dP = -\int_1^2 dh = h_1 - h_2$$

This result could also be obtained from the first-law analysis of an isentropic steady-flow process. Next we determine the enthalpies:

State 1: $$\left.\begin{aligned} P_1 &= 100 \text{ kPa} \\ &\text{(sat. vapor)} \end{aligned}\right\} \quad \begin{aligned} h_1 &= 2675.5 \text{ kJ/kg} \\ s_1 &= 7.3594 \text{ kJ/(kg·K)} \end{aligned} \qquad \text{(Table A-5)}$$

State 2: $$\left.\begin{aligned} P_2 &= 1 \text{ MPa} \\ s_2 &= s_1 \end{aligned}\right\} \quad h_2 = 3195.45 \text{ kJ/kg} \qquad \text{(Table A-6)}$$

Thus, $$w_{rev} = (2675.5 - 3195.45) \text{ kJ/kg} = -519.95 \text{ kJ/kg}$$

That is, compressing steam in the vapor form would require over 500 times more work than compressing it in the liquid form between the same pressure limits.

Proof that Steady-Flow Devices Deliver the Most and Consume the Least Work when the Process Is Reversible

We have shown in Chap. 5 that cyclic devices (heat engines, refrigerators, and heat pumps) deliver the most work and consume the least when reversible processes are used. Now we will demonstrate that this is also the case for discrete devices such as turbines and compressors in steady operation.

Consider two steady-flow devices, one reversible and the other irreversible, operating between the same inlet and exit states. The conservation of energy equation for each of these devices can be expressed in the differential form as

Actual: $\delta q_{act} - \delta w_{act} = dh + d\text{ke} + d\text{pe}$
Reversible: $\delta q_{rev} - \delta w_{rev} = dh + d\text{ke} + d\text{pe}$

The right-hand sides of these two equations are identical since both devices are operating between the same end states. Thus

$$\delta q_{act} - \delta w_{act} = \delta q_{rev} - \delta w_{rev}$$

or

$$\delta w_{rev} - \delta w_{act} = \delta q_{rev} - \delta q_{act}$$

But

$$\delta q_{rev} = T\,ds$$

Substituting this relation into the equation above and dividing each term by T, we obtain

$$\frac{\delta w_{rev} - \delta w_{act}}{T} = ds - \frac{\delta q_{act}}{T} \geqslant 0$$

since

$$ds \geqslant \frac{\delta q_{act}}{T} \qquad \text{(Eq. 6-8}b\text{)}$$

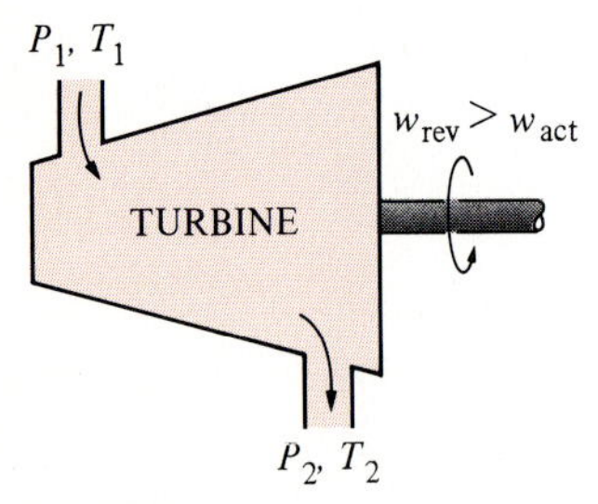

FIGURE 6-56

A reversible turbine delivers more work than an irreversible one if both operate between the same end states.

Also, T is the absolute temperature which is always positive. Thus,

$$\delta w_{rev} \geqslant \delta w_{act} \qquad (6\text{-}55)$$

or

$$w_{rev} \geqslant w_{act} \qquad (6\text{-}56)$$

Thus work-producing devices such as turbines (w is positive) deliver more work, and work-consuming devices such as pumps and compressors (w is negative) require less work when they operate reversibly (Fig. 6-56).

6-12 ■ MINIMIZING THE COMPRESSOR WORK

We have shown in Sec. 6-11 that the work input to a compressor is minimized when the compression process is executed in an internally reversible manner. When the changes in kinetic and potential energies are negligible, the compressor work is given by (Eq. 6-52)

$$w_{rev} = -\int_1^2 v\,dP$$

Obviously one way of minimizing the compressor work is to approach an internally reversible process as much as possible by minimizing the irreversibilities such as friction, turbulence, and non-quasi-equilibrium compression. The extent to which this can be accomplished is limited by economic considerations. A second (and more practical) way of reducing the compressor work is to keep the specific volume of the gas as small as possible during the compression process. This is done by maintaining the temperature of the gas as low as possible during compression since the specific volume of a gas is proportional to temperature. Therefore, reducing the work input to a compressor requires that the gas be cooled as it is compressed.

To have a better understanding of the effect of cooling during the compression process, we compare the work input requirements for three kinds of processes: *an isentropic process* (involves no cooling), *a polytropic process* (involves some cooling), and *an isothermal process* (involves plenty of cooling). Assuming all three processes are executed between the same pressure levels (P_1 and P_2) in an internally reversible manner and the gas behaves as an ideal gas ($Pv = RT$), we see that the compression work is determined by performing the integration in Eq. 6-52 for each case, with the following results:

Isentropic (Pv^k = constant):

$$w_{\text{comp}} = \frac{kR(T_1 - T_2)}{k-1} = \frac{kRT_1}{k-1}\left[1 - \left(\frac{P_2}{P_1}\right)^{(k-1)/k}\right] \qquad (6\text{-}57a)$$

Polytropic (Pv^n = constant):

$$w_{\text{comp}} = \frac{nR(T_1 - T_2)}{n-1} = \frac{nRT_1}{n-1}\left[1 - \left(\frac{P_2}{P_1}\right)^{(n-1)/n}\right] \qquad (6\text{-}57b)$$

Isothermal (Pv = constant):

$$w_{\text{comp}} = RT \ln \frac{P_1}{P_2} \qquad (6\text{-}57c)$$

The three processes are plotted on a P-v diagram in Fig. 6-57 for the same inlet state and exit pressure. On a P-v diagram, the area to the left of the process curve is the integral of $v\,dP$. Thus it is a measure of the steady-flow compression work. It is interesting to observe from this diagram that of the three internally reversible cases considered, the adiabatic compression (Pv^k = constant) requires the maximum work and the isothermal compression (T = constant or Pv = constant) requires the minimum. The work input requirement for the polytropic case (Pv^n = constant) is between these two and decreases as the polytropic exponent n is decreased, by increasing the heat rejection during the compression process. If sufficient heat is removed, the value of n approaches unity and the process becomes isothermal. One common way of cooling the gas during compression is to use cooling jackets around the casing of the compressors.

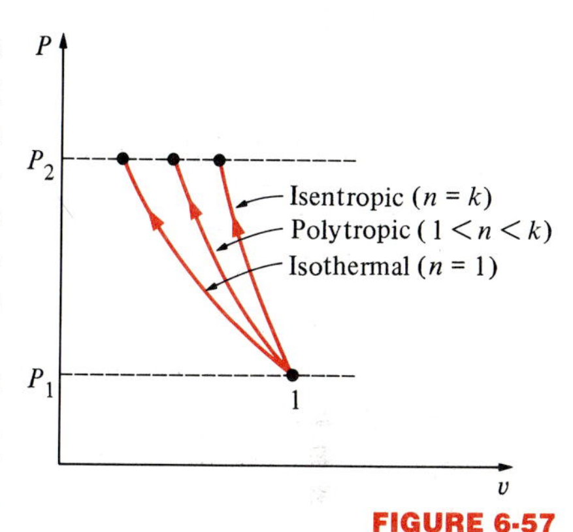

FIGURE 6-57
P-v diagrams of isentropic, polytropic, and isothermal compression processes between the same pressure limits.

Multistage Compression with Intercooling

It is clear from the above arguments that cooling a gas as it is compressed is desirable since this reduces the required work input to the compressor. However, often it is not possible to have effective cooling through the casing of the compressor, and it becomes necessary to use other techniques to achieve effective cooling. One such technique is **multistage compression with intercooling**, where the gas is compressed in stages and cooled between each stage by passing it through a heat exchanger called an *intercooler*. Ideally, the cooling process takes place at constant pressure, and the gas is cooled to the initial temperature T_1 at each intercooler. Multistage compression with intercooling is especially attractive when a gas is to be compressed to very high pressures.

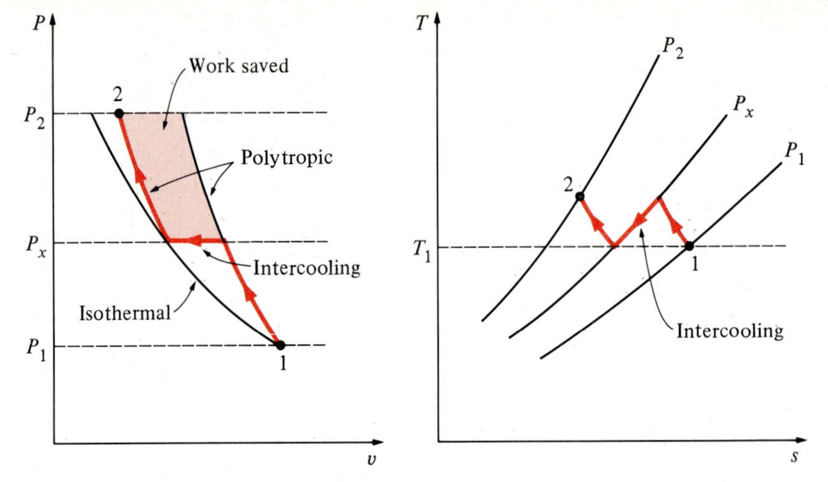

FIGURE 6-58

P-v and *T-s* diagrams for a two-stage steady-flow compression process.

The effect of intercooling on compressor work is graphically illustrated on *P-v* and *T-s* diagrams in Fig. 6-58 for a two-stage compressor. The gas is compressed in the first stage from P_1 to an intermediate pressure P_x, cooled at constant pressure to the initial temperature T_1, and compressed in the second stage to the final pressure P_2. The compression processes, in general, can be modeled as polytropic (Pv^n = constant) where the value of n varies between k and 1. The colored area on the *P-v* diagram represents the work saved as a result of two-stage compression with intercooling. The process paths for single-stage isothermal and polytropic processes are also shown for comparison.

The size of the colored area (the saved work input) varies with the value of the intermediate pressure P_x, and it is of practical interest to determine the conditions under which this area is maximized. The total work input for a two-stage compressor is the sum of the work inputs for each stage of compression as determined from Eq. 6-57*b*:

$$w_{\text{comp}} = w_{\text{comp,I}} + w_{\text{comp,II}}$$
$$= \frac{nRT_1}{n-1}\left[1 - \left(\frac{P_x}{P_1}\right)^{(n-1)/n}\right] + \frac{nRT_1}{n-1}\left[1 - \left(\frac{P_2}{P_x}\right)^{(n-1)/n}\right]$$

The only variable in this equation is P_x. The P_x value which will minimize the total work is determined by differentiating this expression with respect to P_x and setting the resulting expression equal to zero. It yields

$$P_x = (P_1P_2)^{1/2} \qquad \text{or} \qquad \frac{P_x}{P_1} = \frac{P_2}{P_x} \tag{6-58}$$

That is, *for maximum savings from the work input, the pressure ratio across each stage of the compressor must be the same*. When this condition is satisfied, the compression work at each stage becomes identical, that is, $w_{\text{comp,I}} = w_{\text{comp,II}}$.

EXAMPLE 6-17

Air is compressed steadily by a reversible compressor from an inlet state of 100 kPa and 300 K to an exit pressure of 900 kPa. Determine the compressor work per unit mass for (*a*) isentropic compression with $k = 1.4$, (*b*) polytropic compression with $n = 1.3$, (*c*) isothermal compression, and (*d*) ideal two-stage compression with intercooling with a polytropic exponent of 1.3.

Solution A sketch of the system and the *P-v* diagram of the compression processes are given in Fig. 6-59. At the specified conditions, air can be treated as an ideal gas since it is at a high temperature and low pressure relative to its critical-point values ($T_{cr} = -147°C$ and $P_{cr} = 3390$ kPa for nitrogen, the main constituent of air).

The steady-flow compression work for all these four cases is determined by using the relations developed in this section:

(*a*) Isentropic compression with $k = 1.4$ (Eq. 6-57*a*):

$$w_{comp} = \frac{kRT_1}{k-1}\left[1 - \left(\frac{P_2}{P_1}\right)^{(k-1)/k}\right]$$

$$= \frac{(1.4)[0.287\ \text{kJ/(kg·K)}](300\ \text{K})}{1.4-1}\left[1 - \left(\frac{900\ \text{kPa}}{100\ \text{kPa}}\right)^{(1.4-1)/1.4}\right]$$

$$= -263.2\ \text{kJ/kg}$$

(*b*) Polytropic compression with $n = 1.3$ (Eq. 6-57*b*):

$$w_{comp} = \frac{nRT_1}{n-1}\left[1 - \left(\frac{P_2}{P_1}\right)^{(n-1)/n}\right]$$

$$= \frac{(1.3)[(0.287\ \text{kJ/(kg·K)}](300\ \text{K})}{1.3-1}\left[1 - \left(\frac{900\ \text{kPa}}{100\ \text{kPa}}\right)^{(1.3-1)/1.3}\right]$$

$$= -246.4\ \text{kJ/kg}$$

(*c*) Isothermal compression (Eq. 6-57*c*):

$$w_{comp} = RT\ln\frac{P_1}{P_2} = [0.287\ \text{kJ/(kg·K)}](300\ \text{K})\ln\frac{100\ \text{kPa}}{900\ \text{kPa}}$$

$$= -189.2\ \text{kJ/kg}$$

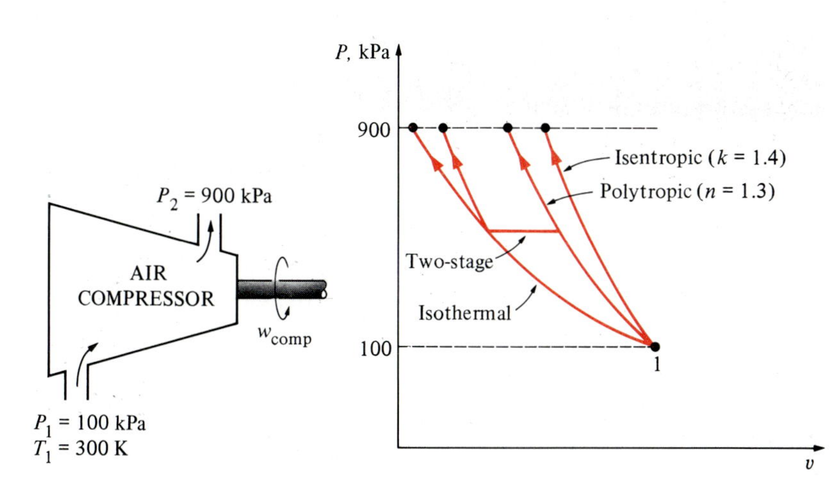

FIGURE 6-59

Schematic and *P-v* diagram for Example 6-17.

(*d*) Ideal two-stage compression with intercooling ($n = 1.3$): In this case, the pressure ratio across each stage is the same, and its value is determined from Eq. 6-58:

$$P_x = (P_1P_2)^{1/2} = [(100\text{ kPa})(900\text{ kPa})]^{1/2} = 300\text{ kPa}$$

The compressor work across each stage is also the same. Thus the total compressor work is twice the compression work for a single stage:

$$\begin{aligned} w_{\text{comp}} &= 2w_{\text{comp,I}} = 2\frac{nRT_1}{n-1}\left[1-\left(\frac{P_x}{P_1}\right)^{(n-1)/n}\right] \\ &= \frac{2(1.3)[0.287\text{ kJ/(kg·K)}](300\text{ K})}{1.3-1}\left[1-\left(\frac{300\text{ kPa}}{100\text{ kPa}}\right)^{(1.3-1)/1.3}\right] \\ &= -215.3\text{ kJ/kg} \end{aligned}$$

Thus, of all four cases considered, the isothermal compression requires the minimum work and the isentropic compression the maximum. The compressor work is decreased when two stages of polytropic compression are utilized instead of just one. As the number of compressor stages is increased, the compressor work approaches the value obtained for the isothermal case.

6-13 ■ ADIABATIC EFFICIENCIES OF SOME STEADY-FLOW DEVICES

We mentioned on several occasions that irreversibilities inherently accompany all actual processes and that their effect is always to downgrade the performance of devices. In engineering analysis, it would be very desirable to have some parameters that would enable us to express the degree of degradation of energy in these devices quantitatively. In Chap. 5 we did this for cyclic devices, such as heat engines and refrigerators, by comparing the actual cycles to the idealized ones, such as the Carnot cycle. A cycle which was composed entirely of reversible processes served as the *model cycle* to which the actual cycles could be compared. This idealized model cycle enabled us to determine the theoretical limits of performance for cyclic devices under specified conditions and to examine how the performance of actual devices suffered as a result of irreversibilities.

Now we extend the analysis to discrete engineering devices working under steady-flow conditions, such as turbines, compressors, and nozzles, and we examine the degree of degradation of energy in these devices as a result of irreversibilities. But first we need to define an ideal process which will serve as a model for the actual processes.

Although some heat transfer between these devices and the surrounding medium is unavoidable, most steady-flow devices are intended to operate under adiabatic conditions. Therefore, the model process for these devices should be an adiabatic one. Furthermore, an ideal process should involve no irreversibilities since the effect of irreversibilities is always to downgrade the performance of engineering devices. Thus the ideal process that can serve as a suitable model for most steady-flow devices is the *isentropic* process (Fig. 6-60).

The more closely the actual process approximates the idealized isentropic process, the better the device will perform. Thus it would be desirable to have a parameter that expresses quantitatively how efficiently an actual device approximates an idealized one. This parameter is called the **isentropic** or **adiabatic efficiency**. It is a measure of the deviation of actual processes from the corresponding idealized ones.

Adiabatic efficiencies are defined differently for different devices since each device is set up to perform different duties. Below we define the adiabatic efficiencies of turbines, compressors, and nozzles by comparing the actual performance of these devices to their performance under isentropic conditions for the same inlet state and exit pressure.

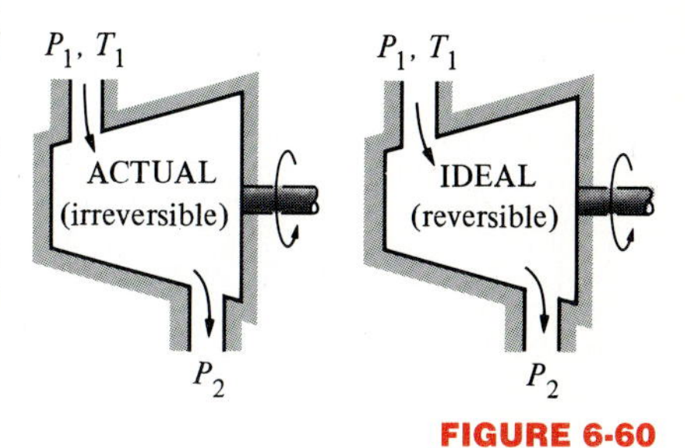

FIGURE 6-60
The isentropic process involves no irreversibilities and serves as the ideal process for adiabatic devices.

Adiabatic Efficiency of Turbines

For a turbine under steady operation, the inlet state of the working fluid and the exhaust pressure are fixed. Therefore, the ideal process for an adiabatic turbine is an isentropic process between the inlet state and the exhaust pressure. The desired output of a turbine is the work produced, and the **adiabatic efficiency of a turbine** is defined as *the ratio of the actual work output of the turbine to the work output that would be achieved if the process between the inlet state and the exit pressure were isentropic:*

$$\eta_T = \frac{\text{actual turbine work}}{\text{isentropic turbine work}} = \frac{w_a}{w_s} \tag{6-59}$$

Usually the changes in kinetic and potential energies associated with a fluid stream flowing through a turbine are small relative to the change in enthalpy, and can be neglected. Then the work output of an adiabatic turbine simply becomes the change in enthalpy, and the above relation for this case can be expressed as

$$\eta_T \cong \frac{h_1 - h_{2a}}{h_1 - h_{2s}} \tag{6-60}$$

where h_{2a} and h_{2s} are the enthalpy values at the exit state for actual and isentropic processes, respectively. The actual and isentropic processes in a turbine are illustrated in Fig. 6-61.

The value of η_T greatly depends on the design of the individual components that make up the turbine. Well-designed, large turbines have adiabatic efficiencies above 90 percent. For small turbines, however, it may drop even below 70 percent. The value of the adiabatic efficiency of a turbine is determined by measuring the actual work output of the turbine and by calculating the isentropic work output for the measured inlet conditions and the exit pressure. This value may then be used conveniently in the design of power plants.

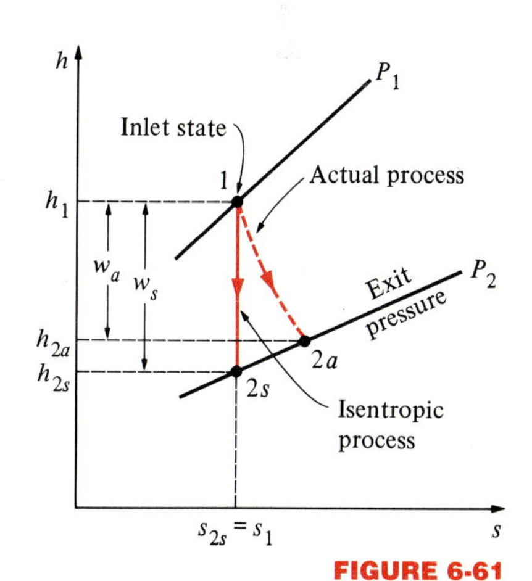

FIGURE 6-61
The *h-s* diagram of the actual and isentropic processes of an adiabatic turbine.

EXAMPLE 6-18

Steam enters an adiabatic turbine steadily at 3 MPa and 400°C and leaves at 50 kPa and 100°C. If the power output of the turbine is 2 MW and the kinetic energy change of the steam is negligible, determine (*a*) the adiabatic efficiency of the turbine and (*b*) the mass flow rate of the steam flowing through the turbine.

Solution (*a*) A sketch of the system and the *T-s* diagram of the process are given in Fig. 6-62. The turbine is adiabatic, and the changes in kinetic and potential energies are negligible; thus the adiabatic efficiency can be determined from Eq. 6-60. But first we need to determine the enthalpies at various states:

State 1: $\left.\begin{array}{l} P_1 = 3 \text{ MPa} \\ T_1 = 400°\text{C} \end{array}\right\}$ $\begin{array}{l} h_1 = 3230.9 \text{ kJ/kg} \\ s_1 = 6.9212 \text{ kJ/(kg·K)} \end{array}$ (Table A-6)

State 2a: $\left.\begin{array}{l} P_{2a} = 50 \text{ kPa} \\ T_{2a} = 100°\text{C} \end{array}\right\}$ $h_{2a} = 2682.5 \text{ kJ/kg}$ (Table A-6)

The exit enthalpy of the steam for the isentropic process h_{2s} is determined from the requirement that the entropy of the steam remain constant ($s_{2s} = s_1$):

State 2s: $\begin{array}{l} P_{2s} = 50 \text{ kPa} \longrightarrow \\ (s_{2s} = s_1) \end{array}$ $\begin{array}{l} s_f = 1.0910 \text{ kJ/(kg·K)} \\ s_g = 7.5939 \text{ kJ/(kg·K)} \end{array}$ (Table A-5)

Obviously, at the end of the isentropic process steam will exist as a saturated liquid–vapor mixture since $s_f < s_{2s} < s_g$. Thus we need to find the quality at state 2s first:

$$x_{2s} = \frac{s_{2s} - s_f}{s_{fg}} = \frac{6.9212 - 1.0910}{6.5029} = 0.897$$

and $h_{2s} = h_f + x_{2s}h_{fg} = 340.49 + 0.897(2305.4) \text{ kJ/kg} = 2407.4 \text{ kJ/kg}$

By substituting these enthalpy values into Eq. 6-60, the adiabatic efficiency of this turbine is determined to be

$$\eta_T \cong \frac{h_1 - h_{2a}}{h_1 - h_{2s}} = \frac{3230.9 - 2682.5}{3230.9 - 2407.4} = 0.667 \quad \text{or } 66.7\%$$

That is, this turbine is delivering only 66.7 percent of the work an isentropic turbine would deliver when operating between the same inlet state and exit pressure.

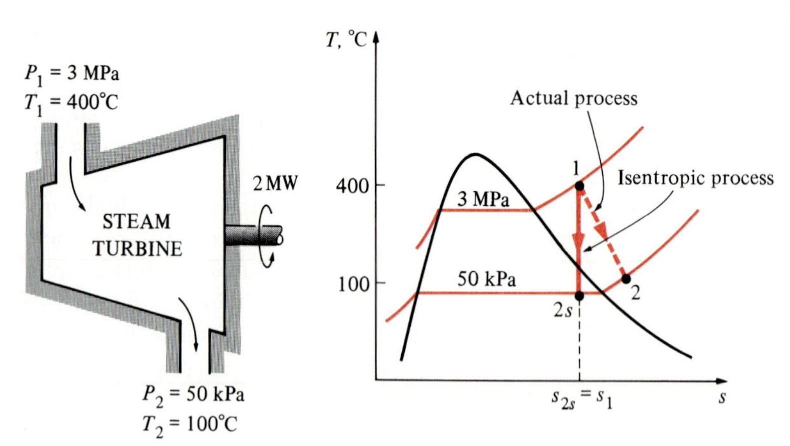

FIGURE 6-62

Schematic and *T-s* diagram for Example 6-18.

(*b*) The mass flow rate of steam through this turbine is determined from the conservation of energy relation for steady-flow systems (Eq. 4-21):

$$\dot{Q}^{\nearrow 0} - \dot{W} = \dot{m}(\Delta h + \Delta ke^{\nearrow 0} + \Delta pe^{\nearrow 0})$$

$$\dot{W} = \dot{m}(h_1 - h_{2a})$$

$$2\text{ MW}\left(\frac{1000\text{ kJ/s}}{1\text{ MW}}\right) = \dot{m}(3230.9 - 2682.5)\text{ kJ/kg}$$

$$\dot{m} = 3.65\text{ kg/s}$$

Adiabatic Efficiency of Compressors

The **adiabatic efficiency of a compressor** is defined as *the ratio of the work input required to raise the pressure of a gas to a specified value in an isentropic manner to the actual work input:*

$$\eta_C = \frac{\text{isentropic compressor work}}{\text{actual compressor work}} = \frac{w_s}{w_a} \tag{6-61}$$

Notice that the adiabatic compressor efficiency is defined with the *isentropic work input in the numerator* instead of in the denominator. This is because w_s is a smaller quantity than w_a, and this definition prevents adiabatic efficiencies from becoming greater than 100 percent, which would falsely imply that the actual compressors performed better than the isentropic ones. Also notice that the inlet conditions and the exit pressure of the gas are the same for both the actual and the isentropic compressor.

When the changes in kinetic and potential energies of the gas being compressed are negligible, the work input to an adiabatic compressor becomes equal to the change in enthalpy, and Eq. 6-61 for this case becomes

$$\eta_C \cong \frac{h_{2s} - h_1}{h_{2a} - h_1} \tag{6-62}$$

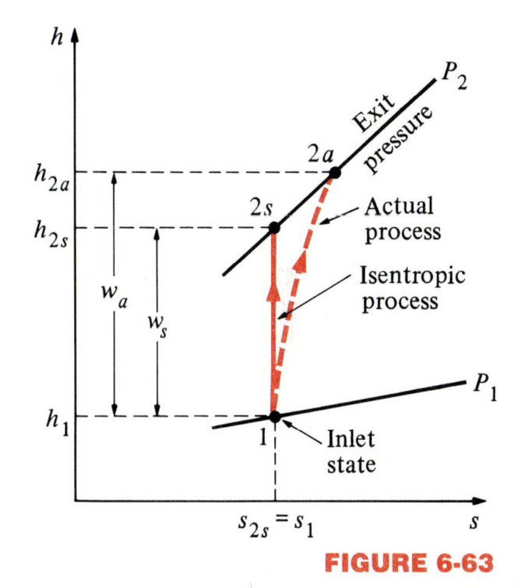

FIGURE 6-63

The *h-s* diagram of the actual and isentropic processes of an adiabatic compressor.

where h_{2a} and h_{2s} are the enthalpy values at the exit state for actual and isentropic compression processes, respectively, as illustrated in Fig. 6-63. Again, the value of η_C greatly depends on the design of the compressor. Well-designed compressors have adiabatic efficiencies that range from 75 to 85 percent.

When no attempt is made to cool the gas as it is compressed, the actual compression process is nearly adiabatic and the reversible adiabatic (i.e., isentropic) process serves well as the ideal process. But sometimes *compressors are cooled intentionally* by utilizing fins or a water jacket placed around the casing to reduce the work input requirements (Fig. 6-64). In this case, the isentropic process is not suitable as the model process since the device is no longer adiabatic and the adiabatic compressor efficiency defined above is meaningless. A realistic model process for compressors that are intentionally cooled during the compression process is the *reversible isothermal process*. Then we can conveniently define

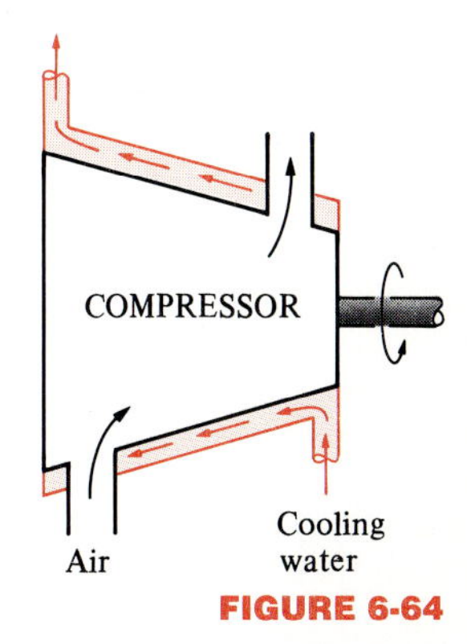

FIGURE 6-64

Compressors are sometimes intentionally cooled to minimize the work input.

an **isothermal efficiency** for such cases by comparing the actual process to a reversible isothermal one:

$$\eta_C = \frac{w_t}{w_a} \tag{6-63}$$

where w_t and w_a are the required work inputs to the compressor for the reversible isothermal and actual cases, respectively. Note that for a fixed inlet state and exit pressure, a reversible isothermal compressor will require less work input than an isentropic one.

EXAMPLE 6-19

Air is compressed by an adiabatic compressor from 100 kPa and 12°C to a pressure of 800 kPa at a steady rate of 0.2 kg/s. If the adiabatic efficiency of the compressor is 80 percent, determine (*a*) the exit temperature of air and (*b*) the required power input to the compressor.

Solution A sketch of the system and the *T*-*s* diagram of the process are given in Fig. 6-65. At the specified conditions, air can be treated as an ideal gas since it is at a high temperature and low pressure relative to its critical-point values ($T_{cr} = -147$°C and $P_{cr} = 3390$ kPa for nitrogen, the main constituent of air).

(*a*) We know only one property (pressure) at the exit state, and we need to know one more to fix the state and thus determine the exit temperature. The property that can be determined with minimal effort in this case is h_{2a} since the adiabatic efficiency of the compressor is given. Neglecting the changes in kinetic and potential energy of the air, we see that the adiabatic efficiency of this compressor is related to the enthalpies by Eq. 6-62.

The enthalpy of an ideal gas is a function of temperature only, and h_1 is easily determined from the air table at the inlet temperature:

$$T_1 = 285 \text{ K} \longrightarrow h_1 = 285.14 \text{ kJ/kg} \quad (P_{r1} = 1.1584) \qquad \text{(Table A-17)}$$

Now we need to determine h_{2s}, the enthalpy of the air at the end of the isentropic compression process. This is done by using one of the isentropic relations of ideal gases, such as Eq. 6-49:

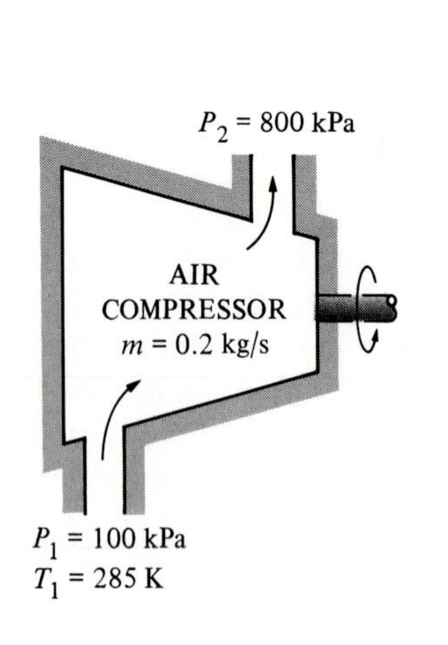

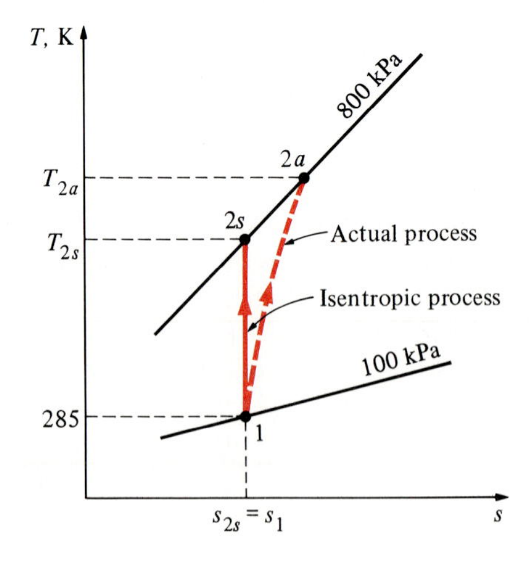

FIGURE 6-65
Schematic and *T*-*s* diagram for Example 6-19.

$$P_{r2} = P_{r1}\left(\frac{P_2}{P_1}\right)_{s=\text{const}} = 1.1584\left(\frac{800\text{ kPa}}{100\text{ kPa}}\right) = 9.2672$$

and $\quad P_{r2} = 9.2672 \longrightarrow h_{2s} = 517.05\text{ kJ/kg} \qquad$ (Table A-17)

(We could also determine T_{2s} from Eq. 6-45 by assuming constant specific heats for air.)

Substituting the known quantities into Eq. 6-62, we have

$$\eta_C \cong \frac{h_{2s} - h_1}{h_{2a} - h_1} \longrightarrow 0.80 = \frac{(517.05 - 285.14)\text{ kJ/kg}}{(h_{2a} - 285.14)\text{ kJ/kg}}$$

Thus, $\quad h_{2a} = 575.03\text{ kJ/kg} \longrightarrow T_{2a} = 569.5\text{ K} \qquad$ (Table A-17)

(*b*) The required power input to the compressor is determined from the conservation of energy relation for steady-flow devices, Eq. 4-21:

$$\dot{Q}_a^{\nearrow 0} - \dot{W}_a = \dot{m}(\Delta h + \Delta ke^{\nearrow 0} + \Delta pe^{\nearrow 0})_a$$

which, under the stated assumptions, reduces to

$$\begin{aligned} \dot{W}_a &= -\dot{m}(h_{2a} - h_1) \\ &= -(0.2\text{ kg/s})[(575.03 - 285.14)\text{ kJ/kg}] \\ &= -58.0\text{ kW} \end{aligned}$$

Notice that in determining the power input to the compressor, we used h_{2a} instead of h_{2s} since h_{2a} is the actual enthalpy of the air as it exits the compressor. The quantity h_{2s} is a hypothetical enthalpy value that the air would have if the process were isentropic.

Adiabatic Efficiency of Nozzles

Nozzles are essentially adiabatic devices and are used to accelerate a fluid. Therefore, the isentropic process serves as a suitable model for nozzles. The **adiabatic efficiency of a nozzle** is defined as *the ratio of the actual kinetic energy of the fluid at the nozzle exit to the kinetic energy value at the exit of an isentropic nozzle for the same inlet state and exit pressure*. That is,

$$\eta_N = \frac{\text{actual KE at nozzle exit}}{\text{isentropic KE at nozzle exit}} = \frac{\mathbf{V}_{2a}^2}{\mathbf{V}_{2s}^2} \tag{6-64}$$

Notice that the exit pressure is the same for both the actual and isentropic processes, but the exit state is different.

Nozzles involve no work interactions, and the fluid experiences little or no change in its potential energy as it flows through the device. If, in addition, the inlet velocity of the fluid is small relative to the exit velocity, the conservation of energy relation (Eq. 4-22) for this steady-flow device reduces to

$$0 = h_{2a} - h_1 + \frac{\mathbf{V}_{2a}^2 - 0}{2}$$

Then the adiabatic efficiency of the nozzle can be expressed in terms of

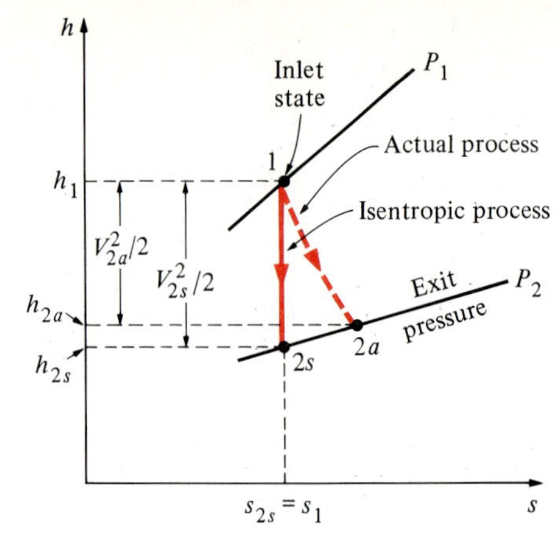

FIGURE 6-66

The *h-s* diagram of the actual and isentropic processes of an adiabatic nozzle.

the enthalpies as

$$\eta_N \cong \frac{h_1 - h_{2a}}{h_1 - h_{2s}} \tag{6-65}$$

where h_{2a} and h_{2s} are the enthalpy values at the nozzle exit for the actual and isentropic processes, respectively (Fig. 6-66). Adiabatic efficiencies of nozzles are typically above 90 percent, and nozzle efficiencies above 95 percent are not uncommon.

EXAMPLE 6-20

Air at 200 kPa and 950 K enters an adiabatic nozzle at low velocity and is discharged at a pressure of 80 kPa. If the adiabatic efficiency of the nozzle is 92 percent, determine (*a*) the maximum possible exit velocity, (*b*) the exit temperature, and (*c*) the actual exit velocity of the air. Assume constant specific heats for air.

Solution A sketch of the system and the *T-s* diagram of the process are given in Fig. 6-67. At the specified conditions, air can be treated as an ideal gas since it is at a high temperature and low pressure relative to its critical-point values ($T_{cr} = -147°C$ and $P_{cr} = 3390$ kPa for nitrogen, the main constituent of air). The temperature of air will drop during this acceleration process because some of its

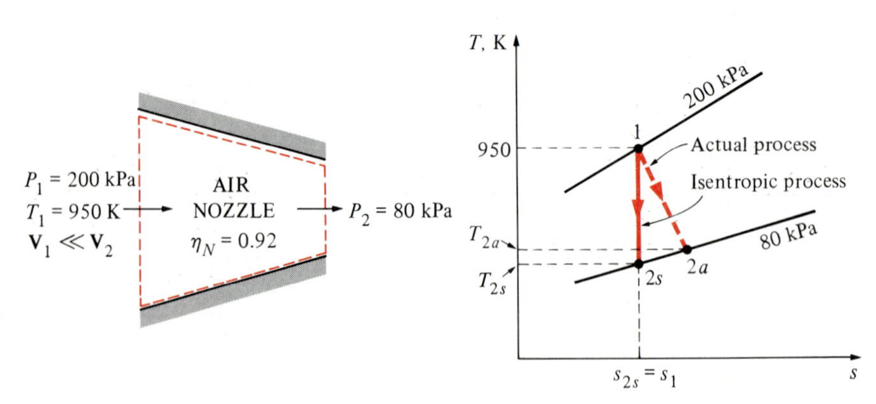

FIGURE 6-67

Schematic and *T-s* diagram for Example 6-20.

internal energy is converted to kinetic energy. This problem can be solved accurately by using property data from the air table. But we will assume constant specific heats (thus sacrifice some accuracy) to demonstrate their use. Let us guess that the average temperature of the air will be about 800 K. Then the average values of C_p and k at this anticipated average temperature are determined, from Table A-2b, to be $C_p = 1.099$ kJ/(kg·K) and $k = 1.354$.

(*a*) The exit velocity of the air will be a maximum when the process in the nozzle involves no irreversibilities. The exit velocity in this case is determined from the steady-flow energy equation. But first we need to determine the exit temperature. For the isentropic process of an ideal gas with constant specific heats, the temperatures and pressures are related by Eq. 6-45:

$$\frac{T_{2s}}{T_1} = \left(\frac{P_{2s}}{P_1}\right)^{(k-1)/k}$$

or

$$T_{2s} = T_1\left(\frac{P_{2s}}{P_1}\right)^{(k-1)/k} = (950\text{ K})\left(\frac{80\text{ kPa}}{200\text{ kPa}}\right)^{0.354/1.354} = 748\text{ K}$$

This will give an average temperature of 749 K, which is somewhat lower than the assumed average temperature (800 K). This result could be refined by reevaluating the k value at 749 K and repeating the calculations, but it is not warranted since the two average temperatures are sufficiently close (doing so would change the temperature by only 1.5 K, which is not significant).

Now we can determine the isentropic exit temperature of the air by simplifying the steady-flow energy relation (Eq. 4-22) for this isentropic process:

$$\cancelto{0}{q_s} - \cancelto{0}{w_s} = \Delta h + \Delta\text{ke} + \cancelto{0}{\Delta\text{pe}}$$

$$0 = C_{p,\text{av}}(T_{2s} - T_1) + \frac{\mathbf{V}_{2s}^2 - 0}{2}$$

or

$$\mathbf{V}_{2s} = \sqrt{2C_{p,\text{av}}(T_1 - T_{2s})}$$

$$= \sqrt{2[1.099\text{ kJ/(kg·K)}][(950 - 748)\text{ K}]\left(\frac{1000\text{ m}^2/\text{s}^2}{1\text{ kJ/kg}}\right)}$$

$$= 666\text{ m/s}$$

(*b*) The actual exit temperature of the air will be higher than the isentropic exit temperature evaluated above, and it is determined from Eq. 6-65. For constant specific heats,

$$\eta_N \cong \frac{h_1 - h_{2a}}{h_1 - h_{2s}} = \frac{C_{p,\text{av}}(T_1 - T_{2a})}{C_{p,\text{av}}(T_1 - T_{2s})}$$

or

$$0.92 = \frac{950 - T_{2a}}{950 - 748} \longrightarrow T_{2a} = 764\text{ K}$$

That is, the temperature will be 16 K higher at the exit of the actual nozzle as a result of irreversibilities such as friction. It represents a loss since this rise in temperature comes at the expense of kinetic energy (Fig. 6-68).

(*c*) The actual exit velocity of air can be determined from the definition of adiabatic efficiency (Eq. 6-64):

$$\eta_N = \frac{\mathbf{V}_{2a}^2}{\mathbf{V}_{2s}^2} \longrightarrow \mathbf{V}_{2a} = \sqrt{\eta_N \mathbf{V}_{2s}^2} = 639\text{ m/s}$$

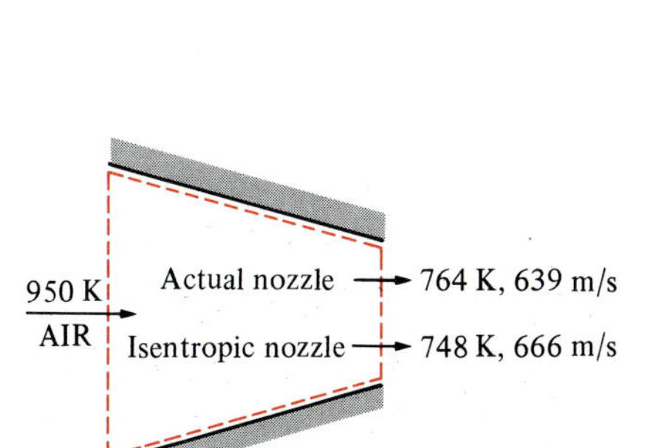

FIGURE 6-68

A substance leaves actual nozzles at a higher temperature (thus a lower velocity) as a result of friction.

6-14 ■ SUMMARY

The second law of thermodynamics leads to the definition of a new property called *entropy,* which is a quantitative measure of microscopic disorder for a system. The definition of entropy is based on the *Clausius inequality,* given by

$$\oint \frac{\delta Q}{T} \leq 0 \qquad \text{(kJ/K)} \tag{6-3}$$

where the equality holds for internally or totally reversible processes and the inequality for irreversible processes. Any quantity whose cyclic integral is zero is a property, and entropy is defined as

$$dS = \left(\frac{\delta Q}{T}\right)_{\text{int rev}} \qquad \text{(kJ/K)} \tag{6-4}$$

The *entropy change* during a process is obtained by integrating this relation:

$$\Delta S = S_2 - S_1 = \int_1^2 \left(\frac{\delta Q}{T}\right)_{\text{int rev}} \qquad \text{(kJ/K)} \tag{6-5}$$

This integration, in general, is not easy to perform since it requires a knowledge of Q as a function of T. For the special case of an internally reversible, isothermal process, this integration can be easily performed to yield

$$\Delta S = \frac{Q}{T_0} \qquad \text{(kJ/K)} \tag{6-6}$$

The inequality part of the Clausius inequality combined with the definition of entropy yields an inequality which is known as the *increase-in-entropy principle:*

$$dS \geq \frac{\delta Q}{T} \tag{6-9}$$

or

$$\Delta S_{\text{adiabatic}} \geq 0 \tag{6-10}$$

or

$$\Delta S_{\text{total}} = \Delta S_{\text{sys}} + \Delta S_{\text{surr}} \geq 0 \tag{6-11}$$

Thus the total entropy change during a process is positive (for actual processes) or zero (for reversible processes). The total entropy change for a process is the amount of entropy generated during that process (S_{gen}), and it is equal to the sum of the entropy change of the system and of the surroundings. The entropy of a system (closed system or control volume) or its surroundings may decrease during a process, but the sum of these two can never decrease.

Entropy change is caused by heat transfer and irreversibilities. Heat transfer to a system increases the entropy, and heat transfer from a system decreases it. The effect of irreversibilities is always to increase the entropy.

The *increase-in-entropy principle for a closed system* is expressed as

$$S_{gen} = \Delta S_{total} = \Delta S_{sys} + \Delta S_{surr} \geqslant 0 \qquad \text{(kJ/K)}$$

where $\Delta S_{sys} = S_2 - S_1 = m(s_2 - s_1) \qquad$ (kJ/K)

$$\Delta S_{surr} = \frac{Q_{surr}}{T_{surr}} \qquad \text{(kJ/K)}$$

The *increase-in-entropy principle for a control volume* is expressed as follows:

1 *General form:*

$$S_{gen} = (S_2 - S_1)_{CV} + \sum \int_{m_e} s_e \, \delta m_e - \sum \int_{m_i} s_i \, \delta m_i + \frac{Q_{surr}}{T_{surr}} \geqslant 0 \quad \text{(kJ/K)} \tag{6-12}$$

2 *Uniform-flow process:*

$$S_{gen} = (m_2 s_2 - m_1 s_1)_{CV} + \sum m_e s_e - \sum m_i s_i + \frac{Q_{surr}}{T_{surr}} \geqslant 0 \quad \text{(kJ/K)} \tag{6-13}$$

3 *Steady-flow process:*

$$\dot{S}_{gen} = \sum \dot{m}_e s_e - \sum \dot{m}_i s_i + \frac{\dot{Q}_{surr}}{T_{surr}} \geqslant 0 \qquad \text{(kW/K)} \tag{6-14}$$

For a single-stream steady-flow device, Eq. 6-14 reduces to

$$\dot{S}_{gen} = \dot{m}(s_e - s_i) + \frac{\dot{Q}_{surr}}{T_{surr}} \geqslant 0 \qquad \text{(kW/K)} \tag{6-15}$$

or

$$s_{gen} = s_e - s_i + \frac{q_{surr}}{T_{surr}} \geqslant 0 \qquad [\text{kJ/(kg·K)}] \tag{6-16}$$

In the above relations $\dot{S}_{gen}$ is the *rate of total entropy change*, or the *rate of entropy generation*. The equality in these relations applies to reversible processes and the inequality to irreversible ones.

The value of ΔS_{total} or S_{gen} can be used to determine whether a process is reversible, irreversible, or impossible:

$$S_{gen} = \Delta S_{total} \begin{cases} > 0 & \text{irreversible process} \\ = 0 & \text{reversible process} \\ < 0 & \text{impossible process} \end{cases}$$

The third law of thermodynamics states that the entropy of a pure crystalline substance at absolute zero temperature is zero. This law provides an absolute reference point for the determination of entropy. The entropy determined relative to this point is called *absolute entropy*.

Entropy is a property, and it can be expressed in terms of more familiar properties through the *T ds* relations, expressed as

$$T\,ds = du + P\,dv \tag{6-23}$$

and

$$T\,ds = dh - v\,dP \tag{6-24}$$

These two relations have many uses in thermodynamics and serve as the starting point in developing entropy-change relations for processes. The successful use of $T\,ds$ relations depends on the availability of property relations. Such relations do not exist for a general pure substance but are available for incompressible substances (solids, liquids) and ideal gases.

The *entropy-change* and *isentropic relations* for a process can be summarized as follows:

1 *Pure substances:*

Any process: $\Delta s = s_2 - s_1$ [kJ/(kg·K)] (6-30)

Isentropic process: $s_2 = s_1$ (6-31)

2 *Incompressible substances:*

Any process: $s_2 - s_1 = C_{av} \ln \dfrac{T_2}{T_1}$ [kJ/(kg·K)] (6-33)

Isentropic process: $T_2 = T_1$ (6-34)

3 *Ideal gases:*

a Constant specific heats (approximate treatment):

Any process:

$$s_2 - s_1 = C_{v,av} \ln \frac{T_2}{T_1} + R \ln \frac{v_2}{v_1} \quad \text{[kJ/(kg·K)]} \tag{6-37}$$

and

$$s_2 - s_1 = C_{p,av} \ln \frac{T_2}{T_1} - R \ln \frac{P_2}{P_1} \quad \text{[kJ/(kg·K)]} \tag{6-38}$$

Or, on a unit-mole basis,

$$\bar{s}_2 - \bar{s}_1 = \bar{C}_{v,av} \ln \frac{T_2}{T_1} + R_u \ln \frac{v_2}{v_1} \quad \text{[kJ/(kmol·K)]} \tag{6-39}$$

and

$$\bar{s}_2 - \bar{s}_1 = \bar{C}_{p,av} \ln \frac{T_2}{T_1} - R_u \ln \frac{P_2}{P_1} \quad \text{[kJ/(kmol·K)]} \tag{6-40}$$

Isentropic process:

$$\left(\frac{T_2}{T_1}\right)_{s=\text{const.}} = \left(\frac{v_1}{v_2}\right)^{k-1} \tag{6-44}$$

$$\left(\frac{T_2}{T_1}\right)_{s=\text{const.}} = \left(\frac{P_2}{P_1}\right)^{(k-1)/k} \tag{6-45}$$

$$\left(\frac{P_2}{P_1}\right)_{s=\text{const.}} = \left(\frac{v_1}{v_2}\right)^{k} \tag{6-46}$$

b Variable specific heats (exact treatment):

Any process:

$$s_2 - s_1 = s_2^\circ - s_1^\circ - R \ln \frac{P_2}{P_1} \qquad [\text{kJ/(kg·K)}] \qquad (6\text{-}42)$$

or $$\bar{s}_2 - \bar{s}_1 = \bar{s}_2^\circ - \bar{s}_1^\circ - R_u \ln \frac{P_2}{P_1} \qquad [\text{kJ/(kmol·K)}] \qquad (6\text{-}43)$$

Isentropic process:

$$s_2^\circ = s_1^\circ + R \ln \frac{P_2}{P_1} \qquad [\text{kJ/(kg·K)}] \qquad (6\text{-}48)$$

$$\left(\frac{P_2}{P_1}\right)_{s=\text{const.}} = \frac{P_{r2}}{P_{r1}} \qquad (6\text{-}49)$$

$$\left(\frac{v_2}{v_1}\right)_{s=\text{const.}} = \frac{v_{r2}}{v_{r1}} \qquad (6\text{-}50)$$

where P_r is the *relative pressure* and v_r is the *relative specific volume*. The function s° depends on temperature only.

The *steady-flow work* for a reversible process can be expressed in terms of the fluid properties as

$$w_{\text{rev}} = -\int_1^2 v\,dP - \Delta\text{ke} - \Delta\text{pe} \qquad (\text{kJ/kg}) \qquad (6\text{-}51)$$

For incompressible substances (v = constant) it simplifies to

$$w_{\text{rev}} = v(P_1 - P_2) - \Delta\text{ke} - \Delta\text{pe} \qquad (\text{kJ/kg}) \qquad (6\text{-}53)$$

The work done during a steady-flow process is proportional to the specific volume. Therefore, v should be kept as small as possible during a compression process to minimize the work input and as large as possible during an expansion process to maximize the work output.

The reversible work inputs to a compressor compressing an ideal gas from T_1, P_1 to P_2 in an isentropic (Pv^k = constant), polytropic (Pv^n = constant), and isothermal (Pv = constant) manner are determined by integration for each case with the following results:

Isentropic:
$$w_{\text{comp}} = \frac{kR(T_1 - T_2)}{k-1} = \frac{kRT_1}{k-1}\left[1 - \left(\frac{P_2}{P_1}\right)^{(k-1)/k}\right] \qquad (\text{kJ/kg}) \qquad (6\text{-}57a)$$

Polytropic:
$$w_{\text{comp}} = \frac{nR(T_1 - T_2)}{n-1} = \frac{nRT_1}{n-1}\left[1 - \left(\frac{P_2}{P_1}\right)^{(n-1)/n}\right] \qquad (\text{kJ/kg}) \qquad (6\text{-}57b)$$

Isothermal:
$$w_{\text{comp}} = RT \ln \frac{P_1}{P_2} \qquad (\text{kJ/kg}) \qquad (6\text{-}57c)$$

The work input to a compressor can be reduced by using multistage compression with intercooling. For maximum savings from the work input, the pressure ratio across each stage of the compressor must be the same.

Most steady-flow devices operate under adiabatic conditions, and the ideal process for these devices is the isentropic process. The parameter that describes how efficiently a device approximates a corresponding isentropic device is called *isentropic* or *adiabatic efficiency*. It is defined for turbines, compressors, and nozzles as follows:

$$\eta_T = \frac{\text{actual turbine work}}{\text{isentropic turbine work}} = \frac{w_a}{w_s} \tag{6-59}$$

or

$$\eta_T \cong \frac{h_1 - h_{2a}}{h_1 - h_{2s}} \qquad \text{when } \Delta\text{ke} \cong \Delta\text{pe} \cong 0 \tag{6-60}$$

$$\eta_C = \frac{\text{isentropic compressor work}}{\text{actual compressor work}} = \frac{w_s}{w_a} \tag{6-61}$$

or

$$\eta_C \cong \frac{h_{2s} - h_1}{h_{2a} - h_1} \qquad \text{when } \Delta\text{ke} \cong \Delta\text{pe} \cong 0 \tag{6-62}$$

$$\eta_N = \frac{\text{actual KE at nozzle exit}}{\text{isentropic KE at nozzle exit}} = \frac{\mathbf{V}_{2a}^2}{\mathbf{V}_{2s}^2} \tag{6-64}$$

or

$$\eta_N \cong \frac{h_1 - h_{2a}}{h_1 - h_{2s}} \qquad \text{when } \mathbf{V}_1 \ll \mathbf{V}_{2a} \tag{6-65}$$

In the relations above, h_{2a} and h_{2s} are the enthalpy values at the exit state for actual and isentropic processes, respectively.

REFERENCES AND SUGGESTED READING

1 A. Bejan, *Entropy Generation through Heat and Fluid Flow,* Wiley-Interscience, New York, 1982.

2 W. Z. Black and J. G. Hartley, *Thermodynamics,* Harper & Row, New York, 1985.

3 J. R. Howell and R. O. Buckius, *Fundamentals of Engineering Thermodynamics,* McGraw-Hill, New York, 1987.

4 J. B. Jones and G. A. Hawkins, *Engineering Thermodynamics,* 2d ed., Wiley, New York, 1986.

5 W. C. Reynolds and H. C. Perkins, *Engineering Thermodynamics,* 2d ed., McGraw-Hill, New York, 1977.

6 G. J. Van Wylen and R. E. Sonntag, *Fundamentals of Classical Thermodynamics,* 3d ed., Wiley, New York, 1985.

7 K. Wark, *Thermodynamics,* 5th ed., McGraw-Hill, New York, 1988.

Clausius Inequality

6-1C Demonstrate the validity of the Clausius inequality, using a reversible and an irreversible refrigerator operating between the same thermal energy reservoirs.

6-2C Does the temperature in the Clausius inequality relation have to be absolute temperature? Why?

6-3C Does a cycle for which $\oint \delta Q > 0$ violate the Clausius inequality? Why?

6-4 A heat engine receives 800 kJ of heat from a high-temperature source at 1100 K and rejects 500 kJ of it to a low-temperature sink at 310 K. Determine if this heat engine violates the second law of thermodynamics on the basis of (*a*) the Clausius inequality and (*b*) the Carnot principle.

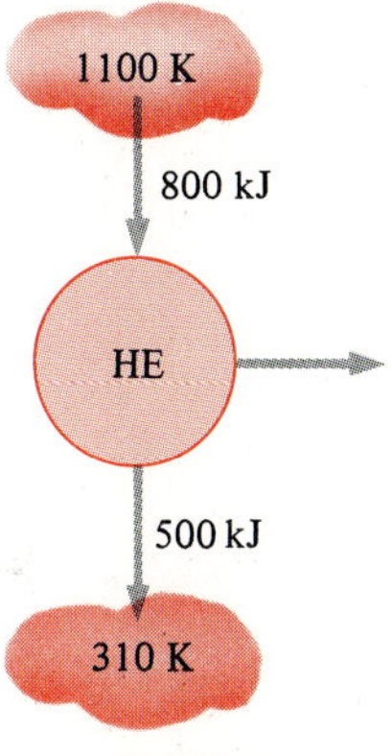

FIGURE P6-4

6-4E A heat engine receives 900 Btu of heat from a high-temperature source at 1800 R and rejects 500 Btu of it to a low-temperature sink at 550 R. Determine if this heat engine violates the second law of thermodynamics on the basis of (*a*) the Clausius inequality and (*b*) the Carnot principle. *Answers:* (*a*) No, (*b*) no

6-5 A heat engine receives heat at a rate of 75 MJ/h from a furnace at 1300 K, rejects heat to river water at 290 K, and has a power output of 8 kW. Determine if this heat engine violates the second law of thermodynamics on the basis of the Clausius inequality.

6-6 A refrigerator removes heat from the refrigerated space at 2°C at a rate of 300 kJ/min and rejects heat to the kitchen air at 26°C at a rate of 345 kJ/min. Determine if this refrigerator violates the second law of thermodynamics on the basis of (*a*) the Clausius inequality and (*b*) the Carnot principle. *Answers:* (*a*) No, (*b*) no

6-6E A refrigerator removes heat from the refrigerated space at 35°F at a rate of 375 Btu/min and rejects heat to the kitchen air at 70°F at a rate of 450 Btu/min. Determine if this refrigerator violates the second law of thermodynamics on the basis of the Clausius inequality.

6-7 An air-based heat pump supplies heat to a house maintained at 21°C at a rate of 75,000 kJ/h while consuming 8 kW of power. If the outside air temperature is −2°C, determine if this heat pump violates the second law of thermodynamics on the basis of the Clausius inequality.

Entropy and the Increase-in-Entropy Principle

6-8C Is a quantity whose cyclic integral is zero necessarily a property?

*Students are encouraged to answer *all* the concept "C" questions.

6-9C Does the cyclic integral of heat have to be zero (i.e., does a system have to reject as much heat as it receives to complete a cycle)? Explain.

6-10C Does the cyclic integral of work have to be zero (i.e., does a system have to produce as much work as it consumes to complete a cycle)? Explain.

6-11C A system undergoes a process between two fixed states first in a reversible manner and then in an irreversible manner. For which case is the entropy change greater? Why?

6-12C Is the value of the integral $\int_1^2 \delta Q/T$ the same for all processes between states 1 and 2? Explain.

6-13C Is the value of the integral $\int_1^2 \delta Q/T$ the same for all reversible processes between states 1 and 2?

6-14C To determine the entropy change for an irreversible process between states 1 and 2, should the integral $\int_1^2 \delta Q/T$ be performed along the actual process path or an imaginary reversible path? Explain.

6-15C Is an isothermal process necessarily internally reversible? Explain your answer with an example.

6-16C How do the values of the integral $\int_1^2 \delta Q/T$ compare for a reversible and irreversible process between the same end states?

6-17C The entropy of a hot baked potato decreases as it cools. Is this a violation of the increase-in-entropy principle? Explain.

6-18C Is it possible to create or destroy entropy?

6-19C Entropy change is defined as

$$\Delta S = \int_1^2 \left(\frac{\delta Q}{T}\right)_{\text{int rev}}$$

Based on this relation, is it reasonable to conclude that the entropy of a closed system remains constant during an adiabatic process?

6-20C During a certain process, it is claimed that the entropies of both the system and the surroundings have increased. Is this a reasonable claim?

6-21C A piston-cylinder device contains helium gas. During a reversible, isothermal process, the entropy of the helium will (*never, sometimes, always*) increase.

6-22C A piston-cylinder device contains nitrogen gas. During a reversible, adiabatic process the entropy of the nitrogen will (*never, sometimes, always*) increase.

6-23C A piston-cylinder device contains superheated steam. During an actual adiabatic process, the entropy of the steam will (*never, sometimes, always*) increase.

6-24C The entropy of steam will (*increase, decrease, remain the same*) as it flows through an actual adiabatic turbine.

6-25C The sum of the entropy changes of a system and its surroundings is (*always, sometimes, never*) negative.

6-26C The entropy of the working fluid of the ideal Carnot cycle (*increases, decreases, remains the same*) during the isothermal heat addition process.

6-27C The entropy of the working fluid of the ideal Carnot cycle (*increases, decreases, remains the same*) during the isothermal heat rejection process.

6-28C During a heat transfer process, the entropy of a system (*always, sometimes, never*) increases.

6-29C During a reversible process, the magnitudes of the entropy changes of a system and its surroundings are (*always, sometimes, never*) equal.

6-30C Is it possible for the entropy change of a closed system to be zero during an irreversible process? Explain.

6-31C What three different mechanisms can cause the entropy of a control volume to change?

6-32C How does mass transfer affect the entropy of a control volume?

6-33C An insulated control volume is initially filled with a hot gas. The intake valve is now opened, and some cooler gas is allowed to enter the control volume. Will the total entropy of the control volume (in kJ/K) increase, decrease, or remain the same as a result of this mass transfer?

6-34C An insulated rigid tank contains a saturated liquid–vapor mixture of refrigerant-12. A leak develops at the bottom of the tank, and some liquid refrigerant leaks out. Will the total entropy (in kJ/K) within the rigid tank increase, decrease, or remain the same as a result of this mass transfer process? How would you answer this question for the specific entropy [in kJ/(kg·K)] of the refrigerant that remains in the tank? Explain.

6-35C Steam is accelerated as it flows through an actual adiabatic nozzle. The entropy of the steam at the nozzle exit will be (*greater than, equal to, less than*) the entropy at the nozzle inlet.

A rigid tank contains an ideal gas at 50°C which is being stirred by a paddle wheel. The paddle wheel does 200 kJ of work on the ideal gas. It is observed that the temperature of the ideal gas remains constant during this process as a result of heat transfer between the system and the surroundings at 25°C. Determine (*a*) the entropy change of the ideal gas and (*b*) the entropy change of the surroundings. Is the increase-in-entropy principle satisfied during this process?
Answers: (*a*) 0, (*b*) 0.671 kJ/K

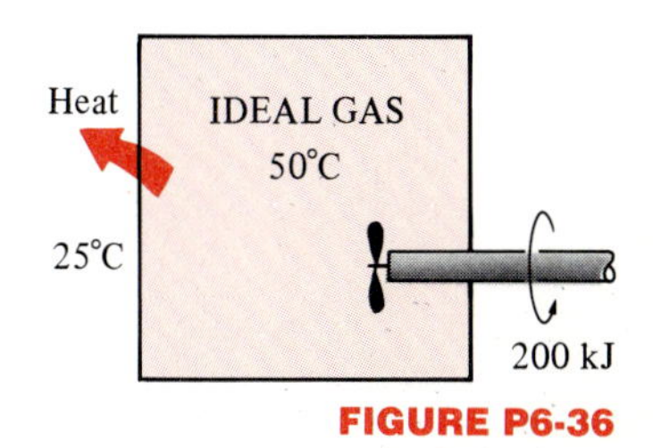

FIGURE P6-36

A rigid tank contains an ideal gas at 90°F which is being stirred

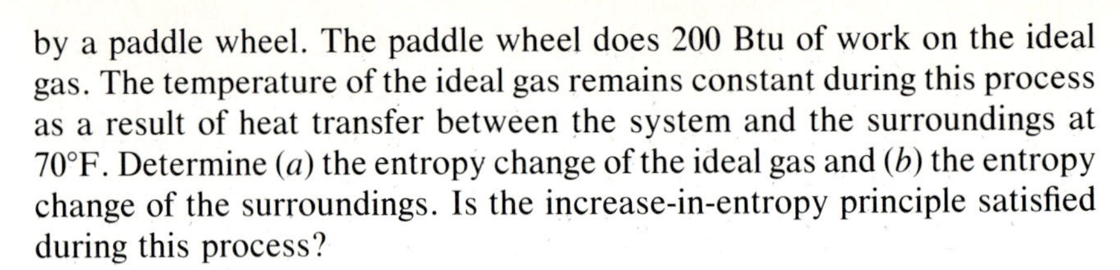

by a paddle wheel. The paddle wheel does 200 Btu of work on the ideal gas. The temperature of the ideal gas remains constant during this process as a result of heat transfer between the system and the surroundings at 70°F. Determine (*a*) the entropy change of the ideal gas and (*b*) the entropy change of the surroundings. Is the increase-in-entropy principle satisfied during this process?

6-37 Air is compressed by a 5-kW compressor from P_1 to P_2. The air temperature is maintained constant at 25°C during this process as a result of heat transfer to the surrounding medium at 10°C. Determine the rate of entropy change of (*a*) the air and (*b*) the surroundings. State the assumptions made in solving this problem. Does this process satisfy the second law of thermodynamics?
Answers: (*a*) −0.0168 kW/K, (*b*) 0.0177 kW/K

6-38 A frictionless piston-cylinder device contains saturated liquid water at 200-kPa pressure. Now 450 kJ of heat is transferred to water from a source at 500°C, and part of the liquid vaporizes at constant pressure. Determine the total entropy change for this process, in kJ/K. Is this process reversible, irreversible, or impossible?
Answers: 0.562 kJ/K, irreversible

6-38E A frictionless piston-cylinder device contains saturated liquid water at 20-psia pressure. Now 600 Btu of heat is transferred to water from a source at 900°F, and part of the liquid vaporizes at constant pressure. Determine the total entropy change for this process, in Btu/R. Is this process reversible, irreversible, or impossible?

6-39 During the isothermal heat addition process of a Carnot cycle, 800 kJ of heat is added to the working fluid from a source at 500°C. Determine (*a*) the entropy change of the working fluid, (*b*) the entropy change of the source, and (*c*) the total entropy change for the process.

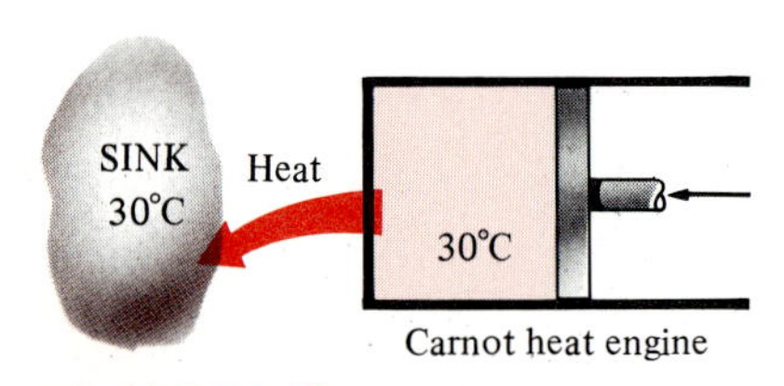

FIGURE P6-40

6-40 During the isothermal heat rejection process of a Carnot cycle, the working fluid experiences an entropy change of −0.6 kJ/K. If the temperature of the energy sink is 30°C, determine (*a*) the amount of heat transfer to the sink, (*b*) the entropy change of the sink, and (*c*) the total entropy change for this process.
Answers: (*a*) 181.8 kJ, (*b*) 0.6 kJ/K, (*c*) 0

6-40E During the isothermal heat rejection process of a Carnot cycle, the working fluid experiences an entropy change of −0.7 Btu/R. If the temperature of the energy sink is 95°F, determine (*a*) the amount of heat transfer, (*b*) the entropy change of the sink, and (*c*) the total entropy change for this process. *Answers:* (*a*) 388.5 Btu, (*b*) 0.7 Btu/R, (*c*) 0

6-41 Refrigerant-12 enters the coils of the evaporator of a refrigeration system as a saturated liquid–vapor mixture at a pressure of 200 kPa. The refrigerant absorbs 70 kJ of heat from the cooled space which is maintained at −5°C and leaves as saturated vapor at the same pressure. Determine (*a*) the entropy change of the refrigerant, (*b*) the entropy change of the cooled space, and (*c*) the total entropy change for this process.

6-42C Are the $T\,ds$ relations developed in Sec. 6-6 limited to reversible processes only, or are they valid for all processes, reversible or irreversible? Explain.

6-43C Are the $T\,ds$ relations developed in Sec. 6-6 limited to closed systems only, or are they also applicable to a fixed mass passing through a control volume? Explain.

6-44C Is a process which is internally reversible and adiabatic necessarily isentropic? Explain.

6-45C Why can we not obtain simple relations for the entropy change of a general pure substance in terms of other properties?

6-46 The radiator of a steam heating system has a volume of 15 L and is filled with superheated water vapor at 200 kPa and 200°C. At this moment both the inlet and the exit valves to the radiator are closed. After a while the temperature of the steam drops to 80°C as a result of heat transfer to the room air. Determine the entropy change of the steam during this process. *Answer:* −0.0605 kJ/K

6-46E The radiator of a steam heating system has a volume of 0.5 ft^3 and is filled with superheated water vapor at 40 psia and 320°F. At this moment both the inlet and the exit valves to the radiator are closed. After a while the temperature of the steam drops to 150°F as a result of heat transfer to the room air. Determine the entropy change of the steam during this process.

6-47 A 0.5-m^3 rigid tank contains refrigerant-12 initially at 200 kPa and 40 percent quality. Heat is added now to the refrigerant from a source at 35°C until the pressure rises to 400 kPa. Determine (*a*) the entropy change of the refrigerant, (*b*) the entropy change of the heat source, and (*c*) the total entropy change for this process.
Answers: (*a*) 3.4835 kJ/K, (*b*) −3.076 kJ/K, (*c*) 0.4075 kJ/K

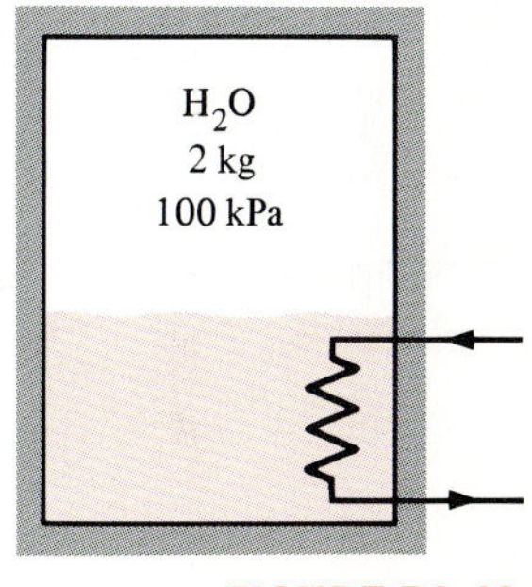

FIGURE P6-48

6-48 A well-insulated rigid tank contains 2 kg of a saturated liquid–vapor mixture of water at 100 kPa. Initially, three-quarters of the mass is in the liquid phase. An electric resistance heater placed in the tank is now turned on and kept on until all the liquid in the tank is vaporized. Determine the entropy change of the steam during this process.
Answer: 8.0962 kJ/K

6-48E A well-insulated rigid tank contains 4 lbm of a saturated liquid–vapor mixture of water at 15 psia. Initially, three-quarters of the mass is in the liquid phase. An electric resistance heater placed in the tank is now turned on and kept on until all the liquid in the tank is vaporized. Determine the entropy change of the steam during this process.

6-49 A rigid tank is divided into two equal parts by a partition. One part of the tank contains 0.5 kg of compressed liquid water at 500 kPa and 60°C while the other part is evacuated. The partition is now removed, and the water expands to fill the entire tank. Determine the entropy change

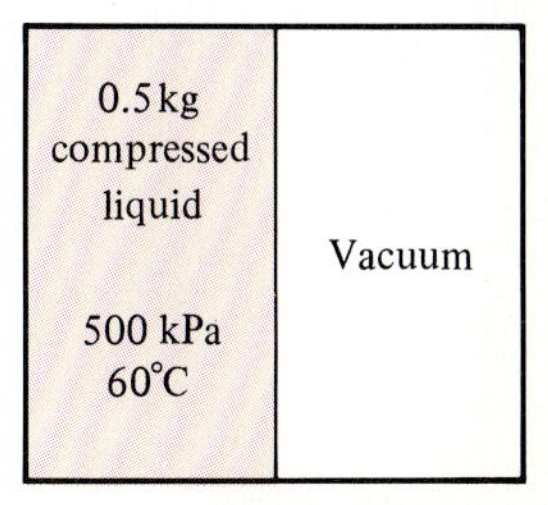

FIGURE P6-49

of the water during this process, if the final pressure in the tank is 15 kPa. *Answer:* −0.0378 kJ/K

6-50 A piston-cylinder device contains 5 kg of refrigerant-12 at 0.8 MPa and 50°C. The refrigerant is now cooled at constant pressure until it exists as a liquid at 24°C. If the temperature of the surroundings is 20°C, determine the total entropy change for this process and whether this process is reversible, irreversible, or impossible.

6-50E A piston-cylinder device contains 3 lbm of refrigerant-12 at 120 psia and 120°F. The refrigerant is now cooled at constant pressure until it exists as a liquid at 90°F. If the temperature of the surroundings is 70°F, determine the total entropy change for this process and whether this process is reversible, irreversible, or impossible.

6-51 An insulated piston-cylinder device contains 2 L of saturated liquid water at a constant pressure of 150 kPa. An electric resistance heater inside the cylinder is now turned on, and electrical work is done on the steam in the amount of 2200 kJ. Determine the entropy change of the water during this process. *Answer:* 5.717 kJ/K

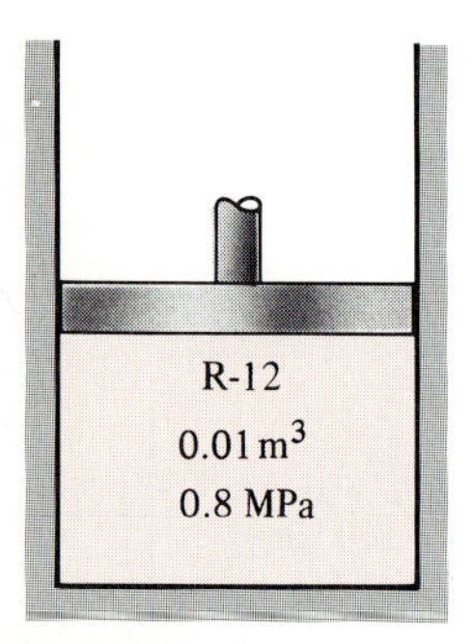

FIGURE P6-52

6-52 An insulated piston-cylinder device contains 0.01 m^3 of saturated refrigerant-12 vapor at 0.8-MPa pressure. The refrigerant is now allowed to expand in a reversible manner until the pressure drops to 0.4 MPa. Determine (*a*) the final temperature in the cylinder and (*b*) the work done by the refrigerant.

6-52E An insulated piston-cylinder device contains 0.4 ft^3 of saturated refrigerant-12 vapor at 120-psia pressure. The refrigerant is now allowed to expand in a reversible manner until the pressure drops to 60 psia. Determine (*a*) the final temperature in the cylinder and (*b*) the work done by the refrigerant.

6-53 Refrigerant-12 enters an adiabatic compressor as saturated vapor at 140 kPa at a rate of 5 m^3/min and is compressed to a pressure of 700 kPa. Determine the minimum power that must be supplied to the compressor.

6-54 Steam enters an adiabatic turbine at 5 MPa and 400°C and leaves at a pressure of 200 kPa. Determine the maximum amount of work that can be delivered by this turbine. *Answer:* 678.4 kJ/kg

6-54E Steam enters an adiabatic turbine at 800 psia and 900°F and leaves at a pressure of 40 psia. Determine the maximum amount of work that can be delivered by this turbine.

6-55 A heavily insulated piston-cylinder device contains 0.05 m^3 of steam at 300 kPa and 150°C. Steam is now compressed in a reversible manner to a pressure of 1 MPa. Determine the work done on the steam during this process.

FIGURE P6-56

6-56 A 2-L well-insulated rigid can initially contains refrigerant-12 at 700

kPa and 20°C. Now a crack develops in the can, and the refrigerant starts to leak out slowly. Assuming the refrigerant remaining in the can has undergone a reversible, adiabatic process, determine the final mass in the can when the pressure drops to 240 kPa. *Answer:* 0.173 kg

6-56E A 0.1-ft^3 well-insulated rigid can initially contains refrigerant-12 at 120 psia and 80°F. Now a crack develops in the can, and the refrigerant starts to leak out slowly. Assuming the refrigerant remaining in the can has undergone a reversible, adiabatic process, determine the final mass in the can when the pressure drops to 30 psia.

6-57 A piston-cylinder device contains 0.5 kg of saturated water vapor at 200°C. Heat is now transferred to steam, and steam expands reversibly and isothermally to a final pressure of 800 kPa. Determine the heat transferred and the work done during this process.

6-58 An insulated tank containing 0.4 m^3 of saturated water vapor at 500 kPa is connected to an initially evacuated, insulated piston-cylinder device. The mass of the piston is such that a pressure of 150 kPa is required to raise it. Now the valve is opened slightly, and part of the steam flows to the cylinder, raising the piston. This process continues until the pressure in the tank drops to 150 kPa. Assuming the steam that remains in the tank to have undergone a reversible adiabatic process, determine the final temperature (*a*) in the rigid tank and (*b*) in the cylinder.

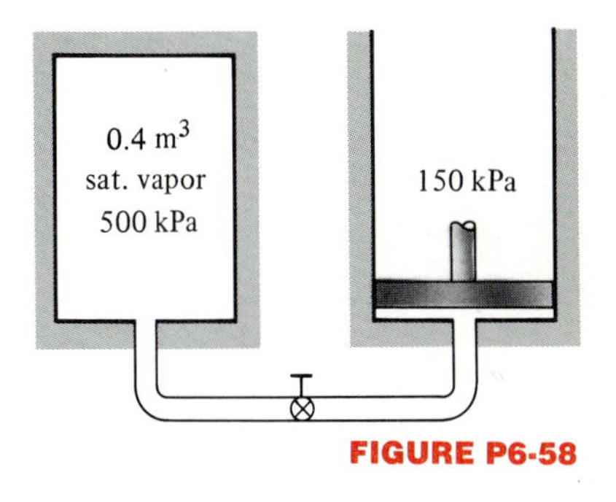

FIGURE P6-58

6-59 Two rigid tanks are connected by a valve. Tank *A* is insulated and contains 0.2 m^3 of steam at 400 kPa and 80 percent quality. Tank *B* is uninsulated and contains 3 kg of steam at 200 kPa and 250°C. The valve is now opened, and steam flows from tank *A* to tank *B* until the pressure in tank *A* drops to 300 kPa. During this process 600 kJ of heat is transferred from tank *B* to the surroundings at 0°C. Assuming the steam remaining inside tank *A* to have undergone a reversible adiabatic process, determine (*a*) the final temperature in each tank and (*b*) the entropy generated during this process. *Answers:* (*a*) 133.55°C, 113.0°C; (*b*) 0.912 kJ/K

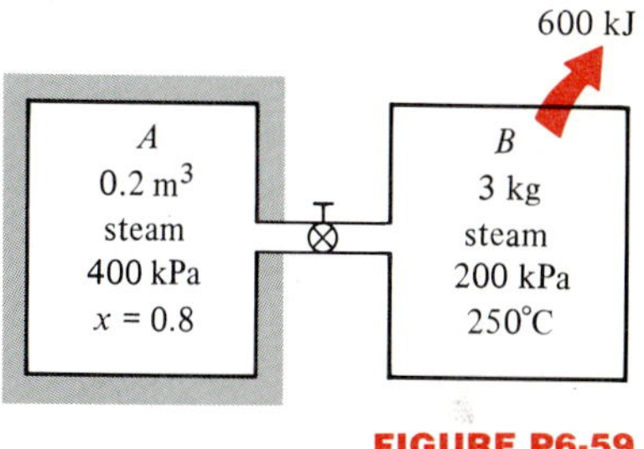

FIGURE P6-59

6-60 Steam enters a diffuser at 10 kPa and 50°C with a velocity of 300 m/s and exits as saturated vapor at 50°C and 50 m/s. The exit area of the diffuser is 2 m^2. Determine (*a*) the mass flow rate of the steam and (*b*) the rate of entropy generation during this process. Assume an ambient temperature of 25°C. *Answers:* (*a*) 8.31 kg/s, (*b*) 0.415 kW/K

6-60E Steam enters a diffuser at 20 psia and 240°F with a velocity of 900 ft/s and exits as saturated vapor at 240°F and 100 ft/s. The exit area of the diffuser is 1 ft^2. Determine (*a*) the mass flow rate of the steam and (*b*) the rate of entropy generation during this process. Assume an ambient temperature of 77°F.

6-61 Steam expands in a turbine steadily at a rate of 25,000 kg/h, entering at 8 MPa and 450°C and leaving at 50 kPa as saturated vapor. If the power generated by the turbine is 4 MW, determine the rate of entropy generation for this process. Assume the surrounding medium is at 25°C.
Answer: 8.38 kW/K

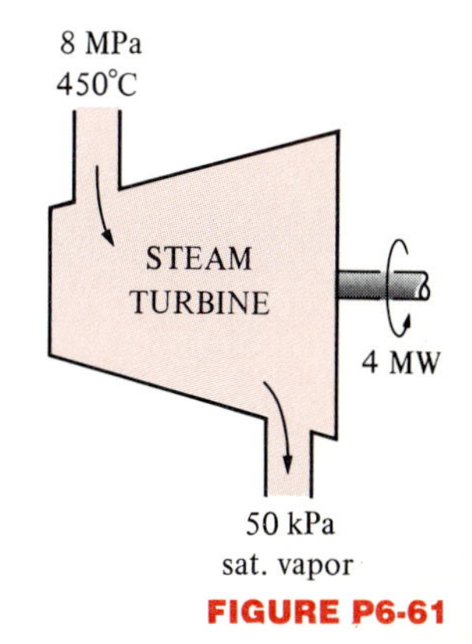

FIGURE P6-61

6-62 A hot-water stream at 70°C enters an adiabatic mixing chamber with a mass flow rate of 0.6 kg/s, where it is mixed with a stream of cold water at 20°C. If the mixture leaves the chamber at 42°C, determine (*a*) the mass flow rate of the cold water and (*b*) the rate of entropy generation during this adiabatic mixing process. Assume all the streams are at a pressure of 200 kPa.

6-62E A hot-water stream at 160°F enters an adiabatic mixing chamber with a mass flow rate of 1.2 lbm/s, where it is mixed with a stream of cold water at 70°F. If the mixture leaves the chamber at 110°F, determine (*a*) the mass flow rate of the cold water and (*b*) the rate of entropy generation during this adiabatic mixing process. Assume all the streams are at a pressure of 30 psia.

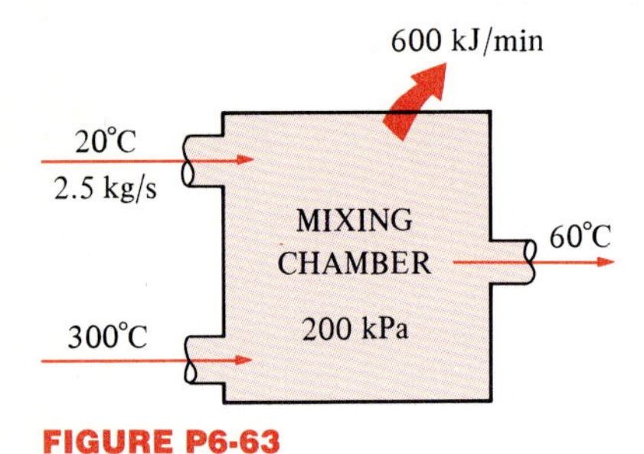

FIGURE P6-63

6-63 Liquid water at 200 kPa and 20°C is heated in a chamber by mixing it with superheated steam at 200 kPa and 300°C. Liquid water enters the mixing chamber at a rate of 2.5 kg/s, and the chamber is estimated to lose heat to the surrounding air at 25°C at a rate of 600 kJ/min. If the mixture leaves the mixing chamber at 200 kPa and 60°C, determine (*a*) the mass flow rate of the superheated steam and (*b*) the rate of entropy generation during this mixing process.
Answers: (*a*) 0.152 kg/s, (*b*) 0.297 kW/K

6-64 In large steam power plants, the feedwater is frequently heated in closed feedwater heaters, which are basically heat exchangers, by steam extracted from the turbine at some stage. Steam enters the feedwater heater at 1 MPa and 200°C and leaves as saturated liquid at the same pressure. Feedwater enters the heater at 2.5 MPa and 50°C and leaves 10°C below the exit temperature of the steam. Neglecting any heat losses from the outer surfaces of the heater, determine (*a*) the ratio of the mass flow rates of the extracted steam and the feedwater heater and (*b*) the total entropy change for this process per unit mass of the feedwater.

6-65 A 0.1-m^3 rigid tank initially contains refrigerant-12 at 1 MPa and 100 percent quality. The tank is connected by a valve to a supply line that carries refrigerant-12 at 1.4 MPa and 30°C. The valve is now opened, allowing the refrigerant to enter the tank, and it is closed when it is observed that the tank contains only saturated liquid at 1.2 MPa. Determine (*a*) the mass of the refrigerant that entered the tank, (*b*) the amount of heat transfer with the surroundings at 50°C, and (*c*) the entropy generated during this process.
Answers: (*a*) 115.67 kg, (*b*) 1564 kJ, (*c*) 0.0907 kJ/K

6-65E A 3-ft^3 rigid tank initially contains refrigerant-12 at 120 psia and 100 percent quality. The tank is connected by a valve to a supply line that carries refrigerant-12 at 160 psia and 80°F. The valve is now opened, allowing the refrigerant to enter the tank, and is closed when it is observed that the tank contains only saturated liquid at 140 psia. Determine (*a*) the mass of the refrigerant that entered the tank, (*b*) the amount of heat transfer with the surroundings at 120°C, and (*c*) the entropy generated during this process.

6-66 A 0.4-m³ rigid tank is filled with saturated liquid water at 200°C. A valve at the bottom of the tank is now opened, and one-half of the total mass is withdrawn from the tank in the liquid form. Heat is transferred to water from a source at 250°C so that the temperature in the tank remains constant. Determine (*a*) the amount of heat transfer and (*b*) the total entropy change for this process.

Entropy Changes of Incompressible Substances

6-67C During a heat transfer process, the entropy change of incompressible substances, such as liquid water, can be determined from Eq. 6-33, that is, $\Delta S = mC_{av} \ln (T_2/T_1)$. Show that for thermal energy reservoirs, such as large lakes, this relation reduces to Eq. 6-6, that is, $\Delta S = Q/T$.

6-68C Equation 6-33 [that is, $\Delta S = mC_{av} \ln (T_2/T_1)$] is developed for incompressible substances. Can this relation be used to determine the entropy changes of ideal gases? If so, under what conditions?

6-69C Consider two solid blocks, one hot and the other cold, brought into contact in an adiabatic container. After a while, thermal equilibrium is established in the container as a result of heat transfer. The first law requires that the amount of energy lost by the hot solid be equal to the energy gained by the cold one. Does the second law require that the decrease in entropy of the hot solid be equal to the increase in entropy of the cold one?

6-70 A 20-kg copper block initially at 80°C is dropped into an insulated tank which contains 150 L of water at 25°C. Determine the final equilibrium temperature and the total entropy change for this process.

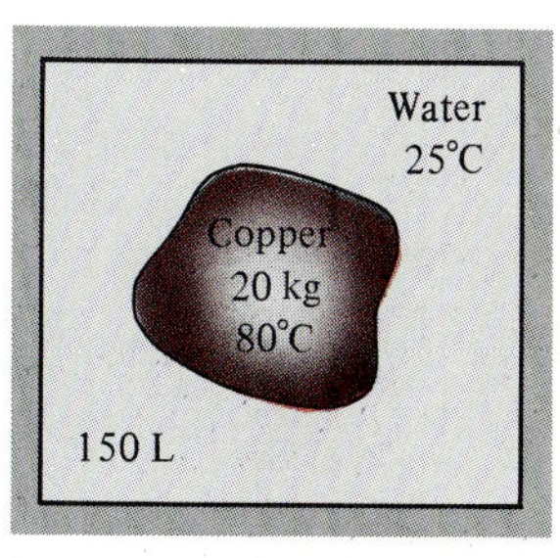

FIGURE P6-70

6-70E A 20-lbm copper block initially at 180°F is dropped into an insulated tank which contains 1 ft³ of water at 75°F. Determine the final equilibrium temperature and the total entropy change for this process.

6-71 A 5-kg iron block initially at 350°C is quenched in an insulated tank which contains 100 kg of water at 22°C. Assuming the water that vaporizes during the process condenses back in the tank, determine the amount of entropy generated during this process.

6-72 A 20-kg aluminum block initially at 200°C is brought into contact with a 20-kg block of iron at 100°C in an insulated enclosure. Determine the final equilibrium temperature and the total entropy change for this process. *Answers:* 168.4°C, 0.169 kJ/K

6-73 An iron block of unknown mass at 85°C is dropped into an insulated tank that contains 100 L of water at 20°C. At the same time, a paddle wheel driven by a 200-W motor is activated to stir the water. It is observed that thermal equilibrium is established after 20 min with a final temperature of 24°C. Determine the mass of the iron block and the entropy generated during this process. *Answers:* 52.2 kg, 1.285 kJ/K

6-73E An iron block of unknown mass at 185°F is dropped into an insulated tank that contains 0.8 ft^3 of water at 70°F. At the same time, a paddle wheel driven by a 200-W motor is activated to stir the water. Thermal equilibrium is established after 10 min with a final temperature of 75°F. Determine the mass of the iron block and the entropy generated during this process.

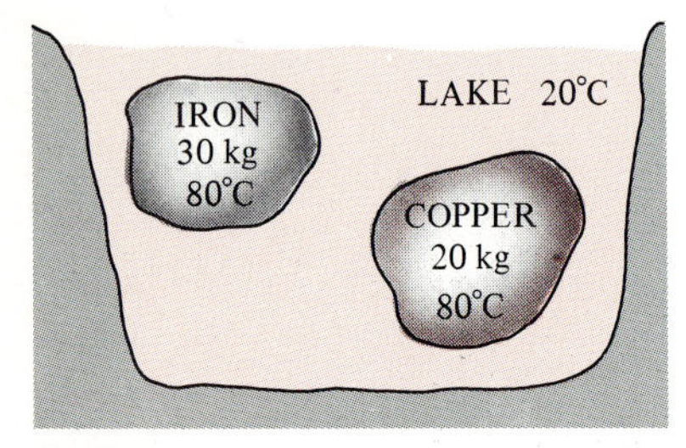

FIGURE P6-74

6-74 A 30-kg iron block and a 20-kg copper block, both initially at 80°C, are dropped into a large lake at 20°C. Thermal equilibrium is established after a while as a result of heat transfer between the blocks and the lake water. Determine the total entropy change for this process.

6-75 A 0.5-m^3 steel container which has a mass of 20 kg when empty is filled with liquid water. Initially, both the steel tank and the water are at 50°C. Now heat is transferred, and the entire system cools to the surrounding air temperature of 25°C. Determine the total entropy change for this process. *Answer* 7.25 kJ/K

6-75E A 15-ft^3 steel container which has a mass of 40 lbm when empty is filled with liquid water. Initially, both the steel tank and the water are at 120°F. Now heat is transferred, and the entire system cools to the surrounding air temperature of 70°F. Determine the total entropy change for this process.

Entropy Changes of Ideal Gases

6-76C Prove that the two relations for entropy changes of ideal gases under the constant-specific-heat assumption (Eqs. 6-37 and 6-38) are equivalent.

6-77C Starting with the second $T\,ds$ relation (Eq. 6-26), obtain Eq. 6-38 for the entropy change of ideal gases under the constant-specific-heat assumption.

6-78C What does the function $s°$ in the ideal-gas tables represent?

6-79C Some properties of ideal gases such as internal energy and enthalpy vary with temperature only [that is, $u = u(T)$ and $h = h(T)$]. Is this also the case for entropy?

6-80C Starting with Eq. 6-38, obtain Eq. 6-45.

6-81C What are P_r and v_r called? Is their use limited to isentropic processes? Explain.

6-82C Can the entropy of an ideal gas change during an isothermal process?

6-83C An ideal gas undergoes a process between two specified temperatures, first at constant pressure and then at constant volume. For which case will the ideal gas experience a larger entropy change? Explain.

6-84 Oxygen gas is compressed in a piston-cylinder device from an initial state of 0.8 m^3/kg and 25°C to a final state of 0.1 m^3/kg and 287°C. Determine the entropy change of the oxygen during this process, assuming (*a*) constant specific heats and (*b*) variable specific heats.

6-85 A 0.2-m^3 insulated rigid tank contains 0.4 kg of carbon dioxide at 100 kPa. Now paddle-wheel work is done on the system until the pressure in the tank rises to 120 kPa. Determine the entropy change of carbon dioxide during this process. Assume constant specific heats.
Answer: 0.0479 kJ/K

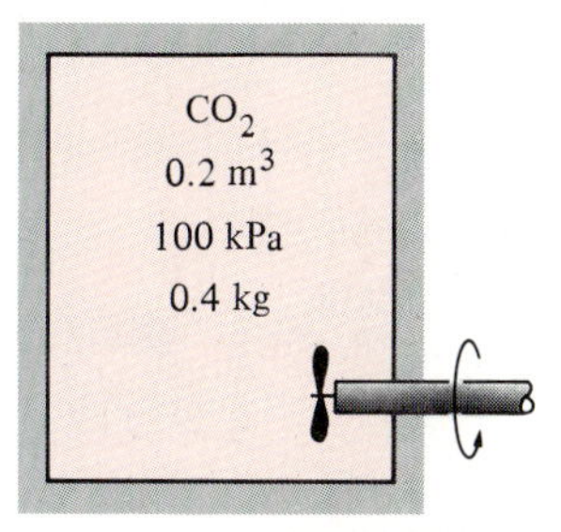

FIGURE P6-85

6-86 An insulated piston-cylinder device initially contains 300 L of air at 120 kPa and 17°C. Air is now heated for 15 min by a 200-W resistance heater placed inside the cylinder. The pressure of air is maintained constant during this process. Determine the entropy change of air, assuming (*a*) constant specific heats and (*b*) variable specific heats.

6-86E An insulated piston-cylinder device initially contains 10 ft^3 of air at 20 psia and 60°F. Air is now heated for 15 min by a 200-W resistance heater placed inside the cylinder. The pressure of air is maintained constant during this process. Determine the entropy change of air, assuming (*a*) constant specific heats and (*b*) variable specific heats.

6-87 A piston-cylinder device initially contains 0.5 m^3 of helium gas at 150 kPa and 20°C. Helium is now compressed in a polytropic process (PV^n = constant) to 400 kPa and 140°C. Determine (*a*) the entropy change of helium, (*b*) the entropy change of the surroundings, and (*c*) whether this process is reversible, irreversible, or impossible. Assume the surroundings are at 20°C.

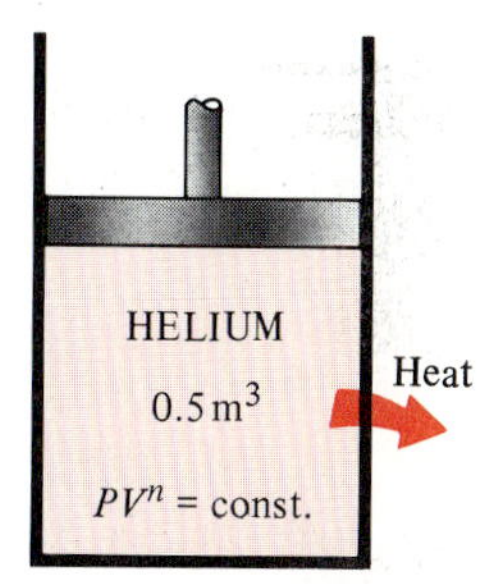

FIGURE P6-87

6-87E A piston-cylinder device initially contains 15 ft^3 of helium gas at 25 psia and 70°F. Helium is now compressed in a polytropic process (PV^n = constant) to 70 psia and 300°F. Determine (*a*) the entropy change of helium, (*b*) the entropy change of the surroundings, and (*c*) whether this process is reversible, irreversible, or impossible. Assume the surroundings are at 70°F.
Answers: (*a*) −0.016 Btu/R, (*b*) 0.019 Btu/R, (*c*) irreversible

6-88 A piston-cylinder device contains 0.8 kg of nitrogen gas at 100 kPa and 27°C. The gas is now compressed slowly in a polytropic process during which $PV^{1.3}$ = constant. The process ends when the volume is reduced by one-half. Determine the entropy change of nitrogen during this process.
Answer: 0.331 kJ/K

6-89 A mass of 2 kg of helium undergoes a process from an initial state of 3 m^3/kg and 15°C to a final state of 0.5 m^3/kg and 80°C. Determine the entropy change of helium during this process, assuming (*a*) the process is reversible and (*b*) the process is irreversible.

6-89E A mass of 5 lbm of helium undergoes a process from an initial state of 50 ft^3/lbm and 60°F to a final state of 10 ft^3/lbm and 140°F. Determine the entropy change of helium during this process, assuming (*a*) the process is reversible and (*b*) the process is irreversible.

6-90 Air is compressed in a piston-cylinder device from 100 kPa and 27°C to 250 kPa in a reversible isothermal process. Determine (*a*) the entropy change of air and (*b*) the work done.

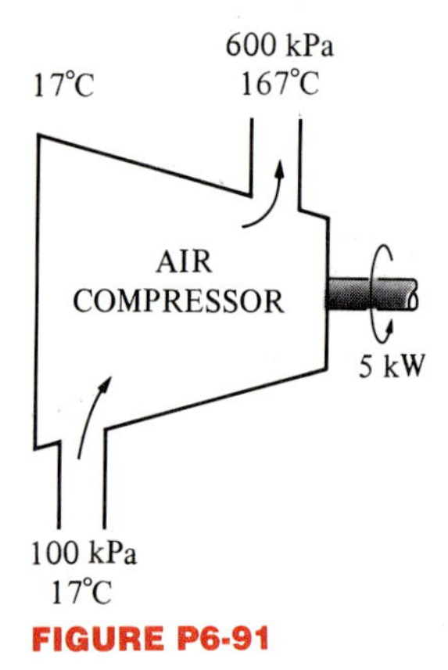

FIGURE P6-91

6-91 Air is compressed steadily by a 5-kW compressor from 100 kPa and 17°C to 600 kPa and 167°C at a rate of 1.6 kg/min. During this process, some heat transfer takes place between the compressor and the surrounding medium at 17°C. Determine (*a*) the rate of entropy change of air, (*b*) the rate of entropy change of the surrounding medium, and (*c*) the rate of entropy generation during this process.
Answers: (*a*) −0.0025 kW/K, (*b*) 0.00331 kW/K, (*c*) 0.00081 kW/K

6-91E Air is compressed steadily by a 25-hp compressor from 15 psia and 60°F to 90 psia and 340°F at a rate of 3.5 lbm/min. During the process, some heat transfer takes place between the compressor and the surrounding medium at 60°F. Determine (*a*) the rate of entropy change of air, (*b*) the rate of entropy change of the surrounding medium, and (*c*) the rate of entropy generation during this process.

6-92 An insulated rigid tank is divided into two equal parts by a partition. Initially, one part contains 2 kmol of an ideal gas at 500 kPa and 70°C, and the other side is evacuated. The partition is now removed, and the gas fills the entire tank. Determine the entropy change during this process. *Answer:* 11.526 kJ/K

6-93 Air is compressed in a piston-cylinder device from 100 kPa and 17°C to 800 kPa in a reversible, adiabatic process. Determine the final temperature and the work done during this process, assuming (*a*) constant specific heats and (*b*) variable specific heats for air.
Answers: (*a*) 525.3 K, 171.1 kJ/kg; (*b*) 522.4 K, 169.3 kJ/kg

6-94 Helium gas is compressed from 100 kPa and 30°C to 500 kPa in a reversible, adiabatic process. Determine the final temperature and the work done, assuming the process takes place (*a*) in a piston-cylinder device and (*b*) in a steady-flow compressor.

6-94E Helium gas is compressed from 15 psia and 90°F to 75 psia in a reversible, adiabatic process. Determine the final temperature and the work done, assuming the process takes place (*a*) in a piston-cylinder device and (*b*) in a steady-flow compressor.

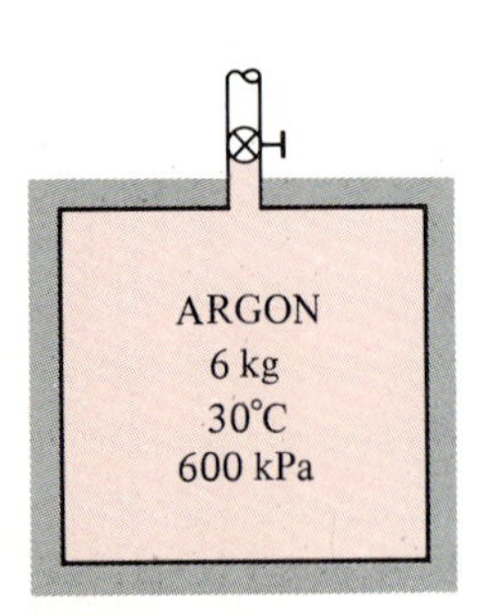

FIGURE P6-95

6-95 An insulated, rigid tank contains 6 kg of argon gas at 600 kPa and 30°C. A valve is now opened, and argon is allowed to escape until the pressure inside drops to 200 kPa. Assuming the argon remaining inside the tank has undergone a reversible, adiabatic process, determine the final mass in the tank. *Answer:* 3.11 kg

6-95E An insulated, rigid tank contains 10 lbm of argon gas at 90 psia and 80°F. A valve is now opened, and argon is allowed to escape until the pressure inside drops to 30 psia. Assuming the argon remaining inside the tank has undergone a reversible, adiabatic process, determine the final mass in the tank.

6-96 Air is compressed steadily by a compressor from 100 kPa and 17°C to 700 kPa at a rate of 2 kg/min. Determine the minimum power input required if the process is (*a*) adiabatic and (*b*) isothermal. Assume air to be an ideal gas with variable specific heats, and neglect the changes in kinetic and potential energies. *Answers:* (*a*) 7.21 kW, (*b*) 5.4 kW

6-97 Air enters an adiabatic nozzle at 400 kPa, 247°C, and 60 m/s and exits at 80 kPa. Assuming air to be an ideal gas with variable specific heats and disregarding any irreversibilities, determine the exit velocity of air.

6-97E Air enters an adiabatic nozzle at 60 psia, 540°F, and 200 ft/s and exits at 12 psia. Assuming air to be an ideal gas with variable specific heats and disregarding any irreversibilities, determine the exit velocity of air.

6-98 Air enters a nozzle steadily at 300 kPa and 77°C with a velocity of 50 m/s and exits at 100 kPa and 320 m/s. The heat losses from the nozzle to the surrounding medium at 20°C are estimated to be 3.2 kJ/kg. Determine (*a*) the exit temperature and (*b*) the total entropy change for this process.

6-99 Air enters a compressor at ambient conditions of 96 kPa and 17°C with a low velocity and exits at 1 MPa, 327°C, and 120 m/s. The compressor is cooled by the ambient air at 17°C at a rate of 1500 kJ/min. The power input to the compressor is 300 kW. Determine (*a*) the mass flow rate of air and (*b*) the rate of entropy generation.
Answers: (*a*) 0.848 kg/s, (*b*) 0.144 kW/K

6-99E Air enters a compressor at ambient conditions of 15 psia and 60°F with a low velocity and exits at 150 psia, 620°F, and 350 ft/s. The compressor is cooled by the ambient air at 60°F at a rate of 1500 Btu/min. The power input to the compressor is 400 hp. Determine (*a*) the mass flow rate of air and (*b*) the rate of entropy generation.

6-100 Air enters the evaporator section of a window air conditioner at 100 kPa and 27°C with a volume flow rate of 6 m^3/min. The refrigerant-12 at 120 kPa with a quality of 0.3 enters the evaporator at a rate of 2 kg/min and leaves as saturated vapor at the same pressure. Determine the exit temperature of the air and the rate of entropy generation for this process, assuming (*a*) the outer surfaces of the air conditioner are insulated and (*b*) heat is transferrred to the evaporator of the air conditioner from the surrounding medium at 32°C at a rate of 30 kJ/min.
Answers: (*a*) −5.7°C, 0.00194 kW/K, (*b*) −1.4°C, 0.0008 kW/K

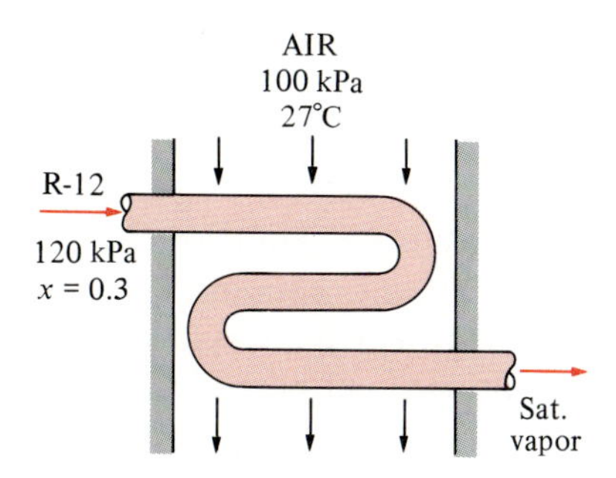

FIGURE P6-100

6-101 An insulated 2-m^3 rigid tank contains air at 500 kPa and 57°C. A valve connected to the tank is now opened, and air is allowed to escape until the pressure inside drops to 200 kPa. The air temperature during this process is maintained constant by an electric resistance heater placed in the tank. Determine (*a*) the electrical work done during this process and (*b*) the total entropy change for this process.
Answers: (*a*) −600.2 kJ, (*b*) 1.76 kJ/K

Reversible Steady-Flow Work

6-102C In large compressors, the gas is frequently cooled while being compressed to reduce the power consumed by the compressor. Explain how cooling the gas during a compression process reduces the power consumption.

6-103C The turbines in steam power plants operate essentially under adiabatic conditions. A plant engineer suggests to end this practice. She proposes to run cooling water through the outer surface of the casing to cool the steam as it flows through the turbine. This way, she reasons, the entropy of the steam will decrease, the performance of the turbine will improve, and as a result the work output of the turbine will increase. How would you evaluate this proposal?

6-104C It is well known that the power consumed by a compressor can be reduced by cooling the gas during compression. Inspired by this, somebody proposes to cool the liquid as it flows through a pump, in order to reduce the power consumption of the pump. Would you support this proposal? Explain.

6-105 Water enters the pump of a steam power plant as saturated liquid at 20 kPa at a rate of 50 kg/s and exits at 8 MPa. Neglecting the changes in kinetic and potential energies and assuming the process to be reversible, determine the power input to the pump.

6-105E Water enters the pump of a steam power plant as saturated liquid at 5 psia at a rate of 100 lbm/s and exits at 1000 psia. Neglecting the changes in kinetic and potential energies and assuming the process to be eversible, determine the power input to the pump.
Answer: 427.4 hp

6-106 Liquid water enters a 10-kW pump at 100-kPa pressure at a rate of 5 kg/s. Determine the highest pressure the liquid water can have at the exit of the pump. Neglect the kinetic and potential energy changes of water, and assume the specific volume of water to be 0.001 m^3/kg.
Answer: 2100 kPa

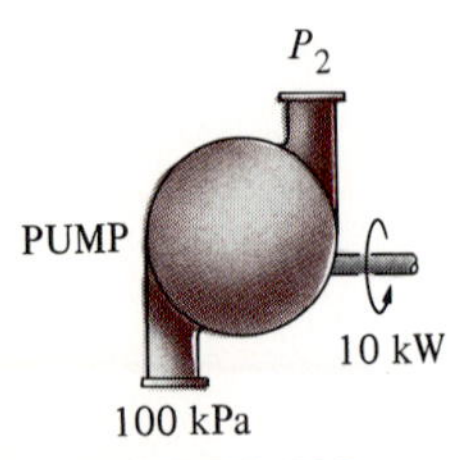

FIGURE P6-106

6-107 Saturated refrigerant-12 vapor at 120 kPa is compressed reversibly in an adiabatic compressor to 800 kPa. Determine the work input to the compressor. What would your answer be if the refrigerant were condensed first at constant pressure before it was compressed?
Answers: 33.49 kJ/kg, 0.50 kJ/kg

6-107E Saturated refrigerant-12 vapor at 20 psia is compressed reversibly in an adiabatic compressor to 120 psia. Determine the work input to the compressor. What would your answer be if the refrigerant were condensed first at constant pressure before it was compressed?

6-108 Consider a steam power plant that operates between the pressure levels of 10 MPa and 20 kPa. Steam enters the pump as saturated liquid and leaves the turbine as saturated vapor. Determine the ratio of the work delivered by the turbine to the work consumed by the pump. Assume the entire cycle to be reversible and the heat losses from the pump and the turbine to be negligible.

6-109 Liquid water at 100 kPa enters a 10-kW pump where its pressure is raised to 5 MPa. If the elevation difference between the exit and the inlet levels is 10 m, determine the highest mass flow rate of liquid water this pump can handle. Neglect the kinetic energy change of water, and assume the specific volume of water to be 0.001 m^3/kg.

6-110 Air enters a two-stage compressor at 100 kPa and 27°C and is compressed to 900 kPa. The pressure ratio across each stage is the same, and the air is cooled to the initial temperature between the two stages. Assuming the compression process to be isentropic, determine the power input to the compressor for a mass flow rate of 0.02 kg/s. What would your answer be if only one stage of compression were used?
Answers: 4.44 kW, 5.26 kW

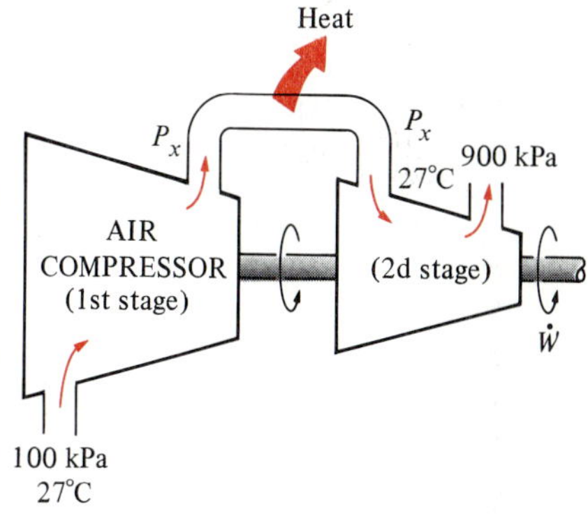

FIGURE P6-110

6-111 Helium gas is compressed from 80 kPa and 20°C to 600 kPa at a rate of 0.2 m^3/s. Determine the power input to the compressor, assuming the compression process to be (*a*) isentropic, (*b*) polytropic with $n = 1.2$, (*c*) isothermal, and (*d*) ideal two-stage polytropic with $n = 1.2$.

6-111E Helium gas is compressed from 14 psia and 70°F to 120 psia at a rate of 5 ft^3/s. Determine the power input to the compressor, assuming the compression process to be (*a*) isentropic, (*b*) polytropic with $n = 1.2$, (*c*) isothermal, and (*d*) ideal two-stage polytropic with $n = 1.2$.

6-112 Nitrogen gas is compressed from 100 kPa and 27°C to 600 kPa by a 5-kW compressor. Determine the mass flow rate of nitrogen through the compressor, assuming the compression process to be (*a*) isentropic, (*b*) polytropic with $n = 1.3$, (*c*) isothermal, and (*d*) ideal two-stage polytropic with $n = 1.3$.
Answers: (*a*) 0.024 kg/s, (*b*) 0.025 kg/s, (*c*) 0.031 kg/s, (*d*) 0.028 kg/s

6-113 Consider a three-stage isentropic compressor with two intercoolers which cool the gas to the initial temperature between the stages. Determine the two intermediate pressures (P_x and P_y) in terms of inlet and exit pressures (P_1 and P_2) that will minimize the work input to the compressor. *Answers:* $P_x = (P_1^2 P_2)^{1/3}$, $P_y = (P_1 P_2^2)^{1/3}$

6-114 Write a computer program to determine the work input to a multistage compressor for a given set of inlet and exit pressures for any number of stages. Assume that the pressure ratio across each stage is identical and the compression process is polytropic. List and plot the compressor work against the number of stages for P_1 = 100 kPa, T_1 = 17°C, P_2 = 800 kPa, and n = 1.3 for air. Based on this chart, can you justify using compressors with many stages?

Adiabatic Efficiencies of Steady-Flow Devices

6-115C Describe the ideal process for an (*a*) adiabatic turbine, (*b*) adiabatic compressor, and (*c*) adiabatic nozzle, and define the adiabatic efficiency for each device.

6-116C Is the isentropic process a suitable model for compressors that are cooled intentionally? Explain.

6-117C On a *T-s* diagram, does the actual exit state (state 2) of an adiabatic turbine have to be on the right-hand side of the isentropic exit state (state 2*s*)? Why?

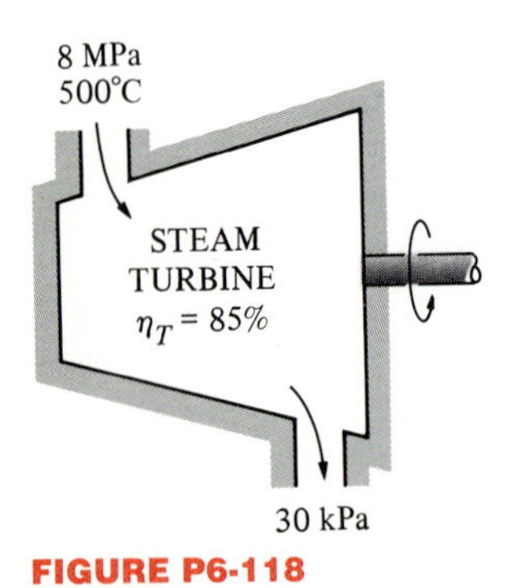

FIGURE P6-118

6-118 Steam enters an adiabatic turbine at 8 MPa and 500°C with a mass flow rate of 5 kg/s and leaves at 30 kPa. The adiabatic efficiency of the turbine is 0.85. Neglecting the kinetic energy change of the steam, determine (*a*) the temperature at the turbine exit and (*b*) the power output of the turbine. *Answers:* (*a*) 69.1°C, (*b*) 4804 kW

6-118E Steam enters an adiabatic turbine at 1000 psia and 900°F with a mass flow rate of 9 lbm/s and leaves at 5 psia. The adiabatic efficiency of the turbine is 0.85. Neglecting the kinetic energy change of the steam, determine (*a*) the temperature of the steam at the turbine exit and (*b*) the power output of the turbine.

6-119 Steam enters an adiabatic turbine at 6 MPa, 600°C, and 80 m/s and leaves at 50 kPa, 100°C, and 140 m/s. If the power output of the turbine is 5 MW, determine (*a*) the mass flow rate of the steam flowing through the turbine and (*b*) the adiabatic efficiency of the turbine. *Answers:* (*a*) 5.16 kg/s, (*b*) 83.7 percent

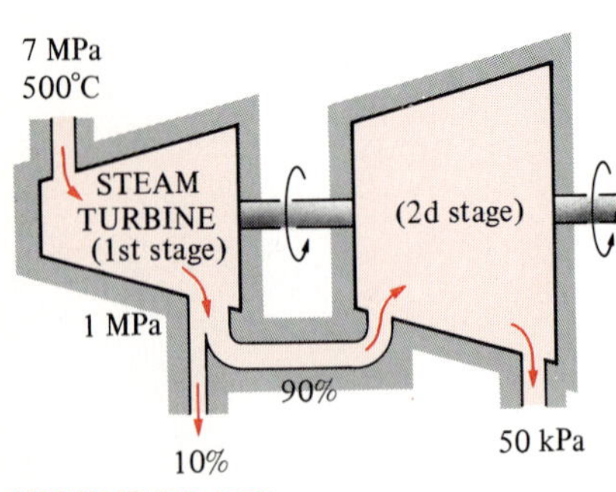

FIGURE P6-120

6-120 Steam at 7 MPa and 500°C enters a two-stage adiabatic turbine at a rate of 15 kg/s. Ten percent of the steam is extracted at the end of the first stage at a pressure of 1 MPa for other use. The remainder of the steam is further expanded in the second stage and leaves the turbine at 50 kPa. Determine the power output of the turbine, assuming (*a*) the process is reversible (*b*) the turbine has an adiabatic efficiency of 88 percent. *Answers:* (*a*) 14,928 kW, (*b*) 13,136 kW

6-121 Steam enters a two-stage adiabatic turbine at 8 MPa and 500°C. It expands in the first stage to a pressure of 2 MPa. Then steam is reheated at constant pressure to 500°C before it is expanded in a second stage to a pressure of 100 kPa. The work output of the turbine is 10 MW. Assuming an adiabatic efficiency of 90 percent for each stage of the turbine, determine the required mass flow rate of steam. Also show the process on a *T-s* diagram with respect to saturation lines. *Answer:* 9.57 kg/s

6-122 Argon gas enters an adiabatic turbine at 800°C and 1.5 MPa at a rate of 20 kg/min and exhausts at 200 kPa. If the power output of the turbine is 100 kW, determine the adiabatic efficiency of the turbine.

6-123 Combustion gases enter an adiabatic gas turbine at 1200 K, 800 kPa, and leave at 400 kPa with a low velocity. Treating the combustion gases as air and assuming an adiabatic efficiency of 86 percent, determine the work output of the turbine. *Answer:* 188.8 kJ/kg

6-123E Combustion gases enter an adiabatic gas turbine at 1540°F, 120 psia, and leave at 60 psia with a low velocity. Treating the combustion gases as air and assuming an adiabatic efficiency of 86 percent, determine the work output of the turbine. *Answer:* 75.6 Btu/lbm

6-124 Refrigerant-12 enters an adiabatic compressor as saturated vapor at 120 kPa at a rate of 2 m^3/min and exits at 1-MPa pressure. If the adiabatic efficiency of the compressor is 80 percent, determine (*a*) the temperature of the refrigerant at the exit of the compressor and (*b*) the power input. Also show the process on a *T-s* diagram with respect to saturation lines. *Answers:* (*a*) 67.1°C, (*b*) 11.6 kW

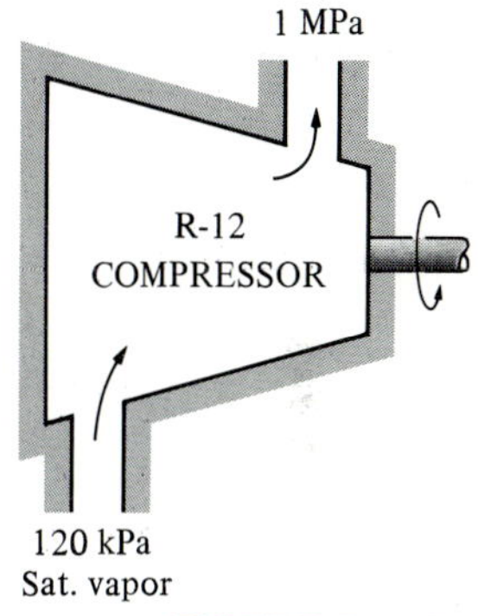

FIGURE P6-124

6-124E Refrigerant-12 enters an adiabatic compressor as saturated vapor at 20 psia at a rate of 50 ft^3/min and exits at 100-psia pressure. If the adiabatic efficiency of the compressor is 80 percent, determine (*a*) the temperature of the refrigerant at the exit of the compressor and (*b*) the power input. Also show the process on a *T-s* diagram with respect to saturation lines.

6-125 Refrigerant-12 at 140 kPa and −20°C is compressed by an adiabatic 0.5-kW compressor to an exit state of 700 kPa and 60°C. Neglecting the changes in kinetic and potential energies, determine (*a*) the adiabatic efficiency of the compressor, (*b*) the volume flow rate of the refrigerant at the compressor inlet, in L/min, and (*c*) the maximum volume flow rate at the inlet conditions that this adiabatic 0.5-kW compressor can handle without violating the second law.
Answers: (*a*) 66.3 percent, (*b*) 82 L/min, (*c*) 124 L/min

6-126 Air enters an adiabatic compressor at 100 kPa and 17°C at a rate of 2 m^3/s, and it exits at 257°C. The compressor has an adiabatic efficiency of 84 percent. Neglecting the changes in kinetic and potential energies, determine (*a*) the exit pressure of air and (*b*) the power required to drive the compressor.

6-126E Air enters an adiabatic compressor at 15 psia and 60°F at a rate of 20 ft^3/s, and it exits at 500°F. The compressor has an adiabatic efficiency of 84 percent. Neglecting the changes in kinetic and potential energies, determine (*a*) the exit pressure of air and (*b*) the power required to drive the compressor.

6-127 Air is compressed by an adiabatic compressor from 95 kPa and 27°C to 600 kPa and 277°C. Assuming variable specific heats and neglecting the changes in kinetic and potential energies, determine (*a*) the adiabatic efficiency of the compressor and (*b*) the exit temperature of air if the process were reversible. *Answers:* (*a*) 81.9 percent, (*b*) 505.5 K

6-128 Argon gas enters an adiabatic compressor at 120 kPa and 30°C with a velocity of 20 m/s, and it exits at 1.2 MPa and 80 m/s. If the adiabatic efficiency of the compressor is 80 percent, determine (*a*) the exit temperature of the argon and (*b*) the work input to the compressor.

6-128E Argon gas enters an adiabatic compressor at 20 psia and 90°F with a velocity of 60 ft/s, and it exits at 200 psia and 240 ft/s. If the adiabatic efficiency of the compressor is 80 percent, determine (*a*) the exit temperature of the argon and (*b*) the work input to the compressor.

6-129 Carbon dioxide enters an adiabatic compressor at 100 kPa and 300 K at a rate of 0.5 kg/s and exits at 600 kPa and 450 K. Neglecting the kinetic energy changes, determine (*a*) the adiabatic efficiency of the compressor and (*b*) the rate of entropy generation during this process.

6-130 Air enters a nozzle at 400 kPa and 547°C with low velocity and exits with a velocity of 290 m/s. If the adiabatic efficiency of the nozzle is 90 percent, determine the exit temperature and pressure of the air.

6-130E Air enters a nozzle at 60 psia and 1020°F with low velocity and exits at 800 ft/s. If the adiabatic efficiency of the nozzle is 90 percent, determine the exit temperature and pressure of the air.

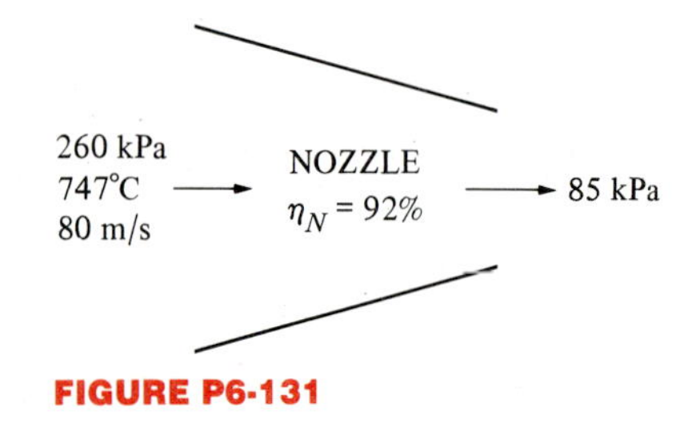

FIGURE P6-131

6-131 Hot combustion gases enter the nozzle of a turbojet engine at 260 kPa, 747°C, and 80 m/s, and they exit at a pressure of 85 kPa. Assuming an adiabatic efficiency of 92 percent and treating the combustion gases as air, determine (*a*) the exit velocity and (*b*) the exit temperature. *Answers:* (*a*) 828.2 m/s, (*b*) 786.3 K

6-132 Helium gas enters a nozzle whose adiabatic efficiency is 94 percent with a low velocity and exits at 95 kPa, 80°C, and 370 m/s. Determine the pressure and temperature at the nozzle inlet.
Answers: 104.7 kPa, 366.2 K

6-132E Helium gas enters a nozzle whose adiabatic efficiency is 94 percent with a low velocity, and it exits at 14 psia, 180°F, and 1000 ft/s. Determine the pressure and temperature at the nozzle inlet.

6-133 Steam enters an adiabatic nozzle at 3 MPa and 400°C with a velocity of 70 m/s and exits at 2 MPa and 320 m/s. If the nozzle has an inlet area of 5 cm^2, determine (*a*) the exit temperature and (*b*) the rate of entropy generation for this process. *Answers:* (*a*) 370.4°C, (*b*) 0.0369 kW/K

Second-Law Analysis of Engineering Systems

CHAPTER

7

The increased awareness that the world's energy resources are limited has caused some governments to reexamine their energy policies and take drastic measures in eliminating waste. It has also sparked interest in the scientific community to take a closer look at the energy conversion devices and to develop new techniques to better utilize the existing limited resources. The first law of thermodynamics deals with the *quantity* of energy and asserts that energy cannot be created or destroyed. This law merely serves as a necessary tool for bookkeeping of energy during a process and offers no challenges to the engineer. The second law, however, deals with the *quality* of energy. More specifically, it is concerned with the degradation of energy during a process, the entropy generation, the lost opportunities to do work; and it offers plenty of space for improvement.

The second law of thermodynamics has proved to be a very powerful tool in the optimization of complex thermodynamic systems. In this chapter, we examine the performance of engineering devices in light of the second law of thermodynamics. We start our discussions with the introduction of *availability,* which is the maximum useful work that could be obtained from a system at a given state, and we continue with the *reversible work,* which is the maximum useful work that can be obtained as a system undergoes a process between two specified states. Next we discuss *irreversibility,* which is the lost work potential during a process as a result of irreversibilities, and we define a *second-law efficiency*. Later in this chapter, these concepts are applied to closed systems and control volumes.

7-1 ■ AVAILABILITY—Maximum Work Potential

When a new energy source, such as a geothermal well, is discovered, the first thing the explorers do is estimate the amount of energy contained in the source. This information alone, however, is of little value in deciding whether to build a power plant on that site. What we really need to know is the *work potential* of the source—that is, the amount of energy that we can extract as useful work and, say, run a generator. The rest of the energy, i.e., the portion that is not available for converting to work, will eventually be discarded as waste energy and is not worthy of our consideration. Thus it would be very desirable to have a property to enable us to determine the useful work potential of a given amount of energy at some specified state. This property is *availability*.

The work potential of the energy contained in a system at a specified state is simply the maximum useful work that can be obtained from the system. You will recall that the work done during a process depends on the initial state, the final state, and the process path. That is,

$$\text{Work} = f(\text{initial state, process path, final state})$$

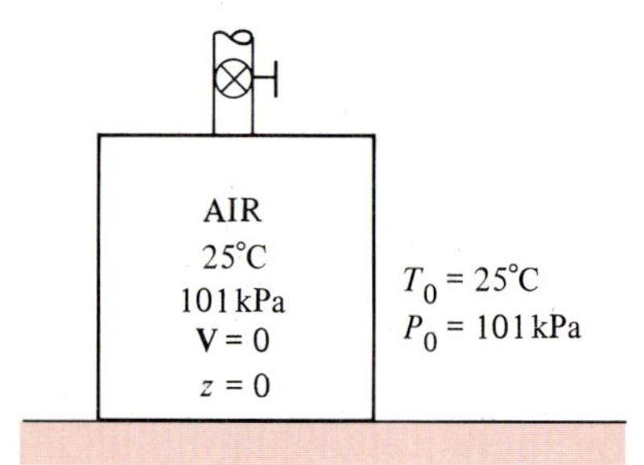

FIGURE 7-1

A system which is in equilibrium with its surroundings is said to be at the dead state.

In an availability analysis, the initial state is specified, and thus it is not a variable. As shown in Sec. 6-10, the work output is maximized when the process between two specified states is executed in a reversible manner. Therefore, all the irreversibilities are disregarded in determining the work potential. Finally, the system must be in the *dead state* at the end of the process to maximize the work output.

A system is said to be in the **dead state** when it is in thermodynamic equilibrium with its surroundings (Fig. 7-1). At the dead state, a system is at the temperature and pressure of its surroundings (in thermal and mechanical equilibrium); it has no kinetic or potential energy relative to its surroundings (zero velocity and zero elevation above a reference level), and it does not react with the surroundings (chemically inert). Also there are no unbalanced magnetic, electrical, and surface tension effects between the system and its surroundings, if these are relevant to the situation at hand. The properties of a system at the dead state are denoted by subscript zero, for example, P_0, T_0, h_0, u_0, and s_0. Unless specified otherwise, the dead-state temperature and pressure are assumed to be T_0 = 25°C (77°F) and P_0 = 1 atm (101.325 kPa or 14.7 psia). A system has zero availability at the dead state (Fig. 7-2).

FIGURE 7-2

At the dead state, the useful work potential (availability) of a system is zero.

The notion that a system must go to the dead state at the end of the process to maximize the work output can be explained as follows: If the system temperature at the final state is greater than (or less than) the temperature of the environment it is in, we can always produce additional work by running a heat engine between these two temperature levels. If the final pressure is greater than (or less than) the pressure of the environment, we can still obtain work by letting the system expand to the pressure of the environment. If the final velocity of the system is not zero, we can catch that extra kinetic energy by a turbine and convert it to rotating shaft work, and so on. No work can be produced from a system that is initially at the dead state. The atmosphere around us contains a tremendous amount of energy. However, the atmosphere is in the dead state, and the energy it contains has no work potential (Fig. 7-3).

Therefore, we conclude that *a system will deliver the maximum possible work as it undergoes a reversible process from the specified initial state to the state of its environment, i.e., the dead state* (Fig. 7-4). This represents the *useful work potential* of the system at the specified state and is called **availability**. It is important to realize that availability does not represent the amount of work that a work-producing device will actually deliver upon installation. Rather, it represents the *upper limit on the amount of work a device can deliver without violating any thermodynamic laws*. There will always be a difference, large or small, between availability and the actual work delivered by a device. This difference represents the room engineers have for improvement.

Note that the availability of a system at a specified state depends on the conditions of the environment (the dead state) as well as the properties of the system. Therefore, availability is a property of the system-surroundings combination and not of the system alone. Altering the environment is another way of increasing availability, but it is definitely not an easy alternative.

The term *availability* was made popular in the United States by the M.I.T. School of Engineering in the 1940s. Today, an equivalent term, *exergy,* introduced in Europe in the 1950s, is finding global acceptance partly because it can be adapted without requiring translation. In this text, these two terms are used interchangeably. The reader should be aware that some authors define them slightly differently.

FIGURE 7-3

The atmosphere contains a tremendous amount of energy, but zero availability.

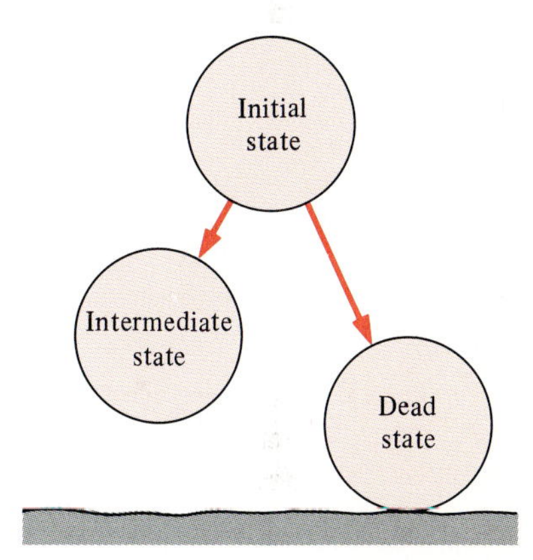

FIGURE 7-4

A system will deliver the most work as it undergoes a reversible process to the dead state.

EXAMPLE 7-1

A windmill with a 12-m-diameter rotor, as shown in Fig. 7-5, is to be installed at a location where the wind is blowing steadily at an average velocity of 10 m/s. Determine the available power for the windmill.

Solution The air flowing with the wind has the same properties as the stagnant atmospheric air except that it possesses a velocity and thus some kinetic energy. This air will reach the dead state when it is brought to a complete stop. Therefore,

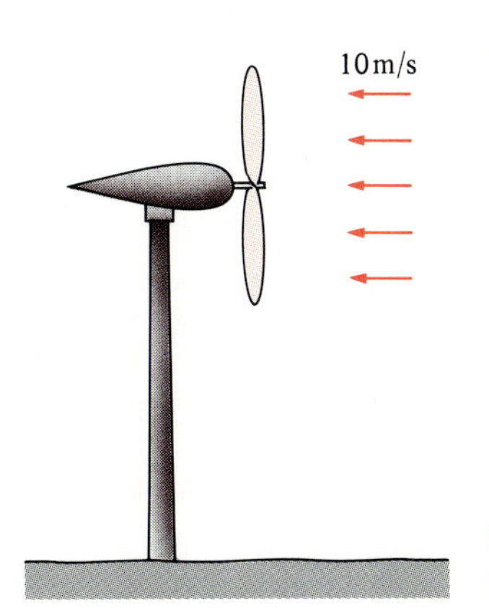

FIGURE 7-5

Schematic for Example 7-1.

the availability of the blowing air is simply the kinetic energy it possesses:

$$\text{Availability} = \text{ke}_1 = \frac{\mathbf{V}_1^2}{2} = \frac{(10\text{ m/s})^2}{2}\left(\frac{1\text{ kJ/kg}}{1000\text{ m}^2/\text{s}^2}\right) = 0.05\text{ kJ/kg}$$

That is, every unit mass of air flowing at a velocity of 10 m/s has a work potential of 0.05 kJ/kg. In other words, a perfect windmill will bring the air to a complete stop and capture that 0.05 kJ/kg of work potential. To determine the available power, we need to know the amount of air passing through the rotor of the windmill per unit time, i.e., the mass flow rate. Assuming standard atmospheric conditions (25°C, 101 kPa), the density of air is 1.18 kg/m^3, and its mass flow rate is

$$\dot{m} = \rho A \mathbf{V}_1 = \rho \frac{\pi D^2}{4} \mathbf{V}_1 = (1.18\text{ kg/m}^3)\left(\frac{3.14}{4}\right)(12\text{ m})^2(10\text{ m/s})$$
$$= 1334\text{ kg/s}$$

Thus,

$$\text{Available power} = \dot{m}(\text{ke}_1) = (1334\text{ kg/s})(0.05\text{ kJ/kg}) = 66.7\text{ kW}$$

This is the maximum power available to the windmill. Assuming a conversion efficiency of 25 percent, an actual windmill will convert 16.7 kW to electricity. Notice that the work potential for this case is equal to the entire kinetic energy of the air. This is because kinetic energy itself is a form of "mechanical" energy.

EXAMPLE 7-2

Consider a large furnace that can supply energy in the form of heat at 2000 R at a steady rate of 3000 Btu/s. Determine the availability of this energy. Assume an environment temperature of 77°F.

Solution The furnace in this example can be modeled as a thermal energy reservoir that supplies heat indefinitely at a constant temperature. The availability of this heat energy is its useful work potential, i.e., the maximum possible amount of work that can be extracted from it. This corresponds to the amount of work that a reversible heat engine operating between the furnace and the environment can produce. First, let us determine the thermal efficiency of this reversible heat engine:

$$\eta_{\text{th,max}} = \eta_{\text{th,rev}} = 1 - \frac{T_0}{T_H} = 1 - \frac{537\text{ R}}{2000\text{ R}} = 0.732 \quad \text{or } 73.2\%$$

That is, a heat engine can convert, at best, 73.2 percent of the heat received from this furnace to work. In other words, only 73.2 percent of this heat energy is available for producing work. Thus the availability of this furnace, in rate form, is equivalent to the power produced by the reversible heat engine:

$$\dot{W}_{\text{max}} = \dot{W}_{\text{rev}} = \eta_{\text{th,rev}}\dot{Q}_{\text{in}} = (0.732)(3000\text{ Btu/s}) = 2196\text{ Btu/s}$$

Notice that 26.8 percent of the energy transferred from the furnace as heat is not available for doing work. The portion of energy which cannot be converted to work is called **unavailable energy** (Fig. 7-6). Unavailable energy is simply the difference between the total energy of a system at a specified state and the availability of that energy.

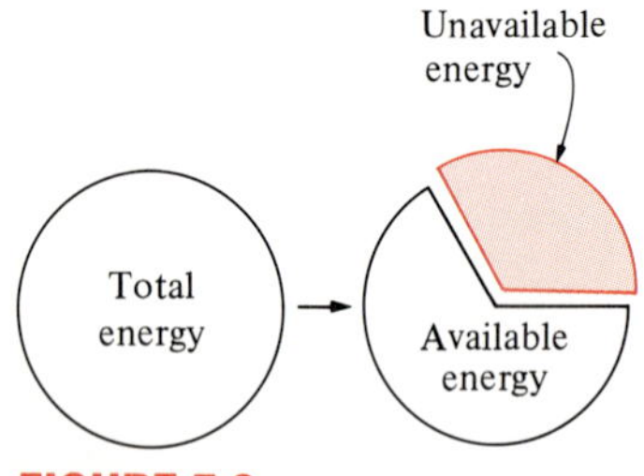

FIGURE 7-6
Unavailable energy is the portion of energy which cannot be converted to work by even a reversible heat engine.

Reversible Work (Work Potential) and Irreversibility (Wasted Work Potential)

The availability analysis described above is useful for determining the useful work potential of the energy of a system at a specified state. It serves as a valuable tool in determining the quality of energy and comparing the work potentials of different energy sources. The availability analysis alone, however, is not of great value for studying engineering devices operating between two fixed states. This is because in an availability analysis, the final state is always assumed to be the *dead state,* which is hardly ever the case for actual engineering systems.

The adiabatic efficiencies discussed in Chap. 6 are also of limited use because the exit state of the model (isentropic) process is not the same as the actual exit state. In this section, we describe two new quantities that are related to the exact initial and final states of the actual process. These two quantities are *reversible work* and *irreversibility,* and they serve as valuable tools in the optimization studies of components in complex thermodynamic systems. But first we examine the **surroundings work**, the work done by or against the surroundings during a process.

The work done by work-producing devices is not always entirely in a usable form. For example, when a gas in a piston-cylinder device is expanding, part of the work done by the gas is used to push the atmospheric air out of the way of the piston (Fig. 7-7). This work, which cannot be recovered and utilized for any useful purpose, is equal to the atmospheric pressure P_0 times the volume change of the system $V_2 - V_1$:

$$W_{\text{surr}} = P_0(V_2 - V_1) = mP_0(v_2 - v_1) \qquad \text{(kJ)} \tag{7-1}$$

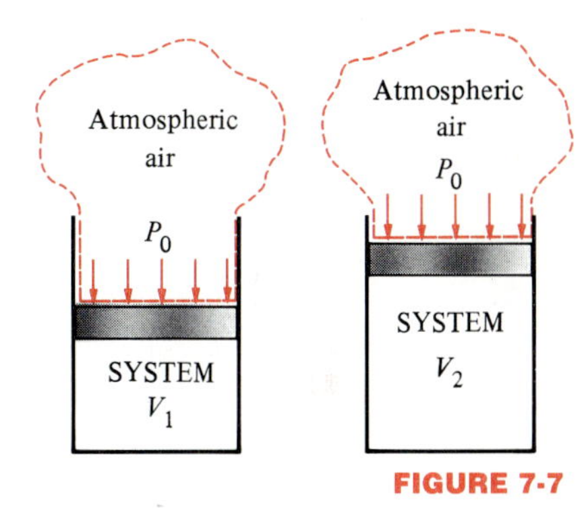

FIGURE 7-7

As a closed system expands, some work needs to be done to push the atmospheric air out of the way (W_{surr}).

The difference between the actual work W and the surroundings work W_{surr} is called the **actual useful work**, or just **useful work** W_u:

$$W_u = W - W_{\text{surr}} \qquad \text{(kJ)} \tag{7-2}$$

When a system is expanding and doing work, part of the work done is used to overcome the atmospheric pressure, and thus W_{surr} represents a loss. When a system is compressed, however, the atmospheric pressure helps the compression process, and thus W_{surr} represents a gain.

Note that the work done by or against the atmospheric pressure has significance only for systems whose volume changes during the process (i.e., systems that involve moving boundary work). It has no significance for cyclic devices and systems whose boundaries remain fixed during a process such as rigid tanks and steady-flow devices (turbines, compressors, nozzles, heat exchangers, etc.), as shown in Fig. 7-8.

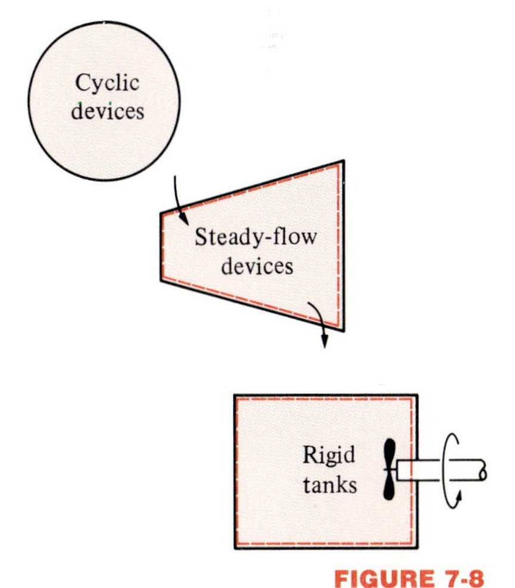

FIGURE 7-8

For constant-volume systems, the total actual and useful works are identical ($W_u = W$).

Reversible work W_{rev} is defined as *the maximum amount of useful work that can be obtained as a system undergoes a process between the specified initial and final states.* This is the useful work output (or input) obtained when the process between the initial and final states is executed in a totally reversible manner. That is, any heat transfer between the system and the surroundings must take place reversibly, and no irreversibilities should be present within the system during the process. When

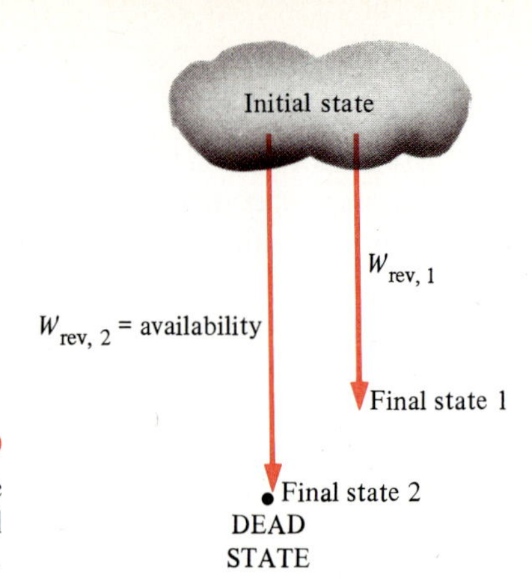

FIGURE 7-9
For a fixed initial state, the reversible work and availability are identical when the final state is the dead state.

the final state is the dead state, the reversible work equals availability (Fig. 7-9). For processes that require work, reversible work represents the minimum amount of work necessary to carry out that process. For convenience in presentation, the term *work* is used to denote both work and power throughout this chapter.

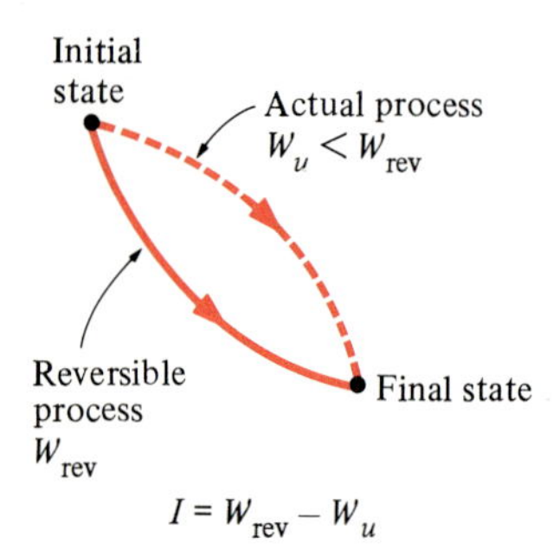

FIGURE 7-10
The difference between reversible work and actual useful work is the irreversibility.

Any difference between the reversible work W_{rev} and the useful work W_u is due to the irreversibilities present during the process, and this difference is called **irreversibility** I. It is expressed as (Fig. 7-10)

$$I = W_{\text{rev}} - W_u \qquad \text{(kJ)} \tag{7-3a}$$

or

$$i = w_{\text{rev}} - w_u \qquad \text{(kJ/kg)} \tag{7-3b}$$

Irreversibility generated during a process per unit time is called the *irreversibility rate* $\dot{I}$ and is given by

$$\dot{I} = \dot{W}_{\text{rev}} - \dot{W}_u \qquad \text{(kW)} \tag{7-3c}$$

For a totally reversible process, the actual and reversible work terms are identical, and thus irreversibility is zero. This is expected since totally reversible processes generate no entropy, which is a measure of irreversibilities occurring during a process. For all actual (irreversible) processes, irreversibility is a positive quantity since the work term is positive and $W_{\text{rev}} > W_u$ for work-producing devices, and the work term is negative and $|W_{\text{rev}}| < |W_u|$ for work-consuming devices.

Irreversibility can be viewed as the *lost opportunity* to do work. It represents the energy that could have been converted to work but was not. The smaller the irreversibility associated with a process, the greater the work that will be produced (or the smaller the work that will be consumed). To improve the performance of complex engineering systems, the primary sources of irreversibility associated with each component in the system should be located, and efforts should be made to minimize them.

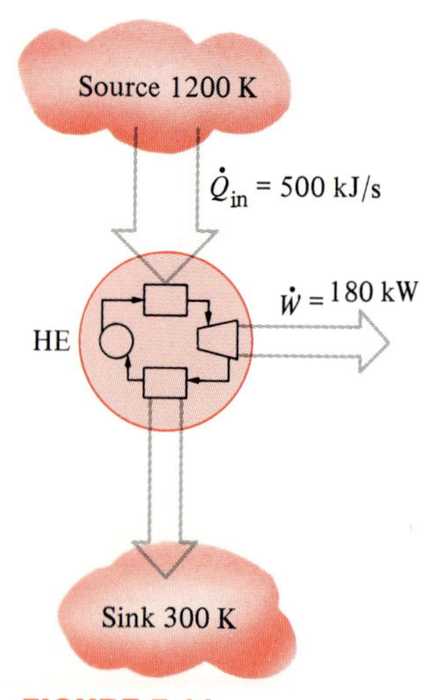

FIGURE 7-11
Schematic for Example 7-3.

EXAMPLE 7-3

A heat engine shown in Fig. 7-11 receives heat from a source at 1200 K at a rate of 500 kJ/s and rejects the waste heat to a medium at 300 K. The power output of the heat engine is 180 kW. Determine the reversible power and the irreversibility rate for this process.

Solution The reversible power for this process is the amount of power that a reversible heat engine, such as a Carnot heat engine, would produce when operating between the same temperature limits. It is determined by using the definition of thermal efficiency for a reversible heat-engine cycle:

$$\dot{W}_{\text{rev}} = \eta_{\text{th,rev}}\dot{Q}_{\text{in}} = \left(1 - \frac{T_{\text{sink}}}{T_{\text{source}}}\right)\dot{Q}_{\text{in}} = \left(1 - \frac{300\text{ K}}{1200\text{ K}}\right)(500\text{ kW}) = 375\text{ kW}$$

This is the maximum power that a heat engine operating between the specified temperature limits and receiving heat at the specified rate can produce. This would also represent the *available power* if 300 K were the lowest temperature available for heat rejection.

The irreversibility rate is the difference between the reversible power (maximum power that could have been produced) and the useful power output:

$$\dot{I} = \dot{W}_{\text{rev}} = \dot{W}_u = (375 - 180)\text{ kW} = 195\text{ kW}$$

That is, 195 kW of power potential is wasted during this process as a result of irreversibilities. Notice that the 500 − 375 = 125 kW of heat rejected to the sink is not available for converting to work and thus is not part of the irreversibility.

EXAMPLE 7-4

A 500-kg iron block shown in Fig. 7-12 is initially at 200°C and is allowed to cool to 27°C by transferring heat to the surrounding air at 27°C. Determine the reversible work and the irreversibility for this process.

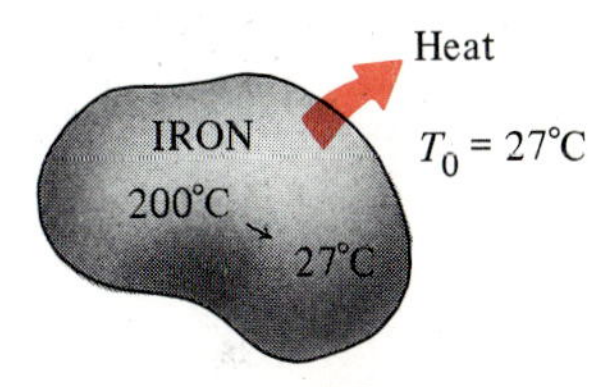

FIGURE 7-12

Schematic for Example 7-4.

Solution It probably came as a surprise to you that we are asking to find the "reversible work" for a process that does not involve any work interactions. Well, even if no attempt is made to produce work during this process, the potential to do work still exists, and the reversible work is a quantitative measure of this potential.

The reversible work in this case is determined by considering a series of imaginary reversible heat engines operating between the source (at a variable temperature T) and the sink (at a constant temperature T_0), as shown in Fig. 7-13, and summing their work output:

$$\delta W_{\text{rev}} = \eta_{\text{th,rev}}\,\delta Q_{\text{in}} = \left(1 - \frac{T_{\text{sink}}}{T_{\text{source}}}\right)\delta Q_{\text{in}} = \left(1 - \frac{T_0}{T}\right)\delta Q_{\text{in}}$$

and

$$W_{\text{rev}} = \int \left(1 - \frac{T_0}{T}\right)\delta Q_{\text{in}}$$

The source temperature T changes from T_i = 200°C = 473 K to T_0 = 27°C = 300 K during this process. The heat input to the heat engine is determined by applying the conservation of energy equation to the iron block. Recognizing that no work crosses the boundaries of the iron block and treating the specific heat of iron as a constant at the average value, we obtain

$$\delta Q - \cancelto{0}{\delta W} = dU = mC_{\text{av}}\,dT$$

$$\delta Q = mC_{\text{av}}\,dT$$

and

$$\delta Q_{\text{in}} = -\delta Q = -mC_{\text{av}}\,dT$$

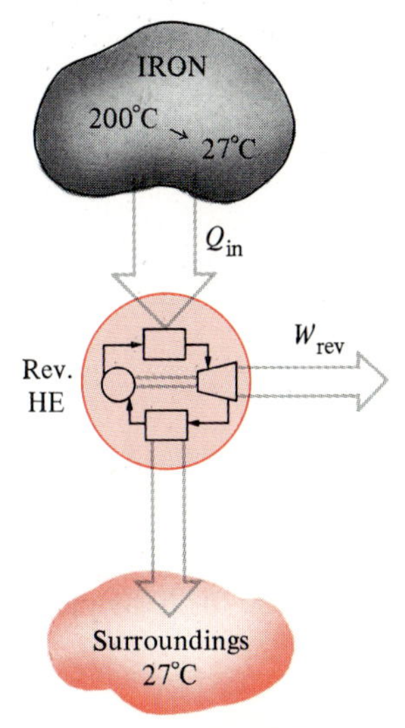

FIGURE 7-13

An irreversible heat transfer process can be made reversible by the use of a reversible heat engine.

since heat transfers from the iron and to the heat engine are equal in magnitude and opposite in sign. Substituting and performing the integration, we see the reversible work is

$$W_{rev} = \int_{T_i}^{T_0} \left(1 - \frac{T_0}{T}\right)(-mC_{av}\,dT) = mC_{av}(T_i - T_0) - mC_{av}T_0 \ln \frac{T_i}{T_0}$$

$$= (500\text{ kg})[0.45\text{ kJ/(kg·K)}]\left[(473 - 300)\text{ K} - (300\text{ K}) \ln \frac{473\text{ K}}{300\text{ K}}\right]$$

$$= 8191\text{ kJ}$$

where the specific heat value is obtained from Table A-3. The first term in the above equation $[Q = mC_{av}(T_i - T_0) = 38{,}925\text{ kJ}]$ is the total heat transfer from the iron block to the heat engines. The reversible work for this problem is found to be 8191 kJ, which means that 8191 (21 percent) of the 38,925 kJ of heat transferred from the iron block to the ambient air *could* have been converted to work. If the specified ambient temperature of 27°C is the lowest available environment temperature, the reversible work determined above also represents the availability, i.e., the maximum work potential of the energy contained in the iron block.

The irreversibility for this process is determined from its definition, Eq. 7-3*a*:

$$I = W_{rev} - W_u = (8191 - 0)\text{ kJ} = 8191\text{ kJ}$$

Notice that the reversible work (the work potential) and irreversibility (the wasted work potential) are the same for this case since the entire work potential is wasted. The source of irreversibility in this process is the heat transfer through a finite temperature difference.

EXAMPLE 7-5

The iron block discussed in Example 7-4 is to be used to maintain a house at 27°C when the outdoor temperature is 5°C. Determine the maximum amount of heat that can be supplied to the house as the iron cools to 27°C.

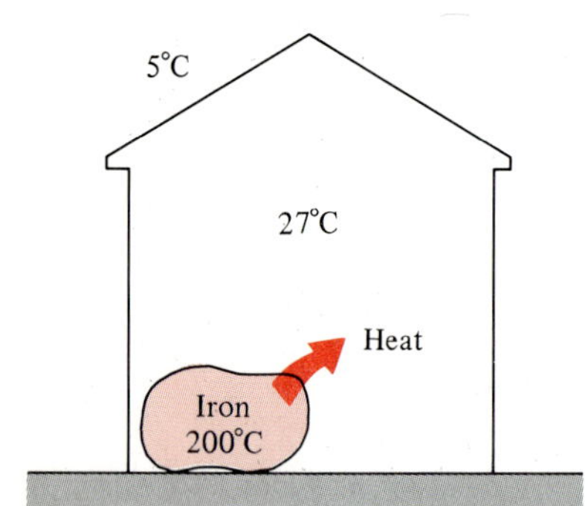

FIGURE 7-14
Schematic for Example 7-5.

Solution Probably the first thought that comes to mind to make the most use of the energy stored in the iron block is to take it inside and let it cool in the house, as shown in Fig. 7-14, transferring its energy as heat to the indoors air (provided that it meets the approval of the household, of course). Since the house is maintained at 27°C, the iron block can keep "losing" heat until its temperature drops to 27°C, transferring a total of 38,925 kJ of heat. Since we utilized the entire energy of the iron block available for heating without wasting a single kilojoule, it seems like we have a 100 percent efficient operation, and nothing can beat this, right? Well, not quite.

In Example 7-4 we learned that this process has an irreversibility of 8191 kJ, which implies that things are not as "perfect" as they seem. A "perfect" process is one that involves "zero" irreversibility. The irreversibility in this process is associated with the heat transfer through a finite temperature difference which can be eliminated by running a reversible heat engine between the iron block and the indoor air. This heat engine will produce (as determined in Example 7-4) 8191 kJ of work and reject the remaining 38,925 − 8191 = 30,734 kJ of heat to the house. Now we managed to eliminate the irreversibility and ended up with 8191 kJ of work. What can we do with this work? Well, at worst we can convert it to heat by running a paddle wheel, for example, creating an equal amount of irreversibility. Or we can supply this work to a heat pump which will transport heat

from the outdoors at 5°C to the indoors at 27°C. Such a heat pump, if reversible, will have a coefficient of performance (COP) of (Eq. 5-23)

$$\mathrm{COP_{HP}} = \frac{1}{1 - T_L/T_H} = \frac{1}{1 - (278\text{ K})/(300\text{ K})} = 13.6$$

That is, this heat pump can supply the house with 13.6 times the energy it consumes as work. In our case, it will consume the 8191 kJ of work and deliver 8191 × 13.6 = 111,398 kJ of heat to the house. Therefore, the hot iron block has the potential of supplying

$$(30{,}734 + 111{,}398)\text{ kJ} = 142{,}132\text{ kJ}$$

of heat to the house. The irreversibility for this process is zero, and this is *the best* we can do under the specified conditions. A similar argument can be given for the electric heating of residential or commercial buildings.

Now try to answer the following question: What would happen if the heat engine were operated between the iron block and the outside air instead of the house until the temperature of the iron block fell to 27°C? Would the amount of heat supplied to the house still be 142,132 kJ? Here is a hint: The initial and final states in both cases are the same, and the irreversibility for both cases is zero.

7-3 ■ SECOND-LAW EFFICIENCY η_{II}

In Chap. 5 we defined the *thermal efficiency* and the *coefficient of performance* for devices, such as heat engines, refrigerators, and heat pumps, as a measure of their performance. They were defined on the basis of the first law only, and they are sometimes referred to as the *first-law efficiencies*.

The first-law efficiency (also known as the *conversion efficiency*), however, makes no reference to the best possible performance, and thus it may be misleading.

Consider two heat engines, for example, both having a thermal efficiency of 30 percent, as shown in Fig. 7-15. One of the engines (engine *A*) is supplied with heat from a source at 600 K, and the other one (engine *B*) from a source at 1000 K. Both engines reject heat to a medium at 300 K. At first glance, both engines seem to convert the same fraction of heat that they receive to work; thus they are performing equally well. When we take a second look at these engines in light of the second law of thermodynamics, however, we see a totally different picture. These engines, at best, can perform as reversible (Carnot) engines, in which case their efficiencies would be

$$\eta_{\text{rev},A} = \left(1 - \frac{T_L}{T_H}\right)_A = 1 - \frac{300\text{ K}}{600\text{K}} = 50\%$$

$$\eta_{\text{rev},B} = \left(1 - \frac{T_L}{T_H}\right)_B = 1 - \frac{300\text{ K}}{1000\text{ K}} = 70\%$$

Now it is becoming apparent that engine *B* has a greater work potential available to it (70 percent of the heat supplied as compared to 50 percent

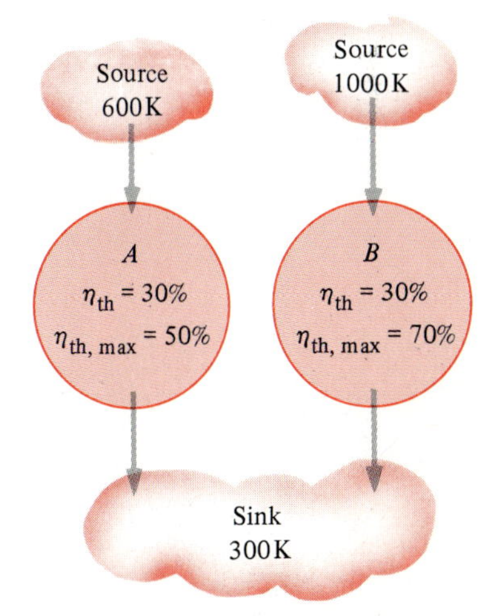

FIGURE 7-15
Two heat engines that have the same thermal efficiency but different maximum thermal efficiencies.

for engine A) and thus should do a lot better than engine A. Therefore, we can say that engine B is performing poorly relative to engine A even though both have the same thermal efficiency.

It is obvious from this example that the first-law efficiency alone is not a realistic measure of the performance of engineering devices. To overcome this deficiency, we define a **second-law efficiency** $\eta_{\rm II}$, which is also called the **effectiveness**, as the ratio of the actual thermal efficiency to the maximum possible (reversible) thermal efficiency under the same conditions (Fig. 7-16):

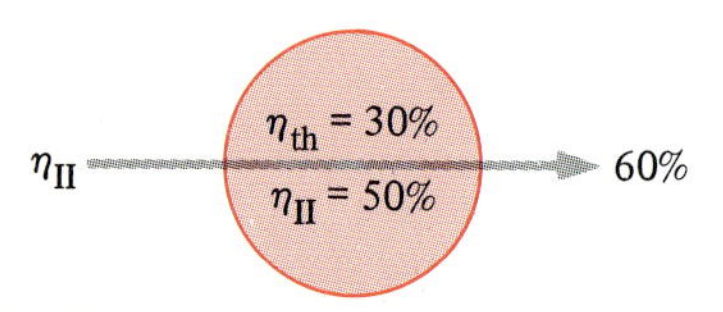

FIGURE 7-16
Second-law efficiency is a measure of the performance of a device relative to its performance under reversible conditions.

$$\eta_{\rm II} = \frac{\eta_{\rm th}}{\eta_{\rm th,rev}} \qquad \text{(heat engines)} \tag{7-4a}$$

Based on this definition, the second-law efficiencies of the two heat engines discussed above are

$$\eta_{{\rm II},A} = \frac{0.30}{0.50} = 0.60 \qquad \text{and} \qquad \eta_{{\rm II},B} = \frac{0.30}{0.70} = 0.43$$

That is, engine A is converting 60 percent of the available work potential to useful work. This ratio is only 43 percent for engine B.

The second-law efficiency can also be expressed as the ratio of the useful work output and the maximum possible (reversible) work output:

$$\eta_{\rm II} = \frac{W_u}{W_{\rm rev}} \qquad \text{(work-producing devices)} \tag{7-4b}$$

This definition is more general since it can be applied to processes (in turbines, piston-cylinder devices, etc.) as well as to cycles. Note that the second-law efficiency cannot exceed 100 percent (Fig. 7-17).

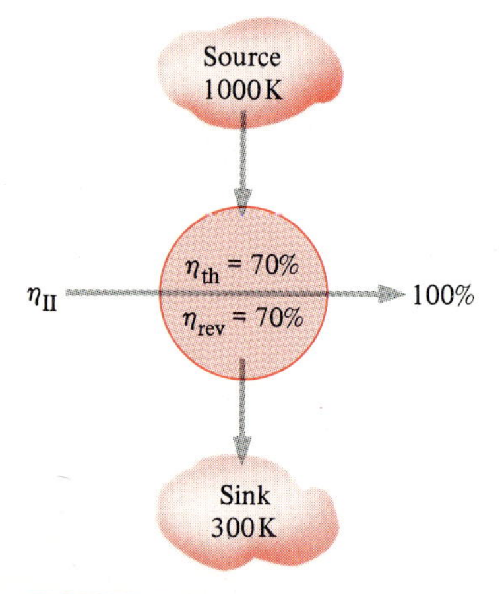

FIGURE 7-17
The second-law efficiency of all reversible devices is 100 percent.

We can also define a second-law efficiency for work-consuming noncyclic (such as compressors) and cyclic (such as refrigerators) devices as the ratio of the minimum (reversible) work input to the useful work input:

$$\eta_{\rm II} = \frac{W_{\rm rev}}{W_u} \qquad \text{(work-consuming devices)} \tag{7-5a}$$

For cyclic devices such as refrigerators and heat pumps, it can also be expressed in terms of the coefficients of performance as

$$\eta_{\rm II} = \frac{\rm COP}{{\rm COP}_{\rm rev}} \qquad \text{(refrigerators and heat pumps)} \tag{7-5b}$$

Again, because of the way we defined the second-law efficiency, its value cannot exceed 100 percent. In the above relations, the reversible work $W_{\rm rev}$ should be determined by using the same initial and final states as in the actual process.

EXAMPLE 7-6

A dealer advertises that she has just received a shipment of electric resistance heaters for residential buildings which have an efficiency of 100 percent, as shown in Fig. 7-18. Assuming an indoor temperature of 21°C and outdoor temperature of 10°C, determine the second-law efficiency of these heaters.

Solution Obviously the efficiency that the dealer is referring to is the first-law efficiency, meaning that for each unit of electric energy (work) consumed, the heater will supply the house with 1 unit of energy (heat). That is, the advertised heater has a COP of 1.

At the specified conditions, a reversible heat pump would have a coefficient of performance of

$$\text{COP}_{\text{HP,rev}} = \frac{1}{1 - T_L/T_H} = \frac{1}{1 - (283\text{ K})/(294\text{ K})} = 26.7$$

That is, it would supply the house with 26.7 units of heat (extracted from the cold outside air) for each unit of electric energy it consumed.

The second-law efficiency of this resistance heater is determined from Eq. 7-5*b* to be

$$\eta_{\text{II}} = \frac{\text{COP}}{\text{COP}_{\text{rev}}} = \frac{1.0}{26.7} = 0.037 \quad \text{or } 3.7\%$$

which does not look so impressive. The dealer definitely will not be happy to see this value. Considering the high price of electricity, a consumer will probably be better off with a "less" efficient gas heater.

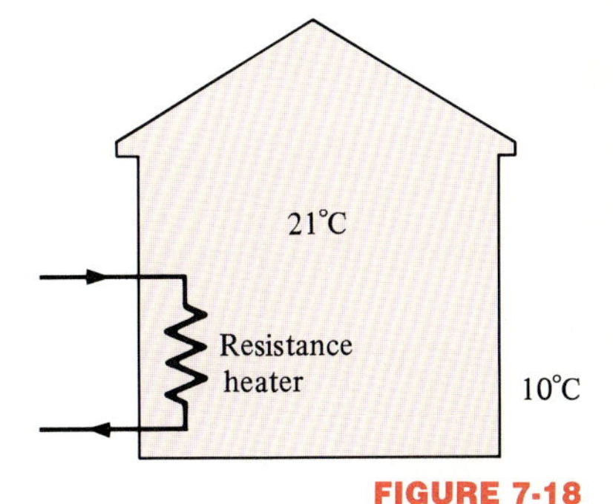

FIGURE 7-18
Schematic for Example 7-6.

Despite their importance, the concepts of availability, irreversibility, and reversible work are often perceived by students as being too difficult and confusing because of the heavy formulation associated with them. These concepts are introduced above in a light context where the emphasis has been on *understanding* them, and they are applied to some simple processes. In the sections that follow, the emphasis will be on *using* them in more challenging engineering problems, and it is advisable at this point to review the material, if necessary, to have a clear understanding of these terms.

To summarize, the maximum useful work output (or the minimum useful work input) as a system undergoes a process between two specified states is called *reversible work;* in the case of the final state being the state of the surroundings, the reversible work is called the *maximum reversible work,* or *availability;* and the difference between the useful work and the reversible work for a process is called *irreversibility*. In the following sections we use these concepts to investigate the second-law aspects of the processes involving closed systems and control volumes.

7-4 ■ SECOND-LAW ANALYSIS OF CLOSED SYSTEMS

Consider a closed system which undergoes a process from state 1 to state 2 in an environment (such as the surrounding atmosphere) at P_0 and T_0, as shown in Fig. 7-19. The system is allowed to exchange heat only with its surroundings and not with any other thermal energy reservoirs (this restriction is lifted later in this section). The closed system is assumed to be stationary relative to its surroundings, and therefore it does not experience any changes in its kinetic or potential energy. The first and the second laws of thermodynamics for this closed system can be expressed as follows:

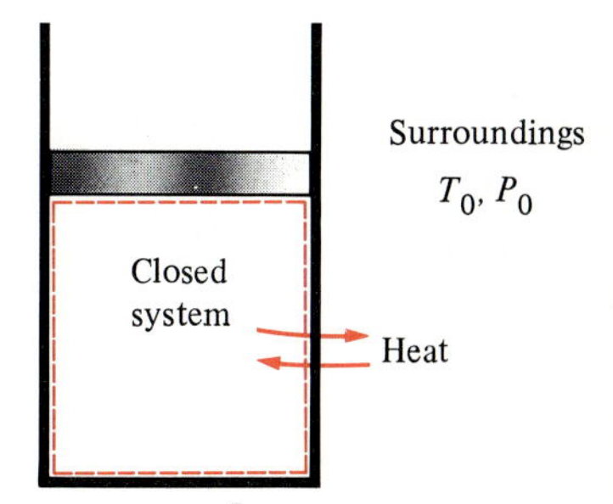

FIGURE 7-19
A closed system exchanging heat with its surroundings only.

First law (Eq. 3-26): $$Q - W = U_2 - U_1$$

Second law (Eq. 6-11): $$S_{gen} = (S_2 - S_1)_{sys} + \frac{Q_{surr}}{T_{surr}}$$

where $T_{surr} = T_0$, $Q_{surr} = -Q$, and S_{gen} is the amount of entropy generated (i.e., the total entropy change) for the process. Eliminating the heat transfer term between these two equations and solving for W, we obtain

$$W = (U_1 - U_2) - T_0(S_1 - S_2) - T_0 S_{gen} \tag{7-6}$$

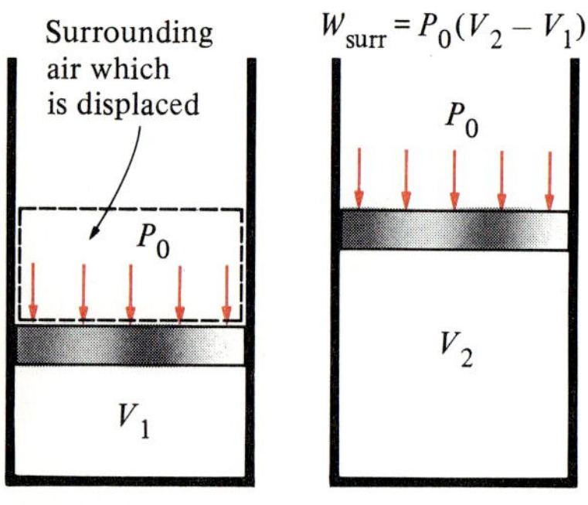

FIGURE 7-20
A closed system, in general, involves surroundings work which cannot be used for any useful purpose.

This is the total *actual work* done during the process. If the volume of the system is changing during the process, part of this work is done by (or against) the surroundings (Fig. 7-20). Then the *useful work*, which is the difference between the actual work and the surroundings work (Eq. 7-2), becomes

$$W_u = W - W_{surr} = W - P_0(V_2 - V_1)$$

or $$W_u = (U_1 - U_2) - T_0(S_1 - S_2) + P_0(V_1 - V_2) - T_0 S_{gen} \tag{7-7}$$

This is the useful work done on or by a closed system during a process 1-2 in terms of the system properties at the initial and the final states, the state of the surrounding medium (P_0 and T_0), and the amount of entropy generated during the process S_{gen}.

Now let us execute this process between the *same* initial and final states, in the *same* environment, in a reversible manner. That is, any heat transfer between the system and the surroundings takes place reversibly, and the system involves no irreversibilities such as friction during the process. Reversible processes generate no entropy, and so $S_{gen} = 0$ in this case. The useful work done during this process is the **reversible work** W_{rev} and is obtained from Eq. 7-7 by setting the entropy generation term S_{gen} equal to zero:

$$W_{rev} = (U_1 - U_2) - T_0(S_1 - S_2) + P_0(V_1 - V_2) \quad \text{(kJ)} \tag{7-8}$$

This is the maximum useful work that can be done as a closed system changes from state 1 to state 2 while exchanging heat only with the surroundings at T_0 and P_0.

Closed-system availability, which is the maximum useful work potential of a closed system at a given state, is denoted by Φ (or ϕ per unit mass) and is obtained from Eq. 7-8 by replacing state 1 by the existing state (no subscript) and state 2 by the dead state ("0" subscript):

$$\Phi = (U - U_0) - T_0(S - S_0) + P_0(V - V_0) \quad \text{(kJ)} \tag{7-9a}$$

or $$\phi = (u - u_0) - T_0(s - s_0) + P_0(v - v_0) \quad \text{(kJ/kg)} \tag{7-9b}$$

It is clear from these equations that the availability of a closed system at the *dead state* ($u = u_0$, $s = s_0$, and $v = v_0$) is zero. That is, no work can be extracted from a system which is in equilibrium with its surroundings. Note that these relations are applicable to nonreacting systems. The availability of chemically reacting systems is discussed in Chap. 14.

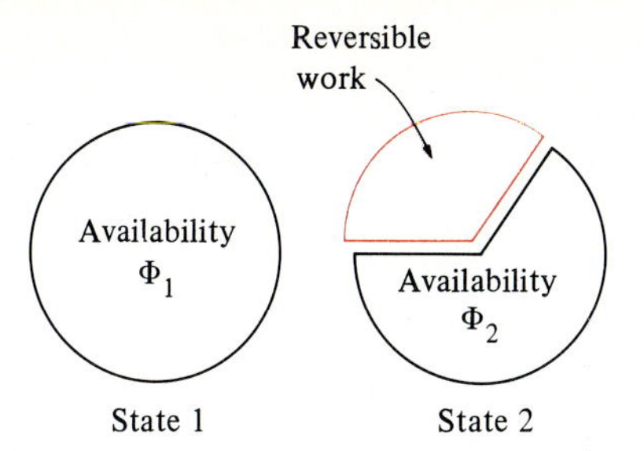

FIGURE 7-21
Reversible work is the *change* in availability.

The reversible work associated with a process between states 1 and 2 can also be expressed in terms of availabilities as

$$W_{rev} = \Phi_1 - \Phi_2 \qquad \text{(kJ)} \tag{7-10a}$$

or

$$w_{rev} = \phi_1 - \phi_2 \qquad \text{(kJ/kg)} \tag{7-10b}$$

That is, the reversible work associated with a closed system is simply the *decrease* (or, the *increase*, for the case of compression) in the availability of the closed system (Fig. 7-21). This can be demonstrated easily by expressing Eq. 7-9*a* at the final and the initial states and taking their difference. In the limiting case of the final state reaching the dead state ($\Phi_2 = 0$), the availability of the closed system at the initial state (Φ_1) and the reversible work for the process (W_{rev}) become identical (Fig. 7-22). Again, notice that availability is associated with a *state* whereas reversible work is associated with a *process*.

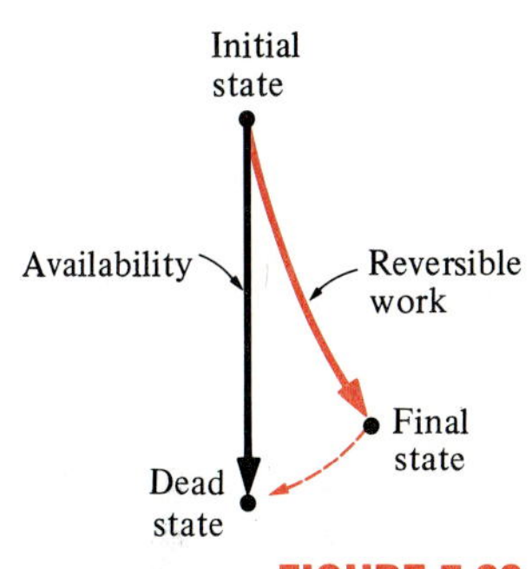

FIGURE 7-22
Reversible work approaches availability as the final state approaches the dead state.

Any changes in kinetic and potential energies of a closed system can properly be accounted for by replacing the internal energy (U) terms in the above equations by the total energy E.

For closed systems, the **irreversibility** associated with a process is determined from its definition (Fig. 7-23):

$$I = W_{rev} - W_u = T_0 S_{gen} \qquad \text{(kJ)} \tag{7-11a}$$

or

$$i = w_{rev} - w_u = T_0 s_{gen} \qquad \text{(kJ/kg)} \tag{7-11b}$$

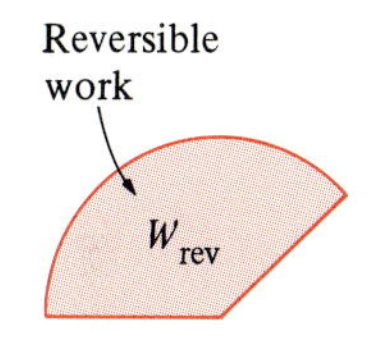

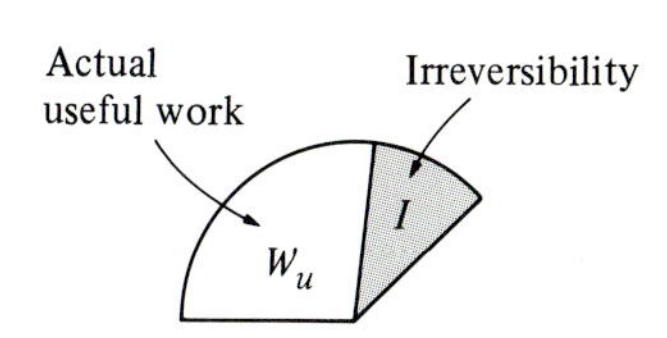

FIGURE 7-23
Irreversibility is the difference between the reversible work and the actual useful work.

EXAMPLE 7-7

A piston-cylinder device contains 0.05 kg of steam at 1 MPa and 300°C. The steam now expands to a final state of 200 kPa and 150°C, doing work. Heat losses from the system to the surroundings are estimated to be 2 kJ during this process. Assuming the surroundings to be at T_0 = 25°C and P_0 = 100 kPa, determine (*a*) the availability of the steam at the initial and the final states, (*b*) the reversible work, (*c*) the irreversibility, and (*d*) the second-law efficiency for this process.

Solution We take the steam contained within the piston-cylinder device as our system, as shown in Fig. 7-24. This is a closed system since no mass crosses the system boundaries, and thus the reversible work, availability, and irreversibility relations developed in this section apply.

(*a*) First let us determine the properties of the steam at the initial and final states as well as the state of the surroundings:

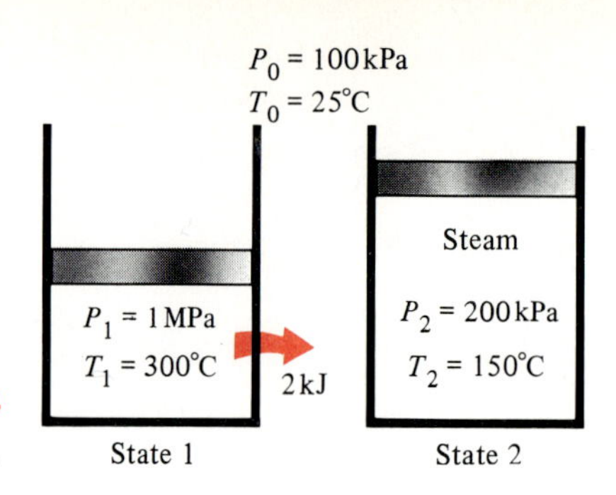

FIGURE 7-24
Schematic for Example 7-7.

State 1: $\left.\begin{aligned} P_1 &= 1 \text{ MPa} \\ T_1 &= 300°\text{C} \end{aligned}\right\}$ $\begin{aligned} u_1 &= 2793.2 \text{ kJ/kg} \\ v_1 &= 0.2579 \text{ m}^3\text{/kg} \\ s_1 &= 7.1229 \text{ kJ/(kg·K)} \end{aligned}$ (Table A-6)

State 2: $\left.\begin{aligned} P_2 &= 200 \text{ kPa} \\ T_2 &= 150°\text{C} \end{aligned}\right\}$ $\begin{aligned} u_2 &= 2576.9 \text{ kJ/kg} \\ v_2 &= 0.9596 \text{ m}^3\text{/kg} \\ s_2 &= 7.2795 \text{ kJ/(kg·K)} \end{aligned}$ (Table A-6)

Dead state: $\left.\begin{aligned} P_0 &= 100 \text{ kPa} \\ T_0 &= 25°\text{C} \end{aligned}\right\}$ $\begin{aligned} u_0 &\cong u_{f\,@\,25°\text{C}} = 104.88 \text{ kJ/kg} \\ v_0 &\cong v_{f\,@\,25°\text{C}} = 0.0010 \text{ m}^3\text{/kg} \\ s_0 &\cong s_{f\,@\,25°\text{C}} = 0.3674 \text{ kJ/(kg·K)} \end{aligned}$ (Table A-4)

The availability of the system at the initial state Φ_1 and at the final state Φ_2 is determined from Eq. 7-9 to be

$$\begin{aligned} \Phi_1 &= m[(u_1 - u_0) - T_0(s_1 - s_0) + P_0(v_1 - v_0)] \\ &= (0.05 \text{ kg})\{(2793.2 - 104.88) \text{ kJ/kg} \\ &\quad - (298 \text{ K})[(7.1229 - 0.3674) \text{ kJ/(kg·K)}] \\ &\quad + (100 \text{ kPa})[(0.2579 - 0.0010) \text{ m}^3\text{/kg}][\text{kJ/(kPa·m}^3)]\} \\ &= 35.0 \text{ kJ} \end{aligned}$$

and

$$\begin{aligned} \Phi_2 &= m[(u_2 - u_0) - T_0(s_2 - s_0) + P_0(v_2 - v_0)] \\ &= (0.05 \text{ kg})\{(2576.9 - 104.88) \text{ kJ/kg} \\ &\quad - (298 \text{ K})[(7.2795 - 0.3674) \text{ kJ/(kg·K)}] \\ &\quad + (100 \text{ kPa})[(0.9596 - 0.0010) \text{ m}^3\text{/kg}][\text{kJ/(kPa·m}^3)]\} \\ &= 25.4 \text{ kJ} \end{aligned}$$

That is, the system (steam) initially has an availability (maximum useful work potential) of 35 kJ, which reduces to 25.4 kJ at the end of the process. In other words, if the system were allowed to undergo a reversible process from the initial state to the state of the surroundings (the dead state), it would produce 35 kJ of useful work.

(*b*) The reversible work for a process is simply the *decrease* in availability of the system during the process, Eq. 7-10*a*:

$$W_{\text{rev}} = \Phi_1 - \Phi_2 = (35.0 - 25.4) \text{ kJ} = 9.6 \text{ kJ}$$

That is, if the process between states 1 and 2 were executed in a reversible manner, the system would deliver 9.6 kJ of useful work. The reversible work could also be determined directly from Eq. 7-8.

(*c*) The irreversibility of a process is simply the difference between the reversible work and useful work for that process. The reversible work for this process is determined above to be 9.6 kJ. The actual work is determined by applying the conservation of energy equation to the actual process:

$$Q - W = \Delta U = m(u_2 - u_1)$$
$$-2 \text{ kJ} - W = (0.05 \text{ kg})[(2576.9 - 2793.2)\text{kJ/kg}]$$
$$W = 8.8 \text{ kJ}$$

This is the *total actual work* done by the system, including the work done against the surrounding atmospheric air to push it out of the way during the expansion process. The *useful work* is the difference between these two:

$$W_u = W - W_{surr} = W - P_0(V_2 - V_1) = W - P_0 m(v_2 - v_1)$$
$$= 8.82 \text{ kJ} - (100 \text{ kPa})(0.05 \text{ kg})[(0.9596 - 0.2579) \text{ m}^3/\text{kg}]\left(\frac{1 \text{ kJ}}{1 \text{ kPa·m}^3}\right)$$
$$= 5.3 \text{ kJ}$$

Thus, $$I = W_{rev} - W_u = (9.6 - 5.3) \text{ kJ} = 4.3 \text{ kJ}$$

That is, 4.3 kJ of work potential is wasted during this process. In other words, an additional 4.3 kJ of energy *could have been* converted to work during this process, but was not.

The irreversibility could also be determined from the second part of Eq. 7-11*a*:

$$I = T_0 S_{gen} = T_0\left[m(s_2 - s_1) + \frac{Q_{surr}}{T_0}\right]$$
$$= (298 \text{ K})\left\{(0.05 \text{ kg})[(7.2795 - 7.1229) \text{ kJ/(kg·K)}] + \frac{2 \text{ kJ}}{298 \text{ K}}\right\}$$
$$= 4.3 \text{ kJ}$$

which is the same result obtained before.

(*d*) The second-law efficiency of a process is the ratio of the useful work to the reversible work, Eq. 7-4*b*:

$$\eta_{II} = \frac{W_u}{W_{rev}} = \frac{5.3 \text{ kJ}}{9.6 \text{ kJ}} = 0.552 \quad \text{or } 55.2\%$$

That is, 44.8 percent of the work potential is wasted during this process.

EXAMPLE 7-8

An insulated rigid tank contains 2 lbm of air at 20 psia and 70°F. A paddle wheel inside the tank is now rotated by an external power source until the temperature in the tank rises to 130°F. If the surrounding air is at $T_0 = 70°F$, determine (*a*) the irreversibility and (*b*) the reversible work for this process.

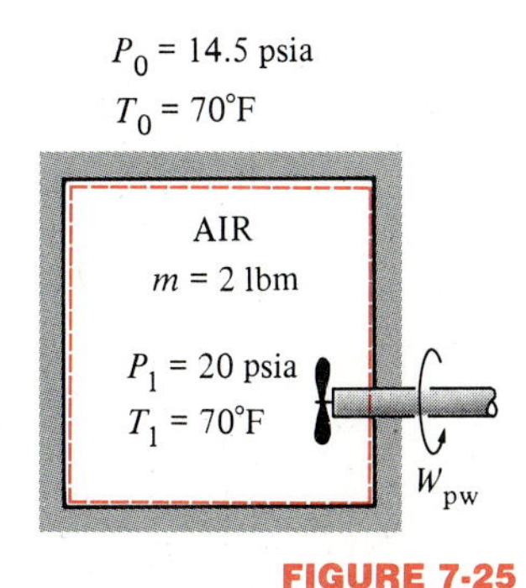

FIGURE 7-25
Schematic for Example 7-8.

Solution We take the air in the rigid tank as our system, as shown in Fig. 7-25. At or about atmospheric temperature and pressure, air can be treated as an ideal gas with negligible error. Furthermore, the air experiences a relatively small temperature change during this process, and thus the constant-specific-heat assumption is appropriate. At the average temperature of 100°F, the constant-volume specific heat of air is determined, from Table A-2E, to be $C_v = 0.172$ Btu/(lbm·°F). Also note that the system is stationary and experiences no changes in kinetic or potential energies.

(*a*) The irreversibility of this process can be determined from Eq. 7-11*a*:

$$I = T_0 S_{gen} = T_0(\Delta S_{sys} + \cancelto{0}{\Delta S_{surr}}) = T_0 \Delta S_{sys} = T_0 m \left(C_v \ln \frac{T_2}{T_1} + R \cancelto{0}{\ln \frac{V_2}{V_1}} \right)$$

$$= T_0 m C_v \ln \frac{T_2}{T_1} = (530 \text{ R})(2 \text{ lbm})[0.172 \text{ Btu/(lbm·R)}] \ln \frac{590 \text{ R}}{530 \text{ R}}$$

$$= 19.6 \text{ Btu}$$

since ΔS_{surr} (= Q_{surr}/T_0) is zero for an adiabatic process and the volume of the system is constant and thus $\ln (V_2/V_1) = 0$.

(*b*) The reversible work for this process can be determined from Eq. 7-8. Notice that $V_1 = V_2$ for this case, and the value of $T_0 \Delta S_{sys}$ is determined above. Thus,

$$\begin{aligned} W_{rev} &= (U_1 - U_2) - T_0(S_1 - S_2) + P_0(V_1 - V_2) \\ &= mC_v(T_1 - T_2) - T_0(-\Delta S_{sys}) + 0 \\ &= (2 \text{ lbm})[0.172 \text{ Btu/(lbm·R)}][(530 - 590) \text{ R}] + 19.6 \text{ Btu} \\ &= (-20.6 + 19.6) \text{ Btu} \\ &= -1.0 \text{ Btu} \end{aligned}$$

Discussion The solution is complete at this point. But to gain some physical insight, we will set the stage for a discussion. First, let us determine the actual work (the paddle-wheel work W_{pw}) done during this process by applying the conservation of energy equation to this closed system:

$$\begin{aligned} \cancelto{0}{Q} - W &= \Delta U \\ -W_{pw} &= mC_{v,av}(T_2 - T_1) \\ W_{pw} &= -20.6 \text{ Btu} \end{aligned}$$

since the system is adiabatic ($Q = 0$) and involves no moving boundaries ($W_b = 0$). Also, the value of ΔU is determined in part (*b*) above.

To put the information into perspective, 20.6 Btu of work is consumed during the process, 19.6 Btu of irreversibility is generated, and the reversible work for the process is −1.0 Btu. What does all this mean? It simply means that we could have created the same effect on the closed system (raising its temperature to 130°F at constant volume) by consuming 1.0 Btu of work only instead of 20.6 Btu, and thus saving 19.6 Btu of work from going to waste. This would have been accomplished by a reversible heat pump.

To prove what we have just said, consider a Carnot heat pump that absorbs heat from the surroundings at $T_0 = 530$ R and transfers it to the air in the rigid tank until the air temperature T rises from 530 to 590 R, as shown in Fig. 7-26. The system (the air in the rigid tank) involves no direct work interactions in this case, and the heat supplied to the system can be expressed in differential form as

$$\delta Q_H = dU = mC_v \, dT$$

The coefficient of performance of a reversible heat pump is given by (Eq. 5-23)

$$\text{COP}_{HP} = \frac{\delta Q_H}{\delta W_{net,in}} = \frac{1}{1 - T_0/T}$$

Thus $$\delta W_{net,in} = \frac{\delta Q_H}{\text{COP}_{HP}} = \left(1 - \frac{T_0}{T}\right) mC_v \, dT$$

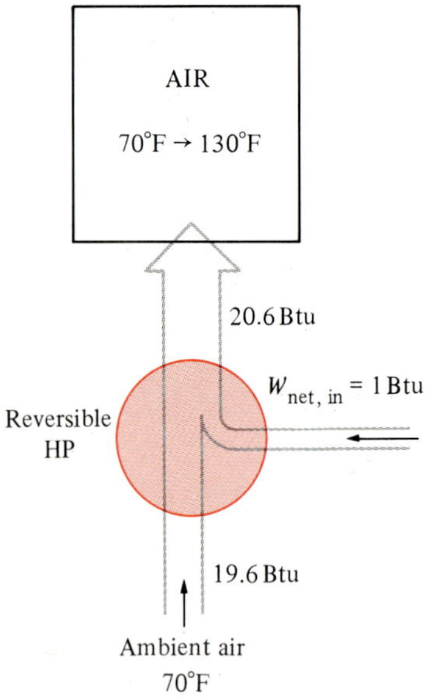

FIGURE 7-26
The same effect on the system can be accomplished by a reversible heat pump which consumes only 1 Btu of work.

Integrating, we get

$$W_{\text{net,in}} = \int_1^2 \left(1 - \frac{T_0}{T}\right) mC_v\, dT$$

$$= mC_{v,\text{av}}(T_2 - T_1) - T_0 mC_{v,\text{av}} \ln \frac{T_2}{T_1}$$

$$= (20.6 - 19.6) \text{ Btu} = 1.0 \text{ Btu}$$

The first term on the right-hand side of the final expression above is recognized as ΔU and the second term as the irreversibility I, whose values were determined earlier. By substituting those values, the total work input to the heat pump is determined to be 1.0 Btu, proving our claim. Notice that the system is still supplied with 20.6 Btu of energy; all we did in the latter case is replace the 19.6 Btu of valuable work by an equal amount of "useless" energy captured from the surroundings.

It is also worth mentioning that the availability of the system as a result of 20.6 Btu of paddle-wheel work done on it has increased by 1.0 Btu only, i.e., by the amount of the reversible work. In other words, if the system were returned to its initial state, it would produce, at most, 1.0 Btu of work.

Heat Transfer with Other Systems or Bodies

To keep complexities at a manageable level, we have limited the analysis so far to processes that involve heat transfer with the surrounding medium only. Now we lift this restriction and allow the system to exchange heat with other bodies at different temperatures as well as with its surroundings.

The easiest way of handling such processes is to treat the main system and the bodies to or from which heat is transferred as separate systems and to superimpose (sum) the results, as illustrated in Fig. 7-27.

The *availability,* as mentioned earlier, is associated with a *state,* and the availability of a system at a specified state depends on the state of the system and of the surrounding medium (the ultimate sink) only, and nothing else. Therefore, the availability of a system is not affected by the existence of other bodies (thermal energy reservoirs or other systems) in the neighborhood which are interested in a *thermal* relationship.

The *reversible work* and *irreversibility,* however, are associated with a *process,* and heat transfer with bodies other than the surrounding medium will affect the reversible work and the irreversibility of a process.

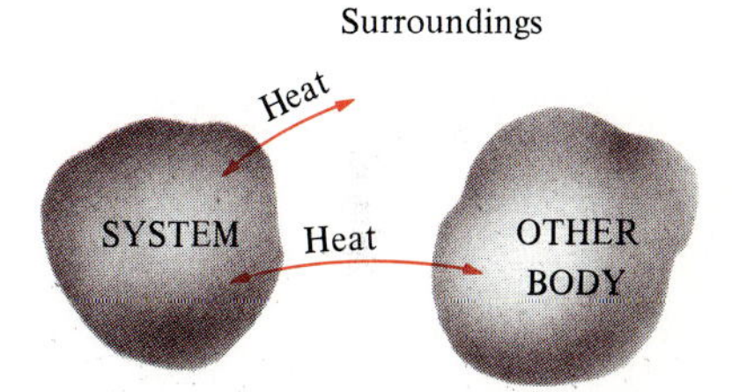

$$\Phi_{\text{total}} = \Phi_{\text{system}} + \Phi_{\text{other body}}$$
$$W_{\text{rev, total}} = W_{\text{rev, system}} + W_{\text{rev, other body}}$$

FIGURE 7-27

Heat transfer with bodies other than the surroundings can be handled with superposition.

EXAMPLE 7-9

A 5-kg iron block initially at 350°C is quenched in an insulated tank which contains 100 kg of water at 30°C. Assuming the water that vaporizes during the process condenses back in the tank and the surroundings are at 20°C and 100 kPa, determine (*a*) the final equilibrium temperature, (*b*) the availability of the combined system at the initial and the final states, and (*c*) the wasted work potential during this process.

Solution We take the iron block and the water in the tank as our system, as

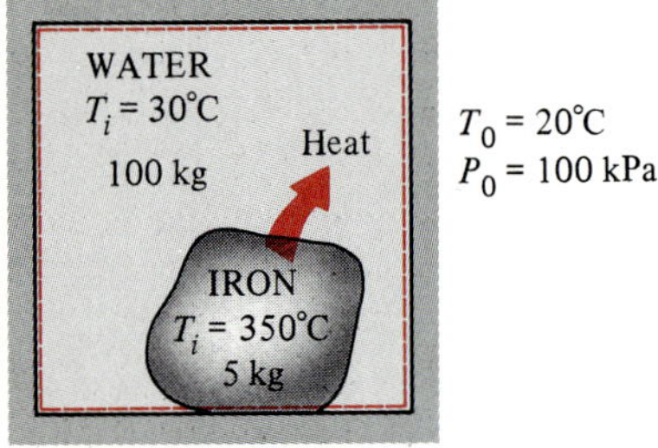

FIGURE 7-28
Schematic for Example 7-9.

shown in Fig. 7-28. Both the water and the iron block can be approximated as incompressible substances, and the kinetic and potential energy changes of the system, if any, are considered insignificant.

(*a*) The final equilibrium temperature of the system can be determined from the conservation of energy equation for closed systems, Eq. 3-20. Noting that the boundaries of the system are fixed, and no heat or work crosses them, we see that Eq. 3-20 reduces to

$$\cancelto{0}{Q} - \cancelto{0}{W} = \Delta U + \cancelto{0}{\Delta KE} + \cancelto{0}{\Delta PE}$$
$$0 = (\Delta U)_{\text{iron}} + (\Delta U)_{\text{water}}$$
$$0 = [mC(T_f - T_i)]_{\text{iron}} + [mC(T_f - T_i)]_{\text{water}}$$

By using the specific-heat values for water and iron at room temperature (from Table A-3*a*), the final equilibrium temperature T_f becomes

$$0 = (5 \text{ kg})[0.45 \text{ kJ/(kg·°C)}](T_f - 350°\text{C}) + (100 \text{ kg})[4.184 \text{ kJ/(kg·°C)}](T_f - 30°\text{C})$$

which yields

$$T_f = 31.7°\text{C}$$

(*b*) Availability Φ is an extensive property, and the availability of a composite system at a specified state is the sum of the availabilities of the components of that system at that state. It is determined from Eq. 7-9*a*, which for an incompressible substance reduces to

$$\Phi = (U - U_0) - T_0(S - S_0) + P_0(\cancelto{0}{V - V_0})$$
$$= mC(T - T_0) - T_0 mC \ln \frac{T}{T_0} + 0$$
$$= mC\left(T - T_0 - T_0 \ln \frac{T}{T_0}\right)$$

where T is the absolute temperature at the specified state and T_0 is the absolute temperature of the surroundings. At the initial state,

$$\Phi_{1,\text{iron}} = (5 \text{ kg})[0.45 \text{ kJ/(kg·K)}][(623 - 293) \text{ K} - (293 \text{ K}) \ln \tfrac{623}{293}]$$
$$= 245.2 \text{ kJ}$$
$$\Phi_{1,\text{water}} = (100 \text{ kg})[4.184 \text{ kJ/(kg·K)}][(303 - 293) \text{ K} - (293 \text{ K}) \ln \tfrac{303}{293}]$$
$$= 69.8 \text{ kJ}$$
$$\Phi_{1,\text{total}} = \Phi_{1,\text{iron}} + \Phi_{1,\text{water}} = (245.2 + 69.8) \text{ kJ} = 315 \text{ kJ}$$

Similarly, the availability at the final state is

$$\Phi_{2,\text{iron}} = 0.5 \text{ kJ}$$
$$\Phi_{2,\text{water}} = 95.2 \text{ kJ}$$
$$\Phi_{2,\text{total}} = \Phi_{2,\text{iron}} + \Phi_{2,\text{water}} = (0.5 + 95.2) \text{ kJ} = 95.7 \text{ kJ}$$

That is, the availability of the combined system (water + iron) decreased from 315 to 95.7 kJ as a result of this irreversible heat transfer process.

(*c*) The wasted work potential for a process is equivalent to the irreversibility for that process. The irreversibility is the difference between the reversible work and the useful work. This process involves no work, and thus the irreversibility is equal to the reversible work, which is the change in the total availability of the system (Eq. 7-10*a*):

$$I = W_{rev} = \Phi_1 - \Phi_2 = (315 - 95.7)\ \text{kJ} = 219.3\ \text{kJ}$$

That is, 219.3 kJ of work could have been produced as the iron was cooled from 350 to 31.7°C and water was heated from 30 to 31.7°C, but was not. The total reversible work for this process could also be determined by applying Eq. 7-8 first to the iron and then to the water and summing the results.

A simple relation for the reversible work can be obtained when the temperatures of the bodies with which the system exchanges heat remain constant during the process. Consider a closed system which is receiving heat in the amount of Q_R from a reservoir at temperature T_R. The system is located in an environment at temperature T_0 and changes from state 1 to state 2 during the process. Writing the first- and the second-law relations for this process and eliminating the heat transfer term between the system and its surroundings, we see that the reversible work is

$$W_{rev} = (\Phi_1 - \Phi_2)_{sys} - Q_R\left(1 - \frac{T_0}{T_R}\right) \quad \text{(kJ)} \tag{7-12}$$

where T_0 = absolute temperature of surroundings
T_R = absolute temperature of reservoir
Q_R = heat transfer between system and reservoir (its sign is determined with respect to reservoir: *positive* if *to* the reservoir and *negative* if *from* the reservoir)

The irreversibility relations remain the same. But in determining the entropy generation S_{gen}, the entropy change of the reservoir should be included:

$$S_{gen} = (S_2 - S_1)_{sys} + \frac{Q_{surr}}{T_0} + \frac{Q_R}{T_R} \quad \text{(kJ/K)} \tag{7-13}$$

Additional thermal energy reservoirs can be handled in a similar manner.

EXAMPLE 7-10

A frictionless piston-cylinder device, shown in Fig. 7-29, initially contains 0.01 m^3 of argon gas at 400 K and 350 kPa. Heat is now transferred to the argon from a furnace at 1200 K, and the argon expands isothermally until its volume is doubled. No heat transfer takes place between the argon and the surrounding atmospheric air which is at T_0 = 300 K and P_0 = 100 kPa. Determine (*a*) the useful work output, (*b*) the reversible work, and (*c*) the irreversibility for this process.

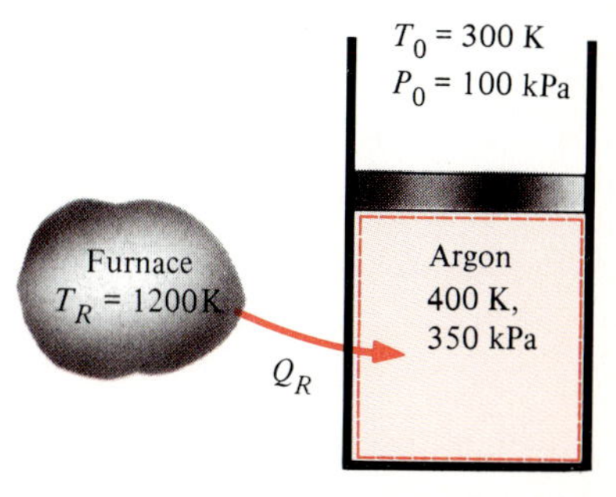

FIGURE 7-29
Schematic for Example 7-10.

Solution We take the argon gas as our system and treat it as an ideal gas since it is well above its critical temperature (151 K). We also assume the process to be quasi-equilibrium and thus internally reversible, and we neglect the changes in kinetic and potential energies.

(*a*) The only work interaction involved during this process is the quasi-equilibrium boundary work. It is determined from Eq. 3-9, which reduces to Eq. 3-12 for an isothermal process:

$$W = W_b = \int_1^2 P\,dV = P_1V_1 \ln\frac{V_2}{V_1} = (350\ \text{kPa})(0.01\ \text{m}^3)\ln\frac{0.02\ \text{m}^3}{0.01\ \text{m}^3}$$
$$= 2.43\ \text{kPa·m}^3 = 2.43\ \text{kJ}$$

This is the total boundary work done by the argon gas. Part of this work is done against the atmospheric pressure P_0 to push the air out of the way, and it cannot be used for any useful purpose. It is determined from Eq. 7-1:

$$W_{surr} = P_0(V_2 - V_1) = (100 \text{ kPa})[(0.02 - 0.01) \text{ m}^3]\left(\frac{1 \text{ kJ}}{1 \text{ kPa}\cdot\text{m}^3}\right) = 1 \text{ kJ}$$

The useful work is the difference between these two:

$$W_u = W - W_{surr} = (2.43 - 1) \text{ kJ} = 1.43 \text{ kJ}$$

That is, 1.43 kJ of the work done is available for creating a useful effect such as rotating a shaft.

(*b*) This process involves heat transfer with a reservoir other than the surrounding medium, and thus the reversible work (the maximum work that *could* be obtained during this process) should be determined from Eq. 7-12:

$$W_{rev} = (\Phi_1 - \Phi_2)_{sys} - Q_R\left(1 - \frac{T_0}{T_R}\right)$$

$$= (U_1 - U_2) - T_0(S_1 - S_2) + P_0(V_1 - V_2) - Q_R\left(1 - \frac{T_0}{T_R}\right)$$

The first term on the right-hand side is the change in internal energy, which is zero for the isothermal process of an ideal gas, and the third term is easily recognized as the work done against the surroundings. Thus

$$W_{rev} = T_0(S_2 - S_1) - W_{surr} - Q_R\left(1 - \frac{T_0}{T_R}\right)$$

The heat transfer between the system and the furnace Q_R is determined by applying the first law to the closed system:

$$Q - W = \Delta U = 0 \longrightarrow Q = W = 2.43 \text{ kJ} \longrightarrow Q_R = -Q = -2.43 \text{ kJ}$$

The entropy change of the argon during this process can be determined by using the entropy-change relations for ideal gases. For the special case of a reversible isothermal process, it can also be determined from Eq. 6-6:

$$\Delta S = S_2 - S_1 = \frac{Q}{T} = \frac{+2.43 \text{ kJ}}{400 \text{ K}} = 0.00608 \text{ kJ/K}$$

Thus, $$W_{rev} = (300 \text{ K})(0.00608 \text{ kJ/K}) - 1 \text{ kJ} - (-2.43 \text{ kJ})\left(1 - \frac{300 \text{ K}}{1200 \text{ K}}\right)$$

$$= 2.64 \text{ kJ}$$

Alternative approach The reversible work could also be determined by applying the basics only, without resorting to any fancy-looking formulas. This is done by replacing the irreversible portions of the process by reversible ones that will create the same effect on the system. The useful work output of this idealized process (between the actual end states) is the reversible work.

The only irreversibility the actual process involves is the heat transfer between the system and the furnace through a finite temperature difference. This irreversibility can be eliminated by operating a reversible heat engine between the furnace at 1200 K and the surroundings at 300 K. When 2.43 kJ of heat is supplied to this heat engine, it will have a work output of

$$W_{HE} = \eta_{rev} Q_H = \left(1 - \frac{T_L}{T_H}\right) Q_H = \left(1 - \frac{300 \text{ K}}{1200 \text{ K}}\right)(2.43 \text{ kJ}) = 1.82 \text{ kJ}$$

The 2.43 kJ of heat that was transferred to the system from the source is now extracted from the surrounding air at 300 K by a reversible heat pump which will require a work input of

$$W_{\text{HP,in}} = \frac{Q_H}{\text{COP}_{\text{HP}}} = \left[\frac{Q_H}{T_H/(T_H - T_L)}\right]_{\text{HP}} = \frac{2.43 \text{ kJ}}{(400 \text{ K})/[(400 - 100)\text{K}]} = 0.61 \text{ kJ}$$

Then the net work output of this reversible process (i.e., the reversible work) becomes

$$W_{\text{rev}} = W_u + W_{\text{HE}} - W_{\text{HP,in}} = (1.43 + 1.82 - 0.61) \text{ kJ} = 2.64 \text{ kJ}$$

which is the same result as obtained earlier.

(c) The irreversibility (the lost potential to do work) is the difference between the reversible work and the useful work:

$$I = W_{\text{rev}} - W_u = (2.65 - 1.43) \text{ kJ} = 1.22 \text{ kJ}$$

It could also be determined from the second part of Eq. 7-11*a*:

$$I = T_0 S_{\text{gen}} = T_0 \left[(S_2 - S_1) + \frac{Q_{\text{surr}}}{T_0} + \frac{Q_R}{T_R}\right]$$

$$= (300 \text{ K}) \left(0.00608 \text{ kJ/K} + 0 + \frac{-2.43 \text{ kJ}}{1200 \text{ K}}\right) = 1.22 \text{ kJ}$$

which is the same result as obtained earlier.

7-5 ■ SECOND-LAW ANALYSIS OF STEADY-FLOW SYSTEMS

In the preceding section we discussed the second-law aspects of the processes involving closed systems. In this section, we repeat the analysis for processes involving steady-flow devices such as nozzles, diffusers, turbines, compressors, and heat exchangers.

Consider a system undergoing a steady-flow process. The system may have multiple inlets and exits and may exchange heat with the surrounding medium at P_0 and T_0 (Fig. 7-30). The first and the second laws of thermodynamics for this steady-flow process can be expressed as follows:

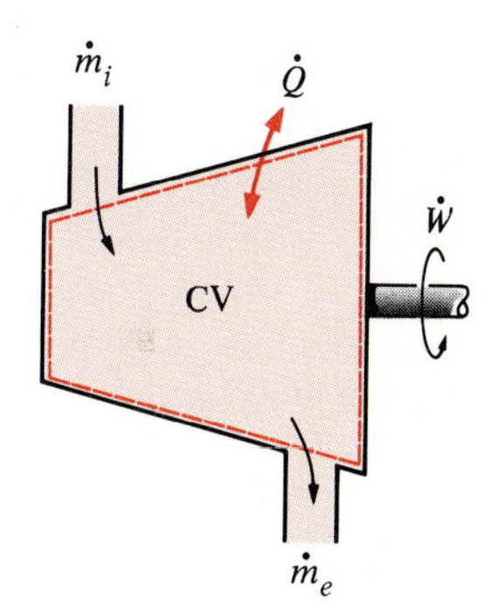

FIGURE 7-30
A steady-flow device which exchanges heat with its surroundings only.

First law (Eq. 4-19): $$\dot{Q} - \dot{W} = \sum \dot{m}_e\left(h_e + \frac{\mathbf{V}_e^2}{2} + gz_e\right) - \sum \dot{m}_i\left(h_i + \frac{\mathbf{V}_i^2}{2} + gz_i\right)$$

Second law (Eq. 6-14) $$\dot{S}_{\text{gen}} = \sum \dot{m}_e s_e - \sum \dot{m}_i s_i + \frac{\dot{Q}_{\text{surr}}}{T_0}$$

where $\dot{Q}_{\text{surr}} = -\dot{Q}$ and $\dot{S}_{\text{gen}}$ is the rate of entropy generation for the process. Eliminating the heat transfer term between these two equations, we obtain

$$\dot{W} = \sum \dot{m}_i \left(h_i + \frac{\mathbf{V}_i^2}{2} + gz_i - T_0 s_i\right) - \sum \dot{m}_e \left(h_e + \frac{\mathbf{V}_e^2}{2} + gz_e - T_0 s_e\right) - T_0 \dot{S}_{\text{gen}} \quad (7\text{-}14)$$

This is the **actual work** done during the process, which is also the **useful work** since steady-flow devices have fixed boundaries and do not involve any work done by or against the surroundings (Fig. 7-31).

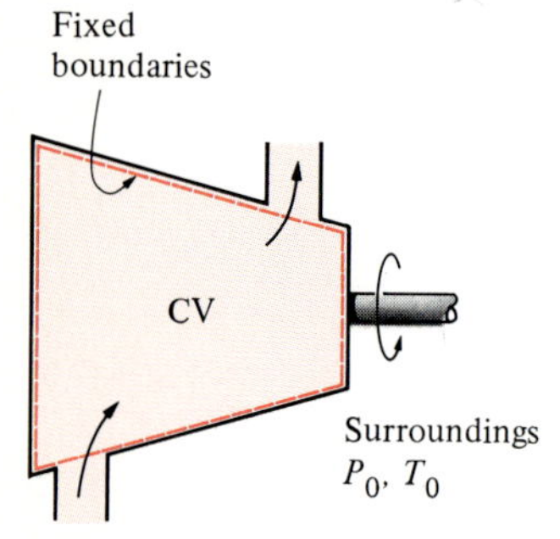

FIGURE 7-31
The total actual work and the useful work for steady-flow devices are the same since they do not involve any surroundings work.

The **reversible work** is obtained by setting the entropy generation term $\dot{S}_{\text{gen}}$ equal to zero:

$$\dot{W}_{\text{rev}} = \sum \dot{m}_i \left(h_i + \frac{\mathbf{V}_i^2}{2} + gz_i - T_0 s_i \right) - \sum \dot{m}_e \left(h_e + \frac{\mathbf{V}_e^2}{2} + gz_e - T_0 s_e \right) \quad \text{(kW)} \quad (7\text{-}15)$$

For a *single fluid stream* entering and leaving the steady-flow device,

$$\dot{W}_{\text{rev}} = \dot{m} \left[(h_i - h_e) - T_0(s_i - s_e) + \frac{\mathbf{V}_i^2 - \mathbf{V}_e^2}{2} + g(z_i - z_e) \right] \quad \text{(kW)} \quad (7\text{-}16a)$$

or

$$w_{\text{rev}} = (h_i - h_e) - T_0(s_i - s_e) + \frac{\mathbf{V}_i^2 - \mathbf{V}_e^2}{2} + g(z_i - z_e) \quad \text{(kJ/kg)} \quad (7\text{-}16b)$$

$$= T_0 \Delta s - \Delta h - \Delta \text{ke} - \Delta \text{pe} \quad \text{(kJ/kg)} \quad (7\text{-}16c)$$

The availability of a fluid stream is called the **stream availability** and is denoted by ψ (Fig. 7-32). It is obtained from Eq. 7-16*b* by replacing the inlet state by the existing state (no subscript) and the exit state by the dead state ("0" subscript with $\mathbf{V}_0 = 0$ and $z_0 = 0$):

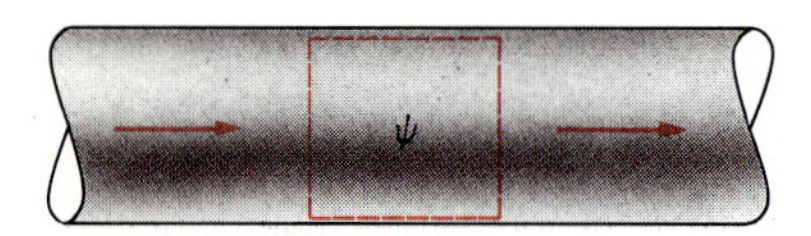

FIGURE 7-32
The maximum work potential of a flowing fluid stream is called the stream availability.

$$\psi = (h - h_0) - T_0(s - s_0) + \frac{\mathbf{V}^2}{2} + gz \quad \text{(kJ/kg)} \quad (7\text{-}17)$$

Then the reversible work for a steady-flow process can be expressed in terms of the stream availability as

$$\dot{W}_{\text{rev}} = \Sigma \dot{m}_i \psi_i - \Sigma \dot{m}_e \psi_e \quad \text{(kW)} \quad (7\text{-}18)$$

or, for a single stream of fluid flowing through a steady-flow device,

$$w_{\text{rev}} = \psi_i - \psi_e \quad \text{(kJ/kg)} \quad (7\text{-}19a)$$

or

$$\dot{W}_{\text{rev}} = \dot{m}(\psi_i - \psi_e) \quad \text{(kW)} \quad (7\text{-}19b)$$

The discussions in Sec. 7-4 on reversible work, availability, and irreversibility in relation to a closed system equally apply to steady-flow devices and other control volumes.

The **rate of irreversibility** $\dot{I}$, or the irreversibility per unit mass i, associated with a steady-flow process is the difference between the useful work and the reversible work:

$$\dot{I} = \dot{W}_{\text{rev}} - \dot{W}_u = T_0 \dot{S}_{\text{gen}} \quad \text{(kW)} \quad (7\text{-}20)$$

It can also be expressed per unit mass as (Eq. 7-11*b*)

$$i = w_{\text{rev}} - w_u = T_0 s_{\text{gen}} \quad \text{(kJ/kg)}$$

Heat transfer between a steady-flow device and bodies other than the surroundings can be handled in the same manner discussed in Sec. 7-4 in relation to a closed system. For example, when a steady-flow device exchanges heat with a thermal reservoir at T_R at a rate of $\dot{Q}_R$, the reversible work relation is obtained by expressing the first and second laws for this process and eliminating the heat transfer term with the surroundings, yielding

$$\dot{W}_{\text{rev}} = \sum \dot{m}_i\psi_i - \sum \dot{m}_e\psi_e - \dot{Q}_R\left(1 - \frac{T_0}{T_R}\right) \qquad \text{(kW)} \qquad (7\text{-}21)$$

where the sign of $\dot{Q}_R$ is determined with respect to the reservoir.

EXAMPLE 7-11

Steam enters a turbine steadily at 3 MPa and 450°C at a rate of 8 kg/s and exits at 0.2 MPa and 150°C. The steam is losing heat to the surrounding air at 100 kPa and 25°C at a rate of 300 kW, and the kinetic and potential energy changes are negligible. Determine (*a*) the actual power output, (*b*) the maximum possible power output (the reversible power), (*c*) the second-law efficiency, (*d*) the irreversibility, and (*e*) the availability of the steam at the inlet conditions.

Solution A sketch of this single-stream steady-flow system is given in Fig. 7-33. Before we attempt to solve this problem, let us determine the steam properties at the inlet and the exit states and the state of the surroundings:

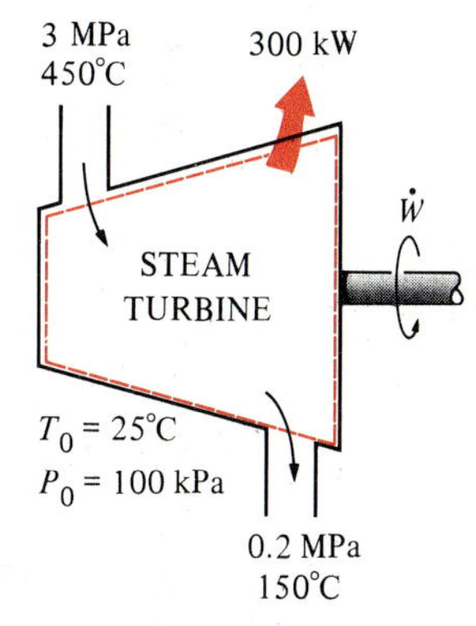

FIGURE 7-33
Schematic for Example 7-11.

Inlet state: $\left.\begin{aligned} P_1 &= 3\text{ MPa} \\ T_1 &= 450°\text{C} \end{aligned}\right\}$ $\begin{aligned} h_1 &= 3344.0\text{ kJ/kg} \\ s_1 &= 7.0834\text{ kJ/(kg·K)} \end{aligned}$ (Table A-6)

Exit state: $\left.\begin{aligned} P_2 &= 0.2\text{ MPa} \\ T_2 &= 150°\text{C} \end{aligned}\right\}$ $\begin{aligned} h_2 &= 2768.8\text{ kJ/kg} \\ s_2 &= 7.2795\text{ kJ/(kg·K)} \end{aligned}$ (Table A-6)

Dead state: $\left.\begin{aligned} P_0 &= 100\text{ kPa} \\ T_0 &= 25°\text{C} \end{aligned}\right\}$ $\begin{aligned} h_0 &\cong h_{f\,@\,25°\text{C}} = 104.89\text{ kJ/kg} \\ s_0 &\cong s_{f\,@\,25°\text{C}} = 0.3674\text{ kJ/(kg·K)} \end{aligned}$ (Table A-4)

(*a*) The actual power output, which is equivalent to the useful power output for steady-flow devices, is determined by applying the energy equation (Eq. 4-21*b*) to the actual process:

$$\begin{aligned} \dot{Q} - \dot{W} &= \dot{m}(\Delta h + \cancelto{0}{\Delta ke} + \cancelto{0}{\Delta pe}) \\ \dot{Q} - \dot{W} &= \dot{m}(h_2 - h_1) \\ -300\text{ kW} - \dot{W} &= (8\text{ kg/s})[(2768.8 - 3344.0)\text{ kJ/kg}] \\ \dot{W} &= 4302\text{ kW} \end{aligned}$$

(*b*) The maximum possible power output will be obtained when the process between the inlet and the exit states is carried out in a reversible manner. That is, it is the reversible power $\dot{W}_{\text{rev}}$ which can be determined from Eq. 7-16*a*:

$$\begin{aligned} \dot{W}_{\text{rev}} &= \dot{m}\left[(h_i - h_e) - T_0(s_i - s_e) + \cancelto{0}{\frac{\mathbf{V}_i^2 - \mathbf{V}_e^2}{2}} + \cancelto{0}{g(z_i - z_e)}\right] \\ &= \dot{m}[(h_1 - h_2) - T_0(s_1 - s_2)] \\ &= (8\text{ kg/s})\{(3344.0 - 2768.8)\text{ kJ/kg} \\ &\qquad - (298\text{ K})[(7.0834 - 7.2795)\text{ kJ/(kg·K)}]\} \\ &= 5069\text{ kW} \end{aligned}$$

(*c*) The second-law efficiency of a process is the ratio of the useful power to the reversible power, Eq. 7-4*b*:

$$\eta_{II} = \frac{\dot{W}_u}{\dot{W}_{rev}} = \frac{4302 \text{ kW}}{5069 \text{ kW}} = 0.849 \quad \text{or } 84.9\%$$

That is, 15.1 percent of the work potential is wasted during this process.

(*d*) Irreversibility, in rate form, is the difference between the reversible power and the useful power output, Eq. 7-20:

$$\dot{I} = \dot{W}_{rev} - \dot{W}_u = (5069 - 4302) \text{ kW} = 767 \text{ kW}$$

That is, the potential to produce useful work is wasted at a rate of 767 kW during this process. The irreversibility could also be determined by first calculating the net entropy generated $\dot{S}_{gen}$ during the process.

(*e*) The availability (maximum work potential) of the steam at the inlet conditions is determined from Eq. 7-17:

$$\begin{aligned}\psi_1 &= (h_1 - h_0) - T_0(s_1 - s_0) + \frac{\mathbf{V}_1^2}{2}^{\nearrow 0} + gz_1^{\nearrow 0} \\ &= (h_1 - h_0) - T_0(s_1 - s_0) \\ &= (3344.0 - 104.89) \text{ kJ/kg} - (298 \text{ K})[(7.0834 - 0.3674) \text{ kJ/(kg·K)}] \\ &= 1238 \text{ kJ/kg}\end{aligned}$$

That is, not counting the kinetic and potential energies, every kilogram of the steam entering the turbine has a work potential of 1238 kJ. This corresponds to a power potential of (8 kg/s)(1238 kJ/kg) = 9904 kW. Obviously, the turbine is converting 4302/9904 = 43.4 percent of the available work potential of the steam to work.

EXAMPLE 7-12

Water at 20 psia and 50°F enters a mixing chamber at a rate of 300 lbm/min, where it is mixed steadily with steam entering at 20 psia and 240°F. The mixture leaves the chamber at 20 psia and 130°F, and heat is being lost to the surrounding air at T_0 = 70°F at a rate of 180 Btu/min. Neglecting the changes in kinetic and potential energies, determine the reversible work and irreversibility for this process.

Solution This is a steady-flow process which was discussed in Example 6-12 with regard to entropy generation. A sketch of the mixing chamber and the system boundaries is given in Fig. 7-34. By applying the conservation of mass and the conservation of energy equations, the mass flow rate of the steam was determined in Example 6-12 to be $\dot{m}_2$ = 22.7 lbm/min. The reversible work for this process can be determined from Eq. 7-15:

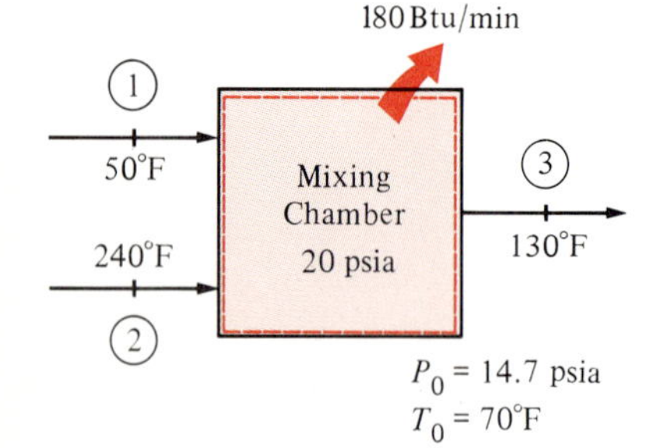

FIGURE 7-34
Schematic for Example 7-12.

$$\begin{aligned}\dot{W}_{rev} &= \sum \dot{m}_i \left(h_i + \frac{\mathbf{V}_i^2}{2}^{\nearrow 0} + gz_i^{\nearrow 0} - T_0 s_i\right) - \sum \dot{m}_e \left(h_e + \frac{\mathbf{V}_e^2}{2}^{\nearrow 0} + gz_e^{\nearrow 0} - T_0 s_e\right) \\ &= \Sigma \dot{m}_i(h_i - T_0 s_i) - \Sigma \dot{m}_e(h_e - T_0 s_e) \\ &= \dot{m}_1(h_1 - T_0 s_1) + \dot{m}_2(h_2 - T_0 s_2) - \dot{m}_3(h_3 - T_0 s_3) \\ &= (300 \text{ lbm/min})\{18.06 \text{ Btu/lbm} - (530 \text{ R})[0.03607 \text{ Btu/(lbm·R)}]\} \\ &\quad + (22.7 \text{ lbm/min})\{1162.3 \text{ Btu/lbm} - (530 \text{ R})[1.7405 \text{ Btu/(lbm·R)}]\} \\ &\quad\quad - (322.7 \text{ lbm/min})\{97.98 \text{ Btu/lbm} - (530 \text{ R})[0.18172 \text{ Btu/(lbm·R)}]\} \\ &= 4588.7 \text{ Btu/min}\end{aligned}$$

That is, we could have produced work at a rate of 4588.7 Btu/min if we ran a heat engine between the hot and the cold fluid streams instead of allowing them to mix directly.

The irreversibility is determined from Eq. 7-20:

$$\dot{I} = \dot{W}_{rev} - \cancelto{0}{\dot{W}_u} = T_0\dot{S}_{gen}$$

Thus,

$$\dot{I} = \dot{W}_{rev} = 4588.9 \text{ Btu/min}$$

since there is no actual work produced during this process (Fig. 7-35).

The entropy generation rate for this process was determined in Example 6-12 to be $\dot{S}_{gen} = 8.65$ Btu/(min·R). Thus the irreversibility could also be determined from the second part of the above equation:

$$\dot{I} = T_0\dot{S}_{gen} = (530 \text{ R})[8.65 \text{ Btu/(min·R)}] = 4584.5 \text{ Btu/min}$$

The slight difference between the two results is due to roundoff error.

FIGURE 7-35
For systems that involve no actual work, the reversible work and irreversibility are identical.

7-6 ■ SECOND-LAW ANALYSIS OF UNSTEADY-FLOW SYSTEMS

In Sec. 7-5 we discussed the second-law aspects of control volumes undergoing steady-flow (time-independent) processes. In this section, we extend the analysis to control volumes undergoing unsteady-flow (time-dependent, or transient) processes. The charging or discharging of storage vessels and the start-up or shutdown of equipment, for example, involve variations with time and thus should be analyzed as unsteady-flow processes.

Below we analyze the unsteady-flow systems first by assuming the fluid properties at any inlet or exit to remain constant (*uniform-flow analysis*) and then by allowing them to vary with time as well as with position over the inlets and the exits (*general unsteady-flow analysis*).

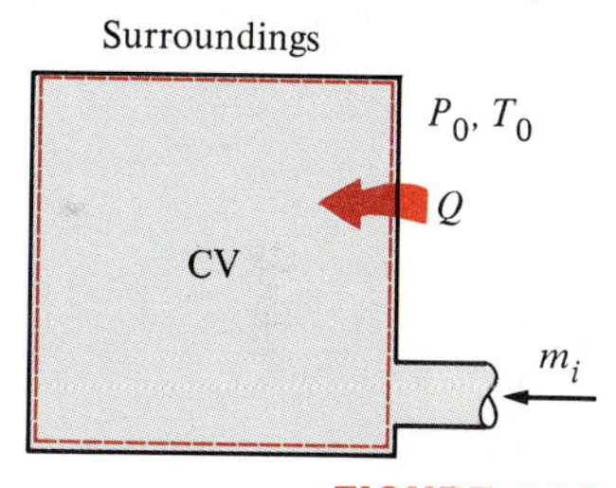

FIGURE 7-36
A uniform-flow system which exchanges heat with its surroundings only.

Uniform-Flow Processes

Consider a stationary control volume which undergoes a uniform-flow process from state 1 to state 2, as shown in Fig. 7-36. Some mass may enter the control volume at a uniform state i, and some may exit at another uniform state e. The control volume may involve multiple inlets and exits and may exchange heat with the surrounding medium at P_0 and T_0. The first and the second laws of thermodynamics for this uniform-flow process (Eqs. 4-35 and 6-13) can be expressed as follows:

$$Q - W = \sum m_e\left(h_e + \frac{\mathbf{V}_e^2}{2} + gz_e\right) - \sum m_i\left(h_i + \frac{\mathbf{V}_i^2}{2} + gz_i\right) + (U_2 - U_1)_{CV}$$

and

$$S_{gen} = (S_2 - S_1)_{CV} + \sum m_e s_e - \sum m_i s_i + \frac{Q_{surr}}{T_0}$$

where $Q_{surr} = -Q$ and S_{gen} is the amount of entropy generated during the process. As before, eliminating the heat transfer term between these two equations and solving for W, we obtain

$$W = \sum m_i\left(h_i + \frac{\mathbf{V}_i^2}{2} + gz_i - T_0 s_i\right) - \sum m_e\left(h_e + \frac{\mathbf{V}_e^2}{2} + gz_e - T_0 s_e\right) + [(U_1 - U_2) - T_0(S_1 - S_2)]_{CV} - T_0 S_{gen} \quad (7\text{-}22)$$

This is the **actual work** done during the process. A uniform-flow system, in general, may involve moving boundaries and thus surroundings work $W_{surr} = P_0(V_2 - V_1)$, as shown in Fig. 7-37. The **reversible work**, which is the maximum possible useful work, is obtained by subtracting W_{surr} from the above equation and setting the entropy generation term S_{gen} equal to zero:

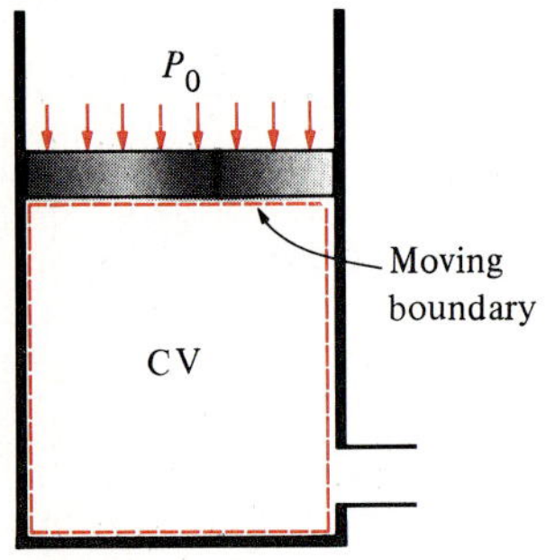

FIGURE 7-37

A uniform-flow system may involve moving boundaries and thus surroundings work.

$$W_{rev} = \sum m_i\left(h_i + \frac{\mathbf{V}_i^2}{2} + gz_i - T_0 s_i\right) - \sum m_e\left(h_e + \frac{\mathbf{V}_e^2}{2} + gz_e - T_0 s_e\right) + [(U_1 - U_2) - T_0(S_1 - S_2) + P_0(V_1 - V_2)]_{CV} \quad \text{(kJ)} \quad (7\text{-}23)$$

Using the definitions of closed-system availability and stream availability, we can also express the reversible work as

$$W_{rev} = \Sigma m_i\psi_i - \Sigma m_e\psi_e + (\Phi_1 - \Phi_2)_{CV} \quad \text{(kJ)} \quad (7\text{-}24)$$

In the above relations, subscript 1 refers to the initial state of the control volume, 2 to the final state, i to the inlet state, and e to the exit state.

The **availability** of a uniform-flow system at specified conditions is the maximum work potential available $W_{rev,max}$ and is determined from Eq. 7-24 by requiring both the control volume and the exit streams to be at the dead state (i.e., the state of zero availability, $\psi_e = 0$ and $\Phi_2 = 0$) at the end of the process:

$$\text{Availability} = W_{rev,max} = \Sigma m_i\psi_i + \Phi_{1,CV} \quad \text{(kJ)} \quad (7\text{-}25)$$

The **irreversibility** I associated with a uniform-flow process is the difference between the reversible work and the useful work (Eq. 7-11a):

$$I = W_{rev} - W_u = T_0 S_{gen} \quad \text{(kJ)}$$

Heat transfer between a uniform-flow system and bodies other than the atmosphere can be handled as discussed in preceding sections.

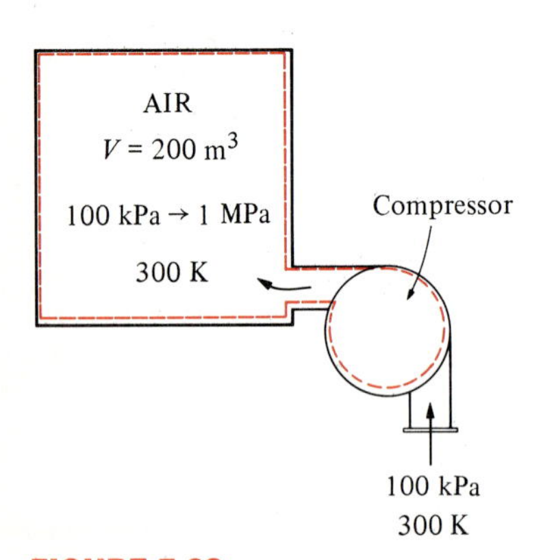

FIGURE 7-38

Schematic for Example 7-13.

EXAMPLE 7-13

A 200-m^3 rigid tank initially contains atmospheric air at 100 kPa and 300 K and is to be used as a storage vessel for compressed air at 1 MPa and 300 K. Compressed air is to be supplied by a compressor which takes in atmospheric air at P_0 = 100 kPa and T_0 = 300 K. Determine the minimum work requirement for this process.

Solution We take the rigid tank combined with the compressor as our control volume, as shown in Fig. 7-38. Overall this is a charging process, and thus the uniform-flow analysis is applicable. Air at specified conditions can be treated as an ideal gas since it is well above the critical temperature of its components. Air enters the control volume at uniform conditions (P_i = 100 kPa and T_i = 300 K). The minimum work required for a process is the reversible work which, in this case, can be determined from Eq. 7-23:

$$W_{rev} = \sum m_i\left(h_i + \frac{\mathbf{V}_i^2}{2} + gz_i - T_0 s_i\right) - \sum m_e\left(h_e + \frac{\mathbf{V}_e^2}{2} + gz_e - T_0 s_e\right) + [(U_1 - U_2) - T_0(S_1 - S_2) + P_0(V_1 - V_2)]_{CV}$$

No mass exits the control volume (m_e = 0), therefore the second term on the right-hand side drops out. The last term, in this case, is zero since the control volume has fixed boundaries ($V_1 = V_2$). By neglecting the kinetic and potential energies, this equation reduces to

$$W_{rev} = m_i(h_i - T_0 s_i) + [(m_1 u_1 - m_2 u_2) - T_0(m_1 s_1 - m_2 s_2)]_{CV}$$

The conservation of mass equation (Eq. 4-28) for this uniform-flow process yields

$$\Sigma m_i - \cancelto{0}{\Sigma m_e} = (m_2 - m_1)_{CV} \quad \longrightarrow \quad m_i = m_2 - m_1$$

Substituting this into the above equation and rearranging it (for convenience in entropy-change calculations), we obtain

$$W_{rev} = [m_2(h_i - u_2) - m_1(h_i - u_1)] - T_0[m_2(s_i - s_2) - m_1(s_i - s_1)]$$

The air temperatures at the inlet, initial, and the final states are given to be the same ($T_i = T_1 = T_2$ = 300 K). As an ideal gas, the enthalpy and internal energy of air depend on temperature only, and at 300 K they are

$$h_i = 300.19 \text{ kJ/kg}$$

and

$$u_1 = u_2 = 214.07 \text{ kJ/kg}$$

The mass of air in the tank at the initial and the final states is determined from the ideal-gas equation:

$$m_1 = \frac{P_1V_1}{RT_1} = \frac{(100 \text{ kPa})(200 \text{ m}^3)}{[0.287 \text{ kPa·m}^3/(\text{kg·K})](300 \text{ K})} = 232.29 \text{ kg}$$

and

$$m_2 = \frac{P_2V_2}{RT_2} = \frac{(1000 \text{ kPa})(200 \text{ m}^3)}{[0.287 \text{ kPa·m}^3/(\text{kg·K})](300 \text{ K})} = 2322.9 \text{ kg}$$

The entropy changes of the air are determined from Eq. 6-38:

$$s_i - s_1 = C_{p,av} \cancelto{0}{\ln \frac{T_i}{T_1}} - R \cancelto{0}{\ln \frac{P_i}{P_1}} = 0$$

and

$$s_i - s_2 = C_{p,av} \cancelto{0}{\ln \frac{T_i}{T_2}} - R \ln \frac{P_i}{P_2} = -R \ln \frac{P_i}{P_2}$$

$$= -[0.287 \text{ kJ/(kg·K)}] \ln \frac{100 \text{ kPa}}{1000 \text{ kPa}} = 0.661 \text{ kJ/(kg·K)}$$

Substituting yields

$$W_{rev} = \{(2322.9 \text{ kg})[(300.19 - 214.07) \text{ kJ/kg}] - (232.29 \text{ kg})[(300.19 - 214.07) \text{ kJ/kg}]\} - (300 \text{ K})\{(2322.9 \text{ kg})[0.661 \text{ kJ/(kg·K)}] - 0\}$$

$$= -280{,}588 \text{ kJ}$$

That is, a minimum of 280,588 kJ of work input is required to fill the tank with compressed air at 300 K and 1 MPa. In reality, the required work input will be greater by an amount equal to the irreversibility of the actual process.

EXAMPLE 7-14

Determine the availability of the 2322.9 kg of compressed air at 1 MPa and 300 K stored in the 200-m^3 rigid tank discussed in Example 7-13.

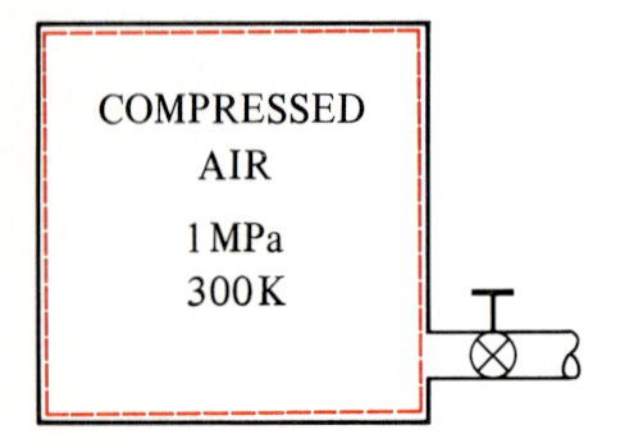

FIGURE 7-39
Schematic for Example 7-14.

Solution This time we have a fixed mass (closed system) at 1 MPa and 300 K (state 2 in Example 7-13) as our system, as shown in Fig. 7-39. Its availability can be determined from Eq. 7-9*a*:

$$\Phi_2 = (U_2 - U_0) - T_0(S_2 - S_0) + P_0(V_2 - V_0)$$

Note that the dead state (P_0 = 100 kPa and T_0 = 300 K) is identical to the inlet state (state *i*) in Example 7-13. Thus,

$$U_2 - U_0 = m(u_2 - u_0) = mC_v(T_2 - T_0) = 0$$

$$s_2 - s_0 = s_2 - s_i = -0.661 \text{ kJ/(kg·K)}$$

and $$\frac{P_0V_0}{T_0} = \frac{P_2V_2}{T_2} \longrightarrow V_0 = \frac{P_2}{P_0}V_2 = \frac{1000 \text{ kPa}}{100 \text{ kPa}}(200 \text{ m}^3) = 2000 \text{ m}^3$$

Substituting, we see that the availability is

$$\begin{aligned}\Phi_2 &= m(u_2 - u_0)^{\nearrow 0} - T_0m(s_2 - s_0) + P_0(V_2 - V_0)\\ &= 0 - (300 \text{ K})(2322.9 \text{ kg})[-0.661 \text{ kJ/(kg·K)}]\\ &\qquad + (100 \text{ kPa})[(200 - 2000) \text{ m}^3][\text{kJ/(kPa·m}^3)]\\ &= 280{,}631 \text{ kJ}\end{aligned}$$

That is, a maximum of 280,631 kJ of useful work could be obtained from this compressed air stored in the tank. This is (supposed to be) the same result as obtained in Example 7-13. The slight difference between the two is due to roundoff error.

General Unsteady-Flow Processes

When the properties of a fluid stream entering or exiting a control volume vary considerably with time or over the cross-sectional area of an inlet or exit, the uniform-flow assumption utilized above may yield unacceptable errors and may no longer be appropriate. In those cases, the variation of fluid properties during a process should be properly accounted for (Fig. 7-40). This is done by replacing the finite quantity

$$\left(h + \frac{\mathbf{V}^2}{2} + gz - T_0s\right)m \tag{7-26}$$

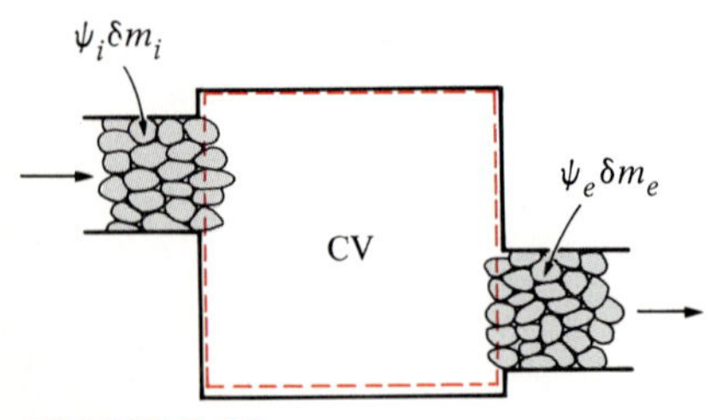

FIGURE 7-40
When the fluid properties at an opening are not uniform, the availability of a fluid stream is obtained by integration.

by the sum (integral) of differential quantities as

$$\int_1^2 \left(h + \frac{\mathbf{V}^2}{2} + gz - T_0s\right)\delta m \tag{7-27}$$

where $\delta m = \rho\mathbf{V}\,dA\,dt$ (in which ρ = density, $\mathbf{V}$ = normal velocity,

dA = differential cross-sectional area of an inlet or exit, dt = differential time interval) is the differential mass entering or leaving the control volume and the integration limits 1 and 2 are the initial and the final states, respectively, of the process. Obviously when the fluid properties remain constant during a process, the quantity in parentheses can be taken out of the integral and Eq. 7-27 reduces to Eq. 7-26.

By following this procedure, the reversible work relations for a control volume exchanging heat with only the surroundings (Eqs. 7-23 and 7-24) can be expressed for the general case as

$$\begin{aligned} W_{\text{rev}} = & \sum \int_1^2 \left(h_i + \frac{\mathbf{V}_i^2}{2} + gz_i - T_0 s_i\right) \delta m_i \\ & - \sum \int_1^2 \left(h_e + \frac{\mathbf{V}_e^2}{2} + gz_e - T_0 s_e\right) \delta m_e \\ & + [(U_1 - U_2) - T_0(S_1 - S_2) + P_0(V_1 - V_2)]_{\text{CV}} \quad \text{(kJ)} \quad (7\text{-}28) \end{aligned}$$

or

$$W_{\text{rev}} = \sum \int_1^2 \psi_i \, \delta m_i - \sum \int_1^2 \psi_e \, \delta m_e + (\Phi_1 - \Phi_2)_{\text{CV}} \quad \text{(kJ)} \quad (7\text{-}29)$$

To perform the integrations, one needs to know how the properties vary with time as well as with position. If such relations are not available, the integrations should be performed numerically.

The reversible work can also be expressed in rate form by dividing each term in Eq. 7-28 by Δt and taking the limit as $\Delta t \rightarrow 0$:

$$\begin{aligned} \dot{W}_{\text{rev}} = & \sum \int_{A_i} \left(h_i + \frac{\mathbf{V}_i^2}{2} + gz_i - T_0 s_i\right) \delta \dot{m}_i \\ & - \sum \int_{A_e} \left(h_e + \frac{\mathbf{V}_e^2}{2} + gz_e - T_0 s_e\right) \delta \dot{m}_e \\ & - \frac{d(E + P_0 V - T_0 S)_{\text{CV}}}{dt} \quad \text{(kJ)} \quad (7\text{-}30) \end{aligned}$$

Equations 7-28 and 7-30 can be modified to handle heat transfer with K other bodies (at temperatures T_k in the amounts of Q_k or $\dot{Q}_k$) by subtracting

$$\sum_{k=1}^{K} Q_k \left(1 - \frac{T_0}{T_k}\right) \qquad \text{from Eq. 7-28}$$

or

$$\sum_{k=1}^{K} \dot{Q}_k \left(1 - \frac{T_0}{T_k}\right) \qquad \text{from Eq. 7-30}$$

Here the sign of Q_k (or $\dot{Q}_k$) is determined with respect to the other bodies, not the system.

The irreversibility associated with the process can be determined from Eq. 7-11*a*, or from Eq. 7-20 in rate form.

The reversible work relations developed above are sufficiently general and can be applied to any thermodynamic system. Relations for closed systems and steady-flow and uniform-flow systems can be obtained from these relations as special cases.

EXAMPLE 7-15

An insulated storage tank with $V = 50\ m^3$ initially contains argon gas at $T_1 = 500$ K and $P_1 = 800$ kPa. The argon is allowed to escape until the pressure in the tank drops to $P_2 = 100$ kPa. The process in the tank is sufficiently slow and adiabatic and thus can be assumed to be isentropic. The surrounding medium is at $T_0 = 25°C$ and $P_0 = 100$ kPa. Determine the reversible work and irreversibility for this process while accounting for the variation of the properties at the tank exit. Disregard any changes in kinetic and potential energies.

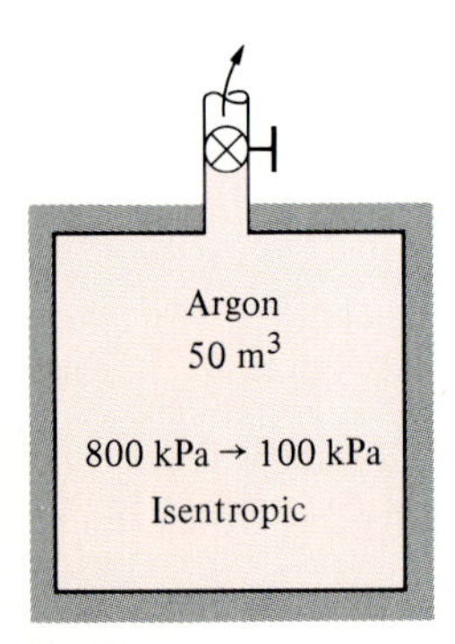

FIGURE 7-41
Schematic for Example 7-15.

Solution We take the volume within the storage tank as the control volume, as shown in Fig. 7-41. This is a transient process since the fluid properties within the control volume are changing with time. The process is assumed to be sufficiently slow, thus the fluid properties within the tank can be taken to be uniform at any instant during the process. The argon is well above its critical temperature (151 K) and can be treated as an ideal gas for this process. The process is isentropic, and the pressure and the temperature in the tank change according to the ideal-gas isentropic relation (Eq. 6-47*b*)

$$TP^{(1-k)/k} = \text{constant}$$

where k is the specific heat ratio C_p/C_v. Taking 0 K as the reference temperature, we can express the internal energy and enthalpy of argon as

$$u = C_v T \qquad \text{and} \qquad h = C_p T$$

where T is the absolute temperature. The control volume involves no actual work interactions, and therefore the reversible work and the irreversibility are identical in this case. The reversible work will be determined from Eq. 7-28 to account for the variation of properties at the tank exit.

Disregarding the kinetic and potential energy changes, Eq. 7-28 reduces to

$$\begin{aligned} W_{rev} &= -\int_1^2 (h_e - T_0 s_e)\,\delta m_e + [(U_1 - U_2) - T_0(S_1 - S_2)]_{CV} \\ &= -\int_1^2 h_e\,\delta m_e + \int_1^2 T_0 s_e\,\delta m_e + [(m_1u_1 - m_2u_2) - T_0(m_1s_1 - m_2s_2)]_{CV} \end{aligned}$$

The specific entropy of a system remains constant during an isentropic process ($s_1 = s_2 = s_e =$ constant). Thus the second integral in the above equation can easily be found to yield $T_0 s_e m_e$. After rearranging, the above equation reduces to

$$\begin{aligned} W_{rev} &= -\int_1^2 h_e\,\delta m_e + (m_1u_1 - m_2u_2)_{CV} + T_0[m_1(\overset{\nearrow 0}{s_e - s_1}) - m_2(\overset{\nearrow 0}{s_e - s_2})]_{CV} \\ &= -\int_1^2 h_e\,\delta m_e + (m_1u_1 - m_2u_2)_{CV} \end{aligned}$$

The integral in the final relation was discussed and performed in Chap. 4 for the polytropic process ($Pv^n =$ constant) of an ideal gas. The isentropic process is a special case of the polytropic process with $n = k$. Replacing the polytropic exponent n by the isentropic exponent k and simplifying, we obtain

$$-\int_1^2 h_e\,\delta m_e = \frac{C_p}{k}\frac{V}{R}(P_2 - P_1) = C_v \frac{V}{R}(P_2 - P_1)$$

since $k = C_p/C_v$. The last two terms of the equation can be expressed as

$$m_1u_1 - m_2u_2 = \left(\frac{P_1V}{RT_1}\right) C_vT_1 - \left(\frac{P_2V}{RT_2}\right) C_vT_2 = C_v \frac{V}{R}(P_1 - P_2)$$

Substituting, we get

$$W_{rev} = C_v \frac{V}{R}(P_2 - P_1) + C_v \frac{V}{R}(P_1 - P_2) = 0$$

That is, the reversible work (and irreversibility) is zero for this process, meaning no work potential is wasted during an isentropic discharging process. This is true given that the fluid is discharged to a work-producing device such as a turbine. If it is throttled by a valve (a highly irreversible process) and discharged to the atmosphere, the entire work potential will be wasted. Even with a valve, the process in the tank (excluding the valve) is still reversible, but the combined process would be irreversible.

7-7 ■ SUMMARY

The energy content of the universe is constant, just as its mass content is. Yet at times of crisis we are bombarded with speeches and articles on how to "conserve" energy. As engineers, we know that energy is already conserved. What is not conserved is the availability, which is the useful work potential of the energy. Once the availability is wasted, it can never be recovered. When we use energy (to heat our homes, for example), we are not destroying any energy; we are merely converting it to a less useful form, a form of less availability.

The maximum useful work potential of a system at the specified state is called *availability*. Availability is a property and is associated with the state of the system and the surroundings. A system which is in equilibrium with its surroundings has zero availability and is said to be at the *dead state*.

The mechanical forms of energy (such as kinetic and potential energies) are entirely available energy. The availability of the thermal energy of thermal reservoirs is equivalent to the work output of a Carnot heat engine operating between the reservoir and the environment.

The *actual work* W done during a process can be determined from the conservation of energy equations discussed in Chaps. 3 and 4. If the volume of the system changes during a process, part of this work, called the *surroundings work* W_{surr}, is used to push the surrounding medium at a constant pressure of P_0 out of the way, and it cannot be used for any useful purpose. The difference between the total actual work and the surroundings work is called the *useful work* W_u, and it is determined from

$$W_u = W - W_{surr} = W - P_0(V_2 - V_1) \qquad \text{(kJ)} \qquad (7\text{-}2)$$

The surroundings work W_{surr} is zero for cyclic devices, steady-flow devices, and systems with fixed boundaries.

The maximum amount of useful work that can be obtained from a system as it undergoes a process between two specified states is called *reversible work* W_{rev}. This is the useful work output (or input) obtained when the process between the initial and final states is executed in a totally reversible manner. For the special case of the final state being the state of the environment (P_0, T_0), the reversible work and availability become identical.

The difference between the reversible work W_{rev} and the useful work W_u for a process is due to the irreversibilities and is called *irreversibility* I. For any system, closed or open, it is expressed as

$$I = W_{rev} - W_u = T_0 S_{gen} \quad \text{(kJ)} \qquad (7\text{-}11a)$$

or

$$i = w_{rev} - w_u = T_0 s_{gen} \quad \text{(kJ/kg)} \qquad (7\text{-}11b)$$

or

$$\dot{I} = \dot{W}_{rev} - \dot{W}_u = T_0 \dot{S}_{gen} \quad \text{(kW)} \qquad (7\text{-}20)$$

where S_{gen} (or $\dot{S}_{gen}$) is the entropy generated during the process. For a totally reversible process, the useful and reversible work terms are identical and thus irreversibility is zero.

The first-law efficiency alone is not a realistic measure of performance for engineering devices. Thus a parameter is defined to express the performance of a device relative to the performance under reversible conditions for the same end states. This parameter is called the *second-law efficiency* η_{II} and is given by

$$\eta_{II} = \frac{\eta_{th}}{\eta_{th,rev}} = \frac{W_u}{W_{rev}} \qquad (7\text{-}4a,b)$$

for heat engines and other work-producing devices, and

$$\eta_{II} = \frac{\text{COP}}{\text{COP}_{rev}} = \frac{W_{rev}}{W_u} \qquad (7\text{-}5a,b)$$

for refrigerators, heat pumps, and other work-consuming devices.

The availabilities of a closed system (ϕ) and a flowing fluid stream (ψ) are given on a unit-mass basis by

$$\phi = (u - u_0) - T_0(s - s_0) + P_0(v - v_0) \quad \text{(kJ/kg)} \qquad (7\text{-}9b)$$

and

$$\psi = (h - h_0) - T_0(s - s_0) + \frac{\mathbf{V}^2}{2} + gz \quad \text{(kJ/kg)} \qquad (7\text{-}17)$$

where properties with "0" subscript are to be evaluated at the state of the surroundings, P_0 and T_0.

The reversible work expressions can be summarized as follows:

Cyclic devices:

$$W_{rev} = \eta_{th,rev} Q_H \qquad \text{heat engines, Eq. 5-20}$$

$$-W_{rev} = \frac{Q_L}{\text{COP}_{R,rev}} \qquad \text{refrigerators, Eq. 5-22}$$

$$-W_{rev} = \frac{Q_H}{\text{COP}_{HP,rev}} \qquad \text{heat pumps, Eq. 5-23}$$

Closed systems:

$$W_{\text{rev}} = (U_1 - U_2) - T_0(S_1 - S_2) + P_0(V_1 - V_2) \qquad (7\text{-}8)$$

$$= m(\phi_1 - \phi_2) \qquad \text{(kJ)} \qquad (7\text{-}10a)$$

Steady-flow systems:

$$\dot{W}_{\text{rev}} = \sum \dot{m}_i \left(h_i + \frac{\mathbf{V}_i^2}{2} + gz_i - T_0 s_i \right) - \sum \dot{m}_e \left(h_e + \frac{\mathbf{V}_e^2}{2} + gz_e - T_0 s_e \right) \qquad (7\text{-}15)$$

$$= \Sigma \dot{m}_i \psi_i - \Sigma \dot{m}_e \psi_e \qquad \text{(kW)} \qquad (7\text{-}18)$$

For a single-stream device it reduces to

$$\dot{W}_{\text{rev}} = \dot{m} \left[(h_i - h_e) - T_0(s_i - s_e) + \frac{\mathbf{V}_i^2 - \mathbf{V}_e^2}{2} + g(z_i - z_e) \right] \qquad \text{(kW)} \qquad (7\text{-}16a)$$

$$= \dot{m}(\psi_i - \psi_e) \qquad \text{(kW)} \qquad (7\text{-}19b)$$

Uniform-flow systems:

$$W_{\text{rev}} = \sum m_i \left(h_i + \frac{\mathbf{V}_i^2}{2} + gz_i - T_0 s_i \right) - \sum m_e \left(h_e + \frac{\mathbf{V}_e^2}{2} + gz_e - T_0 s_e \right) + [(U_1 - U_2) - T_0(S_1 - S_2) + P_0(V_1 - V_2)]_{\text{CV}} \qquad \text{(kJ)} \qquad (7\text{-}23)$$

$$= \Sigma m_i \psi_i - \Sigma m_e \psi_e + (\Phi_1 - \Phi_2)_{\text{CV}} \qquad \text{(kJ)} \qquad (7\text{-}24)$$

In the above relations, subscript 1 refers to the initial state of the system, 2 to the final state, i to inlet state, e to the exit state, and 0 to the state of the environment at T_0 and P_0. It is assumed that any heat transfer takes place between the system and its surroundings only. The reversible work relations above can be modified to handle heat transfer with K other bodies (at temperatures T_k in the amounts of Q_k or $\dot{Q}_k$) by subtracting

$$\sum_{k=1}^{K} Q_k \left(1 - \frac{T_0}{T_k} \right) \quad \text{or} \quad \sum_{k=1}^{K} \dot{Q}_k \left(1 - \frac{T_0}{T_k} \right)$$

from them. Here the sign of Q_k (or $\dot{Q}_k$) is determined with respect to the other bodies, not the system.

The reversible work can also be expressed in a general form as

$$W_{\text{rev}} = \sum \int_1^2 \left(h_i + \frac{\mathbf{V}_i^2}{2} + gz_i - T_0 s_i \right) \delta m_i - \sum \int_1^2 \left(h_e + \frac{\mathbf{V}_e^2}{2} + gz_e - T_0 s_e \right) \delta m_e + [(U_1 - U_2) - T_0(S_1 - S_2) + P_0(V_1 - V_2)]_{\text{CV}} - \sum_{k=1}^{K} Q_k \left(1 - \frac{T_0}{T_k} \right) \qquad \text{(kJ)}$$

REFERENCES AND SUGGESTED READING

1 J. E. Ahern, *The Exergy Method of Energy Systems Analysis,* Wiley, New York, 1980.

2 A Bejan, *Entropy Generation through Heat and Fluid Flow,* Wiley-Interscience, New York, 1982.

3 W. Z. Black and J. G. Hartley, *Thermodynamics,* Harper & Row, New York, 1985.

4 J. B. Jones and G. A. Hawkins, *Engineering Thermodynamics,* 2d ed., Wiley, New York, 1986.

5 J. H. Keenan, *Thermodynamics,* Wiley, New York, 1941.

6 G. J. Van Wylen and R. E. Sonntag, *Fundamentals of Classical Thermodynamics,* 3d ed., Wiley, New York, 1985.

7 K. Wark, *Thermodynamics,* 5th ed., McGraw-Hill, New York, 1988.

PROBLEMS*

Availability, Irreversibility, Reversible Work, and Second-Law Efficiency

7-1C What is the dead state?

7-2C What is the difference between reversible work and availability?

7-3C How does reversible work differ from the useful work?

7-4C Under what conditions does the reversible work equal availability?

7-5C Under what conditions does the reversible work equal irreversibility for a process?

7-6C Under what conditions does the reversible work equal the actual work for a process?

7-7C What final state will maximize the work output of a device?

7-8C Is the availability of a system different in different environments? Explain.

7-9C How does useful work differ from actual work? For what kind of systems are these two identical?

7-10C Are unavailable energy and irreversibility basically the same thing? If not, how do they differ?

7-11C Consider a process that involves no irreversibilities. Will the actual useful work for that process be equal to the reversible work?

---

*Students are encouraged to answer *all* the concept "C" questions.

7-12C Consider two geothermal wells whose energy contents are estimated to be the same. Will the availabilities of these wells necessarily be the same? Explain.

7-13C Consider two systems that are at the same pressure as the environment. The first system is at the same temperature as the environment, whereas the second system is at a lower temperature than the environment. How would you compare the availabilities of these two systems?

7-14C Consider an environment of zero absolute pressure (such as outer space). How will the actual work and the useful work compare in that environment?

7-15C What is the second-law efficiency? How does it differ from the first-law efficiency?

7-16C Does a power plant that has a higher thermal efficiency necessarily have a higher second-law efficiency than one with a lower thermal efficiency? Explain.

7-17C Does a refrigerator that has a higher COP necessarily have a higher second-law efficiency than one with a lower COP? Explain.

7-18C Can a process for which the reversible work is zero be reversible? Can it be irreversible? Explain.

7-19C Consider a steam turbine operating between two specified states. As the design of the turbine improves, will the actual work increase? How about the reversible work? Explain.

7-20C Consider a process during which no entropy is generated ($S_{gen} = 0$). Does the irreversibility for this process have to be zero?

7-21 The electric power needs of a community are to be met by windmills with 10-m-diameter rotors. The windmills are to be located where the wind is blowing steadily at an average velocity of 8 m/s. Determine the minimum number of windmills that need to be installed if the required power output is 300 kW.

7-22 One method of meeting the extra electric power demand at peak periods is to pump some water from a large body of water (such as a lake) to a water reservoir at a higher elevation at times of low demand and to generate electricity at times of high demand by letting this water run down and rotate a turbine (i.e, convert the electric energy to potential energy and then back to electric energy). For an energy storage capacity of 5×10^6 kWh, determine the minimum amount of water that needs to be stored at an average elevation (relative to the ground level) of 75 m. *Answer:* 2.45×10^{10} kg

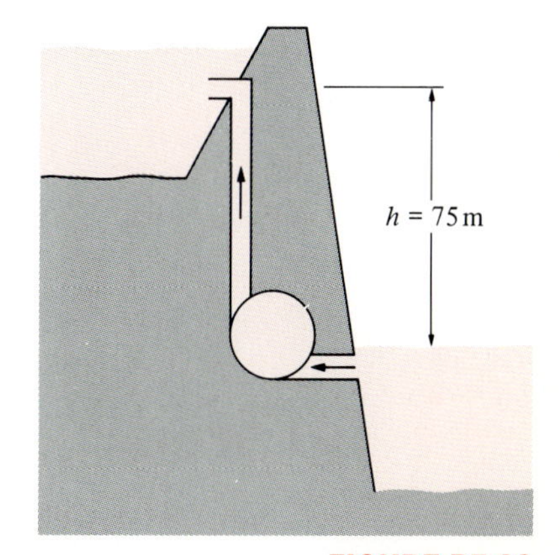

FIGURE P7-22

7-23 A crater lake has a base area of 10,000 m^2, and the water it contains is 12 m deep. The ground surrounding the crater is nearly flat and is 140 m below the base of the lake. Determine the maximum amount of electrical work, in kWh, that can be generated by feeding this water to a hydroelectric power plant. *Answer:* 47,800 kWh

7-23E A crater lake has a base area of 80,000 ft^2, and the water it contains is 35 ft deep. The ground surrounding the crater is nearly flat and is 400 ft below the base of the lake. Determine the maximum amount of electrical work, in kWh, that can be generated by feeding this water to a hydroelectric power plant. *Answer:* 27,318 kWh

7-24 Consider a thermal energy reservoir at 1500 K that can supply heat at a rate of 300,000 kJ/h. Determine the availability of this supplied energy, assuming an environmental temperature of 25°C.

7-25 A heat engine receives heat from a source at 1500 K at a rate of 700 kJ/s, and it rejects the waste heat to a medium at 320 K. The measured power output of the heat engine is 320 kW, and the lowest naturally occurring surrounding temperature is 25°C. Determine (*a*) the reversible power, (*b*) the rate of irreversibility, and (*c*) the second-law efficiency of this heat engine.
Answers: (*a*) 550.7 kW, (*b*) 230.7 kW, (*c*) 58.1 percent

7-26 A heat engine which rejects waste heat to a sink at 310 K has a thermal efficiency of 36 percent and a second-law efficiency of 60 percent. Determine the temperature of the source that supplies heat to this heat engine. *Answer:* 775 K

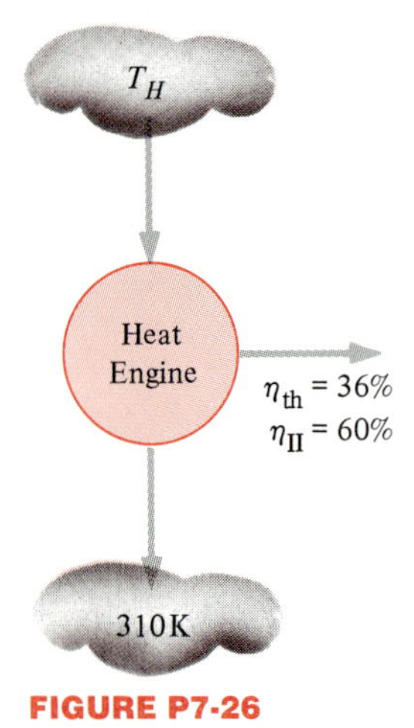

FIGURE P7-26

7-26E A heat engine which rejects waste heat to a sink at 530 R has a thermal efficiency of 36 percent and a second-law efficiency of 60 percent. Determine the temperature of the source that supplies heat to this heat engine.

7-27 How much of the 100 kJ of thermal energy at 850 K can be converted to useful work? Assume the environment to be at 25°C.

7-28 A heat engine that receives heat from a furnace at 1100°C and rejects waste heat to a river at 20°C has a thermal efficiency of 35 percent. Determine the second-law efficiency of this power plant.

7-29 A house which is losing heat at a rate of 60,000 kJ/h when the outside temperature drops to 15°C is to be heated by electric resistance heaters. If the house is to be maintained at 22°C at all times, determine the reversible work for this process and the irreversibility.
Answers: −0.4 kW, 16.27 kW

7-29E A house which is losing heat at a rate of 60,000 Btu/h when the outside temperature drops to 60°F is to be heated by electric resistance heaters. If the house is to be maintained at 80°F at all times, determine the reversible work for this process and the irreversibility.

7-30 A freezer is maintained at −6°C by removing heat from it at a rate of 75 kJ/min. The power input to the freezer is 500 W, and the surrounding air is at 24°C. Determine (*a*) the reversible power, (*b*) the irreversibility, and (*c*) the second-law efficiency of this freezer.

7-30E A freezer is maintained at 20°F by removing heat from it at a rate of 75 Btu/min. The power input to the freezer is 0.70 hp, and the

surrounding air is at 75°F. Determine (*a*) the reversible power, (*b*) the irreversibility, and (*c*) the second-law efficiency of this freezer.
Answers: (*a*) 0.20 hp, (*b*) 0.50 hp, (*c*) 28.9 percent

7-31 A refrigerator has a second-law efficiency of 45 percent, and heat is removed from it at a rate of 180 kJ/min. If the refrigerator is maintained at 3°C while the surrounding air temperature is 27°C, determine the power input to the refrigerator. *Answer:* 0.58 kW

7-31E A refrigerator has a second-law efficiency of 45 percent, and heat is removed from it at a rate of 200 Btu/min. If the refrigerator is maintained at 35°F while the surrounding air temperature is 75°F, determine the power input to the refrigerator.

7-32 One method of passive solar heating is to stack gallons of liquid water inside the buildings, exposing them to the sun. The solar energy stored in the water during the day is released at night to the room air, providing some heating. Consider a house which is maintained at 22°C and whose heating is assisted by a 300-L water storage system. If the water is heated to 45°C during the day, determine the amount of heating that this water will provide to the house at night. Assuming an outside temperature of 5°C, determine the irreversibility associated with this process. *Answers:* 28,842 kJ, 1061 kJ

Second-Law Analysis of Closed Systems

7-33C Is a process during which no entropy is generated ($S_{gen} = 0$) necessarily reversible?

7-34C Writing the first- and the second-law relations and simplifying, obtain the reversible work relation for a closed system which exchanges heat with the surrounding medium at T_0 in the amount of Q_0 as well as a thermal reservoir at T_R in the amount of Q_R. (*Hint:* Eliminate Q_0 between the two equations.)

7-35 A piston-cylinder device initially contains 2 L of air at 100 kPa and 25°C. Air is now compressed to a final state of 600 kPa and 150°C. The useful work input is 1.2 kJ. Assuming the surroundings are at 100 kPa and 25°C, determine (*a*) the availability of the air at the initial and the final states, (*b*) the minimum work that must be supplied to accomplish this compression process, and (*c*) the second-law efficiency of this process. *Answers:* (*a*) 0, 0.171 kJ; (*b*) 0.171 kJ; (*c*) 14.3 percent

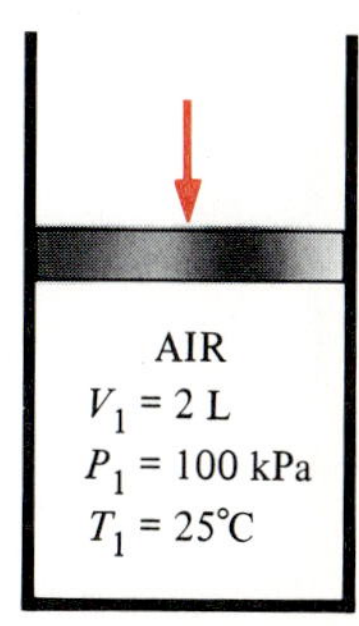

FIGURE P7-35

7-36 A piston-cylinder device contains 5 kg of refrigerant-12 at 0.8 MPa and 50°C. The refrigerant is now cooled at constant pressure until it exists as a liquid at 30°C. If the surroundings are at 100 kPa and 30°C, determine (*a*) the availability of the refrigerant at the initial and the final states and (*b*) the irreversibility for this process.

7-36E A piston-cylinder device contains 3 lbm of refrigerant-12 at 120 psia and 120°F. The refrigerant is now cooled at constant pressure until it exists as a liquid at 90°F. If the surroundings are at 15 psia and 80°F,

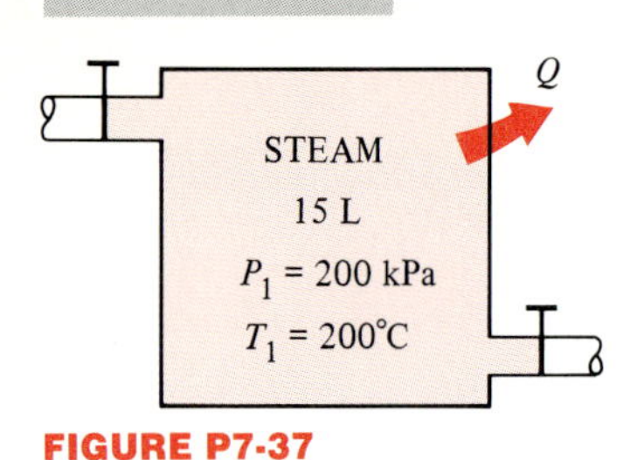

FIGURE P7-37

determine (*a*) the availability of the refrigerant at the initial and final states and (*b*) the irreversibility for this process.

7-37 The radiator of a steam heating system has a volume of 15 L and is filled with superheated water vapor at 200 kPa and 200°C. At this moment both the inlet and the exit valves to the radiator are closed. After a while it is observed that the temperature of the steam drops to 80°C as a result of heat transfer to the room air, which is at 21°C. Assuming the surroundings to be at 0°C, determine (*a*) the amount of heat transfer to the room and (*b*) the maximum amount of heat that can be supplied to the room if this heat from the radiator is supplied to a heat engine which is driving a heat pump. Assume the heat engine operates between the radiator and the surroundings. *Answers:* (*a*) 22.78 kJ, (*b*) 73.4 kJ

7-38 A well-insulated rigid tank contains 2 kg of saturated liquid–vapor mixture of water at 100 kPa. Initially, three-quarters of the mass is in the liquid phase. An electric resistance heater placed in the tank is turned on and kept on until all the liquid in the tank is vaporized. Assuming the surroundings to be at 25°C and 100 kPa, determine (*a*) the irreversibility and (*b*) the second-law efficiency for this process.

7-38E A well-insulated rigid tank contains 4 lbm of saturated liquid–vapor mixture of water at 15 psia. Initially, three-quarters of the mass is in the liquid phase. An electric resistance heater placed in the tank is turned on and kept on until all the liquid in the tank is vaporized. Assuming the surroundings to be at 75°F and 14.7 psia, determine (*a*) the irreversibility and (*b*) the second-law efficiency for this process.

7-39 A rigid tank is divided into two equal parts by a partition. One part of the tank contains 0.5 kg of compressed liquid water at 500 kPa and 60°C and the other side is evacuated. The partition is removed, and the water expands to fill the entire tank. If the final pressure in the tank is 15 kPa, determine the irreversibility for this process. Assume the surroundings to be at 25°C and 100 kPa. *Answer:* 1.22 kJ

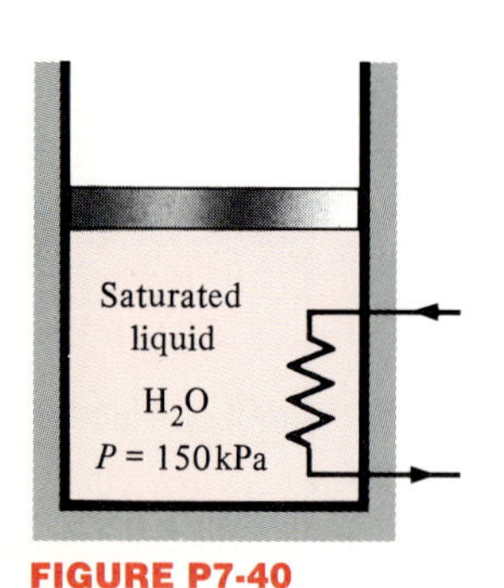

FIGURE P7-40

7-40 An insulated piston-cylinder device contains 2 L of saturated liquid water at a constant pressure of 150 kPa. An electric resistance heater inside the cylinder is turned on, and electrical work is done on the steam in the amount of 2200 kJ. Assuming the surroundings to be at 25°C and 100 kPa, determine (*a*) the minimum work with which this process could be accomplished and (*b*) the irreversibility for this process.
Answers: (*a*) 437.8 kJ, (*b*) 1704.5 kJ

7-40E An insulated piston-cylinder device contains 0.05 ft^3 of saturated liquid water at a constant pressure of 20 psia. An electric resistance heater inside the cylinder is turned on, and electrical work is done on the steam in the amount of 2200 Btu. Assuming the surroundings to be at 77°F and 14.7 psia, determine (*a*) the minimum work with which this process could be accomplished and (*b*) the irreversibility for this process.

7-41 An insulated piston-cylinder device contains 0.01 m^3 of saturated refrigerant-12 vapor at 0.8-MPa pressure. The refrigerant is now allowed to expand in a reversible manner until the pressure drops to 0.4 MPa. Determine the change in the availability of the refrigerant during this process and the reversible work. Assume the surroundings to be at 25°C and 100 kPa.

7-42 Two rigid tanks are connected by a valve. Tank *A* is insulated and contains 0.2 m^3 of steam at 400 kPa and 80 percent quality. Tank *B* is uninsulated and contains 3 kg of steam at 200 kPa and 250°C. The valve is now opened, and steam flows from tank *A* to tank *B* until the pressure in tank *A* drops to 300 kPa. During this process 600 kJ of heat is transferred from tank *B* to the surroundings at 0°C. Assuming the steam remaining inside tank *A* to have undergone a reversible adiabatic process, determine (*a*) the final temperature in each tank and (*b*) the work potential wasted during this process.

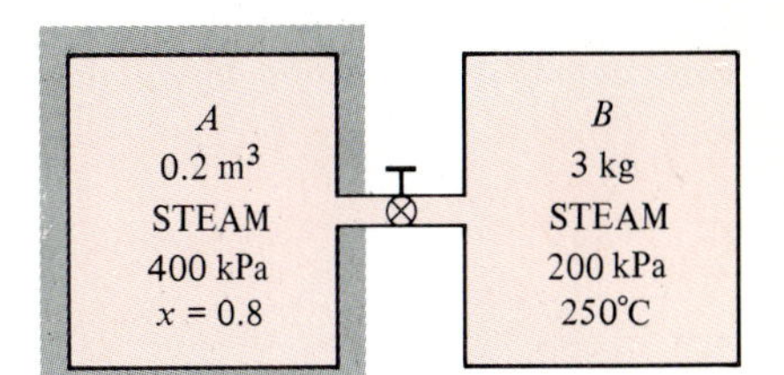

FIGURE P7-42

7-43 Oxygen gas is compressed in a piston-cylinder device from an initial state of 0.8 m^3/kg and 25°C to a final state of 0.1 m^3/kg and 287°C. Determine the reversible work and the increase in the availability of the oxygen during this process. Assume the surroundings to be at 25°C and 100 kPa.

7-43E Oxygen gas is compressed in a piston-cylinder device from an initial state of 12 ft^3/lbm and 75°F to a final state of 1.5 ft^3/lbm and 525°F. Determine the reversible work and the increase in the availability of the oxygen during this process. Assume the surroundings to be at 14.7 psia and 75°F. *Answers:* −60.7 Btu/lbm, 60.7 Btu/lbm

7-44 A 1.2-m^3 insulated rigid tank contains 2.13 kg of carbon dioxide at 100 kPa. Now paddle-wheel work is done on the system until the pressure in the tank rises to 120 kPa. Determine (*a*) the actual paddle-wheel work done during this process and (*b*) the minimum paddle-wheel work with which this process (between the same end states) could be accomplished. Assume the surroundings to be at 25°C and 100 kPa.
Answers: (*a*) −87.0 kJ, (*b*) −7.66 kJ

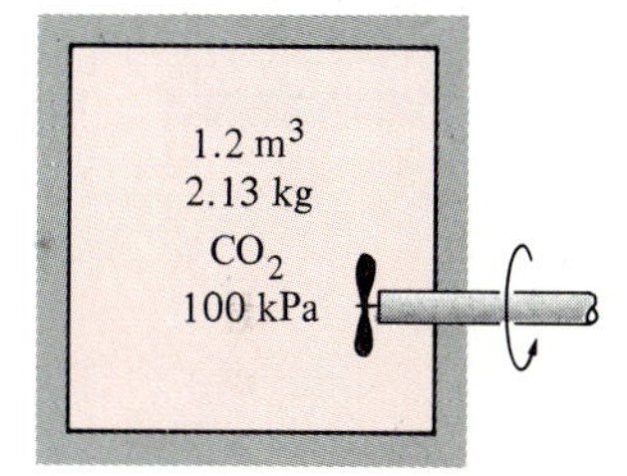

FIGURE P7-44

7-45 An insulated piston-cylinder device initially contains 30 L of air at 120 kPa and 27°C. Air is now heated for 5 min by a 50-W resistance heater placed inside the cylinder. The pressure of air is maintained constant during this process, and the surroundings are at 27°C and 100 kPa. Determine the irreversibility for this process. *Answer:* 9.9 kJ

7-45E An insulated piston-cylinder device initially contains 0.7 ft^3 of air at 15 psia and 80°F. Air is now heated for 5 min by a 50-W resistance heater placed inside the cylinder. The pressure of air is maintained constant during this process, and the surroundings are at 80°F and 14.7 psia. Determine the irreversibility for this process.

7-46 A mass of 2 kg of helium undergoes a process from an initial state of 3 m^3/kg and 15°C to a final state of 0.5 m^3/kg and 80°C. Assuming the surroundings to be at 25°C, determine the increase in the useful work potential of the helium during this process.

7-47 An insulated rigid tank is divided into two equal parts by a partition. Initially, one part contains 2 kg of argon gas at 500 kPa and 70°C, and the other side is evacuated. The partition is now removed, and the gas fills the entire tank. Assuming the surroundings to be at 25°C, determine the irreversibility associated with this process. *Answer:* 85.9 kJ

7-48 A 20-kg copper block initially at 80°C is dropped into an insulated tank which contains 150 L of water at 25°C. Determine (*a*) the final equilibrium temperature and (*b*) the work potential wasted during this process. Assume the surroundings to be at 25°C.
Answers: (*a*) 25.67°C, (*b*) 35 kJ

7-48E A 20-lbm copper block initially at 180°F is dropped into an insulated tank which contains 1 ft^3 of water at 75°F. Determine (*a*) the final equilibrium temperature and (*b*) the work potential wasted during this process. Assume the surroundings to be at 75°F.

7-49 A 20-kg aluminum block initially at 200°C is brought into contact with a 20-kg block of iron at 100°C in an insulated enclosure. Assuming the surroundings to be at 25°C, determine (*a*) the final equilibrium temperature and (*b*) the reversible work for this process.

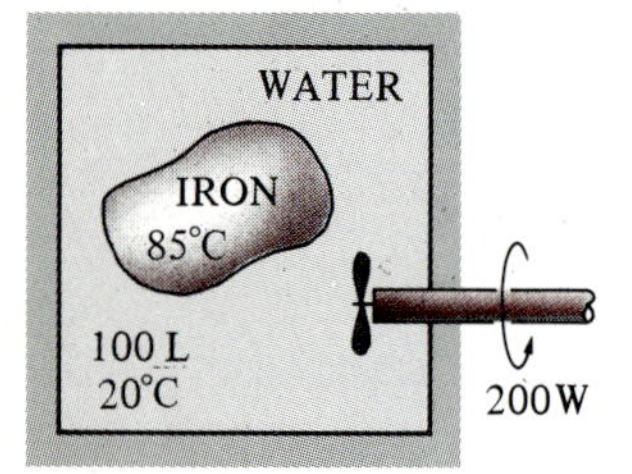

FIGURE P7-50

7-50 An iron block of unknown mass at 85°C is dropped into an insulated tank that contains 100 L of water at 20°C. At the same time, a paddle wheel driven by a 200-W motor is activated to stir the water. It is observed that thermal equilibrium is established after 20 min with a final temperature of 24°C. Assuming the surroundings to be at 20°C, determine (*a*) the mass of the iron block, and (*b*) the irreversibility associated with this process. *Answers:* (*a*) 52.23 kg, (*b*) 376 kJ

7-51 A 30-kg iron block and a 20-kg copper block, both initially at 80°C, are dropped into a large lake at 20°C. Thermal equilibrium is established after a while as a result of heat transfer between the blocks and the lake water. Assuming the surroundings to be at 20°C, determine the amount of work that could have been produced if the entire process was executed in a reversible manner.

7-51E A 60-lbm iron block and a 40-lbm copper block, both initially at 180°F, are dropped into a large lake at 70°F. Thermal equilibrium is established after a while as a result of heat transfer between the blocks and the lake water. Assuming the surroundings to be at 70°F, determine the amount of work that could have been produced if the entire process was executed in a reversible manner. *Answer:* 101.8 Btu

7-52 A 0.5-m^3 rigid tank contains refrigerant-12 initially at 200 kPa and 40 percent quality. Heat is transferred now to the refrigerant from a source at 35°C until the pressure rises to 400 kPa. Assuming the surroundings to be at 25°C, determine (*a*) the amount of heat transfer between the source and the refrigerant and (*b*) the reversible work for this process.
Answers: (*a*) 947.6 kJ, (*b*) 124.3 kJ

7-52E A 20-ft^3 rigid tank contains refrigerant-12 initially at 30 psia and 40 percent quality. Heat is transferred now to the refrigerant from a source at 120°F until the pressure rises to 60 psia. Assuming the surroundings to be at 75°F, determine (*a*) the amount of heat transfer between the source and the refrigerant and (*b*) the reversible work for this process.

7-53 A piston-cylinder device initially contains 0.5 m^3 of helium gas at 150 kPa and 20°C. Helium is now compressed in a polytropic process (PV^n = constant) to 400 kPa and 140°C. Assuming the surroundings to be at 20°C and 100 kPa, determine (*a*) the actual useful work consumed and (*b*) the minimum useful work that could be consumed during this process. *Answers:* (*a*) −33.4 kJ, (*b*) −31.6 kJ

7-53E A piston-cylinder device initially contains 15 ft^3 of helium gas at 25 psia and 70°F. Helium is now compressed in a polytropic process (PV^n = constant) to 70 psia and 300°F. Assuming the surroundings to be at 14.7 psia and 70°F, determine (*a*) the actual useful work consumed and (*b*) the minimum useful work that could be consumed during this process. *Answers:* (*a*) −36 Btu, (*b*) −34.3 Btu

Second-Law Analysis of Steady-Flow Systems

7-54C Writing the first- and second-law relations and simplifying, obtain the reversible work relation for a steady-flow system which exchanges heat with the surrounding medium at T_0 in the amount of Q_0 as well as a thermal reservoir at T_R in the amount of Q_R. (*Hint:* Eliminate Q_0 between the two equations.)

7-55 Steam is throttled from 10 MPa and 500°C to 7 MPa. Determine the wasted work potential during this throttling process. Assume the surroundings to be at 25°C. *Answer:* 45.3 kJ/kg

7-56 Air is compressed steadily by a 5-kW compressor from 100 kPa and 17°C to 600 kPa and 167°C at a rate of 1.6 kg/min. Neglecting the changes in kinetic and potential energies, determine (*a*) the increase in the availability of the air and (*b*) the rate of irreversibility for this process. Assume the surroundings to be at 17°C.

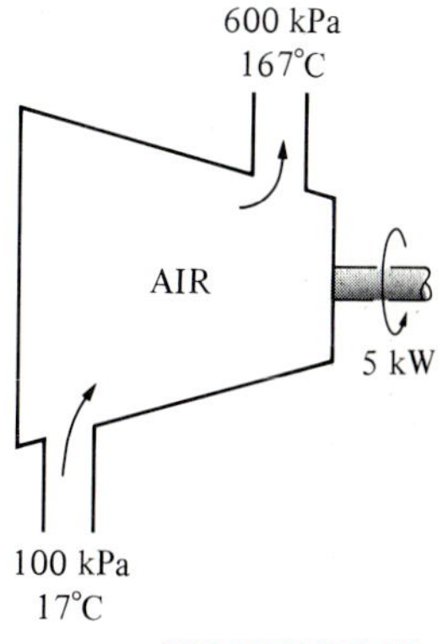

FIGURE P7-56

7-57 Refrigerant-12 at 1 MPa and 80°C is throttled to a pressure of 0.2 MPa. Determine the reversible work and irreversibility for this throttling process. Assume the surroundings to be at 25°C.

7-58 Air enters a nozzle steadily at 300 kPa and 87°C with a velocity of 50 m/s and exits at 95 kPa and 300 m/s. The heat losses from the nozzle to the surrounding medium at 17°C are estimated to be 4 kJ/kg. Determine (*a*) the exit temperature and (*b*) the irreversibility for this process.
Answers: (*a*) 39.5°C, (*b*) 58.4 kJ/kg

7-58E Air enters a nozzle steadily at 50 psia and 180°F with a velocity of 150 ft/s and exits at 15 psia and 880 ft/s. The heat losses from the nozzle to the surrounding medium at 70°F are estimated to be 2 Btu/lbm. Determine (*a*) the exit temperature and (*b*) the irreversibility for this process.

7-59 Steam enters a diffuser at 10 kPa and 50°C with a velocity of 300 m/s and exits as saturated vapor at 50°C and 50 m/s. The exit area of the diffuser is 2 m^2. Determine (*a*) the mass flow rate of the steam and (*b*) the wasted work potential during this process. Assume the surroundings to be at 25°C.

7-60 Air is compressed steadily by a compressor from 100 kPa and 17°C to 700 kPa and 287°C at a rate of 2 kg/min. Assuming the surroundings to be at 17°C, determine the minimum power input to the compressor. Assume air to be an ideal gas with variable specific heats, and neglect the changes in kinetic and potential energies. *Answer:* 8.1 kW

7-60E Air is compressed steadily by a compressor from 14.7 psia and 60°F to 100 psia and 480°F at a rate of 5 lbm/min. Assuming the surroundings to be at 60°F, determine the minimum power input to the compressor. Assume air to be an ideal gas with variable specific heats, and neglect the changes in kinetic and potential energies.

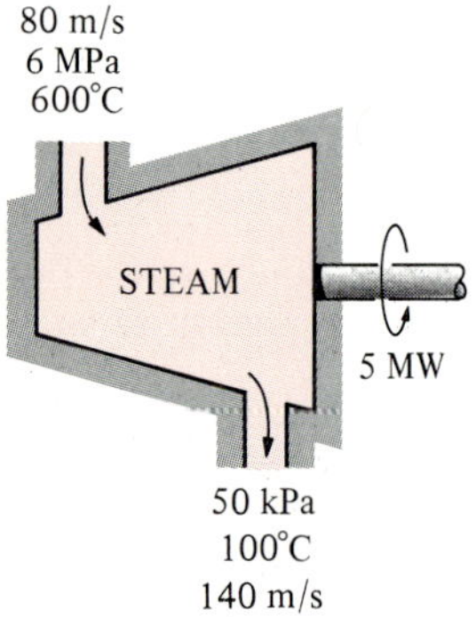

FIGURE P7-61

7-61 Steam enters an adiabatic turbine at 6 MPa, 600°C, and 80 m/s and leaves at 50 kPa, 100°C, and 140 m/s. If the power output of the turbine is 5 MW, determine (*a*) the reversible power and (*b*) the second-law efficiency of the turbine. Assume the surroundings to be at 25°C.
Answers: (*a*) 5.81 MW, (*b*) 86.1 percent

7-61E Steam enters an adiabatic turbine at 1000 psia, 900°F, and 250 ft/s and leaves at 5 psia, 200°F, and 400 ft/s. If the power output of the turbine is 5 MW, determine (*a*) the reversible power and (*b*) the second-law efficiency of the turbine. Assume the surroundings to be at 75°F.

7-62 Steam is throttled from 10 MPa and 700°C to a pressure of 7 MPa. Determine the decrease in availability of the steam during this process. Assume the surroundings to be at 25°C. *Answer:* 47.6 kJ/kg

7-62E Steam is throttled from 1000 psia and 900°F to a pressure of 800 psia. Determine the decrease in availability of the steam during this process. Assume the surroundings to be at 75°F. *Answer:* 12.3 Btu/lbm

7-63 Steam at 7 MPa and 500°C enters a two-stage adiabatic turbine at a rate of 15 kg/s. Ten percent of the steam is extracted at the end of the first stage at a pressure of 1 MPa for other use. The remainder of the steam is further expanded in the second stage and leaves the turbine at 50 kPa. If the turbine has an adiabatic efficiency of 88 percent, determine the wasted power potential during this process as a result of irreversibilities. Assume the surroundings to be at 25°C.

7-64 Steam enters a two-stage adiabatic turbine at 8 MPa and 500°C. It expands in the first stage to a state of 2 MPa and 350°C. Steam is then reheated at constant pressure to a temperature of 500°C before it is routed to the second stage, which it exits at 30 kPa and a quality of 97 percent. The work output of the turbine is 5 MW. Assuming the surroundings to be at 25°C, determine the reversible power and irreversibility for this turbine. *Answers:* 5463 kW, 463 kW

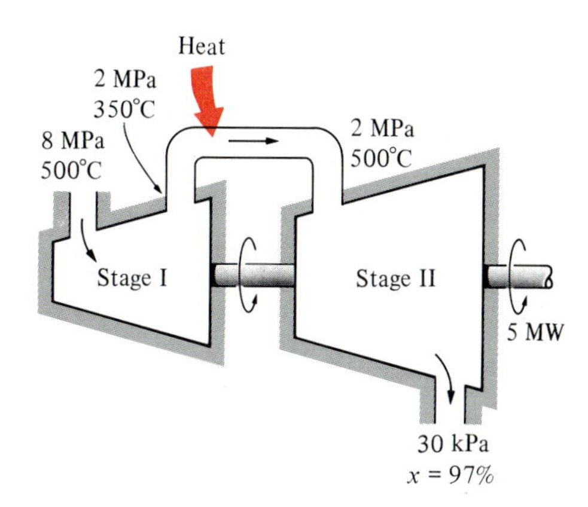

FIGURE P7-64

7-65 Argon gas enters an adiabatic turbine at 800°C and 1.5 MPa at a rate of 20 kg/min and exhausts at 200 kPa. If the power output of the turbine is 90 kW, determine (*a*) the adiabatic efficiency and (*b*) the second-law efficiency of the turbine. Assume the surroundings to be at 25°C. *Answers:* (*a*) 87.4 percent, (*b*) 92.3 percent

7-65E Argon gas enters an adiabatic turbine at 1500°F and 200 psia at a rate of 40 lbm/min and exhausts at 30 psia. If the power output of the turbine is 95 hp, determine (*a*) the adiabatic efficiency and (*b*) the second-law efficiency of the turbine. Assume the surroundings to be at 77°F.

7-66 Combustion gases enter a gas turbine at 900°C, 800 kPa, and 100 m/s and leave at 650°C, 400 kPa, and 220 m/s. Taking $C_p = 1.15$ kJ/(kg·°C) and $k = 1.3$ for the combustion gases, determine (*a*) the availability of the combustion gases at the turbine inlet and (*b*) the work output of the turbine under reversible conditions. Assume the surroundings to be at 25°C.

7-67 Refrigerant-12 enters an adiabatic compressor as saturated vapor at 120 kPa at a rate of 2 m³/min and exits at 1-MPa pressure. If the adiabatic efficiency of the compressor is 80 percent, determine (*a*) the actual power input and (*b*) the second-law efficiency of the compressor. Assume the surroundings to be at 25°C. *Answers:* (*a*) 11.64 kW, (*b*) 82.3 percent

7-67E Refrigerant-12 enters an adiabatic compressor as saturated vapor at 20 psia at a rate of 50 ft³/min and exits at 100-psia pressure. If the adiabatic efficiency of the compressor is 80 percent, determine (*a*) the

actual power input and (*b*) the second-law efficiency of the compressor. Assume the surroundings to be at 75°F.
Answers: (*a*) 7.65 hp, (*b*) 81.3 percent

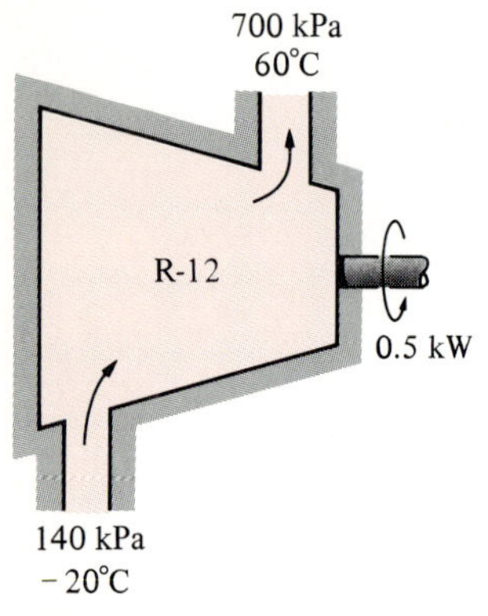

FIGURE P7-68

7-68 Refrigerant-12 at 140 kPa and −20°C is compressed by an adiabatic 0.5-kW compressor to an exit state of 700 kPa and 60°C. Neglecting the changes in kinetic and potential energies and assuming the surroundings to be at 25°C, determine (*a*) the adiabatic efficiency and (*b*) the second-law efficiency of the compressor.

7-69 Air is compressed by a compressor from 95 kPa and 27°C to 600 kPa and 277°C at a rate of 0.02 kg/s. Neglecting the changes in kinetic and potential energies and assuming the surroundings to be at 25°C, determine the reversible power for this process. *Answer:* −4.57 kW

7-69E Air is compressed by a compressor from 14.5 psia and 90°F to 80 psia and 540°F at a rate of 0.04 lbm/s. Neglecting the changes in kinetic and potential energies and assuming the surroundings to be at 77°F, determine the reversible power for this process.

7-70 Argon gas enters an adiabatic compressor at 120 kPa and 30°C with a velocity of 20 m/s and exits at 1.2 MPa, 530°C, and 80 m/s. The inlet area of the compressor is 80 cm^2. Assuming the surroundings to be at 25°C, determine the reversible power and irreversibility for this process. *Answers:* −77.6 kW, 2.53 kW

7-71 Steam expands in a turbine steadily at a rate of 15,000 kg/h, entering at 8 MPa and 450°C and leaving at 50 kPa as saturated vapor. Assuming the surroundings to be at 100 kPa and 25°C, determine (*a*) the power potential of the steam at the inlet conditions and (*b*) the power output of the turbine if there were no irreversibilities present.
Answers: (*a*) 5513 kW, (*b*) 3899 kW

7-72 Air enters a compressor at ambient conditions of 96 kPa and 17°C with a low velocity and exits at 1 MPa, 327°C, and 120 m/s. The compressor is cooled by the ambient air at 17°C at a rate of 1500 kJ/min. The power input to the compressor is 300 kW. Determine (*a*) the mass flow rate of air and (*b*) the portion of the power input that is used just to overcome the irreversibilities.

7-72E Air enters a compressor at ambient conditions of 15 psia and 60°F with a low velocity and exits at 150 psia, 620°F, and 350 ft/s. The compressor is cooled by the ambient air at 60°F at a rate of 1500 Btu/min. The power input to the compressor is 400 hp. Determine (*a*) the mass flow rate of air and (*b*) the portion of the power input that is used just to overcome the irreversibilities.

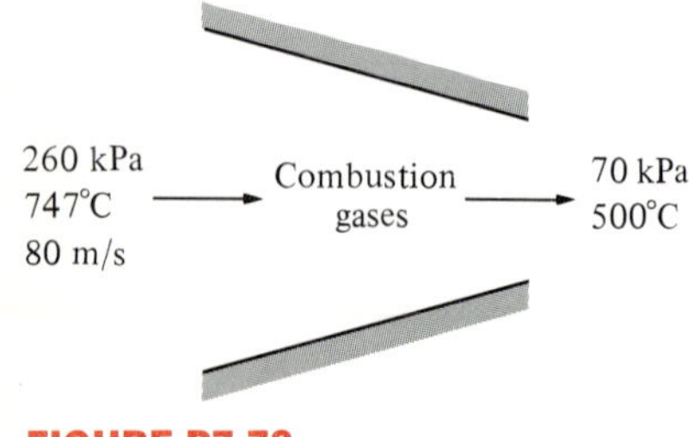

FIGURE P7-73

7-73 Hot combustion gases enter the nozzle of a turbojet engine at 260 kPa, 747°C, and 80 m/s and exit at 70 kPa and 500°C. Assuming the nozzle to be adiabatic and the surroundings to be at 17°C, determine (*a*) the exit velocity and (*b*) the decrease in availability of the gases. Take k = 1.3 and C_p = 1.15 kJ/(kg·°C) for the combustion gases.

7-73E Hot combustion gases enter the nozzle of a turbojet engine at 34 psia, 1600°F, and 250 ft/s and exit at 12 psia and 1200°F. Assuming the nozzle to be adiabatic and the surroundings to be at 65°F, determine (*a*) the exit velocity and (*b*) the decrease in availability of the gases. Take $k = 1.3$ and $C_p = 0.275$ Btu/(lbm·°F) for the combustion gases.

7-74 Steam is usually accelerated in the nozzle of a turbine before it strikes the turbine blades. Steam enters an adiabatic nozzle at 7 MPa and 500°C with a velocity of 70 m/s and exits at 5 MPa and 450°C. Assuming the surroundings to be at 25°C, determine (*a*) the exit velocity of the steam, (*b*) the adiabatic efficiency, and (*c*) the irreversibility associated with the nozzle.

7-75 Carbon dioxide enters a compressor at 100 kPa and 300 K at a rate of 0.5 kg/s and exits at 600 kPa and 450 K. Determine the power input to the compressor if the process involved no irreversibilities. Assume the surroundings to be at 25°C. *Answer:* −49.6 kW

7-76 A hot-water stream at 70°C enters an adiabatic mixing chamber with a mass flow rate of 0.6 kg/s, where it is mixed with a stream of cold water at 20°C. If the mixture leaves the chamber at 55°C, determine (*a*) the mass flow rate of the cold water and (*b*) the irreversibility for this adiabatic mixing process. Assume all the streams are at a pressure of 200 kPa and the surroundings are at 25°C.
Answers: (*a*) 0.257 kg/s, (*b*) 2.66 kW

7-76E A hot-water stream at 160°F enters an adiabatic mixing chamber with a mass flow rate of 1.2 lbm/s, where it is mixed with a stream of cold water at 70°F. If the mixture leaves the chamber at 110°F, determine (*a*) the mass flow rate of the cold water and (*b*) the irreversibility for this adiabatic mixing process. Assume all the streams are at a pressure of 30 psia and the surroundings are at 75°F.
Answers: (*a*) 1.50 lbm/s, (*b*) 4.39 Btu/s

7-77 Liquid water at 200 kPa and 20°C is heated in a chamber by mixing it with superheated steam at 200 kPa and 300°C. Liquid water enters the mixing chamber at a rate of 2.5 kg/s, and the chamber is estimated to lose heat to the surrounding air at 25°C at a rate of 600 kJ/min. If the mixture leaves the mixing chamber at 200 kPa and 60°C, determine (*a*) the mass flow rate of the superheated steam and (*b*) the wasted work potential during this mixing process.

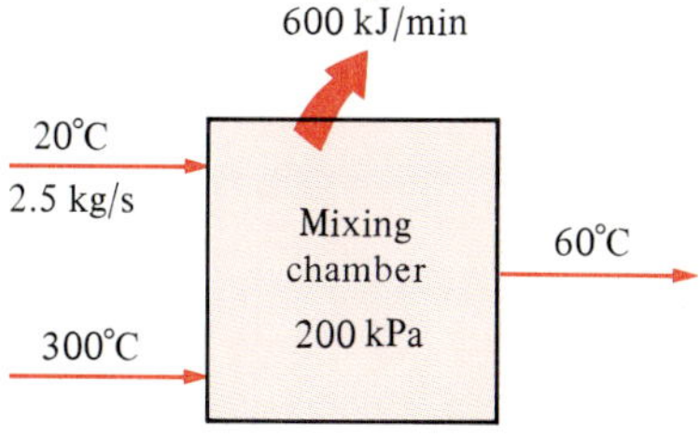

FIGURE P7-77

7-78 Air enters the evaporator section of a window air conditioner at 100 kPa and 27°C with a volume flow rate of 6 m³/min. The refrigerant-12 at 120 kPa with a quality of 0.3 enters the evaporator at a rate of 2 kg/min and leaves as saturated vapor at the same pressure. Determine the exit temperature of the air and the irreversibility for this process, assuming (*a*) heat is transferred to the evaporator of the air conditioner from the surrounding medium at 32°C at a rate of 30 kJ/min and (*b*) the outer surfaces of the air conditioner are insulated.

7-79 In large steam power plants, the feedwater is frequently heated in

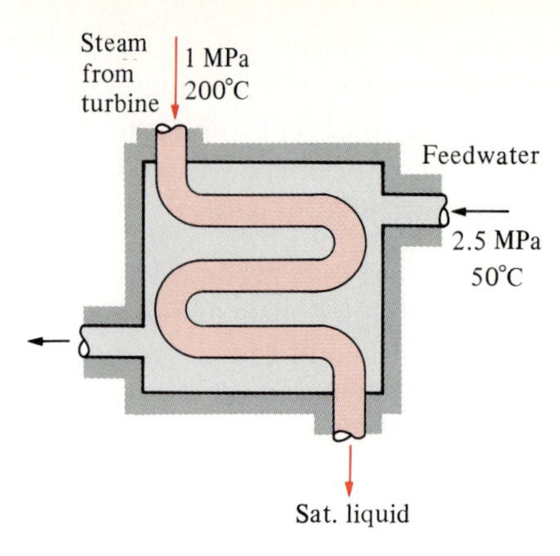

FIGURE P7-79

closed feedwater heaters, which are basically heat exchangers, by steam extracted from the turbine at some stage. Steam enters the feedwater heater at 1 MPa and 200°C and leaves as saturated liquid at the same pressure. Feedwater enters the heater at 2.5 MPa and 50°C and leaves 10°C below the exit temperature of the steam. Neglecting any heat losses from the outer surfaces of the heater, determine (*a*) the ratio of the mass flow rates of the extracted steam and the feedwater heater and (*b*) the reversible work for this process per unit mass of the feedwater. Assume the surroundings to be at 25°C. *Answers:* (*a*) 0.247, (*b*) 63.5 kJ/kg

Second-Law Analysis of Unsteady-Flow Systems

7-80C Writing the first- and second-law relations and simplifying, obtain the reversible work relation for a uniform-flow system which exchanges heat with the surrounding medium at T_0 in the amount of Q_0 as well as a thermal energy reservoir at T_R in the amount of Q_R. (*Hint:* Eliminate Q_0 between the two equations.)

7-81 A 0.1-m^3 rigid tank initially contains refrigerant-12 at 1 MPa and 100 percent quality. The tank is connected by a valve to a supply line that carries refrigerant-12 at 1.4 MPa and 30°C. The valve is now opened, allowing the refrigerant to enter the tank, and it is closed when the tank contains only saturated liquid at 1.2 MPa. The refrigerant exchanges heat with its surroundings at 50°C and 100 kPa during this process. Determine (*a*) the mass of the refrigerant that entered the tank and (*b*) the irreversibility for this process.

7-81E A 3-ft^3 rigid tank initially contains refrigerant-12 at 120 psia and 100 percent quality. The tank is connected by a valve to a supply line that carries refrigerant-12 at 160 psia and 80°F. The valve is opened, allowing the refrigerant to enter the tank, and is closed when the tank contains only saturated liquid at 140 psia. The refrigerant exchanges heat with its surroundings at 120°F and 14.7 psia during this process. Determine (*a*) the mass of the refrigerant that entered the tank and (*b*) the irreversibility for this process.

7-82 A 0.4-m^3 rigid tank is filled with saturated liquid water at 200°C. A valve at the bottom of the tank is now opened, and one-half of the total mass is withdrawn from the tank in liquid form. Heat is transferred to water from a source at 250°C so that the temperature in the tank remains constant. Determine (*a*) the amount of heat transfer and (*b*) the reversible work and irreversibility for this process. Assume the surroundings to be at 25°C and 100 kPa.
Answers: (*a*) 3077 kJ; (*b*) 183.6 kJ, 183.6 kJ

7-83 An insulated 2-m^3 rigid tank contains air at 500 kPa and 52°C. A valve connected to the tank is now opened, and air is allowed to escape until the pressure inside drops to 200 kPa. The air temperature during this process is maintained constant by an electric resistance heater placed in the tank. Determine (*a*) the electrical work done during this process and (*b*) the irreversibility. Assume the surroundings to be at 22°C.

7-83E An insulated 60-ft^3 rigid tank contains air at 75 psia and 140°F. A valve connected to the tank is opened, and air is allowed to escape until the pressure inside drops to 30 psia. The air temperature during this process is maintained constant by an electric resistance heater placed in the tank. Determine (*a*) the electrical work done during this process and (*b*) the irreversibility. Assume the surroundings to be at 70°F.
Answers: (*a*) −499 Btu, (*b*) 427 Btu

7-84 A 0.1-m^3 rigid tank contains saturated refrigerant-12 at 800 kPa. Initially, 20 percent of the volume is occupied by liquid and the rest by vapor. A valve at the bottom of the tank is opened, and liquid is withdrawn from the tank. Heat is transferred to the refrigerant from a source at 50°C so that the pressure inside the tank remains constant. The valve is closed when no liquid is left in the tank and vapor starts to come out. Assuming the surroundings to be at 25°C, determine (*a*) the final mass in the tank and (*b*) the reversible work for this process.
Answers: (*a*) 45.7 kg, (*b*) 62.8 kJ

7-84E A 4-ft^3 rigid tank contains saturated refrigerant-12 at 100 psia. Initially, 20 percent of the volume is occupied by liquid and the rest by vapor. A valve at the bottom of the tank is now opened, and liquid is withdrawn from the tank. Heat is transferred to the refrigerant from a source at 150°F so that the pressure inside the tank remains constant. The valve is closed when no liquid is left in the tank and vapor starts to come out. Assuming the surroundings to be at 75°F and 14.7 psia, determine (*a*) the final mass in the tank and (*b*) the reversible work for this process.

7-85 A vertical piston-cylinder device initially contains 0.1 m^3 of helium at 20°C. The mass of the piston is such that it maintains a constant pressure of 300 kPa inside. A valve is now opened, and helium is allowed to escape until the volume inside the cylinder is decreased by one-half. Heat transfer takes place between the helium and its surroundings at 25°C and 100 kPa so that the temperature of helium in the cylinder remains constant. De-

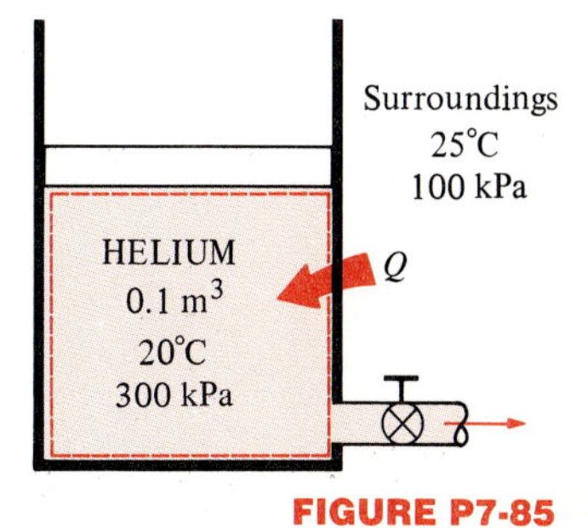

FIGURE P7-85

termine (*a*) the maximum work potential of the helium at the initial state and (*b*) the irreversibility for this process.

7-85E A vertical piston-cylinder device initially contains 3 ft^3 of helium at 70°F. The mass of the piston is such that it maintains a constant pressure of 50 psia inside. A valve is now opened, and helium is allowed to escape until the volume inside the cylinder is decreased by one-half. Heat transfer takes place between the helium and its surroundings at 80°F and 14.7 psia so that the temperature of helium in the cylinder remains constant. Determine (*a*) the maximum work potential of the helium at the initial state and (*b*) the irreversibility for this process.

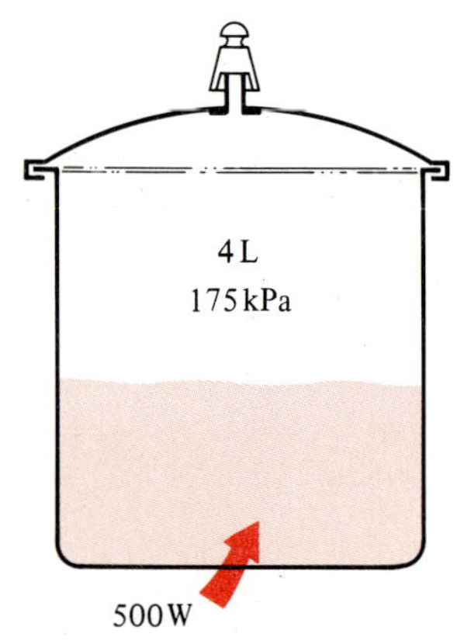

FIGURE P7-86

7-86 A 4-L pressure cooker has an operating pressure of 175 kPa. Initially, one-half of the volume is filled with liquid water and the other half by water vapor. The cooker is now placed on top of a 500-W electrical heating unit which is kept on for 30 min. Assuming the surroundings to be at 25°C and 100 kPa, determine (*a*) the amount of water that remained in the cooker and (*b*) the irreversibility associated with the entire process, including the conversion of electric energy to heat energy.
Answers: (*a*) 1.487 kg, (*b*) 690 kJ

7-87 What would your answer to Prob. 7-86 be if heat were supplied to the pressure cooker from a thermal energy source at 180°C instead of the electrical heating unit?

7-88 A 0.2-m^3 rigid tank initially contains saturated refrigerant-12 vapor at 1 MPa. The tank is connected by a valve to a supply line that carries refrigerant-12 at 1.4 MPa and 60°C. The valve is now opened, and the refrigerant is allowed to enter the tank. The valve is closed when one-half of the volume of the tank is filled with liquid and the rest with vapor at 1.2 MPa. The refrigerant exchanges heat during this process with the surroundings at 25°C. Determine (*a*) the amount of heat transfer and (*b*) the irreversibility associated with this process.

7-89 An insulated vertical piston-cylinder device initially contains 5 kg of water, 4 kg of which is in the vapor phase. The mass of the piston is such that it maintains a constant pressure of 300 kPa inside the cylinder. Now steam at 1 MPa and 400°C is allowed to enter the cylinder from a supply line until all the liquid in the cylinder is vaporized. Assuming the surroundings to be at 25°C and 100 kPa, determine (*a*) the amount of steam that has entered and (*b*) the irreversibility for this process. *Answers:* (*a*) 4.02 kg, (*b*) 1019 kJ

Gas Power Cycles

CHAPTER 8

Two important areas of application for thermodynamics are power generation and refrigeration. Both power generation and refrigeration are usually accomplished by systems that operate on a thermodynamic cycle. Thermodynamic cycles can be divided into two general categories: *power cycles,* which are discussed in this and the next chapter, and *refrigeration cycles,* which are discussed in Chap. 10.

The devices or systems used to produce a net power output are often called *engines,* and the thermodynamic cycles they operate on are called *power cycles*. The devices or systems used to produce refrigeration are called *refrigerators, air conditioners,* or *heat pumps,* and the cycles they operate on are called *refrigeration cycles*.

Thermodynamic cycles can also be categorized as *gas cycles* or *vapor cycles* depending on the *phase* of the working fluid—the substance that circulates through the cyclic device. In gas cycles, the working fluid remains in the gaseous phase throughout the entire cycle, whereas in vapor cycles the working fluid exists in the vapor phase during one part of the cycle and in the liquid phase during another part.

Thermodynamic cycles can be categorized yet another way: *closed* and *open cycles*. In closed cycles, the working fluid (such as the steam in steam power plants) is returned to the initial state at the end of the cycle and is recirculated. In open cycles, the working fluid is renewed at the end of each cycle instead of being recirculated. In automobile engines, for example, the combustion gases are exhausted and replaced by fresh air-fuel mixture at the end of each cycle. The engine operates on a mechanical cycle, but the working fluid in this type of device does not go through a complete thermodynamic cycle.

Heat engines are categorized as *internal combustion* or *external combustion engines,* depending on how the heat is supplied to the working fluid. In external combustion engines (such as steam power plants), energy is supplied to the working fluid from an external source such as a furnace, a geothermal well, a nuclear reactor, or even the sun. In internal combustion engines (such as automobile engines), this is done by burning the fuel within the system boundary. In this chapter, various gas power cycles are analyzed under some simplifying assumptions.

Most power-producing devices operate on cycles, and the study of power cycles is an exciting and important part of thermodynamics. The cycles encounterd in actual devices are difficult to analyze because of the presence of complicating effects, such as friction, and the absence of sufficient time for the establishment of the equilibrium conditions during the cycle. To make an analytical study of a cycle feasible, we have to keep the complexities at a manageable level and utilize some idealizations (Fig. 8-1). When the actual cycle is stripped of all the internal irreversibilities and complexities, we end up with a cycle which resembles the actual cycle closely but is made up entirely of internally reversible processes. Such a cycle is called an **ideal cycle** (Fig. 8-2).

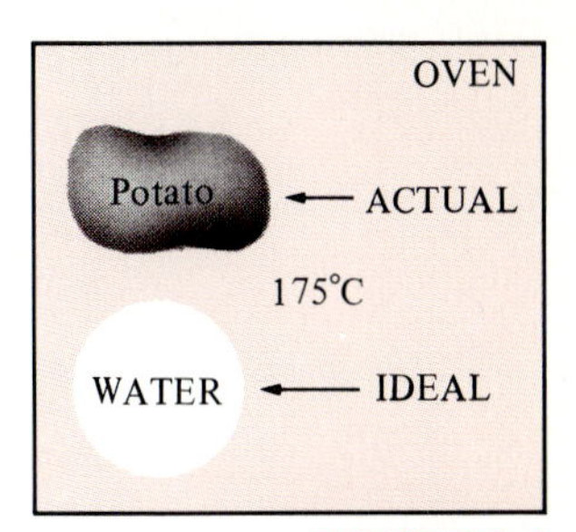

FIGURE 8-1

Modeling is a powerful engineering tool that provides great insight and simplicity at the expense of some loss in accuracy.

A simple idealized model enables engineers to study the effects of the major parameters which dominate the cycle without getting bogged down in the details. The cycles discussed in this chapter are somewhat idealized, but they still retain the general characteristics of the actual cycles they represent. The conclusions reached from the analysis of ideal cycles are often applicable to actual cycles. The thermal efficiency of the Otto cycle, the ideal cycle for spark-ignition automobile engines, for example, increases with the compression ratio. This is also the case for actual automobile engines. The numerical values obtained from the analysis of an ideal cycle, however, are not necessarily representative of the actual cycles, and care should be exercised in their interpretation (Fig. 8-3). The simplified analysis presented in this chapter for various power cycles of practical interest may also serve as the starting point for a more in-depth study.

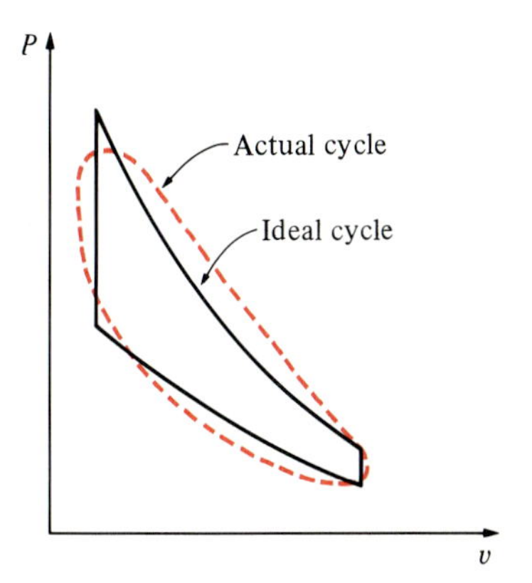

FIGURE 8-2

The analysis of many complex processes can be reduced to a manageable level by utilizing some idealizations.

Heat engines are designed for the purpose of converting other forms of energy (usually in the form of heat) to work, and their performance is expressed in terms of the **thermal efficiency** η_{th}, which is the ratio of the net work produced by the engine to the total heat input:

$$\eta_{th} = \frac{W_{net}}{Q_{in}} \quad \text{or} \quad \eta_{th} = \frac{w_{net}}{q_{in}} \qquad (8\text{-}1a, b)$$

FIGURE 8-3

Care should be exercised in the interpretation of the results from ideal cycles.

It was pointed out in Chap. 6 that heat engines which operate on a totally reversible cycle, such as the Carnot cycle, have the highest thermal efficiency of all heat engines operating between the same temperature levels. That is, nobody can develop a cycle more efficient that the Carnot cycle. Then the following question arises naturally: If the Carnot cycle is the best possible cycle, why do we not use it as the model cycle for all the heat engines instead of bothering with several so-called *ideal* cycles? The answer to this question is hardware-related. Most cycles encountered in practice differ significantly from the Carnot cycle, which makes it unsuitable as a realistic model. Each ideal cycle discussed in this chapter is related to a specific work-producing device and is an *idealized* version of the actual cycle.

The ideal cycles are *internally reversible,* but, unlike the Carnot cycle, they are not necessarily externally reversible. That is, they may involve irreversibilities external to the system such as heat transfer through a finite temperature difference. Therefore, the thermal efficiency of an ideal cycle, in general, is less than that of a totally reversible cycle operating between the same temperature limits. However, it is still considerably higher than the thermal efficiency of an actual cycle because of the idealizations utilized.

The idealizations and simplifications commonly employed in the analysis of power cycles can be summarized as follows:

1 The cycle does not involve any friction. Therefore, the working fluid does not experience any pressure drop as it flows in pipes or devices such as heat exchangers.

2 All expansion and compression processes take place in a quasi-equilibrium manner (Fig. 8-4).

3 The pipes connecting the various components of a system are well insulated, and heat transfer through them is negligible.

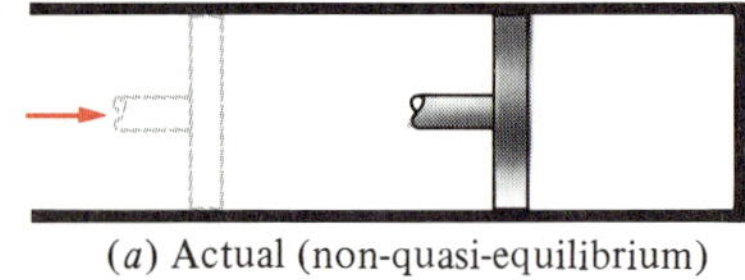
(*a*) Actual (non-quasi-equilibrium) compression

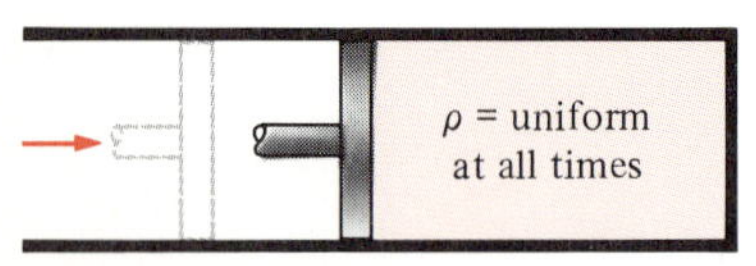

(*b*) Ideal (quasi-equilibrium) compression

FIGURE 8-4

All compression and expansion processes in ideal cycles are assumed to be quasi-equilibrium (internally reversible).

Neglecting the changes in kinetic and potential energies of the working fluid is another commonly utilized simplification in the analysis of power cycles. This is a reasonable assumption since in devices that involve shaft work, such as turbines, compressors, and pumps, the kinetic and potential energy terms are usually very small relative to the other terms in the energy equation. Fluid velocities encountered in devices such as condensers, boilers, and mixing chambers are typically low, and the fluid streams experience little change in their velocities, again making kinetic energy changes negligible. The only devices where the changes in kinetic energy are significant are the nozzles and diffusers which are specifically designed to create large changes in velocity.

In the preceding chapters, property diagrams such as the P-v and T-s diagrams have served as valuable aids in the analysis of thermodynamic processes. On both the P-v and T-s diagrams, the area enclosed by the process curves of a cycle represents the net work produced during the cycle (Fig. 8-5). It is also equivalent to the net heat transfer for that

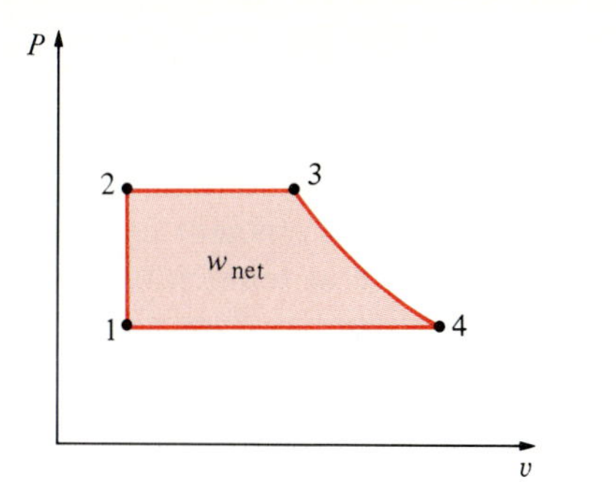

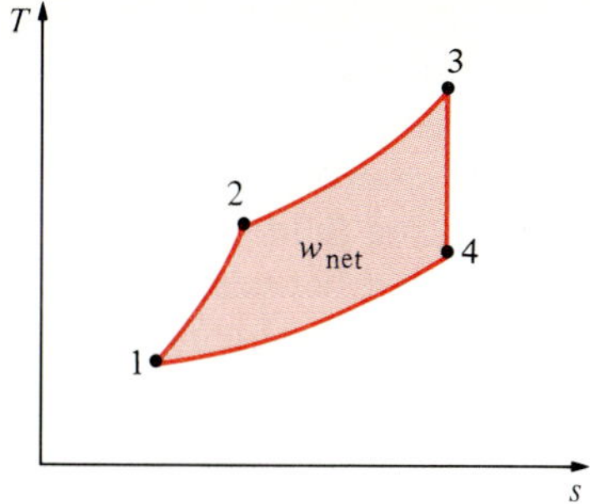

FIGURE 8-5

On both P-v and T-s diagrams, the area enclosed by the process curve represents the net work of the cycle.

cycle. The T-s diagram is particularly useful as a visual aid in the analysis of ideal power cycles. An ideal power cycle does not involve any internal irreversibilities, and so the only effect that can change the entropy of the working fluid during a process is heat transfer.

On a T-s diagram, a heat addition process proceeds in the direction of increasing entropy, a heat rejection process proceeds in the direction of decreasing entropy, and an isentropic (internally reversible, adiabatic) process proceeds at constant entropy. The area under the process curve on a T-s diagram represents the heat transfer for that process. The area under the heat addition process on a T-s diagram is a geometric measure of the total heat added during the cycle q_{in}, and the area under the heat rejection process is a measure of the total heat rejected q_{out}. The difference between these two (the area enclosed by the cyclic curve) is the net heat transfer, which is also the net work produced during the cycle. Therefore, on a T-s diagram, the ratio of the area enclosed by the cyclic curve to the area under the heat-addition process curve represents the thermal efficiency of the cycle. *Any modification that will increase the ratio of these two areas will also improve the thermal efficiency of the cycle.*

Although the working fluid in an ideal power cycle operates on a closed loop, the type of individual processes that compose the cycle depends on the individual devices used to execute the cycle. In the Rankine cycle, which is the ideal cycle for steam power plants, the working fluid flows through a series of steady-flow devices such as the turbine and condenser whereas in the Otto cycle, which is the ideal cycle for the spark-ignition automobile engine, the working fluid is alternately expanded and compressed in a piston-cylinder device. Therefore, equations pertaining to steady-flow systems should be used in the analysis of the Rankine cycle, and equations pertaining to closed systems should be used in the analysis of the Otto cycle.

8-2 ■ THE CARNOT CYCLE AND ITS VALUE IN ENGINEERING

The Carnot cycle, which was introduced and discussed in Chap. 5, is composed of four totally reversible processes: isothermal heat addition, isentropic expansion, isothermal heat rejection, and isentropic compression. The P-v and T-s diagrams of a Carnot cycle are replotted in Fig. 8-6. The Carnot cycle can be executed in a closed system (a piston-cylinder device) or a steady flow system (utilizing two turbines and two compressors, as shown in Fig. 8-7), and either a gas or a vapor can be utilized as the working fluid. The Carnot cycle is the most efficient cycle that can be executed between a thermal energy source at temperature T_H and a sink at temperature T_L, and its thermal efficiency is expressed as

$$\eta_{\text{th,Carnot}} = 1 - \frac{T_L}{T_H} \tag{8-2}$$

Reversible isothermal heat transfer is very difficult to achieve in reality because it would require very large heat exchangers and it would take a very long time (a power cycle in a typical engine is completed in a fraction of a second). Therefore, it is not practical to build an engine that would operate on a cycle which closely approximates the Carnot cycle.

The real value of the Carnot cycle comes from its being a standard against which the actual or other ideal cycles can be compared. The thermal efficiency of the Carnot cycle is a function of the sink and source temperatures only, and the thermal efficiency relation for the Carnot cycle (Eq. 8-2) conveys an important message which is equally applicable to both ideal and actual cycles: *Thermal efficiency increases with an increase in the average temperature at which heat is added to the system or with a decrease in the average temperature at which heat is rejected from the system.*

The source and sink temperatures that can be used in practice are not without limits, however. The highest temperature in the cycle is limited by the maximum temperature that the components of the heat engine, such as the piston or the turbine blades, can withstand. The lowest temperature is limited by the temperature of the cooling medium utilized in the cycle such as a lake, a river, or the atmospheric air.

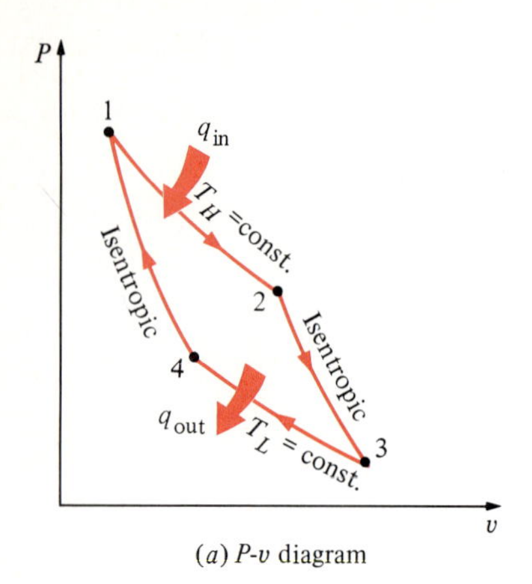

(a) P-v diagram

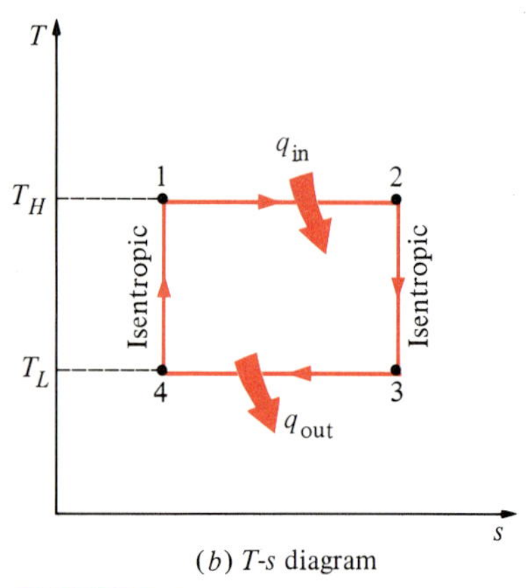

(b) T-s diagram

FIGURE 8-6

P-v and T-s diagrams of a Carnot cycle.

EXAMPLE 8-1

Show that the thermal efficiency of a Carnot cycle operating between the

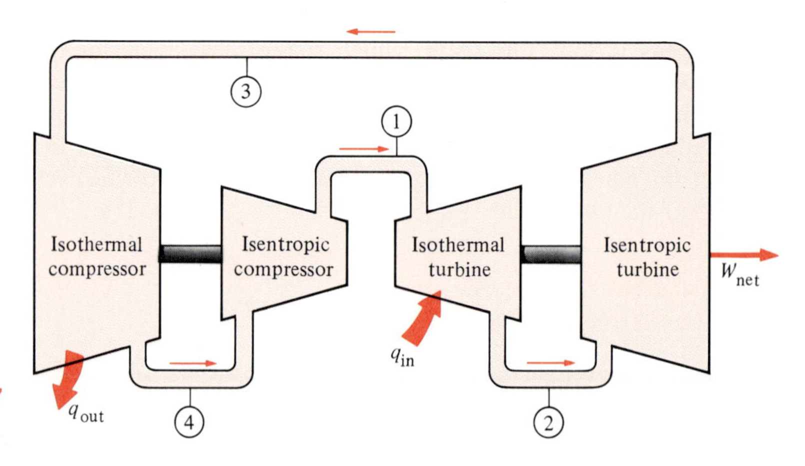

FIGURE 8-7

A steady-flow Carnot engine.

temperature limits of T_H and T_L is solely a function of these two temperatures and is given by Eq. 8-2.

Solution The *T-s* diagram of a Carnot cycle is redrawn in Fig. 8-8. All four processes that compose the Carnot cycle are reversible, and thus the area under each process curve represents the heat transfer for that process. Heat is added to the system during process 1-2 and rejected during process 3-4. Therefore, the amount of heat input and heat output for the cycle can be expressed as

$$q_{in} = T_H(s_2 - s_1) \quad \text{and} \quad q_{out} = T_L(s_3 - s_4) = T_L(s_2 - s_1)$$

since processes 2-3 and 4-1 are isentropic and thus $s_2 = s_3$ and $s_4 = s_1$. Substituting these into Eq. 8-1*b*, we see that the thermal efficiency of a Carnot cycle is

$$\eta_{th} = \frac{w_{net}}{q_{in}} = 1 - \frac{q_{out}}{q_{in}} = 1 - \frac{T_L(s_2 - s_1)}{T_H(s_2 - s_1)} = 1 - \frac{T_L}{T_H}$$

which is the desired result. Notice that the thermal efficiency of a Carnot cycle is independent of the type of the working fluid used (an ideal gas, steam, etc.) or whether the cycle is executed in a closed or steady-flow system.

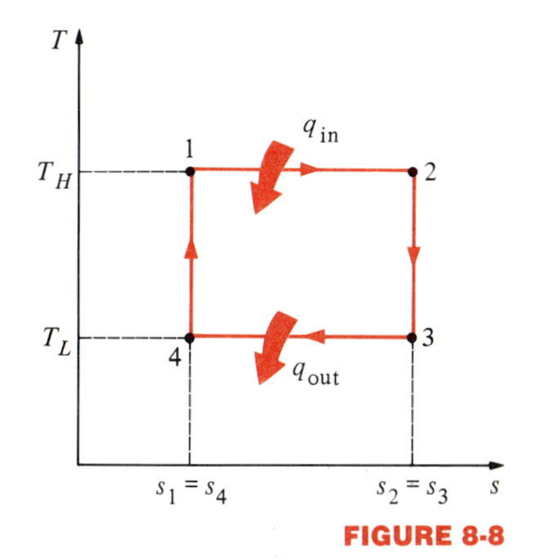

FIGURE 8-8
T-s diagram for Example 8-1.

8-3 ■ AIR-STANDARD ASSUMPTIONS

In gas power cycles, the working fluid remains a gas throughout the entire cycle. Spark-ignition automobile engines, diesel engines, and the conventional gas turbines are familiar examples of devices that operate on gas cycles. In all these engines, energy is provided by burning a fuel within the system boundaries. That is, they are *internal combustion engines*. Because of this combustion process, the composition of the working fluid changes from air and fuel to combustion products during the course of the cycle. However, considering that air is predominantly nitrogen which undergoes hardly any chemical reactions in the combustion chamber, the working fluid closely resembles air at all times.

Even though internal combustion engines operate on a mechanical cycle (the piston returns to its starting position at the end of each revolution), the working fluid does not undergo a complete thermodynamic cycle. It is thrown out of the engine at some point in the cycle (as exhaust gases) instead of being returned to the initial state. Working on an open cycle is the characteristic of all internal combustion engines.

The actual gas power cycles are rather complex. To reduce the analysis to a manageable level, we utilize the following approximations, commonly known as the **air-standard assumptions**

1 The working fluid is air which continuously circulates in a closed loop and always behaves as an ideal gas.

2 All the processes that make up the cycle are internally reversible.

3 The combustion process is replaced by a heat addition process from an external source (Fig. 8-9).

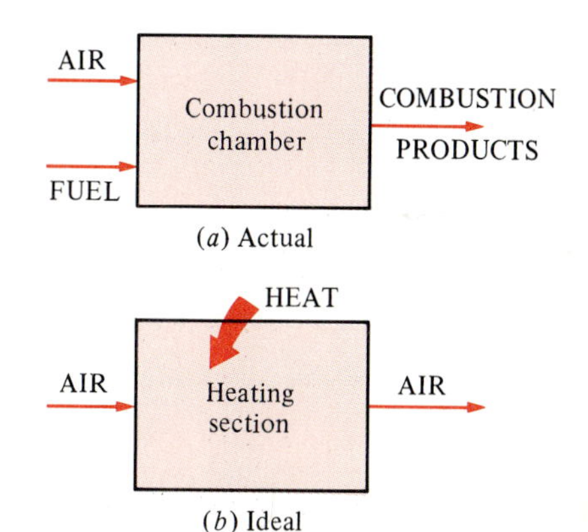

FIGURE 8-9
The combustion process is replaced by a heat addition process in ideal cycles.

4 The exhaust process is replaced by a heat rejection process that restores the working fluid to its initial state.

Another assumption which is often utilized to simplify the analysis even more is that the air has constant specific heats whose values are determined at room temperature (25°C, or 77°F). When this assumption is utilized, the air-standard assumptions are called the **cold-air-standard assumptions**. A cycle for which the air-standard assumptions are applicable is frequently referred to as an **air-standard cycle**.

The air-standard assumptions stated above provide considerable simplification in the analysis without significantly deviating from the actual cycles. This simplified model enables us to study qualitatively the influence of major parameters on the performance of the actual engines.

8-4 BRIEF OVERVIEW OF RECIPROCATING ENGINES

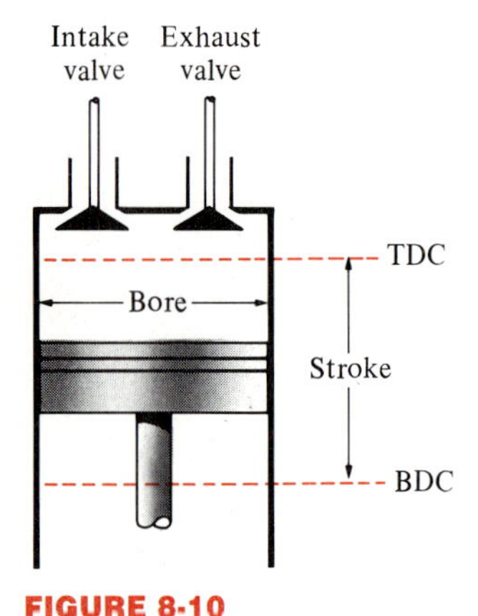

FIGURE 8-10
Nomenclature for reciprocating engines.

Despite its simplicity, the reciprocating engine (basically a piston-cylinder device) is one of the rare inventions that has proved to be very versatile and to have a wide range of applications. It is the powerhouse of the vast majority of automobiles, trucks, light aircraft, ships, and electric power generators as well as many other devices.

The basic components of a reciprocating engine are shown in Fig. 8-10. The piston reciprocates in the cylinder between two fixed positions called the **top dead center** (TDC)—the position of the piston when it forms the smallest volume in the cylinder—and the **bottom dead center** (BDC)—the position of the piston when it forms the largest volume in the cylinder. The distance between the TDC and the BDC is the largest distance that the piston can travel in one direction, and it is called the **stroke** of the engine. The diameter of the piston is called the **bore**. The air or air-fuel mixture is drawn into the cylinder through the **intake valve**, and the combustion products are expelled from the cylinder through the **exhaust valve**.

The minimum volume formed in the cylinder when the piston is at TDC is called the **clearance volume** (Fig. 8-11). The volume displaced by

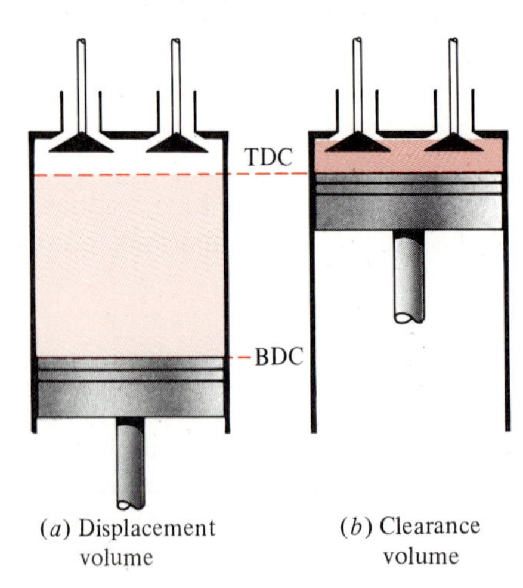

FIGURE 8-11
Displacement and clearance volumes of a reciprocating engine.

the piston as it moves between TDC and BDC is called the **displacement volume**. The ratio of the maximum volume formed in the cylinder to the minimum (clearance) volume is called the **compression ratio** r of the engine:

$$r = \frac{V_{\text{max}}}{V_{\text{min}}} = \frac{V_{\text{BDC}}}{V_{\text{TDC}}} \quad (8\text{-}3)$$

Notice that the compression ratio is a *volume ratio* and should not be confused with the pressure ratio.

Another term frequently used in conjunction with reciprocating engines is the **mean effective pressure** MEP. It is a fictitious pressure which, if it acted on the piston during the entire power stroke, would produce the same amount of net work as that produced during the actual cycle (Fig. 8-12). That is,

$$W_{\text{net}} = \text{MEP(piston area)(stroke)} = \text{MEP(displacement volume)}$$

or

$$\text{MEP} = \frac{W_{\text{net}}}{V_{\text{max}} - V_{\text{min}}} = \frac{w_{\text{net}}}{v_{\text{max}} - v_{\text{min}}} \quad \text{(kPa)} \quad (8\text{-}4)$$

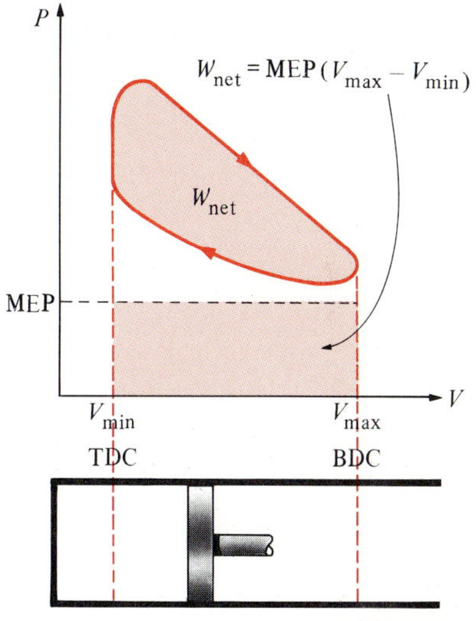

FIGURE 8-12

The net work output of a cycle is equivalent to the product of the mean effective pressure and the displacement volume.

The mean effective pressure can be used as a parameter to compare the performances of reciprocating engines of equal size. The engine that has a larger value of MEP will deliver more net work per cycle and thus will perform better.

Reciprocating engines are classified as **spark-ignition** (SI) **engines** or **compression-ignition** (CI) **engines**, depending on how the combustion process in the cylinder is initiated. In SI engines, the combustion of the air-fuel mixture is initiated by a spark plug. In CI engines, the air-fuel mixture is self-ignited as a result of compressing the mixture above its self-ignition temperature. In the next two sections, we discuss the *Otto* and *Diesel cycles,* which are the ideal cycles for the SI and CI reciprocating engines, respectively.

8-5 ■ OTTO CYCLE—The Ideal Cycle for Spark-Ignition Engines

The Otto cycle is the ideal cycle for spark-ignition reciprocating engines. It is named after Nikolaus A. Otto, who built a successful four-stroke engine in 1876 in Germany. In most spark-ignition engines, the piston executes four complete strokes (two mechanical cycles) within the cylinder, and the crankshaft completes 2 revolutions for each thermodynamic cycle. These engines are called **four-stroke** internal combustion engines. A schematic of each stroke as well as a P-v diagram for an actual four-stroke spark-ignition engine is given in Fig. 8-13*a*.

Initially, both the intake and the exhaust valves are closed, and the piston is at its lowest position (BDC). During the *compression stroke,* the piston moves upward, compressing the air-fuel mixture. Shortly before the piston reaches its highest position (TDC), the spark plug fires and the mixture ignites, increasing the pressure and temperature of the system. The high-pressure gases force the piston down, which in turn forces the

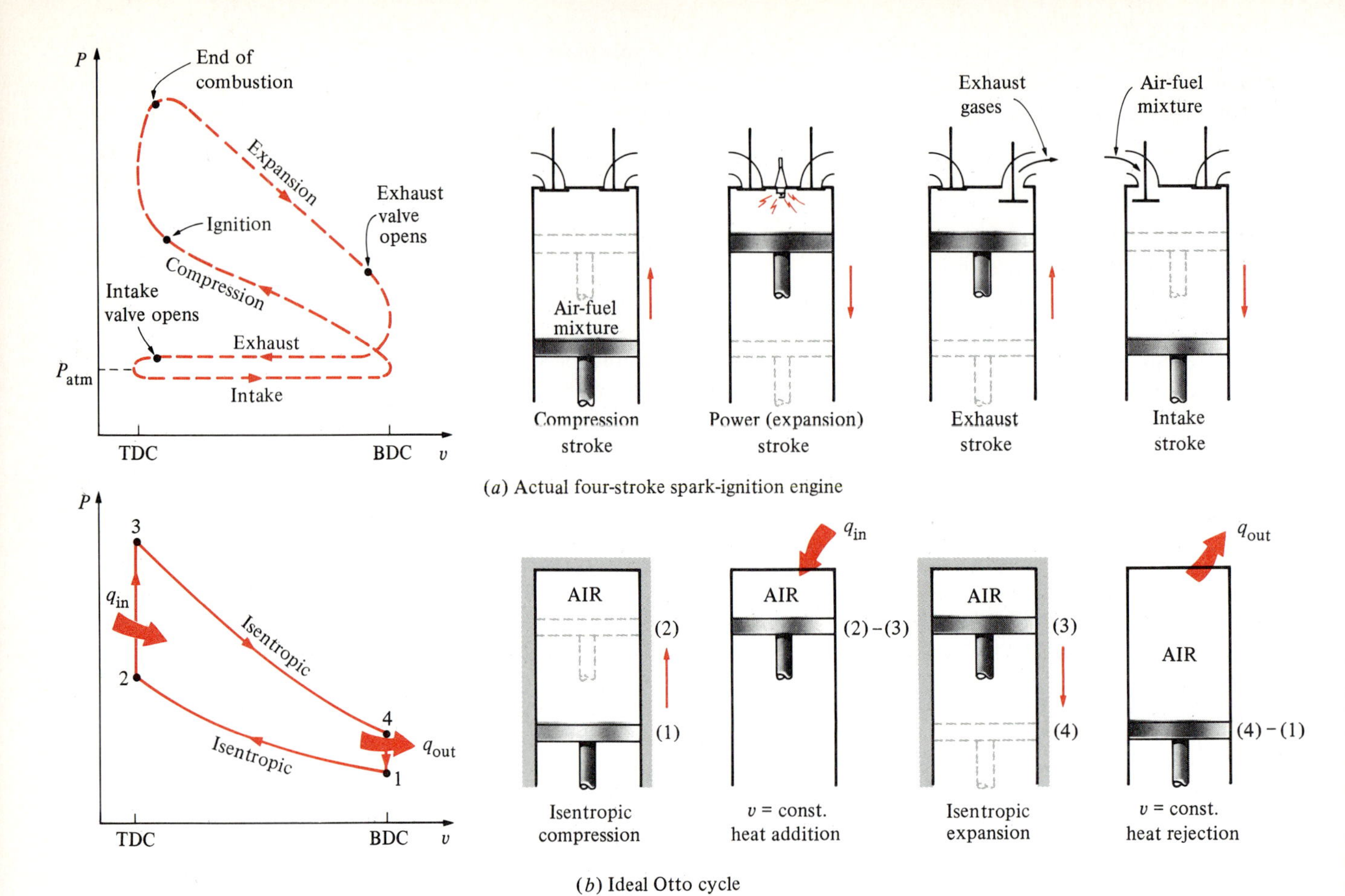

FIGURE 8-13

Actual and ideal cycles in spark-ignition engines and their *P-v* diagrams.

crankshaft to rotate, producing a useful work output during the *expansion* or *power stroke*. At the end of this stroke, the piston is at its lowest position (the completion of the first mechanical cycle), and the cylinder is filled with combustion products. Now the piston moves upward one more time, purging the exhaust gases through the exhaust valve (the *exhaust stroke*), and down a second time, drawing in fresh air-fuel mixture through the intake valve (the *intake stroke*). Notice that the pressure in the cylinder is slightly above the atmospheric value during the exhaust stroke and slightly below during the intake stroke.

In **two-stroke engines**, all four functions described above are executed in just two strokes: the power stroke and the compression stroke. In these engines, the crankcase is sealed, and the outward motion of the piston is used to slightly pressurize the air-fuel mixture in the crankcase, as shown in Fig. 8-14. Also, the intake and exhaust valves are replaced by openings in the lower portion of the cylinder wall. During the latter part of the power stroke, the piston uncovers first the exhaust port, allowing the exhaust gases to be partially expelled, and then the intake port, allowing the fresh air-fuel mixture to rush in and drive most of the remaining exhaust gases out of the cylinder. This mixture is then compressed as the piston moves upward during the compression stroke and is subsequently ignited by a spark plug.

The two-stroke engines are generally less efficient than their four-stroke counterparts because of the incomplete expulsion of the exhaust gases and the partial expulsion of the fresh air-fuel mixture with the exhaust gases. However, they are relatively simple and inexpensive, and they have high power-to-weight and power-to-volume ratios, which make them suitable for applications requiring small size and weight such as for motorcycles, chain saws, and lawn mowers.

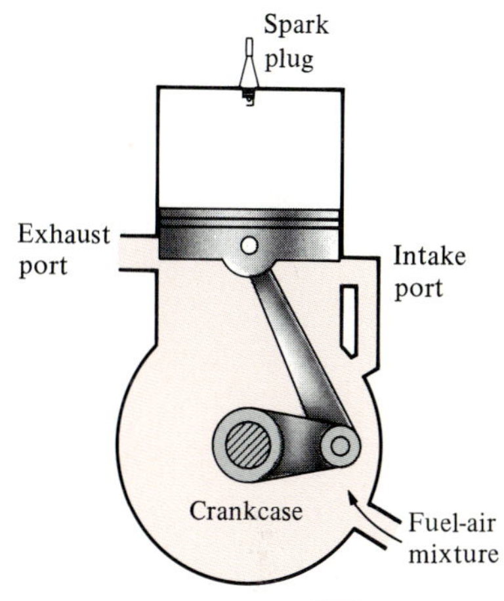

FIGURE 8-14
Schematic of a two-stroke reciprocating engine.

The thermodynamic analysis of the actual four-stroke or two-stroke cycles described above is not a simple task. However, the analysis can be simplified significantly if the air-standard assumptions are utilized. The resulting cycle which closely resembles the actual operating conditions is the ideal **Otto cycle**. It consists of four internally reversible processes:

1-2	Isentropic compression
2-3	v = constant heat addition
3-4	Isentropic expansion
4-1	v = constant heat rejection

The execution of the Otto cycle in a piston-cylinder device together with a *P-v* diagram is illustrated in Fig. 8-13*b*. The *T-s* diagram of the Otto cycle is given in Fig. 8-15.

The Otto cycle is executed in a closed system, and thus the first-law relation for any of the processes is expressed, on a unit-mass basis, as

$$q - w = \Delta u \qquad \text{(kJ/kg)} \tag{8-5}$$

No work is involved during the two heat transfer processes since both take place at constant volume. Therefore, heat transfer to and from the working fluid can be expressed, under the cold-air-standard assumptions, as

$$q_{\text{in}} = q_{23} = u_3 - u_2 = C_v(T_3 - T_2) \tag{8-6a}$$

and

$$q_{\text{out}} = -q_{41} = -(u_1 - u_4) = C_v(T_4 - T_1) \tag{8-6b}$$

Then the thermal efficiency of the ideal air-standard Otto cycle becomes

$$\eta_{\text{th,Otto}} = \frac{w_{\text{net}}}{q_{\text{in}}} = 1 - \frac{q_{\text{out}}}{q_{\text{in}}} = 1 - \frac{T_4 - T_1}{T_3 - T_2} = 1 - \frac{T_1(T_4/T_1 - 1)}{T_2(T_3/T_2 - 1)}$$

Processes 1-2 and 3-4 are isentropic, and $v_2 = v_3$ and $v_4 = v_1$. Thus,

$$\frac{T_1}{T_2} = \left(\frac{v_2}{v_1}\right)^{k-1} = \left(\frac{v_3}{v_4}\right)^{k-1} = \frac{T_4}{T_3} \tag{8-7}$$

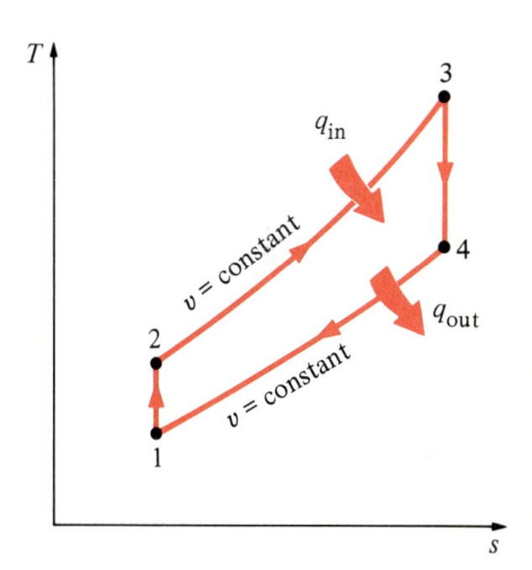

FIGURE 8-15
T-s diagram for the ideal Otto cycle.

Substituting these equations into the thermal efficiency relation and simplifying give

$$\eta_{\text{th,Otto}} = 1 - \frac{1}{r^{k-1}} \tag{8-8}$$

where
$$r = \frac{V_{\max}}{V_{\min}} = \frac{V_1}{V_2} = \frac{v_1}{v_2} \tag{8-9}$$

is the compression ratio and k is the specific heat ratio C_p/C_v.

Equation 8-8 shows that under the cold-air-standard assumptions, the thermal efficiency of an ideal Otto cycle depends on the compression ratio of the engine and the specific heat ratio of the working fluid (if different from air). The thermal efficiency of the ideal Otto cycle increases with both the compression ratio and the specific heat ratio. This is also true for actual spark-ignition internal combustion engines. A plot of thermal efficiency versus the compression ratio is given in Fig. 8-16 for $k = 1.4$, which is the specific-heat-ratio value of air at room temperature. For a given compression ratio, the thermal efficiency of an actual spark-ignition engine will be less than that of an ideal Otto cycle because of the irreversibilities, such as friction, and other factors such as incomplete combustion.

We can observe from Fig. 8-16 that the thermal efficiency curve is rather steep at low compression ratios but flattens out starting with a compression ratio value of about 8. Therefore, the increase in thermal efficiency with the compression ratio is not that pronounced at high compression ratios. Also when high compression ratios are used, the temperature of the air-fuel mixture rises above the autoignition temperature of the fuel (the temperature at which the fuel ignites without the help of a spark) during the combustion process, causing an early and rapid burn of the fuel (Fig. 8-17). This premature ignition of the fuel, called **autoignition**, produces an audible noise, which is called **engine knock**. Autoignition in spark-ignition engines cannot be tolerated because it hurts performance and can cause engine damage. The requirement that autoig-

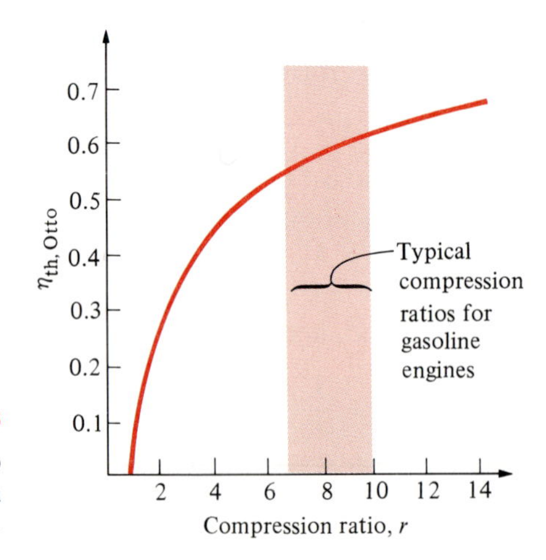

FIGURE 8-16

Thermal efficiency of the ideal Otto cycle as a function of compression ratio ($k = 1.4$).

nition not be allowed places an upper limit on the compression ratios that can be used in spark-ignition internal combustion engines.

Improvement of the thermal efficiency of gasoline engines by utilizing higher compression ratios (up to about 12) without facing the autoignition problem has been made possible by using gasoline blends that have good antiknock characteristics, such as gasoline mixed with tetraethyl lead. Tetraethyl lead has been added to gasoline since the 1920s because it is the cheapest method of raising the octane rating. The leaded gasoline, however, has a very undesirable side effect: It forms compounds during the combustion process that are hazardous to health and pollute the environment. In an effort to combat air pollution, the government adopted a policy in the mid-1970s that would lead to the eventual phase-out of the leaded gasoline. Unable to use lead, the refiners developed other, more elaborate techniques to improve the antiknock characteristics of the gasoline. Most cars made since 1975 have been designed to use unleaded gasoline, and the compression ratios had to be lowered to avoid engine knock. The thermal efficiency of car engines has somewhat decreased as a result of decreased compression ratios. But, owing to the improvements in other areas (reduction in overall automobile weight, improved aerodynamic design, etc.), today's cars have better fuel economy and consequently get more miles per gallon of fuel. This is an example of how engineering decisions involve compromises, and efficiency is only one of the considerations in reaching a final decision.

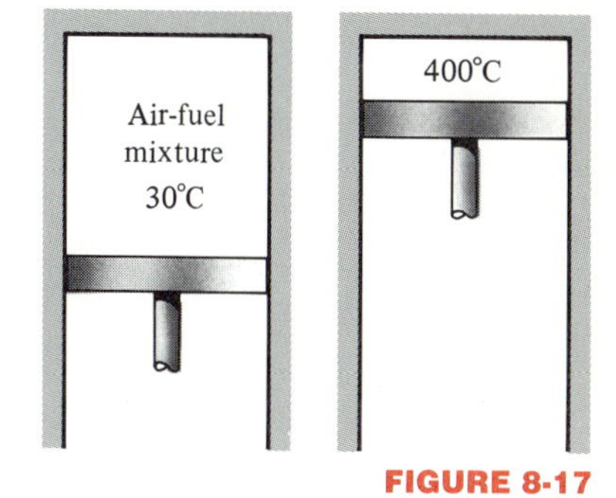

FIGURE 8-17

At high compression ratios, the air-fuel mixture temperature rises above the self-ignition temperature of the fuel during the compression process.

The second parameter that affects the thermal efficiency of an ideal Otto cycle is the specific heat ratio k. For a given compression ratio, an ideal Otto cycle using a monatomic gas (such as argon or helium, $k = 1.667$) as the working fluid will have the highest thermal efficiency. The specific heat ratio k, and thus the thermal efficiency of the ideal Otto cycle, decreases as the molecules of the working fluid get larger (Fig. 8-18). At room temperature, it is 1.4 for air, 1.3 for carbon dioxide, and 1.2 for ethane. The working fluid in actual engines contains larger molecules such as carbon dioxide, and the specific heat ratio decreases with

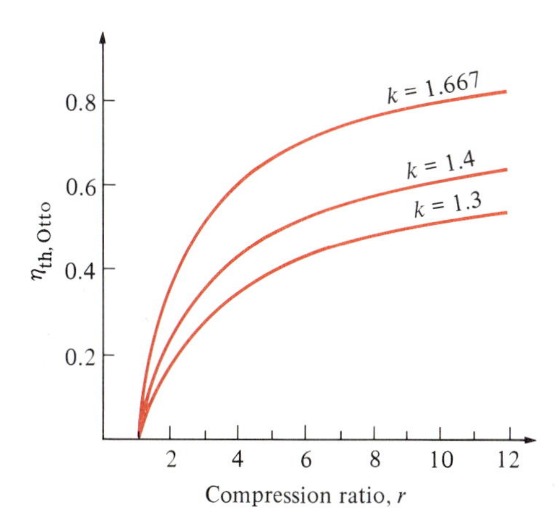

FIGURE 8-18

The thermal efficiency of the Otto cycle increases with the specific heat ratio k of the working fluid.

temperature, which is one of the reasons that the actual cycles have lower thermal efficiencies than the ideal Otto cycle.

EXAMPLE 8-2

An ideal Otto cycle has a compression ratio of 8. At the beginning of the compression process, the air is at 100 kPa and 17°C, and 800 kJ/kg of heat is transferred to air during the constant-volume heat addition process. Accounting for the variation of specific heats of air with temperature, determine (*a*) the maximum temperature and pressure that occur during the cycle, (*b*) the net work output, (*c*) the thermal efficiency, and (*d*) the mean effective pressure for the cycle.

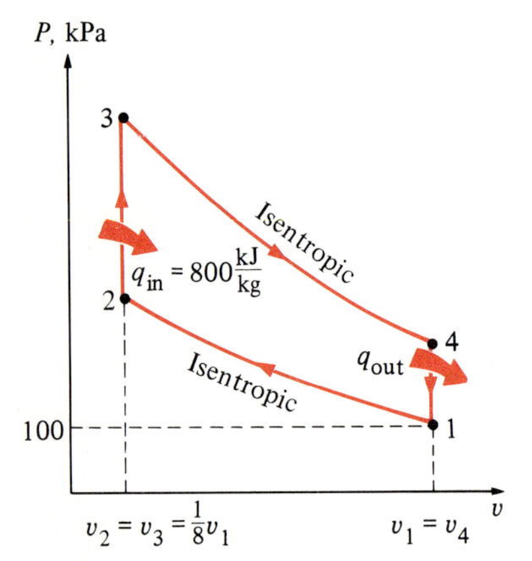

FIGURE 8-19

P-v diagram for the Otto cycle discussed in Example 8-2.

Solution The Otto cycle described is shown on a *P-v* diagram in Fig. 8-19. The air contained in the cylinder forms a closed system, which should be remembered in applying the thermodynamic laws to the individual processes. It is specifically asked in the problem to account for the variation of the specific heats with temperature. Therefore, any ideal-gas relation which utilizes the constant-specific-heat assumption should not be used in the analysis.

(*a*) The maximum temperature and pressure in an Otto cycle occur at the end of the constant-volume heat addition process (state 3). But first we need to determine the temperature and pressure of air at the end of the isentropic compression process (state 2), using data from Table A-17:

$$T_1 = 290 \text{ K} \longrightarrow u_1 = 206.91 \text{ kJ/kg}$$
$$v_{r1} = 676.1$$

Process 1-2 (isentropic compression of an ideal gas):

$$\frac{v_{r2}}{v_{r1}} = \frac{v_2}{v_1} = \frac{1}{r} \longrightarrow v_{r2} = \frac{v_{r1}}{r} = \frac{676.1}{8} = 84.51 \longrightarrow T_2 = 652.4 \text{ K}$$
$$u_2 = 475.11 \text{ kJ/kg}$$

$$\frac{P_2 v_2}{T_2} = \frac{P_1 v_1}{T_1} \longrightarrow P_2 = P_1 \left(\frac{T_2}{T_1}\right)\left(\frac{v_1}{v_2}\right)$$
$$= (100 \text{ kPa})\left(\frac{652.4 \text{ K}}{290 \text{ K}}\right)(8) = 1799.7 \text{ kPa}$$

Process 2-3 (v = constant heat addition):

$$q_{23} - \cancelto{0}{w_{23}} = u_3 - u_2$$
$$800 \text{ kJ/kg} = u_3 - 475.11 \text{ kJ/kg}$$
$$u_3 = 1275.11 \text{ kJ/kg} \longrightarrow T_3 = 1575.1 \text{ K}$$
$$v_{r3} = 4.998$$
$$\frac{P_3 v_3}{T_3} = \frac{P_2 v_2}{T_2} \longrightarrow P_3 = P_2 \left(\frac{T_3}{T_2}\right)\left(\frac{v_2}{v_3}\right)$$
$$= (1.7997 \text{ MPa})\left(\frac{1575.1 \text{ K}}{652.1 \text{ K}}\right)(1) = 4.347 \text{ MPa}$$

(*b*) The net work output for the cycle is determined either by finding the boundary (*P dV*) work involved in each process by integration and adding them or by finding the net heat transfer which is equivalent to the net work done during the cycle. We take the latter approach. But first we need to find the internal energy of the air at state 4:

Process 3-4 (isentropic expansion of an ideal gas):

$$\frac{v_{r4}}{v_{r3}} = \frac{v_4}{v_3} = r \longrightarrow v_{r4} = rv_{r3} = (8)(4.998) = 39.98 \longrightarrow \begin{array}{l} T_4 = 853.7 \text{ K} \\ u_4 = 636.22 \text{ kJ/kg} \end{array}$$

Process 4-1 (v = constant heat rejection):

$$q_{41} - w_{41}^{\nearrow 0} = u_1 - u_4$$

$$q_{41} = (206.91 - 636.22) \text{ kJ/kg} = -429.31 \text{ kJ/kg}$$

or

$$q_{out} = 429.31 \text{ kJ/kg}$$

Thus, $w_{net} = q_{net} = q_{in} - q_{out} = (800 - 429.31) \text{ kJ/kg} = 370.69 \text{ kJ/kg}$

(*c*) The thermal efficiency of the cycle is determined from its definition, Eq. 8-1:

$$\eta_{th} = \frac{w_{net}}{q_{in}} = \frac{370.69 \text{ kJ/kg}}{800 \text{ kJ/kg}} = 0.463 \quad \text{or } 46.3\%$$

Under the cold-air-standard assumptions (constant specific heat values at room temperature), the thermal efficiency would be (Eq. 8-8)

$$\eta_{th,Otto} = 1 - \frac{1}{r^{k-1}} = 1 - r^{1-k} = 1 - (8)^{1-1.4} = 0.565 \quad \text{or } 56.5\%$$

which is considerably different from the value obtained above. Therefore, care should be exercised in utilizing the cold-air-standard assumptions.

(*d*) The mean effective pressure is determined from its definition, Eq. 8-4:

$$\text{MEP} = \frac{w_{net}}{v_1 - v_2} = \frac{w_{net}}{v_1 - v_1/r} = \frac{w_{net}}{v_1(1 - 1/r)}$$

where

$$v_1 = \frac{RT_1}{P_1} = \frac{[0.287 \text{ kPa·m}^3/(\text{kg·K})](290 \text{ K})}{100 \text{ kPa}} = 0.832 \text{ m}^3/\text{kg}$$

Thus,

$$\text{MEP} = \frac{370.69 \text{ kJ/kg}}{(0.832 \text{ m}^3/\text{kg})(1 - \frac{1}{8})}\left(\frac{1 \text{ kPa·m}^3}{1 \text{ kJ}}\right) = 509.2 \text{ kPa}$$

Therefore, a constant pressure of 509.2 kPa during the power stroke would produce the same net work output as the entire cycle.

8-6 ■ DIESEL CYCLE—The Ideal Cycle for Compression-Ignition Engines

The Diesel cycle is the ideal cycle for CI reciprocating engines. The CI engine, first proposed by Rudolph Diesel in the 1890s, is very similar to the SI engine discussed in the last section, differing mainly in the method of initiating combustion. In spark-ignition engines (also known as *gasoline engines*), the air-fuel mixture is compressed to a temperature which is below the autoignition temperature of the fuel, and the combustion process is initiated by firing a spark plug. In CI engines (also known as *diesel engines*) the air is compressed to a temperature which is above the autoignition temperature of the fuel, and combustion starts on contact as the fuel is injected into this hot air. Therefore, the spark plug and carburetor are replaced by a fuel injector in diesel engines (Fig. 8-20).

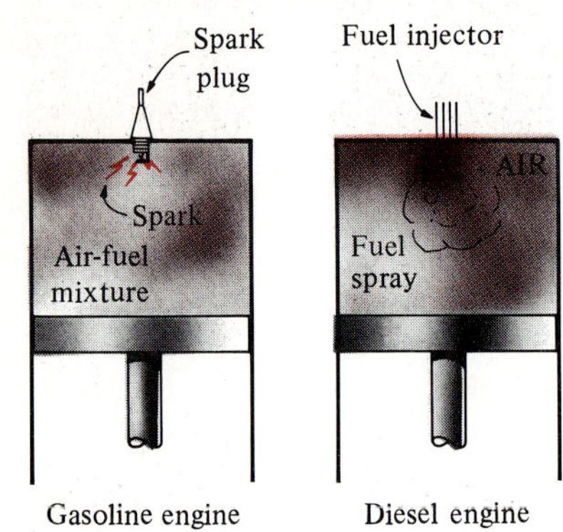

FIGURE 8-20

In diesel engines, the spark plug is replaced by a fuel injector, and only air is compressed during the compression process.

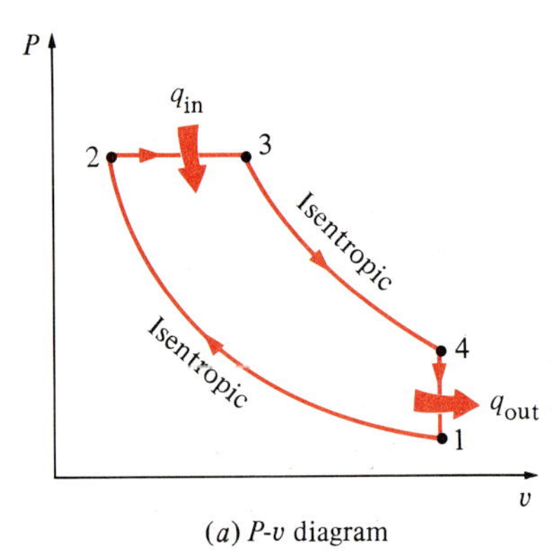

(*a*) *P-v* diagram

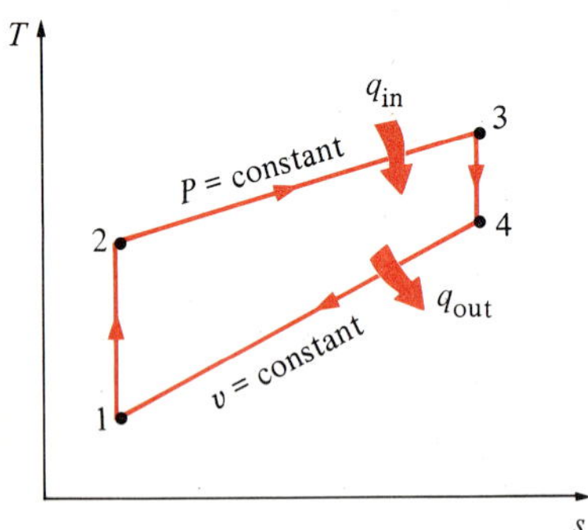

(*b*) *T-s* diagram

FIGURE 8-21

T-s and *P-v* diagrams for the ideal Diesel cycle.

In gasoline engines, a mixture of air and fuel is compressed during the compression stroke, and the compression ratios are limited by the onset of autoignition or engine knock. In diesel engines, only air is compressed during the compression stroke, eliminating the possibility of autoignition. Therefore, diesel engines can be designed to operate at much higher compression ratios, typically between 12 and 24. Not having to deal with the problem of autoignition has another benefit: Many of the stringent requirements placed on the gasoline can now be removed, and fuels that are less refined (thus less expensive) can be used in diesel engines.

The fuel injection process in diesel engines starts when the piston approaches TDC and continues during the first part of the power stroke. Therefore, the combustion process in these engines takes place over a longer interval. Because of this longer duration, the combustion process in the ideal Diesel cycle is approximated as a constant-pressure heat addition process. In fact, this is the only process where the Otto and the Diesel cycles differ. The remaining three processes are the same for both ideal cycles. That is, process 1-2 is isentropic compression, 3-4 is isentropic expansion, and 4-1 is constant-volume heat rejection. The similarity between the two cycles is also apparent from the *P-v* and *T-s* diagrams of the Diesel cycle, shown in Fig. 8-21.

A measure of performance for any power cycle is its thermal efficiency. Below we develop a relation for the thermal efficiency of a Diesel cycle, utilizing the cold-air-standard assumptions. Such a relation will enable us to examine the effects of major parameters on the performance of diesel engines.

The Diesel cycle, like the Otto cycle, is executed in a piston-cylinder device, which forms a closed system. Therefore, equations developed for closed systems should be used in the analysis of individual processes. Under the cold-air-standard assumptions, the amount of heat added to the working fluid at constant pressure and rejected from it at constant volume can be expressed as

$$\begin{aligned} q_{in} &= q_{23} = w_{23} + (\Delta u)_{23} = P_2(v_3 - v_2) + (u_3 - u_2) \\ &= h_3 - h_2 = C_p(T_3 - T_2) \end{aligned} \tag{8-10a}$$

and

$$\begin{aligned} q_{out} &= -q_{41} = -\overset{\nearrow 0}{w_{41}} - (\Delta u)_{41} = u_4 - u_1 \\ &= C_v(T_4 - T_1) \end{aligned} \tag{8-10b}$$

Then the thermal efficiency of the ideal Diesel cycle under the cold-air-standard assumptions becomes

$$\eta_{th,Diesel} = \frac{w_{net}}{q_{in}} = 1 - \frac{q_{out}}{q_{in}} = 1 - \frac{T_4 - T_1}{k(T_3 - T_2)} = 1 - \frac{T_1(T_4/T_1 - 1)}{kT_2(T_3/T_2 - 1)}$$

We now define a new quantity, the **cutoff ratio** r_c, as the ratio of the cylinder volumes after and before the combustion process:

$$r_c = \frac{V_3}{V_2} = \frac{v_3}{v_2} \tag{8-11}$$

Utilizing this definition and the isentropic ideal-gas relations for processes 1-2 and 3-4 (Eqs. 8-7), we see that the thermal efficiency relation reduces to

$$\eta_{\text{th,Diesel}} = 1 - \frac{1}{r^{k-1}}\left[\frac{r_c^k - 1}{k(r_c - 1)}\right] \tag{8-12}$$

where r is the compression ratio defined by Eq. 8-9. Looking at Eq. 8-12 carefully, one would notice that under the cold-air-standard assumptions, the efficiency of a Diesel cycle differs from the efficiency of an Otto cycle by the quantity in the brackets. This quantity is always greater than 1. Therefore,

$$\eta_{\text{th,Otto}} > \eta_{\text{th,Diesel}} \tag{8-13}$$

when both cycles operate on the same compression ratio. Also as the cutoff ratio decreases, the efficiency of the Diesel cycle increases (Fig. 8-22). For the limiting case of $r_c = 1$, the quantity in the brackets becomes unity (can you prove it?), and the efficiencies of the Otto and Diesel cycles become identical. Remember, though, that diesel engines operate at much higher compression ratios and thus are usually more efficient than the spark-ignition (gasoline) engines. The diesel engines also burn the fuel more completely since they usually operate at lower revolutions per minute than spark-ignition engines.

The higher efficiency and lower fuel costs of diesel engines make them the clear choice in applications requiring relatively large amounts of power, such as in locomotive engines, emergency power generation units, large ships, and heavy trucks. As an example of how large diesel engines can be, a 12-cylinder diesel engine built in 1964 by the Fiat Corporation of Italy had a normal power output of 25,200 hp (18.8 MW) at 122 rpm, a cylinder bore of 90 cm, and a stroke of 91 cm.

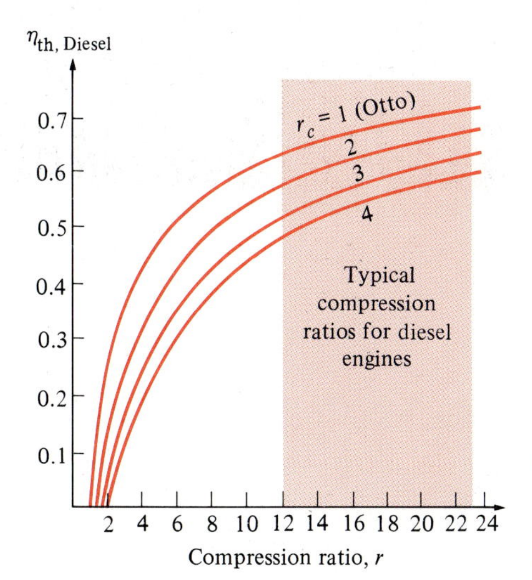

FIGURE 8-22
Thermal efficiency of the ideal Diesel cycle as a function of compression and cutoff ratios ($k = 1.4$).

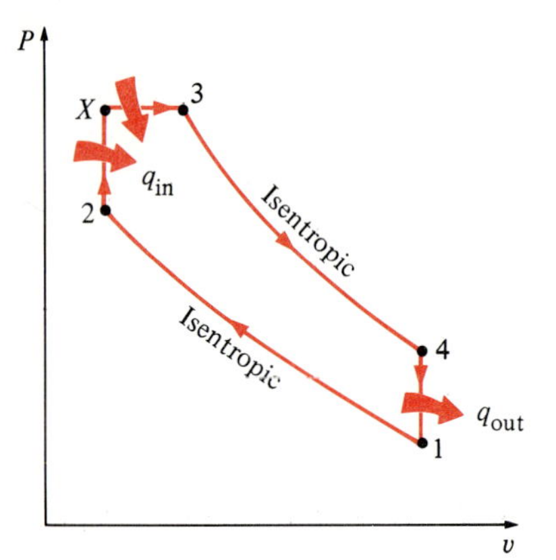

FIGURE 8-23

P-v diagram of an ideal dual cycle.

Approximating the combustion process in internal combustion engines as a constant-volume or a constant-pressure heat addition process is overly simplistic and not quite realistic. Probably a better (but slightly more complex) approach would be to model the combustion process in both gasoline and diesel engines as a combination of two heat transfer processes, one occurring at constant volume and the other at constant pressure. The ideal cycle based on this concept is called the **dual cycle**, and a *P-v* diagram for it is given in Fig. 8-23. The relative amounts of heat added during each process can be adjusted to more closely approximate the actual cycle. Note that both the Otto and the Diesel cycles can be obtained as special cases of the dual cycle.

EXAMPLE 8-3

An ideal Diesel cycle with air as the working fluid has a compression ratio of 18 and a cutoff ratio of 2. At the beginning of the compression process, the working fluid is at 14.7 psia, 80°F, and 117 in³. Utilizing the cold-air-standard assumptions, determine (*a*) the temperature and pressure of the air at the end of each process, (*b*) the net work output and the thermal efficiency, and (*c*) the mean effective pressure.

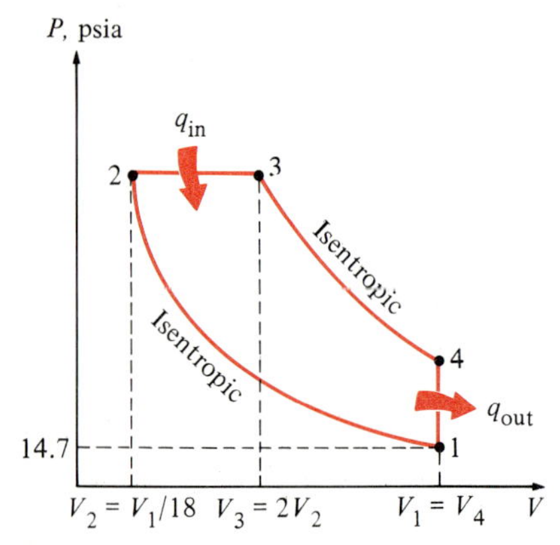

FIGURE 8-24

P-v diagram for the ideal Diesel cycle discussed in Example 8-3.

Solution The *P-v* diagram of the cycle is shown in Fig. 8-24. The ideal Diesel cycle is executed in a piston-cylinder device with a fixed amount of air, and thus all four processes should be analyzed as closed-system processes. Under the cold-air-standard assumptions, the working fluid (air) is assumed to be an ideal gas and to have constant specific heats evaluated at room temperature. The gas constant of air is $R = 0.06855$ Btu/(lbm·R) (Table A-1E), and its specific heats at room temperature are $C_p = 0.240$ Btu/(lbm·R) and $C_v = 0.171$ Btu/(lbm·R) (Table A-3E*a*).

(*a*) The temperature and pressure values at the end of each process can be determined by utilizing the ideal-gas isentropic relations for processes 1-2 and 3-4. But first we determine the volumes at the end of each process from the definitions of the compression ratio and the cutoff ratio:

$$V_2 = \frac{V_1}{r} = \frac{117 \text{ in}^3}{18} = 6.5 \text{ in}^3$$

$$V_3 = r_c V_2 = (2)(6.5 \text{ in}^3) = 13 \text{ in}^3$$

$$V_4 = V_1 = 117 \text{ in}^3$$

Process 1-2 (isentropic compression of an ideal gas, constant specific heats):

$$T_2 = T_1 \left(\frac{V_1}{V_2}\right)^{k-1} = (540 \text{ R})(18)^{1.4-1} = 1716 \text{ R}$$

$$P_2 = P_1 \left(\frac{V_1}{V_2}\right)^{k} = (14.7 \text{ psia})(18)^{1.4} = 841 \text{ psia}$$

Process 2-3 (P = constant heat addition to an ideal gas):

$$P_3 = P_2 = 841 \text{ psia}$$

$$\frac{P_2 V_2}{T_2} = \frac{P_3 V_3}{T_3} \longrightarrow T_3 = T_2 \left(\frac{V_3}{V_2}\right) = (1716 \text{ R})(2) = 3432 \text{ R}$$

Processs 3-4 (isentropic expansion of an ideal gas, constant specific heats):

$$T_4 = T_3\left(\frac{V_3}{V_4}\right)^{k-1} = (3432 \text{ R})\left(\frac{13 \text{ in}^3}{117 \text{ in}^3}\right)^{1.4-1} = 1425 \text{ R}$$

$$P_4 = P_3\left(\frac{V_3}{V_4}\right)^{k} = (841 \text{ psia})\left(\frac{13 \text{ in}^3}{117 \text{ in}^3}\right)^{1.4} = 38.8 \text{ psia}$$

(*b*) The net work for a cycle is equivalent to the net heat transfer, i.e., the difference between the total heat supplied and the total heat rejected. But first we find the mass of air:

$$m = \frac{P_1V_1}{RT_1} = \frac{(14.7 \text{ psia})(117 \text{ in}^3)}{[0.3704 \text{ psia}\cdot\text{ft}^3/(\text{lbm}\cdot\text{R})](540 \text{ R})}\left(\frac{1 \text{ ft}^3}{1728 \text{ in}^3}\right) = 0.00498 \text{ lbm}$$

Process 2-3 is a constant-pressure heat addition process, for which the boundary work and Δu terms can be combined into Δh. Thus,

$$\begin{aligned} Q_{in} &= Q_{23} = m(h_3 - h_2) = mC_p(T_3 - T_2) \\ &= (0.00498 \text{ lbm})[0.240 \text{ Btu}/(\text{lbm}\cdot\text{R})][(3432 - 1716) \text{ R}] \\ &= 2.051 \text{ Btu} \end{aligned}$$

Process 4-1 is a constant-volume heat rejection process (it involves no work interactions), and the amount of heat rejected is

$$\begin{aligned} Q_{out} &= -Q_{41} = m(u_4 - u_1) = mC_v(T_4 - T_1) \\ &= (0.00498 \text{ lbm})[0.171 \text{ Btu}/(\text{lbm}\cdot\text{R})][(1425 - 540) \text{ R}] \\ &= 0.758 \text{ Btu} \end{aligned}$$

Thus, $\quad W_{net} = Q_{in} - Q_{out} = (2.051 - 0.758) \text{ Btu} = 1.293 \text{ Btu}$

Then the thermal efficiency becomes

$$\eta_{th} = \frac{W_{net}}{Q_{in}} = \frac{1.293 \text{ Btu}}{2.051 \text{ Btu}} = 0.630 \quad (\text{or } 63.0\%)$$

The thermal efficiency of this Diesel cycle under the cold-air-standard assumptions could also be determined from Eq. 8-12.

(*c*) The mean effective pressure is determined from its definition, Eq. 8-4:

$$\begin{aligned} \text{MEP} &= \frac{W_{net}}{V_{max} - V_{min}} = \frac{W_{net}}{V_1 - V_2} = \frac{1.293 \text{ Btu}}{(117 - 6.5) \text{ in}^3}\left(\frac{778.17 \text{ lbf}\cdot\text{ft}}{1 \text{ Btu}}\right)\left(\frac{12 \text{ in}}{1 \text{ ft}}\right) \\ &= 109.3 \text{ psia} \end{aligned}$$

Therefore, a constant pressure of 109.3 psia during the power stroke would produce the same net work output as the entire Diesel cycle.

8-7 ■ STIRLING AND ERICSSON CYCLES

The ideal Otto and Diesel cycles discussed in the preceding sections are composed entirely of internally reversible processes and thus are internally reversible cycles. These cycles are not totally reversible, however, since they involve heat transfer through a finite temperature difference during the nonisothermal heat addition and rejection processes, which are irreversible. Therefore, the thermal efficiency of an Otto or Diesel engine will be less than that of a Carnot engine operating between the same temperature limits.

Consider a heat engine operating between a high-temperature reservoir at T_H and a low-temperature reservoir at T_L. For the heat-engine cycle to be totally reversible, the temperature difference between the working fluid and the thermal energy source (or sink) should never exceed a differential amount dT during any heat transfer process. That is, both the heat addition and heat rejection processes during the cycle must take place isothermally, one at a temperature of T_H and the other at a temperature of T_L. This is precisely what happens in a Carnot cycle.

There are two other cycles that involve an isothermal heat addition process at T_H and an isothermal heat rejection process at T_L: the *Stirling cycle* and the *Ericsson cycle*. They differ from the Carnot cycle in that the two isentropic processes are replaced by two constant-volume regeneration processes in the Stirling cycle, and by two constant-pressure regeneration processes in the Ericsson cycle. Both cycles utilize **regeneration**, a process during which heat is transferred to a thermal energy storage device (called a *regenerator*) during one part of the cycle and is transferred back to the working fluid during another part of the cycle (Fig. 8-25).

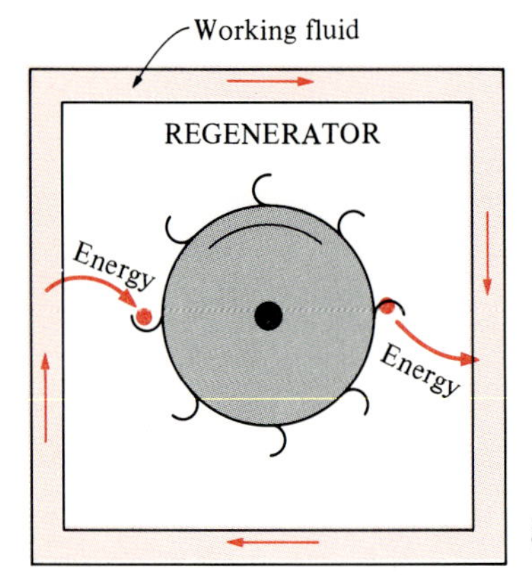

FIGURE 8-25

The regenerator is a device that borrows energy from the working fluid during one part of the cycle and pays it back (without interest) during another part.

Figure 8-26*b* shows the *T*-*s* and *P*-*v* diagrams of the **Stirling cycle**, which is made up of four totally reversible processes:

1-2 $\quad T$ = constant expansion (heat addition from external source)

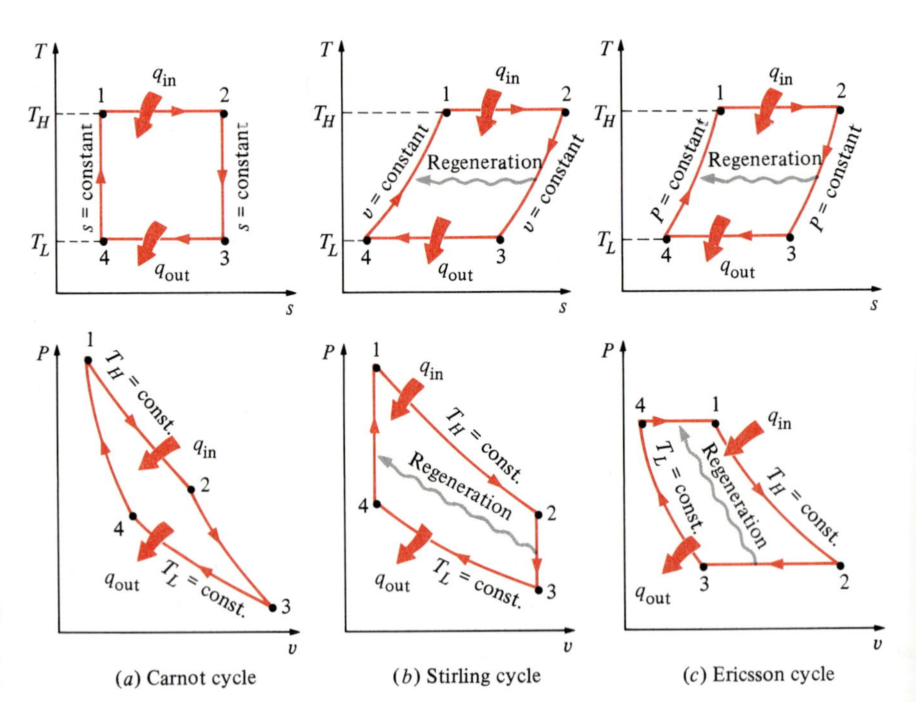

FIGURE 8-26

T-*s* and *P*-*v* diagrams of Carnot, Stirling, and Ericsson cycles.

2-3 v = constant regeneration (internal heat transfer from the working fluid to regenerator)

3-4 T = constant compression (heat rejection to external sink)

4-1 v = constant regeneration (internal heat transfer from regenerator back to the working fluid)

The execution of the Stirling cycle requires rather innovative hardware. The actual Stirling engines, including the original one patented by Robert Stirling, are very heavy and complicated. To spare the reader the complexities, the execution of the Stirling cycle in a closed system is explained with the help of the hypothetical engine shown in Fig. 8-27.

This system consists of a cylinder with two pistons on each side and a regenerator in the middle. The regenerator can be a wire or ceramic mesh or any kind of porous plug that has a high thermal mass (mass times specific heat). It is used for the temporary storage of thermal energy. The mass of the working fluid contained within the regenerator at any instant is considered negligible.

Initially, the left chamber houses the entire working fluid (a gas) which is at a high temperature and pressure. During *process 1-2,* heat is added to the gas at T_H from a source at T_H. As the gas expands isothermally, the left piston moves outward, doing work, and the gas pressure drops. During *process 2-3,* both pistons are moved to the right at the same rate (to keep the volume constant) until the entire gas is forced into the right chamber. As the gas passes through the regenerator, heat is transferred to the regenerator and the gas temperature drops from T_H to T_L. For this heat transfer process to be reversible, the temperature difference between the gas and the regenerator should not exceed a differential amount dT at any point. Thus, the temperature of the regenerator will be T_H at the left end and T_L at the right end of the regenerator when state 3 is reached. During *process 3-4,* the right piston is moved inward, compressing the gas. Heat is transferred from the gas to a sink at temperature T_L, so that the gas temperature remains constant at T_L while the pressure rises. Finally, during *process 4-1,* both pistons are moved to the left at the same rate (to keep the volume constant), forcing the entire gas into the left chamber. The gas temperature rises from T_L to T_H as it passes through the regenerator and picks up the thermal energy stored there during process 2-3. This completes the cycle.

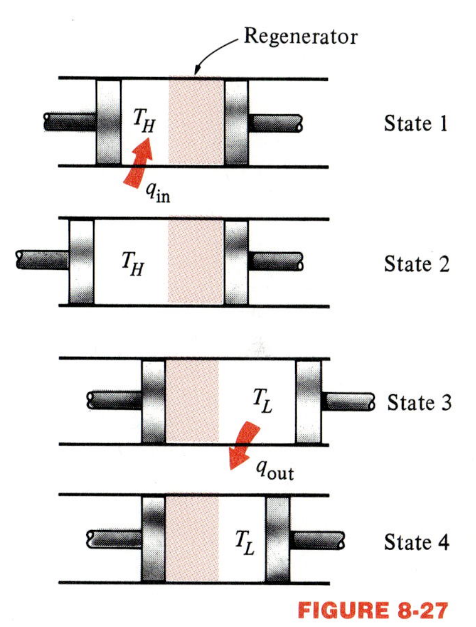

FIGURE 8-27
The execution of the Stirling cycle.

Notice that the second v = constant process takes place at a smaller volume than the first one, and the net heat transfer to the regenerator during a cycle is zero. That is, the amount of thermal energy stored in the regenerator during process 2-3 is equal to the amount picked up by the gas during process 4-1.

The *T-s* and *P-v* diagrams of the **Ericsson cycle** are shown in Fig. 8-25*c*. The Ericsson cycle is very much like the Stirling cycle, except that the two constant-volume processes are replaced by two constant-pressure processes.

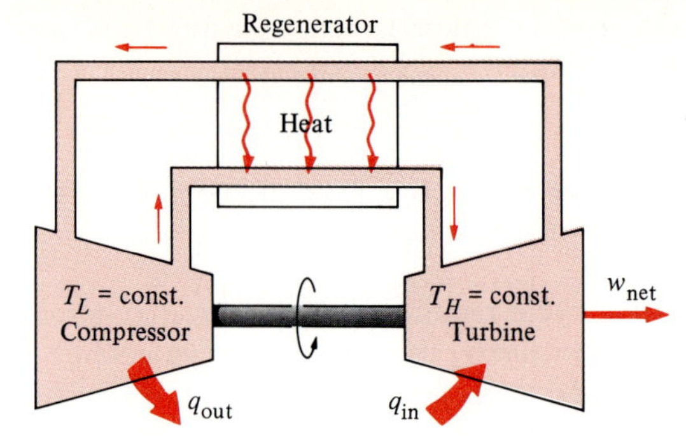

FIGURE 8-28
A steady-flow Ericsson engine.

A steady-flow system operating on an Ericsson cycle is shown in Fig. 8-28. Here the isothermal expansion and compression processes are executed in a compressor and a turbine, respectively, and a counterflow heat exchanger serves as a regenerator. Hot and cold fluid streams enter the heat exchanger from opposite ends, and heat transfer takes place between the two streams. In the ideal case, the temperature difference between the two fluid streams does not exceed a differential amount at any point, and the cold fluid stream leaves the heat exchanger at the inlet temperature of the hot stream.

Both the Stirling and Ericsson cycles are totally reversible as is the Carnot cycle, and so according to the Carnot principle, all three cycles will have the same thermal efficiency when operating between the same temperature limits:

$$\eta_{\text{th,Stirling}} = \eta_{\text{th,Ericsson}} = \eta_{\text{th,Carnot}} = 1 - \frac{T_L}{T_H} \tag{8-14}$$

This is proved for the Carnot cycle in Example 8-1 and can be proved in a similar manner for both the Stirling and Ericsson cycles.

EXAMPLE 8-4

Using an ideal gas as the working fluid, show that the thermal efficiency of an Ericsson cycle is identical to the efficiency of a Carnot cycle operating between the same temperature limits.

Solution Heat is added to the working fluid isothermally from an external source at temperature T_H during process 2-3, and it is rejected again isothermally to an external sink at temperature T_L during process 4-1. For a reversible isothermal process, heat transfer is related to the entropy change by

$$q = T\,\Delta s$$

The entropy change of an ideal gas during an isothermal process is given by

$$\Delta s = C_p \ln \cancelto{0}{\frac{T_e}{T_i}} - R \ln \frac{P_e}{P_i} = -R \ln \frac{P_e}{P_i}$$

Then the amount of heat input and heat output can be expressed, on a unit-mass basis, as

$$q_{in} = q_{23} = T_H(s_3 - s_2) = T_H\left(-R \ln \frac{P_3}{P_2}\right) = RT_H \ln \frac{P_2}{P_3}$$

and
$$q_{out} = -q_{41} = -T_L(s_1 - s_4) = -T_L\left(-R \ln \frac{P_1}{P_4}\right) = RT_L \ln \frac{P_1}{P_4}$$

Then the thermal efficiency of the Ericsson cycle becomes

$$\eta_{th,Ericsson} = 1 - \frac{q_{out}}{q_{in}} = 1 - \frac{RT_L \ln (P_1/P_4)}{RT_H \ln (P_2/P_3)} = 1 - \frac{T_L}{T_H}$$

since $P_2 = P_1$ and $P_4 = P_3$. Notice that this result is independent of whether the cycle is executed in a closed or steady-flow system.

Stirling and Ericsson cycles are difficult to achieve in practice because they involve heat transfer through a differential temperature difference in all components including the regenerator. This would require providing infinitely large surface areas for heat transfer or allowing an infinitely long time for the process. Neither is practical. In actuality, all heat transfer processes will take place through a finite temperature difference, the regenerator will not have an efficiency of 100 percent, and the pressure losses in the regenerator will be considerable. Because of these limitations, both Stirling and Ericsson cycles have long been of only theoretical interest. However, there is renewed interest in engines that operate on these cycles because of their potential for higher efficiency and better emission control. The Ford Motor Company, General Motors Corporation, and the Phillips Research Laboratories of the Netherlands have successfully developed Stirling engines suitable for trucks, buses, and even automobiles. More research and development is needed before these engines can compete with the gasoline or diesel engines.

Both the Stirling and the Ericsson engines are *external combustion* engines. That is, the fuel in these engines is burned outside the system as opposed to gasoline or diesel engines, where the fuel is burned inside the cylinder.

External combustion offers several advantages. First, a variety of fuels can be used as a source of thermal energy. Second, there is more time for combustion, and thus the combustion process is more complete, which means less air pollution and more energy extraction from the fuel. Third, these engines operate on closed cycles, and thus a working fluid that has the most desirable characteristics (stable, chemically inert, high thermal conductivity) can be utilized as the working fluid. Hydrogen and helium are two gases commonly employed in these engines.

Despite the physical limitations and impracticalities associated with them, both the Stirling and Ericsson cycles give a strong message to design engineers: *Regeneration can increase efficiency*. It is no coincidence that modern gas-turbine and steam power plants make extensive use of regeneration. In fact, the Brayton cycle with intercooling, reheating, and regeneration, which is utilized in large gas-turbine power plants and discussed later in this chapter, closely resembles the Ericsson cycle.

8-8 ■ BRAYTON CYCLE—The Ideal Cycle for Gas-Turbine Engines

The Brayton cycle was first proposed by George Brayton for use in the reciprocating oil-burning engine that he developed around 1870. Today, it is used for gas turbines only where both the compression and expansion processes take place in rotating machinery. Gas turbines usually operate on an *open cycle,* as shown in Fig. 8-29. Fresh air at ambient conditions is drawn into the compressor where its temperature and pressure are raised. The high-pressure air proceeds into the combustion chamber where the fuel is burned at constant pressure. The resulting high-temperature gases then enter the turbine where they expand to the atmospheric pressure, thus producing power. The exhaust gases leaving the turbine are thrown out (not recirculated), causing the cycle to be classified as an open cycle.

The open gas-turbine cycle described above can be modeled as a *closed cycle,* as shown in Fig. 8-30, by utilizing the air-standard assumptions. Here the compression and expansion processes remain the same, but the combustion process is replaced by a constant-pressure heat

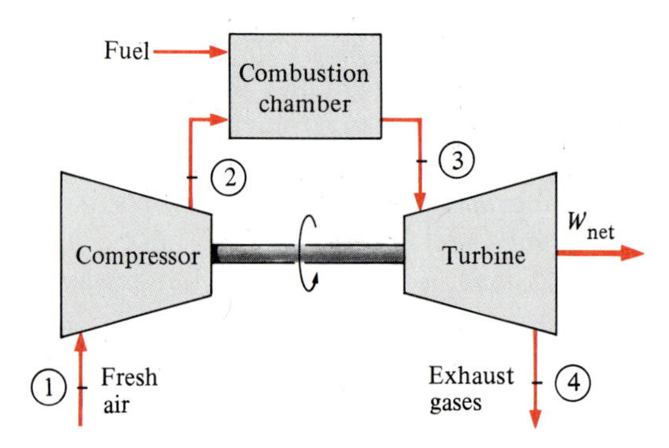

FIGURE 8-29
An open-cycle gas-turbine engine.

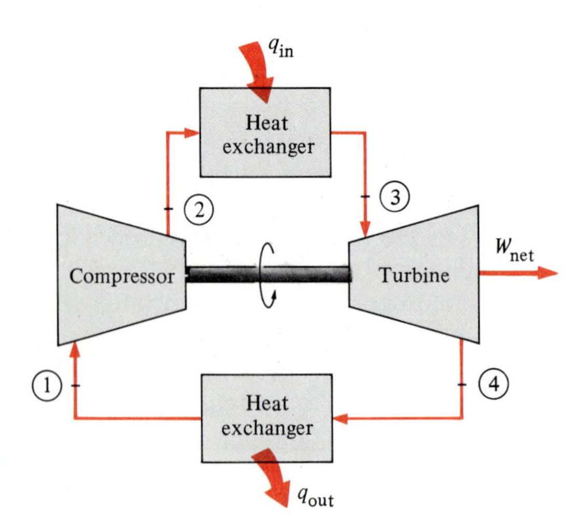

FIGURE 8-30
A closed-cycle gas-turbine engine.

addition process from an external source, and the exhaust process is replaced by a constant-pressure heat rejection process to the ambient air. The ideal cycle which the working fluid undergoes in this closed loop is the **Brayton cycle**, which is made up of four internally reversible processes:

1-2	Isentropic compression (in a compressor)
2-3	P = constant heat addition
3-4	Isentropic expansion (in a turbine)
4-1	P = constant heat rejection

The T-s and P-v diagrams of an ideal Brayton cycle are shown in Fig. 8-31. Notice that all four processes of the Brayton cycle are executed in steady-flow devices; thus they should be analyzed as steady-flow processes. When the changes in kinetic and potential energies are neglected, the conservation of energy equation for a steady-flow process can be expressed, on a unit-mass basis, as

$$q - w = h_{\text{exit}} - h_{\text{inlet}} \tag{8-15}$$

Assuming constant specific heats at room temperature (cold-air-standard assumption), heat transfer to and from the working fluid becomes

$$q_{\text{in}} = q_{23} = h_3 - h_2 = C_p(T_3 - T_2) \tag{8-16a}$$

and

$$q_{\text{out}} = -q_{41} = h_4 - h_1 = C_p(T_4 - T_1) \tag{8-16b}$$

Then the thermal efficiency of the ideal Brayton cycle becomes

$$\eta_{\text{th,Brayton}} = \frac{w_{\text{net}}}{q_{\text{in}}} = 1 - \frac{q_{\text{out}}}{q_{\text{in}}} = 1 - \frac{C_p(T_4 - T_1)}{C_p(T_3 - T_2)} = 1 - \frac{T_1(T_4/T_1 - 1)}{T_2(T_3/T_2 - 1)}$$

Processes 1-2 and 3-4 are isentropic, and $P_2 = P_3$ and $P_4 = P_1$. Thus,

$$\frac{T_2}{T_1} = \left(\frac{P_2}{P_1}\right)^{(k-1)/k} = \left(\frac{P_3}{P_4}\right)^{(k-1)/k} = \frac{T_3}{T_4}$$

Substituting these equations into the thermal efficiency relation and simplifying give

$$\eta_{\text{th,Brayton}} = 1 - \frac{1}{r_p^{(k-1)/k}} \tag{8-17}$$

where

$$r_p = \frac{P_2}{P_1} \tag{8-18}$$

is the **pressure ratio** and k is the specific heat ratio. Equation 8-17 shows that under the cold-air-standard assumptions, the thermal efficiency of an ideal Brayton cycle depends on the pressure ratio of the gas turbine and the specific heat ratio of the working fluid (if different from air). The thermal efficiency increases with both these parameters, which is also the case for actual gas turbines. A plot of thermal efficiency versus the

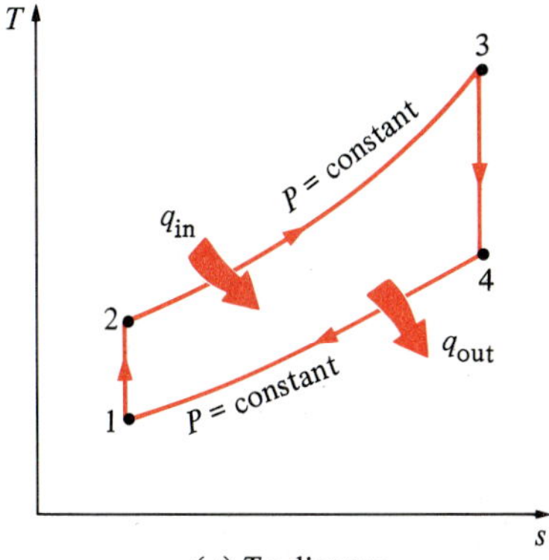

(a) T-s diagram

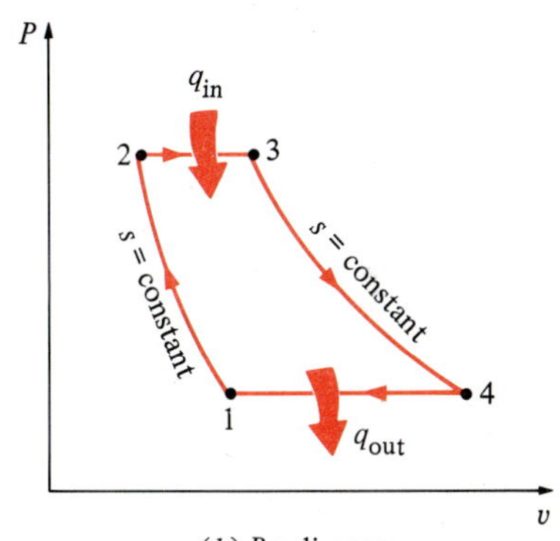

(b) P-v diagram

FIGURE 8-31
T-s and P-v diagrams for the ideal Brayton cycle.

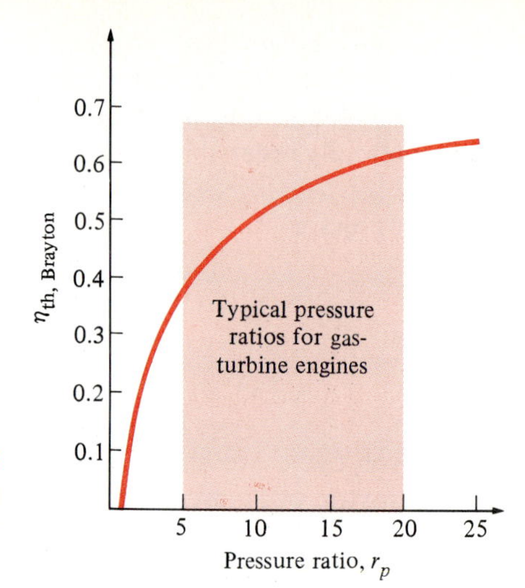

FIGURE 8-32
Thermal efficiency of the ideal Brayton cycle as a function of the pressure ratio.

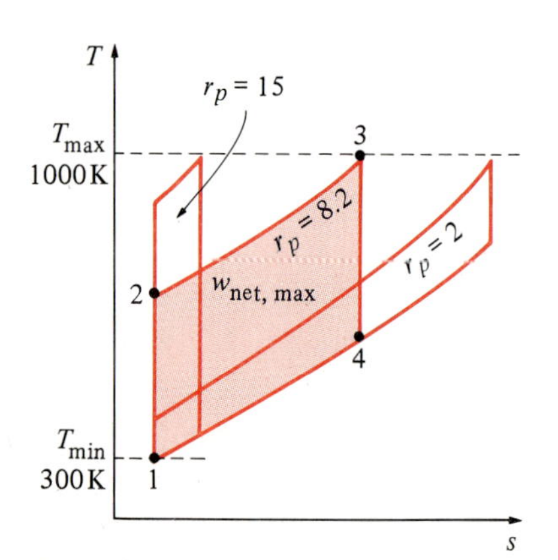

FIGURE 8-33
For fixed values of T_{min} and T_{max}, the net work of the Brayton cycle first increases with the pressure ratio, then reaches a maximum at $r_p = (T_{max}/T_{min})^{k/[2(k-1)]}$, and finally decreases.

pressure ratio is given in Fig. 8-32 for $k = 1.4$, which is the specific-heat-ratio value of air at room temperature.

The highest temperature in the cycle occurs at the end of the combustion process (state 3), and it is limited by the maximum temperature that the turbine blades can withstand. This also limits the pressure ratios that can be used in the cycle. For a fixed turbine inlet temperature T_3, the net work output per cycle increases with the pressure ratio, reaches a maximum, and then starts to decrease, as shown in Fig. 8-33. Therefore, there should be a compromise between the pressure ratio (thus the thermal efficiency) and the net work output. With less work output per cycle, a larger mass flow rate (thus a larger system) is needed to maintain the same power output, which may not be economical.

The air in gas turbines performs two important functions: It supplies the necessary oxidant for the combustion of the fuel, and it serves as a coolant to keep the temperature of various components within safe limits. The second function is accomplished by drawing in more air than is needed for the complete combustion of the fuel. In gas turbines, an air-fuel mass ratio of 50 or above is not uncommon. Therefore, in a cycle analysis, treating the combustion gases as air will not cause any appreciable error. Also, the mass flow rate through the turbine will be greater than that through the compressor, the difference being equal to the mass flow rate of the fuel. Thus, assuming a constant mass flow rate throughout the cycle will yield conservative results for open-loop gas-turbine engines.

The thermal efficiency of a gas-turbine engine depends on the allowable maximum gas temperature at the turbine inlet. Significant advances, such as coating the turbine blades with ceramic layers and cooling the blades with the discharge air from the compressor, have been made during the last two decades. As a result, today's gas turbines can withstand temperatures as high as 1425°C (2600°F) at the turbine inlet, and they have impressive efficiencies.

The two major application areas of gas-turbine engines are aircraft propulsion and electric power generation. When it is used for aircraft propulsion, the gas turbine produces just enough power to drive the compressor and a small generator to power the auxiliary equipment. The high-velocity exhaust gases are responsible for producing the necessary thrust to propel the aircraft. The gas turbines are also used as stationary power plants to generate electricity. Electricity is predominantly generated by large steam power plants which are discussed in the next chapter. Gas-turbine power plants are mostly utilized in the power generation industry to cover emergencies and peak periods because of their relatively low cost and quick response time. Gas turbines are also used in conjunction with steam power plants on the high-temperature side, forming a dual cycle. In these plants, the exhaust gases of the gas turbine serve as the heat source for the steam. The gas-turbine cycle can also be executed as a closed cycle for use in nuclear power plants. This time the working fluid is not limited to air, and a gas with more desirable characteristics (such as helium) can be used.

In gas-turbine power plants, the ratio of the compressor work to the turbine work, called the **back work ratio**, is very high (Fig. 8-34). Usually more than one-half of the turbine work output is used to drive the compressor. The situation is even worse when the adiabatic efficiencies of the compressor and the turbine are low. This is quite in contrast to steam power plants where the back work ratio is only a few percent. This is not surprising, however, since a liquid is compressed in steam power plants instead of a gas, and the reversible steady-flow work is proportional to the specific volume of the working fluid ($w = -\int v\,dP$) when the kinetic and potential energy changes are negligible.

A power plant with a high back work ratio requires a larger turbine to provide the additional power requirements of the compressor. Therefore, the turbines used in gas-turbine power plants are larger than those used in steam power plants of the same power rating.

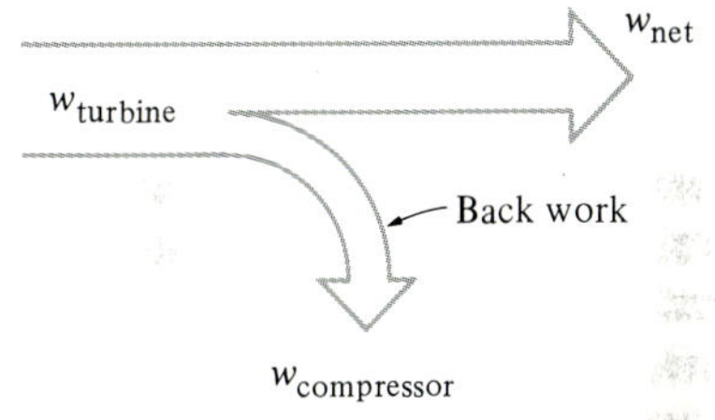

FIGURE 8-34
The fraction of the turbine work used to drive the compressor is called the back work ratio.

EXAMPLE 8-5

A stationary power plant operating on an ideal Brayton cycle has a pressure ratio of 8. The gas temperature is 300 K at the compressor inlet and 1300 K at the turbine inlet. Utilizing the air-standard assumptions and accounting for the variation of specific heats with temperature, determine (*a*) the gas temperature at the exits of the compressor and the turbine, (*b*) the back work ratio, and (*c*) the thermal efficiency.

Solution The cycle is shown on a *T-s* diagram in Fig. 8-35. Under the air-standard assumptions, the working fluid is assumed to be air, which behaves as an ideal gas, and all four processes that make up the cycle are internally reversible. Furthermore, the combustion and exhaust processes are replaced by heat addition and heat rejection processes, respectively. Also, when changes in kinetic and potential energies are neglected, the energy equation for a steady-flow device simplifies to Eq. 8-15.

(*a*) The air temperatures at the compressor and turbine exits are determined by applying the energy equation to processes 1-2 and 3-4:

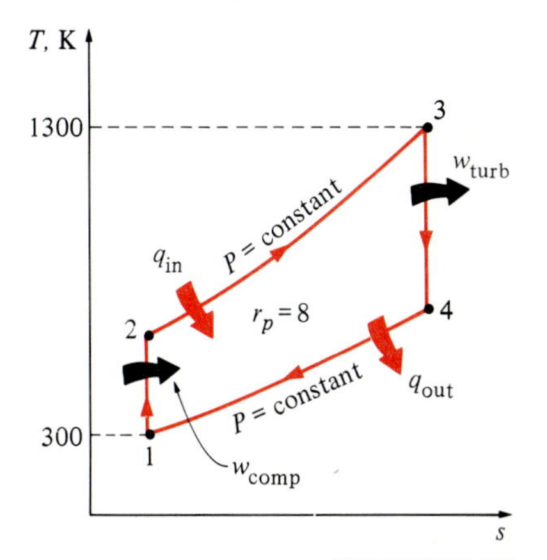

FIGURE 8-35
T-s diagram for the Brayton cycle discussed in Example 8-5.

Process 1-2 (isentropic compression of an ideal gas):

$$T_1 = 300 \text{ K} \longrightarrow h_1 = 300.19 \text{ kJ/kg}$$
$$P_{r_1} = 1.386$$

$$P_{r_2} = \frac{P_2}{P_1} P_{r_1} = (8)(1.386) = 11.09 \longrightarrow T_2 = 540 \text{ K} \quad \text{(at compressor exit)}$$
$$h_2 = 544.35 \text{ kJ/kg}$$

Process 3-4 (isentropic expansion of an ideal gas):

$$T_3 = 1300 \text{ K} \longrightarrow h_3 = 1395.97 \text{ kJ/kg}$$
$$P_{r_3} = 330.9$$

$$P_{r_4} = \frac{P_4}{P_3} P_{r_3} = (\tfrac{1}{8})(330.9) = 41.36 \longrightarrow T_4 = 770 \text{ K} \quad \text{(at turbine exit)}$$
$$h_4 = 789.11 \text{ kJ/kg}$$

(*b*) To find the back work ratio, we need to find the work input to the compressor and the work output of the turbine:

$$w_{\text{comp,in}} = h_2 - h_1 = (544.35 - 300.19) \text{ kJ/kg} = 244.16 \text{ kJ/kg}$$
$$w_{\text{turb,out}} = h_3 - h_4 = (1395.97 - 789.11) \text{ kJ/kg} = 606.86 \text{ kJ/kg}$$

Thus, Back work ratio $r_{bw} = \dfrac{w_{\text{comp,in}}}{w_{\text{turb,out}}} = \dfrac{244.16 \text{ kJ/kg}}{606.86 \text{ kJ/kg}} = 0.402$

That is, 40.2 percent of the turbine work output is used just to drive the compressor.

(*c*) The thermal efficiency of the cycle is the ratio of the net power output to the total heat input:

$$q_{\text{in}} = h_3 - h_2 = (1395.97 - 544.35) \text{ kJ/kg} = 851.62 \text{ kJ/kg}$$
$$w_{\text{net}} = w_{\text{out}} - w_{\text{in}} = (606.86 - 244.16) \text{ kJ/kg} = 362.7 \text{ kJ/kg}$$

Thus, $\eta_{\text{th}} = \dfrac{w_{\text{net}}}{q_{\text{in}}} = \dfrac{362.7 \text{ kJ/kg}}{851.62 \text{ kJ/kg}} = 0.426 \quad \text{(or 42.6\%)}$

The thermal efficiency could also be determined from

$$\eta_{\text{th}} = 1 - \frac{q_{\text{out}}}{q_{\text{in}}}$$

where $q_{\text{out}} = h_4 - h_1 = (789.11 - 300.19) \text{ kJ/kg} = 488.92 \text{ kJ/kg}$

Under the cold-air-standard assumptions (constant specific heats, values at room temperature), the thermal efficiency would be, from Eq. 8-17,

$$\eta_{\text{th,Brayton}} = 1 - \frac{1}{r_p^{(1-k)/k}} = 1 - \frac{1}{8^{(1.4-1)/1.4}} = 0.448$$

which is sufficiently close to the value obtained by accounting for the variation of specific heats with temperature.

Deviation of Actual Gas-Turbine Cycles from Idealized Ones

The actual gas-turbine cycle differs from the ideal Brayton cycle on several accounts. For one thing, some pressure drop during the heat addition and rejection processes is inevitable. More importantly, the actual work input to the compressor will be more, and the actual work output of the turbine will be less because of irreversibilities such as friction and non-quasi-equilibrium operation conditions of these devices. The deviation of actual compressor and turbine behavior from the idealized isentropic behavior can be accurately accounted for, however, by utilizing the adiabatic efficiencies of the turbine and compressor, defined as

$$\eta_C = \frac{w_s}{w_a} \cong \frac{h_{2s} - h_1}{h_{2a} - h_1} \tag{8-19}$$

and

$$\eta_T = \frac{w_a}{w_s} \cong \frac{h_3 - h_{4a}}{h_3 - h_{4s}} \tag{8-20}$$

where states $2a$ and $4a$ are the actual exit states of the compressor and the turbine, respectively, and $2s$ and $4s$ are the corresponding states for the isentropic case, as illustrated in Fig. 8-36. The effect of the turbine and compressor efficiencies on the thermal efficiency of the gas-turbine engine is illustrated below with an example.

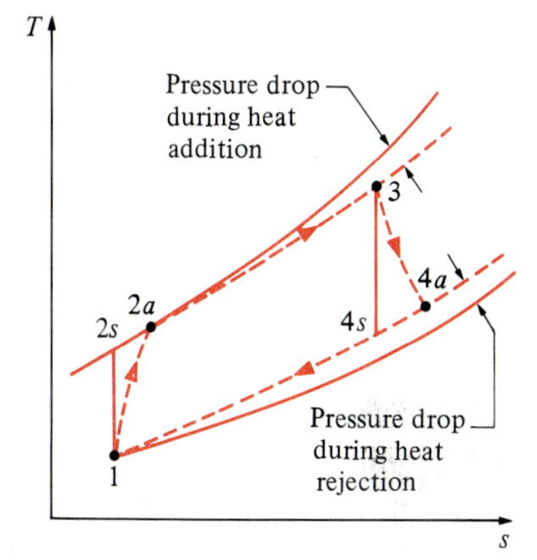

FIGURE 8-36

The deviation of an actual gas-turbine cycle from the ideal Brayton cycle as a result of irreversibilities.

EXAMPLE 8-6

Assuming a compressor efficiency of 80 percent and a turbine efficiency of 85 percent, determine (*a*) the back work ratio, (*b*) the thermal efficiency, and (*c*) the turbine exit temperature of the gas-turbine power plant discussed in Example 8-5.

Solution (*a*) The *T*-*s* diagram of the cycle is shown in Fig. 8-37. The actual compressor work and turbine work are determined by using the definitions of compressor and turbine efficiencies, Eqs. 8-19 and 8-20:

Compressor: $$w_a = \frac{w_s}{\eta_C} = \frac{-244.16 \text{ kJ/kg}}{0.80} = -305.20 \text{ kJ/kg}$$

Turbine: $$w_a = \eta_T w_s = (0.85)(606.86 \text{ kJ/kg}) = 515.83 \text{ kJ/kg}$$

Thus, $$r_{bw} = \frac{w_{comp,in}}{w_{turb,out}} = \frac{305.20 \text{ kJ/kg}}{515.83 \text{ kJ/kg}} = 0.592$$

That is, the compressor is now consuming 59.2 percent of the work produced by the turbine (up from 40.2 percent). This increase is due to the irreversibilities that occur within the compressor and the turbine.

(*b*) In this case, air will leave the compressor at a higher temperature and enthalpy, which are determined to be

$$w_{comp,in} = h_{2a} - h_1 \longrightarrow h_{2a} = h_1 + w_{comp,in}$$
$$= (300.19 + 305.20) \text{ kJ/kg}$$
$$= 605.39 \text{ kJ/kg} \qquad (\text{and } T_{2a} = 598 \text{ K})$$

Thus, $$q_{in} = h_3 - h_{2a} = (1395.97 - 605.39) \text{ kJ/kg} = 790.58 \text{ kJ/kg}$$
$$w_{net} = w_{out} - w_{in} = (515.83 - 305.20) \text{ kJ/kg} = 210.63 \text{ kJ/kg}$$

and $$\eta_{th} = \frac{w_{net}}{q_{in}} = \frac{210.63 \text{ kJ/kg}}{790.58 \text{ kJ/kg}} = 0.266 \quad (\text{or } 26.6\%)$$

That is, the irreversibilities occurring within the turbine and compressor caused the thermal efficiency of the plant to drop from 42.6 to 26.6 percent. This example shows how sensitive the performance of a gas-turbine power plant is to the efficiencies of the compressor and the turbine. In fact, gas-turbine thermal

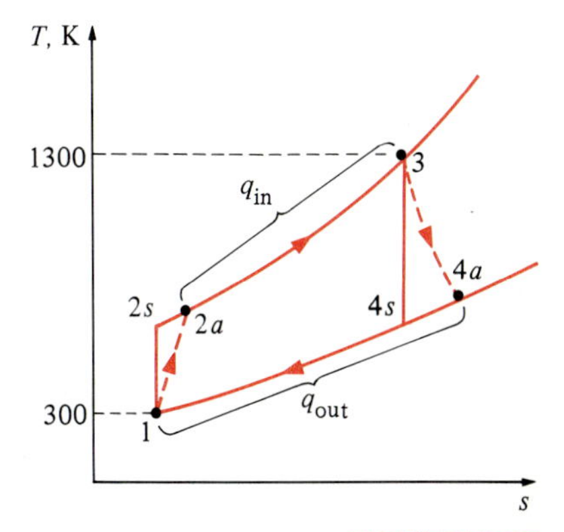

FIGURE 8-37

T-*s* diagram of the gas-turbine cycle discussed in Example 8-6.

efficiencies did not reach competitive values until significant improvements were made in the design of gas turbines and compressors.

(c) The air temperature at the turbine exit is determined from the steady-flow first-law relation for the turbine:

$$\cancelto{0}{q_{34a}} - w_{34a} = h_3 - h_{4a} \longrightarrow h_{4a} = h_3 - w_{\text{turb},a}$$
$$= (1395.97 - 515.83) \text{ kJ/kg}$$
$$= 880.14 \text{ kJ/kg}$$

Then, from Table A-17,

$$T_{4a} = 853 \text{ K}$$

This value is considerably higher than the air temperature at the compressor exit (T_{2a} = 598 K), which suggests the use of regeneration to reduce the heat input requirements.

8-9 ■ THE BRAYTON CYCLE WITH REGENERATION

In gas-turbine engines, the temperature of the exhaust gas leaving the turbine is often considerably higher than the temperature of the air leaving the compressor. Therefore, the high-pressure air leaving the compressor can be heated by transferring heat to it from the hot exhaust gases in a counterflow heat exchanger, which is also known as a *regenerator,* or a *recuperator.* A sketch of the gas-turbine engine utilizing a regenerator and the *T-s* diagram of the new cycle are shown in Figs. 8-38 and 8-39, respectively.

The thermal efficiency of the Brayton cycle increases as a result of regeneration since the portion of energy of the exhaust gases that is normally rejected to the surroundings is now used to preheat the air entering the combustion chamber. This, in turn, decreases the heat input (thus fuel) requirements for the same net work output. Note, however, that the use of a regenerator is recommended only when the turbine exhaust temperature is higher than the compressor exit temperature. Otherwise, heat will flow in the reverse direction (*to* the exhaust gases), decreasing the

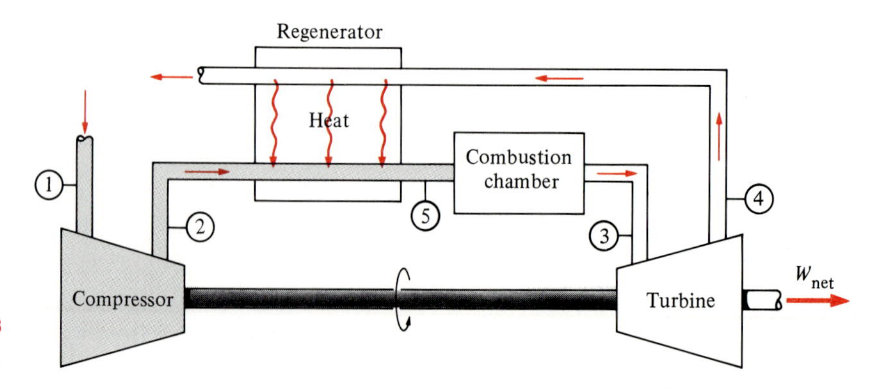

FIGURE 8-38
A gas-turbine engine with regenerator.

efficiency. This situation is encountered in gas-turbine engines operating at very high pressure ratios.

The highest temperature that occurs within the regenerator is T_4, the temperature of the exhaust gases leaving the turbine and entering the regenerator. Under no conditions can the air be preheated in the regenerator to a temperature above this value. Air normally leaves the regenerator at a lower temperature, T_5. In the limiting (ideal) case, the air will exit the regenerator at the inlet temperature of the exhaust gases T_4. Assuming the regenerator to be well insulated and any changes in kinetic and potential energies to be negligible, the actual and maximum heat transfers from the exhaust gases to the air can be expressed as

$$q_{\text{regen,act}} = h_5 - h_2 \tag{8-21}$$

and

$$q_{\text{regen,max}} = h_{5'} - h_2 = h_4 - h_2 \tag{8-22}$$

The extent to which a regenerator approaches an ideal regenerator is called the **effectiveness** ε and is defined as

$$\varepsilon = \frac{q_{\text{regen,act}}}{q_{\text{regen,max}}} = \frac{h_5 - h_2}{h_4 - h_2} \tag{8-23}$$

When the cold-air-standard assumptions are utilized, it reduces to

$$\varepsilon \cong \frac{T_5 - T_2}{T_4 - T_2} \tag{8-24}$$

It is obvious that a regenerator with a higher effectiveness will save a greater amount of fuel since it will preheat the air to a higher temperature prior to combustion. However, achieving a higher effectiveness requires the use of a larger regenerator, which carries a higher price tag and causes a larger pressure drop. Therefore, the use of a regenerator with a very high effectiveness cannot be justified economically unless the savings from the fuel costs exceed the additional expenses involved. Most regenerators used in practice have an effectiveness of 0.7 or below.

Under the cold-air-standard assumptions, the thermal efficiency of an ideal Brayton cycle with regeneration is

$$\eta_{\text{th,regen}} = 1 - \left(\frac{T_1}{T_3}\right)(r_p)^{(k-1)/k} \tag{8-25}$$

Therefore, the thermal efficiency of an ideal Brayton cycle with regeneration depends on the ratio of the minimum to maximum temperatures as well as the pressure ratio. The thermal efficiency is plotted in Fig. 8-40 for various pressure ratios and minimum-to-maximum temperature ratios. This figure shows that regeneration is most effective at low pressure ratios and low minimum-to-maximum temperature ratios.

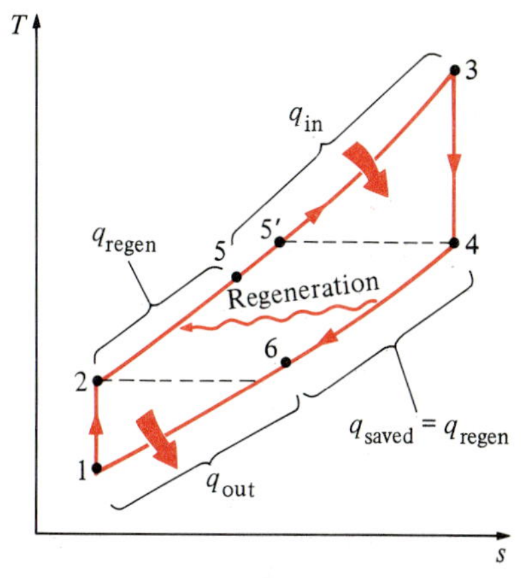

FIGURE 8-39
T-s diagram of a Brayton cycle with regeneration.

EXAMPLE 8-7

Determine the thermal efficiency of the gas-turbine power plant described in Example 8-6 if a regenerator having an effectiveness of 80 percent is installed.

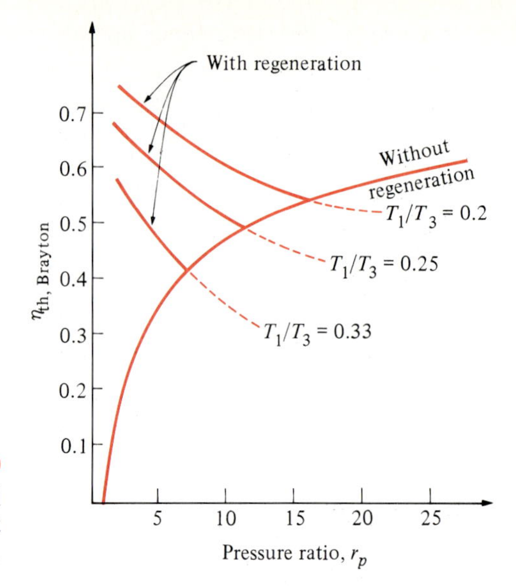

FIGURE 8-40
Thermal efficiency of the ideal Brayton cycle with and without regeneration.

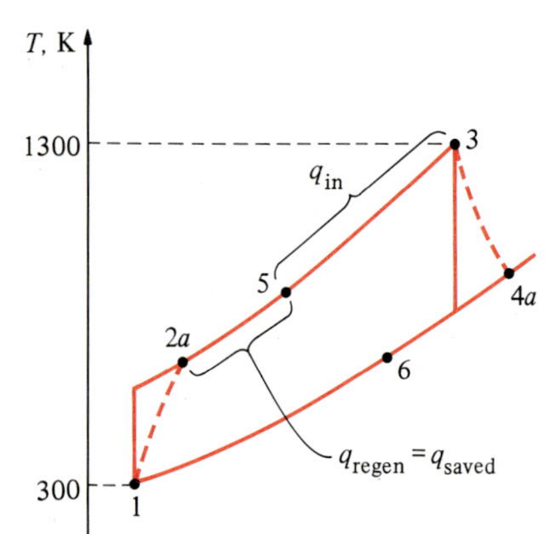

FIGURE 8-41
T-s diagram of the regenerative Brayton cycle described in Example 8-7.

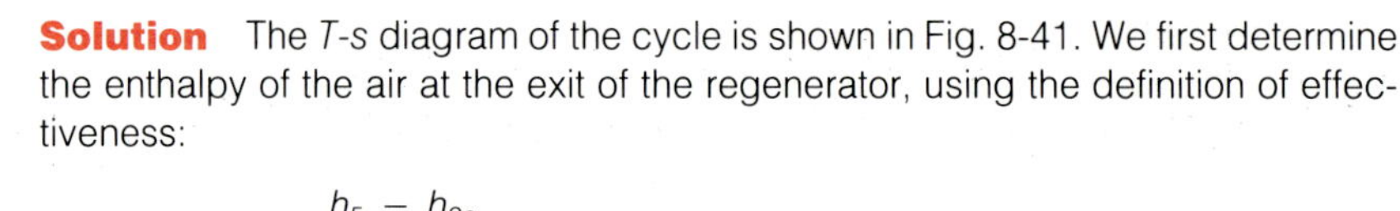

Solution The *T-s* diagram of the cycle is shown in Fig. 8-41. We first determine the enthalpy of the air at the exit of the regenerator, using the definition of effectiveness:

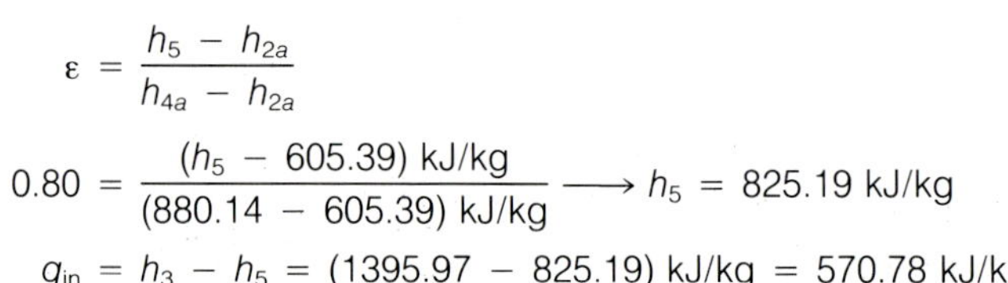

$$\varepsilon = \frac{h_5 - h_{2a}}{h_{4a} - h_{2a}}$$

$$0.80 = \frac{(h_5 - 605.39)\text{ kJ/kg}}{(880.14 - 605.39)\text{ kJ/kg}} \longrightarrow h_5 = 825.19\text{ kJ/kg}$$

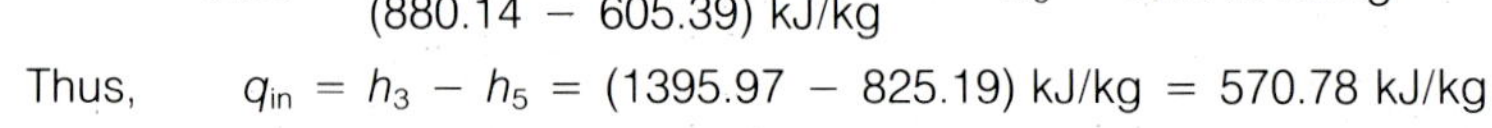

Thus, $q_{in} = h_3 - h_5 = (1395.97 - 825.19)\text{ kJ/kg} = 570.78\text{ kJ/kg}$

This represents a savings of 219.8 kJ/kg from the heat input requirements. The addition of a regenerator (assumed to be frictionless) does not affect the net work output of the plant. Thus,

$$\eta_{th} = \frac{w_{net}}{q_{in}} = \frac{210.63\text{ kJ/kg}}{570.78\text{ kJ/kg}} = 0.369 \quad \text{(or 36.9\%)}$$

That is, the thermal efficiency of the power plant has gone up from 26.6 to 36.9 percent as a result of installing a regenerator which helps to recuperate some of the excess energy of the exhaust gases.

8-10 ■ THE BRAYTON CYCLE WITH INTERCOOLING, REHEATING, AND REGENERATION

The net work of a gas-turbine cycle is the difference between the turbine work output and the compressor work input, and it can be increased by either decreasing the compressor work or increasing the turbine work, or both. It was shown in Chap. 6 that the work required to compress a gas between two specified pressures can be decreased by carrying out the compression process in stages and cooling the gas in between

(Fig. 8-42)—that is, using *multistage compression with intercooling*. As the number of stages is increased, the compression process becomes isothermal at the compressor inlet temperature, and the compression work decreases.

Likewise, the work output of a turbine operating between two pressure levels can be increased by expanding the gas in stages and reheating it in between—that is, utilizing *multistage expansion with reheating*. This is accomplished without raising the maximum temperature in the cycle. As the number of stages is increased, the expansion process becomes isothermal. The foregoing argument is based on a simple principle: The steady-flow compression or expansion work is proportional to the specific volume of the fluid. Therefore, the specific volume of the working fluid should be as low as possible during a compression process and as high as possible during an expansion process. This is precisely what intercooling and reheating accomplish.

The working fluid leaves the compressor at a lower temperature, and the turbine at a higher temperature, when intercooling and reheating are utilized. This makes regeneration more attractive since a greater potential for regeneration exists. Also the gases leaving the compressor can be heated to a higher temperature before they enter the combustion chamber because of the higher temperature of the turbine exhaust.

A schematic of the physical arrangement and the *T-s* diagram of an ideal two-stage gas-turbine cycle with intercooling, reheating, and regeneration are shown in Figs. 8-43 and 8-44, respectively. The gas enters the first stage of the compressor at state 1, is compressed isentropically to an intermediate pressure P_2, is cooled at constant pressure to state 3 ($T_3 = T_1$), and is compressed in the second stage isentropically to the final pressure P_4. At state 4 the gas enters the regenerator, where it is

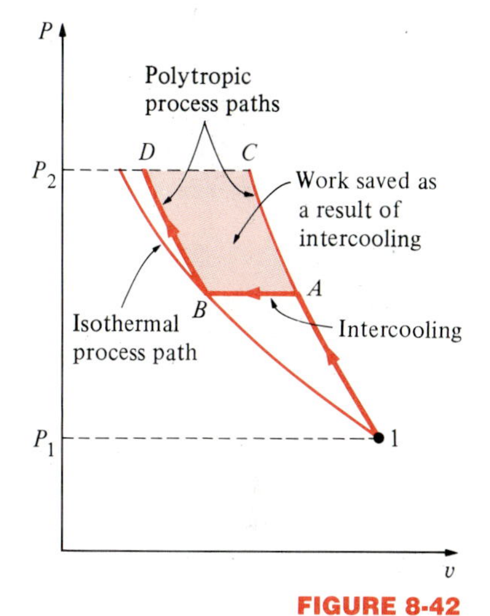

FIGURE 8-42

Comparison of work inputs to a single-stage compressor (1*AC*) and a two-stage compressor with intercooling (1*ABD*).

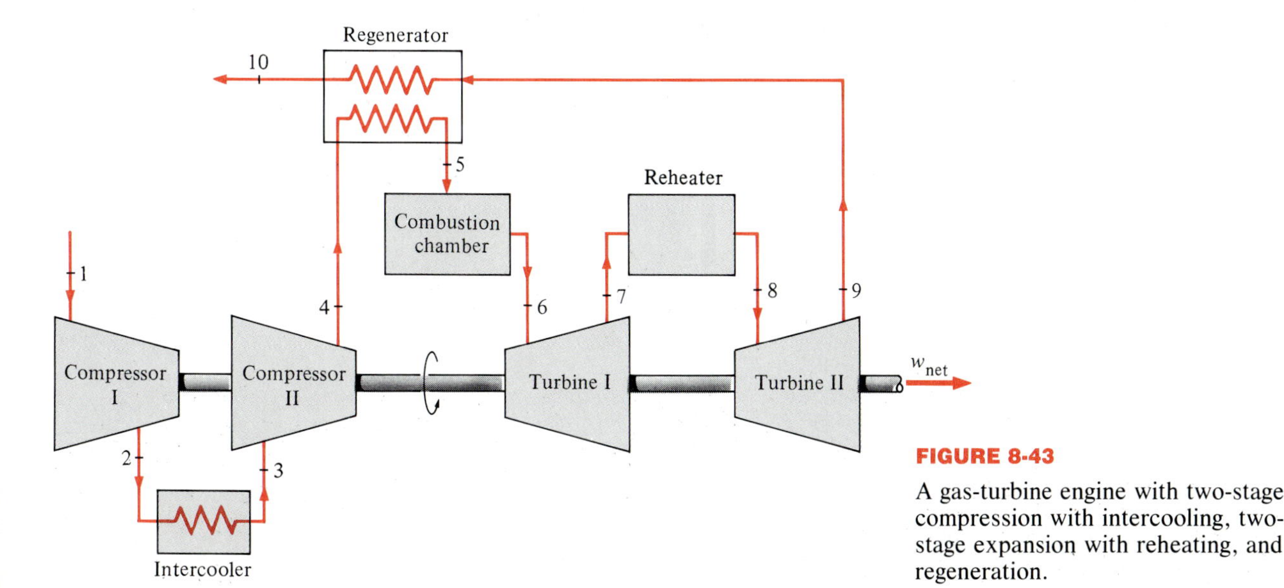

FIGURE 8-43

A gas-turbine engine with two-stage compression with intercooling, two-stage expansion with reheating, and regeneration.

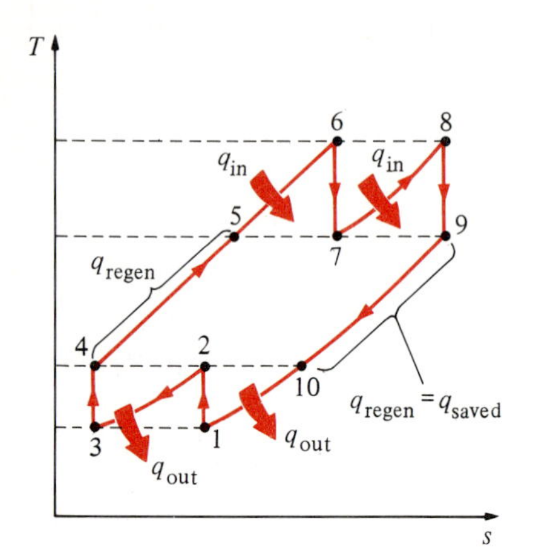

FIGURE 8-44

T-s diagram of an ideal gas-turbine cycle with intercooling, reheating, and regeneration.

heated to T_5 at constant pressure. In an ideal regenerator, the gas will leave the regenerator at the temperature of the turbine exhaust, that is, $T_5 = T_9$. The primary heat addition (or combustion) process takes place between states 5 and 6. The gas enters the first stage of the turbine at state 6 and expands isentropically to state 7, where it enters the reheater. It is reheated at constant pressure to state 8 ($T_8 = T_6$), where it enters the second stage of the turbine. The gas exits the turbine at state 9 and enters the regenerator, where it is cooled to state 10 at constant pressure. The cycle is completed by cooling the gas to the initial state (or purging the exhaust gases).

It was shown in Chap. 6 that the work input to a two-stage compressor is minimized when equal pressure ratios are maintained across each stage. It can be shown that this procedure also maximizes the turbine work output. Thus, for optimum operation we have

$$\frac{P_2}{P_1} = \frac{P_4}{P_3} \quad \text{and} \quad \frac{P_6}{P_7} = \frac{P_8}{P_9} \tag{8-26}$$

In the analysis of the actual gas-turbine cycles, the irreversibilities that are present within the compressor, the turbine, and the regenerator as well as the pressure drops in the heat exchangers should be taken into consideration.

The back work ratio of a gas-turbine cycle improves as a result of intercooling and reheating. However, this does not mean that the thermal efficiency will also improve. The fact is, intercooling and reheating will always decrease the thermal efficiency unless they are accompanied by regeneration. This is because intercooling decreases the average temperature at which heat is added, and reheating increases the average temperature at which heat is rejected. This is also apparent from Fig. 8-44. Therefore, in gas-turbine power plants, intercooling and reheating are always used in conjunction with regeneration.

If the number of compression and expansion stages is increased, the ideal gas-turbine cycle with intercooling, reheating, and regeneration will approach the Ericsson cycle, as illustrated in Fig. 8-45, and the thermal efficiency will approach the theoretical limit (the Carnot efficiency).

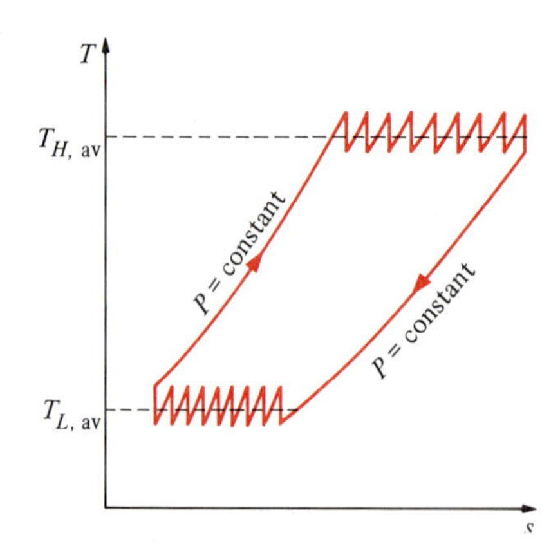

FIGURE 8-45

As the number of compression and expansion stages increases, the gas-turbine cycle with intercooling reheating, and regeneration approaches the Ericsson cycle.

However, the contribution of each additional stage to the thermal efficiency is less and less, and the use of more than two or three stages cannot be justified economically.

EXAMPLE 8-8

An ideal gas-turbine cycle with two stages of compression and two stages of expansion has an overall pressure ratio of 8. Air enters each stage of the compressor at 800 K and each stage of the turbine at 1300 K. Determine the back work ratio and the thermal efficiency of this gas-turbine cycle, assuming (*a*) no regenerators and (*b*) an ideal regenerator with 100 percent effectiveness. Compare the results with those obtained in Example 8-5.

Solution The *T-s* diagram of the gas-turbine cycle described above is shown in Fig. 8-46. The cycle is assumed to be ideal; thus all the processes are internally reversible, and no pressure drops occur during intercooling and reheating ($P_2 = P_3$ and $P_7 = P_8$). For two-stage compression and expansion, the work input is minimized and the work output is maximized when both stages of the compressor and the turbine have the same pressure ratio, as shown in Chap. 6. Thus,

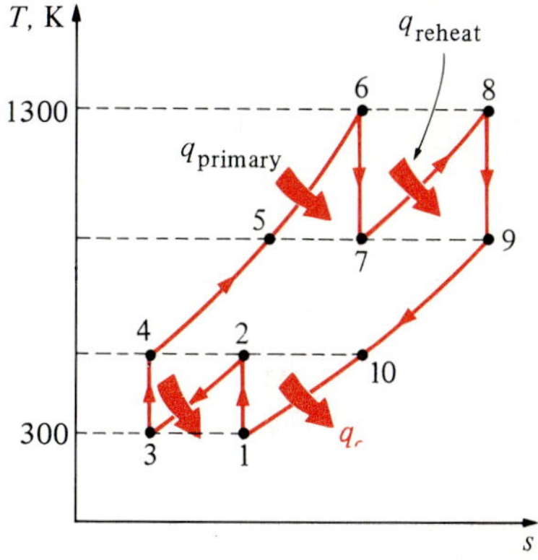

FIGURE 8-46

T-s diagram of the gas-turbine cycle discussed in Example 8-8.

$$\frac{P_2}{P_1} = \frac{P_4}{P_3} = \sqrt{8} = 2.83 \qquad \text{and} \qquad \frac{P_6}{P_7} = \frac{P_8}{P_9} = \sqrt{8} = 2.83$$

Air enters each stage of the compressor at the same temperature, and each stage has the same adiabatic efficiency (100 percent in this case). Therefore, the temperature (and enthalpy) of the air at the exit of each compression stage will be the same. A similar argument can be given for the turbine. Thus,

At inlets: $T_1 = T_3, \quad h_1 = h_3 \quad$ and $\quad T_6 = T_8, \quad h_6 = h_8$

At exits: $T_2 = T_4, \quad h_2 = h_4 \quad$ and $\quad T_7 = T_9, \quad h_7 = h_9$

Under these conditions, the work input to each stage of the compressor will be the same, and so will the work output from each stage of the turbine.

(*a*) In the absence of any regeneration, the back work ratio and the thermal efficiency are determined by using data from Table A-17 as follows:

$$T_1 = 300 \text{ K} \longrightarrow h_1 = 300.19 \text{ kJ/kg}$$
$$P_{r_1} = 1.386$$

$$P_{r_2} = \frac{P_2}{P_1} P_{r_1} = \sqrt{8}\,(1.386) = 3.92 \longrightarrow T_2 = 403.3 \text{ K}$$
$$h_2 = 404.33 \text{ kJ/kg}$$

$$T_6 = 1300 \text{ K} \longrightarrow h_6 = 1395.97 \text{ kJ/kg}$$
$$P_{r_6} = 330.9$$

$$P_{r_7} = \frac{P_7}{P_6} P_{r_6} = \frac{1}{\sqrt{8}}(330.9) = 117.0 \longrightarrow T_7 = 1006.4 \text{ K}$$
$$h_7 = 1053.35 \text{ kJ/kg}$$

Then

$$w_{\text{comp,in}} = 2(w_{\text{comp,in,I}}) = 2(h_2 - h_1) = 2(404.33 - 300.19) \text{ kJ/kg}$$
$$= 208.28 \text{ kJ/kg}$$
$$w_{\text{turb,out}} = 2(w_{\text{turb,out,I}}) = 2(h_6 - h_7) = 2(1395.97 - 1053.35) \text{ kJ/kg}$$
$$= 685.24 \text{ kJ/kg}$$
$$w_{\text{net}} = w_{\text{turb,out}} - w_{\text{comp,in}} = (685.24 - 208.28) \text{ kJ/kg} = 476.96 \text{ kJ/kg}$$
$$q_{\text{in}} = q_{\text{primary}} + q_{\text{reheat}} = (h_6 - h_4) + (h_8 - h_7)$$
$$= (1395.97 - 404.33) \text{ kJ/kg} + (1395.97 - 1053.35) \text{ kJ/kg}$$
$$= 1334.19 \text{ kJ/kg}$$

Thus, $$r_{bw} = \frac{w_{comp,in}}{w_{turb,out}} = \frac{208.28 \text{ kJ/kg}}{685.24 \text{ kJ/kg}} = 0.304 \quad \text{(or 30.4\%)}$$

and $$\eta_{th} = \frac{w_{net}}{q_{in}} = \frac{476.96 \text{ kJ/kg}}{1334.19 \text{ kJ/kg}} = 0.357 \quad \text{(or 35.7\%)}$$

A comparison of these results with those obtained in Example 8-5 (single-stage compression and expansion) reveals that multistage compression with intercooling and multistage expansion with reheating improve the back work ratio (it drops from 40.2 to 30.4 percent) but hurt the thermal efficiency (drops from 42.6 to 35.7 percent). Therefore, intercooling and reheating are not recommended in gas-turbine power plants unless they are accompanied by regeneration.

(*b*) The addition of an ideal regenerator (no pressure drops, 100 percent effectiveness) does not affect the compressor work and the turbine work. Therefore, the net work output and the back work ratio of an ideal gas-turbine cycle will be identical whether there is a regenerator or not. A regenerator, however, reduces the heat input requirements by preheating the air leaving the compressor, using the hot exhaust gases leaving the turbine. In an ideal regenerator, the compressed air is heated to the turbine exit temperature T_9 before it enters the combustion chamber. Thus under the air-standard assumptions, $h_5 = h_7 = h_9$.

The heat input and the thermal efficiency in this case are

$$\begin{aligned} q_{in} &= q_{primary} + q_{reheat} = (h_6 - h_5) + (h_8 - h_7) \\ &= (1395.97 - 1053.35)\,\text{kJ/kg} + (1395.97 - 1053.35)\,\text{kJ/kg} \\ &= 685.24\,\text{kJ/kg} \end{aligned}$$

and $$\eta_{th} = \frac{w_{net}}{q_{in}} = \frac{476.96 \text{ kJ/kg}}{685.24 \text{ kJ/kg}} = 0.696 \quad \text{(or 69.6\%)}$$

That is, the thermal efficiency almost doubles as a result of regeneration compared to the no-regeneration case. The overall effect of two-stage compression and expansion with intercooling, reheating, and regeneration on the thermal efficiency is an increase of over 63 percent. As the number of compression and expansion stages is increased, the cycle will approach the Ericsson cycle, and the thermal efficiency will approach

$$\eta_{th,Ericsson} = \eta_{th,Carnot} = 1 - \frac{T_L}{T_H} = 1 - \frac{300 \text{ K}}{1300 \text{ K}} = 0.769$$

Adding a second stage increases the thermal efficiency from 42.6 to 69.6 percent, an increase of 27 percent. This is a significant increase in efficiency, and usually it is well worth the extra cost associated with the second stage. Adding more stages, however (no matter how many), can increase the efficiency an additional 7.3 percent at most and usually cannot be justified economically.

8-11 ■ IDEAL JET-PROPULSION CYCLES

Gas-turbine engines are widely used to power aircraft because they are light and compact and have a high power-to-weight ratio. Aircraft gas turbines operate on an open cycle called a **jet-propulsion cycle**. The ideal jet-propulsion cycle differs from the simple ideal Brayton cycle in that the gases are not expanded to the ambient pressure in the turbine. Instead, the gases are expanded to a pressure such that the power produced by the turbine is just sufficient to drive the compressor and the auxiliary equipment, such as a small generator and hydraulic pumps. That is, the net work output of a jet-propulsion cycle is zero. The gases that exit the turbine at a relatively high pressure are subsequently accelerated in a nozzle to provide the thrust to propel the aircraft (Fig. 8-47). Also aircraft gas turbines operate at higher pressure ratios (typically between 10 and 25), and the fluid passes through a diffuser first, where it is decelerated and its pressure is increased before it enters the compressor.

Aircraft are propelled by accelerating a fluid in the opposite direction to motion. This is accomplished by either slightly accelerating a large mass of fluid (*propeller-driven engine*) or greatly accelerating a small mass of fluid (*jet* or *turbojet engine*) or both (*turboprop engine*).

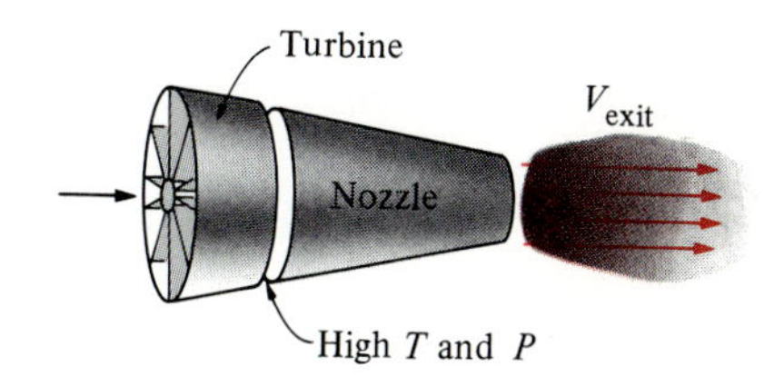

FIGURE 8-47
In jet engines, the high-temperature and high-pressure gases leaving the turbine are accelerated in a nozzle to provide thrust.

A schematic of a turbojet engine and the *T-s* diagram of the ideal turbojet cycle are shown in Fig. 8-48. The pressure of air rises slightly as it is decelerated in the diffuser. Air is compressed in the compressor. It is mixed with fuel in the combustion chamber, where the mixture is burned at constant pressure. The high-pressure and high-temperature combustion gases partially expand in the turbine, producing enough power to drive the compressor and other equipment. Finally, the gases expand in a nozzle to the ambient pressure and leave the aircraft at a high velocity.

In the ideal case, the turbine work is assumed to equal the compressor work. Also the processes in the diffuser, the compressor, the turbine, and the nozzle are assumed to be isentropic. In the analysis of actual cycles, however, the irreversibilities associated with these devices should be considered. The effect of the irreversibilities is to reduce the thrust that can be obtained from a turbojet engine.

The **thrust** developed in a turbojet engine is the unbalanced force which is caused by the difference in the momentum of the low-velocity air entering the engine and the high-velocity exhaust gases leaving the engine, and it is determined from Newton's second law. The pressures at the inlet and the exit of a turbojet engine are identical (the ambient pressure), thus the net thrust developed by the engine is

$$F = (\dot{m}\mathbf{V})_{exit} - (\dot{m}\mathbf{V})_{inlet} = \dot{m}(\mathbf{V}_{exit} - \mathbf{V}_{inlet}) \quad \text{(N)} \qquad (8\text{-}27)$$

where $\mathbf{V}_{exit}$ is the exit velocity of the exhaust gases and $\mathbf{V}_{inlet}$ is the inlet

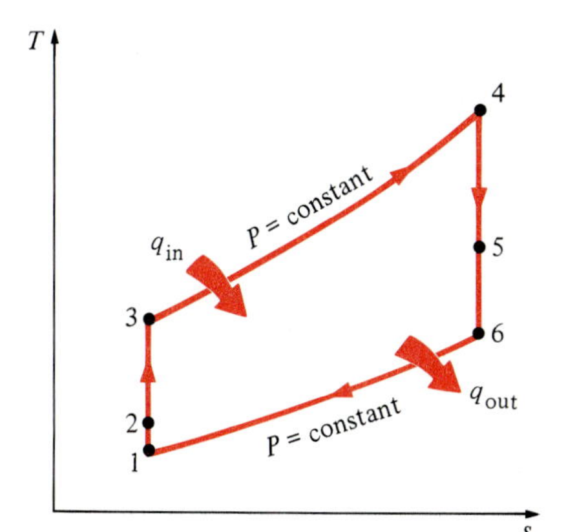

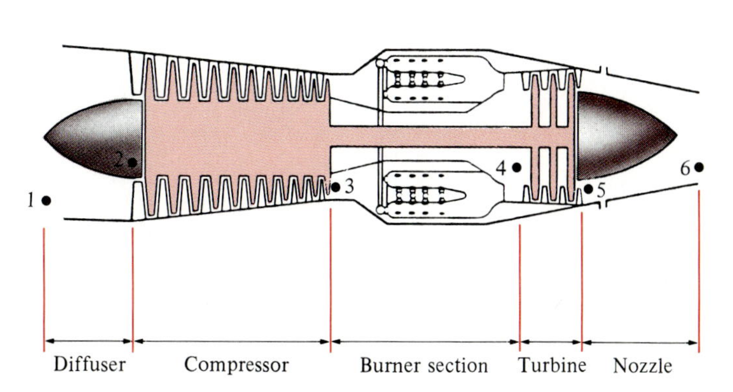

FIGURE 8-48
Basic components of a turbojet engine and the *T-s* diagram for the ideal turbojet cycle. [*Source:* The Aircraft Gas Turbine Engine and Its Operation. © United Aircraft Corporation (Now United Technologies Corp.), 1951, 1974.]

velocity of the air, both relative to the aircraft. Thus for an aircraft cruising in still air, $\mathbf{V}_{\text{inlet}}$ is the aircraft velocity. In reality, the mass flow rates of the gases at the engine exit and the inlet are different, the difference being equal to the combustion rate of the fuel. But the air-fuel mass ratio used in jet-propulsion engines is usually very high, making this difference very small. Thus $\dot{m}$ in Eq. 8-27 is taken as the mass flow rate of air through the engine. For an aircraft cruising at a steady speed, the thrust is used to overcome the fluid drag, and the net force acting on the body of the aircraft is zero. Commercial airplanes save fuel by flying at higher altitudes during long trips since the air at higher altitudes is thinner and exerts a smaller drag force on aircraft.

The power developed from the thrust of the engine is called the **propulsive power** $\dot{W}_P$, which is the *propulsive force* (*thrust*) times the *distance* this force acts on the aircraft per unit time, i.e., the thrust times the aircraft velocity (Fig. 8-49):

$$\dot{W}_P = (F)\mathbf{V}_{\text{aircraft}} = \dot{m}(\mathbf{V}_{\text{exit}} - \mathbf{V}_{\text{inlet}})\mathbf{V}_{\text{aircraft}} \qquad \text{(kW)} \qquad (8\text{-}28)$$

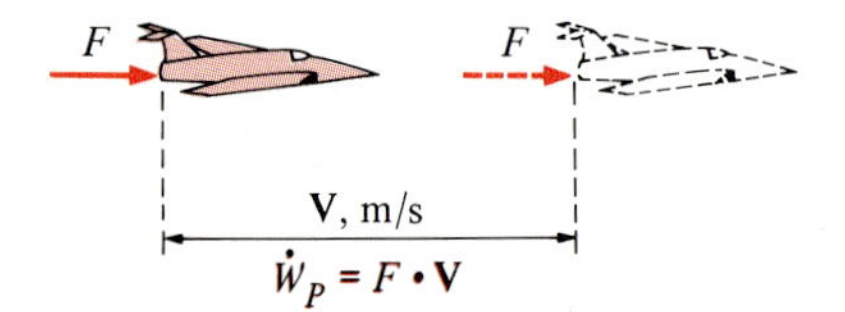

FIGURE 8-49
Propulsive power is the thrust acting on the aircraft through a distance per unit time.

The net work developed by a turbojet engine is zero. Thus we cannot define the efficiency of a turbojet engine in the same way as stationary gas-turbine engines. Instead, we should use the general definition of efficiency, which is the ratio of the desired output to the required input. The desired output in a turbojet engine is the *power produced* to propel the aircraft $\dot{W}_P$, and the required input is the *thermal energy of the fuel* released during the combustion process $\dot{Q}_{\text{in}}$. The ratio of these two quantities is called the **propulsive efficiency** and is given by

$$\eta_P = \frac{\text{propulsive power}}{\text{energy input rate}} = \frac{\dot{W}_P}{\dot{Q}_{\text{in}}} \qquad (8\text{-}29)$$

Propulsive efficiency is a measure of how efficiently the energy released during the combustion process is converted to propulsive energy. The remaining part of the energy released will show up as the kinetic energy of the exhaust gases relative to a fixed point on the ground and as an increase in the enthalpy of the air leaving the engine.

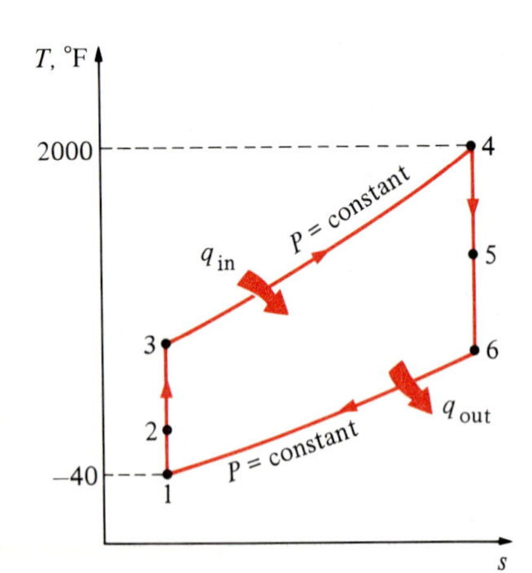

FIGURE 8-50
T-s diagram for the turbojet cycle described in Example 8-9.

EXAMPLE 8-9

A turbojet aircraft flies with a velocity of 850 ft/s at an altitude where the air is at 5 psia and −40°F. The compressor has a pressure ratio of 10, and the temperature of the gases at the turbine inlet is 2000°F. Air enters the compressor at a rate of 100 lbm/s. Utilizing the cold-air-standard assumptions, determine (*a*) the temperature and pressure of the gases at the turbine exit, (*b*) the velocity of the gases at the nozzle exit, and (*c*) the propulsive efficiency of the cycle.

Solution The *T-s* diagram of the cycle is shown in Fig. 8-50. Under the cold-air-standard assumptions, all the processes are assumed to be internally reversible, and the working fluid is air which behaves as an ideal gas and has constant specific heats evaluated at room temperature [C_p = 0.240 Btu/(lbm·R) and k = 1.4, from Table A-2E*a*]. The combustion process is also replaced by a heat addition process.

(*a*) Before we can determine the temperature and pressure at the turbine exit, we need to find the temperatures and pressures at other states:

Process 1-2 (isentropic compression of an ideal gas in a diffuser): For convenience, we can assume that the aircraft is stationary and the air is moving toward the aircraft at a velocity of $\mathbf{V}_1 = 850$ ft/s. Ideally, the air will leave the diffuser with a negligible velocity ($\mathbf{V}_2 \cong 0$):

$$q_{12}^{\nearrow 0} - w_{12}^{\nearrow 0} = h_2 - h_1 + \frac{\mathbf{V}_2^{2 \nearrow 0} - \mathbf{V}_1^2}{2}$$

$$0 = C_p(T_2 - T_1) - \frac{\mathbf{V}_1^2}{2}$$

$$T_2 = T_1 + \frac{\mathbf{V}_1^2}{2\,C_p}$$

$$= 420\text{ R} + \frac{(850\text{ ft/s})^2}{2[0.240\text{ Btu/(lbm·R)}]}\left(\frac{1\text{ Btu/lbm}}{25{,}037\text{ ft}^2/\text{s}^2}\right)$$

$$= 480.1\text{ R}$$

$$P_2 = P_1\left(\frac{T_2}{T_1}\right)^{k/(k-1)} = (5\text{ psia})\left(\frac{480.1\text{ R}}{420\text{ R}}\right)^{1.4/(1.4-1)} = 8.0\text{ psia}$$

Process 2-3 (isentropic compression of an ideal gas in a compressor):

$$P_3 = (r_p)(P_2) = (10)(8.0\text{ psia}) = 80\text{ psia } (= P_4)$$

$$T_3 = T_2\left(\frac{P_3}{P_2}\right)^{(k-1)/k} = (480.1\text{ R})(10)^{(1.4-1)/1.4} = 926.9\text{ R}$$

Process 4-5 (isentropic expansion of an ideal gas in a turbine): Neglecting the kinetic energy changes across the compressor and the turbine and assuming the turbine work to be equal to the compressor work, we find the temperature and pressure at the turbine exit to be

$$w_{\text{comp,in}} = w_{\text{turb,out}}$$

$$h_3 - h_2 = h_4 - h_5$$

$$C_p(T_3 - T_2) = C_p(T_4 - T_5)$$

$$T_5 = T_4 - T_3 + T_2 = (2460 - 926.9 + 480.1)\text{ R} = 2013.2\text{ R}$$

$$P_5 = P_4\left(\frac{T_5}{T_4}\right)^{k/(k-1)} = (80\text{ psia})\left(\frac{2013.2\text{ R}}{2460\text{ R}}\right)^{1.4/(1.4-1)} = 39.7\text{ psia}$$

(*b*) To find the air velocity at the nozzle exit, we need to first determine the nozzle exit temperature and then apply the steady-flow energy equation.

Process 5-6 (isentropic expansion of an ideal gas in a nozzle):

$$T_6 = T_5\left(\frac{P_6}{P_5}\right)^{(k-1)/k} = (2013.2\text{ R})\left(\frac{5\text{ psia}}{39.7\text{ psia}}\right)^{(1.4-1)/1.4} = 1113.8\text{ R}$$

$$q_{56}^{\nearrow 0} - w_{56}^{\nearrow 0} = h_6 - h_5 + \frac{\mathbf{V}_6^2 - \mathbf{V}_5^{2 \nearrow 0}}{2}$$

$$0 = C_p(T_6 - T_5) + \frac{\mathbf{V}_6^2}{2}$$

$$\mathbf{V}_6 = \sqrt{2\,C_p(T - T)}$$

$$= \sqrt{2[0.240\text{ Btu/(lbm·R)}][(2013.2 - 1113.8)\text{ R}]\left(\frac{25{,}037\text{ ft}^2/\text{s}^2}{1\text{ Btu/lbm}}\right)}$$

$$= 3287.7\text{ ft/s}$$

(c) The propulsive efficiency of a turbojet engine is the ratio of the propulsive power developed $\dot{W}_p$ to the total heat transfer rate to the working fluid:

$$\dot{W}_P = \dot{m}(\mathbf{V}_{exit} - \mathbf{V}_{inlet})\mathbf{V}_{aircraft}$$
$$= (100\ \text{lbm/s})[(3287.7 - 850)\ \text{ft/s}](850\ \text{ft/s})\left(\frac{1\ \text{Btu/lbm}}{25{,}037\ \text{ft}^2/\text{s}^2}\right)$$
$$= 8276\ \text{Btu/s} \quad (11{,}707\ \text{hp})$$
$$\dot{Q}_{in} = \dot{m}(h_4 - h_3) = \dot{m}C_p(T_4 - T_3)$$
$$= (100\ \text{lbm/s})[0.240\ \text{Btu/(lbm·R)}][(2460 - 926.9)\ \text{R}]$$
$$= 36{,}794\ \text{Btu/s}$$
$$\eta_P = \frac{\dot{W}_P}{\dot{Q}_{in}} = \frac{8276\ \text{Btu/s}}{36{,}794\ \text{Btu/s}} = \mathbf{22.5\%}$$

That is, 22.5 percent of the energy input is used to propel the aircraft and to overcome the drag force exerted by the air.

For those who are wondering what happened to the rest of the energy, here is a brief account:

$$\dot{KE}_{out} = \dot{m}\frac{\mathbf{V}_g^2}{2} = (100\ \text{lbm/s})\left\{\frac{[(3287.7 - 850)\ \text{ft/s}]^2}{2}\right\}\left(\frac{1\ \text{Btu/lbm}}{25{,}037\ \text{ft}^2/\text{s}^2}\right)$$
$$= 11{,}867\ \text{Btu/s} \quad (32.2\%)$$
$$\dot{Q}_{out} = \dot{m}(h_6 - h_1) = \dot{m}C_p(T_6 - T_1)$$
$$= (100\ \text{lbm/s})[0.24\ \text{Btu/(lbm·R)}][(1113.8 - 420)\ \text{R}]$$
$$= 16{,}651\ \text{Btu/s} \quad (45.3\%)$$

Thus, 32.2 percent of the energy shows up as excess kinetic energy (kinetic energy of the gases relative to a fixed point on the ground). Notice that for the highest propulsion efficiency, the velocity of the exhaust gases relative to the ground $\mathbf{V}_g$ should be zero. That is, the exhaust gases should leave the nozzle at the velocity of the aircraft. The remaining 45.3 percent of the energy shows up as an increase in enthalpy of the gases leaving the engine. These last two forms of energy eventually become part of the internal energy of the atmospheric air (Fig. 8-51).

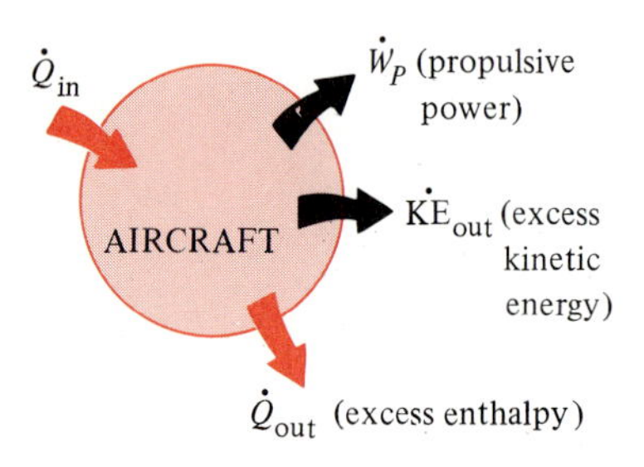

FIGURE 8-51
Energy supplied to an aircraft (from the burning of a fuel) manifests itself in various forms.

Modifications to Turbojet Engines

Both propeller-driven engines and jet-propulsion-driven engines have their own strengths and limitations, and several attempts have been made to combine the desirable characteristics of both in one engine. Two such modifications are the *turboprop engine* and the *turbofan engine*.

In **turboprop** (or **propjet**) engines, shown in Fig. 8-52, the turbine also drives a propeller, which provides most of the thrust. Propellers are more suitable for low-speed operation since their efficiency decreases at high speeds. Therefore, the use of turboprop engines is presently limited to low-speed applications. Research into higher-speed use is under way.

The most widely used engine in aircraft propulsion is the **turbofan** (or **fan-jet**) engine. In these engines, as shown in Fig. 8-53, a large fan driven by the turbine forces a considerable amount of air through a duct surrounding the engine components. The fan exhaust leaves the duct at a higher velocity, enhancing the total thrust of the engine significantly.

Another modification which is popular with military aircraft is the

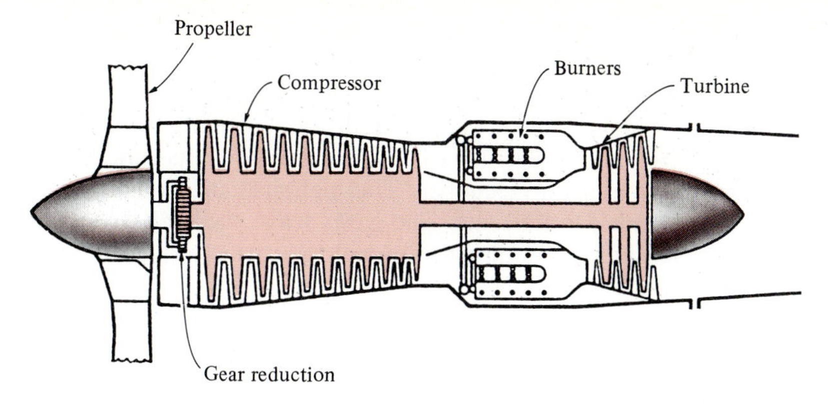

FIGURE 8-52

A turboprop engine. [*Source:* The Aircraft Gas Turbine Engine and Its Operation. © United Aircraft Corporation (Now United Technologies Corporation), 1951, 1974.]

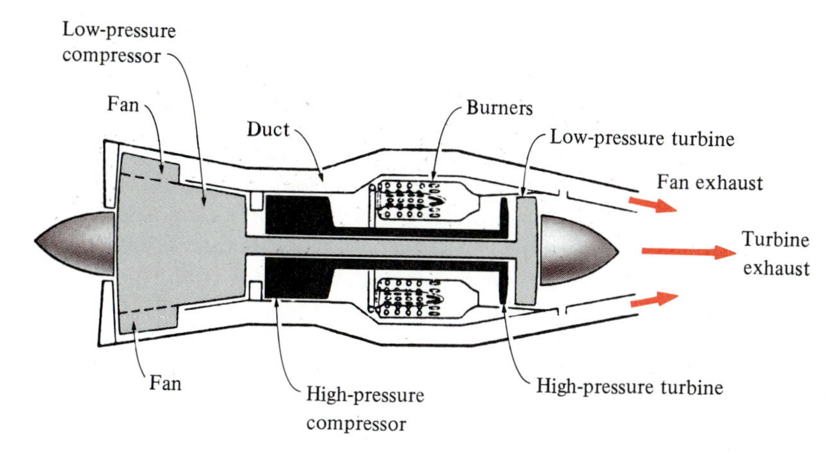

FIGURE 8-53

A turbofan engine. [*Source:* The Aircraft Gas Turbine and Its Operation. © United Aircraft Corporation (Now United Technologies Corporation), 1951, 1974.]

addition of an **afterburner** section between the turbine and the nozzle. Whenever a need for extra thrust arises, such as for short takeoffs or combat conditions, additional fuel is injected into the oxygen-rich combustion gases leaving the turbine. As a result of this added energy, the exhaust gases leave at a higher velocity, providing a greater thrust.

A **ramjet** engine is a properly shaped duct with no compressor or turbine, as shown in Fig. 8-54, and is sometimes used for high-speed

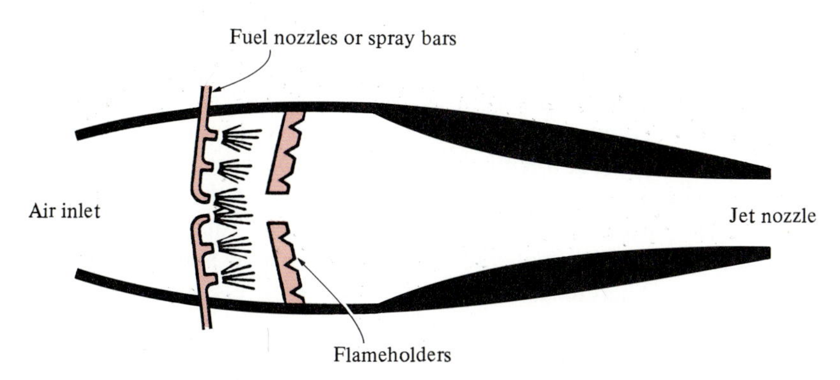

FIGURE 8-54

A ramjet engine. [*Source:* The Aircraft Gas Turbine Engine and Its Operation. © United Aircraft Corporation (Now United Technologies Corporation), 1951, 1974.]

propulsion of missiles and aircraft. The pressure rise in the engine is provided by the ram effect of the incoming high-kinetic energy air. Therefore, a ramjet engine needs to be brought to a sufficiently high speed by an external source before it can be fired.

Finally, a **rocket** is a device where a solid or liquid fuel and an oxidizer react in the combustion chamber. The high-pressure combustion gases are then expanded in a nozzle. The gases leave the rocket at very high velocities, producing the thrust to propel the rocket.

8-12 ■ SECOND-LAW ANALYSIS OF GAS POWER CYCLES

The ideal Carnot, Ericsson, and Stirling cycles are *totally reversible,* thus they do not involve any irreversibilities. The ideal Otto, Diesel, and Brayton cycles, however, are only *internally reversible,* and they may involve irreversibilities external to the system. A second-law analysis of these cycles will reveal where the largest irreversibilities occur and where to start improvements.

Relations for *availability* and *irreversibility* for both closed and steady-flow systems are developed in Chap. 7. The irreversibility for a *process* of a closed system which may be exchanging heat with a reservoir at temperature T_R in the amount of Q_R can be expressed as

$$I = T_0 S_{\text{gen}} = T_0 \left[(S_e - S_i)_{\text{sys}} + \frac{Q_R}{T_R} \right] \quad \text{(kJ)} \qquad \text{(8-30)}$$

where T_0 is the ambient temperature and subscripts i and e denote the initial and final states of a process, respectively. Also the sign of Q_R is to be determined with respect to the reservoir. A similar relation for steady-flow systems can be expressed, in rate form, as

$$\dot{I} = T_0 \dot{S}_{\text{gen}} = T_0 \left(\sum \dot{m}_e s_e - \sum \dot{m}_i s_i + \frac{\dot{Q}_R}{T_R} \right) \quad \text{(kW)} \qquad \text{(8-31)}$$

or, on a unit-mass basis for a one-inlet, one-exit steady-flow device, as

$$i = T_0 s_{\text{gen}} = T_0 \left(s_e - s_i + \frac{q_R}{T_R} \right) \quad \text{(kJ/kg)} \qquad \text{(8-32)}$$

where subscripts i and e denote the inlet and exit states for the process, respectively.

The irreversibility of a *cycle* is the sum of the irreversibilities of the processes that compose that cycle. The irreversibility of a cycle can also be determined without tracing the individual processes by considering the entire cycle as a single process and using one of the relations above. Entropy is a property, and its value depends on the state only. For a cycle, reversible or actual, the initial and the final states are identical, thus $s_e = s_i$. Therefore, the irreversibility of a cycle depends on the magnitude of the heat transfer with the high- and low-temperature reservoirs involved and on their temperatures. It can be expressed on a unit-mass basis as

$$i = T_0 \sum \frac{q_i}{T_i} \qquad \text{(kJ/kg)} \qquad (8\text{-}33)$$

For a cycle which involves heat transfer only with a source at T_H and a sink at T_L, the irreversibility becomes

$$i = T_0 \left(\frac{q_{out}}{T_L} - \frac{q_{in}}{T_H} \right) \qquad \text{(kJ/kg)} \qquad (8\text{-}34)$$

The availabilities of a closed system ϕ and a fluid stream ψ at any state can be determined from Eqs. 7-9*a* and 7-17:

$$\phi = (u - u_0) - T_0(s - s_0) + P_0(v - v_0) \qquad \text{(kJ/kg)} \qquad (8\text{-}35)$$

and

$$\psi = (h - h_0) - T_0(s - s_0) + \frac{\mathbf{V}^2}{2} + gz \qquad \text{(kJ/kg)} \qquad (8\text{-}36)$$

where subscript "0" denotes the state of the surroundings. The reversible work for any process can be determined by finding the change in availability of the working fluid during that process.

EXAMPLE 8-10

Determine the irreversibility of the Otto cycle (all four processes as well as the cycle) discussed in Example 8-2, assuming that heat is transferred to the working fluid from a source at 1700 K and heat is rejected to the surroundings at 290 K. Also determine the availability of the exhaust gases when they are purged.

Solution In Example 8-2, various quantities of interest were given or determined to be

$$r = 8 \qquad P_2 = 1.7997 \text{ MPa}$$
$$T_0 = 290 \text{ K} \qquad P_3 = 4.347 \text{ MPa}$$
$$T_1 = 290 \text{ K} \qquad q_{23} = 800 \text{ kJ/kg}$$
$$T_2 = 652.4 \text{ K} \qquad q_{41} = -429.31 \text{ kJ/kg}$$
$$T_3 = 1575.1 \text{ K} \qquad w_{net} = 370.69 \text{ kJ/kg}$$

Processes 1-2 and 3-4 are isentropic ($s_1 = s_2$, $s_3 = s_4$) and therefore do not involve any internal or external irreversibilities, that is, $i_{12} = 0$ and $i_{34} = 0$.

Processes 2-3 and 4-1 are constant-volume heat addition and rejection processes, respectively, and are internally reversible. But the heat transfer between the working fluid and the source or the sink takes place through a finite temperature difference, rendering both processes irreversible. The irreversibility associated with each process is determined from Eq. 8-32. But first we need to determine the entropy change of air during these processes:

$$s_3 - s_2 = s_3^\circ - s_2^\circ - R \ln \frac{P_3}{P_2}$$

$$= (3.5045 - 2.4975) \text{ kJ/(kg·K)} - [0.287 \text{ kJ/(kg·K)}] \ln \frac{4.347 \text{ MPa}}{1.7997 \text{ MPa}}$$

$$= 0.7539 \text{ kJ/(kg·K)}$$

Also, $\qquad q_{R,23} = -q_{23} = -800 \text{ kJ/kg} \qquad \text{and} \qquad T_R = 1700 \text{ K}$

Thus,
$$i_{23} = T_0\left[(s_3 - s_2)_{sys} + \frac{q_{R,23}}{T_R}\right]$$
$$= (290\ \text{K})\left[0.7539\ \text{kJ/(kg·K)} + \frac{-800\ \text{kJ/kg}}{1700\ \text{K}}\right]$$
$$= 82.2\ \text{kJ/kg}$$

For process 4-1, $s_1 - s_4 = s_2 - s_3 = -0.7539$ kJ/(kg·K), $q_{R,41} = -q_{41} =$ 429.31 kJ/kg, and $T_R = 290$ K. Thus,

$$i_{41} = T_0\left[(s_2 - s_4)_{sys} + \frac{q_{R,41}}{T_R}\right)$$
$$= (290\ \text{K})\left[-0.7539\ \text{kJ/(kg·K)} + \frac{429.31\ \text{kJ/kg}}{290\ \text{K}}\right]$$
$$= 210.7\ \text{kJ/kg}$$

Therefore, the irreversibility of the cycle is

$$i_{cycle} = i_{12} + i_{23} + i_{34} + i_{41}$$
$$= 0 + 82.2\ \text{kJ/kg} + 0 + 210.7\ \text{kJ/kg}$$
$$= 292.9\ \text{kJ/kg}$$

The irreversibility of the cycle could also be determined from Eq. 8-34. Notice that the largest irreversibility in the cycle occurs during the heat rejection process. Therefore, any attempt to reduce the irreversibilities should start with this process.

The availability (work potential) of the working fluid before it is purged (state 4) is determined from Eq. 8-35:

$$\phi_4 = (u_4 - u_0) - T_0(s_4 - s_0) + P_0(v_4 - v_0)$$

where $s_4 - s_0 = s_4 - s_1 = 0.7539$ kJ/(kg·K)
$u_4 - u_0 = u_4 - u_1 = q_{out} = 429.31$ kJ/kg
$v_4 - v_0 = v_4 - v_1 = 0$

Thus, $\phi_4 = 429.31\ \text{kJ/kg} - (290\ \text{K})[0.7539\ \text{kJ/(kg·K)}] + 0 = 210.7\ \text{kJ/kg}$

which is equivalent to the irreversibility for process 4-1. (Why?) That is, 210.7 kJ/kg of work could be obtained from the exhaust gases if they are brought to the state of the surroundings in a reversible manner. Notice that this is 57 percent of the net work output of the cycle.

8-13 ■ SUMMARY

A cycle during which a net amount of work is produced is called a *power cycle*, and a power cycle during which the working fluid remains a gas throughout is called a *gas power cycle*. The most efficient cycle operating between a source at temperature T_H and a sink at temperature T_L is the *Carnot cycle*, and its thermal efficiency is given by

$$\eta_{th,Carnot} = 1 - \frac{T_L}{T_H} \tag{8-2}$$

The actual gas cycles are rather complex. The approximations used to simplify the analysis are known as the *air-standard assumptions*. Under these assumptions, all the processes are assumed to be internally reversible; the working fluid is assumed to be air which behaves as an ideal gas; and the combustion and exhaust processes are replaced by heat addition and heat rejection processes, respectively. The air-standard assumptions are called *cold-air-standard assumptions* if, in addition, air is assumed to have constant specific heats at room temperature.

In reciprocating engines, the *compression ratio* r and the *mean effective pressure* MEP are defined as

$$r = \frac{V_{\text{max}}}{V_{\text{min}}} = \frac{V_{\text{BDC}}}{V_{\text{TDC}}} \tag{8-3}$$

$$\text{MEP} = \frac{w_{\text{net,cycle}}}{v_{\text{max}} - v_{\text{min}}} \quad \text{(kPa)} \tag{8-4}$$

The *Otto cycle* is the ideal cycle for the spark-ignition reciprocating engines, and it consists of four internally reversible processes: isentropic compression, v = constant heat addition, isentropic expansion, and v = constant heat rejection. Under cold-air-standard assumptions, the thermal efficiency of the ideal Otto cycle is

$$\eta_{\text{th,Otto}} = 1 - \frac{1}{r^{k-1}} \tag{8-8}$$

where r is the compression ratio and k is the specific heat ratio C_p/C_v.

The *Diesel cycle* is the ideal cycle for the compression-ignition reciprocating engines. It is very similar to the Otto cycle, except that the v = constant heat addition process is replaced by a P = constant heat addition process. Its thermal efficiency under cold-air-standard assumptions is

$$\eta_{\text{th,diesel}} = 1 - \frac{1}{r^{k-1}}\left[\frac{r_c^k - 1}{k(r_c - 1)}\right] \tag{8-12}$$

where r_c is the *cutoff ratio,* defined as the ratio of the cylinder volumes after and before the combustion process.

Stirling and *Ericsson cycles* are two totally reversible cycles that involve an isothermal heat addition process at T_H and an isothermal heat rejection process at T_L. They differ from the Carnot cycle in that the two isentropic processes are replaced by two v = constant regeneration processes in the Stirling cycle and by two P = constant regeneration processes in the Ericsson cycle. Both cycles utilize *regeneration,* a process during which heat is transferred to a thermal energy storage device (called a *regenerator*) during one part of the cycle which is transferred back to the working fluid during another part of the cycle.

The ideal cycle for the modern gas-turbine engines is the *Brayton cycle,* which is made up of four internally reversible processes: isentropic compression (in a compressor), P = constant heat addition, isentropic expansion (in a turbine), and P = constant heat rejection. Under cold-air-standard assumptions, its thermal efficiency is

$$\eta_{\text{th,Brayton}} = 1 - \frac{1}{r_p^{(k-1)/k}} \tag{8-17}$$

where $r_p = P_{max}/P_{min}$ is the pressure ratio and k is the specific heat ratio. The thermal efficiency of the simple Brayton cycle increases with the pressure ratio.

The deviation of the actual compressor and the turbine from the idealized isentropic ones can be accurately accounted for by utilizing their adiabatic efficiencies, defined as

$$\eta_C = \frac{w_s}{w_a} \cong \frac{h_{2s} - h_1}{h_{2a} - h_1} \qquad (8\text{-}19)$$

and

$$\eta_T = \frac{w_a}{w_s} \cong \frac{h_3 - h_{4a}}{h_3 - h_{4s}} \qquad (8\text{-}20)$$

where states 1 and 3 are the inlet states, $2a$ and $4a$ are the actual exit states, and $2s$ and $4s$ are the isentropic exit states.

In gas-turbine engines, the temperature of the exhaust gas leaving the turbine is often considerably higher than the temperature of the air leaving the compressor. Therefore, the high-pressure air leaving the compressor can be heated by transferring heat to it from the hot exhaust gases in a counterflow heat exchanger, which is also known as a *regenerator*. The extent to which a regenerator approaches an ideal regenerator is called the *effectiveness* ε and is defined as

$$\varepsilon = \frac{q_{\text{regen,act}}}{q_{\text{regen,max}}} \qquad (8\text{-}23)$$

Under cold-air-standard assumptions, the thermal efficiency of an ideal Brayton cycle with regeneration becomes

$$\eta_{\text{th,regen}} = 1 - \left(\frac{T_1}{T_3}\right)(r_p)^{(k-1)/k} \qquad (8\text{-}25)$$

where T_1 and T_3 are the minimum and maximum temperatures, respectively, in the cycle.

The thermal efficiency of the Brayton cycle can also be increased by utilizing *multistage compression with intercooling, regeneration, and multistage expansion with reheating*. The work input to the compressor is minimized when equal pressure ratios are maintained across each stage. This procedure also maximizes the turbine work output.

Gas-turbine engines are widely used to power aircraft because they are light and compact and have a high power-to-weight ratio. The ideal *jet-propulsion cycle* differs from the simple ideal Brayton cycle in that the gases are partially expanded in the turbine. The gases that exit the turbine at a relatively high pressure are subsequently accelerated in a nozzle to provide the thrust needed to propel the aircraft.

The *net thrust* developed by the engine is

$$F = \dot{m}(\mathbf{V}_{\text{exit}} - \mathbf{V}_{\text{inlet}}) \quad \text{(N)} \qquad (8\text{-}27)$$

where $\dot{m}$ is the mass flow rate of gases, $\mathbf{V}_{\text{exit}}$ is the exit velocity of the exhaust gases, and $\mathbf{V}_{inlet}$ is the inlet velocity of the air, both relative to the aircraft.

The power developed from the thrust of the engine is called the *propulsive power* $\dot{W}_P$, and it is given by

$$\dot{W}_P = \dot{m}(\mathbf{V}_{exit} - \mathbf{V}_{inlet})\mathbf{V}_{aircraft} \qquad \text{(kW)} \tag{8-28}$$

Propulsive efficiency is a measure of how efficiently the energy released during the combustion process is converted to propulsive energy, and it is defined as

$$\eta_P = \frac{\text{propulsive power}}{\text{energy input rate}} = \frac{\dot{W}_P}{\dot{Q}_{in}} \tag{8-29}$$

The *irreversibility* for a process involving heat transfer with a reservoir at temperature T_R in the amount of Q_R can be expressed as

$$I = T_0 S_{gen} = T_0 \left[(S_e - S_i)_{sys} + \frac{Q_R}{T_R} \right] \qquad \text{(kJ)} \tag{8-30}$$

for closed systems and as

$$\dot{I} = T_0 \dot{S}_{gen} = T_0 \left(\sum \dot{m}_e s_e - \sum \dot{m}_i s_i + \frac{\dot{Q}_R}{T_R} \right) \qquad \text{(kW)} \tag{8-31}$$

for steady-flow systems. Irreversibility can also be expressed on a unit-mass basis as

$$i = T_0 s_{gen} = T_0 \left(s_e - s_i + \frac{q_R}{T_R} \right) \qquad \text{(kJ/kg)} \tag{8-32}$$

For a cycle which involves heat transfer only with a source at T_H and a sink at T_L, the irreversibility becomes

$$i = T_0 \left(\frac{q_{out}}{T_L} - \frac{q_{in}}{T_H} \right) \qquad \text{(kJ/kg)} \tag{8-34}$$

The *availability* of a closed system ϕ and that of a fluid stream ψ at any state can be determined from Eqs. 7-9*a* and 7-17:

$$\phi = (u - u_0) - T_0(s - s_0) + P_0(v - v_0) \qquad \text{(kJ/kg)} \tag{8-35}$$

and

$$\psi = (h - h_0) - T_0(s - s_0) + \frac{\mathbf{V}^2}{2} + gz \qquad \text{(kJ/kg)} \tag{8-36}$$

where the subscript 0 denotes the state of the surroundings.

REFERENCES AND SUGGESTED READING

1 W. Z. Black and J. G. Hartley, *Thermodynamics,* Harper & Row, New York, 1985.

2 R. C. Fellinger and W. J. Cook, *Introduction to Engineering Thermodynamics,* WCB, Dubuque, IA, 1985.

3 J. P. Holman, *Thermodynamics,* 3d ed., McGraw-Hill, New York, 1980.

4 J. R. Howell and R. O. Buckius, *Fundamentals of Engineering Thermodynamics,* McGraw-Hill, New York, 1987.

5 J. B. Jones and G. A. Hawkins, *Engineering Thermodynamics,* 2d ed., Wiley, New York, 1986.

6 B. V. Karlekar, *Thermodynamics for Engineers,* Prentice-Hall, Englewood Cliffs, NJ, 1983.

7 L. C. Lichty, *Combustion Engine Processes,* McGraw-Hill, New York, 1967.

8 D. C. Look, Jr., and H. J. Sauer, Jr., *Engineering Thermodynamics,* PWS Engineering, Boston, 1986.

9 C. F. Taylor, *The Internal Combustion Engine in Theory and Practice,* M.I.T. Press, Cambridge, MA, 1968.

10 G. J. Van Wylen and R. E. Sonntag, *Fundamentals of Classical Thermodynamics,* 3d ed., Wiley, New York, 1985.

11 K. Wark, *Thermodynamics,* 5th ed., McGraw-Hill, New York, 1988.

12 M. J. Zucrow, *Aircraft and Missile Propulsion,* vol. 1, Wiley, New York, 1958.

PROBLEMS*

Actual and Ideal Cycles, Carnot Cycle, Air-Standard Assumptions, Reciprocating Engines

8-1C How do gas power cycles differ from vapor power cycles?

8-2C Why is the Carnot cycle not suitable as an ideal cycle for all power-producing cyclic devices?

8-3C How does the thermal efficiency of an ideal cycle, in general, compare to that of a Carnot cycle operating between the same temperature limits?

8-4C What does the area enclosed by the cycle represent on a *P-v* diagram? How about on a *T-s* diagram?

8-5C The efficiency of a Carnot cycle is given by $\eta_{th,C} = 1 - T_L/T_H$. Does this relation suggest any ways of improving the thermal efficiency of actual cycles?

*Students are encourged to answer *all* the concept "C" questions.

8-6C What are the air-standard assumptions?

8-7C What is the difference between air-standard assumptions and the cold-air-standard assumptions?

8-8C Do internal combustion engines operate on a closed or an open cycle? Why?

8-9C How are the combustion and exhaust processes modeled under the air-standard assumptions?

8-10C Define the following terms related to reciprocating engines: stroke, bore, top dead center, and clearance volume.

8-11C What is the difference between the clearance volume and the displacement volume of reciprocating engines?

8-12C Define the compression ratio for reciprocating engines.

8-13C How is the mean effective pressure for reciprocating engines defined?

8-14C Can the mean effective pressure of an automobile engine in operation be less than the atmospheric pressure?

8-15C As a car gets older, will its compression ratio change? How about the mean effective pressure?

8-16C What is the difference between spark-ignition and compression-ignition engines?

8-17 An air-standard cycle is executed in a closed system and is composed of the following four processes:

1-2 v = constant heat addition from 100 kPa and 27°C to 300 kPa
2-3 P = constant heat addition to 1027°C
3-4 Isentropic expansion to 100 kPa
4-1 P = constant heat rejection to initial state

(*a*) Show the cycle on P-v and T-s diagrams.
(*b*) Calculate the net work output per unit mass.
(*c*) Determine the thermal efficiency.

8-18 An air-standard cycle is executed in a closed system and is composed of the following four processes:

1-2 Isentropic compression from 100 kPa and 27°C to 800 kPa
2-3 v = constant heat addition to 1800 K
3-4 Isentropic expansion to 100 kPa
4-1 P = constant heat rejection to initial state

(*a*) Show the cycle on P-v and T-s diagrams.
(*b*) Calculate the net work output per unit mass.
(*c*) Determine the thermal efficiency.

8-18E An air-standard cycle is executed in a closed system and is composed of the following four processes:

1-2 Isentropic compression from 14.7 psia and 80°F to 120 psia
2-3 v = constant heat addition to 3000 R
3-4 Isentropic expansion to 14.7 psia
4-1 P = constant heat rejection to initial state

(*a*) Show the cycle on P-v and T-s diagrams.
(*b*) Calculate the net work output per unit mass.
(*c*) Determine the thermal efficiency.

8-19 An air-standard cycle is executed in a closed system and is composed of the following four processes:

1-2 Isentropic compression from 100 kPa and 27°C to 1 MPa
2-3 P = constant heat addition in amount of 2840 kJ/kg
3-4 v = constant heat rejection to 100 kPa
4-1 P = constant heat rejection to initial state

(*a*) Show the cycle on P-v and T-s diagrams.
(*b*) Calculate the maximum temperature in the cycle.
(*c*) Determine the thermal efficiency.
Assume constant specific heats at room temperature.
Answers: (*b*) 3405.1 K, (*c*) 21.1 percent

8-20 An air-standard cycle is executed in a closed system and is composed of the following four processes:

1-2 v = constant heat addition from 100 kPa and 27°C in amount of 701.5 kJ/kg
2-3 P = constant heat addition to 2000 K
3-4 Isentropic expansion to 100 kPa
4-1 P = constant heat rejection to initial state

(*a*) Show the cycle on P-v and T-s diagrams.
(*b*) Calculate the total heat input per unit mass.
(*c*) Determine the thermal efficiency.
Account for the variation of specific heats with temperature.

8-20E An air-standard cycle is executed in a closed system and is composed of the following four processes:

1-2 v = constant heat addition from 14.7 psia and 80°F in amount of 300 Btu/lbm
2-3 P = constant heat addition to 3200 R
3-4 Isentropic expansion to 14.7 psia
4-1 P = constant heat rejection to initial state

(*a*) Show the cycle on P-v and T-s diagrams.
(*b*) Calculate the total heat input per unit mass.
(*c*) Determine the thermal efficiency.
Account for the variation of specific heats with temperature.
Answers: (*b*) 612.4 Btu/lbm, (*c*) 44.1 percent

8-21 An air-standard cycle is executed in a closed system with 1.5 kg of air, and it consists of the following three processes:

1-2 Isentropic compression from 100 kPa and 27°C to 700 kPa
2-3 P = constant heat addition to initial specific volume
3-1 v = constant heat rejection to initial state

(*a*) Show the cycle on *P-v* and *T-s* diagrams.
(*b*) Calculate the maximum temperature in the cycle.
(*c*) Determine the thermal efficiency.

Answers: (*b*) 2100 K, (*c*) 15.8 percent

8-21E An air-standard cycle is executed in a closed system with 4.0 lbm of air and consists of the following three processes:

1-2 Isentropic compression from 14.7 psia and 80°F to 120 psia
2-3 P = constant heat addition to initial specific volume
3-1 v = constant heat rejection to initial state

(*a*) Show the cycle on *P-v* and *T-s* diagrams.
(*b*) Calculate the maximum temperature in the cycle.
(*c*) Determine the thermal efficiency.

8-22 An air-standard cycle is executed in a closed system with 1.2 kg of air and consists of the following three processes:

1-2 Isentropic compression from 100 kPa and 27°C to 1 MPa
2-3 P = constant heat addition in amount of 2208 kJ
3-1 $P = cv$ heat rejection to initial state (c is a constant)

(*a*) Show the cycle on *P-v* and *T-s* diagrams.
(*b*) Calculate the heat rejected.
(*c*) Determine the thermal efficiency.

Assume constant specific heats at room temperature.
Answers: (*b*) 1422 kJ/kg, (*c*) 22.7 percent

8-23 An air-standard cycle is executed in a closed system with 2.0 kg of air and consists of the following three processes:

1-2 v = constant heat addition from 95 kPa and 17°C to 380 kPa
2-3 Isentropic expansion to 95 kPa
3-1 P = constant heat rejection to initial state

(*a*) Show the cycle on *P-v* and *T-s* diagrams.
(*b*) Calculate the net work per cycle, in kJ.
(*c*) Determine the thermal efficiency.

8-24 Consider a Carnot cycle executed in a closed system with 1.4 kg of air. The temperature limits of the cycle are 300 and 1000 K, and the minimum and maximum pressures that occur during the cycle are 20 and 1800 kPa. Assuming constant specific heats, determine the net work output per cycle.

8-25 An air-standard Carnot cycle is executed in a closed system between the temperature limits of 350 and 1200 K. The pressures before and after the isothermal compression are 150 and 300 kPa, respectively. If the net work output per cycle is 400 kJ, determine (*a*) the maximum pressure in the cycle, (*b*) the heat transfer to air, and (*c*) the mass of air. Assume variable specific heats for air.
Answers: (*a*) 30,013 kPa, (*b*) 238.80 kJ/kg, (*c*) 2.36 kg

8-25E An air-standard Carnot cycle is executed in a closed system between the temperature limits of 540 and 2000 R. The pressures before and after the isothermal compression are 20 and 40 psia, respectively. If the net work output per cycle is 450 Btu, determine (*a*) the maximum pressure in the cycle, (*b*) the heat transfer to air, and (*c*) the mass of air. Assume variable specific heats for air.

8-26 Repeat Prob. 8-25, using helium as the working fluid.

8-27 A Carnot cycle is executed in a closed system and uses 0.5 kg of air as the working fluid. The cycle efficiency is 70 percent, and the lowest temperature in the cycle is 300 K. The pressure at the beginning of the isentropic expansion is 700 kPa, and at the end of the isentropic compression it is 1 MPa. Determine the net work output per cycle.

Otto Cycle

8-28C What four processes make up the ideal Otto cycle?

8-29C How do the efficiencies of the ideal Otto cycle and the Carnot cycle compare for the same temperature limits? Explain.

8-30C How is the rpm (revolutions per minute) of an actual four-stroke gasoline engine related to the number of thermodynamic cycles? What would your answer be for a two-stroke engine?

8-31C Are the processes that make up the Otto cycle analyzed as closed-system or steady-flow processes? Why?

8-32C How does the thermal efficiency of an ideal Otto cycle change with the compression ratio of the engine and the specific heat ratio of the working fluid?

8-33C Why are high compression ratios not used in spark-ignition engines?

8-34C An ideal Otto cycle with a specified compression ratio is executed using (*a*) air, (*b*) argon, and (*c*) ethane as the working fluid. For which case will the thermal efficiency be the highest? Why?

8-35 An ideal Otto cycle has a compression ratio of 8. At the beginning of the compression process, air is at 95 kPa and 27°C, and 750 kJ/kg of heat is transferred to air during the constant-volume heat addition process. Taking into account the variation of specific heats with temperature, determine (*a*) the pressure and temperature at the end of the heat addition

process, (*b*) the net work output, (*c*) the thermal efficiency, and (*d*) the mean effective pressure for the cycle.
Answers: (*a*) 3898.1 kPa, 1538.7 K; (*b*) 392.38 kJ/kg; (*c*) 52.3 percent; (*d*) 495 kPa

8-35E An ideal Otto cycle has a compression ratio of 8. At the beginning of the compression process, air is at 14.5 psia and 80°F, and 450 Btu/lbm of heat is transferred to the air during the constant-volume heat addition process. Accounting for the variation of specific heats with temperature, determine (*a*) the pressure and temperature at the end of the heat addition process, (*b*) the net work output, (*c*) the thermal efficiency, and (*d*) the mean effective pressure for the cycle.

8-36 A four-cylinder spark-ignition engine has a compression rato of 8, and each cylinder has a maximum volume of 0.6 L. At the beginning of the compression process, the air is at 98 kPa and 17°C, and the maximum temperature in the cycle is 1800 K. Assuming the engine to operate on the ideal Otto cycle, determine (*a*) the amount of heat supplied per cylinder, (*b*) the thermal efficiency, and (*c*) the number of revolutions per minute required for a net power output of 80 kW. Assume variable specific heats for air.

8-37 The compression ratio of an air-standard Otto cycle is 9.5. Prior to the isentropic compression process, the air is at 100 kPa, 17°C, and 600 cm^3. The temperature at the end of the isentropic expansion process is 800 K. Using specific heat values at room temperature, determine (*a*) the highest temperature and pressure in the cycle, (*b*) the amount of heat added, in kJ, (*c*) the thermal efficiency, and (*d*) the mean effective pressure.
Answers: (*a*) 1968.7 K, 6449.2 kPa; (*b*) 0.65 kJ; (*c*) 59.4 percent; (*d*) 719 kPa

8-37E The compression ratio of an air-standard Otto cycle is 9.5. Prior to the isentropic compression process, the air is at 14.7 psia, 60°F, and 35 in^3. The temperature at the end of the isentropic expansion process is 1400 R. Using specific heat values at room temperature, determine (*a*) the highest temperature and pressure in the cycle, (*b*) the amount of heat added, in Btu, (*c*) the thermal efficiency, and (*d*) the mean effective pressure.

8-38 An ideal Otto cycle with air as the working fluid has a compression ratio of 8. The minimum and maximum temperatures in the cycle are 310 and 1600 K. Accounting for the variation of specific heats with temperature, determine (*a*) the amount of heat transferred to air during the heat addition process, (*b*) the thermal efficiency, and (*c*) the thermal efficiency of a Carnot cycle operating between the same temperature limits.
Answers: (*a*) 790.2 kJ/kg, (*b*) 52.1 percent, (*c*) 80.6 percent

8-38E An ideal Otto cycle with air as the working fluid has a compression ratio of 8. The minimum and maximum temperatures in the cycle are 540 and 2200 R. Accounting for the variation of specific heats with

temperature, determine (*a*) the amount of heat transferred to the air during the heat addition process, (*b*) the thermal efficiency, and (*c*) the thermal efficiency of a Carnot cycle operating between the same temperature limits.
Answers: (*a*) 198.15 Btu/lbm, (*b*) 53.5 percent, (*c*) 75.5 percent

8-39 Repeat Prob. 8-38, using argon as the working fluid.

8-40 An ideal Otto cycle has a compression ratio of 9.2 and uses air as the working fluid. At the beginning of the compression process, air is at 98 kPa and 27°C. The pressure is doubled during the constant-volume heat addition process. Accounting for the variation of specific heats with temperature, determine (*a*) the amount of heat transferred to the air, (*b*) the net work output, (*c*) the thermal efficiency, and (*d*) the mean effective pressure for the cycle.

8-41 Write a computer program to study the effect of variable specific heats on the thermal efficiency of the ideal Otto cycle using air as the working fluid. At the beginning of the compression process, air is at 100 kPa and 300 K. Use the equation in Table A-2*c* to account for the variation of specific heats with temperature. Determine the percentage of error involved in using constant specific heat values at room temperature for the following combinations of compression ratios and maximum cycle temperatures: r = 7, 8, 9, 10, 11, 12 and T_{max} = 1200, 1400, 1600, 1800, 2000, 2500 K.

Diesel Cycle

8-42C How does a diesel engine differ from a gasoline engine?

8-43C How does the ideal Diesel cycle differ from the ideal Otto cycle?

8-44C For a specified compression ratio, is a diesel or gasoline engine more efficient?

8-45C Do diesel or gasoline engines operate at higher compression ratios? Why?

8-46C What is the cutoff ratio? How does it affect the thermal efficiency of a Diesel cycle?

8-47C What is the dual cycle? How does it differ from the Otto and Diesel cycles?

8-48 An air-standard Diesel cycle has a compression ratio of 16 and a cutoff ratio of 2. At the beginning of the compression process, air is at 95 kPa and 27°C. Accounting for the variation of specific heats with temperature, determine (*a*) the temperature after the heat addition process, (*b*) the thermal efficiency, and (*c*) the mean effective pressure.
Answers: (*a*) 1724.8 K, (*b*) 56.3 percent, (*c*) 675.9 kPa

8-48E An air-standard Diesel cycle has a compression ratio of 16 and a cutoff ratio of 2. At the beginning of the compression process, air is at

14.5 psia and 80°F. Accounting for the variation of specific heats with temperature, determine (*a*) the temperature after the heat addition process, (*b*) the thermal efficiency, and (*c*) the mean effective pressure.

8-49 Consider an engine operating on the ideal Diesel cycle with air as the working fluid. The volume of the cylinder is 1200 cm^3 at the beginning of the compression process, 75 cm^3 at the end, and 150 cm^3 after the heat addition process. Air is at 17°C and 100 kPa at the beginning of the compression process. Determine (*a*) the pressure at the beginning of the heat rejection process, (*b*) the net work per cycle, in kJ, and (*c*) the mean effective pressure.

8-50 Repeat Prob. 8-49, using argon as the working fluid.

8-51 An air-standard Diesel cycle has a compression ratio of 18.2. Air is at 27°C and 0.1 MPa at the beginning of the compression process and at 2000 K at the end of the heat addition process. Accounting for the variation of specific heats with temperature, determine (*a*) the cutoff ratio, (*b*) the heat rejection per unit mass, and (*c*) the thermal efficiency.

8-51E An air-standard Diesel cycle has a compression ratio of 18.2. Air is at 80°F and 14.7 psia at the beginning of the compression process and at 3400 R at the end of the heat addition process. Accounting for the variation of specific heats with temperature, determine (*a*) the cutoff ratio, (*b*) the heat rejection per unit mass, and (*c*) the thermal efficiency. *Answers:* (*a*) 2.09, (*b*) 216.3 Btu/lbm, (*c*) 57.3 percent

8-52 A four-cylinder 3.0-L diesel engine which operates on an ideal Diesel cycle has a compression ratio of 17 and a cutoff ratio of 2.2. Air is at 27°C and 97 kPa at the beginning of the compression process. Using the cold-air-standard assumptions, determine how much power the engine will deliver at 1500 rpm.

8-53 Repeat Prob. 8-52, using nitrogen as the working fluid.

8-54 An ideal diesel engine has a compression ratio of 20 and uses air as the working fluid. The state of air at the beginning of the compression process is 95 kPa and 20°C. If the maximum temperature in the cycle is not to exceed 2200 K, determine (*a*) the thermal efficiency and (*b*) the mean effective pressure. Assume constant specific heats for air at room temperature. *Answers:* (*a*) 63.5 percent, (*b*) 933 kPa

8-55 An ideal dual cycle has a compression ratio of 12 and uses air as the working fluid. At the beginning of the compression process, air is 100 kPa and 30°C and occupies a volume of 1.2 L. During the heat addition process, 0.3 kJ of heat is transferred to air at constant volume and 1.1 kJ at constant pressure. Using constant specific heats evaluated at room temperature, determine the thermal efficiency of the cycle.

8-55E An ideal dual cycle has a compression ratio of 12 and uses air as the working fluid. At the beginning of the compression process, air is 14.7 psia and 90°F and occupies a volume of 75 in^3. During the heat addition process, 0.3 Btu of heat is transferred to air at constant volume and 1.1 Btu at constant pressure. Using constant specific heats evaluated at room temperature, determine the thermal efficiency of the cycle.

8-56 The compression ratio of an ideal dual cycle is 14. Air is at 100 kPa and 300 K at the beginning of the compression process and at 2200 K at the end of the heat addition process. Heat transfer to air takes place partly at constant volume and partly at constant pressure, and it amounts to 1520.4 kJ/kg. Assuming variable specific heats for air, determine (*a*) the fraction of heat transferred at constant volume and (*b*) the thermal efficiency of the cycle.

8-57 Write a computer program to study the effect of variable specific heats on the thermal efficiency of the ideal Diesel cycle using air as the working fluid. At the beginning of the compression process, air is at 100 kPa and 300 K. Use the equation in Table A-2*c* to account for the variation of specific heats with temperature. Determine the percentage of error involved in using constant specific heat values at room temperature for the following combinations of compression ratios and maximum cycle temperatures: r = 10, 12, 14, 18, 20 and T_{max} = 1400, 1600, 1800, 2000, 2500 K.

Stirling and Ericsson Cycles

8-58C Consider the ideal Otto, Stirling, and Carnot cycles operating between the same temperature limits. How would you compare the thermal efficiencies of these three cycles?

8-59C Consider the ideal Diesel, Ericsson, and Carnot cycles operating between the same temperature limits. How would you compare the thermal efficiencies of these three cycles?

8-60C What cycle is composed of two isothermal and two constant-volume processes?

8-61C How does the ideal Ericsson cycle differ from the Carnot cycle?

8-62C How is regeneration accomplished in the ideal Ericsson cycle?

8-63C Name three advantages that external combustion engines have over internal combustion engines.

8-64 An ideal Ericsson engine using helium as the working fluid operates between temperature limits of 300 and 1800 K and pressure limits of 150 and 1200 kPa. Assuming a mass flow rate of 2 kg/s, determine (*a*) the thermal efficiency of the cycle, (*b*) the heat transfer rate in the regenerator, and (*c*) the power delivered.

8-64E An ideal Ericsson engine using helium as the working fluid operates between temperature limits of 550 and 3000 R and pressure limits of 25 and 200 psia. Assuming a mass flow rate of 5 lbm/s, determine (*a*) the thermal efficiency of the cycle, (*b*) the heat transfer rate in the regenerator, and (*c*) the power delivered.

8-65 Consider an ideal Ericsson cycle with air as the working fluid executed in a steady-flow system. Air is at 27°C and 120 kPa at the beginning of the isothermal compression process during which 150 kJ/kg of heat is rejected. Heat transfer to air occurs at 1200 K. Determine (*a*) the maximum pressure in the cycle, (*b*) the net work output per unit mass of air, and (*c*) the thermal efficiency of the cycle.
Answers: (*a*) 658.2 kPa, (*b*) 450 kJ/kg, (*c*) 75 percent

8-66 An ideal Stirling engine using helium as the working fluid operates between temperature limits of 300 and 2000 K and pressure limits of 100 kPa and 2 MPa. Assuming the mass of the helium used in the cycle is 1.5 kg, determine (*a*) the thermal efficiency of the cycle, (*b*) the amount of heat transfer in the regenerator, and (*c*) the work output per cycle.

8-67 Consider an ideal Stirling cycle using air as the working fluid. Air is at 350 K and 200 kPa at the beginning of the isothermal compression process, and heat is supplied to air from a source at 1600 K in the amount of 800 kJ/kg. Determine (*a*) the maximum pressure in the cycle and (*b*) the net work output per unit mass of air.
Answers: (*a*) 5218 kPa, (*b*) 625 kJ/kg

Ideal and Actual Gas-Turbine (Brayton) Cycles

8-68C What four processes make up the simple ideal Brayton cycle?

8-69C For fixed maximum and minimum temperatures, what is the effect of the pressure ratio on (*a*) the thermal efficiency and (*b*) the net work output of a simple ideal Brayton cycle?

8-70C Why are gas turbines operated at very high air-fuel mass ratios?

8-71C Should the processes that make up the Brayton cycle be analyzed as closed-system or steady-flow processes? Why?

8-72C What is the back work ratio? What are typical back work ratio values for gas-turbine engines?

8-73C Why are the back work ratios relatively high in gas-turbine engines?

8-74C How can the irreversibilities in the turbine and compressor of gas-turbine engines be properly accounted for?

8-75C How do the inefficiencies of the turbine and the compressor affect (*a*) the back work ratio and (*b*) the thermal efficiency of a gas-turbine engine?

8-76 A simple ideal Brayton cycle with air as the working fluid has a pressure ratio of 10. The air enters the compressor at 300 K and the turbine at 1200 K. Accounting for the variation of specific heats with temperature, determine (*a*) the air temperature at the compressor exit, (*b*) the back work ratio, and (*c*) the thermal efficiency.

8-76E A simple ideal Brayton cycle with air as the working fluid has a pressure ratio of 10. The air enters the compressor at 520 R and the turbine at 2000 R. Accounting for the variation of specific heats with temperature, determine (*a*) the air temperature at the compressor exit, (*b*) the back work ratio, and (*c*) the thermal efficiency.

8-77 A simple Brayton cycle using air as the working fluid has a pressure ratio of 8. The minimum and maximum temperatures in the cycle are 310 and 1160 K. Assuming an adiabatic efficiency of 75 percent for the compressor and 82 percent for the turbine, determine (*a*) the air temperature at the turbine exit, (*b*) the net work output, and (*c*) the thermal efficiency.

8-78 Air is used as the working fluid in a simple Brayton cycle which has a pressure ratio of 12, a compressor inlet temperature of 300 K, and a turbine inlet temperature of 1000 K. Determine the required mass flow rate of air for a net power output of 10 MW, assuming both the compressor and the turbine have an isentropic efficiency of (*a*) 100 percent and (*b*) 80 percent. Assume constant specific heats at room temperature.
Answers: (*a*) 50.2 kg/s, (*b*) 527 kg/s

8-79 A stationary gas-turbine power plant operates on a simple ideal Brayton cycle with air as the working fluid. The air enters the compressor at 95 kPa and 290 K and the turbine at 760 kPa and 1100 K. Heat is transferred to air at a rate of 25,000 kJ/s. Determine the power delivered by this plant, (*a*) assuming constant specific heats at room temperature and (*b*) accounting for the variation of specific heats with temperature.

8-80 Air enters the compressor of a gas-turbine engine at 300 K and 100 kPa, where it is compressed to 700 kPa and 580 K. Heat is transferred to air in the amount of 950 kJ/kg before it enters the turbine. For a turbine efficiency of 86 percent, determine (*a*) the fraction of the turbine work output used to drive the compressor and (*b*) the thermal efficiency. Assume variable specific heats for air.
Answers: (*a*) 64.7 percent, (*b*) 16.4 percent

8-80E Air enters the compressor of a gas-turbine engine at 540 R and 14.5 psia, where it is compressed to 116 psia and 1000 R. Heat is transferred to air in the amount of 420 Btu/lbm before it enters the turbine. For a turbine efficiency of 86 percent, determine (*a*) the fraction of the turbine work output used to drive the compressor and (*b*) the thermal efficiency.

8-81 A gas-turbine power plant operates on a simple Brayton cycle with air as the working fluid. The air enters the turbine at 1 MPa and 1000 K and leaves at 125 kPa and 600 K. Heat is rejected to the surroundings at a rate of 3961 kJ/s, and air flows through the cycle at a rate of 12.5 kg/s.

Assuming a compressor efficiency of 80 percent, determine the net power output of the plant.

8-81E A gas-turbine power plant operates on a simple Brayton cycle with air as the working fluid. The air enters the turbine at 120 psia and 2000 R and leaves at 15 psia and 1200 R. Heat is rejected to the surroundings at a rate of 3200 Btu/s, and air flows through the cycle at a rate of 20 lbm/s. Assuming a compressor efficiency of 80 percent, determine the net power output of the plant. *Answer:* 1687 kW

8-82 For what compressor efficiency will the gas-turbine power plant in Prob. 8-81 produce zero net work?

8-83 A gas-turbine power plant operates on the simple Brayton cycle with air as the working fluid and delivers 15 MW of power. The minimum and maximum temperatures in the cycle are 310 and 900 K, and the pressure of air at the compressor exit is 8 times the value at the compressor inlet. Assuming an adiabatic efficiency of 80 percent for the compressor and 86 percent for the turbine, determine the mass flow rate of air through the cycle.

8-84 Consider a simple ideal Brayton cycle with air as the working fluid. The pressure ratio of the cycle is 6, and the minimum and maximum temperatures are 300 and 1300 K, respectively. Now the pressure ratio is doubled without changing the minimum and maximum temperatures in the cycle. Determine the change in (*a*) the net work output per unit mass and (*b*) the thermal efficiency of the cycle as a result of this modification. Assume variable specific heats for air.
Answers: (*a*) 41.5 kJ/kg, (*b*) 10.6 percent

8-85 Write a computer program to determine the effects of pressure ratio, maximum cycle temperature, and compressor and turbine inefficiencies on the net work output per unit mass and the thermal efficiency of a simple Brayton cycle. Assume the working fluid is air which is at 100 kPa and 300 K at the compressor inlet. Also assume constant specific heats for air at room temperature. Determine the net work output and the thermal efficiency for all combinations of the following parameters:

Pressure ratio: 6, 10, 15
Maximum cycle temperature: 800, 1200, and 1600 K
Compressor adiabatic efficiency: 60, 80, and 100 percent
Turbine adiabatic efficiency: 60, 80, and 100 percent

Draw conclusions from the results.

8-86 Repeat Prob. 8-85, using helium as the working fluid.

8-87 Repeat Prob. 8-85 by considering the variation of specific heats of air with temperature. Use the specific-heat expressions for air given in Table A-2*c*.

8-88 Repeat Prob. 8-85 by considering the variation of specific heats of air with temperature. Use the curve-fitted h and P_r data from Table A-17, each expressed as a fifth-degree polynomial in T for the temperature range of 300 to 1600 K.

Brayton Cycle with Regeneration

8-89C How does regeneration affect the efficiency of a Brayton cycle, and how does it accomplish it?

8-90C Somebody claims that at very high pressure ratios, the use of regeneration actually decreases the thermal efficiency of a gas-turbine engine. Is there any truth in this claim? Explain.

8-91C Define the effectiveness of a regenerator used in gas-turbine cycles.

8-92C In an ideal regenerator, is the air leaving the compressor heated to the temperature at (*a*) turbine inlet, (*b*) turbine exit, (*c*) slightly above turbine exit?

8-93 An ideal Brayton cycle with regeneration has a pressure ratio of 10. The air enters the compressor at 300 K and the turbine at 1200 K. If the effectiveness of the regenerator is 100 percent, determine the net work output and the thermal efficiency of the cycle.

8-93E An ideal Brayton cycle with regeneration has a pressure ratio of 10. The air enters the compressor at 520 R and the turbine at 2000 R. If the effectiveness of the regenerator is 100 percent, determine the net work output and the thermal efficiency.

8-94 A Brayton cycle with regeneration using air as the working fluid has a pressure ratio of 8. The minimum and maximum temperatures in the cycle are 310 and 1150 K. Assuming an adiabatic efficiency of 75 percent for the compressor and 82 percent for the turbine and an effectiveness of 65 percent for the regenerator, determine (*a*) the air temperature at the turbine exit, (*b*) the net work output, and (*c*) the thermal efficiency. Assume variable specific heats for air.
Answers: (*a*) 763.07 kJ/kg, (*b*) 101.64 kJ/kg, (*c*) 21.0 percent

8-95 Helium is used as the working fluid in a Brayton cycle with regeneration. The pressure ratio of the cycle is 8, the compressor inlet temperature is 300 K, and the turbine inlet temperature is 1800 K. The effectiveness of the regenerator is 75 percent. Determine the thermal efficiency and the required mass flow rate of air for a net power output of 10 MW, assuming both the compressor and the turbine have an isentropic efficiency of (*a*) 100 percent and (*b*) 80 percent.

8-96 A stationary gas-turbine power plant operates on an ideal regenerative Brayton cycle (ε = 100 percent) with air as the working fluid. Air enters the compressor at 95 kPa and 290 K and the turbine at 760 kPa and 1100 K. Heat is transferred to air from an external source at a rate of 25,000 kJ/s. Determine the power delivered by this plant, (*a*) assuming constant specific heats for air at room temperature and (*b*) accounting for the variation of specific heats with temperature.

8-97 Air enters the compressor of a regenerative gas-turbine engine at 300 K and 100 kPa, where it is compressed to 800 kPa and 580 K. The regenerator has an effectiveness of 65 percent, and the air enters the turbine at 1200 K. For a turbine efficiency of 86 percent, determine (*a*) the amount of heat transfer in the regenerator and (*b*) the thermal efficiency. Assume variable specific heats for air.
Answers: (*a*) 137.7 kJ/kg, (*b*) 35.0 percent

8-97E Air enters the compressor of a regenerative gas-turbine engine at 540 R and 14.5 psia, where it is compressed to 116 psia and 1080 R. The regenerator has an effectiveness of 65 percent, and the air enters the turbine at 2000 R. For a turbine efficiency of 86 percent, determine (*a*) the amount of heat transfer in the regenerator and (*b*) the thermal efficiency. Assume variable specific heats for air.
Answers: (*a*) 34.6 Btu/lbm, (*b*) 28.0 percent

8-98 Repeat Prob. 8-76, assuming that a regenerator of 75 percent effectiveness is added to the gas-turbine power plant.

8-99 Repeat Prob. 8-77, assuming that a regenerator of 70 percent effectiveness is added to the gas-turbine power plant.

8-100 Repeat Prob. 8-83, assuming that a regenerator of 65 percent effectiveness is added to the gas-turbine power plant.

8-101 Write a computer program to determine the effects of pressure ratio, maximum cycle temperature, regenerator effectiveness, and compressor and turbine efficiencies on the net work output per unit mass and on the thermal efficiency of a regenerative Brayton cycle. Assume the working fluid is air which is at 100 kPa and 300 K at the compressor inlet. Also assume constant specific heats for air at room temperature. Determine the net work output and the thermal efficiency for all combinations of the following parameters:

Pressure ratio: 6, 12, 15
Maximum cycle temperature: 1000, 1400, 1600 K
Compressor adiabatic efficiency: 60, 80, and 100 percent
Turbine adiabatic efficiency: 60, 80, 100 percent
Regenerator effectiveness: 50, 70, 80 percent

8-102 Repeat Prob. 8-101, using helium as the working fluid.

8-103 Repeat Prob. 8-101 by considering the variation of specific heats of air with temperature. Use the specific-heat expressions for air given in Table A-2*c*.

Brayton Cycle with Intercooling, Reheating, and Regeneration

8-104C Under what modifications will the ideal simple gas-turbine cycle approach the Ericsson cycle?

8-105C The single-stage compression process of an ideal Bryaton cycle without regeneration is replaced by a multistage compression process with intercooling between the same pressure limits. As a result of this modification,

(*a*) Does the compressor work increase, decrease, or remain the same?

(*b*) Does the back work ratio increase, decrease, or remain the same?

(*c*) Does the thermal efficiency increase, decrease or remain the same?

8-106C The single-stage expansion process of an ideal Brayton cycle without regeneration is replaced by a multistage expansion process with reheating between the same pressure limits. As a result of this modification,

(*a*) Does the turbine work increase, decrease, or remain the same?

(*b*) Does the back work ratio increase, decrease, or remain the same?

(*c*) Does the thermal efficiency increase, decrease, or remain the same?

8-107C A simple ideal Brayton cycle without regeneration is modified to incorporate multistage compression with intercooling and multistage expansion with reheating, without changing the pressure or temperature limits of the cycle. As a result of these two modifications,

(*a*) Does the net work output increase, decrease, or remain the same?

(*b*) Does the back work ratio increase, decrease, or remain the same?

(*c*) Does the thermal efficiency increase, decrease, or remain the same?

(*d*) Does the heat rejected increase, decrease, or remain the same?

8-108C A simple ideal Brayton cycle is modified to incorporate multistage compression with intercooling, multistage expansion with reheating, and regeneration without changing the pressure limits of the cycle. As a result of these modifications,

(*a*) Does the net work output increase, decrease, or remain the same?

(*b*) Does the back work ratio increase, decrease, or remain the same?

(*c*) Does the thermal efficiency increase, decrease, or remain the same?

(*d*) Does the heat rejected increase, decrease, or remain the same?

8-109C For a specified pressure ratio, why does multistage compression with intercooling decrease the compressor work, and multistage expansion with reheating increase the turbine work?

8-110C In an ideal gas-turbine cycle with intercooling, reheating, and regeneration, as the number of compression and expansion stages is increased, the cycle thermal efficiency approaches (*a*) 100 percent, (*b*) the Otto cycle efficiency, or (*c*) the Carnot cycle efficiency.

8-111 Consider an ideal gas-turbine cycle with two stages of compression and two stages of expansion. The pressure ratio across each stage of compressor and turbine is 3. The air enters each stage of the compressor at 300 K and each stage of the turbine at 1200 K. Determine the back work ratio and the thermal efficiency of the cycle, assuming (*a*) no regenerator is used and (*b*) a regenerator with 75 percent effectiveness is used.

8-111E Consider an ideal gas-turbine cycle with two stages of compression and two stages of expansion. The pressure ratio across each stage of compressor and turbine is 3. The air enters each stage of the compressor at 540 R and each stage of the turbine at 2000 R. Determine the back work ratio and the thermal efficiency of the cycle, assuming (*a*) no regenerator is used and (*b*) a regenerator with 75 percent effectiveness is used.

8-112 Repeat Prob. 8-111, assuming an efficiency of 80 percent for each compressor stage and an efficiency of 85 percent for each turbine stage.

8-113 A gas-turbine engine with regeneration operates with two stages of compression and two stages of expansion. The pressure ratio across each stage of compressor and turbine is 3.5. The air enters each stage of the compressor at 300 K and each stage of the turbine at 1200 K. The compressor and turbine efficiencies are 78 and 86 percent, respectively, and the effectiveness of the regenerator is 72 percent. Determine the back work ratio and the thermal efficiency of the cycle, assuming constant specific heats for air at room temperature.
Answers: 53.2 percent, 39.2 percent

8-114 Repeat Prob. 8-113, using helium as the working fluid.

8-115 Consider a regenerative gas-turbine power plant with two stages of compression and two stages of expansion. The overall pressure ratio of the cycle is 9. The air enters each stage of the compressor at 300 K and each stage of the turbine at 1200 K. Accounting for the variation of specific heats with temperature, determine the minimum mass flow rate of air needed to develop a net power output of 50 MW.
Answer: 113.4 kg/s

8-115E Consider a regenerative gas-turbine power plant with two stages of compression and two stages of expansion. The overall pressure ratio of the cycle is 9. The air enters each stage of the compressor at 540 R and each stage of the turbine at 2200 R. Accounting for the variation of

specific heats with temperature, determine the minimum mass flow rate of air needed to develop a net power output of 50 MW.
Answer: 243 lbm/s

8-116 Repeat Prob. 8-115, using argon as the working fluid.

8-117 Write a computer program to determine the effect of the number of compression and expansion stages on the thermal efficiency of an ideal regenerative Brayton cycle with multistage compression and expansion. Assume that the overall pressure ratio of the cycle is 12, and the air enters each stage of the compressor at 300 K and each stage of the turbine at 1200 K. Using constant specific heats for air at room temperature, determine the thermal efficiency of the cycle by varying the number of stages from 1 to 20. Plot the thermal efficiency versus the number of stages. Compare your results to the efficiency of an Ericsson cycle operating between the same temperature limits.

8-118 Repeat Prob. 8-117, using helium as the working fluid.

Jet-Propulsion Cycles

8-119C How does the ideal jet-propulsion cycle differ from the ideal Brayton cycle?

8-120C What is the function of the nozzle in turbojet engines?

8-121C What is propulsive power? How is it related to thrust?

8-122C What is propulsive efficiency? How is it determined?

8-123C Is the effect of turbine and compressor irreversibilities of a turbojet engine to reduce (*a*) the net work, (*b*) the thrust, or (*c*) the fuel consumption rate?

8-124 A turbojet aircraft is flying with a velocity 280 m/s at an altitude of 6100 m, where the ambient conditions are 48 kPa and −13°C. The pressure ratio across the compressor is 13, and the temperature at the turbine inlet is 1300 K. Assuming ideal operation for all components and constant specific heats for air at room temperature, determine (*a*) the pressure at the turbine exit, (*b*) the velocity of the exhaust gases, and (*c*) the propulsive efficiency.
Answers: (*a*) 374.3 kPa, (*b*) 933.6 m/s, (*c*) 26.9 percent

8-124E A turbojet aircraft is flying with a velocity 900 ft/s at an altitude of 20,000 ft, where the ambient conditions are 7 psia and 10°F. The pressure ratio across the compressor is 13, and the temperature at the turbine inlet is 2400 R. Assuming ideal operation for all components and constant specific heats for air at room temperature, determine (*a*) the pressure at the turbine exit, (*b*) the velocity of the exhaust gases, and (*c*) the propulsive efficiency.

8-125 Repeat Prob. 8-124, accounting for the variation of specific heats with temperature.

8-126 A turbojet aircraft is flying with a velocity 320 m/s at an altitude of 9150 m, where the ambient conditions are 32 kPa and −32°C. The pressure ratio across the compressor is 12, and the temperature at the turbine inlet is 1400 K. Air enters the compressor at a rate of 40 kg/s, and the jet fuel has a heating value of 42,700 kJ/kg. Assuming ideal operation for all components and constant specific heats for air at room temperature, determine (*a*) the velocity of the exhaust gases, (*b*) the propulsive power developed, and (*c*) the rate of fuel consumption.

8-126E A turbojet aircraft is flying with a velocity 950 ft/s at an altitude of 30,000 ft, where the ambient conditions are 5 psia and −26°F. The pressure ratio across the compressor is 12, and the temperature of the gases at the turbine inlet is 2300 R. Air enters the compressor at a rate of 75 lbm/s, and the jet fuel has a heating value of 18,400 Btu/lbm. Assuming ideal operation for all components and constant specific heats for air at room temperature, determine (*a*) the velocity of the exhaust gases, (*b*) the propulsive power developed, and (*c*) the rate of fuel consumption.

8-127 Repeat Prob. 8-126, using a compressor efficiency of 80 percent and a turbine efficiency of 85 percent.

8-128 Consider an aircraft powered by a turbojet engine that has a compression ratio of 12. The aircraft is stationary on the ground, held in position by its brakes. The ambient air is at 27°C and 95 kPa and enters the engine at a rate of 5 kg/s. The jet fuel has a heating value of 42,700 kJ/kg, and it is burned at a rate of 0.1 kg/s. Neglecting the effect of the diffuser and disregarding the slight increase in mass at the engine exit as well as the inefficiencies of engine components, determine the force that must be applied on the brakes to hold the plane stationary.
Answer: 4544 N

8-129 Air at 7°C enters a turbojet engine at a rate of 40 kg/s and at a velocity of 300 m/s (relative to the engine). Air is heated in the combustion chamber at a rate of 40,000 kJ/s and it leaves the engine at 427°C. Determine the thrust produced by this turbojet engine. (*Hint:* Choose the entire engine as your control volume.)

Second-Law Analysis of Gas Power Cycles

8-130 Determine the total irreversibility associated with the Otto cycle described in Prob. 8-35, assuming a source temperature of 2000 K and a sink temperature of 300 K. Also determine the availability at the end of the power stroke. *Answers:* 245.12 kJ/kg, 145.2 kJ/kg

8-130E Determine the irreversibility associated with the Otto cycle described in Prob. 8-35E, assuming a source temperature of 3200 R and a sink temperature of 540 R. Also determine the availability at the end of the power stroke. *Answers:* 142.1 Btu/lbm, 104.1 Btu/lbm

8-131 Determine the irreversibility associated with each of the processes of the Otto cycle described in Prob. 8-38, assuming a source temperature of 1800 K and a sink temperature of 310 K. Also determine the availability at the end of the isentropic expansion process.

8-132 Determine the total irreversibility associated with the Diesel cycle described in Prob. 8-48, assuming a source temperature of 2000 K and a sink temperature of 300 K. Also determine the availability at the end of the isentropic compression process.
Answers: 292.7 kJ/kg, 348.6 kJ/kg

8-133 Determine the irreversibility associated with the heat rejection process of the Diesel cycle described in Prob. 8-51, assuming a source temperature of 2000 K and a sink temperature of 300 K. Also determine the availability at the end of the isentropic expansion process.
Answers: 284.6 kJ/kg, 284.6 kJ/kg

8-133E Determine the irreversibility associated with the heat rejection process of the Diesel cycle described in Prob. 8-51E, assuming a source temperature of 3500 R and a sink temperature of 540 R. Also determine the availability at the end of the isentropic expansion process.

8-134 Determine the total irreversibility associated with the Brayton cycle described in Prob. 8-76, assuming a source temperature of 1400 K and a sink temperature of 300 K. Also find the availability of the working fluid at the turbine exit. Take $P_{exit} = P_0 = 100$ kPa.

8-135 Calculate the irreversibility associated with each of the processes of the Brayton cycle described in Prob. 8-77, assuming a source temperature of 1400 K and a sink temperature of 310 K.

8-136 Determine the total irreversibility associated with the Brayton cycle described in Prob. 8-93, assuming a source temperature of 1400 K and a sink temperature of 300 K. Also determine the availability of the exhaust gases at the exit of the regenerator.

8-136E Determine the total irreversibility associated with the Brayton cycle described in Prob. 8-93E, assuming a source temperature of 3400 R and a sink temperature of 520 R. Also determine the availability of the exhaust gases at the exit of the regenerator.

8-137 Determine the irreversibility associated with each of the processes of the Brayton cycle described in Prob. 8-94, assuming a source temperature of 1260 K and a sink temperature of 310 K. Also determine the availability of the exhaust gases at the exit of the regenerator. Take $P_{exhaust} = P_0 = 100$ kPa.

Vapor and Combined Power Cycles

CHAPTER 9

In Chap. 8 we discussed gas power cycles for which the working fluid remains a gas throughout the entire cycle. In this chapter, we consider *vapor power cycles* for which the working fluid is alternately vaporized and condensed. We also consider power generation coupled with process heating called *cogeneration*.

The continued quest for higher thermal efficiencies has resulted in some innovative modifications to the basic vapor power cycle. Among these, we discuss the *reheat* and *regenerative cycles* as well as power cycles that consist of two separate cycles known as *binary cycles* and *combined cycles* where the heat rejected by one fluid is used as the heat input to another fluid operating at a lower temperature.

Steam is the most common working fluid used in vapor power cycles because of its many desirable characteristics, such as low cost, availability, and high enthalpy of vaporization. Other working fluids used include sodium, potassium, and mercury for high-temperature applications and some organic fluids such as benzene and the freons for low-temperature applications. The majority of this chapter is devoted to the discussion of steam power plants, which produce most of the electric power in the world today.

Steam power plants are commonly referred to as *coal plants*, *nuclear plants*, *geothermal* or *natural gas plants*, depending on the type of fuel used to supply heat to the steam. But the steam goes through the same basic cycle in all of them. Therefore, all can be analyzed in the same manner.

9-1 ■ THE CARNOT VAPOR CYCLE

We have mentioned many times that the Carnot cycle is the most efficient cycle operating between two specified temperature levels. Thus it is natural to look at the Carnot cycle first as a prospective ideal cycle for vapor power plants. If we could, we would certainly adopt it as the ideal cycle. But as explained below, the Carnot cycle is an unsuitable model for vapor power cycles. Throughout the discussions, we assume steam to be the working fluid since it is the working fluid predominantly used in vapor power cycles.

Consider a steady-flow Carnot cycle executed within the saturation dome of a pure substance such as water, as shown in Fig. 9-1*a*. Water is heated reversibly and isothermally in a boiler (process 1-2), expanded isentropically in a turbine (process 2-3), condensed reversibly and isothermally in a condenser (process 3-4), and compressed isentropically by a compressor to the initial state (process 4-1).

Several impracticalities are associated with this cycle:

1 Isothermal heat transfer to or from a two-phase system is not difficult to achieve in practice since maintaining a constant pressure in the device will automatically fix the temperature at the saturation value. Therefore, processes 1-2 and 3-4 can be approached closely in actual boilers and condensers. Limiting the heat transfer processes to two-phase systems, however, severely limits the maximum temperature that can be used in the cycle (it has to remain under the critical-point value, which is 374°C for water). Limiting the maximum temperature in the cycle also limits the thermal efficiency. Any attempt to raise the maximum temperature in the cycle will involve heat transfer to the working fluid in a single phase, which is not easy to accomplish isothermally.

2 The isentropic expansion process (process 2-3) can be approximated closely by a well-designed turbine. However, the quality of the steam decreases during this process, as shown on the *T-s* diagram in Fig. 9-1*a*. Thus the turbine will have to handle steam with low quality, i.e., steam with a high moisture content. The impingement of liquid droplets

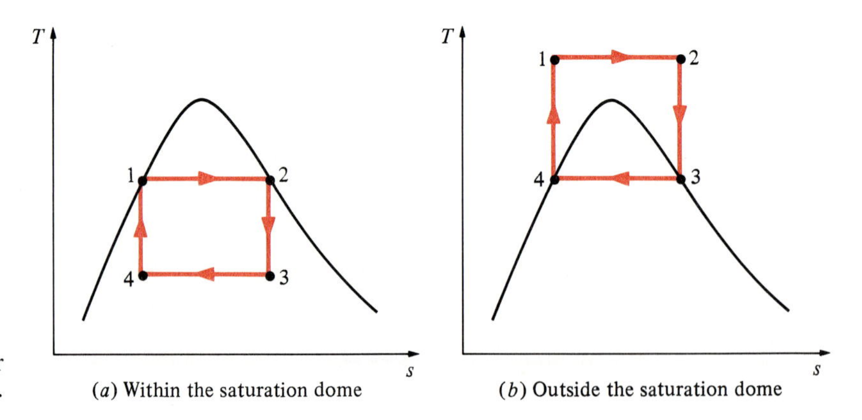

FIGURE 9-1

T-s diagram of two Carnot vapor cycles.

on the turbine blades causes erosion and is a major source of wear. Thus steam with qualities less than about 90 percent cannot be tolerated in the operation of power plants. This problem could be eliminated by using a working fluid with a very steep saturated vapor line.

3 The isentropic compression process (process 4-1) involves the compression of a liquid-vapor mixture to a saturated liquid. There are two difficulties associated with this process. First, it is not easy to control the condensation process so precisely as to end up with the desired quality at state 4. Second, it is not practical to design a compressor that will handle two phases.

Some of these problems could be eliminated by executing the Carnot cycle in a different way, as shown in Fig. 9-1*b*. This cycle, however, presents other problems such as isentropic compression to extremely high pressures and isothermal heat transfer at variable pressures. Thus we conclude that the Carnot cycle cannot be approximated in actual devices and is not a realistic model for vapor power cycles.

9-2 ■ RANKINE CYCLE— The Ideal Cycle for Vapor Power Cycles

Many of the impracticalities associated with the Carnot cycle can be eliminated by superheating the steam in the boiler and condensing it completely in the condenser, as shown schematically on a *T-s* diagram in Fig. 9-2. The cycle that results is the **Rankine cycle**, which is the ideal cycle for vapor power plants. The ideal Rankine cycle does not involve any internal irreversibilities and consists of the following four processes:

1-2 Isentropic compression in a pump

2-3 P = constant heat addition in a boiler

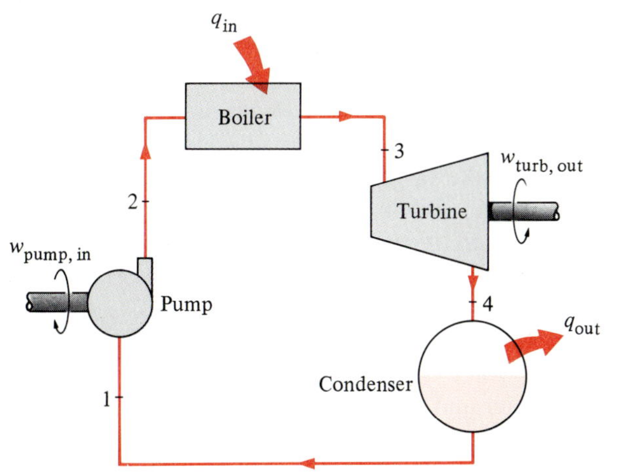

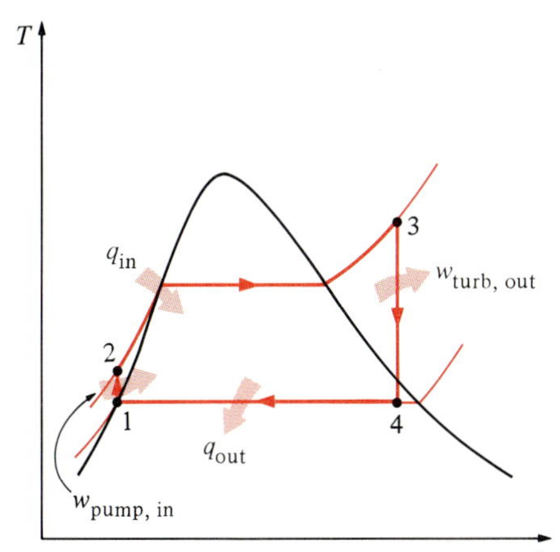

FIGURE 9-2
The simple ideal Rankine cycle.

3-4 Isentropic expansion in a turbine

4-1 P = constant heat rejection in a condenser

Water enters the *pump* at state 1 as saturated liquid and is compressed isentropically to the operating pressure of the boiler. The water temperature increases somewhat during this isentropic compression process due to a slight decrease in the specific volume of the water. The vertical distance between states 1 and 2 on the T-s diagram is greatly exaggerated for clarity. (If water were truly incompressible, would the temperature change at all during this process?)

Water enters the *boiler* as a compressed liquid at state 2 and leaves as a superheated vapor at state 3. The boiler is basically a large heat exchanger where the heat originating from combustion gases, nuclear reactors, or other sources is transferred to the water essentially at constant pressure. The boiler, together with the section where the steam is superheated (the superheater), is often called the *steam generator*.

The superheated vapor at state 3 enters the *turbine* where it expands isentropically and produces work by rotating the shaft connected to an electric generator. The pressure and the temperature of the steam drop during this process to the values at state 4, where steam enters the *condenser*. At this state, steam is usually a saturated liquid–vapor mixture with a high quality. Steam is condensed at constant pressure in the condenser, which is basically a large heat exchanger, by rejecting heat to a cooling medium such as a lake, a river, or the atmosphere. Steam leaves the condenser as saturated liquid and enters the pump, completing the cycle.

Remembering that the area under the process curve on a T-s diagram represents the heat transfer for internally reversible processes, we see that the area under process curve 2-3 represents the heat transferred to the water in the boiler and the area under the process curve 4-1 represents the heat rejected in the condenser. The difference between these two (the area enclosed by the cycle) is the net work produced during the cycle.

Energy Analysis of the Ideal Rankine Cycle

All four components associated with the Rankine cycle (the pump, boiler, turbine, and condenser) are steady-flow devices, and thus all four processes that make up the Rankine cycle can be analyzed as steady-flow processes. The kinetic and potential energy changes of the steam are usually small relative to the work and heat transfer terms and are therefore usually neglected. Then the *steady-flow energy equation* per unit mass of steam reduces to (Eq. 4-26)

$$q - w = h_e - h_i \qquad \text{(kJ/kg)} \tag{9-1}$$

The boiler and the condenser do not involve any work, and the pump and the turbine are assumed to be isentropic. Then the conservation of energy relation for each device can be expressed as follows:

Pump ($q = 0$): $$w_{\text{pump,in}} = h_2 - h_1 \tag{9-2}$$

or, from Eq. 6-53, $$w_{\text{pump,in}} = v(P_2 - P_1) \tag{9-3}$$

where $$h_1 = h_{f@P_1} \quad \text{and} \quad v \cong v_1 = v_{f@P_1} \tag{9-4}$$

Boiler ($w = 0$): $$q_{\text{in}} = h_3 - h_2 \tag{9-5}$$

Turbine ($q = 0$): $$w_{\text{turb,out}} = h_3 - h_4 \tag{9-6}$$

Condenser ($w = 0$): $$q_{\text{out}} = h_4 - h_1 \tag{9-7}$$

The *thermal efficiency* of the Rankine cycle is determined from

$$\eta_{\text{th}} = \frac{w_{\text{net}}}{q_{\text{in}}} = 1 - \frac{q_{\text{out}}}{q_{\text{in}}} \tag{9-8}$$

where $$w_{\text{net}} = q_{\text{in}} - q_{\text{out}} = w_{\text{turb,out}} - w_{\text{pump,in}} \tag{9-9}$$

The thermal efficiency can also be interpreted as the ratio of the area enclosed by the cycle on a *T-s* diagram to the area under the heat addition process. The use of these relations is illustrated in the following example.

EXAMPLE 9-1

Consider a steam power plant operating on the simple ideal Rankine cycle. The steam enters the turbine at 3 MPa and 350°C and is condensed in the condenser at a pressure of 75 kPa. Determine the thermal efficiency of this cycle.

Solution The schematic of the power plant and the *T-s* diagram of the cycle are shown in Fig. 9-3. Since the power plant operates on the ideal Rankine cycle, it is assumed that the turbine and the pump are isentropic, there are no pressure drops in the boiler and the condenser, and steam leaves the condenser and enters the pump as saturated liquid at the condenser pressure.

First, we determine the enthalpies at various points in the cycle, using data from steam tables (Tables A-4, A-5, and A-6):

State 1: $P_1 = 75$ kPa, Sat. liquid $\Big\}$ $h_1 = h_{f@75\text{ kPa}} = 384.39$ kJ/kg, $v_1 = v_{f@75\text{ kPa}} = 0.001037\ \text{m}^3/\text{kg}$

State 2: $P_2 = 3$ MPa $(s_2 = s_1)$

$$w_{\text{pump,in}} = v_1(P_2 - P_1) = (0.001037\ \text{m}^3/\text{kg})[(3000 - 75)\ \text{kPa}]\left(\frac{1\ \text{kJ}}{1\ \text{kPa}\cdot\text{m}^3}\right)$$

$$= 3.03\ \text{kJ/kg}$$

$$h_2 = h_1 + w_{\text{pump,in}} = (384.39 + 3.03)\ \text{kJ/kg} = 387.42\ \text{kJ/kg}$$

State 3: $P_3 = 3$ MPa, $T_3 = 350°\text{C}$ $\Big\}$ $h_3 = 3115.3$ kJ/kg, $s_3 = 6.7428$ kJ/(kg·K)

State 4: $P_4 = 75$ kPa, $s_4 = s_3$ (sat. mixture)

$$x_4 = \frac{s_4 - s_f}{s_{fg}} = \frac{6.7428 - 1.213}{6.2434} = 0.886$$

$$h_4 = h_f + x_4 h_{fg} = 384.39 + 0.886(2278.6) = 2403.2\ \text{kJ/kg}$$

Thus,
$$q_{\text{in}} = h_3 - h_2 = (3115.3 - 387.42)\ \text{kJ/kg} = 2727.88\ \text{kJ/kg}$$
$$q_{\text{out}} = h_4 - h_1 = (2403.2 - 384.39)\ \text{kJ/kg} = 2018.81\ \text{kJ/kg}$$

and
$$\eta_{\text{th}} = 1 - \frac{q_{\text{out}}}{q_{\text{in}}} = 1 - \frac{2018.81\ \text{kJ/kg}}{2727.88\ \text{kJ/kg}} = 0.260 \quad \text{(or 26.0\%)}$$

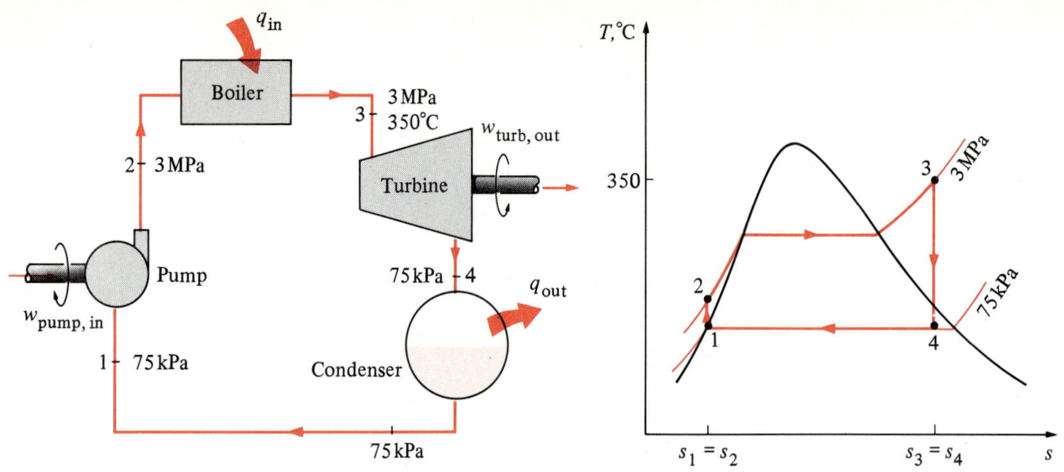

FIGURE 9-3
Schematic and *T-s* diagram for Example 9-1.

The thermal efficiency could also be determined from

$$w_{turb,out} = h_3 - h_4 = (3115.3 - 2403.2) \text{ kJ/kg} = 712.1 \text{ kJ/kg}$$

$$w_{net} = w_{turb,out} - w_{pump,in} = (712.1 - 3.03) \text{ kJ/kg} = 709.07 \text{ kJ/kg}$$

or

$$w_{net} = q_{in} - q_{out} = (2727.88 - 2018.81) \text{ kJ/kg} = 709.07 \text{ kJ/kg}$$

and

$$\eta_{th} = \frac{w_{net}}{q_{in}} = \frac{709.07 \text{ kJ/kg}}{2727.88 \text{ kJ/kg}} = 0.260 \quad \text{(or 26.0\%)}$$

That is, this power plant converts 26 percent of the heat it receives in the boiler to net work. An actual power plant operating between the same temperature and pressure limits will have a lower efficiency because of the irreversibilities such as friction.

Notice that the back work ratio ($r_{bw} = w_{in}/w_{out}$) of this power plant is 0.004. That is, only 0.4 percent of the turbine work output is required to operate the pump. Having low back work ratios (usually under 1 percent) is characteristic of vapor power cycles. This is in contrast to the gas power cycles which typically have very high back work ratios (40 to 80 percent).

It is also interesting to note the thermal efficiency of a Carnot cycle operating between the same temperature limits

$$\eta_{th,Carnot} = 1 - \frac{T_{min}}{T_{max}} = 1 - \frac{(91.78 + 273) \text{ K}}{(350 + 273) \text{ K}} = 0.414$$

The difference between the two efficiencies is due to the large temperature difference between the steam and the combustion gases during the heat addition process.

9-3 ■ DEVIATION OF ACTUAL VAPOR POWER CYCLES FROM IDEALIZED ONES

The actual vapor power cycle differs from the ideal Rankine cycle, as illustrated in Fig. 9-4*a*, as a result of irreversibilities in various

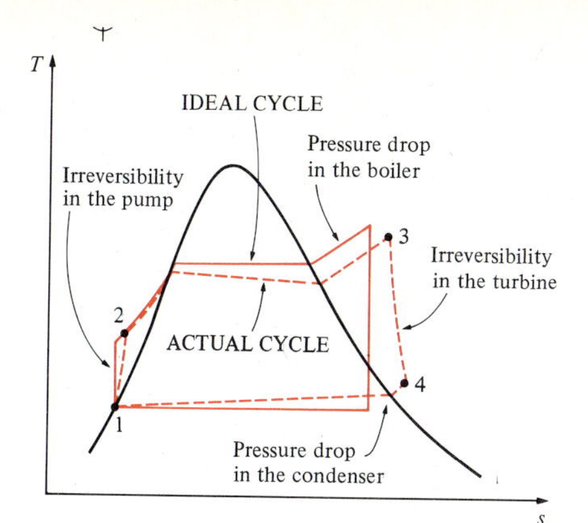

FIGURE 9-4a
Deviation of actual vapor power cycle from the ideal Rankine cycle.

components. Fluid friction and undesired heat loss to the surroundings are the two most common sources of irreversibilities.

Fluid friction causes pressure drops in the boiler, the condenser, and the piping between various components. As a result, steam leaves the boiler at a somewhat lower pressure. Also the pressure at the turbine inlet is somewhat lower than that at the boiler exit due to the pressure drop in the connecting pipes. The pressure drop in the condenser is usually very small. To compensate for these pressure drops, the water must be pumped to a sufficiently higher pressure than the ideal cycle calls for. This requires a larger pump and larger work input to the pump.

The other major source of irreversibility is the *heat loss* from the steam to the surroundings as the steam flows through various components. To maintain the same level of net work output, more heat needs to be transferred to the steam in the boiler to compensate for these undesired heat losses. As a result, cycle efficiency decreases.

Of particular importance are the irreversibilities occurring within the pump and the turbine. A pump requires a greater work input, and a turbine produces a smaller work output as a result of irreversibilities. Under ideal conditions, the flow through these devices is isentropic. The deviation of actual pumps and turbines from the isentropic ones can be accurately accounted for, however, by utilizing *adiabatic efficiencies,* defined as

$$\eta_P = \frac{w_s}{w_a} = \frac{h_{2s} - h_1}{h_{2a} - h_1} \tag{9-10}$$

and

$$\eta_T = \frac{w_a}{w_s} = \frac{h_3 - h_{4a}}{h_3 - h_{4s}} \tag{9-11}$$

where states $2a$ and $4a$ are the actual exit states of the pump and the turbine, respectively, and $2s$ and $4s$ are the corresponding states for the isentropic case (Fig. 9-4*b*).

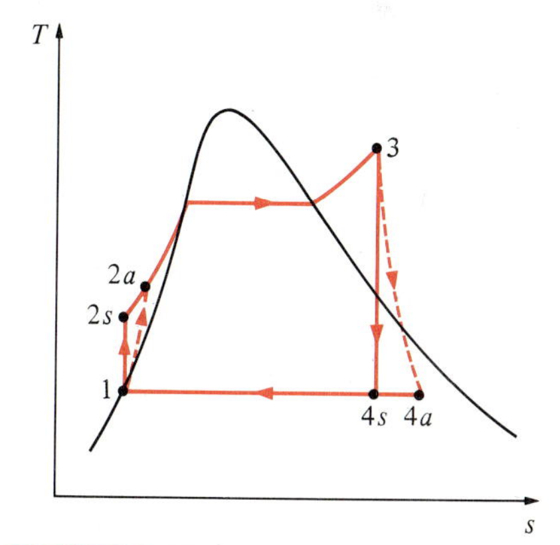

FIGURE 9-4b
The effect of pump and turbine irreversibilities on the ideal Rankine cycle.

Other factors also need to be considered in the analysis of actual vapor power cycles. In actual condensers, for example, the liquid is usually subcooled to prevent the onset of *cavitation,* the rapid vaporization and condensation of the fluid at the low-pressure side of the pump impeller, which may eventually destroy it. Additional losses occur at the bearings between the moving parts as a result of friction. Steam that leaks out during the cycle and air that leaks into the condenser represent two other sources of loss. Finally, the power consumed by the auxiliary equipment such as fans which supply air to the furnace should also be considered in evaluating the performance of actual power plants.

The effect of irreversibilities on the thermal efficiency of a steam power cycle is illustrated below with an example.

EXAMPLE 9-2

A steam power plant operates on the cycle shown in Fig. 9-5. If the adiabatic efficiency of the turbine is 87 percent and the adiabatic efficiency of the pump is 85 percent, determine (*a*) the thermal efficiency of the cycle and (*b*) the net power output of the plant for a mass flow rate of 15 kg/s.

Solution The temperatures and pressures of the steam at various points are indicated in Fig. 9-5. The *T-s* diagram of the cycle is also shown in this figure. All the components are treated as steady-flow devices, and any changes in kinetic and potential energies are assumed to be negligible. The required properties at the specified states are determined from the steam tables.

(*a*) The thermal efficiency of a cycle is the ratio of the net work output to the heat input, and it is determined as follows:

Pump work input:

$$w_{\text{pump,in}} = \frac{w_{s,\text{pump,in}}}{\eta_P} = \frac{v_1(P_2 - P_1)}{\eta_P}$$

$$= \frac{(0.001008\ \text{m}^3/\text{kg})[(16{,}000 - 9)\ \text{kPa}]}{0.85}\left(\frac{1\ \text{kJ}}{1\ \text{kPa}\cdot\text{m}^3}\right)$$

$$= 19.0\ \text{kJ/kg}$$

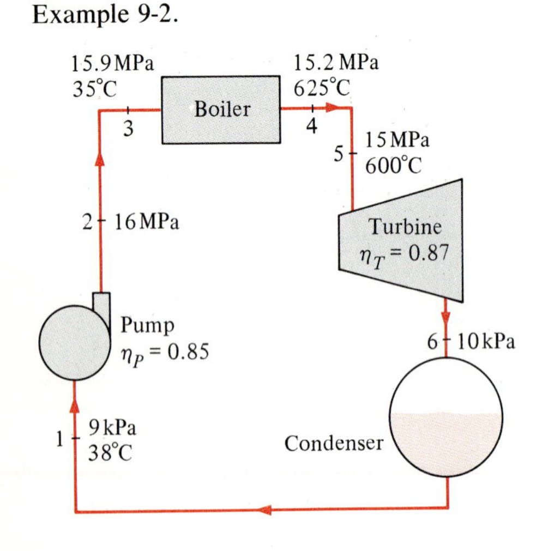

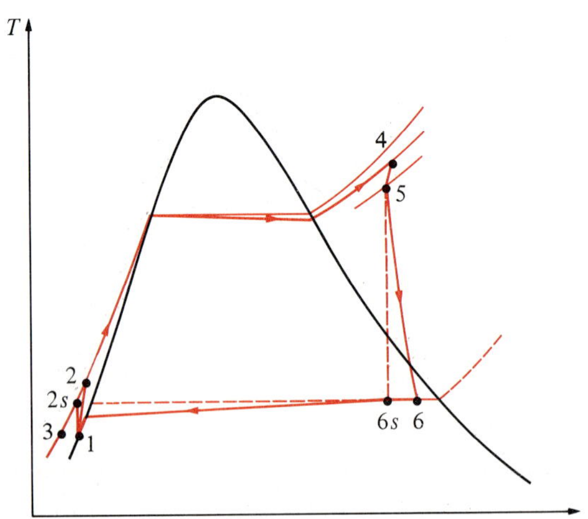

FIGURE 9-5

Schematic and *T-s* diagram for Example 9-2.

Turbine work output:

$$w_{\text{turb,out}} = \eta_T w_{s,\text{turb,out}} = \eta_T(h_5 - h_{6s}) = 0.87(3582.3 - 2115.7)\text{ kJ/kg} = 1275.9\text{ kJ/kg}$$

Boiler heat input: $q_{in} = h_4 - h_3 = (3645.7 - 146.7)\text{ kJ/kg} = 3499.0\text{ kJ/kg}$

Thus, $w_{net} = w_{\text{turb,out}} - w_{\text{pump,in}} = (1275.9 - 19.0)\text{ kJ/kg} = 1256.9\text{ kJ/kg}$

$$\eta_T = \frac{w_{net}}{q_{in}} = \frac{1256.9\text{ kJ/kg}}{3499.0\text{ kJ/kg}} = 0.359 \quad (\text{or } 35.9\%)$$

Without the irreversibilities, the thermal efficiency of this cycle would be 43.0 percent (see Example 9-3*c*).

(*b*) The power produced by this power plant is determined from

$$\dot{W}_{net} = \dot{m}(w_{net}) = (15\text{ kg/s})(1256.9\text{ kJ/kg}) = 18{,}854\text{ kW}$$

9-4 ■ HOW CAN WE INCREASE THE EFFICIENCY OF THE RANKINE CYCLE?

Steam power plants are responsible for the production of most of the electric power in the world, and even small increases in thermal efficiency can mean large savings from the fuel requirements. Therefore, every effort is made to improve the efficiency of the cycle on which steam power plants operate.

The basic idea behind all the modifications to increase the thermal efficiency of a power cycle is the same: *Increase the average temperature at which heat is transferred to the working fluid in the boiler, or decrease the average temperature at which heat is rejected from the working fluid in the condenser.* That is, the average fluid temperature should be as high as possible in the boiler and as low as possible in the condenser. Next we discuss three ways of accomplishing this for the simple ideal Rankine cycle.

1 Lowering the Condenser Pressure (*Lowers* $T_{\text{low,av}}$)

Steam exists as a saturated mixture in the condenser at the saturation temperature corresponding to the pressure inside. Therefore, lowering the operating pressure of the condenser automatically lowers the temperature of the steam and thus the temperature at which heat is rejected.

The effect of lowering the condenser pressure on the Rankine cycle efficiency is illustrated on a *T-s* diagram in Fig. 9-6. For comparison purposes, the turbine inlet state is maintained the same. The colored area on this diagram represents the increase in net work output as a result of lowering the condenser pressure from P_4 to P_4'. The heat input requirements also increase (represented by the area under curve 2′-2), but this increase is very small. Thus the overall effect of lowering the condenser pressure is an increase in the thermal efficiency of the cycle.

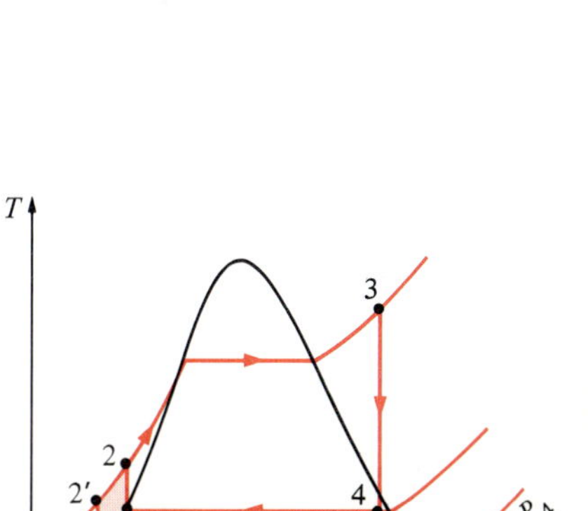

FIGURE 9-6
The effect of lowering the condenser pressure on the ideal Rankine cycle.

To take advantage of the increased efficiencies at low pressures, the condensers of steam power plants usually operate well below the atmospheric pressure. This does not present a major problem since the vapor power cycles operate in a closed loop. However, there is a lower limit on the condenser pressure that can be used. It cannot be lower than the saturation pressure corresponding to the temperature of the cooling medium. Consider, for example, a condenser which is to be cooled by a nearby river at 15°C. Allowing a temperature difference of 10°C for effective heat transfer, the steam temperature in the condenser must be above 25°C, thus the condenser pressure must be above 3.2 kPa, which is the saturation pressure at 25°C.

Lowering the condenser pressure is not without any side effects, however. For one thing, it creates the problem of air leakage into the condenser. More importantly, it increases the moisture content of the steam at the last stages of the turbine, as can be seen from Fig. 9-6. The presence of large quantities of moisture is highly undesirable in turbines because it decreases the turbine efficiency and erodes the turbine blades. Fortunately, this problem can be corrected, as discussed later in this chapter.

2 Superheating the Steam to High Temperatures (*Increases* $T_{\text{high,av}}$)

The average temperature at which heat is added to the steam can be increased without increasing the boiler pressure by superheating the steam to high temperatures. The effect of superheating on the performance of vapor power cycles is illustrated on a *T-s* diagram in Fig. 9-7. The colored area on this diagram represents the increase in the net work. The total area under the process curve 3-3′ represents the increase in the heat input. Thus both the net work and heat input increases as a result of superheating the steam to a higher temperature. The overall effect is an increase in thermal efficiency, however, since the average temperature at which heat is added increases.

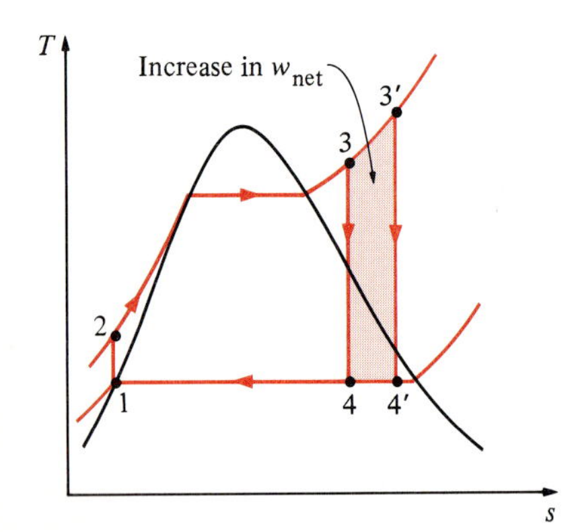

FIGURE 9-7

The effect of superheating the steam to higher temperatures on the ideal Rankine cycle.

Superheating the steam to higher temperatures has another very desirable effect: It decreases the moisture content of the steam at the turbine exit, as can be seen from the *T-s* diagram (the quality at state 4′ is higher than that at state 4).

The temperature to which steam can be superheated is limited, however, by metallurgical considerations. Presently the highest steam temperature allowed at the turbine inlet is about 620°C (1150°F). Any increase in this value depends on improving the present materials or finding new ones that can withstand higher temperatures. Ceramics are very promising in this regard.

3 Increasing the Boiler Pressure (*Increases* $T_{\text{high,av}}$)

Another way of increasing the average temperature during the heat addition process is to increase the operating pressure of the boiler, which automatically raises the temperature at which boiling takes place. This,

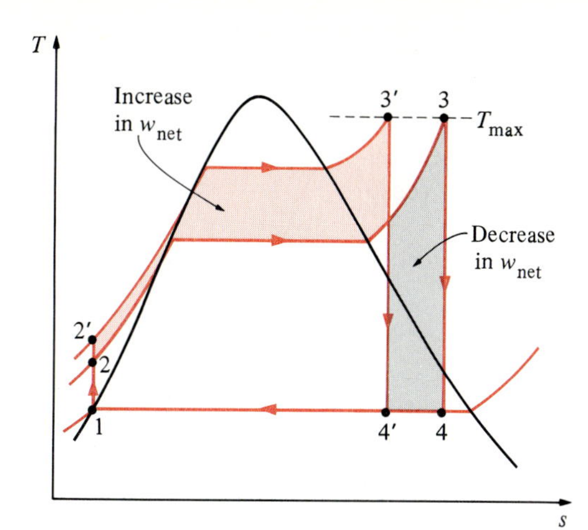

FIGURE 9-8
The effect of increasing the boiler pressure on the ideal Rankine cycle.

in turn, raises the average temperature at which heat is added to the steam and thus raises the thermal efficiency of the cycle.

The effect of increasing the boiler pressure on the performance of vapor power cycles is illustrated on a *T-s* diagram in Fig. 9-8. Notice that for a fixed turbine inlet temperature, the cycle shifts to the left and the moisture content of steam at the turbine exit increases. This undesirable side effect can be corrected, however, by reheating the steam, as discussed in the next section.

Operating pressures of boilers have gradually increased over the years from about 2.7 MPa (400 psia) in 1922 to over 30 MPa (4500 psia) today, generating enough steam to produce a net power output of 1000 MW or more. Today many modern steam power plants operate at supercritical pressures ($P > 22.09$ MPa) and have thermal efficiencies of about 40 percent for fossil-fuel plants and 34 percent for nuclear plants. The lower efficiencies of nuclear power plants is due to the lower maximum temperatures used in those plants for safety reasons. The *T-s* diagram of a supercritical Rankine cycle is shown in Fig. 9-9.

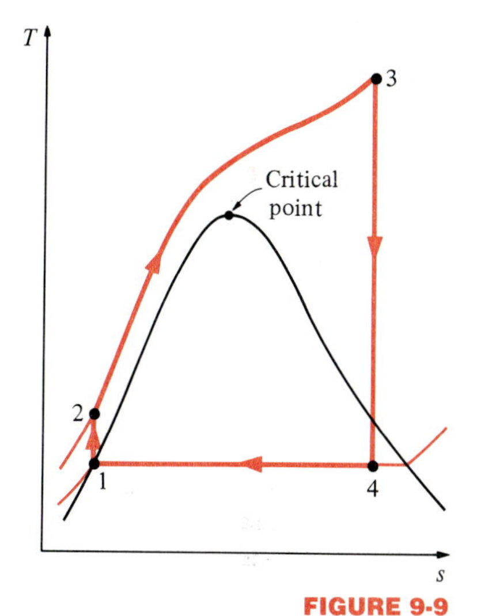

FIGURE 9-9
A supercritical Rankine cycle.

The effects of lowering the condenser pressure, superheating to a higher temperature, and increasing the boiler pressure on the thermal efficiency of the Rankine cycle are illustrated below with an example.

EXAMPLE 9-3

Consider a steam power plant operating on the ideal Rankine cycle. The steam enters the turbine at 3 MPa and 350°C and is condensed in the condenser at a pressure of 10 kPa. Determine (*a*) the thermal efficiency of this power plant, (*b*) the thermal efficiency if steam is superheated to 600°C instead of 350°C, (*c*) the thermal efficiency if the boiler pressure is raised to 15 MPa while the turbine inlet temperature is maintained at 600°C.

Solution The *T-s* diagrams of the cycle for all three cases are given in Fig. 9-10.

(*a*) This is the steam power plant discussed in Example 9-1, except that the

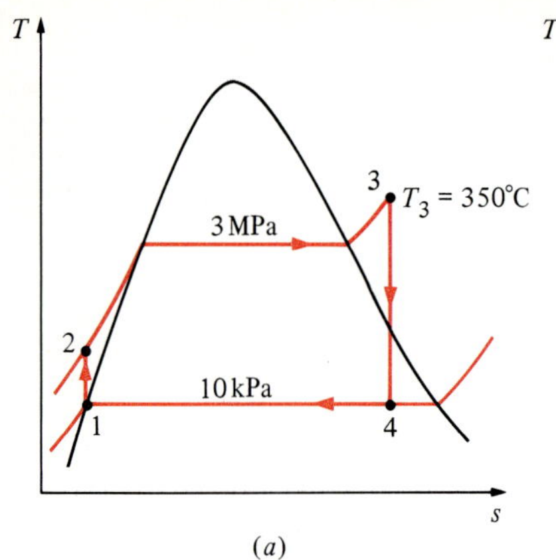

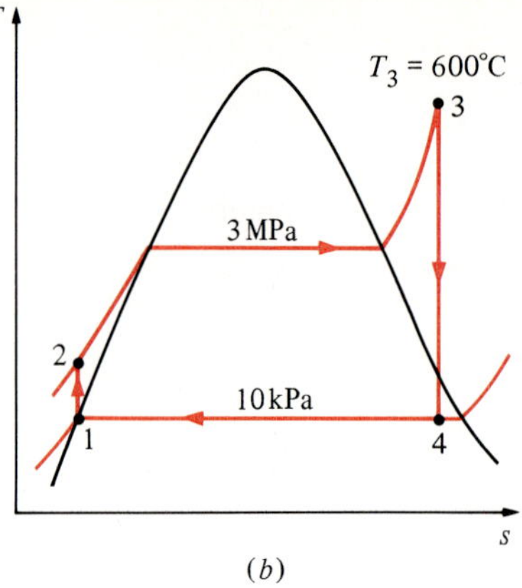

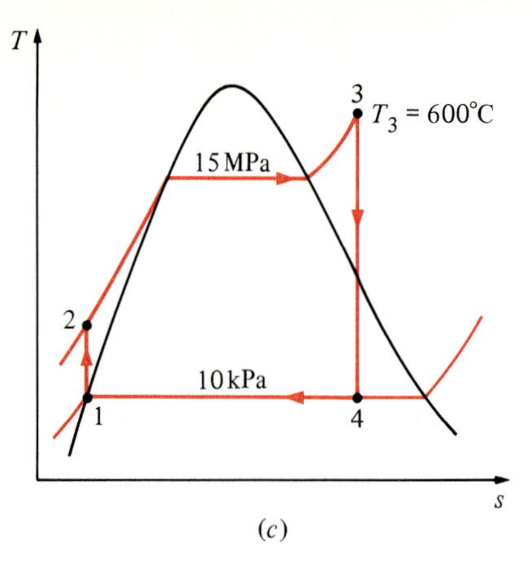

FIGURE 9-10

T-s diagrams of the three cycles discussed in Example 9-3.

condenser pressure is lowered to 10 kPa. The thermal efficiency is determined in a similar manner:

State 1: $\left.\begin{array}{l} P_1 = 10 \text{ kPa} \\ \text{Sat. liquid} \end{array}\right\}$ $\begin{array}{l} h_1 = h_{f@10\text{ kPa}} = 191.83 \text{ kJ/kg} \\ v_1 = v_{f@10\text{ kPa}} = 0.001008 \text{ m}^3\text{/kg} \end{array}$

State 2: $P_2 = 3$ MPa
$(s_2 = s_1)$

$$w_{\text{pump,in}} = v_1(P_2 - P_1) = (0.001008 \text{ m}^3\text{/kg})[(3000 - 10) \text{ kPa}]\left(\frac{1 \text{ kJ}}{1 \text{ kPa}\cdot\text{m}^3}\right)$$
$$= 3.01 \text{ kJ/kg}$$
$$h_2 = h_1 + w_{\text{pump,in}} = (191.83 + 3.01) \text{ kJ/kg} = 194.84 \text{ kJ/kg}$$

State 3: $\left.\begin{array}{l} P_3 = 3 \text{ MPa} \\ T_3 = 350°\text{C} \end{array}\right\}$ $\begin{array}{l} h_3 = 3115.3 \text{ kJ/kg} \\ s_3 = 6.7428 \text{ kJ/(kg}\cdot\text{K)} \end{array}$

State 4: $\begin{array}{l} P_4 = 10 \text{ kPa} \\ s_4 = s_3 \end{array}$ (sat. mixture)

$$x_4 = \frac{s_4 - s_f}{s_{fg}} = \frac{6.7428 - 0.6493}{7.5009} = 0.812$$
$$h_4 = h_f + x_4 h_{fg} = 191.83 + 0.812(2392.8) = 2134.8 \text{ kJ/kg}$$

Thus, $q_{\text{in}} = h_3 - h_2 = (3115.3 - 194.84) \text{ kJ/kg} = 2920.46 \text{ kJ/kg}$
$q_{\text{out}} = h_4 - h_1 = (2134.8 - 191.83) \text{ kJ/kg} = 1942.97 \text{ kJ/kg}$

and $$\eta_{\text{th}} = 1 - \frac{q_{\text{out}}}{q_{\text{in}}} = 1 - \frac{1942.97 \text{ kJ/kg}}{2920.46 \text{ kJ/kg}} = 0.335 \quad \text{(or 33.5\%)}$$

Therefore, the thermal efficiency increases from 26.0 to 33.5 percent as a result of lowering the condenser pressure from 75 to 10 kPa. At the same time, however, the quality of the steam decreases from 0.886 to 0.812 (in other words, the moisture content increases from 11.4 to 18.8 percent).

(*b*) States 1 and 2 remain the same in this case, and the enthalpies at state 3 (3 MPa and 600°C) and state 4 (10 kPa and $s_4 = s_3$) are determined in a similar manner to be

$$h_3 = 3682.3 \text{ kJ/kg}$$
$$h_4 = 2378.8 \text{ kJ/kg} \qquad (x_4 = 0.914)$$

Thus, $q_{in} = h_3 - h_2 = (3682.3 - 194.84)$ kJ/kg $= 3487.46$ kJ/kg

$q_{out} = h_4 - h_1 = (2378.8 - 191.83)$ kJ/kg $= 2186.97$ kJ/kg

and $$\eta_{th} = 1 - \frac{q_{out}}{q_{in}} = 1 - \frac{2186.97 \text{ kJ/kg}}{3487.46 \text{ kJ/kg}} = 0.373 \quad \text{(or 37.3\%)}$$

Therefore, the thermal efficiency increases from 33.5 to 37.3 percent as a result of superheating the steam from 350 to 600°C. At the same time, the quality of the steam increases from 0.812 to 0.914 (the moisture content decreases from 18.8 to 8.6 percent).

(c) State 1 remains the same in this case, but the other states change. The enthalpies at state 2 (15 MPa and $s_2 = s_1$), state 3 (15 MPa and 600°C), and state 4 (10 kPa and $s_4 = s_3$) are determined in a similar manner to be

$$h_2 = 206.94 \text{ kJ/kg}$$
$$h_3 = 3582.3 \text{ kJ/kg}$$
$$h_4 = 2115.7 \text{ kJ/kg} \qquad (x_4 = 0.804)$$

Thus, $q_{in} = h_3 - h_2 = (3582.3 - 206.94)$ kJ/kg $= 3375.36$ kJ/kg

$q_{out} = h_4 - h_1 = (2115.7 - 191.83)$ kJ/kg $= 1923.87$ kJ/kg

and $$\eta_{th} = 1 - \frac{q_{out}}{q_{in}} = 1 - \frac{1923.87 \text{ kJ/kg}}{3375.36 \text{ kJ/kg}} = 0.430 \quad \text{(or 43.0\%)}$$

Therefore, the thermal efficiency increases from 37.3 to 43.0 percent as a result of raising the boiler pressure from 3 to 15 MPa while maintaining the steam temperature at the turbine inlet at 600°C. At the same time, the quality of the steam decreases from 0.914 to 0.804 (the moisture content increases from 8.6 to 19.6 percent).

9-5 ■ THE IDEAL REHEAT RANKINE CYCLE

We noted in the last section that increasing the boiler pressure increases the thermal efficiency of the Rankine cycle, but it also increases the moisture content of the steam to unacceptable levels. Then it is natural to ask the following question:

> *How can we take advantage of the increased efficiencies at higher boiler pressures without facing the problem of excessive moisture at the final stages of the turbine?*

Two possibilities come to mind:

1 Superheat the steam to very high temperatures before it enters the turbine. This would be the desirable solution since the average temperature at which heat is added would also increase, thus increasing the cycle efficiency. This is not a viable solution, however, since it will require raising the steam temperature to metallurgically unsafe levels.

2 Expand the steam in the turbine in two stages, and reheat it in between. In other words, modify the simple ideal Rankine cycle with a **reheat** process. Reheating is a practical solution to the excessive moisture

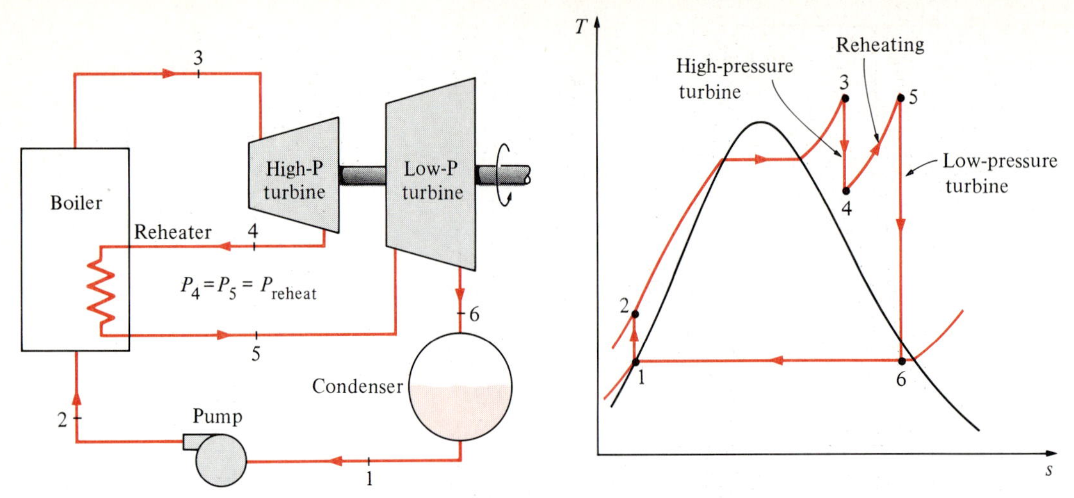

FIGURE 9-11

The ideal reheat Rankine cycle.

problem in turbines, and it is used frequently in modern steam power plants.

The *T-s* diagram of the ideal reheat Rankine cycle and the schematic of the power plant operating on this cycle are shown in Fig. 9-11. The ideal reheat Rankine cycle differs from the simple ideal Rankine cycle in that the expansion process takes place in two stages. In the first stage (the high-pressure turbine), steam is expanded isentropically to an intermediate pressure and sent back to the boiler where it is reheated at constant pressure, usually to the inlet temperature of the first turbine stage. Steam then expands isentropically in the second stage (low-pressure turbine) to the condenser pressure. Thus the total heat input and the total turbine work output for a reheat cycle become

$$q_{in} = q_{primary} + q_{reheat} = (h_3 - h_2) + (h_5 - h_4) \quad \text{(kJ/kg)} \tag{9-12}$$

and

$$w_{turb,out} = w_{turb,I} + w_{turb,II} = (h_3 - h_4) + (h_5 - h_6) \quad \text{(kJ/kg)} \tag{9-13}$$

The reheating process, in general, does not significantly change the average temperature at which heat is added. Therefore, the cycle efficiency is not influenced greatly by the reheating process. The efficiency of the Rankine cycle may increase or decrease somewhat as a result of reheating, depending on the average temperature at which heat is added during the reheat process. This temperature should be maintained as high as possible (without allowing excessive moisture) to prevent any decrease in the cycle efficiency. Also care must be taken to prevent the exit state of the low-pressure turbine from falling into the superheated vapor region, since this will cause the average temperature for heat rejection to increase and thus the cycle efficiency to decrease.

The average temperature during the reheat process can be increased by increasing the number of expansion and reheat stages. As the number of stages is increased, the expansion and reheat processes approach an

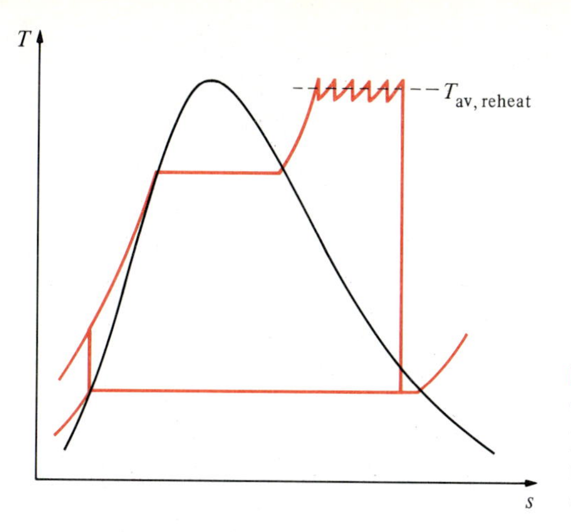

FIGURE 9-12

The average temperature at which heat is added during reheating increases as the number of reheat stages is increased.

isothermal process at the maximum temperature, as shown in Fig. 9-12. The optimum number is determined by economical considerations. The use of more than one or two reheat stages, in general, cannot be justified economically even in large power plants because the savings resulting from increased efficiency are more than offset by the increased cost.

Remember that the sole purpose of the reheat cycle is to reduce the moisture content of the steam at the final stages of the expansion process. If we had materials that could withstand sufficiently high temperatures, there would be no need for the reheat cycle.

EXAMPLE 9-4

Consider a steam power plant operating on the ideal reheat Rankine cycle. The steam enters the high-pressure turbine at 15 MPa and 600°C and is condensed in the condenser at a pressure of 10 kPa. If the moisture content of the steam at the exit of the low-pressure turbine is not to exceed 10.4 percent, determine (*a*) the pressure at which the steam should be reheated and (*b*) the thermal efficiency of the cycle. Assume the steam is reheated to the inlet temperature of the high-pressure turbine.

Solution The schematic of the power plant and the *T*-*s* diagram of the cycle are shown in Fig. 9-13. Since the power plant operates on the ideal reheat Rankine cycle, we assume that both stages of the turbine and the pump are isentropic, there are no pressure drops in the boiler and the condenser, and steam leaves the condenser and enters the pump as a saturated liquid at the condenser pressure.

(*a*) The reheat pressure is determined from the requirement that the entropies at states 5 and 6 be the same:

State 6:

$$P_6 = 10 \text{ kPa}$$

$$x_6 = 0.896 \text{ (sat. mixture)}$$

$$s_6 = s_f + x_6 s_{fg} = 0.6493 + 0.896(7.5009) = 7.370 \text{ kJ/(kg·K)}$$

Also,

$$h_6 = h_f + x_6 h_{fg} = 191.83 + 0.896(2392.8) = 2335.8 \text{ kJ/kg}$$

Thus,

State 5: $\left.\begin{aligned} T_5 &= 600°C \\ s_5 &= s_6 \end{aligned}\right\}$ $\quad P_5 = 4.0 \text{ MPa}$, $\quad h_5 = 3674.4 \text{ kJ/kg}$

Therefore, steam should be reheated at a pressure of 4 MPa or lower to prevent a moisture content above 10.4 percent.

(*b*) To determine the thermal efficiency, we need to know the enthalpies at all the states:

State 1: $\left.\begin{aligned} P_1 &= 10 \text{ kPa} \\ &\text{Sat. liquid} \end{aligned}\right\}$ $\quad h_1 = h_{f@10\text{ kPa}} = 191.83 \text{ kJ/kg}$, $\quad v_1 = v_{f@10\text{ kPa}} = 0.001010 \text{ m}^3\text{/kg}$

State 2: $P_2 = 15 \text{ MPa}$, $\quad s_2 = s_1$

$$w_{\text{pump,in}} = v_1(P_2 - P_1) = (0.001010 \text{ m}^3\text{/kg})[(15{,}000 - 10) \text{ kPa}]\left(\frac{1 \text{ kJ}}{1 \text{ kPa·m}^3}\right)$$
$$= 15.11 \text{ kJ/kg}$$
$$h_2 = h_1 + w_{\text{pump,in}} = (191.83 + 15.11) \text{ kJ/kg} = 206.94 \text{ kJ/kg}$$

State 3: $\left.\begin{aligned} P_3 &= 15 \text{ MPa} \\ T_3 &= 600°C \end{aligned}\right\}$ $\quad h_3 = 3582.3 \text{ kJ/kg}$, $\quad s_3 = 6.6776 \text{ kJ/(kg·K)}$

State 4: $\left.\begin{aligned} P_4 &= 4 \text{ MPa} \\ s_4 &= s_3 \end{aligned}\right\}$ $\quad h_4 = 3154.3 \text{ kJ/kg}$, $\quad (T_4 = 375.5°C)$

Thus,
$$\begin{aligned} q_{\text{in}} &= (h_3 - h_2) + (h_5 - h_4) \\ &= (3582.3 - 206.94) \text{ kJ/kg} + (3674.4 - 3154.3) \text{ kJ/kg} \\ &= 3895.46 \text{ kJ/kg} \\ q_{\text{out}} &= h_6 - h_1 = (2335.8 - 191.83) \text{ kJ/kg} \\ &= 2143.97 \text{ kJ/kg} \end{aligned}$$

and
$$\eta_{\text{th}} = 1 - \frac{q_{\text{out}}}{q_{\text{in}}} = 1 - \frac{2143.97 \text{ kJ/kg}}{3895.46 \text{ kJ/kg}} = 0.450 \quad \text{(or 45.0\%)}$$

This problem was worked out in Example 9-3c for the same pressure and temperature limits but without the reheat process. A comparison of the two results reveals that reheating reduces the moisture content from 19.6 to 10.4 percent while increasing the thermal efficiency from 43.0 to 45.0 percent.

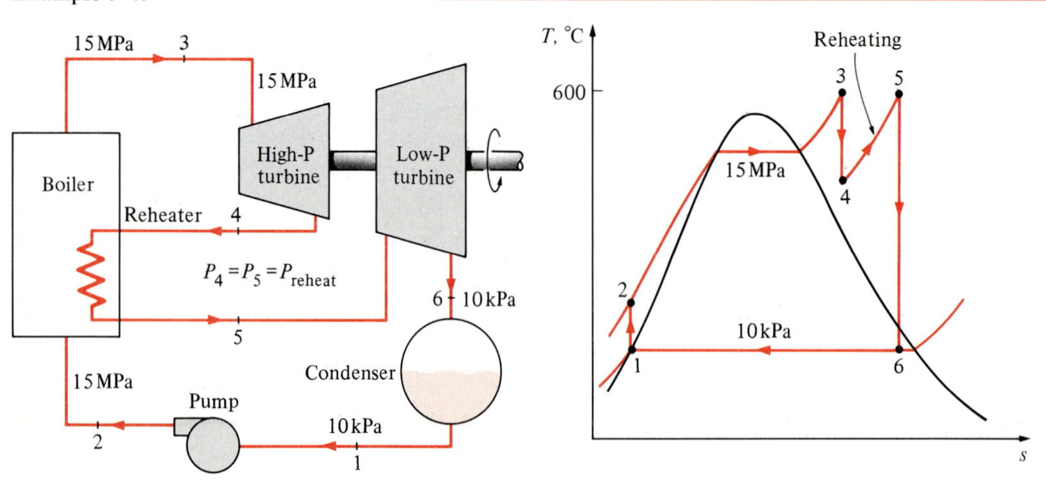

FIGURE 9-13
Schematic and *T-s* diagram for Example 9-4.

A careful examination of the *T-s* diagram of the Rankine cycle redrawn in Fig. 9-14 reveals that heat is added to the working fluid during process 2-2′ at a relatively low temperature. This lowers the average temperature at which heat is added and thus the cycle efficiency.

To overcome this shortcoming, we look for ways to raise the temperature of the liquid (called the *feedwater*) leaving the pump before it enters the boiler. One such possibility is to compress the feedwater isentropically to a high temperature, as in the Carnot cycle. This, however, would involve extremely high pressures and is therefore impractical. Another possibility is to transfer heat to the feedwater from the expanding steam in a counterflow heat exchanger built into the turbine, that is, to use **regeneration**. This solution is also impractical because it is difficult to design such a heat exchanger and because it would increase the moisture content of the steam at the final stages of the turbine.

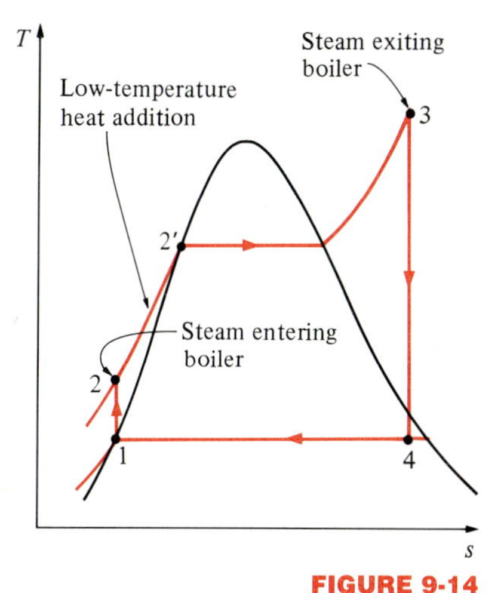

FIGURE 9-14
The first part of the heat addition process in the boiler takes place at relatively low temperatures.

A practical regeneration process in steam power plants is accomplished by extracting, or "bleeding," steam from the turbine at various points. This steam, which could have produced more work by expanding further in the turbine, is used to heat the feedwater instead. The device where the feedwater is heated by regeneration is called a **regenerator**, or a **feedwater heater**.

Regeneration not only improves cycle efficiency, but also provides a convenient means of deaerating the feedwater (removing the air that leaks in at the condenser) to prevent corrosion in the boiler. It also helps control the large volume flow rate of the steam at the final stages of the turbine (due to the large specific volumes at low pressures). Therefore, regeneration is used in all modern steam power plants.

A feedwater heater is basically a heat exchanger where heat is transferred from the steam to the feedwater either by mixing the two fluid streams (open feedwater heaters) or without mixing them (closed feedwater heaters). Regeneration with both types of feedwater heaters is discussed below.

Open Feedwater Heaters

An **open** (or **direct-contact**) **feedwater heater** is basically a *mixing chamber,* where the steam extracted from the turbine mixes with the feedwater exiting the pump. Ideally, the mixture leaves the heater as a saturated liquid at the heater pressure. The schematic of a steam power plant with one open feedwater heater (also called *single-stage regenerative cycle*) and the *T-s* diagram of the cycle are shown in Fig. 9-15.

In an ideal regenerative Rankine cycle, steam enters the turbine at the boiler pressure (state 5) and expands isentropically to an intermediate pressure (state 6). Some steam is extracted at this state and routed to the feedwater heater, while the remaining steam continues to expand isentropically to the condenser pressure (state 7). This steam leaves the condenser as a saturated liquid at the condenser pressure (state 1). The con-

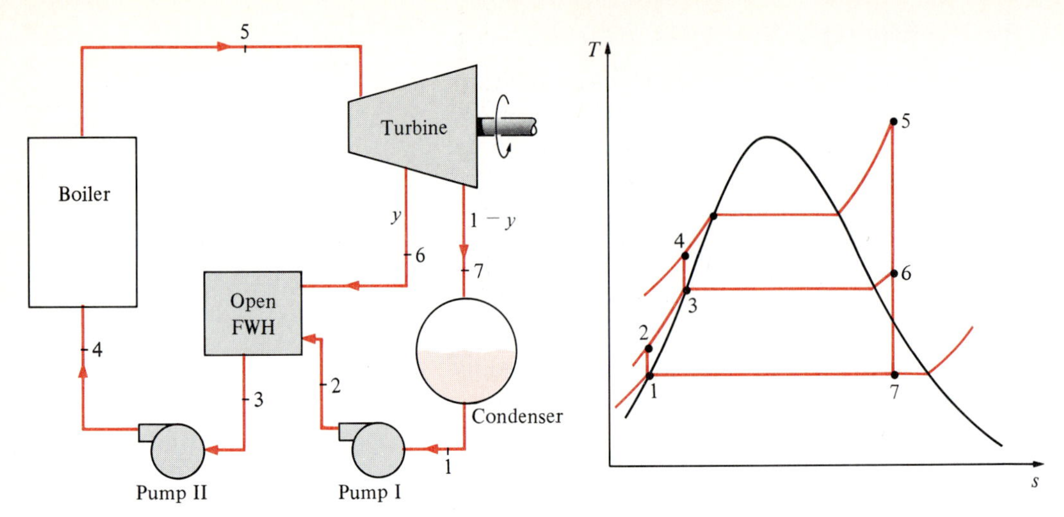

FIGURE 9-15

The ideal regenerative Rankine cycle with an open feedwater heater.

densed water, which is also called the *feedwater,* then enters an isentropic pump, where it is compressed to the feedwater heater pressure (state 2) and is routed to the feedwater heater where it mixes with the steam extracted from the turbine. The fraction of the steam extracted is such that the mixture leaves the heater as a saturated liquid at the heater pressure (state 3). A second pump raises the pressure of the water to the boiler pressure (state 4). The cycle is completed by heating the water in the boiler to the turbine inlet state (state 5).

In the analysis of steam power plants, it is more convenient to work with quantities expressed per unit mass of the steam flowing through the boiler. For each 1 kg of steam leaving the boiler, y kg expands partially in the turbine and is extracted at state 6. The remaining $(1 - y)$ kg expands completely to the condenser pressure. Therefore, the mass flow rates are different in different components. If the mass flow rate through the boiler is $\dot{m}$, for example, it will be $(1 - y)\dot{m}$ through the condenser. This aspect of the regenerative Rankine cycle should be considered in the analysis of the cycle as well as in the interpretation of the areas on the T-s diagram. In light of Fig. 9-15, the heat and work interactions of a regenerative Rankine cycle with one feedwater heater can be expressed per unit mass of steam flowing through the boiler as follows:

$$q_{\text{in}} = h_5 - h_4 \tag{9-14}$$

$$q_{\text{out}} = (1 - y)(h_7 - h_1) \tag{9-15}$$

$$w_{\text{turb,out}} = (h_5 - h_6) + (1 - y)(h_6 - h_7) \tag{9-16}$$

$$w_{\text{pump,in}} = (1 - y)w_{\text{pumpI,in}} + w_{\text{pumpII,in}} \tag{9-17}$$

where $y = \dot{m}_6/\dot{m}_5$ (fraction of steam extracted from turbine)

$$w_{\text{pumpI,in}} = v_1(P_2 - P_1)$$

$$w_{\text{pumpII,in}} = v_3(P_4 - P_3)$$

The thermal efficiency of the Rankine cycle increases as a result of regeneration. This is because regeneration raises the average temperature at which heat is added to the steam in the boiler by raising the temperature of the water before it enters the boiler. The cycle efficiency increases further as the number of feedwater heaters is increased. Many large plants in operation today use as many as eight feedwater heaters. The optimum number of feedwater heaters is determined from economical considerations. The use of an additional feedwater heater cannot be justified unless it saves more from the fuel costs than its own cost.

Closed Feedwater Heaters

Another type of feedwater heater frequently used in steam power plants is the **closed feedwater heater**, in which heat is transferred from the extracted steam to the feedwater without any mixing taking place. The two streams now can be at different pressures, since they do not mix. The schematic of a steam power plant with one closed feedwater heater and the *T-s* diagram of the cycle are shown in Fig. 9-16. In an ideal closed feedwater heater, the feedwater is heated to the exit temperature of the extracted steam, which ideally leaves the heater as a saturated liquid at the extraction pressure. In actual power plants, the feedwater leaves the heater below the exit temperature of the extracted steam because a temperature difference of at least a few degrees is required for any effective heat transfer to take place.

The condensed steam is then either pumped to the feedwater line or routed to another heater or to the condenser through a device called a **trap**. A trap allows the liquid to be throttled to a lower pressure region but *traps* the vapor. The enthalpy of steam remains constant during this throttling process.

The open and closed feedwater heaters can be compared as follows. Open feedwaters are simple and inexpensive and have good heat transfer

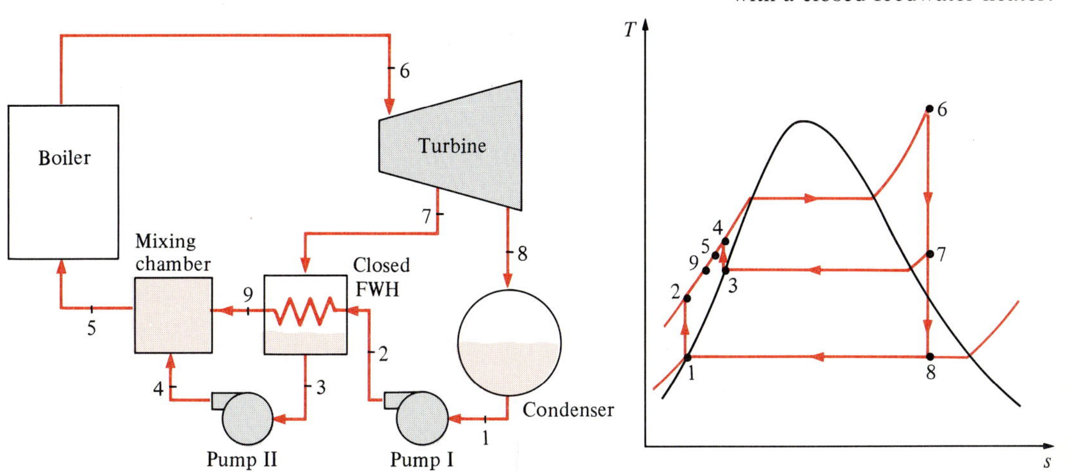

FIGURE 9-16
The ideal regenerative Rankine cycle with a closed feedwater heater.

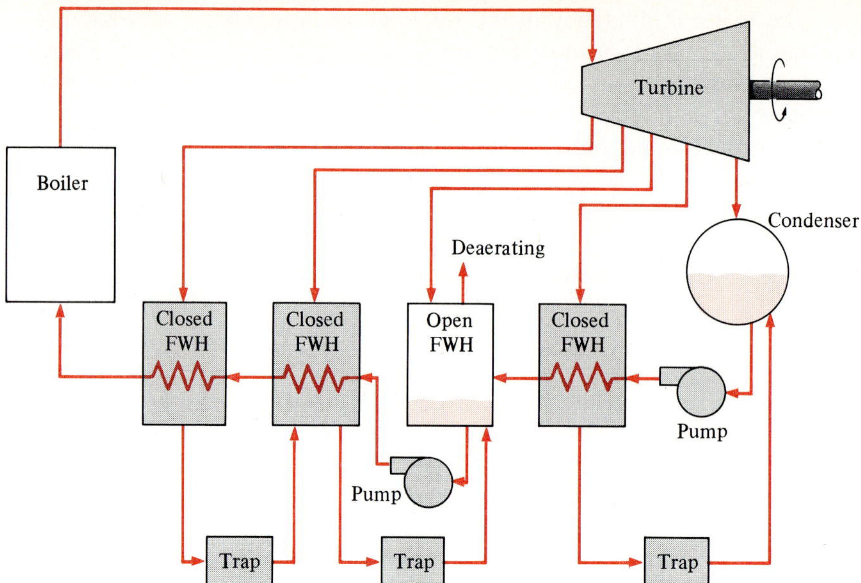

FIGURE 9-17

A steam power plant with one open and three closed feedwater heaters.

characteristics. They also bring the feedwater to the saturation state. But for each heater, a pump is required to handle the feedwater. The closed feedwater heaters are more complex because of the internal piping network, and thus they are more expensive. Heat transfer in closed feedwater heaters is also less effective since the two streams are not allowed to be in direct contact. However, closed feedwater heaters do not require a separate pump for each heater since the extracted steam and the feedwater can be at different pressures. Most steam power plants use a combination of open and closed feedwater heaters, as shown in Fig. 9-17.

EXAMPLE 9-5

Consider a steam power plant operating on the ideal regenerative Rankine cycle with one open feedwater heater. Steam enters the turbine at 15 MPa and 600°C and is condensed in the condenser at a pressure of 10 kPa. Some steam leaves the turbine at a pressure of 1.2 MPa and enters the open feedwater heater. Determine the fraction of steam extracted from the turbine and the thermal efficiency of the cycle.

Solution The schematic of the power plant and the *T-s* diagram of the cycle are given in Fig. 9-18. Since the power plant operates on the ideal regenerative Rankine cycle, we assume that both the turbine and the pump are isentropic; there are no pressure drops in the boiler, feedwater heater, and the condenser; and steam leaves the condenser and the feedwater heater as a saturated liquid.

First, we determine the enthalpies at various states:

State 1: $\left.\begin{array}{l} P_1 = 10 \text{ kPa} \\ \text{Sat. liquid} \end{array}\right\}$ $\begin{array}{l} h_1 = h_{f@10\text{ kPa}} = 191.83 \text{ kJ/kg} \\ v_1 = v_{f@10\text{ kPa}} = 0.001008 \text{ m}^3\text{/kg} \end{array}$

State 2: $P_2 = 1.2$ MPa
$(s_2 = s_1)$

$$w_{\text{pumpI,in}} = v_1(P_2 - P_1) = (0.001008 \text{ m}^3/\text{kg})[(1200 - 10) \text{ kPa}]\left(\frac{1 \text{ kJ}}{1 \text{ kPa}\cdot\text{m}^3}\right)$$
$$= 1.20 \text{ kJ/kg}$$
$$h_2 = h_1 + w_{\text{pumpI,in}} = (191.83 + 1.20) \text{ kJ/kg} = 193.03 \text{ kJ/kg}$$

State 3: $\left.\begin{array}{l} P_3 = 1.2 \text{ MPa} \\ \text{Sat. liquid} \end{array}\right\} \quad h_3 = h_{f@1.2 \text{ MPa}} = 798.65 \text{ kJ/kg}$

State 4: $P_4 = 15$ MPa
$(s_4 = s_3)$

$$w_{\text{pumpII,in}} = v_3(P_4 - P_3)$$
$$= (0.001139 \text{ m}^3/\text{kg})[(15{,}000 - 1200) \text{ kPa}]\left(\frac{1 \text{ kJ}}{1 \text{ kPa}\cdot\text{m}^3}\right)$$
$$= 15.72 \text{ kJ/kg}$$
$$h_4 = h_3 + w_{\text{pumpII,in}} = (798.65 + 15.72) \text{ kJ/kg} = 814.37 \text{ kJ/kg}$$

State 5: $\left.\begin{array}{l} P_5 = 15 \text{ MPa} \\ T_5 = 600°\text{C} \end{array}\right\} \quad \begin{array}{l} h_5 = 3582.3 \text{ kJ/kg} \\ s_5 = 6.6776 \text{ kJ/(kg·K)} \end{array}$

State 6: $\left.\begin{array}{l} P_6 = 1.2 \text{ MPa} \\ s_6 = s_5 \end{array}\right\} \quad \begin{array}{l} h_6 = 2859.5 \text{ kJ/kg} \\ (T_6 = 218.3°\text{C}) \end{array}$

State 7: $\left.\begin{array}{l} P_7 = 10 \text{ kPa} \\ s_7 = s_5 \end{array}\right\} \quad x_7 = \frac{s_7 - s_f}{s_{fg}} = \frac{6.6776 - 0.6493}{7.5009} = 0.804$

$$h_7 = h_f + x_7 h_{fg} = 191.83 + 0.804(2392.8) = 2115.6 \text{ kJ/kg}$$

The energy analysis of open feedwater heaters is identical to the energy analysis of mixing chambers discussed in Chap. 4. The feedwater heaters are

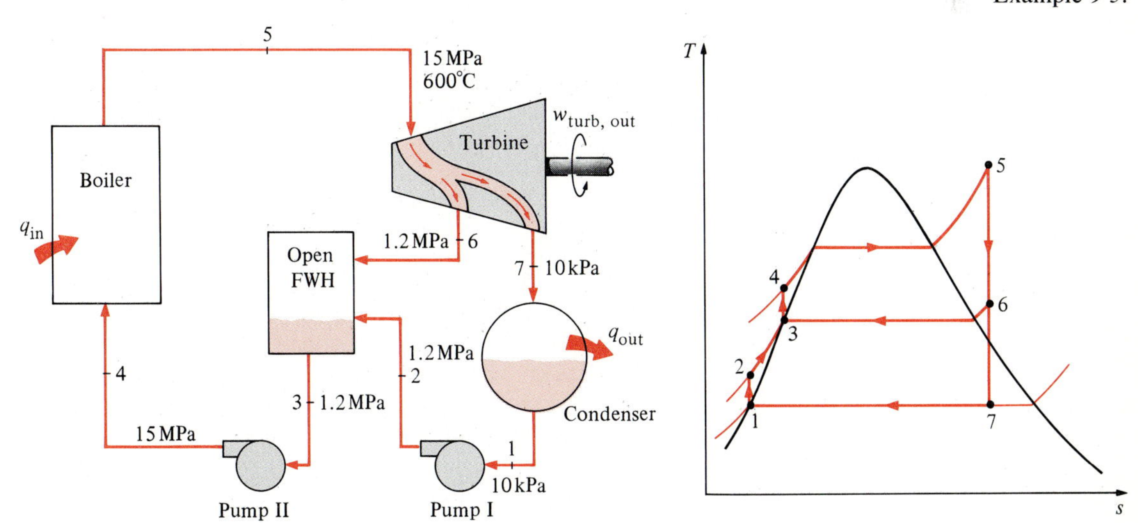

FIGURE 9-18
Schematic and *T-s* diagram for Example 9-5.

generally well insulated ($\dot{Q} = 0$), and they do not involve any work interactions ($\dot{W} = 0$). By neglecting the kinetic and potential energies of the streams the steady-flow conservation of energy equation (Eq. 4-19) reduces for a feedwater heater to

$$\Sigma \dot{m}_i h_i = \Sigma \dot{m}_e h_e$$

or

$$y h_6 + (1 - y) h_2 = 1(h_3)$$

where y is the fraction of steam extracted from the turbine ($= \dot{m}_6/\dot{m}_5$). Solving for y and substituting the enthalpy values, we find

$$y = \frac{h_3 - h_2}{h_6 - h_2} = \frac{798.65 - 193.03}{2859.5 - 193.03} = 0.227$$

Thus,

$$q_{in} = h_5 - h_4 = (3582.3 - 814.37)\ \text{kJ/kg} = 2767.93\ \text{kJ/kg}$$

$$q_{out} = (1 - y)(h_7 - h_1) = (1 - 0.227)(2115.6 - 191.83)\ \text{kJ/kg}$$
$$= 1487.1\ \text{kJ/kg}$$

and

$$\eta_{th} = 1 - \frac{q_{out}}{q_{in}} = 1 - \frac{1487.1\ \text{kJ/kg}}{2767.93\ \text{kJ/kg}} = 0.463 \quad (\text{or } 46.3\%)$$

This problem was worked out in Example 9-3c for the same pressure and temperature limits but without the regeneration process. A comparison of the two results reveals that the thermal efficiency of the cycle has increased from 43.0 to 46.3 percent as a result of regeneration. The net work output decreases by 171 kJ/kg, but the heat input decreases by 607 kJ/kg, which results in a net increase in the thermal efficiency.

EXAMPLE 9-6

Consider a steam power plant that operates on an ideal reheat-regenerative Rankine cycle with one open feedwater heater, one closed feedwater heater, and one reheater. Steam enters the turbine at 15 MPa and 600°C and is condensed in the condenser at a pressure of 10 kPa. Some steam is extracted from the turbine at 4 MPa for the closed feedwater heater, and the remaining steam is reheated at the same pressure to 600°C. The extracted steam is completely condensed in the heater and is pumped to 15 MPa before it mixes with the feedwater at the same pressure. Steam for the open feedwater heater is extracted from the low-pressure turbine at a pressure of 0.5 MPa. Determine the fraction of steam extracted from the turbine each time as well as the thermal efficiency of the cycle.

Solution The schematic of the power plant and the T-s diagram of the cycle are given in Fig. 9-19. The fractions of steam extracted for the closed and open feedwater heaters are assumed to be y and z, respectively. Since the power plant operates on the ideal reheat-regenerative Rankine cycle, it is assumed that both the turbine and the pump are isentropic; there are no pressure drops in the boiler, reheater, feedwater heaters, and the condenser; and steam leaves the condenser and the feedwater heaters as a saturated liquid.

The enthalpies at the various states and the pump work per unit mass of fluid flowing through them are

$$h_1 = 191.83\ \text{kJ/kg} \qquad h_4 = 656.08\ \text{kJ/kg}$$
$$h_2 = 192.32\ \text{kJ/kg} \qquad h_5 = 1087.31\ \text{kJ/kg}$$
$$h_3 = 640.23\ \text{kJ/kg} \qquad h_6 = 1087.31\ \text{kJ/kg}$$

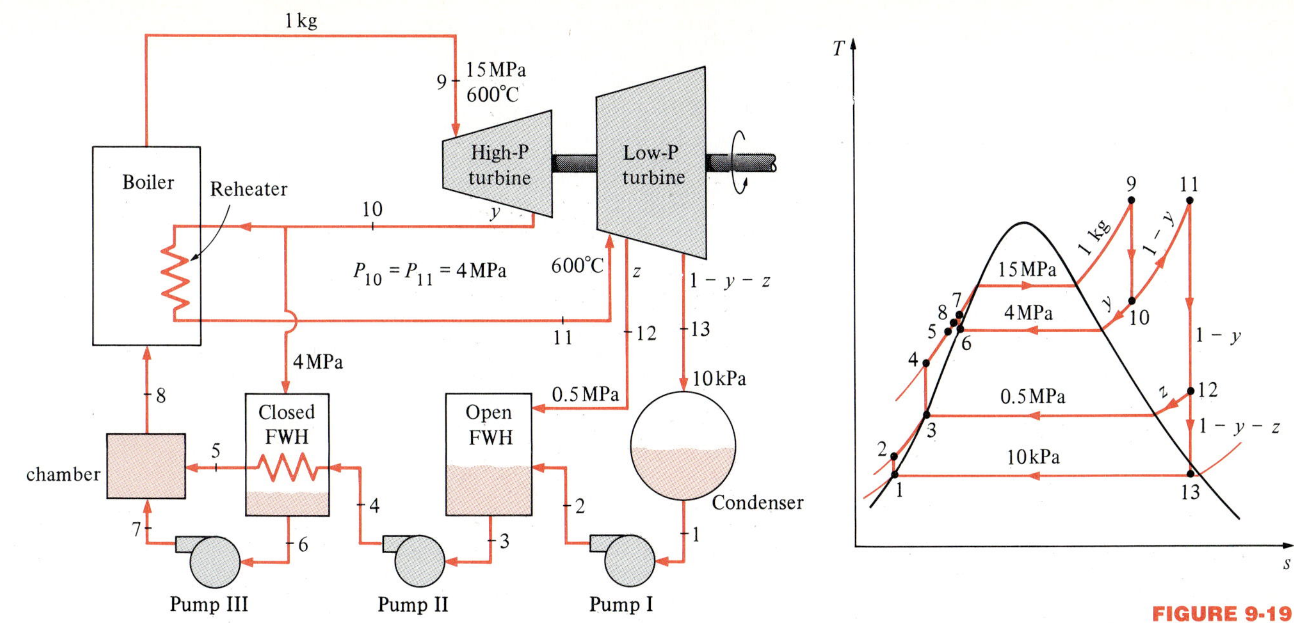

FIGURE 9-19

Schematic and *T-s* diagram for Example 9-6.

$$h_7 = 1101.08 \text{ kJ/kg} \qquad h_{13} = 2335.8 \text{ kJ/kg}$$
$$h_9 = 3582.3 \text{ kJ/kg} \qquad w_{\text{pumpI,in}} = 0.49 \text{ kJ/kg}$$
$$h_{10} = 3154.3 \text{ kJ/kg} \qquad w_{\text{pumpII,in}} = 15.85 \text{ kJ/kg}$$
$$h_{11} = 3674.4 \text{ kJ/kg} \qquad w_{\text{pumpIII,in}} = 13.77 \text{ kJ/kg}$$
$$h_{12} = 3014.3 \text{ kJ/kg}$$

The fractions of steam extracted are determined from the mass and energy balances of the feedwater heaters:

Closed feedwater heater:

$$\Sigma \dot{m}_i h_i = \Sigma \dot{m}_e h_e$$
$$y(h_{10} - h_6) = (1 - y)(h_5 - h_4)$$
$$y = \frac{h_5 - h_4}{(h_{10} - h_6) + (h_5 - h_4)} = \frac{1087.31 - 656.08}{(3154.3 - 1087.31) + (1087.31 - 656.08)} = 0.173$$

Open feedwater heater:

$$\Sigma \dot{m}_i h_i = \Sigma \dot{m}_e h_e$$
$$zh_{12} + (1 - y - z)h_2 = (1 - y)h_3$$
$$z = \frac{(1 - y)(h_3 - h_2)}{h_{12} - h_2} = \frac{(1 - 0.173)(640.23 - 192.32)}{3014.13 - 192.32} = 0.131$$

The enthalpy at state 8 is determined by applying the conservation of mass and energy equations to the mixing chamber which is assumed to be insulated:

$$\Sigma \dot{m}_e h_e = \Sigma \dot{m}_i h_i$$
$$(1)h_8 = (1 - y)h_5 + yh_7$$
$$h_8 = (1 - 0.173)(1087.31) \text{ kJ/kg} + 0.173(1101.08) \text{ kJ/kg}$$
$$= 1089.7 \text{ kJ/kg}$$

Thus, $q_{in} = (h_9 - h_8) + (1 - y)(h_{11} - h_{10})$
$= (3582.5 - 1089.7)$ kJ/kg
$+ (1 - 0.173)(3674.4 - 3154.3)$ kJ/kg
$= 2922.7$ kJ/kg

$q_{out} = (1 - y - z)(h_{13} - h_1)$
$= (1 - 0.173 - 0.131)(2335.8 - 191.83)$ kJ/kg
$= 1492.2$ kJ/kg

and $\eta_{th} = 1 - \frac{q_{out}}{q_{in}} = 1 - \frac{1492.2 \text{ kJ/kg}}{2922.7 \text{ kJ/kg}} = 0.489$ (or 48.9%)

This problem was worked out in Example 9-4 for the same pressure and temperature limits with reheat but without the regeneration process. A comparison of the two results reveals that the thermal efficiency of the cycle has increased from 45.0 to 48.9 percent as a result of regeneration.

The thermal efficiency of this cycle could also be determined from

$$\eta_{th} = \frac{w_{net}}{q_{in}} = \frac{w_{turb,out} - w_{pump,in}}{q_{in}}$$

where $w_{turb,out} = (h_9 - h_{10}) + (1 - y)(h_{11} - h_{12}) + (1 - y - z)(h_{12} - h_{13})$
$w_{pump,in} = (1 - y - z)w_{pumpI,in} + (1 - y)w_{pumpII,in} + (y)w_{pumpIII,in}$

9-7 ■ SECOND-LAW ANALYSIS OF VAPOR POWER CYCLES

The ideal Carnot cycle is a *totally reversible cycle,* and thus it does not involve any irreversibilities. The ideal Rankine cycles (simple, reheat, or regenerative), however, are only *internally reversible,* and they may involve irreversibilities external to the system, such as heat transfer through a finite temperature difference. A second-law analysis of these cycles will reveal where the largest irreversibilities occur and what their magnitudes are.

Relations for availability and irreversibility for steady-flow systems are developed in Chap. 7. The irreversibility for a steady-flow system which may be exchanging heat with a reservoir at temperature T_R in the amount of $\dot{Q}_R$ can be expressed, in rate form, as

$$\dot{I} = T_0\dot{S}_{gen} = T_0\left(\sum \dot{m}_e s_e - \sum \dot{m}_i s_i + \frac{\dot{Q}_R}{T_R}\right) \quad \text{(kW)} \qquad \text{(9-18)}$$

or on a unit-mass basis for a one-inlet, one-exit, steady-flow device as

$$i = T_0 s_{gen} = T_0\left(s_e - s_i + \frac{q_R}{T_R}\right) \quad \text{(kJ/kg)} \qquad \text{(9-19)}$$

where T_0 is the ambient temperature and subscripts i and e are used to denote the inlet and exit states for the process, respectively. Also the sign of $\dot{Q}_R$ is to be determined with respect to the reservoir.

The irreversibility of a cycle depends on the magnitude of the heat transfer with the high- and low-temperature reservoirs involved, and their temperatures, as explained in Chap. 8. It can be expressed on a unit-mass basis as

$$i = T_0 \sum \frac{q_i}{T_i} \qquad \text{(kJ/kg)} \tag{9-20}$$

For a cycle which involves only heat transfer with a source at T_H and a sink at T_L, the irreversibility becomes

$$i = T_0 \left(\frac{q_{\text{out}}}{T_L} - \frac{q_{\text{in}}}{T_H} \right) \qquad \text{(kJ/kg)} \tag{9-21}$$

The availability of a fluid stream ψ at any state can be determined from Eq. 7-17:

$$\psi = (h - h_0) - T_0(s - s_0) + \frac{\mathbf{V}^2}{2} + gz \qquad \text{(kJ/kg)} \tag{9-22}$$

where the subscript "0" denotes the state of the surroundings. The reversible work for any process can be determined by finding the change in availability of the working fluid during that process.

EXAMPLE 9-7

Determine the irreversibility of the Rankine cycle (all four processes as well as the cycle) discussed in Example 9-1, assuming that heat is transferred to the steam in a furnace at 1600 K and heat is rejected to a cooling medium at 290 K and 100 kPa. Also determine the availability of the steam leaving the turbine.

Solution In Example 9-1, the heat input was determined to be 2727.88 kJ/kg, and the heat rejected is 2018.81 kJ/kg.

Processes 1-2 and 3-4 are isentropic ($s_1 = s_2$, $s_3 = s_4$) and therefore do not involve any internal or external irreversibilities, i.e.,

$$i_{12} = 0 \qquad \text{and} \qquad i_{34} = 0$$

Processes 2-3 and 4-1 are constant-pressure heat addition and rejection processes, respectively, and they are internally reversible. But the heat transfer between the working fluid and the source or the sink takes place through a finite temperature difference, rendering both processes irreversible. The irreversibility associated with each process is determined from Eq. 9-19. The entropy of the steam at each state is determined from the steam tables:

$$s_2 = s_1 = s_{f@75\text{ kPa}} = 1.213 \text{ kJ/(kg·K)}$$
$$s_4 = s_3 = 6.7428 \text{ kJ/(kg·K)} \qquad \text{(at 3 MPa, 350°C)}$$

Thus,
$$i_{23} = T_0 \left(s_3 - s_2 + \frac{q_{R,23}}{T_R} \right)$$
$$= (290 \text{ K}) \left[(6.7428 - 1.213) \text{ kJ/(kg·K)} + \frac{-2727.88 \text{ kJ/kg}}{1600 \text{ K}} \right]$$
$$= 1109.2 \text{ kJ/kg}$$

$$i_{41} = T_0 \left(s_1 - s_4 + \frac{q_{R,41}}{T_R} \right)$$
$$= (290 \text{ K}) \left[(1.213 - 6.7428) \text{ kJ/(kg·K)} + \frac{2018.81 \text{ kJ/kg}}{290 \text{ K}} \right]$$
$$= 415.2 \text{ kJ/kg}$$

Therefore, the irreversibility of the cycle is

$$\begin{aligned} i_{\text{cycle}} &= i_{12} + i_{23} + i_{34} + i_{41} \\ &= 0 + 1109.2 \text{ kJ/kg} + 0 + 415.2 \text{ kJ/kg} \\ &= 1524.4 \text{ kJ/kg} \end{aligned}$$

The irreversibility of the cycle could also be determined from Eq. 9-21. Notice that the largest irreversibility in the cycle occurs during the heat addition process. Therefore, any attempt to reduce the irreversibilities should start with this process. Raising the turbine inlet temperature of the steam, for example, would reduce the temperature difference and thus the irreversibility.

The availability (maximum work potential) of the steam leaving the turbine is determined from Eq. 9-22. Disregarding the kinetic and potential energy of the steam, it reduces to

$$\begin{aligned} \psi_4 &= (h_4 - h_0) - T_0(s_4 - s_0) + \frac{\mathbf{V}_4^2}{2}^{\nearrow 0} + gz_4^{\nearrow 0} \\ &= (h_4 - h_0) - T_0(s_4 - s_0) \end{aligned}$$

where

$$h_0 = h_{@290 \text{ K, } 100 \text{ kPa}} \cong h_{f@290 \text{ K}} = 71.34 \text{ kJ/kg}$$

$$s_0 = s_{@290 \text{ K, } 100 \text{ kPa}} \cong s_{f@290 \text{ K}} = 0.2533 \text{ kJ/kg}$$

Thus,

$$\begin{aligned} \psi_4 &= (2403.2 - 71.34) \text{ kJ/kg} - (290 \text{ K})[(6.7428 - 0.2533) \text{ kJ/(kg·K)}] \\ &= 449.9 \text{ kJ/kg} \end{aligned}$$

That is, 449.9 kJ/kg of work could be obtained from the steam leaving the turbine if it is brought to the state of the surroundings in a reversible manner. Notice that this is 70 percent of the net work output of the cycle.

9-8 ■ COGENERATION

In all the cycles discussed so far, the sole purpose was to convert a portion of the heat transferred to the working fluid to work, which is the most valuable form of energy. The remaining portion of the heat is rejected to rivers, lakes, oceans, or the atmosphere as waste heat, because its quality (or grade) is too low to be of any practical use. Wasting a large amount of heat is a price we have to pay to produce work, because electrical or mechanical work is the only form of energy on which many engineering devices (such as a fan) can operate.

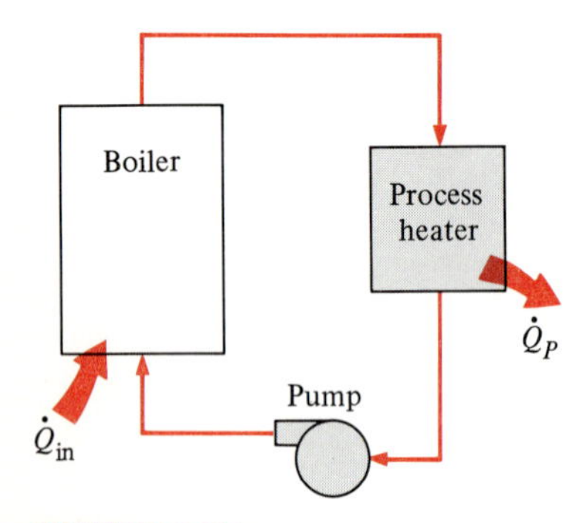

FIGURE 9-20
A simple process heating plant.

Many systems or devices, however, require energy input in the form of heat, called *process heat*. Some industries that heavily rely on process heat are chemical, pulp and paper, oil production and refining, steel making, food processing, and textile industries. Process heat in these industries is usually supplied by steam at 5 to 7 atm and 150 to 200°C (300 to 400°F). Energy is usually transferred to the steam by burning coal, oil, natural gas, or another fuel in a furnace.

Now let us examine the operation of a process heating plant closely. Disregarding any heat losses in the piping, all the heat transferred to the steam in the boiler is used in the process heating units, as shown in Fig. 9-20. Therefore, process heating seems like a perfect operation with practically no waste of energy. From the second-law point of view, however, things do not look so perfect. The temperature in furnaces is typically very high (around 1370°C), and thus the energy in the furnace is of very high quality. This high-quality energy is transferred to water to produce steam at about 200°C or below (a highly irreversible process). Associated with this irreversibility is, of course, a loss in availability or work potential. It is simply not wise to use high-quality energy to accomplish a task that could be accomplished with low-quality energy.

Industries that use large amounts of process heat also consume a large amount of electric power. Therefore, it makes economical as well as engineering sense to use the already existing work potential to produce power, instead of letting it go to waste. The result is a plant which produces electricity while meeting the process heat requirements of certain industrial processes. Such a plant is called a *cogeneration plant*. In general, **cogeneration** is *the production of more than one useful form of energy (such as process heat and electric power) from the same energy source.*

Either a steam-turbine (Rankine) cycle or a gas-turbine (Brayton) cycle or even a combined cycle (discussed later) can be used as the power cycle in a cogeneration plant. The schematic of an ideal steam-turbine cogeneration plant is shown in Fig. 9-21. Let us say this plant is to supply process heat $\dot{Q}_p$ at 500 kPa at a rate of 100 kW. To meet this demand, steam is expanded in the turbine to a pressure of 500 kPa, producing power at a rate of, say, 20 kW. The flow rate of the steam can be adjusted such that steam leaves the process heating section as a saturated liquid at 500 kPa. Steam is then pumped to the boiler pressure and is heated in the boiler to state 3. The pump work is usually very small and can be neglected. Disregarding any heat losses, the rate of heat input in the boiler is determined from an energy balance to be 120 kW.

Probably the most striking feature of the ideal steam-turbine cogeneration plant shown in Fig. 9-21 is the absence of a condenser. Thus no heat is rejected from this plant as waste heat. In other words, all the energy transferred to the steam in the boiler is utilized as either process heat or electric power. Thus it is appropriate to define a **utilization factor** ε_u for a cogeneration plant as

$$\varepsilon_u = \frac{\text{net work output} + \text{process heat delivered}}{\text{total heat input}} = \frac{\dot{W}_{\text{net}} + \dot{Q}_p}{\dot{Q}_{\text{in}}} \tag{9-23}$$

or

$$\varepsilon_u = 1 - \frac{\dot{Q}_{\text{out}}}{\dot{Q}_{\text{in}}} \tag{9-24}$$

where $\dot{Q}_{\text{out}}$ represents the heat rejected in the condenser. Strictly speaking, $\dot{Q}_{\text{out}}$ also includes all the undesirable heat losses from the piping and other components, but they are usually small and thus neglected. The utilization factor of the ideal steam-turbine cogeneration plant is obviously 100 percent. Actual cogeneration plants have utilization factors as high as 70 percent. Future cogeneration plants are expected to have even higher utilization factors.

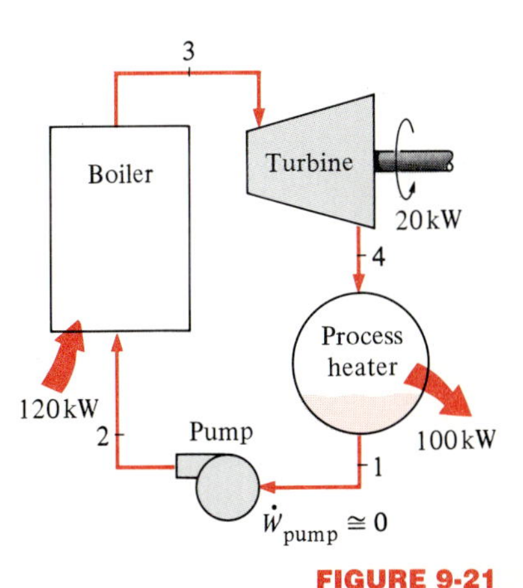

FIGURE 9-21
An ideal cogeneration plant.

Notice that without the turbine, we would need to supply heat to the steam in the boiler at a rate of 100 kW only instead of at 120 kW. The additional 20 kW of heat supplied is converted to work. Therefore, a cogeneration power plant is equivalent to a process heating plant combined with a power plant that has a thermal efficiency of 100 percent.

The ideal steam-turbine cogeneration plant described above is not practical because it cannot adjust to the variations in power and process heat loads. The schematic of a more practical (but more complex) cogeneration plant is shown in Fig. 9-22. Under normal operation, some steam is extracted from the turbine at some predetermined intermediate pressure P_6. The rest of the steam expands to the condenser pressure P_7 and is then cooled at constant pressure. The heat rejected from the condenser represents the waste heat for the cycle.

At times of high demand for process heat, all the steam is routed to the process heating units and none to the condenser ($\dot{m}_7 = 0$). The waste heat is zero in this mode. If this is not sufficient, some steam leaving the boiler is throttled by an expansion or pressure-reducing valve (PRV) to the extraction pressure P_6 and is directed to the process heating unit. Maximum process heating is realized when all the steam leaving the boiler passes through the PRV ($\dot{m}_5 = \dot{m}_4$). No power is produced in this mode. When there is no demand for process heat, all the steam passes through the turbine and the condenser ($\dot{m}_5 = \dot{m}_6 = 0$), and the cogeneration plant operates as an ordinary steam power plant. The rates of heat input, heat rejected, and the process heat supply as well as the power produced for this cogeneration plant can be expressed as follows:

$$\dot{Q}_{in} = \dot{m}_3(h_4 - h_3) \tag{9-25}$$

$$\dot{Q}_{out} = \dot{m}_7(h_7 - h_1) \tag{9-26}$$

$$\dot{Q}_p = \dot{m}_5 h_5 + \dot{m}_6 h_6 - \dot{m}_8 h_8 \tag{9-27}$$

$$\dot{W}_{turb} = (\dot{m}_4 - \dot{m}_5)(h_4 - h_6) + \dot{m}_7(h_6 - h_7) \tag{9-28}$$

Under optimum conditions, a cogeneration plant simulates the ideal cogeneration plant discussed earlier. That is, all the steam expands in the turbine to the extraction pressure and continues to the process heating unit. No steam passes through the PRV or the condenser, thus no waste

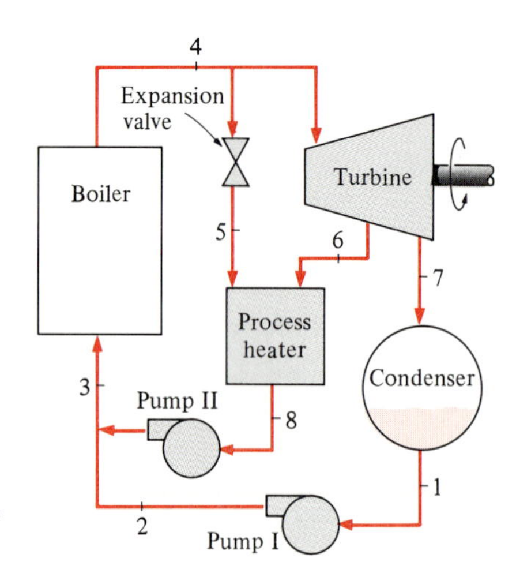

FIGURE 9-22
A cogeneration plant with adjustable loads.

heat is rejected ($\dot{m}_4 = \dot{m}_6$ and $\dot{m}_5 = \dot{m}_7 = 0$). This condition may be difficult to achieve in practice because of the constant variations in the process heat and power loads. But the plant should be designed so that the optimum operating conditions are approximated most of the time.

The use of cogeneration dates to the beginning of this century when power plants were integrated to a community to provide district heating, i.e., space, hot water, and process heating for residential and commercial buildings. The district heating systems lost their popularity in the 1940s owing to low fuel prices. But the rapid rise in the fuel prices in the 1970s brought about renewed interest in district heating.

Cogeneration plants have proved to be economically very attractive. Consequently, more and more such plants are being installed in recent years.

EXAMPLE 9-8

Consider the cogeneration plant shown in Fig. 9-23. Steam enters the turbine at 7 MPa and 500°C. Some steam is extracted from the turbine at 500 kPa for process heating. The remaining steam continues to expand to 5 kPa. Steam is then condensed at constant pressure and pumped to the boiler pressure of 7 MPa. At times of high demand for process heat, some steam leaving the boiler is throttled to 500 kPa and is routed to the process heater. The extraction fractions are adjusted so that steam leaves the process heater as a saturated liquid at 500 kPa. It is subsequently pumped to 7 MPa. The mass flow rate of steam through the boiler is 15 kg/s. Disregarding any pressure drops and heat losses in the piping, and assuming the turbine and the pump to be isentropic, determine (*a*) the maximum rate at which process heat can be supplied, (*b*) the power produced and the utilization factor when no process heat is supplied, and (*c*) the rate of process heat supply when 10 percent of the steam is extracted before it enters the turbine and 70 percent of the steam is extracted from the turbine at 500 kPa for process heating.

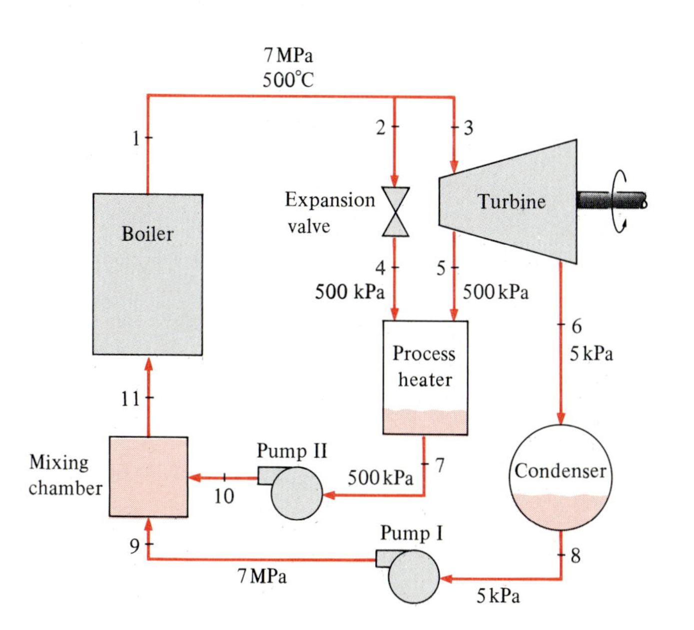

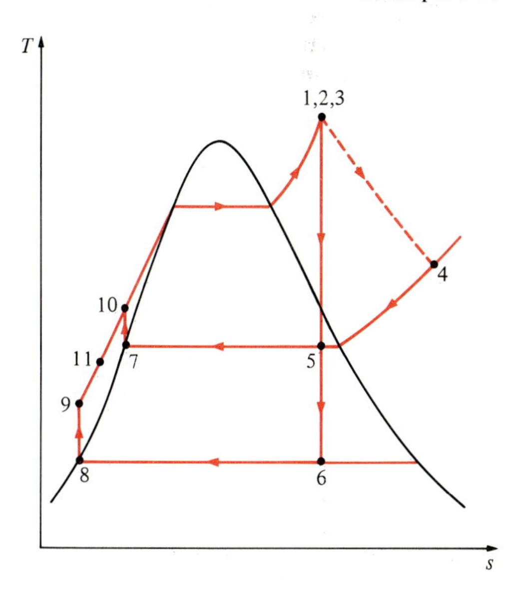

FIGURE 9-23
Schematic and *T-s* diagram for Example 9-8.

Solution The schematic of the cogeneration plant and the *T-s* diagram of the cycle are shown in Fig. 9-23. The work inputs to the pumps and the enthalpies at various states are as follows:

$$w_{\text{pumpI,in}} = v_8(P_9 - P_8) = (0.001005\ \text{m}^3/\text{kg})[(7000 - 5)\ \text{kPa}]\left(\frac{1\ \text{kJ}}{1\ \text{kPa}\cdot\text{m}^3}\right) = 7.03\ \text{kJ/kg}$$

$$w_{\text{pumpII,in}} = v_7(P_{10} - P_7) = (0.001095\ \text{m}^3/\text{kg})[(7000 - 500)\ \text{kPa}]\left(\frac{1\ \text{kJ}}{1\ \text{kPa}\cdot\text{m}^3}\right) = 7.12\ \text{kJ/kg}$$

$$h_1 = h_2 = h_3 = h_4 = 3410.3\ \text{kJ/kg}$$
$$h_5 = 2738.2\ \text{kJ/kg} \qquad (x_5 = 0.995)$$
$$h_6 = 2071.9\ \text{kJ/kg} \qquad (x_6 = 0.798)$$
$$h_7 = h_{f@500\ \text{kPa}} = 640.23\ \text{kJ/kg}$$
$$h_8 = h_{f@5\ \text{kPa}} = 137.82\ \text{kJ/kg}$$
$$h_9 = h_8 + w_{\text{pumpI,in}} = (137.82 + 7.03)\ \text{kJ/kg} = 144.85\ \text{kJ/kg}$$
$$h_{10} = h_7 + w_{\text{pumpII,in}} = (640.23 + 7.12)\ \text{kJ/kg} = 647.35\ \text{kJ/kg}$$

(*a*) The maximum rate of process heat is achieved when all the steam leaving the boiler is throttled and sent to the process heater and none is sent to the turbine (that is, $\dot{m}_4 = \dot{m}_7 = \dot{m}_1 = 15$ kg/s and $\dot{m}_3 = \dot{m}_5 = \dot{m}_6 = 0$). Thus,

$$\dot{Q}_{p,\text{max}} = \dot{m}_1(h_7 - h_4) = (15\ \text{kg/s})[(3410.3 - 640.23)\ \text{kJ/kg}] = 41{,}551\ \text{kW}$$

The utilization factor is 100 percent in this case since no heat is rejected in the condenser and heat losses from the piping and other components are assumed to be negligible.

(*b*) When no process heat is supplied, all the steam leaving the boiler will pass through the turbine and will expand to the condenser pressure of 5 kPa (that is, $\dot{m}_3 = \dot{m}_6 = \dot{m}_1 = 15$ kg/s and $\dot{m}_2 = \dot{m}_5 = 0$). Maximum power will be produced in this mode, which is determined to be

$$\dot{W}_{\text{turb,out}} = \dot{m}_1(h_3 - h_6) = (15\ \text{kg/s})[(3410.3 - 2071.9)\ \text{kJ/kg}] = 20{,}076\ \text{kW}$$

$$\dot{W}_{\text{pump,in}} = \dot{W}_{\text{pumpI,in}} + \cancelto{0}{\dot{W}_{\text{pumpII,in}}} = \dot{m}_1 w_{\text{pumpI,in}} = (15\ \text{kg/s})(7.03\ \text{kJ/kg}) = 105\ \text{kW}$$

$$\dot{W}_{net,out} = \dot{W}_{\text{turb,out}} - \dot{W}_{\text{pump,in}} = (20{,}076 - 105)\ \text{kW} = 19{,}971\ \text{kW}$$

$$\dot{Q}_{\text{in}} = \dot{m}_1(h_1 - h_{11}) = (15\ \text{kg/s})[(3410.3 - 144.85)\ \text{kJ/kg}] = 48{,}982\ \text{kW}$$

Thus,

$$\varepsilon_u = \frac{\dot{W}_{\text{net}} + \dot{Q}_p}{\dot{Q}_{\text{in}}} = \frac{(19{,}971 + 0)\ \text{kJ/kg}}{48{,}982\ \text{kJ/kg}} = 0.408 \quad (\text{or } 40.8\%)$$

That is, 40.8 percent of the energy is utilized for a useful purpose. Notice that the utilization factor is equivalent to the thermal efficiency in this case.

(*c*) Neglecting any kinetic and potential energy changes, an energy balance on the process heater yields

$$\dot{Q} - \cancelto{0}{\dot{W}} = \Sigma\dot{m}_e h_e - \Sigma\dot{m}_i h_i$$
$$\dot{Q} = \dot{m}_7 h_7 - \dot{m}_4 h_4 - \dot{m}_5 h_5$$

where
$$\dot{m}_4 = (0.1)(15\text{ kg/s}) = 1.5\text{ kg/s}$$
$$\dot{m}_5 = (0.7)(15\text{ kg/s}) = 10.5\text{ kg/s}$$
$$\dot{m}_7 = \dot{m}_4 + \dot{m}_5 = 1.5 + 10.5 = 12\text{ kg/s}$$

Thus,
$$\dot{Q} = (12\text{ kg/s})(640.23\text{ kJ/kg}) - (1.5\text{ kg/s})(3410.3\text{ kJ/kg}) - (10.5\text{ kg/s})(2738.2\text{ kJ/kg})$$
$$= -26{,}184\text{ kW}$$

or
$$\dot{Q}_p = 26{,}184\text{ kW}$$

That is, 26,184 kW of the heat added will be utilized in the process heater. We could also show that 10,299 kW of power is produced in this case, and the rate of heat input in the boiler is 42,951 kW. Thus the utilization factor is 84.9 percent.

9-9 ■ BINARY VAPOR CYCLES

With the exception of a few specialized applications, the working fluid predominantly used in vapor power cycles is water. Water is the *best* working fluid presently available, but it is far from being the *ideal* one. The binary cycle is an attempt to overcome some of the shortcomings of water and to approach the *ideal* working fluid by using two fluids. Before we discuss the binary cycle, let us list the characteristics of a working fluid most suitable for vapor power cycles:

1 A high critical temperature and a safe maximum pressure. A critical temperature above the metallurgically allowed maximum temperature (about 620°C) makes it possible to add a considerable portion of the heat isothermally at the maximum temperature as the fluid changes phase. This will make the cycle approach the Carnot cycle. Very high pressures at the maximum temperature are undesirable because they create material-strength problems.

2 Low triple-point temperature. A triple-point temperature below the temperature of the cooling medium will prevent any solidification problems.

3 A condenser pressure which is not too low. Condensers usually operate below atmospheric pressure. Pressures well below the atmospheric pressure create air leakage problems. Therefore, a substance whose saturation pressure at the ambient temperature is too low is not a good candidate.

4 A high enthalpy of vaporization (h_{fg}) so that heat transfer will approach being isothermal and large mass flow rates will not be needed.

5 A saturation dome that resembles an inverted U. This will eliminate the formation of excessive moisture in the turbine and the need for reheating.

6 Good heat transfer characteristics (high thermal conductivity).

7 Other properties such as being inert, inexpensive, readily available, and nontoxic.

Not surprisingly, no fluid possesses all these characteristics. Water comes the closest, although it does not fare well with respect to characteristics 1, 3, and 5. We can cope with its subatmospheric condenser pressure by careful sealing, and with the inverted V-shaped saturation dome by reheating, but there is not much we can do about item 1. Water has a low critical temperature (374°C, well below the 620°C limit) and very high saturation pressures at high temperatures (16.5 MPa and 350°C).

Well, we cannot change the way water behaves during the high-temperature part of the cycle, but we certainly can replace it with a more suitable fluid. The result is a power cycle which is actually a combination of two cycles, one in the high-temperature region and the other in the low-temperature region. Such a cycle is called a **binary vapor cycle**. In binary vapor cycles, the condenser of the high-temperature cycle (also called the *topping cycle*) serves as the boiler of the low-temperature cycle (also called the *bottoming cycle*). That is, the heat output of the high-temperature cycle is used as the heat input to the low-temperature one.

Some working fluids found suitable for the high-temperature cycle are mercury, sodium, potassium, and sodium-potassium mixtures. The schematic and *T-s* diagram for a mercury-water binary vapor cycle are shown in Fig. 9-24. The critical temperature of mercury is 898°C (well above the 620°C metallurgical limit), and its critical pressure is only about

FIGURE 9-24

Mercury-water binary vapor cycle.

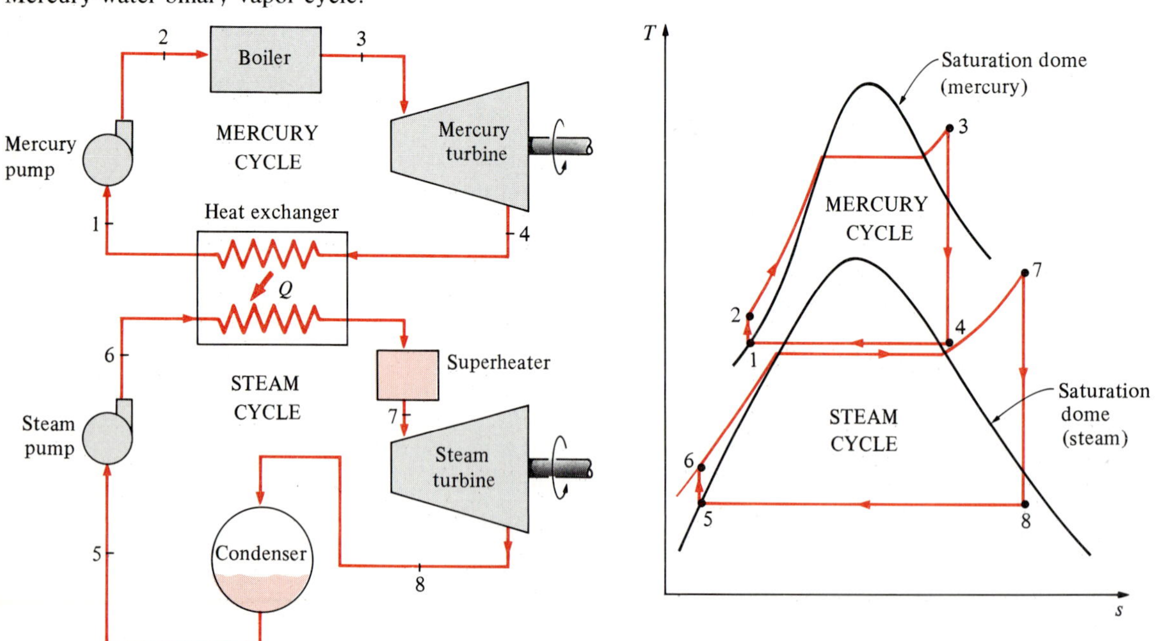

18 MPa. This makes mercury a very suitable working fluid for the topping cycle. Mercury is not suitable as the sole working fluid for the entire cycle, however, since at a condenser temperature of 32°C its saturation pressure is 0.07 Pa. A power plant cannot operate at this vacuum because of air leakage problems. At an acceptable condenser pressure of 7 kPa, the saturation temperature of mercury is 237°C, which is too high as the minimum temperature in the cycle. Therefore, the use of mercury as a working fluid is limited to the high-temperature cycles. Other disadvantages of mercury are its toxicity and high cost. The mass flow rate of mercury in binary vapor cycles is several times that of water because of its low enthalpy of vaporization.

It is evident from the *T-s* diagram in Fig. 9-24 that the binary vapor cycle approximates the Carnot cycle more closely than the steam cycle for the same temperature limits. Therefore, the thermal efficiency of a power plant can be increased by switching to binary cycles. The use of mercury-water binary cycles in the United States dates to 1928. Several such plants have been built since then in the New England area where fuel costs are typically higher. A small (40-MW) mercury-steam power plant which was in service in New Hampshire in 1950 had a higher thermal efficiency than most of the large modern power plants in use at that time.

Studies show that thermal efficiencies of 50 percent or higher are possible with binary vapor cycles. However, binary vapor cycles are not economically attractive because of their high initial cost. This may change in the future, however, as fuel prices escalate.

9-10 ■ COMBINED GAS-VAPOR POWER CYCLES

The continued quest for higher thermal efficiencies has resulted in rather innovative modifications to conventional power plants. The *binary vapor cycle* discussed above is one such modification. A more popular modification involves a gas power cycle topping a vapor power cycle, which is called the **combined gas-vapor cycle**, or just the **combined cycle**. The combined cycle of greatest interest is the gas-turbine (Brayton) cycle topping a steam-turbine (Rankine) cycle, which has a higher thermal efficiency than either of the cycles executed individually.

Gas-turbine cycles typically operate at considerably higher temperatures than steam cycles. The maximum fluid temperature at the turbine inlet is about 620°C (1150°F) for modern steam power plants, but over 1150°C (2100°F) for gas-turbine power plants. The use of higher temperatures in gas turbines is made possible by recent developments in cooling the turbine blades and coating the blades with high-temperature-resistant materials such as ceramics. Because of the higher average temperature at which heat is added, gas-turbine cycles have a greater potential for higher thermal efficiencies. However, the gas-turbine cycles have one inherent disadvantage: The gas leaves the gas turbine at very high temperatures (usually above 500°C), which wipes out any potential gains in the thermal efficiency. The situation can be improved somewhat by using regeneration, but the improvement is limited. Consequently, the thermal efficiency of gas-turbine power plants, in general, is lower than that of steam power plants.

It makes engineering sense to take advantage of ve.y desirable characteristics of the gas-turbine cycle at high temperatures *and* to use the high-temperature exhaust gases as the energy source for a bottoming cycle such as a steam power cycle. The result is a combined gas-steam cycle, as shown in Fig. 9-25. In this cycle, energy is recovered from the exhaust gases by transferring it to the steam in a heat exchanger that serves as the boiler. In general, more than one gas turbine is needed to supply sufficient heat to the steam. Also, the steam cycle may involve regeneration as well as reheating. Energy for the reheating process can be supplied by burning some additional fuel in the oxygen-rich exhaust gases.

Recent developments in gas-turbine technology have made the combined gas-steam cycle economically very attractive. The combined cycle increases the efficiency without appreciably increasing the initial cost. Consequently, many new power plants operate on combined cycles, and many more existing steam- or gas-turbine plants are being converted to combined-cycle power plants. Thermal efficiencies well over 40 percent are reported as a result of conversion.

A 1090-MW Tohoku combined plant which was put in commercial operation in 1985 in Niigata, Japan, is reported to operate at a thermal

FIGURE 9-25

Combined gas-steam power plant.

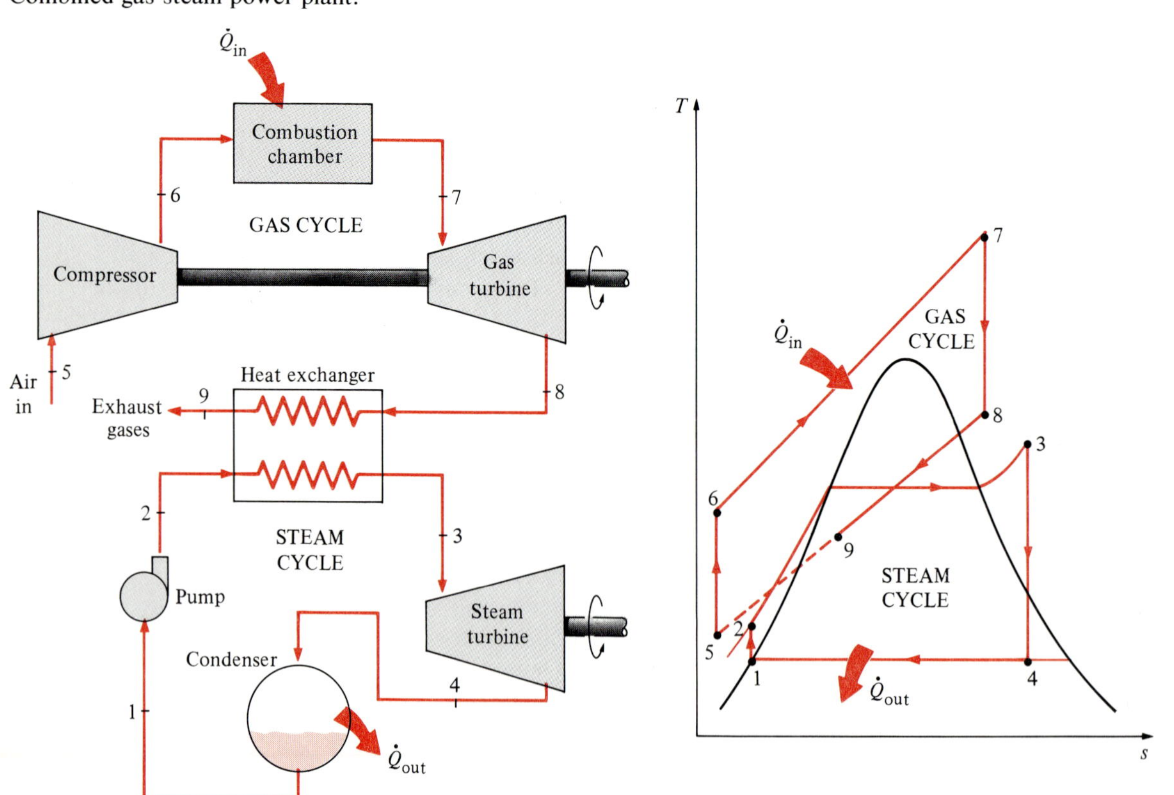

efficiency of 44 percent. This plant has two 191-MW steam turbines and six 118-MW gas turbines. Hot combustion gases enter the gas turbines at 1154°C, and steam enters the steam turbines at 500°C. Steam is cooled in the condenser by cooling water at an average temperature of 15°C. The compressors have a pressure ratio of 14, and the mass flow rate of air through the compressors is 443 kg/s.

EXAMPLE 9-9

Consider the combined gas-steam power cycle shown in Fig. 9-26. The topping cycle is a gas-turbine cycle that has a pressure ratio of 8. Air enters the compressor at 300 K and the turbine at 1300 K. The adiabatic efficiency of the compressor is 80 percent, and that of the gas turbine is 85 percent. The bottoming cycle is a simple ideal Rankine cycle operating between the pressure limits of 7 MPa and 5 kPa. Steam is heated in a heat exchanger by the exhaust gases to a temperature of 500°C. The exhaust gases leave the heat exchanger at 450 K. Determine (*a*) the ratio of the mass flow rates of the steam and the combustion gases and (*b*) the thermal efficiency of the combined cycle.

Solution The *T-s* diagrams of both cycles are given in Fig. 9-26. The gas-turbine cycle alone was analyzed in Example 8-6, and the steam cycle in Example 9-8*b*, with the following results:

Gas cycle: $h'_4 = 880.14$ kJ/kg $\quad (T'_4 = 853$ K$)$
$q_{in} = 790.58$ kJ/kg $\quad w_{net} = 210.63$ kJ/kg $\quad \eta_{th} = 26.6\%$
$h'_5 = h_{@\,450\,K} = 451.80$ kJ/kg

Steam cycle: $h_2 = 144.85$ kJ/kg $\quad (T_2 = 33°C)$
$h_3 = 3410.3$ kJ/kg $\quad (T_3 = 500°C)$
$w_{net} = 1331.4$ kJ/kg $\quad \eta_{th} = 40.8\%$

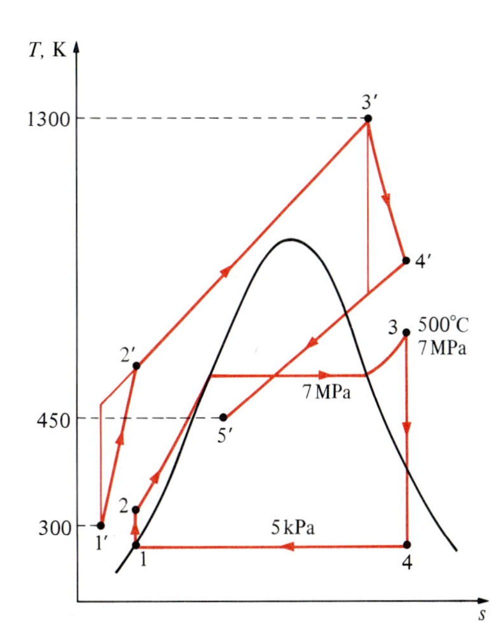

FIGURE 9-26
T-s diagram of the gas-steam combined cycle described in Example 9-9.

(a) The ratio of mass flow rates is determined from an energy balance on the heat exchanger:

$$\dot{Q}^{\nearrow 0} - \dot{W}^{\nearrow 0} = \Sigma \dot{m}_e h_e - \Sigma \dot{m}_i h_i$$
$$\dot{m}_g h_5' + \dot{m}_s h_3 = \dot{m}_g h_4' + \dot{m}_s h_2$$
$$\dot{m}_s(h_3 - h_2) = \dot{m}_g(h_4' - h_5')$$
$$\dot{m}_s(3410.3 - 144.85) = \dot{m}_g(880.14 - 451.80)$$

Thus,
$$\frac{\dot{m}_s}{\dot{m}_g} = y = 0.131$$

That is, 1 kg of exhaust gases can heat only 0.131 kg of steam from 33 to 500°C as they are cooled from 853 to 450 K. Then the total net work output per kilogram of combustion gases becomes

$$\begin{aligned} w_{net} &= w_{net,gas} + y w_{net,steam} \\ &= (210.63 \text{ kJ/kg gas}) + (0.131 \text{ kg steam/kg gas})(1331.4 \text{ kJ/kg steam}) \\ &= 385.04 \text{ kJ/kg gas} \end{aligned}$$

Therefore, for each 1 kg of combustion gases produced, the combined plant will deliver 385.04 kJ of work. The net power output of the plant is determined by multiplying this value by the mass flow rate of the working fluid in the gas-turbine cycle.

(b) The thermal efficiency of the combined cycle is determined from

$$\eta_{th} = \frac{w_{net}}{q_{in}} = \frac{385.04 \text{ kJ/kg gas}}{790.58 \text{ kJ/kg gas}} = 0.487 \quad (\text{or } 48.7\%)$$

Therefore, this combined cycle will convert 48.7 percent of the energy supplied to the gas in the combustion chamber to useful work. Notice that this value is considerably higher than the thermal efficiency of the gas-turbine cycle (26.6 percent) or the steam-turbine cycle (40.7 percent) operating alone.

9-11 ■ SUMMARY

The *Carnot cycle* is not a suitable model for vapor power cycles because it cannot be approximated in practice. The model cycle for vapor power cycles is the *Rankine cycle,* which is composed of four internally reversible processes: constant-pressure heat addition in a boiler, isentropic expansion in a turbine, constant-pressure heat rejection in a condenser, and isentropic compression in a pump. Steam leaves the condenser as a saturated liquid at the condenser pressure.

The thermal efficiency of the Rankine cycle can be increased by increasing the average temperature at which heat is added to the working fluid and/or by decreasing the average temperature at which heat is rejected to the cooling medium, such as a lake or a river. The average temperature during heat rejection can be decreased by lowering the turbine exit pressure. Consequently, the condenser pressure of most vapor power plants is well below the atmospheric pressure. The average temperature during heat addition can be increased by raising the boiler pressure or by superheating the fluid to high temperatures. There is a limit to the degree of superheating, however, since the fluid temperature is not allowed to exceed a metallurgically safe value.

Superheating has the added advantage of decreasing the moisture content of the steam at the turbine exit. Lowering the exhaust pressure or raising the boiler pressure, however, increases the moisture content. To take advantage of the improved efficiencies at higher boiler pressures and lower condenser pressures, steam is usually *reheated* after expanding partially in the high-pressure turbine. This is done by extracting the steam after partial extraction in the high-pressure turbine, sending it back to the boiler where it is reheated at constant pressure, and returning it to the low-pressure turbine for complete expansion to the condenser pressure. The reheat process does not greatly affect the average temperature at which heat is added. Therefore, no significant change in the thermal efficiency should be expected as a result of reheating. The main virtue of the reheating process is to decrease the moisture content at the turbine exit.

Another way of increasing the thermal efficiency of the Rankine cycle is by *regeneration*. During a regeneration process, liquid water (feedwater) leaving the pump is heated by some steam bled off the turbine at some intermediate pressure in devices called *feedwater heaters*. The two streams are mixed in open feedwater heaters, and the mixture leaves as a saturated liquid at the heater pressure. In closed feedwater heaters, heat is transferred from the steam to the feedwater without mixing. Therefore an open feedwater heater is, in essence, a mixing chamber, and a closed feedwater heater is a heat exchanger.

The production of more than one useful form of energy (such as process heat and electric power) from the same energy source is called *cogeneration*. Cogeneration plants produce electric power while meeting the process heat requirements of certain industrial processes. This way, more of the energy transferred to the fluid in the boiler is utilized for a useful purpose. The fraction of energy which is used for either process heat or power generation is called the *utilization factor* of the cogeneration plant.

The overall thermal efficiency of a power plant can be increased by using *binary cycles* or *combined cycles*. A binary cycle is composed of two separate cycles, one at high temperatures (topping cycle) and the other at relatively low temperatures. The most common combined cycle is the gas-steam combined cycle where a gas-turbine cycle operates at the high-temperature range and a steam-turbine cycle at the low-temperature range. Steam is heated by the high-temperature exhaust gases leaving the gas turbine. Combined cycles have a higher thermal efficiency than the steam- or gas-turbine cycles operating alone.

REFERENCES AND SUGGESTED READING

1 W. Z. Black and J. G. Hartley, *Thermodynamics*, Harper & Row, New York, 1985.

2 M. D. Burghard, *Engineering Thermodynamics with Applications,* 3d ed., Harper & Row, New York, 1986.

3 M. M. El-Wakil, *Powerplant Technology,* McGraw-Hill, New York, 1984.

4 R. C. Fellinger and W. J. Cook, *Introduction to Engineering Thermodynamics,* WCB, Dubuque, IA, 1985.

5 J. R. Howell and R. O. Buckius, *Fundamentals of Engineering Thermodynamics,* McGraw-Hill, New York, 1987.

6 J. B. Jones and G. A. Hawkins, *Engineering Thermodynamics,* 2d ed., Wiley, New York, 1986.

7 B. V. Karlekar, *Thermodynamics for Engineers,* Prentice-Hall, Englewood Cliffs, NJ, 1983.

8 J. H. Keenan, *Thermodynamics,* Wiley, New York, 1941.

9 K. W. Li and A. P. Priddy, *Power Plant System Design,* Wiley, New York, 1985.

10 D. C. Look, Jr., and H. J. Sauer, Jr., *Engineering Thermodynamics,* PWS Engineering, Boston, 1986.

11 W. C. Reynolds and H. C. Perkins, *Engineering Thermodynamics,* 2d ed., McGraw-Hill, New York, 1977.

12 H. Sorensen, *Energy Conversion Systems,* Wiley, New York, 1983.

13 *Steam, Its Generation and Use,* 39th ed., Babcock and Wilcox Co., New York, 1978.

14 *Turbomachinery,* vol. 28, no. 2, Business Journals, Inc., Norwalk, CT, March/April 1987.

15 K. Wark, *Thermodynamics,* 5th ed., McGraw-Hill, New York, 1988.

16 J. Weisman and R. Eckart, *Modern Power Plant Engineering,* Prentice-Hall, Englewood Cliffs, NJ, 1985.

17 G. J. Van Wylen and R. E. Sonntag, *Fundamentals of Classical Thermodynamics,* 3d ed., Wiley, New York, 1985.

PROBLEMS*

Carnot Vapor Cycle

9-1C Why is the Carnot cycle not a realistic model for steam power plants?

9-2C Why is excessive moisture in steam undesirable in steam turbines? What is the highest moisture content allowed?

---

*Students are encouraged to answer *all* the concept "C" questions.

9-3C How is moisture content related to quality?

9-4 A steady-flow Carnot cycle uses water as the working fluid. Water changes from saturated liquid to saturated vapor as heat is transferred to it from a source at 200°C. Heat rejection takes place at a pressure of 20 kPa. Show the cycle on a *T-s* diagram relative to the saturation lines, and determine (*a*) the thermal efficiency, (*b*) the amount of heat rejected, in kJ/kg, and (*c*) the net work output.

9-5 Consider a steady-flow Carnot cycle with water as the working fluid. The maximum and minimum temperatures in the cycle are 350 and 60°C. The quality of water is 0.891 at the beginning of the heat rejection process and 0.1 at the end. Show the cycle on a *T-s* diagram relative to the saturation lines, and determine (*a*) the thermal efficiency, (*b*) the pressure at the turbine inlet, and (*c*) the net work output.
Answers: (*a*) 0.465, (*b*) 1.40 MPa, (*c*) 1621.5 kJ/kg

9-6 Water enters the boiler of a steady-flow Carnot engine as a saturated liquid at 1 MPa and leaves with a quality of 0.95. Steam leaves the turbine at a pressure of 100 kPa. Show the cycle on a *T-s* diagram relative to the saturation lines, and determine (*a*) the thermal efficiency, (*b*) the quality at the end of the isothermal heat rejection process, and (*c*) the net work output.

9-6E Water enters the boiler of a steady-flow Carnot engine as a saturated liquid at 120 psia and leaves with a quality of 0.95. Steam leaves the turbine at a pressure of 14.7 psia. Show the cycle on a *T-s* diagram relative to the saturation lines, and determine (*a*) the thermal efficiency, (*b*) the quality at the end of the isothermal heat rejection process, and (*c*) the net work output.
Answers: (*a*) 16.1 percent, (*b*) 0.1245, (*c*) 134.4 Btu/lbm

The Simple Rankine Cycle

9-7C What four processes make up the simple ideal Rankine cycle?

9-8C How does the efficiency of the Rankine cycle compare to that of a Carnot cycle operating between the same temperature limits?

9-9C Does the pressure of steam rise as it is heated in the boiler? How about the temperature?

9-10C Consider a simple ideal Rankine cycle with fixed turbine inlet conditions. What is the effect of lowering the condenser pressure on

Pump work input: (*a*) increases, (*b*) decreases, (*c*) remains the same
Turbine work output: (*a*) increases, (*b*) decreases, (*c*) remains the same
Heat added: (*a*) increases, (*b*) decreases, (*c*) remains the same
Heat rejected: (*a*) increases, (*b*) decreases, (*c*) remains the same
Cycle efficiency: (*a*) increases, (*b*) decreases, (*c*) remains the same
Moisture content at turbine exit: (*a*) increases, (*b*) decreases, (*c*) remains the same

9-11C Consider a simple ideal Rankine cycle with fixed turbine inlet temperature and condenser pressure. What is the effect of increasing the boiler pressure on

Pump work input:	(*a*) increases, (*b*) decreases, (*c*) remains the same
Turbine work output:	(*a*) increases, (*b*) decreases, (*c*) remains the same
Heat added:	(*a*) increases, (*b*) decreases, (*c*) remains the same
Heat rejected:	(*a*) increases, (*b*) decreases, (*c*) remains the same
Cycle efficiency:	(*a*) increases, (*b*) decreases, (*c*) remains the same
Moisture content at turbine exit:	(*a*) increases, (*b*) decreases, (*c*) remains the same

9-12C Consider a simple ideal Rankine cycle with fixed boiler and condenser pressures. What is the effect of superheating the steam to a higher temperature on

Pump work input:	(*a*) increases, (*b*) decreases, (*c*) remains the same
Turbine work output:	(*a*) increases, (*b*) decreases, (*c*) remains the same
Heat added:	(*a*) increases, (*b*) decreases, (*c*) remains the same
Heat rejected:	(*a*) increases, (*b*) decreases, (*c*) remains the same
Cycle efficiency:	(*a*) increases, (*b*) decreases, (*c*) remains the same
Moisture content at turbine exit:	(*a*) increases, (*b*) decreases, (*c*) remains the same

9-13C How do actual vapor power cycles differ from the idealized ones?

9-14C Compare the pressures at the inlet and the exit of the boiler for (*a*) actual and (*b*) ideal cycles.

9-15C The entropy of steam increases in actual steam turbines as a result of irreversibilities. In an effort to control entropy increase, it is proposed to cool the steam in the turbine by running cooling water around the turbine casing. It is argued that this will reduce the entropy and the enthalpy of the steam at the turbine exit and thus increase the work output. How would you evaluate this proposal?

9-16C Is it possible to maintain a pressure of 10 kPa in a condenser which is being cooled by river water entering at 20°C?

9-17 A steam power plant operates on a simple ideal Rankine cycle between the pressure limits of 3 MPa and 50 kPa. The temperature of the steam at the turbine inlet is 400°C, and the mass flow rate of steam through the cycle is 40 kg/s. Show the cycle on a *T-s* diagram with respect to saturation lines, and determine (*a*) the thermal efficiency of the cycle and (*b*) the net power output of the power plant.

9-18 Consider a 200-MW steam power plant that operates on a simple ideal Rankine cycle. Steam enters the turbine at 10 MPa and 500°C and is cooled in the condenser at a pressure of 10 kPa. Show the cycle on a *T-s* diagram with respect to saturation lines, and determine (*a*) the quality

of the steam at the turbine exit, (*b*) the thermal efficiency of the cycle, and (*c*) the mass flow rate of the steam.
Answers: (*a*) 0.793, (*b*) 40.2 percent, (*c*) 156.9 kg/s

9-18E Consider a 200-MW steam power plant that operates on a simple ideal Rankine cycle. Steam enters the turbine at 1000 psia and 1000°F and is cooled in the condenser at a pressure of 2 psia. Show the cycle on a *T-s* diagram with respect to saturation lines, and determine (*a*) the quality of the steam at the turbine exit, (*b*) the thermal efficiency of the cycle, and (*c*) the mass flow rate of the steam.

9-19 Repeat Prob. 9-18, assuming an adiabatic efficiency of 85 percent for both the turbine and the pump.
Answers: (*a*) 0.874, (*b*) 34.1 percent, (*c*) 185.2 kg/s

9-20 Steam enters the turbine of a steam power plant which operates on a simple ideal Rankine cycle at a pressure of 6 MPa, and it leaves as a saturated vapor at 7.5 kPa. Heat is transferred to the steam in the boiler at a rate of 10^5 kJ/s. Steam is cooled in the condenser by the cooling water from a nearby river, which enters the condenser at 18°C. Show the cycle on a *T-s* diagram with respect to saturation lines, and determine (*a*) the turbine inlet temperature, (*b*) the net power output and thermal efficiency, and (*c*) the minimum mass flow rate of the cooling water required.

9-21 A steam power plant operates on a simple ideal Rankine cycle between the pressure limits of 9 MPa and 10 kPa. The mass flow rate of steam through the cycle is 25 kg/s. The moisture content of the steam at the turbine exit is not to exceed 10 percent. Show the cycle on a *T-s* diagram with respect to saturation lines, and determine (*a*) the minimum turbine inlet temperature, (*b*) the rate of heat input in the boiler, and (*c*) the thermal efficiency of the cycle.

9-21E A steam power plant operates on a simple ideal Rankine cycle between the pressure limits of 1250 and 2 psia. The mass flow rate of steam through the cycle is 50 lbm/s. The moisture content of the steam at the turbine exit is not to exceed 10 percent. Show the cycle on a *T-s* diagram with respect to saturation lines, and determine (*a*) the minimum turbine inlet temperature, (*b*) the rate of heat input in the boiler, and (*c*) the thermal efficiency of the cycle.

9-22 Repeat Prob. 9-21, assuming an adiabatic efficiency of 85 percent for both the turbine and the pump.

9-23 Consider a coal-fired steam power plant which produces 300 MW of electric power. The power plant operates on a simple ideal Rankine cycle with turbine inlet conditions of 5 MPa and 450°C and a condenser pressure of 25 kPa. The coal used has a heating value (energy released when the fuel is burned) of 29,300 kJ/kg. Assuming that 75 percent of this energy is transferred to the steam in the boiler and that the electric generator has an efficiency of 96 percent, determine (*a*) the overall plant efficiency (the ratio of net electric power output to the energy input as fuel) and (*b*) the required rate of coal supply, in t/h [1 metric ton (t) = 1000 kg]. *Answers:* (*a*) 24.6 percent, (*b*) 150.1 t/h

9-23E Consider a coal-fired steam power plant which produces 300 MW of electric power. The power plant operates on a simple ideal Rankine cycle with turbine inlet conditions of 700 psia and 800°F and a condenser pressure of 3 psia. The coal used has a heating value (energy released when the fuel is burned) of 12,600 Btu/lbm. Assuming that 75 percent of this energy is transferred to the steam in the boiler and that the electric generator has an efficiency of 96 percent, determine (*a*) the overall plant efficiency (the ratio of net electric power output to the energy input as fuel) and (*b*) the required rate of coal supply, in tons/h (1 ton = 2000 lbm). *Answers:* (*a*) 24.6 percent, (*b*) 183.8 tons/h

9-24 Consider a solar-pond power plant which operates on a simple ideal Rankine cycle with refrigerant-12 as the working fluid. The refrigerant enters the turbine as a saturated vapor at 1.6 MPa and leaves at 0.7 MPa. The mass flow rate of the refrigerant is 2 kg/s. Show the cycle on a *T-s* diagram with respect to saturation lines, and determine (*a*) the thermal efficiency of the cycle and (*b*) the power output of this plant.

9-25 Consider a steam power plant that operates on a simple ideal Rankine cycle and has a net power output of 20 MW. Steam enters the turbine at 7 MPa and 500°C and is cooled in the condenser at a pressure of 10 kPa by running cooling water from a lake through the tubes of the condenser at a rate of 1750 kg/s. Show the cycle on a *T-s* diagram with respect to saturation lines, and determine (*a*) the thermal efficiency of the cycle, (*b*) the mass flow rate of the steam, and (*c*) the temperature rise of the cooling water. *Answers:* (*a*) 38.9 percent, (*b*) 16.0 kg/s, (*c*) 4.3°C

9-25E Consider a steam power plant that operates on a simple Rankine cycle and has a net power output of 20 MW. Steam enters the turbine at 1000 psia and 1000°F and is cooled in the condenser at a pressure of 2 psia by running cooling water from a lake through the tubes of the condenser at a rate of 3400 lbm/s. The turbine and the pump have adiabatic efficiencies of 84 percent. Show the cycle on a *T-s* diagram with respect to saturation lines, and determine (*a*) the thermal efficiency of the cycle, (*b*) the mass flow rate of the steam, and (*c*) the temperature rise of the cooling water.

9-26 Repeat Prob. 9-25, assuming an adiabatic efficiency of 87 percent for both the turbine and the pump.
Answers: (*a*) 33.8 percent, (*b*) 18.4 kg/s, (*c*) 5.35°C

9-27 Write a computer program to determine the effect of condenser pressure on the performance of a simple ideal Rankine cycle. Assume steam enters the turbine at 5 MPa and 500°C, and neglect the pump work.

Determine the thermal efficiency of the cycle, and plot it against condenser pressure for condenser pressures of 100, 75, 50, 25, 10, and 5 kPa.

9-28 Write a computer program to determine the effect of boiler pressure on the performance of a simple ideal Rankine cycle. Assume steam enters the turbine at 500°C and exits at 10 kPa. Neglect the pump work. Determine the thermal efficiency of the cycle, and plot it against boiler pressure for boiler pressures of 0.5, 1, 3, 6, 10, 15, and 20 MPa.

9-29 Write a computer program to determine the effect of superheating the steam on the performance of a simple ideal Rankine cycle. Assume steam enters the turbine at 3 MPa and exits at 10 kPa. Neglect the pump work. Determine the thermal efficiency of the cycle, and plot it against turbine inlet temperature for turbine inlet temperatures of 250, 300, 400, 500, 700, 900, and 1100°C.

The Reheat Rankine Cycle

9-30C How do the following quantities change when a simple ideal Rankine cycle is modified with reheating? Assume the mass flow rate is maintained the same.

Pump work input:	(*a*) increases, (*b*) decreases, (*c*) remains the same
Turbine work output:	(*a*) increases, (*b*) decreases, (*c*) remains the same
Heat added:	(*a*) increases, (*b*) decreases, (*c*) remains the same
Heat rejected:	(*a*) increases, (*b*) decreases, (*c*) remains the same
Moisture content at turbine exit:	(*a*) increases, (*b*) decreases, (*c*) remains the same

9-31C Under what conditions does reheating increase the efficiency of the ideal Rankine cycle?

9-32C Can the efficiency of the ideal Rankine cycle decrease as a result of reheating? Explain.

9-33C Show the ideal Rankine cycle with three stages of reheating on a *T-s* diagram. Assume the turbine inlet temperature is the same for all stages. How does the cycle efficiency vary with the number of reheat stages?

9-34C Consider a simple ideal Rankine cycle and an ideal Rankine cycle with three reheat stages. Both cycles operate between the same pressure limits. The maximum temperature is 700°C in the simple cycle and 500°C in the reheat cycle. Which cycle do you think will have a higher thermal efficiency?

9-35 A steam power plant operates on the ideal reheat Rankine cycle. Steam enters the high-pressure turbine at 8 MPa and 500°C and leaves at 3 MPa. Steam is then reheated at constant pressure to 500°C before it expands to 20 kPa in the low-pressure turbine. Determine the turbine work output, in kJ/kg, and the thermal efficiency of the cycle. Also show the cycle on a *T-s* diagram with respect to saturation lines.

9-36 Consider a steam power plant that operates on a reheat Rankine cycle and has a net power output of 200 MW. Steam enters the high-pressure turbine at 10 MPa and 500°C and the low-pressure turbine at 1 MPa and 500°C. Steam leaves the condenser as a saturated liquid at a pressure of 10 kPa. The adiabatic efficiency of the turbine is 80 percent, and that of the pump is 95 percent. Show the cycle on a *T-s* diagram with respect to saturation lines, and determine (*a*) the quality (or temperature, if superheated) of the steam at the turbine exit, (*b*) the thermal efficiency of the cycle, and (*c*) the mass flow rate of the steam.
Answers: (*a*) 87.5°C, (*b*) 34.1 percent, (*c*) 156.6 kg/s

9-36E Consider a steam power plant that operates on a reheat Rankine cycle and has a net power output of 200 MW. Steam enters the high-pressure turbine at 1500 psia and 1100°F and the low-pressure turbine at 500 psia and 1000°F. Steam leaves the condenser as a saturated liquid at a pressure of 2 psia. The adiabatic efficiency of the turbine is 80 percent, and that of the pump is 95 percent. Show the cycle on a *T-s* diagram with respect to saturation lines, and determine (*a*) the quality (or temperature, if superheated) of the steam at the turbine exit, (*b*) the thermal efficiency of the cycle, and (*c*) the mass flow rate of the steam.

9-37 Repeat Prob. 9-36, assuming both the pump and the turbine are isentropic. *Answers:* (*a*) 0.948, (*b*) 41.4 percent, (*c*) 125.0 kg/s

9-38 Steam enters the high-pressure turbine of a steam power plant which operates on the ideal reheat Rankine cycle at 6 MPa and 450°C and leaves as saturated vapor. Steam is then reheated to 400°C before it expands to a pressure of 7.5 kPa. Heat is transferred to the steam in the boiler at a rate of 10^5 kJ/s. Steam is cooled in the condenser by the cooling water from a nearby river, which enters the condenser at 18°C. Show the cycle on a *T-s* diagram with respect to saturation lines, and determine (*a*) the pressure at which reheating takes place, (*b*) the net power output and thermal efficiency, and (*c*) the minimum mass flow rate of the cooling water required.

9-38E Steam enters the high-pressure turbine of a steam power plant which operates on the ideal reheat Rankine cycle at 800 psia and 900°F and leaves as saturated vapor. Steam is then reheated to 800°F before it expands to a pressure of 1 psia. Heat is transferred to the steam in the boiler at a rate of 10^5 Btu/s. Steam is cooled in the condenser by the cooling water from a nearby river, which enters the condenser at 50°F. Show the cycle on a *T-s* diagram with respect to saturation lines, and determine (*a*) the pressure at which reheating takes place, (*b*) the net power output and thermal efficiency, and (*c*) the minimum mass flow rate of the cooling water required.

9-39 A steam power plant operates on an ideal reheat Rankine cycle between the pressure limits of 9 MPa and 10 kPa. The mass flow rate of steam through the cycle is 25 kg/s. Steam enters both stages of the turbine at 500°C. If the moisture content of the steam at the exit of the low-pressure turbine is not to exceed 10 percent, determine (*a*) the pressure at which reheating takes place, (*b*) the total rate of heat input in the boiler, and (*c*) the thermal efficiency of the cycle. Also show the cycle on a *T-s* diagram with respect to saturation lines.

9-40 A steam power plant operates on an ideal Rankine cycle with two stages of reheat and has a net power output of 500 MW. Steam enters all three stages of the turbine at 500°C. The maximum pressure in the cycle is 15 MPa, and the minimum pressure 5 kPa. Steam is reheated at 5 MPa the first time and at 1 MPa the second time. Show the cycle on a *T-s* diagram with respect to saturation lines, and determine (*a*) the thermal efficiency of the cycle and (*b*) the mass flow rate of the steam.
Answers: (*a*) 45.5 percent, (*b*) 268.5 kg/s

9-40E A steam power plant operates on an ideal Rankine cycle with two stages of reheat and has a net power output of 500 MW. Steam enters all three stages of the turbine at 1000°F. The maximum pressure in the cycle is 1250 psia, and the minimum pressure 3 psia. Steam is reheated at 600 psia the first time and at 160 psia the second time. Show the cycle on a *T-s* diagram with respect to saturation lines, and determine (*a*) the thermal efficiency of the cycle and (*b*) the mass flow rate of the steam.
Answers: (*a*) 41.0 percent, (*b*) 679.7 lbm/s

9-41 Write a computer program to determine the effect of reheat pressure on the performance of an ideal reheat Rankine cycle. The maximum and minimum pressures in the cycle are 15 MPa and 10 kPa, respectively. Steam enters both stages of the turbine at 500°C. Determine the thermal efficiency of the cycle, and plot it against reheat pressure for reheat pressures of 12.5, 10, 7, 5, 2, 1, 0.5, and 0.1 MPa.

9-42 Write a computer program to determine the effect of the number of reheat stages on the performance of an ideal reheat Rankine cycle. The maximum and minimum pressures in the cycle are 15 MPa and 10 kPa, respectively. Steam enters all stages of the turbine at 500°C. Determine the thermal efficiency of the cycle, and plot it against the number of reheat stages of one, two, four, and eight reheat stages. For each case, maintain roughly the same pressure ratio across each turbine stage.

Regenerative Rankine Cycle

9-43C How do the following quantities change when the simple ideal Rankine cycle is modified with regeneration? Assume the mass flow rate is the same.

Turbine work output:	(*a*) increases, (*b*) decreases, (*c*) remains the same
Heat added:	(*a*) increases, (*b*) decreases, (*c*) remains the same
Heat rejected:	(*a*) increases, (*b*) decreases, (*c*) remains the same
Moisture content at turbine exit:	(*a*) increases, (*b*) decreases, (*c*) remains the same

9-44C During a regeneration process, some steam is extracted from the turbine and is used to heat the liquid water leaving the pump. This does not seem like a smart thing to do since the extracted steam could produce some more work in the turbine. How do you justify this action?

9-45C How do open feedwater heaters differ from closed feedwater heaters?

9-46C Consider a simple ideal Rankine cycle and an ideal regenerative Rankine cycle with one open feedwater heater. The two cycles are very much alike, except the feedwater in the regenerative cycle is heated by extracting some steam just before it enters the turbine. How would you compare the efficiencies of these two cycles?

9-47C Devise an ideal regenerative Rankine cycle that has the same thermal efficiency as the Carnot cycle. Show the cycle on a *T-s* diagram.

9-48 A steam power plant operates on an ideal regenerative Rankine cycle. Steam enters the turbine at 6 MPa and 450°C and is condensed in the condenser at 20 kPa. Steam is extracted from the turbine at 0.4 MPa to heat the feedwater in an open feedwater heater. Water leaves the feedwater heater as a saturated liquid. Show the cycle on a *T-s* diagram, and determine (*a*) the net work output per kilogram of steam flowing through the boiler and (*b*) the thermal efficiency of the cycle.
Answers: (*a*) 1016 kJ/kg, (*b*) 37.8 percent

9-49 Repeat Prob. 9-48 by replacing the open feedwater heater with a closed feedwater heater. Assume that the feedwater leaves the heater at the condensation temperature of the extracted steam and that the extracted steam leaves the heater as a saturated liquid and is pumped to the line carrying the feedwater.

9-50 Consider a steam power plant that operates on a regenerative Rankine cycle and has a net power output of 300 MW. Steam enters the turbine at 10 MPa and 500°C and the condenser at 10 kPa. The adiabatic efficiency of the turbine is 80 percent, and that of the pumps is 95 percent. Steam is extracted from the turbine at 0.5 MPa to heat the feedwater in an open feedwater heater. Water leaves the feedwater heater as a saturated liquid. Show the cycle on a *T-s* diagram, and determine (*a*) the mass flow rate of steam through the boiler and (*b*) the thermal efficiency of the cycle.

9-50E Consider a steam power plant that operates on a regenerative Rankine cycle and has a net power output of 300 MW. Steam enters the turbine at 1500 psia and 1000°F and the condenser at 2 psia. The adiabatic efficiency of the turbine is 80 percent, and that of the pumps is 75. Steam is extracted from the turbine at 120 psia to heat the feedwater in an open feedwater heater. Water leaves the feedwater heater as a saturated liquid. Show the cycle on a *T-s* diagram, and determine (*a*) the mass flow rate of steam through the boiler and (*b*) the thermal efficiency of the cycle.
Answers: (*a*) 708.2 lbm/s, (*b*) 34.3 percent

9-51 Repeat Prob. 9-50, assuming both the pump and the turbine are isentropic.

9-51E Repeat Prob. 9-50E, assuming both the pump and the turbine are isentropic.

9-52 A steam power plant operates on an ideal regenerative Rankine cycle with two open feedwater heaters. Steam enters the turbine at 10 MPa and 600°C and exhausts to the condenser at 5 kPa. Steam is extracted from the turbine at 0.6 and 0.2 MPa. Water leaves both feedwater heaters as a saturated liquid. The mass flow rate of steam through the boiler is 6 kg/s. Show the cycle on a *T-s* diagram, and determine (*a*) the net power output of the power plant and (*b*) the thermal efficiency of the cycle. *Answers:* (*a*) 8181 kW, (*b*) 46.3 percent

9-53 Consider an ideal steam regenerative Rankine cycle with two feedwater heaters, one closed and one open. Steam enters the turbine at 12.5 MPa and 550°C and exhausts to the condenser at 10 kPa. Steam is extracted from the turbine at 0.8 MPa for the closed feedwater heater and at 0.3 MPa for the open one. The feedwater is heated to the condensation temperature of the extracted steam in the closed feedwater heater. The extracted steam leaves the closed feedwater heater as a saturated liquid, which is subsequently throttled to the open feedwater heater. Show the cycle on a *T-s* diagram with respect to saturation lines, and determine (*a*) the mass flow rate of steam through the boiler for a net power output of 100 MW and (*b*) the thermal efficiency of the cycle.

9-54 A steam power plant operates on an ideal reheat-regenerative Rankine cycle and has a net power output of 80 MW. Steam enters the high-pressure turbine at 10 MPa and 550°C and leaves at 0.8 MPa. Some steam is extracted at this pressure to heat the feedwater in an open feedwater heater. The rest of the steam is reheated to 500°C and is expanded in the low-pressure turbine to the condenser pressure of 10 kPa. Show the cycle on a *T-s* diagram with respect to saturation lines, and determine (*a*) the mass flow rate of steam through the boiler and (*b*) the thermal efficiency of the cycle. *Answers:* (*a*) 54.56 kg/s, (*b*) 44.4 percent

9-54E A steam power plant operates on an ideal reheat-regenerative Rankine cycle and has a net power output of 80 MW. Steam enters the high-pressure turbine at 1500 psia and 1000°F and leaves at 120 psia. Some steam is extracted at this pressure to heat the feedwater in an open feedwater heater. The rest of the steam is reheated to 1000°F and expanded in the low-pressure turbine to the condenser pressure of 1 psia. Show the cycle on a *T-s* diagram with respect to saturation lines, and determine (*a*) the mass flow rate of steam through the boiler and (*b*) the thermal efficiency of the cycle.

9-55 Repeat Prob. 9-54 by replacing the open feedwater heater with a closed feedwater heater. Assume that the feedwater leaves the heater at the condensation temperature of the extracted steam and that the extracted steam leaves the heater as a saturated liquid and is pumped to the line carrying the feedwater.

9-56 Consider an ideal reheat-regenerative Rankine cycle with one open feedwater heater. The boiler pressure is 10 MPa, the condenser pressure is 15 kPa, the reheater pressure is 1 MPa, and the feedwater pressure is 0.6 MPa. Steam enters both the high- and low-pressure turbines at 500°C. Show the cycle on a *T-s* diagram with respect to saturation lines, and determine (*a*) the fraction of steam extracted for regeneration and (*b*) the thermal efficiency of the cycle. *Answers:* (*a*) 0.174, (*b*) 42.6 percent

9-57 Repeat Prob. 9-56, assuming an adiabatic efficiency of 84 percent for the turbines and 100 percent for the pumps.

9-58 A steam power plant operates on an ideal reheat-regenerative Rankine cycle with one reheater and two open feedwater heaters. Steam enters the high-pressure turbine at 10 MPa and 550°C and leaves the low-pressure turbine at 5 kPa. Steam is extracted from the turbine at 1 and 0.2 MPa, and it is reheated to 500°C at a pressure of 0.8 MPa. Water leaves both feedwater heaters as a saturated liquid. Heat is transferred to the steam in the boiler at a rate of 10^6 kJ/s. Show the cycle on a *T-s* diagram with respect to saturation lines, and determine (*a*) the mass flow rate of steam through the boiler, (*b*) the net power output of the plant, and (*c*) the thermal efficiency of the cycle.
Answers: (*a*) 290.6 kg/s, (*b*) 465.7 MW, (*c*) 46.6 percent

9-58E A steam power plant operates on an ideal reheat-regenerative Rankine cycle with one reheater and two open feedwater heaters. Steam enters the high-pressure turbine at 1500 psia and 1100°F and leaves the low-pressure turbine at 1 psia. Steam is extracted from the turbine at 250 and 40 psia, and it is reheated to 1000°F at a pressure of 140 psia. Water leaves both feedwater heaters as a saturated liquid. Heat is transferred to the steam in the boiler at a rate of 10^6 Btu/s. Show the cycle on a *T-s* diagram with respect to saturation lines, and determine (*a*) the mass flow rate of steam through the boiler, (*b*) the net power output of the plant, and (*c*) the thermal efficiency of the cycle.

9-59 A steam power plant operates on an ideal reheat-regenerative Rankine cycle with one reheater and two feedwater heaters, one open and one closed. Steam enters the high-pressure turbine at 15 MPa and 600°C and the low-pressure turbine at 1 MPa and 500°C. The condenser pressure is 5 kPa. Steam is extracted from the turbine at 0.6 MPa for the closed feedwater heater and at 0.2 MPa for the open feedwater heater. In the closed feedwater heater, the feedwater is heated to the condensation temperature of the extracted steam. The extracted steam leaves the closed feedwater heater as a saturated liquid, which is subsequently throttled to the open feedwater heater. Show the cycle on a *T-s* diagram with respect to saturation lines. Determine (*a*) the fraction of steam extracted from the turbine for the open feedwater heater, (*b*) the thermal efficiency of the cycle, and (*c*) the net power output for a mass flow rate of 20 kg/s through the boiler.

9-60 Write a computer program to determine the effect of extraction pressure on the performance of an ideal regenerative Rankine cycle with one open feedwater heater. Steam enters the turbine at 15 MPa and 600°C and the condenser at 10 kPa. Neglecting the pump work, determine the thermal efficiency of the cycle, and plot it against extraction pressure for extraction pressures of 12.5, 10, 7, 5, 2, 1, 0.5, 0.1, and 0.05 MPa.

9-61 Write a computer program to determine the effect of the number of regeneration stages on the performance of an ideal regenerative Rankine cycle. Steam enters the turbine at 15 MPa and 600°C and the condenser at 5 kPa. Neglecting the pump work, determine the thermal efficiency of the cycle, and plot it against the number of regeneration stages for 1, 2, 3, 4, 5, 6, 8, and 10 regeneration stages. For each case, maintain the temperature difference between any two regeneration stages about the same.

Second-Law Analysis of Vapor Power Cycles

9-62C Starting with Eq. 9-20, show that the irreversibility associated with a simple ideal Rankine cycle can be expressed as $i = q_{in}(\eta_{th,c} - \eta_{th})$, where η_{th} is efficiency of the Rankine cycle and $\eta_{th,c}$ is the efficiency of the Carnot cycle operating between the same temperature limits.

9-63 Determine the irreversibilities associated with each of the processes of the Rankine cycle described in Prob. 9-17, assuming a source temperature of 1500 K and a sink temperature of 290 K.

9-64 Determine the irreversibilities associated with each of the processes of the Rankine cycle described in Prob. 9-18, assuming a source temperature of 1500 K and a sink temperature of 290 K.
Answers: 0, 111.5 kJ/kg, 0, 172.8 kJ/kg

9-64E Determine the irreversibilities associated with each of the processes of the Rankine cycle described in Prob. 9-18E, assuming a source temperature of 2500 R and a sink temperature of 520 R.

9-65 Determine the rate of irreversibility associated with the heat rejection process in Prob. 9-21. Assume a source temperature of 1500 K and a sink temperature of 290 K. Also determine the availability of the steam at the boiler exit. Take P_0 = 100 kPa.

9-66 Determine the irreversibility associated with the heat rejection process in Prob. 9-25. Assume a source temperature of 1500 K and a sink temperature of 290 K. Also determine the availability of the steam at the boiler exit. Take P_0 = 100 kPa.

9-67 Determine the irreversibilities associated with each of the processes of the reheat Rankine cycle described in Prob. 9-35. Assume a source temperature of 1500 K and a sink temperature of 290 K.

9-68 Determine the irreversibilities associated with the heat addition process and the expansion process in Prob. 9-36. Assume a source temperature of 1500 K and a sink temperature of 290 K. Also determine the availability of the steam at the boiler exit. Take P_0 = 100 kPa.
Answers: 997.4 kJ/kg, 249.9 kJ/kg, 1462.2 kJ/kg

9-68E Determine the irreversibilities associated with the heat addition process and the expansion process in Prob. 9-36E. Assume a source temperature of 2500 R and a sink temperature of 520 R. Also determine the availability of the steam at the boiler exit.

9-69 Determine the irreversibility associated with the regenerative cycle described in Prob. 9-48. Assume a source temperature of 1500 K and a sink temperature of 290 K. *Answer:* 1154.7 kJ/kg

9-70 Determine the irreversibility associated with the regeneration process described in Prob. 9-50. Assume a source temperature of 1300 K and a sink temperature of 303 K.

9-70E Determine the irreversibility associated with the regeneration process described in Prob. 9-50E. Assume a source temperature of 2300 R and a sink temperature of 530 R.

9-71 Determine the irreversibilities associated with the reheating and regeneration processes described in Prob. 9-54. Assume a source temperature of 1500 K and a sink temperature of 300 K.

9-71E Determine the irreversibilities associated with the reheating and regeneration processes described in Prob. 9-54E. Assume a source temperature of 2500 R and a sink temperature of 537 R.

9-72 Determine the irreversibilities associated with the reheating and regenerative processes described in Prob. 9-58. Assume a source temperature of 1500 K and a sink temperature of 290 K.
Answers: 193.0 kJ/kg, 104.0 kJ/kg

Cogeneration

9-73C How is the utilization factor ε_u for cogeneration plants defined? Could ε_u be unity for a cogeneration plant that does not produce any power?

9-74C Consider a cogeneration cycle for which the utilization factor is 1. Is the irreversibility associated with this cycle necessarily zero? Explain.

9-75C Consider a cogeneration cycle for which the utilization factor is 0.5. Can the irreversibility associated with this cycle be zero? If yes, under what conditions?

9-76C What is the difference between cogeneraion and regeneration?

9-77 Steam enters the turbine of a cogeneration plant at 7 MPa and 500°C. One-fourth of the steam is extracted from the turbine at 600-kPa pressure for process heating. The remaining steam continues to expand to 10 kPa. The extracted steam is then condensed and mixed with feedwater at constant pressure and the mixture is pumped to the boiler pressure of 7 MPa. The mass flow rate of steam through the boiler is 15 kg/s. Disregarding any pressure drops and heat losses in the piping, and assuming the turbine and the pump to be isentropic, determine the net power produced and the utilization factor of the plant.

9-78 A large food processing plant requires 2 kg/s of saturated or slightly superheated steam at 0.4 MPa, which is extracted from the turbine of a cogeneration plant. The boiler generates steam at 8 MPa and 500°C at a rate of 5 kg/s, and the condenser pressure is 15 kPa. Steam leaves the process heater as a saturated liquid. It is then mixed with the feedwater at the same pressure and this mixture is pumped to the boiler pressure. Assuming both the pumps and the turbine have adiabatic efficiencies of 84 percent, determine (*a*) the rate of heat supply in the boiler and (*b*) the power output of the cogeneration plant.
Answers: (*a*) 15,054 kW, (*b*) 4303 kW

9-78E A large food processing plant requires 4 lbm/s of saturated or slightly superheated steam at 80 psia, which is extracted from the turbine of a cogeneration plant. The boiler generates steam at 1000 psia and 1000°F at a rate of 10 lbm/s, and the condenser pressure is 2 psia. Steam leaves the process heater as a saturated liquid. It is then mixed with the feedwater at the same pressure and this mixture is pumped to the boiler pressure. Assuming both the pumps and the turbine have adiabatic efficiencies of 84 percent, determine (*a*) the rate of heat supply in the boiler and (*b*) the power output of the cogeneration plant.
Answers: (*a*) 13,329 Btu/s, (*b*) 3957 kW

9-79 Steam is generated in the boiler of a cogeneration plant at 10 MPa and 450°C at a steady rate of 15 kg/s. Under normal operation, steam expands in a turbine to a pressure of 0.5 MPa and is then routed to the process heater, where it supplies the process heat. Steam leaves the process heater as a saturated liquid and is pumped to the boiler pressure. In this mode, no steam passes through the condenser, which operates at 20 kPa.

(*a*) Determine the power produced and the rate at which process heat is supplied in this mode,
(*b*) Determine the power produced and the rate of process heat supplied if only 60 percent of the steam is routed to the process heater and the remainder is expanded to the condenser pressure.

9-80 Consider a cogeneration power plant modified with regeneration. Steam enters the turbine at 6 MPa and 450°C and expands to a pressure of 0.4 MPa. At this pressure, 60 percent of the steam is extracted from the turbine, and the remainder expands to 10 kPa. Part of the extracted steam is used to heat the feedwater in an open feedwater heater. The rest of the extracted steam is used for process heating and leaves the process heater as a saturated liquid at 0.4 MPa. It is subsequently mixed with the feedwater leaving the feedwater heater, and the mixture is pumped to the boiler pressure. Show the cycle on a *T-s* diagram with respect to saturation lines, and determine the mass flow rate of steam through the boiler for a net power output of 2 MW. *Answer:* 2.37 kg/s

9-81 Consider a cogeneration power plant which is modified with reheat and which produces 3 MW of power and supplies 7 MW of process heat. Steam enters the high-pressure turbine at 8 MPa and 500°C and expands to a pressure of 1 MPa. At this pressure, part of the steam is extracted from the turbine and routed to the process heater, while the remainder is reheated to 500°C and expanded in the low-pressure turbine to the condenser pressure of 15 kPa. The condensate from the condenser is pumped to 1 MPa and is mixed with the extracted steam, which leaves the process heater as a compressed liquid at 120°C. The mixture is then pumped to the boiler pressure. Show the cycle on a *T-s* diagram with respect to saturation lines, and disregarding pump work, determine (*a*) the rate of heat input in the boiler and (*b*) the fraction of steam extracted for process heating.

9-82 Steam is generated in the boiler of a cogeneration plant at 6 MPa and 500°C at a rate of 6 kg/s. The plant is to produce power while meeting the process steam requiremetns for a certain industrial application. One-third of the steam leaving the boiler is throttled to a pressure of 0.8 MPa and routed to the process heater. The rest of the steam is expanded in an isentropic turbine to a pressure of 0.8 MPa and is also routed to the process heater. Steam leaves the process heater at 120°C. Neglecting the pump work, determine (*a*) the net power produced, (*b*) the rate of process heat supply, and (*c*) the utilization factor of this plant.
Answers: (*a*) 2203 kW, (*b*) 15,308 kW, (*c*) 1.0

9-82E Steam is generated in the boiler of a cogeneration plant at 800 psia and 900°F at a rate of 6 lbm/s. The plant is to produce power while meeting the process steam requirements for a certain industrial application. One-third of the steam leaving the boiler is throttled to a pressure of 120 psia and is routed to the process heater. The rest of the steam is expanded in an isentropic turbine to a pressure of 120 psia and is also routed to the process heater. Steam leaves the process heater at 240°F. Neglecting the pump work, determine (*a*) the net power produced, (*b*) the rate of process heat supply, and (*c*) the utilization factor of this plant.

Binary Vapor and Combined Gas-Vapor Power Cycles

9-83C What is a binary power cycle? What is its purpose?

9-84C By writing an energy balance on the heat exchanger of a binary vapor power cycle, obtain a relation for the ratio of mass flow rates of two fluids in terms of their enthalpies.

9-85C Why is steam not an ideal working fluid for vapor power cycles?

9-86C Why is mercury a suitable working fluid for the topping portion of a binary vapor cycle but not for the bottoming cycle?

9-87C What is the difference between the binary vapor power cycle and the combined gas-steam power cycle?

9-88C In combined gas-steam cycles, what is the energy source for the steam?

9-89C Why is the combined gas-steam cycle more efficient than either of the cycles operated alone?

9-90 The gas-turbine portion of a combined gas-steam power plant has a pressure ratio of 16. Air enters the compressor at 300 K at a rate of 70 kg/s and is heated to 1500 K in the combustion chamber. The combustion gases leaving the gas turbine are used to heat the steam to 400°C at 10 MPa in a heat exchanger. The combustion gases leave the heat exchanger at 420 K. The steam leaving the turbine is condensed at 15 kPa. Assuming all the compression and expansion processes to be isentropic, determine (*a*) the mass flow rate of the steam, (*b*) the net power output, and (*c*) the thermal efficiency of the combined cycle. For air, assume constant specific heats at room temperature.
Answers: (*a*) 6.38 kg/s, (*b*) 39,090 kW, (*c*) 66.3 percent

9-91 Consider a combined gas-steam power plant that has a net power output of 600 MW. The pressure ratio of the gas-turbine cycle is 14. Air enters the compressor at 300 K and the turbine at 1400 K. The combustion gases leaving the gas turbine are used to heat the steam at 8 MPa to 400°C in a heat exchanger. The combustion gases leave the heat exchanger at 460 K. An open feedwater heater is incorporated with the steam cycle which operates at a pressure of 0.6 MPa. The condenser pressure is 20 kPa. Assuming all the compression and expansion processes to be isentropic, determine (*a*) the mass flow rate ratio of air to steam, (*b*) the required rate of heat input in the combustion chamber, and (*c*) the thermal efficiency of the combined cycle.

9-91E Consider a combined gas-steam power plant that has a net power output of 600 MW. The pressure ratio of the gas-turbine cycle is 14. Air enters the compressor at 540 R and the turbine at 2500 R. The combustion gases leaving the gas turbine are used to heat the steam at 1000 psia to 800°F in a heat exchanger. The combustion gases leave the heat exchanger at 840 R. An open feedwater heater is incorporated with the steam cycle which operates at a pressure of 60 psia. The condenser pressure is 2 psia. Assuming all the compression and expansion processes to be isentropic, determine (*a*) the mass flow rate ratio of air to steam, (*b*) the required rate of heat input in the combustion chamber, and (*c*) the thermal efficiency of the combined cycle.

9-92 Repeat Prob. 9-91, assuming adiabatic efficiencies of 100 percent for the pump, 82 percent for the compressor, and 86 percent for the gas and steam turbines.

9-92E Repeat Prob. 9-91E, assuming adiabatic efficiencies of 100 percent for the pump, 82 percent for the compressor, and 86 percent for the gas and steam turbines.

9-93 The gas-turbine cycle of a combined gas-steam power plant has a

pressure ratio of 8. Air enters the compressor at 290 K and the turbine at 1400 K. The combustion gases leaving the gas turbine are used to heat the steam at 15 MPa to 450°C in a heat exchanger. The combustion gases leave the heat exchanger at 247°C. Steam expands in a high-pressure turbine to a pressure of 3 MPa and is reheated in the combustion chamber to 500°C before it expands in a low-pressure turbine to 10 kPa. The mass flow rate of steam is 20 kg/s. Assuming all the compression and expansion processes to be isentropic, determine (*a*) the mass flow rate of air in the gas-turbine cycle, (*b*) the rate of total heat input, and (*c*) the thermal efficiency of the combined cycle.
Answers: (*a*) 175.2 kg/s, (*b*) 186,840 kW, (*c*) 45.9 percent

9-94 Repeat Prob. 9-93, assuming adiabatic efficiencies of 100 percent for the pump, 80 percent for the compressor, and 85 percent for the gas and steam turbines.

Refrigeration Cycles

CHAPTER

10

A major application area of thermodynamics is *refrigeration*, which is the transfer of heat from a lower temperature region to a higher temperature one. Devices that produce refrigeration are called *refrigerators* (or *heat pumps*), and the cycles on which they operate are called *refrigeration cycles*. The most frequently used refrigeration cycle is the *vapor-compression refrigeration cycle* in which the refrigerant is vaporized and condensed alternately and is compressed in the vapor phase. Another well-known refrigeration cycle is the *gas refrigeration cycle* in which the refrigerant remains in the gaseous phase throughout. Other refrigeration cycles discussed in this chapter are *cascade refrigeration,* where more than one refrigeration cycle is used; *absorption refrigeration,* where the refrigerant is dissolved in a liquid before it is compressed; and *thermoelectric refrigeration,* where refrigeration is produced by the passage of electric current through two dissimilar materials.

10-1 ■ REFRIGERATORS AND HEAT PUMPS

We all know from experience that heat flows in the direction of decreasing temperature, i.e., from high-temperature regions to low-temperature ones. This heat transfer process occurs in nature without requiring any devices. The reverse process, however, cannot occur by itself. The transfer of heat from a low-temperature region to a high-temperature one requires special devices called **refrigerators**.

Refrigerators are cyclic devices, and the working fluids used in the refrigeration cycles are called **refrigerants** A refrigerator is shown schematically in Fig. 10-1*a*. Here Q_L is the magnitude of the heat removed from the refrigerated space at temperature T_L, Q_H is the magnitude of the heat rejected to the warm space at temperature T_H, and $W_{net,in}$ is the net work input to the refrigerator. As discussed in Chap. 5, Q_L and Q_H represent magnitudes and thus are positive quantities.

Another device that transfers heat from a low-temperature medium to a high-temperature one is the **heat pump**. Refrigerators and heat pumps are essentially the same devices; they differ in their objectives only. The objective of a refrigerator is to maintain the refrigerated space at a low temperature by removing heat from it. Discharging this heat to a higher temperature medium is merely a necessary part of the operation, not the purpose. The objective of a heat pump, however, is to maintain a heated space at a high temperature. This is accomplished by absorbing heat from a low-temperature source, such as well water or cold outside air in winter, and supplying this heat to a warmer medium such as a house (Fig. 10-1*b*).

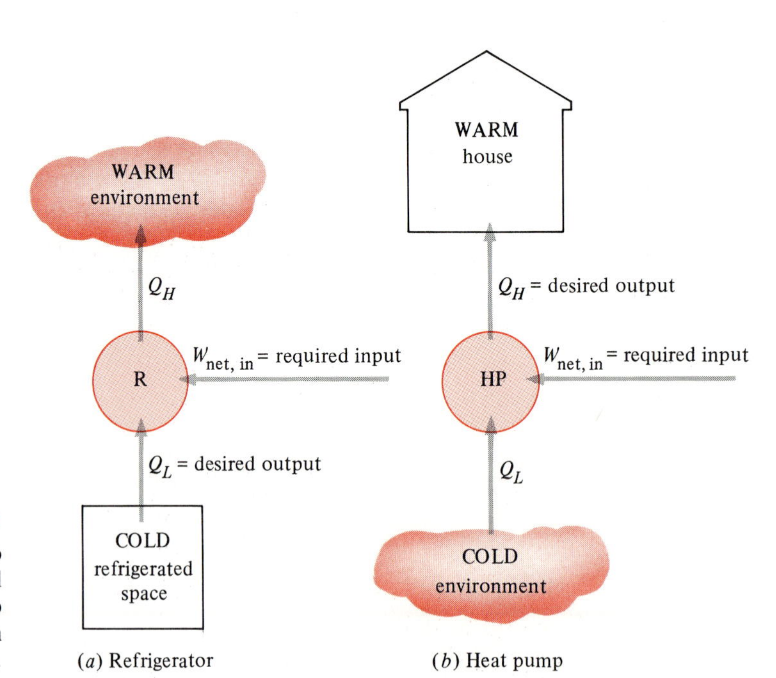

FIGURE 10-1
The objective of a refrigerator is to remove heat (Q_L) from the cold medium; the objective of a heat pump is to supply heat (Q_H) to a warm medium.

The performance of refrigerators and heat pumps is expressed in terms of the **coefficient of performance** (COP), which is defined as

$$\mathrm{COP_R} = \frac{\text{desired output}}{\text{required input}} = \frac{\text{cooling effect}}{\text{work input}} = \frac{Q_L}{W_{\text{net,in}}} \tag{10-1}$$

$$\mathrm{COP_{HP}} = \frac{\text{desired output}}{\text{required input}} = \frac{\text{heating effect}}{\text{work input}} = \frac{Q_H}{W_{\text{net,in}}} \tag{10-2}$$

These relations can also be expressed in rate form by replacing the quantities Q_L, Q_H, and $W_{\text{net,in}}$ by $\dot{Q}_L$, $\dot{Q}_H$, and $\dot{W}_{\text{net,in}}$, respectively. Notice that both $\mathrm{COP_R}$ and $\mathrm{COP_{HP}}$ can be greater than 1. A comparison of Eqs. 10-1 and 10-2 reveals that

$$\mathrm{COP_{HP}} = \mathrm{COP_R} + 1 \tag{10-3}$$

for fixed values of Q_L and Q_H. This relation implies that $\mathrm{COP_{HP}} > 1$ since $\mathrm{COP_R}$ is a positive quantity. That is, a heat pump will function, at worst, as a resistance heater, supplying as much energy to the house as it consumes. In reality, however, part of Q_H is lost to the outside air through piping and other devices, and $\mathrm{COP_{HP}}$ may drop below unity when the outside air temperature is too low. When this happens, the system usually switches to a resistance heating mode.

The cooling capacity of a refrigeration system at some specified conditions is often expressed in terms of **tons of refrigeration**. The capacity of a refrigeration system that can freeze 1 ton (2000 lbm) of liquid water at 0°C (32°F) into ice at 0°C in 24 h is said to be 1 ton. One ton of refrigeration is equivalent to 211 kJ/min or 200 Btu/min.

10-2 ■ THE REVERSED CARNOT CYCLE

You will recall from the preceding chapters that the Carnot cycle is a totally reversible cycle which consists of two reversible isothermal and two isentropic processes. It has the maximum thermal efficiency for given temperature limits, and it serves as a standard against which actual power cycles can be compared. The Carnot cycle proved to be a valuable tool in the study of gas and vapor power cycles discussed in Chaps. 8 and 9.

Since it is a reversible cycle, all four processes that comprise the Carnot cycle can be reversed. Reversing the cycle will also reverse the directions of any heat and work interactions. The result is a cycle which operates in the counterclockwise direction, which is called the **reversed Carnot cycle**. A refrigerator or heat pump which operates on the reversed Carnot cycle is called a **Carnot refrigerator** or a **Carnot heat pump**.

Consider a reversed Carnot cycle executed within the saturation dome of a refrigerant, as shown in Fig. 10-2. The refrigerant absorbs heat isothermally from a low-temperature source at T_L in the amount of Q_L (process 1-2), is compressed isentropically to state 3 (temperature rises to T_H), rejects heat isothermally to a high-temperature sink at T_H in the amount of Q_H (process 3-4), and expands isentropically to state 1 (tem-

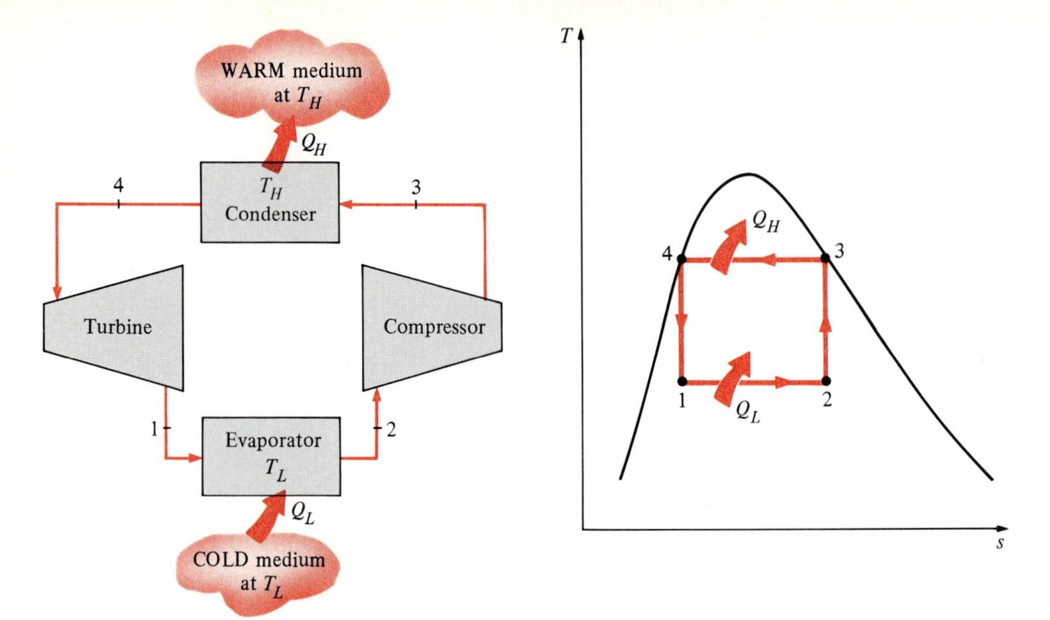

FIGURE 10-2
Schematic of a Carnot refrigerator and *T-s* diagram of the reversed Carnot cycle.

perature drops to T_L). The refrigerant changes from a saturated vapor state to a saturated liquid state in the condenser during process 3-4.

The coefficients of performance of Carnot refrigerators and heat pumps were determined in Sec. 5-11 to be

$$\text{COP}_{\text{R,Carnot}} = \frac{1}{T_H/T_L - 1} \tag{10-4}$$

and

$$\text{COP}_{\text{HP,Carnot}} = \frac{1}{1 - T_L/T_H} \tag{10-5}$$

Notice that both COPs increase as the difference between the two temperatures decreases, i.e., as T_L rises or T_H falls.

The reversed Carnot cycle is the *most efficient* refrigeration cycle operating between two specified temperature levels. Therefore, it is natural to look at it first as a prospective ideal cycle for refrigerators and heat pumps. If we could, we certainly would adapt it as the ideal cycle. But as explained below, the reversed Carnot cycle is an unsuitable model for refrigeration cycles.

The two isothermal heat transfer processes are not difficult to achieve in practice since maintaining a constant pressure automatically fixes the temperature of a two-phase mixture at the saturation value. Therefore, processes 1-2 and 3-4 can be approached closely in actual evaporators and condensers. However, processes 2-3 and 4-1 cannot be approximated closely in practice. This is because process 2-3 involves the compression of a liquid-vapor mixture which requires a compressor that will handle two phases, and process 4-1 involves the expansion of high-moisture-content refrigerant.

It seems as if these problems could be eliminated by executing the reversed Carnot cycle outside the saturation region. But in this case we will have difficulty in maintaining isothermal conditions during the heat absorption and heat rejection processes. Therefore, we conclude that the reversed Carnot cycle cannot be approximated in actual devices and is not a realistic model for refrigeration cycles. However, the reversed Carnot cycle can serve as a standard against which actual refrigeration cycles are compared.

10-3 ■ THE IDEAL VAPOR-COMPRESSION REFRIGERATION CYCLE

Many of the impracticalities associated with the reversed Carnot cycle can be eliminated by vaporizing the refrigerant completely before it is compressed and by replacing the turbine with a throttling device, such as an expansion valve or capillary tube. The cycle that results is called **ideal vapor-compression refrigeration cycle**, and it is shown schematically and on a *T-s* diagram in Fig. 10-3. The vapor-compression refrigeration cycle is the most widely used cycle for refrigerators, air conditioning systems, and heat pumps. It consists of four processes:

1-2 Isentropic compression in a compressor

2-3 P = constant heat rejection in a condenser

3-4 Throttling in an expansion device

4-1 P = constant heat absorption in an evaporator

In an ideal vapor-compression refrigeration cycle, the refrigerant enters the compressor at state 1 as saturated vapor and is compressed isentropically to the condenser pressure. The temperature of the refrigerant

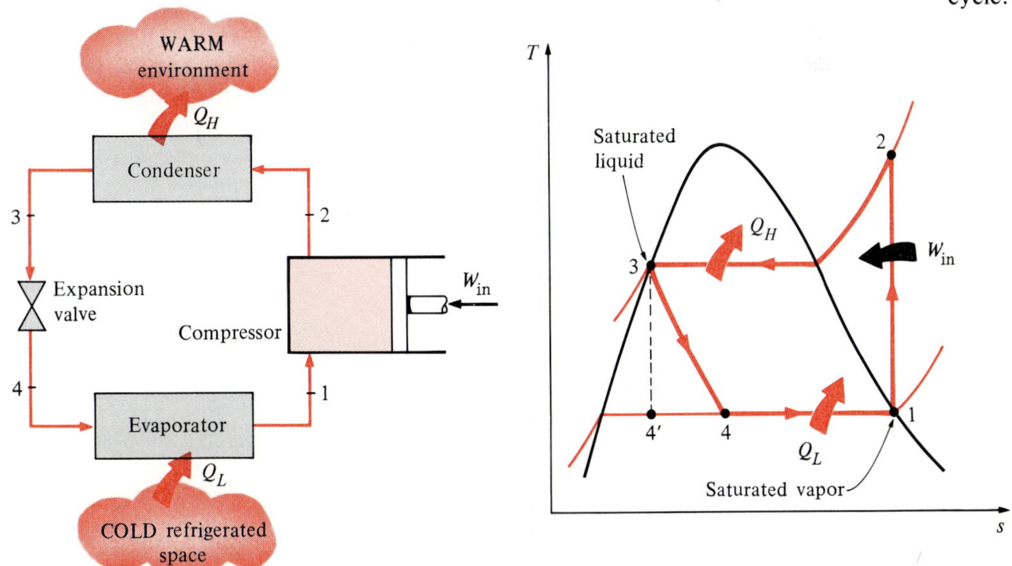

FIGURE 10-3
Schematic and *T-s* diagram for the ideal vapor-compression refrigeration cycle.

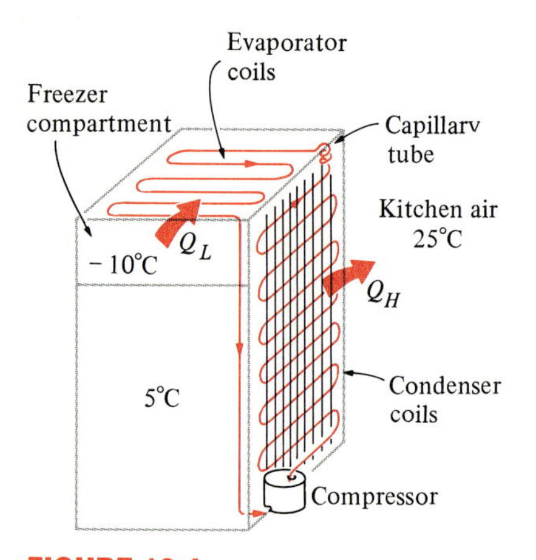

FIGURE 10-4
An ordinary household refrigerator.

increases during this isentropic compression process to well above the temperature of the surrounding medium, such as atmospheric air. The refrigerant then enters the condenser as superheated vapor at state 2 and leaves as saturated liquid at state 3 as a result of heat rejection to the surroundings. The temperature of the refrigerant at this state is still above the temperature of the surroundings.

The saturated liquid refrigerant at state 3 is throttled to the evaporator pressure by passing it through an expansion valve or capillary tube. The temperature of the refrigerant drops below the temperature of the refrigerated space during this process. The refrigerant enters the evaporator at state 4 as a low-quality saturated mixture, and it completely evaporates by absorbing heat from the refrigerated space. The refrigerant leaves the evaporator as saturated vapor and reenters the compressor, completing the cycle.

In a household refrigerator, the freezer compartment where heat is absorbed by the refrigerant, serves as the evaporator. The coils behind the refrigerator where heat is dissipated to the kitchen air serve as the condenser (Fig. 10-4).

Remember that the area under the process curve on a *T-s* diagram represents the heat transfer for internally reversible processes. The area under process curve 4-1 represents the heat absorbed by the refrigerant in the evaporator, and the area under process curve 2-3 represents the heat rejected in the condenser. Another diagram frequently used in the analysis of vapor-compression refrigeration cycles is the *P-h* diagram, as shown in Fig. 10-5. On this diagram, three of the four processes appear as straight lines, and the heat transfer in the condenser and the evaporator is proportional to the lengths of the corresponding process curves.

Notice that unlike the ideal cycles discussed before, the ideal vapor-compression refrigeration cycle is not an internally reversible cycle since it involves an irreversible (throttling) process. This process is maintained in the cycle to make it a more realistic model for the actual vapor-compression refrigeration cycle. If the throttling device were replaced by an isentropic turbine, the refrigerant would enter the evaporator at state 4′ instead of state 4. As a result, the refrigeration capacity would increase (by the area under process curve 4′-4 in Fig. 10-3) and the net work input

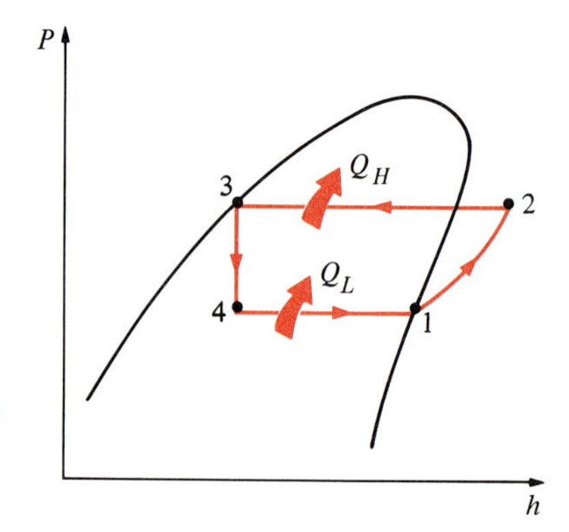

FIGURE 10-5
The *P-h* diagram of an ideal vapor-compression refrigeration cycle.

would decrease (by the amount of work output of the turbine). Replacing the expansion valve by a turbine is not practical, however, since the added benefits cannot justify the added expense and complexity.

All four components associated with the vapor-compression refrigeration cycle are steady-flow devices, and thus all four processes that make up the cycle can be analyzed as steady-flow processes. The kinetic and potential energy changes of the refrigerant are usually small relative to the work and heat transfer terms, and therefore they can be neglected. Then the steady-flow energy equation on a unit-mass basis reduces to (Eq. 4-26)

$$q - w = h_e - h_i \qquad (10\text{-}6)$$

The condenser and the evaporator do not involve any work, and the compressor can be approximated as adiabatic. Then the COPs of refrigerators and heat pumps operating on the vapor-compression refrigeration cycle can be expressed as

$$\text{COP}_\text{R} = \frac{q_L}{w_{\text{net,in}}} = \frac{h_1 - h_4}{h_2 - h_1} \qquad (10\text{-}7)$$

and

$$\text{COP}_\text{HP} = \frac{q_H}{w_{\text{net,in}}} = \frac{h_2 - h_3}{h_2 - h_1} \qquad (10\text{-}8)$$

where $h_1 = h_{g\,@\,P_1}$ and $h_3 = h_{f\,@\,P_3}$ for the ideal case.

EXAMPLE 10-1

A refrigerator uses refrigerant-12 as the working fluid and operates on an ideal vapor-compression refrigeration cycle between 0.14 and 0.8 MPa. If the mass flow rate of the refrigerant is 0.05 kg/s, determine (*a*) the rate of heat removal from the refrigerated space and the power input to the compressor, (*b*) the heat rejection rate to the environment, and (*c*) the COP of the refrigerator.

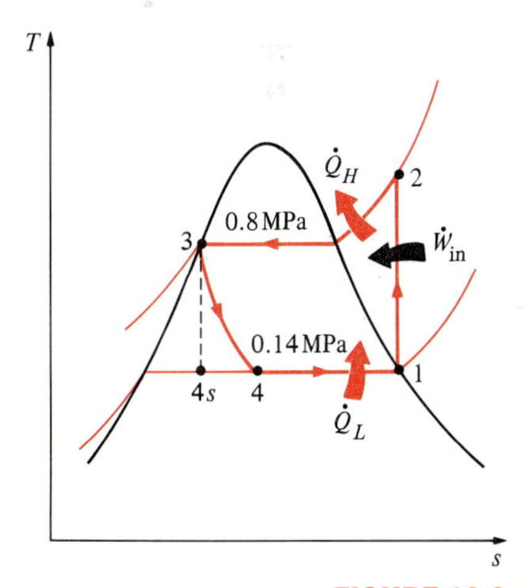

FIGURE 10-6
T-s diagram of the ideal vapor-compression refrigeration cycle described in Example 10-1.

Solution The refrigeration cycle is shown on a *T-s* diagram in Fig. 10-6. In an ideal vapor-compression refrigeration cycle, the compression process is isentropic, and the refrigerant enters the compressor as a saturated vapor at the evaporator pressure. Also, the refrigerant leaves the condenser as saturated liquid at the condenser pressure.

From refrigerant-12 tables, the enthalpies of the refrigerant at all four states are as follows:

$$P_1 = 0.14 \text{ MPa} \longrightarrow \begin{array}{l} h_1 = h_{g\,@\,0.14\text{ MPa}} = 177.87 \text{ kJ/kg} \\ s_1 = s_{g\,@\,0.14\text{ MPa}} = 0.7102 \text{ kJ/(kg·K)} \end{array}$$

$$\left.\begin{array}{l} P_2 = 0.8 \text{ MPa} \\ s_2 = s_1 \end{array}\right\} \longrightarrow h_2 = 208.65 \text{ kJ/kg} \qquad (T_2 = 43.5°\text{C})$$

$$P_3 = 0.8 \text{ MPa} \longrightarrow h_3 = h_{f\,@\,0.8\text{ MPa}} = 67.3 \text{ kJ/kg}$$

$$h_4 \cong h_3 \text{ (throttling)} \longrightarrow h_4 = 67.3 \text{ kJ/kg}$$

(*a*) The rate of heat removal from the refrigerated space and the power input to the compressor are determined from their definitions:

$$\dot{Q}_L = \dot{m}(h_1 - h_4) = (0.05 \text{ kg/s})[(177.87 - 67.30) \text{ kJ/kg}] = 5.53 \text{ kW}$$

and $\dot{W}_{in} = \dot{m}(h_2 - h_1) = (0.05 \text{ kg/s})[(208.65 - 177.87) \text{ kJ/kg}] = 1.54 \text{ kW}$

(*b*) The rate of heat rejection from the refrigerant to the environment is determined from

$$\dot{Q}_H = \dot{m}(h_2 - h_3) = (0.05 \text{ kg/s})[(208.65 - 67.3) \text{ kJ/kg}] = 7.07 \text{ kW}$$

It could also be determined from

$$\dot{Q}_H = \dot{Q}_L + \dot{W}_{in} = 5.53 + 1.54 = 7.07 \text{ kW}$$

(*c*) The coefficient of performance of the refrigerator is determined from its definition (Eq. 10-1):

$$\text{COP}_R = \frac{\dot{Q}_L}{\dot{W}_{in}} = \frac{5.53 \text{ kW}}{1.54 \text{ kW}} = 3.59$$

That is, this refrigerator removes 3.59 units of energy from the refrigerated space for each unit of electric energy it consumes.

Discussion It would be interesting to see what happens if the throttling valve were replaced by an isentropic turbine. The enthalpy at state 4s [the turbine exit, $P_{4s} = 0.14$ MPa, and $s_{4s} = s_3 = 0.2487$ kJ/(kg·K)] in this case would be 61.9 kJ/kg, and the turbine would produce 0.27 kW of power. This would decrease the power input to the refrigerator from 1.54 to 1.27 kW and increase the rate of heat removal from the refrigerated space from 5.53 to 5.80 kW. As a result, the COP of the refrigerator would increase from 3.59 to 4.57, an increase of 27.3 percent.

10-4 ■ ACTUAL VAPOR-COMPRESSION REFRIGERATION CYCLES

An actual vapor-compression refrigeration cycle differs from the ideal one in several ways, owing mostly to the irreversibilities that occur in various components. Two common sources of irreversibilities are fluid friction (causes pressure drops) and heat transfer to or from the surroundings. The *T-s* diagram of an actual vapor-compression refrigeration cycle is shown in Fig. 10-7.

In the ideal cycle, the refrigerant leaves the evaporator and enters the compressor as *saturated vapor*. This cannot be accomplished in practice, however, since it is not possible to control the state of the refrigerant so precisely. Instead, the system is designed so that the refrigerant is slightly superheated at the compressor inlet. This slight overdesign ensures that the refrigerant is completely vaporized when it enters the compressor. Also, the line connecting the evaporator to the compressor is usually very long, thus the pressure drop caused by fluid friction and heat transfer from the surroundings to the refrigerant can be very significant. The result of superheating, heat gain in the connecting line, and pressure drops in the evaporator and the connecting line is an increase in the specific volume, thus an increase in the power input requirements to the compressor since steady-flow work is proportional to the specific volume.

The *compression process* in the ideal cycle is internally reversible

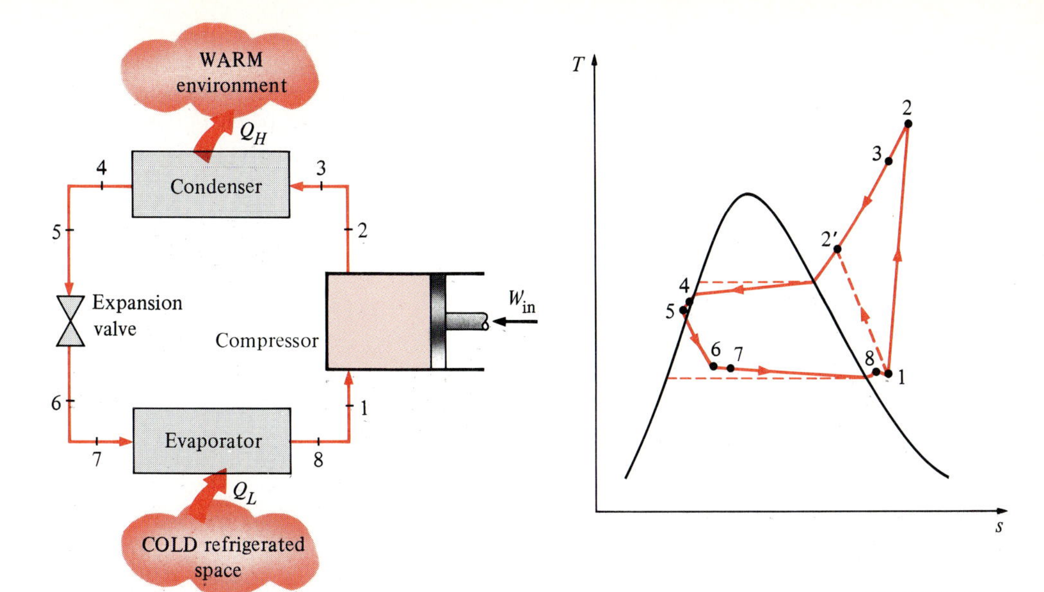

FIGURE 10-7
Schematic and *T-s* diagram for the actual vapor-compression refrigeration cycle.

and adiabatic, thus isentropic. The actual compression process, however, will involve frictional effects, which increases the entropy, and heat transfer, which may increase or decrease the entropy, depending on the direction. Therefore, the entropy of the refrigerant may increase (process 1-2) or decrease (process 1-2′) during an actual compression process, depending on which effects dominate. The compression process 1-2′ may be even more desirable than the isentropic compression process since the specific volume of the refrigerant and thus the work input requirement are smaller in this case. Therefore, the refrigerant should be cooled during the compression process whenever it is practical and economical to do so.

In the ideal case, the refrigerant is assumed to leave the condenser as *saturated liquid* at the compressor exit pressure. In actual situations, however, it is unavoidable to have some pressure drop in the condenser as well as in the lines connecting the condenser to the compressor and to the throttling valve. Also, it is not easy to execute the condensation process with such precision that the refrigerant is a saturated liquid at the end, and it is undesirable to route the refrigerant to the throttling valve before the refrigerant is completely condensed. Therefore, the refrigerant is subcooled somewhat before it enters the throttling valve. We do not mind this at all, however, since the refrigerant in this case enters the evaporator with a lower enthalpy and thus can absorb more heat from the refrigerated space. The throttling valve and the evaporator are usually located very close to each other, so the pressure drop in the connecting line is small.

EXAMPLE 10-2

Refrigerant-12 enters the compressor of a refrigerator as superheated vapor at 0.14 MPa and −20°C at a rate of 0.05 kg/s and leaves at 0.8 MPa and 50°C.

The refrigerant is cooled in the condenser to 26°C and 0.72 MPa, and is throttled to 0.15 MPa. Disregarding any heat transfer and pressure drops in the connecting lines between the components, determine (*a*) the rate of heat removal from the refrigerated space and the power input to the compressor, (*b*) the adiabatic efficiency of the compressor, and (*c*) the coefficient of performance of the refrigerator.

Solution The refrigeration cycle is shown on a *T-s* diagram in Fig. 10-8. From refrigerant-12 tables, the enthalpies of the refrigerant at all four states are

$$\left.\begin{aligned} P_1 &= 0.14 \text{ MPa} \\ T_1 &= -20°\text{C} \end{aligned}\right\} \quad h_1 = 179.01 \text{ kJ/kg}$$

$$\left.\begin{aligned} P_2 &= 0.8 \text{ MPa} \\ T_2 &= 50°\text{C} \end{aligned}\right\} \quad h_2 = 213.45 \text{ kJ/kg}$$

$$\left.\begin{aligned} P_3 &= 0.72 \text{ MPa} \\ T_3 &= 26°\text{C} \end{aligned}\right\} \quad h_3 \cong h_{f\,@\,26°\text{C}} = 60.68 \text{ kJ/kg}$$

$$h_4 \cong h_3 \text{ (throttling)} \longrightarrow h_4 = 60.58 \text{ kJ/kg}$$

(*a*) The rate of heat removal from the refrigerated space and the power input to the compressor are determined from their definitions:

$$\dot{Q}_L = \dot{m}(h_1 - h_4) = (0.05 \text{ kg/s})[(179.01 - 60.68) \text{ kJ/kg}] = 5.92 \text{ kW}$$

and $$\dot{W}_{in} = \dot{m}(h_2 - h_1) = (0.05 \text{ kg/s})[(213.45 - 179.01) \text{ kJ/kg}] = 1.72 \text{ kW}$$

(*b*) The adiabatic efficiency of the compressor is determined from

$$\eta_C \cong \frac{h_{2s} - h_1}{h_2 - h_1}$$

where the enthalpy at state 2*s* [$P_{2s} = 0.8$ MPa and $s_{2s} = s_1 = 0.7147$ kJ/(kg·K)] is 210.08 kJ/kg ($T_{2s} = 45.4°$C). Thus,

$$\eta_C = \frac{210.08 - 179.01}{213.45 - 179.01} = 0.902 \quad \text{(or 90.2\%)}$$

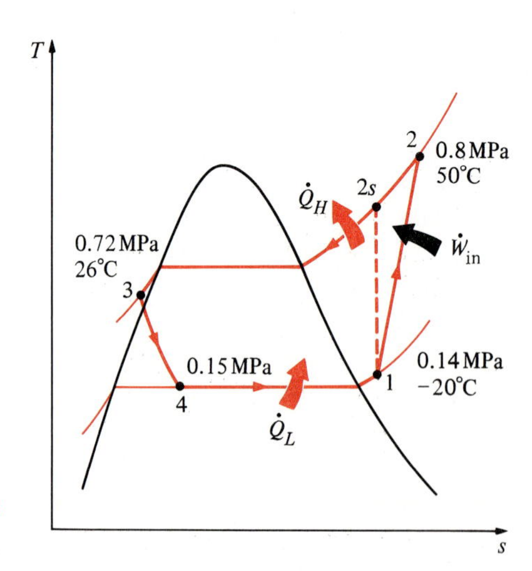

FIGURE 10-8
T-s diagram for Example 10-2.

(c) The coefficient of performance of the refrigerator is determined from its definition (Eq. 10-1):

$$COP_R = \frac{\dot{Q}_L}{\dot{W}_{in}} = \frac{5.92 \text{ kW}}{1.72 \text{ kW}} = 3.44$$

This problem is identical to the one worked out in Example 10-1, except that the refrigerant is slightly superheated at the compressor inlet and subcooled at the condenser exit. Also, the compressor is not isentropic. As a result, the heat removal rate from the refrigerated space increases (by 7.1 percent), but the power input to the compressor increases even more (by 11.7 percent). Consequently, the COP of the refrigerator decreases from 3.59 to 3.44.

10-5 ■ SELECTING THE RIGHT REFRIGERANT

There are several suitable refrigerants to choose from when designing a refrigeration system. Of these, fluorocarbon refrigerants (such as refrigerant-11, refrigerant-12, refrigerant-22, and refrigerant-502) account for over 90 percent of the market in the United States. The right choice for the refrigerant depends on the situation at hand. Two important parameters that need to be considered in the selection of a refrigerant are the temperatures of the two media (the refrigeration space and the environment) with which the refrigerant exchanges heat.

To have heat transfer at a reasonable rate, a temperature difference of 10°C or more should be maintained between the refrigerant and the medium it is exchanging heat with. If a refrigerated space is to be maintained at −10°C, for example, the temperature of the refrigerant should remain at about −20°C while it absorbs heat in the evaporator. The lowest pressure in a refrigeration cycle occurs in the evaporator, and this pressure should be above atmospheric pressure to prevent any air leakage into the refrigeration system. Therefore, a refrigerant should have a saturation pressure of 1 atm or higher at −20°C in this particular case. Ammonia and refrigerant-12 (usually marketed under the trade name Freon-12) are two such substances.

The temperature (and thus the pressure) of the refrigerant on the condenser side depends on the medium to which heat is rejected. Lower temperatures in the condenser (thus higher COPs) can be maintained if the refrigerant is cooled by liquid water instead of air. The use of water cooling cannot be justified economically, however, except in large industrial refrigeration systems. The temperature of the refrigerant in the condenser cannot fall below the temperature of the cooling medium (about 20°C for a household refrigerator), and the saturation pressure of the refrigerant at this temperature should be well below its critical pressure if the heat rejection process is to be approximately isothermal. If no single refrigerant can meet the temperature requirements, then two or more refrigeration cycles with different refrigerants can be used in series. Such a refrigeration system is called a *cascade system,* and it is discussed later in this chapter.

Other desirable characteristics of a refrigerant include being nontoxic, noncorrosive, nonflammable, and chemically stable; having a high enthalpy of vaporization (minimizes the mass flow rate); and, of course, being available at low cost.

In the case of heat pumps, the minimum temperature (and pressure) for the refrigerant may be considerably higher since heat is usually extracted from media that are well above the temperatures encountered in refrigeration systems.

10-6 ■ HEAT PUMP SYSTEMS

Heat pumps are generally more expensive to purchase and install than other heating systems, but they save money in the long run in some areas because they lower the heating bill. Despite their relatively higher initial costs, the popularity of heat pumps is increasing. About one-third of all single-family homes built in the United States in 1984 are heated by heat pumps.

The most common energy source for heat pumps is atmospheric air (air-to-air systems), although water and soil are also used. The major problem with air-source systems is *frosting,* which occurs in humid climates when the temperature falls below 2 to 5°C. The frost accumulation on the evaporator coils is highly undesirable since it seriously disrupts the heat transfer. The coils can be defrosted, however, by reversing the heat pump cycle (running it as an air conditioner). This results in a reduction in the efficiency of the system. Water-source systems usually use well water from depths of up to 80 m in the temperature range of 5 to 18°C, and they do not have a frosting problem. They typically have higher COPs but are more complex and require easy access to a large body of water such as underground water. Soil-source systems are also rather involved since they require long tubing placed deep in the ground where the soil temperature is relatively constant. The COP of heat pumps usually ranges between 1.5 and 4, depending on the particular system used and the temperature of the source.

Both the capacity and the efficiency of a heat pump fall significantly at low temperatures. Therefore, most air-source heat pumps require a supplementary heating system such as electric resistance heaters or an oil or gas furnace. Since water and soil temperatures do not fluctuate much, supplementary heating may not be required for water-source or soil-source systems. But the heat pump system must be large enough to meet the maximum heating load.

Heat pumps and air conditioners have the same mechanical components. Therefore, it is not economical to have two separate systems to meet the heating and cooling requirements of a building or a house. One system can be used as a heat pump in winter and an air conditioner in summer. This is accomplished by adding a reversing valve to the cycle, as shown in Fig. 10-9. As a result of this modification, the condenser of the heat pump (located indoors) functions as the evaporator of the air

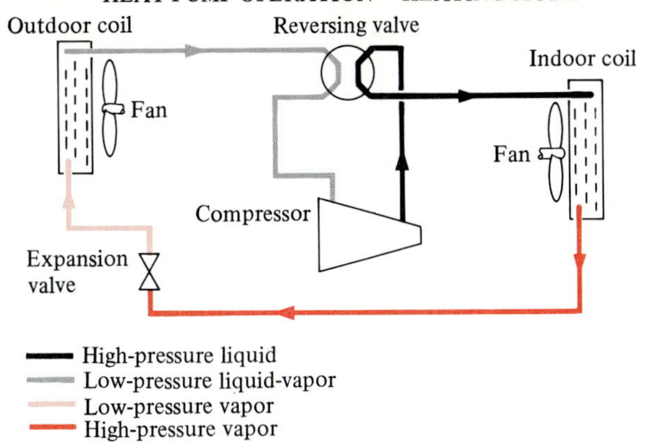

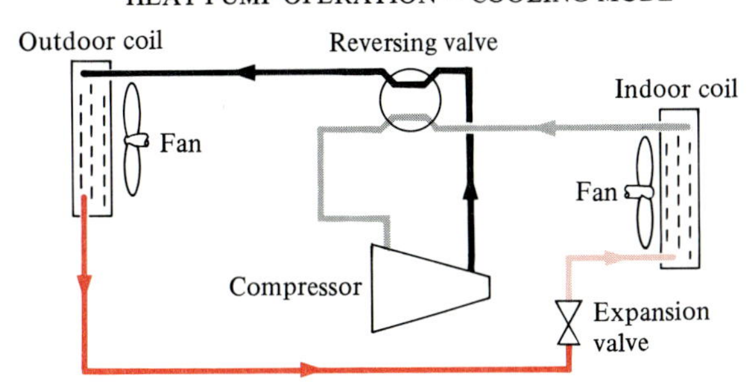

FIGURE 10-9

A heat pump can be used to heat a house in winter and to cool it in summer.

conditioner in summer. Also, the evaporator of the heat pump (located outdoors) serves as the condenser of the air conditioner. This feature increases the competitiveness of the heat pump. Such dual-purpose window units are commonly used in motels.

Heat pumps are most competitive in areas that have a large cooling load during the cooling season and a relatively small heating load during the heating season, such as in the southern parts of the United States. In these areas, the heat pump can meet the entire cooling and heating needs of residential or commercial buildings. The heat pump is least competitive in areas where the heating load is significant and the cooling load is small or nonexistent, such as in the northern parts of the United States.

10-7 ■ INNOVATIVE VAPOR-COMPRESSION REFRIGERATION SYSTEMS

The simple vapor-compression refrigeration cycle discussed above is the most widely used refrigeration cycle, and it is adequate for most refrigeration applications. The ordinary vapor-compression refrigeration systems are simple, inexpensive, reliable, and practically maintenance-free (when was the last time you serviced your household refrigerator?). However, for large industrial applications *efficiency,* not simplicity, is the major concern. Also for some applications the simple vapor-compression refrigeration cycle is inadequate and needs to be modified. We shall now discuss a few such modifications and refinements.

1 Cascade Refrigeration Systems

Some industrial applications require moderately low temperatures, and the temperature range they involve may be too large for a single vapor-compression refrigeration cycle to be practical. A large temperature range also means a large pressure range in the cycle and a poor performance for a reciprocating compressor. One way of dealing with such situations is to perform the refrigeration process in stages, that is, to have two or more refrigeration cycles which operate in series. Such refrigeration cycles are called **cascade refrigeration cycles**.

A two-stage cascade refrigeration cycle is shown in Fig. 10-10. The two cycles are connected through the heat exchanger in the middle, which serves as the evaporator for the topping cycle (cycle A) and the condenser for the bottoming cycle (cycle B). Assuming the heat exchanger is well insulated and the kinetic and potential energies are negligible, the heat transfer from the fluid in the bottoming cycle should be equal to the heat transfer to the fluid in the topping cycle. Thus the ratio of mass flow rates

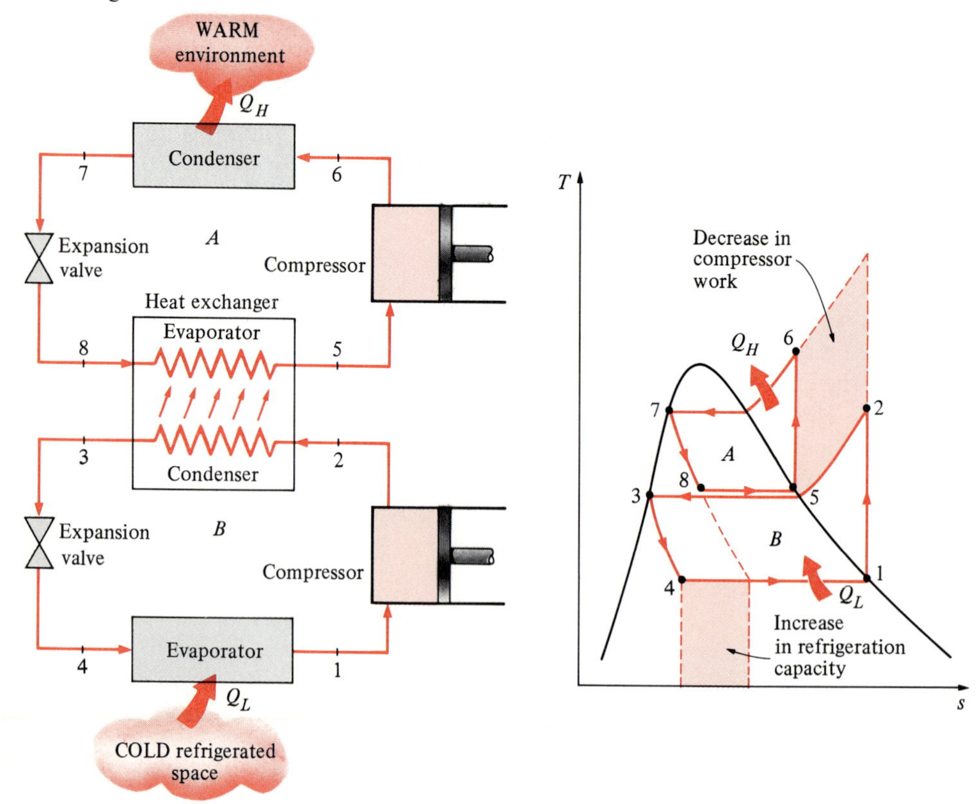

FIGURE 10-10
A two-stage cascade refrigeration system with the same refrigerant in both stages.

through each cycle should be

$$\dot{m}_A(h_5 - h_8) = \dot{m}_B(h_2 - h_3) \longrightarrow \frac{\dot{m}_A}{\dot{m}_B} = \frac{h_2 - h_3}{h_5 - h_8} \qquad (10\text{-}9)$$

Also,

$$\text{COP}_{\text{R,cascade}} = \frac{\dot{Q}_L}{\dot{W}_{\text{net,in}}} = \frac{\dot{m}_B(h_1 - h_4)}{\dot{m}_A(h_6 - h_5) + \dot{m}_B(h_2 - h_1)} \qquad (10\text{-}10)$$

In the cascade system shown in the figure, the refrigerants in both cycles are assumed to be the same. This is not necessary, however, since there is no mixing taking place in the heat exchanger. Therefore, refrigerants with more desirable characteristics can be used in each cycle. In this case, there would be a separate saturation dome for each fluid, and the *T-s* diagram for one of the cycles would be different. Also, in actual cascade refrigeration systems the two cycles would overlap somewhat since a temperature difference between the two fluids is needed for any heat transfer to take place.

It is evident from the *T-s* diagram in Fig. 10-10 that the compressor work decreases and the amount of heat absorbed from the refrigerated space increases as a result of cascading. Therefore, cascading improves the COP of a refrigeration system. Some refrigeration systems use three or four stages of cascading.

EXAMPLE 10-3

Consider a two-stage cascade refrigeration system operating between the pressure limits of 0.8 and 0.14 MPa. Each stage operates on an ideal vapor-compression refrigeration cycle with refrigerant-12 as the working fluid. Heat rejection from the lower cycle to the upper cycle takes place in an adiabatic counterflow heat exchanger where both streams enter at about 0.32 MPa. (In practice, the working fluid of the lower cycle will be at a higher pressure and temperature in the heat exchanger for effective heat transfer.) If the mass flow rate of the refrigerant through the upper cycle is 0.05 kg/s, determine (*a*) the mass flow rate of the refrigerant through the lower cycle, (*b*) the rate of heat removal from the refrigerated space and the power input to the compressor, and (*c*) the coefficient of performance of this cascade refrigerator.

Solution The two-stage cascade refrigeration cycle is shown on a *T-s* diagram in Fig. 10-11. Here cycle *A* is the upper cycle, and cycle *B* is the lower cycle. In ideal vapor-compression refrigeration cycles, the compression process is isentropic, and the refrigerant enters the compressor as a saturated vapor at the evaporator pressure. Also, the refrigerant leaves the condenser as a saturated liquid at the condenser pressure.

The enthalpies of the refrigerant at all eight states are determined from the refrigerant-12 tables and are indicated on the *T-s* diagram.

(*a*) The mass flow rate of the refrigerant through the lower cycle is determined from an energy balance on the heat exchanger:

$$\cancel{\dot{Q}}^{\,0} - \cancel{\dot{W}}^{\,0} = \Sigma\dot{m}_e h_e - \Sigma\dot{m}_i h_i$$

$$\dot{m}_A(h_5 - h_8) = \dot{m}_B(h_2 - h_3)$$

$$(0.05 \text{ kg/s})[(188.00 - 67.3) \text{ kJ/kg}] = \dot{m}_B[(191.97 - 37.08) \text{ kJ/kg}]$$

$$\dot{m}_B = 0.039 \text{ kg/s}$$

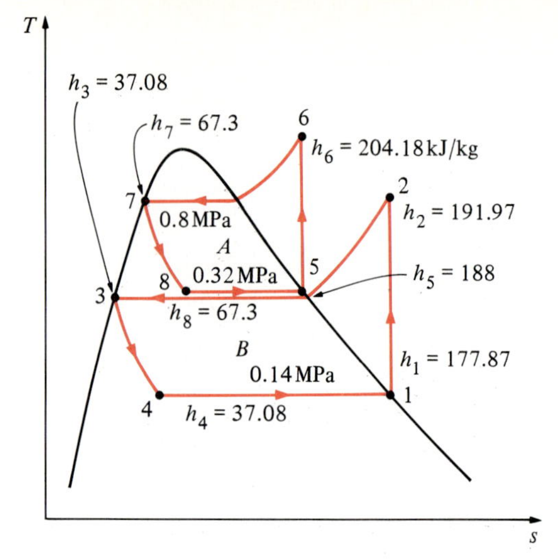

FIGURE 10-11

T-s diagram of the cascade refrigeration cycle described in Example 10-3.

(*b*) The rate of heat removal by a cascade cycle is the rate of heat absorption in the evaporator of the lowest stage. The power input to a cascade cycle is the sum of the power inputs to all the compressors:

$$\dot{Q}_L - \dot{m}_B(h_1 - h_4) = (0.039\,\text{kg/s})[(177.87 - 37.08)\,\text{kJ/kg}] = 5.49\,\text{kW}$$

$$\begin{aligned}\dot{W}_{\text{in}} &= \dot{W}_{\text{compI,in}} + \dot{W}_{\text{compII,in}} = \dot{m}_A(h_6 - h_5) + \dot{m}_B(h_2 - h_1) \\ &= (0.05\,\text{kg/s})[(204.18 - 188)\,\text{kJ/kg}] + (0.039\,\text{kg/s})[(191.97 - 177.87)\,\text{kJ/kg}] \\ &= 1.36\,\text{kW}\end{aligned}$$

(*c*) The COP of a refrigeration system is the ratio of the refrigeration rate to the net power input:

$$\text{COP}_R = \frac{\dot{Q}_L}{\dot{W}_{\text{net,in}}} = \frac{5.49\text{ kW}}{1.36\text{ kW}} = 4.04$$

This problem was worked out in Example 10-1 for a single-stage refrigeration system. Notice that the COP of the refrigeration system increases from 3.59 to 4.04 as a result of cascading. The COP of the system can be increased even more by increasing the number of cascade stages.

2 Multistage Compression Refrigeration Systems

When the fluid used throughout the cascade refrigeration system is the same, the heat exchanger between the stages can be replaced by a mixing chamber (called a *flash chamber*), since it has better heat transfer characteristics. Such systems are called **multistage compression refrigeration systems**. A two-stage compression refrigeration system is shown in Fig. 10-12.

In this system, the liquid refrigerant expands in the first expansion valve to the flash chamber pressure, which is the same as the compressor interstage pressure. Part of the liquid vaporizes during this process. This saturated vapor (state 7) is mixed with the superheated vapor from the

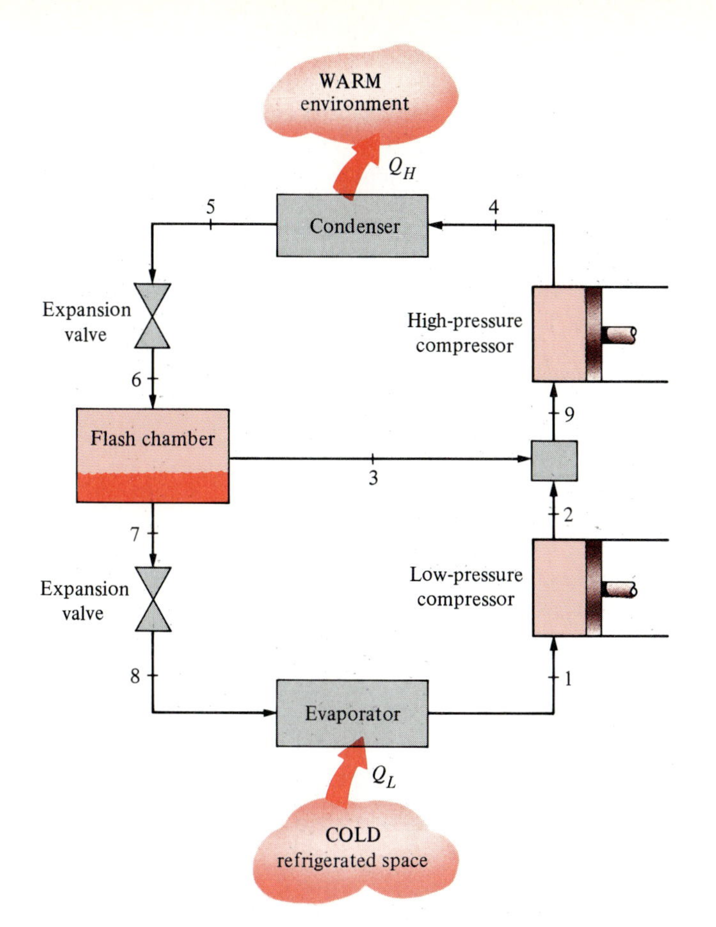

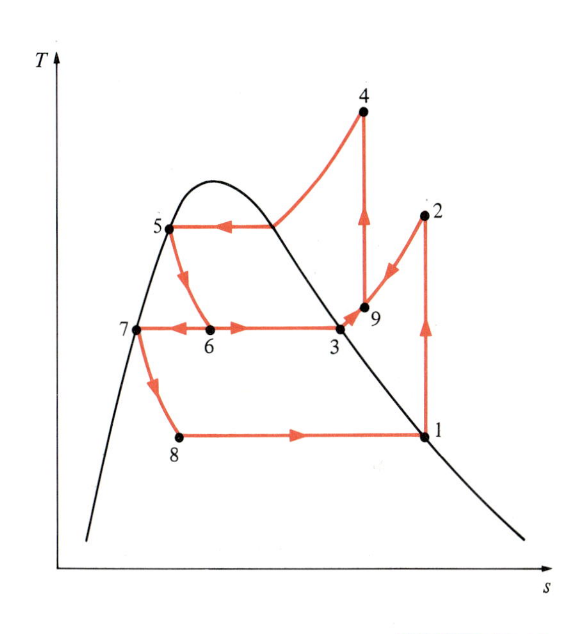

FIGURE 10-12

A two-stage compression refrigeration system with a flash chamber.

low-pressure compressor (state 2), and the mixture enters the high-pressure compressor at state 3. This is, in essence, a regeneration process. The saturated liquid (state 8) expands through the second expansion valve into the evaporator, where it picks up heat from the refrigerated space.

The compression process in this system resembles a two-stage compression with intercooling, and the compressor work decreases. Care should be exercised in the interpretations of the areas on the *T-s* diagram since the mass flow rates are different in different parts of the cycle.

EXAMPLE 10-4

Consider a two-stage compression refrigeration system operating between the pressure limits of 0.8 and 0.14 MPa. The working fluid is refrigerant-12. The refrigerant leaves the condenser as a saturated liquid and is throttled to a flash chamber operating at 0.32 MPa. Part of the refrigerant evaporates during this flashing process, and this vapor is mixed with the refrigerant leaving the low-pressure compressor. The mixture is then compressed to the condenser pressure by the high-pressure compressor. The liquid in the flash chamber is throttled to the evaporator pressure and cools the refrigerated space as it vaporizes in the evaporator. Assuming the refrigerant leaves the evaporator as a saturated vapor and both compressors are isentropic, determine (*a*) the fraction of the refrigerant which evaporates as it is throttled to the flash chamber, (*b*) the amount of heat removed from the refrigerated space and the compressor work per unit mass of refrigerant flowing through the condenser, and (*c*) the coefficient of performance.

Solution The two-stage compression refrigeration cycle is shown on a *T-s* diagram in Fig. 10-13. The enthalpies of the refrigerant at various states are determined from the refrigerant-12 tables and are indicated on this *T-s* diagram.

(*a*) The fraction of the refrigerant which evaporates as it is throttled to the flash chamber is simply the quality at state 6, which is

$$x_6 = \frac{h_6 - h_f}{h_{fg}} = \frac{67.3 - 37.08}{150.92} = 0.200$$

(*b*) The amount of heat removed from the refrigerated space and the compressor work input per unit mass of refrigerant flowing through the condenser are

$$q_L = (1 - x_6)(h_1 - h_8)$$
$$= (1 - 0.2)[(177.87 - 37.08)\text{ kJ/kg}] = 112.63\text{ kJ/kg}$$

and $$w_{in} = w_{compI,in} + w_{compII,in} = (1 - x_6)(h_2 - h_1) + (1)(h_4 - h_9)$$

The enthalpy at state 9 is determined from an energy balance on the mixing chamber:

$$\dot{Q}^{\nearrow 0} - \dot{W}^{\nearrow 0} = \Sigma\dot{m}_e h_e - \Sigma\dot{m}_i h_i$$
$$(1)h_9 = x_6 h_3 + (1 - x_6)h_2$$
$$h_9 = (0.2)(188.00) + (1 - 0.2)(191.97) = 191.18\text{ kJ/kg}$$

Also, $s_9 = 0.7074$ kJ/(kg·K). Thus the enthalpy at state 4 (0.8 MPa, $s_4 = s_9$) is $h_4 = 207.76$ kJ/kg. Substituting,

$$w_{in} = (1 - 0.2)[(191.97 - 177.87)\text{ kJ/kg}] + (207.76 - 191.18)\text{ kJ/kg}$$
$$27.86\text{ kJ/kg}$$

(*c*) The coefficient of performance is determined from

$$\text{COP}_R = \frac{q_L}{w_{in}} = \frac{112.63\text{ kJ/kg}}{27.86\text{ kJ/kg}} = 4.04$$

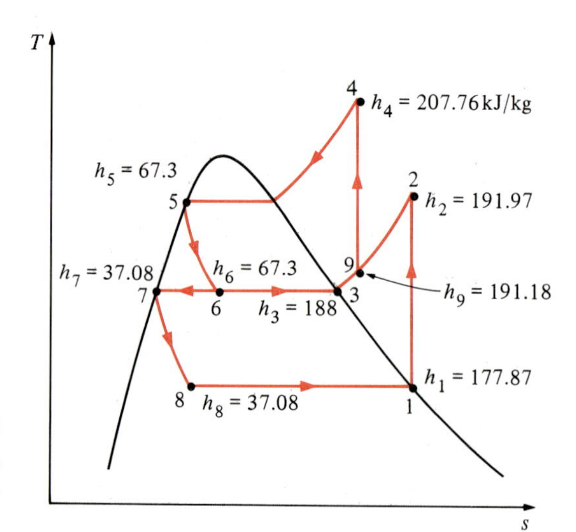

FIGURE 10-13

T-s diagram of the two-stage compression refrigeration cycle described in Example 10-4.

This problem was worked out in Example 10-1 for a single-stage refrigeration system (COP = 3.59) and in Example 10-3 for a two-stage cascade refrigeration system (COP = 4.04). Notice that the COP of the refrigeration system increased considerably relative to the single-stage compression, but did not change at all relative to the two-stage cascade compression.

3 Multipurpose Refrigeration Systems with a Single Compressor

Some applications require refrigeration at more than one temperature. This could be accomplished by using a separate throttling valve and a separate compressor for each evaporator operating at different temperatures. However, such a system will be bulky and probably uneconomical. A more practical and economical approach would be to route all the exit streams from the evaporators to a single compressor and let it handle the compression process for the entire system.

Consider, for example, an ordinary refrigerator-freezer unit. A simplified schematic of the unit and the *T-s* diagram of the cycle are shown in Fig. 10-14. Most refrigerated goods have a high water content, and the refrigerated space must be maintained above the ice point (at about 5°C) to prevent freezing. The freezer compartment, however, is maintained at about −15°C. Therefore, the refrigerant should enter the freezer at about −25°C to have heat transfer at a reasonable rate in the freezer. If a single expansion valve and evaporator were used, the refrigerant would have to circulate in both compartments at about −25°C, which would cause ice formation in the neighborhood of the evaporator coils and dehydration of the produce. This would not be acceptable to a household. This problem can be eliminated by throttling the refrigerant to a higher pressure (hence temperature) for use in the refrigerated space and then throttling it to the minimum pressure for use in the freezer. The entire refrigerant leaving

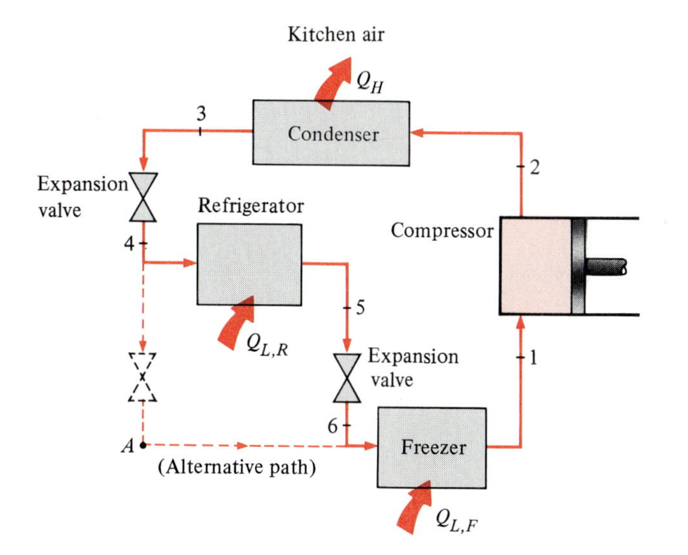

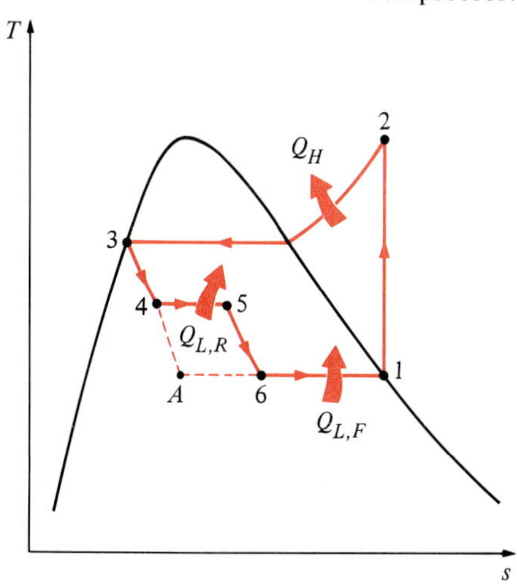

FIGURE 10-14
Schematic and *T-s* diagram for a refrigeration-freezer unit with one compressor.

the freezer compartment is subsequently compressed by a single compressor to the condenser pressure.

4 Liquefaction of Gases

The liquefaction of gases has always been an important area of refrigeration since many important scientific and engineering processes at cryogenic temperatures (temperatures below about $-100°C$) depend on liquefied gases. Some examples of such processes are the separation of oxygen and nitrogen from air, preparation of liquid propellants for rockets, study of material properties at low temperatures, and study of some exciting phenomena such as superconductivity.

At temperatures above the critical-point value, a substance exists in the gas phase only. The critical temperatures of helium, hydrogen, and nitrogen (three commonly used liquefied gases) are -268, -240, and $-147°C$, respectively. Therefore, none of these substances will exist in liquid form at atmospheric conditions. Furthermore, low temperatures of this magnitude cannot be obtained with ordinary refrigeration techniques. Then the question that needs to be answered in the liquefaction of gases is this: *How can we lower the temperature of a gas below its critical-point value?*

Several cycles, some complex and others simple, are used successfully for the liquefaction of gases. Below we discuss the Linde-Hampson cycle which is shown schematically and on a *T-s* diagram in Fig. 10-15.

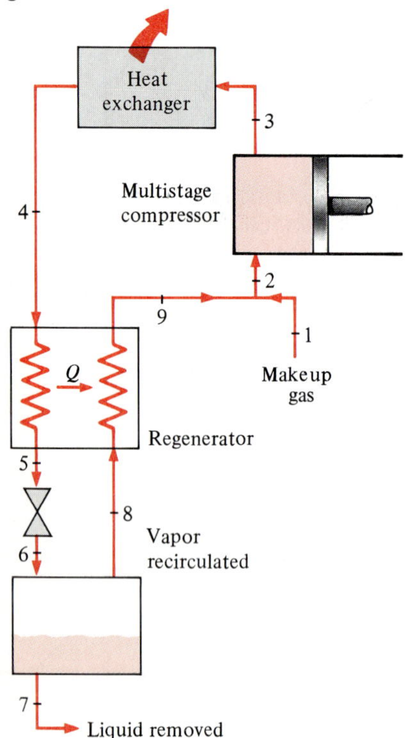

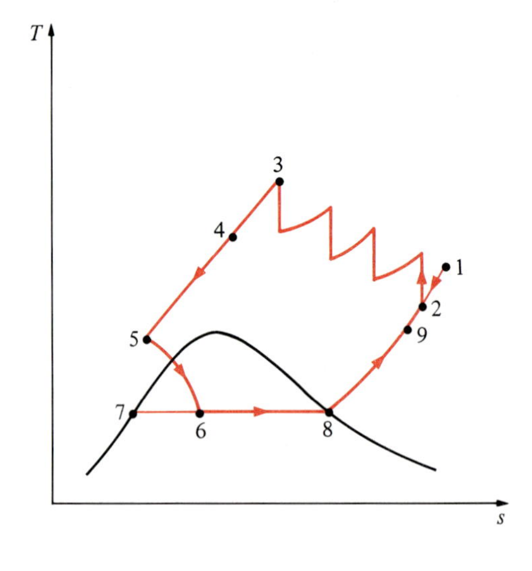

FIGURE 10-15
Linde-Hampson system for liquefying gases.

Makeup gas is mixed with the uncondensed portion of the gas from the previous cycle, and the mixture at state 2 is compressed by a multistage compressor to state 3. The compression process approaches an isothermal process due to intercooling. The high-pressure gas is cooled in an after-cooler by a cooling medium or by a separate external refrigeration system to state 4. The gas is further cooled in a regenerative counterflow heat exchanger by the uncondensed portion of gas from the previous cycle to state 5, and it is throttled to state 6, which is a saturated liquid–vapor mixture state. The liquid (state 7) is collected as the desired product, and the vapor (state 8) is routed through the regenerator to cool the high-pressure gas approaching the throttling valve. Finally, the gas is mixed with fresh makeup gas, and the cycle is repeated.

This and other refrigeration cycles used for the liquefaction of gases can also be used for the solidification of gases.

10-8 ■ GAS REFRIGERATION CYCLES

As explained in Sec. 10-2, the Carnot cycle (the standard of comparison for power cycles) and the reversed Carnot cycle (the standard of comparison for refrigeration cycles) are identical, except that the reversed Carnot cycle operates "backward." This suggests that the power cycles discussed in earlier chapters can be used as refrigeration cycles by simply reversing them. In fact, the vapor-compression refrigeration cycle is essentially a modified Rankine cycle operating in reverse. Another example is the reversed Stirling cycle, which is the cycle on which Stirling refrigerators operate. In this section, we discuss the *reversed Brayton cycle,* better known as the **gas refrigeration cycle**.

Consider the gas refrigeration cycle shown in Fig. 10-16. The sur-

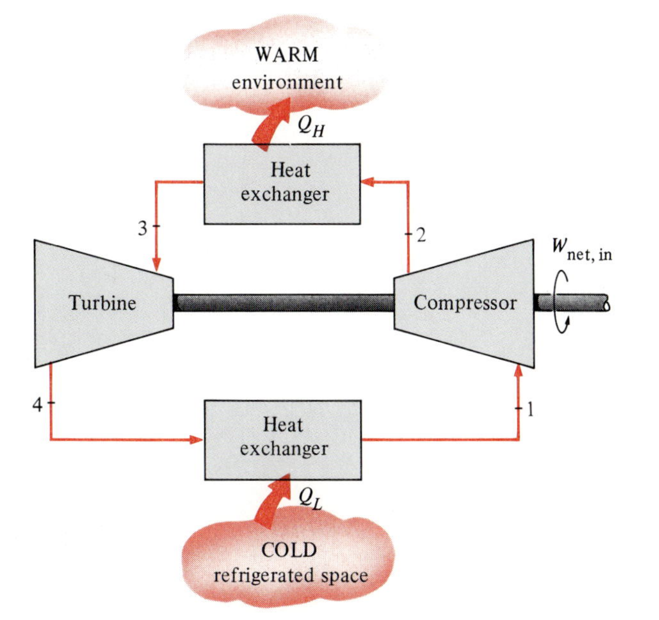

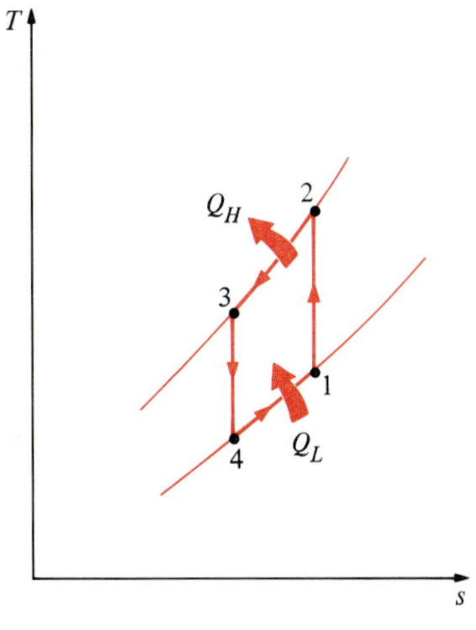

FIGURE 10-16
Simple gas refrigeration cycle.

roundings are at T_0, and the refrigerated space is to be maintained at T_L. The gas is compressed during process 1-2. The high-pressure, high-temperature gas at state 2 is then cooled at constant pressure to T_0 by rejecting heat to the surroundings. This is followed by an expansion process in a turbine, during which the gas temperature drops to T_4. (Can we achieve the cooling effect by using a throttling valve instead of a turbine?) Finally, the cool gas absorbs heat from the refrigerated space until its temperature rises to T_1.

All the processes described above are internally reversible, and the cycle executed is the *ideal* gas refrigeration cycle. In actual gas refrigeration cycles, the compression and expansion processes will deviate from the isentropic ones, and T_3 will be higher than T_0 unless the heat exchanger is infinitely large.

On a *T-s* diagram, the area under process curve 4-1 represents the heat removed from the refrigerated space; the enclosed area 1-2-3-4-1 represents the net work input. The ratio of these areas is the COP for the cycle, which may be expressed as

$$\text{COP}_\text{R} = \frac{q_L}{w_{\text{net,in}}} = \frac{q_L}{w_{\text{comp,in}} - w_{\text{turb,out}}} \tag{10-11}$$

where

$$q_L = h_1 - h_4$$
$$w_{\text{turb,out}} = h_3 - h_4$$
$$w_{\text{comp,in}} = h_2 - h_1$$

The gas refrigeration cycle deviates from the reversed Carnot cycle because the heat transfer processes are not isothermal. In fact, the gas temperature varies considerably during heat transfer processes. Consequently, the gas refrigeration cycles have lower COPs relative to the vapor-compression refrigeration cycles or the reversed Carnot cycle. This is also evident from the *T-s* diagram in Fig. 10-17. The reversed Carnot cycle consumes a fraction of the net work (rectangular area 1*A*3*B*1), but produces a greater amount of refrigeration (rectangular area under *B*1).

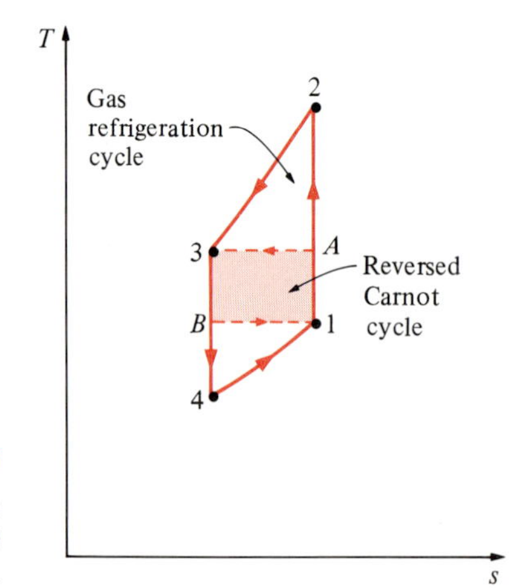

FIGURE 10-17

A reversed Carnot cycle produces more refrigeration (area under *B*1) with less work input (area 1*A*3*B*).

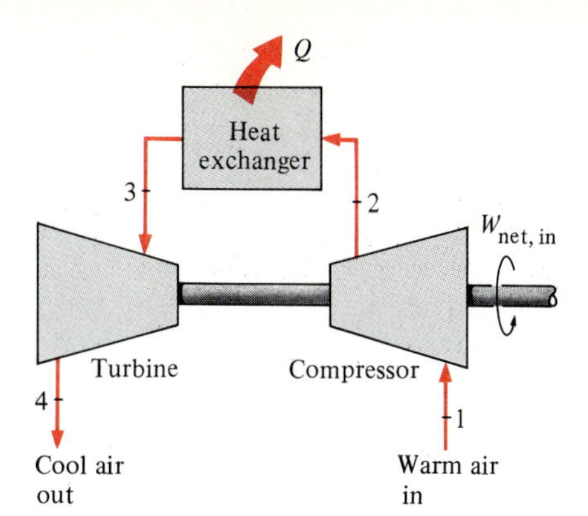

FIGURE 10-18
An open-cycle aircraft cooling system.

Despite their relatively low COPs, the gas refrigeration cycles have two desirable characteristics: They involve simple, lighter components, which make them suitable for aircraft cooling, and they can incorporate regeneration, which makes them suitable for liquefaction of gases and cryogenic applications. An aircraft cooling system, which operates on an open cycle, is shown in Fig. 10-18. Atmospheric air is compressed by a compressor, cooled by the surrounding air, and expanded in a turbine. The cool air leaving the turbine is then directly routed to the cabin.

The regenerative gas cycle is shown in Fig. 10-19. Regenerative cooling is achieved by inserting a counterflow heat exchanger into the cycle. Without regeneration, the lowest turbine inlet temperature is T_0, the temperature of the surroundings or any other cooling medium. With regeneration, the high-pressure gas is further cooled to T_4 before expanding in the turbine. Lowering the turbine inlet temperature automatically lowers the turbine exit temperature, which is the minimum temperature in the

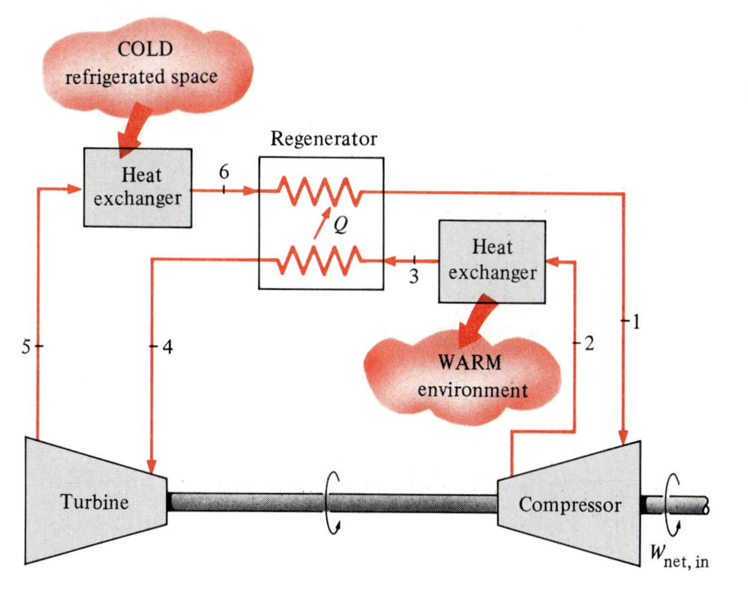

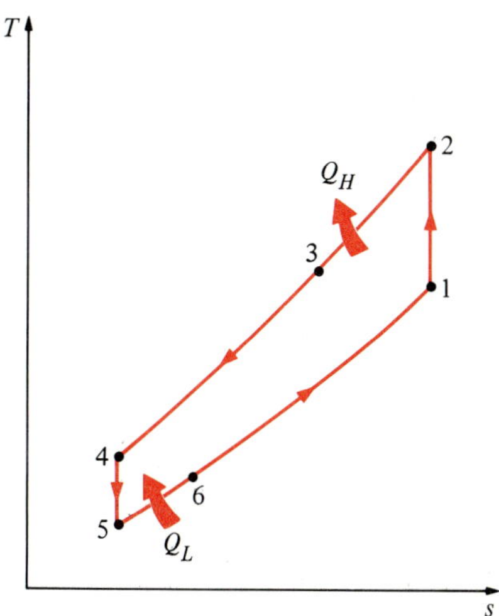

FIGURE 10-19
Gas refrigeration cycle with regeneration.

cycle. Extremely low temperatures can be achieved by repeating this procedure.

EXAMPLE 10-5

An ideal-gas refrigeration cycle using air as the working medium is to maintain a refrigerated space at 0°F while rejecting heat to the surrounding medium at 80°F. The pressure ratio of the compressor is 4. Determine (*a*) the maximum and minimum temperatures in the cycle, (*b*) the coefficient of performance, and (*c*) the rate of refrigeration for a mass flow rate of 0.1 lbm/s.

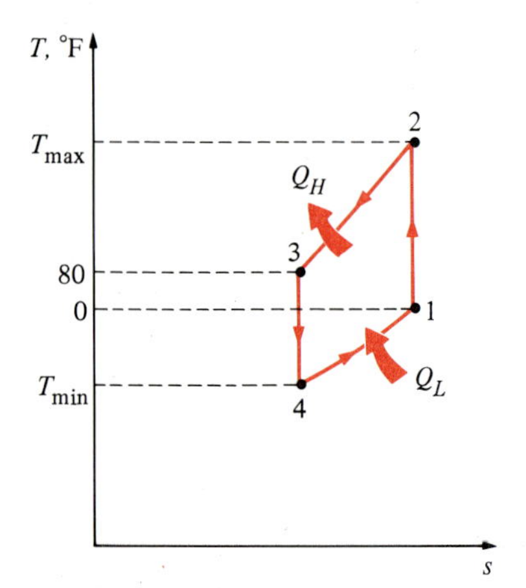

FIGURE 10-20

T-s diagram of the ideal-gas refrigeration cycle described in Example 10-5.

Solution The *T-s* diagram of the cycle is shown in Fig. 10-20. Since this is an ideal cycle, we assume that both the turbine and the compressor are isentropic, and the turbine inlet temperature is 80°F, the temperature of the surroundings. At the specified conditions, air can be assumed to be an ideal gas.

(*a*) The maximum and minimum temperatures in the cycle are determined from the isentropic relations of ideal gases for the compression and expansion processes:

$$T_1 = 460 \text{ R} \longrightarrow h_1 = 109.90 \text{ Btu/lbm} \quad \text{and} \quad P_{r1} = 0.7913$$

$$P_{r2} = \frac{P_2}{P_1} P_{r1} = (4)(0.7913) = 3.165 \longrightarrow \begin{cases} h_2 = 163.5 \text{ Btu/lbm} \\ T_2 = 683 \text{ R} \quad (\text{or } 223°\text{F}) \end{cases}$$

$$T_3 = 540 \text{ R} \longrightarrow h_3 = 129.06 \text{ Btu/lbm} \quad \text{and} \quad P_{r3} = 1.3860$$

$$P_{r4} = \frac{P_4}{P_3} P_{r3} = (0.25)(1.386) = 0.3465 \longrightarrow \begin{cases} h_4 = 86.7 \text{ Btu/lbm} \\ T_4 = 363 \text{ R} \quad (\text{or } -97°\text{F}) \end{cases}$$

Therefore, the highest and the lowest temperatures in the cycle are 223 and −97°F, respectively.

(*b*) The COP of this ideal gas refrigeration cycle is determined from Eq. 10-11:

$$\text{COP}_\text{R} = \frac{q_L}{w_{\text{net,in}}} = \frac{q_L}{w_{\text{comp,in}} - w_{\text{turb,out}}}$$

where

$$q_L = h_1 - h_4 = 109.9 - 86.7 = 23.2 \text{ Btu/lbm}$$

$$w_{\text{turb,out}} = h_3 - h_4 = 129.06 - 86.7 = 42.36 \text{ Btu/lbm}$$

$$w_{\text{comp,in}} = h_2 - h_1 = 163.5 - 109.9 = 53.6 \text{ Btu/lbm}$$

Thus,

$$\text{COP}_\text{R} = \frac{23.2}{53.6 - 42.36} = 2.06$$

It is worth noting that an ideal vapor-compression cycle working under similar conditions would have a COP greater than 3.

(*c*) The rate of refrigeration is

$$\dot{Q}_{\text{refrig}} = \dot{m}(q_L) = (0.1 \text{ lbm/s})(23.2 \text{ Btu/lbm}) = 2.32 \text{ Btu/s}$$

10-9 ABSORPTION REFRIGERATION SYSTEMS

Another form of refrigeration which becomes economically attractive when there is a source of inexpensive heat energy at a temperature of 100 to

200°C is **absorption refrigeration**. Some examples of inexpensive heat energy sources include geothermal energy, solar energy, and waste heat from cogeneration or process steam plants; that is, heat energy that otherwise would be wasted.

As the name implies, absorption refrigeration systems involve the absorption of a *refrigerant* by a *transport medium*. The most widely used absorption refrigeration system is the ammonia-water system, where ammonia (NH_3) serves as the refrigerant and water (H_2O) as the transport medium. Other absorption refrigeration systems include water–lithium bromide and water–lithium chloride systems, where water serves as the refrigerant. The latter two systems are limited to applications such as air conditioning where the minimum temperature is above the freezing point of water (0°C).

To understand the basic principles involved in absorption refrigeration, we examine the NH_3-H_2O system shown in Fig. 10-21. You will immediately notice that this system looks very much like the vapor-compression system, except that the compressor has been replaced by a complex absorption mechanism consisting of an absorber, a pump, a generator, a regenerator, a valve, and a rectifier. Once the pressure of NH_3 is raised by the components in the box (this is the only thing they are set up to do), it is cooled and condensed in the condenser by rejecting heat to the surroundings, is throttled to the evaporator pressure, and picks up heat from the refrigerated space as it flows through the evaporator. So, there is nothing new there. Here is what happens in the box:

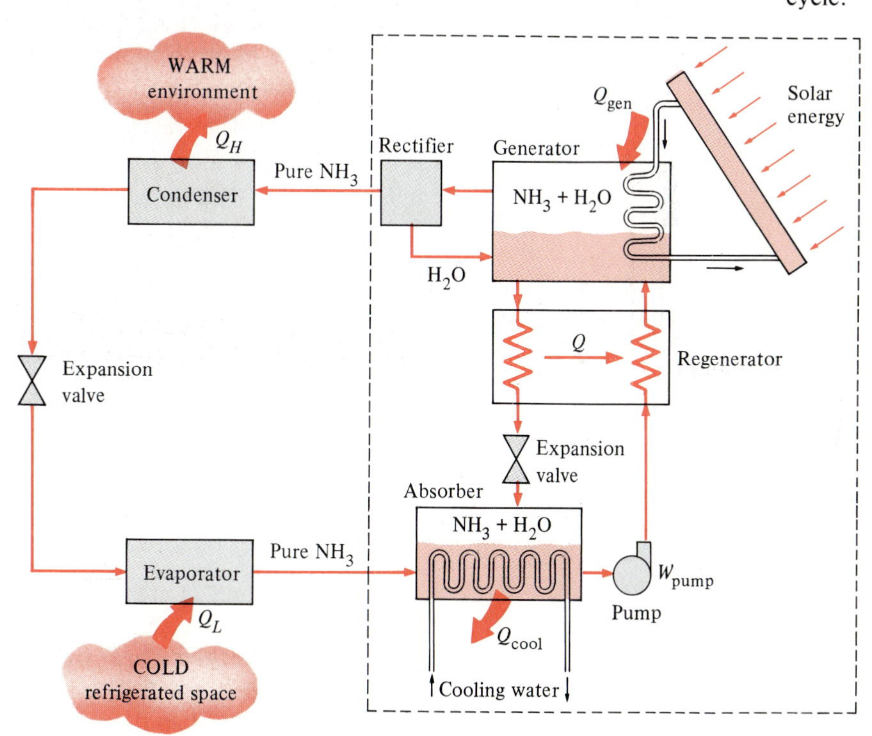

FIGURE 10-21
Ammonia-absorption refrigeration cycle.

Ammonia vapor leaves the evaporator and enters the absorber, where it dissolves and chemically reacts with water to form $NH_3 \cdot H_2O$. This is an exothermic reaction, thus heat is released during this process. The amount of NH_3 that can be dissolved in H_2O is inversely proportional to the temperature. Therefore, it is necessary to cool the absorber to maintain its temperature as low as possible, hence to maximize the amount of NH_3 dissolved in water. The liquid $NH_3 + H_2O$ solution, which is rich in NH_3, is then pumped to the generator. Heat is transferred to the solution from a source to vaporize some of the solution. The vapor which is rich in NH_3 passes through a rectifier, which separates the water and returns it to the generator. The high-pressure pure NH_3 vapor then continues its journey through the rest of the cycle. The hot $NH_3 + H_2O$ solution, which is weak in NH_3, then passes through a regenerator, where it transfers some heat to the rich solution leaving the pump, and is throttled to the absorber pressure.

Compared with vapor-compression systems, absorption refrigeration systems have one major advantage: A liquid is compressed instead of a vapor. The steady-flow work is proportional to the specific volume, and thus the work input for absorption refrigeration systems is very small and often neglected in the cycle analysis. The operation of these systems is based on heat transfer from an external source. Therefore, absorption refrigeration systems are often classified as *heat-driven systems*. These systems are not without disadvantages, however; they are bulky, complex, and, of course, expensive. They are economically competitive only when there is an available source of energy that would otherwise be wasted. They are generally used in industrial applications.

The COP of absorption refrigeration systems is defined as

$$\mathrm{COP_R} = \frac{\text{desired output}}{\text{required input}} = \frac{Q_L}{Q_{gen} + W_{pump,in}} \cong \frac{Q_L}{Q_{gen}} \tag{10-12}$$

The maximum COP of an absorption refrigeration system is determined by assuming that the entire cycle is totally reversible (i.e., the cycle involves no irreversibilities and any heat transfer is through a differential temperature difference). The refrigeration system would be reversible if the heat from the source (Q_{gen}) were transferred to a Carnot heat engine, and the work output of this heat engine ($W = \eta_{th,C} Q_{gen}$) is supplied to a Carnot refrigerator to remove heat from the refrigerated space. Note that $Q_L = W \times \mathrm{COP_{R,C}} = \eta_{th,C} Q_{gen} \mathrm{COP_{R,C}}$. Then the overall COP of an absorption refrigeration system under reversible conditions becomes

$$\mathrm{COP_{R,rev}} = \frac{Q_L}{Q_{gen}} = \eta_{th,C} \mathrm{COP_{R,C}} = \left(1 - \frac{T_0}{T_s}\right)\left(\frac{T_L}{T_0 - T_L}\right) \tag{10-13}$$

where T_L, T_0, and T_s are the temperatures of the refrigerated space, environment, and heat source, respectively. Any absorption refrigeration system which receives heat from a source at T_s and removes heat from the refrigerated space at T_L while operating in an environment at T_0 will have a lower COP than the one determined from Eq. 10-13. For example, when the source is at 120°C, the refrigerated space is at −10°C, and the environment is at 25°C, the maximum COP that an absorption refrigeration system can have is 1.8. The COP of actual absorption refrigeration systems is usually less than 1.

Another absorption refrigeration system which is quite popular with campers is a propane-fired system invented by two Swedish undergraduate students. In this system, the pump is replaced by a third fluid (hydrogen), which makes it a truly portable unit.

10-10 THERMOELECTRIC POWER GENERATION AND REFRIGERATION SYSTEMS

All the refrigeration systems discussed above involve many moving parts and bulky, complex components. Then this question comes to mind: Is it really necessary for a refrigeration system to be so complex? Can we not achieve the same effect in a more direct way? The answer to this question is *yes*. It is possible to use electric energy more directly to produce cooling without involving any refrigerants and moving parts. Below we discuss one such system, called a *thermoelectric refrigerator*.

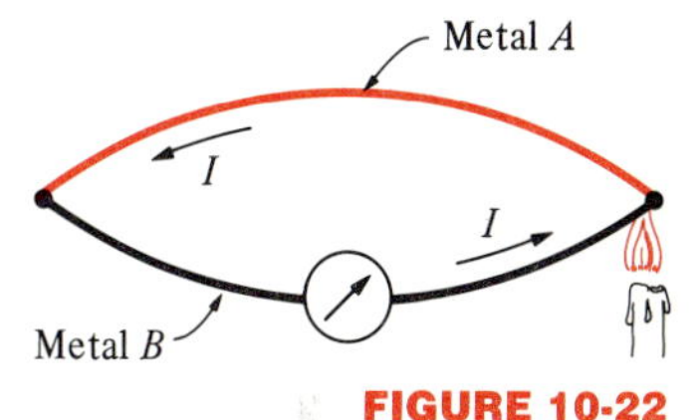

FIGURE 10-22

When one of the junctions of two dissimilar metals is heated, a current I flows through the closed circuit.

Consider two wires made from different metals joined at both ends (junctions), forming a closed circuit. Ordinarily, nothing will happen. But when one of the ends is heated, something interesting happens: A current flows continuously in the circuit, as shown in Fig. 10-22. This is called the **Seebeck effect**, in honor of Thomas Seebeck, who made this discovery in 1821. The circuit which incorporates both thermal and electrical effects is called a **thermoelectric circuit**, and a device that operates on this circuit is called a **thermoelectric device**.

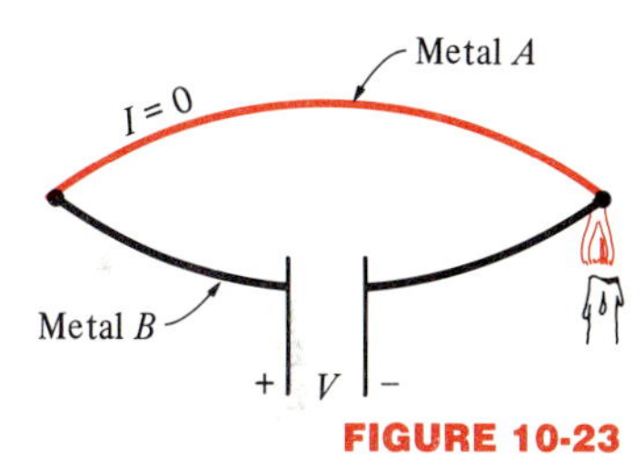

FIGURE 10-23

When a thermoelectric circuit is broken, a potential difference is generated.

The Seebeck effect has two major applications: temperature measurement and power generation. When the thermoelectric circuit is broken, as shown in Fig. 10-23, the current ceases to flow, and we can measure the driving force (the electromotive force) or the voltage generated in the circuit by a voltmeter. The voltage generated is a function of the temperature difference and the materials of the two wires used. Therefore, temperature can be measured by simply measuring voltages. The two wires used to measure the temperature in this manner form a *thermocouple,* which is the most versatile and most widely used temperature measurement device. A common T-type thermocouple, for example, consists of copper and constantan wires, and it produces about 40 μV per degree Celsius temperature difference.

The Seebeck effect also forms the basis for thermoelectric power generation. The schematic diagram of a **thermoelectric generator** is shown in Fig. 10-24. Heat is transferred from a high-temperature source to the hot junction in the amount of Q_H, and it is rejected to a low-temperature sink from the cold junction in the amount of Q_L. The difference between these two quantities is the net electrical work produced, that is, $W_e = Q_H - Q_L$. It is evident from Fig. 10-24 that the thermoelectric power cycle closely resembles an ordinary heat engine cycle, with electrons

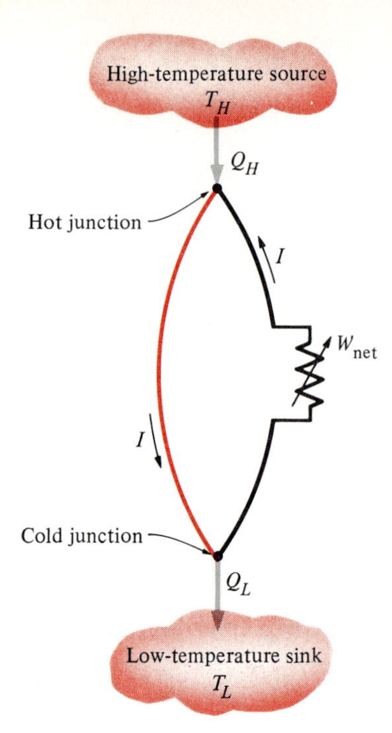

FIGURE 10-24
Schematic of a simple thermoelectric power generator.

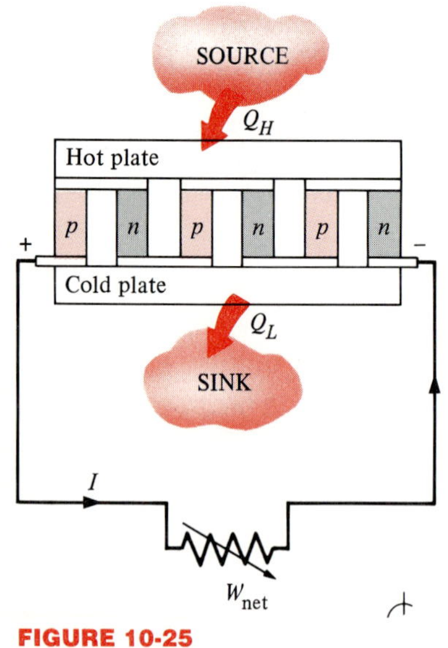

FIGURE 10-25
A thermoelectric power generator.

serving as the working fluid. Therefore, the thermal efficiency of a thermoelectric generator operating between the temperature limits of T_H and T_L is limited by the efficiency of a Carnot cycle operating between the same temperature limits. Thus in the absence of any irreversibilites (such as I^2R heating, where R is the total electrical resistance of the wires), the thermoelectric generator will have the Carnot efficiency.

The major drawback of thermoelectric generators is their low efficiency. The future success of these devices depends on finding materials with more desirable characteristics. For example, the voltage output of thermoelectric devices has been increased several times by switching from metal pairs to semiconductors. A practical thermoelectric generator using *n*-type (heavily doped to create excess electrons) and *p*-type (heavily doped to create a deficiency of electrons) materials connected in series is shown in Fig. 10-25. Despite their low efficiencies, thermoelectric generators have a definite weight and reliability advantage and are presently used in rural areas and in space applications.

If Seebeck had been fluent in thermodynamics, he would probably have tried reversing the direction of flow of electrons in the thermoelectric circuit (by externally applying a potential difference in the reverse direction), to create a refrigeration effect. But this honor belongs to Jean Charles Athanase Peltier, who discovered this phenomenon in 1834. He noticed during his experiments that when a small current was passed through the junction of two dissimilar wires, the junction was cooled, as shown in Fig. 10-26. This is called the **Peltier effect**, and it forms the basis for **thermoelectric refrigeration**. A practical thermoelectric refrigeration circuit using semiconductor materials is shown in Fig. 10-27. Heat is absorbed from the refrigerated space in the amount of Q_L and rejected to the warmer environment in the amount of Q_H. The difference between these two quantities is the net electrical work that needs to be supplied, that is, $W_e = Q_H - Q_L$. Thermoelectric refrigerators presently cannot compete with vapor-compression refrigeration systems because of their low coefficient of performance. They are available in the market, however, and are preferred in some applications, because of their small size, simplicity, quietness, and reliability.

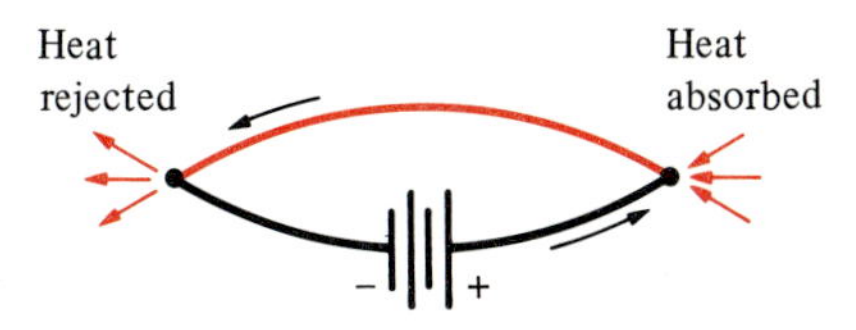

FIGURE 10-26
When a current is passed through the junction of two dissimilar materials, the junction is cooled.

10-11 ■ SUMMARY

The transfer of heat from lower temperature regions to higher temperature ones is called *refrigeration*. Devices that produce refrigeration are called *refrigerators,* and the cycles on which they operate are called *refrigeration cycles*. The working fluids used in the refrigeration cycles are called *refrigerants*. Refrigerators used for the purpose of heating a space by transferring heat from a cooler medium are called *heat pumps*.

The performance of refrigerators and heat pumps is expressed in terms of the *coefficient of performance* (COP), defined as

$$\text{COP}_\text{R} = \frac{\text{desired output}}{\text{required input}} = \frac{\text{cooling effect}}{\text{work input}} = \frac{Q_L}{W_{\text{net,in}}} \tag{10-1}$$

$$\text{COP}_\text{HP} = \frac{\text{desired output}}{\text{required input}} = \frac{\text{heating effect}}{\text{work input}} = \frac{Q_H}{W_{\text{net,in}}} \tag{10-2}$$

The standard of comparison for refrigeration cycles is the *reversed Carnot cycle*. A refrigerator or heat pump which operates on the reversed Carnot cycle is called a *Carnot refrigerator* or a *Carnot heat pump,* and their COPs are

$$\text{COP}_{\text{R,Carnot}} = \frac{1}{T_H/T_L - 1} \tag{10-4}$$

and

$$\text{COP}_{\text{HP,Carnot}} = \frac{1}{1 - T_L/T_H} \tag{10-5}$$

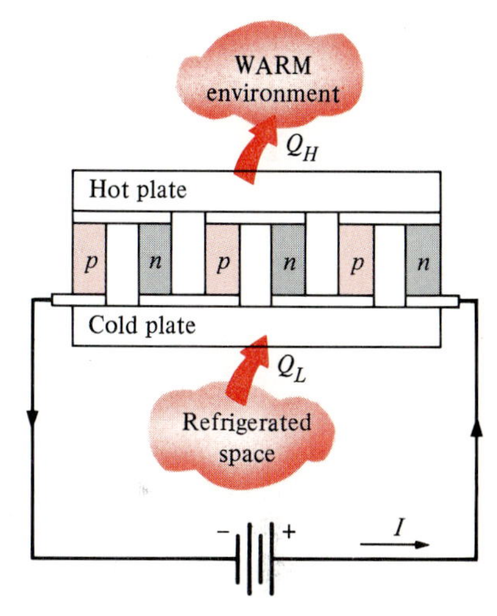

FIGURE 10-27
A thermoelectric refrigerator.

The most widely used refrigeration cycle is the *vapor-compression refrigeration cycle*. In an ideal vapor-compression refrigeration cycle, the refrigerant enters the compressor as a saturated vapor and is cooled to the saturated liquid state in the condenser. It is then throttled to the evaporator pressure and vaporizes as it absorbs heat from the refrigerated space.

Very low temperatures can be achieved by operating two or more vapor-compression systems in series, called *cascading*. The COP of a refrigeration system also increases as a result of cascading. Another way of improving the performance of a vapor-compression refrigeration system is by using *multistage compression with regenerative cooling*. A refrigerator with a single compressor can provide refrigeration at several temperatures by throttling the refrigerant in stages. The vapor-compression refrigeration cycle can also be used to liquefy gases after some modifications.

The power cycles can be used as refrigeration cycles by simply reversing them. Of these, the *reversed Brayton cycle,* which is also known as the *gas refrigeration cycle,* is extensively used in aircraft cooling. It is also used to obtain very low (cryogenic) temperatures after it is modified with regeneration. The work output of the turbine can be used to reduce the work input requirements to the compressor. Thus the COP of a gas refrigeration cycle is

$$\mathrm{COP_R} = \frac{q_L}{w_{\mathrm{net,in}}} = \frac{q_L}{w_{\mathrm{comp,in}} - w_{\mathrm{turb,out}}} \tag{10-11}$$

Another form of refrigeration which becomes economically attractive when there is a source of inexpensive heat energy at a temperature of 100 to 200°C is *absorption refrigeration,* where the refrigerant is absorbed by a transport medium and compressed in liquid form. The most widely used absorption refrigeration system is the ammonia-water system, where ammonia serves as the refrigerant and water as the transport medium. The work input to the pump is usually very small, and the COP of absorption refrigeration systems is defined as

$$\mathrm{COP_R} = \frac{\text{desired output}}{\text{required input}} = \frac{Q_L}{Q_{\mathrm{gen}} + W_{\mathrm{pump,in}}} \cong \frac{Q_L}{Q_{\mathrm{gen}}} \tag{10-12}$$

The maximum COP an absorption refrigeration system can have is determined by assuming totally reversible operation, which yields

$$\mathrm{COP_{R,max}} = \eta_{\mathrm{th,C}}\mathrm{COP_{R,C}} = \left(1 - \frac{T_0}{T_s}\right)\left(\frac{T_L}{T_0 - T_L}\right) \tag{10-13}$$

where T_0, T_L, and T_s are the absolute temperatures of the environment, refrigerated space, and heat source, respectively.

A refrigeration effect can also be achieved without using any moving parts by simply passing a small current through a closed circuit made up of two dissimilar materials. This effect is called the *Peltier effect,* and a refrigerator that works on this principle is called a *thermoelectric refrigerator.*

REFERENCES AND SUGGESTED READING

1 W. Z. Black and J. G. Hartley, *Thermodynamics,* Harper & Row, New York, 1985.

2 R. C. Fellinger and W. J. Cook, *Introduction to Engineering Thermodynamics,* WCB, Dubuque, IA, 1985.

3 J. P. Holman, *Thermodynamics,* 3d ed., McGraw-Hill, New York, 1980.

4 J. B. Jones and G. A. Hawkins, *Engineering Thermodynamics,* 2d ed., Wiley, New York, 1986.

5 D. C. Look, Jr., and H. J. Sauer, Jr., *Engineering Thermodynamics,* PWS Engineering, Boston, 1986.

6 W. F. Stoecker and J. W. Jones, *Refrigeration and Air Conditioning,* 2d ed., McGraw-Hill, New York, 1982.

7 G. J. Van Wylen and R. E. Sonnntag, *Fundamentals of Classical Thermodynamics,* 3d ed., Wiley, New York, 1985.

8 K. Wark, *Thermodynamics,* 5th ed., McGraw-Hill, New York, 1988.

9 *ASHRAE Fundamentals Handbook,* American Society of Heating, Refrigerating, and Air-Conditioning Engineers, Atlanta, 1985.

10 *Heat Pump Systems—A Technology Review,* OECD Report, Paris, 1982.

PROBLEMS*

The Reversed Carnot Cycle

10-1C Why is the reversed Carnot cycle executed within the saturation dome not a realistic model for refrigeration cycles?

10-2C Draw the *T-s* diagrams of a reversed Carnot cycle and a Carnot cycle operating between the same temperature limits. How do they differ?

10-3C Consider a refrigerator and a heat pump operating between the same temperature limits. For which device will the COP be higher?

10-4C What is the difference between a refrigerator and a heat pump?

10-5C Consider two identical refrigerators operating in different environments. Which refrigerator will consume more work, the one that operates in the warmer environment or the one that operates in the cooler environment?

10-6 A steady-flow Carnot refrigeration cycle uses refrigerant-12 as the working fluid. The refrigerant changes from saturated vapor to saturated liquid at 30°C in the condenser as it rejects heat. The evaporator pressure is 120 kPa. Show the cycle on a *T-s* diagram relative to saturation lines, and determine (*a*) the coefficient of performance, (*b*) the amount of heat absorbed from the refrigerated space, and (*c*) the net work input.
Answers: (*a*) 4.44, (*b*) 110.2 kJ/kg, (*c*) 24.8 kJ/kg

10-6E A steady-flow Carnot refrigeration cycle uses refrigerant-12 as the working fluid. The refrigerant changes from saturated vapor to saturated

*Students are encouraged to answer *all* the concept "C" questions.

liquid at 90°F in the condenser as it rejects heat. The evaporator pressure is 20 psia. Show the cycle on a *T-s* diagram relative to saturation lines, and determine (*a*) the coefficient of performance, (*b*) the amount of heat absorbed from the refrigerated space, and (*c*) the net work input.
Answers: (*a*) 4.61, (*b*) 47.2 Btu/lbm, (*c*) 10.2 Btu/lbm

10-7 Consider a steady-flow Carnot refrigeration cycle which uses refrigerant-12 as the working fluid. The maximum and minimum temperatures in the cycle are 20 and −20°C, respectively. The quality of the refrigerant is 0.2 at the beginning of the heat absorption process and 0.85 at the end. Show the cycle on a *T-s* diagram relative to saturation lines, and determine (*a*) the coefficient of performance, (*b*) the condenser and evaporator pressures, and (*c*) the net work input.

10-8 Refrigerant-12 enters the condenser of a steady-flow Carnot refrigerator as a saturated vapor at 800 kPa and leaves with a quality of 0.05. The heat absorption from the refrigerated space takes place at 140 kPa. Show the cycle on a *T-s* diagram relative to saturation lines, and determine (*a*) the coefficient of performance, (*b*) the quality at the beginning of the heat absorption process, and (*c*) the net work input.
Answers: (*a*) 4.59, (*b*) 0.3317, (*c*) 22.7 kJ/kg

10-8E Refrigerant-12 enters the condenser of a steady-flow Carnot refrigerator as a saturated vapor at 90 psia, and it leaves with a quality of 0.05. The heat absorption from the refrigerated space takes place at a pressure of 30 psia. Show the cycle on a *T-s* diagram relative to saturation lines, and determine (*a*) the coefficient of performance, (*b*) the quality at the beginning of the heat absorption process, and (*c*) the net work input.

Ideal and Actual Vapor-Compression Refrigeration Cycles

10-9C Does the ideal vapor-compression refrigeration cycle involve any internal irreversibilities?

10-10C Why is the throttling valve not replaced by an isentropic turbine in the ideal vapor-compression refrigeration cycle?

10-11C It is proposed to use water instead of refrigerant-12 as the working fluid in air conditioning applications, where the minimum temperature never falls below the freezing point. Would you support this proposal? Explain.

10-12C In a refrigeration system, would you recommend condensing the refrigerant-12 at a pressure of 0.7 or 1.0 MPa if heat is to be rejected to a cooling medium at 15°C? Why?

10-13C Does the area enclosed by the cycle on a *T-s* diagram represent the net work input for the reversed Carnot cycle? How about for the ideal vapor-compression refrigeration cycle?

10-14C Consider two vapor-compression refrigeration cycles. The refrigerant enters the throttling valve as a saturated liquid at 30°C in one cycle and as subcooled liquid at 30°C in the other one. The evaporator pressure for both cycles is the same. Which cycle do you think will have a higher COP?

10-15C The COP of vapor-compression refrigeration cycles improves when the refrigerant is subcooled before it enters the throttling valve. Can the refrigerant be subcooled indefinitely to maximize this effect, or is there a lower limit? Explain.

10-16 A refrigerator uses refrigerant-12 as the working fluid and operates on an ideal vapor-compression refrigeraton cycle between 0.12 and 0.7 MPa. The mass flow rate of the refrigerant is 0.1 kg/s. Show the cycle on a *T-s* diagram with respect to saturation lines. Determine (*a*) the rate of heat removal from the refrigerated space and the power input to the compressor, (*b*) the rate of heat rejection to the environment, and (*c*) the coefficient of performance.
Answers: (*a*) 11.4 kW, 3.1 kW; (*b*) 14.5 kW; (*c*) 3.68

10-17 If the throttling valve in Prob. 10-16 is replaced by an isentropic turbine, determine the percentage increase in the COP and in the rate of heat removal from the refrigerated space.
Answers: 4.3 percent, 4.4 percent

10-18 Consider a 12-kW refrigeration system which operates on an ideal vapor-compression refrigeration cycle with refrigerant-12 as the working fluid. The refrigerant enters the compressor as saturated vapor at 140 kPa and is compressed to 800 kPa. Show the cycle on a *T-s* diagram with respect to saturation lines, and determine (*a*) the quality of the refrigerant at the end of the throttling process, (*b*) the coefficient of performance, and (*c*) the power input to the compressor

10-18E Consider a 5-ton refrigeration system which operates on an ideal vapor-compression refrigeration cycle with refrigerant-12 as the working fluid. The refrigerant enters the compressor as saturated vapor at 20 psia and is compressed to a pressure of 140 psia. Show the cycle on a *T-s* diagram with respect to saturation lines, and determine (*a*) the quality of the refrigerant at the end of the throttling process, (*b*) coefficient of performance, and (*c*) the power input to the compressor.
Answers: (*a*) 0.364, (*b*) 2.98, (*c*) 7.91 hp

10-19 Repeat Prob. 10-18, assuming an adiabatic efficiency of 85 percent for the compressor. Also determine the irreversibility rate associated with the compression process in this case.

10-19E Repeat Prob. 10-18E by replacing the throttling valve with an isentropic turbine.

10-20 Refrigerant-12 enters the compressor of a refrigerator as superheated vapor at 0.14 MPa and −20°C at a rate of 0.1 kg/s, and it leaves at 0.7 MPa and 50°C. The refrigerant is cooled in the condenser to 24°C and 0.65 MPa, and it is throttled to 0.15 MPa. Disregarding any heat transfer and pressure drops in the connecting lines between the compo-

nents, show the cycle on a *T-s* diagram with respect to saturation lines, and determine (*a*) the rate of heat removal from the refrigerated space and the power input to the compressor, (*b*) the adiabatic efficiency of the compressor, and (*c*) the COP of the refrigerator.
Answers: (*a*) 12.03 kW, 3.59 kW; (*b*) 79.4 percent; (*c*) 3.35

10-21 An ice-making machine operates on the ideal vapor-compression cycle, using refrigerant-12. The refrigerant enters the compressor as saturated vapor at 160 kPa and leaves the condenser as saturated liquid at 700 kPa. Water enters the ice machine at 15°C and leaves as ice at −5°C. For an ice production rate of 8 kg/h, determine the power input to the ice maker (384 kJ of heat needs to be removed from each kilogram of water at 15°C to turn it into ice at −5°C). *Answer:* 0.19 kW

10-21E An ice-making machine operates on the ideal vapor-compression cycle, using refrigerant-12. The refrigerant enters the compressor as saturated vapor at 20 psia and leaves the condenser as saturated liquid at 100 psia. Water enters the ice machine at 55°F and leaves as ice at 25°F. For an ice production rate of 15 lbm/h, determine the power input to the ice machine (169 Btu of heat needs to be removed from each lbm of water at 55°F to turn it into ice at 25°F).

10-22 Refrigerant-12 enters the compressor of a refrigerator at 140 kPa and −10°C at a rate of 1 m^3/min, and leaves at 1 MPa. The adiabatic efficiency of the compressor is 78 percent. The refrigerant enters the throttling valve at 0.95 MPa and 30°C and leaves the evaporator as saturated vapor at −18.5°C. Show the cycle on a *T-s* diagram with respect to saturation lines, and determine (*a*) the power input to the compresssor, (*b*) the rate of heat removal from the refrigerated space, and (*c*) the pressure drop and rate of heat gain in the line between the evaporator and the compressor.
Answers: (*a*) 6.42 kW; (*b*) 15.5 kW; (*c*) 20 kPa, 0.75 kW

10-23 A large refrigeration plant is to be maintained at −15°C, and it requires refrigeration at a rate of 100 kW. The condenser of the plant is to be cooled by liquid water, which experiences a temperature rise of 8°C as it flows over the coils of the condenser. Assuming the plant operates on the ideal vapor-compression cycle using refrigerant-12 between the pressure limits of 120 and 700 kPa, determine (*a*) the mass flow rate of the refrigerant, (*b*) the power input to the compressor, and (*c*) the mass flow rate of the cooling water.

10-24 Repeat Prob. 10-23, assuming the compressor has an adiabatic efficiency of 75 percent. Also determine the irreversibility rate associated with the compression process in this case.

10-25 Write a computer program to determine the effect of the evaporator pressure on the COP of an ideal vapor-compression refrigeration cycle. Assume the condenser pressure is kept constant at 1 MPa. Calculate the COP of the refrigeration cycle for the following evaporator pressures: 100, 120, 140, 160, 200, 280, 320, 400, and 500 kPa. Plot the COPs against the evaporator pressure.

10-26 Write a computer program to determine the effect of the condenser pressure on the COP of an ideal vapor-compression refrigeration cycle. Assume the evaporator pressure is maintained constant at 120 kPa. Calculate the COP of the refrigeration cycle for the following condenser pressures: 400, 500, 600, 700, 800, 900, 1000, and 1400 kPa. Plot the COPs against the condenser pressure.

Selecting the Right Refrigerant

10-27C When selecting a refrigerant for a certain application, what qualities would you look for in the refrigerant?

10-28C Consider a refrigeration system using refrigerant-12 as the working fluid. If this refrigerator is to operate in an environment at 30°C, what is the minimum pressure to which the refrigerant should be compressed? Why?

10-29C A refrigerant-12 refrigerator is to maintain the refrigerated space at −10°C. Would you recommend an evaporator pressure of 0.12 or 0.14 MPa for this system? Why?

10-30 A refrigerator which operates on the ideal vapor-compression cycle with refrigerant-12 is to maintain the refrigerated space at −10°C while rejecting heat to the environment at 25°C. Select reasonable pressures for the evaporator and the condenser, and explain why you chose those values.

10-31 A heat pump which operates on the ideal vapor-compression cycle with refrigerant-12 is used to heat a house and maintain it at 20°C by using underground water at 10°C as the heat source. Select reasonable pressures for the evaporator and the condenser, and explain why you chose those values.

Heat Pump Systems

10-32C What are the advantages and disadvantages of heat pumps? How do they compare to other heating systems?

10-33C Do you think a heat pump system will be more cost-effective in New York or in Miami? Why?

10-34C What is a water-source heat pump? How does the COP of a water-source heat pump system compare to that of an air-source system?

10-35 A heat pump which operates on the ideal vapor-compression cycle with refrigerant-12 is used to heat a house and maintain it at 20°C, using underground water at 10°C as the heat source. The house is losing heat at a rate of 60,000 kJ/h. The evaporator and condenser pressures are 320 and 800 kPa, respectively. Determine the power input to the heat pump and the electric power saved by using a heat pump instead of a resistance heater. *Answers:* 1.97 kW, 14.7 kW

10-35E A heat pump which operates on the ideal vapor-compression cycle with refrigerant-12 is used to heat a house and maintain it at 75°F by using underground water at 50°F as the heat source. The house is losing heat at a rate of 60,000 Btu/h. The evaporator and condenser pressures are 50 and 120 psia, respectively. Determine the power input to the heat pump and the electric power saved by using a heat pump instead of a resistance heater. *Answers:* 3.675 hp, 19.905 hp

10-36 A heat pump which operates on the ideal vapor-compression cycle with refrigerant-12 is used to heat water from 15 to 54°C at a rate of 0.25 kg/s. The condenser and evaporator pressures are 1.4 and 0.32 MPa, respectively. Determine the power input to the heat pump.

10-37 A heat pump which operates on the ideal vapor-compression cycle with refrigerant-12 is used to heat a house. The mass flow rate of the refrigerant is 0.15 kg/s. The condenser and evaporator pressures are 900 and 240 kPa, respectively. Show the cycle on a *T-s* diagram with respect to saturation lines, and determine (*a*) the rate of heat supply to the house, (*b*) the volume flow rate of the refrigerant at the compressor inlet, and (*c*) the COP for this heat pump.

10-38 A heat pump using refrigerant-12 heats a house by using underground water at 8°C as the heat source. The house is losing heat at a rate of 75,000 kJ/h. The refrigerant enters the compressor at 280 kPa and 0°C, and it leaves at 1 MPa and 60°C. The refrigerant leaves the condenser at 30°C. Determine (*a*) the power input to the heat pump, (*b*) the rate of heat absorption from the water, and (*c*) the increase in electric power input if an electric resistance heater is used instead of a heat pump.
Answers: (*a*) 4.06 kW, (*b*) 16.8 kW, (*c*) 16.77 kW

10-38E A heat pump using refrigerant-12 heats a house by using underground water at 45°F as the heat source. The house is losing heat at a rate of 70,000 Btu/h. The refrigerant enters the compressor at 30 psia and 20°F and leaves at 120 psia and 140°F. The refrigerant leaves the condenser at 90°F. Determine (*a*) the power input to the heat pump, (*b*) the rate of heat absorption from the water, and (*c*) the increase in electric power input if an electric resistance heater is used instead of a heat pump.

Innovative Refrigeration Systems

10-39C What is cascade refrigeration? What are the advantages and disadvantages of cascade refrigeration?

10-40C How does the COP of a cascade refrigeration system compare to that of a simple vapor-compression cycle operating between the same pressure limits?

10-41C A certain application requires maintaining the refrigerated space at −35°C. Would you recommend a simple refrigeration cycle with refrigerant-12 or a two-stage cascade refrigeration cycle with a different refrigerant at the bottoming cycle? Why?

10-42C Derive a relation for the COP of the two-stage refrigeration system with a flash chamber shown in Fig. 10-12 in terms of enthalpies and the quality at state 6. Consider a unit mass in the condenser.

10-43C Consider a two-stage cascade refrigeration cycle and a two-stage compression refrigeration cycle with a flash chamber. Both cycles operate between the same pressure limits and use the same refrigerant. Which system would you favor? Why?

10-44C Can a vapor-compression refrigeration system with a single compressor handle several evaporators operating at different pressures? How?

10-45C Is it possible to have liquid helium at room temperature?

10-46C In the liquefaction process, why are gases compressed to very high pressures?

10-47 Consider a two-stage cascade refrigeration system operating between the pressure limits of 0.8 and 0.14 MPa. Each stage operates on the ideal vapor-compression refrigeration cycle with refrigerant-12 as the working fluid. Heat rejection from the lower cycle to the upper cycle takes place in an adiabatic counterflow heat exchanger where both streams enter at about 0.4 MPa. If the mass flow rate of the refrigerant through the upper cycle is 0.2 kg/s, determine (*a*) the mass flow rate of the refrigerant through the lower cycle, (*b*) the rate of heat removal from the refrigerated space and the power input to the compressor, and (*c*) the coefficient of performance of this cascade refrigerator.
Answers: (*a*) 0.161 kg/s; (*b*) 21.5 kW, 5.35 kW; (*c*) 4.0

10-47E Consider a two-stage cascade refrigeration system operating between the pressure limits of 160 and 20 psia. Each stage operates on an ideal vapor-compression refrigeration cycle with refrigerant-12 as the working fluid. Heat rejection from the lower cycle to the upper cycle takes place in an adiabatic counterflow heat exchanger where both streams enter at about 70 psia. If the mass flow rate of the refrigerant through the upper cycle is 0.5 lbm/s, determine (*a*) the mass flow rate of the refrigerant through the lower cycle, (*b*) the rate of heat removal from the refrigerated space and the power input to the compressor, and (*c*) the COP of this cascade refrigerator.

10-48 Repeat Prob. 10-47 for a heat exchanger pressure of 0.5 MPa.

10-49 Consider a two-stage compression refrigeration system operating between the pressure limits of 0.8 and 0.14 MPa. The working fluid is refrigerant-12. The refrigerant leaves the condenser as a saturated liquid and is throttled to a flash chamber operating at 0.4 MPa. Part of the refrigerant evaporates during this flashing process, and this vapor is mixed with the refrigerant leaving the low-pressure compressor. The mixture is then compressed to the condenser pressure by the high-pressure compressor. The liquid in the flash chamber is throttled to the evaporator pressure, and it cools the refrigerated space as it vaporizes in the evaporator. Assuming the refrigerant leaves the evaporator as saturated vapor

and both compressors are isentropic, determine (*a*) the fraction of the refrigerant which evaporates as it is throttled to the flash chamber, (*b*) the amount of heat removed from the refrigerated space and the compressor work per unit mass of refrigerant flowing through the condenser, and (*c*) the coefficient of performance.
Answers: (*a*) 0.161; (*b*) 112.6 kJ/kg, 27.8 kJ/kg; (*c*) 4.0

10-50 A two-stage compression refrigeration system operates with refrigerant-12 between the pressure limits of 1 and 0.14 MPa. The refrigerant leaves the condenser as a saturated liquid and is throttled to a flash chamber operating at 0.5 MPa. The refrigerant leaving the low-pressure compressor at 0.5 MPa is also routed to the flash chamber. The vapor in the flash chamber is then compressed to the condenser pressure by the high-pressure compressor, and the liquid is throttled to the evaporator pressure. Assuming the refrigerant leaves the evaporator as saturated vapor and both compressors are isentropic, determine (*a*) the fraction of the refrigerant which evaporates as it is throttled to the flash chamber, (*b*) the amount of heat removed from the refrigerated space for a mass flow rate of 0.42 kg/s through the condenser, and (*c*) the coefficient of performance.

10-50E A two-stage compression refrigeration system operates with refrigerant-12 between the pressure limits of 20 and 160 psia. The refrigerant leaves the condenser as a saturated liquid and is throttled to a flash chamber operating at 70 psia. The refrigerant leaving the low-pressure compressor at 70 psia is also routed to the flash chamber. The vapor in the flash chamber is then compressed to the condenser pressure by the high-pressure compressor, and the liquid is throttled to the evaporator pressure. Assuming the refrigerant leaves the evaporator as saturated vapor and both compressors are isentropic, determine (*a*) the fraction of the refrigerant which evaporates as it is throttled to the flash chamber, (*b*) the amount of heat removed from the refrigerated space for a mass flow rate of 1 lbm/s through the condenser, and (*c*) the coefficient of performance.

10-51 Repeat Prob. 10-50 for a flash chamber pressure of 0.32 MPa.

Gas Refrigeration Cycle

10-52C How does the ideal-gas refrigeration cycle differ from the Brayton cycle?

10-53C Devise a refrigeration cycle which works on the reversed Stirling cycle. Also determine the COP for this cycle.

10-54C How does the ideal-gas refrigeration cycle differ from the Carnot refrigeration cycle?

10-55C How is the ideal-gas refrigeration cycle modified for aircraft cooling?

10-56C In gas refrigeration cycles, can we replace the turbine by an expansion valve as we did in vapor-compression refrigeration cycles? Why?

10-57C How do we achieve very low temperatures with gas refrigeration cycles?

10-58 An ideal-gas refrigeration cycle using air as the working fluid is to maintain a refrigerated space at −23°C while rejecting heat to the surrounding medium at 27°C. If the pressure ratio of the compressor is 3, determine (*a*) the maximum and minimum temperatures in the cycle, (*b*) the coefficient of performance, and (*c*) the rate of refrigeration for a mass flow rate of 0.2 kg/s.

10-59 Air enters the compressor of an ideal-gas refrigeration cycle at 12°C and 50 kPa and the turbine at 47°C and 250 kPa. The mass flow rate of air through the cycle is 0.2 kg/s. Assuming variable specific heats for air, determine (*a*) the rate of refrigeration, (*b*) the net power input, and (*c*) coefficient of performance.
Answers: (*a*) 16.7 kW, (*b*) 9.7 kW, (*c*) 1.72

10-59E Air enters the compressor of an ideal-gas refrigeration cycle at 40°F and 10 psia and the turbine at 120°F and 30 psia. The mass flow rate of air through the cycle is 0.5 lbm/s. Determine (*a*) the rate of refrigeration, (*b*) the coefficient of performance, and (*c*) the net power input.

10-60 Repeat Prob. 10-59 for a compressor adiabatic efficiency of 80 percent and a turbine adiabatic efficiency of 85 percent.

10-61 An aircraft on the ground is to be cooled by a gas refrigeration cycle operating with air on an open cycle. Air enters the compressor at 30°C and 100 kPa and is compressed to 300 kPa. Air is cooled to 70°C before it enters the turbine. Assuming both the turbine and the compressor to be isentropic, determine the temperature of the air leaving the turbine and entering the cabin. *Answer:* −22.4°C

10-62 A gas refrigeration cycle with a pressure ratio of 3 uses helium as the working fluid. The temperature of the helium is −10°C at the compressor inlet and 50°C at the turbine inlet. Assuming adiabatic efficiencies of 82 percent for both the turbine and the compressor, determine (*a*) the minimum temperature in the cycle, (*b*) the coefficient of performance, and (*c*) the mass flow rate of the helium for a refrigeration rate of 8 kW.

10-63 A gas refrigeration system using air as the working fluid has a pressure ratio of 4. Air enters the compressor at −7°C. The high-pressure air is cooled to 27°C by rejecting heat to the surroundings. It is further cooled to −15°C by regenerative cooling before it enters the turbine. Assuming both the turbine and the compressor to be isentropic and using constant specific heats at room temperature, determine (*a*) the lowest temperature that can be obtained by this cycle, (*b*) the coefficient of performance of the cycle, and (*c*) the mass flow rate of air for a refrigeration rate of 5 kW. *Answers:* (*a*) −99.4°C, (*b*) 1.12, (*c*) 0.0987 kg/s

10-63E A gas refrigeration system using air as the working fluid has a pressure ratio of 4. Air enters the compressor at 10°F. The high-pressure air is cooled to 70°F by rejecting heat to the surroundings. It is further cooled to 5°F by regenerative cooling before it enters the turbine. Assuming both the turbine and the compressor to be isentropic and using constant specific heats for air at room temperature, determine (*a*) the lowest temperature that can be obtained by this cycle, (*b*) the coefficient of performance of the cycle, and (*c*) the mass flow rate of air for a refrigeration rate of 5 tons.
Answers: (*a*) −147.1°F, (*b*) 1.21, (*c*) 0.754 lbm/s

10-64 Repeat Prob. 10-63, assuming adiabatic efficiencies of 75 percent for the compressor and 80 percent for the turbine.

10-65 Consider a regenerative gas refrigeration cycle using helium as the working fluid. Helium enters the compressor at 100 kPa and −10°C and is compressed to 300 kPa. Helium is then cooled to 20°C by water. It then enters the regenerator where it is cooled further before it enters the turbine. Helium leaves the refrigerated space at −25°C and enters the regenerator. Assuming both the turbine and the compressor to be isentropic, determine (*a*) the temperature of the helium at the turbine inlet, (*b*) the coefficient of performance of the cycle, and (*c*) the net power input required for a mass flow rate of 0.4 kg/s.
Answers: (*a*) −5°C, (*b*) 1.51, (*c*) 103.6 kW

Absorption Refrigeration Systems

10-66C What is absorption refrigeration? How does an absorption refrigeration system differ from a vapor-compression refrigeration system?

10-67C What are the advantages and disadvantages of absorption refrigeration?

10-68C Can water be used as a refrigerant in air conditioning applications? Explain.

10-69C In absorption refrigeration systems, why is the fluid in the absorber cooled, and the fluid in the generator heated?

10-70C How is the coefficient of performance of an absorption refrigeration system defined?

10-71C What are the functions of the rectifier and the regenerator in an absorption refrigeration system?

10-72 An absorption refrigeration system which receives heat from a source at 130°C and maintains the refrigerated space at −7°C is claimed to have a COP of 2. If the environment temperature is 30°C, can this claim be valid? *Answer:* No

10-73 An absorption refrigeration system receives heat from a source at 100°C and maintains the refrigerated space at −20°C. If the temperature of the environment is 25°C, what is the maximum COP this absorption refrigeration system can have?

10-74 Heat is supplied to an absorption refrigeration system from a geothermal well at 130°C at a rate of 10^5 kJ/h. The environment is at 25°C, and the refrigerated space is maintained at −30°C. Determine the maximum rate at which this system can remove heat from the refrigerated space. *Answer:* 7.7×10^{-5} kJ/kg

10-74E Heat is supplied to an absorption refrigeration system from a geothermal well at 250°F at a rate of 10^5 Btu/h. The environment is at 80°F, and the refrigerated space is maintained at 0°F. Determine the maximum rate at which this system can remove heat from the refrigerated space. *Answer:* 1.38×10^5 Btu/h

10-75 An absorption refrigeration system is to remove heat from the refrigerated space at −10°C at a rate of 15 kW while operating in an environment at 25°C. Heat is to be supplied from a solar pond at 85°C. What is the minimum rate of heat supply required?

Thermoelectric Power Generation and Refrigeration Systems

10-76C What is a thermoelectric circuit?

10-77C Describe the Seebeck and the Peltier effects.

10-78C Consider a circular copper wire formed by connecting the two ends of a copper wire. The connection point is now heated by a burning candle. Do you expect any current to flow through the wire?

10-79C An iron and a constantan wire are formed into a closed circuit by connecting the ends. Now both junctions are heated and are maintained at the same temperature. Do you expect any current to flow through this circuit?

10-80C A copper and a constantan wire are formed into a closed circuit by connecting the ends. Now one junction is heated by a burning candle while the other is maintained at room temperature. Do you expect any current to flow through this circuit?

10-81C How does a thermocouple work as a temperature measurement device?

10-82C Why are semiconductor materials preferable to metals in thermoelectric refrigerators?

10-83C Is the efficiency of a thermoelectric generator limited by the Carnot efficiency? Why?

10-84 A thermoelectric generator receives heat from a source at 100°C and rejects the waste heat to the environment at 25°C. What is the maximum thermal efficiency this thermoelectric generator can have? *Answer:* 20.1 percent

10-84E A thermoelectric generator receives heat from a source at 200°F and rejects the waste heat to the environment at 80°F. What is the maximum thermal efficiency this thermoelectric generator can have? *Answer:* 18.2 percent

10-85 It is proposed to run a thermoelectric generator in conjunction with a solar pond which can supply heat at a rate of 10^6 kJ/h at 80°C. The waste heat is to be rejected to the environment at 30°C. What is the maximum power this thermoelectric generator can produce?

10-86 A thermoelectric refrigerator removes heat from a refrigerated space at −5°C at a rate of 200 W and rejects it to an environment at 20°C. Determine the maximum coefficient of performance this thermoelectric refrigerator can have and the minimum required power input. *Answers:* 10.72, 16.7 kW

10-87 A thermoelectric cooler has a COP of 0.6 and removes heat from a refrigerated space at a rate of 250 W. Determine the required power input to the thermoelectric cooler.

10-87E A thermoelectric cooler has a COP of 0.6 and removes heat from a refrigerated space at a rate of 20 Btu/min. Determine the required power input to the thermoelectric cooler.

Thermodynamic Property Relations

CHAPTER

11

In the preceding chapters we made extensive use of the property tables. We tend to take the tables for granted, but thermodynamic laws and principles are of little use to engineers without them. In this chapter, we focus our attention on how the property tables are prepared and how some unknown properties can be determined from limited available data.

It will come as no surprise that some properties such as temperature, pressure, volume, and mass can be measured directly. Other properties such as density and specific volume can be determined from these using some simple relations. But properties such as internal energy, enthalpy, and entropy are not so easy to determine because they cannot be measured directly or related to easily measurable properties through some simple relations. Therefore, it is essential that we develop some fundamental relations between commonly encountered thermodynamic properties and express the properties that cannot be measured directly in terms of easily measurable properties.

By the nature of the material, this chapter makes extensive use of partial derivatives. Therefore, we start by reviewing them. Then we develop the Maxwell relations which form the basis for many thermodynamic relations. Next we discuss the Clapeyron equation which enables us to determine the enthalpy of vaporization from P, v, and T measurements alone, and we develop general relations for C_v, C_p, du, dh, and ds which are valid for all pure substances under all conditions. Then we discuss the Joule-Thomson coefficient which is a measure of the temperature change with pressure during a throttling process. Finally, we develop a method of evaluating the Δh, Δu, and Δs of real gases through the use of generalized enthalpy and entropy departure charts.

11-1 ■ A LITTLE MATH—Partial Derivatives and Associated Relations

Many of the expressions developed in this chapter are based on the state postulate which says that the state of a simple, compressible substance is completely specified by any two independent, intensive properties. All other properties at that state can be expressed in terms of those two properties. Mathematically speaking,

$$z = z(x,y)$$

where x and y are the two independent properties that fix the state and z represents any other property. Most basic thermodynamic relations involve differentials. Therefore, we start by reviewing the derivatives and various relations among derivatives to the extent necessary in this chapter.

Consider a function f which depends on a single variable x, that is, $f = f(x)$. Figure 11-1 shows such a function which starts out flat but gets rather steep as x increases. The steepness of the curve is a measure of the degree of dependence of f and x. In our case, the function f depends on x more strongly at larger x values. The steepness of a curve at a point is measured by the slope of a line tangent to the curve at that point, and it is equivalent to the **derivative** of the function at that point:

$$\frac{df}{dx} = \lim_{\Delta x \to 0} \frac{\Delta f}{\Delta x} = \lim_{\Delta x \to 0} \frac{f(x + \Delta x) - f(x)}{\Delta x} \tag{11-1}$$

Therefore, *the derivative of a function $f(x)$ with respect to x represents the rate of change of $f(x)$ with x.*

EXAMPLE 11-1

The C_p of ideal gases depends on temperature only, and it is expressed as $C_p(T) = dh(T)/dT$. Determine the C_p of air at 300 K, using the enthalpy data from Table A-17, and compare it to the value listed in Table A-2*b*.

Solution The C_p values of air at various temperatures are listed in Table A-2*b*. The listed value at 300 K is 1.005 kJ/(kg·K).

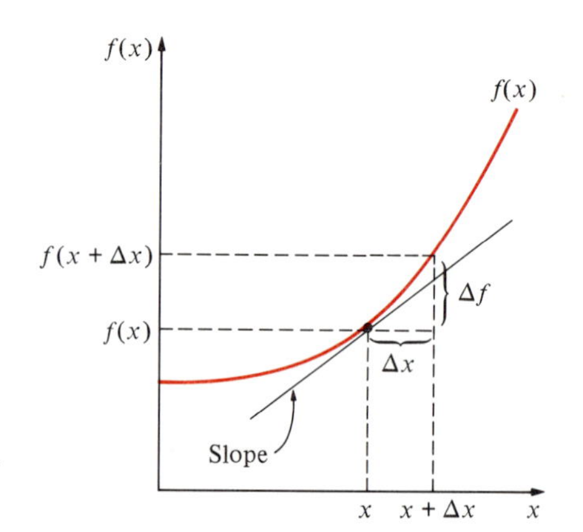

FIGURE 11-1

The derivative of a function at a specified point represents the slope of the function at that point.

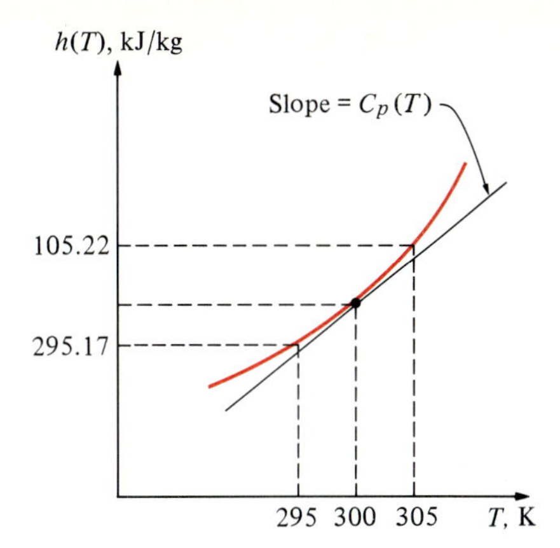

FIGURE 11-2

Schematic for Example 11-1.

This value could also be determined by differentiating the function $h(T)$ with respect to T and evaluating the result at $T = 300$ K. However, the function $h(T)$ is not available. But we can still determine the C_p value approximately by replacing the differentials in the $C_p(T)$ relation by the corresponding differences in the neighborhood of the specified point (Fig. 11-2):

$$C_p(300\text{ K}) = \left[\frac{dh(T)}{dT}\right]_{T=300\text{ K}} \cong \left[\frac{\Delta h(T)}{\Delta T}\right]_{T\cong 300\text{ K}} = \frac{h(305\text{ K}) - h(295\text{ K})}{(305 - 295)\text{ K}}$$

$$= \frac{(305.22 - 295.17)\text{ kJ/kg}}{(305 - 295)\text{ K}} = 1.005\text{ kJ/(kg·K)}$$

which is identical to the listed value. Therefore, differential quantities can be viewed as differences. They can even be replaced by differences, whenever necessary, to obtain approximate results. The widely used finite difference numerical method is based on this simple principle.

Partial Differentials

Now consider a function that depends on two (or more) variables, such as $z = z(x,y)$. This time the value of z depends on both x and y. It is sometimes desired to examine the dependence of z on only one of the variables. This is done by allowing one variable to change while holding the others constant and observing the change in the function. The variation of $z(x,y)$ with x when y is held constant is called the **partial derivative** of z with respect to x, and it is expressed as

$$\left(\frac{\partial z}{\partial x}\right)_y = \lim_{\Delta x\to 0}\left(\frac{\Delta z}{\Delta x}\right)_y = \lim_{\Delta x\to 0}\frac{z(x+\Delta x, y) - z(x,y)}{\Delta x} \qquad (11\text{-}2)$$

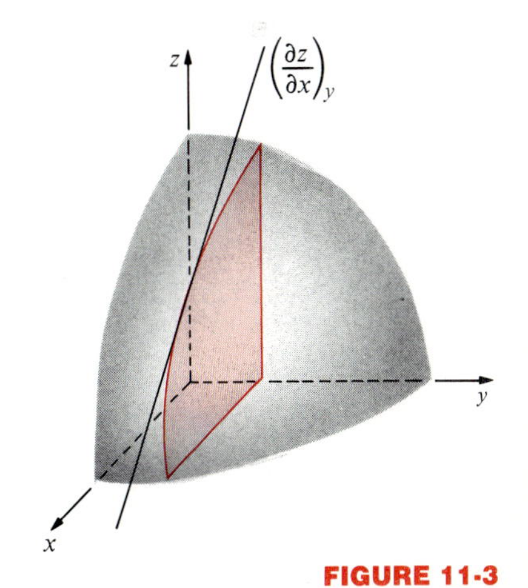

FIGURE 11-3

Geometric representation of partial derivative $(\partial z/\partial x)_y$.

This is illustrated in Fig. 11-3. The symbol ∂ represents differential changes, just like the symbol d. They differ in that the symbol d represents the *total* differential change and reflects the influence of all variables, whereas ∂ represents the *partial* differential change due to the variation of a single variable.

Note that the changes indicated by d and ∂ are identical for indepen-

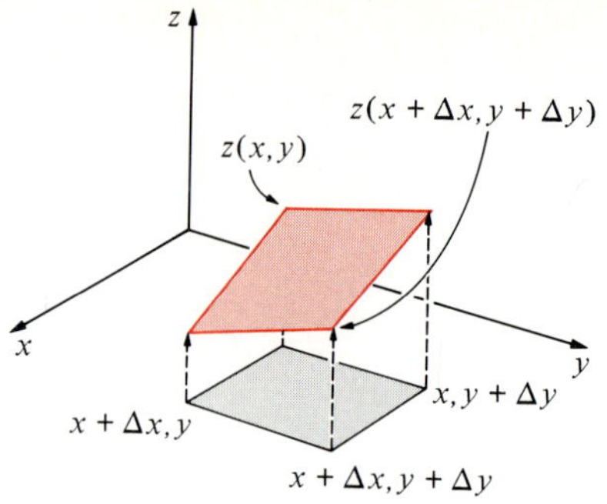

FIGURE 11-4
Geometric representation of total derivative dz for a function $z(x,y)$.

dent variables, but not for dependent variables. For example, $(\partial x)_y = dx$ but $(\partial z)_y \neq dz$. [In our case, $dz = (\partial z)_x + (\partial z)_y$.] Also note that the value of the partial derivative $(\partial z/\partial x)_y$, in general, will be different at different y values.

To obtain a relation for the total differential change in $z(x,y)$ for simultaneous changes in x and y, consider a small portion of the surface $z(x,y)$ shown in Fig. 11-4. When the independent variables x and y change by Δx and Δy, respectively, the dependent variable z changes by Δz, which can be expressed as

$$\Delta z = z(x + \Delta x, y + \Delta y) - z(x,y)$$

Adding and subtracting $z(x, y + \Delta y)$, we get

$$\Delta z = z(x + \Delta x, y + \Delta y) - z(x, y + \Delta y) + z(x, y + \Delta y) - z(x,y)$$

or

$$\Delta z = \frac{z(x + \Delta x, y + \Delta y) - z(x, y + \Delta y)}{\Delta x}\Delta x + \frac{z(x, y + \Delta y) - z(x,y)}{\Delta y}\Delta y$$

Taking the limits as $\Delta x \to 0$ and $\Delta y \to 0$ and using the definition of partial derivatives, we obtain

$$dz = \left(\frac{\partial z}{\partial x}\right)_y dx + \left(\frac{\partial z}{\partial y}\right)_x dy \tag{11-3}$$

Equation 11-3 is the fundamental relation for the **total differential** of a dependent variable in terms of its partial derivatives with respect to the independent variables. This relation can easily be extended to include more independent variables.

EXAMPLE 11-2

Consider an ideal gas at 300 K and 0.86 m^3/kg. As a result of some disturbance, the state of the gas changes to 302 K and 0.87 m^3/kg. Estimate the change in the pressure of the gas as a result of this disturbance, using Eq. 11-3.

Solution Strictly speaking, Eq. 11-3 is valid for differential changes in variables.

But it can also be used with reasonable accuracy if these changes are small. The changes in T and v, respectively, can be expressed as

$$dT \cong \Delta T = (302 - 300)\ \text{K} = 2\ \text{K}$$

and

$$dv \cong \Delta v = (0.87 - 0.86)\ \text{m}^3/\text{kg} = 0.01\ \text{m}^3/\text{kg}$$

An ideal gas obeys the relation $Pv = RT$. Solving for P yields

$$P = \frac{RT}{v}$$

Note that R is a constant and $P = P(T,v)$. Applying Eq. 11-3 and using average values for T and v, we find

$$\begin{aligned} dP &= \left(\frac{\partial P}{\partial T}\right)_v dT + \left(\frac{\partial P}{\partial v}\right)_T dv \\ &= \frac{R\,dT}{v} - \frac{RT\,dv}{v^2} \\ &= [0.287\ \text{kPa}\cdot\text{m}^3/(\text{kg}\cdot\text{K})]\left[\frac{2\ \text{K}}{0.865\ \text{m}^3/\text{kg}} - \frac{(301\ \text{K})(0.01\ \text{m}^3/\text{kg})}{(0.865\ \text{m}^3/\text{kg})^2}\right] \\ &= 0.664\ \text{kPa} - 1.155\ \text{kPa} \\ &= -0.491\ \text{kPa} \end{aligned}$$

Therefore, the pressure will decrease by 0.491 kPa as a result of this disturbance. Notice that if the temperature had remained constant ($dT = 0$), the pressure would decrease by 1.155 kPa as a result of the 0.01-m³/kg increase in specific volume. However, if the specific volume had remained constant ($dv = 0$), the pressure would increase by 0.664 kPa as a result of the 2-K rise in temperature (Fig. 11-5). That is,

$$\left(\frac{\partial P}{\partial T}\right)_v dT = (\partial P)_v = 0.664\ \text{kPa}$$

$$\left(\frac{\partial P}{\partial v}\right)_T dv = (\partial P)_T = -1.155\ \text{kPa}$$

and

$$dP = (\partial P)_v + (\partial P)_T = 0.664 - 1.155 = -0.491\ \text{kPa}$$

Of course, we could solve this problem easily (and exactly) by evaluating

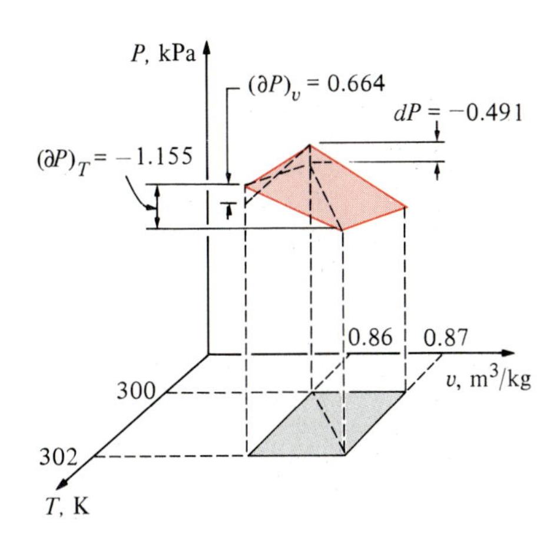

FIGURE 11-5
Geometric representation of the disturbance discussed in Example 11-2.

the pressure from the ideal-gas relation $P = RT/v$ at the final state (302 K and 0.87 m^3/kg) and the initial state (300 K and 0.86 m^3/kg) and taking their difference. This yields -0.491 kPa, which is exactly the value obtained above. Thus the small finite quantities (2 K, 0.01 m^3/kg) can be approximated as differential quantities with reasonable accuracy.

Partial Differential Relations

Now let us rewrite Eq. 11-3 as

$$dz = M\,dx + N\,dy \tag{11-4}$$

where $M = \left(\frac{\partial z}{\partial x}\right)_y$ and $N = \left(\frac{\partial z}{\partial y}\right)_x$

Taking the partial derivative of M with respect to y, and of N with respect to x yields

$$\left(\frac{\partial M}{\partial y}\right)_x = \frac{\partial^2 z}{\partial x\,\partial y} \quad \text{and} \quad \left(\frac{\partial N}{\partial x}\right)_y = \frac{\partial^2 z}{\partial y\,\partial x}$$

The order of differentiation is immaterial for properties since they are continuous point functions and have exact differentials. Therefore, the two relations above are identical:

$$\left(\frac{\partial M}{\partial y}\right)_x = \left(\frac{\partial N}{\partial x}\right)_y \tag{11-5}$$

This is an important relation for partial derivatives, and it is used in calculus to test whether a differential dz is exact or inexact [i.e., whether $z(x,y)$ is a point function or a path function]. In thermodynamics, this relation forms the basis for the development of the Maxwell relations discussed in the next section.

Finally, we develop two important relations for partial derivatives—the reciprocity and the cyclic relations. The function $z = z(x,y)$ can also be expressed as $x = x(y,z)$ if y and z are taken to be the independent variables. Then the total differential of x becomes, from Eq. 11-3,

$$dx = \left(\frac{\partial x}{\partial y}\right)_z dy + \left(\frac{\partial x}{\partial z}\right)_y dz \tag{11-6}$$

Eliminating dx by combining Eqs. 11-3 and 11-6, we have

$$dz = \left[\left(\frac{\partial z}{\partial x}\right)_y \left(\frac{\partial x}{\partial y}\right)_z + \left(\frac{\partial z}{\partial y}\right)_x\right] dy + \left(\frac{\partial x}{\partial z}\right)_y \left(\frac{\partial z}{\partial x}\right)_y dz$$

Rearranging,

$$\left[\left(\frac{\partial z}{\partial x}\right)_y \left(\frac{\partial x}{\partial y}\right)_z + \left(\frac{\partial z}{\partial y}\right)_x\right] dy = \left[1 - \left(\frac{\partial x}{\partial z}\right)_y \left(\frac{\partial z}{\partial x}\right)_y\right] dz \tag{11-7}$$

The variables y and z are independent of each other and thus can be varied independently. For example, y can be held constant ($dy = 0$), and z can be varied over a range of values ($dz \neq 0$). For this equation to be valid at all times, the terms in the brackets must equal zero, regardless of the values of y and z. Setting the terms in each bracket equal to zero, we obtain

$$\left(\frac{\partial x}{\partial z}\right)_y \left(\frac{\partial z}{\partial x}\right)_y = 1 \longrightarrow \left(\frac{\partial x}{\partial z}\right)_y = \frac{1}{(\partial z/\partial x)_y} \quad (11\text{-}8)$$

$$\left(\frac{\partial z}{\partial x}\right)_y \left(\frac{\partial x}{\partial y}\right)_z = -\left(\frac{\partial z}{\partial y}\right)_x \longrightarrow \left(\frac{\partial x}{\partial y}\right)_z \left(\frac{\partial y}{\partial z}\right)_x \left(\frac{\partial z}{\partial x}\right)_y = -1 \quad (11\text{-}9)$$

The first relation is called the **reciprocity relation**, and it shows that the inverse of a partial derivative is equal to its reciprocal (Fig. 11-6). The second relation is called the **cyclic relation**, and it is frequently used in thermodynamics (Fig. 11-7).

Function: $z + 2xy - 3y^2z = 0$

1. $z = \dfrac{2xy}{3y^2 - 1} \rightarrow \left(\dfrac{\partial z}{\partial x}\right)_y = \dfrac{2y}{3y^2 - 1}$

2. $x = \dfrac{3y^2z - z}{2y} \rightarrow \left(\dfrac{\partial x}{\partial z}\right)_y = \dfrac{3y^2 - 1}{2y}$

Thus, $\left(\dfrac{\partial z}{\partial x}\right)_y = \dfrac{1}{\left(\dfrac{\partial x}{\partial z}\right)_y}$

FIGURE 11-6
Demonstration of the reciprocity relation for the function $z + 2xy - 3y^2z = 0$.

EXAMPLE 11-3

Using the ideal-gas equation of state, verify (*a*) the cyclic relation and (*b*) the reciprocity relation at constant P.

Solution The ideal-gas equation of state $Pv = RT$ involves the three variables P, v, and T. Any two of these can be taken as the independent variables, with the remaining one being the dependent variable.

(*a*) Replacing x, y, and z in Eq. 11-9 by P, v, and T, respectively, we can express the cyclic relation for an ideal gas as

$$\left(\frac{\partial P}{\partial v}\right)_T \left(\frac{\partial v}{\partial T}\right)_P \left(\frac{\partial T}{\partial P}\right)_v = -1$$

where $P = P(v,T) = \dfrac{RT}{v} \longrightarrow \left(\dfrac{\partial P}{\partial v}\right)_T = -\dfrac{RT}{v^2}$

$$v = v(P,T) = \frac{RT}{P} \longrightarrow \left(\frac{\partial v}{\partial T}\right)_P = \frac{R}{P}$$

$$T = T(P,v) = \frac{Pv}{R} \longrightarrow \left(\frac{\partial T}{\partial P}\right)_v = \frac{v}{R}$$

Substituting yields

$$\left(-\frac{RT}{v^2}\right)\left(\frac{R}{P}\right)\left(\frac{v}{R}\right) = -\frac{RT}{Pv} = -1$$

which is the desired result.

(*b*) The reciprocity rule for an ideal gas at P = constant can be expressed as

$$\left(\frac{\partial v}{\partial T}\right)_P = \frac{1}{(\partial T/\partial v)_P}$$

Performing the differentiations and substituting, we have

$$\frac{R}{P} = \frac{1}{P/R} \longrightarrow \frac{R}{P} = \frac{R}{P}$$

Thus the proof is complete.

FIGURE 11-7
Partial differentials are powerful tools that are supposed to make life easier, not harder.

11-2 ■ THE MAXWELL RELATIONS

The equations that relate the partial derivatives of properties P, v, T, and s of a simple compressible substance to each other are called the *Maxwell relations*. They are obtained from the four Gibbs equations by exploiting the exactness of the differentials of thermodynamic properties.

Two of the Gibbs relations were derived in Chap. 7 and are expressed as

$$du = T\,ds - P\,dv \tag{11-10}$$

$$dh = T\,ds + v\,dP \tag{11-11}$$

The other two Gibbs relations are based on two new combination properties—the **Helmholtz function** a and the **Gibbs function** g, defined as

$$a = u - Ts \tag{11-12}$$

$$g = h - Ts \tag{11-13}$$

Differentiating, we get

$$da = du - T\,ds - s\,dT$$

$$dg = dh - T\,ds - s\,dT$$

Simplifying the above relations by using Eqs. 11-10 and 11-11, we obtain the other two Gibbs relations for simple compressible systems:

$$da = -s\,dT - P\,dv \tag{11-14}$$

$$dg = -s\,dT + v\,dP \tag{11-15}$$

A careful examination of the four Gibbs relations presented above reveals that they are of the form

$$dz = M\,dx + N\,dy \tag{11-4}$$

with

$$\left(\frac{\partial M}{\partial y}\right)_x = \left(\frac{\partial N}{\partial x}\right)_y \tag{11-5}$$

since u, h, a, and g are properties and thus have exact differentials. Applying Eq. 11-5 to each of them, we obtain

$$\left(\frac{\partial T}{\partial v}\right)_s = -\left(\frac{\partial P}{\partial s}\right)_v \tag{11-16}$$

$$\left(\frac{\partial T}{\partial P}\right)_s = \left(\frac{\partial v}{\partial s}\right)_P \tag{11-17}$$

$$\left(\frac{\partial s}{\partial v}\right)_T = \left(\frac{\partial P}{\partial T}\right)_v \tag{11-18}$$

$$\left(\frac{\partial s}{\partial P}\right)_T = -\left(\frac{\partial v}{\partial T}\right)_P \tag{11-19}$$

$$\left(\frac{\partial T}{\partial v}\right)_s = -\left(\frac{\partial P}{\partial s}\right)_v$$

$$\left(\frac{\partial T}{\partial P}\right)_s = \left(\frac{\partial v}{\partial s}\right)_P$$

$$\left(\frac{\partial s}{\partial v}\right)_T = \left(\frac{\partial P}{\partial T}\right)_v$$

$$\left(\frac{\partial s}{\partial P}\right)_T = -\left(\frac{\partial v}{\partial T}\right)_P$$

FIGURE 11-8
Maxwell relations are extremely valuable in thermodynamic analysis.

These are called the **Maxwell relations** (Fig. 11-8). They are extremely valuable in thermodynamics because they provide a means of determining the change in entropy, which cannot be measured directly, by simply measuring the changes in properties P, v, and T. Note that the Maxwell relations given above are limited to simple compressible systems. However, other similar relations can be written just as easily for nonsimple systems such as those involving electrical, magnetic, and other effects.

EXAMPLE 11-4

Verify the validity of the last Maxwell relation (Eq. 11-19) for steam at 250°C and 300 kPa.

Solution The last Maxwell relation says that for a simple compressible substance, the change in entropy with pressure at constant temperature is equal to the negative of the change in specific volume with temperature at constant pressure.

If we had explicit analytical relations for the entropy and specific volume of steam in terms of other properties, we could easily verify this by performing the indicated derivations. However, all we have for steam is tables of properties listed at certain intervals. Therefore, the only course we can take to solve this problem, without taking a trip to the library, is to replace the differential quantities in Eq. 11-19 with corresponding finite quantities, using property values from the tables (Table A-6 in this case) at or about the specified state.

$$\left(\frac{\partial s}{\partial P}\right)_T \overset{?}{=} -\left(\frac{\partial v}{\partial T}\right)_P$$

$$\left(\frac{\Delta s}{\Delta P}\right)_{T=250^\circ\text{C}} \overset{?}{\cong} -\left(\frac{\Delta v}{\Delta T}\right)_{P=300\text{ kPa}}$$

$$\left[\frac{s_{400\text{ kPa}} - s_{200\text{ kPa}}}{(400-200)\text{ kPa}}\right]_{T=250^\circ\text{C}} \overset{?}{\cong} -\left[\frac{v_{300^\circ\text{C}} - v_{200^\circ\text{C}}}{(300-200)\text{ °C}}\right]_{P=300\text{ kPa}}$$

$$\frac{(7.3789 - 7.7086)\text{ kJ/(kg·K)}}{(400-200)\text{ kPa}} \overset{?}{\cong} -\frac{(0.8753 - 0.7163)\text{ m}^3\text{/kg}}{(300-200)\text{ °C}}$$

$$-0.00165\text{ m}^3\text{/(kg·K)} \cong -0.00159\text{ m}^3\text{/(kg·K)}$$

since kJ = kPa·m^3 and K ≡ °C for temperature differences. The two values are within 4 percent of each other. This difference is due to replacing the differential quantities by relatively large finite quantities. Based on the close agreement between the two values, the steam seems to satisfy Eq. 11-19 at the specified state.

This example shows that the entropy change of a simple compressible substance during an isothermal process can be determined from a knowledge of the easily measurable properties P, v, and T alone.

11-3 ■ THE CLAPEYRON EQUATION

The Maxwell relations have far-reaching implications in thermodynamics and are frequently used to derive useful thermodynamic relations. The Clapeyron equation is one such relation, and it enables us to determine the enthalpy change associated with a phase change (such as the enthalpy of vaporization h_{fg}) from a knowledge of P, v, and T data alone.

Consider the third Maxwell relation, Eq. 11-18:

$$\left(\frac{\partial P}{\partial T}\right)_v = \left(\frac{\partial s}{\partial v}\right)_T$$

During a phase-change process, the pressure is the saturation pressure which depends on the temperature only and is independent of the specific volume. That is, $P_{sat} = f(T_{sat})$. Therefore, the partial derivative $(\partial P/\partial T)_v$ can be expressed as a total derivative $(dP/dT)_{sat}$, which is the slope of the saturation curve on a *P-T* diagram at a specified saturation state (Fig. 11-9). This slope is independent of the specific volume, and thus it can be treated as a constant during the integration of Eq. 11-18 between two saturation states at the same temperature. For an isothermal liquid-vapor phase-change process, for example, the integration yields

$$s_g - s_f = \left(\frac{dP}{dT}\right)_{sat} (v_g - v_f) \tag{11-20}$$

or

$$\left(\frac{dP}{dT}\right)_{sat} = \frac{s_{fg}}{v_{fg}} \tag{11-21}$$

During this process the pressure also remains constant. Therefore, from Eq. 11-11,

$$dh = T\,ds + v\,\cancelto{0}{dP} \longrightarrow \int_f^g dh = \int_f^g T\,ds \longrightarrow h_{fg} = Ts_{fg}$$

Substituting this result into Eq. 11-21, we obtain

$$\left(\frac{dP}{dT}\right)_{sat} = \frac{h_{fg}}{Tv_{fg}} \tag{11-22}$$

which is called the **Clapeyron equation**. This is an important thermodynamic relation since it enables us to determine the enthalpy of vaporization h_{fg} at a given temperature by simply measuring the slope of the saturation curve on a *P-T* diagram and the specific volume of saturated liquid and saturated vapor at the given temperature.

The Clapeyron equation is applicable to any phase-change process that occurs at constant temperature and pressure. It can be expressed in a general form as

$$\left(\frac{dP}{dT}\right)_{sat} = \frac{h_{12}}{Tv_{12}} \tag{11-23}$$

where the subscripts 1 and 2 indicate the two phases.

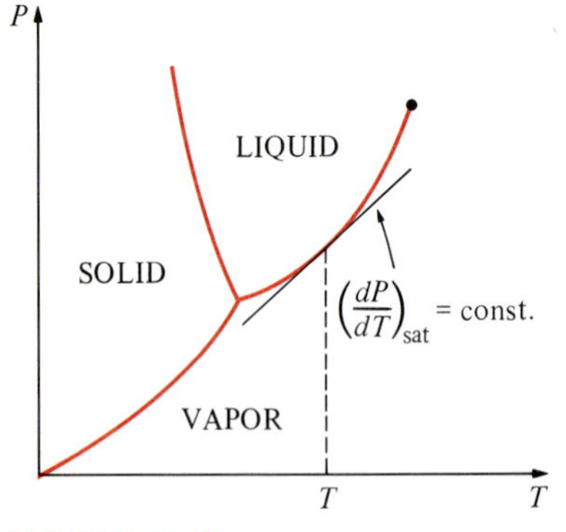

FIGURE 11-9
The slope of the saturation curve on a *P-T* diagram is constant at a constant *T* or *P*.

EXAMPLE 11-5

Using the Clapeyron equation, estimate the value of the enthalpy of vaporization of refrigerant-12 at 20°C, and compare it with the tabulated value.

Solution From Eq. 11-22,

$$h_{fg} = Tv_{fg}\left(\frac{dP}{dT}\right)_{sat}$$

where, from Table A-11,

$$v_{fg} = (v_g - v_f)_{@\,20^\circ\text{C}} = (0.03078 - 0.0007525)\ \text{m}^3/\text{kg}$$
$$= 0.03003\ \text{m}^3/\text{kg}$$

$$\left(\frac{dP}{dT}\right)_{\text{sat},20^\circ\text{C}} \cong \left(\frac{\Delta P}{\Delta T}\right)_{\text{sat},20^\circ\text{C}} = \frac{P_{\text{sat}\,@\,24^\circ\text{C}} - P_{\text{sat}\,@\,16^\circ\text{C}}}{24^\circ\text{C} - 16^\circ\text{C}}$$
$$= \frac{(634.05 - 505.91)\ \text{kPa}}{8^\circ\text{C}} = 16.02\ \text{kPa/K}$$

since $\Delta T(^\circ\text{C}) \equiv \Delta T(\text{K})$. Substituting, we get

$$h_{fg} = (293\ \text{K})(0.03003\ \text{m}^3/\text{kg})(16.02\ \text{kPa/K})\left(\frac{1\ \text{kJ}}{1\ \text{kPa}\cdot\text{m}^3}\right)$$
$$= 140.96\ \text{kJ/kg}$$

The tabulated value of h_{fg} at 20°C is 140.91 kJ/kg. The small difference between the two values is due to the approximation used in determining the slope of the saturation curve at 20°C.

The Clapeyron equation can be simplified for liquid-vapor and solid-vapor phase changes by utilizing some approximations. At low pressures $v_g >> v_f$, and thus $v_{fg} \cong v_g$. By treating the vapor as an ideal gas, we have $v_g = RT/P$. Substituting these approximations into Eq. 11-22, we find

$$\left(\frac{dP}{dT}\right)_{\text{sat}} = \frac{Ph_{fg}}{RT^2}$$

or

$$\left(\frac{dP}{P}\right)_{\text{sat}} = \frac{h_{fg}}{R}\left(\frac{dT}{T^2}\right)_{\text{sat}}$$

For small temperature intervals h_{fg} can be treated as a constant at some average value. Then integrating this equation between two saturation states yields

$$\ln\left(\frac{P_2}{P_1}\right)_{\text{sat}} \cong \frac{h_{fg}}{R}\left(\frac{1}{T_1} - \frac{1}{T_2}\right)_{\text{sat}} \tag{11-24}$$

This equation is called the **Clapeyron-Clausius equation**, and it can be used to determine the variation of saturation pressure with temperature. It can also be used in the solid-vapor region by replacing h_{fg} by h_{ig} (the enthalpy of sublimation) of the substance.

EXAMPLE 11-6

Estimate the saturation pressure of refrigerant-12 at −50°F, using the data available in the refrigerant tables.

Solution Table A-11E lists saturation data at temperatures −40°F and above. Therefore, we should either resort to other sources or use extrapolation to obtain saturation data at lower temperatures. Equation 11-24 provides an intelligent way to extrapolate:

$$\ln\left(\frac{P_2}{P_1}\right)_{\text{sat}} \cong \frac{h_{fg}}{R}\left(\frac{1}{T_1} - \frac{1}{T_2}\right)_{\text{sat}}$$

In our case $T_1 = -40°F$ and $T_2 = -50°F$. For refrigerant-12, $R = 0.01643$ Btu/(lbm·R). Also from Table A-11E at $-40°F$, we read $h_{fg} = 72.91$ Btu/lbm and $P_1 = P_{sat\,@\,-40°F} = 9.308$ psia. Substituting these values into Eq. 11-24 gives

$$\ln\left(\frac{P_2}{9.308\text{ psia}}\right) \cong \frac{72.91\text{ Btu/lbm}}{0.01643\text{ Btu/(lbm·R)}}\left(\frac{1}{420\text{ R}} - \frac{1}{410\text{ R}}\right)$$

$$P_2 \cong 7.19\text{ psia}$$

Therefore, according to Eq. 11-24, the saturation pressure of refrigerant-12 at $-50°F$ is 7.19 psia. The actual value, obtained from another source, is 7.12 psia. Thus the value predicted by Eq. 11-24 is in error by about 1 percent, which is quite acceptable for most purposes. (If we had used linear extrapolation instead, we would have obtained 6.62 psia, which is in error by 7 percent.

11-4 ■ GENERAL RELATIONS FOR du, dh, ds, C_v, AND C_p

The state postulate established that the state of a simple compressible system is completely specified by two independent, intensive properties. Therefore, at least theoretically, we should be able to calculate all the properties of a system at any state once two independent, intensive properties are available. This is certainly good news for properties that cannot be measured directly such as the internal energy, enthalpy, and entropy. But the successful calculation of these properties from measurable ones depends on the availability of simple and accurate relations between the two groups.

In this section we develop general relations for changes in internal energy, enthalpy, and entropy in terms of pressure, specific volume, temperature, and specific heats alone. We also develop some general relations involving specific heats. The relations developed will enable us to determine the *changes* in these properties. The property values at specified states can be determined only after a reference state is selected, the choice of which is quite arbitrary.

1 Internal Energy Changes

We choose the internal energy to be a function of T and v, $u = u(T,v)$, and take its total differential (Eq. 11-3),

$$du = \left(\frac{\partial u}{\partial T}\right)_v dT + \left(\frac{\partial u}{\partial v}\right)_T dv$$

Using the definition of C_v, we have

$$du = C_v\,dT + \left(\frac{\partial u}{\partial v}\right)_T dv \tag{11-25}$$

Now we choose the entropy to be a function of T and v, $s = s(T,v)$, and take its total differential,

$$ds = \left(\frac{\partial s}{\partial T}\right)_v dT + \left(\frac{\partial s}{\partial v}\right)_T dv \tag{11-26}$$

Substituting this into the $T\,ds$ relation $du = T\,ds - P\,dv$ yields

$$du = T\left(\frac{\partial s}{\partial T}\right)_v dT + \left[T\left(\frac{\partial s}{\partial v}\right)_T - P\right] dv \qquad (11\text{-}27)$$

Equating the coefficients of dT and dv in Eqs. 11-25 and 11-27 gives

$$\left(\frac{\partial s}{\partial T}\right)_v = \frac{C_v}{T}$$

$$\left(\frac{\partial u}{\partial v}\right)_T = T\left(\frac{\partial s}{\partial v}\right)_T - P \qquad (11\text{-}28)$$

Using the third Maxwell relation (Eq. 11-18), we get

$$\left(\frac{\partial u}{\partial v}\right)_T = T\left(\frac{\partial P}{\partial T}\right)_v - P$$

Substituting this into Eq. 11-25, we obtain the desired relation for du:

$$du = C_v\,dT + \left[T\left(\frac{\partial P}{\partial T}\right)_v - P\right] dv \qquad (11\text{-}29)$$

The change in internal energy of a simple compressible system associated with a change of state from (T_1,v_1) to (T_2,v_2) is determined by integration:

$$u_2 - u_1 = \int_{T_1}^{T_2} C_v\,dT + \int_{v_1}^{v_2} \left[T\left(\frac{\partial P}{\partial T}\right)_v - P\right] dv \qquad (11\text{-}30)$$

2 Enthalpy Changes

The general relation for dh is determined in exactly the same manner. This time we choose the enthalpy to be a function of T and P ' = $h(T,P)$, and take its total differential,

$$dh = \left(\frac{\partial h}{\partial T}\right)_P dT + \left(\frac{\partial h}{\partial P}\right)_T dP$$

Using the definition of C_p, we have

$$dh = C_p\,dT + \left(\frac{\partial h}{\partial P}\right)_T dP \qquad (11\text{-}31)$$

Now we choose the entropy to be a function of T and P, $s = s(T,P)$, and take its total differential,

$$ds = \left(\frac{\partial s}{\partial T}\right)_P dT + \left(\frac{\partial s}{\partial P}\right)_T dP \qquad (11\text{-}32)$$

Substituting this into the $T\,ds$ relation $dh = T\,ds + v\,dP$ gives

$$dh = T\left(\frac{\partial s}{\partial T}\right)_P dT + \left[T\left(\frac{\partial s}{\partial P}\right)_T + v\right] dP \qquad (11\text{-}33)$$

Equating the coefficients of dT and dP in Eqs. 11-31 and 11-33, we obtain

$$\left(\frac{\partial s}{\partial T}\right)_P = \frac{C_P}{T}$$

$$\left(\frac{\partial h}{\partial P}\right)_T = T\left(\frac{\partial s}{\partial P}\right)_T + v \qquad (11\text{-}34)$$

Using the fourth Maxwell relation (Eq. 11-19), we have

$$\left(\frac{\partial h}{\partial P}\right)_T = v - T\left(\frac{\partial v}{\partial T}\right)_P$$

Substituting this into Eq. 11-31, we obtain the desired relation for dh:

$$dh = C_p\, dT + \left[v - T\left(\frac{\partial v}{\partial T}\right)_P\right] dP \qquad (11\text{-}35)$$

The change in enthalpy of a simple compressible system associated with a change of state from (T_1,P_1) to (T_2,P_2) is determined by integration:

$$h_2 - h_1 = \int_{T_1}^{T_2} C_p\, dT + \int_{P_1}^{P_2} \left[v - T\left(\frac{\partial v}{\partial T}\right)_P\right] dP \qquad (11\text{-}36)$$

In reality, one only needs to determine either $u_2 - u_1$ from Eq. 11-30 or $h_2 - h_1$ from Eq. 11-36, depending on which is more suitable to the data at hand. The other can easily be determined by using the definition of enthalpy $h = u + Pv$:

$$h_2 - h_1 = u_2 - u_1 + (P_2v_2 - P_1v_1) \qquad (11\text{-}37)$$

3 Entropy Changes

We develop two general relations for the entropy change of a simple compressible system, using the relations developed earlier in this section.

The first relation is obtained by replacing the first partial derivative in the total differential of ds (Eq. 11-26) by Eq. 11-29 and the second partial derivative by the third Maxwell relation (Eq. 11-18), yielding

$$ds = \frac{C_v}{T}\, dT + \left(\frac{\partial P}{\partial T}\right)_v dv \qquad (11\text{-}38)$$

and

$$s_2 - s_1 = \int_{T_1}^{T_2} \frac{Cv}{T}\, dT + \int_{v_1}^{v_2} \left(\frac{\partial P}{\partial T}\right)_v dv \qquad (11\text{-}39)$$

The second relation is obtained by replacing the first partial derivative in the total differential of ds (Eq. 11-32) by Eq. 11-34, and the second partial derivative by the fourth Maxwell relation (Eq. 11-19), yielding

$$ds = \frac{C_p}{T}\, dT - \left(\frac{\partial v}{\partial T}\right)_P dP \qquad (11\text{-}40)$$

and
$$s_2 - s_1 = \int_{T_1}^{T_2} \frac{C_p}{T}\,dT - \int_{P_1}^{P_2} \left(\frac{\partial v}{\partial T}\right)_P dP \qquad (11\text{-}41)$$

Either relation can be used to deterine the entropy change. The proper choice will depend on the suitability of the available data to a particular relation.

4 Specific Heats C_v and C_p

We mentioned in Chap. 3 that the specific heats of an ideal gas depend on temperature only. For a general pure substance, however, the specific heats depend on specific volume or pressure as well as on the temperature. Below we develop some general relations to relate the specific heats of a substance to pressure, specific volume, and temperature.

At low pressures gases behave as ideal gases, and their specific heats essentially depend on temperature only. These specific heats are called *zero-pressure,* or *ideal-gas, specific heats* (denoted C_{v0} and C_{p0}), and they are relatively easier to measure. Thus it is desirable to have some general relations which will enable us to calculate the specific heats at higher pressures (or lower specific volumes) from a knowledge of C_{v0} or C_{p0} and the P-v-T behavior of the substance. Such relations are obtained by applying the test of exactness (Eq. 11-5) on Eqs. 11-38 and 11-40, which yields

$$\left(\frac{\partial C_v}{\partial v}\right)_T = T\left(\frac{\partial^2 P}{\partial T^2}\right)_v \qquad (11\text{-}42)$$

and
$$\left(\frac{\partial C_p}{\partial P}\right)_T = -T\left(\frac{\partial^2 v}{\partial T^2}\right)_P \qquad (11\text{-}43)$$

The deviation of C_p from C_{p0} with increasing pressure, for example, is determined by integrating Eq. 11-43 from zero pressure to any pressure P along an isothermal path:

$$(C_p - C_{p0})_T = -T\int_0^P \left(\frac{\partial^2 v}{\partial T^2}\right)_P dP \qquad (11\text{-}44)$$

The integration on the right-hand side requires a knowledge of the P-v-T behavior of the substance alone. The notation indicates that v should be differentiated twice with respect to T while P is held constant. The resulting expression should be integrated with respect to P while T is held constant.

Another desirable general relation involving specific heats is one that relates the two specific heats C_P and C_v. The advantage of such a relation is obvious: We will need to measure only one specific heat (usually C_p) and calculate the other one using that relation and the P-v-T data of the substance. We start the development of such a relation by equating the two ds relations (Eqs. 11-38 and 11-40) and solving for dT:

$$dT = \frac{T(\partial P/\partial T)_v}{C_p - C_v}\, dv + \frac{T(\partial v/\partial T)_P}{C_p - C_v}\, dP$$

Choosing $T = T(v,P)$ and differentiating, we get

$$dT = \left(\frac{\partial T}{\partial v}\right)_P dv + \left(\frac{\partial T}{\partial P}\right)_v dP$$

Equating the coefficients of either dv or dP of the above two equations gives the desired result:

$$C_p - C_v = T\left(\frac{\partial v}{\partial T}\right)_P \left(\frac{\partial P}{\partial T}\right)_v \tag{11-45}$$

An alternative form of this relation is obtained by using the cyclic relation:

$$\left(\frac{\partial P}{\partial T}\right)_v \left(\frac{\partial T}{\partial v}\right)_P \left(\frac{\partial v}{\partial P}\right)_T = -1 \longrightarrow \left(\frac{\partial P}{\partial T}\right)_v = -\left(\frac{\partial v}{\partial T}\right)_P \left(\frac{\partial P}{\partial v}\right)_T$$

Substituting the result into Eq. 11-45 gives

$$C_p - C_v = -T\left(\frac{\partial v}{\partial T}\right)_P^2 \left(\frac{\partial P}{\partial v}\right)_T \tag{11-46}$$

This relation can be expressed in terms of two other thermodynamic properties called the **volume expansivity** β and the **isothermal compressibility** α, which are defined as (Fig. 11-10)

$$\beta = \frac{1}{v}\left(\frac{\partial v}{\partial T}\right)_P \tag{11-47}$$

and

$$\alpha = -\frac{1}{v}\left(\frac{\partial v}{\partial P}\right)_T \tag{11-48}$$

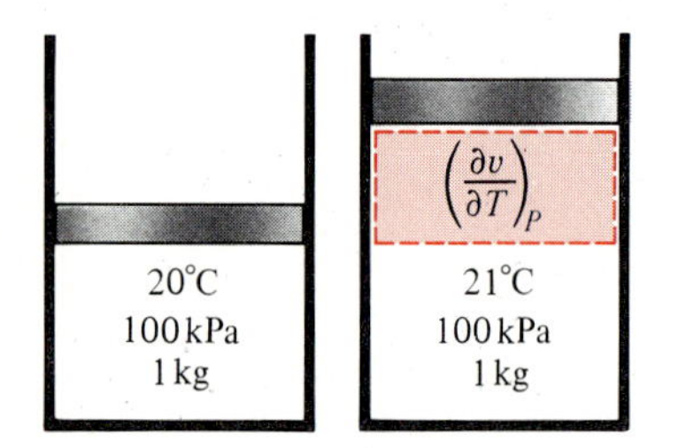

(a) A substance with a large β

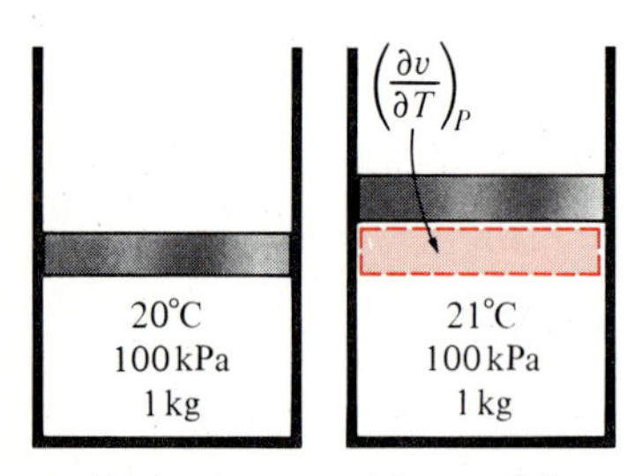

(b) A substance with a small β

FIGURE 11-10

The volume expansivity (also called the *coefficient of volumetric expansion*) is a measure of the change in volume with temperature at constant pressure.

Substituting these two relations into Eq. 11-46, we obtain a third general relation for $C_p - C_v$:

$$C_p - C_v = \frac{vT\beta^2}{\alpha} \tag{11-49}$$

We can draw several conclusions from this equation:

1 The isothermal compressibility α is a positive quantity for all substances in all phases. The volume expansivity could be negative for some substances (such as liquid water below 4°C), but its square is always positive or zero. The temperature T in this relation is absolute temperature, which is also positive. Therefore we conclude that *the constant-pressure specific heat is always greater than or equal to the constant-volume specific heat:*

$$C_p \geqslant C_v \tag{11-50}$$

2 The difference between C_p and C_v approaches zero as the absolute temperature approaches zero.

3 The two specific heats are identical for truly incompressible substances since v = constant. The difference between the two specific heats is very small and is usually disregarded for substances that are *almost* incompressible, such as liquids and solids.

EXAMPLE 11-7

Derive a relation for the internal energy change of a gas which obeys the van der Waals equation of state. Assume that in the range of interest C_v varies according to the relation $C_v = c_1 + c_2T$, where c_1 and c_2 are constants.

Solution The change in internal energy of any simple compressible substance in any phase during any process can be determined from Eq. 11-30:

$$u_2 - u_1 = \int_{T_1}^{T_2} C_v\, dT + \int_{v_1}^{v_2} \left[T\left(\frac{\partial P}{\partial T}\right)_v - P\right] dv$$

The van der Waals equation of state was discussed in Chap. 2. It can be expressed as

$$P = \frac{RT}{v - b} - \frac{a}{v^2}$$

Then
$$\left(\frac{\partial P}{\partial T}\right)_v = \frac{R}{v - b}$$

Thus,
$$T\left(\frac{\partial P}{\partial T}\right)_v - P = \frac{RT}{v - b} - \frac{RT}{v - b} + \frac{a}{v^2} = \frac{a}{v^2}$$

Substituting gives

$$u_2 - u_1 = \int_{T_1}^{T_2} (c_1 + c_2T)\, dT + \int_{v_1}^{v_2} \frac{a}{v^2}\, dv$$

Integrating yields

$$u_2 - u_1 = c_1(T_2 - T_1) + \frac{c_2}{2}(T_2^2 - T_1^2) + a\left(\frac{1}{v_1} - \frac{1}{v_2}\right)$$

which is the desired relation.

EXAMPLE 11-8

Show that the internal energy of (*a*) an ideal gas and (*b*) an incompressible substance is a function of temperature only, $u = u(T)$.

Solution The differential change in the internal energy of a general simple compressible substance is given by Eq. 11-29:

$$du = C_v\, dT + \left[T\left(\frac{\partial P}{\partial T}\right)_v - P\right] dv$$

(*a*) For an ideal gas $Pv = RT$. Then

$$T\left(\frac{\partial P}{\partial T}\right)_v - P = T\left(\frac{R}{v}\right) - P = P - P = 0$$

Thus, $$du = C_v\,dT$$

To complete the proof, we need to show that C_v is not a function of v either. This is done with the help of Eq. 11-42:

$$\left(\frac{\partial C_v}{\partial v}\right)_T = T\left(\frac{\partial^2 P}{\partial T^2}\right)_v$$

For an ideal gas $P = RT/v$. Then

$$\left(\frac{\partial P}{\partial T}\right)_v = \frac{R}{v} \quad \text{and} \quad \left(\frac{\partial^2 P}{\partial T^2}\right)_v = \left[\frac{\partial(R/v)}{\partial T}\right]_v = 0$$

Thus, $$\left(\frac{\partial C_v}{\partial v}\right)_T = 0$$

which says that C_v does not change with specific volume. That is, C_v is not a function of specific volume either. Therefore we conclude that the internal energy of an ideal gas is a function of temperature only (Fig. 11-11).

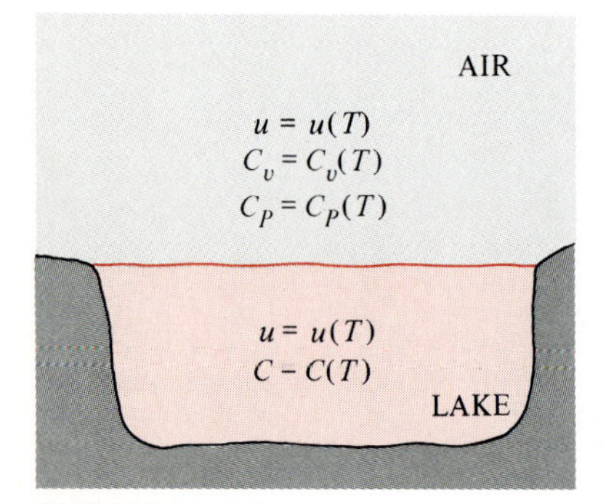

FIGURE 11-11
The internal energies and specific heats of ideal gases and incompressible substances depend on temperature only.

(*b*) For an incompressible substance, v = constant and thus $dv = 0$. Also from Eq. 11-49, $C_p = C_v = C$ since $\alpha = \beta = 0$ for incompressible substances. Then Eq. 11-29 reduces to

$$du = C\,dT$$

Again we need to show that the specific heat C depends on temperature only and not on the pressure or the specific volume. This is easily done with the help of Eq. 11-43:

$$\left(\frac{\partial C_p}{\partial P}\right)_T = -T\left(\frac{\partial^2 v}{\partial T^2}\right)_P = 0$$

since v = constant. Therefore we conclude that the internal energy of a truly incompressible substance depends on temperature only.

EXAMPLE 11-9

Show that $C_p - C_v = R$ for an ideal gas.

Solution This relation is easily proved by showing that the right-hand side of Eq. 11-46 is equivalent to the gas constant R of the ideal gas:

$$C_p - C_v = -T\left(\frac{\partial v}{\partial T}\right)_P^2\left(\frac{\partial P}{\partial v}\right)_T$$

$$P = \frac{RT}{v} \longrightarrow \left(\frac{\partial P}{\partial v}\right)_T = -\frac{RT}{v^2} = -\frac{P}{v}$$

$$v = \frac{RT}{P} \longrightarrow \left(\frac{\partial v}{\partial T}\right)_P^2 = \left(\frac{R}{P}\right)^2$$

Substituting, $$-T\left(\frac{\partial v}{\partial T}\right)_P^2\left(\frac{\partial P}{\partial v}\right)_T = -T\left(\frac{R}{P}\right)^2\left(-\frac{P}{v}\right) = R$$

Therefore, $$C_p - C_v = R$$

The Joule-Thomson Coefficient

When a fluid passes through a restriction such as a porous plug, a capillary tube, or an ordinary valve, its pressure decreases. As we have shown in Chap. 4, the enthalpy of the fluid remains approximately constant during such a process, called a throttling process. You will remember that a fluid may experience a large drop in its temperature as a result of throttling, which forms the basis of operation for refrigerators and most air conditioners. This is not always the case, however. The temperature of the fluid may remain unchanged, or it may even increase during a throttling process (Fig. 11-12).

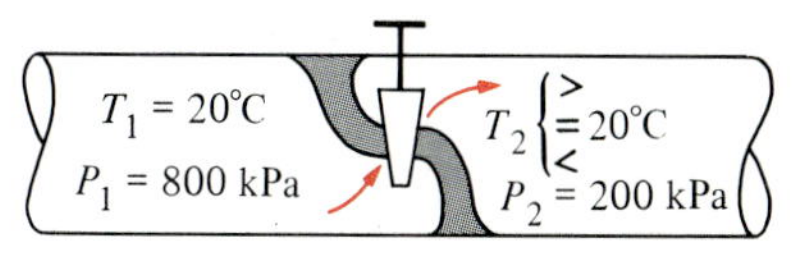

FIGURE 11-12

The temperature of a fluid may increase, decrease, or remain constant during a throttling process.

The temperature behavior of a fluid during a throttling (h = constant) process is described by the **Joule-Thomson coefficient**, defined as

$$\mu = \left(\frac{\partial T}{\partial P}\right)_h \tag{11-51}$$

Thus the Joule-Thomson coefficient is a measure of the change in temperature with pressure during a constant-enthalpy process. Notice that if

$$\mu \begin{cases} < 0 & \text{temperature increases} \\ = 0 & \text{temperature remains constant} \\ > 0 & \text{temperature decreases} \end{cases}$$

during a throttling process.

A careful look at its defining equation reveals that the Joule-Thomson coefficient represents the slope of h = constant lines on a T-P diagram. Such diagrams can be easily constructed from temperature and pressure measurements alone during throttling processes. A fluid at a fixed temperature and pressure T_1 and P_1 (thus fixed enthalpy) is forced to flow through a porous plug, and its temperature and pressure downstream (T_2 and P_2) are measured. The experiment is repeated for different sizes of porous plugs, each giving a different set of T_2 and P_2. Plotting the temperatures against the pressures gives us an h = constant line on a T-P diagram, as shown in Fig. 11-13. Repeating the experiment for different

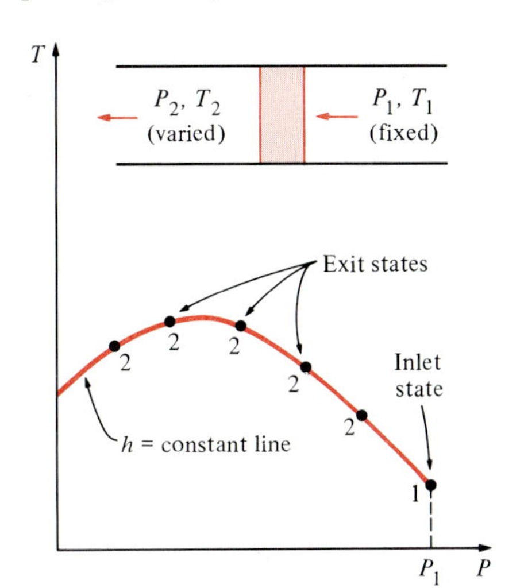

FIGURE 11-13

The development of an h = constant line on a P-T diagram.

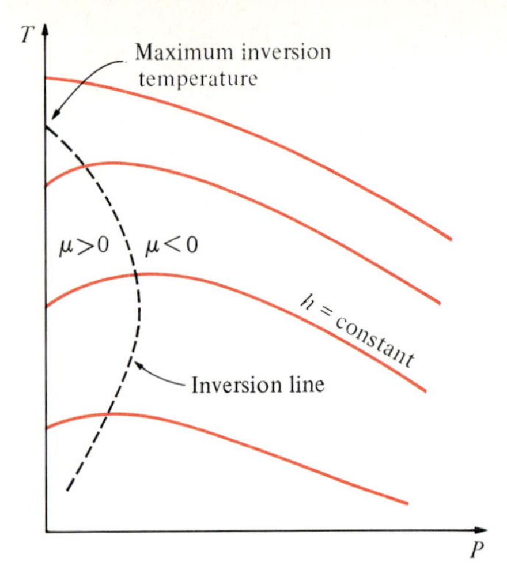

FIGURE 11-14
Constant-enthalpy lines of a substance on a *T-P* diagram.

sets of inlet pressure and temperature and plotting the results, we can construct a *T-P* diagram for a substance with several h = constant lines, as shown in Fig. 11-14.

Some constant-enthalpy lines on the *T-P* diagram pass through a point of zero slope or zero Joule-Thomson coefficient. The line that passes through these points is called the **inversion line**, and the temperature at a point where a constant-enthalpy line intersects the inversion line is called the **inversion temperature**. The temperature at the intersection of the $P = 0$ line (ordinate) and the upper part of the inversion line is called the **maximum inversion temperature**. Notice that the slopes of the h = constant lines are negative ($\mu < 0$) at states to the right of the inversion line and positive ($\mu > 0$) to the left of the inversion line.

A throttling process proceeds along a constant-enthalpy line in the direction of decreasing pressure, i.e., from right to left. Therefore, the temperature of a fluid will increase during a throttling process that takes place on the right-hand side of the inversion line. However, the fluid temperature will decrease during a throttling process that takes place on the left-hand side of the inversion line. It is clear from this diagram that a cooling effect cannot be achieved by throttling unless the fluid is below its maximum inversion temperature. This presents a problem for substances whose maximum inversion temperature is well below room temperature. For hydrogen, for example, the maximum inversion temperature is -68°C. Thus hydrogen must be cooled below this temperature if any further cooling is to be achieved by throttling.

Next we would like to develop a general relation for the Joule-Thomson coefficient in terms of the specific heats, pressure, specific volume, and temperature. This is easily accomplished by modifying the generalized relation for enthalpy change (Eq. 11-35)

$$dh = C_p\, dT + \left[v - T\left(\frac{\partial v}{\partial T}\right)_P\right] dP$$

Taking the partial derivative of each term with respect to P holding h constant and rearranging, we obtain

$$\frac{1}{C_p}\left[v - T\left(\frac{\partial v}{\partial T}\right)_P\right] = \left(\frac{\partial T}{\partial P}\right)_h = \mu \tag{11-52}$$

which is the desired relation. Thus the Joule-Thomson coefficient can be determined from a knowledge of the constant-pressure specific heat and the P-v-T behavior of the substance. Of course, it is also possible to predict the constant-pressure specific heat of a substance by using the Joule-Thomson coefficient, which is relatively easy to measure, together with the P-v-T data for the substance.

EXAMPLE 11-10

Show that the Joule-Thomson coefficient of an ideal gas is zero.

Solution For an ideal gas $v = RT/P$, and thus

$$\left(\frac{\partial v}{\partial T}\right)_P = \frac{R}{P}$$

Substituting into Eq. 11-52 yields

$$\mu = \frac{1}{C_p}\left[v - T\left(\frac{\partial v}{\partial T}\right)_P\right] = \frac{1}{C_p}\left[v - T\frac{R}{P}\right] = \frac{1}{C_p}(v - v) = 0$$

This result is not surprising since the enthalpy of an ideal gas is a function of temperature only, $h = h(T)$, which requires that the temperature remain constant unless the enthalpy changes. Therefore, a throttling process cannot be used to cool an ideal gas (Fig. 11-15).

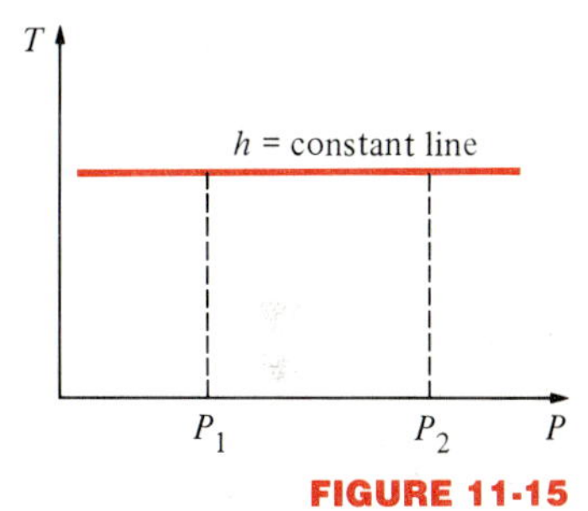

FIGURE 11-15

The temperature of an ideal gas remains constant during a throttling process since h = constant and T = constant lines on a T-P diagram coincide.

11-6 THE Δh, Δu, AND Δs OF REAL GASES

We have mentioned several times that gases at low pressures behave as ideal gases and obey the relation $Pv = RT$. The properties of ideal gases are relatively easy to evaluate since the properties u, h, C_v, and C_p depend on temperature only. At high pressures, however, gases deviate considerably from ideal-gas behavior, and it becomes necessary to account for this deviation. In Chap. 2 we accounted for the deviation in properties P, v, and T by either using more complex equations of state or evaluating the compressibility factor Z from the compressibility charts. Now we extend the analysis to evaluate the changes in the enthalpy, internal energy, and entropy of nonideal (real) gases, using the general relations for du, dh, and ds developed earlier.

1 Enthalpy Changes of Real Gases

The enthalpy of a real gas, in general, depends on the pressure as well as on the temperature. Thus the enthalpy change of a real gas during a

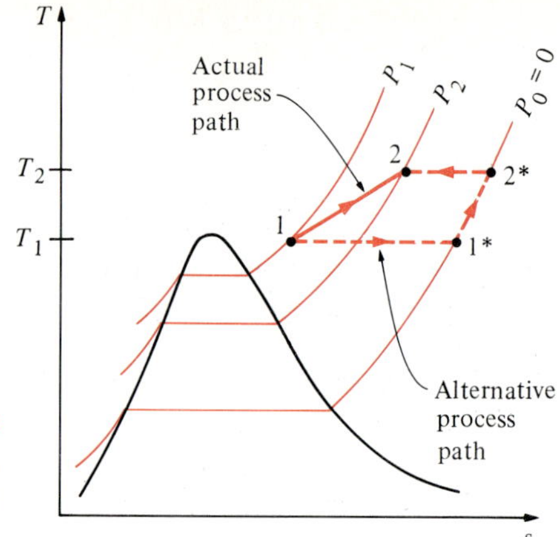

FIGURE 11-16

An alternative process path to evaluate the enthalpy changes of real gases.

process can be evaluated from the general relation for dh (Eq. 11-36)

$$h_2 - h_1 = \int_{T_1}^{T_2} C_p\, dT + \int_{P_1}^{P_2} \left[v - T\left(\frac{\partial v}{\partial T}\right)_P \right] dP$$

where P_1, T_1 and P_2, T_2 are the pressures and temperatures of the gas at the initial and the final states, respectively. For an isothermal process $dT = 0$, and the first term vanishes. For a constant-pressure process, $dP = 0$ and the second term vanishes.

Properties are point functions, and thus the change in a property between two specified states is the same no matter which process path is followed. This fact can be exploited to greatly simplify the integration of Eq. 11-36. Consider, for example, the process shown on a T-s diagram in Fig. 11-16. The enthalpy change during this process $h_2 - h_1$ can be determined by performing the integrations in Eq. 11-36 along a path that consists of two isothermal (T_1 = constant and T_2 = constant) lines and one isobaric (P_0 = constant) line instead of the actual process path, as shown in Fig. 11-16.

Although this approach increases the number of integrations, it also simplifies them since one property remains constant now during each part of the process. The pressure P_0 can be chosen to be very low or zero, so that the gas can be treated as an ideal gas during the P_0 = constant process. Using a superscript asterisk (*) to denote an ideal-gas state, we can express the enthalpy change of a real gas during process 1-2 as

$$h_2 - h_1 = (h_2 - h_2^*) + (h_2^* - h_1^*) + (h_1^* - h_1) \qquad (11\text{-}53)$$

where, from Eq. 11-36,

$$h_2 - h_2^* = 0 + \int_{P_2^*}^{P_2} \left[v - T\left(\frac{\partial v}{\partial T}\right)_P \right]_{T=T_2} dP = \int_{0}^{P_2} \left[v - T\left(\frac{\partial v}{\partial T}\right)_P \right]_{T=T_2} dP \qquad (11\text{-}54)$$

$$h_2^* - h_1^* = \int_{T_1}^{T_2} C_p\, dT + 0 = \int_{T_1}^{T_2} C_{p0}(T)\, dT \qquad (11\text{-}55)$$

$$h_1^* - h_1 = 0 + \int_{P_1}^{P_1^*} \left[v - T\left(\frac{\partial v}{\partial T}\right)_P \right]_{T=T_1} dP = -\int_0^{P_1} \left[v - T\left(\frac{\partial v}{\partial T}\right)_P \right]_{T=T_1} dP \quad (11\text{-}56)$$

The difference between h and h^* is called the **enthalpy departure**, and it represents the variation of the enthalpy of a gas with pressure at a fixed temperature. The calculation of enthalpy departure requires a knowledge of the P-v-T behavior of the gas. In the absence of such data, we can use the relation $Pv = ZRT$, where Z is the compressibility factor. Substituting $v = ZRT/P$ and simplifying Eq. 11-56, we can write the enthalpy departure at any temperature T and pressure P as

$$(h^* - h)_T = -RT^2 \int_0^P \left(\frac{\partial Z}{\partial T}\right)_P \frac{dP}{P}$$

The above equation can be generalized by expressing it in terms of the reduced coordinates, using $T = T_{cr}T_R$ and $P = P_{cr}P_R$. After some manipulations, the enthalpy departure can be expressed in a nondimensionalized form as

$$Z_h = \frac{(\bar{h}^* - \bar{h})_T}{R_u T_{cr}} = T_R^2 \int_0^{P_R} \left(\frac{\partial Z}{\partial T_R}\right)_{P_R} d(\ln P_R) \quad (11\text{-}57)$$

where Z_h is called the **enthalpy departure factor**. The integral in the above equation can be performed graphically or numerically by employing data from the compressibility charts for various values of P_R and T_R. The values of Z_h are presented in graphical form as a function of P_R and T_R in Fig. A-31. This graph is called the **generalized enthalpy departure chart**, and it is used to determine the deviation of the enthalpy of a gas at a given P and T from the enthalpy of an ideal gas at the same T. By replacing h^* by h_{ideal} for clarity, Eq. 11-53 for the enthalpy change of a gas during a process 1-2 can be rewritten as

$$\bar{h}_2 - \bar{h}_1 = R_u T_{cr}(Z_{h_1} - Z_{h_2}) + (\bar{h}_2 - \bar{h}_1)_{ideal} \quad (11\text{-}58)$$

or

$$h_2 - h_1 = RT_{cr}(Z_{h_1} - Z_{h_2}) + (h_2 - h_1)_{ideal} \quad (11\text{-}59)$$

where the values of Z_h are determined from the generalized enthalpy chart, and $(\bar{h}_2 - \bar{h}_1)_{ideal}$ is determined from the ideal-gas tables. Notice that the first terms on the right-hand side are zero for an ideal gas.

2 Internal Energy Changes of Real Gases

The internal energy change of a real gas is determined by relating it to the enthalpy change through the definition $\bar{h} = \bar{u} + P\bar{v} = \bar{u} + ZR_uT$:

$$\bar{u}_2 - \bar{u}_1 = (\bar{h}_2 - \bar{h}_1) - R_u(Z_2T_2 - Z_1T_1) \quad (11\text{-}60)$$

3 Entropy Changes of Real Gases

The entropy change of a real gas is determined by following an approach similar to that used above for the enthalpy change. There is some difference in derivation, however, owing to the dependence of the ideal-gas entropy on pressure as well as the temperature.

The general relation for ds was expressed as (Eq. 11-41)

$$s_2 - s_1 = \int_{T_1}^{T_2} \frac{C_p}{T}\, dT - \int_{P_1}^{P_2} \left(\frac{\partial v}{\partial T}\right)_P dP$$

where P_1, T_1 and P_2, T_2 are the pressures and temperatures of the gas at the initial and the final states, respectively. The thought that comes to mind at this point is to perform the integrations in the above equation first along a T_1 = constant line to zero pressure, then along the $P = 0$ line to T_2, and finally along the T_2 = constant line to P_2, as we did for the enthalpy. This approach is not suitable for entropy-change calculations, however, since it involves the value of entropy at zero pressure, which is infinity. We can avoid this difficulty by choosing a different (but more complex) path between the two states, as shown in Fig. 11-17. Then the entropy change can be expressed as

$$s_2 - s_1 = (s_2 - s_b^*) + (s_b^* - s_2^*) + (s_2^* - s_1^*) + (s_1^* - s_a^*) + (s_a^* - s_1) \quad (11\text{-}61)$$

States 1 and 1* are identical ($T_1 = T_1^*$ and $P_1 = P_1^*$), so are states 2 and 2*. States 1* and 2* exist only in the imagination, and the gas is assumed to behave as an ideal gas at these two states as well as at the states between the two. Therefore, the entropy change during process 1*-2* can be determined from the entropy-change relations for ideal gases. The calculation of entropy change between an actual state and the corresponding imaginary ideal-gas state is more involved, however, and requires the use of generalized entropy charts, as explained below.

Consider a gas at a pressure P and temperature T. To determine how

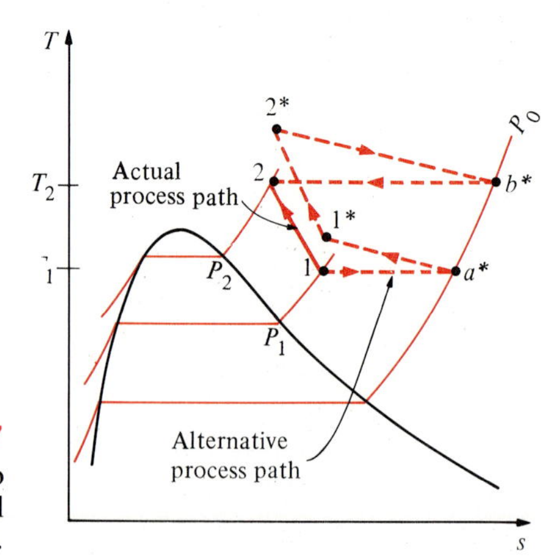

FIGURE 11-17

An alternative process path to evaluate the entropy changes of real gases during process 1-2.

much different the entropy of this gas would be if it were an ideal gas at the same temperature and pressure, we consider an isothermal process from the actual state P, T to zero (or close to zero) pressure and back to the imaginary ideal-gas state P^*, T^* (denoted by superscript *), as shown in Fig. 11-17. The entropy change during this isothermal process can be expressed as

$$\begin{aligned}(s_P - s_P^*)_T &= (s_P - s_0^*)_T + (s_0^* - s_P^*)_T \\ &= -\int_0^P \left(\frac{\partial v}{\partial T}\right)_P dP - \int_P^0 \left(\frac{\partial v^*}{\partial T}\right)_P dP\end{aligned}$$

where $v = ZRT/P$ and $v^* = v_{\text{ideal}} = RT/P$. Performing the differentiations and rearranging, we obtain

$$(s_P - s_P^*)_T = \int_0^P \left[\frac{(1 - Z)R}{P} - \frac{RT}{P}\left(\frac{\partial Z}{\partial T}\right)_P\right] dP$$

By substituting $T = T_{\text{cr}}T_R$ and $P = P_{\text{cr}}P_R$ and rearranging, the entropy departure can be expressed in a nondimensionalized form as

$$Z_s = \frac{(\bar{s}^* - \bar{s})_{T,P}}{R_u} = \int_0^{P_R} \left[Z - 1 + T_R\left(\frac{\partial Z}{\partial T_R}\right)_{P_R}\right] d(\ln P_R) \qquad (11\text{-}62)$$

The difference $(\bar{s}^* - \bar{s})_{T,P}$ is called the **entropy departure** and Z_s is called the **entropy departure factor**. The integral in the above equation can be performed by using data from the compressibility charts. The values of Z_s are presented in graphical form as a function of P_R and T_R in Fig. A-32. This graph is called the **generalized entropy departure chart**, and it is used to determine the deviation of the entropy of a gas at a given P and T from the entropy of an ideal gas at the same P and T. Replacing s^* by s_{ideal} for clarity, we can rewrite Eq. 11-61 for the entropy change of a gas during a process 1-2 as

$$\bar{s}_2 - \bar{s}_1 = R_u(Z_{s_1} - Z_{s_2}) + (\bar{s}_2 - \bar{s}_1)_{\text{ideal}} \qquad (11\text{-}63)$$

or

$$s_2 - s_1 = R(Z_{s_1} - Z_{s_2}) + (s_2 - s_1)_{\text{ideal}} \qquad (11\text{-}64)$$

where the values of Z_s are determined from the generalized entropy chart, and the entropy change $(\bar{s}_2 - \bar{s}_1)_{\text{ideal}}$ is determined from the ideal-gas relations for entropy change. Notice that the first terms on the right-hand side are zero for an ideal gas.

EXAMPLE 11-11

Determine the enthalpy change and the entropy change of oxygen per unit mole as it undergoes a change of state from 220 K and 5 MPa to 300 K and 10 MPa (*a*) by assuming ideal-gas behavior and (*b*) by accounting for the deviation from ideal-gas behavior.

Solution The critical temperature and pressure of oxygen are $T_{\text{cr}} = 154.8$ K and $P_{\text{cr}} = 5.08$ MPa (Table A-1), respectively. The oxygen remains above its critical temperature; therefore it is in the gas phase, but its pressure is quite high.

Therefore, the oxygen will deviate from ideal-gas behavior and should be treated as a real gas.

(*a*) If the O_2 is assumed to behave as an ideal gas, its enthalpy will depend on temperature only, and the enthalpy values at the initial and the final temperatures can be determined from the ideal-gas table of O_2 (Table A-19) at the specified temperatures:

$$\begin{aligned}(\bar{h}_2 - \bar{h}_1)_{\text{ideal}} &= \bar{h}_{2,\text{ideal}} - \bar{h}_{1,\text{ideal}} \\ &= (8736 - 6404)\ \text{kJ/kmol} \\ &= 2332\ \text{kJ/kmol}\end{aligned}$$

The entropy depends on both temperature and pressure even for ideal gases. Under the ideal-gas assumption, the entropy change of oxygen is determined from

$$\begin{aligned}(\bar{s}_2 - \bar{s}_1)_{\text{ideal}} &= \bar{s}_2^\circ - \bar{s}_1^\circ - R_u \ln\frac{P_2}{P_1} \\ &= (205.213 - 196.171)\ \text{kJ/(kmol·K)} - [8.314\ \text{kJ/(kmol·K)}] \times \ln\frac{10\ \text{MPa}}{5\ \text{MPa}} \\ &= 3.28\ \text{kJ/(kmol·K)}\end{aligned}$$

(*b*) The deviation from the ideal-gas behavior can be accounted for by determining the enthalpy and entropy departures from the generalized charts at each state:

$$\left.\begin{aligned}T_{R_1} &= \frac{T_1}{T_{\text{cr}}} = \frac{220}{154.8} = 1.42 \\ P_{R_1} &= \frac{P_1}{P_{\text{cr}}} = \frac{5}{5.08} = 0.98\end{aligned}\right\} \quad Z_{h_1} = 0.53,\ Z_{s_1} = 0.25$$

and

$$\left.\begin{aligned}T_{R_2} &= \frac{T_2}{T_{\text{cr}}} = \frac{300}{154.8} = 1.94 \\ P_{R_2} &= \frac{P_2}{P_{\text{cr}}} = \frac{10}{5.08} = 1.97\end{aligned}\right\} \quad Z_{h_2} = 0.48,\ Z_{s_2} = 0.20$$

Then the enthalpy and entropy changes of oxygen during this process are determined by substituting the values above into Eqs. 11-58 and 11-63,

$$\begin{aligned}\bar{h}_2 - \bar{h}_1 &= R_u T_{\text{cr}}(Z_{h_1} - Z_{h_2}) + (\bar{h}_2 - \bar{h}_1)_{\text{ideal}} \\ &= [8.314\ \text{kJ/(kmol·K)}](154.8\ \text{K})(0.53 - 0.48) + 2332\ \text{kJ/kmol} \\ &= 2396\ \text{kJ/kmol}\end{aligned}$$

and

$$\begin{aligned}\bar{s}_2 - \bar{s}_1 &= R_u(Z_{s_1} - Z_{s_2}) + (\bar{s}_2 - \bar{s}_1)_{\text{ideal}} \\ &= [8.314\ \text{kJ/(kmol·K)}](0.25 - 0.20) + 3.28\ \text{kJ/(kmol·K)} \\ &= 3.68\ \text{kJ/(kmol·K)}\end{aligned}$$

Therefore, in this case, the ideal-gas assumption would underestimate the enthalpy change of the oxygen by 2.7 percent and the entropy change by 10.9 percent.

11-7 ■ SUMMARY

Some thermodynamic properties can be measured directly, but many others cannot. Therefore it is necessary to develop some relations between these two groups so that the properties that cannot be measured directly can be evaluated. The derivations are based on the fact that properties are point functions, and the state of a simple, compressible system is completely specified by any two independent, intensive properties.

The equations that relate the partial derivatives of properties P, v, T, and s of a simple compressible substance to each other are called the *Maxwell relations*. They are obtained from the four *Gibbs equations*, expressed as

$$du = T\,ds - P\,dv \tag{11-10}$$

$$dh = T\,ds + v\,dP \tag{11-11}$$

$$da = -s\,dT - P\,dv \tag{11-14}$$

$$dg = -s\,dT + v\,dP \tag{11-15}$$

The *Maxwell relations* are

$$\left(\frac{\partial T}{\partial v}\right)_s = -\left(\frac{\partial P}{\partial s}\right)_v \tag{11-16}$$

$$\left(\frac{\partial T}{\partial P}\right)_s = \left(\frac{\partial v}{\partial s}\right)_P \tag{11-17}$$

$$\left(\frac{\partial s}{\partial v}\right)_T = \left(\frac{\partial P}{\partial T}\right)_v \tag{11-18}$$

$$\left(\frac{\partial s}{\partial P}\right)_T = -\left(\frac{\partial v}{\partial T}\right)_P \tag{11-19}$$

The *Clapeyron equation* enables us to determine the enthalpy change associated with a phase change from a knowledge of P, v, and T data alone. It is expressed as

$$\left(\frac{dP}{dT}\right)_{\text{sat}} = \frac{h_{fg}}{Tv_{fg}} \tag{11-22}$$

For liquid-vapor and solid-vapor phase-change processes at low pressures, it can be approximated as

$$\ln \left(\frac{P_2}{P_1}\right)_{\text{sat}} = \frac{h_{fg}}{R}\left(\frac{T_2 - T_1}{T_1 T_2}\right)_{\text{sat}} \tag{11-24}$$

The changes in internal energy, enthalpy, and entropy of a simple, compressible substance can be expressed in terms of pressure, specific volume, temperature, and specific heats alone as

$$du = C_v\, dT + \left[T\left(\frac{\partial P}{\partial T}\right)_v - P\right] dv \tag{11-29}$$

$$dh = C_p\, dT + \left[v - T\left(\frac{\partial v}{\partial T}\right)_P\right] dP \tag{11-35}$$

$$ds = \frac{C_v}{T}\, dT + \left(\frac{\partial P}{\partial T}\right)_v dv \tag{11-38}$$

or

$$ds = \frac{C_p}{T}\, dT - \left(\frac{\partial v}{\partial T}\right)_P dP \tag{11-40}$$

For specific heats, we have the following general relations:

$$\left(\frac{\partial C_v}{\partial v}\right)_T = T\left(\frac{\partial^2 P}{\partial T^2}\right)_v \tag{11-42}$$

$$\left(\frac{\partial C_p}{\partial P}\right)_T = -T\left(\frac{\partial^2 v}{\partial T^2}\right)_P \tag{11-43}$$

$$C_{p,T} - C_{p0,T} = -T \int_0^P \left(\frac{\partial^2 v}{\partial T^2}\right)_P dP \tag{11-44}$$

$$C_p - C_v = -T\left(\frac{\partial v}{\partial T}\right)_P^2 \left(\frac{\partial P}{\partial v}\right)_T \tag{11-46}$$

$$C_p - C_v = \frac{vT\beta^2}{\alpha} \tag{11-49}$$

where β is the *volume expansivity* and α is the *isothermal compressibility*, defined as

$$\beta = \frac{1}{v}\left(\frac{\partial v}{\partial T}\right)_P \quad \text{and} \quad \alpha = -\frac{1}{v}\left(\frac{\partial v}{\partial P}\right)_T \tag{11-47, 11-48}$$

The difference $C_p - C_v$ is equal to R for ideal gases and to zero for incompressible substances.

The temperature behavior of a fluid during a throttling (h = constant) process is described by the *Joule-Thomson coefficient,* defined as

$$\mu = \left(\frac{\partial T}{\partial P}\right)_h \tag{11-51}$$

The Joule-Thomson coefficient is a measure of the change in temperature of a substance with pressure during a constant-enthalpy process, and it can also be expressed as

$$\mu = \frac{1}{C_p}\left[v - T\left(\frac{\partial v}{\partial T}\right)_P\right] \tag{11-52}$$

The enthalpy, internal energy, and entropy changes of real gases can be determined accurately by utilizing *generalized enthalpy* or *entropy departure charts* to account for the deviation from the ideal-gas behavior by using the following relations:

$$\bar{h}_2 - \bar{h}_1 = R_u T_{cr}(Z_{h_1} - Z_{h_2}) + (\bar{h}_2 - \bar{h}_1)_{ideal} \tag{11-58}$$

$$\bar{u}_2 - \bar{u}_1 = (\bar{h}_2 - \bar{h}_1) - R_u(Z_2 T_2 - Z_1 T_1) \tag{11-60}$$

$$\bar{s}_2 - \bar{s}_1 = R_u(Z_{s_1} - Z_{s_2}) + (\bar{s}_2 - \bar{s}_1)_{ideal} \tag{11-63}$$

where the values of Z_h and Z_s are determined from the generalized charts.

REFERENCES AND SUGGESTED READING

1 W. Z. Black and J. G. Hartley, *Thermodynamics,* Harper & Row, New York, 1985.

2 J. R. Howell and R. O. Buckius, *Fundamentals of Engineering Thermodynamics,* McGraw-Hill, New York, 1987.

3 J. B. Jones and G. A. Hawkins, *Engineering Thermodynamics,* 2d ed., Wiley, New York, 1986.

4 G. J. Van Wylen and R. E. Sonntag, *Fundamentals of Classical Thermodynamics,* 3d ed., Wiley, New York, 1985.

5 K. Wark, *Thermodynamics,* 5th ed., McGraw-Hill, New York, 1988.

PROBLEMS*

Partial Derivatives and Associated Relations

11-1C What is the difference between partial differentials and ordinary differentials?

11-2C Consider the function $z(x,y)$, its partial derivatives $(\partial z/\partial x)_y$ and $(\partial z/\partial y)_x$, and the total derivative dz/dx.

(*a*) How do the magnitudes $(\partial x)_y$ and dx compare?
(*b*) How do the magnitudes $(\partial z)_y$ and dz compare?
(*c*) Is there any relation among dz, $(\partial z)_x$, and $(\partial z)_y$?

*Students are encouraged to answer *all* the concept "C" questions.

11-3C Consider a function $z(x,y)$ and its partial derivative $(\partial z/\partial y)_x$. Under what conditions is this partial derivative equal to the total derivative dz/dy?

11-4C Consider a function $z(x,y)$ and its partial derivative $(\partial z/\partial y)_x$. If this partial derivative is equal to zero for all values of x, what does it indicate?

11-5C Consider a function $z(x,y)$ and its partial derivative $(\partial z/\partial y)_x$. Can this partial derivative still be a function of x?

11-6C Consider a function $f(x)$ and its derivative df/dx. Can this derivative be determined by evaluating dx/df and taking its inverse?

11-7C Consider the function $z(x,y)$. Plot a differential surface on x-y-z coordinates, and indicate ∂x, dx, ∂y, dy, $(\partial z)_x$, $(\partial z)_y$, and dz.

11-8 Prove for an ideal gas that (*a*) the P = constant lines on a T-v diagram are straight lines and (*b*) the high-pressure lines are steeper than the low-pressure lines.

11-9 Derive a relation for the slope of the v = constant lines on a T-P diagram for a gas that obeys the van der Waals equation of state.
Answer: $(v - b)/R$

11-10 Nitrogen gas at 400 K and 100 kPa behaves as an ideal gas. Estimate the C_p and C_v of the nitrogen at this state, using enthalpy and internal energy data from Table A-18, and compare them to the values listed in Table A-2*b*.

11-10E Nitrogen gas at 600 R and 15 psia behaves as an ideal gas. Estimate the C_p and C_v of the nitrogen at this state, using enthalpy and internal energy data from Table A-18E, and compare them to the values listed in Table A-2E*b*. *Answers:* 0.249 Btu/(lbm·R), 0.178 Btu/(lbm·R)

11-11 Consider an ideal gas at 400 K and 100 kPa. As a result of some disturbance, the conditions of the gas change to 404 K and 98 kPa. Estimate the change in the specific volume of the gas as a result of this disturbance, using (*a*) Eq. 11-3 and (*b*) the ideal-gas relation at each state.

11-12 Using the equation of state $P(v - a) = RT$, verify (*a*) the cyclic relation and (*b*) the reciprocity relation at constant v.

11-13 Consider air at 350 K and 0.90 m^3/kg. Using Eq. 11-3, determine the change in pressure corresponding to an increase of (*a*) 1 percent in temperature at constant specific volume, (*b*) 1 percent in specific volume at constant temperature, and (*c*) 1 percent in both the temperature and specific volume.

Maxwell Relations

11-14 Verify the validity of the last Maxwell relation (Eq. 11-19) for steam at 300°C and 600 kPa.

11-14E Verify the validity of the last Maxwell relation (Eq. 11-19) for steam at 900°F and 450 psia.

11-15 Verify the validity of the last Maxwell relation (Eq. 11-19) for refrigerant-12 at 80°C and 1.2 MPa.

11-16 Using the cyclic relation and the first Maxwell relation, derive the other three Maxwell relations.

11-17 Using the Maxwell relations, determine a relation for $(\partial s/\partial P)_T$ for a gas whose equation of state is $P(v - b) = RT$. *Answer:* $-R/P$

11-18 Using the Maxwell relations, determine a relation for $(\partial s/\partial v)_T$ for a gas whose equation of state is $(P - a/v^2)(v - b) = RT$.

11-19 Using the Maxwell relations and the ideal-gas equation of state, determine a relation for $(\partial s/\partial v)_T$ for an ideal gas. *Answer:* R/v

11-20 Starting with the relation $dh = T\,ds + v\,dP$, show that the slope of a constant-pressure line on an *h-s* diagram (*a*) is constant in the saturation region and (*b*) increases with temperature in the superheat region.

The Clapeyron Equation

11-21C What is the value of the Clapeyron equation in thermodynamics?

11-22C Does the Clapeyron equation involve any approximations, or is it exact?

11-23C What approximations are involved in the Clapeyron-Clausius equation?

11-24 Using the Clapeyron equation, estimate the enthalpy of vaporization of refrigerant-12 at 30°C, and compare it with the tabulated value.
Answers: 135.04 kJ/kg, 135.03 kJ/kg

11-24E Using the Clapeyron equation, estimate the enthalpy of vaporization of refrigerant-12 at 25°F, and compare it to the tabulated value.
Answers: 65.98 Btu/lbm, 65.96 Btu/lbm

11-25 Using the Clapeyron equation, estimate the enthalpy of vaporization of steam at 200 kPa, and compare it to the tabulated value.

11-26 Calculate the h_{fg} and s_{fg} of steam at 150°C from the Clapeyron equation, and compare them to the tabulated values.

11-27 Determine the h_{fg} of refrigerant-12 at 0°C on the basis of (*a*) the Clapeyron equation and (*b*) the Clapeyron-Clausius equation. Compare your results to the tabulated h_{fg} value.
Answers: (*a*) 149.81 kJ/kg; (*b*) 168.69 kJ/kg, 151.48 kJ/kg

11-27E Determine the h_{fg} of refrigerant-12 at 50°F on the basis of (*a*) the Clapeyron equation and (*b*) the Clapeyron-Clausius equation. Compare your results to the tabulated h_{fg} value.

General Relations for du, dh, ds, C_v, and C_p

11-28C Can the variation of specific heat C_p with pressure at a given temperature be determined from a knowledge of P-v-T data alone?

11-29 Show that the enthalpy of an ideal gas is a function of temperature only and that for an incompressible substance it also depends on pressure.

11-30 Derive expressions for (*a*) Δu, (*b*) Δh, and (*c*) Δs for a gas that obeys the van der Waals equation of state for an isothermal process.

11-31 Derive expressions for (*a*) Δu, (*b*) Δh, and (*c*) Δs for a gas whose equation of state is $P(v - a) = RT$ for an isothermal process.
Answers: (*a*) 0, (*b*) $a(P_2 - P_1)$, (*c*) $R \ln(P_2/P_1)$

11-32 Derive relations for (*a*) Δu, (*b*) Δh, and (*c*) Δs of a gas which obeys the equation of state $(P + a/v^2)v = RT$ for an isothermal process.

11-33 Derive expressions for $(\partial u/\partial P)_T$ and $(\partial h/\partial v)_T$ in terms of P, v, and T only.

11-34 Derive an expression for the specific-heat difference $C_p - C_v$ for (*a*) an ideal gas, (*b*) a van der Waals gas, and (*c*) an incompressible substance.

11-35 Estimate the specific-heat difference $C_p - C_v$ for liquid water at 10 MPa and 40°C. *Answer:* 0.1094 kJ/(kg·K)

11-35E Estimate the specific-heat difference $C_p - C_v$ for liquid water at 1000 psia and 150°F. *Answer:* 0.0560 Btu/(lbm·R)

11-36 Derive a relation for the volume expansivity β and the isothermal compressibility α (*a*) for an ideal gas and (*b*) for a gas whose equation of state is $P(v - a) = RT$.

11-37 Estimate the volume expansivity β and the isothermal compressibility α of refrigerant-12 at 140 kPa and 20°C.
Answers: 0.00379 K^{-1}, 0.00803 kPa^{-1}

11-38 Show that

$$C_v = -T\left(\frac{\partial v}{\partial T}\right)_s\left(\frac{\partial P}{\partial T}\right)_v \quad \text{and} \quad C_p = T\left(\frac{\partial P}{\partial T}\right)_s\left(\frac{\partial v}{\partial T}\right)_P$$

11-39 Estimate the C_p of nitrogen at 200 kPa and 400 K, using (*a*) the relation in the above problem and (*b*) its definition. Compare your results to the value listed in Table A-2*b*.

Joule-Thomson Coefficient

11-40C What does the Joule-Thomson coefficient represent?

11-41C Describe the inversion line and the maximum inversion temperature.

11-42C The pressure of a fluid always decreases during an adiabatic throttling process. Is this also the case for the temperature?

11-43C Does the Joule-Thomson coefficient of a substance change with temperature at a fixed pressure?

11-44C Will the temperature of helium change if it is throttled adiabatically from 300 K and 500 kPa to 100 kPa?

11-45 Consider a gas whose equation of state is $P(v - a) = RT$, where a is a positive constant. Is it possible to cool this gas by throttling?

11-46 Derive a relation for the Joule-Thomson coefficient and the inversion temperature for a gas whose equation of state is $(P + a/v^2)v = RT$.

11-47 Estimate the Joule-Thomson coefficient of nitrogen at (*a*) 2 MPa and 250 K and (*b*) 6 MPa and 275 K.

11-47E Estimate the Joule-Thomson coefficient of nitrogen at (*a*) 200 psia and 500 R and (*b*) 2000 psia and 400 R.

11-48 Estimate the Joule-Thomson coefficient of refrigerant-12 at 0.6 MPa and 100°C. *Answer:* 12.25 °C/MPa

11-49 Steam is throttled slightly from 1 MPa and 300°C. Will the temperature of the steam increase, decrease, or remain the same during this process?

11-50 Steam is throttled from 4.5 MPa and 400°C to 3.5 MPa. Estimate the temperature change of the steam during this process and the average Joule-Thomson coefficient. *Answers:* −7.44°C, 7.44°C/MPa

The Δh, Δu, and Δs of Real Gases

11-51C What is the enthalpy departure?

11-52C On the generalized enthalpy departure chart, the normalized enthalpy departure values seem to approach zero as the reduced pressure P_R approaches zero. How do you explain this behavior?

11-53C Why is the generalized enthalpy departure chart prepared by using P_R and T_R as the parameters instead of P and T?

11-54 Determine the enthalpy of nitrogen, in kJ/kg, at 175 K and 8 MPa using (*a*) data from superheated nitrogen table, (*b*) data from the ideal-gas nitrogen table, and (*c*) the generalized enthalpy departure chart. *Answers:* (*a*) 125.536 kJ/kg, (*b*) 181.48 kJ/kg, (*c*) 121.6 kJ/kg

11-54E Determine the enthalpy of nitrogen, in Btu/lbm, at 400 R and 2000 psia using (*a*) data from the superheated nitrogen table, (*b*) data from the ideal-gas nitrogen table, and (*c*) the generalized enthalpy chart.

11-55 What is the error involved in the (*a*) enthalpy and (*b*) internal energy of CO_2 at 350 K and 10 MPa if it is assumed to be an ideal gas? *Answers:* (*a*) 50%, (*b*) 49%

11-56 Determine the enthalpy change and the entropy change of nitrogen per unit mole as it undergoes a change of state from 225 K and 6 MPa to 300 K and 10 MPa, (*a*) by assuming ideal-gas behavior, (*b*) by accounting for the deviation from ideal-gas behavior through the use of generalized charts, and (*c*) by using data from the superheated nitrogen table.

11-57 Determine the enthalpy change and the entropy change of CO_2 per unit mass as it undergoes a change of state from 250 K and 7 MPa to 280 K and 12 MPa, (*a*) by assuming ideal-gas behavior and (*b*) by accounting for the deviation from ideal-gas behavior.

11-57E Determine the enthalpy change and the entropy change of CO_2 per unit mass as it undergoes a change of state from 300 R and 2000 psia to 400 R and 1500 psia, (*a*) by assuming ideal-gas behavior and (*b*) by accounting for the deviation from ideal-gas behavior.

11-58 A rigid tank contains 5 m^3 of argon at $-100°C$ and 1 MPa. Heat is now transferred to argon until the temperature in the tank rises to 0°C. Using the generalized charts, determine (*a*) the mass of the argon in the tank, (*b*) the final pressure, and (*c*) the heat transfer.
Answers: (*a*) 146.2 kg, (*b*) 1531 kPa, (*c*) 5212 kJ

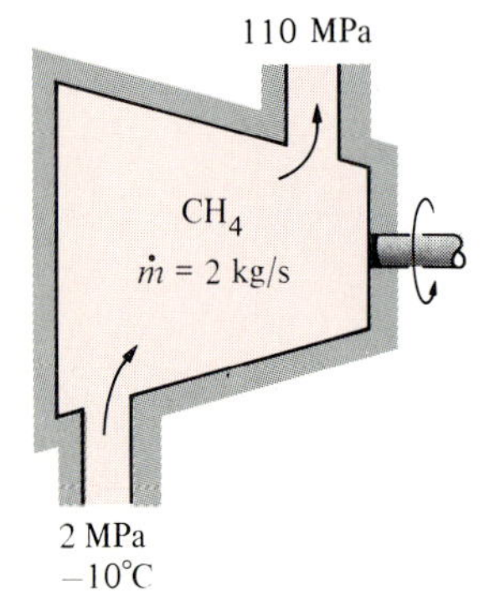

FIGURE P11-59

11-59 Methane is compressed adiabatically by a steady-flow compressor from 2 MPa and $-10°C$ to 10 MPa and 110°C at a rate of 2 kg/s. Using the generalized charts, determine the required power input to the compressor. *Answer:* -462 kW

11-60 Propane is compressed isothermally by a piston-cylinder device from 100°C and 1 MPa to 4 MPa. Using the generalized charts, determine the work done and the heat transfer per unit mass of propane.

11-60E Propane is compressed isothermally by a piston-cylinder device from 200°F and 200 psia to 800 psia. Using the generalized charts, determine the work done and the heat transfer per unit mass of the propane.
Answers: -45.3 Btu/lbm, -132.1 Btu/lbm

11-61 Determine the irreversibility associated with the process described in Prob. 11-60. Assume $T_0 = 25°C$.

11-62 Carbon dioxide enters an adiabatic nozzle at 8 MPa and 450 K with a low velocity and leaves at 2 MPa and 350 K. Using the generalized enthalpy departure chart, determine the exit velocity of the carbon dioxide. *Answer:* 384 m/s

11-63 Argon gas enters a turbine at 7 MPa and 600 K with a velocity of 100 m/s and leaves at 1 MPa and 280 K with a velocity of 150 m/s at a rate of 3 kg/s. Heat is being lost to the surroundings at 25°C at a rate of 25 kW. Using the generalized charts, determine (*a*) the power output of the turbine and (*b*) the irreversibility associated with the process.

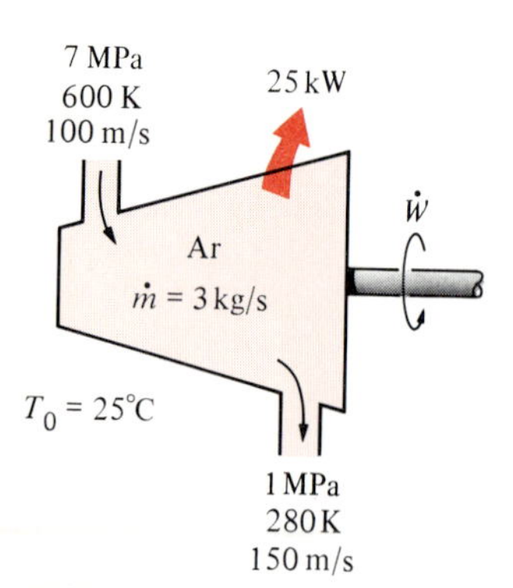

FIGURE P11-63

11-63E Argon gas enters a turbine at 1000 psia and 1000 R with a velocity of 300 ft/s and leaves at 150 psia and 500 R with a velocity of 450 ft/s at a rate of 5 lbm/s. Heat is being lost to the surroundings at 75°F at a rate of 25 Btu/s. Using the generalized charts, determine (*a*) the power output of the turbine and (*b*) the irreversibility associated with the process. *Answers:* (*a*) 284.3 Btu/s, (*b*) 47.7 Btu/s

11-64 A 1-m^3 well-insulated rigid tank contains oxygen at 220 K and 10 MPa. A paddle wheel placed in the tank is turned on, and the temperature of the oxygen rises to 250 K. Using the generalized charts, determine (*a*) the final pressure in the tank and (*b*) the paddle-wheel work done during this process. *Answers:* (*a*) 12,190 kPa, (*b*) 4702 kJ

11-65 An adiabatic 0.5-m^3 storage tank which is initially evacuated is connected to a supply line that carries nitrogen at 225 K and 10 MPa. A valve is opened, and nitrogen flows into the tank from tne supply line. The valve is closed when the pressure in the tank reaches 10 MPa. Determine the final temperature in the tank (*a*) using data from the superheated nitrogen table, (*b*) treating nitrogen as an ideal gas, and (*c*) using generalized charts.

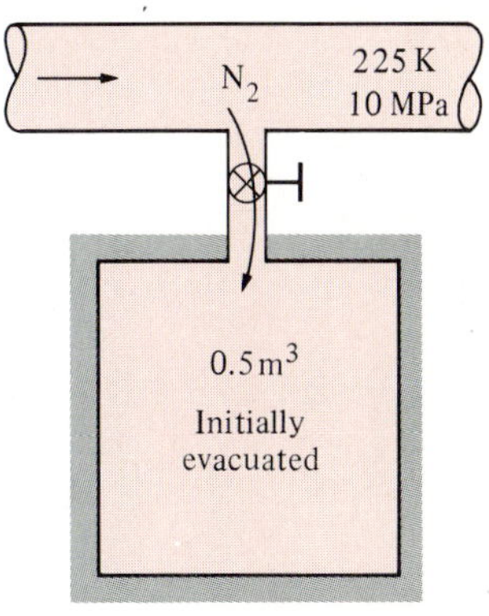

FIGURE P11-65

Gas Mixtures

CHAPTER 12

Up to this point, we have limited our consideration to thermodynamic systems which involve a single pure substance such as water, refrigerant-12, or nitrogen. Many important thermodynamic applications, however, involve *mixtures* of several pure substances rather than a single pure substance. Therefore, it is important to develop an understanding of mixtures and learn how to handle them.

In this chapter, we deal with nonreacting gas mixtures. A nonreacting gas mixture can be treated as a pure substance since it is usually a homogeneous mixture of different gases. The properties of a gas mixture obviously will depend on the properties of the individual gases (called *components,* or *constituents*) as well as on the amount of each gas in the mixture. Therefore, it is possible to prepare tables of properties for mixtures. This has been done for common mixtures such as air. It is not practical to prepare property tables for every conceivable mixture composition, however, since the number of possible compositions is endless. Therefore, we need to develop rules for determining mixture properties from a knowledge of mixture composition and the properties of the individual components. We do this first for ideal-gas mixtures and then for real-gas mixtures. The basic principles involved are also applicable to liquid or solid mixtures, called *solutions*.

12-1 ■ COMPOSITION OF A GAS MIXTURE— Mass and Mole Fractions

To determine the properties of a mixture, we need to know the *composition* of the mixture as well as the properties of the individual components. There are two ways to describe the composition of a mixture: either by specifying the number of moles of each component, called **molar analysis**, or by specifying the mass of each component, called **gravimetric analysis**.

Consider a gas mixture composed of k components. You will agree that the mass of the mixture m_m is the sum of the masses of the individual components, and the mole number of the mixture N_m is the sum of the mole numbers of the individual components* (Figs. 12-1 and 12-2). That is,

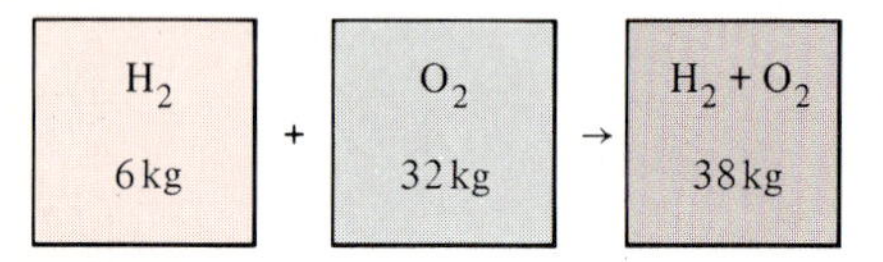

FIGURE 12-1

The mass of a mixture is equal to the sum of the masses of its components.

$$m_m = \sum_{i=1}^{k} m_i \qquad \text{and} \qquad N_m = \sum_{i=1}^{k} N_i \qquad (12\text{-}1a, b)$$

The ratio of the mass of a component to the mass of the mixture is called the **mass fraction** (mf), and the ratio of the mole number of a component to the mole number of the mixture is called the **mole fraction** y:

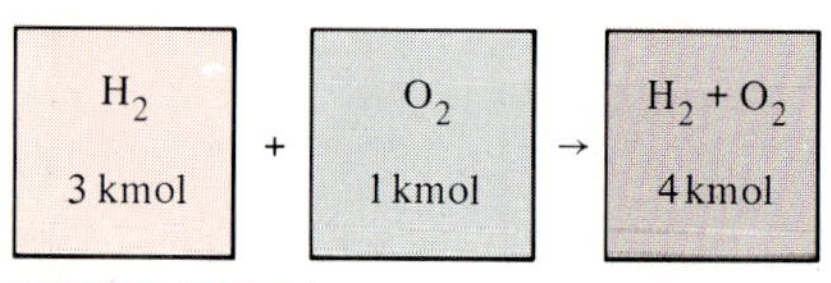

FIGURE 12-2

The number of moles of a nonreacting mixture is equal to the sum of the number of moles of its components.

$$\text{mf}_i = \frac{m_i}{m_m} \qquad \text{and} \qquad y_i = \frac{N_i}{N_m} \qquad (12\text{-}2a, b)$$

Dividing Eq. 12-1a by m_m or Eq. 12-1b by N_m, we can easily show that the sum of the mass fractions or mole fractions for a mixture is equal to 1 (Fig. 12-3):

$$\sum_{i=1}^{k} \text{mf}_i = 1 \qquad \text{and} \qquad \sum_{i=1}^{k} y_i = 1 \qquad (12\text{-}3a, b)$$

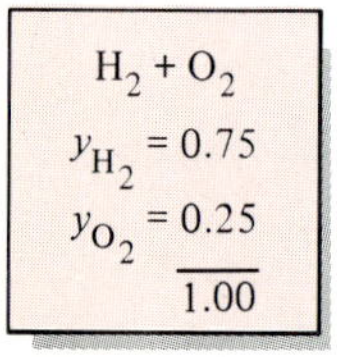

$H_2 + O_2$
$y_{H_2} = 0.75$
$y_{O_2} = 0.25$
1.00

FIGURE 12-3

The sum of the mole fractions of a mixture is equal to 1.

The mass of a substance can be expressed in terms of the mole number N and molar mass M of the substance as

$$m = NM \qquad \text{(kg)}$$

Then the **apparent** (or **average**) **molar mass** of a mixture can be expressed as

$$M_m = \frac{m_m}{N_m} = \frac{\Sigma m_i}{N_i} = \frac{\Sigma N_i M_i}{N_m} = \sum_{i=1}^{k} y_i M_i \qquad \text{(kg/kmol)} \qquad (12\text{-}4)$$

Then the **average** (or **apparent**) **gas constant** of the mixture can be determined from

$$R_m = \frac{R_u}{M_m} \qquad [\text{kJ/(kg·K)}] \qquad (12\text{-}5)$$

*Throughout this chapter, the subscript m will denote the gas mixture, and the subscript i will denote any single component of the mixture.

EXAMPLE 12-1

Consider a gas mixture which consists of 3 kg of O_2, 5 kg of N_2, and 12 kg of CH_4, as shown in Fig. 12-4. Determine (*a*) the mass fraction of each component, (*b*) the mole fraction of each component, and (*c*) the average molar mass and gas constant of the mixture.

3 kg O_2
5 kg N_2
12 kg CH_4

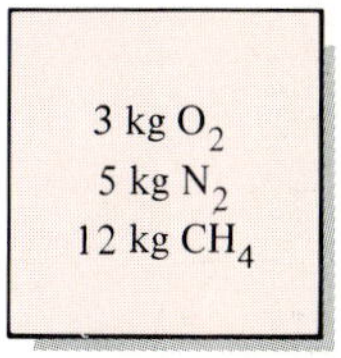

FIGURE 12-4
Schematic for Example 12-1.

Solution (*a*) The total mass of the mixture is

$$m_m = m_{O_2} + m_{N_2} + m_{CH_4} = (3 + 5 + 12)\text{ kg} = 20\text{ kg}$$

Then the mass fraction of each component becomes

$$\text{mf}_{O_2} = \frac{m_{O_2}}{m_m} = \frac{3\text{ kg}}{20\text{ kg}} = 0.15$$

$$\text{mf}_{N_2} = \frac{m_{N_2}}{m_m} = \frac{5\text{ kg}}{20\text{ kg}} = 0.25$$

$$\text{mf}_{CH_4} = \frac{m_{CH_4}}{m_m} = \frac{12\text{ kg}}{20\text{ kg}} = 0.60$$

(*b*) To find the mole fractions, we need to determine the mole numbers of each component first:

$$N_{O_2} = \frac{m_{O_2}}{M_{O_2}} = \frac{3\text{ kg}}{32\text{ kg/kmol}} = 0.094\text{ kmol}$$

$$N_{N_2} = \frac{m_{N_2}}{M_{N_2}} = \frac{5\text{ kg}}{28\text{ kg/kmol}} = 0.179\text{ kmol}$$

$$N_{CH_4} = \frac{m_{CH_4}}{M_{CH_4}} = \frac{12\text{ kg}}{16\text{ kg/kmol}} = 0.750\text{ kmol}$$

Thus, $N_m = N_{O_2} + N_{N_2} + N_{CH_4} = 0.094 + 0.179 + 0.750 = 1.023\text{ kmol}$

and

$$y_{O_2} = \frac{N_{O_2}}{N_m} = \frac{0.094\text{ kmol}}{1.023\text{ kmol}} = 0.092$$

$$y_{N_2} = \frac{N_{N_2}}{N_m} = \frac{0.179\text{ kmol}}{1.023\text{ kmol}} = 0.175$$

$$y_{CH_4} = \frac{N_{CH_4}}{N_m} = \frac{0.750\text{ kmol}}{1.023\text{ kmol}} = 0.733$$

(*c*) The average molar mass and gas constant of the mixture are determined from their definitions:

$$M_m = \frac{m_m}{N_m} = \frac{20\text{ kg}}{1.023\text{ kmol}} = 19.6\text{ kg/kmol}$$

or

$$\begin{aligned} M_m = \Sigma y_i M_i &= y_{O_2}M_{O_2} + y_{N_2}M_{N_2} + y_{CH_4}M_{CH_4} \\ &= (0.092)(32) + (0.175)(28) + (0.733)(16) \\ &= 19.6\text{ kg/kmol} \end{aligned}$$

Also,

$$R_m = \frac{R_u}{M_m} = \frac{8.314\text{ kJ/(kmol·K)}}{19.6\text{ kg/kmol}} = 0.424\text{ kJ/(kg·K)}$$

12-2 ■ *P-v-T* BEHAVIOR OF GAS MIXTURES—Ideal and Real Gases

An ideal gas is defined in Chap. 2 as a gas whose molecules are spaced far apart so that the behavior of a molecule is not influenced by the presence of other molecules—a situation encountered at low densities. We also mentioned that real gases approximate this behavior closely when they are at a low pressure or high temperature relative to their critical-point values. The P-v-T behavior of an ideal gas is expressed by the simple relation $Pv = RT$, which is called the *ideal-gas equation of state*. The P-v-T behavior of real gases is expressed by more complex equations of state or by $Pv = ZRT$, where Z is the compressibility factor.

When two or more ideal gases are mixed, the behavior of a molecule normally is not influenced by the presence of other similar or dissimilar molecules, and therefore a nonreacting mixture of ideal gases also behaves as an ideal gas. Air, for example, is conveniently treated as an ideal gas in the range where nitrogen and oxygen behave as an ideal gas. When a gas mixture consists of real (nonideal) gases, however, the prediction of the P-v-T behavior of the mixture becomes rather involved and difficult.

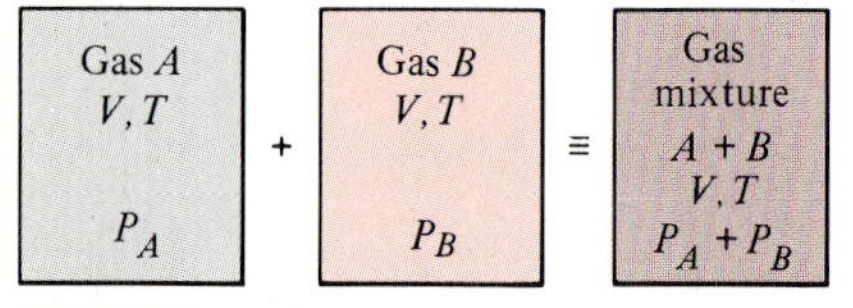

FIGURE 12-5

Dalton's law of additive pressures for a mixture of two ideal gases.

The prediction of the P-v-T behavior of gas mixtures is usually based on two models: *Dalton's law of additive pressures* and *Amagat's law of additive volumes*. Both models are described and discussed below.

> **Dalton's law of additive pressures:** *The pressure of a gas mixture is equal to the sum of the pressures each gas would exert if it existed alone at the mixture temperature and volume* (Fig. 12-5).

> **Amagat's law of additive volumes:** *The volume of a gas mixture is equal to the sum of the volumes each gas would occupy if it existed alone at the mixture temperature and pressure* (Fig. 12-6).

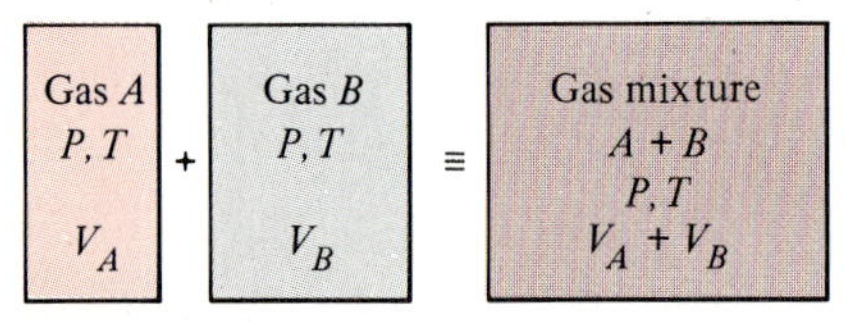

FIGURE 12-6

Amagat's law of additive volumes for a mixture of two ideal gases.

Dalton's and Amagat's laws hold exactly for ideal-gas mixtures, but only approximately for real-gas mixtures. This is due to intermolecular forces which may be significant for real gases at high densities. For ideal gases, these two laws are identical and give identical results.

Dalton's and Amagat's laws can be expressed as follows:

Dalton's law: $$P_m = \sum_{i=1}^{k} P_i(T_m, V_m) \qquad (12\text{-}6)$$

Amagat's law: $$V_m = \sum_{i=1}^{k} V_i(T_m, P_m) \qquad (12\text{-}7)$$

(exact for ideal gases and approximate for real gases)

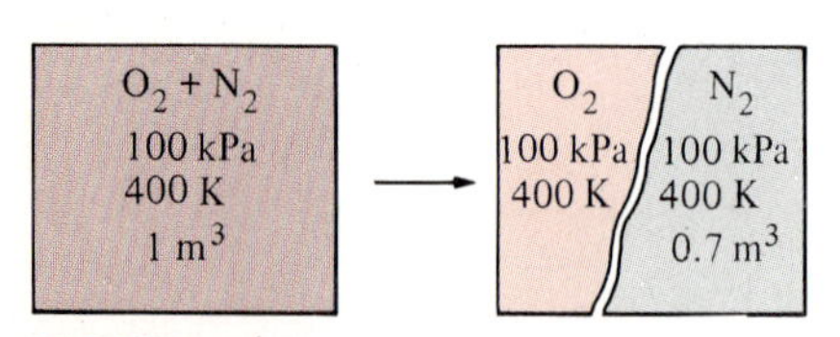

FIGURE 12-7

The volume a component would occupy if it existed alone at the mixture T and P is called the *component volume* (for ideal gases, it is equal to the partial volume y_iV_m).

In these relations, P_i is called the **component pressure**, and V_i is called the **component volume** (Fig. 12-7). Note that V_i is the volume a component *would* occupy if it existed alone at T_m and P_m, not the actual volume occupied by the component in the mixture. (In a vessel that holds a gas mixture, each component fills the entire volume of the vessel. Therefore, the volume of each component is equal to the volume of the vessel.) Also the ratio P_i/P_m is called the **pressure fraction**, and the ratio V_i/V_m is called the **volume fraction** of component i.

1 Ideal-Gas Mixtures

For ideal gases, P_i and V_i can be related to y_i by using the ideal-gas relation for both the components and the gas mixture:

$$\frac{P_i(T_m,V_m)}{P_m} = \frac{N_iR_uT_m/V_m}{N_mR_uT_m/V_m} = \frac{N_i}{N_m} = y_i$$

$$\frac{V_i(T_m,P_m)}{V_m} = \frac{N_iR_uT_m/P_m}{N_mR_uT_m/P_m} = \frac{N_i}{N_m} = y_i$$

Therefore,

$$\frac{P_i}{P_m} = \frac{V_i}{V_m} = \frac{N_i}{N_m} = y_i \qquad (12\text{-}8)$$

Equation 12-8 is strictly valid for ideal-gas mixtures since it is derived by assuming ideal-gas behavior for the gas mixture and each of its components. The quantity y_iP_m is called the **partial pressure** (identical to the *component pressure* for ideal gases), and the quantity y_iV_m is called the **partial volume** (identical to the *component volume* for ideal gases). For an ideal-gas mixture, the mole fraction, the pressure fraction, and the volume fraction of a component are identical.

The composition of an ideal-gas mixture (such as the exhaust gases leaving a combustion chamber) is frequently determined by a volumetric analysis and Eq. 12-8. A sample gas at a known volume, pressure, and temperature is passed into a vessel containing reagents that absorb one of the gases. The volume of the remaining gas is then measured at the original pressure and temperature. The ratio of the reduction in volume to the original volume (volume fraction) represents the mole fraction of that particular gas.

2 Real-Gas Mixtures

Dalton's law of additive pressures and Amagat's law of additive volumes can also be used for real gases, often with reasonable accuracy. This time, however, the component pressures or component volumes should be evaluated from relations that take into account the deviation of each component from ideal-gas behavior. One way of doing that is to use more exact equations of state (van der Waals, Beattie-Bridgeman, Benedict-Webb-Rubin, etc.) instead of the ideal-gas equation of state. Another way is to use the compressibility factor (Fig. 12-8) as

$$PV = ZNR_uT \qquad (12\text{-}9)$$

The compressibility factor of the mixture Z_m can be expressed in terms of the compressibility factors of the individual gases Z_i by applying Eq. 12-9 to both sides of Dalton's law or Amagat's law expression and simplifying. We obtain

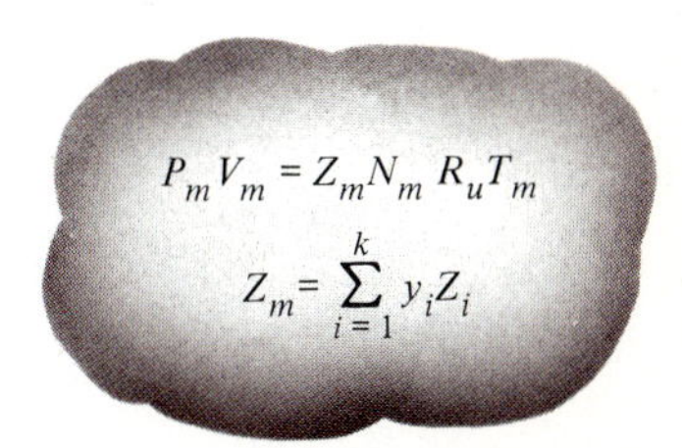

FIGURE 12-8
One way of predicting the *P-v-T* behavior of a real-gas mixture is to use compressibility factors.

$$Z_m = \sum_{i=1}^{k} y_i Z_i \tag{12-10}$$

where Z_i is determined either at T_m and V_m (Dalton's law) or at T_m and P_m (Amagat's law) for each individual gas. It may seem that using either law will give the same result, but it does not.

The compressibility-factor approach, in general, gives more accurate results when the Z_i's in Eq. 12-10 are evaluated by using Amagat's law instead of Dalton's law. This is because Amagat's law involves the use of mixture pressure P_m, which accounts for the influence of intermolecular forces between the molecules of different gases. Dalton's law disregards the influence of dissimilar molecules in a mixture on each other. As a result, it tends to underpredict the pressure of a gas mixture for a given V_m and T_m. Therefore, Dalton's law is more appropriate for gas mixtures at low pressures. Amagat's law is more appropriate at high pressures.

Note that there is a significant difference between using the compressibility factor for a single gas and for a mixture of gases. The compressibility factor predicts the *P-v-T* behavior of single gases rather accurately, as discussed in Chap. 2, but not for mixtures of gases. When we use compressibility factors for the components of a gas mixture, we account for the influence of like molecules on each other; the influence of dissimilar molecules largely remains unaccounted for. Consequently, a property value predicted by this approach may be considerably different from the experimentally determined value.

Pseudopure substance

$$P'_{cr} = \sum_{i=1}^{k} y_i P_{cr,i}$$

$$T'_{cr} = \sum_{i=1}^{k} y_i T_{cr,i}$$

FIGURE 12-9

Another way of predicting the *P-v-T* behavior of a real-gas mixture is to treat it as a pseudopure substance with critical properties P'_{cr} and T'_{cr}.

Another approach for predicting the *P-v-T* behavior of a gas mixture is to treat the gas mixture as a pseudopure substance (Fig. 12-9). One such method, proposed by W. B. Kay in 1936 and called **Kay's rule**, involves the use of a *pseudocritical pressure* $P'_{cr,m}$ and *pseudocritical temperature* $T'_{cr,m}$ for the mixture, defined in terms of the critical pressures and temperatures of the mixture components as

$$P'_{cr,m} = \sum_{i=1}^{k} y_i P_{cr,i} \quad \text{and} \quad T'_{cr,m} = \sum_{i=1}^{k} y_i T_{cr,i} \tag{12-11a, b}$$

The compressibility factor of the mixture Z_m is then easily determined by using these pseudocritical properties. The result obtained by using Kay's rule is accurate to within about 10 percent over a wide range of temperatures and pressures, which is acceptable for most engineering purposes.

Another way of treating a gas mixture as a pseudopure substance is to use a more accurate equation of state such as the van der Waals, Beattie-Bridgeman, or Benedict-Webb-Rubin equations for the mixture, and to determine the constant coefficients in terms of the coefficients of the components. In the van der Waals equation, for example, the two constants for the mixture are determined from

$$a_m = \left(\sum_{i=1}^{k} y_i a_i^{1/2} \right)^2 \quad \text{and} \quad b_m = \sum_{i=1}^{k} y_i b_i \tag{12-12a, b}$$

where expressions for a_i and b_i are given by Eq. 2-24.

EXAMPLE 12-2

A rigid tank contains 2 kmol of N_2 and 6 kmol of CO_2 gases at 300 K and 15 MPa. Estimate the volume of the tank on the basis of (*a*) the ideal-gas equation of state, (*b*) Kay's rule, (*c*) compressibility factors and Amagat's law, and (*d*) compressibility factors and Dalton's law.

Solution A sketch of the tank containing the N_2 and CO_2 mixture is shown in Fig. 12-10.

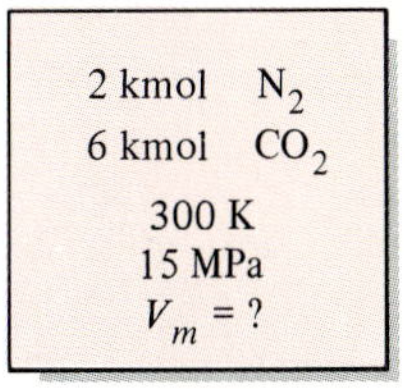

2 kmol N_2
6 kmol CO_2
300 K
15 MPa
$V_m = ?$

FIGURE 12-10

Schematic for Example 12-2.

(*a*) When the mixture is assumed to behave as an ideal gas, the volume of the mixture is easily determined from the ideal-gas relation for the mixture:

$$V_m = \frac{N_m R_u T_m}{P_m} = \frac{(8\text{ kmol})[8.314\text{ kPa}\cdot\text{m}^3/(\text{kmol}\cdot\text{K})](300\text{ K})}{15{,}000\text{ kPa}} = 1.330\text{ m}^3$$

since $N_m = N_{N_2} + N_{CO_2} = (2 + 6)\text{ kmol} = 8\text{ kmol}$.

(*b*) To use Kay's rule, we need to determine the pseudocritical temperature and pseudocritical pressure of the mixture by using the critical-point properties of N_2 and CO_2 from Table A-1. But first we need to determine the mole fraction of each component:

$$y_{N_2} = \frac{N_{N_2}}{N_m} = \frac{2\text{ kmol}}{8\text{ kmol}} = 0.25 \quad \text{and} \quad y_{CO_2} = \frac{N_{CO_2}}{N_m} = \frac{6\text{ kmol}}{8\text{ kmol}} = 0.75$$

$$T'_{cr,m} = \Sigma y_i T_{cr,i} = y_{N_2} T_{cr,N_2} + y_{CO_2} T_{cr,CO_2}$$
$$= (0.25)(126.2\text{ K}) + (0.75)(304.2\text{ K}) = 259.7\text{ K}$$

$$P'_{cr,m} = \Sigma y_i P_{cr,i} = y_{N_2} P_{cr,N_2} + y_{CO_2} P_{cr,CO_2}$$
$$= (0.25)(3.39\text{ MPa}) + (0.75)(7.39\text{ MPa}) = 6.39\text{ MPa}$$

Then

$$\left.\begin{aligned} T_R &= \frac{T_m}{T'_{cr,m}} = \frac{300\text{ K}}{259.7\text{ K}} = 1.16 \\ P_R &= \frac{P_m}{P'_{cr,m}} = \frac{15\text{ MPa}}{6.39\text{ MPa}} = 2.35 \end{aligned}\right\} \quad Z_m = 0.49 \quad \text{(Fig. A-30}b\text{)}$$

Thus,

$$V_m = \frac{Z_m N_m R_u T_m}{P_m} = Z_m V_{ideal} = (0.49)(1.330\text{ m}^3) = 0.652\text{ m}^3$$

(*c*) When Amagat's law is used in conjunction with compressibility factors, Z_m is determined from Eq. 12-10. But first we need to determine the Z of each component on the basis of Amagat's law:

N_2:

$$\left.\begin{aligned} T_{R,N_2} &= \frac{T_m}{T_{cr,N_2}} = \frac{300\text{ K}}{126.2\text{ K}} = 2.38 \\ P_{R,N_2} &= \frac{P_m}{P_{cr,N_2}} = \frac{15\text{ MPa}}{3.39\text{ MPa}} = 4.42 \end{aligned}\right\} \quad Z_{N_2} = 1.02 \quad \text{(Fig. A-30}b\text{)}$$

CO_2:

$$\left.\begin{aligned} T_{R,CO_2} &= \frac{T_m}{T_{cr,CO_2}} = \frac{300\text{ K}}{304.2\text{ K}} = 0.99 \\ P_{R,CO_2} &= \frac{P_m}{P_{cr,CO_2}} = \frac{15\text{ MPa}}{7.39\text{ MPa}} = 2.03 \end{aligned}\right\} \quad Z_{CO_2} = 0.30 \quad \text{(Fig. A-30}b\text{)}$$

Mixture:

$$Z_m = \Sigma y_i Z_i = y_{N_2} Z_{N_2} + y_{CO_2} Z_{CO_2}$$
$$= (0.25)(1.02) + (0.75)(0.30) = 0.48$$

Thus, $$V_m = \frac{Z_m N_m R_u T_m}{P_m} = Z_m V_{\text{ideal}} = (0.48)(1.330\ \text{m}^3) = 0.638\ \text{m}^3$$

The compressibility factor in this case turned out to be almost the same as the one determined by using Kay's rule. This is not always the case, however.

(*d*) When Dalton's law is used in conjunction with compressibility factors, Z_m is again determined from Eq. 12-10. But this time the Z of each component is to be determined at the mixture temperature and volume, which is not known. Therefore, an iterative solution is required. We start the calculations by assuming that the volume of the gas mixture is 1.330 m^3, the value determined by assuming ideal-gas behavior.

The T_R values in this case are identical to those obtained in part (*c*) and remain constant. The pseudoreduced volume is determined from its definition, Eq. 2-21:

$$v_{R,N_2} = \frac{\bar{v}_{N_2}}{R_u T_{cr,N_2}/P_{cr,N_2}} = \frac{V_m/N_{N_2}}{R_u T_{cr,N_2}/P_{cr,N_2}}$$

$$= \frac{1.33\ \text{m}^3/(2\ \text{kmol})}{[8.314\ \text{kPa·m}^3/(\text{kmol·K})](126.2\ \text{K})/(3390\ \text{kPa})} = 2.15$$

Similarly $$v_{R,CO_2} = \frac{1.33\ \text{m}^3/(6\ \text{kmol})}{[8.314\ \text{kPa·m}^3/(\text{kmol·K})](304.2\ \text{K})/(7390\ \text{kPa})} = 0.648$$

From Fig. A-30*b*, we read $Z_{N_2} = 0.99$ and $Z_{CO_2} = 0.56$. Thus,

$$Z_m = y_{N_2} Z_{N_2} + y_{CO_2} Z_{CO_2} = (0.25)(0.99) + (0.75)(0.56) = 0.67$$

and $$V_m = \frac{Z_m N_m R_u T_m}{P_m} = Z_m V_{\text{ideal}} = (0.67)(1.330\ \text{m}^3) = 0.891\ \text{m}^3$$

This is 33 percent lower than the assumed value. Therefore, we should repeat the calculations, using the new value of V_m. When the calculations are repeated, we obtain 0.738 m^3 after the second iteration, 0.678 m^3 after the third iteration, and 0.648 m^3 after the fourth iteration. This value does not change with more iterations. Therefore,

$$V_m = 0.648\ \text{m}^3$$

Notice that the results obtained in parts (*b*), (*c*), and (*d*) are very close. But they are very different from the values obtained from the ideal-gas relation. Therefore, treating a mixture of gases as an ideal gas may yield unacceptable errors at high pressures.

12-3 ■ PROPERTIES OF GAS MIXTURES—Ideal and Real Gases

Consider a gas mixture that consists of 2 kg of N_2 and 3 kg of CO_2. The mass (an *extensive* property) of this mixture is, to nobody's surprise, 5 kg. How did we do it? Well, we simply added the mass of each component. This example suggests a simple way of evaluating the **extensive properties** of a nonreacting ideal- or real-gas mixture: *Just add the contributions of each component of the mixture* (Fig. 12-11). Then the total

2 kmol A
3 kmol B
$U_A = 1000$ kJ
$U_B = 1800$ kJ
⇓
$U_m = 2800$ kJ

FIGURE 12-11

The extensive properties of a mixture are determined by simply adding the properties of the components.

internal energy, enthalpy, and entropy of a gas mixture can be expressed, respectively, as

$$U_m = \sum_{i=1}^{k} U_i = \sum_{i=1}^{k} m_i u_i = \sum_{i=1}^{k} N_i \bar{u}_i \qquad \text{(kJ)} \tag{12-13}$$

$$H_m = \sum_{i=1}^{k} H_i = \sum_{i=1}^{k} m_i h_i = \sum_{i=1}^{k} N_i \bar{h}_i \qquad \text{(kJ)} \tag{12-14}$$

$$S_m = \sum_{i=1}^{k} S_i = \sum_{i=1}^{k} m_i s_i = \sum_{i=1}^{k} N_i \bar{s}_i \qquad \text{(kJ/K)} \tag{12-15}$$

By following a similar logic, the changes in internal energy, enthalpy, and entropy of a gas mixture during a process can be expressed, respectively, as

$$\Delta U_m = \sum_{i=1}^{k} \Delta U_i = \sum_{i=1}^{k} m_i\, \Delta u_i = \sum_{i=1}^{k} N_i\, \Delta \bar{u}_i \qquad \text{(kJ)} \tag{12-16}$$

$$\Delta H_m = \sum_{i=1}^{k} \Delta H_i = \sum_{i=1}^{k} m_i\, \Delta h_i = \sum_{i=1}^{k} N_i\, \Delta \bar{h}_i \qquad \text{(kJ)} \tag{12-17}$$

$$\Delta S_m = \sum_{i=1}^{k} \Delta S_i = \sum_{i=1}^{k} m_i\, \Delta s_i = \sum_{i=1}^{k} N_i\, \Delta \bar{s}_i \qquad \text{(kJ/K)} \tag{12-18}$$

Now reconsider the same mixture, and assume that both N_2 and CO_2 are at 25°C. The temperature (an *intensive* property) of the mixture is, as you would expect, also 25°C. Notice that we did not add the component temperatures to determine the mixture temperature. Instead, we used some kind of averaging scheme, a characteristic approach for determining the **intensive properties** of a gas mixture. The internal energy, enthalpy, and entropy of a gas mixture *per unit mass* or *per unit mole* of the mixture can be determined by dividing the equations above by the mass or the mole number of the mixture (m_m or N_m). We obtain (Fig. 12-12)

2 kmol A
3 kmol B
$\bar{u}_A$ = 500 kJ/kmol
$\bar{u}_B$ = 600 kJ/kmol
⇓
$\bar{u}_m$ = 560 kJ/kmol

FIGURE 12-12
The intensive properties of a mixture are determined by weighed averaging.

$$u_m = \sum_{i=1}^{k} \text{mf}_i u_i \quad \text{and} \quad \bar{u}_m = \sum_{i=1}^{k} y_i \bar{u}_i \qquad \text{(kJ/kg or kJ/kmol)} \tag{12-19}$$

$$h_m = \sum_{i=1}^{k} \text{mf}_i h_i \quad \text{and} \quad \bar{h}_m = \sum_{i=1}^{k} y_i \bar{h}_i \qquad \text{(kJ/kg or kJ/kmol)} \tag{12-20}$$

$$s_m = \sum_{i=1}^{k} \text{mf}_i s_i \quad \text{and} \quad \bar{s}_m = \sum_{i=1}^{k} y_i \bar{s}_i \qquad [\text{kJ/(kg·K) or kJ/(kmol·K)}] \tag{12-21}$$

Similarly, the specific heats of a gas mixture can be expressed as

$$C_{v,m} = \sum_{i=1}^{k} \text{mf}_i C_{v,i} \quad \text{and} \quad \bar{C}_{v,m} = \sum_{i=1}^{k} y_i \bar{C}_{v,i} \quad [\text{kJ/(kg·°C) or kJ/(kmol·°C)}] \tag{12-22}$$

$$C_{p,m} = \sum_{i=1}^{k} \text{mf}_i C_{p,i} \quad \text{and} \quad \bar{C}_{p,m} = \sum_{i=1}^{k} y_i \bar{C}_{p,i} \quad [\text{kJ/(kg·°C) or kJ/(kmol·°C)}] \tag{12-23}$$

Notice that *properties per unit mass involve mass fractions* (mf_i) *and properties per unit mole involve mole fractions* (y_i).

The relations given above are generally valid and are applicable to both ideal- and real-gas mixtures. (In fact, they are also applicable to nonreacting liquid and solid solutions.) The only major difficulty associated with these relations is the determination of properties for each individual gas in the mixture. The analysis can be simplified greatly, however, by treating the individual gases as an ideal gas, if doing so does not introduce a significant error.

1 Ideal-Gas Mixtures

The gases that comprise a mixture are often at a high temperature and low pressure relative to the critical-point values of individual gases. In such cases, the gas mixture and its components can be treated as ideal gases with negligible error. Under the ideal-gas approximation, the properties of a gas are not influenced by the presence of other gases, and each gas component in the mixture behaves as if it exists alone at the mixture temperature T_m and mixture volume V_m. This principle is known as the **Gibbs-Dalton law**, which is an extension of Dalton's law of additive pressures. Also, the h, u, C_v, and C_p of an ideal gas depend on temperature only and are independent of the pressure or the volume of the ideal-gas mixture. The partial pressure of a component in an ideal-gas mixture is simply $P_i = y_i P_m$, where P_m is the mixture pressure.

Evaluation of Δu or Δh of the components of an ideal-gas mixture during a process is relatively easy since it requires only a knowledge of the initial and final temperatures. Care should be exercised, however, in evaluating the Δs of the components since the entropy of an ideal gas depends on the pressure or volume of the component as well as on its temperature. The entropy change of individual gases in an ideal-gas mixture during a process can be determined from

$$\Delta s_i = s^\circ_{i,2} - s^\circ_{i,1} - R_i \ln \frac{P_{i,2}}{P_{i,1}} \qquad [\text{kJ/(kg·K)}] \qquad (12\text{-}24)$$

$$\cong C_{p,i} \ln \frac{T_{i,2}}{T_{i,1}} - R_i \ln \frac{P_{i,2}}{P_{i,1}} \qquad [\text{kJ/(kg·K)}] \qquad (12\text{-}25)$$

or

$$\Delta \bar{s}_i = \bar{s}^\circ_{i,2} - \bar{s}^\circ_{i,1} - R_u \ln \frac{P_{i,2}}{P_{i,1}} \qquad [\text{kJ/(kmol·K)}] \qquad (12\text{-}26)$$

$$\cong \bar{C}_{p,i} \ln \frac{T_{i,2}}{T_{i,1}} - R_u \ln \frac{P_{i,2}}{P_{i,1}} \qquad [\text{kJ/(kmol·K)}] \qquad (12\text{-}27)$$

where $P_{i,2} = y_{i,2} P_{m,2}$ and $P_{i,1} = y_{i,1} P_{m,1}$. Notice that the partial pressure P_i of each component is used in the evaluation of the entropy change, not the mixture pressure P_m (Fig. 12-13).

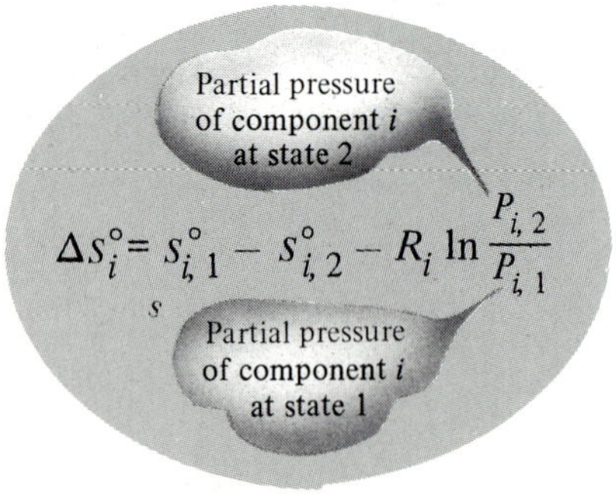

FIGURE 12-13

Partial pressures (not the mixture pressure) are used in the evaluation of entropy changes of ideal-gas mixtures.

EXAMPLE 12-3

An insulated rigid tank is divided into two compartments by a partition. One compartment contains 7 kg of oxygen gas at 40°C and 100 kPa, and the other compartment contains 4 kg of nitrogen gas at 20°C and 150 kPa. Now the partition is removed, and the two gases are allowed to mix. Determine (*a*) the mixture temperature and (*b*) the mixture pressure after equilibrium has been established.

Solution Both the oxygen and nitrogen are at much higher temperatures than their critical temperature and much lower pressures than their critical pressure. Therefore, both gases can be treated as ideal gases, and their mixture as an ideal-gas mixture.

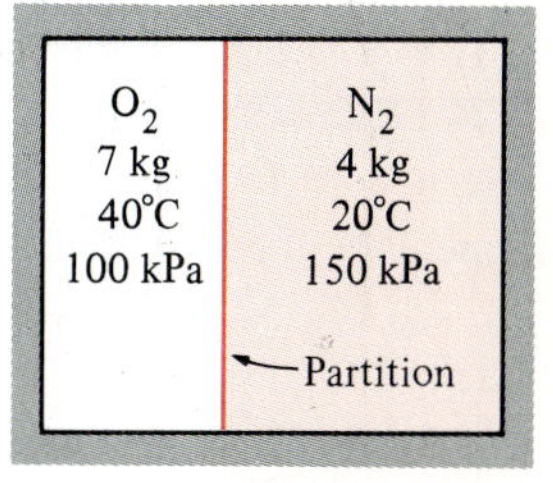

FIGURE 12-14
Schematic for Example 12-3.

(*a*) We take both gases as our system. The boundaries of the system are indicated in Fig. 12-14. No heat, work, or mass crosses the system boundary; therefore this is a closed system with $Q = 0$ and $W = 0$. Then the conservation of energy equation for this closed system reduces to

$$\cancel{Q}^{0} - \cancel{W}^{0} = \Delta U = \Delta U_{N_2} + \Delta U_{O_2}$$

$$[mC_v(T_m - T_1)]_{N_2} + [mC_v(T_m - T_1)]_{O_2} = 0$$

By using C_v values at room temperature (from Table A-2*a*), the final temperature of the mixture is determined to be

$$(4\text{ kg})[0.743\text{ kJ/(kg·°C)}](T_m - 20°\text{C}) + (7\text{ kg})[0.658\text{ kJ/(kg·°C)}](T_m - 40°\text{C}) = 0$$

$$T_m = 32.2°\text{C}$$

(*b*) The final pressure of the mixture is determined from the ideal-gas relation

$$P_m V_m = N_m R_u T_m$$

where

$$N_{O_2} = \frac{m_{O_2}}{M_{O_2}} = \frac{7\text{ kg}}{32\text{ kg/kmol}} = 0.219\text{ kmol}$$

$$N_{N_2} = \frac{m_{N_2}}{M_{N_2}} = \frac{4\text{ kg}}{28\text{ kg/kmol}} = 0.143\text{ kmol}$$

$$N_m = N_{O_2} + N_{N_2} = (0.219 + 0.143)\text{ kmol} = 0.362\text{ kmol}$$

and

$$V_{O_2} = \left(\frac{NR_uT_1}{P_1}\right)_{O_2} = \frac{(0.219\text{ kmol})[8.314\text{ kPa·m}^3\text{/(kmol·K)}](313\text{ K})}{100\text{ kPa}} = 5.70\text{ m}^3$$

$$V_{N_2} = \left(\frac{NR_uT_1}{P_1}\right)_{N_2} = \frac{(0.143\text{ kmol})[8.314\text{ kPa·m}^3\text{/(kmol·K)}](293\text{ K})}{150\text{ kPa}} = 2.32\text{ m}^3$$

$$V_m = V_{O_2} + V_{N_2} = (5.70 + 2.32)\text{ m}^3 = 8.02\text{ m}^3$$

Thus,

$$P_m = \frac{N_m R_u T_m}{V_m} = \frac{(0.362\text{ kmol})[8.314\text{ kPa·m}^3\text{/(kmol·K)}](305.2\text{ K})}{8.02\text{ m}^3}$$

$$= 114.5\text{ kPa}$$

We could also determine the mixture pressure by using $P_m V_m = m_m R T_m$, where R is the apparent gas constant of the mixture. This would not be any easier, however, since the evaluation of R requires the determination of the mole fractions of the components.

EXAMPLE 12-4

An insulated rigid tank is divided into two compartments by a partition, as shown in Fig. 12-15. One compartment contains 3 kmol of O_2, and the other compartment contains 5 kmol of CO_2. Both gases are initially at 25°C and 200 kPa. Now the partition is removed, and the two gases are allowed to mix. Assuming the surroundings are at 25°C and both gases behave as ideal gases, determine the entropy change and irreversibility associated with this process.

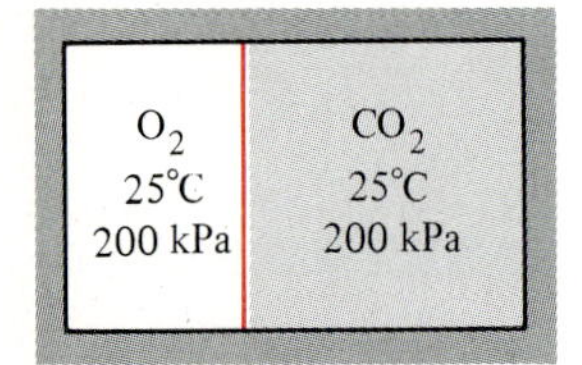

FIGURE 12-15
Schematic for Example 12-4.

Solution When two ideal gases initially at the same temperature and pressure are mixed by removing a partition between them, the mixture will also be at the same temperature and pressure. (Can you prove it? Will this be true for nonideal gases?) Therefore, the temperature and pressure in the tank will still be 25°C and 200 kPa, respectively, after the mixing. The entropy change of each component gas can be determined from Eqs. 12-18 and 12-27:

$$\Delta S_m = \sum \Delta S_i = \sum N_i \, \Delta \bar{s}_i = \sum N_i \left(\bar{C}_{p,i} \ln \cancelto{0}{\frac{T_{i,2}}{T_{i,1}}} - R_u \ln \frac{P_{i,2}}{P_{i,1}} \right)$$

$$= -R_u \sum N_i \ln \frac{y_i P_{m,2}}{P_{i,1}} = -R_u \sum N_i \ln y_i \qquad (12\text{-}28)$$

since $P_{m,2} = P_{i,1} = 200$ kPa. It is obvious that the entropy change is independent of the composition of the mixture in this case and depends on only the mole fraction of the gases in the mixture. What is not so obvious is that if the same gas in two different chambers is mixed at constant temperature and pressure, the entropy change is zero.

Substituting the known values, we see that the entropy change is

$$N_m = N_{O_2} + N_{CO_2} = (3 + 5) \text{ kmol} = 8 \text{ kmol}$$

$$y_{O_2} = \frac{N_{O_2}}{N_m} = \frac{3 \text{ kmol}}{8 \text{ kmol}} = 0.375$$

$$y_{CO_2} = \frac{N_{CO_2}}{N_m} = \frac{5 \text{ kmol}}{8 \text{ kmol}} = 0.625$$

$$\begin{aligned} \Delta S_m &= -R_u(N_{O_2} \ln y_{O_2} + N_{CO_2} \ln y_{CO_2}) \\ &= -[8.314 \text{ kJ/(kmol·K)}][(3 \text{ kmol})(\ln 0.375) + (5 \text{ kmol})(\ln 0.625)] \\ &= \mathbf{44.0 \text{ kJ/K}} \end{aligned}$$

The irreversibility associated with this mixing process is determined from

$$\begin{aligned} I &= T_0 S_{gen} = T_0(\Delta S_{sys} + \cancelto{0}{\Delta S_{surr}}) = T_0 \, \Delta S_{sys} \\ &= (298 \text{ K})(44.0 \text{ kJ/K}) \\ &= \mathbf{13{,}112 \text{ kJ}} \end{aligned}$$

This result shows that mixing processes are highly irreversible.

2 Real-Gas Mixtures

When the components of a gas mixture do not behave as ideal gases, the analysis becomes more complex because the properties of real (nonideal) gases such as u, h, C_v, and C_p depend on the pressure (or specific volume) as well as on the temperature. In such cases, the effects of deviation from ideal-gas behavior on the mixture properties should be accounted for.

Consider two nonideal gases contained in two separate compartments of an adiabatic rigid tank at 100 kPa and 25°C. The partition separating the two gases is removed, and the two gases are allowed to mix. What do you think the final pressure in the tank will be? You are probably tempted to say 100 kPa, which would be true for ideal gases. However, this is not true for nonideal gases because of the influence of the molecules of different gases on each other (deviation from Dalton's law, Fig. 12-16).

When real-gas mixtures are involved, it may be necessary to account for the effect of nonideal behavior on the mixture properties such as enthalpy and entropy. One way of doing that is to use compressibility factors in conjunction with generalized equations and charts developed in Chap. 11 for real gases.

Consider the following $T\,ds$ relation for a gas mixture:

$$dh_m = T_m\,ds_m + v_m\,dP_m$$

It can also be expressed as

$$d(\Sigma \text{mf}_i h_i) = T_m\,d(\Sigma \text{mf}_i s_i) + (\Sigma \text{mf}_i v_i)\,dP_m$$

or

$$\Sigma \text{mf}_i (dh_i - T_m\,ds_i - v_i\,dP_m) = 0$$

which yields

$$dh_i = T_m\,ds_i + v_i\,dP_m \tag{12-29}$$

This is an important result because Eq. 12-29 is the starting equation in the development of the generalized relations and charts for enthalpy and entropy. It suggests that the generalized property relations and charts for real gases developed in Chap. 11 can also be used for the components of real-gas mixtures. But the reduced temperature T_R and reduced pressure P_R for each component should be evaluated by using the mixture temperature T_m and mixture pressure P_m. This is because Eq. 12-29 involves the mixture pressure P_m, not the component pressure P_i.

The approach described above is somewhat analogous to Amagat's law of additive volumes (evaluating mixture properties at the mixture pressure and temperature), which holds exactly for ideal-gas mixtures and approximately for real-gas mixtures. Therefore, the mixture properties determined with this approach will not be exact, but they will be sufficiently accurate for most purposes.

What if the mixture volume and temperature are specified instead of the mixture pressure and temperature? Well, there is no need to panic. Just evaluate the mixture pressure, using Dalton's law of additive pressures, and then use this value (which is only approximate) as the mixture pressure.

Another way of evaluating the properties of a real-gas mixture is to treat the mixture as a pseudopure substance having pseudocritical properties, determined in terms of the critical properties of the component gases by using Kay's rule. The approach is quite simple, and the accuracy is usually acceptable.

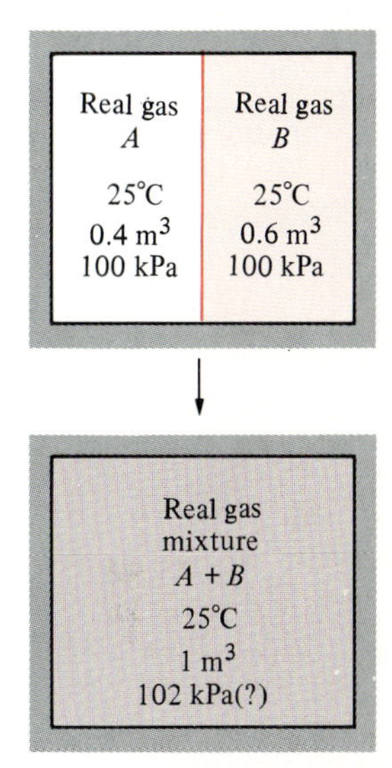

FIGURE 12-16

It is difficult to predict the behavior of nonideal-gas mixtures because of the influence of dissimilar gas molecules on each other.

EXAMPLE 12-5

Air is a mixture of N_2, O_2, and small amounts of other gases, and it can be approximated as 79 percent N_2 and 21 percent O_2 on a mole basis. During a steady-flow process, the air is cooled from 220 to 160 K at a constant pressure of 10 MPa (Fig. 12-17). Determine the heat transfer during this process per mole of air, using (*a*) the ideal-gas approximation, (*b*) Kay's rule, and (*c*) Amagat's law.

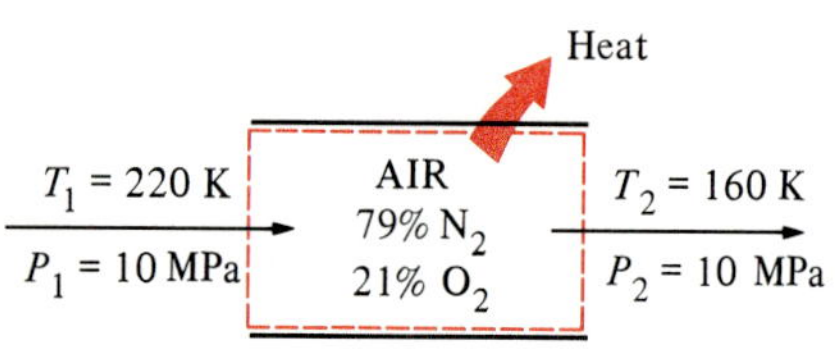

FIGURE 12-17
Schematic for Example 12-5.

Solution The critical properties are T_{cr} = 126.2 K and P_{cr} = 3.39 MPa for N_2 and T_{cr} = 154.8 and P_{cr} = 5.08 MPa for O_2. Both gases remain above their critical temperatures, but they are also above their critical pressures. Therefore, air will probably deviate from ideal-gas behavior, and thus it should be treated as a real-gas mixture.

Disregarding any changes in kinetic and potential energies, the conservation of energy equation for this gas mixture can be written, on a unit-mole basis, as

$$\bar{q} - \bar{w}^{\nearrow 0} = \Delta\bar{h} = \Sigma y_i\,\Delta\bar{h}_i$$

$$\bar{q} = y_{N_2}(\bar{h}_2 - \bar{h}_1)_{N_2} - y_{O_2}(\bar{h}_2 - \bar{h}_1)_{O_2}$$

where the enthalpy change for either component can be determined from the generalized enthalpy chart (Fig. A-31) and Eq. 11-58:

$$\bar{h}_2 - \bar{h}_1 = R_uT_{cr}\,(Z_{h1} - Z_{h2}) + \bar{h}_{2,\text{ideal}} - \bar{h}_{1,\text{ideal}}$$

The last term in this equation represents the ideal-gas enthalpy change of the component. The terms in parentheses represent the deviation from the ideal-gas behavior, and its evaluation requires a knowledge of reduced pressure P_R and reduced temperature T_R, which are calculated at the mixture temperature T_m and mixture pressure P_m.

(*a*) If the N_2 and O_2 mixture is assumed to behave as an ideal gas, the enthalpy of the mixture will depend on temperature only, and the enthalpy values at the initial and the final temperatures can be determined from the ideal-gas tables of N_2 and O_2 (Tables A-18 and A-19):

$$T_1 = 220\text{ K} \longrightarrow \bar{h}_{1,\text{ideal},N_2} = 6391\text{ kJ/kmol}$$
$$\bar{h}_{1,\text{ideal},O_2} = 6404\text{ kJ/kmol}$$
$$T_2 = 160\text{ K} \longrightarrow \bar{h}_{2,\text{ideal},N_2} = 4648\text{ kJ/kmol}$$
$$\bar{h}_{2,\text{ideal},O_2} = 4657\text{ kJ/kmol}$$

$$\begin{aligned}\bar{q} &= y_{N_2}(\bar{h}_2 - \bar{h}_1)_{N_2} + y_{O_2}(\bar{h}_2 - \bar{h}_1)_{O_2}\\ &= (0.79)(4648 - 6391)\text{ kJ/kmol} + (0.21)(4657 - 6404)\text{ kJ/kmol}\\ &= \mathbf{-1744\ kJ/kmol}\end{aligned}$$

(*b*) Kay's rule is based on treating a gas mixture as a pseudopure substance whose critical temperature and pressure are, respectively,

$$\begin{aligned}T'_{cr,m} &= \Sigma y_iT_{cr,i} = y_{N_2}T_{cr,N_2} + y_{O_2}T_{cr,O_2}\\ &= (0.79)(126.2\text{ K}) + (0.21)(154.8\text{ K}) = 132.2\text{ K}\end{aligned}$$

and

$$\begin{aligned}P'_{cr,m} &= \Sigma y_iP_{cr,i} = y_{N_2}P_{cr,N_2} + y_{O_2}P_{cr,O_2}\\ &= (0.79)(3.39\text{ MPa}) + (0.21)(5.08\text{ MPa}) = 3.74\text{ MPa}\end{aligned}$$

Then

$$T_{R,1} = \frac{T_{m,1}}{T'_{cr,m}} = \frac{220\text{ K}}{132.2\text{ K}} = 1.66$$

$$P_R = \frac{P_m}{P'_{cr,m}} = \frac{10\text{ MPa}}{3.74\text{ MPa}} = 2.67$$

$$T_{R,2} = \frac{T_{m,2}}{T'_{cr,m}} = \frac{160\text{ K}}{132.2\text{ K}} = 1.21$$

$$\left.\begin{matrix}T_{R,1}, P_R\end{matrix}\right\} Z_{h1,m} = 1.0 \qquad \left.\begin{matrix}P_R, T_{R,2}\end{matrix}\right\} Z_{h2,m} = 2.6$$

Also,

$$\begin{aligned}\bar{h}_{m1,\text{ideal}} &= y_{N_2}\bar{h}_{1,\text{ideal},N_2} + y_{O_2}\bar{h}_{1,\text{ideal},O_2}\\ &= (0.79)(6391\text{ kJ/kmol}) + (0.21)(6404\text{ kJ/mol})\\ &= 6394\text{ kJ/kmol}\\ \bar{h}_{m2,\text{ideal}} &= y_{N_2}\bar{h}_{2,\text{ideal},N_2} + y_{O_2}\bar{h}_{2,\text{ideal},O_2}\\ &= (0.79)(4648\text{ kJ/kmol}) + (0.21)(4657\text{ kJ/kmol})\\ &= 4650\text{ kJ/kmol}\end{aligned}$$

Therefore, from Eq. 11-58,

$$\begin{aligned}\bar{q} &= R_uT_{cr}(Z_{h1} - Z_{h2})_m + (\bar{h}_{m2,\text{ideal}} - \bar{h}_{m1,\text{ideal}})\\ &= [8.314\text{ kJ/(kmol·K)}](132.2\text{ K})(1.0 - 2.6) + [(4650 - 6394)\text{ kJ/(kmol·K)}]\\ &= -3503\text{ kJ/kmol}\end{aligned}$$

(*c*) The reduced temperatures and pressures for both N_2 and O_2 at the initial and final states and the corresponding enthalpy departure factors are, from Fig. A-31,

N_2:

$$T_{R_1,N_2} = \frac{T_{m,1}}{T_{cr,N_2}} = \frac{220\text{ K}}{126.2\text{ K}} = 1.74$$

$$P_{R_1,N_2} = P_{R_2,N_2} = \frac{P_m}{P_{cr,N_2}} = \frac{10\text{ MPa}}{3.39\text{ MPa}} = 2.95$$

$$T_{R_2,N_2} = \frac{T_{m,2}}{T_{cr,N_2}} = \frac{160\text{ K}}{126.2\text{ K}} = 1.27$$

$$Z_{h1,N_2} = 0.9 \qquad Z_{h2,N_2} = 2.4$$

O_2:

$$T_{R_1,O_2} = \frac{T_{m,1}}{T_{cr,O_2}} = \frac{220\text{ K}}{154.8\text{ K}} = 1.42$$

$$P_{R_1,O_2} = P_{R_2,O_2} = \frac{P_m}{P_{cr,O_2}} = \frac{10\text{ MPa}}{5.08\text{ MPa}} = 1.97$$

$$T_{R_2,O_2} = \frac{T_{m,2}}{T_{cr,O_2}} = \frac{160\text{ K}}{154.8\text{ K}} = 1.03$$

$$Z_{h1,O_2} = 1.3 \qquad Z_{h2,O_2} = 4.0$$

From Eq. 11-58,

$$\begin{aligned}(\bar{h}_2 - \bar{h}_1)_{N_2} &= R_uT_{cr}(Z_{h1} - Z_{h2})_{N_2} + (\bar{h}_{2,\text{ideal}} - \bar{h}_{1,\text{ideal}})_{N_2}\\ &= [8.314\text{ kJ/(kmol·K)}](126.2\text{ K})(0.9 - 2.4)\\ &\qquad + [(4648 - 6391)\text{ kJ/(kmol·K)}]\\ &= -3317\text{ kJ/kmol}\\ (\bar{h}_2 - \bar{h}_1)_{O_2} &= R_uT_{cr}(Z_{h1} - Z_{h2})_{O_2} + (\bar{h}_{2,\text{ideal}} - \bar{h}_{1,\text{ideal}})_{O_2}\\ &= [8.314\text{ kJ/(kmol·K)}](154.8\text{ K})(1.3 - 4.0)\\ &\qquad + [(4657 - 6404)\text{ kJ/(kmol·K)}]\\ &= -5222\text{ kJ/kmol}\end{aligned}$$

Therefore,

$$\begin{aligned}\bar{q} &= y_{N_2}(\bar{h}_2 - \bar{h}_1)_{N_2} + y_{O_2}(\bar{h}_2 - \bar{h}_1)_{O_2}\\ &= (0.79)(-3317\text{ kJ/kmol}) + (0.21)(-5222\text{ kJ/kmol})\\ &= -3717\text{ kJ/kmol}\end{aligned}$$

This result is about 6 percent greater than the result obtained in part (*b*) by using Kay's rule. But it is more than twice the result obtained by assuming the mixture to be an ideal gas.

12-4 ■ SUMMARY

A mixture of two or more gases of fixed chemical composition is called a *nonreacting gas mixture*. The composition of a gas mixture is described by specifying either the *mole fraction* or the *mass fraction* of each component, defined as

$$\mathrm{mf}_i = \frac{m_i}{m_m} \quad \text{and} \quad y_i = \frac{N_i}{N_m} \tag{12-2a, b}$$

where

$$m_m = \sum_{i=1}^{k} m_i \quad \text{and} \quad N_m = \sum_{i=1}^{k} N_i \tag{12-1a, b}$$

The *apparent* (or average) *molar mass* and *gas constant* of a mixture are expressed as

$$M_m = \frac{m_m}{N_m} = \sum_{i=1}^{k} y_i M_i \qquad \text{(kg/kmol)} \tag{12-4}$$

and

$$R_m = \frac{R_u}{M_m} \qquad [\text{kJ/(kg·K)}] \tag{12-5}$$

Dalton's law of additive pressures states that the pressure of a gas mixture is equal to the sum of the pressures each gas would exert if it existed alone at the mixture temperature and volume. *Amagat's law of additive volumes* states that the volume of a gas mixture is equal to the sum of the volumes each gas would occupy if it existed alone at the mixture temperature and pressure. Dalton's and Amagat's laws hold exactly for ideal-gas mixtures, but only approximately for real-gas mixtures. They can be expressed as

Dalton's law:

$$P_m = \sum_{i=1}^{k} P_i(T_m, V_m) \tag{12-6}$$

Amagat's law:

$$V_m = \sum_{i=1}^{k} V_i(T_m, P_m) \tag{12-7}$$

Here P_i is called the *component pressure,* and V_i is called the *component volume*. Also the ratio P_i/P_m is called the *pressure fraction,* and the ratio V_i/V_m is called the *volume fraction* of component i. For ideal gases, P_i and V_i can be related to y_i by

$$\frac{P_i}{P_m} = \frac{V_i}{V_m} = \frac{N_i}{N_m} = y_i \tag{12-8}$$

The quantity y_iP_m is called the *partial pressure,* and the quantity y_iV_m is called the *partial volume*. The *P-v-T* behavior of real-gas mixtures can be predicted by using generalized compressibility charts. The compressibility factor of the mixture can be expressed in terms of the compressibility factors of the individual gases as

$$Z_m = \sum_{i=1}^{k} y_i Z_i \tag{12-10}$$

where Z_i is determined either at T_m and V_m (Dalton's law) or at T_m and P_m (Amagat's law) for each individual gas. The P-v-T behavior of a gas mixture can also be predicted approximately by *Kay's rule,* which involves treating a gas mixture as a pure substance with pseudocritical properties determined from

$$P'_{\mathrm{cr},m} = \sum_{i=1}^{k} y_i P_{\mathrm{cr},i} \quad \text{and} \quad T'_{\mathrm{cr},m} = \sum_{i=1}^{k} y_i T_{\mathrm{cr},i} \qquad (12\text{-}11a, b)$$

The *extensive properties* of a gas mixture, in general, can be determined by summing the contributions of each component of the mixture. The evaluation of *intensive properties* of a gas mixture, however, involves averaging in terms of mass or mole fractions:

$$U_m = \sum_{i=1}^{k} U_i = \sum_{i=1}^{k} m_i u_i = \sum_{i=1}^{k} N_i \bar{u}_i \qquad \text{(kJ)} \qquad (12\text{-}13)$$

$$H_m = \sum_{i=1}^{k} H_i = \sum_{i=1}^{k} m_i h_i = \sum_{i=1}^{k} N_i \bar{h}_i \qquad \text{(kJ)} \qquad (12\text{-}14)$$

$$S_m = \sum_{i=1}^{k} S_i = \sum_{i=1}^{k} m_i s_i = \sum_{i=1}^{k} N_i \bar{s}_i \qquad \text{(kJ/K)} \qquad (12\text{-}15)$$

and

$$u_m = \sum_{i=1}^{k} \mathrm{mf}_i u_i \quad \text{and} \quad \bar{u}_m = \sum_{i=1}^{k} y_i \bar{u}_i \qquad \text{(kJ/kg or kJ/kmol)} \qquad (12\text{-}19)$$

$$h_m = \sum_{i=1}^{k} \mathrm{mf}_i h_i \quad \text{and} \quad \bar{h}_m = \sum_{i=1}^{k} y_i \bar{h}_i \qquad \text{(kJ/kg or kJ/kmol)} \qquad (12\text{-}20)$$

$$s_m = \sum_{i=1}^{k} \mathrm{mf}_i s_i \quad \text{and} \quad \bar{s}_m = \sum_{i=1}^{k} y_i \bar{s}_i \quad [\text{kJ/(kg·K) or kJ/(kmol·K)}] \qquad (12\text{-}21)$$

$$C_{v,m} = \sum_{i=1}^{k} \mathrm{mf}_i C_{v,i} \quad \text{and} \quad \bar{C}_{v,m} = \sum_{i=1}^{k} y_i \bar{C}_{v,i} \quad [\text{kJ/(kg·°C) or kJ/(kmol·°C)}] \qquad (12\text{-}22)$$

$$C_{p,m} = \sum_{i=1}^{k} \mathrm{mf}_i C_{p,i} \quad \text{and} \quad \bar{C}_{p,m} = \sum_{i=1}^{k} y_i \bar{C}_{p,i} \quad [\text{kJ/(kg·°C) or kJ/(kmol·°C)}] \qquad (12\text{-}23)$$

These relations are applicable to both ideal- and real-gas mixtures. The properties or property changes of individual components can be determined by using ideal-gas or real-gas relations developed in earlier chapters.

REFERENCES AND SUGGESTED READING

1 W. Z. Black and J. G. Hartley, *Thermodynamics,* Harper & Row, New York, 1985.

2 J. P. Holman, *Thermodynamics,* 3d ed., McGraw-Hill, New York, 1980.

3 J. R. Howell and R. O. Buckius, *Fundamentals of Engineering Thermodynamics,* McGraw-Hill, New York, 1987.

4 J. B. Jones and G. A. Hawkins, *Engineering Thermodynamics,* 2d ed., Wiley, New York, 1986.

5 G. J. Van Wylen and R. E. Sonntag, *Fundamentals of Classical Thermodynamics,* 3d ed., Wiley, New York, 1985.

6 K. Wark, *Thermodynamics,* 5th ed., McGraw-Hill, New York, 1988.

PROBLEMS*

Composition of Gas Mixtures

12-1C What are the mass fraction and the mole fraction?

12-2C Using the definitions of mass and mole fractions, derive a relation between them.

12-3C Somebody claims that the mass and mole fractions for a mixture of CO_2 and N_2O gases are identical. Is this true? Why?

12-4C The sum of the mass fractions for an ideal-gas mixture is equal to 1. Is this also true for a real-gas mixture?

12-5C The sum of the mole fractions for an ideal-gas mixture is equal to 1. Is this also true for a real-gas mixture?

12-6C Consider a mixture of several gases of identical masses. Will all the mass fractions be identical? How about the mole fractions?

12-7C What is the *apparent molar mass* for a gas mixture? Does the mass of every molecule in the mixture equal the apparent molar mass?

12-8C Consider a mixture of two gases. Can the apparent molar mass of this mixture be determined by simply taking the arithmetic average of the molar masses of the individual gases? When will this be the case?

12-9C What is the *apparent gas constant* for a gas mixture? Can it be larger than the largest gas constant in the mixture?

12-10 A gas mixture consists of 5 kg of O_2, 8 kg of N_2, and 10 kg of CO_2. Determine (*a*) the mass fraction of each component, (*b*) the mole fraction of each component, and (*c*) the average molar mass and gas constant of the mixture.

12-10E A gas mixture consists of 5 lbm of O_2, 8 lbm of N_2, and 10 lbm of CO_2. Determine (*a*) the mass fraction of each component, (*b*) the mole fraction of each component, and (*c*) the average molar mass and gas constant of the mixture.

---

*Students are encouraged to answer *all* the concept "C" questions.

12-11 Air has the following composition on a mole basis: 21 percent O_2, 78 percent N_2, and 1 percent Ar. Determine the gravimetric analysis of air and its molar mass.
Answers: 23.2 percent O_2, 75.4 percent N_2, 1.4 percent Ar; 28.96 kg/kmol

12-12 A gas mixture has the following composition on a mole basis: 65 percent N_2 and 35 percent CO_2. Determine the gravimetric analysis of the mixture, its molar mass, and gas constant.

12-13 Determine the mole fractions of a gas mixture which consists of 85 percent CH_4 and 15 percent CO_2 by mass. Also determine the gas constant of the mixture.

12-14 A gas mixture consists of 5 kmol of H_2 and 2 kmol of N_2. Determine the mass of each gas and the apparent gas constant of the mixture.
Answers: 10 kg, 56 kg, 9.43 kg/kmol

12-14E A gas mixture consists of 5 lbmol of H_2 and 2 lbmol of N_2. Determine the mass of each gas and the apparent gas constant of the mixture.

P-v-T Behavior of Gas Mixtures

12-15C Is a mixture of ideal gases also an ideal gas? Give an example.

12-16C Express Dalton's law of additive pressures. Does this law hold exactly for ideal-gas mixtures? How about nonideal-gas mixtures?

12-17C Express Amagat's law of additive volumes. Does this law hold exactly for ideal-gas mixtures? How about nonideal-gas mixtures?

12-18C How is the *P-v-T* behavior of a component in an ideal-gas mixture expressed? How is the *P-v-T* behavior of a component in a real-gas mixture expressed?

12-19C What is the difference between the *component pressure* and the *partial pressure?* When are these two equivalent?

12-20C What is the difference between the *component volume* and the *partial volume?* When are these two equivalent?

12-21C In a gas mixture, which component will have the higher partial pressure—the one with the highest mole number or the one with the largest molar mass?

12-22C Consider a rigid tank which contains a mixture of two ideal gases. A valve is opened, and some gas escapes. As a result, the pressure in the tank drops. Will the partial pressure of each component change? How about the pressure fraction of each component?

12-23C Consider a rigid tank which contains a mixture of two ideal gases. The gas mixture is heated, and the pressure and temperature in the tank rise. Will the partial pressure of each component change? How about the pressure fraction of each component?

12-24C Is this statement correct? *The volume of an ideal-gas mixture is equal to the sum of the volumes of each individual gas in the mixture*. If not, how would you correct it?

12-25C Is this statement correct? *The temperature of an ideal-gas mixture is equal to the sum of the temperatures of each individual gas in the mixture*. If not, how would you correct it?

12-26C Is this statement correct? *The pressure of an ideal-gas mixture is equal to the sum of the partial pressures of each individual gas in the mixture*. If not, how would you correct it?

12-27C Using Amagat's law, show that

$$Z_m = \sum_{i=1}^{k} y_i Z_i$$

for a real-gas mixture of k gases, where Z is the compressibility factor.

12-28C Using Dalton's law, show that

$$Z_m = \sum_{i=1}^{k} y_i Z_i$$

for a real-gas mixture of k gases, where Z is the compressibility factor.

12-29C Explain how a real-gas mixture can be treated as a pseudopure substance using Kay's rule.

12-30 A rigid tank contains 3 kmol of O_2 and 6 kmol of CO_2 gases at 300 K and 150 kPa. Estimate the volume of the tank.
Answer: 149.7 m^3

12-31 A rigid tank contains 0.5 kmol of Ar and 2 kmol of N_2 at 100 kPa and 280 K. The mixture is now heated to 400 K. Determine the volume of the tank and the final pressure of the mixture.

12-31E A rigid tank contains 0.5 lbmol of Ar and 2 lbmol of N_2 at 14.7 psia and 580 R. The mixture is now heated to 800 K. Determine the volume of the tank and the final pressure of the mixture.
Answers: 1058.4 ft^3, 20.3 psia

12-32 A gas mixture at 400 K and 150 kPa consists of 1 kg of CO_2 and 3 kg of CH_4. Determine the partial pressure of each gas and the apparent molar mass of the gas mixture.

12-32E A gas mixture at 600 R and 20 psia consists of 1 lbm of CO_2 and 3 lbm of CH_4. Determine the partial pressure of each gas and the apparent molar mass of the gas mixture.

12-33 A 0.1-m^3 rigid tank contains 0.6 kg of N_2 and 0.4 kg of O_2 at 300 K. Determine the partial pressure of each gas and the total pressure of the mixture. *Answers:* 534.2 kPa, 311.8 kPa, 846.0 kPa

12-34 A gas mixture at 290 K and 120 kPa has the following volumetric analysis: 65 percent N_2, 20 percent O_2, and 15 percent CO_2. Determine the mass fraction and partial pressure of each gas.

12-35 A rigid tank which contains 2 kg of N_2 at 25°C and 100 kPa is connected to another rigid tank which contains 3 kg of O_2 at 25°C and 350 kPa. The valve connecting the two tanks is opened, and the two gases are allowed to mix. If the final mixture temperature is 25°C, determine the volume of each tank and the final mixture pressure.
Answers: 1.77 m³, 0.66 m³, 168.4 kPa

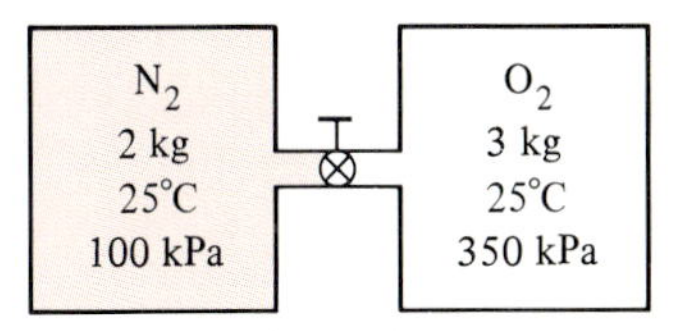

FIGURE P12-35

12-35E A rigid tank which contains 2 lbm of N_2 at 70°F and 15 psia is connected to another rigid tank which contains 3 lbm of O_2 at 70°F and 50 psia. The valve connecting the two tanks is now opened, and the two gases are allowed to mix. If the final mixture temperature is 70°F, determine the volume of each tank and the final mixture pressure.

12-36 A rigid tank contains 2 kmol of N_2 and 6 kmol of CH_4 gases at 200 K and 10 MPa. Estimate the volume of the tank, using (*a*) the ideal-gas equation of state, (*b*) Kay's rule, and (*c*) the compressibility chart and Amagat's law.

12-37 A volume of 0.3 m³ of O_2 at 200 K and 8 MPa is mixed with 0.5 m³ of N_2 at the same temperature and pressure, forming a mixture at 200 K and 8 MPa. Determine the volume of the mixture, using (*a*) the ideal-gas equation of state, (*b*) Kay's rule, and (*c*) the compressibility chart and Amagat's law. *Answers:* (*a*) 0.8 m³, (*b*) 0.79 m³, (*c*) 0.80 m³

12-38 A rigid tank contains 1 kmol of Ar gas at 220 K and 5 MPa. A valve is now opened, and 3 kmol of N_2 gas is allowed to enter the tank at 190 K and 8 MPa. The final mixture temperature is 200 K. Determine the pressure of the mixture, using (*a*) the ideal-gas equation of state and (*b*) the compressiblity chart and Dalton's law.

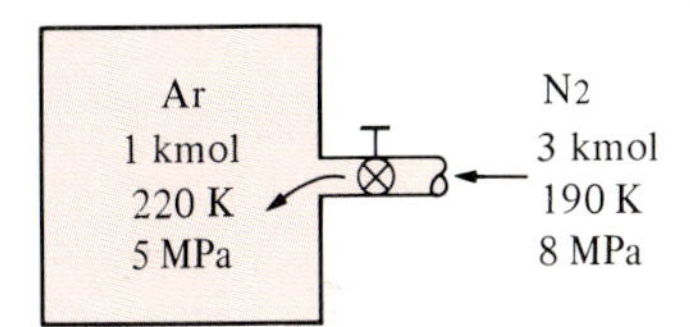

FIGURE P12-38

12-38E A rigid tank contains 1 lbmol of argon gas at 400 R and 750 psia. A valve is now opened, and 3 lbmol of N_2 gas is allowed to enter the tank at 340 R and 1200 psia. The final mixture temperature is 360 R. Determine the pressure of the mixture, using (*a*) the ideal-gas equation of state and (*b*) the compressibility chart and Dalton's law.
Answers: (*a*) 2700 psia, (*b*) 2472 psia

Properties of Gas Mixtures

12-39C Is the total internal energy of an ideal-gas mixture equal to the sum of the internal energies of each individual gas in the mixture? Answer the same question for a real-gas mixture.

12-40C Is the specific internal energy of a gas mixture equal to the sum of the specific internal energy of each individual gas in the mixture?

12-41C Is the total internal energy change of an ideal-gas mixture equal to the sum of the internal energy changes of each individual gas in the mixture? Answer the same question for a real-gas mixture.

12-42C Answer Probs. 12-39C and 12-40C for entropy.

12-43C When evaluating the entropy change of the components of an ideal-gas mixture, do we have to use the partial pressure of each component or the total pressure of the mixture?

12-44C Suppose we want to determine the enthalpy change of a real-gas mixture undergoing a process. The enthalpy change of each individual gas is determined by using the generalized enthalpy chart, and the enthalpy change of the mixture is determined by summing them. Is this an exact approach? Explain.

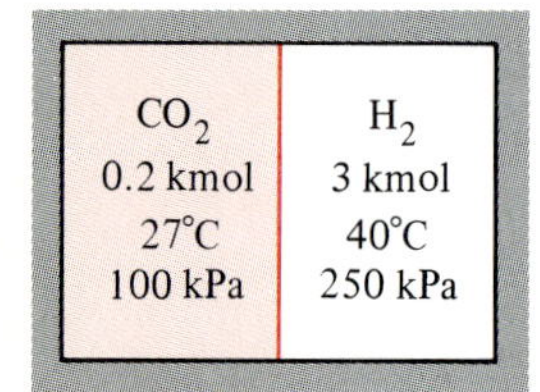

FIGURE P12-45

12-45 An insulated rigid tank is divided into two compartments by a partition. One compartment contains 0.2 kmol of CO_2 at 27°C and 100 kPa, and the other compartment contains 3 kmol of H_2 gas at 40°C and 250 kPa. Now the partition is removed, and the two gases are allowed to mix. Determine (*a*) the mixture temperature and (*b*) the mixture pressure after equilibrium has been established. Assume constant specific heats at room temperature for both gases.

12-46 A 0.6-m^3 rigid tank is divided into two equal compartments by a partition. One compartment contains Ne at 20°C and 150 kPa, and the other compartment contains Ar at 50°C and 300 kPa. Now the partition is removed, and the two gases are allowed to mix. Heat is lost to the surrounding air at 20°C during this process in the amount of 15 kJ. Determine (*a*) the final mixture temperature and (*b*) the final mixture pressure. *Answers:* (*a*) 26.4°C, (*b*) 215.7 kPa

12-46E A 20-ft^3 rigid tank is divided into two equal compartments by a partition. One compartment contains Ne at 70°F and 20 psia, and the other compartment contains Ar at 130°F and 50 psia. Now the partition is removed, and the two gases are allowed to mix. Heat is lost to the surrounding air at 70°F during this process in the amount of 15 Btu. Determine (*a*) the final mixture temperature and (*b*) the final mixture pressure.

12-47 Ethane (C_2H_6) at 20°C and 200 kPa and methane (CH_4) at 45°C and 200 kPa enter an adiabatic mixing chamber. The mass flow rate of ethane is 6 kg/s, which is twice the mass flow rate of methane. Determine (*a*) the mixture temperature and (*b*) the rate of entropy change during this process.

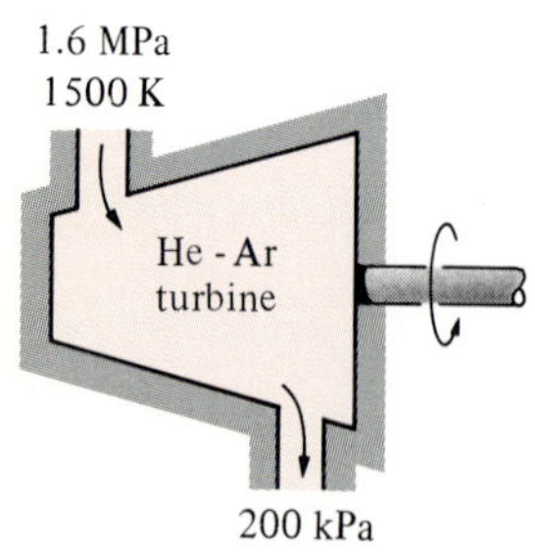

FIGURE P12-48

12-48 An equimolar mixture of helium and argon gases is to be used as the working fluid in a closed-loop gas-turbine cycle. The mixture enters the turbine at 1.6 MPa and 1500 K and expands isentropically to a pressure of 200 kPa. Determine the work output of the turbine per unit mass of the mixture.

12-48E An equimolar mixture of helium and argon gases is to be used as the working fluid in a closed-loop gas-turbine cycle. The mixture enters the turbine at 225 psia and 2800 R and expands isentropically to a pressure of 30 psia. Determine the work output of the turbine per unit mass of the mixture. *Answer:* 352.6 Btu/lbm

12-49 A mixture of 80 percent N_2 and 20 percent CO_2 gases (on a mass basis) enters the nozzle of a turbojet engine at 600 kPa and 1000 K with a low velocity, and it expands to a pressure of 80 kPa. If the adiabatic efficiency of the nozzle is 92 percent, determine (*a*) the exit temperature and (*b*) the exit velocity of the mixture. Assume constant specific heats at room temperature. *Answers:* (*a*) 609.4 K, (*b*) 883.9 m/s

12-49E A mixture of 80 percent N_2 and 20 percent CO_2 gases (on a mass basis) enters the nozzle of a turbojet engine at 90 psia and 1800 R with a low velocity, and it expands to a pressure of 12 psia. If the adiabatic efficiency of the nozzle is 92 percent, determine (*a*) the exit temperature and (*b*) the exit velocity of the mixture. Assume constant specific heats at room temperature.

12-50 A piston-cylinder device contains a mixture of 0.1 kg of H_2 and 0.8 kg of N_2 at 100 kPa and 300 K. Heat is now transferred to the mixture at constant pressure until the volume is doubled. Assuming constant specific heats at the average temperature, determine (*a*) the heat transfer and (*b*) the entropy change of the mixture.

12-51 An insulated tank which contains 2 kg of O_2 at 15°C and 300 kPa is connected to a 4-m^3 uninsulated tank which contains N_2 at 50°C and 500 kPa. The valve connecting the two tanks is opened, and the two gases form a homogeneous mixture at 25°C. Determine (*a*) the final pressure in the tank, (*b*) the heat transfer, and (*c*) the entropy generated during this process. Assume T_0 = 25°C.
Answers: (*a*) 444.6 kPa, (*b*) −374.3 kJ, (*c*) 1.923 kJ/K

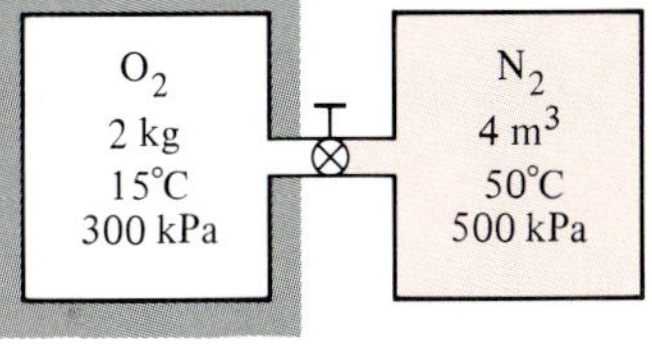

FIGURE P12-51

12-51E An insulated tank which contains 4 lbm of O_2 at 60°F and 50 psia is connected to a 120-ft^3 uninsulated tank which contains N_2 at 140°F and 75 psia. The valve connecting the two tanks is opened, and the two gases form a homogeneous mixture at 70°F. Determine (*a*) the final pressure in the tank, (*b*) the heat transfer, and (*c*) the entropy generated during this process. Assume T_0 = 70°F.

12-52 A steady stream of equimolar N_2-and-CO_2 mixture at 100 kPa and 27°C is to be separated into N_2 and CO_2 gases at 100 kPa and 27°C. Determine the minimum work required per unit mass of mixture to accomplish this separation process. Assume T_0 = 27°C.

12-53 A gas mixture consists of O_2 and N_2. The ratio of the mole numbers of N_2 to O_2 is 3:1. This mixture is heated during a steady-flow process from 180 to 210 K at a constant pressure of 8 MPa. Determine the heat transfer during this process per mole of the mixture, using (*a*) the ideal-gas approximation and (*b*) Kay's rule.

12-54 Determine the total entropy change and irreversibility associated with the process described in Prob. 12-53, using (*a*) the ideal-gas approximation and (*b*) Kay's rule. Assume constant specific heats and T_0 = 25°C.

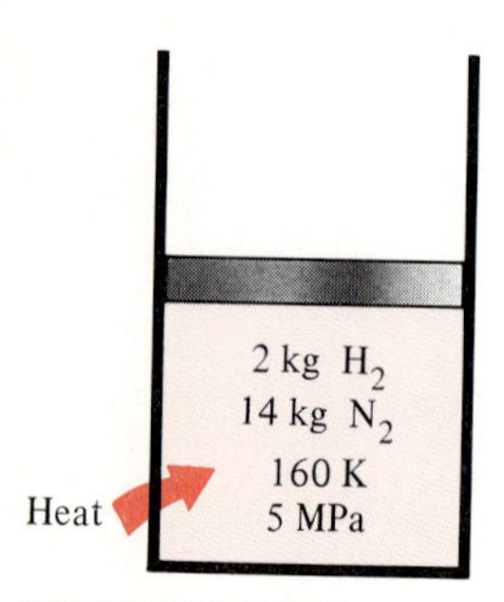

FIGURE P12-55

12-55 A piston-cylinder device contains 2 kg of H_2 and 14 kg of N_2 at 160 K and 5 MPa. Heat is now transferred to the device, and the mixture expands at constant pressure until the temperature rises to 200 K. Determine the heat transfer during this process by treating the mixture (*a*) as an ideal gas and (*b*) as a nonideal gas and using Amagat's law.
Answers: (*a*) 1715 kJ, (*b*) 2030 kJ

12-55E A piston-cylinder device contains 2 lbm of H_2 and 14 lbm of N_2 at 300 R and 800 psia. Heat is now transferred to the device, and the mixture expands at constant pressure until the temperature rises to 360 R. Determine the heat transfer during this process by treating the mixture (*a*) as an ideal gas and (*b*) as a nonideal gas and using Amagat's law.

12-56 Determine the total entropy change and irreversibility associated with the process described in Prob. 12-55 by treating the mixture (*a*) as an ideal gas and (*b*) as a nonideal gas and using Amagat's law. Assume constant specific heats at room temperature and take T_0 = 25°C.

12-56E Determine the total entropy change and irreversibility associated with the process described in Prob. 12-55E by treating the mixture (*a*) as an ideal gas and (*b*) as a nonideal gas and using Amagat's law. Assume constant specific heats at room temperature and take T_0 = 77°F.
Answers: (*a*) 0.491 Btu/R, 264 Btu; (*b*) 0.834 Btu/R, 448 Btu

12-57 A rigid tank contains a mixture of 4 kg of He and 8 kg of O_2 at 170 K and 7 MPa. Heat is now transferred to the tank, and the mixture temperature rises to 220 K. Treating the He as an ideal gas and the O_2 as a nonideal gas, determine (*a*) the final pressure of the mixture and (*b*) the heat transfer.

12-58 Air, which may be considered as a mixture of 79 percent N_2 and 21 percent O_2 by mole numbers, is compressed isothermally at 200 K from 4 to 8 MPa in a steady-flow device. The compression process is internally reversible, and the mass flow rate of air is 1.45 kg/s. Determine the power input to the compressor and the rate of heat transfer by treating the mixture (*a*) as an ideal gas and (*b*) as a nonideal gas and using Amagat's law. *Answers:* (*a*) 63.4 kW, −63.4 kW; (*b*) 54.5 kW, −79.0 kW

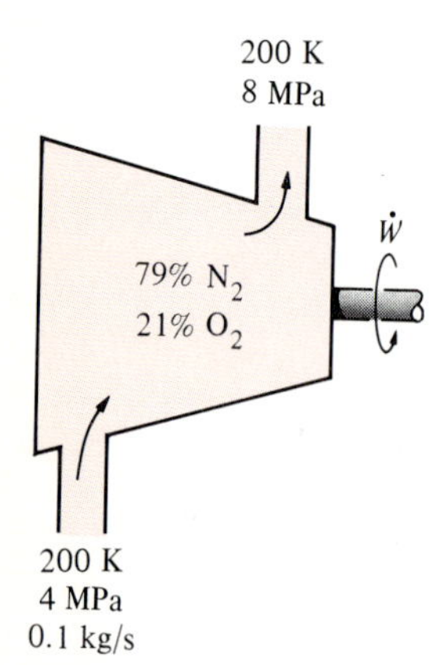

FIGURE P12-58

Gas-Vapor Mixtures and Air Conditioning

CHAPTER 13

At temperatures below the critical temperature, the gas phase of a substance is frequently referred to as a *vapor*. The term *vapor* implies a gaseous state which is close to the saturation region of the substance, raising the possibility of condensation during a process.

In Chap. 12, we discussed mixtures of gases which were usually above their critical temperatures. Therefore, we were not concerned about any of the gases condensing during a process. Not having to deal with two phases greatly simplified the analysis. When we are dealing with a gas-vapor mixture, however, the vapor may condense out of the mixture during a process, forming a two-phase mixture. This may complicate the analysis considerably. Therefore, a gas-vapor mixture needs to be treated differently from an ordinary gas mixture.

Several gas-vapor mixtures are encountered in engineering. In this chapter, we consider the *air–water-vapor mixture,* which is the most commonly used gas-vapor mixture in practice. We also discuss *air conditioning,* which is the primary application area of air–water-vapor mixtures.

13-1 ■ DRY AND ATMOSPHERIC AIR

Air is a mixture of nitrogen, oxygen, and small amounts of some other gases. The air in the atmosphere normally contains some water vapor (or *moisture*) and is referred to as **atmospheric air**. By contrast, air that contains no water vapor is called **dry air**. It is often convenient to treat air as a mixture of water vapor and dry air since the composition of dry air remains relatively constant but the amount of water vapor changes as a result of condensation and evaporation from oceans, lakes, rivers, showers, and even the human body. Although the amount of water vapor in the air is small, it plays a major role in human comfort. Therefore, it is an important consideration in air conditioning applications.

The temperature of air in air conditioning applications ranges from about −10 to about 50°C. In this range, the dry air can be treated as an ideal gas with a constant C_p value of 1.005 kJ/(kg·K) [0.240 Btu/(lbm·R)] with negligible error (under 0.2 percent), as illustrated in Fig. 13-1. Taking 0°C as the reference temperature, the enthalpy and the enthalpy change of dry air can be determined from

DRY AIR	
T, °C	C_p, kJ/(kg · °C)
−10	1.0038
0	1.0041
10	1.0045
20	1.0049
30	1.0054
40	1.0059
50	1.0065

FIGURE 13-1

The C_p of air can be assumed to be constant at 1.005 kJ/(kg·°C) in the temperature range −10 to 50°C with an error under 0.2 percent.

$$h_{\text{dry air}} = C_p T = [1.005 \text{ kJ/(kg·°C)}]T \qquad \text{(kJ/kg)} \qquad (13\text{-}1a)$$

and

$$\Delta h_{\text{dry air}} = C_p \,\Delta T = [1.005 \text{ kJ/(kg·°C)}]\,\Delta T \qquad \text{(kJ/kg)} \qquad (13\text{-}1b)$$

where T is the air temperature in °C and the ΔT is the change in temperature. In air conditioning processes we are concerned with the *changes* in enthalpy Δh, which is independent of the reference point selected.

Would it not be convenient to also treat the water vapor in the air as an ideal gas? You would probably be willing to sacrifice some accuracy for such convenience. It turns out that we can have the convenience without much sacrifice. At 50°C, the saturation pressure of water is 12.3 kPa. As can be seen from Fig. 2-54, at pressures below this value, water vapor can be treated as an ideal gas with negligible error (under 0.2 percent), even when it is a saturated vapor. Therefore, the water vapor in the air behaves as if it existed alone and obeys the ideal-gas relation $Pv = RT$. Then the atmospheric air can be treated as an ideal-gas mixture whose pressure is the sum of the partial pressure of dry air* P_a and that of the water vapor P_v:

$$P = P_a + P_v \qquad \text{(kPa)} \qquad (13\text{-}2)$$

The partial pressure of water vapor is usually referred to as the **vapor pressure**. It is the pressure the water vapor would exert if it existed alone at the temperature and volume of the mixture.

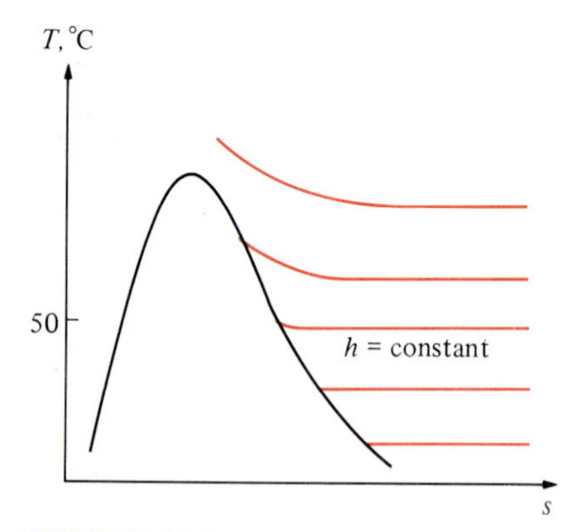

FIGURE 13-2

At temperatures below 50°C, the h = constant lines coincide with the T = constant lines in the superheated vapor region of water.

Since water vapor is an ideal gas, the enthalpy of water vapor is a function of temperature only, that is, $h = h(T)$. This can also be observed from the T-s diagram of water given in Fig. A-9 and Fig. 13-2 where the constant-enthalpy lines coincide with constant-temperature lines at tem-

*Throughout this chapter, the subscript a will denote dry air, and the subscript v will denote water vapor.

	WATER VAPOR		
T, °C	h_g, kJ/kg (Table A-4)	h_g, kJ/kg (Eq. 13-4)	Difference
−10	2482.9	2483.1	−0.2
0	2501.3	2501.3	0.0
10	2519.8	2519.5	0.3
20	2538.1	2537.7	0.4
30	2556.3	2555.9	0.4
40	2574.3	2574.1	0.2
50	2592.1	2592.3	−0.2

FIGURE 13-3
In the temperature range −10 to 50°C, the h_g of water can be determined from Eq. 13-4 with negligible error.

peratures below 50°C. Therefore, *the enthalpy of water vapor in air can be taken to be equal to the enthalpy of the saturated vapor at the same temperature*. That is,

$$h_v(T, \text{low } P) \cong h_g(T) \tag{13-3}$$

The enthalpy of water vapor at 0°C is 2501.3 kJ/kg. The average C_p value of water vapor in the temperature range −20 to 50°C can be taken to be 1.82 kJ/(kg·°C). Then the enthalpy of water vapor can be determined approximately from

$$h_g(T) \cong 2501.3 + 1.82T \qquad \text{(kJ/kg)}, \qquad T \text{ in °C} \tag{13-4}$$

or

$$h_g(T) \cong 1061.5 + 0.435T \qquad \text{(Btu/lbm)}, \qquad T \text{ in °F} \tag{13-5}$$

in the temperature range −10 to 50°C (or 15 to 120°F), with negligible error, as shown in Fig. 13-3.

13-2 ■ SPECIFIC AND RELATIVE HUMIDITY OF AIR

The amount of water vapor in the air can be specified in various ways. Probably the most logical way is to specify directly the mass of water vapor present in a unit mass of dry air. This is called **absolute** or **specific humidity** (also called *humidity ratio*) and is denoted by ω:

$$\omega = \frac{m_v}{m_a} \qquad \text{(kg water vapor/kg dry air)} \tag{13-6}$$

The specific humidity can also be expressed as

$$\omega = \frac{m_v}{m_a} = \frac{P_vV/(R_vT)}{P_aV/(R_aT)} = \frac{P_v/R_v}{P_a/R_a} = 0.622\frac{P_v}{P_a} \tag{13-7}$$

or

$$\omega = \frac{0.622P_v}{P - P_v} \qquad \text{(kg water vapor/kg dry air)} \tag{13-8}$$

where P is the total pressure.

Consider 1 kg of dry air. By definition, dry air contains no water vapor, and thus its specific humidity is zero. Now let us add some water vapor to this dry air. The specific humidity will increase. As more vapor

AIR
25°C, 100 kPa

($P_{sat,\ H_2O\ @\ 25°C}$ = 3.169 kPa)

$P_v = 0$ → dry air
$P_v < 3.169$ kPa → unsaturated air
$P_v = 3.169$ kPa → saturated air

FIGURE 13-4
For saturated air, the vapor pressure is equal to the saturation pressure of water.

AIR
25°C, 1 atm
$m_a = 1$ kg
$m_v = 0.01$ kg
$m_{v,\ max} = 0.02$ kg
Specific humidity: $\omega = 0.01 \dfrac{\text{kgH}_2\text{O}}{\text{kg dry air}}$
Relative humidity: $\phi = 50\%$

FIGURE 13-5
Specific humidity is the actual amount of water vapor in 1 kg of dry air, whereas relative humidity is the ratio of the actual amount of moisture in the air to the maximum amount of moisture air can hold at that temperature.

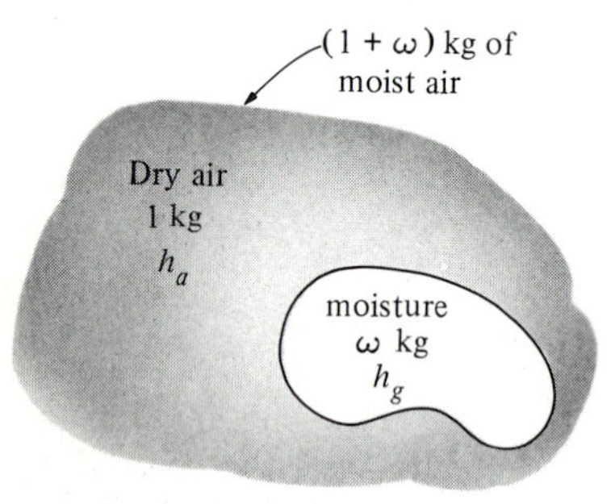

$h = h_a + \omega h_g$, kJ/kg dry air

FIGURE 13-6
The enthalpy of moist (atmospheric) air is expressed per unit mass of dry air, not per unit mass of moist air.

or moisture is added, the specific humidity will keep increasing until the air can hold no more moisture. At this point, the air is said to be saturated with moisture, and it is called **saturated air**. Any moisture introduced into saturated air will condense. The amount of water vapor in saturated air at a specified temperature and pressure can be determined from Eq. 13-8 by replacing P_v by P_g, the saturation pressure of water at that temperature (Fig. 13-4).

The amount of moisture in the air has a definite effect on how comfortable we feel in an environment. However, the comfort level depends more on the amount of moisture the air holds (m_v) relative to the maximum amount of moisture the air can hold at the same temperature (m_g). The ratio of these two quantities is called the **relative humidity** ϕ (Fig. 13-5)

$$\phi = \frac{m_v}{m_g} = \frac{P_v V/(R_v T)}{P_g V/(R_v T)} = \frac{P_v}{P_g} \tag{13-9}$$

where

$$P_g = P_{\text{sat @ }T} \tag{13-10}$$

Combining Eqs. 13-8 and 13-9, we can also express the relative humidity as

$$\phi = \frac{\omega P}{(0.622 + \omega)P_g} \quad \text{and} \quad \omega = \frac{0.622\phi P_g}{P - \phi P_g} \tag{13-11a, b}$$

The relative humidity ranges from 0 for dry air to 1 for saturated air. Note that the amount of moisture air can hold depends on its temperature. Therefore, the relative humidity of air changes with temperature even when its specific humidity remains constant.

Atmospheric air is a mixture of dry air and water vapor, and thus the enthalpy of air is expressed in terms of the enthalpies of the dry air and the water vapor. In most practical applications, the amount of dry air in the air–water-vapor mixture remains constant, but the amount of water vapor changes. Therefore, the enthalpy of atmospheric air is expressed *per unit mass of dry air* instead of per unit mass of the air–water-vapor mixture.

The total enthalpy (an extensive property) of atmospheric air is the sum of the enthalpies of the dry air and the water vapor:

$$H = H_a + H_v = m_a h_a + m_v h_v$$

Dividing by m_a gives

$$h = \frac{H}{m_a} = h_a + \frac{m_v}{m_a} h_v = h_a + \omega h_v$$

or

$$h = h_a + \omega h_g \qquad \text{(kJ/kg dry air)} \tag{13-12}$$

since $h_v \cong h_g$ (Fig. 13-6).

Also note that the ordinary temperature of atmospheric air is frequently referred to as the **dry-bulb temperature** to differentiate it from other forms of temperatures that shall be discussed.

EXAMPLE 13-1

A 5 m × 5 m × 3 m room shown in Fig. 13-7 contains air at 25°C and 100 kPa at a relative humidity of 75 percent. Determine (*a*) the partial pressure of dry air, (*b*) the specific humidity of the air, (*c*) the enthalpy per unit mass of the dry air, and (*d*) the masses of the dry air and water vapor in the room.

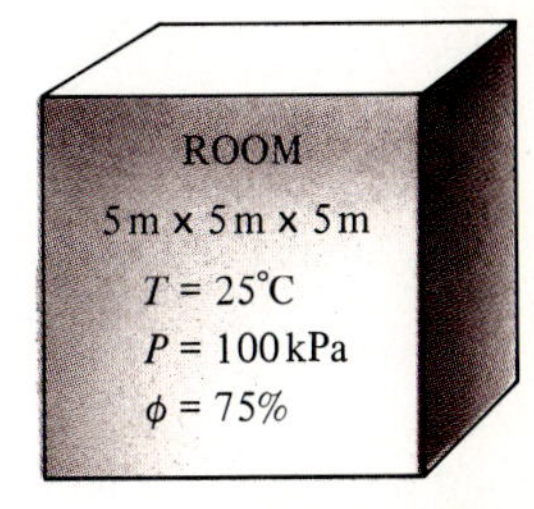

FIGURE 13-7

Solution (*a*) The partial pressure of dry air can be determined from Eq. 13-2:

$$P_a = P - P_v$$

where $P_v = \phi P_g = \phi P_{\text{sat @ 25°C}} = (0.75)(3.169 \text{ kPa}) = 2.38 \text{ kPa}$

Thus, $P_a = (100 - 2.38) \text{ kPa} = 97.62 \text{ kPa}$

(*b*) The specific humidity of air is determined from Eq. 13-8:

$$\omega = \frac{0.622 P_v}{P - P_v} = \frac{(0.622)(2.38 \text{ kPa})}{(100 - 2.38) \text{ kPa}} = 0.0152 \text{ kg H}_2\text{O/kg dry air}$$

(*c*) The enthalpy of air per unit mass of dry air is determined from Eq. 13-12, where h_g is taken from Table A-4:

$$\begin{aligned} h &= h_a + \omega h_v \cong C_p T + \omega h_g \\ &= [1.005 \text{ kJ/(kg·°C)}](25°\text{C}) + (0.0152)(2547.2 \text{ kJ/kg}) \\ &= 63.8 \text{ kJ/kg dry air} \end{aligned}$$

The enthalpy of water vapor (2547.2 kJ/kg) could also be determined from the approximation given by Eq. 13-4:

$$h_{g \text{ @ 25°C}} \cong 2501.3 + 1.82(25) = 2546.8 \text{ kJ/kg}$$

which is very close to the value obtained from Table A-4.

(*d*) Both the dry air and the water vapor fill the entire room completely. Therefore, the volume of each of them is equal to the volume of the room:

$$V_a = V_v = V_{\text{room}} = (5)(5)(3) = 75 \text{ m}^3$$

The masses of the dry air and the water vapor are determined from the ideal-gas relation applied to each gas separately:

$$m_a = \frac{P_a V_a}{R_a T} = \frac{(97.62 \text{ kPa})(75 \text{ m}^3)}{[0.287 \text{ kPa·m}^3\text{/(kg·K)}](298 \text{ K})} = 85.61 \text{ kg}$$

$$m_v = \frac{P_v V_v}{R_v T} = \frac{(2.38 \text{ kPa})(75 \text{ m}^3)}{[0.4619 \text{ kPa·m}^3\text{/(kg·K)}](298 \text{ K})} = 1.3 \text{ kg}$$

The mass of the water vapor in the air could also be determined from Eq. 13-6:

$$m_v = \omega m_a = (0.0152)(85.61 \text{ kg}) = 1.3 \text{ kg}$$

13-3 ■ DEW-POINT TEMPERATURE

If you live in humid weather, you are probably used to waking up most summer mornings and finding the grass wet. You know it did not rain the night before. So what happened? Well, the excess moisture in the air simply condensed on the cool surfaces, forming what we call *dew*. In summer, a considerable amount of water vaporizes during the day. As

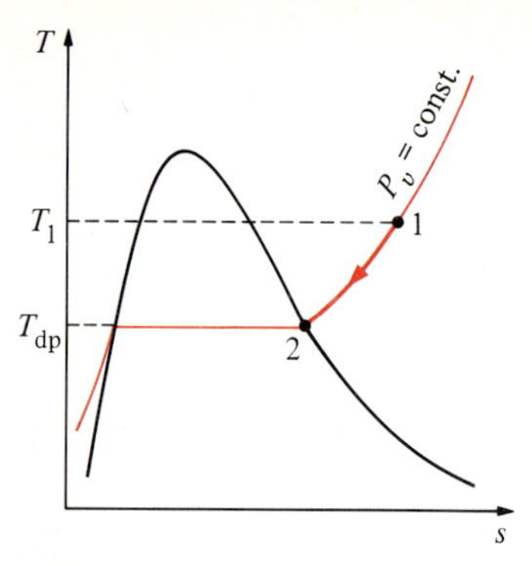

FIGURE 13-8

Constant-pressure cooling of moist air and the dew-point temperature on the *T-s* diagram of water.

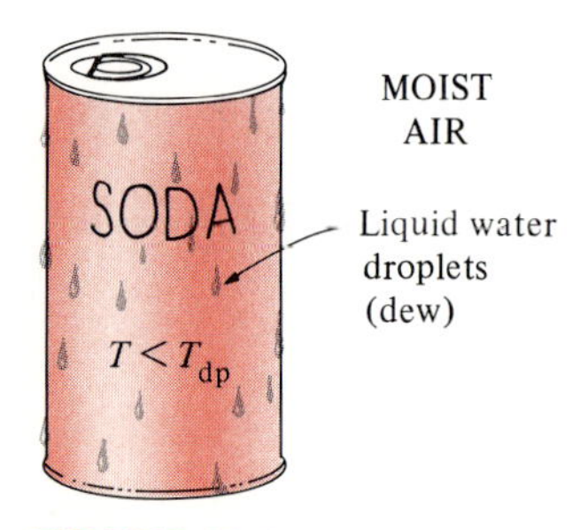

FIGURE 13-9

When the temperature of a cold drink is below the dew-point temperature of the surrounding air, it "sweats."

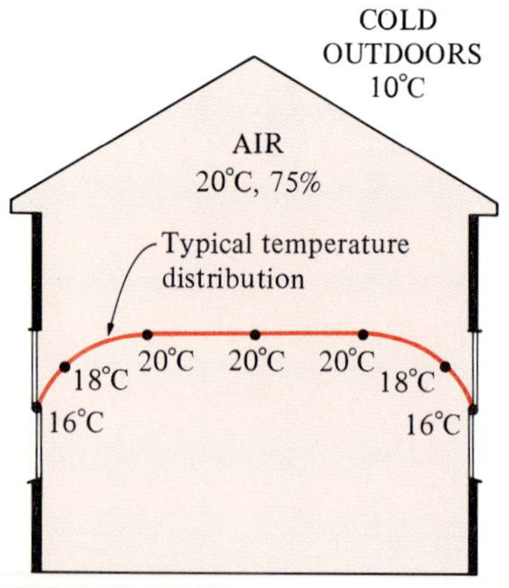

FIGURE 13-10

Schematic for Example 13-2.

the temperature falls during the night, so does the "moisture capacity" of air, which is the maximum amount of moisture air can hold. (What happens to the relative humidity during this process?) After a while, the moisture capacity of the air equals the moisture content of the air. At this point, the air is saturated, and its relative humidity is 100 percent. Any further drop in air temperature results in the condensation of some of the moisture, and this is the beginning of dew formation.

The **dew-point temperature** T_{dp} is defined as *the temperature at which condensation begins if the air is cooled at constant pressure*. In other words, T_{dp} is the saturation temperature of water corresponding to the vapor pressure:

$$T_{dp} = T_{sat @ P_v} \tag{13-13}$$

This is also illustrated in Fig. 13-8. As the air cools at constant pressure, the vapor pressure P_v remains constant. Therefore, the vapor in the air (state 1) undergoes a constant-pressure cooling process until it strikes the saturated vapor line (state 2). The temperature at this point is T_{dp}, and if the temperature drops any further, some vapor condenses out. As a result, the amount of vapor in the air decreases, which results in a decrease in P_v. The air remains saturated during the condensation process and thus follows a path of 100 percent relative humidity (the saturated vapor line). The ordinary temperature and the dew-point temperature of saturated air are identical.

You probably noticed that when you buy a cold canned drink from a vending machine on a hot and humid day, dew forms on the can. The formation of dew on the can indicates that the temperature of the drink is below the dew-point temperature of the surrounding air (Fig. 13-9).

The dew-point temperature of room air can be determined easily by cooling some water in a metal cup by adding small amounts of ice and stirring. The temperature of the outer surface of the cup when dew starts to form on the surface is the dew-point temperature of the air.

EXAMPLE 13-2

In cold weather, condensation frequently occurs on the inner surfaces of the windows due to the lower air temperatures near the window surface. Consider a house, shown in Fig. 13-10, which contains air at 20°C and 75 percent relative humidity. At what window temperature will the moisture in the air start condensing on the inner surfaces of the windows?

Solution The temperature distribution in a house, in general, is not uniform. When the outdoors temperature drops in winter, so does the indoors temperature near the walls and the windows. Therefore, the air near the walls and the windows remains at a lower temperature than at the inner parts of a house even though the total pressure and the vapor pressure remain constant throughout the house. As a result, the air near the walls and the windows will undergo a P_v = constant cooling process until the moisture in the air starts condensing. This will happen when the air reaches its dew point temperature T_{dp}. The dew point is determined from Eq. 13-13:

$$T_{dp} = T_{sat @ P_v}$$

where $P_v = \phi P_{g\,@\,20°C} = (0.75)(2.339 \text{ kPa}) = 1.754 \text{ kPa}$

Thus, $T_{dp} = T_{sat\,@\,1.754\text{ kPa}} = 15.3°C$

Therefore, the inner surface of the window should be maintained above 15.3°C if condensation on the window surfaces is to be avoided.

13-4 ■ ADIABATIC SATURATION AND WET-BULB TEMPERATURES

Relative humidity and specific humidity are frequently used in engineering and atmospheric sciences, but neither is easy to measure directly. Therefore, we must relate them to easily measurable quantities such as temperature and pressure.

One way of determining the relative humidity is to determine the dew-point temperature of air, as discussed in the last section. Knowing the dew-point temperature, we can determine the vapor pressure P_v and thus the relative humidity. This approach is simple, but not quite practical.

Another way of determining the absolute or relative humidity is related to an *adiabatic saturation process,* shown schematically and on a *T-s* diagram in Fig. 13-11. The system consists of a long insulated channel that contains a pool of water. A steady stream of unsaturated air which has a specific humidity of ω_1 (unknown) and a temperature of T_1 is passed through this channel. As the air flows over the water, some water will evaporate and mix with the airstream. The moisture content of air will increase during this process, and its temperature will decrease, since part of the latent heat of vaporization of the water that evaporates will come from the air. If the channel is long enough, the airstream will exit as

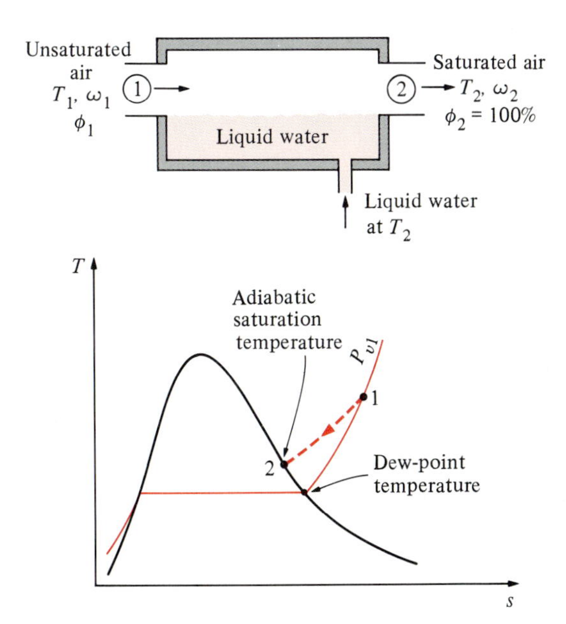

FIGURE 13-11

The adiabatic saturation process and its representation on a *T-s* diagram.

saturated air ($\phi = 100$ percent) at temperature T_2, which is called the **adiabatic saturation temperature**.

If makeup water is supplied to the channel at the rate of evaporation at temperature T_2, the adiabatic saturation process described above can be analyzed as a steady-flow process. The process involves no heat or work interactions, and the kinetic and potential energy changes can be neglected. Then the conservation of mass and conservation of energy relations for this two-inlet, one-exit steady-flow system reduces to the following:

Conservation of mass:

$\dot{m}_{a_1} = \dot{m}_{a_2} = \dot{m}_a$ (The mass flow rate of dry air remains constant)

$\dot{m}_{w_1} + \dot{m}_f = \dot{m}_{w_2}$

or $\dot{m}_a\omega_1 + \dot{m}_f = \dot{m}_a\omega_2$ (The mass flow rate of vapor in the air increases by an amount equal to the rate of evaporation $\dot{m}_f$)

Thus, $\dot{m}_f = \dot{m}_a(\omega_2 - \omega_1)$

Conservation of energy:

$$\Sigma\dot{m}_i h_i = \Sigma\dot{m}_e h_e \qquad (\text{since } \dot{Q} = 0 \text{ and } \dot{W} = 0)$$

$$\dot{m}_{a_1}h_1 + \dot{m}_f h_{f_2} = \dot{m}_{a_2}h_2$$

or $$\dot{m}_a h_1 + \dot{m}_a(\omega_2 - \omega_1)h_{f_2} = \dot{m}_a h_2$$

Dividing by $\dot{m}_a$ gives

$$h_1 + (\omega_2 - \omega_1)h_{f_2} = h_2$$

or $$(C_pT_1 + \omega_1 h_{g_1}) + (\omega_2 - \omega_1)h_{f_2} = (C_pT_2 + \omega_2 h_{g_2})$$

which yields

$$\omega_1 = \frac{C_p(T_2 - T_1) + \omega_2 h_{fg_2}}{h_{g_1} - h_{f_2}} \tag{13-14}$$

where, from Eq. 13-11*b*,

$$\omega_2 = \frac{0.622P_{g_2}}{P_2 - P_{g_2}} \tag{13-15}$$

since $\phi_2 = 100$ percent. Thus we conclude that the specific humidity (and relative humidity) of air can be determined from Eqs. 13-14 and 13-15 by measuring the pressure and temperature of the air at the inlet and the exit of an adiabatic saturator.

If the air entering the channel is already saturated, then the adiabatic saturation temperature T_2 will be identical to the inlet temperature T_1, in which case Eq. 13-14 yields $\omega_1 = \omega_2$. In general, the adiabatic saturation temperature will be between the inlet and dew-point temperatures.

The adiabatic saturation process discussed above provides a means of determining the absolute or relative humidity of air, but it requires a long channel or a spray mechanism to achieve saturation conditions at the exit. A more practical approach is to use a thermometer whose bulb is covered with a cotton wick saturated with water and to blow air over the wick, as shown in Fig. 13-12. The temperature measured in this manner is called the **wet-bulb temperature** T_{wb}, and it is commonly used in air conditioning applications.

The basic principle involved is similar to that in adiabatic saturation. When unsaturated air passes over the wet wick, some of the water in the wick evaporates. As a result, the temperature of the water drops, creating a temperature difference (which is the driving force for heat transfer) between the air and the water. After a while, the heat loss from the water by evaporation equals the heat gain from the air, and the water temperature stabilizes. The thermometer reading at this point is the wet-bulb temperature. The wet-bulb temperature can also be measured by placing the wet wicked thermometer in a holder attached to a handle and rotating the holder rapidly, that is, by moving the thermometer instead of the air. A device that works on this principle is called a *sling psychrometer* and is shown in Fig. 13-13. Usually a dry-bulb thermometer is also mounted on the frame of this device so that both the wet- and dry-bulb temperatures can be read simultaneously.

In general, the adiabatic saturation temperature and the wet-bulb temperature are not the same. But for air–water-vapor mixtures at atmospheric pressure, the wet-bulb temperature happens to be approximately equal to the adiabatic saturation temperature. Therefore, the wet-bulb temperature T_{wb} can be used in Eq. 13-14 in place of T_2 to determine the specific humidity of air.

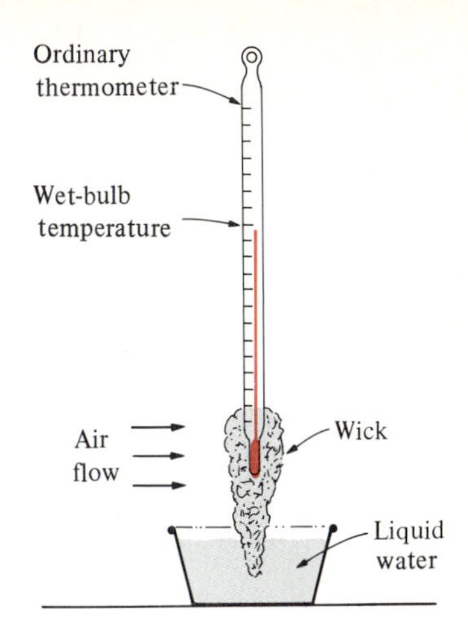

FIGURE 13-12
A simple arrangement to measure the wet-bulb temperature.

EXAMPLE 13-3

The dry- and the wet-bulb temperatures of atmospheric air at 1-atm (101.325-kPa) pressure are measured with a sling psychrometer and determined to be 25 and 15°C, respectively. Determine (*a*) the specific humidity, (*b*) the relative humidity, and (*c*) the enthalpy of the air.

Solution (*a*) The specific humidity ω_1 is determined from Eq. 13-14:

$$\omega_1 = \frac{C_p(T_2 - T_1) + \omega_2 h_{fg_2}}{h_{g_1} - h_{f_2}}$$

where T_2 is the wet-bulb temperature, and ω_2 is determined from Eq. 13-15 to be

$$\omega_2 = \frac{0.622P_{g_2}}{P_2 - P_{g_2}} = \frac{(0.622)(1.705 \text{ kPa})}{(101.325 - 1.705) \text{ kPa}}$$

$$= 0.01065 \text{ kg H}_2\text{O/kg dry air}$$

$$\text{Thus,} \quad \omega_1 = \frac{[1.005 \text{ kJ/(kg·°C)}][(15 - 25)\text{ °C}] + (0.01065)(2465.9 \text{ kJ/kg})}{(2547.2 - 62.99) \text{ kJ/kg}}$$

$$= 0.00653 \text{ kg H}_2\text{O/kg dry air}$$

(*b*) The relative humidity ϕ_1 is determined from Eq. 13-11*a*:

$$\phi_1 = \frac{\omega_1 P_2}{(0.622 + \omega_1)P_{g_1}} = \frac{(0.00653)(101.325 \text{ kPa})}{(0.622 + 0.00653)(3.169 \text{ kPa})} = 0.332 \quad \text{(or 33.2\%)}$$

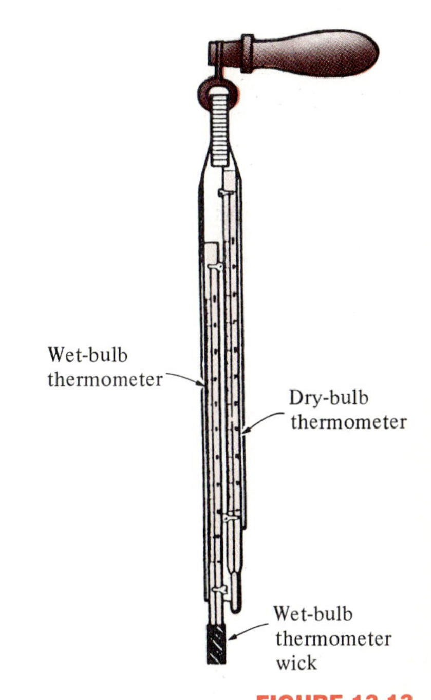

FIGURE 13-13
Sling psychrometer.

(c) The enthalpy of air per unit mass of dry air is determined from Eq. 13-12:

$$h_1 = h_{a_1} + \omega_1 h_{v_1} \cong C_p T_1 + \omega_1 h_{g_1}$$
$$= [1.005 \text{ kJ/(kg·°C)}](25°\text{C}) + (0.00653)(2547.2 \text{ kJ/kg})$$
$$= 41.8 \text{ kJ/kg dry air}$$

13-5 ■ THE PSYCHROMETRIC CHART

The state of the atmospheric air at a specified pressure is completely specified by two independent intensive properties. The rest of the properties can be calculated easily from the relations above. The sizing of a typical air conditioning system involves numerous such calculations, which may eventually get on the nerves of even the most patient engineers. Therefore, there is clear motivation to do these calculations once and to present the data in the form of easily readable charts. Such charts are called **psychrometric charts**, and they are used extensively in air conditioning applications. A psychrometric chart for a pressure of 1 atm (101.325 kPa or 14.696 psia) is given in Fig. A-33 in SI units and in Fig. A-33E in English units. Psychrometric charts at other pressures (for use at considerably higher elevations than sea level) are also available.

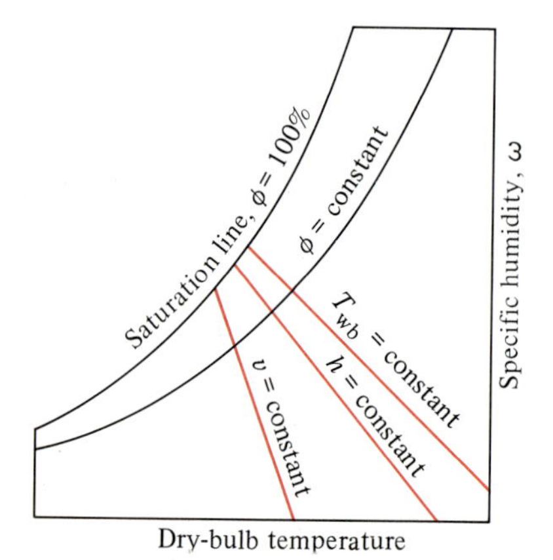

FIGURE 13-14
Schematic of a psychrometric chart.

The basic features of the psychrometric chart are illustrated in Fig. 13-14. The dry-bulb temperatures are shown on the horizontal axis, and the specific humidity is shown on the vertical axis. (Some charts also show the vapor pressure on the vertical axis since at a fixed total pressure P there is a one-to-one correspondence between the specific humidity ω and the vapor pressure P_v, as can be seen from Eq. 13-8.) On the left end of the chart, there is a curve (called the *saturation line*) instead of a straight line. All the saturated air states are located on this curve. Therefore, it is also the curve of 100 percent relative humidity. Other constant relative-humidity curves have the same general shape.

Lines of constant wet-bulb temperature have a downhill appearance to the right. Lines of constant specific volume (in m³/kg dry air) look similar, except they are steeper. Lines of constant enthalpy (in kJ/kg dry air) lie very nearly parallel to the lines of constant wet-bulb temperature. Therefore, the constant-wet-bulb-temperature lines are used as constant-enthalpy lines in some charts.

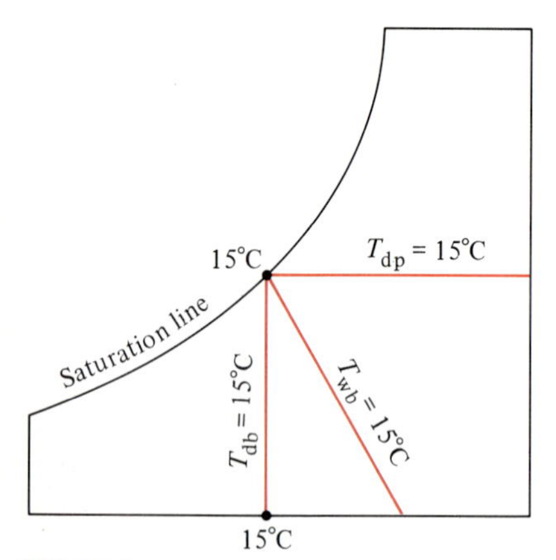

FIGURE 13-15
For saturated air, the dry-bulb, wet-bulb, and dew-point temperatures are identical.

For saturated air, the dry-bulb, wet-bulb, and dew-point temperatures are identical (Fig. 13-15). Therefore, the dew-point temperature of atmospheric air at any point on the chart can be determined by drawing a horizontal line (a line of ω = constant or P_v = constant) from the point to the saturation curve. The temperature value at the intersection point is the dew-point temperature.

The psychrometric chart also serves as a valuable aid in visualizing the air conditioning processes. An ordinary heating or cooling process, for example, will appear as a horizontal line on this chart if no humidification or dehumidification is involved (that is, ω = constant). Any deviation from a horizontal line indicates that moisture is added or removed from the air during the process.

EXAMPLE 13-4

Consider a room which contains air at 1 atm, 35°C, and 40 percent relative humidity. Using the psychrometric chart, determine (*a*) the specific humidity, (*b*) the enthalpy (in kJ/kg dry air), (*c*) the wet-bulb temperature, (*d*) the dew-point temperature, and (*e*) the specific volume of the air (in m^3/kg dry air).

Solution At a given total pressure, the state of atmospheric air is completely specified by two independent properties such as the dry-bulb temperature and the relative humidity. Other properties are determined by directly reading their values at the specified state.

(*a*) The specific humidity is determined by drawing a horizontal line from the specified state to the right until it intersects with the ω axis, as shown in Fig. 13-16. At the intersection point we read

$$\omega = 0.0142 \text{ kg } H_2O/\text{kg dry air}$$

(*b*) The enthalpy of air per unit mass of dry air is determined by drawing a line parallel to the h = constant lines from the specified state until it intersects the enthalpy scale. At the intersection point we read

$$h = 71.5 \text{ kJ/kg dry air}$$

(*c*) The wet-bulb temperature is determined by drawing a line parallel to the T_{wb} = constant lines from the specified state until it intersects the saturation line. At the intersection point we read

$$T_{wb} = 24°C$$

(*d*) The dew-point temperature is determined by drawing a horizontal line from the specified state to the left until it intersects the saturation line. At the intersection point we read

$$T_{dp} = 19.4°C$$

(*e*) The specific volume per unit mass of dry air is determined by noting the distances between the specified state and the v = constant lines on both sides of the point. The specific volume is determined by visual interpolation to be

$$v = 0.893 \text{ m}^3/\text{kg dry air}$$

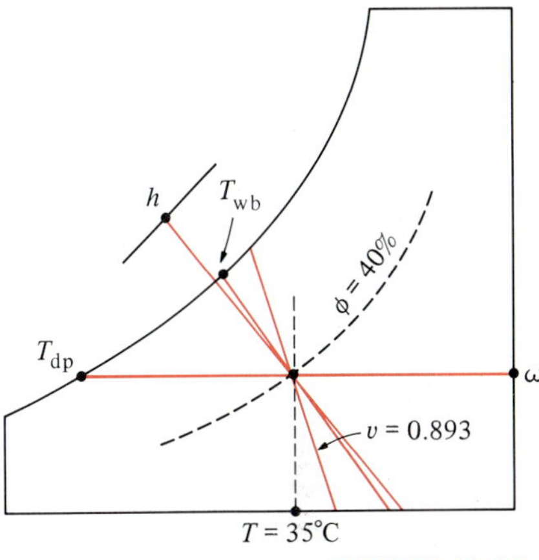

FIGURE 13-16
Schematic for Example 13-4.

13-6 ■ HUMAN COMFORT AND AIR CONDITIONING

Human beings have an inherent weakness—they want to feel comfortable. They want to live in an environment which is neither hot nor cold, neither very humid nor very dry. But comfort does not come easily since the desires of the human body and the weather usually are not quite compatible. Achieving comfort requires a constant struggle against the factors that cause discomfort, such as high or low temperatures and high or low humidity. As engineers, it is our duty to help people feel comfortable. (Besides, it keeps us employed.)

It did not take long for people to realize that they could not change the weather in an area. All they can do is change it in a confined space such as a house or a workplace (Fig. 13-17). In the past, this was partially accomplished by fire and simple indoor heating systems. Today, modern

FIGURE 13-17
We cannot change the weather, but we can change the climate in a confined space by air conditioning.

air conditioning systems can heat, cool, humidify, dehumidify, clean, and even deodorize the air—in other words, *condition* the air to peoples' desires. Air conditioning systems are designed to *satisfy* the needs of the human body; therefore, it is essential that we understand the thermodynamic aspects of it.

The human body can be viewed as a heat engine whose energy input is food. As with any other heat engine, the human body generates waste heat which must be rejected to the environment if the body is to continue operating. The rate of heat generation depends on the level of the activity. For an average adult male, it is about 87 W when sleeping, 115 W when resting or doing office work, 230 W when bowling, and 440 W when doing heavy physical work. The corresponding numbers for an adult female are about 15 percent less. (This difference is due to the body size, not the body temperature. The deep-body temperature of a healthy person is maintained constant at 37°C.) A body will feel comfortable in environments in which it can dissipate this waste heat comfortably (Fig. 13-18).

FIGURE 13-18

A body feels comfortable when it can freely dissipate its waste heat, and no more.

Heat transfer is proportional to the temperature difference. Therefore in cold environments, a body will lose more heat than it normally generates, which results in a feeling of discomfort. The body tries to minimize the energy deficit by cutting down the blood circulation near the skin (causing a pale look). This lowers the skin temperature and thus the heat transfer rate. We can also reduce the heat loss from the body either by putting barriers (additional clothes, blankets, etc.) in the path of heat or by increasing the rate of heat generation within the body by exercising. Or we can just cuddle up and put our hands between our legs to reduce the surface area through which heat flows.

In hot environments, we have the opposite problem—we do not seem to be dissipating enough heat from our bodies, and we feel as if we are going to burst (Fig. 13-19). We dress lightly to make it easier for heat to get away from our bodies, and we reduce the level of activity to minimize the rate of waste heat generation in the body. We also turn on the fan to continuously replace the warmer air layer that forms around our bodies as a result of body heat by the cooler air in other parts of the room. The body also helps out by perspiring or sweating more. As this sweat evaporates, it absorbs heat from the body and cools it. In general, most of the heat dissipation from the body takes place through evaporation. Perspiration is not much help, however, if the relative humidity of the environment is close to 100 percent. Prolonged sweating without any fluid intake will cause dehydration and reduced sweating, which may lead to a rise in body temperature and a heat stroke.

FIGURE 13-19

Hot and humid environments make it very difficult for the body to dissipate its waste heat.

Another important factor that affects human comfort is heat transfer by radiation between the body and the surrounding surfaces such as walls and windows. The sun's rays travel through space by radiation. You warm up in front of a fire even if the air between you and the fire is quite cold. Likewise, in a warm room you will feel chilly if the ceiling or the wall surfaces are at a considerably lower temperature. This is due to direct heat transfer between your body and the surrounding surfaces by radiation. Radiant heaters are commonly used for heating hard-to-heat places such as car repair shops.

The comfort of the human body primarily depends on three factors: the (dry-bulb) temperature, relative humidity, and air motion (Fig. 13-20). The temperature of the environment is the single most important index of comfort. Most people feel comfortable when the environment temperature is between 22 and 27°C (72 and 80°F). The relative humidity also has a considerable effect on comfort since it affects the amount of heat a body can dissipate through evaporation. Relative humidity is a measure of air's ability to absorb more moisture. High relative humidity slows down heat rejection by evaporation, and low relative humidity speeds it up. Most people prefer a relative humidity of 40 to 60 percent.

Air motion also plays an important role in human comfort. It removes the warm, moist air that builds up around the body and replaces it with fresh air. Therefore, air motion improves heat rejection by both convection and evaporation. Air motion should be strong enough to remove heat and moisture from the vicinity of the body, but gentle enough to be unnoticed. Most people feel comfortable at an airspeed of about 15 m/min. Other factors that affect comfort are air cleanliness, odor, noise, and radiation effect.

FIGURE 13-20
A comfortable environment.

13-7 ■ AIR CONDITIONING PROCESSES

Maintaining a living space or an industrial facility at the desired temperature and humidity requires some processes called air conditioning processes. These processes include *simple heating* (raising the temperature), *simple cooling* (lowering the temperature), *humidifying* (adding moisture), and *dehumidifying* (removing moisture). Sometimes two or more of these processes are needed to bring the air to a desired temperature and humidity level.

Various air conditioning processes are illustrated on the psychrometric chart in Fig. 13-21. Notice that simple heating and cooling processes appear as horizontal lines on this chart since the moisture content of the air remains constant (ω = constant) during these processes. Air is commonly heated and humidified in winter and cooled and dehumidified in summer. Notice how these processes appear on the psychrometric chart.

Most air conditioning processes can be modeled as steady-flow processes, and therefore they can be analyzed by applying the steady-flow conservation of mass (for both dry air and water) and conservation of energy principles:

Dry air mass: $$\Sigma \dot{m}_{a,i} = \Sigma \dot{m}_{a,e} \tag{13-16}$$

Water mass: $$\Sigma \dot{m}_{w,i} = \Sigma \dot{m}_{w,e} \quad \text{or} \quad \Sigma \dot{m}_{a,i}\omega_i = \Sigma \dot{m}_{a,e}\omega_e \tag{13-17}$$

Energy: $$\dot{Q} - \dot{W} = \Sigma \dot{m}_e h_e - \Sigma \dot{m}_i h_i \tag{13-18}$$

Here subscripts i and e denote inlet and exit states, respectively. The

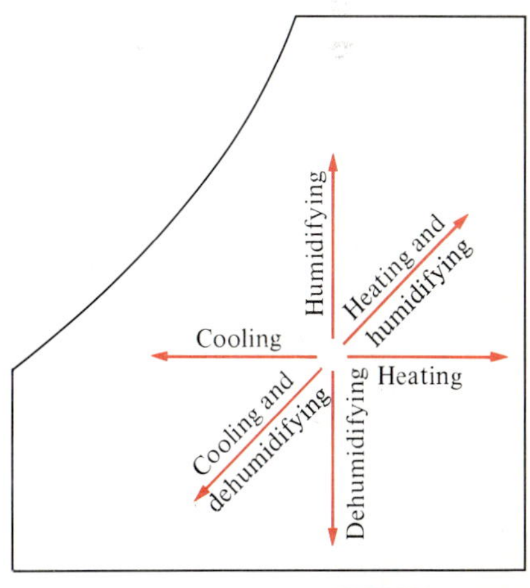

FIGURE 13-21
Various air conditioning processes.

changes in kinetic and potential energies are assumed to be negligible. The work term usually consists of the fan work, which is very small relative to the other terms in the energy equation. Next we examine some commonly encountered processes in air conditioning.

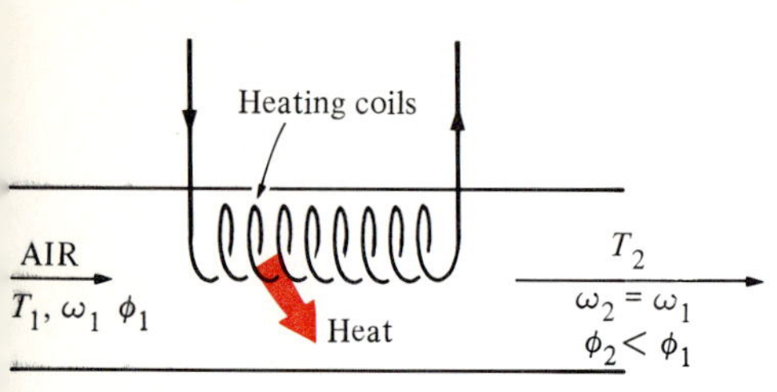

FIGURE 13-22

During simple heating, specific humidity remains constant, but relative humidity decreases.

1 Simple Heating and Cooling (ω = constant)

Many residential heating systems consist of a stove, a heat pump, or an electric resistance heater. The air in these systems is heated by circulating it through a duct that contains the tubing for the hot gases or the electric resistance wires, as shown in Fig. 13-22. The amount of moisture in the air remains constant during this process since no moisture is added to or removed from the air. That is, the specific humidity of the air remains constant (ω = constant) during a heating (or cooling) process with no humidification or dehumidification. Such a heating process will proceed in the direction of increasing dry-bulb temperature following a line of constant specific humidity on the psychrometric chart, which appears as a horizontal line.

Notice that the relative humidity of air decreases during a heating process even if the specific humidity ω remains constant. This is because the relative humidity is the ratio of the moisture content to the moisture capacity of air at the same temperature, and moisture capacity increases with temperature. Therefore, the relative humidity of heated air may be well below comfortable levels, causing dry skin, respiratory difficulties, and an increase in static electricity.

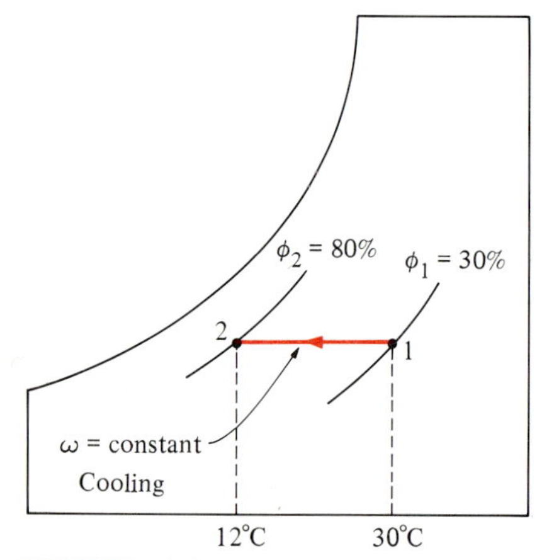

FIGURE 13-23

During simple cooling, specific humidity remains constant, but relative humidity increases.

A cooling process at constant specific humidity is similar to the heating process discussed above, except the dry-bulb temperature decreases and the relative humidity increases during such a process, as shown in Fig. 13-23. Cooling can be accomplished by passing the air over some coils through which a refrigerant or cool water flows.

The conservation of mass equations for a heating or cooling process that involves no humidification or dehumidification reduce to $\dot{m}_{a_1} = \dot{m}_{a_2} = \dot{m}_a$ for dry air and $\omega_1 = \omega_2$ for water. Neglecting any fan work that may be present, the conservation of energy equation in this case reduces to $\dot{Q} = \dot{m}_a(h_2 - h_1)$ or $q = h_2 - h_1$, where h_1 and h_2 are enthalpies per unit mass of dry air at the inlet and the exit of the heating or cooling section, respectively.

2 Heating with Humidification

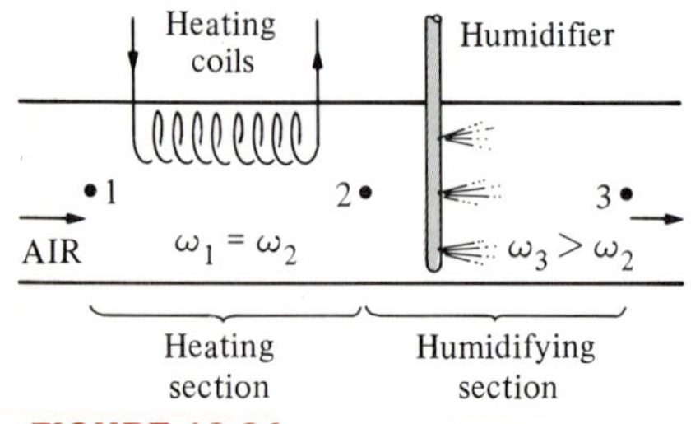

FIGURE 13-24

Heating and humidification.

Problems associated with low relative humidity resulting from simple heating can be eliminated by humidifying the heated air. This is accomplished by passing the air first through a heating section (process 1-2) and then through a humidifying section (process 2-3), as shown in Fig. 13-24.

The location of state 3 depends on how the humidification is accomplished. If steam is introduced in the humidification section, this will result in humidification with additional heating ($T_3 > T_2$). If humidification is accomplished by spraying water into the airstream instead, part of the latent heat of vaporization will come from the air, which will result in the cooling of the heated airstream ($T_3 > T_2$). Air should be heated to a higher temperature in the heating section in this case to make up for the cooling during the humidification process.

EXAMPLE 13-5

An air conditioning system is to take in outdoor air at 10°C and 30 percent relative humidity at a steady rate of 45 m^3/min and to condition it to 25°C and 60 percent relative humidity. The outdoor air is first heated to 22°C in the heating section and then humidified by the injection of hot steam in the humidifying section. Assuming the entire process takes place at a pressure of 100 kPa, determine (*a*) the rate of heat supply in the heating section and (*b*) the mass flow rate of the steam required in the humidifying section.

Solution The schematic of the system and the psychrometric chart of the process are shown in Fig. 13-25. The mass flow rate of the dry air remains constant during the entire process. The amount of moisture in the air remains constant as it flows through the heating section ($\omega_1 = \omega_2$), but increases in the humidifying section ($\omega_3 > \omega_2$).

(*a*) The conservation of mass and the conservation of energy equations for the heating section reduce to these:

Dry air mass: $\Sigma \dot{m}_{a,i} = \Sigma \dot{m}_{a,e} \longrightarrow \dot{m}_{a_1} = \dot{m}_{a_2} = \dot{m}_a$

Water mass: $\Sigma \dot{m}_{w,i} = \Sigma \dot{m}_{w,e} \longrightarrow \dot{m}_{a_1}\omega_1 = \dot{m}_{a_2}\omega_2$ or $\omega_1 = \omega_2$

Energy: $\dot{Q} - \dot{W}^{\nearrow 0} = \Sigma \dot{m}_e h_e - \Sigma \dot{m}_i h_i \longrightarrow \dot{Q} = \dot{m}_a(h_2 - h_1)$

The psychrometric chart offers great convenience in determining the properties

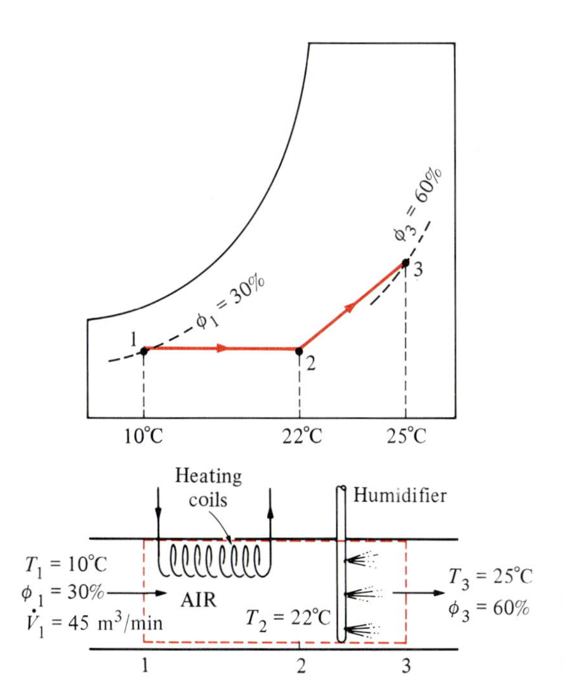

FIGURE 13-25

Schematic and psychrometric chart for Example 13-5.

of moist air. However, its use is limited to a specified pressure only, which is 1 atm (101.325 kPa) for the one given in the Appendix. At pressures other than 1 atm, either other charts for that pressure or the relations developed earlier should be used. In our case, the choice is clear:

$$P_{v_1} = \phi_1 P_{g_1} = \phi_1 P_{\text{sat @ 10°C}} = (0.3)(1.2276\text{ kPa}) = 0.368\text{ kPa}$$
$$P_{a_1} = P_1 - P_{v_1} = (100 - 0.368)\text{ kPa} = 99.632\text{ kPa}$$

$$v_1 = \frac{R_a T_1}{P_{a_1}} = \frac{[0.287\text{ kPa·m}^3/(\text{kg·K})](283\text{ K})}{99.632\text{ kPa}} = 0.815\text{ m}^3/\text{kg dry air}$$

$$\dot{m}_a = \frac{\dot{V}_1}{v_1} = \frac{45\text{ m}^3/\text{min}}{0.815\text{ m}^3/\text{kg}} = 55.2\text{ kg/min}$$

$$\omega_1 = \frac{0.622P_{v_1}}{P_1 - P_{v_1}} = \frac{0.622(0.368\text{ kPa})}{(100 - 0.368)\text{ kPa}} = 0.0023\text{ kg H}_2\text{O/kg dry air}$$

$$h_1 = C_pT_1 + \omega_1 h_{g_1} = [1.005\text{ kJ/(kg·°C)}](10°\text{C}) + (0.0023)(2519.8\text{ kJ/kg})$$
$$= 15.8\text{ kJ/kg dry air}$$
$$h_2 = C_pT_2 + \omega_2 h_{g_2} = [1.005\text{ kJ/(kg·°C)}](22°\text{C}) + (0.0023)(2541.7\text{ kJ/kg})$$
$$= 28.0\text{ kJ/kg dry air}$$

since $\omega_2 = \omega_1$. Then the rate of heat transfer to the air in the heating section becomes

$$\dot{Q} = \dot{m}_a(h_2 - h_1) = (55.2\text{ kg/min})[(28.0 - 15.8)\text{ kJ/kg}]$$
$$= 673.4\text{ kJ/min}$$

(*b*) The conservation of mass equation for water in the humidifying section can be expressed as

$$\dot{m}_{a_2}\omega_2 + \dot{m}_w = \dot{m}_{a_3}\omega_3$$

or

$$\dot{m}_w = \dot{m}_a(\omega_3 - \omega_2)$$

where

$$\omega_3 = \frac{0.622\phi_3 P_{g_3}}{P_3 - \phi_3 P_{g_3}} = \frac{0.622(0.60)(3.169\text{ kPa})}{[100 - (0.60)(3.169)]\text{ kPa}}$$
$$= 0.01206\text{ kg H}_2\text{O/kg dry air}$$

Thus,

$$\dot{m}_w = (55.2\text{ kg/min})(0.01206 - 0.0023)$$
$$= 0.539\text{ kg/min}$$

3 Cooling with Dehumidification

The specific humidity of air remains constant during a simple cooling process, but its relative humidity increases. If the relative humidity reaches undesirably high levels, it may be necessary to remove some moisture from the air, i.e., to dehumidify it. This requires cooling the air below its dew-point temperature.

The cooling process with dehumidifying is illustrated schematically and on the psychrometric chart in Fig. 13-26 in conjunction with Example 13-6. Hot, moist air enters the cooling section at state 1. As it passes through the cooling coils, its temperature decreases and its relative humidity increases at constant specific humidity. If the cooling section is sufficiently long, air will reach its dew point (state 2, saturated air). Further

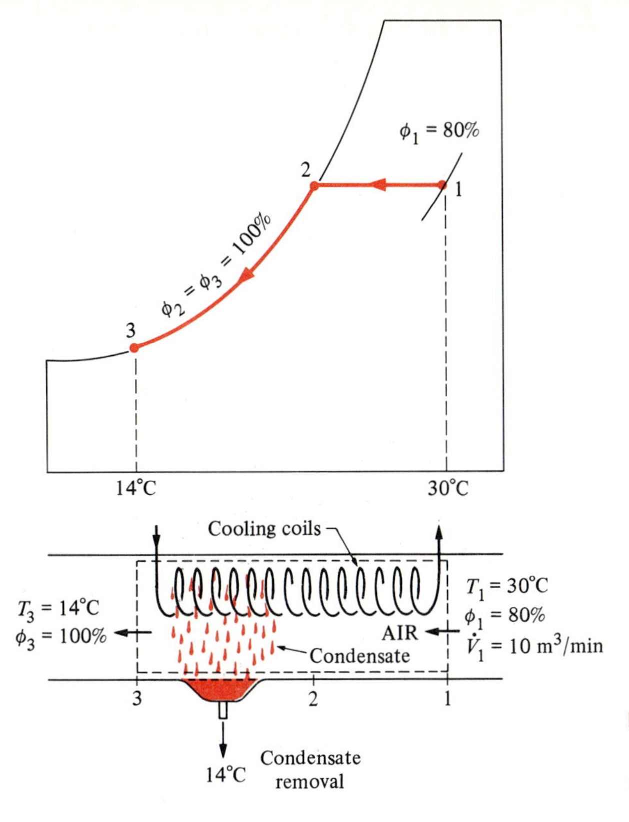

FIGURE 13-26

Schematic and psychrometric chart for Example 13-6.

cooling of air results in the condensation of part of the moisture in the air. Air remains saturated during the entire condensation process, which follows a line of 100 percent relative humidity until the final state (state 3) is reached. The water vapor that condenses out of the air during this process is removed from the cooling section through a separate channel. The condensate is usually assumed to leave the cooling section at T_3.

The cool, saturated air at state 3 is usually routed directly to the room, where it mixes with the room air. In some cases, however, the air at state 3 may be at the right specific humidity but at a very low temperature. In such cases, the air is passed through a heating section where its temperature is raised to a more comfortable level before it is routed to the room.

EXAMPLE 13-6

Air enters a window air conditioner at 1 atm, 30°C, and 80 percent relative humidity at a rate of 10 m^3/min, and it leaves as saturated air at 14°C. Part of the moisture in the air which condenses during the process is also removed at 14°C. Determine the rates of heat and moisture removal from the air.

Solution The schematic of the system and the psychrometric chart of the process are shown in Fig. 13-26. The mass flow rate of dry air remains constant during the entire process, but the amount of moisture in the air decreases due to dehumidification ($\omega_2 < \omega_1$). The conservation of mass and the conservation of energy equations for the combined cooling and dehumidification section reduce to these:

Dry air mass: $\Sigma \dot{m}_{a,i} = \Sigma \dot{m}_{a,e} \longrightarrow \dot{m}_{a_1} = \dot{m}_{a_2} = \dot{m}_a$

Water mass: $\Sigma \dot{m}_{w,i} = \Sigma \dot{m}_{w,e} \longrightarrow \dot{m}_{a_1}\omega_1 = \dot{m}_{a_2}\omega_2 + \dot{m}_w$

or $\dot{m}_w = \dot{m}_a(\omega_1 - \omega_2)$

Energy: $\dot{Q} - \dot{W}^{\nearrow 0} = \Sigma \dot{m}_e h_e - \Sigma \dot{m}_i h_i \longrightarrow \dot{Q} = \dot{m}_{a_2}h_2 - \dot{m}_{a_1}h_1 + \dot{m}_w h_w$

$$= \dot{m}_a(h_2 - h_1) + \dot{m}_w h_w$$

The inlet and the exit states of the air are completely specified, and the total pressure is 1 atm. Therefore, we can determine the properties of the air at both states from the psychrometric chart:

$$h_1 = 85.4 \text{ kJ/kg dry air}$$
$$\omega_1 = 0.0216 \text{ kg } H_2O\text{/kg dry air}$$
$$v_1 = 0.889 \text{ m}^3\text{/kg dry air}$$

and
$$h_2 = 39.3 \text{ kJ/kg dry air}$$
$$\omega_2 = 0.0100 \text{ kg } H_2O\text{/kg dry air}$$

Also,
$$h_w \cong h_{f\,@\,14°C} = 58.8 \text{ kJ/kg} \qquad \text{(Table A-4)}$$

Then,
$$\dot{m}_{a_1} = \frac{\dot{V}_1}{v_1} = \frac{10 \text{ m}^3\text{/min}}{0.889 \text{ m}^3\text{/kg dry air}} = 11.3 \text{ kg/min}$$

$$\dot{m}_w = (11.3 \text{ kg/min})(0.0216 - 0.0100) = 0.131 \text{ kg/min}$$

$$\dot{Q} = (11.3 \text{ kg/min})[(39.3 - 85.4) \text{ kJ/kg}] + (0.131 \text{ kg/min})(58.8 \text{ kJ/kg})$$
$$= -513 \text{ kJ/min}$$

Therefore, this air conditioning unit removes moisture and heat from the air at rates of 0.131 kg/min and 513 kJ/min, respectively.

4 Evaporative Cooling

Conventional cooling systems operate on a refrigeration cycle, and they can be used in any part of the world. But they have a high initial and operating cost. In desert (hot and dry) climates, we can avoid the high cost of cooling by using *evaporative coolers,* also known as *swamp coolers.*

Evaporative cooling is based on a simple principle: As water evaporates, the latent heat of vaporization is absorbed from the water body and the surrounding air. As a result, both the water and the air are cooled during the process. This approach has been used for thousands of years to cool water. A porous jug or pitcher filled with water is left in an open, shaded area. A small amount of water leaks out through the porous holes, and the pitcher "sweats." In a dry environment, this water evaporates and cools the remaining water in the pitcher (Fig. 13-27).

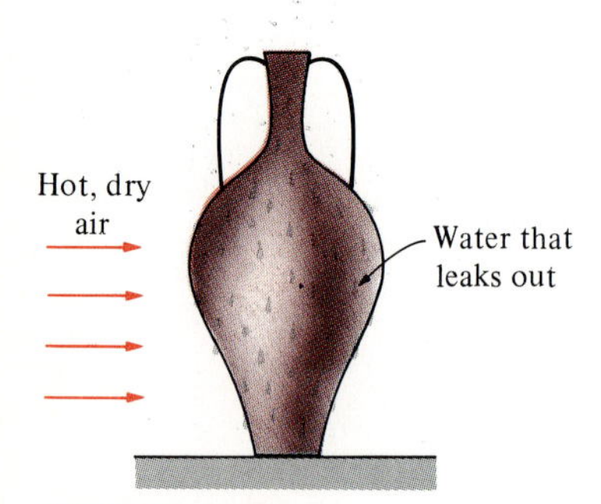

FIGURE 13-27
The water in a porous jug left in an open, breezy area cools as a result of evaporative cooling.

You have probably noticed that on a hot, dry day the air feels a lot cooler when the yard is watered. This is because water absorbs heat from

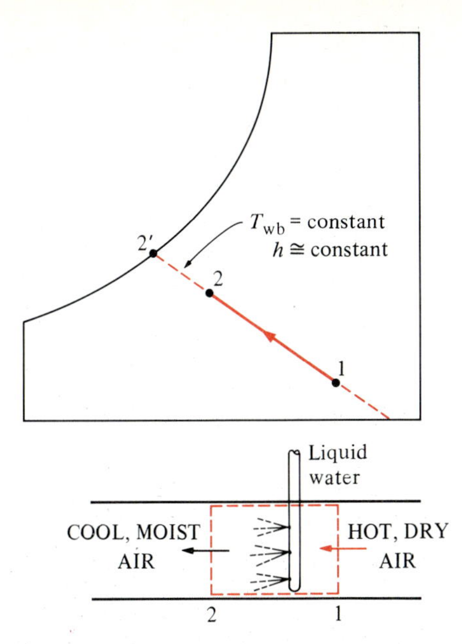

FIGURE 13-28
Evaporative cooling.

the air as it evaporates. An evaporative cooler works on the same principle. The evaporative cooling process is shown schematically and on a psychrometric chart in Fig. 13-28. Hot, dry air at state 1 enters the evaporative cooler, where it is sprayed with liquid water. Part of the water evaporates during this process by absorbing heat from the airstream. As a result, the temperature of the airstream decreases and its humidity increases (state 2). In the limiting case, the air will leave the cooler saturated at state 2′. This is the lowest temperature that can be achieved by this process.

The evaporative cooling process is essentially identical to the adiabatic saturation process since the heat transfer between the airstream and the surroundings is usually negligible. Therefore, the evaporative cooling process follows a line of constant wet-bulb temperature on the psychrometric chart. (Note that this will not exactly be the case if the liquid water is supplied at a temperature different from the exit temperature of the airstream.) Since the constant-wet-bulb-temperature lines almost coincide with the constant-enthalpy lines, the enthalpy of the airstream can also be assumed to remain constant. That is,

$$T_{wb} \cong \text{constant} \tag{13-19}$$

and

$$h \cong \text{constant} \tag{13-20}$$

during an evaporative cooling process. This is a reasonably accurate approximation, and it is commonly used in air conditioning calculations.

EXAMPLE 13-7

Air enters an evaporative cooler at 14.7 psi, 95°F, and 20 percent relative humidity, and it exits at 80 percent relative humidity. Determine (*a*) the exit temperature of the air and (*b*) the lowest temperature to which the air can be cooled by this evaporative cooler.

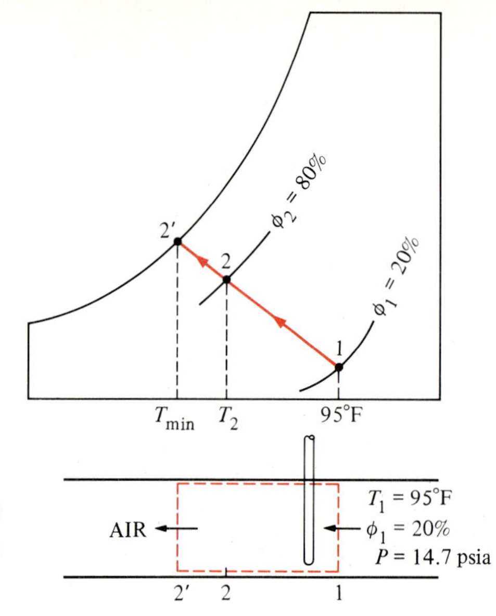

FIGURE 13-29
Schematic and psychrometric chart for Example 13-7.

The schematic of the evaporative cooler and the psychrometric chart of the process are shown in Fig. 13-29.

(*a*) If we assume the liquid water is supplied at a temperature not much different from the exit temperature of the airstream, the evaporative cooling process follows a line of constant wet-bulb temperature on the psychrometric chart. That is,

$$T_{wb} \cong \text{constant}$$

The wet-bulb temperature at 95°F and 20 percent relative humidity is determined from the psychrometric chart to be 66.0°F. The interseciton point of the T_{wb} = 66.0°F and the ϕ = 80 percent lines is the exit state of the air. The temperature at this point is the exit temperature of the air, and it is determined from the psychrometric chart to be

$$T_2 = 70.4°\text{F}$$

(*b*) In the limiting case, the air will leave the evaporative cooler saturated (ϕ = 100 percent), and the exit state of the air in this case will be the state where the T_{wb} = 66.0°F line intersects the saturation line. For saturated air, the dry- and the wet-bulb temperatures are identical. Therefore, the lowest temperature to which the air can be cooled is the wet-bulb temperature, which is

$$T_{min} = T_2' = 66.0°\text{F}$$

5 Adiabatic Mixing of Airstreams

Many air conditioning applications require the mixing of two airstreams. This is particularly true for large buildings, most production and process plants, and hospitals, which require that the conditioned air be mixed with a certain fraction of fresh outside air before it is routed into the living

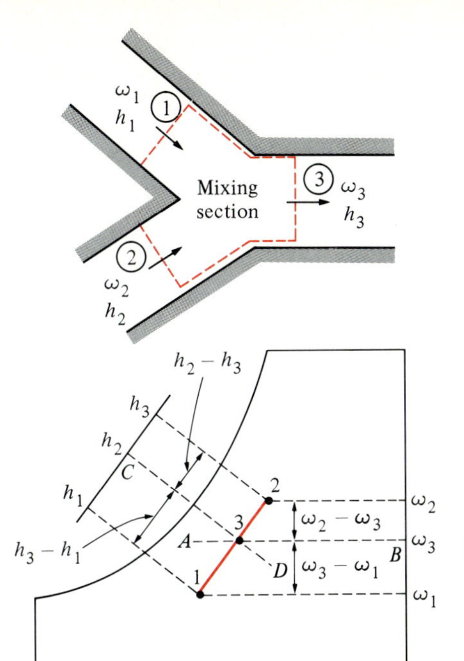

FIGURE 13-30

When two airstreams at states 1 and 2 are mixed adiabatically, the state of the mixture lies on the straight line connecting the two states.

space. The mixing is accomplished by simply merging the two airstreams, as shown in Fig. 13-30.

The heat transfer with the surroundings is usually small, and thus the mixing processes can be assumed to be adiabatic. Mixing processes normally involve no work interactions, and the changes in kinetic and potential energies, if any, are negligible. Then the conservation of mass and conservation of energy equations for the adiabatic mixing of two airstreams reduce to

Mass of dry air: $$\dot{m}_{a_1} + \dot{m}_{a_2} = \dot{m}_{a_3} \tag{13-21}$$

Mass of water vapor: $$\omega_1\dot{m}_{a_1} + \omega_2\dot{m}_{a_2} = \omega_3\dot{m}_{a_3} \tag{13-22}$$

Energy: $$\dot{m}_{a_1}h_1 + \dot{m}_{a_2}h_2 = \dot{m}_{a_3}h_3 \tag{13-23}$$

Eliminating $\dot{m}_{a_3}$ from the relations above, we obtain

$$\frac{\dot{m}_{a_1}}{\dot{m}_{a_2}} = \frac{\omega_2 - \omega_3}{\omega_3 - \omega_1} = \frac{h_2 - h_3}{h_3 - h_1} \tag{13-24}$$

This equation has an instructive geometric interpretation on the psychrometric chart. It shows that the ratio of $\omega_2 - \omega_3$ to $\omega_3 - \omega_1$ is equal to the ratio of $\dot{m}_{a_1}$ to $\dot{m}_{a_2}$. The states that satisfy this condition are indicated by the dashed line *AB*. The ratio of $h_2 - h_3$ to $h_3 - h_1$ is also equal to the ratio of $\dot{m}_{a_1}$ to $\dot{m}_{a_2}$, and the states that satisfy this condition are indicated by the dashed line *CD*. The only state that satisfies both conditions is the intersection point of these two dashed lines, which is located on the straight line connecting states 1 and 2. Thus we conclude that *when two airstreams at two different states (states 1 and 2) are mixed adiabatically, the state of the mixture (state 3) will lie on the straight line connecting states 1 and 2 on the psychrometric chart, and the ratio of the distances 2-3 and 3-1 is equal to the ratio of mass flow rates* m_{a_1} *and* m_{a_2}.

The concave nature of the saturation curve and the conclusion above lead to an interesting possibility. When states 1 and 2 are located close to the saturation curve, the straight line connecting the two states will cross the saturation curve, and state 3 may lie to the left of the saturation curve. In this case, some water will inevitably condense during the mixing process.

EXAMPLE 13-8

Saturated air leaving the cooling section of an air conditioning system at 14°C at a rate of 50 m³/min is mixed adiabatically with the outside air at 32°C and 60 percent relative humidity at a rate of 20 m³/min. Assuming that the mixing process occurs at a pressure of 1 atm, determine the specific humidity, the relative humidity, the dry-bulb temperature, and the volume flow rate of the mixture.

Solution The schematic of the system and the psychrometric chart of the process are shown in Fig. 13-31. The properties of each inlet stream are determined from the psychrometric chart to be

$$h_1 = 39.4 \text{ kJ/kg dry air}$$
$$\omega_1 = 0.010 \text{ kg H}_2\text{O/kg dry air}$$
$$v_1 = 0.826 \text{ m}^3\text{/kg dry air}$$

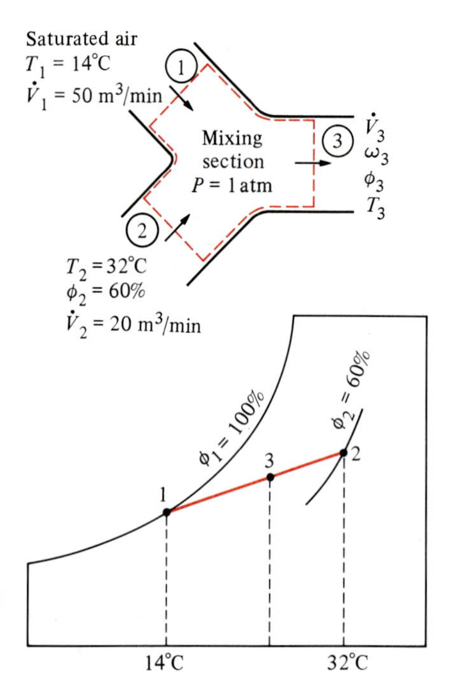

FIGURE 13-31
Schematic and psychrometric chart for Example 13-8.

and
$$h_2 = 79.0 \text{ kJ/kg dry air}$$
$$\omega_2 = 0.0182 \text{ kg } H_2O\text{/kg dry air}$$
$$v_2 = 0.889 \text{ m}^3\text{/kg dry air}$$

Then the mass flow rates of dry air in each stream are

$$\dot{m}_{a_1} = \frac{\dot{V}_1}{v_1} = \frac{50 \text{ m}^3\text{/min}}{0.826 \text{ m}^3\text{/kg dry air}} = 60.5 \text{ kg/min}$$

$$\dot{m}_{a_2} = \frac{\dot{V}_2}{v_2} = \frac{20 \text{ m}^3\text{/min}}{0.889 \text{ m}^3\text{/kg dry air}} = 22.5 \text{ kg/min}$$

From the conservation of mass,

$$\dot{m}_{a_3} = \dot{m}_{a_1} + \dot{m}_{a_2} = (60.5 + 22.5) \text{ kg/min} = 83 \text{ kg/min}$$

The specific humidity and the enthalpy of the mixture can be determined from Eqs. 13-24, which are obtained by combining the conservation of mass and conservation of energy equations for the adiabatic mixing of two streams:

$$\frac{\dot{m}_{a_1}}{\dot{m}_{a_2}} = \frac{\omega_2 - \omega_3}{\omega_3 - \omega_1} = \frac{h_2 - h_3}{h_3 - h_1}$$

$$\frac{60.5}{22.5} = \frac{0.0182 - \omega_3}{\omega_3 - 0.010} = \frac{79.0 - h_3}{h_3 - 39.4}$$

which yield

$$\omega_3 = 0.0122 \text{ kg } H_2O\text{/kg dry air}$$
$$h_3 = 50.1 \text{ kJ/kg dry air}$$

These two properties fix the state of the mixture. Other properties of the mixture are determined from the psychrometric chart:

$$T_3 = 19.0°\text{C}$$
$$\phi_3 = 89\%$$
$$v_3 = 0.844 \text{ m}^3\text{/kg dry air}$$

Finally, the volume flow rate of the mixture is determined from

$$\dot{V}_3 = \dot{m}_{a_3} v_3 = (83 \text{ kg/min})(0.844 \text{ m}^3\text{/kg}) = 70.1 \text{ m}^3\text{/min}$$

Notice that the volume flow rate of the mixture is approximately equal to the sum of the volume flow rates of the two incoming streams. This is typical in air conditioning applications.

6 Wet Cooling Towers

Power plants, large air conditioning systems, and some industries generate large quantities of waste heat which is often rejected to cooling water from nearby lakes or rivers. In some cases, however, the water supply is limited or thermal pollution is a serious concern. In such cases, the waste heat must be rejected to the atmosphere, with cooling water recirculating and serving as a transport medium for heat between the source and the sink (the atmosphere). One way of achieving this is through the use of wet cooling towers.

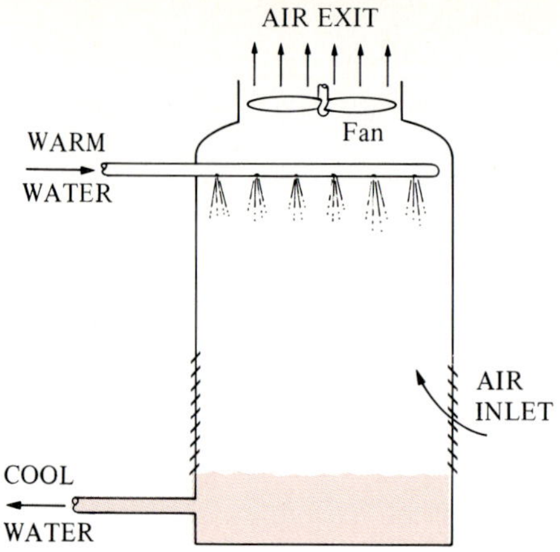

FIGURE 13-32
An induced-draft counterflow cooling tower.

A **wet cooling tower** is essentially a semienclosed evaporative cooler. An induced-draft counterflow wet cooling tower is shown schematically in Fig. 13-32. Air is drawn into the tower from the bottom and leaves at the top. Warm water from the condenser is pumped to the top of the tower and is sprayed into this airstream. The purpose of spraying is to expose a large surfacc arca of water to the air. As the water droplets fall under the influence of gravity, a small fraction of water (usually a few percent) evaporates and cools the remaining water. The temperature and the moisture content of the air increase during this process. The cooled water collects at the bottom of the tower and is pumped back to the condenser to pick up additional waste heat. Makeup water must be added to the cycle to replace the water lost by evaporation and air draft. To minimize water carried away by the air, drift eliminators are installed in the wet cooling towers above the spray section.

The air circulation in the cooling tower described above is provided by a fan, and therefore it is classified as a forced-draft cooling tower. Another popular type of cooling tower is the **natural-draft cooling tower**, which looks like a large chimney and works as an ordinary chimney. The air in the tower has a high water vapor content, and thus it is lighter than the outside air. Consequently, the light air in the tower rises, and the heavier outside air fills the vacant space, creating an airflow from the bottom of the tower to the top. The flow rate of air is controlled by the conditions of the atmospheric air. Natural-draft cooling towers do not require any external power to induce the air, but they cost a lot more to build than forced-draft cooling towers. The natural-draft cooling towers are hyperbolic in profile, as shown in Fig. 13-33, and some are over 100 m high. The hyperbolic profile is for greater structural strength, not for any thermodynamic reason.

The idea of a cooling tower started with the **spray pond**, where the warm water is sprayed into the air and is cooled by the air as it falls into the pond, as shown in Fig. 13-34. Some spray ponds are still in use today.

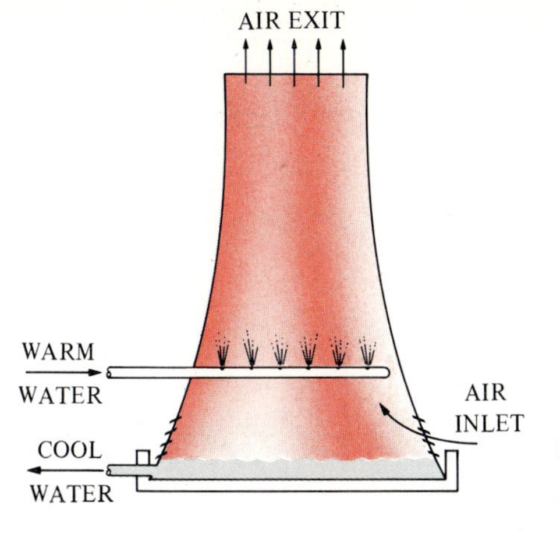

FIGURE 13-33
A natural-draft cooling tower.

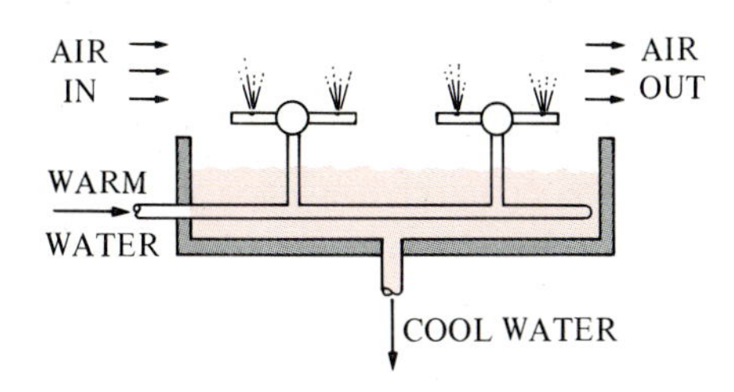

FIGURE 13-34
A spray pond.

But they require 25 to 50 times the area of a cooling tower, water loss due to air drift is high, and they are unprotected against dust and dirt.

We could also dump the waste heat into a still **cooling pond**, which is basically a large lake open to the atmosphere. But heat transfer from the pond surface to the atmosphere is very slow, and we would need about 20 times the area of a spray pond in this case to achieve the same cooling.

EXAMPLE 13-9

Cooling water leaves the condenser of a power plant and enters a wet cooling tower at 35°C at a rate of 100 kg/s. The water is cooled to 22°C in the cooling tower by air which enters the tower at 1 atm, 20°C, 60 percent and leaves saturated at 30°C. Neglecting the power input to the fan, determine (*a*) the volume flow rate of air into the cooling tower and (*b*) the mass flow rate of the required makeup water.

Solution The cooling tower is shown schematically in Fig. 13-35. The process that the air undergoes is illustrated on the psychrometric chart. The mass flow rate of dry air through the tower remains constant ($\dot{m}_{a_1} = \dot{m}_{a_2} = \dot{m}_a$), but the mass flow rate of liquid water decreases by an amount equal to the amount of water that vaporizes in the tower during the cooling process. The water lost through evaporation must be made up later in the cycle to maintain steady operation.

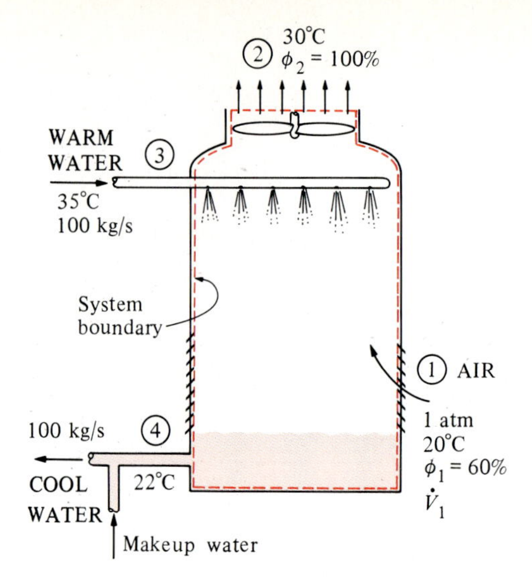

FIGURE 13-35
Schematic for Example 13-9.

(*a*) Applying the conservation of mass and the conservation of energy equations yields

Dry air mass: $\quad \Sigma\dot{m}_{a,i} = \Sigma\dot{m}_{a,e} \longrightarrow \dot{m}_{a_1} = \dot{m}_{a_2} = \dot{m}_a$

Water mass: $\quad \Sigma\dot{m}_{w,i} = \Sigma\dot{m}_{w,e} \longrightarrow \dot{m}_3 + \dot{m}_{a_1}\omega_1 = \dot{m}_4 + \dot{m}_{a_2}\omega_2$

$$\dot{m}_3 - \dot{m}_4 = \dot{m}_a(\omega_2 - \omega_1) = \dot{m}_{\text{makeup}}$$

Energy: $\quad \dot{Q}^{\nearrow 0} - \dot{W}^{\nearrow 0} = \Sigma\dot{m}_e h_e - \Sigma\dot{m}_i h_i$

$$0 = \dot{m}_{a_2}h_2 + \dot{m}_4 h_4 - \dot{m}_{a_1}h_1 - \dot{m}_3 h_3$$

$$0 = \dot{m}_a(h_2 - h_1) + (\dot{m}_3 - \dot{m}_{\text{makeup}})h_4 - \dot{m}_3 h_3$$

Solving for $\dot{m}_a$ gives

$$\dot{m}_a = \frac{\dot{m}_3(h_3 - h_4)}{(h_2 - h_1) - (\omega_2 - \omega_1)h_4}$$

From the psychrometric chart,

$$h_1 = 42.2 \text{ kJ/kg dry air}$$
$$\omega_1 = 0.0087 \text{ kg H}_2\text{O/kg dry air}$$
$$v_1 = 0.842 \text{ m}^3\text{/kg dry air}$$

and

$$h_2 = 100.0 \text{ kJ/kg dry air}$$
$$\omega_2 = 0.0273 \text{ kg H}_2\text{O/kg dry air}$$

From Table A-4,

$$h_3 \cong h_{f\,@\,35^\circ\text{C}} = 146.68 \text{ kJ/kg H}_2\text{O}$$
$$h_4 \cong h_{f\,@\,22^\circ\text{C}} = 92.33 \text{ kJ/kg H}_2\text{O}$$

Substituting,

$$\dot{m}_a = \frac{(100 \text{ kg/s})[(146.68 - 92.33) \text{ kJ/kg}]}{[(100.0 - 42.2) \text{ kJ/kg}] - [(0.0273 - 0.0087)(92.33) \text{ kJ/kg}]} = 96.9 \text{ kg/s}$$

Then the volume flow rate of air into the cooling tower becomes

$$\dot{V}_1 = \dot{m}_a v_1 = (96.9 \text{ kg/s})(0.842 \text{ m}^3\text{/kg}) = \mathbf{81.6 \text{ m}^3\text{/s}}$$

(*b*) The mass flow rate of the required makeup water is determined from

$$\dot{m}_{makeup} = \dot{m}_a(\omega_2 - \omega_1) = (96.9 \text{ kg/s})(0.0273 - 0.0087) = 1.80 \text{ kg/s}$$

Therefore, over 98 percent of the cooling water is saved and recirculated in this case.

13-8 ■ SUMMARY

In this chapter we discussed the air–water-vapor mixture, which is the most commonly encountered gas-vapor mixture in practice. The air in the atmosphere normally contains some water vapor, and it is referred to as *atmospheric air*. By contrast, air that contains no water vapor is called *dry air*. In the temperature range encountered in air conditioning applications, both the dry air and the water vapor can be treated as ideal gases. The enthalpy change of dry air during a process can be determined from

$$\Delta h_{\text{dry air}} = C_p\,\Delta T = [1.005 \text{ kJ/(kg·K)}]\,\Delta T \qquad \text{(kJ/kg)} \qquad (13\text{-}1b)$$

The atmospheric air can be treated as an ideal-gas mixture whose pressure is the sum of the partial pressure of dry air P_a and that of the water vapor P_v:

$$P = P_a + P_v \qquad \text{(kPa)} \qquad (13\text{-}2)$$

The enthalpy of water vapor in the air can be taken to be equal to the enthalpy of the saturated vapor at the same temperature:

$$h_v(T, \text{low } P) \cong h_g(T) \qquad (13\text{-}3)$$
$$\cong 2501.3 + 1.82T \qquad \text{(kJ/kg)}, \quad T \text{ in °C} \qquad (13\text{-}4)$$
$$\cong 1061.5 + 0.435T \qquad \text{(Btu/lbm)}, \quad T \text{ in °F} \qquad (13\text{-}5)$$

in the temperature range −10 to 50°C (15 to 120°F).

The mass of water vapor present in 1 unit mass of dry air is called the *specific* or *absolute humidity* ω:

$$\omega = \frac{m_v}{m_a} = \frac{0.622P_v}{P - P_v} \qquad \text{(kg } H_2O\text{/kg dry air)} \qquad (13\text{-}6, 13\text{-}8)$$

where P is the total pressure of air and P_v is the vapor pressure. There is a limit on the amount of vapor the air can hold at a given temperature. The air which is holding as much moisture as it can is called *saturated air*. The ratio of the amount of moisture air holds (m_v) to the maximum amount of moisture air can hold at the same temperature (m_g) is called the *relative humidity* ϕ

$$\phi = \frac{m_v}{m_g} = \frac{P_vV/(R_vT)}{P_gV/(R_vT)} = \frac{P_v}{P_g} \qquad (13\text{-}9)$$

where $$P_g = P_{\text{sat @ } T} \tag{13-10}$$

By combining Eqs. 13-8 and 13-9, the relative and specific humidities can also be expressed as

$$\phi = \frac{\omega P}{(0.622 + \omega)P_g} \quad \text{and} \quad \omega = \frac{0.622\phi P_g}{P - \phi P_g} \tag{13-11a, b}$$

Relative humidity ranges from 0 for dry air to 1 for saturated air.

The enthalpy of atmospheric air is expressed *per unit mass of dry air,* instead of per unit mass of the air–water-vapor mixture, as

$$h = h_a + \omega h_g \qquad \text{(kJ/kg dry air)} \tag{13-12}$$

The ordinary temperature of atmospheric air is frequently referred to as the *dry-bulb temperature* to differentiate it from other forms of temperatures. The temperature at which condensation begins if the air is cooled at constant pressure is called the *dew-point temperature* T_{dp}:

$$T_{\text{dp}} = T_{\text{sat @ } P_v} \tag{13-13}$$

Relative humidity and specific humidity of air cannot be measured directly. But they can be determined by measuring the *adiabatic saturation temperature* of air, which is the temperature the air attains after flowing over water in a long channel until it is saturated:

$$\omega_1 = \frac{C_p(T_2 - T_1) + \omega_2 h_{fg_2}}{h_{g_1} - h_{f_2}} \tag{13-14}$$

where $$\omega_2 = \frac{0.622 P_{g_2}}{P_2 - P_{g_2}} \tag{13-15}$$

and T_2 is the adiabatic saturation temperature. A more practical approach in air conditioning applications is to use a thermometer whose bulb is covered with a cotton wick saturated with water and to blow air over the wick. The temperature measured in this manner is called the *wet-bulb temperature* T_{wb}, and it is used in place of the adiabatic saturation temperature. The properties of atmospheric air at a specified total pressure are presented in the form of easily readable charts, called *psychrometric charts*. The lines of constant enthalpy and the lines of constant wet-bulb temperature are very nearly parallel on these charts.

The needs of the human body and the conditions of the environment are not quite compatible. Therefore, it often becomes necessary to change the conditions of a living space to make it more comfortable. Maintaining a living space or an industrial facility at the desired temperature and humidity requires some air conditioning processes. These processes include simple heating (raising the temperature), simple cooling (lowering the temperature), humidifying (adding moisture), and dehumidifying (removing moisture). Sometimes two or more of these processes are needed to bring the air to a desired temperature and humidity level.

Most air conditioning processes can be modeled as steady-flow processes, and therefore they can be analyzed by applying the steady-flow conservation of mass (for both dry air and water) and conservation of energy principles:

Dry air mass: $$\Sigma \dot{m}_{a,i} = \Sigma \dot{m}_{a,e} \tag{13-16}$$

Water mass: $$\Sigma \dot{m}_{w,i} = \Sigma \dot{m}_{w,e} \tag{13-17}$$

Energy: $$\dot{Q} - \dot{W} = \Sigma \dot{m}_e h_e - \Sigma \dot{m}_i h_i \tag{13-18}$$

where subscripts i and e denote inlet and exit states, respectively. The changes in kinetic and potential energies are assumed to be negligible.

During a simple heating or cooling process, the specific humidity remains constant, but the temperature and the relative humidity change. Sometimes air is humidified after it is heated, and some cooling processes include dehumidification. In dry climates, the air can be cooled via evaporative cooling by passing it through a section where it is sprayed with water. In locations with limited water supply, large amounts of waste heat can be rejected to the atmosphere with minimum water loss through the use of cooling towers.

REFERENCES AND SUGGESTED READING

1 *ASHRAE 1981 Handbook of Fundamentals,* American Society of Heating, Refrigerating, and Air Conditioning Engineers, Atlanta, GA, 1981.

2 W. Z. Black and J. G. Hartley, *Thermodynamics,* Harper & Row, New York, 1985.

3 Elonka, "Cooling Towers," *Power,* March 1963.

4 J. B. Jones and G. A. Hawkins, *Engineering Thermodynamics,* 2d ed., Wiley, New York, 1986.

5 D. C. Look, Jr., and H. J. Sauer, Jr., *Engineering Thermodynamics,* PWS Engineering, Boston, 1986.

6 W. F. Stoecker and J. W. Jones, *Refrigeration and Air Conditioning,* 2d ed., McGraw-Hill, New York, 1982.

7 K. Wark, *Thermodynamics,* 5th ed., McGraw-Hill, New York, 1988.

8 L. D. Winiarski and B. A. Tichenor, "Model of Natural Draft Cooling Tower Performance," *Journal of the Sanitary Engineering Division, Proceedings of the American Society of Civil Engineers,* August 1970.

PROBLEMS*

Dry and Atmospheric Air; Specific and Relative Humidity

13-1C What is the difference between dry air and atmospheric air?

*Students are encouraged to answer *all* the concept "C" questions.

13-2C Can the water vapor in air be treated as an ideal gas? Explain.

13-3C What is vapor pressure?

13-4C How would you compare the enthalpy of water vapor at 20°C and 2 kPa with the enthalpy of water vapor at 20°C and 5 kPa?

13-5C What is the difference between the specific humidity and the relative humidity?

13-6C How will (*a*) the specific humidity and (*b*) the relative humidity of the air contained in a well-sealed room change if it is heated?

13-7C How will (*a*) the specific humidity and (*b*) the relative humidity of the air contained in a well-sealed room change if the air is cooled?

13-8C Is it possible to obtain saturated air from unsaturated air without adding any moisture? Explain.

13-9C Is the relative humidity of saturated air necessarily 100 percent?

13-10C Moist air is passed through a cooling section where it is cooled and dehumidified. How do (*a*) the specific humidity and (*b*) the relative humidity of air change during this process?

13-11 A room contains air at 20°C and 98 kPa at a relative humidity of 85 percent. Determine (*a*) the partial pressure of dry air, (*b*) the specific humidity of the air, and (*c*) the enthalpy per unit mass of dry air.

13-11E A room contains air at 70°F and 14.6 psia at a relative humidity of 85 percent. Determine (*a*) the partial pressure of dry air, (*b*) the specific humidity of the air, and (*c*) the enthalpy per unit mass of dry air. *Answers:* (*a*) 14.291 psia, (*b*) 0.0134 lbm H_2O/lbm dry air, (*c*) 31.43 Btu/lbm dry air

13-12 Determine the mass of dry air and the water vapor contained in a 70-m^3 room at 98 kPa, 23°C, and 50 percent relative humidity. *Answers:* 79.6 kg, 0.727 kg

13-13 A tank contains 7 kg of dry air and 0.1 kg of water vapor at 30°C and 100 kPa total pressure. Determine (*a*) the specific humidity, (*b*) the relative humidity, and (*c*) the volume of the tank.

13-13E A tank contains 15 lbm of dry air and 0.2 lbm of water vapor at 90°F and 14.5 psia total pressure. Determine (*a*) the specific humidity, (*b*) the relative humidity, and (*c*) the volume of the tank.

13-14 A 5-m^3 tank contains saturated air at 25°C and 97 kPa. Determine (*a*) the mass of the dry air, (*b*) the specific humidity, and (*c*) the enthalpy of the air per unit mass of the dry air. *Answers:* (*a*) 5.49 kg, (*b*) 0.0210 kg H_2O/kg dry air, (*c*) 78.62 kJ/kg dry air

Dew-Point, Adiabatic Saturation, and Wet-Bulb Temperatures

13-15C What is the dew-point temperature?

13-16C Andy and Wendy both wear glasses. On a cold winter day, Andy comes from the cold outside and enters the warm house while Wendy leaves the house and goes outside. Whose glasses are more likely to be fogged? Explain.

13-17C In summer, the outer surface of a glass filled with iced water frequently "sweats." How can you explain this sweating?

13-18C In some climates, cleaning the ice off the car windows is a common chore on winter mornings. Explain how ice forms on the car windows during some nights even when there is no rain or snow.

13-19C When are the dry-bulb and dew-point temperatures identical?

13-20C When are the adiabatic saturation and wet-bulb temperatures equivalent for atmospheric air?

13-21 A house contains air at 25°C and 65 percent relative humidity. Will any moisture condense on the inner surfaces of the windows when the temperature of the window drops to 10°C?

13-21E A house contains air at 80°F and 65 percent relative humidity. Will any moisture condense on the inner surfaces of the windows when the temperature of the window drops to 50°F?

13-22 After a long walk in the 12°C outdoors, a person wearing glasses enters a room at 25°C and 40 percent relative humidity. Determine whether the glasses will become fogged.

13-23 Repeat Prob. 13-22 for a relative humidity of 70 percent.

13-24 A thirsty man opens the refrigerator and picks up a cool canned drink at 5°C. Do you think the can will "sweat" as the man enjoys the drink in a room at 25°C and 40 percent relative humidity?
Answer: Yes

13-24E A thirsty woman opens the refrigerator and picks up a cool canned drink at 45°F. Do you think the can will "sweat" as the woman enjoys the drink in a room at 80°F and 40 percent relative humidity?

13-25 The dry- and the wet-bulb temperatures of atmospheric air at 95 kPa are 25 and 20°C, respectively. Determine (*a*) the specific humidity, (*b*) the relative humidity, and (*c*) the enthalpy of the air in kJ/kg dry air.

13-26 The air in a room has a dry-bulb temperature of 22°C and a wet-bulb temperature of 16°C. Assuming a pressure of 100 kPa, determine (*a*) the specific humidity, (*b*) the relative humidity, and (*c*) the dew-point temperature.
Answers: (*a*) 0.0091 kg H_2O/kg dry air, (*b*) 54.1 percent, (*c*) 12.4°C

13-26E The air in a room has a dry-bulb temperature of 73°F and a wet-bulb temperature of 65°F. Assuming a pressure of 14.7 psia, determine (*a*) the specific humidity, (*b*) the relative humidity, and (*c*) the dew-point temperature.
Answers: (*a*) 0.0087 lbm H_2O/lbm dry air, (*b*) 55.8 percent, (*c*) 53.2°F

13-27 Air at 105 kPa, 15°C, and 50 percent relative humidity flows in a 25-cm-diameter duct at a velocity of 20 m/s. Determine (*a*) the dew-point temperature, (*b*) the volume flow rate of air, and (*c*) the mass flow rate of the dry air.

13-27E Air at 15 psia, 60°F, and 50 percent relative humidity flows in a 10-in-diameter duct at a velocity of 60 ft/s. Determine (*a*) the dew-point temperature, (*b*) the volume flow rate of air, and (*c*) the mass flow rate of the dry air.

Psychrometric Chart

13-28C How do constant-enthalpy and constant wet-bulb temperature lines compare on the psychrometric chart?

13-29C At what states on the psychrometric chart are the dry-bulb, wet-bulb, and dew-point temperatures identical?

13-30C How is the dew-point temperature at a specified state determined on the psychrometric chart?

13-31C Can the enthalpy values determined from a psychrometric chart at sea level (1 atm pressure) be used at higher elevations?

13-32 The air in a room is at 1 atm, 32°C, and 60 percent relative humidity. Using the psychrometric chart, determine (*a*) the specific humidity, (*b*) the enthalpy (in kJ/kg dry air), (*c*) the wet-bulb temperature, (*d*) the dew-point temperature, and (*e*) the specific volume of the air (in m^3/kg dry air).

13-33 A room contains air at 1 atm, 26°C, and 70 percent relative humidity. Using the psychrometric chart, determine (*a*) the specific humidity, (*b*) the enthalpy (in kJ/kg dry air), (*c*) the wet-bulb temperature, (*d*) the dew-point temperature, and (*e*) the specific volume of the air (in m^3/kg dry air).

13-33E A room contains air at 1 atm, 82°F, and 70 percent relative humidity. Using the psychrometric chart, determine (*a*) the specific humidity, (*b*) the enthalpy (in Btu/lbm dry air), (*c*) the wet-bulb temperature, (*d*) the dew-point temperature, and (*e*) the specific volume of the air (in ft^3/lbm dry air).

13-34 The air in a room has a pressure of 1 atm, a dry-bulb temperature of 24°C, and a wet-bulb temperature of 17°C. Using the psychrometric chart, determine (*a*) the specific humidity, (*b*) the enthalpy (in kJ/kg dry air), (*c*) the relative humidity, (*d*) the dew-point temperature, and (*e*) the specific volume of the air (in m^3/kg dry air).

13-34E The air in a room has a pressure of 14.7 psia, a dry-bulb temperature of 75°F, and a wet-bulb temperature of 65°F. Using the psychrometric chart, determine (*a*) the specific humidity, (*b*) the enthalpy (in Btu/lbm dry air), (*c*) the relative humidity, (*d*) the dew-point temperature, and (*e*) the specific volume of the air (in ft^3/lbm dry air).

13-35C What does a modern air conditioning system do besides heating or cooling the air?

13-36C How does the human body respond to (*a*) hot weather, (*b*) cold weather, and (*c*) hot and humid weather?

13-37C What is the radiation effect? How does it affect human comfort?

13-38C How does the air motion in the vicinity of the human body affect human comfort?

13-39C Consider a tennis match in cold weather where both players and spectators wear the same clothes. Which group of people will feel colder? Why?

13-40C Why do you think little babies are most susceptible to cold?

13-41C How does humidity affect human comfort?

13-42C What are humidification and dehumidification?

Simple Heating and Cooling

13-43C How do relative and specific humidities change during a simple heating process? Answer the same question for a simple cooling process.

13-44C Why does a simple heating or cooling process appear as a horizontal line on the psychrometric chart?

13-45 Air enters a heating section at 95 kPa, 15°C, and 30 percent relative humidity at a rate of 15 m^3/min, and it leaves at 25°C. Determine (*a*) the rate of heat transfer in the heating section and (*b*) the relative humidity of the air at the exit. *Answers:* (*a*) 172.9 kJ/min, (*b*) 16.1 percent

13-46 A heating section consists of a 35-cm-diameter duct which houses an 8-kW electric resistance heater. Air enters the heating section at 1 atm, 13°C, and 40 percent relative humidity at a velocity of 15 m/s. Determine (*a*) the exit temperature, (*b*) the exit relative humidity of the air, and (*c*) the exit velocity.

13-46E A heating section consists of a 15-in-diameter duct which houses an 8-kW electric resistance heater. Air enters the heating section at 14.7 psia, 50°F, and 40 percent relative humidity at a velocity of 50 ft/s. Determine (*a*) the exit temperature, (*b*) the exit relative humidity of the air, and (*c*) the exit velocity.
Answers: (*a*) 56.8°F, (*b*) 30.8 percent, (*c*) 50.8 m/s

13-47 Air enters a cooling section at 97 kPa, 35°C, and 20 percent relative humidity at a rate of 20 m^3/min, where it is cooled until the moisture in the air starts condensing. Determine (*a*) the temperature of the air at the exit and (*b*) the rate of heat transfer in the cooling section.

13-48 Air enters a 40-cm-diameter cooling section at 1 atm, 32°C, and 30 percent relative humidity at 18 m/s. Heat is removed from the air at

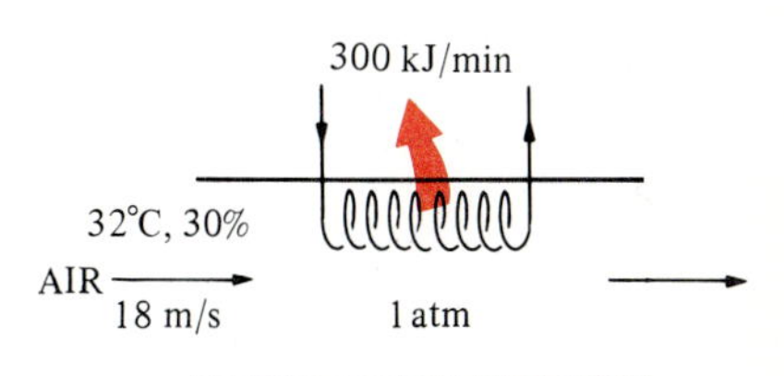

FIGURE P13-48

a rate of 1200 kJ/min. Determine (*a*) the exit temperature, (*b*) the exit relative humidity of the air, and (*c*) the exit velocity.
Answers: (*a*) 24.4°C, (*b*) 46.6 percent, (*c*) 17.6 m/s

13-48E Air enters a 15-in-diameter cooling section at 1 atm, 90°F, and 30 percent relative humidity at 55 ft/s. Heat is removed from the air at a rate of 2400 Btu/min. Determine (*a*) the exit temperature, (*b*) the exit relative humidity of the air, and (*c*) the exit velocity.

Heating with Humidification

13-49C Why is heated air sometimes humidified?

13-50 Outdoor air enters an air conditioning system at 10°C and 40 percent relative humidity at a steady rate of 30 m^3/min, and it leaves at 25°C and 55 percent relative humidity. The outdoor air is first heated to 22°C in the heating section and then humidified by the injection of hot steam in the humidifying section. Assuming the entire process takes place at a pressure of 1 atm, determine (*a*) the rate of heat supply in the heating section and (*b*) the mass flow rate of the steam required in the humidifying section.

13-51 Air at 1 atm, 15°C, and 60 percent relative humidity is first heated to 20°C in a heating section and then humidified by introducing water vapor. The air leaves the humidifying section at 25°C and 65 percent relative humidity. Determine (*a*) the amount of steam added to the air, in kg H_2O/kg dry air, and (*b*) the amount of heat transfer to the air in the heating section, in kJ/kg dry air.
Answers: (*a*) 0.0065 kg H_2O/kg dry air, (*b*) 5.1 kJ/kg dry air

13-51E Air at 14.7 psia, 55°F, and 60 percent relative humidity is first heated to 72°F in a heating section and then humidified by introducing water vapor. The air leaves the humidifying section at 75°F and 65 percent relative humidity. Determine (*a*) the amount of steam added to the air, in lbm H_2O/lbm dry air, and (*b*) the amount of heat transfer to the air in the heating section, in Btu/lbm dry air.

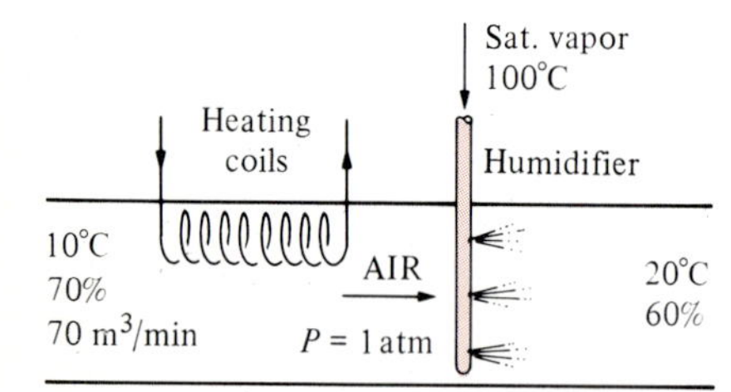

FIGURE P13-52

13-52 An air conditioning system operates at a total pressure of 1 atm and consists of a heating section and a humidifier which supplies wet steam (saturated water vapor) at 100°C. Air enters the heating section at 10°C and 70 percent relative humidity at a rate of 70 m^3/min, and it leaves the humidifying section at 20°C and 60 percent relative humidity. Determine (*a*) the temperature and relative humidity of air when it leaves the heating section, (*b*) the rate of heat transfer in the heating section, and (*c*) the rate at which water is added to the air in the humidifying section.

13-52E An air conditioning system operates at a total pressure of 1 atm and consists of a heating section and a humidifier which supplies wet steam (saturated water vapor) at 212°F. Air enters the heating section at 50°F and 70 percent relative humidity at a rate of 2500 ft^3/min, and it

leaves the humidifying section at 70°F and 60 percent relative humidity. Determine (*a*) the temperature and relative humidity of air when it leaves the heating section, (*b*) the rate of heat transfer in the heating section, and (*c*) the rate at which water is added to the air in the humidifying section.
Answers: (*a*) 69.2°F, 35.0 percent; (*b*) 885 Btu/min; (*c*) 0.79 lbm/min

13-53 Repeat Prob. 13-52 for a total pressure of 95 kPa for the airstream.
Answers: (*a*) 19.5°C, 37.7 percent; (*b*) 782 kJ/min; (*c*) 0.29 kg/min

Cooling with Dehumidification

13-54C Why is cooled air sometimes reheated in summer before it is discharged to a room?

13-55 Air enters a window air conditioner at 1 atm, 32°C, and 70 percent relative humidity at a rate of 15 m^3/min, and it leaves as saturated air at 12°C. Part of the moisture in the air which condenses during the process is also removed at 12°C. Determine the rates of heat and moisture removal from the air. *Answers:* 868.1 kJ/min, 0.21 kg/min

13-56 An air conditioning system is to take in air at 1 atm, 34°C, and 70 percent relative humidity and deliver it at 22°C and 50 percent relative humidity. The air flows first over the cooling coils, where it is cooled and dehumidified, and then over the resistance heating wires, where it is heated to the desired temperature. Assuming that the condensate is removed from the cooling section at 10°C, determine (*a*) the temperature of air before it enters the heating section, (*b*) the amount of heat removed in the cooling section, and (*c*) the amount of heat transferred in the heating section, both in kJ/kg dry air.

13-56E An air conditioning system is to take in air at 14.7 psia, 92°F, and 70 percent relative humidity and deliver it at 72°F and 50 percent relative humidity. The air flows first over the cooling coils, where it is cooled and dehumidified, and then over the resistance heating wires, where it is heated to the desired temperature. Assuming that the condensate is removed from the cooling section at 50°F, determine (*a*) the temperature of air before it enters the heating section, (*b*) the amount of heat removed in the cooling section, and (*c*) the amount of heat transferred in the heating section, both in Btu/lbm dry air.
Answers: (*a*) 52.4°F, (*b*) 21.4 Btu/lbm dry air, (*c*) 4.8 Btu/lbm dry air

13-57 Air enters an air conditioning system which uses refrigerant-12 at 30°C and 70 percent relative humidity at a rate of 10 m^3/min. The refrigerant enters the cooling section at 700 kPa with a quality of 20 percent and leaves as saturated vapor. The air is cooled to 20°C at a pressure of 1 atm. Determine (*a*) the rate of dehumidification, (*b*) the rate of heat transfer, and (*c*) the mass flow rate of the refrigerant.
Answers: (*a*) 0.046 kg/min, (*b*) −231.2 kJ/min, (*c*) 2.12 kg/min

13-58 Repeat Prob. 13-57 for a total pressure of 95 kPa for air.

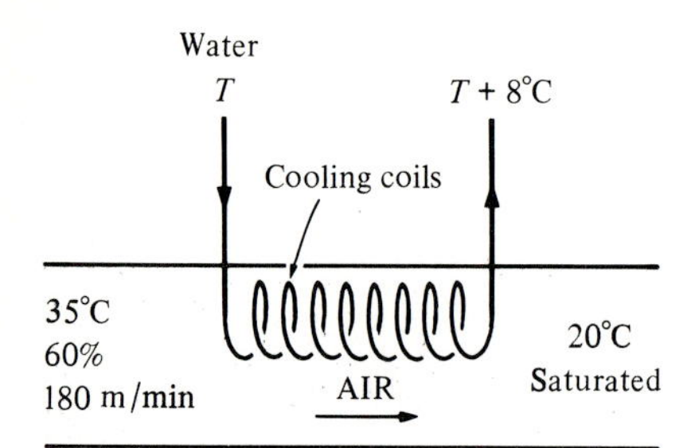

FIGURE P13-59

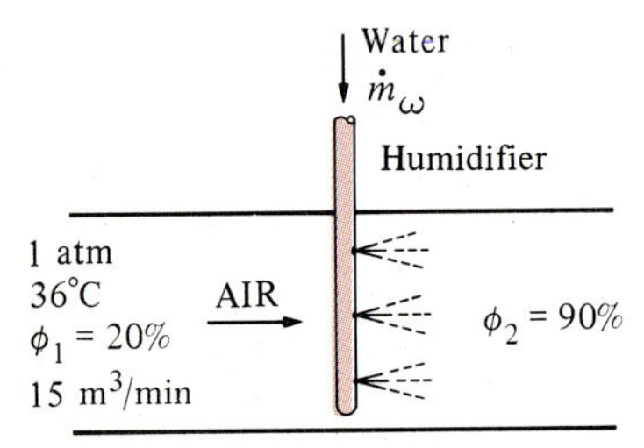

FIGURE P13-62

13-59 Air enters a 30-cm-diameter cooling section at 1 atm, 35°C, and 60 percent relative humidity at 180 m/min. The air is cooled by passing it over a cooling coil through which cold water flows. The water experiences a temperature rise of 8°C. The air leaves the cooling section saturated at 20°C. Determine (*a*) the rate of heat transfer, (*b*) the mass flow rate of the water, and (*c*) the exit velocity of the airstream.

13-59E Air enters a 1-ft-diameter cooling section at 14.7 psia, 90°F, and 60 percent relative humidity at 600 ft/min. The air is cooled by passing it over a cooling coil through which cold water flows. The water experiences a temperature rise of 14°F. The air leaves the cooling section saturated at 70°F. Determine (*a*) the rate of heat transfer, (*b*) the mass flow rate of the water, and (*c*) the exit velocity of the airstream.

13-60 Repeat Prob. 13-59 for a total pressure of 95 kPa for air. *Answers:* (*a*) −413.4 kJ/min, (*b*) 12.4 kg/min, (*c*) 169.4 m/min

13-60E Repeat Prob. 13-59E for a total pressure of 14.4 psia for air.

Evaporative Cooling

13-61C What is evaporative cooling? Will it work in humid climates?

13-62 Air enters an evaporative cooler at 1 atm, 36°C, and 20 percent relative humidity at a rate of 15 m^3/min, and it leaves with a relative humidity of 90 percent. Determine (*a*) the exit temperature of the air and (*b*) the required rate of water supply to the evaporative cooler.

13-62E Air enters an evaporative cooler at 14.7 psia, 90°F, and 20 percent relative humidity at a rate of 500 ft^3/min, and it leaves with a relative humidity of 90 percent. Determine (*a*) the exit temperature of the air and (*b*) the required rate of water supply to the evaporative cooler. *Answers:* (*a*) 64°F, (*b*) 0.20 lbm/min

13-63 Air enters an evaporative cooler at 95 kPa, 35°C, and 30 percent relative humidity and exits saturated. Determine the exit temperature of the air. *Answer:* 21.1°C

13-63E Air enters an evaporative cooler at 14.5 psia, 93°F, and 30 percent relative humidity and exits saturated. Determine the exit temperature of the air.

13-64 Air enters an evaporative cooler at 1 atm, 32°C, and 30 percent relative humidity at a rate of 10 m^3/min and leaves at 22°C. Determine (*a*) the final relative humidity and (*b*) the amount of water added to the air.

13-65 What is the lowest temperature that air can attain in an evaporative cooler if it enters at 1 atm, 29°C, and 40 percent relative humidity? *Answer:* 19.3°C

13-66 Air at 1 atm, 15°C, and 60 percent relative humidity first is heated to 30°C in a heating section and then is passed through an evaporative cooler where its temperature drops to 25°C. Determine (*a*) the exit relative humidity and (*b*) the amount of water added to the air in kg H_2O/kg dry air.

13-66E Air at 14.7 psia, 55°F, and 60 percent relative humidity is first heated to 86°F in a heating section and then passed through an evaporative cooler where its temperature drops to 75°F. Determine (*a*) the exit relative humidity and (*b*) the amount of water added to the air in lbm H_2O/lbm dry air.

13-67 An air conditioning system operates at a total pressure of 1 atm and consists of a heating section and an evaporative cooler. Air enters the heating section at 10°C and 70 percent relative humidity at a rate of 70 m^3/min, and it leaves the evaporative cooler at 20°C and 60 percent relative humidity. Determine (*a*) the temperature and relative humidity of the air when it leaves the heating section, (*b*) the rate of heat transfer in the heating section, and (*c*) the rate of water added to the air in the evaporative cooler.
Answers: (*a*) 28.3°C, 23.0 percent; (*b*) 1624 kJ/min; (*c*) 0.30 kg/min

13-68 Repeat Prob. 13-67 for a total pressure of 96 kPa.

Adiabatic Mixing of Airstreams

13-69C Two unsaturated airstreams are mixed adiabatically. It is observed that some moisture condenses during the mixing process. Under what conditions will this be the case?

13-70C Consider the adiabatic mixing of two airstreams. Does the state of the mixture on the psychrometric chart have to be on the straight line connecting the two states?

13-71 Two airstreams are mixed steadily and adiabatically. The first stream enters at 32°C and 40 percent relative humidity at a rate of 60 m^3/min, while the second stream enters at 12°C and 90 percent relative humidity at a rate of 75 m^3/min. Assuming that the mixing process occurs at a pressure of 1 atm, determine the specific humidity, the relative humidity, the dry-bulb temperature, and the volume flow rate of the mixture.
Answers: 0.0096 kg H_2O/kg dry air, 63.4 percent, 20.6°C, 134.9 m^3/min

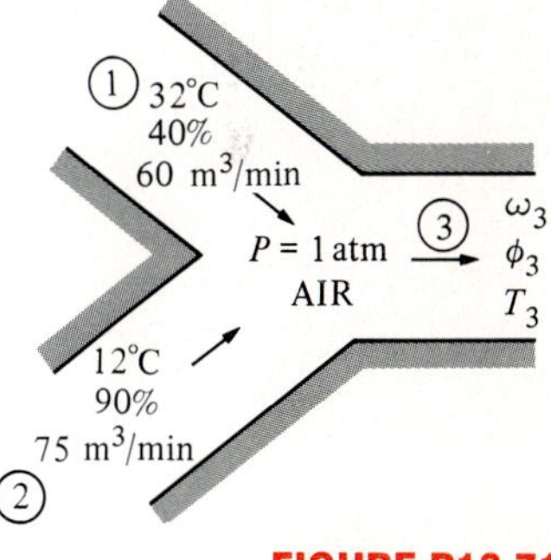

FIGURE P13-71

13-71E Two airstreams are mixed steadily and adiabatically. The first stream enters at 90°F and 40 percent relative humidity at a rate of 2000 ft^3/min, while the second stream enters at 55°F and 90 percent relative humidity at a rate of 2500 ft^3/min. Assuming that the mixing process occurs at a pressure of 1 atm, determine the specific humidity, the relative humidity, the dry-bulb temperature, and the volume flow rate of the mixture.

13-72 Repeat Prob. 13-71 for a total mixing chamber pressure of 95 kPa.

13-73 Conditioned air at 13°C and 90 percent relative humidity is to be mixed with outside air at 34°C and 40 percent relative humidity at 1 atm. If it is desired that the mixture have a relative humidity of 60 percent, determine (*a*) the ratio of the dry air mass flow rates of the conditioned air to the outside air and (*b*) the temperature of the mixture. Use the psychrometric chart.

13-74 During an air conditioning process 60 m^3/min of conditioned air at 17°C and 30 percent relative humidity is mixed adiabatically with 20 m^3/min of outside air at 30°C and 90 percent relative humidity at a pressure of 1 atm. Determine (*a*) the temperature, (*b*) the specific humidity, and (*c*) the relative humidity of the mixture.
Answers: (*a*) 20.2°C, (*b*) 0.0085 kg H_2O/kg dry air, (*c*) 57.8 percent

13-74E During an air conditioning process, 900 ft^3/min of conditioned air at 65°F and 30 percent relative humidity is mixed adiabatically with 300 ft^3/min of outside air at 80°F and 90 percent relative humidity at a pressure of 1 atm. Determine (*a*) the temperature, (*b*) the specific humidity, and (*c*) the relative humidity of the mixture.
Answers: (*a*) 68.7°F, (*b*) 0.0085 lbm H_2O/lbm dry air, (*c*) 52.1 percent

13-75 A stream of warm air with a dry-bulb temperature of 40°C and a wet-bulb temperature of 32°C is mixed adiabatically with a stream of saturated cool air at 18°C. The dry air mass flow rates of the warm and cool airstreams are 2 and 1.5 kg/s, respectively. Assuming a total pressure of 1 atm, determine (*a*) the temperature, (*b*) the specific humidity, and (*c*) the relative humidity of the mixture.

Wet Cooling Towers

13-76C How does a natural-draft wet cooling tower work?

13-77C What is a spray pond? How does its performance compare to the performance of a wet cooling tower?

13-78 The cooling water from the condenser of a power plant enters a wet cooling tower at 40°C at a rate of 130 kg/s. The water is cooled to 25°C in the cooling tower by air which enters the tower at 1 atm, 23°C, and 60 percent relative humidity and leaves saturated at 32°C. Neglecting the power input to the fan, determine (*a*) the volume flow rate of air into the cooling tower and (*b*) the mass flow rate of the required makeup water.

13-78E The cooling water from the condenser of a power plant enters a wet cooling tower at 110°F at a rate of 275 lbm/s. The water is cooled to 80°F in the cooling tower by air which enters the tower at 1 atm, 76°F, and 60 percent relative humidity and leaves saturated at 95°F. Neglecting the power input to the fan, determine (*a*) the volume flow rate of air into the cooling tower and (*b*) the mass flow rate of the required makeup water.
Answers: (*a*) 3608.9 ft^3/s, (*b*) 6.58 lbm/s

13-79 A wet cooling tower is to cool 50 kg/s of water from 40 to 26°C. Atmospheric air enters the tower at 1 atm with dry- and wet-bulb temperatures of 22 and 16°C, respectively, and leaves at 34°C with a relative

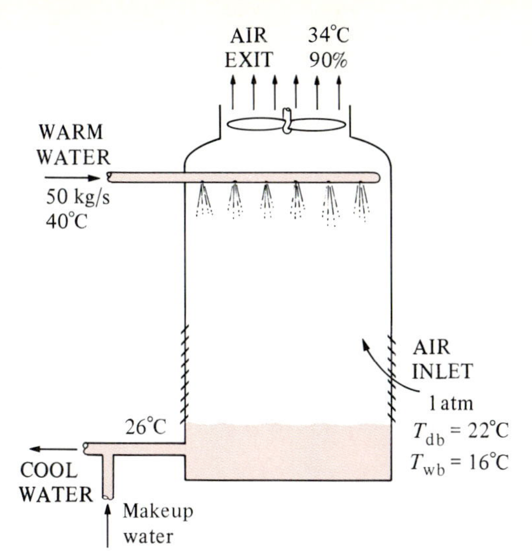

FIGURE P13-79

humidity of 90 percent. Using the psychrometric chart, determine (*a*) the volume flow rate of air into the cooling tower and (*b*) the mass flow rate of the required makeup water. *Answers:* (*a*) 37.4 m³/s, (*b*) 0.97 kg/s

13-79E A wet cooling tower is to cool 100 lbm/s of water from 105 to 80°F. Atmospheric air enters the tower at 14.7 psia with dry- and wet-bulb temperatures of 75 and 60°F, respectively, and leaves at 95°F with a relatively humidity of 90 percent. Using the psychrometric chart, determine (*a*) the volume flow rate of air into the cooling tower and (*b*) the mass flow rate of the required makeup water.

13-80 A natural-draft cooling tower is to remove 50 MW of waste heat from the cooling water which enters the tower at 42°C and leaves at 27°C. Atmospheric air enters the tower at 1 atm with dry- and wet-bulb temperatures of 23 and 18°C, respectively, and leaves saturated at 37°C. Determine (*a*) the mass flow rate of the cooling water, (*b*) the volume flow rate of air into the cooling tower, and (*c*) the mass flow rate of the required makeup water.

13-81 A wet cooling tower is to cool 180 kg/s of cooling water from 40 to 25°C at a location where the atmospheric pressure is 96 kPa. Atmospheric air enters the tower at 20°C and 70 percent relative humidity and leaves saturated at 35°C. Neglecting the power input to the fan, determine (*a*) the volume flow rate of air into the cooling tower and (*b*) the mass flow rate of the required makeup water.
Answers: (*a*) 119.6 m³/s, (*b*) 3.74 kg/s

Chemical Reactions

CHAPTER 14

In the preceding chapters we limited our consideration to nonreacting thermodynamic systems. That is, the chemical composition of all the systems considered remained unchanged during a process. This was the case even with mixing processes during which a homogeneous mixture is formed from two or more fluids without the occurrence of any chemical reactions. In this chapter, we specifically deal with systems whose chemical composition changes during a process, i.e., systems that involve chemical reactions.

When dealing with nonreacting systems, we need to consider only the *sensible internal energy* (associated with temperature and pressure changes) and the *latent internal energy* (associated with phase changes). When dealing with reacting systems, however, we also need to consider the *chemical internal energy,* which is the energy associated with the destruction and the formation of chemical bonds between the atoms. The conservation of energy relations developed for nonreacting systems are equally applicable to reacting systems, but the energy terms in the latter case should be modified to include the chemical energy of the system.

In this chapter we focus on a particular type of chemical reaction, known as *combustion,* because of its importance in engineering. The reader should keep in mind, however, that the principles developed are equally applicable to any chemical reaction.

We start with a general discussion of fuels and combustion. Then we apply the conservation of mass and conservation of energy principles to reacting systems. In this regard we discuss the adiabatic flame temperature, which is the highest temperature a reacting mixture can attain. Finally, we examine the second-law aspects of chemical reactions.

14-1 ■ FUELS AND COMBUSTION

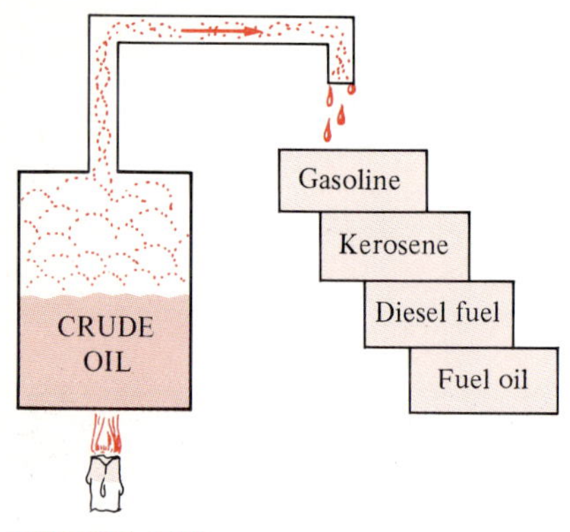

FIGURE 14-1

Most liquid hydrocarbon fuels are obtained from crude oil by distillation.

Any material that can be burned to release energy is called a **fuel**. Most familiar fuels consist primarily of hydrogen and carbon. They are called **hydrocarbon fuels** and are denoted by the general formula C_nH_m. Hydrocarbon fuels exist in all phases, some examples being coal, gasoline, and natural gas.

The main constituent of coal is carbon. Coal also contains varying amounts of oxygen, hydrogen, nitrogen, sulfur, moisture, and ash. It is difficult to give an exact mass analysis for coal since its composition varies considerably from one geographical location to the next and even within the same geographical location. Most liquid hydrocarbon fuels are a mixture of numerous hydrocarbons and are obtained from crude oil by distillation (Fig. 14-1). The most volatile hydrocarbons vaporize first, forming what we know as gasoline. The less volatile fuels obtained during distillation are kerosene, diesel fuel, and fuel oil. The composition of a particular fuel depends on the source of the crude oil as well as on the refinery.

FIGURE 14-2

Combustion is a chemical reaction during which a fuel is oxidized and a large quantity of energy is released.

Although liquid hydrocarbon fuels are mixtures of many different hydrocarbons, they are usually considered to be a single hydrocarbon for convenience. For example, gasoline is treated as **octane**, C_8H_{18}, and the diesel fuel as **dodecane**, $C_{12}H_{26}$. Another common liquid hydrocarbon fuel is **methyl alcohol**, CH_3OH, which is also called *methanol* and is used in some gasoline blends. The gaseous hydrocarbon fuel natural gas, which is a mixture of methane and smaller amounts of other gases, is sometimes treated as **methane**, CH_4, for simplicity.

A chemical reaction during which a fuel is oxidized and a large quantity of energy is released is called **combustion** (Fig. 14-2). The oxidizer most often used in combustion processes is air, for obvious reasons—it is free and readily available. Pure oxygen O_2 is used as an oxidizer only in some specialized applications where air cannot be used. Therefore, a few words about the composition of air are in order.

On a mole or a volume basis, dry air is composed of 20.9 percent oxygen, 78.1 percent nitrogen, 0.9 percent argon, and small amounts of carbon dioxide, helium, neon, and hydrogen. In the analysis of combustion processes, the argon in the air is treated as nitrogen, and the gases that exist in trace amounts are disregarded. Then dry air can be approximated as 21 percent oxygen and 79 percent nitrogen by mole numbers. Therefore, each mole of oxygen entering a combustion chamber will be accompanied by 0.79/0.21 = 3.76 mol of nitrogen (Fig. 14-3). That is,

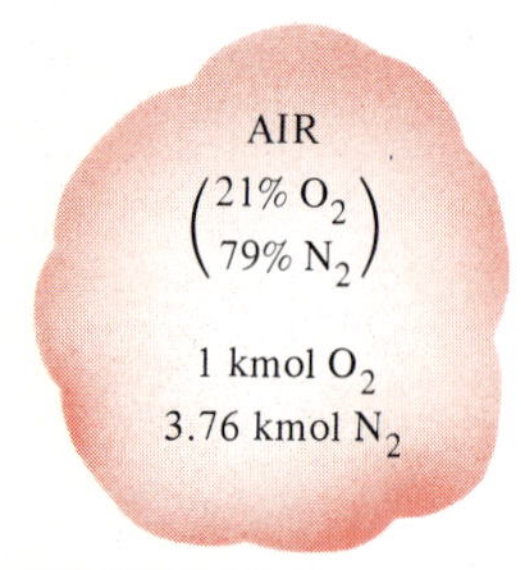

FIGURE 14-3

Each kmol of O_2 in air is accompanied by 3.76 mol of N_2.

$$1 \text{ kmol } O_2 + 3.76 \text{ kmol } N_2 = 4.76 \text{ kmol air} \tag{14-1}$$

At ordinary combustion temperatures, nitrogen behaves as an inert gas and does not react with other chemical elements. But even then the presence of nitrogen greatly affects the outcome of a combustion process since nitrogen usually enters a combustion chamber in large quantities at low temperatures and exits at considerably higher temperatures, absorbing a large proportion of the chemical energy released during combustion. Throughout this chapter, nitrogen is assumed to remain perfectly inert.

Keep in mind, however, that at very high temperatures, such as those encountered in internal combustion engines, a small fraction of nitrogen reacts with oxygen, forming hazardous gases such as nitric oxide.

The air which enters a combustion chamber normally contains some water vapor (or moisture), which also deserves consideration. For most combustion processes, the moisture in the air can also be treated as an inert gas, like nitrogen. At very high temperatures, however, some water vapor dissociates into H_2 and O_2 as well as into H, O, and OH. When the combustion gases are cooled below the dew-point temperature of the water vapor, some moisture will condense out. It is important to be able to predict the dew-point temperature since the water droplets often combine with the sulfur dioxide that may be present in the combustion gases, forming sulfuric acid, which is highly corrosive.

During a combustion process, the components that exist before the reaction are called **reactants**, and the components that exist after the reaction are called **products** (Fig. 14-4). Consider, for example, the combustion of 1 kmol of carbon with 1 kmol of pure oxygen, forming carbon dioxide:

$$C + O_2 \longrightarrow CO_2 \tag{14-2}$$

Here C and O_2 are the reactants since they exist before combustion, and CO_2 is the product since it exists after combustion. Note that a reactant does not have to chemically react in the combustion chamber. For example, if carbon is burned with air instead of pure oxygen, both sides of the combustion equation will include N_2. That is, the N_2 will appear both as a reactant and as a product.

FIGURE 14-4

In a steady-flow combustion process, the components that enter the reaction chamber are called reactants, and the components that exit are called products.

We should also mention that bringing a fuel into intimate contact with oxygen is not sufficient to start a combustion process. (Thank goodness it is not. Otherwise, the whole world would be on fire now.) The fuel must be brought above its **ignition temperature** to start the combustion. The minimum ignition temperatures of various substances in atmospheric air are approximately 260°C for gasoline, 400°C for carbon, 580°C for hydrogen, 610°C for carbon monoxide, and 630°C for methane. Moreover, the proportions of the fuel and air must be in the proper range for combustion to begin.

As you will recall from your chemistry courses, chemical equations are balanced on the basis of the **conservation of mass principle**, which can be stated as follows: *The total mass of each element is conserved during a chemical reaction* (Fig. 14-5). That is, the total mass of each element on the right-hand side of the reaction equation (the products) must be equal to the total mass of that element on the left-hand side (the reactants). Also the total number of atoms of each element is conserved during a chemical reaction since the total number of atoms of an element is equal to total mass of the element divided by its atomic mass.

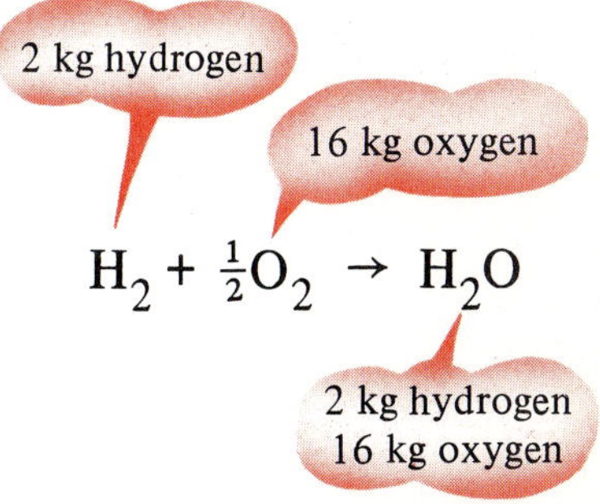

FIGURE 14-5

The mass (and number of atoms) of each element is conserved during a chemical reaction.

For example, both sides of Eq. 14-2 contain 12 kg of carbon and 32 kg of oxygen, even though the carbon and the oxygen exist as elements in the reactants and as a compound in the product. Also the total mass of reactants is equal to the total mass of products, each being 44 kg. (It is common practice to round the molar masses to the nearest integer if great accuracy is not required.) However, notice that the total mole number of the reactants (2 kmol) is not equal to the total mole number of the products (1 kmol). That is, *the total number of moles is not conserved during a chemical reaction.*

A frequently used quantity in the analysis of combustion processes is the **air-fuel ratio** AF. It is usually expressed on a mass basis and is defined as *the ratio of the mass of air to the mass of fuel during a combustion process* (Fig. 14-6). That is,

$$AF = \frac{m_{air}}{m_{fuel}} \tag{14-3}$$

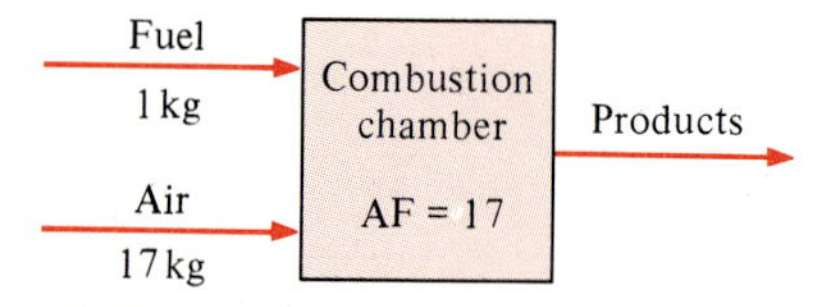

FIGURE 14-6

The air-fuel ratio (AF) represents the amount of air used per unit mass of fuel during a combustion process.

The mass m of a substance is related to the number of moles N through the relation $m = NM$, where M is the molar mass.

The air-fuel ratio can also be expressed on a mole basis as the ratio of the mole numbers of air to the mole numbers of fuel. But we will use the former definition. The reciprocal of air-fuel ratio is called the **fuel-air ratio**

EXAMPLE 14-1

One kmol of octane (C_8H_{18}) is burned with air which contains 20 kmol of O_2, as shown in Fig. 14-7. Assuming the products contain only CO_2, H_2O, O_2, and N_2, determine the mole number of each gas in the products and the air-fuel ratio for this combustion process.

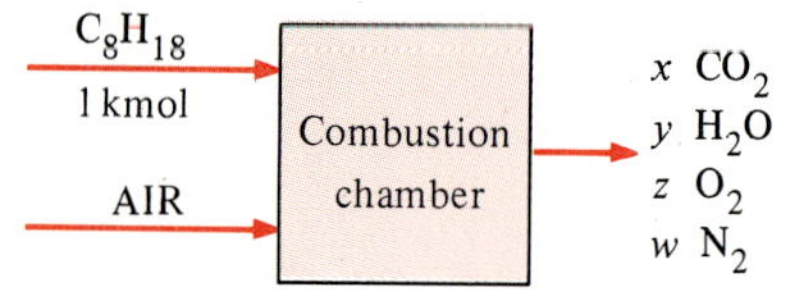

FIGURE 14-7

Schematic for Example 14-1.

Solution The chemical equation for this combustion process can be written as

$$C_8H_{18} + 20(O_2 + 3.76N_2) \longrightarrow xCO_2 + yH_2O + zO_2 + wN_2$$

where the terms in the parantheses represents the composition of dry air which contains 1 kmol of O_2, and x, y, z, and w represent the unknown mole numbers of the gases in the products. These unknowns are determined by applying the conservation of mass principle to each of the elements—that is, by requiring that the total mass or mole number of each element in the reactants be equal to that in the products:

$$\begin{aligned} &\text{C:} & 8 &= x \longrightarrow x = 8 \\ &\text{H:} & 18 &= 2y \longrightarrow y = 9 \\ &\text{O:} & 40 &= 2x + y + 2z \longrightarrow z = 7.5 \\ &\text{N}_2\text{:} & (20)(3.76) &= w \longrightarrow w = 75.2 \end{aligned}$$

Substituting yields

$$C_8H_{18} + 20(O_2 + 3.76N_2) \longrightarrow 8CO_2 + 9H_2O + 7.5O_2 + 75.2N_2$$

The air-fuel (AF) ratio is determined by taking the ratio of the mass of the fuel and the mass of the air (Eq. 14-3):

$$\begin{aligned} AF = \frac{m_{air}}{m_{fuel}} &= \frac{(NM)_{air}}{(NM)_C + (NM)_{H_2}} \\ &= \frac{(20 \times 4.76 \text{ kmol})(29 \text{ kg/kmol})}{(8 \text{ kmol})(12 \text{ kg/kmol}) + (9 \text{ kmol})(2 \text{ kg/kmol})} \\ &= 24.2 \text{ kg air/kg fuel} \end{aligned}$$

That is, 24.2 kg of air is used to burn each kilogram of fuel during this combustion process.

14-2 ■ THEORETICAL AND ACTUAL COMBUSTION PROCESSES

It is often instructive to study the combustion of a fuel by assuming that the combustion is complete. A combustion process is **complete** if all the carbon in the fuel burns to CO_2, all the hydrogen burns to H_2O, and all the sulfur (if any) burns to SO_2. That is, all the combustible components of a fuel are burned to completion during a complete combustion process (Fig. 14-8). Conversely, a combustion process is **incomplete** if the combustion products contain any unburned fuel or components such as C, H_2, CO, or OH.

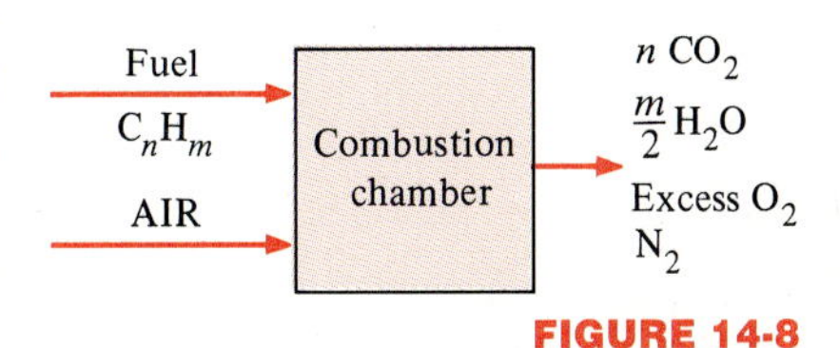

FIGURE 14-8

A combustion process is complete if all the combustible components of the fuel are burned to completion.

Insufficient oxygen is an obvious reason for incomplete combustion, but it is not the only one. Incomplete combustion occurs even when more oxygen is present in the combustion chamber than is needed for complete combustion. This may be attributed to insufficient mixing in the combustion chamber during the limited time that the fuel and the oxygen are in contact. Another cause of incomplete combustion is dissociation, which becomes important at high temperatures.

Oxygen is more strongly attracted to hydrogen than it is to carbon. Therefore, the hydrogen in the fuel normally burns to completion, forming H_2O, even when there is less oxygen than needed for complete combustion. Some of the carbon, however, ends up as CO or just as plain C particles in the product gases.

The minimum amount of air needed for the complete combustion of a fuel is called the **stoichiometric** or **theoretical air**. Thus when a fuel is completely burned with theoretical air, no uncombined oxygen will be present in the product gases. The theoretical air is also referred to as the *chemically correct* amount of air, or 100 percent theoretical air. A combustion process with less than the theoretical air is bound to be incomplete. The ideal combustion process during which a fuel is burned completely with theoretical air is called the **stoichiometric** or **theoretical combustion** of that fuel (Fig. 14-9). For example, the theoretical combustion of methane is

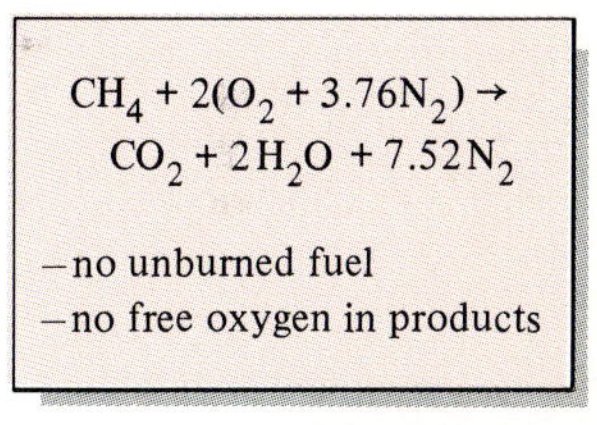

FIGURE 14-9

The complete combustion process with no free oxygen in the products is called theoretical combustion.

$$CH_4 + 2(O_2 + 3.76N_2) \longrightarrow CO_2 + 2H_2O + 7.52N_2$$

Notice that the products of the theoretical combustion contain no unburned methane, C, H_2, CO, OH, or free O_2.

In actual combustion processes, it is common practice to use more air than the stoichiometric amount, to increase the chances of complete combustion or to control the temperature of the combustion chamber. The amount of air in excess of the stoichiometric amount is called **excess air**. The amount of excess air is usually expressed in terms of the stoichiometric air as **percentage of excess air** or **percentage of theoretical air**. For example, 50 percent excess air is equivalent to 150 percent theoretical air, and 200 percent excess air is equivalent to 300 percent theoretical air. Of course, the stoichiometric air can be expressed as 0 percent excess air or 100 percent theoretical air. The amount of air used in combustion processes is also expressed in terms of the **equivalence ratio**, which is the ratio of the actual amount of air used to the stoichiometric amount of air.

Predicting the composition of the products is relatively easy when the combustion process is assumed to be complete and the exact amounts of the fuel and air used are known. All one needs to do in this case is simply apply the conservation of mass principle to each element that appears in the combustion equation, without needing to take any measurements. Things are not so simple, however, when one is dealing with actual combustion processes. For one thing, actual combustion processes are hardly ever complete, even in the presence of considerable excess air. Therefore, it is impossible to predict the composition of the products on the basis of the conservation of mass principle alone. Then the only alternative we have is to measure the amount of each component in the products directly.

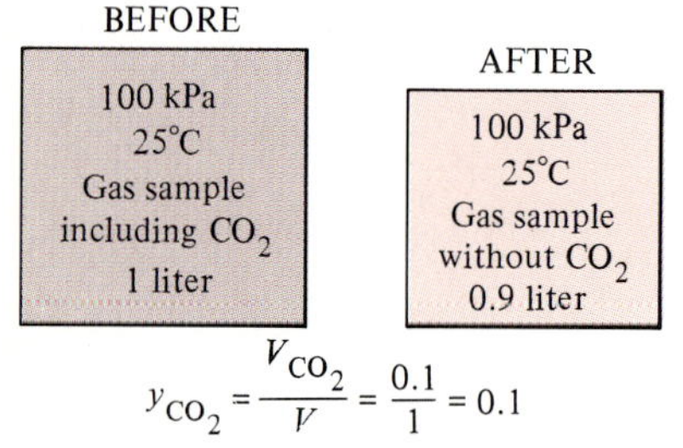

$$y_{CO_2} = \frac{V_{CO_2}}{V} = \frac{0.1}{1} = 0.1$$

FIGURE 14-10

Determining the mole fraction of the CO_2 in combustion gases by using the Orsat gas analyzer.

A commonly used device to analyze the composition of combustion gases is the **Orsat gas analyzer**. In this device, a sample of the combustion gases is collected and cooled to room temperature and pressure, at which point its volume is measured. The sample is then brought into contact with a chemical that absorbs the CO_2. The remaining gases are returned to the room temperature and pressure, and the new volume they occupy is measured. The ratio of the reduction in volume to the original volume is the volume fraction of the CO_2, which is equivalent to the mole fraction if ideal-gas behavior is assumed (Fig. 14-10). The volume fractions of the other gases are determined by repeating this procedure. In Orsat analysis the gas sample is collected over water and is maintained saturated at all times. Therefore, the vapor pressure of water remains constant during the entire test. For this reason the presence of water vapor in the test chamber is ignored and data is reported on a dry basis. But the amount of H_2O formed during combustion is easily determined by balancing the combustion equation.

EXAMPLE 14-2

Ethane (C_2H_6) is burned with 20 percent excess air during a combustion process, as shown in Fig. 14-11. Assuming complete combustion and a total pressure of 100 kPa, determine (*a*) the air-fuel ratio and (*b*) the dew-point temperature of the products.

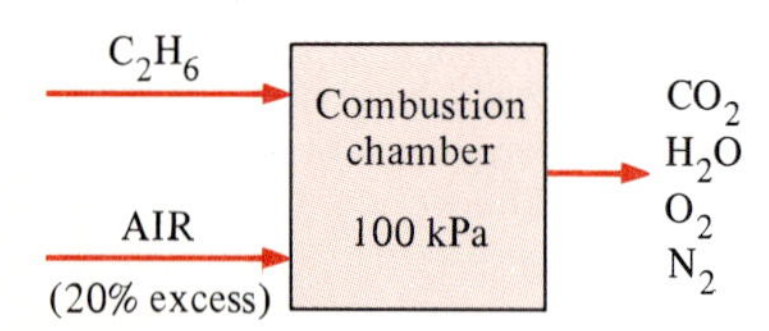

FIGURE 14-11

Schematic for Example 14-2.

Solution It is specified that the ethane is burned completely. Therefore, the products will contain only CO_2, H_2O, N_2, and the unused O_2. The combustion equation in this case can be written as

$$C_2H_6 + 1.2a_{th}(O_2 + 3.76N_2) \longrightarrow 2CO_2 + 3H_2O + 0.2a_{th}O_2 + (1.2 \times 3.76)a_{th}N_2$$

where a_{th} is the stoichiometric coefficient for air. We have automatically accounted for the 20 percent excess air by using the factor $1.2a_{th}$ instead of a_{th} for air. The stoichiometric amount of oxygen ($a_{th}O_2$) will be used to oxidize the fuel, and the remaining excess amount ($0.2a_{th}O_2$) will appear in the products as unused oxygen. Notice that the coefficient of N_2 is the same on both sides of the equation. You will also notice that we did the C and H_2 balance in our heads as we wrote the combustion equation because it is so obvious. The coefficient a_{th} is determined from the O_2 balance:

$$O_2: \quad 1.2a_{th} = 2 + 1.5 + 0.2a_{th} \longrightarrow a_{th} = 3.5$$

Substituting gives

$$C_2H_6 + 4.2(O_2 + 3.76N_2) \longrightarrow 2CO_2 + 3H_2O + 0.7O_2 + 15.79N_2$$

(*a*) The air-fuel ratio is determined by taking the ratio of the mass of the air to the mass of the fuel (Eq. 14-3):

$$AF = \frac{m_{air}}{m_{fuel}} = \frac{(4.2 \times 4.76 \text{ kmol})(29 \text{ kg/kmol})}{(2 \text{ kmol})(12 \text{ kg/kmol}) + (3 \text{ kmol})(2 \text{ kg/kmol})}$$
$$= 19.3 \text{ kg air/kg fuel}$$

That is, 19.3 kg of air is supplied for each kilogram of fuel during this combustion process.

(*b*) The dew-point temperature of the products is the temperature at which the water vapor in the products starts to condense as the products are cooled. You will recall from Chap. 13 that the dew-point temperature of a gas-vapor mixture is the saturation temperature of the water vapor corresponding to its partial pressure. Therefore, we need to determine the partial pressure of the water vapor P_v in the products first. Assuming ideal-gas behavior for the combustion gases, we have

$$P_v = \left(\frac{N_v}{N_{prod}}\right)(P_{prod}) = \left(\frac{3 \text{ kmol}}{21.49 \text{ kmol}}\right)(100 \text{ kPa}) = 13.96 \text{ kPa}$$

Thus, $$T_{dp} = T_{sat\,@\,13.96\,kPa} = 52.3°C \qquad \text{(Table A-5)}$$

EXAMPLE 14-3

A certain natural gas has the following volumetric analysis: 72 percent CH_4, 9 percent H_2, 14 percent N_2, 2 percent O_2, and 3 percent CO_2. This gas is now burned with the stoichiometric amount of air which enters the combustion chamber at 20°C, 1 atm, and 80 percent relative humidity, as shown in Fig. 14-12. Assuming complete combustion and a total pressure of 1 atm, determine the dew-point temperature of the products.

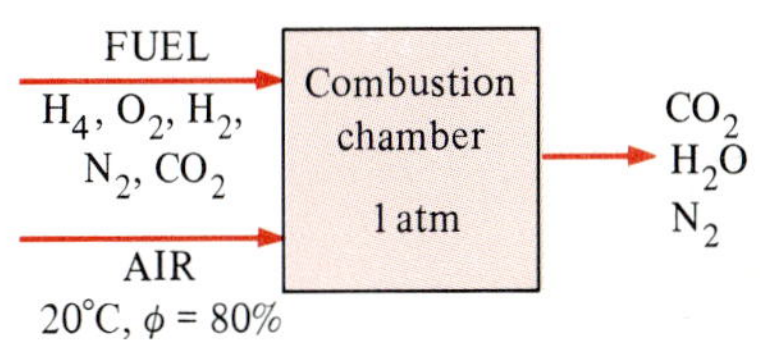

FIGURE 14-12
Schematic for Example 14-3.

Solution The combustion process is assumed to be complete; therefore, all the carbon in the fuel will burn to CO_2, and all the hydrogen to H_2O. Also the fuel is burned with the stoichiometric amount of air; therefore, there will be no free O_2 in the products. The moisture in the air does not react with anything; it simply shows up as additional H_2O in the products. Therefore, for simplicity, we will balance the combustion equation by using dry air and then add the moisture later to both sides of the equation. Considering 1 kmol of fuel,

$$\overbrace{(0.72CH_4 + 0.09H_2 + 0.14N_2 + 0.02O_2 + 0.03CO_2)}^{\text{fuel}} + \overbrace{a_{th}(O_2 + 3.76N_2)}^{\text{dry air}}$$
$$\longrightarrow xCO_2 + yH_2O + zN_2$$

The unknown coefficients in the above equation are determined from a mass balance on various elements:

$$\text{C:} \quad 0.72 + 0.03 = x \longrightarrow x = 0.75$$

$$\text{H:} \quad 0.72 \times 4 + 0.09 \times 2 = 2y \longrightarrow y = 1.53$$

$$\text{O}_2\text{:} \quad 0.02 + 0.03 + a_{th} = x + \frac{y}{2} \longrightarrow a_{th} = 1.465$$

$$\text{N}_2\text{:} \quad 0.14 + 3.76a_{th} = z \longrightarrow z = 5.648$$

Next we determine the amount of moisture that accompanies $4.76a_{th} = (4.76)(1.465) = 6.97$ kmol of dry air. The partial pressure of the moisture in the air is

$$P_{v,\text{air}} = \phi_{\text{air}} P_{\text{sat @ 20°C}} = (0.80)(2.339 \text{ kPa}) = 1.871 \text{ kPa}$$

Assuming ideal-gas behavior, the number of moles of the moisture in the air $N_{v,\text{air}}$ is

$$N_{v,\text{air}} = \left(\frac{P_{v,\text{air}}}{P_{\text{total}}}\right) N_{\text{total}} = \left(\frac{1.871 \text{ kPa}}{101.325 \text{ kPa}}\right)(6.97 + N_{v,\text{air}})$$

which yields

$$N_{v,\text{air}} = 0.131 \text{ kmol}$$

The balanced combustion equation is obtained by substituting the coefficients determined earlier and adding 0.131 kmol of H_2O to both sides of the equation:

$$\overbrace{(0.72\text{CH}_4 + 0.09\text{H}_2 + 0.14\text{N}_2 + 0.02\text{O}_2 + 0.03\text{CO}_2)}^{\text{fuel}} + \overbrace{1.465(\text{O}_2 + 3.76\text{N}_2)}^{\text{dry air}}$$

$$+ \overbrace{0.131\text{H}_2\text{O}}^{\text{moisture}} \longrightarrow 0.75\text{CO}_2 + \overbrace{1.661\text{H}_2\text{O}}^{\text{includes moisture}} + 5.648\text{N}_2$$

The dew-point temperature of the products is the temperature at which the water vapor in the products starts to condense as the products are cooled. Again, assuming ideal-gas behavior, the partial pressure of the water vapor in the combustion gases is

$$P_{v,\text{prod}} = \left(\frac{N_{v,\text{prod}}}{N_{\text{prod}}}\right) P_{\text{prod}} = \left(\frac{1.661 \text{ kmol}}{8.059 \text{ kmol}}\right)(101.325 \text{ kPa}) = 20.88 \text{ kPa}$$

Thus,

$$T_{\text{dp}} = T_{\text{sat @ 20.88 kPa}} = 60.9°\text{C}$$

If the combustion process were achieved with dry air instead of moist air, the products would contain less moisture, and the dew-point temperature in this case would be 59.5°C.

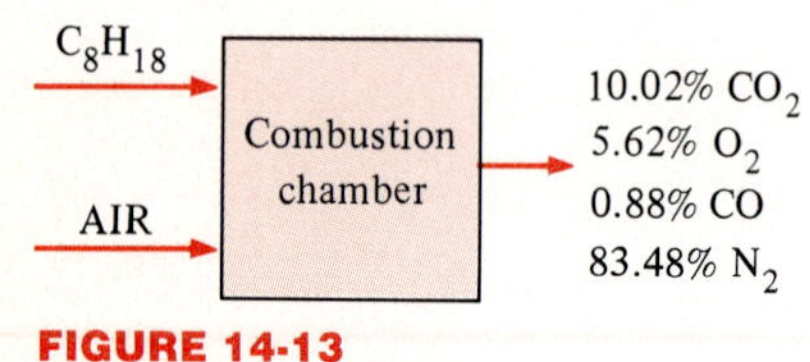

FIGURE 14-13
Schematic for Example 14-4.

EXAMPLE 14-4

Octane (C_8H_{18}) is burned with dry air. The volumetric analysis of the products on a dry basis is (Fig. 14-13)

CO_2	10.02 percent
O_2	5.62 percent
CO	0.88 percent
N_2	83.48 percent

Determine (*a*) the air-fuel ratio, (*b*) the percentage of theoretical air used, and (*c*) the fraction of the H_2O that condenses as the products are cooled to 25°C at 100 kPa.

Solution This time we know the relative composition of the products, but we do not know how much fuel or air is used during the combustion process. But the amounts of fuel and air can be determined from mass balances. If the combustion gases are assumed to be ideal gases, the volume fractions given above become equivalent to mole fractions. By considering 100 kmol of dry products for convenience, the combustion equation can be written as

$$xC_8H_{18} + a(O_2 + 3.76N_2) \longrightarrow 10.02CO_2 + 0.88CO + 5.62O_2 + 83.48N_2 + bH_2O$$

The unknown coefficients *x*, *a*, and *b* are determined from mass balances:

$$N_2: \quad 3.76a = 83.48 \longrightarrow a = 22.20$$

$$C: \quad 8x = 10.02 + 0.88 \longrightarrow x = 1.36$$

$$H: \quad 18x = 2b \longrightarrow b = 12.24$$

$$O_2: \quad a = 10.02 + 0.44 + 5.62 + \frac{b}{2} \longrightarrow 22.20 = 22.20$$

The O_2 balance is not necessary, but it can be used to check the values obtained from the other mass balances, as we did above. Substituting, we get

$$1.36C_8H_{18} + 22.2(O_2 + 3.76N_2) \longrightarrow 10.02CO_2 + 0.88CO + 5.62O_2 + 83.48N_2 + 12.24H_2O$$

The combustion equation for 1 kmol of fuel is obtained by dividing the above equation by 1.36:

$$C_8H_{18} + 16.32(O_2 + 3.76N_2) \longrightarrow 7.37CO_2 + 0.65CO + 4.13O_2 + 61.38N_2 + 9H_2O$$

(*a*) The air-fuel ratio is determined by taking the ratio of the mass of the air to the mass of the fuel (Eq. 14-3):

$$\text{AF} = \frac{m_{\text{air}}}{m_{\text{fuel}}} = \frac{(16.32 \times 4.76 \text{ kmol})(29 \text{ kg/kmol})}{(8 \text{ kmol})(12 \text{ kg/kmol}) + (9 \text{ kmol})(2 \text{ kg/kmol})}$$

$$= 19.76 \text{ kg air/kg fuel}$$

(*b*) To find the percentage of theoretical air used, we need to know the theoretical amount of air, which is determined from the theoretical combustion equation of the fuel:

$$C_8H_{18} + a_{\text{th}}(O_2 + 3.76N_2) \longrightarrow 8CO_2 + 9H_2O + 61.27N_2$$

$$O_2: \quad a_{\text{th}} = 8 + 4.5 \longrightarrow a_{\text{th}} = 12.5$$

Then,

$$\text{Percentage of theoretical air} = \frac{m_{\text{air,act}}}{m_{\text{air,th}}} = \frac{N_{\text{air,act}}}{N_{\text{air,th}}} = \frac{(16.32)(4.76) \text{ kmol}}{(12.50)(4.76) \text{ kmol}} = 131\%$$

That is, 31 percent excess air was used during this combustion process. Notice that some carbon formed carbon monoxide even though there was considerably more oxygen than needed for complete combustion.

(c) For each kmol of fuel burned, 7.37 + 0.65 + 4.13 + 61.38 + 9 = 82.53 kmol of products are formed, including 9 kmol of H_2O. Assuming that the dew-point temperature of the products is above 25°C, some of the water vapor will condense as the products are cooled to 25°C. If N_w kmol of H_2O condenses, there will be $9 - N_w$ kmol of water vapor left in the products. The mole number of the products in the gas phase will also decrease to $80.94 - N_w$ as a result. By treating the product gases (including the remaining water vapor) as ideal gases, N_w is determined by equating the mole fraction of the water vapor to its pressure fraction:

$$\frac{N_v}{N_{\text{prod,gas}}} = \frac{P_v}{P_{\text{prod}}}$$

$$\frac{9 - N_w}{82.53 - N_w} = \frac{3.169\text{ kPa}}{100\text{ kPa}}$$

$$N_w = 6.59\text{ kmol}$$

since $P_v = P_{\text{sat @ 25°C}} = 3.169$ kPa. Therefore, the majority of the water vapor in products (73 percent of it) will condense as the product gases are cooled to 25°C.

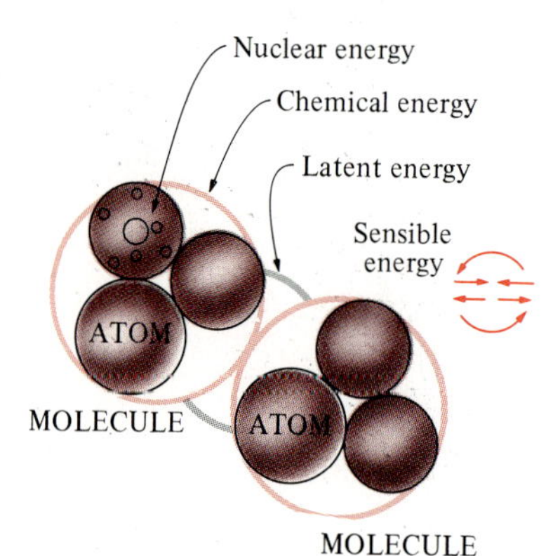

FIGURE 14-14
The microscopic form of energy of a substance consists of sensible, latent, chemical, and nuclear energies.

14-3 ■ ENTHALPY OF FORMATION AND ENTHALPY OF COMBUSTION

We mentioned in Chap. 1 that the molecules of a system possess energy in various forms such as *sensible and latent energy* (associated with a change of state), *chemical energy* (associated with the molecular structure), and *nuclear energy* (associated with the atomic structure), as illustrated in Fig. 14-14. In this text we do not intend to deal with nuclear energy, so we leave that sleeping giant alone. We also ignored chemical energy until now since the systems considered in previous chapters involved no changes in their chemical structure, and thus no changes in chemical energy. Consequently, all we needed to deal with were the sensible and latent energies.

During a chemical reaction, some chemical bonds which bind the atoms into molecules are broken, and other new ones are formed. The chemical energy associated with these bonds, in general, is different for the reactants and the products. Therefore, a process which involves chemical reactions will involve changes in chemical energies, which must be accounted for in an energy balance (Fig. 14-15). Assuming the atoms of each reactant remain intact (no nuclear reactions) and disregarding any changes in kinetic and potential energies, the energy change of a system during a chemical reaction will be due to a change in state and a change in chemical composition. That is,

$$\Delta E_{\text{sys}} = \Delta E_{\text{state}} + \Delta E_{\text{chem}} \tag{14-4}$$

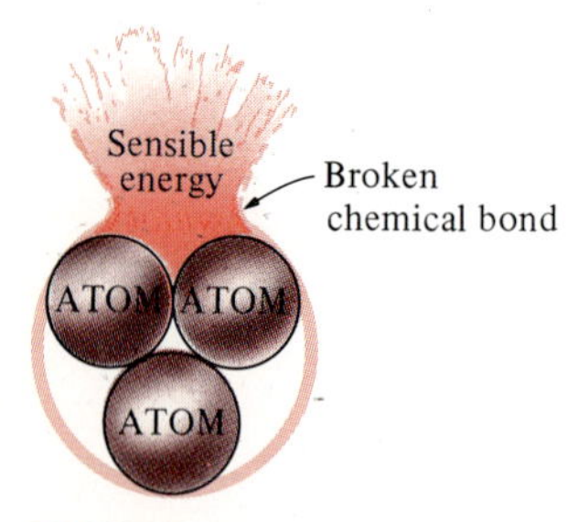

FIGURE 14-15
When the existing chemical bonds are destroyed and new ones are formed during a combustion process, usually a large amount of sensible energy is absorbed or released.

Therefore, when the products formed during a chemical reaction exit the reaction chamber at the inlet state of the reactants, we have $\Delta E_{state} = 0$ and the energy change of the system in this case is due to the changes in its chemical composition only.

In thermodynamics we are concerned with the *changes* in the energy of a system during a process, and not the energy values at the particular states. Therefore, we can choose any state as the reference state and assign a value of zero to the internal energy or enthalpy of a substance at that state. When a process involves no changes in chemical composition, the reference state chosen has no effect on the results. When the process involves chemical reactions, however, the composition of the system at the end of a process is no longer the same as that at the beginning of the process. In this case it becomes necessary to have a common reference state for all substances. The chosen reference state is 25°C (77°F) and 1 atm, which is known as the **standard reference state**. Property values at the standard reference state are indicated by a superscript "°" (such as $h°$ and $u°$).

When analyzing reacting systems, we must use property values relative to the standard reference state. However, it is not necessary to prepare a new set of property tables for this purpose. We can use the existing tables by subtracting the property values at the standard reference state from the values at the specified state. The ideal-gas enthalpy of N_2 at 500 K relative to the standard reference state, for example, is $\bar{h}_{500\ K} - \bar{h}° = 14{,}581 - 8669 = 5912$ kJ/kmol.

Consider the formation of CO_2 from its elements, carbon and oxygen, during a steady-flow combustion process (Fig. 14-16). Both the carbon and the oxygen enter the combustion chamber at 25°C and 1 atm. The CO_2 formed during this process leaves the combustion chamber also at 25°C and 1 atm. The combustion of carbon is an *exothermic reaction* (a reaction during which chemical energy is released in the form of heat). Therefore, some heat will be transferred from the combustion chamber to the surroundings during this process, which is 393,520 kJ/kmol CO_2 formed. (When one is dealing with chemical reactions, it is more convenient to work with quantities per unit mole than per unit time, even for steady-flow processes.)

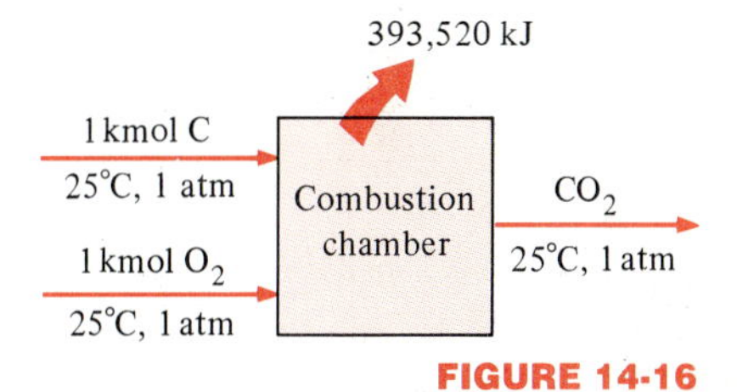

FIGURE 14-16

The formation of CO_2 during a steady-flow combustion process at 25°C and 1 atm.

The process described above involves no work interactions. Therefore, from the steady-flow conservation of energy equation, the heat transfer during this process must be equal to the difference between the enthalpy of the products and the enthalpy of the reactants. That is,

$$Q = H_p - H_R = -393{,}520 \text{ kJ/kmol} \tag{14-5}$$

Since both the reactants and the products are at the same state, the enthalpy change during this process is solely due to the changes in the chemical composition of the system. This enthalpy change will be different for different reactions, and it would be very desirable to have a property to represent the changes in chemical energy during a reaction. This property is the **enthalpy of reaction** h_R, which is defined as *the difference between the enthalpy of the products at a specified state and the enthalpy of the reactants at the same state for a complete reaction* (Fig. 14-17).

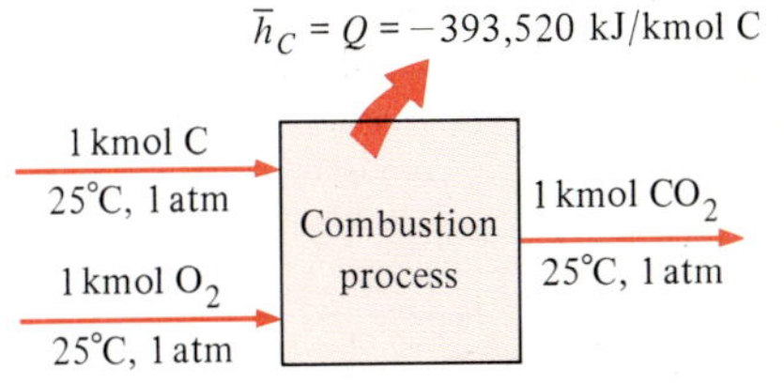

FIGURE 14-17

The enthalpy of combustion represents the amount of energy released as a fuel is burned during a steady-flow process at a specified state.

For combustion processes the enthalpy of reaction is usually referred to as the **enthalpy of combustion** h_C, which represents the amount of heat released during a steady-flow combustion process when 1 kmol (or 1 kg) of fuel is burned completely at a specified temperature and pressure. It is expressed as

$$h_C = H_P - H_R \tag{14-6}$$

which is $-393{,}520$ kJ/kmol for C. The enthalpy of combustion of a particular fuel will be different at different temperatures and pressures. Table A-27 lists h_C values for various fuels at the standard reference state of 25°C and 1 atm.

The enthalpy of combustion is obviously a very useful property for analyzing the combustion processes of fuels. However, there are so many different fuels and fuel mixtures that it is not practical to list h_C values for all possible cases. Besides, the enthalpy of combustion is not of much use when the combustion is incomplete. Therefore a more practical approach would be to have a more fundamental property to represent the chemical energy of an element or a compound at some reference state. This property is the **enthalpy of formation** $\bar{h}_f$, which can be viewed as *the enthalpy of a substance at a specified state due to its chemical composition.*

To establish a starting point, we assign the enthalpy of formation of all stable elements (such as O_2, N_2, H_2, and C) a value of zero at the standard reference state of 25°C and 1 atm. That is, $\bar{h}_f^\circ = 0$ for all stable elements. (This is no different from assigning the internal energy of saturated liquid water a value of zero at 0.01°C.) Perhaps we should clarify what we mean by *stable*. The stable form of an element is simply the chemically stable form of that element at 25°C and 1 atm. Nitrogen, for example, exists in diatomic form (N_2) at 25°C and 1 atm. Therefore, the stable form of nitrogen at the standard reference state is diatomic nitrogen N_2, not monatomic nitrogen N. If an element exists in more than one stable form at 25°C and 1 atm, one of the forms should be specified as the stable form. For carbon, for example, the stable form is assumed to be graphite, not diamond.

Now reconsider the formation of CO_2 (a compound) from its elements C and O_2 at 25°C and 1 atm during a steady-flow process. The enthalpy change during this process was determined to be (Eq. 14-5)

$$H_P - H_R = -393{,}520 \text{ kJ/kmol}$$

But $H_R = 0$ since both reactants are elements at the standard reference state, and the products consist of 1 kmol of CO_2 at the same state. Therefore, the enthalpy of formation of CO_2 at the standard reference state is $-393{,}520$ kJ/kmol (Fig. 14-18). That is,

$$\bar{h}^\circ_{f,CO_2} = -393{,}520 \text{ kJ/kmol}$$

The negative sign is due to the fact that the enthalpy of 1 kmol of CO_2 at 25°C and 1 atm is 393,520 kJ less than the enthalpy of 1 kmol of C and 1 kmol of O_2 at the same state. In other words, 393,520 kJ of chemical

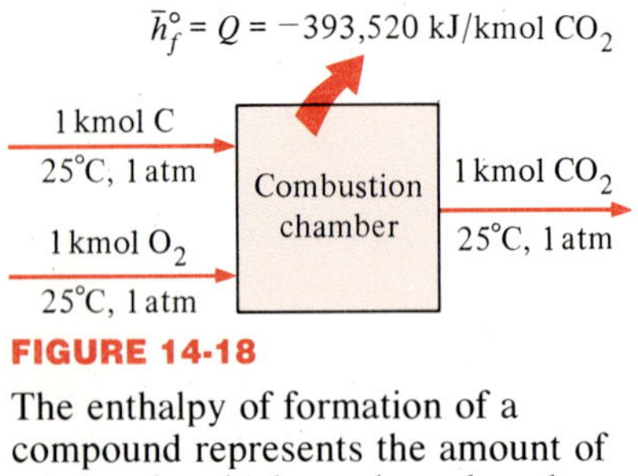

FIGURE 14-18

The enthalpy of formation of a compound represents the amount of energy absorbed or released as the component is formed from its stable elements during a steady-flow process at a specified state.

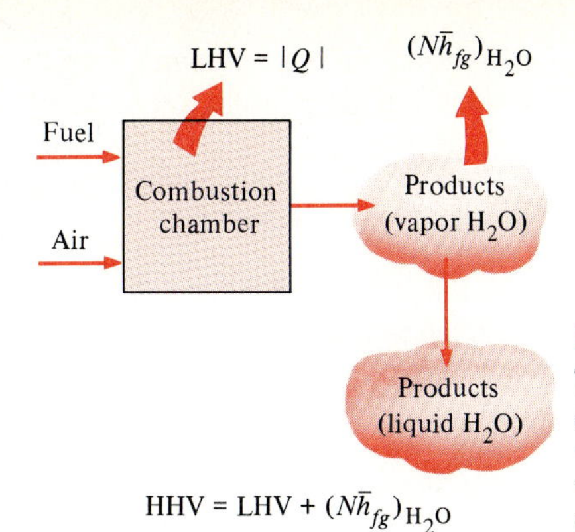

FIGURE 14-19

The higher heating value of a fuel is equal to the sum of the lower heating value of the fuel and the latent heat of vaporization of the H_2O in the products.

energy is released (leaving the system as heat) when C and O_2 combine to form 1 kmol of CO_2. Therefore, a negative enthalpy of formation for a compound indicates that heat is released during the formation of that compound from its stable elements. A positive value indicates heat is absorbed.

You will notice that two $\bar{h}_f^\circ$ values are given for H_2O in Table A-26, one for liquid water and the other for water vapor. This is because both phases of H_2O are encountered at 25°C, and the effect of pressure on the enthalpy of formation is small. The difference between the two enthalpies of formation is approximately equal to the h_{fg} of water at 25°C, which is 2442.3 kJ/kg or 44,000 kJ/kmol.

Another term commonly used in conjunction with the combustion of fuels is the **heating value** of the fuel, which is defined as the amount of energy released when a fuel is burned completely in a steady-flow process and the products are returned to the state of the reactants. In other words, the heating value of a fuel is equal to the absolute value of the enthalpy of combustion of the fuel. That is,

$$\text{Heating value} = |h_C| \qquad \text{(kJ/kg fuel)}$$

The heating value depends on the *phase* of the H_2O in the products. The heating value is called the **higher heating value** (HHV) when the H_2O in the products is in the liquid form, and it is called the **lower heating value** (LHV) when the H_2O in the products is in the vapor form (Fig. 14-19). The two heating values are related by

$$\text{HHV} = \text{LHV} + (N\bar{h}_{fg})_{H_2O} \qquad \text{(kJ/kg fuel)} \qquad (14\text{-}7)$$

where N is the number of moles of H_2O in the products and $\bar{h}_{fg}$ is the enthalpy of vaporization of water at 25°C.

The heating value or enthalpy of combustion of a fuel can be determined from a knowledge of the enthalpy of formation for the compounds involved. This is illustrated with the following example.

EXAMPLE 14-5

Determine the enthalpy of combustion of gaseous octane (C_8H_{18}) at 25°C and 1 atm, using enthalpy-of-formation data from Table A-26. Assume the water in the products is in the liquid form.

Solution The combustion of C_8H_{18} is illustrated in Fig. 14-20. The stoichiometric equation for this reaction is

$$C_8H_{18} + a_{th}(O_2 + 3.76N_2) \longrightarrow 8CO_2 + 9H_2O(\ell) + 3.76a_{th}N_2$$

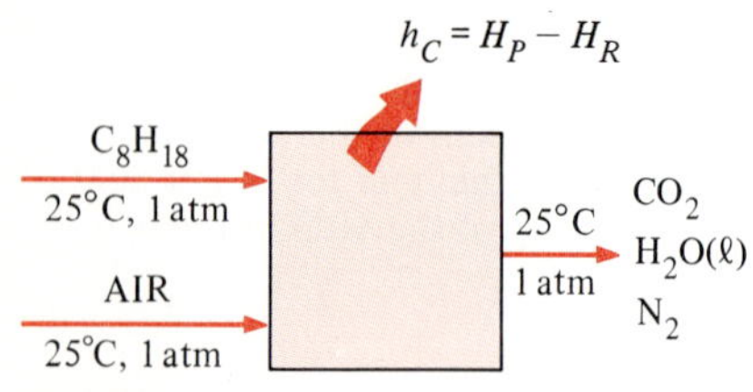

FIGURE 14-20
Schematic for Example 14-5.

Both the reactants and the products are at the standard reference state of 25°C and 1 atm. Also, N_2 and O_2 are stable elements, and thus their enthalpy of formation is zero. Then the enthalpy of combustion of C_8H_{18} becomes (Eq. 14-6)

$$\begin{aligned}\bar{h}_C &= H_P - H_R \\ &= \Sigma N_P\bar{h}^\circ_{f,P} - \Sigma N_R\bar{h}^\circ_{f,R} = (N\bar{h}^\circ_f)_{CO_2} + (N\bar{h}^\circ_f)_{H_2O} - (N\bar{h}^\circ_f)_{C_8H_{18}}\end{aligned}$$

Using $\bar{h}^\circ_f$ values from Table A-26, we get

$$\begin{aligned}\bar{h}_C &= (8\text{ kmol})(-393{,}520\text{ kJ/kmol}) + (9\text{ kmol})(-285{,}830\text{ kJ/kmol}) \\ &\quad - (1\text{ kmol})(-208{,}450\text{ kJ/kmol}) \\ &= -5{,}512{,}180\text{ kJ/kmol } C_8H_{18}\end{aligned}$$

which is practically identical to the listed value of −5,512,200 kJ in Table A-27. Since the water in the products is assumed to be in the liquid phase, this h_C value corresponds to the HHV of C_8H_{18}.

14-4 FIRST-LAW ANALYSIS OF REACTING SYSTEMS

The conservation of energy (or first-law) relations developed in Chaps. 3 and 4 are applicable to both reacting and nonreacting systems. However, chemically reacting systems involve changes in their chemical energy, and thus it is more convenient to rewrite the first-law relations so that the changes in chemical energies are explicitly expressed. We do this first for steady-flow systems and then for closed systems.

Steady-Flow Systems

When the changes in kinetic and potential energies are negligible, the conservation of energy relation for a chemically reacting steady-flow system can be expressed as

$$Q - W = H_P - H_R \qquad \text{(kJ)} \tag{14-8}$$

where H_P represents the enthalpy of the products (the exit streams) and H_R represents the enthalpy of the reactants (the inlet streams). Then H_P and H_R can be expressed in terms of enthalpies and mole numbers of their components as (Fig. 14-21)

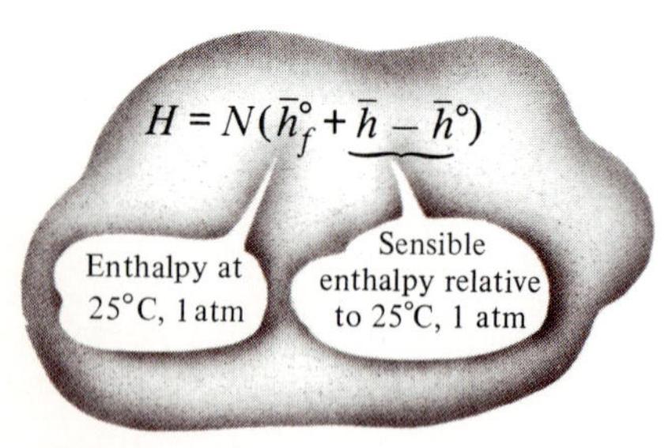

FIGURE 14-21
The enthalpy of a chemical component at a specified state is the sum of the enthalpy of the component at 25°C, 1 atm ($\bar{h}^\circ_f$), and the sensible enthalpy of the component relative to 25°C, 1 atm.

$$H_P = \Sigma N_P(\bar{h}^\circ_f + \bar{h} - \bar{h}^\circ)_P \qquad \text{(kJ)} \tag{14-9}$$

$$H_R = \Sigma N_R(\bar{h}^\circ_f + \bar{h} - \bar{h}^\circ)_R \qquad \text{(kJ)} \tag{14-10}$$

where the superscript ° represents properties at the standard reference state of 25°C and 1 atm. Substituting into Eq. 14-8, we obtain the conservation of energy relation for chemically reacting steady-flow systems:

$$Q - W = \Sigma N_P(\bar{h}_f^\circ + \bar{h} - \bar{h}^\circ)_P - \Sigma N_R(\bar{h}_f^\circ + \bar{h} - \bar{h}^\circ)_R \quad \text{(kJ)} \quad (14\text{-}11)$$

where $\bar{h}_f^\circ$ represents the *chemical enthalpy* of a substance at the standard reference state and $\bar{h} - \bar{h}^\circ$ represents the *sensible and latent enthalpy* relative to the standard reference state. If the enthalpy of combustion for a particular reaction is available, Eq. 14-11 can also be expressed as

$$Q - W = N_F\bar{h}_C^\circ + \Sigma N_P(\bar{h} - \bar{h}^\circ)_P - \Sigma N_R(\bar{h} - \bar{h}^\circ)_R \quad \text{(kJ)} \quad (14\text{-}12)$$

where N_F is the number of moles of the fuel. The two equations above are sometimes written without the work term since most steady-flow combustion processes do not involve any work interactions.

Closed Systems

The conservation of energy relation for a chemically reacting stationary closed system can be expressed as

$$Q - W = U_P - U_R \quad \text{(kJ)} \quad (14\text{-}13)$$

where U_P represents the internal energy of the products and U_R represents the internal energy of the reactants. To avoid using another property—*the internal energy of formation* $\bar{u}_f^\circ$—we utilize the definition of enthalpy ($\bar{u} = \bar{h} - P\bar{v}$ or $\bar{u}_f^\circ + \bar{u} - \bar{u}^\circ = \bar{h}_f^\circ + \bar{h} - \bar{h}^\circ - P\bar{v}$) and express the above equation as (Fig. 14-22).

$$Q - W = \Sigma N_P(\bar{h}_f^\circ + \bar{h} - \bar{h}^\circ - P\bar{v})_P - \Sigma N_R(\bar{h}_f^\circ + \bar{h} - \bar{h}^\circ - P\bar{v})_R \quad \text{(kJ)} \quad (14\text{-}14)$$

The $P\bar{v}$ terms are negligibly small for solids and liquids, and can be replaced by R_uT for gases which behave as an ideal gas. Also, if desired, the $\bar{h} - P\bar{v}$ terms in Eq. 14-14 can be replaced by $\bar{u}$.

The work term in Eq. 14-14 represents all forms of work, including the boundary work. It was shown in Chap. 3 that $\Delta U + W_b = \Delta H$ for nonreacting closed systems undergoing a quasi-equilibrium P = constant expansion or compression process. This is also the case for chemically reacting systems.

$$U = H - PV$$
$$= N(h_f^\circ + h - h^\circ) - PV$$
$$= N(h_f^\circ + h - h^\circ - P\bar{v})$$

FIGURE 14-22
An expression for the internal energy of a chemical component in terms of the enthalpy.

EXAMPLE 14-6

Liquid propane (C_3H_8) enters a combustion chamber at 25°C at a rate of 0.05 kg/min where it is mixed and burned with 50 percent excess air which enters the combustion chamber at 7°C, as shown in Fig. 14-23. An analysis of the combustion gases reveals that all the hydrogen in the fuel burns to H_2O but only 90 percent of the carbon burns to CO_2, with the remaining 10 percent forming CO. If the exit temperature of the combustion gases is 1500 K, determine (*a*) the mass flow rate of air and (*b*) the rate of heat transfer from the combustion chamber.

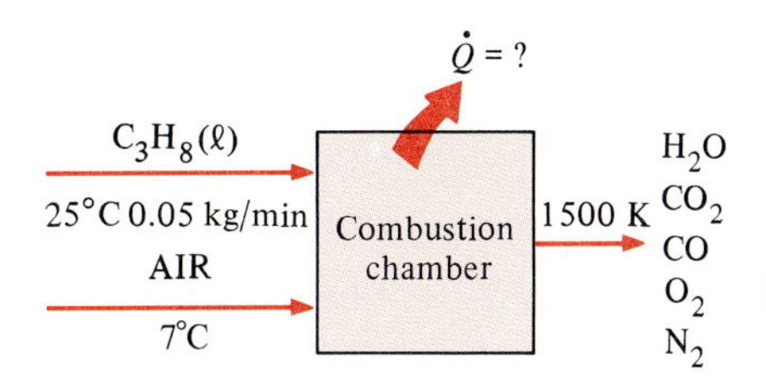

FIGURE 14-23
Schematic for Example 14-6.

Solution The amount of air is specified relative to theoretical requirements, which is determined by an oxygen balance on the theoretical reaction:

$$C_3H_8(\ell) + a_{th}(O_2 + 3.76N_2) \longrightarrow 3CO_2 + 4H_2O + 3.76a_{th}N_2$$

O_2 balance: $a_{th} = 3 + 2 = 5$

Then the balanced equation for the actual combustion process with 50 percent excess air and some CO in the products becomes

$$C_3H_8(\ell) + 7.5(O_2 + 3.76N_2) \longrightarrow 2.7CO_2 + 0.3CO + 4H_2O + 2.65O_2 + 28.2N_2$$

(*a*) The air-fuel ratio for this combustion process is

$$AF = \frac{m_{air}}{m_{fuel}} = \frac{(7.5 \times 4.76\text{ kmol})(29\text{ kg/kmol})}{(3\text{ kmol})(12\text{ kg/kmol}) + (4\text{ kmol})(2\text{ kg/kmol})}$$
$$= 23.53\text{ kg air/kg fuel}$$

Thus,

$$\dot{m}_{air} = (AF)(\dot{m}_{fuel})$$
$$= (23.53\text{ kg air/kg fuel})(0.05\text{ kg fuel/min})$$
$$= 1.18\text{ kg air/min}$$

(*b*) The heat transfer for this steady-flow combustion process is determined from Eq. 14-11 with $W = 0$:

$$Q = \Sigma N_P(\bar{h}_f^\circ + \bar{h} - \bar{h}^\circ)_P - \Sigma N_R(\bar{h}_f^\circ + \bar{h} - \bar{h}^\circ)_R$$

Assuming the air and the combustion products to be ideal gases, we have $h = h(T)$ and using data from the property tables we form the following minitable:

Substance	$\bar{h}_f^\circ$, kJ/kmol	$\bar{h}_{280\text{ K}}$, kJ/kmol	$\bar{h}_{298\text{ K}}$, kJ/kmol	$\bar{h}_{1500\text{ K}}$, kJ/kmol
$C_3H_8(\ell)$	−118,910	—	—	—
O_2	0	8150	8682	49,292
N_2	0	8141	8669	47,073
$H_2O(g)$	−241,820	—	9904	57,999
CO_2	−393,520	—	9364	71,078
CO	−110,530	—	8669	47,517

The $\bar{h}_f^\circ$ of liquid propane is obtained by adding the $\bar{h}_{fg}$ of propane at 25°C to the $\bar{h}_f^\circ$ of gas propane. Substituting gives

$$\begin{aligned}
Q = {} & (2.7\text{ kmol }CO_2)[(-393{,}520 + 71{,}078 - 9364)\text{ kJ/kmol }CO_2] \\
& + (0.3\text{ kmol CO})[(-110{,}530 + 47{,}517 - 8669)\text{ kJ/kmol CO}] \\
& + (4\text{ kmol }H_2O)[(-241{,}820 + 57{,}999 - 9904)\text{ kJ/kmol }H_2O] \\
& + (2.65\text{ kmol }O_2)[(0 + 49{,}292 - 8682)\text{ kJ/kmol }O_2] \\
& + (28.2\text{ kmol }N_2)[(0 + 47{,}073 - 8669)\text{ kJ/kmol }N_2] \\
& - (1\text{ kmol }C_3H_8)[(-118{,}910 + h_{298} - h_{298})\text{ kJ/kmol }C_3H_8] \\
& - (7.5\text{ kmol }O_2)[(0 + 8150 - 8682)\text{ kJ/kmol }O_2] \\
& - (28.2\text{ kmol }N_2)[(0 + 8141 - 8669)\text{ kJ/kmol }N_2] \\
= {} & -363{,}882\text{ kJ/kmol of }C_3H_8
\end{aligned}$$

Thus 363,882 kJ of heat is transferred from the combustion chamber for each kmol (44 kg) of propane. This corresponds to 363,882/44 = 8270.0 kJ of heat loss per kilogram of propane. Then the rate of heat transfer for a mass flow rate of 0.05 kg/min for the propane becomes

$$\dot{Q} = \dot{m}q = (0.05 \text{ kg/min})(-8270.0 \text{ kJ/kg}) = -413.5 \text{ kJ/min}$$
$$= -6.89 \text{ kW}$$

EXAMPLE 14-7

The constant-volume tank shown in Fig. 14-24 contains 1 lbmol of methane (CH_4) gas and 3 lbmol of O_2 at 77°F and 1 atm. The contents of the tank are ignited, and the methane gas burns completely. If the final temperature is 1800 R, determine (*a*) the final pressure in the tank and (*b*) the heat transfer during this process.

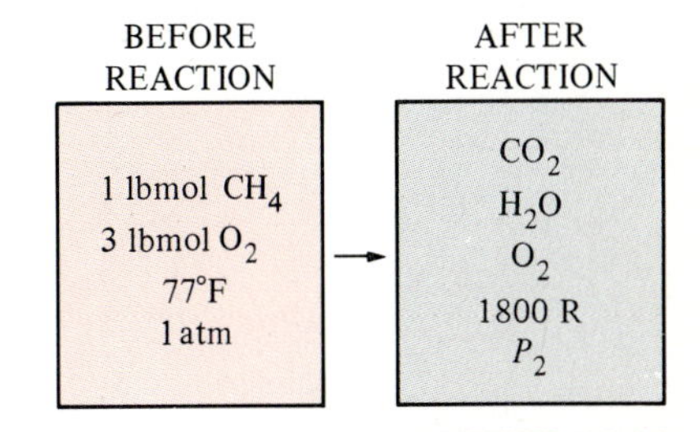

FIGURE 14-24
Schematic for Example 14-7.

Solution Since the combustion is assumed to be complete, all the carbon in the methane burns to CO_2 and all the hydrogen to H_2O. Then the balanced combustion equation becomes

$$CH_4(g) + 3O_2 \longrightarrow CO_2 + 2H_2O + O_2$$

(*a*) At 1800 R, water exists in the gas phase. Assuming both the reactants and the products to be ideal gases, the final pressure in the tank is

$$\left.\begin{aligned} P_R V &= N_R R_u T_R \\ P_P V &= N_P R_u T_P \end{aligned}\right\} \qquad P_P = P_R \left(\frac{N_P}{N_R}\right)\left(\frac{T_P}{T_R}\right)$$

Substituting, we get

$$P_P = (1 \text{ atm})\left(\frac{4 \text{ lbmol}}{4 \text{ lbmol}}\right)\left(\frac{1800 \text{ R}}{537 \text{ R}}\right) = 3.35 \text{ atm}$$

which is relatively low. Therefore, the ideal-gas assumption utilized earlier is appropriate.

(*b*) The heat transfer during this constant-volume combustion process can be determined from Eq. 14-14:

$$Q - W = \Sigma N_P(\bar{h}_f^\circ + \bar{h} - \bar{h}^\circ - P\bar{v})_P - \Sigma N_R(\bar{h}_f^\circ + \bar{h} - \bar{h}^\circ - P\bar{v})_R$$

Since both the reactants and the products are assumed to be ideal gases, all the internal energy and enthalpies depend on temperature only, and the $P\bar{v}$ terms in this equation can be replaced by R_uT. It yields

$$Q = \Sigma N_P(\bar{h}_f^\circ + \bar{h}_{1800 \text{ R}} - \bar{h}_{537 \text{ R}} - R_uT)_P - \Sigma N_R(\bar{h}_f^\circ - R_uT)_R$$

since $W = 0$ and the reactants are at the standard reference temperature of 537 R. From $\bar{h}_f^\circ$ and ideal-gas tables,

Substance	$\bar{h}_f^\circ$, Btu/lbmol	$\bar{h}_{537 \text{ R}}$, Btu/lbmol	$\bar{h}_{1800 \text{ R}}$, Btu/lbmol
CH_4	−32,210	—	—
O_2	0	3725.1	13,485.8
CO_2	−169,300	4027.5	18,391.5
$H_2O(g)$	−104,040	4258.0	15,433.0

Substituting, we have

$$\begin{aligned} Q = {} & (1 \text{ lbmol } CO_2)[(-169{,}300 + 18{,}391.5 - 4027.5 - 1.986 \times 1800) \text{ Btu/lbmol } CO_2] \\ & + (2 \text{ lbmol } H_2O)[(-104{,}040 + 15{,}433.0 - 4258.0 - 1.986 \times 1800) \text{ Btu/lbmol } H_2O] \\ & + (1 \text{ lbmol } O_2)[(0 + 13{,}485.8 - 3725.1 - 1.986 \times 1800) \text{ Btu/lbmol } O_2] \\ & - (1 \text{ lbmol } CH_4)[(-32{,}210 - 1.986 \times 537) \text{ Btu/lbmol } CH_4] \\ & - (3 \text{ lbmol } O_2)[(0 - 1.986 \times 537) \text{ Btu/lbmol } O_2] \\ = {} & -308{,}729 \text{ Btu/lbmol } CH_4 \end{aligned}$$

The negative sign indicates that heat is transferred from the combustion chamber to the surroundings. On a mass basis, the heat transfer would be $-308{,}729/16 = -19{,}296$ Btu/lbm.

14-5 ■ ADIABATIC FLAME TEMPERATURE

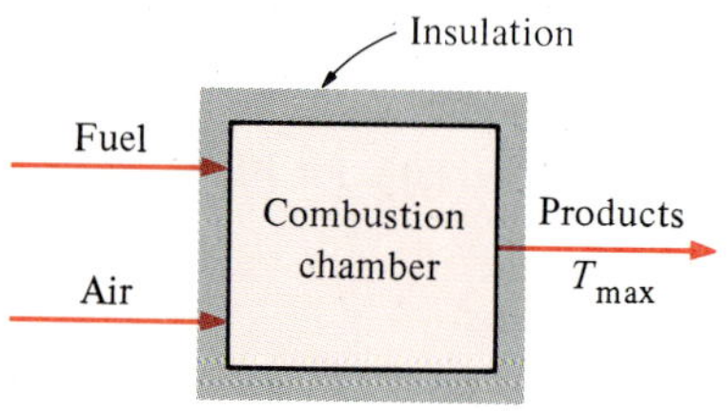

FIGURE 14-25
The temperature of a combustion chamber will be maximum when combustion is complete and no heat is lost to the surroundings ($Q = 0$).

In the absence of any work interactions and any changes in kinetic or potential energies, the chemical energy released during a combustion process either is lost as heat to the surroundings or is used internally to raise the temperature of the combustion products. The smaller the heat loss, the larger the temperature rise. In the limiting case of no heat loss to the surroundings ($Q = 0$), the temperature of the products will reach a maximum, which is called the **adiabatic flame temperature** of the reaction (Fig. 14-25).

The adiabatic flame temperature of a steady-flow combustion process is determined from Eq. 14-8 by setting $Q = 0$ and $W = 0$. This yields

$$H_P = H_R \tag{14-15}$$

or

$$\Sigma N_P(\overline{h}_f^\circ + \overline{h} - \overline{h}^\circ)_P = \Sigma N_R(\overline{h}_f^\circ + \overline{h} - \overline{h}^\circ)_R \tag{14-16}$$

Once the reactants and their states are specified, the enthalpy of the reactants H_R can be easily determined. The calculation of the enthalpy of the products H_P is not so straightforward, however, because the temperature of the products is not known prior to the calculations. Therefore, the determination of the adiabatic flame temperature requires the use of an iterative technique unless equations for the sensible enthalpy changes of the combustion products are available. A temperature is assumed for the product gases, and the H_P is determined for this temperature. If it is not equal to H_R, calculations are repeated with another temperature. The adiabatic flame temperature is then determined from these two results by interpolation. When the oxidant is air, the product gases mostly consist of N_2, and a good first guess for the adiabatic flame temperature is obtained by treating the entire product gases as N_2.

In combustion chambers, the highest temperature to which a material can be exposed is limited by metallurgical considerations. Therefore, the adiabatic flame temperature is an important consideration in the design of combustion chambers, gas turbines, and nozzles. The maximum temperatures that occur in these devices are considerably lower than the adiabatic flame temperature; however, since the combustion is usually incomplete, some heat loss takes place, and some combustion gases dissociate at high temperatures (Fig. 14-26). The maximum temperature in a combustion chamber can be controlled by adjusting the amount of excess air, which serves as a coolant.

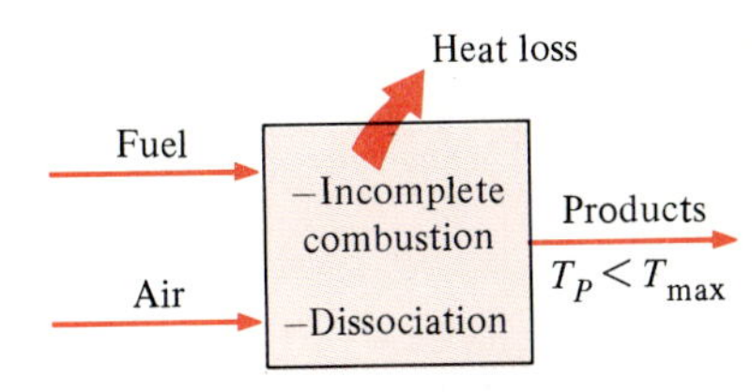

FIGURE 14-26
The maximum temperature encountered in a combustion chamber is lower than the theoretical adiabatic flame temperature.

EXAMPLE 14-8

Liquid octane (C_8H_{18}) enters the combustion chamber of a gas turbine steadily at 1 atm and 25°C, and it is burned with air which enters the combustion chamber at the same state, as shown in Fig. 14-27. Disregarding any changes in kinetic and potential energies, determine the adiabatic flame temperature for (*a*) complete combustion with 100 percent theoretical air, (*b*) complete combustion with 400 percent theoretical air, and (*c*) incomplete combustion (some CO in the products) with 90 percent theoretical air.

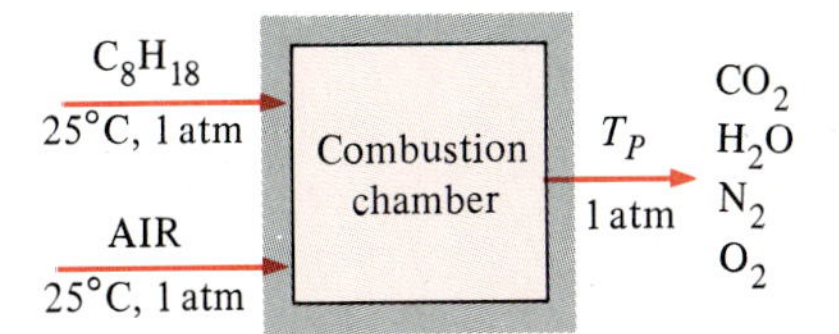

FIGURE 14-27
Schematic for Example 14-8.

Solution Under adiabatic conditions, there will be no heat transfer from the combustion chamber, and the products will exit at the highest possible temperature, which is the adiabatic flame temperature.

(*a*) The balanced equation for the combustion process with the theoretical amount of air is

$$C_8H_{18}(\ell) + 12.5(O_2 + 3.76N_2) \longrightarrow 8CO_2 + 9H_2O + 47N_2$$

The adiabatic flame temperature relation (Eq. 14-16) in this case reduces to

$$\Sigma N_P(\bar{h}_f^\circ + \bar{h} - \bar{h}^\circ)_P = \Sigma N_R \bar{h}_{f,R}^\circ = (N\bar{h}_f^\circ)_{C_8H_{18}}$$

since all the reactants are at the standard reference state and $\bar{h}_f^\circ = 0$ for O_2 and N_2. The $\bar{h}_f^\circ$ and h values of various components at 298 K are

Substance	$\bar{h}_f^\circ$, kJ/kmol	$\bar{h}_{298}$, kJ/kmol
$C_8H_{18}(\ell)$	−249,950	—
O_2	0	8682
N_2	0	8669
$H_2O(g)$	−241,820	9904
CO_2	−393,520	9364

Substituting, we have

$$\begin{aligned}&(8 \text{ kmol } CO_2)[(-393{,}520 + \bar{h}_{CO_2} - 9364) \text{ kJ/kmol } CO_2]\\ &+ (9 \text{ kmol } H_2)[(-241{,}820 + \bar{h}_{H_2O} - 9904) \text{ kJ/kmol } H_2O]\\ &+ (47 \text{ kmol } N_2)[(0 + \bar{h}_{N_2} - 8669) \text{ kJ/kmol } N_2]\\ &\qquad = (1 \text{ kmol } C_8H_{18})(-249{,}950 \text{ kJ/kmol } C_8H_{18})\end{aligned}$$

which yields

$$8\bar{h}_{CO_2} + 9\bar{h}_{H_2O} + 47\bar{h}_{N_2} = 5{,}646{,}081 \text{ kJ}$$

It appears that we have one equation with three unknowns. But actually we have only one unknown—the temperature of the products T_P—since $h = h(T)$ for ideal gases. Therefore, we will have to use a trial-and-error solution to determine the temperature of the products.

A first guess is obtained by dividing the right-hand side of the equation by the total number of moles, which yields 5,646,081/(8 + 9 + 47) = 88,220 kJ/kmol. This enthalpy value will correspond to about 2650 K for N_2, 2100 K for H_2O, and 1800 K for CO_2. Noting that the majority of the moles are N_2, we see that T_P will be close to 2650 K, but somewhat under it. Therefore, a good first guess is 2400 K. At this temperature,

$$8\bar{h}_{CO_2} + 9\bar{h}_{H_2O} + 47\bar{h}_{N_2} = 8 \times 125{,}152 + 9 \times 103{,}508 + 47 \times 79{,}320$$
$$= 5{,}660{,}828 \text{ kJ}$$

This value is higher than 5,646,081 kJ. Therefore, the actual temperature will be slightly under 2400 K. Next we choose 2350 K. It yields

$$8 \times 122{,}091 + 9 \times 100{,}508 + 47 \times 77{,}496 = 5{,}523{,}612 \text{ kJ}$$

which is lower than 5,646,081 kJ. Therefore, the actual temperature of the products is between 2350 and 2400 K. By interpolation, it is found to be $T_P = 2395$ K

(*b*) The balanced equation for the complete combustion process with 400 percent theoretical air is

$$C_8H_{18}(\ell) + 50(O_2 + 3.76N_2) \longrightarrow 8CO_2 + 9H_2O + 37.5O_2 + 188N_2$$

By following the procedure used in (*a*), the adiabatic flame temperature in this case is determined to be $T_P = 962$ K

Notice that the temperature of the products decreases significantly as a result of using excess air.

(*c*) The balanced equation for the incomplete combustion process with 90 percent theoretical air is

$$C_8H_{18}(\ell) + 11.25(O_2 + 3.76N_2) \longrightarrow 5.5CO_2 + 2.5CO + 9H_2O + 42.3N_2$$

Following the procedure used in (*a*), we find the adiabatic flame temperature in this case to be $T_P = 2236$ K

Notice that the adiabatic flame temperature decreases as a result of incomplete combustion or using excess air. Also, *the maximum adiabatic flame temperature is achieved when complete combustion occurs with the theoretical amount of air.*

14-6 ■ ENTROPY CHANGE OF REACTING SYSTEMS

So far we have analyzed combustion processes from the conservation of mass and the conservation of energy points of view. The thermodynamic analysis of a process is not complete, however, without the examination of the second-law aspects. Of particular interest are the irreversibility and reversible work, both of which are related to entropy.

As discussed in Chap. 7, entropy *can* be generated but it *cannot* be destroyed. The entropy generated S_{gen} during a process is equal to the total entropy change associated with that process, which is the sum of the entropy change of the system and of the surroundings. The second law of thermodynamics requires that

$$S_{gen} = \Delta S_{sys} + \Delta S_{surr} \geq 0 \qquad \text{(kJ/K)} \qquad (14\text{-}17)$$

whether a system is chemically reacting or not. For reacting systems, ΔS_{sys} can be taken to represent the entropy change associated with the reaction within the reaction chamber boundaries, which is equal to the difference between the entropy of the products S_P and the entropy of the reactants S_R, as illustrated in Fig. 14-28. That is,

$$\Delta S_{sys} = S_P - S_R = \Sigma N_P \bar{s}_P - \Sigma N_R \bar{s}_R \qquad \text{(kJ/K)} \qquad (14\text{-}18)$$

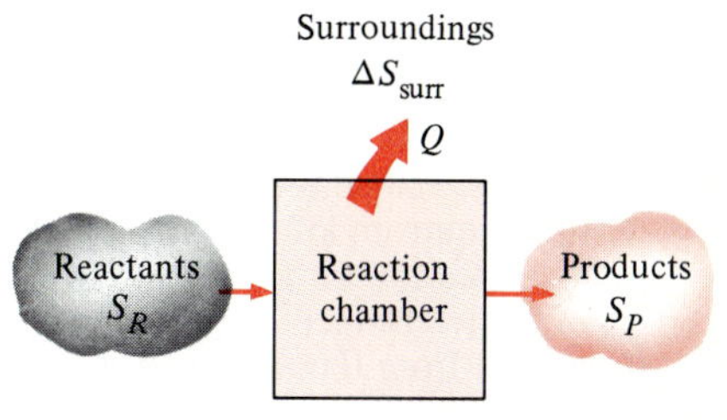

FIGURE 14-28

The total entropy change S_{gen} associated with a chemical reaction.

In general, ΔS_{surr} represents the entropy change of everything external to the system under the influence of the chemical reaction. If the only interaction between a reacting system and its surroundings is heat transfer, then

$$\Delta S_{surr} = \frac{Q_{surr}}{T_0} \qquad \text{(kJ/K)} \qquad (14\text{-}19)$$

where Q_{surr} is the heat transfer from the system to the surroundings and T_0 is the temperature of the surroundings, which is assumed to remain constant.

The determination of the entropy change associated with a chemical reaction seems to be straightforward, except for one thing: The entropy relations for the reactants and the products (Eq. 14-18) involve the *entropies* of the components, *not entropy changes* which was the case for nonreacting systems. Thus we are faced with the problem of finding a common base for the entropy of all substances, as we did with enthalpy. The search for such a common base led to the establishment of the **third law of thermodynamics** in the early part of this century. The third law was expressed in Chap. 6 as follows: *The entropy of a pure crystalline substance at absolute zero temperature is zero.*

Therefore, the third law of thermodynamics provides an absolute base for the entropy values for all substances. Entropy values relative to this base are called the **absolute entropy**. The $\bar{s}°$ values listed in Tables A-18 through A-25 for various gases such as N_2, O_2, CO, CO_2, H_2, H_2O, OH, and O are the ideal-gas *absolute entropy values* at the specified temperature and *at a pressure of 1 atm*. The absolute entropy values for various fuels are listed in Table A-26 together with the $\bar{h}_f°$ values at the standard reference state of 25°C and 1 atm.

Equation 14-18 is a general relation for the entropy change of a reacting system. It requires the determination of the entropy of each individual component of the reactants and the products, which in general is not very easy to do. The entropy calculations can be simplified somewhat if the gaseous components of the reactants and the products are approximated as ideal gases. However, entropy calculations are never as easy as enthalpy or internal energy calculations, since entropy is a function of both temperature and pressure even for ideal gases.

When evaluating the entropy of a component of an ideal-gas mixture, we should use the temperature and the partial pressure of the component. Note that the temperature of a component is the same as the temperature of the mixture, and the partial pressure of a component is equal to the mixture pressure multiplied by the mole fraction of the component.

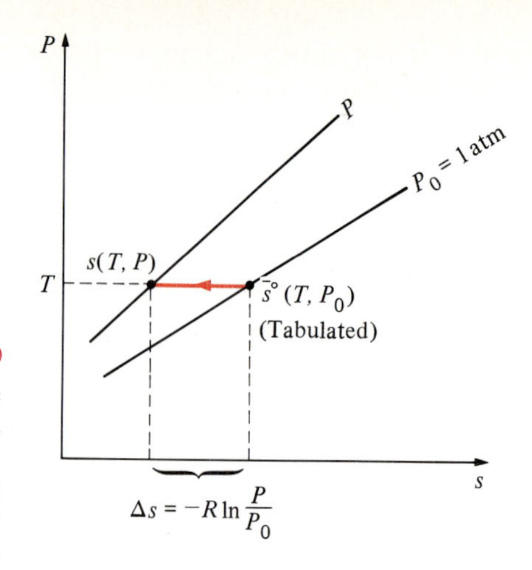

FIGURE 14-29
At a specified temperature, the absolute entropy of an ideal gas at pressures other than $P_0 = 1$ atm can be determined by subtracting $R \ln(P/P_0)$ from the tabulated value at 1 atm.

Absolute entropy values at pressures other than $P_0 = 1$ atm for any temperature T can easily be obtained from the ideal-gas entropy change relation written for an imaginary isothermal process between states (T, P_0) and (T, P), as illustrated in Fig. 14-29:

$$\bar{s}(T, P) = \bar{s}°(T, P_0) - R_u \ln \frac{P}{P_0} \qquad \text{[kJ/(kmol·K)]} \qquad (14\text{-}20)$$

For the component i of an ideal-gas mixture, this relation can be written

$$\bar{s}_i(T, P_i) = \bar{s}_i°(T, P_0) - R_u \ln y_i P_m \qquad \text{[kJ/(kmol·K)]} \qquad (14\text{-}21)$$

where P_i is the partial pressure, y_i is the mole fraction of the component, and P_m is the total pressure of the mixture in atmospheres.

If a gas mixture is at a relatively high pressure or low temperature, the deviation from the ideal-gas behavior should be accounted for by incorporating more accurate equations of state or the generalized entropy charts.

14-7 ■ SECOND-LAW ANALYSIS OF REACTING SYSTEMS

Once the total entropy change or the entropy generation is evaluated, the **irreversibility** (wasted work potential) I associated with a chemical reaction can be determined from

$$I = T_0 S_{gen} \qquad \text{(kJ)} \qquad (14\text{-}22)$$

where T_0 is the absolute temperature of the surroundings.

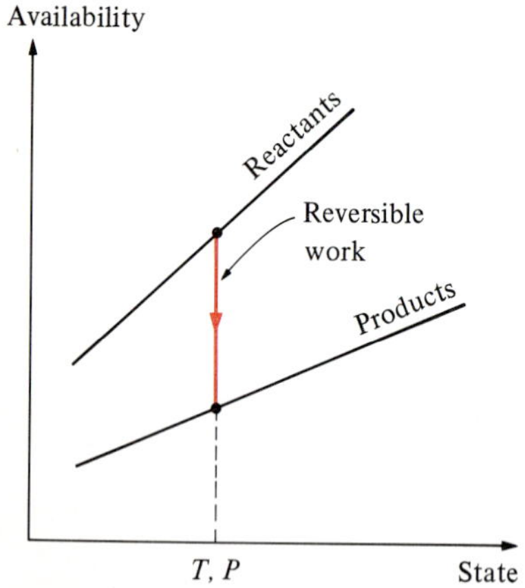

Availability
Reactants
Reversible work
Products
T, P
State

FIGURE 14-30
The difference between the availability of the reactants and the products during a chemical reaction is the reversible work associated with that reaction.

When analyzing reacting systems, we are more concerned with the changes in the availability of reacting systems than with the values of availability at various states (Fig. 14-30). You will recall from Chap. 7 that the change in the availability of a system during a process is called the **reversible work** W_{rev}, and it represents the maximum work that can be done during a process. In the absence of any changes in kinetic and potential energies, the reversible work relation for a steady-flow combustion process which involves heat transfer only with the surroundings at T_0 can be obtained by replacing the enthalpy terms in Eq. 7-15 by $\overline{h}_f^\circ + \overline{h} - \overline{h}^\circ$, yielding

$$W_{\text{rev}} = \Sigma N_R(\overline{h}_f^\circ + \overline{h} - \overline{h}^\circ - T_0\overline{s})_R - \Sigma N_P(\overline{h}_f^\circ + \overline{h} - \overline{h}^\circ - T_0\overline{s})_P \quad \text{(kJ)} \quad (14\text{-}23)$$

If the combustion chamber involves heat transfer with a reservoir at temperature T_R in the amount of Q_R, the reversible work in this case can be determined by subtracting $Q_R(1 - T_0/T_R)$ from the above relation, as explained in Chap. 7. The sign of Q_R is determined with respect to the reservoir.

An interesting situation arises when both the reactants and the products are at the temperature of the surroundings T_0. In this case, $\overline{h} - T_0\overline{s} = (\overline{h} - T_0\overline{s})_{T_0} = \overline{g}_0$, which is, by definition, the **Gibbs function** of a unit mole of a substance at temperature T_0. The W_{rev} relation in this case can be written as

$$W_{\text{rev}} = \Sigma N_R\overline{g}_{0,R} - \Sigma N_P\overline{g}_{0,P} \quad (14\text{-}24)$$

or $$W_{\text{rev}} = \Sigma N_R(\overline{g}_f^\circ + \overline{g}_{T_0} - \overline{g}^\circ)_R - \Sigma N_P(\overline{g}_f^\circ + \overline{g}_{T_0} - \overline{g}^\circ)_P \quad \text{(kJ)} \quad (14\text{-}25)$$

where $\overline{g}_f^\circ$ is the Gibbs function of formation ($\overline{g}_f^\circ = 0$ for stable elements like N_2 and O_2 at the standard reference state of 25°C and 1 atm, just like the enthalpy of formation) and $\overline{g}_{T_0} - \overline{g}^\circ$ represents the value of the sensible Gibbs function of a substance at temperature T_0 relative to the standard reference state.

For the very special case of $T_R = T_P = T_0 = 25°\text{C}$ (i.e., the reactants, the products, and the surroundings are at 25°C) and the partial pressure $P_i = 1$ atm for each component of the reactants and the products, Eq. 14-25 reduces to

$$W_{\text{rev}} = \Sigma N_R\overline{g}_{f,R}^\circ - \Sigma N_P\overline{g}_{f,P}^\circ \quad \text{(kJ)} \quad (14\text{-}26)$$

We can conclude from the above equation that the $-\overline{g}_f^\circ$ value (the negative of the Gibbs function of formation at 25°C and 1 atm) of a compound represents the *reversible work* associated with the formation of that compound from its stable elements at 25°C and 1 atm in an environment at 25°C and 1 atm (Fig. 14-31). The $\overline{g}_f^\circ$ values of several substances are listed in Table A-26.

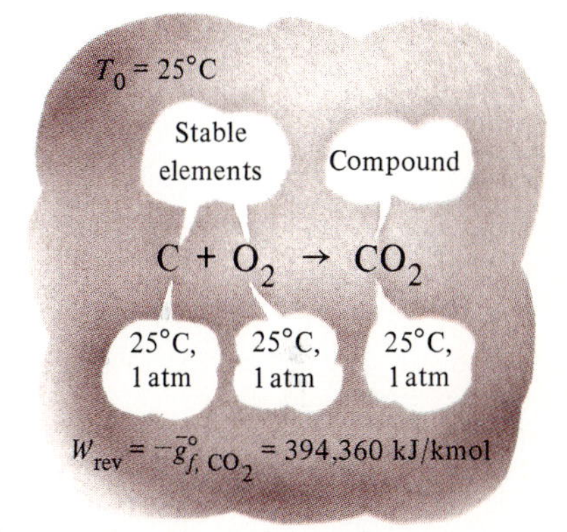

FIGURE 14-31

The negative of the Gibbs function of formation of a compound at 25°C, 1 atm represents the reversible work associated with the formation of that compound from its stable elements at 25°C, 1 atm in an environment which is at 25°C, 1 atm.

EXAMPLE 14-9

One lbmol of carbon at 77°F and 1 atm is burned steadily with 1 lbmol of oxygen at the same state as shown in Fig. 14-32. The CO_2 formed during the process is then brought to 77°F and 1 atm, the conditions of the surroundings. Assuming the combustion is complete, determine the reversible work for this process.

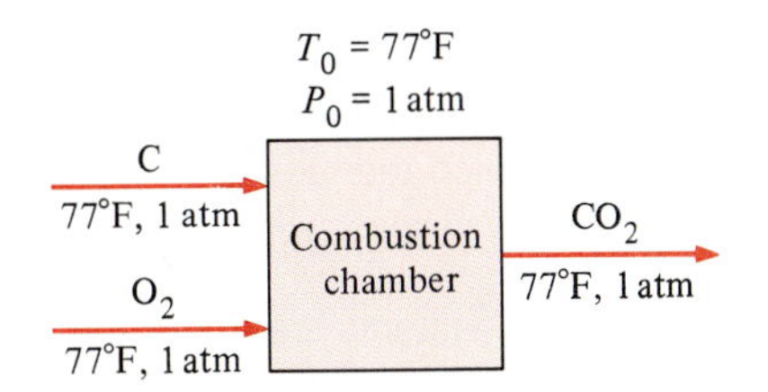

FIGURE 14-32

Schematic for Example 14-9.

Solution The combustion equation is

$$C + O_2 \longrightarrow CO_2$$

The C, O_2, and CO_2 are at 77°F and 1 atm, which is the standard reference state and also the state of the surroundings. Therefore, the reversible work in this case is simply the difference between the Gibbs function of formation of the reactants and that of the products (Eq. 14-26):

$$\begin{aligned} W_{rev} &= \Sigma N_R \bar{g}^\circ_{f,R} - \Sigma N_P \bar{g}^\circ_{f,P} \\ &= N_C \cancelto{0}{\bar{g}^\circ_{f,C}} + N_{O_2} \cancelto{0}{\bar{g}^\circ_{f,O_2}} - N_{CO_2}\bar{g}^\circ_{f,CO_2} \\ &= -N_{CO_2}\bar{g}^\circ_{f,CO_2} = (-1 \text{ lbmol})(-169{,}680 \text{ Btu/lbmol}) \\ &= 169{,}680 \text{ Btu} \end{aligned}$$

since the $\bar{g}^\circ_f$ of stable elements at 77°F and 1 atm is zero. Therefore, 169,680 Btu of work could be done as 1 lbmol of C is burned with 1 lbmol of O_2 at 77°F and 1 atm in an environment at the same state. The reversible work in this case represents the availability of the reactants since the product (the CO_2) is at the state of the surroundings.

We could also determine the reversible work without involving the Gibbs function by using Eq. 14-23:

$$\begin{aligned} W_{rev} &= \Sigma N_R(\bar{h}^\circ_f + \bar{h} - \bar{h}^\circ - T_0\bar{s})_R - \Sigma N_P(\bar{h}^\circ_f + \bar{h} - \bar{h}^\circ - T_0\bar{s})_P \\ &= \Sigma N_R(\bar{h}^\circ_f - T_0\bar{s})_R - \Sigma N_P(\bar{h}^\circ_f - T_0\bar{s})_P \\ &= N_C(\bar{h}^\circ_f - T_0\bar{s}^\circ)_C + N_{O_2}(\bar{h}^\circ_f - T_0\bar{s}^\circ)_{O_2} - N_{CO_2}(\bar{h}^\circ_f - T_0\bar{s}^\circ)_{CO_2} \end{aligned}$$

Substituting the enthalpy of formation and absolute entropy values from Table A-26E, we obtain

$$\begin{aligned} W_{rev} = &(1 \text{ lbmol C})\{0 - (537 \text{ R})[1.36 \text{ Btu/(lbmol·R)}]\} \\ &+ (1 \text{ lbmol } O_2)\{0 - (537 \text{ R})[49.00 \text{ Btu/(lbmol/R)}]\} \\ &- (1 \text{ lbmol } CO_2)\{-169{,}300 \text{ Btu/lbmol} - (537 \text{ R})[51.07 \text{ Btu/(lbmol·R)}]\} \\ = &\ 169{,}681 \text{ Btu} \end{aligned}$$

which is practically identical to the result obtained before.

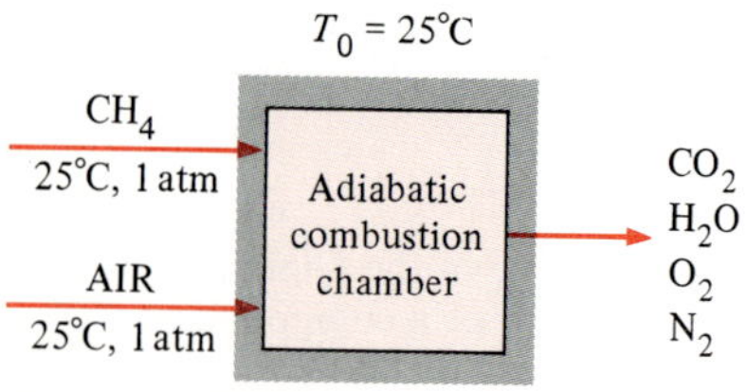

$T_0 = 25°C$

CH_4 25°C, 1 atm

AIR 25°C, 1 atm

Adiabatic combustion chamber

CO_2 H_2O O_2 N_2

FIGURE 14-33

Schematic for Example 14-10.

EXAMPLE 14-10

Methane (CH_4) gas enters a steady-flow adiabatic combustion chamber at 25°C and 1 atm. It is burned with 50 percent excess air which also enters at 25°C and 1 atm, as shown in Fig. 14-33. Assuming complete combustion, determine (*a*) the temperature of the products, (*b*) the entropy generation, and (*c*) the reversible work and irreversibility. Assume that T_0 = 298 K and the products leave the combustion chamber at 1 atm pressure.

Solution (*a*) The balanced equation for the complete combustion process with 20 percent excess air is

$$CH_4(g) + 4.5(O_2 + 3.76N_2) \longrightarrow CO_2 + 2H_2O + 1.5O_2 + 16.92N_2$$

Under steady-flow conditions, the adiabatic flame temperature is determined from Eq. 14-16, which reduces to

$$\Sigma N_P(\bar{h}^\circ_f + \bar{h} - \bar{h}^\circ)_P = \Sigma N_R \bar{h}^\circ_{f,R} = (N\bar{h}^\circ_f)_{CH_4}$$

since all the reactants are at the standard reference state and $\bar{h}^\circ_f = 0$ for O_2 and N_2. Assuming ideal-gas behavior for air and for the products, the $\bar{h}^\circ_f$ and h values of various components at 298 K can be listed as

Substance	$\overline{h}_f^\circ$, kJ/kmol	$\overline{h}_{298\text{ K}}$, kJ/kmol
$CH_4(g)$	−74,850	—
O_2	0	8682
N_2	0	8669
$H_2O(g)$	−241,820	9904
CO_2	−393,520	9364

Substituting, we have

$$\begin{aligned}&(1\text{ kmol }CO_2)[(-393{,}520 + \overline{h}_{CO_2} - 9364)\text{ kJ/kmol }CO_2]\\&+ (2\text{ kmol }H_2O)[(-241{,}820 + \overline{h}_{H_2O} - 9904)\text{ kJ/kmol }H_2O]\\&+ (16.92\text{ kmol }N_2)[(0 + \overline{h}_{N_2} - 8669)\text{ kJ/kmol }N_2]\\&+ (1.5\text{ kmol }O_2)[(0 + \overline{h}_{O_2} - 8682)\text{ kJ/kmol }O_2]\\&= (1\text{ kmol }CH_4)(-74{,}850\text{ kJ/kmol }CH_4)\end{aligned}$$

which yields

$$\overline{h}_{CO_2} + 2\overline{h}_{H_2O} + 1.5\overline{h}_{O_2} + 16.92\overline{h}_{N_2} = 991{,}184.5\text{ kJ}$$

By trial and error, the temperature of the products is found to be

$$T_P = 1416.4\text{ K}$$

(*b*) The entropy generation during this process is determined from Eq. 14-17:

$$S_{gen} = \Delta S_{sys} + \Delta S_{surr}$$

But $\Delta S_{surr} = 0$ since the process is adiabatic. Thus,

$$S_{gen} = \Delta S_{sys} = S_P - S_R = \Sigma N_P \overline{s}_P - \Sigma N_R \overline{s}_R$$

The CH_4 is at 25°C and 1 atm, and thus its absolute entropy is $\overline{s}_{CH_4} = 186.16$ kJ/(kmol·K) (Table A-26). The entropy values listed in the ideal-gas tables are for a 1-atm pressure. Both the air and the product gases are at a total pressure of 1 atm, but the entropies are to be calculated at the partial pressure of the components which is equal to $P_i = y_i P_{total}$, where y_i is the mole fraction of component i from Eq. 14-21:

$$S_i = N_i \overline{s}_i(T, P_i) = N_i[\overline{s}_i^\circ(T, P_0) - R_u \ln y_i P_m]$$

The entropy calculations can be presented in tabular form as follows:

	N_i	y_i	$\overline{s}_i^\circ(T, 1\text{ atm})$	$-R_u \ln y_i P_m$	$N_i \overline{s}_i$
CH_4	1	1.00	186.16	—	186.16
O_2	4.5	0.21	205.04	12.98	981.09
N_2	16.92	0.79	191.61	1.96	3275.20
					S_R = 4442.45
CO_2	1	0.0467	288.785	25.474	314.26
H_2O	2	0.0934	247.777	19.711	534.98
O_2	1.5	0.0700	255.875	22.109	416.98
N_2	16.92	0.7899	239.777	1.961	4090.21
					S_P = 5356.43

Thus,

$$S_{gen} = S_P - S_R = (5356.43 - 4442.45) \text{ kJ/(kmol·K) } CH_4$$
$$= 913.98 \text{ kJ/(kmol·K) } CH_4$$

(*c*) The irreversibility associated with this process is determined from Eq. 14-22:

$$I = T_0 S_{gen} = (298 \text{ K})[913.98 \text{ kJ/(kmol·K) } CH_4]$$
$$= 272{,}366 \text{ kJ/kmol } CH_4$$

That is, 272,366 kJ of work potential is wasted during this combustion process for each kmol of methane burned. This example shows that even complete combustion processes are highly irreversible.

This process involves no actual work. Therefore, the reversible work and irreversibility are identical:

$$W_{rev} = 272{,}366 \text{ kJ/kmol } CH_4$$

That is, 272,366 kJ of work could be done during this process, but is not. Instead, the entire work potential is wasted.

EXAMPLE 14-11

Methane (CH_4) gas enters a steady-flow combustion chamber at 25°C and 1 atm and is burned with 50 percent excess air, which also enters at 25°C and 1 atm, as shown in Fig. 14-34. After combustion, the products are allowed to cool to 25°C. Assuming complete combustion, determine (*a*) the heat transfer per kmol of CH_4, (*b*) the entropy generation, and (*c*) the reversible work and irreversibility. Assume that T_0 = 298 K and the products leave the combustion chamber at a 1-atm pressure.

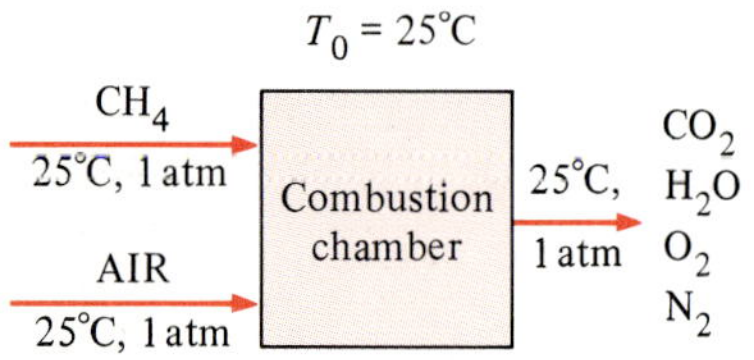

FIGURE 14-34
Schematic for Example 14-11.

Solution This is the same combustion process we discussed in Example 14-10, except that the combustion products are brought to the state of the surroundings by transferring heat from them. Thus the combustion equation remains the same:

$$CH_4(g) + 4.5(O_2 + 3.76N_2) \longrightarrow CO_2 + 2H_2O + 1.5O_2 + 16.92N_2$$

At 25°C part of the water will condense, and the amount of water vapor that remains in the products is determined from (see Example 14-3)

$$\frac{N_v}{N_{gas}} = \frac{P_v}{P_{total}} = \frac{3.169 \text{ kPa}}{101.325 \text{ kPa}} = 0.03128$$

and $$N_v = \left(\frac{P_v}{P_{total}}\right) N_{gas} = (0.03128)(19.42 + N_v) \longrightarrow N_v = 0.63 \text{ kmol}$$

Therefore, 1.37 kmol of the H_2O formed will be in the liquid form, which will be removed at 25°C and 1 atm. When one is evaluating the partial pressures of the components in the product gases, the only water molecules that need to be considered are those that are in the vapor phase. As before, all the gaseous reactants and products will be treated as ideal gases.

Heat transfer during this steady-flow combustion process is determined from Eq. 14-11, which reduces to

$$Q = \Sigma N_P \bar{h}^\circ_{f,P} - \Sigma N_R \bar{h}^\circ_{f,R}$$

since all the reactants and components are at the standard reference of 25°C and the enthalpy of ideal gases depends on temperature only. Substituting the $\bar{h}_f^\circ$ values, we have

$$\begin{aligned} Q &= (1 \text{ kmol } CO_2)(-393{,}520 \text{ kJ/kmol } CO_2) \\ &\quad + [0.63 \text{ kmol } H_2O(g)][-241{,}820 \text{ kJ/kmol } H_2O(g)] \\ &\quad + [1.37 \text{ kmol } H_2O(\ell)][-285{,}830 \text{ kJ/kmol } H_2O(\ell)] \\ &\quad - (1 \text{ kmol } CH_4)(-74{,}850 \text{ kJ/kmol } CH_4) \\ &= -862{,}604 \text{ kJ/kmol } CH_4 \end{aligned}$$

(*b*) The entropy of the reactants was evaluated in Example 14-10 and was determined to be $S_R = 4442.45$ kJ/(kmol·K) CH_4. By following a similar approach, the entropy of products is determined to be the following:

	N_i	y_i	$\bar{s}_i^\circ(T, 1 \text{ atm})$	$-R_u \ln y_i P_m$	$N_i \bar{s}_i$
$H_2O(\ell)$	1.37	1.0000	69.92	—	95.79
H_2O	0.63	0.0314	188.83	28.77	137.09
CO_2	1	0.0499	213.80	24.92	238.72
O_2	1.5	0.0748	205.04	21.56	339.45
N_2	16.92	0.8439	191.61	1.41	3265.90
					$S_P = 4076.95$

Thus,

$$\begin{aligned} \Delta S_{sys} &= S_P - S_R = (4076.95 - 4442.45) \text{ kJ/(kmol·K) } CH_4 \\ &= -365.50 \text{ kJ/(kmol·K) } CH_4 \end{aligned}$$

$$\Delta S_{surr} = \frac{Q_{surr}}{T_0} = \frac{862{,}604 \text{ kJ/kmol } CH_4}{298 \text{ K}} = 2894.64 \text{ kJ/(kmol·K) } CH_4$$

$$\begin{aligned} S_{gen} &= \Delta S_{sys} + \Delta S_{surr} = (-365.50 + 2894.64) \text{ kJ/(kmol·K) } CH_4 \\ &= 2529.14 \text{ kJ/(kmol·K) } CH_4 \end{aligned}$$

(*c*) The irreversibility and reversible work associated with this process are determined from

$$\begin{aligned} I = T_0 S_{gen} &= (298 \text{ K})[2529.14 \text{ kJ/(kmol·K) } CH_4] \\ &= 753{,}683 \text{ kJ/kmol } CH_4 \end{aligned}$$

and $$W_{rev} = I = 753{,}683 \text{ kJ/kmol } CH_4$$

since this process involves no actual work. Therefore, 753,683 kJ of work could be done during this process, but is not. Instead, the entire work potential is wasted. The reversible work in this case represents the availability of the reactants before the reaction starts since the products are in equilibrium with the surroundings, i.e., they are at the dead state.

There is an important observation to be made from the preceding two examples. Fuels like methane are commonly burned to provide thermal energy at high temperatures for use in heat engines. However, a comparison of the reversible works obtained in the two examples reveals that the availability of the reactants (753,683 kJ/kmol CH_4) decreases by 272,366 kJ/kmol as a result of the irreversible adiabatic combustion process alone.

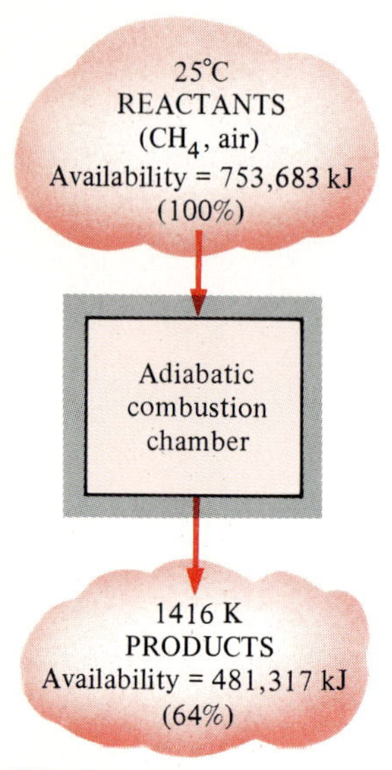

FIGURE 14-35

The availability of methane decreases by 36 percent as a result of irreversible combustion process.

That is, the availability of the hot combustion gases at the end of the adiabatic combustion process is 753,683 − 272,366 = 481,317 kJ/kmol CH_4. In other words, the work potential of the hot combustion gases is about 64 percent of the work potential of the reactants. It seems that when methane is burned, 36 percent of the work potential is lost before we even start using the thermal energy (Fig. 14-35).

Thus the second law of thermodynamics suggests that there should be a better way of converting the chemical energy to work. The better way is, of course, the less irreversible way, the best being the reversible case. In chemical reactions, the irreversibility is due to uncontrolled electron exchange between the reacting components. The electron exchange can be controlled by replacing the combustion chamber by electrolytic cells, like car batteries. In electrolytic cells, the electrons are exchanged through conductor wires connected to a load, and the chemical energy is directly converted to electric energy. The energy conversion devices that work on this principle are called **fuel cells**.

The operation of a hydrogen-oxygen fuel cell is illustrated in Fig. 14-36. Hydrogen is ionized at the surface of the anode, and hydrogen ions flow through the electrolyte to the cathode. There is a potential difference between the anode and the cathode, and free electrons flow from the anode to the cathode through an external circuit (such as a generator). Hydrogen ions combine with oxygen and the free electrons at the surface of the cathode, forming water. In steady operation, hydrogen and oxygen continuously enter the fuel cell as reactants, and water leaves as the product.

Fuel cells are not heat engines, and thus their efficiencies are not limited by the Carnot efficiency. They convert chemical energy to electric energy essentially in an isothermal manner. Despite the irreversible effects such as internal resistance to electron flow, fuel cells have a great potential for higher conversion efficiencies, and they have been used successfully in some small-scale applications. But more research and development is needed before large-scale fuel-cell power plants can be realized.

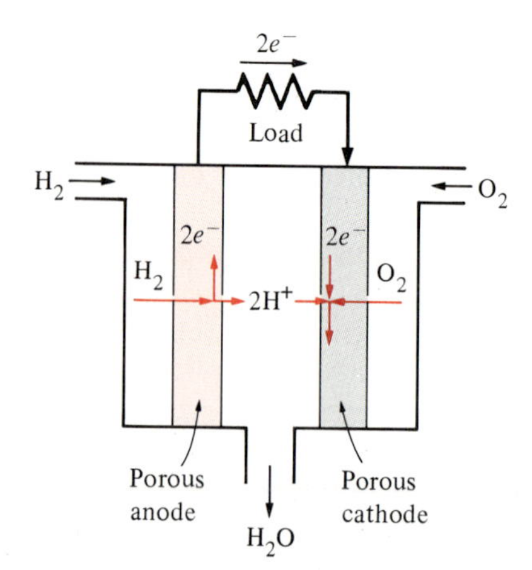

FIGURE 14-36

The operation of a hydrogen-oxygen fuel cell.

14-7 SUMMARY

Any material that can be burned to release energy is called a *fuel,* and a chemical reaction during which a fuel is oxidized and a large quantity of energy is released is called *combustion*. The oxidizer most often used in combustion processes is air. The dry air can be approximated as 21 percent oxygen and 79 percent nitrogen by mole numbers. Therefore,

$$1 \text{ kmol } O_2 + 3.76 \text{ kmol } N_2 = 4.76 \text{ kmol air} \tag{14-1}$$

At ordinary combustion temperatures, nitrogen behaves as an inert gas and does not react with other chemical elements.

During a combustion process, the components that exist before the reaction are called *reactants,* and the components that exist after the reaction are called *products*. Chemical equations are balanced on the basis of the *conservation of mass principle,* which states that the total mass of each element is conserved during a chemical reaction. The ratio of the mass of air to the mass of fuel during a combustion process is called the *air-fuel ratio* AF

$$\text{AF} = \frac{m_{\text{air}}}{m_{\text{fuel}}} \tag{14-3}$$

where $m_{\text{air}} = (NM)_{\text{air}}$ and $m_{\text{fuel}} = \Sigma(N_iM_i)_{\text{fuel}}$.

A combustion process is *complete* if all the carbon in the fuel burns to CO_2, all the hydrogen burns to H_2O, and all the sulfur (if any) burns to SO_2. The minimum amount of air needed for the complete combustion of a fuel is called the *stoichiometric* or *theoretical air*. The theoretical air is also referred to as the chemically correct amount of air or 100 percent theoretical air. The ideal combustion process during which a fuel is burned completely with theoretical air is called the *stoichiometric* or *theoretical combustion* of that fuel. The air in excess of the stoichiometric amount is called the *excess air*. The amount of excess air is usually expressed in terms of the stoichiometric air as *percent excess air* or *percent theoretical air*.

During a chemical reaction, some chemical bonds are broken and others are formed. Therefore, a process which involves chemical reactions will involve changes in chemical energies. Because of the changed composition, it is necessary to have a *standard reference state* for all substances, which is chosen to be 25°C (77°F) and 1 atm.

The difference between the enthalpy of the products at a specified state and the enthalpy of the reactants at the same state for a complete reaction is called the *enthalpy of reaction* h_R. For combustion processes, the enthalpy of reaction is usually referred to as the *enthalpy of combustion* h_C, which represents the amount of heat released during a steady-flow combustion process when 1 kmol (or 1 kg) of fuel is burned completely at a specified temperature and pressure. The enthalpy of a substance at a specified state due to its chemical composition is called the *enthalpy of formation* $\overline{h}_f$. The enthalpy of formation of all stable elements is assigned a value of zero at the standard reference state of

25°C and 1 atm. The *heating value* of a fuel is defined as the amount of energy released when a fuel is burned completely in a steady-flow process and the products are returned to the state of the reactants. The heating value of a fuel is equal to the absolute value of the enthalpy of combustion of the fuel:

$$\text{Heating value} = |h_C| \qquad \text{(kJ/kg fuel)}$$

The heating value is called the *higher heating value* (HHV) when the H_2O in the products is in the liquid form, and it is called the *lower heating value* (LHV) when the H_2O in the products is in the vapor form. The two heating values are related by

$$\text{HHV} = \text{LHV} + (N\bar{h}_{fg})_{H_2O} \qquad \text{(kJ/kg fuel)} \qquad (14\text{-}7)$$

where N is the number of moles of H_2O in the products and $\bar{h}_{fg}$ is the enthalpy of vaporization of water at 25°C.

The conservation of energy relation for chemically reacting steady-flow systems can be expressed as

$$Q - W = \Sigma N_P(\bar{h}_f^\circ + \bar{h} - \bar{h}^\circ)_P - \Sigma N_R(\bar{h}_f^\circ + \bar{h} - \bar{h}^\circ)_R \qquad \text{(kJ)} \qquad (14\text{-}11)$$

where the superscript ° represents properties at the standard reference state of 25°C and 1 atm. For a closed system, it becomes

$$Q - W = \Sigma N_P(\bar{h}_f^\circ + \bar{h} - \bar{h}^\circ - P\bar{v})_P - \Sigma N_R(\bar{h}_f^\circ + \bar{h} - \bar{h}^\circ - P\bar{v})_R \qquad \text{(kJ)} \qquad (14\text{-}14)$$

The $P\bar{v}$ terms are negligibly small for solids and liquids and can be replaced by R_uT for gases which behave as ideal gases.

In the absence of any heat loss to the surroundings ($Q = 0$), the temperature of the products will reach a maximum, which is called the *adiabatic flame temperature* of the reaction. The adiabatic flame temperature of a steady-flow combustion process is determined from Eq. 14-11 by setting $Q = 0$ and $W = 0$:

$$\Sigma N_P(\bar{h}_f^\circ + \bar{h} - \bar{h}^\circ)_P = \Sigma N_R(\bar{h}_f^\circ + \bar{h} - \bar{h}^\circ)_R \qquad (14\text{-}16)$$

The increase-in-entropy principle for reacting systems can be expressed as

$$S_{\text{gen}} = \Delta S_{\text{sys}} + \Delta S_{\text{surr}} \geq 0 \qquad \text{(kJ/K)} \qquad (14\text{-}17)$$

where $$\Delta S_{\text{sys}} = S_P - S_R = \Sigma N_P\bar{s}_P - \Sigma N_R\bar{s}_R \qquad \text{(kJ/K)} \qquad (14\text{-}18)$$

and $$\Delta S_{\text{surr}} = \frac{Q_{\text{surr}}}{T_0} \qquad \text{(kJ/K)} \qquad (14\text{-}19)$$

Here T_0 is the temperature of the surroundings, which is assumed to remain constant.

The *third law of thermodynamics* states that the entropy of a pure crystalline substance at absolute zero temperature is zero. The third law provides a common base for the entropy of all substances, and the entropy values relative to this base are called the *absolute entropy*. The ideal-gas tables list the absolute entropy values over a wide range of temperatures but at a fixed pressure of $P_0 = 1$ atm. Absolute entropy values at other pressures P for any temperature T are determined from

$$\bar{s}(T, P) = \bar{s}^\circ(T, P_0) - R_u \ln \frac{P}{P_0} \qquad [\text{kJ/(kmol·K)}] \qquad (14\text{-}20)$$

For component i of an ideal-gas mixture, this relation can be written as

$$\bar{s}_i(T, P_i) = \bar{s}_i^\circ(T, P_0) - R_u \ln y_i P_m \qquad [\text{kJ/(kmol·K)}] \qquad (14\text{-}21)$$

where P_i is the partial pressure, y_i is the mole fraction of the component, and P_m is the total pressure of the mixture in atmospheres.

The *irreversibility* and the *reversible work* associated with a chemical reaction are determined from

$$I = W_{\text{rev}} - W_{\text{act}} = T_0 S_{\text{gen}} \qquad (\text{kJ}) \qquad (14\text{-}22)$$

and

$$W_{\text{rev}} = \Sigma N_R(\bar{h}_f^\circ + \bar{h} - \bar{h}^\circ - T_0\bar{s})_R - \Sigma N_P(\bar{h}_f^\circ + \bar{h} - \bar{h}^\circ - T_0\bar{s})_P \qquad (\text{kJ}) \qquad (14\text{-}23)$$

When both the reactants and the products are at the temperature of the surroundings T_0, the reversible work can be expressed in terms of the Gibbs functions as

$$W_{\text{rev}} = \Sigma N_R(\bar{g}_f^\circ + \bar{g}_{T_0} - \bar{g}^\circ)_R - \Sigma N_P(\bar{g}_f^\circ + \bar{g}_{T_0} - \bar{g}^\circ)_P \qquad (\text{kJ}) \qquad (14\text{-}25)$$

where $\bar{g}_f^\circ$ is the Gibbs function of formation.

REFERENCES AND SUGGESTED READING

1 S. W. Angrist, *Direct Energy Conversion,* 4th ed., Allyn and Bacon, Boston, 1982.

2 W. Z. Black and J. G. Hartley, *Thermodynamics,* Harper & Row, New York, 1985.

3 I. Glassman, *Combustion,* Academic Press, New York, 1977.

4 J. B. Jones and G. A. Hawkins, *Engineering Thermodynamics,* 2d ed., Wiley, New York, 1986.

5 R. Strehlow, *Fundamentals of Combustion,* International Textbook Co., Scranton, 1968.

6 G. J. Van Wylen and R. E. Sonntag, *Fundamentals of Classical Thermodynamics,* 3d ed., Wiley, New York, 1986.

7 K. Wark, *Thermodynamics,* 5th ed., McGraw-Hill, New York, 1988.

PROBLEMS*

Fuels and Combustion

14-1C How do reacting systems differ from nonreacting systems?

14-2C Is it possible to invent a car that runs on H_2O instead of gasoline?

14-3C What are the approximate chemical compositions of gasoline, diesel fuel, and natural gas?

14-4C How does the presence of N_2 in the air affect the outcome of a combustion process?

14-5C How does the presence of moisture in the air affect the outcome of a combustion process?

14-6C What does the dew-point temperature of the product gases represent? How is it determined?

14-7C Will a fuel start burning when it is brought into intimate contact with oxygen?

14-8C Write three different statements which express the conservation of mass principle for a chemical reaction.

14-9C Is the number of atoms of each element conserved during a chemical reaction? How about the total number of moles?

14-10C What is the air-fuel ratio? How is it related to the fuel-air ratio?

14-11C Is the air-fuel ratio expressed on a mole basis identical to the air-fuel ratio expressed on a mass basis?

Theoretical and Actual Combustion Processes

14-12C What are the causes of incomplete combustion?

14-13C Which is more likely to be found in the products of an incomplete combustion of a hydrocarbon fuel, CO or OH? Why?

14-14C What does 100 percent theoretical air represent?

14-15C Are complete combustion and theoretical combustion identical? If not, how do they differ?

14-16C Consider a fuel which is burned with (*a*) 130 percent theoretical air and (*b*) 70 percent excess air. In which case is the fuel burned with more air?

14-17C What is the operation principle of the Orsat gas analyzer?

14-18 Methane (CH_4) is burned with the stoichiometric amount of air

---

*Students are encouraged to answer *all* the concept "C" questions.

during a combustion process. Assuming complete combustion, determine the air-fuel and fuel-air ratios.

14-19 Propane (C_3H_8) is burned with 50 percent excess air during a combustion process. Assuming complete combustion, determine the air-fuel ratio. *Answer:* 23.5 kg air/kg fuel

14-20 Acetylene (C_2H_2) is burned with stoichiometric air during a combustion process. Assuming complete combustion, determine the air-fuel ratio on a mass and on a mole basis.

14-21 One kmol of ethane (C_2H_6) is burned with an unknown amount of air during a combustion process. An analysis of the combustion products reveals that the combustion is complete, and there is 3 kmol of free O_2 in the products. Determine (*a*) the air-fuel ratio and (*b*) the percentage of theoretical air used during this process.

14-22 Ethylene (C_2H_4) is burned with 200 percent theoretical air during a combustion process. Assuming complete combustion and a total pressure of 98 kPa, determine (*a*) the air-fuel ratio and (*b*) the dew-point temperature of the products. *Answers:* (*a*) 29.6 kg air/kg fuel, (*b*) 37.7°C

14-22E Ethylene (C_2H_4) is burned with 200 percent theoretical air during a combustion process. Assuming complete combustion and a total pressure of 14.5 psia, determine (*a*) the air-fuel ratio and (*b*) the dew-point temperature of the products.
Answers: (*a*) 29.6 lbm air/lbm fuel, (*b*) 100.9°F

14-23 Propylene (C_3H_6) is burned with 50 percent excess air during a combustion process. Assuming complete combustion and a total pressure of 100 kPa, determine (*a*) the air-fuel ratio and (*b*) the temperature at which the water vapor in the products will start condensing.

14-24 Octane (C_8H_{18}) is burned with 250 percent theoretical air, which enters the combustion chamber at 25°C. Assuming complete combustion and a total pressure of 1 atm, determine (*a*) the air-fuel ratio and (*b*) the dew-point temperature of the products.

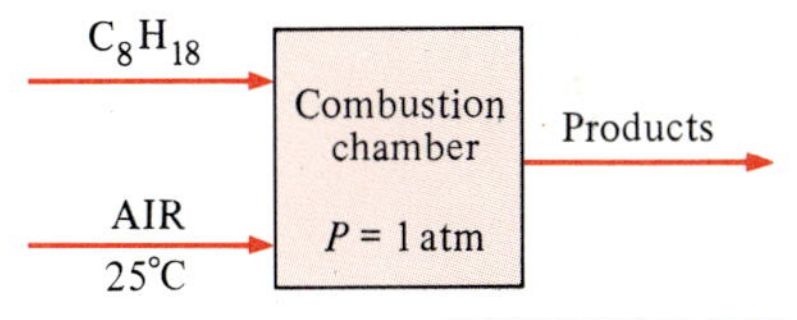

FIGURE P14-24

14-25 Hydrogen (H_2) is burned with 100 percent excess air, which enters the combustion chamber at 30°C, 97 kPa, and 60 percent relative humidity. Assuming complete combustion, determine (*a*) the air-fuel ratio and (*b*) the volume flow rate of air required to burn the hydrogen at a rate of 6 kg/h. *Answers:* (*a*) 70.2 kg air/kg fuel, (*b*) 381.2 m^3/h

14-25E Hydrogen (H_2) is burned with 100 percent excess air, which enters the combustion chamber at 90°F, 14.5 psia, and 60 percent relative humidity. Assuming complete combustion, determine (*a*) the air-fuel ratio and (*b*) the volume flow rate of air required to burn the hydrogen at a rate of 10 lbm/h.

14-26 Gasoline (assumed C_8H_{18}) is burned steadily with air in a jet engine. If the air-fuel ratio is 30 kg air/kg fuel, determine the percentage of theoretical air used during this process.

14-27 In a combustion chamber, ethane (C_2H_6) is burned at a rate of 5 kg/h with air which enters the combustion chamber at a rate of 100 kg/h. Determine the percentage of excess air used during this process. *Answer:* 24 percent

14-28 One kilogram of butane (C_4H_{10}) is burned with 25 kg of air which is at 35°C and 100 kPa. Assuming that the combustion is complete and the pressure of the products is 100 kPa, determine (*a*) the percentage of theoretical air used and (*b*) the dew-point temperature of the products.

14-28E One lbm of butane (C_4H_{10}) is burned with 25 lbm of air which is at 90°F and 14.7. Assuming that the combustion is complete and the pressure of the products is 14.7 psia, determine (*a*) the percentage of theoretical air used and (*b*) the dew-point temperature of the products. *Answers:* (*a*) 161 percent, (*b*) 111.4°F

14-29 A certain natural gas has the following volumetric analysis: 65 percent CH_4, 8 percent H_2, 18 percent N_2, 3 percent O_2 and 6 percent CO_2. This gas is now burned completely with the stoichiometric amount of dry air. What is the air-fuel ratio for this combustion process?

14-30 Repeat Prob. 14-29 by replacing the dry air by moist air which enters the combustion chamber at 25°C, 1 atm, and 85 percent relative humidity.

14-31 A gaseous fuel with a volumetric analysis of 60 percent CH_4, 30 percent H_2, and 10 percent N_2 is burned to completion with 130 percent theoretical air. Determine (*a*) the air-fuel ratio and (*b*) the fraction of water vapor which would condense if the product gases were cooled to 20°C at 1 atm. *Answers:* (*a*) 18.6 kg air/kg fuel, (*b*) 88 percent

14-31E A gaseous fuel with a volumetric analysis of 60 percent CH_4, 30 percent H_2, and 10 percent N_2 is burned to completion with 130 percent theoretical air. Determine (*a*) the air-fuel ratio and (*b*) the fraction of water vapor which would condense if the product gases were cooled to 70°F at 14.7 psia.

14-32 A gaseous fuel with 80 percent CH_4, 15 percent N_2, and 5 percent O_2 (on a mole basis) is burned to completion with 120 percent theoretical air, which enters the combustion chamber at 30°C, 100 kPa, and 60 percent relative humidity. Determine (*a*) the air-fuel ratio and (*b*) the volume flow rate of air required to burn fuel at a rate of 1 kg/min.

14-33 A certain coal has the following analysis on a mass basis: 82 percent C, 5 percent H_2O, 2 percent H_2, 1 percent O_2 and 10 percent ash. The coal is burned with 20 percent excess air. Determine the air-fuel ratio. *Answer:* 12.3 kg air/kg coal

14-34 Octane (C_8H_{18}) is burned with dry air. The volumetric analysis of the products on a dry basis is 9.21 percent CO_2, 0.61 percent CO, 7.06 percent O_2, and 83.12 percent N_2. Determine (*a*) the air-fuel ratio and (*b*) the percentage of theoretical air used.

14-35 Carbon is burned with dry air. The volumetric analysis of the products is 10.06 percent CO_2, 0.42 percent CO, 10.69 percent O_2, and 78.83 percent N_2. Determine (*a*) the air-fuel ratio and (*b*) the percentage of theoretical air used.

14-36 Methane (CH_4) is burned with dry air. The volumetric analysis of the products on a dry basis is 5.20 percent CO_2, 0.33 percent CO, 11.24 percent O_2, and 83.23 percent N_2. Determine (*a*) the air-fuel ratio and (*b*) the percentage of theoretical air used.
Answers: (*a*) 34.5 kg air/kg fuel, (*b*) 200%

14-37 A gaseous fuel with 80 percent CH_4, 15 percent N_2, and 5 percent O_2 (on a mole basis) is burned with dry air which enters the combustion chamber at 25°C and 100 kPa. The volumetric analysis of the products on a dry basis is 3.36 percent CO_2, 0.09 percent CO, 14.91 percent O_2, and 81.64 percent N_2. Determine (*a*) the air-fuel ratio, (*b*) the percent theoretical air used, and (*c*) the volume flow rate of air used to burn fuel at a rate of 1 kg/min.

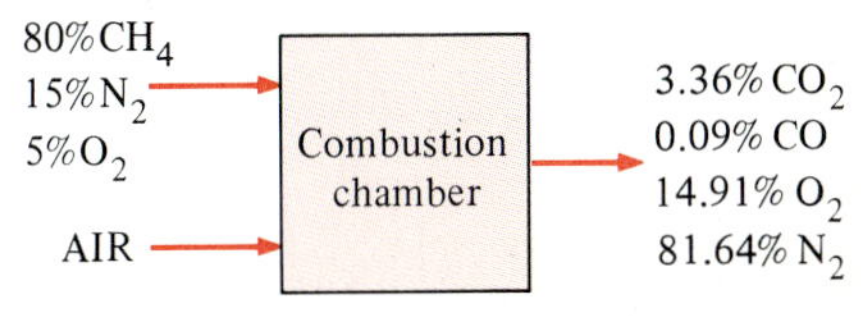

FIGURE P14-37

Enthalpy of Formation and Enthalpy of Combustion

14-38C Do nonreacting systems contain any chemical energy? Do they involve any changes in chemical energy during a process?

14-39C What is enthalpy of combustion? How does it differ from the enthalpy of reaction?

14-40C What is enthalpy of formation? How does it differ from the enthalpy of combustion?

14-41C What are the higher and the lower heating values of a fuel? How do they differ? How is the heating value of a fuel related to the enthalpy of combustion of that fuel?

14-42C When are the enthalpy of formation and the enthalpy of combustion identical?

14-43C Does the enthalpy of formation of a substance change with temperature?

14-44C Is it possible to analyze a chemical reaction by using property tables that are prepared using different reference states? How?

14-45C The $\overline{h}_f^\circ$ of N_2 is listed as zero. Does this mean that N_2 contains no chemical energy at the standard reference state?

14-46C Which contains more chemical energy, 1 kmol of H_2 or 1 kmol of H_2O?

14-47 Determine the enthalpy of combustion of methane (CH_4) at 25°C and 1 atm, using the enthalpy-of-formation data from Table A-26. Assume that the water in the products is in the liquid form. Compare your result to the value listed in Table A-27. *Answer:* −890,330 kJ

14-48 Repeat Prob. 14-47 for gaseous ethane (C_2H_6).

14-49 Repeat Prob. 14-47 for liquid octane (C_8H_{18}).

14-49E Repeat Prob. 14-47 for gaseous propane (C_3H_8), using English units and a temperature of 77°F.

First-Law Analysis of Reacting Systems

14-50C Are the conservation of energy relations different for reacting systems and nonreacting systems?

14-51C Derive a relation for the conservation of energy relation for a reacting closed system undergoing a quasi-equilibrium P = constant expansion or compression process.

14-52C Consider a complete combustion process during which both the reactants and the products are maintained at the same state. Combustion is achieved with (*a*) 100 percent theoretical air, (*b*) 200 percent theoretical air, and (*c*) the chemically correct amount of pure oxygen. For which case will the amount of heat transfer be the highest? Explain.

14-53C Consider a complete combustion process during which the reactants enter the combustion chamber at 20°C and the products leave at 500°C. Combustion is achieved with (*a*) 100 percent theoretical air, (*b*) 200 percent theoretical air, and (*c*) the chemically correct amount of pure oxygen. For which case will the amount of heat transfer be the lowest? Explain.

14-54 Methane (CH_4) is burned completely with the stoichiometric amount of air during a steady-flow combustion process. If both the reactants and the products are maintained at 25°C and 1 atm and the water in the products exists in the liquid form, determine the heat transfer for this process. What would your answer be if combustion were achieved with 50 percent excess air? *Answer:* −890,330 kJ/kmol

14-55 Hydrogen (H_2) is burned completely with the stoichiometric amount of air during a steady-flow combustion process. If both the reactants and the products are maintained at 25°C and 1 atm and the water in the products exists in the liquid form, determine the heat transfer for this process. What would your answer be if combustion were achieved with 80 percent excess air?

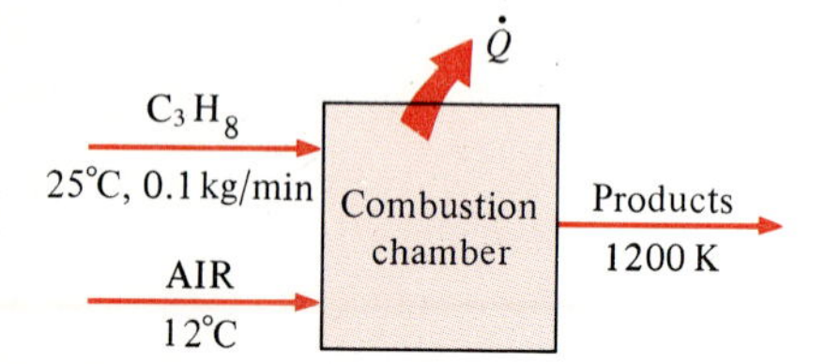

FIGURE P14-56

14-56 Liquid propane (C_3H_8) enters a combustion chamber at 25°C at a rate of 0.1 kg/min where it is mixed and burned with 150 percent excess air, which enters the combustion chamber at 12°C. If the combustion is complete and the exit temperature of the combustion gases is 1200 K, determine (*a*) the mass flow rate of air and (*b*) the rate of heat transfer from the combustion chamber.
Answers: (*a*) 3.92 kg/min, (*b*) −422.2 kJ/min

14-56E Liquid propane (C_3H_8) enters a combustion chamber at 77°F at a rate of 0.1 lbm/min where it is mixed and burned with 150 percent excess air, which enters the combustion chamber at 40°F. If the combustion is complete and the exit temperature of the combustion gases is 1800 R, determine (*a*) the mass flow rate of air and (*b*) the rate of heat transfer from the combustion chamber.
Answers: (*a*) 3.92 lbm/min, (*b*) −522 Btu/min

14-57 Acetylene gas (C_2H_2) is burned completely with 20 percent excess air during a steady-flow combustion process. The fuel and the air enter the combustion chamber at 25°C, and the products leave at 1500 K. Determine (*a*) the air-fuel ratio and (*b*) the heat transfer for this process.

14-58 Liquid octane (C_8H_{18}) at 25°C is burned completely during a steady-flow combustion process with 180 percent theoretical air which enters the combustion chamber at 25°C. If the products leave at 1600 K, determine (*a*) the air-fuel ratio and (*b*) the heat transfer for this process.

14-58E Liquid octane (C_8H_{18}) at 77°F is burned completely during a steady-flow combustion process with 180 percent theoretical air, which enters the combustion chamber at 77°F. If the products leave at 2500 R, determine (*a*) the air-fuel ratio and (*b*) the heat transfer for this process.

14-59 Benzene gas (C_6H_6) at 25°C is burned during a steady-flow combustion process with 95 percent theoretical air, which enters the combustion chamber at 25°C. All the hydrogen in the fuel burns to H_2O, but part of the carbon burns to CO. If the products leave at 1000 K, determine (*a*) the mole fraction of the CO in the products and (*b*) the heat transfer for this process.
Answers: (*a*) 2.1 percent, (*b*) −2,112,779 kJ/kmol C_6H_6

14-59E Benzene gas (C_6H_6) at 77°F is burned during a steady-flow combustion process with 95 percent theoretical air which enters the combustion chamber at 77°F. All the hydrogen in the fuel burns to H_2O, but part of the carbon burns to CO. If the products leave at 1500 R, determine (*a*) the mole fraction of the CO in the products and (*b*) the heat transfer for this process.

14-60 A steady-flow combustion chamber is supplied with CO gas at 37°C and 110 kPa at a rate of 0.8 m^3/min and air at 25°C and 110 kPa at a rate of 3 kg/min. The combustion products leave the combustion chamber at 900 K. Assuming combustion is complete, determine the rate of heat transfer from the combustion chamber.

14-61 Diesel fuel ($C_{12}H_{26}$) at 25°C is burned in a steady-flow combustion chamber with 20 percent excess air which also enters at 25°C. The products leave the combustion chamber at 500 K. Assuming combustion is complete, determine the required mass flow rate of the diesel fuel to supply heat at a rate of 700 kJ/s. *Answer:* 17.3 g/s

14-61E Diesel fuel ($C_{12}H_{26}$) at 77°F is burned in a steady-flow combustion chamber with 20 percent excess air which also enters at 77°F. The products leave the combustion chamber at 800 R. Assuming combustion is complete, determine the required mass flow rate of the diesel fuel to supply heat at a rate of 700 Btu/s. *Answer:* 0.00391 lbm/s

14-62 Octane gas (C_8H_{18}) at 25°C is burned steadily with 30 percent excess air at 25°C, 1 atm, and 60 percent relative humidity. Assuming combustion is complete and the products leave the combustion chamber at 600 K, determine the heat transfer for this process per unit mass of octane.

14-62E Octane gas (C_8H_{18}) at 77°F is burned steadily with 30 percent excess air at 77°F, 1 atm, and 60 percent relative humidity. Assuming combustion is complete and the products leave the combustion chamber at 1000 R, determine the heat transfer for this process per unit mass of octane.

14-63 Methane gas (CH_4) at 25°C is burned steadily with dry air which enters the combustion chamber at 17°C. The volumetric analysis of the products on a dry basis is 5.20 percent CO_2, 0.33 percent CO, 11.24 percent O_2, and 83.23 percent N_2. Determine (*a*) the percentage of theoretical air used and (*b*) the heat transfer from the combustion chamber per kmol of CH_4 if the combustion products leave at 700 K.

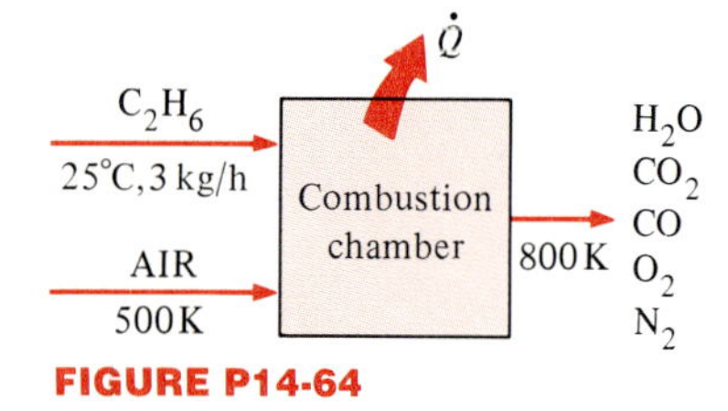

FIGURE P14-64

14-64 Ethane gas (C_2H_6) at 25°C is burned in a steady-flow combustion chamber at a rate of 3 kg/h with the stoichiometric amount of air which is preheated to 500 K before entering the combustion chamber. An analysis of the combustion gases reveals that all the hydrogen in the fuel burns to H_2O but only 95 percent of the carbon burns to CO_2, the remaining 5 percent forming CO. If the products leave the combustion chamber at 800 K, determine the rate of heat transfer from the combustion chamber. *Answer:* −107,652 kJ/h

14-65 A constant-volume tank contains a mixture of 100 g of methane (CH_4) gas and 500 g of O_2 at 25°C and 150 kPa. The contents of the tank are ignited, and the methane gas burns completely. If the final temperature is 1200 K, determine (*a*) the final pressure in the tank and (*b*) the amount of heat transfer during this process.

14-65E A constant-volume tank contains a mixture of 0.25 lbm of methane (CH_4) gas and 1.25 lbm of O_2 at 77°F and 20 psia. The contents of the tank are ignited, and the methane gas burns completely. If the final temperature is 1500 R, determine (*a*) the final pressure in the tank and (*b*) the amount of heat transfer during this process.

14-66 A closed combustion chamber is designed so that it maintains a constant pressure of 200 kPa during a combustion process. The combustion chamber has an initial volume of 1 m^3 and contains a stoichiometric mixture of octane (C_8H_{18}) gas and air at 25°C. The mixture is now ignited, and the product gases are observed to be at 1000 K at the end of the combustion process. Assuming complete combustion, and treating both the reactants and the products as ideal gases, determine the heat transfer from the combustion chamber during this process. *Answer:* 4798 kJ

14-67 A constant-volume tank contains a mixture of 1 kmol of benzene (C_6H_6) gas and 30 percent excess air at 25°C and 1 atm. The contents of the tank are now ignited, and all the hydrogen in the fuel burns to H_2O but only 92 percent of the carbon burns to CO_2, the remaining 8 percent forming CO. If the final temperature in the tank is 1000 K, determine the heat transfer from the combustion chamber during this process.

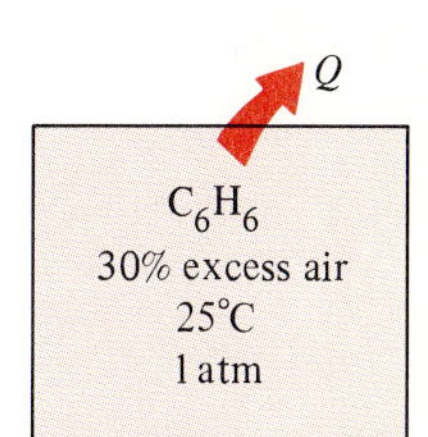

FIGURE P14-67

14-67E A constant-volume tank contains a mixture of 1 lbmol of benzene (C_6H_6) gas and 30 percent excess air at 77°F and 1 atm. The contents of the tank are now ignited, and all the hydrogen in the fuel burns to H_2O but only 92 percent of the carbon burns to CO_2, the remaining 8 percent forming CO. If the final temperature in the tank is 1800 R, determine the heat transfer from the combustion chamber during this process.
Answer: −921,768 Btu

14-68 A 20-m^3 rigid tank initially contains a mixture of 1 kmol of hydrogen (H_2) gas and the stoichiometric amount of air at 25°C. The contents of the tank are ignited, and all the hydrogen in the fuel burns to H_2O. If the combustion products are cooled to 25°C, determine (*a*) the fraction of the H_2O that condenses and (*b*) the heat transfer from the combustion chamber during this process.

Adiabatic Flame Temperature

14-69C A fuel is completely burned first with the stoichiometric amount of air and then with the stoichiometric amount of pure oxygen. For which case will the adiabatic flame temperature be higher?

14-70C A fuel at 25°C is burned in a well-insulated steady-flow combustion chamber with air which is also at 25°C. Under what conditions will the adiabatic flame temperature of the combustion process be a maximum?

14-71 Propane gas (C_3H_8) enters a steady-flow combustion chamber at 1 atm and 25°C and is burned with air which enters the combustion chamber at the same state. Disregarding any changes in kinetic and potential energies, determine the adiabatic flame temperature for (*a*) complete combustion with 100 percent theoretical air, (*b*) complete combustion with 300 percent theoretical air, and (*c*) incomplete combustion (some CO in the products) with 95 percent theoretical air.

14-72 Hydrogen (H_2) at 7°C is burned with 20 percent excess air which is also at 7°C during an adiabatic steady-flow combustion process. Assuming complete combustion, determine the exit temperature of the product gases for this process. *Answer:* 2251.4 K

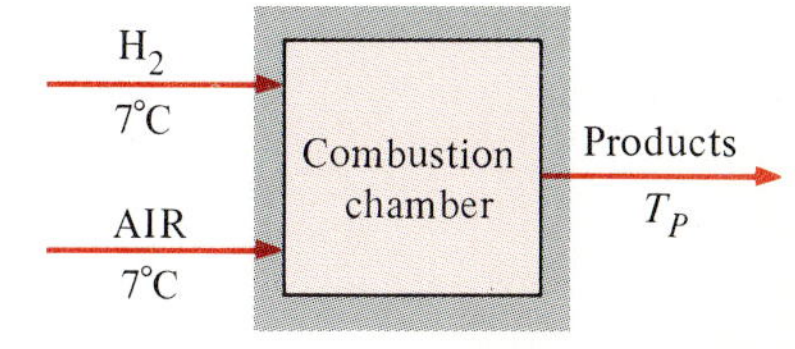

FIGURE P14-72

14-72E Hydrogen (H_2) at 40°F is burned with 20 percent excess air which is also at 40°F during an adiabatic steady-flow combustion process. Assuming complete combustion, find the exit temperature of the product gases for this process.

14-73 Determine the highest possible temperature that can be obtained when liquid gasoline (assumed C_8H_{18}) at 25°C is burned steadily with air at 25°C and 1 atm. What would your answer be if pure oxygen at 25°C were used to burn the fuel instead of air?

14-74 Acetylene gas (C_2H_2) at 25°C is burned during a steady-flow combustion process with 30 percent excess air at 27°C. It is observed that 75,000 kJ of heat is being lost from the combustion chamber to the surroundings per kmol of acetylene. Assuming combustion is complete, determine the exit temperature of the product gases.
Answer: 2303 K

14-75 An adiabatic constant-volume tank contains a mixture of 1 kmol of hydrogen (H_2) gas and the stoichiometric amount of air at 25°C and 1 atm. The contents of the tank are now ignited. Assuming complete combustion, determine the final temperature in the tank.

14-75E An adiabatic constant-volume tank contains a mixture of 1 lbmol of hydrogen (H_2) gas and the stoichiometric amount of air at 77°F and 1 atm. The contents of the tank are now ignited. Assuming complete combustion, determine the final temperature in the tank.
Answer: 5483 R

14-76 Octane gas (C_8H_{18}) at 25°C is burned steadily with 30 percent excess air at 25°C, 1 atm, and 60 percent relative humidity. Assuming combustion is complete and adiabatic, calculate the exit temperature of the product gases.

14-77 Write a computer program to determine the effect of the amount of air on the adiabatic combustion temperature of liquid octane (C_8H_{18}). Assume both the air and the octane are initially at 25°C. Determine the adiabatic combustion temperature for 75, 90, 100, 120, 150, 200, 300, 500, and 800 percent theoretical air. Assume the hydrogen in the fuel always burns to H_2O and the carbon to CO_2, except when there is a deficiency of air. In the latter case, assume that part of the carbon forms CO.

Entropy Change and Second-Law Analysis of Reacting Systems

14-78C Express the increase-in-entropy principle for chemically reacting systems.

14-79C What is the importance of the third law of thermodynamics?

14-80C How are the absolute entropy values of ideal gases at pressures different from 1 atm determined?

14-81C Is it a waste of time to calculate the reversible work associated with chemical reactions since most chemical reactions do not involve any work interactions?

14-82C What does the Gibbs function of formation $\bar{g}_f^\circ$ of a compound represent?

14-83 One kmol of H_2 at 25°C and 1 atm is burned steadily with 0.5 kmol of O_2 at the same state. The H_2O formed during the process is then brought to 25°C and 1 atm, the conditions of the surroundings. Assuming combustion is complete, determine the reversible work and irreversibility for this process.

14-84 Ethylene (C_2H_4) gas enters an adiabatic combustion chamber at 25°C and 1 atm and is burned with 20 percent excess air, which enters at 25°C and 1 atm. The combustion is complete, and the products leave the combustion chamber at 1 atm pressure. Assuming T_0 = 25°C, determine (*a*) the temperature of the products, (*b*) the entropy generation, and (*c*) the irreversibility.
Answers: (*a*) 2269.6 K, (*b*) 1457.45 kJ/(kmol·K), (*c*) 434,320 kJ/Kmol

14-85 Liquid octane (C_8H_{18}) enters a steady-flow combustion chamber at 25°C and 1 atm at a rate of 0.2 kg/min. It is burned with 50 percent excess air which also enters at 25°C and 1 atm. After combustion, the products are allowed to cool to 25°C. Assuming complete combustion and that all the H_2O in the products is in liquid form, determine (*a*) the heat transfer rate from the combustion chamber, (*b*) the entropy generation rate, and (*c*) the reversible work and irreversibility. Assume that T_0 = 298 K and the products leave the combustion chamber at 1 atm pressure.

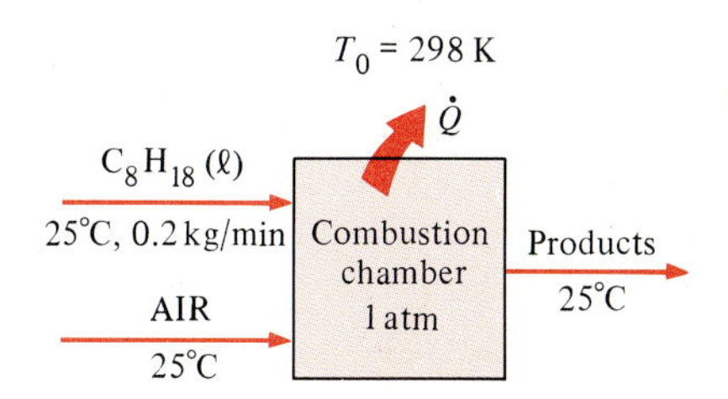

FIGURE P14-85

14-85E Liquid octane (C_8H_{18}) enters a steady-flow combustion chamber at 77°F and 1 atm at a rate of 0.2 lbm/min. It is burned with 50 percent excess air which also enters at 77°F and 1 atm. After combustion, the products are allowed to cool to 77°F. Assuming complete combustion and that all the H_2O in the products is in liquid form, determine (*a*) the heat transfer rate from the combustion chamber, (*b*) the entropy generation rate, and (*c*) the reversible work and irreversibility. Assume that T_0 = 537 R and the products leave the combustion chamber at 1 atm pressure.
Answers: (*a*) 4118.8 Btu/min, (*b*) 8.59 Btu/(min·R), (*c*) 4613 Btu/min

14-86 Acetylene gas (C_2H_2) is burned completely with 20 percent excess air during a steady-flow combustion process. The fuel and the air enter the combustion chamber separately at 25°C and 1 atm, and heat is being lost from the combustion chamber to the surroundings at 25°C at a rate of 300,000 kJ/kmol C_2H_2. The combustion products leave the combustion chamber at 1 atm pressure. Determine (*a*) the temperature of the products, (*b*) the total entropy change per kmol of C_2H_2, and (*c*) the reversible work and irreversibility during this process.

14-87 A steady-flow combustion chamber is supplied with CO gas at 37°C and 110 kPa at a rate of 0.8 m³/min, and air at 25°C and 110 kPa at a rate of 3 kg/min. Heat is transferred to a medium at 800 K, and the combustion products leave the combustion chamber at 900 K. Assuming the combustion is complete and T_0 = 25°C, determine (*a*) the rate of heat transfer from the combustion chamber, (*b*) the reversible work, and (*c*) the rate of irreversibility.
Answers: (*a*) 7133 kJ/min, (*b*) 3228 kJ/min, (*c*) 3228 kJ/min

14-88 Benzene gas (C_6H_6) at 1 atm and 25°C is burned during a steady-flow combustion process with 95 percent theoretical air, which enters the combustion chamber at 25°C and 1 atm. All the hydrogen in the fuel burns to H_2O, but part of the carbon burns to CO. Heat is lost to the surroundings at T_0 = 25°C, and the products leave the combustion chamber at 1 atm and 1000 K. Determine (*a*) the heat transfer for this process and (*b*) the irreversibility.

14-88E Benzene gas (C_6H_6) at 1 atm and 77°F is burned during a steady-flow combustion process with 95 percent theoretical air, which enters the combustion chamber at 77°F and 1 atm. All the hydrogen in the fuel burns to H_2O, but part of the carbon burns to CO. Heat is lost to the surroundings at 77°F, and the products leave the combustion chamber at 1 atm and 1500 R. Determine (*a*) the heat transfer for this process and (*b*) the irreversibility.

14-89 Liquid propane (C_3H_8) enters a combustion chamber at 25°C and 1 atm at a rate of 0.1 kg/min where it is mixed and burned with 150 percent excess air which enters the combustion chamber at 12°C. If the combustion products leave at 1200 K and 1 atm, determine (*a*) the mass flow rate of air, (*b*) the rate of heat transfer from the combustion chamber, and (*c*) the rate of entropy generation during this process. Assume T_0 = 25°C.
Answers: (*a*) 3.92 kg/min, (*b*) 456.4 kJ/min, (*c*) 7.45 kJ/(min·K)

14-90 Determine the work potential of 1 kmol of diesel fuel ($C_{12}H_{26}$) at 25°C and 1 atm in an environment at the same state.

14-90E Determine the work potential of 1 lbmol of diesel fuel ($C_{12}H_{26}$) at 77°F and 1 atm in an environment at the same state.
Answer: 3,315,224 Btu

14-91 Liquid octane (C_8H_{18}) enters a steady-flow combustion chamber at 25°C and 8 atm at a rate of 0.5 kg/min. It is burned with 200 percent excess air which is compressed and preheated to 500 K and 8 atm before entering the combustion chamber. After combustion, the products enter an adiabatic turbine at 1300 K and 8 atm and leave at 950 K and 2 atm. Assuming complete combustion and T_0 = 25°C, determine (*a*) the heat transfer rate from the combustion chamber, (*b*) the power output of the turbine, and (*c*) the reversible work and irreversibility for the entire process.
Answers: (*a*) 481.0 kJ/min; (*b*) 387.5 kW; (*c*) 544.6 kW, 157.1 kW

Chemical and Phase Equilibrium

CHAPTER 15

In Chap. 14 we analyzed combustion processes under the assumption that combustion is complete if there is sufficient time and oxygen. Often this is not the case, however. A chemical reaction may reach a state of equilibrium before reaching completion even if there is sufficient time and oxygen.

A system is said to be in equilibrium if no changes occur within the system when it is isolated from its surroundings. An isolated system is in *mechanical equilibrium* if no changes occur in pressure, in *thermal equilibrium* if no changes occur in temperature, in *phase equilibrium* if no transformations occur from one phase to another, and in *chemical equilibrium* if no changes occur in the chemical composition of the system. The conditions of mechanical and thermal equilibrium are straightforward, but the conditions of chemical and phase equilibrium can be rather involved.

The equilibrium criterion for reacting systems is based on the second law of thermodynamics, more specifically, the increase-in-entropy principle. For adiabatic systems, chemical equilibrium is established when the entropy of the reacting system reaches a maximum. Most reacting systems encountered in practice are not adiabatic, however. Therefore, we need to develop a general equilibrium criterion applicable to any reacting system.

In this chapter, we develop a general criterion for chemical equilibrium and apply it to reacting ideal-gas mixtures. We then extend the analysis to simultaneous reactions. Finally, we study phase equilibrium for nonreacting systems.

15-1 ■ CRITERION FOR CHEMICAL EQUILIBRIUM

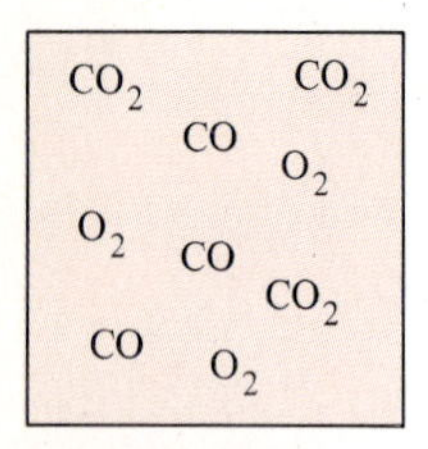

FIGURE 15-1

A reaction chamber that contains a mixture of CO_2, CO, and O_2 at a specified temperature and pressure.

Consider a reaction chamber which contains a mixture of CO, O_2, and CO_2 at a specified temperature and pressure. Let us try to predict what will happen in this combustion chamber (Fig. 15-1). Probably the first thing that comes to mind is a chemical reaction between CO and O_2 to form more CO_2:

$$CO + \tfrac{1}{2}O_2 \longrightarrow CO_2$$

This reaction is certainly a possibility, but it is not the only one. It is also possible that some CO_2 in the combustion chamber dissociated into CO and O_2. Yet a third possibility would be to have no reactions among the three components at all, i.e., for the system to be in chemical equilibrium. Although we know the temperature, pressure, and composition (thus the state) of the system, we are unable to predict whether the system is in chemical equilibrium. In this chapter we develop the necessary tools to change that.

Assume that the CO, O_2, and CO_2 mixture discussed above is in chemical equilibrium at the specified temperature and pressure. The chemical composition of this mixture will not change unless the temperature or the pressure of the mixture is changed. That is, a reacting mixture, in general, will have different equilibrium compositions at different pressures and temperatures. Therefore, when developing a general criterion for chemical equilibrium, we consider a reacting system at a fixed temperature and pressure.

The increase-in-entropy principle for a reacting or nonreacting system was expressed in Chap. 6 as

$$dS_{sys} \geqslant \frac{\delta Q}{T} \tag{15-1}$$

For adiabatic systems it reduces to $dS_{sys} \geqslant 0$. That is, a chemical reaction in an adiabatic chamber proceeds in the direction of increasing entropy. When the entropy reaches a maximum, the reaction stops (Fig. 15-2). Therefore, entropy is a very useful property in the analysis of reacting adiabatic systems.

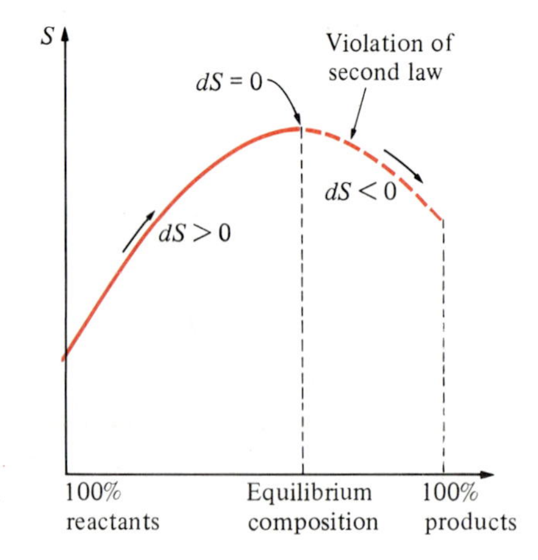

FIGURE 15-2

Equilibrium criteria for a chemical reaction that takes place adiabatically.

When a reacting system involves heat transfer, the increase-in-entropy principle relation (Eq. 15-1) becomes impractical to use, however, since it requires a knowledge of heat transfer between the system and its surroundings. A more practical approach would be to develop a relation for the equilibrium criterion in terms of the properties of the reacting system only. Such a relation is developed below.

Consider a reacting (or nonreacting) simple compressible system of fixed mass with only quasi-equilibrium work modes at a specified temperature T and pressure P (Fig. 15-3). Combining the first- and the second-law relations for this system gives

$$\left.\begin{aligned} \delta Q - P\,dV &= dU \\ dS &\geq \frac{\delta Q}{T} \end{aligned}\right\} \quad dU + P\,dV - T\,dS \leq 0 \tag{15-2}$$

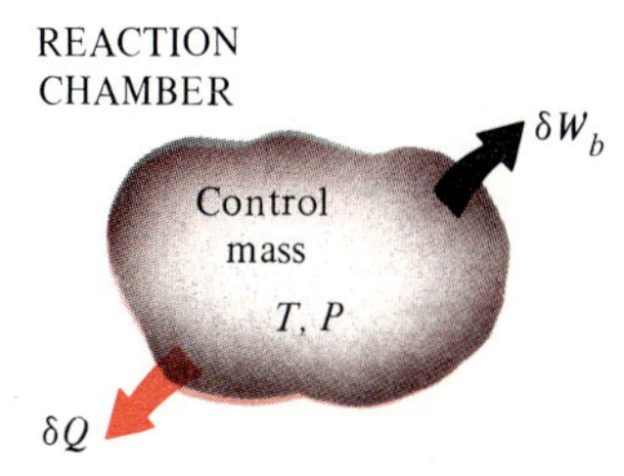

FIGURE 15-3

A control mass undergoing a chemical reaction at a specified temperature and pressure.

The differential of the Gibbs function ($G = H - TS$) at constant temperature and pressure is

$$\begin{aligned} (dG)_{T,P} &= dH - T\,dS - S\,dT \\ &= (dU + P\,dV + V\,\cancelto{0}{dP}) - T\,dS - S\,\cancelto{0}{dT} \\ &= dU + P\,dV - T\,dS \end{aligned} \tag{15-3}$$

From Eqs. 15-2 and 15-3, we have $(dG)_{T,P} \leq 0$. Therefore, a chemical reaction at a specified temperature and pressure will proceed in the direction of a decreasing Gibbs function. The reaction will stop and chemical equilibrium will be established when the Gibbs function attains a minimum value (Fig. 15-4). Therefore, the criterion for chemical equilibrium can be expressed as

$$(dG)_{T,P} = 0 \tag{15-4}$$

A chemical reaction at a specified temperature and pressure cannot proceed in the direction of an increasing Gibbs function since this will be a violation of the second law of thermodynamics. Notice that if the temperature or the pressure is changed, the reacting system will assume a

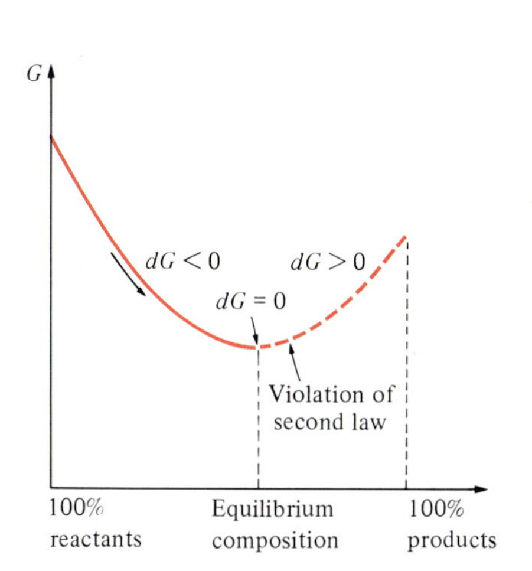

FIGURE 15-4

Criteria for chemical equilibrium at a specified temperature and pressure.

different equilibrium state, which is the state of minimum Gibbs function at the new temperature or pressure.

To obtain a relation for chemical equilibrium in terms of the properties of the individual components, consider a mixture of four chemical components A, B, C, and D which exist in equilibrium at a specified temperature and pressure. Let the number of moles of the respective components be N_A, N_B, N_C, and N_D. Now consider a reaction that occurs to an infinitesimal extent during which differential amounts of A and B (reactants) are converted to C and D (products) while the temperature and the pressure remain constant (Fig. 15-5):

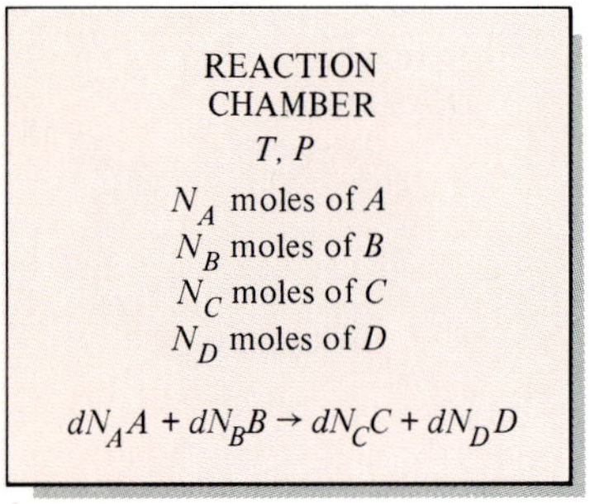

$$dN_A\, A + dN_B\, B \longrightarrow dN_C\, C + dN_D\, D$$

REACTION CHAMBER
T, P
N_A moles of A
N_B moles of B
N_C moles of C
N_D moles of D

$dN_A A + dN_B B \rightarrow dN_C C + dN_D D$

FIGURE 15-5
An infinitesimal reaction in a chamber at constant temperature and pressure.

The equilibrium criterion (Eq. 15-4) requires that the change in the Gibbs function of the mixture during this process be equal to zero. That is,

$$(dG)_{T,P} = \Sigma(dG_i)_{T,P} = \Sigma(\bar{g}_i\, dN_i)_{T,P} = 0 \qquad (15\text{-}5)$$

or

$$\bar{g}_C\, dN_C + \bar{g}_D\, dN_D + \bar{g}_A\, dN_A + \bar{g}_B\, dN_B = 0 \qquad (15\text{-}6)$$

where the $\bar{g}$'s are the molar Gibbs functions (also called the *chemical potentials*) at the specified temperature and pressure and the dN's are the differential changes in the number of moles of the components.

To find a relation between the dN's, we write the corresponding stoichiometric (theoretical) reaction

$$\nu_A A + \nu_B B \rightleftharpoons \nu_C C + \nu_D D \qquad (15\text{-}7)$$

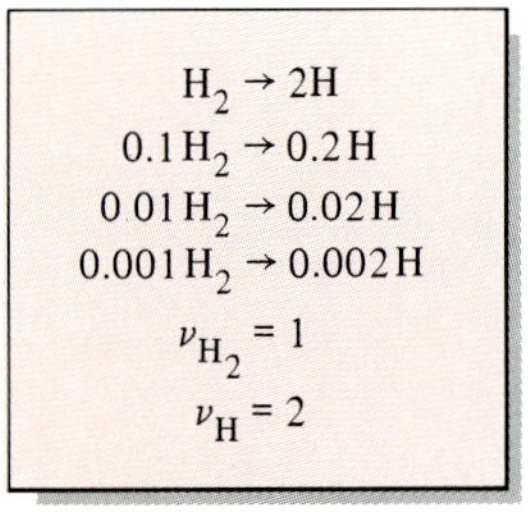

where the ν's are the stoichiometric coefficients, which are evaluated easily once the reaction is specified. The stoichiometric reaction plays an important role in the determination of the equilibrium composition of reacting mixtures because the changes in the number of moles of the components are proportional to the stoichiometric coefficients (Fig. 15-6). That is,

$$\begin{aligned} dN_A &= -\varepsilon\nu_A \qquad & dN_C &= \varepsilon\nu_C \\ dN_B &= -\varepsilon\nu_B \qquad & dN_D &= \varepsilon\nu_D \end{aligned} \qquad (15\text{-}8)$$

$H_2 \rightarrow 2H$
$0.1\,H_2 \rightarrow 0.2\,H$
$0.01\,H_2 \rightarrow 0.02\,H$
$0.001\,H_2 \rightarrow 0.002\,H$
$\nu_{H_2} = 1$
$\nu_H = 2$

FIGURE 15-6
The changes in the number of moles of the components during a chemical reaction are proportional to the stoichiometric coefficients regardless of the extent of the reaction.

where ε is the proportionality constant and represents the extent of a reaction. A minus sign is added to the first two terms because the number of moles of the reactants A and B decreases as the reaction progresses.

For example, if the reactants are C_2H_6 and O_2 and the products are CO_2 and H_2O, the reaction of 1 μmol (10^{-6} mol) of C_2H_6 will result in a 2-μmol increase in CO_2, a 3-μmol increase in H_2O, and a 3.5-μmol decrease in O_2 in accordance with the stoichiometric equation

$$C_2H_6 + 3.5O_2 \longrightarrow 2CO_2 + 3H_2O$$

That is, the change in the number of moles of a component is one-millionth ($\varepsilon = 10^{-6}$) of the stoichiometric coefficient of that component in this case.

Substituting the relations in Eq. 15-8 into Eq. 15-6 and canceling ε, we have

$$\nu_C\bar{g}_C + \nu_D\bar{g}_D - \nu_A\bar{g}_A - \nu_B\bar{g}_B = 0 \qquad (15\text{-}9)$$

This equation involves the stoichiometric coefficients and the molar Gibbs functions of the reactants and the products, and it is known as the **criterion for chemical equilibrium**. It is valid for any chemical reaction regardless of the phases involved.

Equation 15-9 is developed for a chemical reaction that involves two reactants and two products for simplicity, but it can easily be modified to handle chemical reactions with any number of reactants and products. Next we analyze the equilibrium criterion for ideal-gas mixtures.

15-2 ■ THE EQUILIBRIUM CONSTANT FOR IDEAL-GAS MIXTURES

Consider a mixture of ideal gases which exists in equilibrium at a specified temperature and pressure. Like entropy, the Gibbs function of an ideal gas depends on both the temperature and the pressure. The Gibbs function values are usually listed versus temperature at a fixed reference pressure P_0, which is taken to be 1 atm. The variation of the Gibbs function of an ideal gas with pressure at a fixed temperature is determined by using the definition of the Gibbs function ($\overline{g} = \overline{h} - T\overline{s}$) and the entropy-change relation for isothermal processes [$\Delta\overline{s} = -R_u \ln (P_2/P_1)$]. It yields

$$(\Delta\overline{g})_T = \Delta\overline{h}^{\,0} - T\,\Delta\overline{s} = -T\,\Delta\overline{s} = R_uT \ln \frac{P_2}{P_1}$$

Thus the Gibbs function of component i of an ideal-gas mixture at its partial pressure P_i and mixture temperature T can be expressed as

$$\overline{g}_i(T,P_i) = \overline{g}_i^*(T) + R_uT \ln P_i \qquad (15\text{-}10)$$

where $\overline{g}_i^*(T)$ represents the Gibbs function of component i at 1 atm pressure and temperature T, and P_i represents the partial pressure of component i in atmospheres. Substituting the Gibbs function expression for each component into Eq. 15-9, we obtain

$$\nu_C[\overline{g}_C^*(T) + R_uT \ln P_C] + \nu_D[\overline{g}_D^*(T) + R_uT \ln P_D] - \nu_A[\overline{g}_A^*(T) + R_uT \ln P_A] - \nu_B[\overline{g}_B^*(T) + R_uT \ln P_B] = 0$$

For convenience, we define the **standard-state Gibbs function change** as

$$\Delta G^*(T) = \nu_C\overline{g}_C^*(T) + \nu_D\overline{g}_D^*(T) - \nu_A\overline{g}_A^*(T) - \nu_B\overline{g}_B^*(T) \qquad (15\text{-}11)$$

Substituting, we get

$$\Delta G^*(T) = -R_uT(\nu_C \ln P_C + \nu_D \ln P_D - \nu_A \ln P_A - \nu_B \ln P_B)$$

$$= -R_uT \ln \frac{P_C^{\nu_C}P_D^{\nu_D}}{P_A^{\nu_A}P_B^{\nu_B}} \qquad (15\text{-}12)$$

Now we define the **equilibrium constant** K_P for the chemical equilibrium of ideal-gas mixtures as

$$K_P = \frac{P_C^{\nu_C}P_D^{\nu_D}}{P_A^{\nu_A}P_B^{\nu_B}} \qquad (15\text{-}13)$$

Substituting into Eq. 15-12 and rearranging, we obtain

$$K_P = e^{-\Delta G^*(T)/(R_u T)} \tag{15-14}$$

Therefore, the equilibrium constant K_P of an ideal-gas mixture at a specified temperature can be determined from a knowledge of the standard-state Gibbs function change at the same temperature. The K_P values for several reactions are given in Table A-28.

Once the equilibrium constant is available, it can be used to determine the equilibrium composition of reacting ideal-gas mixtures. This is accomplished by expressing the partial pressures of the components in terms of their mole fractions

$$P_i = y_i P = \frac{N_i}{N_{\text{total}}} P$$

where P is the total pressure and N_{total} is the total number of moles present in the reaction chamber, including any inert gases. Replacing the partial pressures in Eq. 15-13 by the above relation and rearranging, we obtain (Fig. 15-7)

$$K_P = \frac{N_C^{\nu_C} N_D^{\nu_D}}{N_A^{\nu_A} N_B^{\nu_B}} \left(\frac{P}{N_{\text{total}}}\right)^{\Delta\nu} \tag{15-15}$$

where $\Delta\nu = \nu_C + \nu_D - \nu_A - \nu_B$

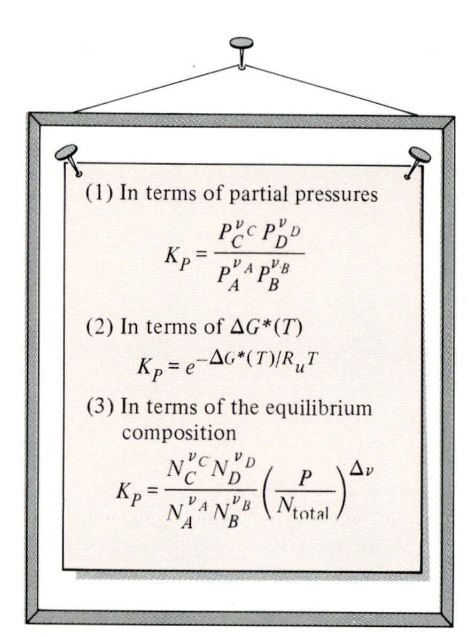

FIGURE 15-7

Three equivalent K_P relations for reacting ideal-gas mixtures.

Equation 15-15 is written for a reaction involving two reactants and two products, but it can be extended to reactions involving any number of reactants and products.

EXAMPLE 15-1

Using Eq. 15-14 and the Gibbs function data, determine the equilibrium constant K_P for the dissociation process $N_2 \rightarrow 2N$ at 25°C. Compare your result with the K_P value listed in Table A-28.

Solution The natural logarithm of the equilibrium constant at 298 K for this reaction is listed in Table A-28 as $\ln K_P = -367.5$. In the absence of such tables, K_P can be determined from the Gibbs function data and Eq. 15-14

$$K_P = e^{-\Delta G^*(T)/(R_u T)}$$

where, from Eq. 15-11,

$$\begin{aligned}\Delta G^*(T) &= \nu_N \bar{g}_N^*(T) - \nu_{N_2} \bar{g}_{N_2}^*(T) \\ &= (2)(455{,}510 \text{ kJ/kmol}) - 0 \\ &= 911{,}020 \text{ kJ/kmol}\end{aligned}$$

Substituting, we find

$$\begin{aligned}\ln K_P &= -\frac{911{,}020 \text{ kJ/kmol}}{[8.314 \text{ kJ/(kmol·K)}](298.15 \text{ K})} \\ &= -367.5\end{aligned}$$

or

$$K_P \cong 2 \times 10^{-160}$$

The calculated K_P value is in agreement with the value listed in Table

A-28. The K_P value for this reaction is practically zero, indicating that this reaction will not occur at this temperature.

Note that this reaction involves one product (N) and one reactant (N_2), and the stoichiometric coefficients for this reaction are $\nu_N = 2$ and $\nu_{N_2} = 1$. Also note that the Gibbs function of all stable elements (such as N_2) is assigned a value of zero at the standard reference state at 25°C and 1 atm. The Gibbs function values at other temperatures can be calculated from the enthalpy and absolute entropy data by using the definition of the Gibbs function, $\bar{g}^*(T) = \bar{h}(T) - T\bar{s}^*(T)$, where $\bar{h}(T) = \bar{h}_f^* + \bar{h}_T - \bar{h}_{298\,K}$.

EXAMPLE 15-2

Determine the temperature at which 10 percent of diatomic hydrogen (H_2) dissociates into monatomic hydrogen (H) at a pressure of 10 atm.

Solution For simplicity we consider 1 kmol of H_2 as shown in Fig. 15-8. The stoichiometric and actual reactions in this case are as follows:

Stoichiometric: $H_2 \rightleftharpoons 2H$ (thus $\nu_{H_2} = 1$ and $\nu_H = 2$)

Actual: $H_2 \longrightarrow \underbrace{0.9H_2}_{\substack{\text{reactants}\\\text{(leftover)}}} + \underbrace{0.2H}_{\text{products}}$

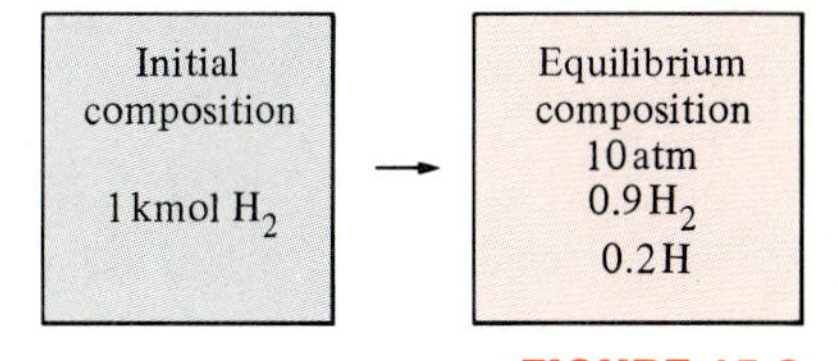

FIGURE 15-8
Schematic for Example 15-2.

A double-headed arrow is used for the stoichiometric reaction to differentiate it from the actual reaction. This reaction involves one reactant (H_2) and one product (H). The equilibrium composition consists of 0.9 kmol of H_2 (the leftover reactant) and 0.2 kmol of H (the newly formed product). Therefore, $N_{H_2} = 0.9$ and $N_H = 0.2$. Assuming ideal-gas behavior for both H_2 and H, the equilibrium constant K_P can be determined from Eq. 15-15:

$$K_P = \frac{N_H^{\nu_H}}{N_{H_2}^{\nu_{H_2}}}\left(\frac{P}{N_{total}}\right)^{\nu_H - \nu_{H_2}} = \frac{(0.2)^2}{0.9}\left(\frac{10}{0.9 + 0.2}\right)^{2-1} = 0.404$$

From Table A-28, the temperature corresponding to this K_P value is

$$T = 3535 \text{ K}$$

Therefore, 10 percent of H_2 will dissociate into H when the temperature is raised to 3535 K. If the temperature is increased further, the percentage of H_2 that dissociates into H will also increase.

15-3 ■ SOME REMARKS ABOUT THE K_P OF IDEAL-GAS MIXTURES

In the last section we developed three equivalent expressions for the equilibrium constant K_P of reacting ideal-gas mixtures: Eq. 15-13 which expresses K_P in terms of the partial pressures, Eq. 15-14 which expresses K_P in terms of the standard-state Gibbs function change $\Delta G^*(T)$, and Eq. 15-15 which expresses K_P in terms of the number of moles of the components. All three relations are equivalent, but sometimes one is more convenient to use than the others. For example, Eq. 15-15 is best suited for determining the equilibrium composition of a reacting ideal-gas mixture at a specified temperature and pressure. On the basis of these relations, we may draw the following conclusions about the equilibrium constant K_P of ideal-gas mixtures:

1 *The K_P of a reaction depends on temperature only*. It is independent of the pressure of the equilibrium mixture and is not affected by the presence of inert gases. This is because K_P depends on $\Delta G^*(T)$ which depends on temperature only, and the $\Delta G^*(T)$ of inert gases is zero (see Eq. 15-14). Thus at a specified temperature the following four reactions have the same K_P value:

$$H_2 + \tfrac{1}{2}O_2 \rightleftharpoons H_2O \qquad \text{at 1 atm}$$
$$H_2 + \tfrac{1}{2}O_2 \rightleftharpoons H_2O \qquad \text{at 5 atm}$$
$$H_2 + \tfrac{1}{2}O_2 + 3N_2 \rightleftharpoons H_2O + 3N_2 \qquad \text{at 3 atm}$$
$$H_2 + 2O_2 + 5N_2 \rightleftharpoons H_2O + 1.5O_2 + 5N_2 \qquad \text{at 2 atm}$$

2 *The K_P of the reverse reaction is $1/K_P$*. This is easily seen from Eq. 15-13. For reverse reactions, the products and reactants switch places, and thus the terms in the numerator move to the denominator and vice versa. Consequently, the equilibrium constant of the reverse reaction becomes $1/K_P$. For example, from Table A-28,

$$K_P = 0.1147 \times 10^{11} \quad \text{for} \quad H_2 + \tfrac{1}{2}O_2 \rightleftharpoons H_2O \quad \text{at 1000 K}$$
$$K_P = 8.718 \times 10^{-11} \quad \text{for} \quad H_2O \rightleftharpoons H_2 + \tfrac{1}{2}O_2 \quad \text{at 1000 K}$$

	$H_2 \rightarrow 2H$ $P = 1$ atm	
T, K	K_P	%mol H
1000	5.17×10^{-18}	0.00
2000	2.65×10^{-6}	0.16
3000	0.025	14.63
4000	2.545	76.80
5000	41.47	97.70
6000	267.7	99.63

FIGURE 15-9

The larger the K_P, the more complete the reaction.

3 *The larger the K_P, the more complete the reaction*. This is also apparent from Fig. 15-9 and Eq. 15-13. If the equilibrium composition consists largely of product gases, the partial pressures of the products (P_C and P_D) will be considerably larger than the partial pressures of the reactants (P_A and P_B), which will result in a large value of K_P. In the limiting case of a complete reaction (no leftover reactants in the equilibrium mixture), K_P will approach infinity. Conversely, very small values of K_P indicate that a reaction will not proceed to any appreciable degree. Thus reactions with very small K_P values at a specified temperature can be neglected.

A reaction with $K_P > 1000$ (or $\ln K_P > 7$) is usually assumed to proceed to completion, and a reaction with $K_P < 0.001$ (or $\ln K_P < -7$) is assumed not to occur at all. For example, $\ln K_P = -6.8$ for the reaction $N_2 \rightleftharpoons 2N$ at 5000 K. Therefore, the dissociation of N_2 into monatomic nitrogen (N) can be disregarded at temperatures below 5000 K.

4 *The mixture pressure affects the equilibrium composition* (although it does not affect the equilibrium constant K_P). This can be seen from Eq. 15-15 which involves the term $P^{\Delta\nu}$, where $\Delta\nu = \Sigma\nu_P - \Sigma\nu_R$ (the difference between the number of moles of products and the number of moles of reactants in the stoichiometric reaction). At a specified temperature, the K_P value of the reaction, and thus the right-hand side of Eq. 15-15, remains constant. Therefore, the mole numbers of the reactants and the products must change to counteract any changes in the pressure term. The direction of the change depends on the sign of $\Delta\nu$. An increase in pressure at a

specified temperature will increase the number of moles of the reactants and decrease the number of moles of products if $\Delta\nu$ is positive, have the opposite effect if $\Delta\nu$ is negative, and have no effect if $\Delta\nu$ is zero.

5 *The presence of inert gases affects the equilibrium composition* (although it does not affect the equilibrium constant K_P). This can be seen from Eq. 15-15 which involves the term $(1/N_{\text{total}})^{\Delta\nu}$, where N_{total} is the total number of moles of the ideal-gas mixture at equilibrium, including inert gases. The sign of $\Delta\nu$ determines how the presence of inert gases will influence the equilibrium composition (Fig. 15-10). An increase in the number of moles of inert gases at a specified temperature and pressure will decrease the number of moles of the reactants and increase the number of moles of products if $\Delta\nu$ is positive, have the opposite effect if $\Delta\nu$ is negative, and have no effect if $\Delta\nu$ is zero.

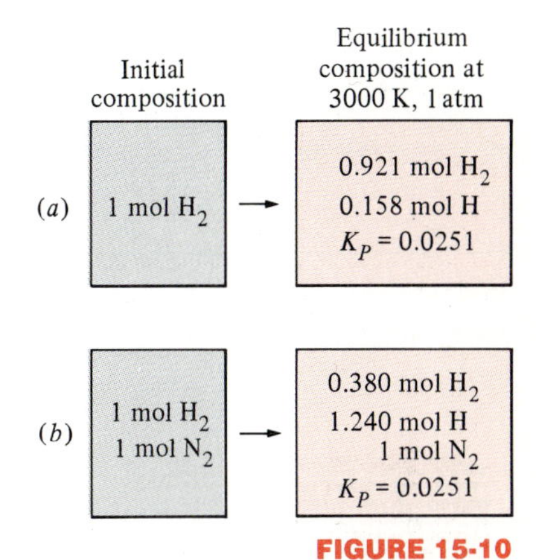

FIGURE 15-10

The presence of inert gases does not affect the equilibrium constant, but it does affect the equilibrium composition.

6 *When the stoichiometric coefficients are doubled, the value of K_P is squared.* Therefore, when one is using K_P values from a table, the stoichiometric coefficients (the ν's) used in a problem must be exactly the same ones appearing in the table from which the K_P values are selected. Multiplying all the coefficients of a stoichiometric equation does not affect the mass balance, but it does affect the equilibrium constant calculations since the stoichiometric coefficients appear as exponents of partial pressures in Eq. 15-13. For example,

For $\qquad H_2 + \tfrac{1}{2}O_2 \rightleftharpoons H_2O \qquad K_{P_1} = \dfrac{P_{H_2O}}{P_{H_2}P_{O_2}^{1/2}}$

But for $\qquad 2H_2 + O_2 \rightleftharpoons 2H_2O \qquad K_{P_2} = \dfrac{P_{H_2O}^2}{P_{H_2}^2 P_{O_2}} = (K_{P_1})^2$

7 *Free electrons in the equilibrium composition can be treated as an ideal gas.* At high temperatures (usually above 2500 K), gas molecules start to dissociate into unattached atoms (such as $H_2 \rightleftharpoons 2H$), and at even higher temperatures atoms start to lose electrons and ionize, e.g.,

$$H \rightleftharpoons H^+ + e^- \tag{15-16}$$

The dissociation and ionization effects are more pronounced at low pressures. Ionization occurs to an appreciable extent only in very high temperature environments, and the mixture of electrons, ions, and neutral atoms can be treated as an ideal gas in such environments. Therefore, the equilibrium composition of ionized gas mixtures can be determined from Eq. 15-15 (Fig. 15-11). This treatment may not be adequate in the presence of strong electric fields, however, since the electrons may be at a different temperature than the ions in this case.

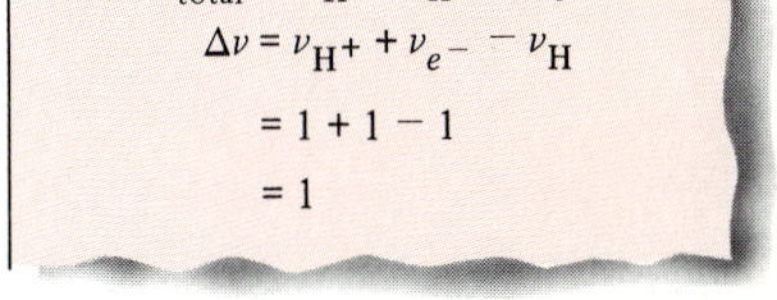

$$H \rightarrow H^+ + e^-$$

$$K_P = \frac{N_{H^+}^{\nu_{H^+}} N_{e^-}^{\nu_{e^-}}}{N_H^{\nu_H}} \left(\frac{P}{N_{\text{total}}}\right)^{\Delta\nu}$$

where

$$N_{\text{total}} = N_H + N_{H^+} + N_{e^-}$$
$$\Delta\nu = \nu_{H^+} + \nu_{e^-} - \nu_H = 1 + 1 - 1 = 1$$

FIGURE 15-11

Equilibrium-constant relation for the ionization reaction of hydrogen.

8 *Equilibrium calculations provide information on the equilibrium composition of a reaction, not on the reaction rate.* Sometimes it may even take years to achieve the indicated equilibrium composition. For example, the equilibrium constant of the reaction $H_2 + \tfrac{1}{2}O_2 \rightleftharpoons H_2O$ at 298 K is about 10^{40}, which suggests that a stoichiometric mixture of H_2 and O_2 at room temperature should react to form H_2O, and the reaction should go

to completion. However, the rate of this reaction is so slow that it practically does not occur. But when the right catalyst is used, the reaction goes to completion in a reasonable time to the predicted value.

EXAMPLE 15-3

A mixture of 2 kmol of CO and 3 kmol of O_2 is heated to 2600 K at a pressure of 304 kPa. Determine the equilibrium composition, assuming the mixture consists of CO_2, CO, and O_2 (Fig. 15-12).

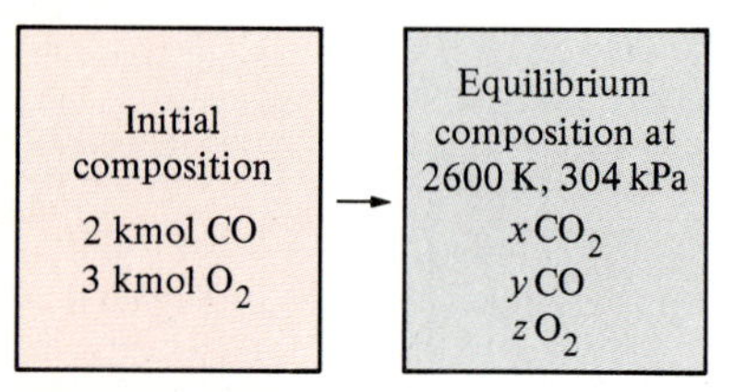

FIGURE 15-12
Schematic for Example 15-3.

Solution The stoichiometric and actual reactions in this case are as follows:

Stoichiometric: $CO + \frac{1}{2}O_2 \rightleftharpoons CO_2$ (thus $\nu_{CO_2} = 1$, $\nu_{CO} = 1$, and $\nu_{O_2} = \frac{1}{2}$)

Actual: $2CO + 3O_2 \longrightarrow \underbrace{xCO_2}_{\text{products}} + \underbrace{yCO + zO_2}_{\text{reactants (leftover)}}$

C balance: $2 = x + y$ or $y = 2 - x$

O balance: $8 = 2x + y + 2z$ or $z = 3 - \dfrac{x}{2}$

Total number of moles: $N_{total} = x + y + z = 5 - \dfrac{x}{2}$

Pressure: $P = 304\,\text{kPa} = 3.0\,\text{atm}$

The closest reaction listed in Table A-28 is $CO_2 \rightleftharpoons CO + \frac{1}{2}O_2$, for which $\ln K_P = -2.801$ at 2600 K. The reaction we have is the inverse of this, and thus $\ln K_P = +2.801$ or $K_P = 16.461$ in our case.

Assuming ideal-gas behavior for all components, the equilibrium constant relation (Eq. 15-15) becomes

$$K_P = \frac{N_{CO_2}^{\nu_{CO_2}}}{N_{CO}^{\nu_{CO}} N_{O_2}^{\nu_{O_2}}} \left(\frac{P}{N_{total}}\right)^{\nu_{CO_2} - \nu_{CO} - \nu_{O_2}}$$

Substituting, we get

$$16.461 = \frac{x}{(2 - x)(3 - x/2)^{1/2}} \left(\frac{3}{5 - x/2}\right)^{-1/2}$$

Solving for x yields

$$x = 1.906$$

Then

$$y = 2 - x = 0.094$$

$$z = 3 - \frac{x}{2} = 2.047$$

Therefore, the equilibrium composition of the mixture at 2600 K and 304 kPa is

$$1.906CO_2 + 0.094CO + 2.047O_2$$

In solving this problem, we disregarded the dissociation of O_2 into O according to the reaction $O_2 \rightarrow 2O$, which is a real possibility at high temperatures. This is because $\ln K_P = -7.521$ at 2600 K for this reaction, which indicates that the amount of O_2 that dissociates into O in this case is negligible. (Besides, we have not learned how to deal with simultaneous reactions yet. We will do so in the next section.)

EXAMPLE 15-4

A mixture of 3 kmol of CO, 2.5 kmol of O_2, and 8 kmol of N_2 is heated to 2600 K at a pressure of 5 atm. Determine the equilibrium composition of the mixture (Fig. 15-13).

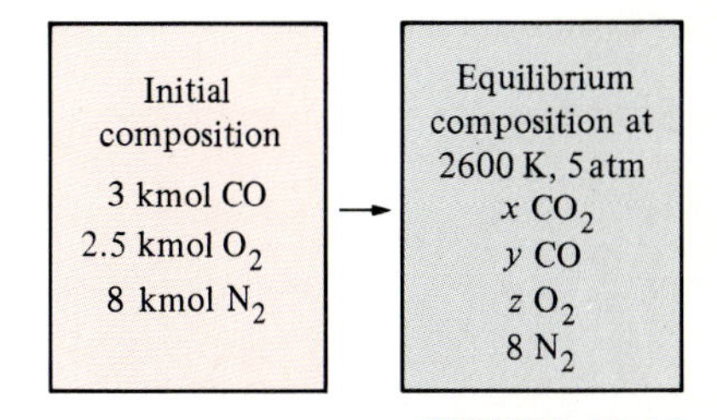

FIGURE 15-13
Schematic for Example 15-4.

Solution This problem is similar to Example 15-3, except that it involves an inert gas N_2. At 2600 K, some possible reactions are $O_2 \rightleftharpoons 2O$ ($\ln K_P = -7.521$), $N_2 \rightleftharpoons 2N$ ($\ln K_P = -28.304$), $\frac{1}{2}O_2 + \frac{1}{2}N_2 \rightleftharpoons NO$ ($\ln K_P = -2.671$), and $CO + \frac{1}{2}O_2 \rightleftharpoons CO_2$ ($\ln K_P = 2.801$ or $K_P = 16.461$). Based on these K_P values, we conclude that the O_2 and N_2 will not dissociate to any appreciable degree, but a small amount will combine to form some oxides of nitrogen. (We disregard the oxides of nitrogen in this example, but they should be considered in a more refined analysis.) We also conclude that most of the CO will combine with O_2 to form CO_2. Notice that despite the changes in pressure, the number of moles of CO and O_2, and the presence of an inert gas, the K_P value of the reaction is the same as that used in Example 15-3.

The stoichiometric and actual reactions in this case are

Stoichiometric: $CO + \frac{1}{2}O_2 \rightleftharpoons CO_2$ (thus $\nu_{CO_2} = 1$, $\nu_{CO} = 1$, and $\nu_{O_2} = \frac{1}{2}$)

Actual: $$3CO + 2.5O_2 + 8N_2 \longrightarrow \underbrace{xCO_2}_{\text{products}} + \underbrace{yCO + zO_2}_{\text{reactants (leftover)}} + \underbrace{8N_2}_{\text{inert}}$$

C balance: $3 = x + y$ or $y = 3 - x$

O balance: $8 = 2x + y + 2z$ or $z = 2.5 - \dfrac{x}{2}$

Total number of moles: $N_{\text{total}} = x + y + z + 8 = 13.5 - \dfrac{x}{2}$

Assuming ideal-gas behavior for all components, the equilibrium constant relation (Eq. 15-15) becomes

$$K_P = \frac{N_{CO_2}^{\nu_{CO_2}}}{N_{CO}^{\nu_{CO}} N_{O_2}^{\nu_{O_2}}} \left(\frac{P}{N_{\text{total}}} \right)^{\nu_{CO_2} - \nu_{CO} - \nu_{O_2}}$$

Substituting, we get

$$16.461 = \frac{x}{(3 - x)(2.5 - x/2)^{1/2}} \left(\frac{5}{13.5 - x/2} \right)^{-1/2}$$

Solving for x gives

$$x = 2.142$$

Then

$$y = 3 - x = 0.858$$

$$z = 2.5 - \frac{x}{2} = 1.429$$

Therefore, the equilibrium composition of the mixture at 2600 K and 5 atm is

$$2.142CO_2 + 0.858CO + 1.429O_2 + 8N_2$$

Note that the inert gases do not affect the K_P value or the K_P relation for a reaction, but they do affect the equilibrium composition.

15-4 ■ CHEMICAL EQUILIBRIUM FOR SIMULTANEOUS REACTIONS

The reacting mixtures we have considered so far involved only one reaction, and writing a K_P relation for that reaction was sufficient to determine the equilibrium composition of the mixture. But most practical chemical reactions involve two or more reactions that occur simultaneously, which makes them more difficult to deal with. In such cases, it becomes necessary to apply the equilibrium criterion to all possible reactions that may occur in the reaction chamber. When a chemical species appears in more than one reaction, the application of the equilibrium criterion, together with the conservation of mass principle for each chemical species, results in a system of simultaneous equations from which the equilibrium composition can be determined.

We have shown earlier that a reacting system at a specified temperature and pressure will achieve chemical equilibrium when its Gibbs function reaches a minimum value, that is, $(dG)_{T,P} = 0$. This is true regardless of the number of reactions that may be occurring. When two or more reactions are involved, this condition is satisfied only when $(dG)_{T,P} = 0$ for each reaction. Assuming ideal-gas behavior, the K_P of each reaction can be determined from Eq. 15-15, with N_{total} being the total number of moles present in the equilibrium mixture.

Composition: CO_2, CO, O_2, O

No. of components: 4

No. of elements: 2

No. of K_P relations needed: 4 − 2 = 2

FIGURE 15-14

The number of K_P relations needed to determine the equilibrium composition of a reacting mixture is the difference between the number of species and the number of elements.

The determination of the equilibrium composition of a reacting mixture requires that we have as many equations as unknowns, where the unknowns are the number of moles of each chemical species present in the equilibrium mixture. The mass balance of each element involved provides one equation. The rest of the equations must come from the K_P relations written for each reaction. Thus we conclude that *the number of K_P relations needed to determine the equilibrium composition of a reacting mixture is equal to the number of chemical species minus the number of elements present in equilibrium.* For an equilibrium mixture that consists of CO_2, CO, O_2, and O, for example, two K_P relations are needed to determine the equilibrium composition since it involves four chemical species and two elements (Fig. 15-14).

The determination of the equilibrium composition of a reacting mixture in the presence of two simultaneous reactions is illustrated below with an example.

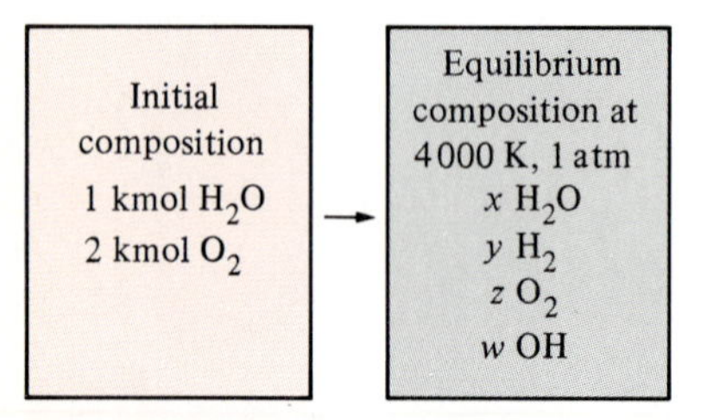

FIGURE 15-15

Schematic for Example 15-5.

EXAMPLE 15-5

A mixture of 1 kmol of H_2O and 2 kmol of O_2 is heated to 4000 K at a pressure of 1 atm. Determine the equilibrium composition of the mixture, assuming that only H_2O, OH, O_2, and H_2 are present (Fig. 15-15).

Solution The chemical reaction during this process can be expressed as

$$H_2O + 2O_2 \longrightarrow xH_2O + yH_2 + zO_2 + wOH$$

Mass balances for hydrogen and oxygen yield

H balance: $$2 = 2x + 2y + w \qquad (1)$$

O balance: $$5 = x + 2z + w \qquad (2)$$

The mass balances provide us with only two equations with four unknowns, and thus we need to have two more equations (to be obtained from the K_P relations) to determine the equilibrium composition of the mixture. It seems that part of the H_2O in the products is dissociated into H_2 and OH during this process, according to the stoichiometric reactions

$$H_2O \rightleftharpoons H_2 + \tfrac{1}{2}O_2 \qquad \text{(reaction 1)}$$
$$H_2O \rightleftharpoons \tfrac{1}{2}H_2 + OH \qquad \text{(reaction 2)}$$

The equilibrium constants for these two reactions at 4000 K are determined from Table A-28 to be

$$\ln K_{P_1} = -0.542 \longrightarrow K_{P_1} = 0.5816$$
$$\ln K_{P_2} = -0.044 \longrightarrow K_{P_2} = 0.9570$$

The K_P relations for these two simultaneous reactions are

$$K_{P_1} = \frac{N_{H_2}^{\nu_{H_2}} H_{O_2}^{\nu_{O_2}}}{N_{H_2O}^{\nu_{H_2O}}} \left(\frac{P}{N_{\text{total}}}\right)^{\nu_{H_2} + \nu_{O_2} - \nu_{H_2O}}$$

$$K_{P_2} = \frac{N_{H_2}^{\nu_{H_2}} N_{OH}^{\nu_{OH}}}{N_{H_2O}^{\nu_{H_2O}}} \left(\frac{P}{N_{\text{total}}}\right)^{\nu_{H_2} + \nu_{OH} - \nu_{H_2O}}$$

where $\qquad N_{\text{total}} = N_{H_2O} + N_{H_2} + N_{O_2} + N_{OH} = x + y + z + w$

Substituting yields

$$0.5816 = \frac{(y)(z)^{1/2}}{x} \left(\frac{1}{x + y + z + w}\right)^{1/2} \tag{3}$$

$$0.9570 = \frac{(w)(y)^{1/2}}{x} \left(\frac{1}{x + y + z + w}\right)^{1/2} \tag{4}$$

Solving Eqs. (1), (2), (3), and (4) simultaneously for the four unknowns x, y, z, and w yields

$$x = 0.271 \qquad z = 1.849$$
$$y = 0.213 \qquad w = 1.032$$

Therefore, the equilibrium composition of 1 kmol H_2O and 2 kmol O_2 at 1 atm and 4000 K is

$$0.271H_2O + 0.213H_2 + 1.849O_2 + 1.032OH$$

We could also solve this problem by using the K_P relation for the stoichiometric reaction $O_2 \rightleftharpoons 2O$ as one of the two equations.

Solving a system of simultaneous nonlinear equations is extremely tedious and time-consuming if it is done by hand. Thus it is often necessary to solve these kinds of problems by using an iterative algorithm in a computer or a programmable calculator.

15-5 VARIATION OF K_P WITH TEMPERATURE

It was shown in Sec. 15-2 that the equilibrium constant K_P of an ideal gas depends on temperature only, and it is related to the standard-state Gibbs function change $\Delta G^*(T)$ through the relation (Eq. 15-14)

$$\ln K_P = -\frac{\Delta G^*(T)}{R_u T}$$

In this section we develop a relation for the variation of K_P with temperature in terms of other thermochemical properties.

Substituting $\Delta G^*(T) = \Delta H^*(T) - T\,\Delta S^*(T)$ into the above relation and differentiating with respect to temperature, we get

$$\frac{d(\ln K_P)}{dT} = \frac{\Delta H^*(T)}{R_u T^2} - \frac{d[\Delta H^*(T)]}{R_u T\,dT} + \frac{d[\Delta S^*(T)]}{R_u\,dT}$$

At constant pressure, the second $T\,ds$ relation, $T\,ds = dh - v\,dP$, reduces to $T\,ds = dh$. Also $T\,d(\Delta S^*) = d(\Delta H^*)$ since ΔS^* and ΔH^* consist of entropy and enthalpy terms of the reactants and the products. Therefore, the last two terms in the above relation cancel, and it reduces to

$$\frac{d(\ln K_P)}{dT} = \frac{\Delta H(T)}{R_u T^2} = \frac{\overline{h}_R(T)}{R_u T^2} \qquad (15\text{-}17)$$

where $\overline{h}_R(T)$ is the enthalpy of reaction at temperature T. Notice that we dropped the superscript (*) which indicates a constant pressure of 1 atm from $\Delta H(T)$, since the enthalpy of an ideal gas depends on temperature only and is independent of pressure. Equation 15-17 is an expression of the variation of K_P with temperature in terms of $\overline{h}_R(T)$, and it is known as the **van't Hoff equation**. To integrate it, we need to know how $\overline{h}_R$ varies with T. For small temperature intervals, $\overline{h}_R$ can be treated as a constant and Eq. 15-17 can be integrated to yield

$$\ln \frac{K_{P_2}}{K_{P_1}} \cong \frac{\overline{h}_R}{R_u}\left(\frac{1}{T_1} - \frac{1}{T_2}\right) \qquad (15\text{-}18)$$

This equation has two important implications. First, it provides a means of calculating the $\overline{h}_R$ of a reaction from a knowledge of K_P, which is easier to determine. Second, it shows that exothermic reactions ($h_R < 0$) such as combustion processes will be less complete at higher temperatures since K_P decreases with temperature for such reactions (Fig. 15-16).

Reaction: $C + O_2 \rightarrow CO_2$

T, K	K_P
1000	4.78×10^{20}
2000	2.25×10^{10}
3000	7.80×10^{6}
4000	1.41×10^{5}

FIGURE 15-16

Exothermic reactions are less complete at higher temperatures.

EXAMPLE 15-6

Estimate the enthalpy of reaction $\overline{h}_R$ for the combustion process of hydrogen $H_2 + 0.5O_2 \rightarrow H_2O$ at 2000 K, using (*a*) enthalpy data and (*b*) K_P data.

Solution (*a*) The $\overline{h}_R$ of the combustion process of H_2 at 2000 K is the amount of energy released as 1 kmol of H_2 is burned in a steady-flow combustion chamber at a temperature of 2000 K. It can be determined from Eq. 14-11:

$$\overline{h}_R = Q = \Sigma N_P(\overline{h}_f^\circ + \overline{h} - \overline{h}^\circ)_P - \Sigma N_R(\overline{h}_f^\circ + \overline{h} - \overline{h}^\circ)_R$$

$$= N_{H_2O}(\overline{h}_f^\circ + \overline{h}_{2000\,K} - \overline{h}_{298\,K})_{H_2O} - N_{H_2}(\overline{h}_f^\circ + \overline{h}_{2000\,K} - \overline{h}_{298\,K})_{H_2}$$

$$- N_{O_2}(\overline{h}_f^\circ + \overline{h}_{2000\,K} - \overline{h}_{298\,K})_{O_2}$$

Substituting yields

$$\begin{aligned}\bar{h}_R &= (1\,\text{kmol}\,H_2O)[(-241{,}820 + 82{,}593 - 9904)\,\text{kJ/kmol}\,H_2O] \\ &\quad - (1\,\text{kmol}\,H_2)[(0 + 61{,}400 - 8468)\,\text{kJ/kmol}\,H_2] \\ &\quad - (0.5\,\text{kmol}\,O_2)[(0 + 67{,}881 - 8682)\,\text{kJ/kmol}\,O_2] \\ &= -251{,}663\,\text{kJ/kmol}\end{aligned}$$

(*b*) The $\bar{h}_R$ value at 2000 K can be estimated by using K_P values at 1800 and 2200 K (the closest two temperatures to 2000 K for which K_P data are available) from Table A-28. They are $K_{P_1} = 18{,}509$ at $T_1 = 1800$ K and $K_{P_2} = 869.6$ at $T_2 = 2200$ K. By substituting these values into Eq. 15-18, the $\bar{h}_R$ value is determined to be

$$\ln\frac{K_{P_2}}{K_{P_1}} \cong \frac{\bar{h}_R}{R_u}\left(\frac{1}{T_1} - \frac{1}{T_2}\right)$$

$$\ln\frac{869.6}{18{,}509} \cong \frac{\bar{h}_R}{8.314\,\text{kJ/(kmol·K)}}\left(\frac{1}{1800\,\text{K}} - \frac{1}{2200\,\text{K}}\right)$$

$$\bar{h}_R \cong -251{,}698\,\text{kJ/kmol}$$

Despite the large temperature difference between T_1 and T_2 (400 K), the two results are almost identical. The agreement between the two results would be even better if a smaller temperature interval were used.

15-6 ■ PHASE EQUILIBRIUM

We showed at the beginning of this chapter that the equilibrium state of a system at a specified temperature and pressure is the state of minimum Gibbs function, and the equilibrium criterion for a reacting or nonreacting system was expressed as (Eq. 15-4)

$$(dG)_{T,P} = 0$$

In the preceding sections we applied the equilibrium criterion to reacting systems. In this section, we apply it to nonreacting multiphase systems.

We know from experience that a wet T-shirt hanging in an open area eventually dries, a small amount of water left in a glass evaporates, and the aftershave in an open bottle quickly disappears (Fig. 15-17). These examples suggest that there is a driving force between the two phases of a substance which forces the mass to transform from one phase to another. The magnitude of this force depends, among other things, on the relative concentrations of the two phases. A wet T-shirt will dry much quicker in dry air than it would in humid air. In fact, it will not dry at all if the relative humidity of the environment is 100 percent. In this case, there will be no transformation from the liquid phase to the vapor phase, and the two phases will be in **phase equilibrium**. The conditions of phase equilibrium will change, however, if the temperature or the pressure is changed. Therefore, we examine at a specified temperature and pressure.

FIGURE 15-17
A wet T-shirt hung in an open area eventually dries as a result of mass tranfer from the liquid phase to the vapor phase.

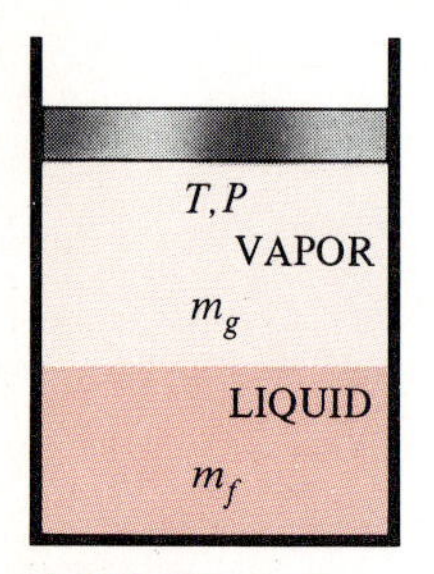

FIGURE 15-18

A liquid-vapor mixture in equilibrium at a constant temperature and pressure.

Phase Equilibrium for a Single-Component System

The equilibrium criterion for two phases of a pure substance such as water is easily developed by considering a mixture of saturated liquid and saturated vapor in equilibrium at a specified temperature and pressure, such as that shown in Fig. 15-18. The total Gibbs function of this mixture is

$$G = m_f g_f + m_g g_g$$

where g_f and g_g are the Gibbs functions of the liquid and vapor phases per unit mass, respectively. Now imagine a disturbance during which a differential amount of liquid dm_f evaporates at constant temperature and pressure. The change in the total Gibbs function during this disturbance is

$$(dG)_{T,P} = g_f\,dm_f + g_g\,dm_g$$

since g_f and g_g remain constant at constant temperature and pressure. At equilibrium, $(dG)_{T,P} = 0$. Also from the conservation of mass, $dm_g = -dm_f$. Substituting, we obtain

$$(dG)_{T,P} = (g_f - g_g)\,dm_f$$

which must be equal to zero at equilibrium. It yields

$$g_f = g_g \tag{15-19}$$

Therefore, *the two phases of a pure substance are in equilibrium when each phase has the same value of specific Gibbs function.* Also at the triple point (the state at which all three phases coexist in equilibrium), the specific Gibbs function of each one of the three phases is equal.

What happens if $g_f > g_g$? Obviously the two phases will not be in equilibrium at that moment. The second law requires that $(dG)_{T,P} = (g_f - g_g)\,dm_f \leq 0$. Thus dm_f must be negative, which means that some liquid must vaporize until $g_f = g_g$. Therefore, the Gibbs function difference is the driving force for phase change, just as the temperature difference is the driving force for heat transfer.

EXAMPLE 15-7

Show that a mixture of saturated liquid water and saturated water vapor at 120°C satisfies the criterion for phase equilibrium.

Solution Using the definition of Gibbs function and enthalpy and entropy data from Table A-4, we have

$$\begin{aligned} g_f = h_f - Ts_f &= 503.71\ \text{kJ/kg} - (393.15\ \text{K})[1.5276\ \text{kJ/(kg·K)}] \\ &= -96.9\ \text{kJ/kg} \end{aligned}$$

and

$$\begin{aligned} g_g = h_g - Ts_g &= 2706.3\ \text{kJ/kg} - (393.15\ \text{K})[7.1296\ \text{kJ/(kg·K)}] \\ &= -96.7\ \text{kJ/kg} \end{aligned}$$

The two results are in close agreement. They would match exactly if more accurate property data were used. Therefore, the criterion for phase equilibrium is satisfied.

Notice that a single-component two-phase system may exist in equilibrium at different temperatures (or pressures). But once the temperature is fixed, the system will be locked into an equilibrium state and all intensive properties of each phase (except their relative amounts) will be fixed. Therefore, a single-component two-phase system has one independent property, which may be taken to be the temperature or the pressure.

In general, the number of independent variables associated with a multicomponent, multiphase system is given by the **Gibbs phase rule**, expressed as

$$\text{IV} = C - \text{PH} + 2 \tag{15-20}$$

where IV = the number of independent variables, C = the number of components, and PH = the number of phases present in equilibrium. For the single-component ($C = 1$) two-phase (PH = 2) system discussed above, for example, one independent intensive property needs to be specified (IV = 1, Fig. 15-19). At the triple point, however, PH = 3 and thus IV = 0. That is, none of the properties of a pure substance at the triple point can be varied. Also based on this rule, a pure substance which exists in a single phase (PH = 1) will have two independent variables. In other words, two independent intensive properties need to be specified to fix the equilibrium state of a pure substance in a single phase.

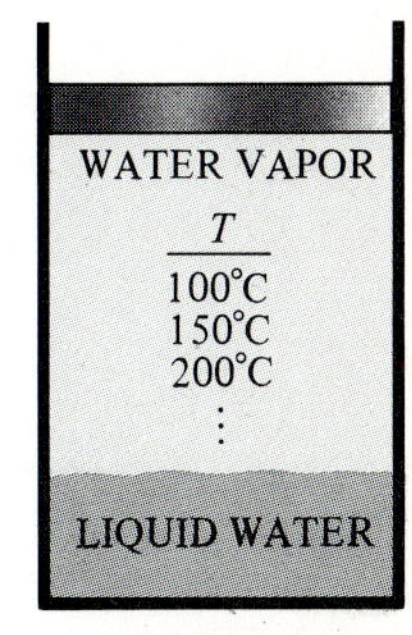

FIGURE 15-19
According to the Gibbs phase rule, a single-component two-phase system can have only one independent variable.

Phase Equilibrium for Multicomponent Systems

Many multiphase systems encountered in practice involve two or more components. A multicomponent multiphase system at a specified temperature and pressure will be in phase equilibrium when there is no driving force between the different phases of each component. Thus for phase equilibrium, the specific Gibbs function (or chemical potential) of each component must be the same in all phases (Fig. 15-20). That is,

$$g_{f,1} = g_{g,1} = g_{s,1} \qquad \text{for component 1}$$
$$g_{f,2} = g_{g,2} = g_{s,2} \qquad \text{for component 2}$$
$$\cdots\cdots\cdots\cdots$$
$$g_{f,N} = g_{g,N} = g_{s,N} \qquad \text{for component } N$$

We could also derive these relations by using mathematical vigor instead of physical arguments.

Some components may exist in more than one solid phase at the specified temperature and pressure. In this case, the specific Gibbs function of each solid phase of a component must also be the same for phase equilibrium.

In this section we examine the phase equilibrium of two-component systems which involve two phases (liquid and vapor) in equilibrium. For such systems, $C = 2$, PH = 2, and thus IV = 2. That is, a two-component, two-phase system has two independent variables, and such a system will

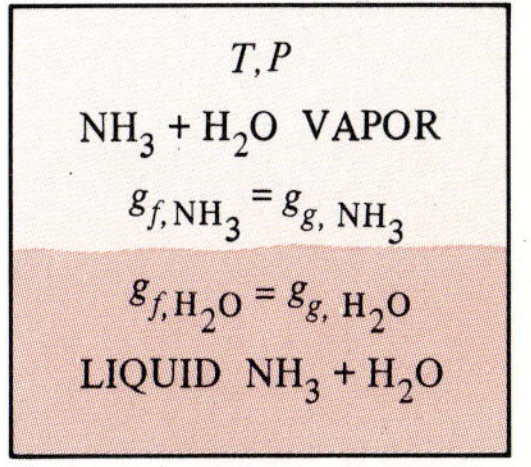

FIGURE 15-20
A multicomponent multiphase system is in phase equilibrium when the specific Gibbs function of each component is the same in all phases.

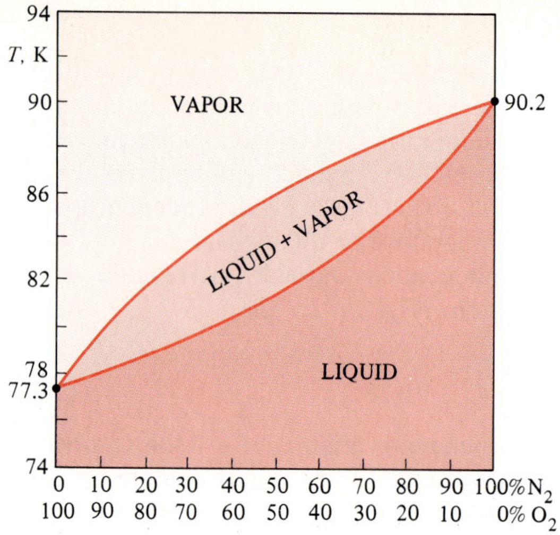

FIGURE 15-21

Equilibrium diagram for the two-phase mixture of oxygen and nitrogen at 0.1 MPa.

not be in equilibrium unless two independent intensive properties are fixed.

In general, the two phases of a two-component system will not have the same composition in each phase. That is, the mole fraction of a component will be different in different phases. This is illustrated in Fig. 15-21 for the two-phase mixture of oxygen and nitrogen at a pressure of 0.1 MPa. On this diagram, the vapor line represents the equilibrium composition of the vapor phase at various temperatures, and the liquid line does the same for the liquid phase. At 84 K, for example, the mole fractions are 70 percent nitrogen and 30 percent oxygen in the liquid phase and 34 percent nitrogen and 66 percent oxygen in the vapor phase. Notice that

$$y_{l,N_2} + y_{l,O_2} = 0.70 + 0.30 = 1$$
$$y_{g,N_2} + y_{g,O_2} = 0.34 + 0.66 = 1$$

Therefore, once the temperature and pressure (two independent variables) of a two-component, two-phase mixture are specified, the equilibrium composition of each phase can be determined from the phase diagram, which is based on experimental measurements.

In some cases, it is possible to calculate the equilibrium composition of a two-phase mixture if each phase is assumed to be an ideal solution. Under this assumption, the gas phase is assumed to behave as an ideal-gas mixture, and thus

$$P_i = y_{g,i}P_{\text{total}} \tag{15-21}$$

Also the vapor pressure P_i is related to the saturation pressure of component i at the mixture temperature by

$$P_i = y_{f,i}P_{\text{sat},i}(T) \tag{15-22}$$

where $y_{f,i}$ and $y_{g,i}$ are the mole fractions in the liquid and the vapor phases and P_{total} is the total pressure of the mixture. Equation 15-22 is called **Raoult's law**, and it gives reasonably good results at low pressures.

EXAMPLE 15-8

In absorption refrigeration systems, a two-phase equilibrium mixture of liquid ammonia (NH_3) and water (H_2O) is frequently used. Consider one such mixture at 40°C, shown in Fig. 15-22. If the composition of the liquid phase is 70 percent NH_3 and 30 percent H_2O by mole numbers, determine the composition of the vapor phase of this mixture.

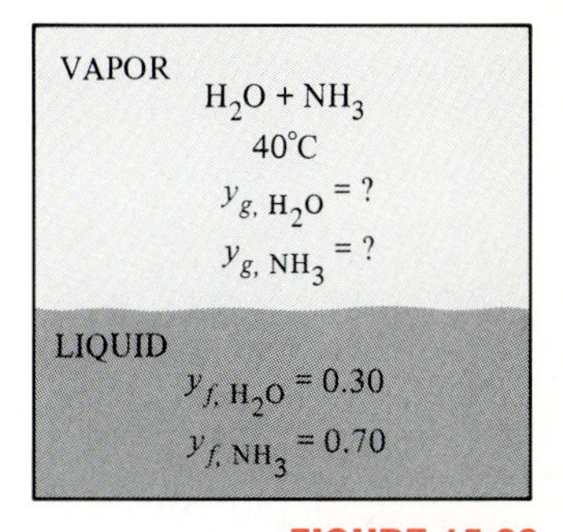

FIGURE 15-22
Schematic for Example 15-8.

Solution The saturation pressures of H_2O and NH_3 at 40°C are $P_{sat,\,H_2O} = 7.384$ kPa and $P_{sat,\,NH_3} = 1554.33$ kPa. Then the vapor pressures are determined from Eq. 15-22:

$$P_{H_2O} = y_{f,H_2O}P_{sat,\,H_2O}(T) = 0.30(7.384\text{ kPa}) = 2.22\text{ kPa}$$

$$P_{NH_3} = y_{f,NH_3}P_{sat,\,NH_3}(T) = 0.70(1554.33\text{ kPa}) = 1088.03\text{ kPa}$$

Thus the total pressure of the mixture is $P_{total} = P_{H_2O} + P_{NH_3} = 2.22 + 1088.03 = 1090.25$ kPa. Then the mole fractions in the vapor phase are

$$y_{g,H_2O} = \frac{P_{H_2O}}{P_{total}} = \frac{2.22\text{ kPa}}{1090.25\text{ kPa}} = 0.0020$$

and

$$y_{g,NH_3} = \frac{P_{NH_3}}{P_{total}} = \frac{1088.03\text{ kPa}}{1090.25\text{ kPa}} = 0.9980$$

Notice that the vapor phase consists almost entirely of ammonia, making this mixture very suitable for absorption refrigeration.

15-7 ■ SUMMARY

An isolated system is said to be in *chemical equilibrium* if no changes occur in the chemical composition of the system. The criterion for chemical equilibrium is based on the second law of thermodynamics, and for a system at a specified temperature and pressure it can be expressed as

$$(dG)_{T,P} = 0 \qquad (15\text{-}4)$$

For the reaction

$$\nu_A A + \nu_B B \longrightarrow \nu_C C + \nu_D D \qquad (15\text{-}7)$$

where the ν's are the stoichiometric coefficients, the *equilibrium criterion* can be expressed in terms of the Gibbs functions as

$$\nu_C \bar{g}_C + \nu_D \bar{g}_D - \nu_A \bar{g}_A - \nu_B \bar{g}_B = 0 \qquad (15\text{-}9)$$

which is valid for any chemical reaction regardless of the phases involved.

For reacting systems that consist of ideal gases only, the equilibrium constant K_P can be expressed as

$$K_P = e^{-\Delta G^*(T)/(R_u T)} \qquad (15\text{-}14)$$

where the *standard-state Gibbs function change* $\Delta G^*(T)$ and the equilibrium constant K_P are defined as

$$\Delta G^*(T) = \nu_C \bar{g}_C^*(T) + \nu_D \bar{g}_D^*(T) - \nu_A \bar{g}_A^*(T) - \nu_B \bar{g}_B^*(T) \qquad (15\text{-}11)$$

and

$$K_P = \frac{P_C^{\nu_C} P_D^{\nu_D}}{P_A^{\nu_A} P_B^{\nu_B}} \qquad (15\text{-}13)$$

Here, P_i's are the partial pressures of the components. The K_P of ideal-gas mixtures can also be expressed in terms of the mole numbers of the components as

$$K_P = \frac{N_C^{\nu_C} N_D^{\nu_D}}{N_A^{\nu_A} N_B^{\nu_B}} \left(\frac{P}{N_{\text{total}}}\right)^{\Delta \nu} \qquad (15\text{-}15)$$

where $\Delta \nu = \nu_C + \nu_D - \nu_A - \nu_B$, P is the total pressure, and N_{total} is the total number of moles present in the reaction chamber, including any inert gases. Equation 15-15 is written for a reaction involving two reactants and two products, but it can be extended to reactions involving any number of reactants and products.

The equilibrium constant K_P of ideal-gas mixtures depends on temperature only. It is independent of the pressure of the equilibrium mixture, and it is not affected by the presence of inert gases. The larger the K_P, the more complete the reaction. Very small values of K_P indicate that a reaction will not proceed to any appreciable degree. A reaction with $K_P > 1000$ is usually assumed to proceed to completion, and a reaction with $K_P < 0.001$ is assumed not to occur at all. The mixture pressure affects the equilibrium composition, although it does not affect the equilibrium constant K_P.

The variation of K_P with temperature is expressed in terms of other thermochemical properties through the *van't Hoff equation*

$$\frac{d(\ln K_P)}{dT} = \frac{\bar{h}_R(T)}{R_u T^2} \qquad (15\text{-}17)$$

where $\bar{h}_R(T)$ is the enthalpy of reaction at temperature T. For small temperature intervals, Eq. 15-17 can be integrated to yield

$$\ln \frac{K_{P_2}}{K_{P_1}} \cong \frac{\bar{h}_R}{R_u}\left(\frac{1}{T_1} - \frac{1}{T_2}\right) \qquad (15\text{-}18)$$

This equation shows that combustion processes will be less complete at higher temperatures since K_P decreases with temperature for exothermic reactions.

Two phases are said to be in *phase equilibrium* when there is no transformation from one phase to the other. Two phases of a pure substance are in equilibrium when each phase has the same value of specific Gibbs function. That is,

$$g_f = g_g \qquad (15\text{-}19)$$

In general, the number of independent variables associated with a multicomponent, multiphase system is given by the *Gibbs phase rule,* expressed as

$$\text{IV} = C - \text{PH} + 2 \tag{15-20}$$

where IV = the number of independent variables, C = the number of components, and PH = the number of phases present in equilibrium.

A multicomponent, multiphase system at a specified temperature and pressure will be in phase equilibrium when the specific Gibbs function of each component is the same in all phases. The equilibrium composition of a two-phase mixture at low pressures can be determined from

$$P_i = y_{g,i}P_{\text{total}} \tag{15-21}$$

$$P_i = y_{f,i}P_{\text{sat},i}(T) \tag{15-22}$$

where P_i is the vapor pressure, $P_{\text{sat},i}(T)$ is the saturation pressure of component i at the mixture temperature, $y_{f,i}$ and $y_{g,i}$ are the mole fractions in the liquid and the vapor phases, respectively, and P_{total} is the total pressure of the mixture.

REFERENCES AND SUGGESTED READING

1 W. Z. Black and J. G. Hartley, *Thermodynamics,* Harper & Row, New York, 1985.

2 I. Glassman, *Combustion,* Academic Press, New York, 1977.

3 J. B. Jones and G. A. Hawkins, *Engineering Thermodynamics,* 2d ed., Wiley, New York, 1986.

4 A. M. Kanury, *Introduction to Combustion Phenomena,* Gordon and Breach, New York, 1975.

5 J. M. Smith and H. C. Van Ness, *Introduction to Chemical Engineering Thermodynamics,* 2d ed., McGraw-Hill, New York, 1975.

6 G. J. Van Wylen and R. E. Sonntag, *Fundamentals of Classical Thermodynamics,* 3d ed., Wiley, New York, 1986.

7 K. Wark, *Thermodynamics,* 5th ed., McGraw-Hill, New York, 1988.

PROBLEMS*

Equilibrium Criterion

15-1C How is chemical equilibrium characterized?

15-2C Why is the criterion for chemical equilibrium expressed in terms of the Gibbs function instead of entropy?

*Students are encouraged to answer *all* the concept "C" questions.

15-3C Express the criterion for chemical equilibrium in terms of the chemical potentials of the components.

K_P and the Equilibrium Composition of Ideal Gases

15-4C Write three different K_P relations for reacting ideal-gas mixtures, and state when each relation should be used.

15-5C The equilibrium constant of the reaction $CO + \frac{1}{2}O_2 \rightarrow CO_2$ at 1000 K and 1 atm is K_{P_1}. Express the equilibrium constants of the following reactions at 1000 K in terms of K_{P_1}:

(*a*)	$CO + \frac{1}{2}O_2 \rightleftharpoons CO_2$	at 3 atm
(*b*)	$CO_2 \rightleftharpoons CO + \frac{1}{2}O_2$	at 1 atm
(*c*)	$CO + O_2 \rightleftharpoons CO_2 + \frac{1}{2}O_2$	at 1 atm
(*d*)	$CO + 2O_2 + 5N_2 \rightleftharpoons CO_2 + 1.5O_2 + 5N_2$	at 4 atm
(*e*)	$2CO + O_2 \rightleftharpoons 2CO_2$	at 1 atm

15-6C The equilibrium constant of the dissociation reaction $H_2 \rightarrow 2H$ at 3000 K and 1 atm is K_{P_1}. Express the equilibrium constants of the following reactions at 3000 K in terms of K_{P_1}:

(*a*)	$H_2 \rightleftharpoons 2H$	at 2 atm
(*b*)	$2H \rightleftharpoons H_2$	at 1 atm
(*c*)	$2H_2 \rightleftharpoons 4H$	at 1 atm
(*d*)	$H_2 + 2N_2 \rightleftharpoons 2H + 2N_2$	at 2 atm
(*e*)	$6H \rightleftharpoons 3H_2$	at 4 atm

15-7C Consider a mixture of CO_2, CO, and O_2 in equilibrium at a specified temperature and pressure. Now the pressure is doubled.

(*a*) Will the equilibrium constant K_P change?
(*b*) Will the number of moles of CO_2, CO, and O_2 change? How?

15-8C Consider a mixture of NO, O_2, and N_2 in equilibrium at a specified temperature and pressure. Now the pressure is tripled.

(*a*) Will the equilibrium constant K_P change?
(*b*) Will the number of moles of NO, O_2, and N_2 change? How?

15-9C A reaction chamber contains a mixture of CO_2, CO, and O_2 in equilibrium at a specified temperature and pressure. How will (*a*) increasing the temperature at constant pressure and (*b*) increasing the pressure at constant temperature affect the number of moles of CO_2?

15-10C A reaction chamber contains a mixture of N_2 and N in equilibrium at a specified temperature and pressure. How will (*a*) increasing the temperature at constant pressure and (*b*) increasing the pressure at constant temperature affect the number of moles of N_2?

15-11C A reaction chamber contains a mixture of CO_2, CO, and O_2 in equilibrium at a specified temperature and pressure. Now some N_2 is added to the mixture while the mixture temperature and pressure are kept constant. Will this affect the number of moles of CO_2? How?

15-12C Which element is more likely to dissociate into its monatomic form at 3000 K, H_2 or N_2? Why?

15-13 Using the Gibbs function data, determine the equilibrium constant K_P for the dissociation process $O_2 \rightleftharpoons 2O$ at 2000 K. Compare your result with the K_P value listed in Table A-28. *Answer:* 4.4×10^{-7}

15-14 Using the Gibbs function data, determine the equilibrium constant K_P for the reaction $H_2 + \frac{1}{2}O_2 \rightleftharpoons H_2O$ at (*a*) 25°C and (*b*) 1800 K. Compare your results with the K_P values listed in Table A-28.

15-14E Using Gibbs function data, determine the equilibrium constant K_P for the reaction $H_2 + \frac{1}{2}O_2 \rightleftharpoons H_2O$ at (*a*) 537 R and (*b*) 3240 R. Compare your results with the K_P values listed in Table A-28.
Answers: (*a*) 1.12×10^{40}, (*b*) 1.80×10^{4}

15-15 Determine the equilibrium constant K_P for the reaction $CH_4 + 2O_2 \rightleftharpoons CO_2 + 2H_2O$ at 25°C. *Answer:* 1.96×10^{140}

15-16 Using the Gibbs function data, determine the equilibrium constant K_P for the dissociation process $CO_2 \rightleftharpoons CO + \frac{1}{2}O_2$ at (*a*) 298 K and (*b*) 2000 K. Compare your results with the K_P values listed in Table A-28.

15-17 Using the Gibbs function data, determine the equilibrium constant K_P for the reaction $H_2O \rightleftharpoons \frac{1}{2}H_2 + OH$ at 25°C. Compare your result with the K_P value listed in Table A-28.

15-17E Using the Gibbs function data, determine the equilibrium constant K_P for the reaction $H_2O \rightleftharpoons \frac{1}{2}H_2 + OH$ at 77°F. Compare your result with the K_P value listed in Table A-28.

15-18 Determine the temperature at which 5 percent of diatomic oxygen (O_2) dissociates into monatomic oxygen (O) at a pressure of 5 atm.
Answer: 3214 K

15-19 A mixture of 1 mol of H_2 and 1 mol of Ar is heated at a constant pressure of 1 atm until 15 percent of H_2 dissociates into monatomic hydrogen (H). Determine the final temperature of the mixture.

15-20 Carbon monoxide is burned with 100 percent excess air during a steady-flow process at a pressure of 4 atm. At what temperature will 97 percent of CO burn to CO_2? Assume the equilibrium mixture consists of CO_2, CO, O_2, and N_2. *Answer:* 2276 K

15-20E Repeat Prob. 15-20, using data in English units.

15-21 Hydrogen is burned with 150 percent theoretical air during a steady-flow process at a pressure of 1 atm. At what temperature will 98 percent of H_2 burn to H_2O? Assume the equilibrium mixture consists of H_2O, H_2, O_2, and N_2.

15-22 Air (79 percent N_2 and 21 percent O_2) is heated to 2000 K at a constant pressure of 2 atm. Assuming the equilibrium mixture consists of N_2, O_2, and NO, determine the equilibrium composition at this state. Is it realistic to assume that no monatomic oxygen or nitrogen will be present in the equilibrium mixture? Will the equilibrium composition change if the pressure is doubled at constant temperature?

15-23 Hydrogen (H_2) is heated to 3200 K at a constant pressure of 10 atm. Determine the percentage of H_2 that will dissociate into H during this process. *Answer:* 13.7 percent

15-23E Hydrogen (H_2) is heated to 5760 R at a constant pressure of 10 atm. Determine the percentage of H_2 that will dissociate into H during this process.

15-24 Carbon dioxide (CO_2) is heated to 2800 K at a constant pressure of 3 atm. Determine the percentage of CO_2 that will dissociate into CO and O_2 during this process.

15-25 A mixture of 1 mol of CO and 3 mol of O_2 is heated to 2200 K at a pressure of 2 atm. Determine the equilibrium composition, assuming the mixture consists of CO_2, CO, and O_2.
Answer: $0.995CO_2$, $0.005CO$, $2.5025O_2$

15-26 A mixture of 2 mol of CO, 2 mol of O_2, and 6 mol of N_2 is heated to 2400 K at a pressure of 3 atm. Determine the equilibrium composition of the mixture.

15-26E A mixture of 2 mol of CO, 2 mol of O_2, and 6 mol of N_2 is heated to 4320 R at a pressure of 3 atm. Determine the equilibrium composition of the mixture. *Answer:* $1.93CO_2$, $0.07CO$, $1.035O_2$, $6N_2$

15-27 A mixture of 1 mol of H_2O, 2 mol of O_2, and 5 mol of N_2 is heated to 2200 K at a pressure of 5 atm. Assuming the equilibrium mixture consists of H_2O, O_2, N_2, and H_2, determine the equilibrium composition at this state. Is it realistic to assume that no OH will be present in the equilibrium mixture?

15-28 A mixture of 3 mol of N_2, 1 mol of O_2, and 0.1 mol of Ar is heated to 2400 K at a constant pressure of 10 atm. Assuming the equilibrium mixture consists of N_2, O_2, Ar, and NO, determine the equilibrium composition. *Answer:* $0.0823NO$, $2.9589N_2$, $0.9589O_2$

15-28E A mixture of 3 mol of N_2, 1 mol of O_2 and 0.1 mol of Ar is heated to 4320 R at a constant pressure of 10 atm. Assuming the equilibrium mixture consists of N_2, O_2, Ar, and NO, determine the equilibrium composition.

15-29 Determine the mole fraction of argon that ionizes according to the reaction $Ar \rightleftharpoons Ar^+ + e^-$ at 10,000 K and 0.1 atm ($K_P = 0.00042$ for this reaction).

15-30 Determine the mole fraction of sodium that ionizes according to the reaction $Na \rightleftharpoons Na^+ + e^-$ at 2000 K and 0.5 atm ($K_P = 0.668$ for this reaction). *Answer:* 75.6 percent

15-31 Methane gas (CH_4) at 25°C is burned with the stoichiometric amount of air at 25°C during an adiabatic steady-flow combustion process at 1

atm. Assuming the product gases consist of CO_2, H_2O, CO, N_2, and O_2, determine (*a*) the equilibrium composition of the product gases and (*b*) the exit temperature.

15-32 Liquid propane (C_3H_8) enters a combustion chamber at 25°C at a rate of 0.1 kg/min where it is mixed and burned with 150 percent excess air which enters the combustion chamber at 12°C. If the combustion gases consist of CO_2, H_2O, CO, O_2, and N_2, which exit at 1200 K and 2 atm, determine (*a*) the equilibrium composition of the product gases and (*b*) the rate of heat transfer from the combustion chamber. Is it realistic to disregard the presence of NO in the product gases?
Answers: (*a*) $3CO_2$, $7.5O_2$, $4H_2O$, $47N_2$; (*b*) 422.2 kJ/min

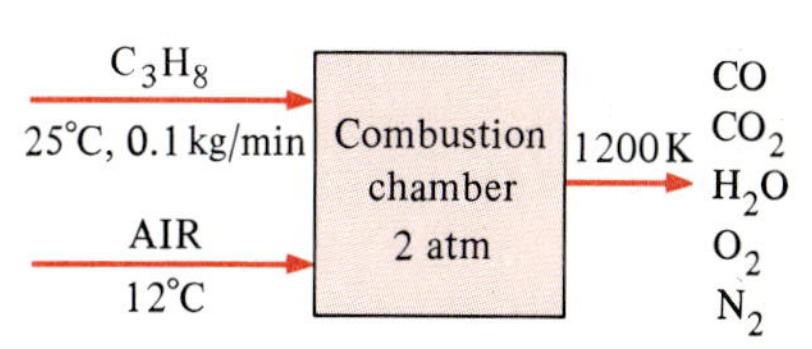

FIGURE P15-32

15-33 A steady-flow combustion chamber is supplied with CO gas at 37°C and 110 kPa at a rate of 0.8 m³/min and with oxygen (O_2) at 25°C and 110 kPa at a rate of 0.7 kg/min. The combustion products leave the combustion chamber at 2000 K and 110 kPa. If the combustion gases consist of CO_2, CO, O_2, and N_2, determine (*a*) the equilibrium composition of the product gases and (*b*) the rate of heat transfer from the combustion chamber.

15-33E A steady-flow combustion chamber is supplied with CO gas at 560 R and 16 psia at a rate of 25 ft³/min and with oxygen (O_2) at 537 R and 16 psia at a rate of 1.4 lbm/min. The combustion products leave the combustion chamber at 3600 R and 16 psia. If the combustion gases consist of CO_2, CO, O_2, and N_2, determine (*a*) the equilibrium composition of the product gases and (*b*) the rate of heat transfer from the combustion chamber.

15-34 A constant-volume tank contains a mixture of 1 mol of H_2 and 0.5 mol of O_2 at 25°C and 1 atm. The contents of the tank are ignited, and the final temperature and pressure in the tank are 2800 K and 5 atm, respectively. If the combustion gases consist of H_2O, H_2, and O_2, determine (*a*) the equilibrium composition of the product gases and (*b*) the amount of heat transfer from the combustion chamber. Is it realistic to assume that no OH will be present in the equilibrium mixture?
Answers: (*a*) $0.944H_2O$, $0.066H_2$, $0.028O_2$; (*b*) 131,993 J

15-35 Oxygen (O_2) is heated during a steady-flow process at 1 atm from 298 to 3000 K at a rate of 0.1 kg/min. Determine the rate of heat supply needed during this process, assuming (*a*) some O_2 dissociates into O and (*b*) no dissociation takes place.

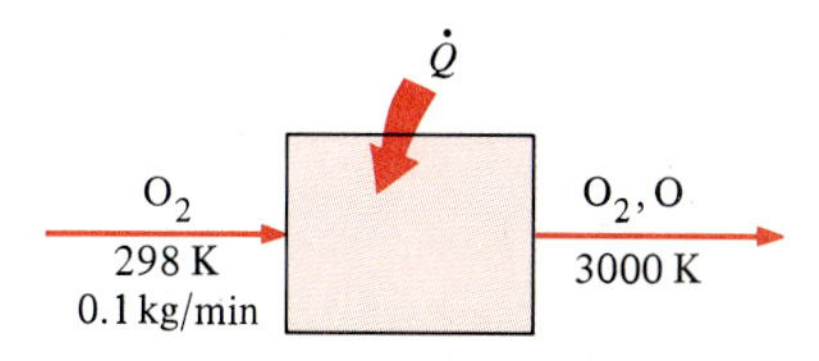

FIGURE P15-35

15-35E Oxygen (O_2) is heated during a steady-flow process at 1 atm from 537 to 5400 R at a rate of 0.2 lbm/min. Determine the rate of heat supply needed during this process, assuming (*a*) some O_2 dissociates into O and (*b*) no dissociation takes place.
Answers: (*a*) 342.2 Btu/min, (*b*) 263.4 Btu/min

15-36 Write a computer program to determine the real roots of the nonlinear equation $54.193(5 - x)(3 - x)^2 = x^2(27 - x)$. Compare your result with the x value in Example 15-4.

Simultaneous Reactions

15-37C What is the equilibrium criterion for systems that involve two or more simultaneous chemical reactions?

15-38C When determining the equilibrium composition of a mixture involving simultaneous reactions, how would you determine the number of K_P relations needed?

15-39 A mixture of 2 mol of H_2O and 3 mol of O_2 is heated to 3600 K at a pressure of 5 atm. Determine the equilibrium composition of the mixture, assuming that only H_2O, OH, O_2, and H_2 are present.

15-40 One mole of H_2O is heated to 3400 K at a pressure of 1 atm. Determine the equilibrium composition, assuming that only H_2O, OH, O_2, and H_2 are present. *Answer:* $0.574H_2O$, $0.308H_2$, $0.095O_2$, 0.236OH

15-40E One mole of H_2O is heated to 6120 R at a pressure of 1 atm. Determine the equilibrium composition, assuming that only H_2O, OH, O_2, and H_2 are present.

15-41 A mixture of 2 mol of CO_2 and 1 mol of O_2 is heated to 3200 K at a pressure of 1 atm. Determine the equilibrium composition of the mixture, assuming that only CO_2, CO, O_2, and O are present.

15-42 A mixture of 3 mol of CO_2 and 2 mol of O_2 is heated to 3400 K at a pressure of 2 atm. Determine the equilibrium composition of the mixture, assuming that only CO_2, CO, O_2, and O are present.
Answer: $1.56CO_2$, 1.44CO, $3.08O_2$, 1.28O

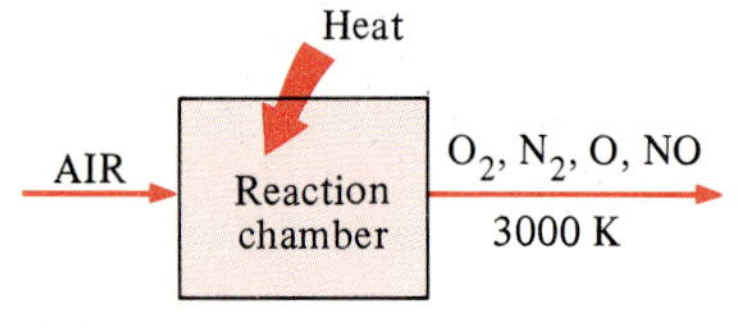

FIGURE P15-43

15-43 Air (21 percent O_2, 79 percent N_2) is heated to 3000 K at a pressure of 1 atm. Determine the equilibrium composition, assuming that only O_2, N_2, O, and NO are present. Is it realistic to assume that no N will be present in the final equilibrium mixture?

15-43E Air (21 percent O_2, 79 percent N_2) is heated to 5400 R at a pressure of 1 atm. Determine the equilibrium composition, assuming that only O_2, N_2, O, and NO are present. Is it realistic to assume that no N will be present in the final equilibrium mixture?
Answer: $3.731N_2$, $0.855O_2$, 0.058NO, 0.232O

15-44 Water vapor (H_2O) is heated during a steady-flow process at 1 atm from 298 to 3000 K at a rate of 0.5 kg/min. Determine the rate of heat supply needed during this process, assuming (*a*) some H_2O dissociates into H_2, O_2, and OH and (*b*) no dissociation takes place.
Answers: (*a*) 5137 kJ/min, (*b*) 3510 kJ/min

15-45 Write a computer program to solve two nonlinear equations with two unknowns simultaneously. Use Eqs. (3) and (4) in Example 15-5 after eliminating two of the unknowns using relations (1) and (2).

Variations of K_P with Temperature

15-46C What is the importance of the van't Hoff equation?

15-47C At what temperature will a fuel burn more completely, 2000 or 2500 K?

15-48 Estimate the enthalpy of reaction $\bar{h}_R$ for the combustion process of hydrogen at 2400 K, using (*a*) enthalpy data and (*b*) K_P data.
Answers: (*a*) −252,377 kJ/kmol, (*b*) −252,047 kJ/kmol

15-49 Estimate the enthalpy of reaction $\bar{h}_R$ for the combustion process of carbon monoxide at 2200 K, using (*a*) enthalpy data and (*b*) K_P data.

15-49E Estimate the enthalpy of reaction $\bar{h}_R$ for the combustion process of carbon monoxide at 3960 R, using (*a*) enthalpy data and (*b*) K_P data.
Answers: (*a*) −119,030 Btu/lbmol, (*b*) −119,041 Btu/lbmol

15-50 Using the enthalpy of reaction $\bar{h}_R$ data and the K_P value at 2400 K, estimate the K_P value of the combustion process $H_2 + \frac{1}{2}O_2 \rightleftharpoons H_2O$ at 2600 K. *Answer:* 104.1

15-51 Estimate the enthalpy of reaction $\bar{h}_R$ for the dissociation process $CO_2 \rightleftharpoons CO + \frac{1}{2}O_2$ at 2000 K, using (*a*) enthalpy data and (*b*) K_P data.

15-51E Estimate the enthalpy of reaction $\bar{h}_R$ for the dissociation process $CO_2 \rightleftharpoons CO + \frac{1}{2}O_2$ at 3600 R, using (*a*) enthalpy data and (*b*) K_P data.

15-52 Estimate the enthalpy of reaction $\bar{h}_R$ for the dissociation process $O_2 \rightleftharpoons 2O$ at 2900 K, using (*a*) enthalpy data and (*b*) K_P data.
Answers: (*a*) 513,229 kJ/kmol, (*b*) 512,957 kJ/kmol

15-53 Using the enthalpy of reaction $\bar{h}_R$ data and the K_P value at 2800 K, estimate the K_P value of the dissociation process $O_2 \rightleftharpoons 2O$ at 3000 K.

Phase Equilibrium

15-54C Consider a cylinder which contains a saturated liquid–vapor mixture of water in equilibrium. Some vapor is now allowed to escape the cylinder at constant temperature and pressure. Will this disturb the phase equilibrium and cause some of the liquid to evaporate?

15-55C Consider a two-phase mixture of ammonia and water in equilibrium. Can this mixture exist in two phases at the same temperature but at a different pressure?

15-56 Show that when the three phases of a pure substance are in equilibrium, the specific Gibbs function of each phase is the same.

15-57 Show that a mixture of saturated liquid water and saturated water vapor at 100°C satisfies the criterion for phase equilibrium.

15-58 Show that a mixture of saturated liquid water and saturated water vapor at 500 kPa satisfies the criterion for phase equilibrium.

15-59 Show that a saturated liquid–vapor mixture of refrigerant-12 at 0°C satisfies the criterion for phase equilibrium.

15-60 Consider a mixture of oxygen and nitrogen in the gas phase. How many independent properties are needed to fix the state of the system? *Answer:* 3

15-61 Show that when the two phases of a two-component system are in equilibrium, the specific Gibbs function of each phase of each component is the same.

15-62 In absorption refrigeration systems, a two-phase equilibrium mixture of liquid ammonia (NH_3) and water (H_2O) is frequently used. Consider a liquid-vapor mixture of ammonia and water in equilibrium at 30°C. If the composition of the liquid phase is 60 percent NH_3 and 40 percent H_2O by mole numbers, determine the composition of the vapor phase of this mixture. Saturation pressure of NH_3 at 30°C is 1116.5 kPa.

15-63 Consider a liquid-vapor mixture of ammonia and water in equilibrium at 20°C. If the composition of the liquid phase is 50 percent NH_3 and 50 percent H_2O by mole numbers, determine the composition of the vapor phase of this mixture. Saturation pressure of NH_3 at 20°C is 857.1 kPa. *Answer:* 0.27 percent, 99.73 percent

15-64 A two-phase mixture of ammonia and water is in equilibrium at 50°C. If the composition of the vapor phase is 99 percent NH_3 and 1 percent H_2O by mole numbers, determine the composition of the liquid phase of this mixture. Saturation pressure of NH_3 at 50°C is 2032.6 kPa.

15-65 Using the liquid-vapor equilibrium diagram of an oxygen-nitrogen mixture, determine the composition of each phase at 80 K and 100 kPa.

15-66 Using the liquid-vapor equilibrium diagram of an oxygen-nitrogen mixture, determine the composition of each phase at 88 K and 100 kPa.

15-67 Using the liquid-vapor equilibrium diagram of an oxygen-nitrogen mixture at 100 kPa, determine the temperature at which the composition of the vapor phase is 79 percent N_2 and 21 percent O_2. *Answer:* 89 K

15-68 Using the liquid-vapor equilibrium diagram of an oxygen-nitrogen mixture at 100 kPa, determine the temperature at which the composition of the liquid phase is 30 percent N_2 and 70 percent O_2.

Thermodynamics of High-Speed Fluid Flow

CHAPTER 16

In this chapter we develop the general relations for the thermodynamics of high-speed fluid flow. We introduce the concepts of *stagnation state, velocity of sound,* and *Mach number* for a compressible fluid. The relationships between the static and stagnation fluid properties are developed for isentropic flows of ideal gases, and they are expressed as functions of specific heat ratios and flow Mach number. The effects of area changes for one-dimensional isentropic subsonic and supersonic flows are discussed. These effects are illustrated with an introduction to the isentropic flow through *converging* and *converging-diverging nozzles*. The concept of the *normal shock wave* and the variation of flow properties across the shock wave are discussed. Finally, we take another look at the influence of nozzle and diffuser efficiencies on the flow parameters for the flow of ideal gases and vapors through these devices.

16-1 ■ STAGNATION PROPERTIES

When analyzing control volumes, we found it very convenient to combine the internal energy and the flow energy of a fluid into a single term, enthalpy, defined as $h = u + Pv$. Whenever the kinetic and potential energies of the fluid are negligible, as is often the case, the enthalpy represents the *total* energy of a fluid. For high-speed flows, the potential energy of the fluid is still negligible, but the kinetic energy is not. In such cases, it is convenient to combine the enthalpy and the kinetic energy of the fluid into a single term called **stagnation** (or **total**) **enthalpy** h_0, defined as

$$h_0 = h + \frac{\mathbf{V}^2}{2} \qquad \text{(kJ/kg)} \tag{16-1}$$

When the potential energy of the fluid is negligible, the stagnation enthalpy represents the *total energy of a flowing fluid stream* per unit mass. Thus it simplifies the thermodynamic analysis of high-speed flows.

Throughout this chapter the ordinary enthalpy h is referred to as the **static enthalpy**, whenever necessary, to distinguish it from the stagnation enthalpy. Notice that the stagnation enthalpy is a combination property of a fluid, just like the static enthalpy, and these two enthalpies become identical when the kinetic energy of the fluid is negligible.

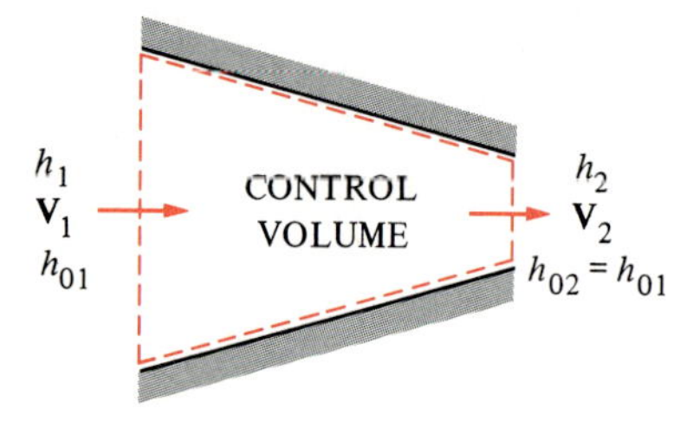

FIGURE 16-1
Steady flow of a fluid through an adiabatic duct.

Consider the steady flow of a fluid through a duct such as a nozzle, diffuser, or some other flow passage where the flow takes place adiabatically and with no shaft or electrical work, as shown in Fig. 16-1. Assuming the fluid experiences little or no change in its elevation and its potential energy, the conservation of energy relation (Eq. 4-23) for this single-stream steady-flow system reduces to

$$\cancelto{0}{q} - \cancelto{0}{w} = \Delta h + \Delta \text{ke} + \cancelto{0}{\Delta \text{pe}}$$

$$0 = (h_2 - h_1) + \left(\frac{\mathbf{V}_2^2}{2} - \frac{\mathbf{V}_1^2}{2}\right)$$

or

$$h_1 + \frac{\mathbf{V}_1^2}{2} = h_2 + \frac{\mathbf{V}_2^2}{2} \tag{16-2}$$

or

$$h_{01} = h_{02} \tag{16-3}$$

That is, in the absence of any heat and work interactions and any changes in potential energy, the stagnation enthalpy of a fluid remains constant during a steady-flow process. Flows through nozzles and diffusers usually satisfy these conditions, and any increase in fluid velocity in these devices will create an equivalent decrease in the static enthalpy of the fluid.

If the fluid were brought to a complete stop, then the velocity at state 2 would be zero and Eq. 16-2 would become

$$h_1 + \frac{\mathbf{V}_1^2}{2} = h_2 = h_{02}$$

Thus the stagnation enthalpy represents the enthalpy of a fluid when it is brought to rest adiabatically.

FIGURE 16-2
Kinetic energy is converted to enthalpy during a stagnation process.

During a stagnation process, the kinetic energy of a fluid is converted to enthalpy (internal energy + flow energy) which results in an increase in the fluid temperature and pressure (Fig. 16-2). The properties of a fluid at the stagnation state are called **stagnation properties** (stagnation temperature, stagnation pressure, stagnation density, etc.). The stagnation state and the stagnation properties are indicated by the subscript 0.

The stagnation state is called the **isentropic stagnation state** when the stagnation process is reversible as well as adiabatic (i.e., isentropic). The entropy of a fluid remains constant during an isentropic stagnation process. The actual (irreversible) and isentropic stagnation processes are shown on the *h-s* diagram in Fig. 16-3. Notice that the stagnation enthalpy of the fluid (and the stagnation temperature if the fluid is an ideal gas) is the same for both cases. But the actual stagnation pressure is lower than the isentropic stagnation pressure since the entropy increases during the actual stagnation process as a result of fluid friction. The stagnation processes are often approximated to be isentropic, and the isentropic stagnation properties are simply referred to as stagnation properties.

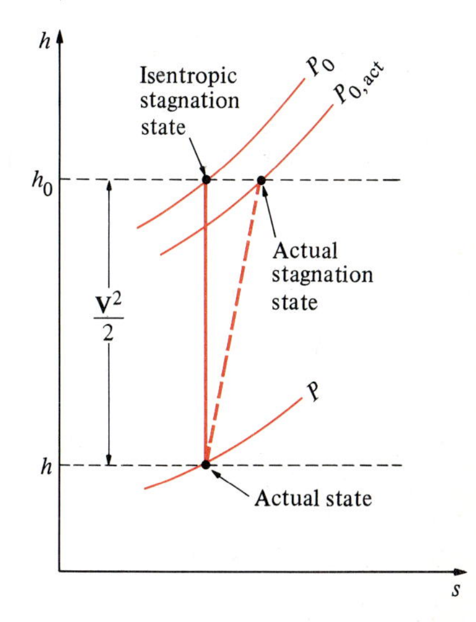

FIGURE 16-3
The actual state, actual stagnation state, and isentropic stagnation state of a fluid on an *h-s* diagram.

EXAMPLE 16-1

Steam at 400°C and 1 MPa is flowing through a device with a velocity of 300 m/s. Determine the stagnation enthalpy, temperature, pressure, and density of the steam.

Solution The enthalpy of the steam at the specified state is determined from Table A-6 to be 3263.9 kJ/kg. Then the stagnation enthalpy of the steam is easily determined from Eq. 16-1:

$$
\begin{aligned}
h_0 &= h + \frac{\mathbf{V}^2}{2} \\
&= 3263.9 \text{ kJ/kg} + \frac{(300 \text{ m/s})^2}{2}\left(\frac{1 \text{ kJ/kg}}{1000 \text{ m}^2/\text{s}^2}\right) \\
&= 3263.9 \text{ kJ/kg} + 45.0 \text{ kJ/kg} \\
&= 3308.9 \text{ kJ/kg}
\end{aligned}
$$

The entropy of the steam at the specified temperaure and pressure is 7.4561 kJ/(kg·K). Assuming the stagnation process to be isentropic, the entropy of the steam at the stagnation state is also this value. Thus the stagnation state is completely specified by the two independent properties, enthalpy and entropy.

T = 400°C
P = 1 MPa
h = 3263.9 kJ/kg
V = 300 m/s → ke = 45.0 kJ/kg

(*a*) Actual state

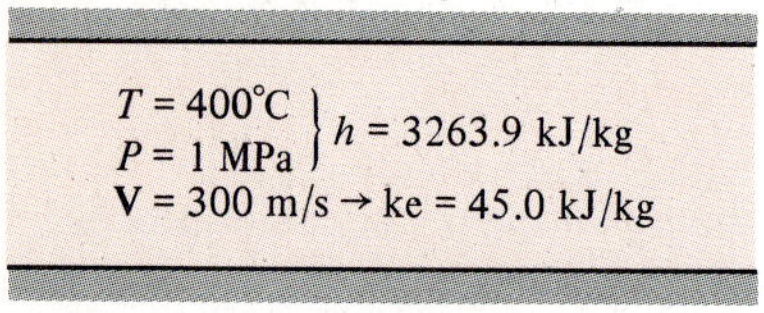

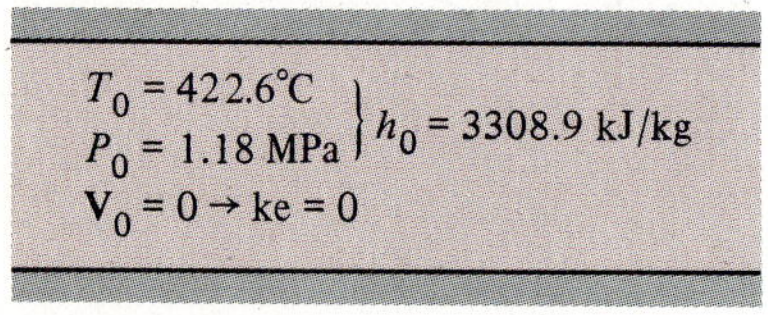

T0 = 422.6°C
P0 = 1.18 MPa
h0 = 3308.9 kJ/kg
V0 = 0 → ke = 0

(*b*) Stagnation state

FIGURE 16-4

The properties of a high-speed fluid change significantly during an adiabatic stagnation process (values from Example 16-1).

Then the temperature, pressure, and density ($\rho_0 = 1/v_0$) at the stagnation state are determined to be (Fig. 16-4)

$$T_0 = 422.6°C$$
$$P_0 = 1.18 \text{ MPa}$$
$$\rho_0 = 3.719 \text{ kg/m}^3$$

Note that if the steam velocity had been 50 m/s, the kinetic energy term would have been only 1.25 kJ/kg and the stagnation enthalpy would have been 3265.15 kJ/kg, which is essentially the same as the enthalpy of the steam. Thus for low-speed flow, the stagnation properties and the (static) fluid properties are approximately the same.

When the fluid is an ideal gas, its enthalpy can be replaced by C_pT. Then Eq. 16-1 can be expressed as

$$C_pT_0 = C_pT + \frac{\mathbf{V}^2}{2}$$

or

$$T_0 = T + \frac{\mathbf{V}^2}{2C_p} \tag{16-4}$$

Here T_0 is called the **stagnation** (or **total**) **temperature**, and it represents *the temperature an ideal gas will attain when it is brought to rest adiabatically*. The term $\mathbf{V}^2/2C_p$ corresponds to the temperature rise during such a process and is called the **dynamic temperature**. For example, the dynamic temperature of air flowing at 100 m/s is $(100 \text{ m/s})^2/[2 \times 1.005 \text{ kJ/(kg·K)}] = 5.0$ K. Therefore, when air at 300 K and 100 m/s is brought to rest adiabatically (at the tip of a temperature probe, for example), its temperature will rise to the stagnation value of 305 K (Fig. 16-5). Note that for low-speed flows, the stagnation and static (or ordinary) temperatures are practically identical. But for high-speed flows, the temperature measured by a stationary probe placed in the fluid (the stagnation temperature) may be significantly higher than the static temperature of the fluid.

Temperature rise during stagnation
305 K
300 K
STOP
AIR
100 m/s

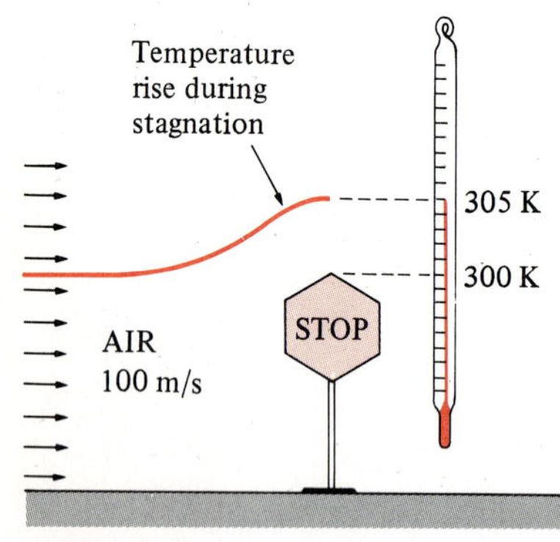

FIGURE 16-5

The temperature of an ideal gas flowing at a velocity **V** rises by $\mathbf{V}^2/(2C_p)$ when it is brought to a complete stop.

The pressure a fluid attains when brought to rest isentropically is called the **stagnation pressure** P_0. For ideal gases with constant specific heats, P_0 is related to the static pressure of the fluid through Eq. 6-45,

$$\frac{P_0}{P} = \left(\frac{T_0}{T}\right)^{k/(k-1)} \tag{16-5}$$

By noting that $\rho = 1/v$ and using Eq. 6-44, the ratio of the stagnation density to static density can be expressed as

$$\frac{\rho_0}{\rho} = \left(\frac{T_0}{T}\right)^{1/(k-1)} \tag{16-6}$$

When stagnation enthalpies are used, there is no need to refer explicitly to kinetic energy. Then the conservation of energy equation (Eq. 4-23) for a single-stream, steady-flow device can be expressed as

$$q - w = (h_{02} - h_{01}) + g(z_2 - z_1) \qquad (16\text{-}7)$$

where h_{01} and h_{02} are the stagnation enthalpies at states 1 and 2, respectively. When the fluid is an ideal gas with constant specific heats, Eq. 16-7 becomes

$$q - w = C_p(T_{02} - T_{01}) + g(z_2 - z_1) \qquad (16\text{-}8)$$

where T_{01} and T_{02} are the stagnation temperatures.

Notice that kinetic energy terms do not explicitly appear in the above relations, but the stagnation enthalpy terms account for their contribution.

EXAMPLE 16-2

An aircraft is flying at a cruising speed of 250 m/s at an altitude of 5000 m where the atmospheric pressure is 54.05 kPa and ambient air temperature is 255.7 K. The ambient air is first decelerated in a diffuser before it enters the compressor (Fig. 16-6). Assuming both the diffuser and the compressor to be isentropic, determine (*a*) the stagnation pressure at the compressor inlet and (*b*) the required compressor work per unit mass if the stagnation pressure ratio of the compressor is 8.

Solution We assume air to be an ideal gas with constant specific heats at room temperature. The constant-pressure specific heat C_p and the specific heat ratio k of air are determined from Table A-2*a* to be

$$C_p = 1.005 \text{ kJ/(kg·K)} \qquad \text{and} \qquad k = 1.4$$

(*a*) Under isentropic conditions, the stagnation pressure at the compressor inlet (diffuser exit) can be determined from Eq. 16-5. But first we need to find the stagnation temperature at the compressor inlet T_{01}. Under stated assumptions, T_{01} can be determined from Eq. 16-4:

$$T_{01} = T_1 + \frac{\mathbf{V}_1^2}{2C_p} = 255.7 \text{ K} + \frac{(250 \text{ m/s})^2}{(2)[1.005 \text{ kJ/(kg·K)}]}\left(\frac{1 \text{ kJ/kg}}{1000 \text{ m}^2/\text{s}^2}\right)$$
$$= 286.8 \text{ K}$$

Then from Eq. 16-5,

$$P_{01} = P_1\left(\frac{T_{01}}{T_1}\right)^{k/(k-1)} = (54.05 \text{ kPa})\left(\frac{286.8 \text{ K}}{255.7 \text{ K}}\right)^{1.4/(1.4-1)}$$
$$= 80.77 \text{ kPa}$$

That is, the temperature of air would increase by 31.1°C and the pressure by

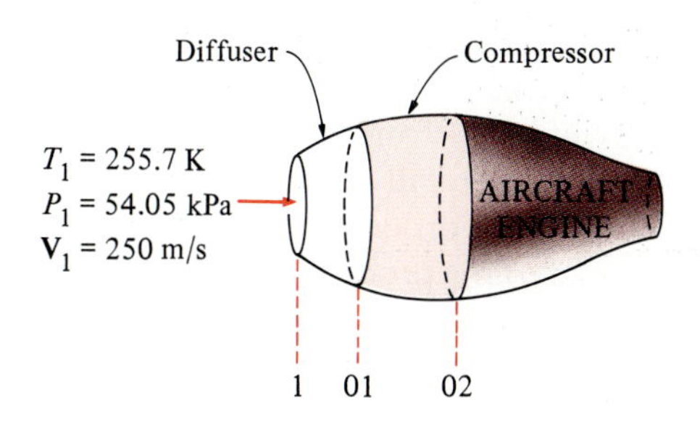

FIGURE 16-6
Schematic for Example 16-2.

26.72 kPa as air were decelerated from 250 m/s to zero velocity. These increases in the temperature and pressure of air are due to the conversion of the kinetic energy into enthalpy.

(*b*) To determine the compressor work, we need to know the stagnation temperature of air at the compressor exit T_{02}. The stagnation pressure ratio across the compressor P_{02}/P_{01} is specified to be 8. Since the compression process is assumed to be isentropic, T_{02} can be determined from the ideal-gas isentropic relation Eq. 16-5:

$$T_{02} = T_{01}\left(\frac{P_{02}}{P_{01}}\right)^{(k-1)/k} = (286.8\text{ K})(8)^{(1.4-1)/1.4} = 519.5\text{ K}$$

Then disregarding potential energy changes, the compressor work per unit mass of air is determined from Eq. 16-8:

$$\begin{aligned} w &= C_p(T_{01} - T_{02}) \\ &= [1.005\text{ kJ/(kg·K)}](286.8\text{ K} - 519.5\text{ K}) \\ &= -233.9\text{ kJ/kg} \end{aligned}$$

Thus the work supplied to the compressor is 233.9 kJ/kg. Notice that using stagnation properties automatically accounts for any changes in the kinetic energy of a fluid stream.

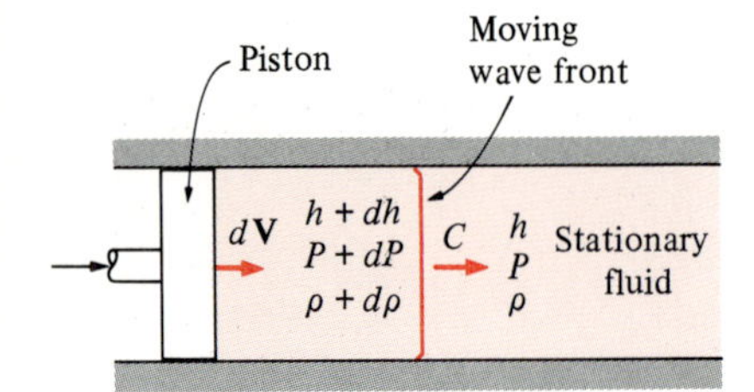

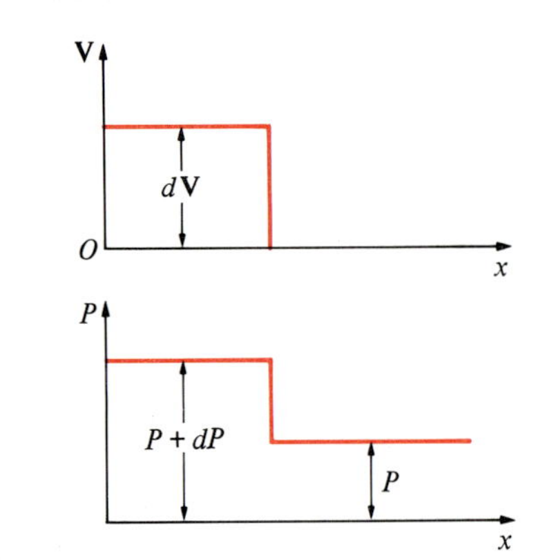

FIGURE 16-7

Propagation of a small pressure wave along a duct.

16-2 ■ VELOCITY OF SOUND AND MACH NUMBER

An important parameter in the study of compressible flow is the **velocity of sound** (or the **sonic velocity**), which is the velocity at which an infinitesimally small pressure wave travels through a medium. The pressure wave may be caused by a small disturbance, which creates a slight rise in local pressure.

To obtain a relation for the velocity of sound in a medium, consider a pipe which is filled with a fluid at rest, as shown in Fig. 16-7. A piston fitted in the pipe is now moved to the right with a constant incremental velocity $d\mathbf{V}$, creating a sonic wave. The wave front moves to the right through the fluid at the velocity of sound C and separates the moving fluid adjacent to the piston from the fluid still at rest. The fluid to the left of the wave front experiences an incremental change in its thermodynamic properties while the fluid on the right of the wave front maintains its original thermodynamic properties, as shown on the figure.

To simplify the analysis, consider a control volume that encloses the wave front and moves with it, as shown in Fig. 16-8. To an observer traveling with the wave front, the fluid to the right will appear to be moving toward the wave front with a velocity of C and the fluid to the left to be moving away from the wave front with a velocity of $C - d\mathbf{V}$. Of course, the observer will think the control volume which encloses the wave front (and herself or himself) is stationary, and the observer will be witnessing a steady-flow process. The conservation of mass principle for this single-stream, steady-flow process can be expressed as

$$\dot{m}_{\text{right}} = \dot{m}_{\text{left}}$$

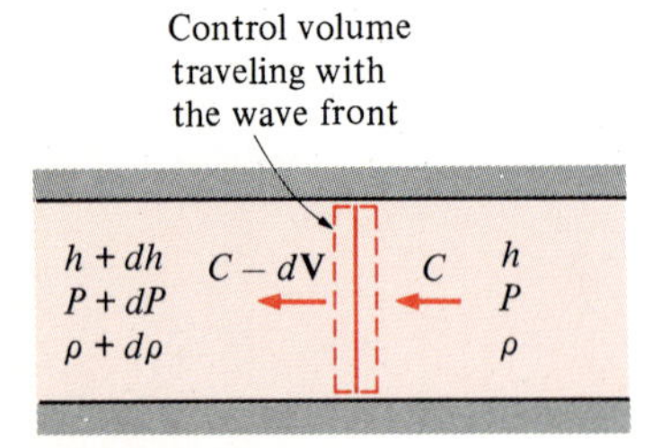

FIGURE 16-8

Control volume moving with the small pressure wave along a duct.

or
$$\rho AC = (\rho + d\rho)A(C - d\mathbf{V})$$

By canceling the cross-sectional (or flow) area A and neglecting the higher-order terms, this equation reduces to

$$C\,d\rho - \rho\,d\mathbf{V} = 0 \qquad (a)$$

No heat or work crosses the boundaries of the control volume during this steady-flow process, and the potential energy change, if any, can be neglected. Then the conservation of energy equation becomes

$$h + \frac{C^2}{2} = h + dh + \frac{(C - d\mathbf{V})^2}{2}$$

which yields

$$dh - C\,d\mathbf{V} = 0 \qquad (b)$$

The amplitude of the ordinary sonic wave is very small and does not cause any appreciable change in the pressure and temperature of the fluid. Therefore, the propagation of a sonic wave is not only adiabatic but also very nearly isentropic. Then the second $T\,ds$ relation (Eq. 6-24) reduces to

$$T\,\cancelto{0}{ds} = dh - \frac{dP}{\rho}$$

or
$$dh = \frac{dP}{\rho} \qquad (c)$$

Combining Eqs. a, b, and c yields the desired expression for the velocity of sound as

$$C^2 = \frac{dP}{d\rho} \quad \text{at } s = \text{constant}$$

or
$$C^2 = \left(\frac{\partial P}{\partial \rho}\right)_s \qquad (16\text{-}9)$$

It is left as an exercise for the reader to show, by using the relations presented in Chap. 11, that Eq. 16-9 can also be written as

$$C^2 = k\left(\frac{\partial P}{\partial \rho}\right)_T \qquad (16\text{-}10)$$

where k is the specific heat ratio of the fluid. Note that the velocity of sound in a fluid is a function of the thermodynamic properties of that fluid.

When the fluid is an ideal gas ($P = \rho RT$), the differentiation in Eq. 16-10 can easily be performed, to yield the following relation

$$C^2 = k\left(\frac{\partial P}{\partial \rho}\right)_T = k\left[\frac{\partial(\rho RT)}{\partial \rho}\right]_T = kRT$$

or
$$C = \sqrt{kRT} \qquad (16\text{-}11)$$

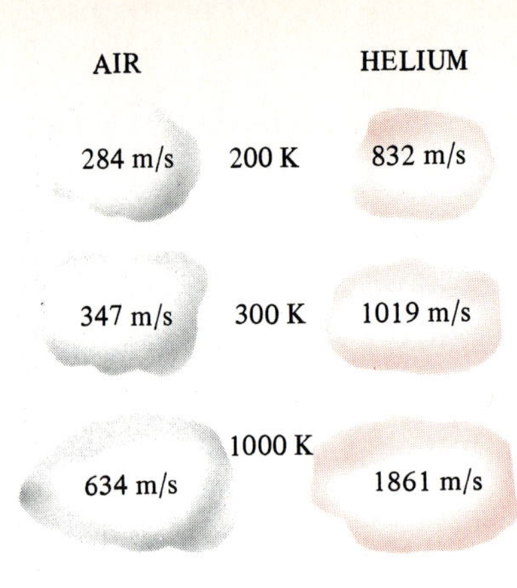

FIGURE 16-9
The velocity of sound changes with temperature.

Noting that the gas constant R has a fixed value for a specified ideal gas and the specific heat ratio k of an ideal gas is, at most, a function of temperature, we see that the velocity of sound in a specified ideal gas is a function of temperature alone (Fig. 16-9).

A second important parameter in the analysis of compressible fluid flow is the **Mach number** M, named after the Austrian physicist Ernst Mach (1838–1916). It is the ratio of the actual velocity of the fluid (or an object in still air) to the velocity of sound in the same fluid at the same state:

$$M = \frac{\mathbf{V}}{C} \tag{16-12}$$

Note that the Mach number depends on the velocity of sound, which depends on the state of the fluid. Therefore, the Mach number of an IFO (identified flying object) cruising at constant velocity in still air may be different at different locations (Fig. 16-10).

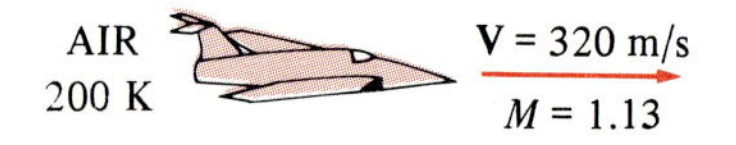

FIGURE 16-10
The Mach number can be different at different temperatures even if the velocity is the same.

Fluid flow regimes are often described in terms of the flow Mach number. The flow is called **sonic** when $M = 1$, **subsonic** when $M < 1$, **supersonic** when $M > 1$, **hypersonic** when $M >> 1$, and **transonic** when $M \cong 1$.

EXAMPLE 16-3

Air enters a diffuser shown in Fig. 16-11 with a velocity of 200 m/s. Determine (*a*) the velocity of sound and (*b*) the flow Mach number at the diffuser inlet when the air temperature is 30°C.

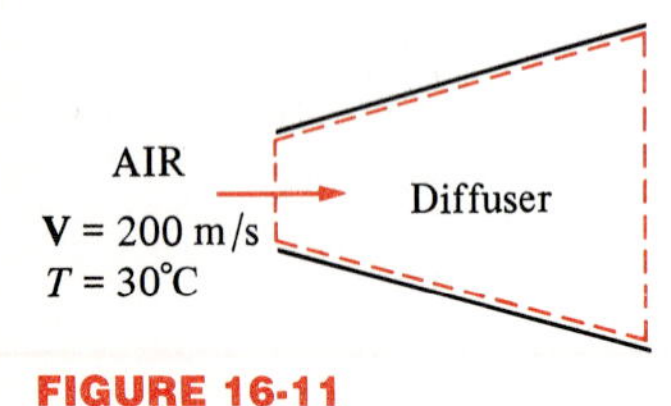

FIGURE 16-11
Schematic for Example 16-3.

Solution At specified conditions, air can be treated as an ideal gas. The gas constant of air is $R = 0.287$ kJ/(kg·K), and its specific heat ratio at 30°C is determined from Table A-2*a* to be 1.4.

(*a*) The velocity of sound in air at 30°C is determined from Eq. 16-11:

$$C = \sqrt{kRT} = \sqrt{(1.4)[0.287 \text{ kJ/(kg·K)}](303 \text{ K})\left(\frac{1000 \text{ m}^2/\text{s}^2}{1 \text{ kJ/kg}}\right)} = 349 \text{ m/s}$$

(*b*) The Mach number is then readily determined from Eq. 16-12:

$$M = \frac{\mathbf{V}}{C} = \frac{200 \text{ m/s}}{349 \text{ m/s}} = 0.573$$

Thus the flow of air at the diffuser inlet is subsonic.

EXAMPLE 16-4

Steam at 1.0 MPa and 350°C flows through a pipe shown in Fig. 16-12 with a velocity of 300 m/s. Determine (*a*) the velocity of sound and (*b*) the flow Mach number for the steam. Compare the result with that obtained by treating the superheated steam as an ideal gas with $k = 1.3$.

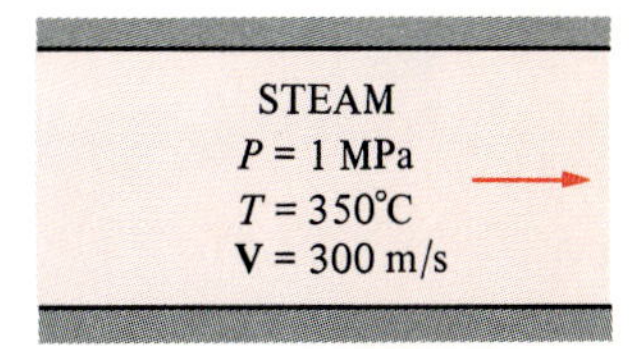

FIGURE 16-12
Schematic for Example 16-4.

Solution (*a*) The velocity of sound in steam at the given state can be determined exactly from Eq. 16-9, but we need to have a relation for pressure P in terms of density ρ. In the absence of such relations, a reasonably accurate value for the velocity of sound can be obtained by replacing the differentials in that equation by differences about the specified pressure:

$$C^2 = \left(\frac{\partial P}{\partial \rho}\right)_s \cong \left[\frac{\Delta P}{\Delta(1/v)}\right]_s$$

The entropy of steam at the given state (1 MPa and 350°C) is determined from Table A-6 to be $s = 7.3011$ kJ/(kg·K). By interpolation, the specific volumes of the steam at this entropy and the listed pressures just below and just above the specified pressure (0.8 and 1.2 MPa) are 0.3359 and 0.2454 m³/kg, respectively. Substituting yields

$$C = \sqrt{\frac{(1200 - 800) \text{ kPa}}{(1/0.2454 - 1/0.3359) \text{ kg/m}^3}\left(\frac{1000 \text{ m}^2/\text{s}^2}{1 \text{ kPa·m}^3/\text{kg}}\right)} = 603.6 \text{ m/s}$$

(*b*) The Mach number is then readily determined from Eq. 16-12:

$$M = \frac{\mathbf{V}}{C} = \frac{300 \text{ m/s}}{603.6 \text{ m/s}} = 0.497$$

Thus the flow of steam through the pipe is subsonic.

If ideal-gas behavior is assumed for the steam, the velocity of sound can be determined from Eq. 16-11:

$$C = \sqrt{kRT} = \sqrt{(1.3)[0.4615 \text{ kJ/(kg·K)}](623 \text{ K})\left(\frac{1000 \text{ m}^2/\text{s}^3}{1 \text{ kJ/kg}}\right)}$$
$$= 611.4 \text{ m/s}$$

and $$M = \frac{\mathbf{V}}{C} = \frac{300 \text{ m/s}}{611.4 \text{ m/s}}$$
$$= 0.491$$

Thus the ideal-gas assumption is a reasonable one for steam at the specified state.

16-3 ■ ONE-DIMENSIONAL ISENTROPIC FLOW

The fluid flow through several thermodynamic devices such as nozzles, diffusers, and turbine blade passages can be approximated as one-dimensional isentropic flow with good accuracy. Therefore, it merits special consideration. Before presenting a formal discussion of one-dimensional isentropic flow, we illustrate some important aspects of it with an example.

EXAMPLE 16-5

Carbon dioxide flows steadily through a varying-cross-sectional-area duct such as a nozzle shown in Fig. 16-13 at a mass flow rate of 3 kg/s. The carbon dioxide enters the duct at a pressure of 1400 kPa and 200°C with a low velocity, and it expands in the nozzle to a pressure of 200 kPa. The duct is designed so that the flow can be approximated as isentropic. Determine the density, velocity, flow area, and Mach number at each location along the nozzle which corresponds to a pressure drop of 200 kPa.

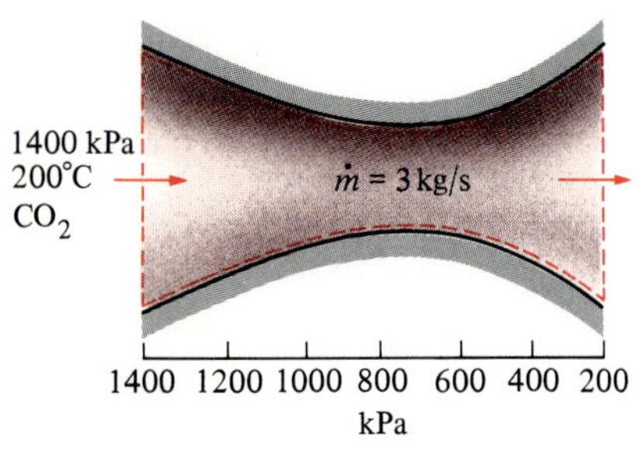

FIGURE 16-13
Schematic for Example 16-5.

We treat carbon dioxide as an ideal gas with constant specific heats, and we approximate the flow through the nozzle as one-dimensional and isentropic. Then it follows from Eq. 16-3 that the stagnation enthalpy h_0 (and thus the stagnation temperature T_0) will remain constant throughout the nozzle. The inlet temperature in this case is equivalent to the stagnation temperature since the inlet velocity of the carbon dioxide is said to be negligible. The stagnation pressure P_0 will also remain constant throughout the duct since the flow is isentropic. That is,

$$T_0 = T_i = 200°\text{C} = 473 \text{ K}$$

and

$$P_0 = P_i = 1400 \text{ kPa}$$

To illustrate the solution procedure, we calculate the desired properties at the location where the pressure is 1200 kPa, the first location that corresponds to a pressure drop of 200 kPa. For simplicity we use $C_p = 0.846$ kJ/(kg·K) and $k = 1.289$ throughout the calculations, which are the constant-pressure specific heat and specific heat ratio values of carbon dioxide at room temperature.

From Eq. 16-5,

$$T = T_0 \left(\frac{P}{P_0}\right)^{(k-1)/k} = (473 \text{ K}) \left(\frac{1200 \text{ kPa}}{1400 \text{ kPa}}\right)^{(1.289-1)/1.289} = 457 \text{ K}$$

From Eq. 16-4,

$$\mathbf{V} = \sqrt{2C_p(T_0 - T)}$$
$$= \sqrt{2[0.846 \text{ kJ/(kg·K)}](473 \text{ K} - 457 \text{ K}) \left(\frac{1000 \text{ m}^2/\text{s}^2}{1 \text{ kJ/kg}}\right)}$$
$$= 164.5 \text{ m/s}$$

From the ideal-gas relation,

$$\rho = \frac{P}{RT} = \frac{1200 \text{ kPa}}{[0.1889 \text{ kPa·m}^3/(\text{kg·K})](457 \text{ K})} = 13.9 \text{ kg/m}^3$$

From the mass flow relation,

$$A = \frac{\dot{m}}{\rho \mathbf{V}} = \frac{3 \text{ kg/s}}{(13.9 \text{ kg/m}^3)(164.5 \text{ m/s})} = 13.1 \times 10^{-4} \text{ m}^2 = 13.1 \text{ cm}^2$$

From Eq. 16-11,

$$C = \sqrt{kRT} = \sqrt{(1.289)[0.1889 \text{ kJ/(kg·K)}](457 \text{ K}) \left(\frac{1000 \text{ m}^2/\text{s}^2}{1 \text{ kJ/kg}}\right)}$$
$$= 333.6 \text{ m/s}$$

TABLE 16-1

Variation of fluid properties in flow direction in duct described in Example 16-5 for $\dot{m}$ = 3 kg/s = constant

P, kPa	T, K	V, m/s	ρ, kg/m³	C m/s	A, cm²	M
1400	473	0	15.7	0	∞	0
1200	457	164.5	13.9	333.6	13.1	0.493
1000	439	240.7	12.1	326.9	10.3	0.736
800	417	306.6	10.1	318.8	9.64	0.962
767*	414	317.2	9.82	317.2	9.63	1.000
600	391	371.4	8.12	308.7	10.0	1.203
400	357	441.9	5.93	295.0	11.5	1.498
200	306	530.9	3.46	272.9	16.3	1.946

*767 kPa is the critical pressure where the local Mach number is unity.

From Eq. 16-12,

$$M = \frac{\mathbf{V}}{C} = \frac{164.5 \text{ m/s}}{333.6 \text{ m/s}} = 0.493$$

The results for the other pressure steps are summarized in Table 16-1.

We note from the above example that as the pressure decreases, the temperature and velocity of sound decrease while the fluid velocity and Mach number increase in the flow direction. The density decreases slowly at first and rapidly later as the fluid velocity increases. The flow area decreases with decreasing pressure up to a critical-pressure value where the Mach number is unity, and then it begins to increase with further reductions in pressure. The Mach number is unity at the location of smallest flow area, called the **throat** (Fig. 16-14). Note that the velocity of the fluid keeps increasing after passing the throat although the flow area increases rapidly in that region. This increase in velocity past the throat is due to the rapid decrease in the fluid density. The flow area of the duct considered in this example first decreases and then increases. Such ducts are called **converging-diverging nozzles**

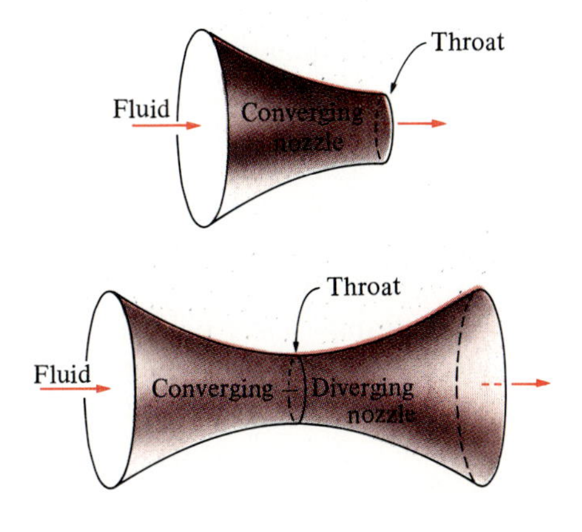

FIGURE 16-14

The cross section of a nozzle at the smallest flow area is called the *throat*.

Variation of Fluid Velocity with Flow Area

It is clear from this example that the couplings among the velocity, density, and flow areas for isentropic duct flow are rather complex. In the remainder of this section we investigate these couplings more thoroughly, and we develop relations for the variations of static to stagnation property ratios with Mach number for pressure, temperature, and density.

We begin our investigation by seeking relationships among the pressure, temperature, density, velocity, flow area, and the Mach number for one-dimensional isentropic flow. Consider the conservation of mass equation for a steady-flow process:

$$\dot{m} = \rho A \mathbf{V} = \text{constant}$$

Differentiating and dividing the resultant equation by the mass flow rate, we obtain

$$\frac{d\rho}{\rho} + \frac{dA}{A} + \frac{d\mathbf{V}}{\mathbf{V}} = 0 \tag{16-13}$$

Neglecting the potential energy, the conservation of energy equation for an isentropic flow with no work interactions can be expressed in the differential form as (Fig. 16-15)

$$\frac{dP}{\rho} + \mathbf{V}\, d\mathbf{V} = 0 \tag{16-14}$$

CONSERVATION OF ENERGY
(steady flow, $\omega = 0$, $q = 0$, $\Delta\text{pe} = 0$)

$$h_i + \frac{\mathbf{V}_i^2}{2} = h_e + \frac{\mathbf{V}_e^2}{2}$$

or

$$h + \frac{\mathbf{V}^2}{2} = \text{constant}$$

Differentiate,

$$dh + \mathbf{V}\, d\mathbf{V} = 0$$

From Eq. 6-24, 0 (isentropic)

$$T\, ds = dh - v\, dP$$

$$dh = v\, dP = \frac{1}{\rho}\, dP$$

Substitute,

$$\frac{dP}{\rho} + \mathbf{V}\, d\mathbf{V} = 0$$

FIGURE 16-15

Derivation of the differential form of the energy equation for steady isentropic flow.

This relation is also the differential form of Bernoulli's equation, Eq. 6-54, which is a form of the conservation of momentum principle for steady-flow control volumes. Combining the two equations above gives

$$\frac{dA}{A} = \frac{dP}{\rho}\left(\frac{1}{\mathbf{V}^2} - \frac{d\rho}{dP}\right) \tag{16-15}$$

Rearranging Eq. 16-9 as $(\partial\rho/\partial P)_s = 1/C^2$ and substituting into Eq. 16-15 yield

$$\frac{dA}{A} = \frac{dP}{\rho \mathbf{V}^2}(1 - M^2) \tag{16-16}$$

This is an important relation for the isentropic flow in ducts since it describes the variation of pressure with flow area. We note that A, ρ, and $\mathbf{V}$ are positive quantities. For *subsonic* flow ($M < 1$), the term $1 - M^2$ is positive; and thus dA and dP must have the same sign. That is, the pressure of the fluid must increase as the flow area of the duct increases and must decrease as the flow area of the duct decreases. Thus at subsonic velocities, the pressure decreases in converging ducts (subsonic nozzles) and increases in diverging ducts (subsonic diffusers).

In *supersonic* flow ($M > 1$), the term $1 - M^2$ is negative, and thus dA and dP must have opposite signs. That is, the pressure of the fluid must increase as the flow area of the duct decreases and must decrease as the flow area of the duct increases. Thus at supersonic velocities, the pressure decreases in diverging ducts (supersonic nozzles) and increases in converging ducts (supersonic diffusers).

Another important relation for the isentropic flow of a fluid is obtained by substituting $\rho\mathbf{V} = -dP/d\mathbf{V}$ from Eq. 16-14 into Eq. 16-16:

$$\frac{dA}{A} = -\frac{d\mathbf{V}}{\mathbf{V}}(1 - M^2) \tag{16-17}$$

This equation governs the shape that a nozzle or a diffuser must have in subsonic or supersonic isentropic flow. Noting that A and $\mathbf{V}$ are positive quantities, we conclude the following:

For subsonic flow ($M < 1$), $\quad \dfrac{dA}{d\mathbf{V}} < 0$

For supersonic flow ($M > 1$), $\quad \dfrac{dA}{d\mathbf{V}} > 0$

For sonic flow ($M = 1$), $\quad \dfrac{dA}{d\mathbf{V}} = 0$

Thus the proper shape of a nozzle depends on the highest velocity desired relative to the sonic velocity. To accelerate a fluid, we must use a converging nozzle at subsonic velocities and a diverging nozzle at supersonic velocities. The velocities encountered in most familiar applications are well below the sonic velocity, and thus it is natural that we visualize a nozzle as a converging duct. But the highest velocity we can achieve by a converging nozzle is the sonic velocity, which will occur at the exit of the nozzle. If we extend the converging nozzle by further decreasing the flow area, in hopes of accelerating the fluid to supersonic velocities, as shown in Fig. 16-16, we are up for disappointment. Now the sonic velocity will occur at the exit of the converging extension, instead of the exit of the original nozzle, and the mass flow rate through the nozzle will decrease because of the reduced exit area.

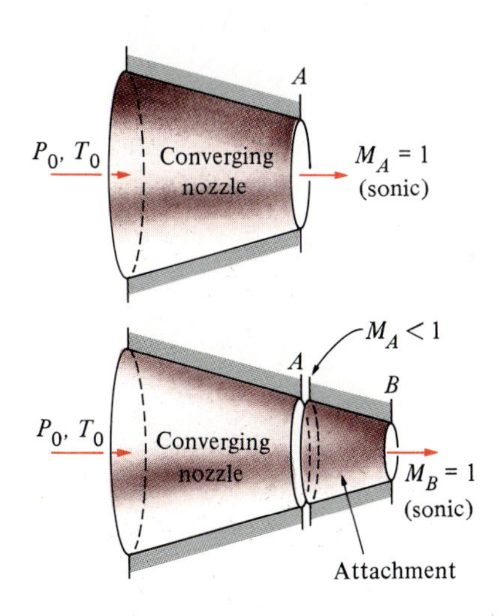

FIGURE 16-16
We cannot obtain supersonic velocities by attaching a converging section to a converging nozzle. Doing so will only move the sonic cross section farther downstream.

Based on Eq. 16-16, which is an expression of the conservation of mass and energy principles, we must add a diverging section to a converging nozzle to accelerate a fluid to supersonic velocities. The result is a converging-diverging nozzle. The fluid first passes through a subsonic (converging) section, where the Mach number increases as the flow area of the nozzle decreases, and then reaches the value of unity at the nozzle throat. The fluid continues to accelerate as it passes through a supersonic (diverging) section. Noting that $\dot{m} = \rho A \mathbf{V}$ for steady flow, we see that the large decrease in density makes acceleration in the diverging section possible. An example of this type of flow is the flow of hot combustion gases through a nozzle in a gas turbine.

The opposite process occurs in the engine inlet of a supersonic aircraft. The fluid is decelerated by passing it first through a supersonic diffuser which has a flow area that decreases in the flow direction. The flow reaches a Mach number of unity at the diffuser throat. The fluid is further decelerated in a subsonic diffuser which has a flow area that increases in the flow direction, as shown in Fig. 16-17.

Property Relations for Isentropic Flow of Ideal Gases

Next we develop relations between the static properties and stagnation properties of an ideal gas in terms of the specific heat ratio k and the Mach number M. We assume the flow is isentropic, and the gas has constant specific heats.

The temperature T of an ideal gas anywhere in the flow is related to the stagnation temperature T_0 through Eq. 16-4:

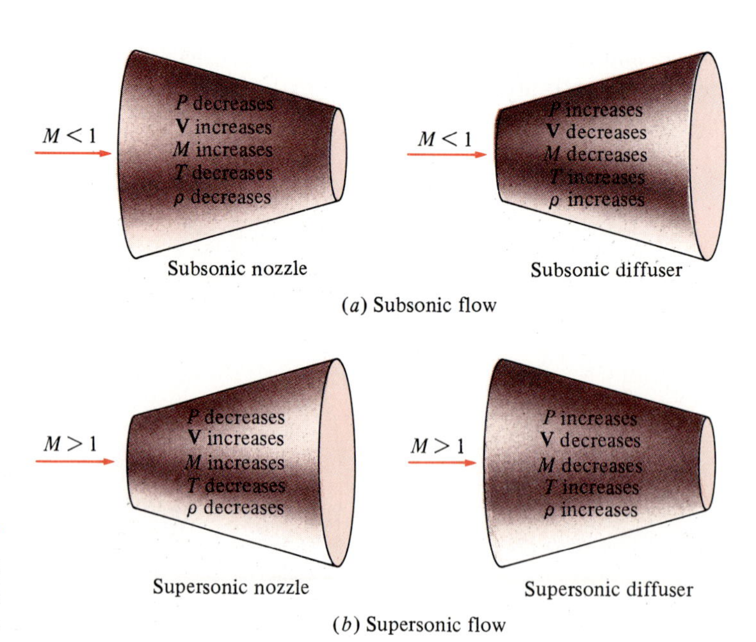

FIGURE 16-17
Variation of flow properties in subsonic and supersonic nozzles and diffusers.

$$T_0 = T + \frac{\mathbf{V}^2}{2C_p}$$

or
$$\frac{T_0}{T} = 1 + \frac{\mathbf{V}^2}{2C_pT}$$

Noting that $C_p = kR/(k - 1)$, $C^2 = kRT$, and $M = \mathbf{V}/C$, we see that

$$\frac{\mathbf{V}^2}{2C_pT} = \frac{\mathbf{V}^2}{2\,[kR/(k - 1)]T} = \left(\frac{k-1}{2}\right)\frac{\mathbf{V}^2}{C^2} = \left(\frac{k-1}{2}\right)M^2$$

Substituting yields
$$\frac{T_0}{T} = 1 + \left(\frac{k-1}{2}\right)M^2 \tag{16-18}$$

which is the desired relation between T_0 and T.

The ratio of the stagnation to static pressure is obtained by substituting Eq. 16-18 above into Eq. 16-5:

$$\frac{P_0}{P} = \left[1 + \left(\frac{k-1}{2}\right)M^2\right]^{k/(k-1)} \tag{16-19}$$

The ratio of the stagnation to static density is obtained by substituting Eq. 16-18 into Eq. 16-6:

$$\frac{\rho_0}{\rho} = \left[1 + \left(\frac{k-1}{2}\right)M^2\right]^{1/(k-1)} \tag{16-20}$$

Numerical values of T/T_0, P/P_0, and ρ/ρ_0 are listed versus the Mach number in Table A-34 for $k = 1.4$.

The properties of a fluid at a location where the Mach number is unity (the throat) are called **critical properties**, and the above ratios are called **critical ratios** (Fig. 16-18). Let the superscript asterisk (*) represent the critical values. Setting $M = 1$ in Eqs. 16-18 through 16-20 yields

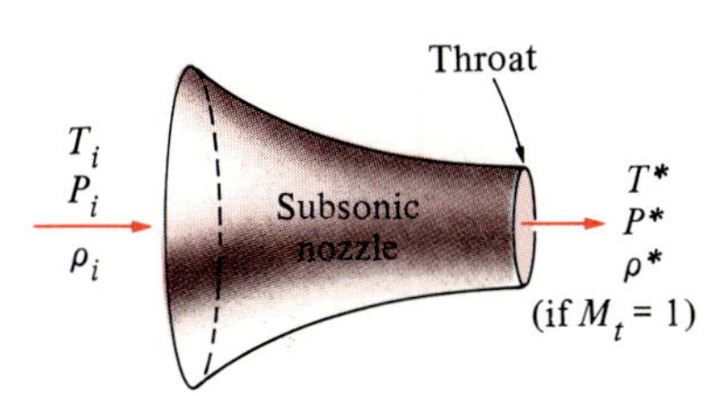

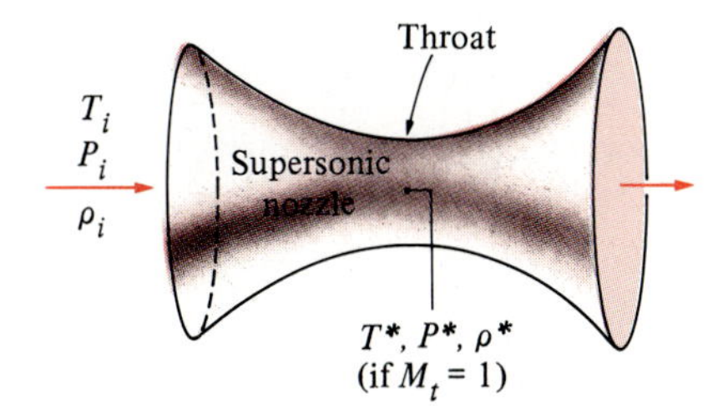

FIGURE 16-18
When $M_t = 1$, the properties at the nozzle throat become the critical properties.

TABLE 16-2
The critical-pressure, critical-temperature, and critical-density ratios for isentropic flow of some ideal gases

	Superheated steam, $k = 1.3$	Hot products of combustion, $k = 1.33$	Air, $k = 1.4$	Monatomic gases, $k = 1.667$
$\frac{P^*}{P_0}$	0.5457	0.5404	0.5283	0.4871
$\frac{T^*}{T_0}$	0.8696	0.8584	0.8333	0.7499
$\frac{\rho^*}{\rho_0}$	0.6276	0.6295	0.6340	0.6495

$$\frac{T^*}{T_0} = \frac{2}{k + 1} \tag{16-21}$$

$$\frac{P^*}{P_0} = \left(\frac{2}{k + 1}\right)^{k/(k-1)} \tag{16-22}$$

$$\frac{\rho^*}{\rho_0} = \left(\frac{2}{k + 1}\right)^{1/(k-1)} \tag{16-23}$$

These ratios are evaluated for various values of k and are listed in Table 16-2.

EXAMPLE 16-6

Calculate the critical pressure and temperature of carbon dioxide for the flow conditions described in Example 16-5 (Fig. 16-19).

Solution Using the specific-heat-ratio value of carbon dioxide at room temperature ($k = 1.289$), we find the ratios of critical to stagnation temperature and pressure, respectively, from Eqs. 16-21 and 16-22 to be

$$\frac{T^*}{T_0} = \frac{2}{k + 1} = \frac{2}{1.289 + 1} = 0.874$$

$$\frac{P^*}{P_0} = \left(\frac{2}{k + 1}\right)^{k/(k-1)} = \left(\frac{2}{1.289 + 1}\right)^{1.289/(1.289-1)} = 0.548$$

Noting that the stagnation temperature and pressure are, from Example 16-5,

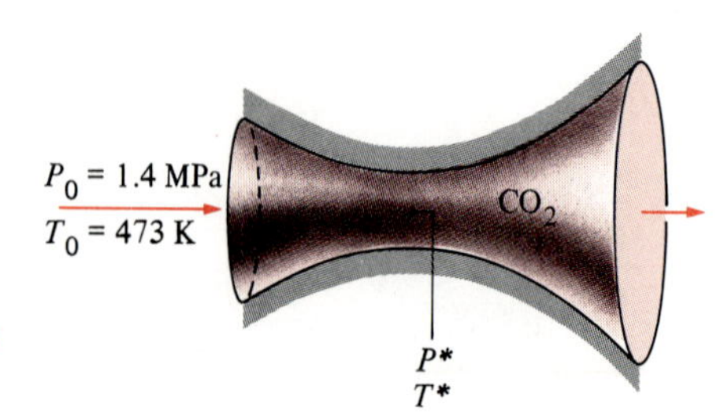

FIGURE 16-19
Schematic for Example 16-6.

T_0 = 473 K and P_0 = 1400 kPa, we see that the critical temperature and pressure in this case are

$$T^* = 0.874T_0 = (0.874)(473 \text{ K}) = 413.4 \text{ K}$$

$$P^* = 0.548P_0 = (0.548)(1400 \text{ kPa}) = 767.2 \text{ kPa}$$

16-4 ■ ISENTROPIC FLOW THROUGH NOZZLES

Converging or converging-diverging nozzles are found in many engineering applications including steam and gas turbines, aircraft propulsion systems, and even ordinary garden hoses, among others. In this section we consider the effects of **back pressure** (i.e., the pressure applied at the nozzle discharge region) on the exit velocity, the mass flow rate, and the pressure distribution along the nozzle.

Converging Nozzles

Consider the subsonic flow through a converging nozzle shown in Fig. 16-20. The nozzle inlet is attached to a reservoir at pressure P_r and temperature T_r. The reservoir is sufficiently large so that the nozzle inlet velocity is negligible. Since the fluid velocity in the reservoir is zero and the flow through the nozzle is isentropic, the stagnation pressure and stagnation temperature of the fluid at any cross section through the nozzle are equal to the reservoir pressure and temperature, respectively.

Now we begin to reduce the back pressure and observe the resulting effects on the pressure distribution along the length of the nozzle, as shown in Fig. 16-20. If the back pressure P_b is equal to P_1, which is equal

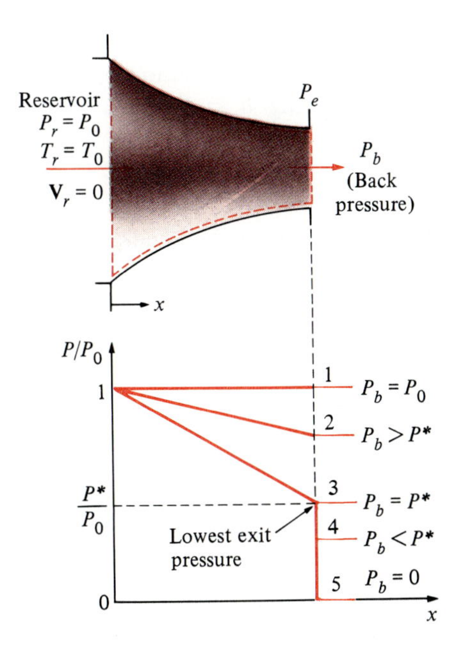

FIGURE 16-20
The effect of back pressure on the pressure distribution along a converging nozzle.

to P_r, there will be no flow and the pressure distribution will be uniform along the nozzle. When the back pressure is reduced to P_2, the exit plane pressure P_e also drops to P_2. This causes the pressure along the nozzle to decrease in the flow direction.

When the back pressure is reduced to P_3 (= P^*, the pressure required to increase the fluid velocity to the velocity of sound at the exit plane or throat), the mass flow reaches a maximum value and the flow is said to be **choked**. Further reduction of the back pressure to level P_4 or below does not result in additional changes in the pressure distribution along the nozzle length.

Under steady-flow conditions, the mass flow rate through the nozzle is constant and can be expressed as

$$\dot{m} = \rho A \mathbf{V} = \left(\frac{P}{RT}\right) A(M\sqrt{kRT}) = PAM\sqrt{\frac{k}{RT}}$$

Solving for T from Eq. 16-18 and for P from Eq. 16-19 and substituting, we have

$$\dot{m} = \frac{AMP_0\sqrt{k/(RT_0)}}{[1 + (k - 1)M^2/2]^{(k+1)/[2(k-1)]}} \qquad (16\text{-}24)$$

Thus the mass flow rate of a particular fluid through a nozzle is a function of the stagnation properties of the fluid, the flow area, and the Mach number. The relation above is valid at any cross section, and thus $\dot{m}$ can be evaluated at any location along the length of the nozzle.

For a specified flow area A and stagnation properties T_0 and P_0, the maximum mass flow rate can be determined by differentiating Eq. 16-24 with respect to M and setting the result equal to zero. It yields $M = 1$. Since the only location in a nozzle where the Mach number can be unity is the location of minimum flow area (the throat), the mass flow rate through a nozzle will be maximum when $M = 1$ at the throat. Denoting this area by A^*, we obtain an expression for the maximum mass flow rate by substituting $M = 1$ in Eq. 16-24:

$$\dot{m}_{\max} = A^*P_0\sqrt{\frac{k}{RT_0}}\left(\frac{2}{k+1}\right)^{(k+1)/[2(k-1)]} \qquad (16\text{-}25)$$

Thus for a particular ideal gas, the maximum mass flow rate through a nozzle with a given throat area is fixed by the stagnation pressure and temperature of the inlet flow. The flow rate can be controlled by changing the stagnation pressure or temperature, and thus a convergent nozzle can be used as a flow meter. The flow rate can also be controlled, of course, by varying the throat area.

A plot of $\dot{m}$ versus P_b/P_0 for a converging nozzle is shown in Fig. 16-21. Notice that the mass flow rate increases with decreasing P_b/P_0, reaches a maximum at $P_b = P^*$, and remains constant for P_b/P_0 values less than this critical ratio. Also illustrated on this figure is the effect of back pressure on the nozzle exit pressure P_e. We observe that

$$P_e = \begin{cases} P_b & \text{for } P_b \geq P^* \\ P^* & \text{for } P_b < P^* \end{cases}$$

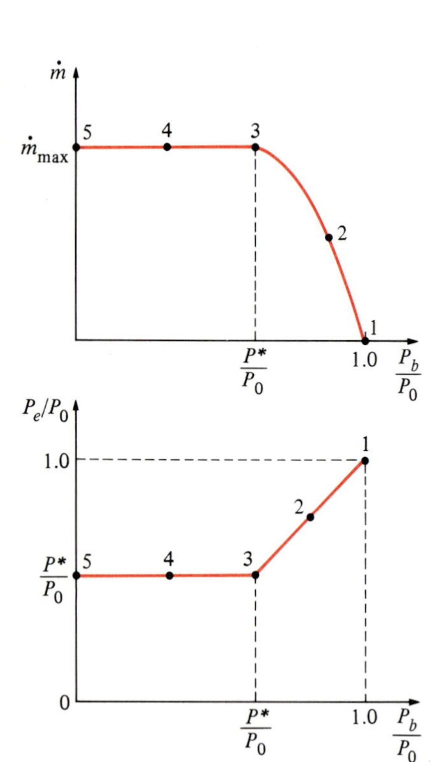

FIGURE 16-21
The effect of back pressure P_b on the mass flow rate $\dot{m}$ and the exit pressure P_e of a converging nozzle.

To summarize, for all back pressures lower than the critical pressure P^*, the pressure at the exit plane of the converging nozzle P_e is equal to P^*, the Mach number at the exit plane is unity, and the mass flow rate is the maximum (or choked) flow rate. Because the velocity of the flow is sonic at the throat for the maximum flow rate, a back pressure lower than the critical pressure cannot be sensed in the nozzle upstream flow and does not affect the flow rate.

The effects of the stagnation temperature T_0 and stagnation pressure P_0 on the mass flow rate through a converging nozzle are illustrated in Fig. 16-22 where the mass flux (mass flow rate per unit area) is plotted against the static-to-stagnation pressure ratio at the throat P_t/P_0. An increase in P_0 (or a decrease of T_0) will increase the mass flux through the converging nozzle; a decrease in P_0 (or an increase in T_0) will decrease it. We could also conclude this by carefully observing Eqs. 16-24 and 16-25.

A relation for the variation of flow area A through the nozzle relative to throat area A^* can be obtained by combining Eqs. 16-24 and 16-25 for the same mass flow rate and stagnation properties of a particular fluid. This yields

$$\frac{A}{A^*} = \frac{1}{M}\left[\left(\frac{2}{k+1}\right)\left(1 + \frac{k-1}{2}M^2\right)\right]^{(k+1)/[2(k-1)]} \tag{16-26}$$

Table A-34 gives values of A/A^* as a function of the Mach number for $k = 1.4$. There is one value of A/A^* for each value of the Mach number, but there are two possible values of Mach number for each value of A/A^*—one for subsonic flow and another for supersonic flow.

Another parameter sometimes used in the analysis of one-dimensional isentropic flow of ideal gases is M^*, which is the ratio of the local velocity to the velocity of sound at the throat:

$$M^* = \frac{\mathbf{V}}{C^*} \tag{16-27}$$

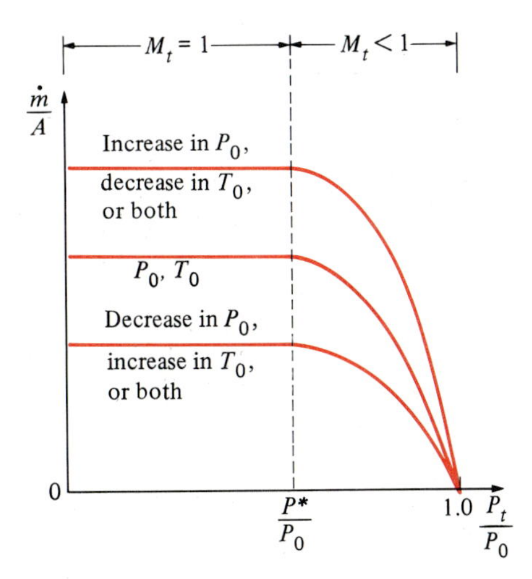

FIGURE 16-22
The variation of the mass flow rate through a nozzle with inlet stagnation properties.

It can also be expressed as

$$M^* = \frac{MC}{C^*} = \frac{M\sqrt{kRT}}{\sqrt{kRT^*}} = M\sqrt{\frac{T}{T^*}}$$

where M is the local Mach number, T is the local temperature, and T^* is the critical temperature. Solving for T from Eq. 16-18 and for T^* from Eq. 16-21 and substituting, we get

$$M^* = M\sqrt{\frac{k+1}{2+(k-1)M^2}} \tag{16-28}$$

Values of M^* are also listed in Table A-34 versus the Mach number for $k = 1.4$ (Fig. 16-23). Note that the parameter M^* differs from the Mach number M in that M^* is the local velocity nondimensionalized with respect to the sonic velocity at the *throat*, whereas M is the local velocity nondimensionalized with respect to the *local* sonic velocity. (Recall that the sonic velocity in a nozzle varies with temperature and thus with location.)

M	M^*	$\frac{A}{A^*}$	$\frac{P}{P_0}$	$\frac{\rho}{\rho_0}$	$\frac{T}{T_0}$
⋮	⋮	⋮	⋮	⋮	⋮
0.90	0.9146	1.0087	0.5913	⋮	⋮
1.00	1.0000	1.0000	0.5283	⋮	⋮
1.10	1.0812	1.0079	0.4684	⋮	⋮
⋮	⋮	⋮	⋮	⋮	⋮

FIGURE 16-23

Various property ratios for isentropic flow through nozzles and diffusers are listed in Table A-34 for $k = 1.4$ for convenience.

EXAMPLE 16-7

Air at 1 MPa and 600°C enters a converging nozzle, shown in Fig. 16-24, with a velocity of 150 m/s. Determine the mass flow rate through the nozzle for a nozzle throat area of 50 cm² when the back pressure is (*a*) 0.7 MPa and (*b*) 0.4 MPa.

Solution We assume air to be an ideal gas having properties found in Table A-2*a* and undergoing an isentropic process. Subscripts i and t are used to represent the properties at the nozzle inlet and the throat, respectively.

The stagnation temperature and pressure at the nozzle inlet are determined from Eqs. 16-4 and 16-5:

$$T_{0i} = T_i + \frac{\mathbf{V}_i^2}{2C_p} = 873 \text{ K} + \frac{(150 \text{ m/s})^2}{2[1.005 \text{ kJ/(kg·K)}]}\left(\frac{1 \text{ kJ/kg}}{1000 \text{ m}^2/\text{s}^2}\right) = 884 \text{ K}$$

$$P_{0i} = P_i\left(\frac{T_{0i}}{T_i}\right)^{k/(k-1)} = (1 \text{ MPa})\left(\frac{884 \text{ K}}{873 \text{ K}}\right)^{1.4/(1.4-1)} = 1.045 \text{ MPa}$$

These stagnation temperature and pressure values remain constant throughout the nozzle since the flow is assumed to be isentropic. That is,

$$T_0 = T_{0i} = 884 \text{ K} \qquad \text{and} \qquad P_0 = P_{0i} = 1.045 \text{ MPa}$$

The critical-pressure ratio is determined from Table 16-2 (or Eq. 16-22) to be $P^*/P_0 = 0.5283$.

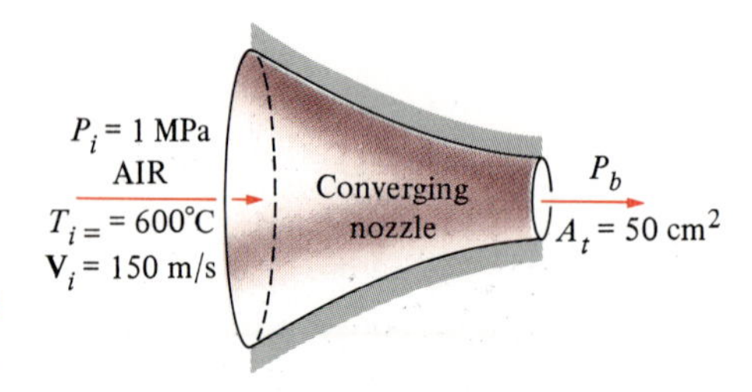

FIGURE 16-24

Schematic for Example 16-7.

(a) The back pressure ratio for this case is

$$\frac{P_b}{P_0} = \frac{0.7 \text{ MPa}}{1.045 \text{ MPa}} = 0.670$$

which is greater than the critical-pressure ratio, 0.5283. Thus the exit plane pressure (or throat pressure, P_t) is equal to the back pressure in this case. That is, $P_t = P_b = 0.7$ MPa, and $P_t/P_0 = 0.670$. Therefore, the flow is not choked. From Table A-34 at $P_t/P_0 = 0.670$ we read $M_t = 0.778$ and $T_t/T_0 = 0.892$.

The mass flow rate through the nozzle can be calculated from Eq. 16-24. But it can also be determined in a step-by-step manner as follows:

$$T_t = 0.892T_0 = 0.892(884 \text{ K}) = 788.5 \text{ K}$$

$$\rho_t = \frac{P_t}{RT_t} = \frac{700 \text{ kPa}}{[0.287 \text{ kPa·m}^3/(\text{kg·K})](788.5 \text{ K})} = 3.093 \text{ kg/m}^3$$

$$\begin{aligned}\mathbf{V}_t &= M_tC_t = M_t\sqrt{kRT_t} \\ &= (0.778)\sqrt{(1.4)[0.287 \text{ kJ/(kg·K)}](788.5 \text{ K})\left(\frac{1000 \text{ m}^2/\text{s}^2}{1 \text{ kJ/kg}}\right)} \\ &= 437.9 \text{ m/s}\end{aligned}$$

Thus, $\dot{m} = \rho_t A_t \mathbf{V}_t = (3.093 \text{ kg/m}^3)(50 \times 10^{-4} \text{ m}^2)(437.9 \text{ m/s}) = 6.766 \text{ kg/s}$

(b) The back pressure ratio for this case is

$$\frac{P_b}{P_0} = \frac{0.4 \text{ MPa}}{1.045 \text{ MPa}} = 0.383$$

which is less than the critical-pressure ratio, 0.5283. Therefore, sonic conditions exist at the exit plane (throat) of the nozzle, and $M = 1$. The flow is choked in this case, and the mass flow rate through the nozzle can be calculated from Eq. 16-25:

$$\begin{aligned}\dot{m} &= A^*P_0\sqrt{\frac{k}{RT_0}}\left(\frac{2}{k+1}\right)^{(k+1)/[2(k-1)]} \\ &= (50 \times 10^{-4} \text{ m}^2)(1045 \text{ kPa}) \\ &\quad \times \sqrt{\frac{1.4}{[0.287 \text{ kJ/(kg·K)}](884 \text{ K})}}\left(\frac{2}{1.4+1}\right)^{2.4/0.8} \\ &= 7.103 \text{ kg/s}\end{aligned}$$

since $\text{kPa·m}^2/\sqrt{\text{kJ/kg}} = \sqrt{1000}$ kg/s. This is the maximum mass flow rate through the nozzle for the specified inlet conditions and nozzle throat area.

EXAMPLE 16-8

Nitrogen enters a duct with varying flow area at $T_1 = 400$ K, $P_1 = 100$ kPa, and $M_1 = 0.3$. Assuming steady isentropic flow, determine T_2, P_2, and M_2 at a location where the flow area has been reduced by 20 percent.

Solution A schematic of the duct is shown in Fig. 16-25. We assume nitrogen to be an ideal gas with $k = 1.4$. For isentropic flow through a duct, the area ratio A/A^* (the flow area over the area of the throat where $M = 1$) is also listed in Table A-34. At the initial Mach number of $M = 0.3$, we read

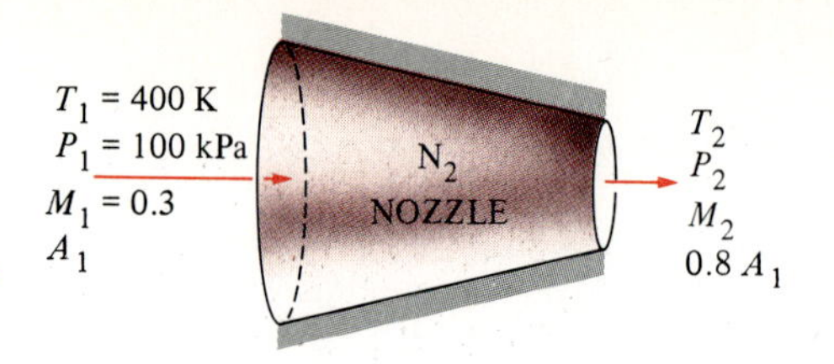

FIGURE 16-25
Schematic for Example 16-8.

$$\frac{A_1}{A^*} = 2.0351 \qquad \frac{T_1}{T_0} = 0.98232 \qquad \frac{P_1}{P_0} = 0.93947$$

With a 20 percent reduction in flow area, $A_2 = 0.8A_1$, and

$$\frac{A_2}{A^*} = \frac{A_2}{A_1}\frac{A_1}{A^*} = (0.8)(2.0351) = 1.6281$$

For this value of A_2/A^* from Table A-34 we read

$$\frac{T_2}{T_0} = 0.97033$$

$$\frac{P_2}{P_0} = 0.89995$$

$$M_2 = 0.391$$

Here we chose the subsonic Mach number for the calculated A_2/A^* instead of the supersonic one because the duct is converging in the flow direction and the initial flow is subsonic. Since the stagnation properties are constant for isentropic flow, we can write

$$\frac{T_2}{T_1} = \frac{T_2/T_0}{T_1/T_0} \longrightarrow T_2 = T_1\left(\frac{T_2/T_0}{T_1/T_0}\right) = (400\text{ K})\left(\frac{0.97033}{0.98232}\right) = 395.1\text{ K}$$

$$\frac{P_2}{P_1} = \frac{P_2/P_0}{P_1/P_0} \longrightarrow P_2 = P_1\left(\frac{P_2/P_0}{P_1/P_0}\right) = (100\text{ kPa})\left(\frac{0.89995}{0.93947}\right) = 95.8\text{ kPa}$$

which are the temperature and pressure at the desired location.

Converging-Diverging Nozzles

When we think of nozzles, we ordinarily think of flow passages whose cross-sectional area decreases in the flow direction. However, the highest velocity to which a fluid can be accelerated in a converging nozzle is limited to the sonic velocity ($M = 1$), which occurs at the exit plane (throat) of the nozzle. Accelerating a fluid to supersonic velocities ($M > 1$) can be accomplished only by attaching a diverging flow section to the subsonic nozzle at the throat. The resulting combined flow section is a converging-diverging nozzle, which is a standard equipment in supersonic aircraft.

Forcing a fluid through a converging-diverging nozzle is no guarantee that the fluid will be accelerated to a supersonic velocity. In fact, the fluid may find itself decelerating in the diverging section instead of accelerating

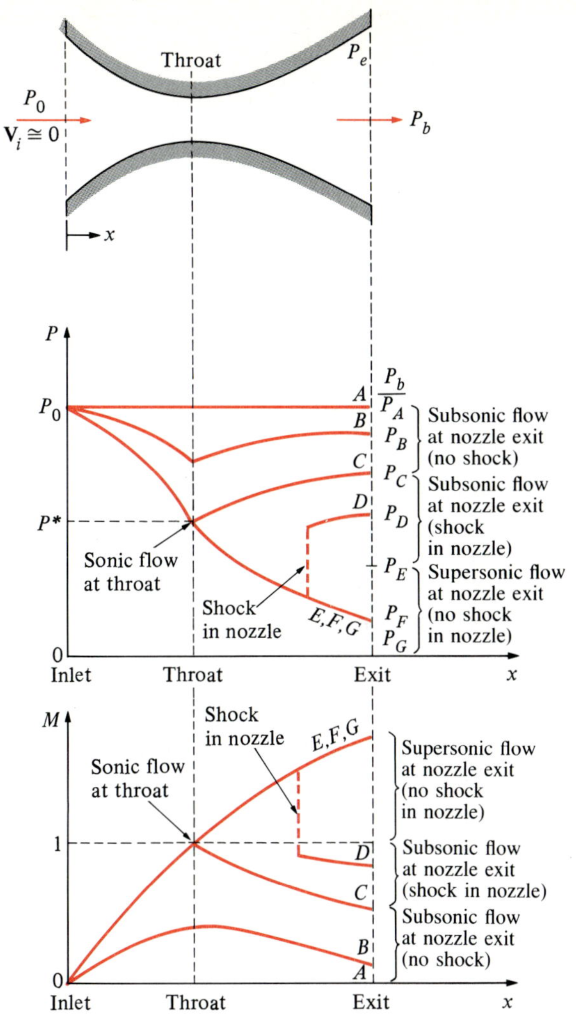

FIGURE 16-26

The effects of back pressure on the flow through a converging-diverging nozzle.

if the back pressure is not in the right range. For given inlet conditions, the flow through a converging-diverging nozzle is governed by the back pressure P_b, as explained below.

Consider the converging-diverging nozzle shown in Fig. 16-26. A fluid enters the nozzle with a low velocity at stagnation pressure P_0. When $P_b = P_0$ (case A), there will be no flow through the nozzle. This is expected since the flow in a nozzle is driven by the pressure difference between the nozzle inlet and the exit. Now let us examine what will happen as the back pressure is lowered.

1. When $P_0 > P_b > P_C$, the flow remains subsonic throughout the nozzle, and the mass flow is less than that for choked flow. The fluid velocity increases in the first (converging) section and reaches a maximum at the throat (but $M < 1$). However, most of the gain in velocity is lost in the

second (diverging) section of the nozzle which acts as a diffuser. The pressure decreases in the converging section, reaches a minimum at the throat, and increases at the expense of velocity in the diverging section.

2. When $P_b = P_C$, the throat pressure becomes P^* and the fluid achieves sonic velocity at the throat. But the diverging section of the nozzle still acts as diffuser, slowing the fluid to subsonic velocities. The mass flow rate which was increasing with decreasing P_b also reaches its maximum value.

Recall that P^* is the lowest pressure that can be obtained at the throat, and the sonic velocity is the highest velocity that can be achieved with a converging nozzle. Thus lowering P_b further will have no influence on the fluid flow in the converging part of the nozzle or the mass flow rate through the nozzle. But it will influence the character of the flow in the diverging section.

3. When $P_C > P_b > P_E$, the fluid which achieved a sonic velocity at the throat continues accelerating to supersonic velocities in the diverging section as the pressure decreases. This acceleration comes to a sudden stop, however, as a **normal shock** develops at a section between the throat and the exit plane, which causes a sudden drop in velocity to subsonic levels and a sudden increase in pressure. The fluid then continues to decelerate further in the remaining part of the converging-diverging nozzle. Flow through the shock is highly irreversible, and thus it cannot be approximated as isentropic. The normal shock moves downstream away from the throat as P_b is decreased, and it approaches the nozzle exit plane as P_b approaches P_E.

When $P_b = P_E$, the normal shock forms at the exit plane of the nozzle. The flow is supersonic through the entire diverging section in this case, and it can be approximated as isentropic. However, the fluid velocity drops to subsonic levels just before leaving the nozzle as it crosses the normal shock. Normal shock waves are discussed in the next section.

4. When $P_E > P_b > 0$, the flow in the diverging section is supersonic, and the fluid expands to P_F at the nozzle exit with no normal shock forming within the nozzle. Thus the flow through the nozzle can be approximated as isentropic. When $P_b = P_F$, no shocks occur within or outside the nozzle. When $P_b < P_F$, irreversible mixing and expansion waves occur downstream of the exit plane of the nozzle. When $P_b > P_F$, however, the pressure of the fluid increases from P_F to P_b irreversibly in the wake of the nozzle exit, creating what are called *oblique shocks*.

EXAMPLE 16-9

Air enters a converging-diverging nozzle, shown in Fig. 16-27, at 1.0 MPa and 800 K with a negligible velocity. The flow is steady, one-dimensional, and isentropic with $k = 1.4$. For an exit Mach number of $M = 2$ and a throat area of 20 cm^2, determine (*a*) the throat conditions, (*b*) the exit plane conditions, including the exit area, and (*c*) the mass flow rate through the nozzle.

Solution Since the exit Mach number is 2, the flow must be sonic at the throat and supersonic in the diverging section of the nozzle. Since the inlet velocity is

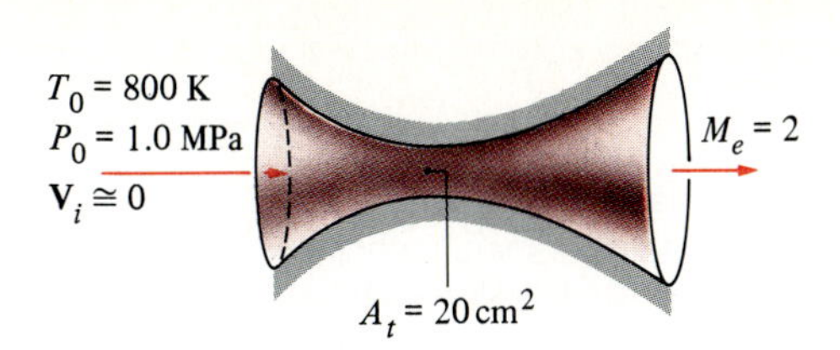

FIGURE 16-27
Schematic for Example 16-9.

negligible, the stagnation pressure and stagnation temperature are the same as the inlet temperature and pressure, $P_0 = 1.0$ MPa and $T_0 = 800$ K. Assuming ideal-gas behavior, the stagnation density is

$$\rho_0 = \frac{P_0}{RT_0} = \frac{1000 \text{ kPa}}{[0.287 \text{ kPa·m}^3/(\text{kg·K})](800 \text{ K})} = 4.355 \text{ kg/m}^3$$

(*a*) At the throat of the nozzle $M = 1$, and from Table A-34 we read

$$\frac{P^*}{P_0} = 0.5283 \qquad \frac{T^*}{T_0} = 0.8333 \qquad \frac{\rho^*}{\rho_0} = 0.6339$$

Thus,

$$P^* = 0.5283P_0 = (0.5283)(1.0 \text{ MPa}) = 0.5283 \text{ MPa}$$
$$T^* = 0.8333T_0 = (0.8333)(800 \text{ K}) = 666.6 \text{ K}$$
$$\rho^* = 0.6339\rho_0 = (0.6339)(4.355 \text{ kg/m}^3) = 2.761 \text{ kg/m}^3$$

Also,

$$\mathbf{V}^* = C^* = \sqrt{kRT^*} = \sqrt{(1.4)[0.287 \text{ kJ/(kg·K)}](666.6 \text{ K})\left(\frac{1000 \text{ m}^2/\text{s}^2}{1 \text{ kJ/kg}}\right)}$$
$$= 517.5 \text{ m/s}$$

(*b*) Since the flow is isentropic, the properties at the exit plane can also be calculated by using data from Table A-34. For $M = 2$ we read

$$\frac{P_e}{P_0} = 0.1278 \qquad \frac{T_e}{T_0} = 0.5556 \qquad \frac{\rho_e}{\rho_0} = 0.2301 \qquad M_e^* = 1.6330 \qquad \frac{A_e}{A^*} = 1.6875$$

Thus,

$$P_e = 0.1278P_0 = (0.1278)(1.0 \text{ MPa}) = 0.1278 \text{ MPa}$$
$$T_e = 0.5556T_0 = (0.5556)(800 \text{ K}) = 444.5 \text{ K}$$
$$\rho_e = 0.2301\rho_0 = (0.2301)(4.355 \text{ kg/m}^3) = 1.002 \text{ kg/m}^3$$
$$A_e = 1.6875A^* = (1.6875)(20 \text{ cm}^2) = 33.75 \text{ cm}^2$$

and

$$\mathbf{V}_e = M^*C^* = (1.6330)(517.5 \text{ m/s}) = 845.1 \text{ m/s}$$

The nozzle exit velocity could also be determined from $\mathbf{V}_e = M_eC_e$, where C_e is the velocity of sound at the exit conditions:

$$\mathbf{V}_e = M_eC_e = M_e\sqrt{kRT_e} = 2\sqrt{(1.4)[0.287 \text{ kJ/(kg·K)}](444.5 \text{ K})\left(\frac{1000 \text{ m}^2/\text{s}^2}{1 \text{ kJ/kg}}\right)}$$
$$= 845.2 \text{ m/s}$$

(*c*) Since the flow is steady, the mass flow rate of the fluid is the same at all sections of the nozzle. Thus it may be calculated by using properties at any cross section of the nozzle. Using the properties at the throat, we find that the mass flow rate is

$$\dot{m} = \rho^* A^* \mathbf{V}^* = (2.761 \text{ kg/m}^3)(20 \times 10^{-4} \text{ m}^2)(517.5 \text{ m/s}) = 2.858 \text{ kg/s}$$

Note that this is the highest possible mass flow rate that can flow through this nozzle for the specified inlet conditions.

16-5 ■ NORMAL SHOCKS IN NOZZLE FLOW

We have seen that sound waves are caused by infinitesimally small pressure disturbances, and they travel through a medium at the speed of sound. We have also seen that for some values of back pressure values, abrupt changes in fluid properties occur in a very thin section of a converging-diverging nozzle under supersonic flow conditions, creating a **shock wave**. It is of interest to study the conditions under which shock waves develop and how they affect the flow. We limit our discussion to shock waves which occur in a plane normal to the direction of flow, called **normal shock waves**. The flow process through the shock wave is highly irreversible and cannot be approximated as being isentropic.

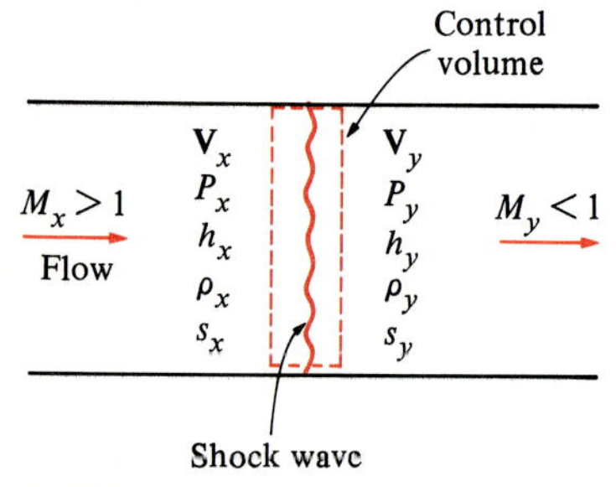

FIGURE 16-28

Control volume for flow across a shock wave.

Next we develop some relations which relate the flow properties before the shock to those after it. We do this by applying the conservation of mass, momentum, and energy relations as well as some property relations to a stationary control volume which contains the shock, as shown in Fig. 16-28. The normal shock waves are extremely thin, so the entrance and exit flow areas for the control volume are equal.

We assume steady flow with no heat and work interactions, and no potential energy changes. Denoting the properties upstream of the shock by the subscript x and those downstream of the shock by y, we have the following:

Conservation of mass:

$$\rho_x A \mathbf{V}_x = \rho_y A \mathbf{V}_y$$

or

$$\rho_x \mathbf{V}_x = \rho_y \mathbf{V}_y \tag{16-29}$$

Conservation of energy:

$$h_x + \frac{\mathbf{V}_x^2}{2} = h_y + \frac{\mathbf{V}_y^2}{2} \tag{16-30}$$

or

$$h_{0_x} = h_{0_y} \tag{16-31}$$

Conservation of momentum: Rearranging Eq. 16-14 and integrating yield

$$A(P_x - P_y) = \dot{m}(\mathbf{V}_y - \mathbf{V}_x) \tag{16-32}$$

The second law:

$$s_y - s_x \geqslant 0 \tag{16-33}$$

We can combine the conservation of mass and energy relations into a single equation and plot it on the h-s diagram, using property relations. The resultant curve is called the **Fanno line**, and it is the locus of states

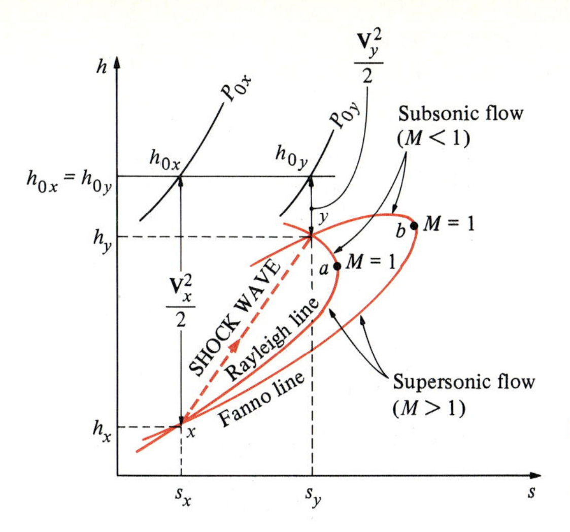

FIGURE 16-29

The *h-s* diagram for flow across a normal shock.

which have the same value of stagnation enthalpy and mass flux (mass flow per unit flow area). Likewise, combining the conservation of mass and momentum equations into a single equation and plotting it on the *h-s* diagram yield a curve called the **Rayleigh line**. Both these lines are shown on the *h-s* diagram in Fig. 16-29. As proved later in Example 16-10, the points of maximum entropy on these lines (points *a* and *b*) correspond to $M = 1$. The state on the upper part of each curve is subsonic and on the lower part supersonic.

The Fanno and Rayleigh lines intersect at two points (points *x* and *y*) which represent the two states at which all three conservation equations are satisfied. One of these (state *x*) corresponds to the state before the shock, and the other (state *y*) corresponds to the state after the shock. Note that the flow is supersonic before the shock and subsonic afterward. Therefore, the flow must change from supersonic to subsonic if a shock is to occur. The larger the Mach number before the shock, the stronger the shock will be. In the limiting case of $M = 1$, the shock wave simply becomes the sonic wave. Notice from Fig. 16-29 that $s_y > s_x$. This is expected since the flow through the shock is adiabatic and irreversible.

The conservation of energy principle (Eq. 16-31) requires that the stagnation enthalpy remain constant across the shock: $h_{0x} = h_{0y}$. For ideal gases $h = h(T)$, and thus

$$T_{0x} = T_{0y} \tag{16-34}$$

That is, the stagnation temperature of an ideal gas also remains constant across the shock. Note, however, that the stagnation pressure decreases across the shock because of the irreversibilities.

We now wish to develop relations between various properties before and after the shock for an ideal gas with constant specific heats. A relation for the ratio of the static temperatures T_y/T_x can be obtained by applying

Eq. 16-18 twice:

$$\frac{T_{0x}}{T_x} = 1 + \left(\frac{k-1}{2}\right) M_x^2 \qquad \text{and} \qquad \frac{T_{0y}}{T_y} = 1 + \left(\frac{k-1}{2}\right) M_y^2$$

Dividing the first equation by the second one and noting that $T_{0x} = T_{0y}$, we have

$$\frac{T_y}{T_x} = \frac{1 + M_x^2(k-1)/2}{1 + M_y^2(k-1)/2} \tag{16-35}$$

From the ideal-gas equation of state,

$$\rho_x = \frac{P_x}{RT_x} \qquad \text{and} \qquad \rho_y = \frac{P_y}{RT_y}$$

Substituting these into the conservation of mass relation $\rho_x \mathbf{V}_x = \rho_y \mathbf{V}_y$ and noting that $M = \mathbf{V}/C$ and $C = \sqrt{kRT}$, we have

$$\frac{T_y}{T_x} = \frac{P_y \mathbf{V}_y}{P_x \mathbf{V}_x} = \frac{P_y M_y C_y}{P_x M_x C_x} = \frac{P_y M_y \sqrt{T_y}}{P_x M_x \sqrt{T_x}} = \left(\frac{P_y}{P_x}\right)^2 \left(\frac{M_y}{M_x}\right)^2 \tag{16-36}$$

Combining Eqs. 16-35 and 16-36 gives the pressure ratio across the shock:

$$\frac{P_y}{P_x} = \frac{M_x\sqrt{1 + M_x^2(k-1)/2}}{M_y\sqrt{1 + M_y^2(k-1)/2}} \tag{16-37}$$

Equation 16-37 is a combination of the conservation of mass and energy equations; thus it is also the equation of the Fanno line for an ideal gas with constant specific heats. A similar relation for the Rayleigh line can be obtained by combining the conservation of mass and momentum equations. From Eq. 16-32,

$$P_x - P_y = \frac{\dot{m}}{A}(\mathbf{V}_y - \mathbf{V}_x) = \rho_y \mathbf{V}_y^2 - \rho_x \mathbf{V}_x^2$$

But $\qquad \rho \mathbf{V}^2 = \left(\frac{P}{RT}\right)(MC)^2 = \left(\frac{P}{RT}\right)(M\sqrt{kRT})^2 = PkM^2$

Thus $\qquad P_x(1 + kM_x^2) = P_y(1 + kM_y^2)$

or

$$\frac{P_y}{P_x} = \frac{1 + kM_x^2}{1 + kM_y^2} \tag{16-38}$$

Combining Eqs. 16-37 and 16-38 yields

$$M_y^2 = \frac{M_x^2 + 2/(k-1)}{2M_x^2 k/(k-1) - 1} \tag{16-39}$$

This represents the intersections of Fanno and Rayleigh lines and relates the Mach number upstream of the shock to that downstream of the shock.

The occurrence of shock waves is not limited to hypersonic nozzles only. Shock waves also occur ahead of supersonic aircraft or bullets, as shown in Fig. 16-30. The shock waves can be understood better if the aircraft is assumed stationary and the air to be moving toward the aircraft

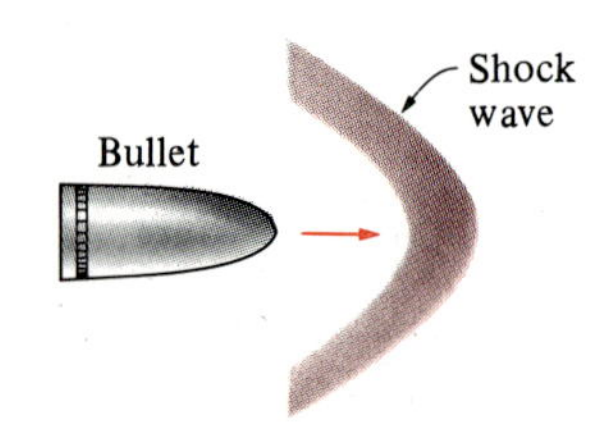

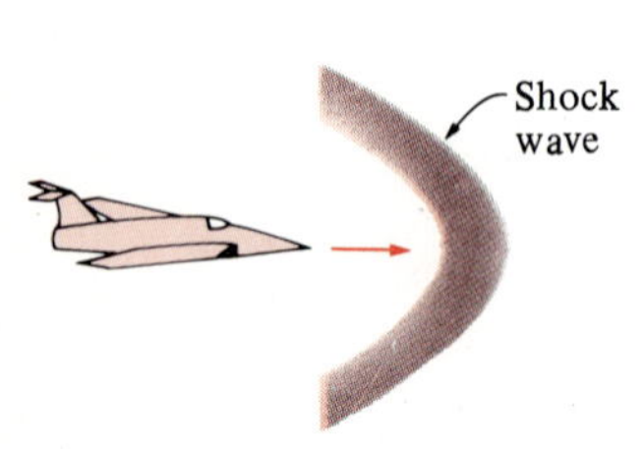

FIGURE 16-30
Shock wave in front of a supersonic aircraft and a bullet.

at the speed of the aircraft. Obviously the air will undergo a stagnation process toward the nose of aircraft and will stop when it reaches the nose. The shock will occur at a location where the air is suddenly decelerated to a subsonic velocity. It creates a sonic boom we all are familiar with. This phenomenon is also observed at the engine inlet of a supersonic aircraft, where the air passes through a shock and decelerates to subsonic velocities before entering the diffuser of the engine.

Various flow property ratios across the shock are listed in Table A-35 for an ideal gas with $k = 1.4$. Inspection of this table reveals that M_y (the Mach number after the shock) is always less than 1 and that the larger the supersonic Mach number before the shock, the smaller the subsonic Mach number after the shock. Also, we see that the static pressure, temperature, and density all increase after the shock while the stagnation pressure decreases.

The entropy change across the shock is obtained by applying the entropy-change equation for an ideal gas, Eq. 6-38, across the shock:

$$s_y - s_x = C_p \ln \frac{T_y}{T_x} - R \ln \frac{P_y}{P_x} \tag{16-40}$$

which can be expressed in terms of k, R, and M_x by using the relations developed earlier in this section. A plot of nondimensional entropy change across the normal shock $(s_y - s_x)/R$ versus M_x is shown in Fig. 16-31. Since the flow across the shock is adiabatic and irreversible, the second law requires that the entropy increase across the shock wave. Thus, a shock wave cannot exist for values of M_x less than unity where the entropy change would be negative. For adiabatic flows, shock waves can only exist for supersonic flows, $M_x > 1$.

EXAMPLE 16-10

Show that the point of maximum entropy on the Fanno line (point *a*) for the adiabatic steady flow of a fluid in a duct corresponds to the sonic velocity, $M = 1$.

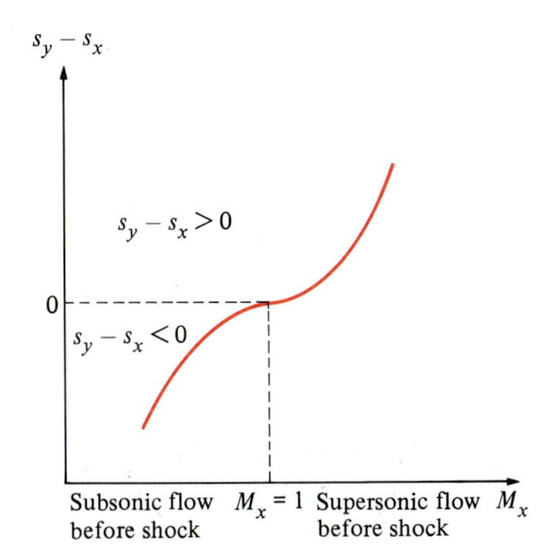

FIGURE 16-31
Entropy change across the normal shock.

Solution In the absence of any heat and work interactions and potential energy changes, the steady-flow energy equation reduces to

$$h + \frac{\mathbf{V}^2}{2} = \text{constant}$$

Differentiating yields

$$dh + \mathbf{V}\, d\mathbf{V} = 0$$

For a constant-flow area, the steady-flow continuity (conservation of mass) equation can be expressed as

$$\rho \mathbf{V} = \text{constant}$$

Differentiating, we have

$$\rho\, d\mathbf{V} + \mathbf{V}\, d\rho = 0$$

Solving for $d\mathbf{V}$ gives

$$d\mathbf{V} = -\mathbf{V}\frac{d\rho}{\rho}$$

Combining this with the energy equation, we have

$$dh - \mathbf{V}^2 \frac{d\rho}{\rho} = 0$$

which is the equation for the Fanno line in differential form. At point a (the point of maximum entropy) $ds = 0$. Then from the second $T\,ds$ relation ($T\,ds = dh - v\,dP$) we have $dh = v\,dP = dP/\rho$. Substituting yields

$$\frac{dP}{\rho} - \mathbf{V}^2 \frac{d\rho}{\rho} = 0 \qquad \text{at } s = \text{const.}$$

Solving for $\mathbf{V}$, we have

$$\mathbf{V} = \left(\frac{\partial P}{\partial \rho}\right)_s^{1/2}$$

which is the relation for the velocity of sound, Eq. 16-9. Thus the proof is complete.

EXAMPLE 16-11

If the air flowing through the converging-diverging nozzle of Example 16-9 experiences a normal shock wave at the nozzle exit plane (Fig. 16-32), determine the following after the shock: (*a*) the stagnation pressure, static pressure, static temperature, and static density; (*b*) the entropy change during shock; (*c*) the exit velocity; and (*d*) the mass flow rate through the nozzle. Assume steady, one-

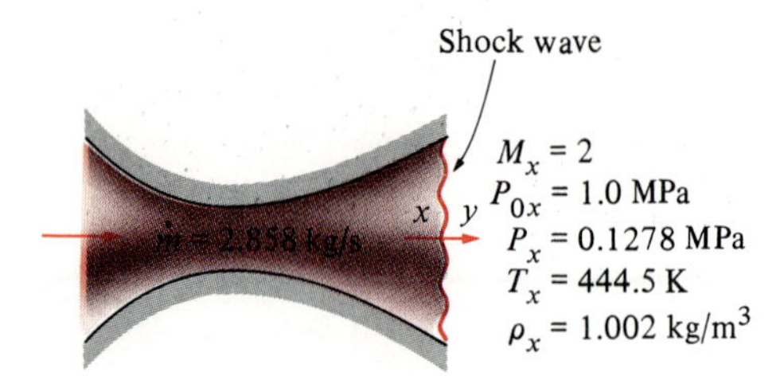

FIGURE 16-32
Schematic for Example 16-11.

dimensional, and isentropic flow with $k = 1.4$ from the nozzle inlet to the shock location.

Solution (*a*) The fluid properties at the exit of the nozzle just before the shock (denoted by subscript *x*) are those evaluated in Example 16-9 at the nozzle exit:

$$P_{0x} = 1.0 \text{ MPa} \qquad P_x = 0.1278 \text{ MPa} \qquad T_x = 444.5 \text{ K} \qquad \rho_x = 1.002 \text{ kg/m}^3$$

The fluid properties after the shock (denoted by subscript *y*) are related to those before the shock through the functions listed in Table A-35. For $M_x = 2.0$, we read

$$M_y = 0.5774 \quad \frac{P_{0y}}{P_{0x}} = 0.7209 \quad \frac{P_y}{P_x} = 4.5000 \quad \frac{T_y}{T_x} = 1.6875 \quad \frac{\rho_y}{\rho_x} = 2.6666$$

Then the stagnation pressure P_{0y}, static pressure P_y, static temperature T_y, and static density ρ_y after the shock are

$$\begin{aligned} P_{0y} &= 0.7209P_{0x} = (0.7209)(1.0 \text{ MPa}) = 0.7209 \text{ MPa} \\ P_y &= 4.5000P_x = (4.5000)(0.1278 \text{ MPa}) = 0.5751 \text{ MPa} \\ T_y &= 1.6875T_x = (1.6875)(444.5 \text{ K}) = 750.1 \text{ K} \\ \rho_y &= 2.6666\rho_x = (2.6666)(1.002 \text{ kg/m}) = 2.672 \text{ kg/m} \end{aligned}$$

(*b*) Assuming constant specific heats and using values from Table A-2*a*, the entropy change across the shock becomes

$$\begin{aligned} s_y - s_x &= C_p \ln \frac{T_y}{T_x} - R \ln \frac{P_y}{P_x} \\ &= [1.005 \text{ kJ/(kg·K)}] \ln (1.6875) - [0.287 \text{ kJ/(kg·K)}] \ln (4.5000) \\ &= 0.0942 \text{ kJ/(kg·K)} \end{aligned}$$

Thus the entropy of the air increases as it experiences a normal shock, which is highly irreversible.

(*c*) The air velocity after the shock can be determined from $\mathbf{V}_y = M_y C_y$, where C_y is the velocity of sound at the exit conditions after the shock:

$$\begin{aligned} \mathbf{V}_y &= M_y C_y = M_y \sqrt{kRT_y} \\ &= (0.5774)\sqrt{(1.4)[0.287 \text{ kJ/(kg·K)}](750.1 \text{ K}) \left(\frac{1000 \text{ m}^2/\text{s}^2}{1 \text{ kJ/kg}}\right)} \\ &= 317.0 \text{ m/s} \end{aligned}$$

(*d*) The mass flow rate through a converging-diverging nozzle with sonic conditions at the throat is not affected by the presence of shock waves in the nozzle. Therefore, the mass flow rate in this case is the same as that determined in Example 16-9:

$$\dot{m} = 2.858 \text{ kg/s}$$

This result can easily be verified by using property values at the nozzle exit after the shock.

This example illustrates that the stagnation pressure and velocity decrease while the static pressure, temperature, density, and entropy increase across the shock. The rise in temperature of the fluid downstream of a shock wave is of major concern to the aerospace engineer because it creates serious heat transfer problems on the leading edges of wings and nose cones of space reentry vehicles and the newly proposed hypersonic space planes.

16-6 ■ FLOW THROUGH ACTUAL NOZZLES AND DIFFUSERS

For the most part, in this chapter we have approximated the flow through the nozzles and diffusers as being isentropic, and it is only human to wonder how realistic this approximation is. We can easily satisfy our curiosity by simply measuring the quantities of interest at the exit of the device and comparing them to those calculated under the isentropic assumption for the same inlet conditions. In Chap. 6 we have done this by comparing the kinetic energies at the nozzle exit, and we defined the **nozzle efficiency** as

$$\eta_N = \frac{\mathbf{V}_2^2/2}{\mathbf{V}_{2s}^2/2} = \frac{\text{actual kinetic energy at nozzle exit}}{\text{kinetic energy at nozzle exit for isentropic flow from same inlet state to same exit pressure}} \quad (16\text{-}41)$$

Writing the energy equation for both the actual and the isentropic processes and using the definition of stagnation enthalpy, we can express the nozzle efficiency as

$$\eta_N = \frac{h_{01} - h_2}{h_{01} - h_{2s}} \quad (16\text{-}42)$$

where h_{01} is the stagnation enthalpy of the fluid at the nozzle inlet, h_2 is the actual enthalpy at the nozzle exit, and h_{2s} is the exit enthalpy under isentropic conditions for the same exit pressure. The various terms that appear in the η_N relation are illustrated on an h-s diagram in Fig. 16-33.

Nozzle efficiencies range from about 90 to 99 percent, with larger nozzles with straight-line axes having the higher efficiencies. Thus actual nozzles closely resemble isentropic (reversible, adiabatic) conditions. This is not surprising since the fluid flow in actual nozzles is very nearly adiabatic, and the irreversibilities are minimized by carefully streamlining the nozzle contour.

The primary cause of irreversibility in nozzles (and diffusers) is the frictional effects which are mostly confined to a region close to the wall surface called the *boundary layer*. Large nozzles have a higher efficiency because the boundary layer in large nozzles occupies a smaller portion of the total flow volume. Another cause of irreversibility in nozzles and diffusers is flow separation, which induces strong turbulence near the nozzle wall. Flow separation occurs when the flow area increases faster than the fluid expands, a situation that must be avoided in the design of a nozzle or a diffuser. It is not uncommon for converging-diverging nozzles

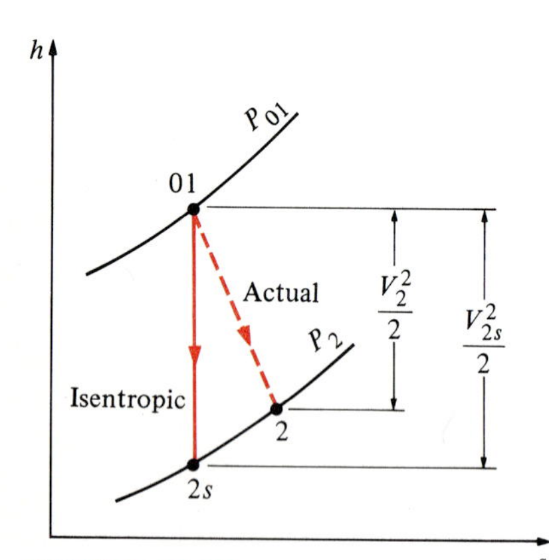

FIGURE 16-33

Isentropic and actual (irreversible) flow in a nozzle between the same inlet state and the exit pressure.

to have efficiencies of less than 90 percent because of the increased irreversibilities in the diverging section.

A second parameter which is used to express the performance of a nozzle is the **velocity coefficient** C_V, defined as

$$C_V = \frac{\mathbf{V}_2}{\mathbf{V}_{2s}} = \frac{\text{actual velocity at nozzle exit}}{\text{velocity at nozzle exit for isentropic flow from same inlet state to same exit pressure}} \qquad (16\text{-}43)$$

Notice that the velocity coefficient is equal to the square root of the nozzle efficiency:

$$C_V = \sqrt{\eta_N} \qquad (16\text{-}44)$$

Another quantity of interest in nozzle design is the mass flow rate of the fluid through the nozzle, which is also affected by the irreversibilities. The effect of irreversibilities on the mass flow rate is expressed in terms of the **discharge coefficient** C_D, which is defined as

$$C_D = \frac{\dot{m}}{\dot{m}_s}$$

$$= \frac{\text{actual mass flow rate through nozzle}}{\text{mass flow rate through nozzle for isentropic flow from same inlet state to same exit pressure}} \qquad (16\text{-}45)$$

If the nozzle is choked, the mass flow rate for the isentropic case is determined by using the sonic velocity and critical density at the nozzle throat.

Using the relations developed in this chapter, we can show that the critical-pressure ratio for which an actual (irreversible) nozzle is choked is smaller than that for an ideal (isentropic) nozzle. The back pressure required to choke the actual nozzle is also smaller than that required to choke an isentropic nozzle.

EXAMPLE 16-12

Hot combustion gases enter a converging nozzle with a stagnation temperature of 600 K and stagnation pressures of (*a*) 180 kPa and (*b*) 100 kPa. Determine the velocity, temperature, and pressure at the nozzle exit for a nozzle back pressure of 60 kPa and a nozzle efficiency of 95 percent. Assume the combustion gases have the properties of air with constant specific heats and $k = 1.33$.

Solution A schematic of the nozzle and the *T-s* diagram of the process are given in Fig. 16-34. Before we attempt to determine the velocity at the nozzle exit, we need to know whether the nozzle is choked. If it is, the velocity at the nozzle exit will be independent of the back pressure, and it will be equal to the velocity of sound (Eq. 16-13). If the nozzle is not choked, then the exit velocity can be determined from Eq. 16-4. We assume the nozzle is adiabatic, so the stagnation enthalpy of the fluid remains constant throughout the nozzle

$$h_{01} = h_{02}$$

or

$$T_{01} = T_{02} = 600 \text{ K}$$

since the specific heats are assumed to be constant.

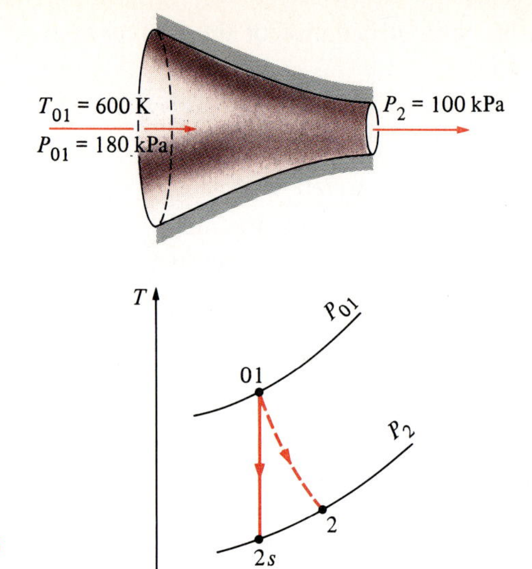

FIGURE 16-34
Schematic and *T-s* diagram for Example 16-12.

We can find out whether the nozzle is choked by calculating the critical pressure P^* at the nozzle exit and comparing it to the specified back pressure P_b. If $P_b < P^*$, then the nozzle is choked; otherwise, it is not.

The critical pressure P^* of a converging nozzle is the actual pressure at the nozzle exit when the sonic velocity is reached at the nozzle exit ($M_2 = 1$). For this condition the actual temperature T_2 and pressure P_2 at the nozzle exit are determined as follows:

From Eq. 16-21 for $M_2 = 1$,

$$\frac{T_2}{T_{02}} = \frac{2}{k+1} \longrightarrow T_2 = T_{02}\left(\frac{2}{k+1}\right) = (600 \text{ K})\left(\frac{2}{1.33+1}\right) = 515.0 \text{ K}$$

For an ideal gas with constant specific heats, the nozzle efficiency can be expressed as

$$\eta_N = \frac{h_{01} - h_2}{h_{01} - h_{2s}} = \frac{C_p(T_{01} - T_2)}{C_p(T_{01} - T_{2s})} = \frac{T_{01} - T_2}{T_{01} - T_{2s}}$$

Thus,

$$0.95 = \frac{600 - 515.0}{600 - T_{2s}}$$

$$T_{2s} = 510.5 \text{ K}$$

Here $T^* = T_2 = 515.0$ K is the actual temperature at the nozzle exit at sonic conditions, and $T_{2s} = 510.5$ K is the exit temperature that would be achieved at sonic conditions if the process were isentropic. The exit pressure for both the actual and the isentropic cases is the same and is determined from

$$\frac{P_2}{P_{01}} = \left(\frac{T_{2s}}{T_{01}}\right)^{k/(k-1)} = \left(\frac{510.5 \text{ K}}{600 \text{ K}}\right)^{1.33/(1.33-1)} = 0.5215$$

$$P^* = P_2 = 0.5215P_{01}$$

(*a*) For an inlet stagnation pressure of $P_{01} = 180$ kPa, the critical pressure is

$$P^* = 0.5215P_{01} = (0.5215)(180 \text{ kPa}) = 93.9 \text{ kPa}$$

which is larger than the back pressure $P_b = 60$ kPa. Therefore, the nozzle is choked in this case, the actual pressure at the nozzle exit is equal to the critical pressure, and the actual temperature at the nozzle exit is equal to the critical temperature:

$$P_2 = P^* = 93.9 \text{ kPa}$$
$$T_2 = T^* = 515.0 \text{ K}$$

Of course, the exit velocity is the sonic velocity at the actual nozzle exit conditions, and it is determined from Eq. 16-4:

$$\mathbf{V}_2 = C_2 = \sqrt{kRT_2} = \sqrt{(1.33)[0.287 \text{ kJ/(kg·K)}](515.0 \text{ K})\left(\frac{1000 \text{ m}^2/\text{s}^2}{1 \text{ kJ/kg}}\right)}$$
$$= 443.4 \text{ m/s}$$

(*b*) For an inlet stagnation pressure of $P_{01} = 100$ kPa, the critical pressure is

$$P^* = 0.5215P_{01} = (0.5215)(100 \text{ kPa}) = 52.15 \text{ kPa}$$

which is less than the back pressure $P_b = 60$ kPa. Therefore, the nozzle is not choked in this case, and the pressure at the nozzle exit is equal to the specified back pressure

$$P_2 = P_b = 60 \text{ kPa}$$

The actual temperature at the nozzle exit is found by first determining the isentropic exit temperature and then using the η_N relation:

$$\frac{T_{2s}}{T_{01}} = \left(\frac{P_2}{P_{01}}\right)^{(k-1)/k} \longrightarrow T_{2s} = (600 \text{ K})\left(\frac{60 \text{ kPa}}{100 \text{ kPa}}\right)^{(1.33-1)/1.33} = 528.6 \text{ K}$$

and

$$\eta_N = \frac{T_{01} - T_2}{T_{01} - T_{2s}}$$

Thus,

$$0.95 = \frac{600 - T_2}{600 - 528.6} \longrightarrow T_2 = 532.2 \text{ K}$$

The actual exit velocity is determined from Eq. 16-4 to be

$$\mathbf{V}_2 = \sqrt{2C_p(T_{01} - T_2)} = \sqrt{2[1.005 \text{ kJ/(kg·K)}][(600 - 532.2) \text{ K}]\left(\frac{1000 \text{ m}^2/\text{s}^2}{1 \text{ kJ/kg}}\right)}$$
$$= 369.2 \text{ m/s}$$

Diffusers are designed to perform the opposite task of a nozzle—to increase the pressure of a fluid by decelerating it. Thus diffusers are judged on their ability to convert the kinetic energy of the fluid to an increase in fluid pressure. The performance of a diffuser is evaluated by comparing it to an ideal diffuser which decelerates the fluid isentropically (an isentropic stagnation process).

You will recall from Sec. 16-1 that the enthalpy of a fluid (and its temperature, if the fluid is an ideal gas) remains constant during an adiabatic stagnation process whether the process is isentropic or not. The pressure of the fluid increases during such a process, and the pressure

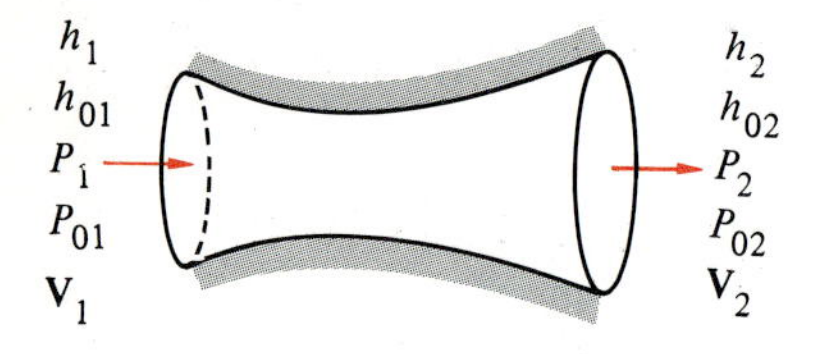

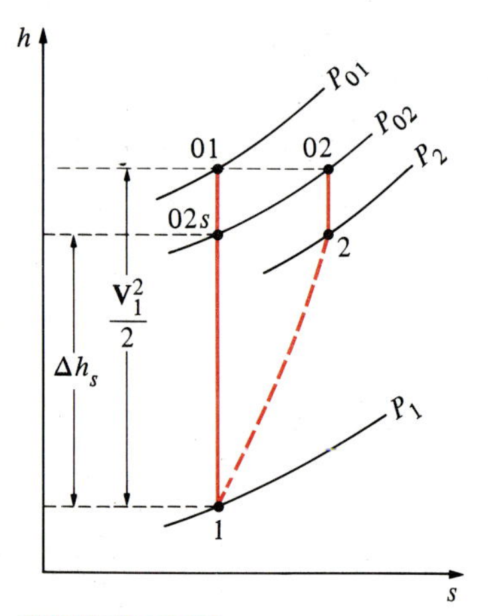

FIGURE 16-35

Schematic and *h-s* diagram for the definition of the diffuser efficiency.

rise is a maximum when the process is isentropic. Thus the isentropic flow through a diffuser can be used as a standard for evaluating the performance of an actual diffuser.

The performance of a diffuser is usually expressed in terms of the **diffuser efficiency** η_D, defined in an awkward way as

$$\eta_D = \frac{\Delta h_s}{\mathbf{V}_1^2/2} = \frac{h_{02s} - h_1}{h_{01} - h_1} \tag{16-46}$$

where $\mathbf{V}_1^2/2$ is the maximum kinetic energy available for converting to pressure rise and Δh_s represents the kinetic energy which can be converted to pressure rise if the fluid is discharged at the actual exit stagnation pressure. These two quantities are identical for an isentropic diffuser since the actual exit stagnation pressure in this case becomes equal to the inlet stagnation pressure, yielding an efficiency of 100 percent.

This definition is better understood with the help of the *h-s* diagram in Fig. 16-35. On this diagram states 1 and 01 are the actual and stagnation states at the nozzle inlet, states 2 and 02 are the actual and stagnation states at the nozzle exit, and state 02*s* is a fictitious exit state which would be attained if the fluid underwent an isentropic process in the diffuser to exit stagnation pressure. The states 2 and 02 become identical when the exit velocity of the diffuser is close to zero.

Diffuser efficiencies vary from about 90 percent to close to 100 percent for subsonic diffusers, and they decrease with increasing Mach numbers.

Another measure of the diffuser's ability to increase the pressure of the fluid is the **pressure recovery factor** F_P, which represents the actual stagnation pressure of a fluid at diffuser exit relative to the maximum possible stagnation pressure. It is defined as

$$F_P = \frac{\text{actual stagnation pressure at diffuser exit}}{\text{isentropic stagnation pressure}} = \frac{P_{02}}{P_{01}} \tag{16-47}$$

For an isentropic diffuser $P_{02} = P_{01}$, and thus $F_P = 1$. It will be less for actual diffusers because the stagnation pressure drops as a result of irreversibilities.

A third parameter frequently used to evaluate diffuser performance is the **pressure rise coefficient** C_{PR}. This parameter is based on the idea that a diffuser is used to increase the static pressure of the flow, and it is defined as the ratio of the actual pressure rise in the diffuser to the pressure rise that would be realized if the process were isentropic:

$$C_{PR} = \frac{\text{actual pressure rise}}{\text{isentropic pressure rise}} = \frac{P_2 - P_1}{P_{01} - P_1} \tag{16-48}$$

The value of the pressure rise coefficient is highly dependent on the character of the flow and the shape of the diffuser, and it is less than 0.8 for most diffusers. Diffusers having highly separated boundary layers experience significant losses in the pressure rise.

EXAMPLE 16-13

Air enters a diffuser with a velocity of 200 m/s, a stagnation pressure of 0.4 MPa, and a stagnation temperature of 500 K. The diffuser exit velocity is 100 m/s. For

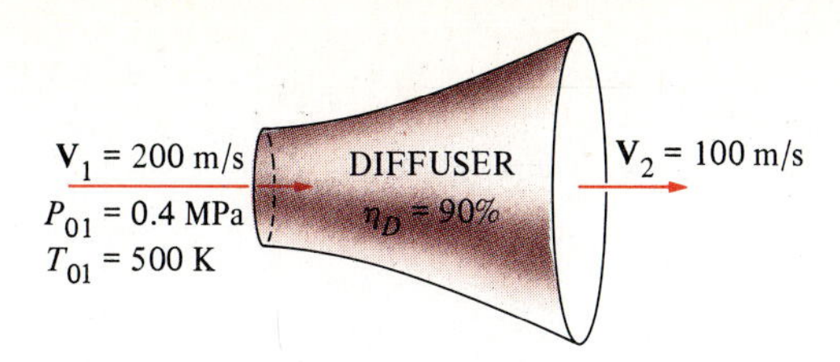

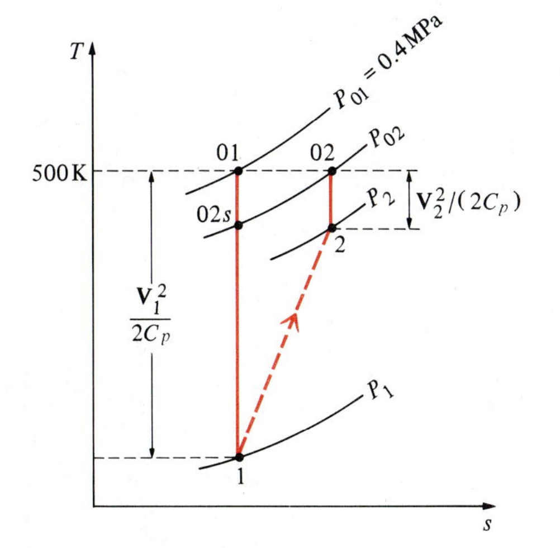

FIGURE 16-36
Schematic and *T-s* diagram for Example 16-13.

a diffuser efficiency of 90 percent, determine (*a*) the pressure rise coefficient and (*b*) the required exit-to-inlet area ratio.

Solution A sketch of the diffuser and the *T-s* diagram of the process are given in Fig. 16-36. We assume air to be an ideal gas with constant specific heats and $k = 1.4$. Assuming the diffuser to be adiabatic, the stagnation enthalpy of the air remains constant as it flows steadily through the diffuser:

$$T_{02} = T_{01} = 500 \text{ K}$$

(*a*) The static temperature at the diffuser inlet is found from Eq. 16-4:

$$T_1 = T_{01} - \frac{\mathbf{V}_1^2}{2C_p} = (500 \text{ K}) - \frac{(200 \text{ m/s})^2}{2[1.005 \text{ kJ/(kg·K)}]}\left(\frac{1 \text{ kJ/kg}}{1000 \text{ m}^2/\text{s}^2}\right) = 480.1 \text{ K}$$

Then the static inlet pressure is

$$P_1 = P_{01}\left(\frac{T_1}{T_{01}}\right)^{k/(k-1)} = (400 \text{ kPa})\left(\frac{480.1 \text{ K}}{500 \text{ K}}\right)^{1.4/(1.4-1)} = 347.0 \text{ kPa}$$

For an ideal gas with constant specific heats, the diffuser efficiency relation can be expressed as

$$\eta_D = \frac{h_{02s} - h_1}{h_{01} - h_1} = \frac{T_{02s} - T_1}{T_{01} - T_1}$$

Then $$0.90 = \frac{T_{02s} - 480.1}{500 - 480.1} \longrightarrow T_{02s} = 498.0 \text{ K}$$

Noting that $P_{02} = P_{02s}$, we see that the isentropic relation between states 1 and 02*s* gives

$$P_{02} = P_{02s} = P_1 \left(\frac{T_{02s}}{T_1}\right)^{k/(k-1)} = (347.0 \text{ kPa}) \left(\frac{498.0 \text{ K}}{480.1 \text{ K}}\right)^{1.4/(1.4-1)}$$
$$= 394.4 \text{ kPa}$$

Thus,

$$T_2 = T_{02} - \frac{\mathbf{V}_2^2}{2C_p} = 500 \text{ K} - \frac{(100 \text{ m/s})^2}{2[1.005 \text{ kJ/(kg·K)}]} \left(\frac{1 \text{ kJ/kg}}{1000 \text{ m}^2/\text{s}^2}\right) = 495.0 \text{ K}$$

Then the static exit pressure becomes

$$P_2 = P_{02} \left(\frac{T_2}{T_{02}}\right)^{k/(k-1)} = (394.4 \text{ kPa}) \left(\frac{495.0 \text{ K}}{500 \text{ K}}\right)^{1.4/(1.4-1)} = 380.8 \text{ kPa}$$

By substituting these values into Eq. 16-48, the pressure rise coefficient is

$$C_{PR} = \frac{P_2 - P_1}{P_{01} - P_1} = \frac{380.8 - 347.0}{400 - 347.0} = 0.638$$

(*b*) The exit-to-inlet area ratio A_2/A_1 is obtained by applying the steady-flow conservation of mass relation and using the ideal-gas relation $\rho = P/(RT)$:

$$\dot{m} = \rho_1 A_1 \mathbf{V}_1 = \rho_2 A_2 \mathbf{V}_2$$

$$\frac{A_2}{A_1} = \frac{\rho_1 \mathbf{V}_1}{\rho_2 \mathbf{V}_2} = \frac{[P_1/(RT_1)]\mathbf{V}_1}{[P_2/(RT_2)]\mathbf{V}_2} = \frac{P_1 T_2 \mathbf{V}_1}{P_2 T_1 \mathbf{V}_2}$$
$$= \frac{(347.0 \text{ kPa})(495.0 \text{ K})(200 \text{ m/s})}{(380.8 \text{ kPa})(480.1 \text{ K})(100 \text{ m/s})}$$
$$= 1.879$$

16-7 ■ STEAM NOZZLES

We have seen in Chap. 2 that water vapor at moderate or high pressures deviates considerably from ideal-gas behavior, and thus most of the relations developed in this chapter are not applicable to the flow of steam through the nozzles or blade passages encountered in steam turbines. Given that the steam properties such as enthalpy are functions of pressure as well as temperature and that no simple property relations exist, an accurate analysis of steam flow through the nozzles is no easy matter. Often it becomes necessary to use steam tables, an *h-s* diagram, or a computer program for the properties of steam.

A further complication in the expansion of steam through nozzles occurs as the steam expands into the saturation region, as shown in Fig. 16-37. As the steam expands in the nozzle, its pressure and temperature drop, and ordinarily one would expect the steam to start condensing when it strikes the saturation line. But this is not always what happens. Owing to the high velocities, the residence time of the steam in the nozzle is small, and there may not be sufficient time for the necessary heat transfer and the formation of liquid droplets. Consequently, the condensation of the steam may be delayed for a little while. This phenomenon is known as **supersaturation**, and the steam which exists in the wet region without

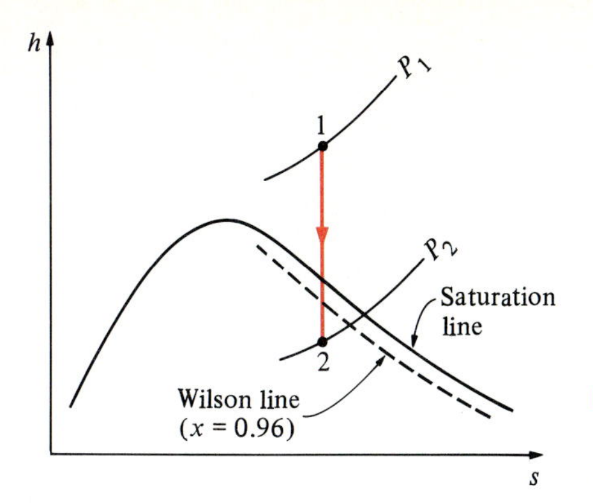

FIGURE 16-37
The h-s diagram for the isentropic expansion of steam in a nozzle.

containing any liquid is called **supersaturated steam**. Supersaturation states are nonequilibrium (or metastable) states.

During the expansion process, the steam reaches a temperature lower than that normally required for the condensation process to begin. Once the temperature drops a sufficient amount below the saturation temperature corresponding to the local pressure, groups of steam moisture droplets of sufficient size are formed, and condensation occurs rapidly. The locus of points where condensation will take place regardless of the initial temperature and pressure at the nozzle entrance is called the **Wilson line**. The Wilson line lies between the 4 and 5 percent moisture curves in the saturation region on the h-s diagram for steam, and it is often approximated by the 4 percent moisture line. Therefore, steam flowing through a high-velocity nozzle is assumed to begin condensation when the 4 percent moisture line is crossed.

The critical-pressure ratio P^*/P_0 for steam depends on the nozzle inlet state as well as on whether the steam is superheated or saturated at the nozzle inlet. However, the ideal-gas relation for the critical-pressure ratio, Eq. 16-22, gives reasonably good results over a wide range of inlet states. As indicated in Table 16-2, the specific heat ratio of superheated steam is approximated as $k = 1.3$. Then the critical-pressure ratio becomes

$$\frac{P^*}{P_0} = \left(\frac{2}{k+1}\right)^{k/(k-1)} = 0.546$$

When steam enters the nozzle as a saturated vapor instead of superheated vapor (a common occurrence in the lower stages of a steam turbine), the critical-pressure ratio is taken to be 0.576, which corresponds to a specific heat ratio of $k = 1.14$.

EXAMPLE 16-14

Steam enters a converging-diverging nozzle at 2 MPa and 400°C with a negligible velocity and a mass flow rate of 2.5 kg/s, and it exits at a pressure of 300 kPa. The flow is isentropic between the nozzle entrance and throat, and the overall nozzle efficiency is 93 percent. Determine (*a*) the throat and exit areas and (*b*) the Mach number at the throat and the nozzle exit.

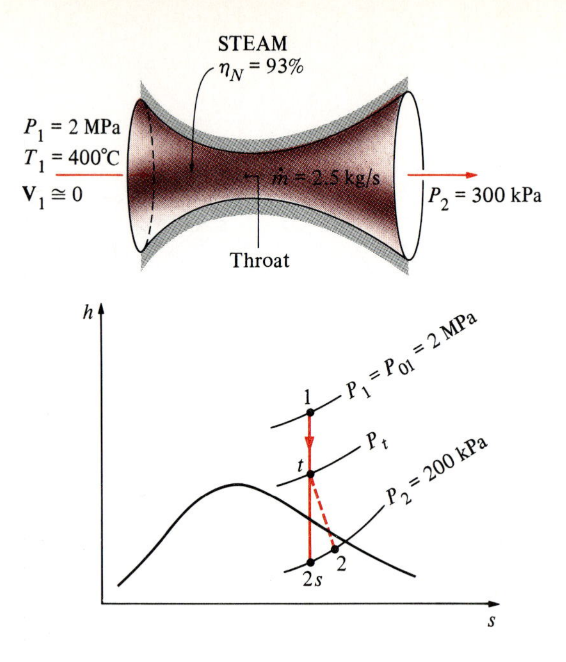

FIGURE 16-38

Schematic and h-s diagram for Example 16-14.

Solution Let the entrance, throat, and exit states be states 1, t, and 2, respectively, as illustrated in Fig. 16-38. Notice that the flow is isentropic between the inlet and the throat and is adiabatic and irreversible between the throat and the exit.

(a) Since the inlet velocity is negligible, the inlet stagnation and static states are identical. The ratio of the exit-to-inlet stagnation pressure is

$$\frac{P_2}{P_{01}} = \frac{300 \text{ kPa}}{2000 \text{ kPa}} = 0.15$$

It is much smaller than critical-pressure ratio, which is taken to be $P^*/P_{01} = 0.546$ since the steam is superheated at the nozzle inlet. Therefore, the flow surely is supersonic at the exit. Then the velocity at the throat is the sonic velocity, and the throat pressure is

$$P_t = 0.546P_{01} = (0.546)(2 \text{ MPa}) = 1.09 \text{ MPa}$$

At the inlet

$$\left.\begin{array}{l} P_1 = P_{01} = 2 \text{ MPa} \\ T_1 = T_{01} = 400°\text{C} \end{array}\right\} \quad \begin{array}{l} h_1 = h_{01} = 3247.6 \text{ kJ/kg} \\ s_1 = s_t = s_{2_s} = 7.1271 \text{ kJ/(kg·K)} \end{array}$$

Also at the throat,

$$\left.\begin{array}{l} P_t = 1.09 \text{ MPa} \\ s_t = 7.1271 \text{ kJ/(kg·K)} \end{array}\right\} \quad \begin{array}{l} h_t = 3076.1 \text{ kJ/kg} \\ v_t = 0.2420 \text{ m}^3\text{/kg} \end{array}$$

Then the throat velocity is determined from Eq. 16-3:

$$\mathbf{V}_t = \sqrt{2(h_{01} - h_t)} = \sqrt{[2(3247.6 - 3076.1) \text{ kJ/kg}]\left(\frac{1000 \text{ m}^2/\text{s}^2}{1 \text{ kJ/kg}}\right)}$$

$$= 585.7 \text{ m/s}$$

The flow area at the throat is determined from the mass flow rate relation:

$$A_t = \frac{\dot{m}v_t}{\mathbf{V}_t} = \frac{(2.5\text{ kg/s})(0.2420\text{ m}^3\text{/kg})}{585.7\text{ m/s}} = 10.33 \times 10^{-4}\text{ m}^2$$
$$= 10.33\text{ cm}^2$$

At state 2s,

$$\left.\begin{aligned} P_{2s} &= P_2 = 300\text{ kPa} \\ s_{2s} &= s_1 = 7.1271\text{ kJ/(kg·K)} \end{aligned}\right\} \quad h_{2s} = 2781.9\text{ kJ/kg}$$

The enthalpy of steam at the actual exit state is determined from

$$\eta_N = \frac{h_{01} - h_2}{h_{01} - h_{2s}}$$

$$0.93 = \frac{3247.6 - h_2}{3247.6 - 2781.9} \longrightarrow h_2 = 2814.5\text{ kJ/kg}$$

Therefore,

$$\left.\begin{aligned} P_2 &= 300\text{ kPa} \\ h_2 &= 2814.5\text{ kJ/kg} \end{aligned}\right\} \quad \begin{aligned} v_2 &= 0.6764\text{ m}^3\text{/kg} \\ s_2 &= 7.2009\text{ kJ/(kg·K)} \end{aligned}$$

The exit velocity is then

$$\mathbf{V}_2 = \sqrt{2(h_{01} - h_t)} = \sqrt{[2(3247.6 - 2814.5)\text{ kJ/kg}]\left(\frac{1000\text{ m}^2\text{/s}^2}{1\text{ kJ/kg}}\right)}$$
$$= 930.7\text{ m/s}$$

And the exit area becomes

$$A_2 = \frac{\dot{m}v_2}{\mathbf{V}_2} = \frac{(2.5\text{ kg/s})(0.6764\text{ m}^3\text{/kg})}{930.7\text{ m/s}} = 18.17 \times 10^{-4}\text{ m}^2 = 18.17\text{ cm}^2$$

(*b*) The velocity of sound and the Mach numbers at the throat and the exit of the nozzle are determined by following the procedure outlined in Example 16-5 and replacing differential quantities with differences:

$$C = \left(\frac{\partial P}{\partial \rho}\right)_s^{1/2} \cong \left[\frac{\Delta P}{\Delta(1/v)}\right]_s^{1/2}$$

The velocity of sound at the throat is determined by evaluating the specific volume at $s_t = 7.1271$ kJ/(kg·K) and at pressures of 1.115 and 1.065 MPa ($P_t \pm 25$ kPa):

$$C = \sqrt{\frac{(1115 - 1065)\text{ kPa}}{(1/0.2378 - 1/0.2464)\text{ kg/m}^3}\left(\frac{1000\text{ m}^2\text{/s}^2}{1\text{ kPa·m}^3}\right)} = 583.7\text{ m/s}$$

The Mach number at the throat is determined from Eq. 16-12:

$$M = \frac{\mathbf{V}}{C} = \frac{585.7\text{ m/s}}{583.7\text{ m/s}} = 1.003$$

Thus the flow at the throat is sonic, as expected. The slight deviation of the Mach number from unity is due to replacing the derivatives by differences.

The velocity of sound and the Mach number at the nozzle exit are determined by evaluating the specific volume at $s_2 = 7.2009$ kJ/(kg·K) and at pressures of 325 and 275 kPa ($P_2 \pm 25$ kPa):

$$C = \sqrt{\frac{(325 - 275)\text{ kPa}}{(1/0.6363 - 1/0.7228)\text{ kg/m}^3}\left(\frac{1000\text{ m}^2\text{/s}^2}{1\text{ kPa·m}^3\text{/kg}}\right)} = 515.6\text{ m/s}$$

and $$M = \frac{\mathbf{V}}{C} = \frac{930.7 \text{ m/s}}{515.7 \text{ m/s}} = 1.805$$

Thus the flow of steam at the nozzle exit is supersonic.

16-8 ■ SUMMARY

In this chapter the thermodynamic aspects of high-speed fluid flow are examined. For high-speed flows, it is convenient to combine the enthalpy and the kinetic energy of the fluid into a single term called *stagnation* (or *total*) *enthalpy* h_0, defined as

$$h_0 = h + \frac{\mathbf{V}^2}{2} \qquad \text{(kJ/kg)} \tag{16-1}$$

The properties of a fluid at the stagnation state are called *stagnation properties* and are indicated by the subscript zero. The *stagnation temperature* of an ideal gas with constant specific heats is

$$T_0 = T + \frac{\mathbf{V}^2}{2C_p} \tag{16-5}$$

which represents the temperature an ideal gas will attain when it is brought to rest adiabatically.

The (isentropic) stagnation properties of an ideal gas are related to the static properties of the fluid by

$$\frac{P_0}{P} = \left(\frac{T_0}{T}\right)^{k/(k-1)} \tag{16-6}$$

$$\frac{\rho_0}{\rho} = \left(\frac{T_0}{T}\right)^{1/(k-1)} \tag{16-7}$$

When stagnation enthalpies are used, the conservation of energy equation for a single-stream, steady-flow device can be expressed as

$$q - w = (h_{02} - h_{01}) + g(z_2 - z_1) \tag{16-8}$$

where h_{01} and h_{02} are the stagnation enthalpies at states 1 and 2, respectively. The velocity at which an infinitesimally small pressure wave travels through a medium is the *velocity of sound* (or the *sonic velocity*). It is expressed as

$$C^2 = \left(\frac{\partial P}{\partial \rho}\right)_s \tag{16-9}$$

For an ideal gas it becomes

$$C = \sqrt{kRT} \tag{16-11}$$

The *Mach number* is the ratio of the actual velocity of the fluid to the velocity of sound at the same state:

$$M = \frac{\mathbf{V}}{C} \tag{16-12}$$

The flow is called *sonic* when $M = 1$, *subsonic* when $M < 1$, *supersonic* when $M > 1$, *hypersonic* when $M >> 1$, and *transonic* when $M \cong 1$.

The nozzles whose flow area decreases in the flow direction are called *converging* nozzles. Nozzles whose flow area first decreases and then increases are called *converging-diverging nozzles*. The location of smallest flow area of a nozzle is called the *throat*. The highest velocity to which a fluid can be accelerated in a convergent nozzle is the sonic velocity. Accelerating a fluid to supersonic velocities is only possible in converging-diverging nozzles. In all supersonic converging-diverging nozzles, the flow velocity at the throat is the velocity of sound.

The ratios of the stagnation to static properties for ideal gases with constant specific heats can be expressed in terms of the Mach number as

$$\frac{T_0}{T} = 1 + \left(\frac{k-1}{2}\right) M^2 \tag{16-18}$$

$$\frac{P_0}{P} = \left[1 + \left(\frac{k-1}{2}\right) M^2\right]^{k/(k-1)} \tag{16-19}$$

$$\frac{\rho_0}{\rho} = \left[1 + \left(\frac{k-1}{2}\right) M^2\right]^{1/(k-1)} \tag{16-20}$$

When $M = 1$, the resulting static-to-stagnation property ratios for the temperature, pressure, and density are called *critical ratios* and are denoted by the superscript asterisk:

$$\frac{T^*}{T_0} = \frac{2}{k+1} \tag{16-21}$$

$$\frac{P^*}{P_0} = \left(\frac{2}{k+1}\right)^{k/(k-1)} \tag{16-22}$$

$$\frac{\rho^*}{\rho_0} = \left(\frac{2}{k+1}\right)^{1/(k-1)} \tag{16-23}$$

The pressure outside the exit plane of a nozzle is called the *back pressure*. For all back pressures lower than P^*, the pressure at the exit plane of the converging nozzle is equal to P^*, the Mach number at the exit plane is unity, and the mass flow rate is the maximum (or choked) flow rate.

Under steady-flow conditions, the mass flow rate through the nozzle is constant and can be expressed as

$$\dot{m} = \frac{AMP_0\sqrt{k/(RT_0)}}{[1 + (k-1)M^2/2]^{(k+1)/[2(k-1)]}} \tag{16-24}$$

The variation of flow area A through the nozzle relative to the throat area

A^* for the same mass flow rate and stagnation properties of a particular ideal gas is

$$\frac{A}{A^*} = \frac{1}{M}\left[\left(\frac{2}{k+1}\right)\left(1 + \frac{k-1}{2}M^2\right)\right]^{(k+1)/[2(k-1)]} \tag{16-26}$$

The parameter M^* is defined as the ratio of the local velocity to the velocity of sound at the throat ($M = 1$):

$$M^* = \frac{\mathbf{V}}{C^*} \tag{16-27}$$

It can also be expressed as

$$M^* = M\sqrt{\frac{k+1}{2 + (k-1)M^2}} \tag{16-28}$$

In some range of back pressure, the fluid which achieved a sonic velocity at the throat of a converging-diverging nozzle and is accelerating to supersonic velocities in the diverging section experiences a *normal shock,* which causes a sudden rise in pressure and temperature, and a sudden drop in velocity to subsonic levels. Flow through the shock is highly irreversible, and thus it cannot be approximated as isentropic. The properties of an ideal gas with constant specific heats before (subscript x) and after (subscript y) a shock are related by

$$T_{0x} = T_{0y} \tag{16-34}$$

$$\frac{T_y}{T_x} = \frac{1 + M_x^2(k-1)/2}{1 + M_y^2(k-1)/2} \tag{16-35}$$

$$\frac{P_y}{P_x} = \frac{M_x\sqrt{1 + M_x^2(k-1)/2}}{M_y\sqrt{1 + M_y^2(k-1)/2}} \tag{16-37}$$

$$M_y^2 = \frac{M_x^2 + 2/(k-1)}{2M_x^2k/(k-1) - 1} \tag{16-39}$$

The entropy change across the shock is obtained by applying the entropy-change equation for an ideal gas across the shock:

$$s_y - s_x = C_p \ln\frac{T_y}{T_x} - R\ln\frac{P_y}{P_x} \tag{16-40}$$

The deviation of actual nozzles from isentropic ones is expressed in terms of the *nozzle efficiency* η_N, *nozzle velocity coefficient* C_V, and the *coefficient of discharge* C_D, which are defined as

$$\eta_N = \frac{\mathbf{V}_2^2/2}{\mathbf{V}_{2s}^2/2} = \frac{\text{actual kinetic energy at nozzle exit}}{\text{kinetic energy at nozzle exit for isentropic flow from same inlet state to same exit pressure}} \tag{16-41}$$

$$= \frac{h_{01} - h_2}{h_{01} - h_{2s}} \tag{16-42}$$

$$C_V = \frac{\mathbf{V}_2}{\mathbf{V}_{2s}}$$

$$= \frac{\text{actual velocity at nozzle exit}}{\text{velocity at nozzle exit for isentropic flow from same inlet state to same exit pressure}} \tag{16-43}$$

$$= \sqrt{\eta_N} \tag{16-44}$$

$$C_D = \frac{\dot{m}}{\dot{m}_s}$$

$$= \frac{\text{actual mass flow rate through nozzle}}{\text{mass flow rate through nozzle for isentropic flow from same inlet state to same exit pressure}} \tag{16-45}$$

where h_{01} is the stagnation enthalpy of the fluid at the nozzle inlet, h_2 is the actual enthalpy at the nozzle exit, and h_{2s} is the exit enthalpy under isentropic conditions for the same exit pressure.

The performance of a diffuser is expressed in terms of the *diffuser efficiency* η_D, the *pressure recovery factor* F_P, and the *pressure rise coefficient* C_{PR}. They are defined as

$$\eta_D = \frac{\Delta h_s}{\mathbf{V}_1^2/2} = \frac{h_{02s} - h_1}{h_{01} - h_1} \tag{16-46}$$

$$F_P = \frac{\text{actual stagnation pressure at diffuser exit}}{\text{isentropic stagnation pressure}} = \frac{P_{02}}{P_{01}} \tag{16-47}$$

$$C_{PR} = \frac{\text{actual pressure rise}}{\text{isentropic pressure rise}} = \frac{P_2 - P_1}{P_{01} - P_1} \tag{16-48}$$

Steam often deviates considerably from ideal-gas behavior, and no simple property relations are available for it. Thus it is often necessary to use steam tables instead of ideal-gas relations. The critical-pressure ratio of steam is often taken to be 0.546, which corresponds to a specific heat ratio of $k = 1.3$ for superheated steam. At high velocities, steam does not start condensing when it encounters the saturation line, and it exists as a *supersaturated substance*. Supersaturation states are non-equilibrium (or metastable) states, and care should be exercised in dealing with them.

REFERENCES AND SUGGESTED READING

1 H. Cohen, G. F. C. Rogers, and H. I. H. Saravanamuttoo, *Gas Turbine Theory,* 3d ed., Wiley, New York, 1987.

2 J. P. Holman, *Thermodynamics,* 4th ed., McGraw-Hill, New York, 1988.

3 J. E. A. John, *Gas Dynamics,* 2d ed., Allyn and Bacon, Boston, 1984.

4 J. B. Jones and G. A. Hawkins, *Engineering Thermodynamics,* 2d ed., Wiley, New York, 1986.

5 J. F. Lee, *Theory and Design of Steam and Gas Turbines,* McGraw-Hill, New York, 1954.

6 A. H. Shapiro, *The Dynamics and Thermodynamics of Compressible Fluid Flow,* vol. 1, Ronald Press Company, New York, 1953.

7 G. J. Van Wylen and R. E. Sonntag, *Fundamentals of Classical Thermodynamics,* English/SI Version, 3d ed., Wiley, New York, 1986.

8 *The Aircraft Gas Turbine Engine and Its Operation,* United Technologies Corporation, 1982.

PROBLEMS*

Stagnation Properties

16-1C Why did we define the stagnation enthalpy h_0? How does it differ from ordinary (static) enthalpy?

16-2C What is dynamic temperature?

16-3C In air conditioning applications, the temperature of air is measured by inserting a probe into the flow stream. Thus the probe actually measures the stagnation temperature. Does this cause any significant error?

16-4C A high-speed aircraft is cruising in still air. How will the temperature of the air at the nose of the aircraft differ from the temperature of the air at some distance from the aircraft?

16-5 Air at 300 K is flowing in a duct at a velocity of (*a*) 1, (*b*) 10, (*c*) 100, and (*d*) 1000 m/s. Determine the temperature that a stationary probe inserted into the duct will read for each case.

16-6 A stationary temperature probe inserted into a duct where air is flowing at 250 m/s reads 85°C. What is the actual temperature of air?
Answer: 53.9°C

16-7 Determine the stagnation temperature and stagnation pressure of air which is flowing at 44 kPa, 245.9 K, and 470 m/s.
Answers: 355.8 K, 160.3 kPa

16-8 Calculate the stagnation temperature and pressure for the following substances flowing through a duct: (*a*) helium at 0.15 MPa, 50°C, and

*Students are encouraged to answer *all* the concept "C" questions.

300 m/s; (*b*) nitrogen at 0.15 MPa, 50°C, and 300 m/s; and (*c*) steam at 2.0 MPa, 350°C, and 480 m/s.

16-9 Nitrogen enters a steady-flow heat exchanger at 150 kPa, 10°C, and 100 m/s, and it receives heat in the amount of 80 kJ/kg as it flows through it. The nitrogen leaves the heat exchanger at 100 kPa with a velocity of 200 m/s. Determine the stagnation pressure and temperature of the nitrogen at the inlet and exit states.

16-10 Air enters a compressor with a stagnation pressure of 100 kPa and a stagnation temperature of 27°C, and it is compressed to a stagnation pressure of 900 kPa. Assuming the compression process to be isentropic, determine the power input to the compressor for a mass flow rate of 0.02 kg/s. *Answer:* 5.27 kW

16-11 Steam flows through a device with a stagnation pressure of 0.8 MPa, a stagnation temperature of 350°C, and a velocity of 300 m/s. Determine the static pressure and temperature of the steam at this state.

16-11E Steam flows through a device with a stagnation pressure of 120 psia, a stagnation temperature of 700°F, and a velocity of 900 ft/s. Determine the static pressure and temperature of the steam at this state. *Answers:* 665.9°F, 105.2 psia

16-12 Products of combustion enter a gas turbine with a stagnation pressure of 1.0 MPa and a stagnation temperature of 600°C, and they expand to a stagnation pressure of 100 kPa. Taking $k = 1.33$ and $R = 0.287$ kJ/(kg·K) for the products of combustion, and assuming the expansion process to be isentropic, determine the power output of the turbine per unit mass flow.

16-13 Air flows through a device such that the stagnation pressure is 0.6 MPa, the stagnation temperature is 400°C, and the velocity is 570 m/s. Determine the static pressure and temperature of air at this state. *Answers:* 518.6 K, 0.23 MPa

16-13E Air flows through a device such that the stagnation pressure is 90 psia, the stagnation temperature is 750°F, and the velocity is 1800 ft/s. Determine the static pressure and temperature of air at this state.

Velocity of Sound and Mach Number

16-14C What is sound? How is it generated? How does it travel?

16-15C Can sound waves travel in a vacuum?

16-16C Is it realistic to assume that the propagation of sound waves is an isentropic process?

16-17C Is the sonic velocity in a specified medium a fixed quantity, or does it change as the properties of the medium changes?

16-18C In which medium does a sound wave travel faster, in cool air or in warm air?

16-19C In which medium will sound travel fastest for a given temperature: air, helium, or argon?

16-20C In which medium does a sound wave travel faster, in air at 20°C and 1 atm or in air at 20°C and 5 atm?

16-21C Does the Mach number of a gas flowing at a constant velocity remain constant? Explain.

16-22 Determine the velocity of sound in air at (*a*) 300 K and (*b*) 1000 K. Also determine the Mach number of an aircraft moving in the air at a velocity of 350 m/s for both cases.

16-23 Carbon dioxide enters an adiabatic nozzle at 1200 K with a velocity of 50 m/s and leaves at 400 K. Determine the Mach number (*a*) at the inlet and (*b*) the exit of the nozzle. *Answers:* (*a*) 0.095, (*b*) 4.41

16-23E Carbon dioxide enters an adiabatic nozzle at 1800 R with a velocity of 150 ft/s, and it leaves at 800 R. Determine the Mach number (*a*) at the inlet and (*b*) the exit of the nozzle. *Answers:* (*a*) 0.097, (*b*) 3.47

16-24 Nitrogen enters a steady-flow heat exchanger at 150 kPa, 10°C, and 100 m/s, and it receives heat in the amount of 80 kJ/kg as it flows through it. Nitrogen leaves the heat exchanger at 100 kPa with a velocity of 200 m/s. Determine the Mach number of the nitrogen at the inlet and the exit of the heat exchanger.

16-25 Determine the velocity of sound in refrigerant-12 at 1 MPa and 60°C.

16-26 Steam flows through a device with a pressure of 0.8 MPa, a temperature of 350°C, and a velocity of 300 m/s. Determine the Mach number of the steam at this state by (*a*) using data from the steam table and (*b*) assuming ideal-gas behavior with $k = 1.3$.
Answers: (*a*) 0.496, (*b*) 0.491

16-26E Steam flows through a device with a pressure of 120 psia, a temperature of 700°F, and a velocity of 900 ft/s. Determine the Mach number of the steam at this state by (*a*) using data from the steam table and (*b*) assuming ideal-gas behavior with $k = 1.3$.
Answers: (*a*) 0.446, (*b*) 0.441

16-27 Derive an expression for the velocity of sound based on van der Waals equation of state. Using this relation, determine the velocity of sound in carbon dioxide at 50°C and 200 kPa, and compare your result to that obtained by assuming ideal-gas behavior. The van der Waals constants for carbon dioxide are $a = 364.3$ kPa·m^6/kmol^2 and $b = 0.0427$ m^3/kmol.

16-28 Starting with Eq. 16-9, verify Eq. 16-10.

16-29 The isentropic process for an ideal gas is expressed as $Pv^k =$

constant. Using this process equation and the definition of the velocity of sound (Eq. 16-9), obtain the expression for the velocity of sound for an ideal gas (Eq. 16-11).

16-30 Air expands isentropically from 1.5 MPa and 90°C to 0.5 MPa. Calculate the ratio of the initial to final velocity of sound.
Answer: 1.17

16-30E Air expands isentropically from 200 psia and 200°F to 60 psia. Calculate the ratio of the initial to final velocity of sound.
Answer: 1.19

One-Dimensional Isentropic Flow

16-31C Consider a converging nozzle with sonic velocity at the exit plane. Now the nozzle exit area is reduced while the nozzle inlet conditions are maintained constant. What will happen to (*a*) the exit velocity and (*b*) the mass flow rate through the nozzle?

16-32C A gas initially at a supersonic velocity enters an adiabatic converging duct. Discuss how this will affect (*a*) the velocity, (*b*) the temperature, (*c*) the pressure, and (*d*) the density of the fluid.

16-33C A gas initially at a supersonic velocity enters an adiabatic diverging duct. Discuss how this will affect (*a*) the velocity, (*b*) the temperature, (*c*) the pressure, and (*d*) the density of the fluid.

16-34C A gas initially at a subsonic velocity enters an adiabatic converging duct. Discuss how this will affect (*a*) the velocity, (*b*) the temperature, (*c*) the pressure, and (*d*) the density of the fluid.

16-35C A gas initially at a subsonic velocity enters an adiabatic diverging duct. Discuss how this will affect (*a*) the velocity, (*b*) the temperature, (*c*) the pressure, and (*d*) the density of the fluid.

16-36C A gas at a specified stagnation temperature and pressure is accelerated to $M = 2$ in a converging-diverging nozzle and to $M = 3$ in another nozzle. What can you say about the pressures at the throats of these two nozzles?

16-37C Is it possible to accelerate a gas to a supersonic velocity in a converging nozzle?

16-38 For ideal gases undergoing isentropic flows, obtain expressions for P/P^*, T/T^*, and ρ/ρ^* as functions of k and M.

16-39 Using Eqs. 16-4, 16-13, and 16-14, verify that for the steady flow of ideal gases $dT_0/T = dA/A + (1 - M^2)\, d\mathbf{V}/\mathbf{V}$. Explain the effect of heating and area changes on the velocity of an ideal gas in steady flow for (*a*) subsonic flow and (*b*) supersonic flow.

16-40 Air enters a converging-diverging nozzle at a pressure of 1 MPa with negligible velocity. What is the lowest pressure that can be obtained at the throat of the nozzle? *Answer:* 528.3 kPa

16-41 Helium enters a converging-diverging nozzle at 1 MPa, 800 K, and 100 m/s. What are the lowest temperature and pressure that can be obtained at the throat of the nozzle?

16-42 Calculate the critical temperature, pressure, and density of (*a*) air at 200 kPa, 100°C, and 250 m/s, and (*b*) helium at 200 kPa, 40°C, and 300 m/s.

16-43 Stationary carbon dioxide at 500 kPa and 400 K is accelerated isentropically to a Mach number of 0.6. Determine the temperature and pressure of the carbon dioxide after acceleration.
Answers: 382.6 K, 400.9 kPa

16-44 Air at 200 kPa, 100°C, and Mach number $M = 0.8$ flows through a duct. Find the velocity and the stagnation pressure, temperature, and density of the air.

16-44E Air at 30 psia, 212°F, and Mach number $M = 0.8$ flows through a duct. Find the velocity and the stagnation pressure, temperature, and density of the air.
Answers: 1016.3 ft/s, 757.7 R, 45.7 psia, 0.163 lbm/ft^3

16-45 A subsonic airplane is flying at 3000-m altitude where the atmospheric conditions are 70.109 kPa and 268.65 K. A pitot tube measures the difference between the static and stagnation pressures to be 14 kPa. Calculate the speed of the airplane and the flight Mach number.
Answers: 169.9 m/s, 0.52

16-46 An aircraft is designed to cruise at Mach number $M = 2.0$ at 8000 m where the atmospheric temperature is 236.15 K. Determine the stagnation temperature on the leading edge of the wing.

16-46E An aircraft is designed to cruise at Mach number $M = 2.0$ at 24,000 ft where the atmospheric temperature is 425 R. Determine the stagnation temperature on the leading edge of the wing.

Isentropic Flow through Nozzles

16-47C Consider subsonic flow in a converging nozzle with fixed inlet conditions. What is the effect of dropping the back pressure to the critical pressure on (*a*) the exit velocity, (*b*) the exit pressure, and (*c*) the mass flow rate through the nozzle?

16-48C Consider subsonic flow in a converging nozzle with specified conditions at the nozzle inlet and critical pressure at the nozzle exit. What is the effect of dropping the back pressure well below the critical pressure on (*a*) the exit velocity, (*b*) the exit pressure, and (*c*) the mass flow rate through the nozzle?

16-49C Consider a converging nozzle and a converging-diverging nozzle having the same throat areas. For the same inlet conditions, how would you compare the mass flow rates through these two nozzles?

16-50C Consider gas flow through a converging nozzle with specified inlet conditions. We know that the highest velocity the fluid can have at the nozzle exit is the sonic velocity, at which point the mass flow rate through the nozzle is a maximum. If it were possible to achieve hypersonic velocities at the nozzle exit, how would it affect the mass flow rate through the nozzle?

16-51C How does the parameter M^* differ from the Mach number M?

16-52C What would happen if we attempted to decelerate a supersonic fluid with a diverging diffuser?

16-53C What would happen if we tried to further accelerate a supersonic fluid with a converging nozzle?

16-54C Consider the isentropic flow of a fluid through a converging-diverging nozzle with a subsonic velocity at the throat. How does the diverging section affect (*a*) the velocity, (*b*) the pressure, and (*c*) the mass flow rate of the fluid?

16-55C Is it possible to accelerate a fluid to supersonic velocities with a velocity other than the sonic velocity at the throat?

16-56 Plot the mass flow parameter $\dot{m}\sqrt{RT_0}/(AP_0)$ versus the Mach number for $k = 1.2$, 1.4, and 1.6 in the range of $0 \leq M \leq 1$.

16-57 Explain why the maximum flow rate per unit area for a given gas depends only on $P_0/\sqrt{T_0}$. For an ideal gas with $k = 1.4$ and $R = 0.287$ kJ/(kg·K), find the constant a such that $\dot{m}/A^* = aP_0/\sqrt{T_0}$.

16-58 For an ideal gas obtain an expression for the ratio of the velocity of sound where $M = 1$ to the velocity of sound based on the stagnation temperature, C^*/C_0.

16-59 An ideal gas flows through a passage that first converges and then diverges during an adiabatic, reversible, steady-flow process. When the inlet flow is subsonic, sketch the variation of pressure, velocity, and Mach number along the length of the nozzle when the Mach number at the minimum flow area is equal to unity.

16-60 Repeat Prob. 16-59 for supersonic flow at the inlet.

16-61 Air enters a nozzle at 0.2 MPa, 350 K, and a velocity of 150 m/s. Assuming isentropic flow, determine the pressure and temperature of the air at a location where the air velocity equals the velocity of sound. What is the ratio of the area at this location to the entrance area?
Answers: 301.0 K, 0.118 MPa, 0.629

16-61E Air enters a nozzle at 30 psia, 630 R, and a velocity of 450 ft/s. Assuming isentropic flow, determine the pressure and temperature of air at a location where the air velocity equals the velocity of sound. What is the ratio of the area at this location to the entrance area?
Answers: 539.8 R, 17.38 psia, 0.574

16-62 Repeat Prob. 16-61, assuming the entrance velocity is negligible.

16-62E Repeat Prob. 16-61E, assuming the entrance velocity is negligible.

16-63 Helium enters a nozzle at 0.8 MPa, 500 K, and a velocity of 120 m/s. Assuming isentropic flow, determine the pressure and temperature of helium at a location where the velocity equals the velocity of sound. What is the ratio of the area at this location to the entrance area?

16-64 Repeat Prob. 16-63, assuming the entrance velocity is negligible.

16-65 Air enters a converging-diverging nozzle at 0.8 MPa with a negligible velocity. Assuming the flow to be isentropic, determine the back pressure that will result in an exit Mach number of 1.5.
Answer: 0.218 MPa

16-66 Nitrogen enters a converging-diverging nozzle at 700 kPa and 310 K with a negligible velocity. Determine the critical velocity, pressure, temperature, and density in the nozzle.

16-67 An ideal gas with $k = 1.4$ is flowing through a nozzle such that the Mach number is 2.5 where the flow area is 15 cm^2. Assuming the flow to be isentropic, determine the flow area at the location where the Mach number is 1.2.

16-67E An ideal gas with $k = 1.4$ is flowing through a nozzle such that the Mach number is 2.5 where the flow area is 0.05 ft^2. Assuming the flow to be isentropic, determine the flow area at the location where the Mach number is 1.2. *Answer:* 0.0195 ft^2

16-68 Repeat Prob. 16-67 for an ideal gas with $k = 1.2$.

16-69 Air at 900 kPa and 400 K enters a converging nozzle with a negligible velocity. The throat area of the nozzle is 10 cm^2. Assuming isentropic flow, calculate and plot the exit pressure, the exit velocity, and the mass flow rate versus the back pressure P_b for $0.9 \geq P_b \geq 0.1$ MPa.

16-70 Air at 0.9 MPa and 400 K enters a converging nozzle with a velocity of 180 m/s. The throat area is 10 cm^2. Assuming isentropic flow, calculate and plot the mass flow rate through the nozzle, the exit velocity, the exit Mach number, and the exit pressure–stagnation pressure ratio versus the back pressure–stagnation pressure ratio for a back pressure range of $0.9 \geq P_b \geq 0.1$ MPa.

16-71 Steam at 6.0 MPa and 700 K enters a converging nozzle with a negligible velocity. The nozzle throat area is 8 cm^2. Assuming isentropic flow, plot the exit pressure, the exit velocity, and the mass flow rate through the nozzle versus the back pressure P_b for $6.0 \geq P_b \geq 3.0$ MPa. Treat the steam as an ideal gas with $k = 1.3$, $C_p = 1.872$ kJ/(kg·K), and $R = 0.462$ kJ/(kg·K).

16-72 Air enters a converging-diverging nozzle of a supersonic wind tunnel at 1 MPa and 300 K with a low velocity. The flow area of the test section is equal to the exit area of the nozzle, which is 0.15 m^2. Calculate the pressure, temperature, velocity, and mass flow rate in the test section for a Mach number $M = 2$. Explain why the air must be very dry for this application. *Answers:* 0.1278 MPa, 166.7 K, 517.6 m/s, 207.4 kg/s

16-72E Air enters a converging-diverging nozzle of a supersonic wind tunnel at 150 psia and 100°F with a low velocity. The flow area of the test section is equal to the exit area of the nozzle, which is 5 ft^2. Calculate the pressure, temperature, velocity, and mass flow rate in the test section for a Mach number $M = 2$. Explain why the air must be very dry for this application. *Answers:* 19.1 psia, 311 R, 1729 ft/s, 1435 lbm/s

16-73 Write a computer program to determine the shape of a converging-diverging nozzle for air for a mass flow rate of 3 kg/s and inlet stagnation conditions of 1400 kPa and 200°C. Assume the flow is isentropic. Repeat calculations for 50-kPa increments of pressure drops. Plot the nozzle to scale. Also calculate and plot the Mach number along the nozzle.

Normal Shocks in Nozzle Flow

16-74C Can a shock wave develop in the converging section of a converging-diverging nozzle?

16-75C What do the states on the Fanno line and the Rayleigh line represent? What do the intersection points of these two curves represent?

16-76C Can the Mach number of a fluid be greater than 1 after a shock wave?

16-77C How does the normal shock affect (*a*) the fluid velocity, (*b*) the static temperature, (*c*) the stagnation temperature, (*d*) the static pressure, and (*e*) the stagnation pressure?

16-78 Find the expression for the ratio of the stagnation pressure after a shock wave to the static pressure before the shock wave as a function of k and the Mach number upstream of the shock wave M_x.

16-79 For an ideal gas flowing through a normal shock, develop a relation for $\mathbf{V}_y/\mathbf{V}_x$ in terms of k, M_x, and M_y.

16-80 Air enters a converging-diverging nozzle of a supersonic wind tunnel at 1 MPa and 300 K with a low velocity. If a normal shock wave occurs at the exit plane of the nozzle at $M = 2$, determine the pressure, temperature, Mach number, velocity, and stagnation pressure after the shock wave.
Answers: 0.5751 MPa, 281.3 K, 0.5774, 194.1 m/s, 0.7209 MPa

16-81 Air enters a converging-diverging nozzle with low velocity at 1.0 MPa and 100°C. If the exit area of the nozzle is twice the throat area, what must the back pressure be to produce a normal shock at the exit plane of the nozzle? *Answer:* 0.512 MPa

16-82 What must the back pressure be in Prob. 16-81 for a normal shock to occur at a location where the cross-sectional area is 3.5 times the throat area?

16-83 Air flowing steadily in a nozzle experiences a normal shock at a Mach number $M = 2.5$. If the pressure and temperature of air are 61.64 kPa and 262.15 K, respectively, upstream of the shock, calculate the pressure, temperature, velocity, Mach number, and stagnation pressure downstream of the shock. Compare these results to those for helium undergoing a normal shock under the same conditions.

16-83E Air flowing steadily in a nozzle experiences a normal shock at a Mach number $M = 2.5$. If the pressure and temperature of air, respectively, are 10.0 psia and 440.5 R upstream of the shock, calculate the pressure, temperature, velocity, Mach number, and stagnation downstream of the shock. Compare these results to those for helium undergoing a normal shock at the same conditions.

16-84 Calculate the entropy change of air across the normal shock wave in Prob. 16-83.

16-84E Calculate the entropy change of air across the normal shock wave in Prob. 16-83E.

16-85 Nitrogen enters a converging-diverging nozzle at 700 kPa and 300 K with a negligible velocity, and it experiences a normal shock at a location where the Mach number is $M = 3.0$. Calculate the pressure, temperature, velocity, Mach number, and stagnation pressure downstream of the shock. Compare these results to those of air undergoing a normal shock at the same conditions.

16-86 Air enters a normal shock at 22.6 kPa, 217 K, and 738 m/s. Calculate the stagnation pressure and Mach number upstream of the shock, as well as pressure, temperature, velocity, Mach number, and stagnation pressure downstream of the shock.
Answers: 125.3 kPa, 2.50, 161.0 kPa, 463.8 K, 227.6 m/s, 0.513, 192.7 kPa

16-87 Calculate the entropy change of air across the normal shock wave in Prob. 16-86. *Answer:* 0.200 kJ/(kg·K)

16-88 Calculate and plot the entropy change of air across the shock for upstream Mach numbers between 0.5 and 1.5. Explain why normal shock waves can only occur for upstream Mach numbers greater than $M = 1$.

Nozzle and Diffuser Efficiencies

16-89C Which is greater, the efficiency or the velocity coefficient of a nozzle?

16-90C What is the difference between the pressure rise coefficient and the pressure recovery factor?

16-91 Atmospheric air at 90 kPa and 260 K enters the diffuser of an aircraft engine at 250 m/s at a rate of 15 kg/s. If the diffuser has an efficiency of 95 percent, find the diffuser exit temperature, pressure, and area for an exit velocity of 80 m/s.
Answers: 287.9 K, 126.1 kPa, 0.123 m^2

16-91E Atmospheric air at 14.0 psia and 450 R enters the diffuser of an aircraft engine at 800 ft/s at a rate of 30 lbm/s. If the diffuser has an efficiency of 95 percent, determine the diffuser exit temperature, pressure, and area for an exit velocity of 250 ft/s and a mass flow rate of 30 lbm/s.

16-92 Determine the pressure rise coefficient and the pressure recovery factor for Prob. 16-91. *Answers:* 0.826, 0.981

16-93 An aircraft flies with a Mach number $M_1 = 0.8$ at an altitude of 7000 m where the pressure is 41.1 kPa and the temperature is 242.7 K. The diffuser has an efficiency of 92 percent and an exit Mach number of $M_2 = 0.3$. For a mass flow rate of 50 kg/s, determine the static pressure rise across the diffuser and the exit area.

16-94 Air at 54.0 kPa and 255.7 K approaches a diffuser with a velocity of 640 m/s. A normal shock wave occurs at the diffuser entrance. For a diffuser efficiency of 90 percent for the flow downstream of the shock and a diffuser exit velocity of 150 m/s, determine the static pressure rise across the diffuser.

16-95 Products of combustion enter the nozzle of a gas turbine at the design conditions of 400 kPa, 1000 K, and 200 m/s, and they exit at a pressure of 270 kPa at a rate of 3 kg/s. Assuming isentropic flow, determine whether the nozzle is converging or converging-diverging. Also find the exit velocity and exit area. Take $k = 1.34$ and $C_p = 1.16$ kJ/(kg·K) for the combustion products.

16-96 Repeat Prob. 16-95 for a nozzle efficiency of 95 percent. Assume no shocks occur in the nozzle.

16-97 Helium expands in a nozzle from 1 MPa, 500 K, and negligible velocity to 0.1 MPa. Calculate the throat and exit areas for a mass flow rate of 0.1 kg/s, assuming the nozzle (*a*) is isentropic and (*b*) has an efficiency of 97 percent. Why must this nozzle be converging-diverging?
Answers: (*a*) 1.40 cm^2, 2.34 cm^2; (*b*) 1.44 cm^2, 4.49 cm^2

16-97E Helium expands in a nozzle from 150 psia, 900 R, and negligible velocity to 15 psia. Calculate the throat and exit areas for a mass flow rate of 0.2 lbm/s, assuming the nozzle (*a*) is isentropic and (*b*) has an efficiency of 97 percent. Why must this nozzle be converging-diverging?
Answers: (*a*) 0.00132 ft^2, 0.00221 ft^2; (*b*) 0.00136 ft^2, 0.00234 ft^2

Steam Nozzles

16-98C What is supersaturation? Under what conditions does it occur?

16-99 Steam enters a convergent nozzle at 3.0 MPa, 600°C, and negligible velocity, and it exits at 1.8 MPa. For a nozzle exit area of 24 cm^2, de-

termine the exit velocity, mass flow rate, and exit Mach number if the nozzle (*a*) is isentropic and (*b*) has an efficiency of 90 percent.
Answers: (*a*) 618.2 m/s, 7.49 kg/s, 0.914; (*b*) 586.5 m/s, 7.03 kg/s, 0.863

16-99E Steam enters a convergent nozzle at 450 psia, 900°F, and negligible velocity and exits at 275 psia. For a nozzle exit area of 3.75 in^2, determine the exit velocity, mass flow rate, and exit Mach number if the nozzle (*a*) is isentropic and (*b*) has an efficiency of 90 percent.
Answers: (*a*) 1845.3 ft/s, 18.7 lbm/s, 0.900; (*b*) 1743.4 ft/s, 17.4 lbm/s, 0.846

16-100 Saturated steam enters a convergent-divergent nozzle at 3.0 MPa, 5 percent moisture, and negligible velocity, and it exits at 1.2 MPa. For a nozzle exit area of 24 cm^2, determine the throat area, exit velocity, mass flow rate, and exit Mach number if the nozzle (*a*) is isentropic and (*b*) has an efficiency of 90 percent.

16-101 Steam enters a converging-diverging nozzle at 1 MPa and 500°C with a negligible velocity at a mass flow rate of 1.7 kg/s, and it exits at a pressure of 200 kPa. Assuming the flow through the nozzle to be isentropic, determine the exit area and the exit Mach number.
Answers: 21.4 cm^2, 1.735

16-102 Repeat Prob. 16-101 for a nozzle efficiency of 95 percent.

Property Tables, Figures, and Charts (SI Units)

APPENDIX 1

TABLE A-1

Molar mass, gas constant, and critical-point properties

Substance	Formula	Molar mass kg/kmol	R kJ/(kg·K)*	Temperature K	Pressure MPa	Volume m³/kmol
Ammonia	NH_3	17.03	0.4882	405.5	11.28	0.0724
Argon	Ar	39.948	0.2081	151	4.86	0.0749
Bromine	Br_2	159.808	0.0520	584	10.34	0.1355
Carbon dioxide	CO_2	44.01	0.1889	304.2	7.39	0.0943
Carbon monoxide	CO	28.011	0.2968	133	3.50	0.0930
Chlorine	Cl_2	70.906	0.1173	417	7.71	0.1242
Deuterium (normal)	D_2	4.00	2.0785	38.4	1.66	—
Helium	He	4.003	2.0769	5.3	0.23	0.0578
Hydrogen (normal)	H_2	2.016	4.1240	33.3	1.30	0.0649
Krypton	Kr	83.80	0.09921	209.4	5.50	0.0924
Neon	Ne	20.183	0.4119	44.5	2.73	0.0417
Nitrogen	N_2	28.013	0.2968	126.2	3.39	0.0899
Nitrous oxide	N_2O	44.013	0.1889	309.7	7.27	0.0961
Oxygen	O_2	31.999	0.2598	154.8	5.08	0.0780
Sulfur dioxide	SO_2	64.063	0.1298	430.7	7.88	0.1217
Water	H_2O	18.015	0.4615	647.3	22.09	0.0568
Xenon	Xe	131.30	0.06332	289.8	5.88	0.1186
Benzene	C_6H_6	78.115	0.1064	562	4.92	0.2603
n-Butane	C_4H_{10}	58.124	0.1430	425.2	3.80	0.2547
Carbon tetrachloride	CCl_4	153.82	0.05405	556.4	4.56	0.2759
Chloroform	$CHCl_3$	119.38	0.06964	536.6	5.47	0.2403
Dichlorodifluoromethane (R-12)	CCl_2F_2	120.91	0.06876	384.7	4.01	0.2179
Dichlorofluoromethane	$CHCl_2F$	102.92	0.08078	451.7	5.17	0.1973
Ethane	C_2H_6	30.070	0.2765	305.5	4.88	0.1480
Ethyl alcohol	C_2H_5OH	46.07	0.1805	516	6.38	0.1673
Ethylene	C_2H_4	28.054	0.2964	282.4	5.12	0.1242
n-Hexane	C_6H_{14}	86.178	0.09647	507.9	3.03	0.3677
Methane	CH_4	16.043	0.5182	191.1	4.64	0.0993
Methyl alcohol	CH_3OH	32.042	0.2595	513.2	7.95	0.1180
Methyl chloride	CH_3Cl	50.488	0.1647	416.3	6.68	0.1430
Propane	C_3H_8	44.097	0.1885	370	4.26	0.1998
Propene	C_3H_6	42.081	0.1976	365	4.62	0.1810
Propyne	C_3H_4	40.065	0.2075	401	5.35	—
Trichlorofluoromethane	CCl_3F	137.37	0.06052	471.2	4.38	0.2478
Air	—	28.97	0.2870	—	—	—

*The unit kJ/(kg·K) is equivalent to kPa·m³/(kg·K). The gas constant is calculated from $R = R_u/M$, where R_u = 8.314 kJ/(kmol·K) and M is the molar mass.

Source: Gordon J. Van Wylen and Richard E. Sonntag, *Fundamentals of Classical Thermodynamics*, English/SI Version, 3d ed., Wiley, New York, 1986, p. 685, table A.6SI. Originally published in K. A. Kobe and R. E. Lynn, Jr., *Chemical Review*, vol. 52, pp. 117–236, 1953.

TABLE A-2

Ideal-gas specific heats of various common gases

(*a*) At 300 K

Gas	Formula	C_{p0} kJ/(kg·K)	C_{v0} kJ/(kg·K)	k
Air	—	1.005	0.718	1.400
Argon	Ar	0.5203	0.3122	1.667
Butane	C_4H_{10}	1.7164	1.5734	1.091
Carbon dioxide	CO_2	0.846	0.657	1.289
Carbon monoxide	CO	1.040	0.744	1.400
Ethane	C_2H_6	1.7662	1.4897	1.186
Ethylene	C_2H_4	1.5482	1.2518	1.237
Helium	He	5.1926	3.1156	1.667
Hydrogen	H_2	14.307	10.183	1.405
Methane	CH_4	2.2537	1.7354	1.299
Neon	Ne	1.0299	0.6179	1.667
Nitrogen	N_2	1.039	0.743	1.400
Octane	C_8H_{18}	1.7113	1.6385	1.044
Oxygen	O_2	0.918	0.658	1.395
Propane	C_3H_8	1.6794	1.4909	1.126
Steam	H_2O	1.8723	1.4108	1.327

Source: Gordon J. Van Wylen and Richard E. Sonntag, *Fundamentals of Classical Thermodynamics*, English/SI Version, 3d ed., Wiley, New York, 1986, p. 687, table A.8SI.

TABLE A-2
(Continued)

(b) At various temperatures

Temperature K	C_{p0} kJ/(kg·K)	C_{v0} kJ/(kg·K)	k	C_{p0} kJ/(kg·K)	C_{v0} kJ/(kg·K)	k	C_{p0} kJ/(kg·K)	C_{v0} kJ/(kg·K)	k
	Air			**Carbon dioxide, CO_2**			**Carbon monoxide, CO**		
250	1.003	0.716	1.401	0.791	0.602	1.314	1.039	0.743	1.400
300	1.005	0.718	1.400	0.846	0.657	1.288	1.040	0.744	1.399
350	1.008	0.721	1.398	0.895	0.706	1.268	1.043	0.746	1.398
400	1.013	0.726	1.395	0.939	0.750	1.252	1.047	0.751	1.395
450	1.020	0.733	1.391	0.978	0.790	1.239	1.054	0.757	1.392
500	1.029	0.742	1.387	1.014	0.825	1.229	1.063	0.767	1.387
550	1.040	0.753	1.381	1.046	0.857	1.220	1.075	0.778	1.382
600	1.051	0.764	1.376	1.075	0.886	1.213	1.087	0.790	1.376
650	1.063	0.776	1.370	1.102	0.913	1.207	1.100	0.803	1.370
700	1.075	0.788	1.364	1.126	0.937	1.202	1.113	0.816	1.364
750	1.087	0.800	1.359	1.148	0.959	1.197	1.126	0.829	1.358
800	1.099	0.812	1.354	1.169	0.980	1.193	1.139	0.842	1.353
900	1.121	0.834	1.344	1.204	1.015	1.186	1.163	0.866	1.343
1000	1.142	0.855	1.336	1.234	1.045	1.181	1.185	0.888	1.335
	Hydrogen, H_2			**Nitrogen, N_2**			**Oxygen, O_2**		
250	14.051	9.927	1.416	1.039	0.742	1.400	0.913	0.653	1.398
300	14.307	10.183	1.405	1.039	0.743	1.400	0.918	0.658	1.395
350	14.427	10.302	1.400	1.041	0.744	1.399	0.928	0.668	1.389
400	14.476	10.352	1.398	1.044	0.747	1.397	0.941	0.681	1.382
450	14.501	10.377	1.398	1.049	0.752	1.395	0.956	0.696	1.373
500	14.513	10.389	1.397	1.056	0.759	1.391	0.972	0.712	1.365
550	14.530	10.405	1.396	1.065	0.768	1.387	0.988	0.728	1.358
600	14.546	10.422	1.396	1.075	0.778	1.382	1.003	0.743	1.350
650	14.571	10.447	1.395	1.086	0.789	1.376	1.017	0.758	1.343
700	14.604	10.480	1.394	1.098	0.801	1.371	1.031	0.771	1.337
750	14.645	10.521	1.392	1.110	0.813	1.365	1.043	0.783	1.332
800	14.695	10.570	1.390	1.121	0.825	1.360	1.054	0.794	1.327
900	14.822	10.698	1.385	1.145	0.849	1.349	1.074	0.814	1.319
1000	14.983	10.859	1.380	1.167	0.870	1.341	1.090	0.830	1.313

Source: Kenneth Wark, *Thermodynamics*, 4th ed., McGraw-Hill, New York, 1983, p. 783, table A-4M. Originally published in *Tables of Thermal Properties of Gases*, NBS Circ. 564, 1955.

TABLE A-2
(Continued)

(*c*) As a function of temperature

$$\bar{C}_{p0} = a + bT + cT^2 + dT^3$$
[T in K, $\bar{C}_{p0}$ in kJ/(kmol·K)]

Substance	Formula	a	b	c	d	Temperature range K	% error Max.	% error Avg.
Nitrogen	N_2	28.90	-0.1571×10^{-2}	0.8081×10^{-5}	-2.873×10^{-9}	273–1800	0.59	0.34
Oxygen	O_2	25.48	1.520×10^{-2}	-0.7155×10^{-5}	1.312×10^{-9}	273–1800	1.19	0.28
Air		28.11	0.1967×10^{-2}	0.4802×10^{-5}	-1.966×10^{-9}	273–1800	0.72	0.33
Hydrogen	H_2	29.11	-0.1916×10^{-2}	0.4003×10^{-5}	-0.8704×10^{-9}	273–1800	1.01	0.26
Carbon monoxide	CO	28.16	0.1675×10^{-2}	0.5372×10^{-5}	-2.222×10^{-9}	273–1800	0.89	0.37
Carbon dioxide	CO_2	22.26	5.981×10^{-2}	-3.501×10^{-5}	7.469×10^{-9}	273–1800	0.67	0.22
Water vapor	H_2O	32.24	0.1923×10^{-2}	1.055×10^{-5}	-3.595×10^{-9}	273–1800	0.53	0.24
Nitric oxide	NO	29.34	-0.09395×10^{-2}	0.9747×10^{-5}	-4.187×10^{-9}	273–1500	0.97	0.36
Nitrous oxide	N_2O	24.11	5.8632×10^{-2}	-3.562×10^{-5}	10.58×10^{-9}	273–1500	0.59	0.26
Nitrogen dioxide	NO_2	22.9	5.715×10^{-2}	-3.52×10^{-5}	7.87×10^{-9}	273–1500	0.46	0.18
Ammonia	NH_3	27.568	2.5630×10^{-2}	0.99072×10^{-5}	-6.6909×10^{-9}	273–1500	0.91	0.36
Sulfur	S_2	27.21	2.218×10^{-2}	-1.628×10^{-5}	3.986×10^{-9}	273–1800	0.99	0.38
Sulfur dioxide	SO_2	25.78	5.795×10^{-2}	-3.812×10^{-5}	8.612×10^{-9}	273–1800	0.45	0.24
Sulfur trioxide	SO_3	16.40	14.58×10^{-2}	-11.20×10^{-5}	32.42×10^{-9}	273–1300	0.29	0.13
Acetylene	C_2H_2	21.8	9.2143×10^{-2}	-6.527×10^{-5}	18.21×10^{-9}	273–1500	1.46	0.59
Benzene	C_6H_6	−36.22	48.475×10^{-2}	-31.57×10^{-5}	77.62×10^{-9}	273–1500	0.34	0.20
Methanol	CH_4O	19.0	9.152×10^{-2}	-1.22×10^{-5}	-8.039×10^{-9}	273–1000	0.18	0.08
Ethanol	C_2H_6O	19.9	20.96×10^{-2}	-10.38×10^{-5}	20.05×10^{-9}	273–1500	0.40	0.22
Hydrogen chloride	HCl	30.33	-0.7620×10^{-2}	1.327×10^{-5}	-4.338×10^{-9}	273–1500	0.22	0.08
Methane	CH_4	19.89	5.024×10^{-2}	1.269×10^{-5}	-11.01×10^{-9}	273–1500	1.33	0.57
Ethane	C_2H_6	6.900	17.27×10^{-2}	-6.406×10^{-5}	7.285×10^{-9}	273–1500	0.83	0.28
Propane	C_3H_8	−4.04	30.48×10^{-2}	-15.72×10^{-5}	31.74×10^{-9}	273–1500	0.40	0.12
n-Butane	C_4H_{10}	3.96	37.15×10^{-2}	-18.34×10^{-5}	35.00×10^{-9}	273–1500	0.54	0.24
i-Butane	C_4H_{10}	−7.913	41.60×10^{-2}	-23.01×10^{-5}	49.91×10^{-9}	273–1500	0.25	0.13
n-Pentane	C_5H_{12}	6.774	45.43×10^{-2}	-22.46×10^{-5}	42.29×10^{-9}	273–1500	0.56	0.21
n-Hexane	C_6H_{14}	6.938	55.22×10^{-2}	-28.65×10^{-5}	57.69×10^{-9}	273–1500	0.72	0.20
Ethylene	C_2H_4	3.95	15.64×10^{-2}	-8.344×10^{-5}	17.67×10^{-9}	273–1500	0.54	0.13
Propylene	C_3H_6	3.15	23.83×10^{-2}	-12.18×10^{-5}	24.62×10^{-9}	273–1500	0.73	0.17

Source: B. G. Kyle, *Chemical and Process Thermodynamics,* Prentice-Hall, Englewood Cliffs, N.J., 1984. Used with permission.

TABLE A-3

Specific heats of common solids and liquids
(*a*) At 25°C

Solid	C_p kJ/(kg·K)	ρ kg/m³	Liquid	C_p kJ/(kg·K)	ρ kg/m³
Aluminum	0.900	2,700	Ammonia	4.800	602
Copper	0.386	8,900	Ethanol	2.456	783
Granite	1.017	2,700	Refrigerant-12	0.977	1,310
Graphite	0.711	2,500	Mercury	0.139	13,560
Iron	0.450	7,840	Methanol	2.550	787
Lead	0.128	11,310	Oil (light)	1.800	910
Rubber (soft)	1.840	1,100	Water	4.184	997
Silver	0.235	10,470			
Tin	0.217	5,730			
Wood (most)	1.760	350–700			

Source: Gordon J. Van Wylen and Richard E. Sonntag, *Fundamentals of Classical Thermodynamics,* English/SI Version, 3d ed., Wiley, New York, 1986, p. 686, table A-7SI.

TABLE A-3
(Continued)

(*b*) At various temperatures

Solids

Substance	Temp.	C_p kJ/(kg·K)	Substance	Temp.	C_p kJ/(kg·K)
Ice	200 K	1.56	Silver	20°C	0.233
	220 K	1.71		200°C	0.243
	240 K	1.86	Lead	−173°C	0.118
	260 K	2.01		−50°C	0.126
	270 K	2.08		27°C	0.129
	273 K	2.11		100°C	0.131
Aluminum	200 K	0.797		200°C	0.136
	250 K	0.859	Copper	−173°C	0.254
	300 K	0.902		−100°C	0.342
	350 K	0.929		−50°C	0.367
	400 K	0.949		0°C	0.381
	450 K	0.973		27°C	0.386
	500 K	0.997		100°C	0.393
Iron	20°C	0.448		200°C	0.403

Liquids

Substance	State	C_p kJ/(kg·K)	Substance	State	C_p kJ/(kg·K)
Water	1 atm, 273 K	4.217	Benzene	1 atm, 15°C	1.80
	1 atm, 280 K	4.198		1 atm, 65°C	1.92
	1 atm, 300 K	4.179	Glycerin	1 atm, 10°C	2.32
	1 atm, 320 K	4.180		1 atm, 50°C	2.58
	1 atm, 340 K	4.188	Mercury	1 atm, 10°C	0.138
	1 atm, 360 K	4.203		1 atm, 315°C	0.134
	1 atm, 373 K	4.218	Sodium	1 atm, 95°C	1.38
Ammonia	Sat., −20°C	4.52		1 atm, 540°C	1.26
	Sat., 50°C	5.10	Propane	1 atm, 0°C	2.41
Refrigerant-12	Sat., −40°C	0.883	Bismuth	1 atm, 425°C	0.144
	Sat., −20°C	0.908		1 atm, 760°C	0.164
	Sat., 50°C	1.02	Ethyl alcohol	1 atm, 25°C	2.43

Source: Adapted from Kenneth Wark, *Thermodynamics*, 4th ed., McGraw-Hill, New York, 1983, p. 813, table A-19M.

TABLE A-4

Saturated water–Temperature table

		Specific volume m³/kg		Internal energy kJ/kg			Enthalpy kJ/kg			Entropy kJ/(kg·K)		
Temp. °C T	Sat. press. kPa P_{sat}	Sat. liquid v_f	Sat. vapor v_g	Sat. liquid u_f	Evap. u_{fg}	Sat. vapor u_g	Sat. liquid h_f	Evap. h_{fg}	Sat. vapor h_g	Sat. liquid s_f	Evap. s_{fg}	Sat. vapor s_g
0.01	0.6113	0.001 000	206.14	0.00	2375.3	2375.3	0.01	2501.3	2501.4	0.0000	9.1562	9.1562
5	0.8721	0.001 000	147.12	20.97	2361.3	2382.3	20.98	2489.6	2510.6	0.0761	8.9496	9.0257
10	1.2276	0.001 000	106.38	42.00	2347.2	2389.2	42.01	2477.7	2519.8	0.1510	8.7498	8.9008
15	1.7051	0.001 001	77.93	62.99	2333.1	2396.1	62.99	2465.9	2528.9	0.2245	8.5569	8.7814
20	2.339	0.001 002	57.79	83.95	2319.0	2402.9	83.96	2454.1	2538.1	0.2966	8.3706	8.6672
25	3.169	0.001 003	43.36	104.88	2304.9	2409.8	104.89	2442.3	2547.2	0.3674	8.1905	8.5580
30	4.246	0.001 004	32.89	125.78	2290.8	2416.6	125.79	2430.5	2556.3	0.4369	8.0164	8.4533
35	5.628	0.001 006	25.22	146.67	2276.7	2423.4	146.68	2418.6	2565.3	0.5053	7.8478	8.3531
40	7.384	0.001 008	19.52	167.56	2262.6	2430.1	167.57	2406.7	2574.3	0.5725	7.6845	8.2570
45	9.593	0.001 010	15.26	188.44	2248.4	2436.8	188.45	2394.8	2583.2	0.6387	7.5261	8.1648
50	12.349	0.001 012	12.03	209.32	2234.2	2443.5	209.33	2382.7	2592.1	0.7038	7.3725	8.0763
55	15.758	0.001 015	9.568	230.21	2219.9	2450.1	230.23	2370.7	2600.9	0.7679	7.2234	7.9913
60	19.940	0.001 017	7.671	251.11	2205.5	2456.6	251.13	2358.5	2609.6	0.8312	7.0784	7.9096
65	25.03	0.001 020	6.197	272.02	2191.1	2463.1	272.06	2346.2	2618.3	0.8935	6.9375	7.8310
70	31.19	0.001 023	5.042	292.95	2176.6	2569.6	292.98	2333.8	2626.8	0.9549	6.8004	7.7553
75	38.58	0.001 026	4.131	313.90	2162.0	2475.9	313.93	2321.4	2635.3	1.0155	6.6669	7.6824
80	47.39	0.001 029	3.407	334.86	2147.4	2482.2	334.91	2308.8	2643.7	1.0753	6.5369	7.6122
85	57.83	0.001 033	2.828	355.84	2132.6	2488.4	355.90	2296.0	2651.9	1.1343	6.4102	7.5445
90	70.14	0.001 036	2.361	376.85	2117.7	2494.5	376.92	2283.2	2660.1	1.1925	6.2866	7.4791
95	84.55	0.001 040	1.982	397.88	2102.7	2500.6	397.96	2270.2	2668.1	1.2500	6.1659	7.4159
	Sat. press. MPa P_{sat}											
100	0.101 35	0.001 044	1.6729	418.94	2087.6	2506.5	419.04	2257.0	2676.1	1.3069	6.0480	7.3549
105	0.120 82	0.001 048	1.4194	440.02	2072.3	2512.4	440.15	2243.7	2683.8	1.3630	5.9328	7.2958
110	0.143 27	0.001 052	1.2102	461.14	2057.0	2518.1	461.30	2230.2	2691.5	1.4185	5.8202	7.2387
115	0.169 06	0.001 056	1.0366	482.30	2041.4	2523.7	482.48	2216.5	2699.0	1.4734	5.7100	7.1833
120	0.198 53	0.001 060	0.8919	503.50	2025.8	2529.3	503.71	2202.6	2706.3	1.5276	5.6020	7.1296
125	0.2321	0.001 065	0.7706	524.74	2009.9	2534.6	524.99	2188.5	2713.5	1.5813	5.4962	7.0775
130	0.2701	0.001 070	0.6685	546.02	1993.9	2539.9	546.31	2174.2	2720.5	1.6344	5.3925	7.0269
135	0.3130	0.001 075	0.5822	567.35	1977.7	2545.0	567.69	2159.6	2727.3	1.6870	5.2907	6.9777
140	0.3613	0.001 080	0.5089	588.74	1961.3	2550.0	589.13	2144.7	2733.9	1.7391	5.1908	6.9299
145	0.4154	0.001 085	0.4463	610.18	1944.7	2554.9	610.63	2129.6	2740.3	1.7907	5.0926	6.8833
150	0.4758	0.001 091	0.3928	631.68	1927.9	2559.5	632.20	2114.3	2746.5	1.8418	4.9960	6.8379
155	0.5431	0.001 096	0.3468	653.24	1910.8	2564.1	653.84	2098.6	2752.4	1.8925	4.9010	6.7935
160	0.6178	0.001 102	0.3071	674.87	1893.5	2568.4	675.55	2082.6	2758.1	1.9427	4.8075	6.7502
165	0.7005	0.001 108	0.2727	696.56	1876.0	2572.5	697.34	2066.2	2763.5	1.9925	4.7153	6.7078
170	0.7917	0.001 114	0.2428	718.33	1858.1	2576.5	719.21	2049.5	2768.7	2.0419	4.6244	6.6663
175	0.8920	0.001 121	0.2168	740.17	1840.0	2580.2	741.17	2032.4	2773.6	2.0909	4.5347	6.6256
180	1.0021	0.001 127	0.194 05	762.09	1821.6	2583.7	763.22	2015.0	2778.2	2.1396	4.4461	6.5857
185	1.1227	0.001 134	0.174 09	784.10	1802.9	2587.0	785.37	1997.1	2782.4	2.1879	4.3586	6.5465
190	1.2544	0.001 141	0.156 54	806.19	1783.8	2590.0	807.62	1978.8	2786.4	2.2359	4.2720	6.5079
195	1.3978	0.001 149	0.141 05	828.37	1764.4	2592.8	829.98	1960.0	2790.0	2.2835	4.1863	6.4698
200	1.5538	0.001 157	0.127 36	850.65	1744.7	2595.3	852.45	1940.7	2793.2	2.3309	4.1014	6.4323
205	1.7230	0.001 164	0.115 21	873.04	1724.5	2597.5	875.04	1921.0	2796.0	2.3780	4.0172	6.3952
210	1.9062	0.001 173	0.104 41	895.53	1703.9	2599.5	897.76	1900.7	2798.5	2.4248	3.9337	6.3585
215	2.104	0.001 181	0.094 79	918.14	1682.9	2601.1	920.62	1879.9	2800.5	2.4714	3.8507	6.3221
220	2.318	0.001 190	0.086 19	940.87	1661.5	2602.4	943.62	1858.5	2802.1	2.5178	3.7683	6.2861
225	2.548	0.001 199	0.078 49	963.73	1639.6	2603.3	966.78	1836.5	2803.3	2.5639	3.6863	6.2503

TABLE A-4

(Continued)

Temp. °C T	Sat. press. MPa P_{sat}	Specific volume m^3/kg: Sat. liquid v_f	Specific volume m^3/kg: Sat. vapor v_g	Internal energy kJ/kg: Sat. liquid u_f	Internal energy kJ/kg: Evap. u_{fg}	Internal energy kJ/kg: Sat. vapor u_g	Enthalpy kJ/kg: Sat. liquid h_f	Enthalpy kJ/kg: Evap. h_{fg}	Enthalpy kJ/kg: Sat. vapor h_g	Entropy kJ/(kg·K): Sat. liquid s_f	Entropy kJ/(kg·K): Evap. s_{fg}	Entropy kJ/(kg·K): Sat. vapor s_g
230	2.795	0.001 209	0.071 58	986.74	1617.2	2603.9	990.12	1813.8	2804.0	2.6099	3.6047	6.2146
235	3.060	0.001 219	0.065 37	1009.89	1594.2	2604.1	1013.62	1790.5	2804.2	2.6558	3.5233	6.1791
240	3.344	0.001 229	0.059 76	1033.21	1570.8	2604.0	1037.32	1766.5	2803.8	2.7015	3.4422	6.1437
245	3.648	0.001 240	0.054 71	1056.71	1546.7	2603.4	1061.23	1741.7	2803.0	2.7472	3.3612	6.1083
250	3.973	0.001 251	0.050 13	1080.39	1522.0	2602.4	1085.36	1716.2	2801.5	2.7927	3.2802	6.0730
255	4.319	0.001 263	0.045 98	1104.28	1596.7	2600.9	1109.73	1689.8	2799.5	2.8383	3.1992	6.0375
260	4.688	0.001 276	0.042 21	1128.39	1470.6	2599.0	1134.37	1662.5	2796.9	2.8838	3.1181	6.0019
265	5.081	0.001 289	0.038 77	1152.74	1443.9	2596.6	1159.28	1634.4	2793.6	2.9294	3.0368	5.9662
270	5.499	0.001 302	0.035 64	1177.36	1416.3	2593.7	1184.51	1605.2	2789.7	2.9751	2.9551	5.9301
275	5.942	0.001 317	0.032 79	1202.25	1387.9	2590.2	1210.07	1574.9	2785.0	3.0208	2.8730	5.8938
280	6.412	0.001 332	0.030 17	1227.46	1358.7	2586.1	1235.99	1543.6	2779.6	3.0668	2.7903	5.8571
285	6.909	0.001 348	0.027 77	1253.00	1328.4	2581.4	1262.31	1511.0	2773.3	3.1130	2.7070	5.8199
290	7.436	0.001 366	0.025 57	1278.92	1297.1	2576.0	1289.07	1477.1	2766.2	3.1594	2.6227	5.7821
295	7.993	0.001 384	0.023 54	1305.2	1264.7	2569.9	1316.3	1441.8	2758.1	3.2062	2.5375	5.7437
300	8.581	0.001 404	0.021 67	1332.0	1231.0	2563.0	1344.0	1404.9	2749.0	3.2534	2.4511	5.7045
305	9.202	0.001 425	0.019 948	1359.3	1195.9	2555.2	1372.4	1366.4	2738.7	3.3010	2.3633	5.6643
310	9.856	0.001 447	0.018 350	1387.1	1159.4	2546.4	1401.3	1326.0	2727.3	3.3493	2.2737	5.6230
315	10.547	0.001 472	0.016 867	1415.5	1121.1	2536.6	1431.0	1283.5	2714.5	3.3982	2.1821	5.5804
320	11.274	0.001 499	0.015 488	1444.6	1080.9	2525.5	1461.5	1238.6	2700.1	3.4480	2.0882	5.5362
330	12.845	0.001 561	0.012 996	1505.3	993.7	2498.9	1525.3	1140.6	2665.9	3.5507	1.8909	5.4417
340	14.586	0.001 638	0.010 797	1570.3	894.3	2464.6	1594.2	1027.9	2622.0	3.6594	1.6763	5.3357
350	16.513	0.001 740	0.008 813	1641.9	776.6	2418.4	1670.6	893.4	2563.9	3.7777	1.4335	5.2112
360	18.651	0.001 893	0.006 945	1725.2	626.3	2351.5	1760.5	720.3	2481.0	3.9147	1.1379	5.0526
370	21.03	0.002 213	0.004 925	1844.0	384.5	2228.5	1890.5	441.6	2332.1	4.1106	0.6865	4.7971
374.14	22.09	0.003 155	0.003 155	2029.6	0	2029.6	2099.3	0	2099.3	4.4298	0	4.4298

Source: Gordon J. Van Wylen and Richard E. Sonntag, *Fundamentals of Classical Thermodynamics*, English/SI Version, 3d ed., Wiley, New York, 1986, pp. 635–637, table A.1.1. Originally published in Joseph H. Keenan, Frederick G. Keyes, Philip G. Hill, and Joan G. Moore, *Steam Tables,* SI Units, Wiley, New York, 1978.

TABLE A-5

Saturated water–Pressure table

		Specific volume m³/kg		Internal energy kJ/kg			Enthalpy kJ/kg			Entropy kJ/(kg·K)		
Press. kPa P	Sat. Temp. °C T_{sat}	Sat. liquid v_f	Sat. vapor v_g	Sat. liquid u_f	Evap. u_{fg}	Sat. vapor u_g	Sat. liquid h_f	Evap. h_{fg}	Sat. vapor h_g	Sat. liquid s_f	Evap. s_{fg}	Sat. vapor s_g
0.6113	0.01	0.001 000	206.14	0.00	2375.3	2375.3	0.01	2501.3	2501.4	0.0000	9.1562	9.1562
1.0	6.98	0.001 000	129.21	29.30	2355.7	2385.0	29.30	2484.9	2514.2	0.1059	8.8697	8.9756
1.5	13.03	0.001 001	87.98	54.71	2338.6	2393.3	54.71	2470.6	2525.3	0.1957	8.6322	8.8279
2.0	17.50	0.001 001	67.00	73.48	2326.0	2399.5	73.48	2460.0	2533.5	0.2607	8.4629	8.7237
2.5	21.08	0.001 002	54.25	88.48	2315.9	2404.4	88.49	2451.6	2540.0	0.3120	8.3311	8.6432
3.0	24.08	0.001 003	45.67	101.04	2307.5	2408.5	101.05	2444.5	2545.5	0.3545	8.2231	8.5776
4.0	28.96	0.001 004	34.80	121.45	2293.7	2415.2	121.46	2432.9	2554.4	0.4226	8.0520	8.4746
5.0	32.88	0.001 005	28.19	137.81	2282.7	2420.5	137.82	2423.7	2561.5	0.4764	7.9187	8.3951
7.5	40.29	0.001 008	19.24	168.78	2261.7	2430.5	168.79	2406.0	2574.8	0.5764	7.6750	8.2515
10	45.81	0.001 010	14.67	191.82	2246.1	2437.9	191.83	2392.8	2584.7	0.6493	7.5009	8.1502
15	53.97	0.001 014	10.02	225.92	2222.8	2448.7	225.94	2373.1	2599.1	0.7549	7.2536	8.0085
20	60.06	0.001 017	7.649	251.38	2205.4	2456.7	251.40	2358.3	2609.7	0.8320	7.0766	7.9085
25	64.97	0.001 020	6.204	271.90	2191.2	2463.1	271.93	2346.3	2618.2	0.8931	6.9383	7.8314
30	69.10	0.001 022	5.229	289.20	2179.2	2468.4	289.23	2336.1	2625.3	0.9439	6.8247	7.7686
40	75.87	0.001 027	3.993	317.53	2159.5	2477.0	317.58	2319.2	2636.8	1.0259	6.6441	7.6700
50	81.33	0.001 030	3.240	340.44	2143.4	2483.9	340.49	2305.4	2645.9	1.0910	6.5029	7.5939
75	91.78	0.001 037	2.217	384.31	2112.4	2496.7	384.39	2278.6	2663.0	1.2130	6.2434	7.4564
Press. MPa P												
0.100	99.63	0.001 043	1.6940	417.36	2088.7	2506.1	417.46	2258.0	2675.5	1.3026	6.0568	7.3594
0.125	105.99	0.001 048	1.3749	444.19	2069.3	2513.5	444.32	2241.0	2685.4	1.3740	5.9104	7.2844
0.150	111.37	0.001 053	1.1593	466.94	2052.7	2519.7	467.11	2226.5	2693.6	1.4336	5.7897	7.2233
0.175	116.06	0.001 057	1.0036	486.80	2038.1	2524.9	486.99	2213.6	2700.6	1.4849	5.6868	7.1717
0.200	120.23	0.001 061	0.8857	504.49	2025.0	2529.5	504.70	2201.9	2706.7	1.5301	5.5970	7.1271
0.225	124.00	0.001 064	0.7933	520.47	2013.1	2533.6	520.72	2191.3	2712.1	1.5706	5.5173	7.0878
0.250	127.44	0.001 067	0.7187	535.10	2002.1	2537.2	535.37	2181.5	2716.9	1.6072	5.4455	7.0527
0.275	130.60	0.001 070	0.6573	548.59	1991.9	2540.5	548.89	2172.4	2721.3	1.6408	5.3801	7.0209
0.300	133.55	0.001 073	0.6058	561.15	1982.4	2543.6	561.47	2163.8	2725.3	1.6718	5.3201	6.9919
0.325	136.30	0.001 076	0.5620	572.90	1973.5	2546.4	573.25	2155.8	2729.0	1.7006	5.2646	6.9652
0.350	138.88	0.001 079	0.5243	583.95	1965.0	2548.9	584.33	2148.1	2732.4	1.7275	5.2130	6.9405
0.375	141.32	0.001 081	0.4914	594.40	1956.9	2551.3	594.81	2140.8	2735.6	1.7528	5.1647	6.9175
0.40	143.63	0.001 084	0.4625	604.31	1949.3	2553.6	604.74	2133.8	2738.6	1.7766	5.1193	6.8959
0.45	147.93	0.001 088	0.4140	622.77	1934.9	2557.6	623.25	2120.7	2743.9	1.8207	5.0359	6.8565
0.50	151.86	0.001 093	0.3749	639.68	1921.6	2561.2	640.23	2108.5	2748.7	1.8607	4.9606	6.8213
0.55	155.48	0.001 097	0.3427	655.32	1909.2	2564.5	655.93	2097.0	2753.0	1.8973	4.8920	6.7893
0.60	158.85	0.001 101	0.3157	669.90	1897.5	2567.4	670.56	2086.3	2756.8	1.9312	4.8288	6.7600
0.65	162.01	0.001 104	0.2927	683.56	1886.5	2570.1	684.28	2076.0	2760.3	1.9627	4.7703	6.7331
0.70	164.97	0.001 108	0.2729	696.44	1876.1	2572.5	697.22	2066.3	2763.5	1.9922	4.7158	6.7080
0.75	167.78	0.001 112	0.2556	708.64	1866.1	2574.7	709.47	2057.0	2766.4	2.0200	4.6647	6.6847
0.80	170.43	0.001 115	0.2404	720.22	1856.6	2576.8	721.11	2048.0	2769.1	2.0462	4.6166	6.6628
0.85	172.96	0.001 118	0.2270	731.27	1847.4	2578.7	732.22	2039.4	2771.6	2.0710	4.5711	6.6421
0.90	175.38	0.001 121	0.2150	741.83	1838.6	2580.5	742.83	2031.1	2773.9	2.0946	4.5280	6.6226
0.95	177.69	0.001 124	0.2042	751.95	1830.2	2582.1	753.02	2023.1	2776.1	2.1172	4.4869	6.6041
1.00	179.91	0.001 127	0.194 44	761.68	1822.0	2583.6	762.81	2015.3	2778.1	2.1387	4.4478	6.5865
1.10	184.09	0.001 133	0.177 53	780.09	1806.3	2586.4	781.34	2000.4	2781.7	2.1792	4.3744	6.5536
1.20	187.99	0.001 139	0.163 33	797.29	1791.5	2588.8	798.65	1986.2	2784.8	2.2166	4.3067	6.5233
1.30	191.64	0.001 144	0.151 25	813.44	1777.5	2591.0	814.93	1972.7	2787.6	2.2515	4.2438	6.4953

TABLE A-5

(Continued)

Press. MPa P	Sat. temp. °C T_{sat}	Specific volume m³/kg Sat. liquid v_f	Sat. vapor v_g	Internal energy kJ/kg Sat. liquid u_f	Evap. u_{fg}	Sat. vapor u_g	Enthalpy kJ/kg Sat. liquid h_f	Evap. h_{fg}	Sat. vapor h_g	Entropy kJ/(kg·K) Sat. liquid s_f	Evap. s_{fg}	Sat. vapor s_g
1.40	195.07	0.001 149	0.140 84	828.70	1764.1	2592.8	830.30	1959.7	2790.0	2.2842	4.1850	6.4693
1.50	198.32	0.001 154	0.131 77	843.16	1751.3	2594.5	844.89	1947.3	2792.2	2.3150	4.1298	6.4448
1.75	205.76	0.001 166	0.113 49	876.46	1721.4	2597.8	878.50	1917.9	2796.4	2.3851	4.0044	6.3896
2.00	212.42	0.001 177	0.099 63	906.44	1693.8	2600.3	908.79	1890.7	2799.5	2.4474	3.8935	6.3409
2.25	218.45	0.001 187	0.088 75	933.83	1668.2	2602.0	936.49	1865.2	2801.7	2.5035	3.7937	6.2972
2.5	223.99	0.001 197	0.079 98	959.11	1644.0	2603.1	962.11	1841.0	2803.1	2.5547	3.7028	6.2575
3.0	233.90	0.001 217	0.066 68	1004.78	1599.3	2604.1	1008.42	1795.7	2804.2	2.6457	3.5412	6.1869
3.5	242.60	0.001 235	0.057 07	1045.43	1558.3	2603.7	1049.75	1753.7	2803.4	2.7253	3.4000	6.1253
4	250.40	0.001 252	0.049 78	1082.31	1520.0	2602.3	1087.31	1714.1	2801.4	2.7964	3.2737	6.0701
5	263.99	0.001 286	0.039 44	1147.81	1449.3	2597.1	1154.23	1640.1	2794.3	2.9202	3.0532	5.9734
6	275.64	0.001 319	0.032 44	1205.44	1384.3	2589.7	1213.35	1571.0	2784.3	3.0267	2.8625	5.8892
7	285.88	0.001 351	0.027 37	1257.55	1323.0	2580.5	1267.00	1505.1	2772.1	3.1211	2.6922	5.8133
8	295.06	0.001 384	0.023 52	1305.57	1264.2	2569.8	1316.64	1441.3	2758.0	3.2068	2.5364	5.7432
9	303.40	0.001 418	0.020 48	1350.51	1207.3	2557.8	1363.26	1378.9	2742.1	3.2858	2.3915	5.6722
10	311.06	0.001 452	0.018 026	1393.04	1151.4	2544.4	1407.56	1317.1	2724.7	3.3596	2.2544	5.6141
11	318.15	0.001 489	0.015 987	1433.7	1096.0	2529.8	1450.1	1255.5	2705.6	3.4295	2.1233	5.5527
12	324.75	0.001 527	0.014 263	1473.0	1040.7	2513.7	1491.3	1193.6	2684.9	3.4962	1.9962	5.4924
13	330.93	0.001 567	0.012 780	1511.1	985.0	2496.1	1531.5	1130.7	2662.2	3.5606	1.8718	5.4323
14	336.75	0.001 611	0.011 485	1548.6	928.2	2476.8	1571.1	1066.5	2637.6	3.6232	1.7485	5.3717
15	342.24	0.001 658	0.010 337	1585.6	869.8	2455.5	1610.5	1000.0	2610.5	3.6848	1.6249	5.3098
16	347.44	0.001 711	0.009 306	1622.7	809.0	2431.7	1650.1	930.6	2580.6	3.7461	1.4994	5.2455
17	352.37	0.001 770	0.008 364	1660.2	744.8	2405.0	1690.3	856.9	2547.2	3.8079	1.3698	5.1777
18	357.06	0.001 840	0.007 489	1698.9	675.4	2374.3	1732.0	777.1	2509.1	3.8715	1.2329	5.1044
19	361.54	0.001 924	0.006 657	1739.9	598.1	2338.1	1776.5	688.0	2464.5	3.9388	1.0839	5.0228
20	365.81	0.002 036	0.005 834	1785.6	507.5	2293.0	1826.3	583.4	2409.7	4.0139	0.9130	4.9269
21	369.89	0.002 207	0.004 952	1842.1	388.5	2230.6	1888.4	446.2	2334.6	4.1075	0.6938	4.8013
22	373.80	0.002 742	0.003 568	1961.9	125.2	2087.1	2022.2	143.4	2165.6	4.3110	0.2216	4.5327
22.09	374.14	0.003 155	0.003 155	2029.6	0	2029.6	2099.3	0	2099.3	4.4298	0	4.4298

Source: Gordon J. Van Wylen and Richard E. Sonntag, *Fundamentals of Classical Thermodynamics,* English/SI Version, 3d ed., Wiley, New York, 1986, pp. 638–640, table A.1.2. Originally published in Joseph H. Keenan, Frederick G. Keyes, Philip G. Hill, and Joan G. Moore, *Steam Tables,* SI Units, Wiley, New York, 1978.

TABLE A-6

Superheated water

T °C	*v* m³/kg	*u* kJ/kg	*h* kJ/kg	*s* kJ/(kg·K)	*v* m³/kg	*u* kJ/kg	*h* kJ/kg	*s* kJ/(kg·K)	*v* m³/kg	*u* kJ/kg	*h* kJ/kg	*s* kJ/(kg·K)
	P = **0.01 MPa (45.81°C)**				*P* = **0.05 MPa (81.33°C)**				*P* = **0.10 MPa (99.63°C)**			
Sat.	14.674	2437.9	2584.7	8.1502	3.240	2483.9	2645.9	7.5939	1.6940	2506.1	2675.5	7.3594
50	14.869	2443.9	2592.6	8.1749								
100	17.196	2515.5	2687.5	8.4479	3.418	2511.6	2682.5	7.6947	1.6958	2506.7	2676.2	7.3614
150	19.512	2587.9	2783.0	8.6882	3.889	2585.6	2780.1	7.9401	1.9364	2582.8	2776.4	7.6134
200	21.825	2661.3	2879.5	8.9038	4.356	2659.9	2877.7	8.1580	2.172	2658.1	2875.3	7.8343
250	24.136	2736.0	2977.3	9.1002	4.820	2735.0	2976.0	8.3556	2.406	2733.7	2974.3	8.0333
300	26.445	2812.1	3076.5	9.2813	5.284	2811.3	3075.5	8.5373	2.639	2810.4	3074.3	8.2158
400	31.063	2968.9	3279.6	9.6077	6.209	2968.5	3278.9	8.8642	3.103	2967.9	3278.2	8.5435
500	35.679	3132.3	3489.1	9.8978	7.134	3132.0	3488.7	9.1546	3.565	3131.6	3488.1	8.8342
600	40.295	3302.5	3705.4	10.1608	8.057	3302.2	3705.1	9.4178	4.028	3301.9	3704.4	9.0976
700	44.911	3479.6	3928.7	10.4028	8.981	3479.4	3928.5	9.6599	4.490	3479.2	3928.2	9.3398
800	49.526	3663.8	4159.0	10.6281	9.904	3663.6	4158.9	9.8852	4.952	3663.5	4158.6	9.5652
900	54.141	3855.0	4396.4	10.8396	10.828	3854.9	4396.3	10.0967	5.414	3854.8	4396.1	9.7767
1000	58.757	4053.0	4640.6	11.0393	11.751	4052.9	4640.5	10.2964	5.875	4052.8	4640.3	9.9764
1100	63.372	4257.5	4891.2	11.2287	12.674	4257.4	4891.1	10.4859	6.337	4257.3	4891.0	10.1659
1200	67.987	4467.9	5147.8	11.4091	13.597	4467.8	5147.7	10.6662	6.799	4467.7	5147.6	10.3463
1300	72.602	4683.7	5409.7	11.5811	14.521	4683.6	5409.6	10.8382	7.260	4683.5	5409.5	10.5183
	P = **0.20 MPa (120.23°C)**				*P* = **0.30 MPa (133.55°C)**				*P* = **0.40 MPa (143.63°C)**			
Sat.	0.8857	2529.5	2706.7	7.1272	0.6058	2543.6	2725.3	6.9919	0.4625	2553.6	2738.6	6.8959
150	0.9596	2576.9	2768.8	7.2795	0.6339	2570.8	2761.0	7.0778	0.4708	2564.5	2752.8	6.9299
200	1.0803	2654.4	2870.5	7.5066	0.7163	2650.7	2865.6	7.3115	0.5342	2646.8	2860.5	7.1706
250	1.1988	2731.2	2971.0	7.7086	0.7964	2728.7	2967.6	7.5166	0.5951	2726.1	2964.2	7.3789
300	1.3162	2808.6	3071.8	7.8926	0.8753	2806.7	3069.3	7.7022	0.6548	2804.8	3066.8	7.5662
400	1.5493	2966.7	3276.6	8.2218	1.0315	2965.6	3275.0	8.0330	0.7726	2964.4	3273.4	7.8985
500	1.7814	3130.8	3487.1	8.5133	1.1867	3130.0	3486.0	8.3251	0.8893	3129.2	3484.9	8.1913
600	2.013	3301.4	3704.0	8.7770	1.3414	3300.8	3703.2	8.5892	1.0055	3300.2	3702.4	8.4558
700	2.244	3478.8	3927.6	9.0194	1.4957	3478.4	3927.1	8.8319	1.1215	3477.9	3926.5	8.6987
800	2.475	3663.1	4158.2	9.2449	1.6499	3662.9	4157.8	9.0576	1.2372	3662.4	4157.3	8.9244
900	2.705	3854.5	4395.8	9.4566	1.8041	3854.2	4395.4	9.2692	1.3529	3853.9	4395.1	9.1362
1000	2.937	4052.5	4640.0	9.6563	1.9581	4052.3	4639.7	9.4690	1.4685	4052.0	4639.4	9.3360
1100	3.168	4257.0	4890.7	9.8458	2.1121	4256.8	4890.4	9.6585	1.5840	4256.5	4890.2	9.5256
1200	3.399	4467.5	5147.5	10.0262	2.2661	4467.2	5147.1	9.8389	1.6996	4467.0	5146.8	9.7060
1300	3.630	4683.2	5409.3	10.1982	2.4201	4683.0	5409.0	10.0110	1.8151	4682.8	5408.8	9.8780
	P = **0.50 MPa (151.86°C)**				*P* = **0.60 MPa (158.85°C)**				*P* = **0.80 MPa (170.43°C)**			
Sat.	0.3749	2561.2	2748.7	6.8213	0.3157	2567.4	2756.8	6.7600	0.2404	2576.8	2769.1	6.6628
200	0.4249	2642.9	2855.4	7.0592	0.3520	2638.9	2850.1	6.9665	0.2608	2630.6	2839.3	6.8158
250	0.4744	2723.5	2960.7	7.2709	0.3938	2720.9	2957.2	7.1816	0.2931	2715.5	2950.0	7.0384
300	0.5226	2802.9	3064.2	7.4599	0.4344	2801.0	3061.6	7.3724	0.3241	2797.2	3056.5	7.2328
350	0.5701	2882.6	3167.7	7.6329	0.4742	2881.2	3165.7	7.5464	0.3544	2878.2	3161.7	7.4089
400	0.6173	2963.2	3271.9	7.7938	0.5137	2962.1	3270.3	7.7079	0.3843	2959.7	3267.1	7.5716
500	0.7109	3128.4	3483.9	8.0873	0.5920	3127.6	3482.8	8.0021	0.4433	3126.0	3480.6	7.8673
600	0.8041	3299.6	3701.7	7.3522	0.6697	3299.1	3700.9	8.2674	0.5018	3297.9	3699.4	8.1333
700	0.8969	3477.5	3925.9	8.5952	0.7472	3477.0	3925.3	8.5107	0.5601	3476.2	3924.2	8.3770
800	0.9896	3662.1	4156.9	8.8211	0.8245	3661.8	4156.5	8.7367	0.6181	3661.1	4155.6	8.6033
900	1.0822	3853.6	4394.7	9.0329	0.9017	3853.4	4394.4	8.9486	0.6761	3852.8	4393.7	8.8153
1000	1.1747	4051.8	4639.1	9.2328	0.9788	4051.5	4638.8	9.1485	0.7340	4051.0	4638.2	9.0153
1100	1.2672	4256.3	4889.9	9.4224	1.0559	4256.1	4889.6	9.3381	0.7919	4255.6	4889.1	9.2050
1200	1.3596	4466.8	5146.6	9.6029	1.1330	4466.5	5146.3	9.5185	0.8497	4466.1	5145.9	9.3855
1300	1.4521	4682.5	5408.6	9.7749	1.2101	4682.3	5408.3	9.6906	0.9076	4681.8	5407.9	9.5575

TABLE A-6
(Continued)

T °C	v m^3/kg	u kJ/kg	h kJ/kg	s kJ/(kg·K)	v m^3/kg	u kJ/kg	h kJ/kg	s kJ/(kg·K)	v m^3/kg	u kJ/kg	h kJ/kg	s kJ/(kg·K)
	P = 1.00 MPa (179.91°C)				**P = 1.20 MPa (187.99°C)**				**P = 1.40 MPa (195.07°C)**			
Sat.	0.194 44	2583.6	2778.1	6.5865	0.163 33	2588.8	2784.8	6.5233	0.140 84	2592.8	2790.0	6.4693
200	0.2060	2621.9	2827.9	6.6940	0.169 30	2612.8	2815.9	6.5898	0.143 02	2603.1	2803.3	6.4975
250	0.2327	2709.9	2942.6	6.9247	0.192 34	2704.2	2935.0	6.8294	0.163 50	2698.3	2927.2	6.7467
300	0.2579	2793.2	3051.2	7.1229	0.2138	2789.2	3045.8	7.0317	0.182 28	2785.2	3040.4	6.9534
350	0.2825	2875.2	3157.7	7.3011	0.2345	2872.2	3153.6	7.2121	0.2003	2869.2	3149.5	7.1360
400	0.3066	2957.3	3263.9	7.4651	0.2548	2954.9	3260.7	7.3774	0.2178	2952.5	3257.5	7.3026
500	0.3541	3124.4	3478.5	7.7622	0.2946	3122.8	3476.3	7.6759	0.2521	3121.1	3474.1	7.6027
600	0.4011	3296.8	3697.9	8.0290	0.3339	3295.6	3696.3	7.9435	0.2860	3294.4	3694.8	7.8710
700	0.4478	3475.3	3923.1	8.2731	0.3729	3474.4	3922.0	8.1881	0.3195	3473.6	3920.8	8.1160
800	0.4943	3660.4	4154.7	8.4996	0.4118	3659.7	4153.8	8.4148	0.3528	3659.0	4153.0	8.3431
900	0.5407	3852.2	4392.9	8.7118	0.4505	3851.6	4392.2	8.6272	0.3861	3851.1	4391.5	8.5556
1000	0.5871	4050.5	4637.6	8.9119	0.4892	4050.0	4637.0	8.8274	0.4192	4049.5	4636.4	8.7559
1100	0.6335	4255.1	4888.6	9.1017	0.5278	4254.6	4888.0	9.0172	0.4524	4254.1	4887.5	8.9457
1200	0.6798	4465.6	5145.4	9.2822	0.5665	4465.1	5144.9	9.1977	0.4855	4464.7	5144.4	9.1262
1300	0.7261	4681.3	5407.4	9.4543	0.6051	4680.9	5407.0	9.3698	0.5186	4680.4	5406.5	9.2984
	P = 1.60 MPa (201.41°C)				**P = 1.80 MPa (207.15°C)**				**P = 2.00 MPa (212.42°C)**			
Sat.	0.123 80	2596.0	2794.0	6.4218	0.110 42	2598.4	2797.1	6.3794	0.099 63	2600.3	2799.5	6.3409
225	0.132 87	2644.7	2857.3	6.5518	0.116 73	2636.6	2846.7	6.4808	0.103 77	2628.3	2835.8	6.4147
250	0.141 84	2692.3	2919.2	6.6732	0.124 97	2686.0	2911.0	6.6066	0.111 44	2679.6	2902.5	6.5453
300	0.158 62	2781.1	3034.8	6.8844	0.140 21	2776.9	3029.2	6.8226	0.125 47	2772.6	3023.5	6.7664
350	0.174 56	2866.1	3145.4	7.0694	0.154 57	2863.0	3141.2	7.0100	0.138 57	2859.8	3137.0	6.9563
400	0.190 05	2950.1	3254.2	7.2374	0.168 47	2947.7	3250.9	7.1794	0.151 20	2945.2	3247.6	7.1271
500	0.2203	3119.5	3472.0	7.5390	0.195 50	3117.9	3469.8	7.4825	0.175 68	3116.2	3467.6	7.4317
600	0.2500	3293.3	3693.2	7.8080	0.2220	3292.1	3691.7	7.7523	0.199 60	3290.9	3690.1	7.7024
700	0.2794	3472.7	3919.7	8.0535	0.2482	3471.8	3918.5	7.9983	0.2232	3470.9	3917.4	7.9487
800	0.3086	3658.3	4152.1	8.2808	0.2742	3657.6	4151.2	8.2258	0.2467	3657.0	4150.3	8.1765
900	0.3377	3850.5	4390.8	8.4935	0.3001	3849.9	4390.1	8.4386	0.2700	3849.3	4389.4	8.3895
1000	0.3668	4049.0	4635.8	8.6938	0.3260	4048.5	4635.2	8.6391	0.2933	4048.0	4634.6	8.5901
1100	0.3958	4253.7	4887.0	8.8837	0.3518	4253.2	4886.4	8.8290	0.3166	4252.7	4885.9	8.7800
1200	0.4248	4464.2	5143.9	9.0643	0.3776	4463.7	5143.4	9.0096	0.3398	4463.3	5142.9	8.9607
1300	0.4538	4679.9	5406.0	9.2364	0.4034	4679.5	5405.6	9.1818	0.3631	4679.0	5405.1	9.1329
	P = 2.50 MPa (223.99°C)				**P = 3.00 MPa (233.90°C)**				**P = 3.50 MPa (242.60°C)**			
Sat.	0.079 98	2603.1	2803.1	6.2575	0.066 68	2604.1	2804.2	6.1869	0.057 07	2603.7	2803.4	6.1253
225	0.080 27	2605.6	2806.3	6.2639								
250	0.087 00	2662.6	2880.1	6.4085	0.070 58	2644.0	2855.8	6.2872	0.058 72	2623.7	2829.2	6.1749
300	0.098 90	2761.6	3008.8	6.6438	0.081 14	2750.1	2993.5	6.5390	0.068 42	2738.0	2977.5	6.4461
350	0.109 76	2851.9	3126.3	6.8403	0.090 53	2843.7	3115.3	6.7428	0.076 78	2835.3	3104.0	6.6579
400	0.120 10	2939.1	3239.3	7.0148	0.099 36	2932.8	3230.9	6.9212	0.084 53	2926.4	3222.3	6.8405
450	0.130 14	3025.5	3350.8	7.1746	0.107 87	3020.4	3344.0	7.0834	0.091 96	3015.3	3337.2	7.0052
500	0.139 93	3112.1	3462.1	7.3234	0.116 19	3108.0	3456.5	7.2338	0.099 18	3103.0	3450.9	7.1572
600	0.159 30	3288.0	3686.3	7.5960	0.132 43	3285.0	3682.3	7.5085	0.113 24	3282.1	3678.4	7.4339
700	0.178 32	3468.7	3914.5	7.8435	0.148 38	3466.5	3911.7	7.7571	0.126 99	3464.3	3908.8	7.6837
800	0.197 16	3655.3	4148.2	8.0720	0.164 14	3653.5	4145.9	7.9862	0.140 56	3651.8	4143.7	7.9134
900	0.215 90	3847.9	4387.6	8.2853	0.179 80	3846.5	4385.9	8.1999	0.154 02	3845.0	4384.1	8.1276
1000	0.2346	4046.7	4633.1	8.4861	0.195 41	4045.4	4631.6	8.4009	0.167 43	4044.1	4630.1	8.3288
1100	0.2532	4251.5	4884.6	8.6762	0.210 98	4250.3	4883.3	8.5912	0.180 80	4249.2	4881.9	8.5192
1200	0.2718	4462.1	5141.7	8.8569	0.226 52	4460.9	5140.5	8.7720	0.194 15	4459.8	5139.3	8.7000
1300	0.2905	4677.8	5404.0	9.0291	0.242 06	4676.6	5402.8	8.9442	0.207 49	4675.5	5401.7	8.8723

TABLE A-6

(Continued)

T °C	*v* m^3/kg	*u* kJ/kg	*h* kJ/kg	*s* kJ/(kg·K)	*v* m^3/kg	*u* kJ/kg	*h* kJ/kg	*s* kJ/(kg·K)	*v* m^3/kg	*u* kJ/kg	*h* kJ/kg	*s* kJ/(kg·K)
	***P* = 4.0 MPa (250.40°C)**				***P* = 4.5 MPa (257.49°C)**				***P* = 5.0 MPa (263.99°C)**			
Sat.	0.049 78	2602.3	2801.4	6.0701	0.044 06	2600.1	2798.3	6.0198	0.039 44	2597.1	2794.3	5.9734
275	0.054 57	2667.9	2886.2	6.2285	0.047 30	2650.3	2863.2	6.1401	0.041 41	2631.3	2838.3	6.0544
300	0.058 84	2725.3	2960.7	6.3615	0.051 35	2712.0	2943.1	6.2828	0.045 32	2698.0	2924.5	6.2084
350	0.066 45	2826.7	3092.5	6.5821	0.058 40	2817.8	3080.6	6.5131	0.051 94	2808.7	3068.4	6.4493
400	0.073 41	2919.9	3213.6	6.7690	0.064 75	2913.3	3204.7	6.7047	0.057 81	2906.6	3195.7	6.6459
450	0.080 02	3010.2	3330.3	6.9363	0.070 74	3005.0	3323.3	6.8746	0.063 30	2999.7	3316.2	6.8186
500	0.086 43	3099.5	3445.3	7.0901	0.076 51	3095.3	3439.6	7.0301	0.068 57	3091.0	3433.8	6.9759
600	0.098 85	3279.1	3674.4	7.3688	0.087 65	3276.0	3670.5	7.3110	0.078 69	3273.0	3666.5	7.2589
700	0.110 95	3462.1	3905.9	7.6198	0.098 47	3459.9	3903.0	7.5631	0.088 49	3457.6	3900.1	7.5122
800	0.122 87	3650.0	4141.5	7.8502	0.109 11	3648.3	4139.3	7.7942	0.098 11	3646.6	4137.1	7.7440
900	0.134 69	3843.6	4382.3	8.0647	0.119 65	3842.2	4380.6	8.0091	0.107 62	3840.7	4378.8	7.9593
1000	0.146 45	4042.9	4628.7	8.2662	0.130 13	4041.6	4627.2	8.2108	0.117 07	4040.4	4625.7	8.1612
1100	0.158 17	4248.0	4880.6	8.4567	0.140 56	4246.8	4879.3	8.4015	0.126 48	4245.6	4878.0	8.3520
1200	0.169 87	4458.6	5138.1	8.6376	0.150 98	4457.5	5136.9	8.5825	0.135 87	4456.3	5135.7	8.5331
1300	0.181 56	4674.3	5400.5	8.8100	0.161 39	4673.1	5399.4	8.7549	0.145 26	4672.0	5398.2	8.7055
	***P* = 6.0 MPa (275.64°C)**				***P* = 7.0 MPa (285.88°C)**				***P* = 8.0 MPa (295.06°C)**			
Sat.	0.032 44	2589.7	2784.3	5.8892	0.027 37	2580.5	2772.1	5.8133	0.023 52	2569.8	2758.0	5.7432
300	0.036 16	2667.2	2884.2	6.0674	0.029 47	2632.2	2838.4	5.9305	0.024 26	2590.9	2785.0	5.7906
350	0.042 23	2789.6	3043.0	6.3335	0.035 24	2769.4	3016.0	6.2283	0.029 95	2747.7	2987.3	6.1301
400	0.047 39	2892.9	3177.2	6.5408	0.039 93	2878.6	3158.1	6.4478	0.034 32	2863.8	3138.3	6.3634
450	0.052 14	2988.9	3301.8	6.7193	0.044 16	2978.0	3287.1	6.6327	0.038 17	2966.7	3272.0	6.5551
500	0.056 65	3082.2	3422.2	6.8803	0.048 14	3073.4	3410.3	6.7975	0.041 75	3064.3	3398.3	6.7240
550	0.061 01	3174.6	3540.6	7.0288	0.051 95	3167.2	3530.9	6.9486	0.045 16	3159.8	3521.0	6.8778
600	0.065 25	3266.9	3658.4	7.1677	0.055 65	3260.7	3650.3	7.0894	0.048 45	3254.4	3642.0	7.0206
700	0.073 52	3453.1	3894.2	7.4234	0.062 83	3448.5	3888.3	7.3476	0.054 81	3443.9	3882.4	7.2812
800	0.081 60	3643.1	4132.7	7.6566	0.069 81	3639.5	4128.2	7.5822	0.060 97	3636.0	4123.8	7.5173
900	0.089 58	3837.8	4375.3	7.8727	0.076 69	3835.0	4371.8	7.7991	0.067 02	3832.1	4368.3	7.7351
1000	0.097 49	4037.8	4622.7	8.0751	0.083 50	4035.3	4619.8	8.0020	0.073 01	4032.8	4616.9	7.9384
1100	0.105 36	4243.3	4875.4	8.2661	0.090 27	4240.9	4872.8	8.1933	0.078 96	4238.6	4870.3	8.1300
1200	0.113 21	4454.0	5133.3	8.4474	0.097 03	4451.7	5130.9	8.3747	0.084 89	4449.5	5128.5	8.3115
1300	0.121 06	4669.6	5396.0	8.6199	0.103 77	4667.3	5393.7	8.5475	0.090 80	4665.0	5391.5	8.4842
	***P* = 9.0 MPa (303.40°C)**				***P* = 10.0 MPa (311.06°C)**				***P* = 12.5 MPa (327.89°C)**			
Sat.	0.020 48	2557.8	2742.1	5.6772	0.018 026	2544.4	2724.7	5.6141	0.013 495	2505.1	2673.8	5.4624
325	0.023 27	2646.6	2856.0	5.8712	0.019 861	2610.4	2809.1	5.7568				
350	0.025 80	2724.4	2956.6	6.0361	0.022 42	2699.2	2923.4	5.9443	0.016 126	2624.6	2826.2	5.7118
400	0.029 93	2848.4	3117.8	6.2854	0.026 41	2832.4	3096.5	6.2120	0.020 00	2789.3	3039.3	6.0417
450	0.033 50	2955.2	3256.6	6.4844	0.029 75	2943.4	3240.9	6.4190	0.022 99	2912.5	3199.8	6.2719
500	0.036 77	3055.2	3386.1	6.6576	0.032 79	3045.8	3373.7	6.5966	0.025 60	3021.7	3341.8	6.4618
550	0.039 87	3152.2	3511.0	6.8142	0.035 64	3144.6	3500.9	6.7561	0.028 01	3125.0	3475.2	6.6290
600	0.042 85	3248.1	3633.7	6.9589	0.038 37	3241.7	3625.3	6.9029	0.030 29	3225.4	3604.0	6.7810
650	0.045 74	3343.6	3755.3	7.0943	0.041 01	3338.2	3748.2	7.0398	0.032 48	3324.4	3730.4	6.9218
700	0.048 57	3439.3	3876.5	7.2221	0.043 58	3434.7	3870.5	7.1687	0.034 60	3422.9	3855.3	7.0536
800	0.054 09	3632.5	4119.3	7.4596	0.048 59	3628.9	4114.8	7.4077	0.038 69	3620.0	4103.6	7.2965
900	0.059 50	3829.2	4364.8	7.6783	0.053 49	3826.3	4361.2	7.6272	0.042 67	3819.1	4352.5	7.5182
1000	0.064 85	4030.3	4614.0	7.8821	0.058 32	4027.8	4611.0	7.8315	0.046 58	4021.6	4603.8	7.7237
1100	0.070 16	4236.3	4867.7	8.0740	0.063 12	4234.0	4865.1	8.0237	0.050 45	4228.2	4858.8	7.9165
1200	0.075 44	4447.2	5126.2	8.2556	0.067 89	4444.9	5123.8	8.2055	0.054 30	4439.3	5118.0	8.0937
1300	0.080 72	4662.7	5389.2	8.4284	0.072 65	4460.5	5387.0	8.3783	0.058 13	4654.8	5381.4	8.2717

TABLE A-6
(Continued)

T °C	v m³/kg	u kJ/kg	h kJ/kg	s kJ/(kg·K)	v m³/kg	u kJ/kg	h kJ/kg	s kJ/(kg·K)	v m³/kg	u kJ/kg	h kJ/kg	s kJ/(kg·K)
	P = **15.0 MPa (342.24°C)**				P = **17.5 MPa (354.75°C)**				P = **20.0 MPa (365.81°C)**			
Sat.	0.010 337	2455.5	2610.5	5.3098	0.007 920	2390.2	2528.8	5.1419	0.005 834	2293.0	2409.7	4.9269
350	0.011 470	2520.4	2692.4	5.4421								
400	0.015 649	2740.7	2975.5	5.8811	0.012 447	2685.0	2902.9	5.7213	0.009 942	2619.3	2818.1	5.5540
450	0.018 445	2879.5	3156.2	6.1404	0.015 174	2844.2	3109.7	6.0184	0.012 695	2806.2	3060.1	5.9017
500	0.020 80	2996.6	3308.6	6.3443	0.017 358	2970.3	3274.1	6.2383	0.014 768	2942.9	3238.2	6.1401
550	0.022 93	3104.7	3448.6	6.5199	0.019 288	3083.9	3421.4	6.4230	0.016 555	3062.4	3393.5	6.3348
600	0.024 91	3208.6	3582.3	6.6776	0.021 06	3191.5	3560.1	6.5866	0.018 178	3174.0	3537.6	6.5048
650	0.026 80	3310.3	3712.3	6.8224	0.022 74	3296.0	3693.9	6.7357	0.019 693	3281.4	3675.3	6.6582
700	0.028 61	3410.9	3840.1	6.9572	0.024 34	3398.7	3824.6	6.8736	0.021 13	3386.4	3809.0	6.7993
800	0.032 10	3610.9	4092.4	7.2040	0.027 38	3601.8	4081.1	7.1244	0.023 85	3592.7	4069.7	7.0544
900	0.035 46	3811.9	4343.8	7.4279	0.030 31	3804.7	4335.1	7.3507	0.026 45	3797.5	4326.4	7.2830
1000	0.038 75	4015.4	4596.6	7.6348	0.033 16	4009.3	4589.5	7.5589	0.028 97	4003.1	4582.5	7.4925
1100	0.042 00	4222.6	4852.6	7.8283	0.035 97	4216.9	4846.4	7.7531	0.031 45	4211.3	4840.2	7.6874
1200	0.045 23	4433.8	5112.3	8.0108	0.038 76	4428.3	5106.6	7.9360	0.033 91	4422.8	5101.0	7.8707
1300	0.048 45	4649.1	5376.0	8.1840	0.041 54	4643.5	5370.5	8.1093	0.036 36	4638.0	5365.1	8.0442
	P = **25.0 MPa**				P = **30.0 MPa**				P = **35.0 MPa**			
375	0.001 973 1	1798.7	1848.0	4.0320	0.001 789 2	1737.8	1791.5	3.9305	0.001 700 3	1702.9	1762.4	3.8722
400	0.006 004	2430.1	2580.2	5.1418	0.002 790	2067.4	2151.1	4.4728	0.002 100	1914.1	1987.6	4.2126
425	0.007 881	2609.2	2806.3	5.4723	0.005 303	2455.1	2614.2	5.1504	0.003 428	2253.4	2373.4	4.7747
450	0.009 162	2720.7	2949.7	5.6744	0.006 735	2619.3	2821.4	5.4424	0.004 961	2498.7	2672.4	5.1962
500	0.011 123	2884.3	3162.4	5.9592	0.008 678	2820.7	3081.1	5.7905	0.006 927	2751.9	2994.4	5.6282
550	0.012 724	3017.5	3335.6	6.1765	0.010 168	2970.3	3275.4	6.0342	0.008 345	2921.0	3213.0	5.9026
600	0.014 137	3137.9	3491.4	6.3602	0.011 446	3100.5	3443.9	6.2331	0.009 527	3062.0	3395.5	6.1179
650	0.015 433	3251.6	3637.4	6.5229	0.012 596	3221.0	3598.9	6.4058	0.010 575	3189.8	3559.9	6.3010
700	0.016 646	3361.3	3777.5	6.6707	0.013 661	3335.8	3745.6	6.5606	0.011 533	3309.8	3713.5	6.4631
800	0.018 912	3574.3	4047.1	6.9345	0.015 623	3555.5	4024.2	6.8332	0.013 278	3536.7	4001.5	6.7450
900	0.021 045	3783.0	4309.1	7.1680	0.017 448	3768.5	4291.9	7.0718	0.014 883	3754.0	4274.9	6.9386
1000	0.023 10	3990.9	4568.5	7.3802	0.019 196	3978.8	4554.7	7.2867	0.016 410	3966.7	4541.1	7.2064
1100	0.025 12	4200.2	4828.2	7.5765	0.020 903	4189.2	4816.3	7.4845	0.017 895	4178.3	4804.6	7.4037
1200	0.027 11	4412.0	5089.9	7.7605	0.022 589	4401.3	5079.0	7.6692	0.019 360	4390.7	5068.3	7.5910
1300	0.029 10	4626.9	5354.4	7.9342	0.024 266	4616.0	5344.0	7.8432	0.020 815	4605.1	5333.6	7.7653
	P = **40.0 MPa**				P = **50.0 MPa**				P = **60.0 MPa**			
375	0.001 640 7	1677.1	1742.8	3.8290	0.001 559 4	1638.6	1716.6	3.7639	0.001 502 8	1609.4	1699.5	3.7141
400	0.001 907 7	1854.6	1930.9	4.1135	0.001 730 9	1788.1	1874.6	4.0031	0.001 633 5	1745.4	1843.4	3.9318
425	0.002 532	2096.9	2198.1	4.5029	0.002 007	1959.7	2060.0	4.2734	0.001 816 5	1892.7	2001.7	4.1626
450	0.003 693	2365.1	2512.8	4.9459	0.002 486	2159.6	2284.0	4.5884	0.002 085	2053.9	2179.0	4.4121
500	0.005 622	2678.4	2903.3	5.4700	0.003 892	2525.5	2720.1	5.1726	0.002 956	2390.6	2567.9	4.9321
550	0.006 984	2869.7	3149.1	5.7785	0.005 118	2763.6	3019.5	5.5485	0.003 956	2658.8	2896.2	5.3441
600	0.008 094	3022.6	3346.4	6.0114	0.006 112	2942.0	3247.6	5.8178	0.004 834	2861.1	3151.2	5.6452
650	0.009 063	3158.0	3520.6	6.2054	0.006 966	3093.5	3441.8	6.0342	0.005 595	3028.8	3364.5	5.8829
700	0.009 941	3283.6	3681.2	6.3750	0.007 727	3230.5	3616.8	6.2189	0.006 272	3177.2	3553.5	6.0824
800	0.011 523	3517.8	3978.7	6.6662	0.009 076	3479.8	3933.6	6.5290	0.007 459	3441.5	3889.1	6.4109
900	0.012 962	3739.4	4257.9	6.9150	0.010 283	3710.3	4224.4	6.7882	0.008 508	3681.0	4191.5	6.6805
1000	0.014 324	3954.6	4527.6	7.1356	0.011 411	3930.5	4501.1	7.0146	0.009 480	3906.4	4475.2	6.9127
1100	0.015 642	4167.4	4793.1	7.3364	0.012 496	4145.7	4770.5	7.2184	0.010 409	4124.1	4748.6	7.1195
1200	0.016 940	4380.1	5057.7	7.5224	0.013 561	4359.1	5037.2	7.4058	0.011 317	4338.2	5017.2	7.3083
1300	0.018 229	4594.3	5323.5	7.6969	0.014 616	4572.8	5303.6	7.5808	0.012 215	4551.4	5284.3	7.4837

Source: Gordon J. Van Wylen and Richard E. Sonntag, *Fundamentals of Classical Thermodynamics*, English/SI Version, 3d ed., Wiley, New York, 1986, pp. 641–648, table A.1.3. Originally published in Joseph H. Keenan, Frederick G. Keyes, Philip G. Hill, and Joan G. Moore, *Steam Tables*, SI Units, Wiley, New York, 1978.

TABLE A-7

Compressed liquid water

T °C	v m^3/kg	u kJ/kg	h kJ/kg	s kJ/(kg·K)	v m^3/kg	u kJ/kg	h kJ/kg	s kJ/(kg·K)	v m^3/kg	u kJ/kg	h kJ/kg	s kJ/(kg·K)
	P = 5 MPa (263.99°C)				P = 10 MPa (311.06°C)				P = 15 MPa (342.24°C)			
Sat.	0.001 285 9	1147.8	1154.2	2.9202	0.001 452 4	1393.0	1407.6	3.3596	0.001 658 1	1585.6	1610.5	3.6848
0	0.000 997 7	0.04	5.04	0.0001	0.000 995 2	0.09	10.04	0.0002	0.000 992 8	0.15	15.05	0.0004
20	0.000 999 5	83.65	88.65	0.2956	0.000 997 2	83.36	93.33	0.2945	0.000 995 0	83.06	97.99	0.2934
40	0.001 005 6	166.95	171.97	0.5705	0.001 003 4	166.35	176.38	0.5686	0.001 001 3	165.76	180.78	0.5666
60	0.001 014 9	250.23	255.30	0.8285	0.001 012 7	249.36	259.49	0.8258	0.001 010 5	248.51	263.67	0.8232
80	0.001 026 8	333.72	338.85	1.0720	0.001 024 5	332.59	342.83	1.0688	0.001 022 2	331.48	346.81	1.0656
100	0.001 041 0	417.52	422.72	1.3030	0.001 038 5	416.12	426.50	1.2992	0.001 036 1	414.74	430.28	1.2955
120	0.001 057 6	501.80	507.09	1.5233	0.001 054 9	500.08	510.64	1.5189	0.001 052 2	498.40	514.19	1.5145
140	0.001 076 8	586.76	592.15	1.7343	0.001 073 7	584.68	595.42	1.7292	0.001 070 7	582.66	598.72	1.7242
160	0.001 098 8	672.62	678.12	1.9375	0.001 095 3	670.13	681.08	1.9317	0.001 091 8	667.71	684.09	1.9260
180	0.001 124 0	759.63	765.25	2.1341	0.001 119 9	756.65	767.84	2.1275	0.001 115 9	753.76	770.50	2.1210
200	0.001 153 0	848.1	853.9	2.3255	0.001 148 0	844.5	856.0	2.3178	0.001 143 3	841.0	858.2	2.3104
220	0.001 186 6	938.4	944.4	2.5128	0.001 180 5	934.1	945.9	2.5039	0.001 174 8	929.9	947.5	2.4953
240	0.001 226 4	1031.4	1037.5	2.6979	0.001 218 7	1026.0	1038.1	2.6872	0.001 211 4	1020.8	1039.0	2.6771
260	0.001 274 9	1127.9	1134.3	2.8830	0.001 264 5	1121.1	1133.7	2.8699	0.001 255 0	1114.6	1133.4	2.8576
280					0.001 321 6	1220.9	1234.1	3.0548	0.001 308 4	1212.5	1232.1	3.0393
300					0.001 397 2	1328.4	1342.3	3.2469	0.001 377 0	1316.6	1337.3	3.2260
320									0.001 472 4	1431.1	1453.2	3.4247
340									0.001 631 1	1567.5	1591.9	3.6546

T °C	v m^3/kg	u kJ/kg	h kJ/kg	s kJ/(kg·K)	v m^3/kg	u kJ/kg	h kJ/kg	s kJ/(kg·K)	v m^3/kg	u kJ/kg	h kJ/kg	s kJ/(kg·K)
	P = 20 MPa (365.81°C)				P = 30 MPa				P = 50 MPa			
Sat.	0.002 036	1785.6	1826.3	4.0139								
0	0.000 990 4	0.19	20.01	0.0004	0.000 985 6	0.25	29.82	0.0001	0.000 976 6	0.20	49.03	0.0014
20	0.000 992 8	82.77	102.62	0.2923	0.000 988 6	82.17	111.84	0.2899	0.000 980 4	81.00	130.02	0.2848
40	0.000 999 2	165.17	185.16	0.5646	0.000 995 1	164.04	193.89	0.5607	0.000 987 2	161.86	211.21	0.5527
60	0.001 008 4	247.68	267.85	0.8206	0.001 004 2	246.06	276.19	0.8154	0.000 996 2	242.98	292.79	0.8052
80	0.001 019 9	330.40	350.80	1.0624	0.001 015 6	328.30	358.77	1.0561	0.001 007 3	324.34	374.70	1.0440
100	0.001 033 7	413.39	434.06	1.2917	0.001 029 0	410.78	441.66	1.2844	0.001 020 1	405.88	456.89	1.2703
120	0.001 049 6	496.76	517.76	1.5102	0.001 044 5	493.59	524.93	1.5018	0.001 034 8	487.65	539.39	1.4857
140	0.001 067 8	580.69	602.04	1.7193	0.001 062 1	576.88	608.75	1.7098	0.001 051 5	569.77	622.35	1.6915
160	0.001 088 5	665.35	687.12	1.9204	0.001 082 1	660.82	693.28	1.9096	0.001 070 3	652.41	705.92	1.8891
180	0.001 112 0	750.95	773.20	2.1147	0.001 104 7	745.59	778.73	2.1024	0.001 091 2	735.69	790.25	2.0794
200	0.001 138 8	837.7	860.5	2.3031	0.001 130 2	831.4	865.3	2.2893	0.001 114 6	819.7	875.5	2.2634
220	0.001 169 5	925.9	949.3	2.4870	0.001 159 0	918.3	953.1	2.4711	0.001 140 8	904.7	961.7	2.4419
240	0.001 204 6	1016.0	1040.0	2.6674	0.001 192 0	1006.9	1042.6	2.6490	0.001 170 2	990.7	1049.2	2.6158
260	0.001 246 2	1108.6	1133.5	2.8459	0.001 230 3	1097.4	1134.3	2.8243	0.001 203 4	1078.1	1138.2	2.7860
280	0.001 296 5	1204.7	1230.6	3.0248	0.001 275 5	1190.7	1229.0	2.9986	0.001 241 5	1167.2	1229.3	2.9537
300	0.001 359 6	1306.1	1333.3	3.2071	0.001 330 4	1287.9	1327.8	3.1741	0.001 286 0	1258.7	1323.0	3.1200
320	0.001 443 7	1415.7	1444.6	3.3979	0.001 399 7	1390.7	1432.7	3.3539	0.001 338 8	1353.3	1420.2	3.2868
340	0.001 568 4	1539.7	1571.0	3.6075	0.001 492 0	1501.7	1546.5	3.5426	0.001 403 2	1452.0	1522.1	3.4557
360	0.001 822 6	1702.8	1739.3	3.8772	0.001 626 5	1626.6	1675.4	3.7494	0.001 483 8	1556.0	1630.2	3.6291
380					0.001 869 1	1781.4	1837.5	4.0012	0.001 588 4	1667.2	1746.6	3.8101

Source: Gordon J. Van Wylen and Richard E. Sonntag, *Fundamentals of Classical Thermodynamics,* English/SI Version, 3d ed., Wiley, New York, 1986, pp. 649–650, table A.1.4. Originally published in Joseph H. Keenan, Frederick G. Keyes, Philip G. Hill, and Joan G. Moore, *Steam Tables,* SI Units, Wiley, New York, 1978.

TABLE A-8

Saturated ice-water vapor

Temp. T °C	Sat. press. P_{sat} kPa	Specific volume m³/kg: Sat. ice $v_i \times 10^3$	Sat. vapor v_g	Internal energy kJ/kg: Sat. ice u_i	Subl. u_{ig}	Sat. vapor u_g	Enthalpy kJ/kg: Sat. ice h_i	Subl. h_{ig}	Sat. vapor h_g	Entropy kJ/(kg·K): Sat. ice s_i	Subl. s_{ig}	Sat. vapor s_g
0.01	0.6113	1.0908	206.1	−333.40	2708.7	2375.3	−333.40	2834.8	2501.4	−1.221	10.378	9.156
0	0.6108	1.0908	206.3	−333.43	2708.8	2375.3	−333.43	2834.8	2501.3	−1.221	10.378	9.157
−2	0.5176	1.0904	241.7	−337.62	2710.2	2372.6	−337.62	2835.3	2497.7	−1.237	10.456	9.219
−4	0.4375	1.0901	283.8	−341.78	2711.6	2369.8	−341.78	2835.7	2494.0	−1.253	10.536	9.283
−6	0.3689	1.0898	334.2	−345.91	2712.9	2367.0	−345.91	2836.2	2490.3	−1.268	10.616	9.348
−8	0.3102	1.0894	394.4	−350.02	2714.2	2364.2	−350.02	2836.6	2486.6	−1.284	10.698	9.414
−10	0.2602	1.0891	466.7	−354.09	2715.5	2361.4	−354.09	2837.0	2482.9	−1.299	10.781	9.481
−12	0.2176	1.0888	553.7	−358.14	2716.8	2358.7	−358.14	2837.3	2479.2	−1.315	10.865	9.550
−14	0.1815	1.0884	658.8	−362.15	2718.0	2355.9	−362.15	2837.6	2475.5	−1.331	10.950	9.619
−16	0.1510	1.0881	786.0	−366.14	2719.2	2353.1	−366.14	2837.9	2471.8	−1.346	11.036	9.690
−18	0.1252	1.0878	940.5	−370.10	2720.4	2350.3	−370.10	2838.2	2468.1	−1.362	11.123	9.762
−20	0.1035	1.0874	1128.6	−374.03	2721.6	2347.5	−374.03	2838.4	2464.3	−1.377	11.212	9.835
−22	0.0853	1.0871	1358.4	−377.93	2722.7	2344.7	−377.93	2838.6	2460.6	−1.393	11.302	9.909
−24	0.0701	1.0868	1640.1	−381.80	2723.7	2342.0	−381.80	2838.7	2456.9	−1.408	11.394	9.985
−26	0.0574	1.0864	1986.4	−385.64	2724.8	2339.2	−385.64	2838.9	2453.2	−1.424	11.486	10.062
−28	0.0469	1.0861	2413.7	−389.45	2725.8	2336.4	−389.45	2839.0	2449.5	−1.439	11.580	10.141
−30	0.0381	1.0858	2943	−393.23	2726.8	2333.6	−393.23	2839.0	2445.8	−1.455	11.676	10.221
−32	0.0309	1.0854	3600	−396.98	2727.8	2330.8	−396.98	2839.1	2442.1	−1.471	11.773	10.303
−34	0.0250	1.0851	4419	−400.71	2728.7	2328.0	−400.71	2839.1	2438.4	−1.486	11.872	10.386
−36	0.0201	1.0848	5444	−404.40	2729.6	2325.2	−404.40	2839.1	2434.7	−1.501	11.972	10.470
−38	0.0161	1.0844	6731	−408.06	2730.5	2322.4	−408.06	2839.0	2430.9	−1.517	12.073	10.556
−40	0.0129	1.0841	8354	−411.70	2731.3	2319.6	−411.70	2839.9	2427.2	−1.532	12.176	10.644

Source: Gordon J. Van Wylen and Richard E. Sonntag, *Fundamentals of Classical Thermodynamics,* English/SI Version, 3d ed., Wiley, New York, 1986, p. 651, table A.1.5. Originally published in Joseph H. Keenan, Frederick G. Keyes, Philip G. Hill, and Joan G. Moore, *Steam Tables,* SI Units, Wiley, New York, 1978.

FIGURE A-9

T-s diagram for water. (*Source*: Lester Haar, John S. Gallagher, and George S. Kell, *NBS/NRC Steam Tables,* 1984. With permission from Hemisphere Publishing Corporation, New York.)

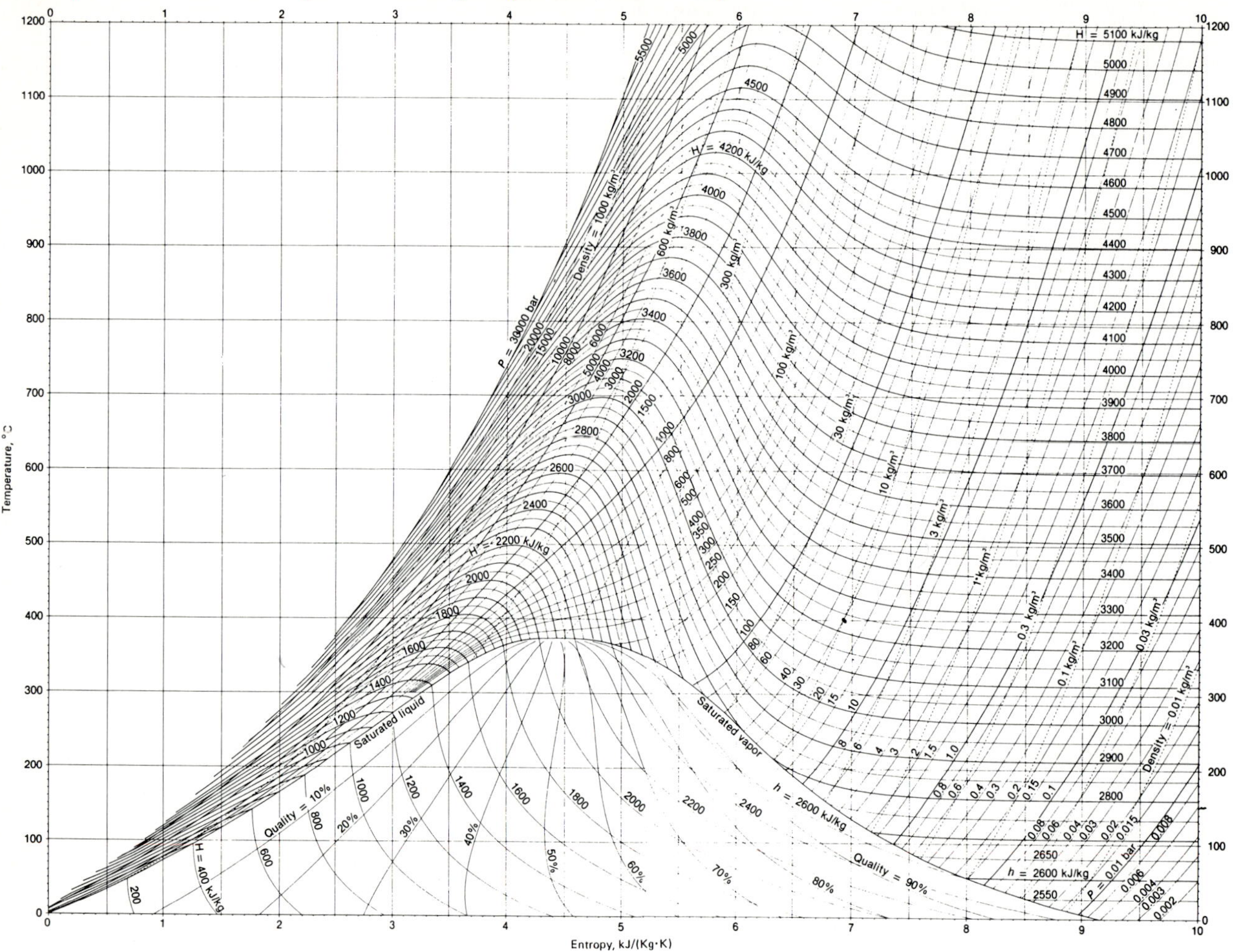

FIGURE A-10

Mollier diagram for water. (*Source*: Lester Haar, John S. Gallagher, and George S. Kell, *NBS/NRC Steam Tables,* 1984. With permission from Hemisphere Publishing Corporation, New York.)

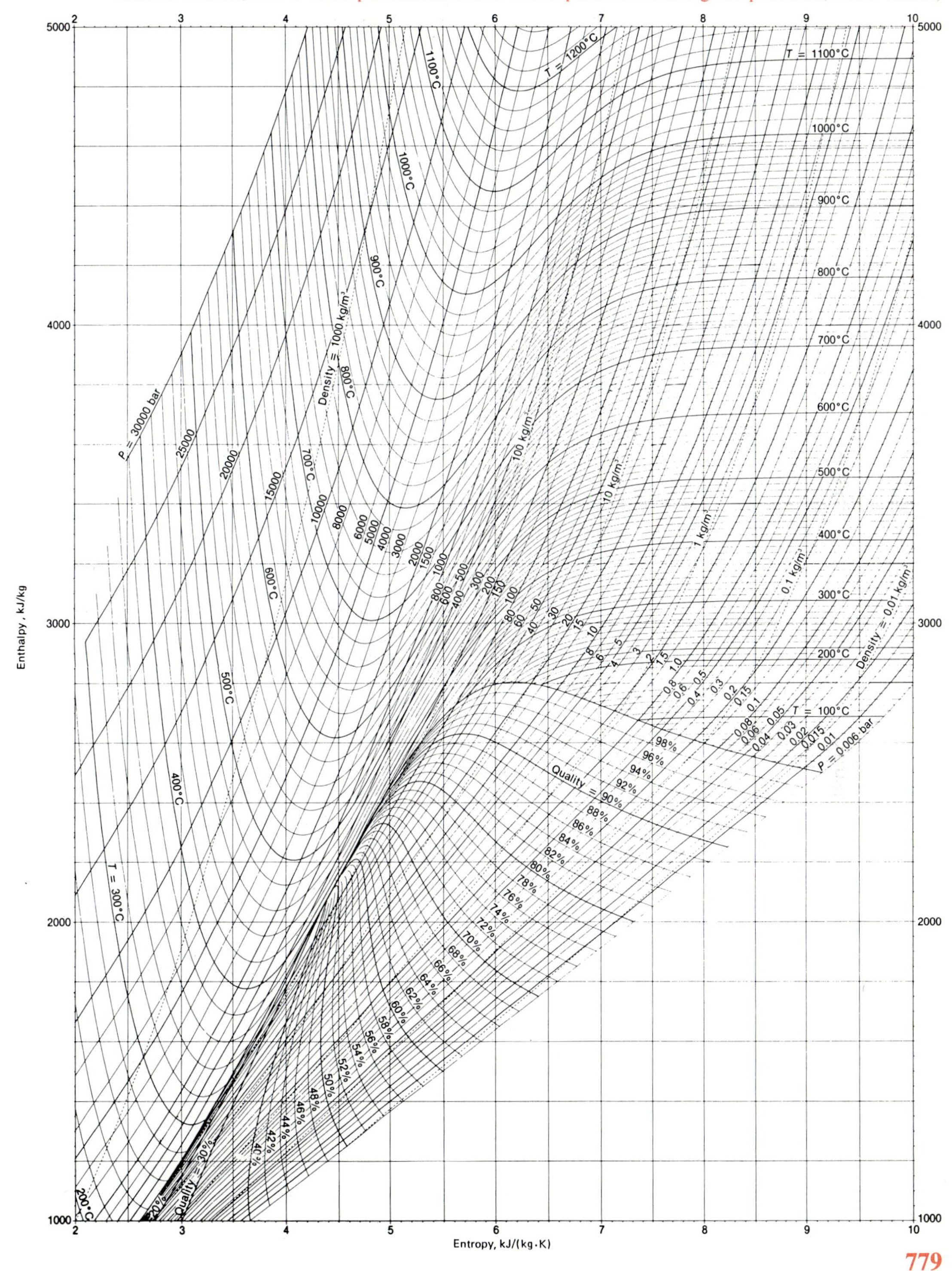

TABLE A-11

Saturated refrigerant-12–Temperature table

Temp. °C T	Sat. press. MPa P_{sat}	Specific volume m³/kg Sat. liquid v_f	Sat. vapor v_g	Internal energy kJ/kg Sat. liquid u_f	Sat. vapor u_g	Enthalpy kJ/kg Sat. liquid h_f	Evap. h_{fg}	Sat. vapor h_g	Entropy kJ/(kg·K) Sat. liquid s_f	Sat. vapor s_g
−40	0.06417	0.0006 595	0.241 91	−0.04	154.07	0	169.59	169.59	0	0.7274
−35	0.08071	0.0006 656	0.195 40	4.37	156.13	4.42	167.48	171.90	0.0187	0.7219
−30	0.10041	0.0006 720	0.159 38	8.79	158.20	8.86	165.33	174.20	0.0371	0.7170
−28	0.10927	0.0006 746	0.147 28	10.58	159.02	10.65	164.46	175.11	0.0444	0.7153
−26	0.11872	0.0006 773	0.136 28	12.35	159.84	12.43	163.59	176.02	0.0517	0.7135
−25	0.12368	0.0006 786	0.131 17	13.25	160.26	13.33	163.15	176.48	0.0552	0.7126
−24	0.12880	0.0006 800	0.126 28	14.13	160.67	14.22	162.71	176.93	0.0589	0.7119
−22	0.13953	0.0006 827	0.117 17	15.92	161.48	16.02	161.82	177.83	0.0660	0.7103
−20	0.15093	0.0006 855	0.108 85	17.72	162.31	17.82	160.92	178.74	0.0731	0.7087
−18	0.16304	0.0006 883	0.101 24	19.51	163.12	19.62	160.01	179.63	0.0802	0.7073
−15	0.18260	0.0006 926	0.091 02	22.20	164.35	22.33	158.64	180.97	0.0906	0.7051
−10	0.21912	0.0007 000	0.076 65	26.72	166.39	26.87	156.31	183.19	0.1080	0.7019
−5	0.26096	0.0007 078	0.064 96	31.27	168.42	31.45	153.93	185.37	0.1251	0.6991
0	0.30861	0.0007 159	0.055 39	35.83	170.44	36.05	151.48	187.53	0.1420	0.6965
4	0.35124	0.0007 227	0.048 95	39.51	172.04	39.76	149.47	189.23	0.1553	0.6946
8	0.39815	0.0007 297	0.043 40	43.21	173.63	43.50	147.41	190.91	0.1686	0.6929
12	0.44962	0.0007 370	0.038 60	46.93	175.20	47.26	145.30	192.56	0.1817	0.6913
16	0.50591	0.0007 446	0.034 42	50.67	176.78	51.05	143.14	194.19	0.1948	0.6898
20	0.56729	0.0007 525	0.030 78	54.44	178.32	54.87	140.91	195.78	0.2078	0.6884
24	0.63405	0.0007 607	0.027 59	58.25	179.85	58.73	138.61	197.34	0.2207	0.6871
26	0.66954	0.0007 650	0.026 14	60.17	180.61	60.68	137.44	198.11	0.2271	0.6865
28	0.70648	0.0007 694	0.024 78	62.09	181.36	62.63	136.24	198.87	0.2335	0.6859
30	0.74490	0.0007 739	0.023 51	64.01	182.11	64.59	135.03	199.62	0.2400	0.6853
32	0.78485	0.0007 785	0.022 31	65.96	182.85	66.57	133.79	200.36	0.2463	0.6847
34	0.82636	0.0007 832	0.021 18	67.90	183.59	68.55	132.53	201.09	0.2527	0.6842
36	0.86948	0.0007 880	0.020 12	69.86	184.31	70.55	131.25	201.80	0.2591	0.6836
38	0.91423	0.0007 929	0.019 12	71.84	185.03	72.56	129.94	202.51	0.2655	0.6831
40	0.96065	0.0007 980	0.018 17	73.82	185.74	74.59	128.61	203.20	0.2718	0.6825
42	1.0088	0.0008 033	0.017 28	75.82	186.45	76.63	127.25	203.88	0.2782	0.6820
44	1.0587	0.0008 086	0.016 44	77.82	187.13	78.68	125.87	204.54	0.2845	0.6814
48	1.1639	0.0008 199	0.014 88	81.88	188.51	82.83	123.00	205.83	0.2973	0.6802
52	1.2766	0.0008 318	0.013 49	86.00	189.83	87.06	119.99	207.05	0.3101	0.6791
56	1.3972	0.0008 445	0.012 24	90.18	191.10	91.36	116.84	208.20	0.3229	0.6779
60	1.5259	0.0008 581	0.011 11	94.43	192.31	95.74	113.52	209.26	0.3358	0.6765
112	4.1155	0.0017 92	0.001 79	175.98	175.98	183.35	0	183.35	0.5687	0.5687

Source: Adapted from Kenneth Wark, *Thermodynamics*, 4th ed., McGraw-Hill, New York, 1983, pp. 807–808, table A-16M. Originally published by E. I. du Pont de Nemours & Company, Inc., 1969.

TABLE A-12

Saturated refrigerant-12–Pressure table

Press. MPa P	Sat. Temp. °C T_{sat}	Specific volume m³/kg Sat. liquid v_f	Sat. vapor v_g	Internal energy kJ/kg Sat. liquid u_f	Sat. vapor u_g	Enthalpy kJ/kg Sat. liquid h_f	Evap. h_{fg}	Sat. vapor h_g	Entropy kJ/(kg·K) Sat. liquid s_f	Sat. vapor s_g
0.06	−41.42	0.000 657 8	0.2575	−1.29	153.49	−1.25	170.19	168.94	−0.0054	0.7290
0.10	−30.10	0.000 671 9	0.1600	8.71	158.15	8.78	165.37	174.15	0.0368	0.7171
0.12	−25.74	0.000 677 6	0.1349	12.58	159.95	12.66	163.48	176.14	0.0526	0.7133
0.14	−21.91	0.000 682 8	0.1168	15.99	161.52	16.09	161.78	177.87	0.0663	0.7102
0.16	−18.49	0.000 687 6	0.1031	19.07	162.91	19.18	160.23	179.41	0.0784	0.7076
0.18	−15.38	0.000 692 1	0.092 25	21.86	164.19	21.98	158.82	180.80	0.0893	0.7054
0.20	−12.53	0.000 696 2	0.083 54	24.43	165.36	24.57	157.50	182.07	0.0992	0.7035
0.24	−7.42	0.000 704 0	0.070 33	29.06	167.44	29.23	155.09	184.32	0.1168	0.7004
0.28	−2.93	0.000 711 1	0.060 76	33.15	169.26	33.35	152.92	186.27	0.1321	0.6980
0.32	1.11	0.000 717 7	0.053 51	36.85	170.88	37.08	150.92	188.00	0.1457	0.6960
0.40	8.15	0.000 729 9	0.043 21	43.35	173.69	43.64	147.33	190.97	0.1691	0.6928
0.50	15.60	0.000 743 8	0.034 82	50.30	176.61	50.67	143.35	194.02	0.1935	0.6899
0.60	22.00	0.000 756 6	0.029 13	56.35	179.09	56.80	139.77	196.57	0.2142	0.6878
0.70	27.65	0.000 768 6	0.025 01	61.75	181.23	62.29	136.45	198.74	0.2324	0.6860
0.80	32.74	0.000 780 2	0.021 88	66.68	183.13	67.30	133.33	200.63	0.2487	0.6845
0.90	37.37	0.000 791 4	0.019 42	71.22	184.81	71.93	130.36	202.29	0.2634	0.6832
1.0	41.64	0.000 802 3	0.017 44	75.46	186.32	76.26	127.50	203.76	0.2770	0.6820
1.2	49.31	0.000 823 7	0.014 41	83.22	188.95	84.21	122.03	206.24	0.3015	0.6799
1.4	56.09	0.000 844 8	0.012 22	90.28	191.11	91.46	116.76	208.22	0.3232	0.6778
1.6	62.19	0.000 866 0	0.010 54	96.80	192.95	98.19	111.62	209.81	0.3329	0.6758

Source: Adapted from Kenneth Wark, *Thermodynamics*, 4th ed., McGraw-Hill, New York, 1983, p. 809. table A-17M. Originally published by E. I. du Pont de Nemours & Company, Inc., 1969.

TABLE A-13

Superheated refrigerant-12

Temp. °C	v m³/kg	u kJ/kg	h kJ/kg	s kJ/(kg·K)	v m³/kg	u kJ/kg	h kJ/kg	s kJ/(kg·K)
	0.060 MPa (T_{sat} = −41.42°C)				**0.10 MPa (T_{sat} = −30.10°C)**			
Sat.	0.257 5	153.49	168.94	0.7290	0.160 0	158.15	174.15	0.7171
−40	0.259 3	154.16	169.72	0.7324				
−20	0.283 8	163.91	180.94	0.7785	0.167 7	163.22	179.99	0.7406
0	0.307 9	174.05	192.52	0.8225	0.182 7	173.50	191.77	0.7854
10	0.319 8	179.26	198.45	0.8439	0.190 0	178.77	197.77	0.8070
20	0.331 7	184.57	204.47	0.8647	0.197 3	184.12	203.85	0.8281
30	0.343 5	189.96	210.57	0.8852	0.204 5	189.57	210.02	0.8488
40	0.355 2	195.46	216.77	0.9053	0.211 7	195.09	216.26	0.8691
50	0.367 0	201.02	223.04	0.9251	0.218 8	200.70	222.58	0.8889
60	0.378 7	206.69	229.41	0.9444	0.226 0	206.38	228.98	0.9084
80	0.402 0	218.25	242.37	0.9822	0.240 1	218.00	242.01	0.9464
	0.14 MPa (T_{sat} = −21.91°C)				**0.18 MPa (T_{sat} = −15.38°C)**			
Sat.	0.116 8	161.52	177.87	0.7102	0.092 2	164.20	180.80	0.7054
−20	0.117 9	162.50	179.01	0.7147				
−10	0.123 5	167.69	184.97	0.7378	0.092 5	164.39	181.03	0.7181
0	0.128 9	172.94	190.99	0.7602	0.099 1	172.37	190.21	0.7408
10	0.134 3	178.28	197.08	0.7821	0.103 4	177.77	196.38	0.7630
20	0.139 7	183.67	203.23	0.8035	0.107 6	183.23	202.60	0.7846
30	0.144 9	189.17	209.46	0.8243	0.111 8	188.77	208.89	0.8057
40	0.150 2	194.72	215.75	0.8447	0.116 0	194.35	215.23	0.8263
50	0.155 3	200.38	222.12	0.8648	0.120 1	200.02	221.64	0.8464
60	0.160 5	206.08	228.55	0.8844	0.124 1	205.78	228.12	0.8662
80	0.170 7	217.74	241.64	0.9225	0.132 2	217.47	241.27	0.9045
100	0.180 9	229.67	255.00	0.9593	0.140 2	229.45	254.69	0.9414
	0.20 MPa (T_{sat} = −12.53°C)				**0.24 MPa (T_{sat} = −7.42°C)**			
Sat.	0.083 5	165.37	182.07	0.7035	0.070 3	167.45	184.32	0.7004
0	0.088 6	172.08	189.08	0.7325	0.072 9	171.49	188.99	0.7177
10	0.092 6	177.50	196.02	0.7548	0.076 3	176.98	195.29	0.7404
20	0.096 4	183.00	202.28	0.7766	0.079 6	182.53	201.63	0.7624
30	0.100 2	188.56	208.60	0.7978	0.082 8	188.14	208.01	0.7838
40	0.104 0	194.17	214.97	0.8184	0.086 0	193.80	214.44	0.8047
50	0.107 7	199.86	221.40	0.8387	0.089 2	199.51	220.92	0.8251
60	0.111 4	205.62	227.90	0.8585	0.092 3	205.31	227.46	0.8450
80	0.118 7	217.35	241.09	0.8969	0.098 5	217.07	240.71	0.8836
100	0.125 9	229.35	254.53	0.9339	0.104 5	229.12	254.20	0.9208
120	0.133 1	241.59	268.21	0.9696	0.110 5	241.41	267.93	0.9566

TABLE A-13

(Continued)

Temp. °C	v m³/kg	u kJ/kg	h kJ/kg	s kJ/(kg·K)	v m³/kg	u kJ/kg	h kJ/kg	s kJ/(kg·K)
	0.28 MPa (T_{sat} = −2.93°C)				**0.32 MPa (T_{sat} = 1.11°C)**			
Sat.	0.060 76	169.26	186.27	0.6980	0.053 51	170.88	188.00	0.6960
0	0.061 66	170.89	188.15	0.7049				
10	0.064 64	176.45	194.55	0.7279	0.055 90	175.90	193.79	0.7167
20	0.067 55	182.06	200.97	0.7502	0.058 52	181.57	200.30	0.7393
30	0.070 40	187.71	207.42	0.7718	0.061 06	187.28	206.82	0.7612
40	0.073 19	193.42	213.91	0.7928	0.063 55	193.02	213.36	0.7824
50	0.075 94	199.18	220.44	0.8134	0.066 00	198.82	219.94	0.8031
60	0.078 65	205.00	227.02	0.8334	0.068 41	204.68	226.57	0.8233
80	0.083 99	216.82	240.34	0.8722	0.073 14	216.55	239.96	0.8623
100	0.089 24	228.29	253.88	0.9095	0.077 78	228.66	253.55	0.8997
120	0.094 43	241.21	267.65	0.9455	0.082 36	241.00	267.36	0.9358
	0.40 MPa (T_{sat} = 8.15°C)				**0.50 MPa (T_{sat} = 15.60°C)**			
Sat.	0.043 21	173.69	190.97	0.6928	0.034 82	176.61	194.02	0.6899
10	0.043 63	174.76	192.21	0.6972				
20	0.045 84	180.57	198.91	0.7204	0.035 65	179.26	197.08	0.7004
30	0.047 97	186.39	205.58	0.7428	0.037 46	185.23	203.96	0.7235
40	0.050 05	192.23	212.25	0.7645	0.039 22	191.20	210.81	0.7457
50	0.052 07	198.11	218.94	0.7855	0.040 91	197.19	217.64	0.7672
60	0.054 06	204.03	225.65	0.8060	0.042 57	203.20	224.48	0.7881
80	0.057 91	216.03	239.19	0.8454	0.045 78	215.32	238.21	0.8281
100	0.061 73	228.20	252.89	0.8831	0.048 89	227.61	252.05	0.8662
120	0.065 46	240.61	266.79	0.9194	0.051 93	240.10	266.06	0.9028
140	0.069 13	253.23	280.88	0.9544	0.054 92	252.77	280.23	0.9379
	0.60 MPa (T_{sat} = 22.00°C)				**0.70 MPa (T_{sat} = 27.65°C)**			
Sat.	0.029 13	179.09	196.57	0.6878	0.025 01	181.23	198.74	0.6860
30	0.030 42	184.01	202.26	0.7068	0.025 35	182.72	200.46	0.6917
40	0.031 97	190.13	209.31	0.7297	0.026 76	189.00	207.73	0.7153
50	0.033 45	196.23	216.30	0.7516	0.028 10	195.23	214.90	0.7378
60	0.034 89	202.34	223.27	0.7729	0.029 39	201.45	222.02	0.7595
80	0.037 65	214.61	237.20	0.8135	0.031 84	213.88	236.17	0.8008
100	0.040 32	227.01	251.20	0.8520	0.034 19	226.40	250.33	0.8398
120	0.042 91	239.57	265.32	0.8889	0.036 46	239.05	264.57	0.8769
140	0.045 45	252.31	279.58	0.9243	0.038 67	251.85	278.92	0.9125
160	0.047 94	265.25	294.01	0.9584	0.040 85	264.83	293.42	0.9468

TABLE A-13

(Continued)

Temp. °C	v m³/kg	u kJ/kg	h kJ/kg	s kJ/(kg·K)	v m³/kg	u kJ/kg	h kJ/kg	s kJ/(kg·K)
	0.80 MPa (T_{sat} = 32.74°C)				**0.90 MPa (T_{sat} = 37.37°C)**			
Sat.	0.021 88	183.13	200.63	0.6845	0.019 42	184.81	202.29	0.6832
40	0.022 83	187.81	206.07	0.7021	0.019 74	186.55	204.32	0.6897
50	0.024 07	194.19	213.45	0.7253	0.020 91	193.10	211.92	0.7136
60	0.025 25	200.52	220.72	0.7474	0.022 01	199.56	219.37	0.7363
80	0.027 48	213.13	235.11	0.7894	0.024 07	212.37	234.03	0.7790
100	0.029 59	225.77	249.44	0.8289	0.026 01	225.13	248.54	0.8190
120	0.031 62	238.51	263.81	0.8664	0.027 85	237.97	263.03	0.8569
140	0.033 59	251.39	278.26	0.9022	0.029 64	250.90	277.58	0.8930
160	0.035 52	264.41	292.83	0.9367	0.031 38	263.99	292.23	0.9276
180	0.037 42	277.60	307.54	0.9699	0.033 09	277.23	307.01	0.9609
	1.0 MPa (T_{sat} = 41.64°C)				**1.2 MPa (T_{sat} = 49.31°C)**			
Sat.	0.017 44	186.32	203.76	0.6820	0.014 41	188.95	206.24	0.6799
50	0.018 37	191.95	210.32	0.7026	0.014 48	189.43	206.81	0.6816
60	0.019 41	198.56	217.97	0.7259	0.015 46	196.41	214.96	0.7065
80	0.021 34	211.57	232.91	0.7695	0.017 22	209.91	230.57	0.7520
100	0.023 13	224.48	247.61	0.8100	0.018 81	223.13	245.70	0.7937
120	0.024 84	237.41	262.25	0.8482	0.020 30	236.27	260.53	0.8326
140	0.026 47	250.43	276.90	0.8845	0.021 72	249.45	275.51	0.8696
160	0.028 07	263.56	291.63	0.9193	0.023 09	263.70	290.41	0.9048
180	0.029 63	276.84	306.47	0.9528	0.024 43	276.05	305.37	0.9385
200	0.031 16	290.26	321.42	0.9851	0.025 74	289.55	320.44	0.9711
	1.4 MPa (T_{sat} = 56.09°C)				**1.6 MPa (T_{sat} = 62.19°C)**			
Sat.	0.012 22	191.11	208.22	0.6778	0.010 54	192.95	209.81	0.6758
60	0.012 58	194.00	211.61	0.6881				
80	0.014 25	208.11	228.06	0.7360	0.011 98	206.17	225.34	0.7209
100	0.015 71	221.70	243.69	0.7791	0.013 37	220.19	241.58	0.7656
120	0.017 05	235.09	258.96	0.8189	0.014 61	233.84	257.22	0.8065
140	0.018 32	248.43	274.08	0.8564	0.015 77	247.38	272.61	0.8447
160	0.019 54	261.80	289.16	0.8921	0.016 86	260.90	287.88	0.8808
180	0.020 71	275.27	304.26	0.9262	0.017 92	274.47	303.14	0.9152
200	0.021 86	288.84	319.44	0.9589	0.018 95	288.11	318.43	0.9482
220	0.022 99	302.51	334.70	0.9905	0.019 96	301.84	333.78	0.9800

Source: Adapted from Kenneth Wark, *Thermodynamics*, 4th ed., McGraw-Hill, New York, 1983, pp. 810–812, table A-18M. Originally published by E. I. du Pont de Nemours & Company, Inc., 1969.

FIGURE A-14

P-h diagram for refrigerant-12. (Freon 12 is the DuPont trademark for refrigerant-12. Copyright E. I. du Pont de Nemours & Company; used with permission.)

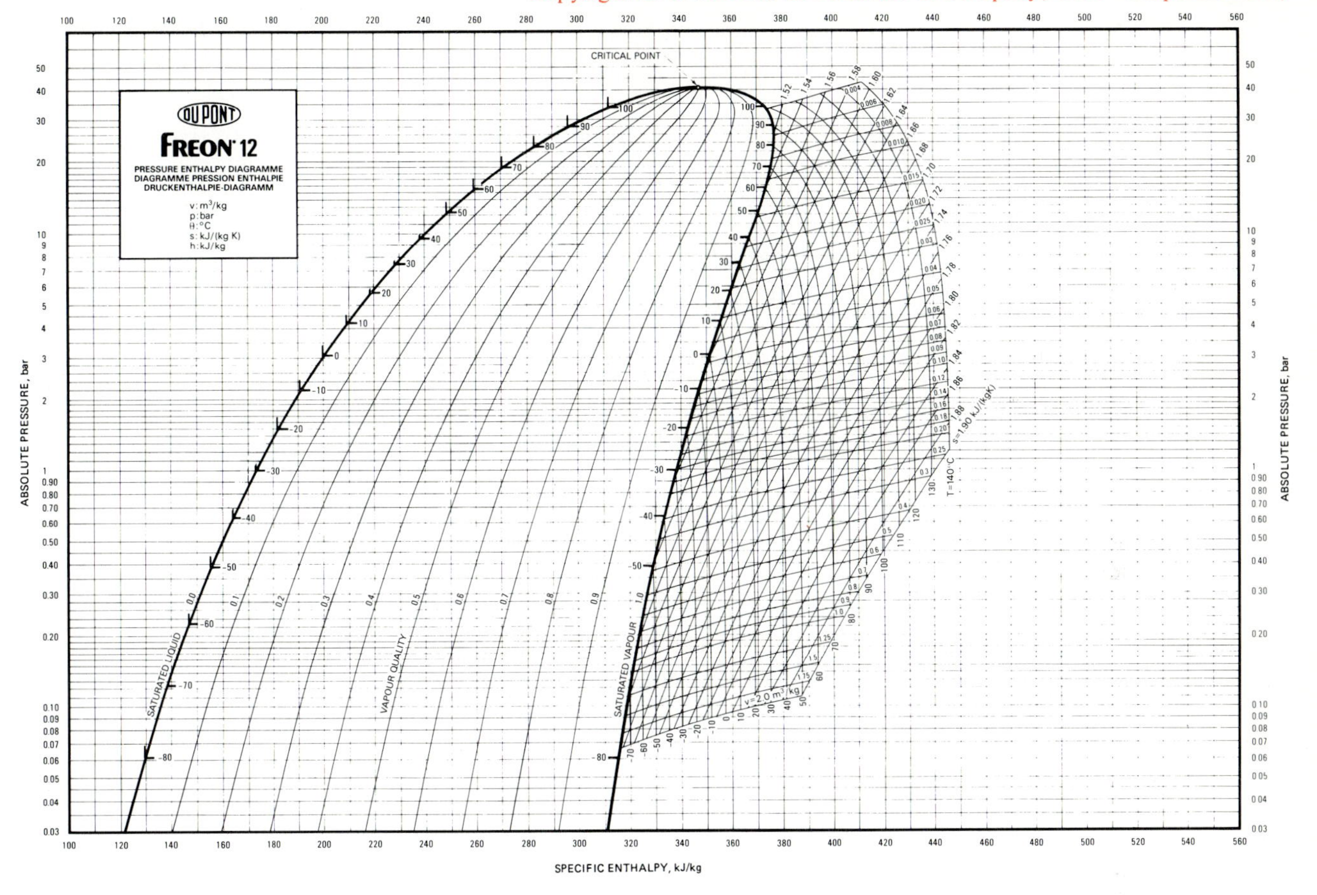

E-52455 *FREON is Du Pont's registered trademark for its fluorocarbon refrigerants

TABLE A-15

Saturated nitrogen–Temperature table

Temp. K	Sat. Press., MPa P_{sat}	Specific volume m³/kg Sat. liquid v_f	Evap. v_{fg}	Sat. vapor v_g	Enthalpy kJ/kg Sat. liquid h_f	Evap. h_{fg}	Sat. vapor h_g	Entropy kJ/(kg·K) Sat. liquid s_f	Evap. s_{fg}	Sat. vapor s_g
63.143	0.012 53	0.001 152	1.480 060	1.481 212	−150.348	215.188	64.840	2.4310	3.4076	5.8386
65	0.017 42	0.001 162	1.093 173	1.094 335	−146.691	213.291	66.600	2.4845	3.2849	5.7694
70	0.038 58	0.001 189	0.525 785	0.526 974	−136.569	207.727	71.158	2.6345	2.9703	5.6048
75	0.076 12	0.001 221	0.280 970	0.282 191	−126.287	201.662	75.375	2.7755	2.6915	5.4670
77.347	0.101 325	0.001 237	0.215 504	0.216 741	−121.433	198.645	77.212	2.8390	2.5706	5.4096
80	0.1370	0.001 256	0.162 794	0.164 050	−115.926	195.089	79.163	2.9083	2.4409	5.3492
85	0.2291	0.001 296	0.100 434	0.101 730	−105.461	187.892	82.431	3.0339	2.2122	5.2461
90	0.3608	0.001 340	0.064 950	0.066 290	−94.817	179.894	85.077	3.1535	2.0001	5.1536
95	0.5411	0.001 392	0.043 504	0.044 896	−83.895	170.877	86.982	3.2688	1.7995	5.0683
100	0.7790	0.001 452	0.029 861	0.031 313	−72.571	160.562	87.991	3.3816	1.6060	4.9876
105	1.0843	0.001 524	0.020 745	0.022 269	−60.691	148.573	87.882	3.4930	1.4150	4.9080
110	1.4673	0.001 613	0.014 402	0.016 015	−48.027	134.319	86.292	3.6054	1.2209	4.8263
115	1.9395	0.001 797	0.009 696	0.011 493	−34.157	116.701	82.544	3.7214	1.0145	4.7359
120	2.5135	0.001 904	0.006 130	0.008 034	−18.017	93.092	75.075	3.8450	0.7803	4.6253
125	3.2079	0.002 323	0.002 568	0.004 891	+6.202	50.114	56.316	4.0356	0.3989	4.4345
126.1	3.4000	0.003 184	0.000 000	0.003 184	+30.791	0.000	30.791	4.2269	0.0000	4.2269

Source: Gordon J. Van Wylen and Richard E. Sonntag, *Fundamentals of Classical Thermodynamics,* English/SI Version, 3d ed., Wiley, New York, 1986, p. 677, table A-4.1SI. Originally published in R. T. Jacobsen and R. R. Stewart, "Thermodynamic Properties of Nitrogen Including Liquid and Vapor Phases from 63 K to 2000 K with Pressures to 10,000 Bar," *Journal of Physical and Chemical Reference Data,* vol. 2, pp. 757–922, 1973.

TABLE A-16

Superheated nitrogen

Temp. K	v m³/kg	h kJ/kg	s kJ/(kg·K)	v m³/kg	h kJ/kg	s kJ/(kg·K)	v m³/kg	h kJ/kg	s kJ/(kg·K)
	0.1 MPa			**0.2 MPa**			**0.5 MPa**		
100	0.290 978	101.965	5.6944	0.142 475	100.209	5.4767	0.055 520	94.345	5.1706
125	0.367 217	128.505	5.9313	0.181 711	127.371	5.7194	0.073 422	123.824	5.4343
150	0.442 619	154.779	6.1228	0.220 014	153.962	5.9132	0.090 150	151.470	5.6361
175	0.517 576	180.935	6.2841	0.257 890	180.314	6.0760	0.106 394	178.434	5.8025
200	0.592 288	207.029	6.4234	0.295 531	206.537	6.2160	0.122 394	205.063	5.9447
225	0.666 552	233.085	6.5460	0.332 841	232.690	6.3388	0.138 173	231.459	6.0690
250	0.741 375	259.122	6.6561	0.370 418	258.796	6.4491	0.154 006	257.828	6.1801
275	0.815 563	285.144	6.7550	0.407 619	284.876	6.5485	0.169 642	284.076	6.2800
300	0.890 205	311.158	6.8457	0.445 047	310.937	6.6393	0.185 346	310.273	6.3715
	1.0 MPa			**2.0 MPa**			**4.0 MPa**		
125	0.033 065	117.422	5.1872	0.014 021	101.489	4.8878			
150	0.041 884	147.176	5.4042	0.019 546	137.916	5.1547	0.008 234	115.716	4.8384
175	0.050 125	175.255	5.5779	0.024 155	168.709	5.3449	0.011 186	154.851	5.0804
200	0.058 096	202.596	5.7237	0.028 436	197.609	5.4992	0.013 648	187.521	5.2553
225	0.065 875	229.526	5.8502	0.032 697	225.578	5.6309	0.015 894	217.757	5.3976
250	0.073 634	256.220	5.9632	0.036 557	253.032	5.7469	0.018 060	246.793	5.5202
275	0.081 260	282.720	6.0639	0.040 485	280.132	5.8501	0.020 133	275.056	5.6277
300	0.088 899	309.173	6.1563	0.044 398	307.014	5.9436	0.022 178	302.848	5.7248
	6.0 MPa			**8.0 MPa**			**10.0 MPa**		
150	0.004 413	87.090	4.5667	0.002 917	61.903	4.3518	0.002 388	48.687	4.2287
175	0.006 913	140.183	4.8966	0.004 863	125.536	4.7470	0.003 750	112.489	4.6239
200	0.008 772	177.447	5.0961	0.006 390	167.680	4.9726	0.005 016	158.578	4.8709
225	0.010 396	210.139	5.2410	0.007 691	202.867	5.1384	0.006 104	196.079	5.0474
250	0.011 934	240.806	5.3796	0.008 903	235.141	5.2750	0.007 112	229.861	5.1900
275	0.013 383	270.222	5.4917	0.010 034	265.676	5.3910	0.008 046	261.450	5.3103
300	0.014 800	298.907	5.5916	0.011 133	295.219	5.4942	0.008 950	291.800	5.4163
	15.0 MPa			**20.0 MPa**					
150	0.001 956	36.922	4.0798	0.001 781	33.637	3.9956			
175	0.002 603	92.284	4.4213	0.002 186	83.453	4.3029			
200	0.003 369	140.886	4.6813	0.002 685	130.291	4.5535			
225	0.004 106	182.034	4.8752	0.003 208	172.307	4.7511			
250	0.004 808	218.710	5.0303	0.003 728	210.456	4.9127			
275	0.005 461	252.465	5.1845	0.004 223	245.640	5.0467			
300	0.006 091	284.523	5.2707	0.004 704	278.942	5.1629			

Source: Gordon J. Van Wylen and Richard E. Sonntag, *Fundamentals of Classical Thermodynamics*, English/SI Version, 3d ed., Wiley, New York, 1986, pp. 678–679, table A-4.2SI. Originally published in R. T. Jacobsen and R. R. Stewart, "Thermodynamic Properties of Nitrogen Including Liquid and Vapor Phases from 63 K to 2000 K with Pressures to 10,000 Bar," *Journal of Physical and Chemical Reference Data,* vol. 2, pp. 757–922, 1973.

TABLE A-17

Ideal-gas properties of air

T K	h kJ/kg	P_r	u kJ/kg	v_r	s° kJ/(kg·K)	T K	h kJ/kg	P_r	u kJ/kg	v_r	s° kJ/(kg·K)
200	199.97	0.3363	142.56	1707.	1.295 59	580	586.04	14.38	419.55	115.7	2.373 48
210	209.97	0.3987	149.69	1512.	1.344 44	590	596.52	15.31	427.15	110.6	2.391 40
220	219.97	0.4690	156.82	1346.	1.391 05	600	607.02	16.28	434.78	105.8	2.409 02
230	230.02	0.5477	164.00	1205.	1.435 57	610	617.53	17.30	442.42	101.2	2.426 44
240	240.02	0.6355	171.13	1084.	1.478 24	620	628.07	18.36	450.09	96.92	2.443 56
250	250.05	0.7329	178.28	979.	1.519 17	630	638.63	19.84	457.78	92.84	2.460 48
260	260.09	0.8405	185.45	887.8	1.558 48	640	649.22	20.64	465.50	88.99	2.477 16
270	270.11	0.9590	192.60	808.0	1.596 34	650	659.84	21.86	473.25	85.34	2.493 64
280	280.13	1.0889	199.75	738.0	1.632 79	660	670.47	23.13	481.01	81.89	2.509 85
285	285.14	1.1584	203.33	706.1	1.650 55	670	681.14	24.46	488.81	78.61	2.525 89
290	290.16	1.2311	206.91	676.1	1.668 02	680	691.82	25.85	496.62	75.50	2.541 75
295	295.17	1.3068	210.49	647.9	1.685 15	690	702.52	27.29	504.45	72.56	2.557 31
300	300.19	1.3860	214.07	621.2	1.702 03	700	713.27	28.80	512.33	69.76	2.572 77
305	305.22	1.4686	217.67	596.0	1.718 65	710	724.04	30.38	520.23	67.07	2.588 10
310	310.24	1.5546	221.25	572.3	1.734 98	720	734.82	32.02	528.14	64.53	2.603 19
315	315.27	1.6442	224.85	549.8	1.751 06	730	745.62	33.72	536.07	62.13	2.618 03
320	320.29	1.7375	228.42	528.6	1.766 90	740	756.44	35.50	544.02	59.82	2.632 80
325	325.31	1.8345	232.02	508.4	1.782 49	750	767.29	37.35	551.99	57.63	2.647 37
330	330.34	1.9352	235.61	489.4	1.797 83	760	778.18	39.27	560.01	55.54	2.661 76
340	340.42	2.149	242.82	454.1	1.827 90	780	800.03	43.35	576.12	51.64	2.690 13
350	350.49	2.379	250.02	422.2	1.857 08	800	821.95	47.75	592.30	48.08	2.717 87
360	360.58	2.626	257.24	393.4	1.885 43	820	843.98	52.59	608.59	44.84	2.745 04
370	370.67	2.892	264.46	367.2	1.913 13	840	866.08	57.60	624.95	41.85	2.771 70
380	380.77	3.176	271.69	343.4	1.940 01	860	888.27	63.09	641.40	39.12	2.797 83
390	390.88	3.481	278.93	321.5	1.966 33	880	910.56	68.98	657.95	36.61	2.823 44
400	400.98	3.806	286.16	301.6	1.991 94	900	932.93	75.29	674.58	34.31	2.848 56
410	411.12	4.153	293.43	283.3	2.016 99	920	955.38	82.05	691.28	32.18	2.873 24
420	421.26	4.522	300.69	266.6	2.041 42	940	977.92	89.28	708.08	30.22	2.897 48
430	431.43	4.915	307.99	251.1	2.065 33	960	1000.55	97.00	725.02	28.40	2.921 28
440	441.61	5.332	315.30	236.8	2.088 70	980	1023.25	105.2	741.98	26.73	2.944 68
450	451.80	5.775	322.62	223.6	2.111 61	1000	1046.04	114.0	758.94	25.17	2.967 70
460	462.02	6.245	329.97	211.4	2.134 07	1020	1068.89	123.4	776.10	23.72	2.990 34
470	472.24	6.742	337.32	200.1	2.156 04	1040	1091.85	133.3	793.36	22.39	3.012 60
480	482.49	7.268	344.70	189.5	2.177 60	1060	1114.86	143.9	810.62	21.14	3.034 49
490	492.74	7.824	352.08	179.7	2.198 76	1080	1137.89	155.2	827.88	19.98	3.056 08
500	503.02	8.411	359.49	170.6	2.219 52	1100	1161.07	167.1	845.33	18.896	3.077 32
510	513.32	9.031	366.92	162.1	2.239 93	1120	1184.28	179.7	862.79	17.886	3.098 25
520	523.63	9.684	374.36	154.1	2.259 97	1140	1207.57	193.1	880.35	16.946	3.118 83
530	533.98	10.37	381.84	146.7	2.279 67	1160	1230.92	207.2	897.91	16.064	3.139 16
540	544.35	11.10	389.34	139.7	2.299 06	1180	1254.34	222.2	915.57	15.241	3.159 16
550	554.74	11.86	396.86	133.1	2.318 09	1200	1277.79	238.0	933.33	14.470	3.178 88
560	565.17	12.66	404.42	127.0	2.336 85	1220	1301.31	254.7	951.09	13.747	3.198 34
570	575.59	13.50	411.97	121.2	2.355 31	1240	1324.93	272.3	968.95	13.069	3.217 51

TABLE A-17

(Continued)

T K	h kJ/kg	P_r	u kJ/kg	v_r	$s°$ kJ/(kg·K)	T K	h kJ/kg	P_r	u kJ/kg	v_r	$s°$ kJ/(kg·K)
1260	1348.55	290.8	986.90	12.435	3.236 38	1600	1757.57	791.2	1298.30	5.804	3.523 64
1280	1372.24	310.4	1004.76	11.835	3.255 10	1620	1782.00	834.1	1316.96	5.574	3.538 79
1300	1395.97	330.9	1022.82	11.275	3.273 45	1640	1806.46	878.9	1335.72	5.355	3.553 81
1320	1419.76	352.5	1040.88	10.747	3.291 60	1660	1830.96	925.6	1354.48	5.147	3.568 67
1340	1443.60	375.3	1058.94	10.247	3.309 59	1680	1855.50	974.2	1373.24	4.949	3.583 35
1360	1467.49	399.1	1077.10	9.780	3.327 24	1700	1880.1	1025	1392.7	4.761	3.5979
1380	1491.44	424.2	1095.26	9.337	3.344 74	1750	1941.6	1161	1439.8	4.328	3.6336
1400	1515.42	450.5	1113.52	8.919	3.362 00	1800	2003.3	1310	1487.2	3.944	3.6684
1420	1539.44	478.0	1131.77	8.526	3.379 01	1850	2065.3	1475	1534.9	3.601	3.7023
1440	1563.51	506.9	1150.13	8.153	3.395 86	1900	2127.4	1655	1582.6	3.295	3.7354
1460	1587.63	537.1	1168.49	7.801	3.412 47	1950	2189.7	1852	1630.6	3.022	3.7677
1480	1611.79	568.8	1186.95	7.468	3.428 92	2000	2252.1	2068	1678.7	2.776	3.7994
1500	1635.97	601.9	1205.41	7.152	3.445 16	2050	2314.6	2303	1726.8	2.555	3.8303
1520	1660.23	636.5	1223.87	6.854	3.461 20	2100	2377.4	2559	1775.3	2.356	3.8605
1540	1684.51	672.8	1242.43	6.569	3.477 12	2150	2440.3	2837	1823.8	2.175	3.8901
1560	1708.82	710.5	1260.99	6.301	3.492 76	2200	2503.2	3138	1872.4	2.012	3.9191
1580	1733.17	750.0	1279.65	6.046	3.508 29	2250	2566.4	3464	1921.3	1.864	3.9474

Source: Kenneth Wark, *Thermodynamics,* 4th ed., McGraw-Hill, New York, 1983, pp. 785–786, table A-5M. Originally published in J. H. Keenan and J. Kays, *Gas Tables,* Wiley, New York, 1948.

TABLE A-18

Ideal-gas properties of nitrogen, N_2

T K	$\bar{h}$ kJ/kmol	$\bar{u}$ kJ/kmol	$\bar{s}^\circ$ kJ/(kmol·K)	T K	$\bar{h}$ kJ/kmol	$\bar{u}$ kJ/kmol	$\bar{s}^\circ$ kJ/(kmol·K)
0	0	0	0	600	17,563	12,574	212.066
220	6,391	4,562	182.639	610	17,864	12,792	212.564
230	6,683	4,770	183.938	620	18,166	13,011	213.055
240	6,975	4,979	185.180	630	18,468	13,230	213.541
250	7,266	5,188	186.370	640	18,772	13,450	214.018
260	7,558	5,396	187.514	650	19,075	13,671	214.489
270	7,849	5,604	188.614	660	19,380	13,892	214.954
280	8,141	5,813	189.673	670	19,685	14,114	215.413
290	8,432	6,021	190.695	680	19,991	14,337	215.866
298	8,669	6,190	191.502	690	20,297	14,560	216.314
300	8,723	6,229	191.682	700	20,604	14,784	216.756
310	9,014	6,437	192.638	710	20,912	15,008	217.192
320	9,306	6,645	193.562	720	21,220	15,234	217.624
330	9,597	6,853	194.459	730	21,529	15,460	218.059
340	9,888	7,061	195.328	740	21,839	15,686	218.472
350	10,180	7,270	196.173	750	22,149	15,913	218.889
360	10,471	7,478	196.995	760	22,460	16,141	219.301
370	10,763	7,687	197.794	770	22,772	16,370	219.709
380	11,055	7,895	198.572	780	23,085	16,599	220.113
390	11,347	8,104	199.331	790	23,398	16,830	220.512
400	11,640	8,314	200.071	800	23,714	17,061	220.907
410	11,932	8,523	200.794	810	24,027	17,292	221.298
420	12,225	8,733	201.499	820	24,342	17,524	221.684
430	12,518	8,943	202.189	830	24,658	17,757	222.067
440	12,811	9,153	202.863	840	24,974	17,990	222.447
450	13,105	9,363	203.523	850	25,292	18,224	222.822
460	13,399	9,574	204.170	860	25,610	18,459	223.194
470	13,693	9,786	204.803	870	25,928	18,695	223.562
480	13,988	9,997	205.424	880	26,248	18,931	223.927
490	14,285	10,210	206.033	890	26,568	19,168	224.288
500	14,581	10,423	206.630	900	26,890	19,407	224.647
510	14,876	10,635	207.216	910	27,210	19,644	225.002
520	15,172	10,848	207.792	920	27,532	19,883	225.353
530	15,469	11,062	208.358	930	27,854	20,122	225.701
540	15,766	11,277	208.914	940	28,178	20,362	226.047
550	16,064	11,492	209.461	950	28,501	20,603	226.389
560	16,363	11,707	209.999	960	28,826	20,844	226.728
570	16,662	11,923	210.528	970	29,151	21,086	227.064
580	16,962	12,139	211.049	980	29,476	21,328	227.398
590	17,262	12,356	211.562	990	29,803	21,571	227.728

TABLE A-18

(Continued)

T K	$\bar{h}$ kJ/kmol	$\bar{u}$ kJ/kmol	$\bar{s}°$ kJ/(kmol·K)	T K	$\bar{h}$ kJ/kmol	$\bar{u}$ kJ/kmol	$\bar{s}°$ kJ/(kmol·K)
1000	30,129	21,815	228.057	1760	56,227	41,594	247.396
1020	30,784	22,304	228.706	1780	56,938	42,139	247.798
1040	31,442	22,795	229.344	1800	57,651	42,685	248.195
1060	32,101	23,288	229.973	1820	58,363	43,231	248.589
1080	32,762	23,782	230.591	1840	59,075	43,777	248.979
1100	33,426	24,280	231.199	1860	59,790	44,324	249.365
1120	34,092	24,780	231.799	1880	60,504	44,873	249.748
1140	34,760	25,282	232.391	1900	61,220	45,423	250.128
1160	35,430	25,786	232.973	1920	61,936	45,973	250.502
1180	36,104	26,291	233.549	1940	62,654	46,524	250.874
1200	36,777	26,799	234.115	1960	63,381	47,075	251.242
1220	37,452	27,308	234.673	1980	64,090	47,627	251.607
1240	38,129	27,819	235.223	2000	64,810	48,181	251.969
1260	38,807	28,331	235.766	2050	66,612	49,567	252.858
1280	39,488	28,845	236.302	2100	68,417	50,957	253.726
1300	40,170	29,361	236.831	2150	70,226	52,351	254.578
1320	40,853	29,378	237.353	2200	72,040	53,749	255.412
1340	41,539	30,398	237.867	2250	73,856	55,149	256.227
1360	42,227	30,919	238.376	2300	75,676	56,553	257.027
1380	42,915	31,441	238.878	2350	77,496	57,958	257.810
1400	43,605	31,964	239.375	2400	79,320	59,366	258.580
1420	44,295	32,489	239.865	2450	81,149	60,779	259.332
1440	44,988	33,014	240.350	2500	82,981	62,195	260.073
1460	45,682	33,543	240.827	2550	84,814	63,613	260.799
1480	46,377	34,071	241.301	2600	86,650	65,033	261.512
1500	47,073	34,601	241.768	2650	88,488	66,455	262.213
1520	47,771	35,133	242.228	2700	90,328	67,880	262.902
1540	48,470	35,665	242.685	2750	92,171	69,306	263.577
1560	49,168	36,197	243.137	2800	94,014	70,734	264.241
1580	49,869	36,732	243.585	2850	95,859	72,163	264.895
1600	50,571	37,268	244.028	2900	97,705	73,593	265.538
1620	51,275	37,806	244.464	2950	99,556	75,028	266.170
1640	51,980	38,344	244.896	3000	101,407	76,464	266.793
1660	52,686	38,884	245.324	3050	103,260	77,902	267.404
1680	53,393	39,424	245.747	3100	105,115	79,341	268.007
1700	54,099	39,965	246.166	3150	106,972	80,782	268.601
1720	54,807	40,507	246.580	3200	108,830	82,224	269.186
1740	55,516	41,049	246.990	3250	110,690	83,668	269.763

Source: Kenneth Wark, *Thermodynamics,* 4th ed., McGraw-Hill, New York, 1983, pp. 787–788, table A-6M. Originally published in JANAF, *Thermochemical Tables,* NSRDS-NBS-37, 1971.

TABLE A-19

Ideal-gas properties of oxygen, O_2

T K	$\bar{h}$ kJ/kmol	$\bar{u}$ kJ/kmol	$\bar{s}^\circ$ kJ/(kmol·K)	T K	$\bar{h}$ kJ/kmol	$\bar{u}$ kJ/kmol	$\bar{s}^\circ$ kJ/(kmol·K)
0	0	0	0	600	17,929	12,940	226.346
220	6,404	4,575	196.171	610	18,250	13,178	226.877
230	6,694	4,782	197.461	620	18,572	13,417	227.400
240	6,984	4,989	198.696	630	18,895	13,657	227.918
250	7,275	5,197	199.885	640	19,219	13,898	228.429
260	7,566	5,405	201.027	650	19,544	14,140	228.932
270	7,858	5,613	202.128	660	19,870	14,383	229.430
280	8,150	5,822	203.191	670	20,197	14,626	229.920
290	8,443	6,032	204.218	680	20,524	14,871	230.405
298	8,682	6,203	205.033	690	20,854	15,116	230.885
300	8,736	6,242	205.213	700	21,184	15,364	231.358
310	9,030	6,453	206.177	710	21,514	15,611	231.827
320	9,325	6,664	207.112	720	21,845	15,859	232.291
330	9,620	6,877	208.020	730	22,177	16,107	232.748
340	9,916	7,090	208.904	740	22,510	16,357	233.201
350	10,213	7,303	209.765	750	22,844	16,607	233.649
360	10,511	7,518	210.604	760	23,178	16,859	234.091
370	10,809	7,733	211.423	770	23,513	17,111	234.528
380	11,109	7,949	212.222	780	23,850	17,364	234.960
390	11,409	8,166	213.002	790	24,186	17,618	235.387
400	11,711	8,384	213.765	800	24,523	17,872	235.810
410	12,012	8,603	214.510	810	24,861	18,126	236.230
420	12,314	8,822	215.241	820	25,199	18,382	236.644
430	12,618	9,043	215.955	830	25,537	18,637	237.055
440	12,923	9,264	216.656	840	25,877	18,893	237.462
450	13,228	9,487	217.342	850	26,218	19,150	237.864
460	13,535	9,710	218.016	860	26,559	19,408	238.264
470	13,842	9,935	218.676	870	26,899	19,666	238.660
480	14,151	10,160	219.326	880	27,242	19,925	239.051
490	14,460	10,386	219.963	890	27,584	20,185	239.439
500	14,770	10,614	220.589	900	27,928	20,445	239.823
510	15,082	10,842	221.206	910	28,272	20,706	240.203
520	15,395	11,071	221.812	920	28,616	20,967	240.580
530	15,708	11,301	222.409	930	28,960	21,228	240.953
540	16,022	11,533	222.997	940	29,306	21,491	241.323
550	16,338	11,765	223.576	950	29,652	21,754	241.689
560	16,654	11,998	224.146	960	29,999	22,017	242.052
570	16,971	12,232	224.708	970	30,345	22,280	242.411
580	17,290	12,467	225.262	980	30,692	22,544	242.768
590	17,609	12,703	225.808	990	31,041	22,809	243.120

TABLE A-19

(Continued)

T K	$\bar{h}$ kJ/kmol	$\bar{u}$ kJ/kmol	$\bar{s}^\circ$ kJ/(kmol·K)	T K	$\bar{h}$ kJ/kmol	$\bar{u}$ kJ/kmol	$\bar{s}^\circ$ kJ/(kmol·K)
1000	31,389	23,075	243.471	1760	58,880	44,247	263.861
1020	32,088	23,607	244.164	1780	59,624	44,825	264.283
1040	32,789	24,142	244.844	1800	60,371	45,405	264.701
1060	33,490	24,677	245.513	1820	61,118	45,986	265.113
1080	34,194	25,214	246.171	1840	61,866	46,568	265.521
1100	34,899	25,753	246.818	1860	62,616	47,151	265.925
1120	35,606	26,294	247.454	1880	63,365	47,734	266.326
1140	36,314	26,836	248.081	1900	64,116	48,319	266.722
1160	37,023	27,379	248.698	1920	64,868	48,904	267.115
1180	37,734	27,923	249.307	1940	65,620	49,490	267.505
1200	38,447	28,469	249.906	1960	66,374	50,078	267.891
1220	39,162	29,018	250.497	1980	67,127	50,665	268.275
1240	39,877	29,568	251.079	2000	67,881	51,253	268.655
1260	40,594	30,118	251.653	2050	69,772	52,727	269.588
1280	41,312	30,670	252.219	2100	71,668	54,208	270.504
1300	42,033	31,224	252.776	2150	73,573	55,697	271.399
1320	42,753	31,778	253.325	2200	75,484	57,192	272.278
1340	43,475	32,334	253.868	2250	77,397	58,690	273.136
1360	44,198	32,891	254.404	2300	79,316	60,193	273.981
1380	44,923	33,449	254.932	2350	81,243	61,704	274.809
1400	45,648	34,008	255.454	2400	83,174	63,219	275.625
1420	46,374	34,567	255.968	2450	85,112	64,742	276.424
1440	47,102	35,129	256.475	2500	87,057	66,271	277.207
1460	47,831	35,692	256.978	2550	89,004	67,802	277.979
1480	48,561	36,256	257.474	2600	90,956	69,339	278.738
1500	49,292	36,821	257.965	2650	92,916	70,883	279.485
1520	50,024	37,387	258.450	2700	94,881	72,433	280.219
1540	50,756	37,952	258.928	2750	96,852	73,987	280.942
1560	51,490	38,520	259.402	2800	98,826	75,546	281.654
1580	52,224	39,088	259.870	2850	100,808	77,112	282.357
1600	52,961	39,658	260.333	2900	102,793	78,682	283.048
1620	53,696	40,227	260.791	2950	104,785	80,258	283.728
1640	54,434	40,799	261.242	3000	106,780	81,837	284.399
1660	55,172	41,370	261.690	3050	108,778	83,419	285.060
1680	55,912	41,944	262.132	3100	110,784	85,009	285.713
1700	56,652	42,517	262.571	3150	112,795	86,601	286.355
1720	57,394	43,093	263.005	3200	114,809	88,203	286.989
1740	58,136	43,669	263.435	3250	116,827	89,804	287.614

Source: Kenneth Wark, *Thermodynamics,* 4th ed., McGraw-Hill, New York, 1983, pp. 789–790, table A-7M. Originally published in JANAF, *Thermochemical Tables,* NSRDS-NBS-37, 1971.

TABLE A-20

Ideal-gas properties of carbon dioxide, CO_2

T K	$\bar{h}$ kJ/kmol	$\bar{u}$ kJ/kmol	$\bar{s}^\circ$ kJ/(kmol·K)	T K	$\bar{h}$ kJ/kmol	$\bar{u}$ kJ/kmol	$\bar{s}^\circ$ kJ/(kmol·K)
0	0	0	0	600	22,280	17,291	243.199
220	6,601	4,772	202.966	610	22,754	17,683	243.983
230	6,938	5,026	204.464	620	23,231	18,076	244.758
240	7,280	5,285	205.920	630	23,709	18,471	245.524
250	7,627	5,548	207.337	640	24,190	18,869	246.282
260	7,979	5,817	208.717	650	24,674	19,270	247.032
270	8,335	6,091	210.062	660	25,160	19,672	247.773
280	8,697	6,369	211.376	670	25,648	20,078	248.507
290	9,063	6,651	212.660	680	26,138	20,484	249.233
298	9,364	6,885	213.685	690	26,631	20,894	249.952
300	9,431	6,939	213.915	700	27,125	21,305	250.663
310	9,807	7,230	215.146	710	27,622	21,719	251.368
320	10,186	7,526	216.351	720	28,121	22,134	252.065
330	10,570	7,826	217.534	730	28,622	22,552	252.755
340	10,959	8,131	218.694	740	29,124	22,972	253.439
350	11,351	8,439	219.831	750	29,629	23,393	254.117
360	11,748	8,752	220.948	760	30,135	23,817	254.787
370	12,148	9,068	222.044	770	30,644	24,242	255.452
380	12,552	9,392	223.122	780	31,154	24,669	256.110
390	12,960	9,718	224.182	790	31,665	25,097	256.762
400	13,372	10,046	225.225	800	32,179	25,527	257.408
410	13,787	10,378	226.250	810	32,694	25,959	258.048
420	14,206	10,714	227.258	820	33,212	26,394	258.682
430	14,628	11,053	228.252	830	33,730	26,829	259.311
440	15,054	11,393	229.230	840	34,251	27,267	259.934
450	15,483	11,742	230.194	850	34,773	27,706	260.551
460	15,916	12,091	231.144	860	35,296	28,125	261.164
470	16,351	12,444	232.080	870	35,821	28,588	261.770
480	16,791	12,800	233.004	880	36,347	29,031	262.371
490	17,232	13,158	233.916	890	36,876	29,476	262.968
500	17,678	13,521	234.814	900	37,405	29,922	263.559
510	18,126	13,885	235.700	910	37,935	30,369	264.146
520	18,576	14,253	236.575	920	38,467	30,818	264.728
530	19,029	14,622	237.439	930	39,000	31,268	265.304
540	19,485	14,996	238.292	940	39,535	31,719	265.877
550	19,945	15,372	239.135	950	40,070	32,171	266.444
560	20,407	15,751	239.962	960	40,607	32,625	267.007
570	20,870	16,131	240.789	970	41,145	33,081	267.566
580	21,337	16,515	241.602	980	41,685	33,537	268.119
590	21,807	16,902	242,405	990	42,226	33,995	268.670

TABLE A-20
(Continued)

T K	$\bar{h}$ kJ/kmol	$\bar{u}$ kJ/kmol	$\bar{s}^\circ$ kJ/(kmol·K)	T K	$\bar{h}$ kJ/kmol	$\bar{u}$ kJ/kmol	$\bar{s}^\circ$ kJ/(kmol·K)
1000	42,769	34,455	269.215	1760	86,420	71,787	301.543
1020	43,859	35,378	270.293	1780	87,612	72,812	302.217
1040	44,953	36,306	271.354	1800	88,806	73,840	302.884
1060	46,051	37,238	272.400	1820	90,000	74,868	303.544
1080	47,153	38,174	273.430	1840	91,196	75,897	304.198
1100	48,258	39,112	274.445	1860	92,394	76,929	304.845
1120	49,369	40,057	275.444	1880	93,593	77,962	305.487
1140	50,484	41,006	276.430	1900	94,793	78,996	306.122
1160	51,602	41,957	277.403	1920	95,995	80,031	306.751
1180	52,724	42,913	278.361	1940	97,197	81,067	307.374
1200	53,848	43,871	279.307	1960	98,401	82,105	307.992
1220	54,977	44,834	280.238	1980	99,606	83,144	308.604
1240	56,108	45,799	281.158	2000	100,804	84,185	309.210
1260	57,244	46,768	282.066	2050	103,835	86,791	310.701
1280	58,381	47,739	282.962	2100	106,864	89,404	312.160
1300	59,522	48,713	283.847	2150	109,898	92,023	313.589
1320	60,666	49,691	284.722	2200	112,939	94,648	314.988
1340	61,813	50,672	285.586	2250	115,984	97,277	316.356
1360	62,963	51,656	286.439	2300	119,035	99,912	317.695
1380	64,116	52,643	287.283	2350	122,091	102,552	319.011
1400	65,271	53,631	288.106	2400	125,152	105,197	320.302
1420	66,427	54,621	288.934	2450	128,219	107,849	321.566
1440	67,586	55,614	289.743	2500	131,290	110,504	322.808
1460	68,748	56,609	290.542	2550	134,368	113,166	324.026
1480	69,911	57,606	291.333	2600	137,449	115,832	325.222
1500	71,078	58,606	292.114	2650	140,533	118,500	326.396
1520	72,246	59,609	292.888	2700	143,620	121,172	327.549
1540	73,417	60,613	292.654	2750	146,713	123,849	328.684
1560	74,590	61,620	294.411	2800	149,808	126,528	329.800
1580	76,767	62,630	295.161	2850	152,908	129,212	330.896
1600	76,944	63,741	295.901	2900	156,009	131,898	331.975
1620	78,123	64,653	296.632	2950	159,117	134,589	333.037
1640	79,303	65,668	297.356	3000	162,226	137,283	334.084
1660	80,486	66,592	298.072	3050	165,341	139,982	335.114
1680	81,670	67,702	298.781	3100	168,456	142,681	336.126
1700	82,856	68,721	299.482	3150	171,576	145,385	337.124
1720	84,043	69,742	300.177	3200	174,695	148,089	338.109
1740	85,231	70,764	300.863	3250	177,822	150,801	339.069

Source: Kenneth Wark, *Thermodynamics,* 4th ed., McGraw-Hill, New York, 1983, pp. 793–794, table A-9M. Originally published in JANAF, *Thermochemical Tables,* NSRDS-NBS-37, 1971.

TABLE A-21

Ideal-gas properties of carbon monoxide, CO

T K	$\bar{h}$ kJ/kmol	$\bar{u}$ kJ/kmol	$\bar{s}^{\circ}$ kJ/(kmol·K)	T K	$\bar{h}$ kJ/kmol	$\bar{u}$ kJ/kmol	$\bar{s}^{\circ}$ kJ/(kmol·K)
0	0	0	0	600	17,611	12,622	218.204
220	6,391	4,562	188.683	610	17,915	12,843	218.708
230	6,683	4,771	189.980	620	18,221	13,066	219.205
240	6,975	4,979	191.221	630	18,527	13,289	219.695
250	7,266	5,188	192.411	640	18,833	13,512	220.179
260	7,558	5,396	193.554	650	19,141	13,736	220.656
270	7,849	5,604	194.654	660	19,449	13,962	221.127
280	8,140	5,812	195.713	670	19,758	14,187	221.592
290	8,432	6,020	196.735	680	20,068	14,414	222.052
298	8,669	6,190	197.543	690	20,378	14,641	222.505
300	8,723	6,229	197.723	700	20,690	14,870	222.953
310	9,014	6,437	198.678	710	21,002	15,099	223.396
320	9,306	6,645	199.603	720	21,315	15,328	223.833
330	9,597	6,854	200.500	730	21,628	15,558	224.265
340	9,889	7,062	201.371	740	21,943	15,789	224.692
350	10,181	7,271	202.217	750	22,258	16,022	225.115
360	10,473	7,480	203.040	760	22,573	16,255	225.533
370	10,765	7,689	203.842	770	22,890	16,488	225.947
380	11,058	7,899	204.622	780	23,208	16,723	226.357
390	11,351	8,108	205.383	790	23,526	16,957	226.762
400	11,644	8,319	206.125	800	23,844	17,193	227.162
410	11,938	8,529	206.850	810	24,164	17,429	227.559
420	12,232	8,740	207.549	820	24,483	17,665	227.952
430	12,526	8,951	208.252	830	24,803	17,902	228.339
440	12,821	9,163	208.929	840	25,124	18,140	228.724
450	13,116	9,375	209.593	850	25,446	18,379	229.106
460	13,412	9,587	210.243	860	25,768	18,617	229.482
470	13,708	9,800	210.880	870	26,091	18,858	229.856
480	14,005	10,014	211.504	880	26,415	19,099	230.227
490	14,302	10,228	212.117	890	26,740	19,341	230.593
500	14,600	10,443	212.719	900	27,066	19,583	230.957
510	14,898	10,658	213.310	910	27,392	19,826	231.317
520	15,197	10,874	213.890	920	27,719	20,070	231.674
530	15,497	11,090	214.460	930	28,046	20,314	232.028
540	15,797	11,307	215.020	940	28,375	20,559	232.379
550	16,097	11,524	215.572	950	28,703	20,805	232.727
560	16,399	11,743	216.115	960	29,033	21,051	233.072
570	16,701	11,961	216.649	970	29,362	21,298	233.413
580	17,003	12,181	217.175	980	29,693	21,545	233.752
590	17,307	12,401	217.693	990	30,024	21,793	234.088

TABLE A-21

(Continued)

T K	$\bar{h}$ kJ/kmol	$\bar{u}$ kJ/kmol	$\bar{s}^\circ$ kJ/(kmol·K)
1000	30,355	22,041	234.421
1020	31,020	22,540	235.079
1040	31,688	23,041	235.728
1060	32,357	23,544	236.364
1080	33,029	24,049	236.992
1100	33,702	24,557	237.609
1120	34,377	25,065	238.217
1140	35,054	25,575	238.817
1160	35,733	26,088	239.407
1180	36,406	26,602	239.989
1200	37,095	27,118	240.663
1220	37,780	27,637	241.128
1240	38,466	28,426	241.686
1260	39,154	28,678	242.236
1280	39,844	29,201	242.780
1300	40,534	29,725	243.316
1320	41,226	30,251	243.844
1340	41,919	30,778	244.366
1360	42,613	31,306	244.880
1380	43,309	31,836	245.388
1400	44,007	32,367	245.889
1420	44,707	32,900	246.385
1440	45,408	33,434	246.876
1460	46,110	33,971	247.360
1480	46,813	34,508	247.839
1500	47,517	35,046	248.312
1520	48,222	35,584	248.778
1540	48,928	36,124	249.240
1560	49,635	36,665	249.695
1580	50,344	37,207	250.147
1600	51,053	37,750	250.592
1620	51,763	38,293	251.033
1640	52,472	38,837	251.470
1660	53,184	39,382	251.901
1680	53,895	39,927	252.329
1700	54,609	40,474	252.751
1720	55,323	41,023	253.169
1740	56,039	41,572	253.582
1760	56,756	42,123	253.991
1780	57,473	42,673	254.398
1800	58,191	43,225	254.797
1820	58,910	43,778	255.194
1840	59,629	44,331	255.587
1860	60,351	44,886	255.976
1880	61,072	45,441	256.361
1900	61,794	45,997	256.743
1920	62,516	46,552	257.122
1940	63,238	47,108	257.497
1960	63,961	47,665	257.868
1980	64,684	48,221	258.236
2000	65,408	48,780	258.600
2050	67,224	50,179	259.494
2100	69,044	51,584	260.370
2150	70,864	52,988	261.226
2200	72,688	54,396	262.065
2250	74,516	55,809	262.887
2300	76,345	57,222	263.692
2350	78,178	58,640	264.480
2400	80,015	60,060	265.253
2450	81,852	61,482	266.012
2500	83,692	62,906	266.755
2550	85,537	64,335	267.485
2600	87,383	65,766	268.202
2650	89,230	67,197	268.905
2700	91,077	68,628	269.596
2750	92,930	70,066	270.285
2800	94,784	71,504	270.943
2850	96,639	72,945	271.602
2900	98,495	74,383	272.249
2950	100,352	75,825	272.884
3000	102,210	77,267	273.508
3050	104,073	78,715	274.123
3100	105,939	80,164	274.730
3150	107,802	81,612	275.326
3200	109,667	83,061	275.914
3250	111,534	84,513	276.494

Source: Kenneth Wark, *Thermodynamics,* 4th ed., McGraw-Hill, New York, 1983, pp. 791–792, table A-8M. Originally published in JANAF, *Thermochemical Tables,* NSRDS-NBS-37, 1971.

TABLE A-22

Ideal-gas properties of hydrogen, H_2

T K	$\bar{h}$ kJ/kmol	$\bar{u}$ kJ/kmol	$\bar{s}^\circ$ kJ/(kmol·K)	T K	$\bar{h}$ kJ/kmol	$\bar{u}$ kJ/kmol	$\bar{s}^\circ$ kJ/(kmol·K)
0	0	0	0	1440	42,808	30,835	177.410
260	7,370	5,209	126.636	1480	44,091	31,786	178.291
270	7,657	5,412	127.719	1520	45,384	32,746	179.153
280	7,945	5,617	128.765	1560	46,683	33,713	179.995
290	8,233	5,822	129.775	1600	47,990	34,687	180.820
298	8,468	5,989	130.574	1640	49,303	35,668	181.632
300	8,522	6,027	130.754	1680	50,622	36,654	182.428
320	9,100	6,440	132.621	1720	51,947	37,646	183.208
340	9,680	6,853	134.378	1760	53,279	38,645	183.973
360	10,262	7,268	136.039	1800	54,618	39,652	184.724
380	10,843	7,684	137.612	1840	55,962	40,663	185.463
400	11,426	8,100	139.106	1880	57,311	41,680	186.190
420	12,010	8,518	140.529	1920	58,668	42,705	186.904
440	12,594	8,936	141.888	1960	60,031	43,735	187.607
460	13,179	9,355	143.187	2000	61,400	44,771	188.297
480	13,764	9,773	144.432	2050	63,119	46,074	189.148
500	14,350	10,193	145.628	2100	64,847	47,386	189.979
520	14,935	10,611	146.775	2150	66,584	48,708	190.796
560	16,107	11,451	148.945	2200	68,328	50,037	191.598
600	17,280	12,291	150.968	2250	70,080	51,373	192.385
640	18,453	13,133	152.863	2300	71,839	52,716	193.159
680	19,630	13,976	154.645	2350	73,608	54,069	193.921
720	20,807	14,821	156.328	2400	75,383	55,429	194.669
760	21,988	15,669	157.923	2450	77,168	56,798	195.403
800	23,171	16,520	159.440	2500	78,960	58,175	196.125
840	24,359	17,375	160.891	2550	80,755	59,554	196.837
880	25,551	18,235	162.277	2600	82,558	60,941	197.539
920	26,747	19,098	163.607	2650	84,368	62,335	198.229
960	27,948	19,966	164.884	2700	86,186	63,737	198.907
1000	29,154	20,839	166.114	2750	88,008	65,144	199.575
1040	30,364	21,717	167.300	2800	89,838	66,558	200.234
1080	31,580	22,601	168.449	2850	91,671	67,976	200.885
1120	32,802	23,490	169.560	2900	93,512	69,401	201.527
1160	34,028	24,384	170.636	2950	95,358	70,831	202.157
1200	35,262	25,284	171.682	3000	97,211	72,268	202.778
1240	36,502	26,192	172.698	3050	99,065	73,707	203.391
1280	37,749	27,106	173.687	3100	100,926	75,152	203.995
1320	39,002	28,027	174.652	3150	102,793	76,604	204.592
1360	40,263	28,955	175.593	3200	104,667	78,061	205.181
1400	41,530	29,889	176.510	3250	106,545	79,523	205.765

Source: Kenneth Wark, *Thermodynamics*, 4th ed., McGraw-Hill, New York, 1983, p. 797, table A-11M. Originally published in JANAF, *Thermochemical Tables*, NSRDS-NBS-37, 1971.

TABLE A-23

Ideal-gas properties of water vapor, H_2O

T K	$\bar{h}$ kJ/kmol	$\bar{u}$ kJ/kmol	$\bar{s}^\circ$ kJ/(kmol·K)	T K	$\bar{h}$ kJ/kmol	$\bar{u}$ kJ/kmol	$\bar{s}^\circ$ kJ/(kmol·K)
0	0	0	0	600	20,402	15,413	212.920
220	7,295	5,466	178.576	610	20,765	15,693	213.529
230	7,628	5,715	180.054	620	21,130	15,975	214.122
240	7,961	5,965	181.471	630	21,495	16,257	214.707
250	8,294	6,215	182.831	640	21,862	16,541	215.285
260	8,627	6,466	184.139	650	22,230	16,826	215.856
270	8,961	6,716	185.399	660	22,600	17,112	216.419
280	9,296	6,968	186.616	670	22,970	17,399	216.976
290	9,631	7,219	187.791	680	23,342	17,688	217.527
298	9,904	7,425	188.720	690	23,714	17,978	218.071
300	9,966	7,472	188.928	700	24,088	18,268	218.610
310	10,302	7,725	190.030	710	24,464	18,561	219.142
320	10,639	7,978	191.098	720	24,840	18,854	219.668
330	10,976	8,232	192.136	730	25,218	19,148	220.189
340	11,314	8,487	193.144	740	25,597	19,444	220.707
350	11,652	8,742	194.125	750	25,977	19,741	221.215
360	11,992	8,998	195.081	760	26,358	20,039	221.720
370	12,331	9,255	196.012	770	26,741	20,339	222.221
380	12,672	9,513	196.920	780	27,125	20,639	222.717
390	13,014	9,771	197.807	790	27,510	20,941	223.207
400	13,356	10,030	198.673	800	27,896	21,245	223.693
410	13,699	10,290	199.521	810	28,284	21,549	224.174
420	14,043	10,551	200.350	820	28,672	21,855	224.651
430	14,388	10,813	201.160	830	29,062	22,162	225.123
440	14,734	11,075	201.955	840	29,454	22,470	225.592
450	15,080	11,339	202.734	850	29,846	22,779	226.057
460	15,428	11,603	203.497	860	30,240	23,090	226.517
470	15,777	11,869	204.247	870	30,635	23,402	226.973
480	16,126	12,135	204.982	880	31,032	23,715	227.426
490	16,477	12,403	205.705	890	31,429	24,029	227.875
500	16,828	12,671	206.413	900	31,828	24,345	228.321
510	17,181	12,940	207.112	910	32,228	24,662	228.763
520	17,534	13,211	207.799	920	32,629	24,980	229.202
530	17,889	13,482	208.475	930	33,032	25,300	229.637
540	18,245	13,755	209.139	940	33,436	25,621	230.070
550	18,601	14,028	209.795	950	33,841	25,943	230.499
560	18,959	14,303	210.440	960	34,247	26,265	230.924
570	19,318	14,579	211.075	970	34,653	26,588	231.347
580	19,678	14,856	211.702	980	35,061	26,913	231.767
590	20,039	15,134	212.320	990	35,472	27,240	232.184

TABLE A-23
(Continued)

T K	$\bar{h}$ kJ/kmol	$\bar{u}$ kJ/kmol	$\bar{s}^\circ$ kJ/(kmol·K)	T K	$\bar{h}$ kJ/kmol	$\bar{u}$ kJ/kmol	$\bar{s}^\circ$ kJ/(kmol·K)
1000	35,882	27,568	232.597	1760	70,535	55,902	258.151
1020	36,709	28,228	233.415	1780	71,523	56,723	258.708
1040	37,542	28,895	234.223	1800	72,513	57,547	259.262
1060	38,380	29,567	235.020	1820	73,507	58,375	259.811
1080	39,223	30,243	235.806	1840	74,506	59,207	260.357
1100	40,071	30,925	236.584	1860	75,506	60,042	260.898
1120	40,923	31,611	237.352	1880	76,511	60,880	261.436
1140	41,780	32,301	238.110	1900	77,517	61,720	261.969
1160	42,642	32,997	238.859	1920	78,527	62,564	262.497
1180	43,509	33,698	239.600	1940	79,540	63,411	263.022
1200	44,380	34,403	240.333	1960	80,555	64,259	263.542
1220	45,256	35,112	241.057	1980	81,573	65,111	264.059
1240	46,137	35,827	241.773	2000	82,593	65,965	264.571
1260	47,022	36,546	242.482	2050	85,156	68,111	265.838
1280	47,912	37,270	243.183	2100	87,735	70,275	267.081
1300	48,807	38,000	243.877	2150	90,330	72,454	268.301
1320	49,707	38,732	244.564	2200	92,940	74,649	269.500
1340	50,612	39,470	245.243	2250	95,562	76,855	270.679
1360	51,521	40,213	245.915	2300	98,199	79,076	271.839
1380	52,434	40,960	246.582	2350	100,846	81,308	272.978
1400	53,351	41,711	247.241	2400	103,508	83,553	274.098
1420	54,273	42,466	247.895	2450	106,183	85,811	275.201
1440	55,198	43,226	248.543	2500	108,868	88,082	276.286
1460	56,128	43,989	249.185	2550	111,565	90,364	277.354
1480	57,062	44,756	249.820	2600	114,273	92,656	278.407
1500	57,999	45,528	250.450	2650	116,991	94,958	279.441
1520	58,942	46,304	251.074	2700	119,717	97,269	280.462
1540	59,888	47,084	251.693	2750	122,453	99,588	281.464
1560	60,838	47,868	252.305	2800	125,198	101,917	282.453
1580	61,792	48,655	252.912	2850	127,952	104,256	283.429
1600	62,748	49,445	253.513	2900	130,717	106,605	284.390
1620	63,709	50,240	254.111	2950	133,486	108,959	285.338
1640	64,675	51,039	254.703	3000	136,264	111,321	286.273
1660	65,643	51,841	255.290	3050	139,051	113,692	287.194
1680	66,614	52,646	255.873	3100	141,846	116,072	288.102
1700	67,589	53,455	256.450	3150	144,648	118,458	288.999
1720	68,567	54,267	257.022	3200	147,457	120,851	289.884
1740	69,550	55,083	257.589	3250	150,272	123,250	290.756

Source: Kenneth Wark, *Thermodynamics,* 4th ed., McGraw-Hill, New York, 1983, pp. 795–796, table A-10M. Originally published in JANAF, *Thermochemical Tables,* NSRDS-NBS-37, 1971.

TABLE A-24

Ideal-gas properties of monatomic oxygen, O

T K	$\bar{h}$ kJ/kmol	$\bar{u}$ kJ/kmol	$\bar{s}°$ kJ/(kmol·K)	T K	$\bar{h}$ kJ/kmol	$\bar{u}$ kJ/kmol	$\bar{s}°$ kJ/(kmol·K)
0	0	0	0	2400	50,894	30,940	204.932
298	6,852	4,373	160.944	2450	51,936	31,566	205.362
300	6,892	4,398	161.079	2500	52,979	32,193	205.783
500	11,197	7,040	172.088	2550	54,021	32,820	206.196
1000	21,713	13,398	186.678	2600	55,064	33,447	206.601
1500	32,150	19,679	195.143	2650	56,108	34,075	206.999
1600	34,234	20,931	196.488	2700	57,152	34,703	207.389
1700	36,317	22,183	197.751	2750	58,196	35,332	207.772
1800	38,400	23,434	198.941	2800	59,241	35,961	208.148
1900	40,482	24,685	200.067	2850	60,286	36,590	208.518
2000	42,564	25,935	201.135	2900	61,332	37,220	208.882
2050	43,605	26,560	201.649	2950	62,378	37,851	209.240
2100	44,646	27,186	202.151	3000	63,425	38,482	209.592
2150	45,687	27,811	202.641	3100	65,520	39,746	210.279
2200	46,728	28,436	203.119	3200	67,619	41,013	210.945
2250	47,769	29,062	203.588	3300	69,720	42,283	211.592
2300	48,811	29,688	204.045	3400	71,824	43,556	212.220
2350	49,852	30,314	204.493	3500	73,932	44,832	212.831

Source: Kenneth Wark, *Thermodynamics,* 4th ed., McGraw-Hill, New York, 1983, p. 798, table A-11M. Originally published in JANAF, *Thermochemical Tables,* NSRDS-NBS-37, 1971.

TABLE A-25

Ideal-gas properties of hydroxyl, OH

T K	$\bar{h}$ kJ/kmol	$\bar{u}$ kJ/kmol	$\bar{s}^\circ$ kJ/(kmol·K)	T K	$\bar{h}$ kJ/kmol	$\bar{u}$ kJ/kmol	$\bar{s}^\circ$ kJ/(kmol·K)
0	0	0	0	2400	77,015	57,061	248.628
298	9,188	6,709	183.594	2450	78,801	58,431	249.364
300	9,244	6,749	183.779	2500	80,592	59,806	250.088
500	15,181	11,024	198.955	2550	82,388	61,186	250.799
1000	30,123	21,809	219.624	2600	84,189	62,572	251.499
1500	46,046	33,575	232.506	2650	85,995	63,962	252.187
1600	49,358	36,055	234.642	2700	87,806	65,358	252.864
1700	52,706	38,571	236.672	2750	89,622	66,757	253.530
1800	56,089	41,123	238.606	2800	91,442	68,162	254.186
1900	59,505	43,708	240.453	2850	93,266	69,570	254.832
2000	62,952	46,323	242.221	2900	95,095	70,983	255.468
2050	64,687	47,642	243.077	2950	96,927	72,400	256.094
2100	66,428	48,968	243.917	3000	98,763	73,820	256.712
2150	68,177	50,301	244.740	3100	102,447	76,673	257.919
2200	69,932	51,641	245.547	3200	106,145	79,539	259.093
2250	71,694	52,987	246.338	3300	109,855	82,418	260.235
2300	73,462	54,339	247.116	3400	113,578	85,309	261.347
2350	75,236	55,697	247.879	3500	117,312	88,212	262.429

Source: Kenneth Wark, *Thermodynamics,* 4th ed., McGraw-Hill, New York, 1983, p. 798, table A-11M. Originally published in JANAF, *Thermochemical Tables,* NSRDS-NBS-37, 1971.

TABLE A-26

Enthalpy of formation, Gibbs function of formation, and absolute entropy at 25°C, 1 atm

Substance	Formula	$\bar{h}_f^\circ$ kJ/kmol	$\bar{g}_f^\circ$ kJ/kmol	$\bar{s}^\circ$ kJ/(kmol·K)
Carbon	$C(s)$	0	0	5.74
Hydrogen	$H_2(g)$	0	0	130.68
Nitrogen	$N_2(g)$	0	0	191.61
Oxygen	$O_2(g)$	0	0	205.04
Carbon monoxide	$CO(g)$	−110,530	−137,150	197.65
Carbon dioxide	$CO_2(g)$	−393,520	−394,360	213.80
Water	$H_2O(g)$	−241,820	−228,590	188.83
Water	$H_2O(l)$	−285,830	−237,180	69.92
Hydrogen peroxide	$H_2O_2(g)$	−136,310	−105,600	232.63
Ammonia	$NH_3(g)$	−46,190	−16,590	192.33
Methane	$CH_4(g)$	−74,850	−50,790	186.16
Acetylene	$C_2H_2(g)$	+226,730	+209,170	200.85
Ethylene	$C_2H_4(g)$	+52,280	+68,120	219.83
Ethane	$C_2H_6(g)$	−84,680	−32,890	229.49
Propylene	$C_3H_6(g)$	+20,410	+62,720	266.94
Propane	$C_3H_8(g)$	−103,850	−23,490	269.91
n-Butane	$C_4H_{10}(g)$	−126,150	−15,710	310.12
n-Octane	$C_8H_{18}(g)$	−208,450	+16,530	466.73
n-Octane	$C_8H_{18}(l)$	−249,950	+6,610	360.79
n-Dodecane	$C_{12}H_{26}(g)$	−291,010	+50,150	622.83
Benzene	$C_6H_6(g)$	+82,930	+129,660	269.20
Methyl alcohol	$CH_3OH(g)$	−200,670	−162,000	239.70
Methyl alcohol	$CH_3OH(l)$	−238,660	−166,360	126.80
Ethyl alcohol	$C_2H_5OH(g)$	−235,310	−168,570	282.59
Ethyl alcohol	$C_2H_5OH(l)$	−277,690	−174,890	160.70
Oxygen	$O(g)$	+249,190	+231,770	161.06
Hydrogen	$H(g)$	+218,000	+203,290	114.72
Nitrogen	$N(g)$	+472,650	+455,510	153.30
Hydroxyl	$OH(g)$	+39,460	+34,280	183.70

Source: From JANAF, *Thermochemical Tables*, Dow Chemical Co., 1971; *Selected Values of Chemical Thermodynamic Properties*, NBS Technical Note 270-3, 1968; and *API Research Project 44*, Carnegie Press, 1953.

TABLE A-27

Enthalpy of combustion and enthalpy of vaporization at 25°C, 1 atm
(Water appears as a liquid in the products of combustion)

Substance	Formula	$\Delta \bar{h}_c^\circ$ = −HHV kJ/kmol	$\bar{h}_{fg}$ kJ/kmol
Hydrogen	$H_2(g)$	−285,840	
Carbon	$C(s)$	−393,520	
Carbon monoxide	$CO(g)$	−282,990	
Methane	$CH_4(g)$	−890,360	
Acetylene	$C_2H_2(g)$	−1,299,600	
Ethylene	$C_2H_4(g)$	−1,410,970	
Ethane	$C_2H_6(g)$	−1,559,900	
Propylene	$C_3H_6(g)$	−2,058,500	
Propane	$C_3H_8(g)$	−2,220,000	15,060
n-Butane	$C_4H_{10}(g)$	−2,877,100	21,060
n-Pentane	$C_5H_{12}(g)$	−3,536,100	26,410
n-Hexane	$C_6H_{14}(g)$	−4,194,800	31,530
n-Heptane	$C_7H_{16}(g)$	−4,853,500	36,520
n-Octane	$C_8H_{18}(g)$	−5,512,200	41,460
Benzene	$C_6H_6(g)$	−3,301,500	33,830
Toluene	$C_7H_8(g)$	−3,947,900	39,920
Methyl alcohol	$CH_3OH(g)$	−764,540	37,900
Ethyl alcohol	$C_2H_5OH(g)$	−1,409,300	42,340

Source: Kenneth Wark, *Thermodynamics,* 3d ed., McGraw-Hill, New York, 1977, pp. 834–835, table A-23M.

TABLE A-28

Logarithms to base e of the equilibrium constant K_p

The equilibrium constant K_p for the reaction $\nu_A A + \nu_B B \rightleftharpoons \nu_C C + \nu_D D$ is defined as $K_p \equiv \dfrac{P_C^{\nu_C} P_D^{\nu_D}}{P_A^{\nu_A} P_B^{\nu_B}}$

Temp. K	$H_2 \rightleftharpoons 2H$	$O_2 \rightleftharpoons 2O$	$N_2 \rightleftharpoons 2N$	$H_2O \rightleftharpoons H_2 + \frac{1}{2}O_2$	$H_2O \rightleftharpoons \frac{1}{2}H_2 + OH$	$CO_2 \rightleftharpoons CO + \frac{1}{2}O_2$	$\frac{1}{2}N_2 + \frac{1}{2}O_2 \rightleftharpoons NO$
298	−164.005	−186.975	−367.480	−92.208	−106.208	−103.762	−35.052
500	−92.827	−105.630	−213.372	−52.691	−60.281	−57.616	−20.295
1000	−39.803	−45.150	−99.127	−23.163	−26.034	−23.529	−9.388
1200	−30.874	−35.005	−80.011	−18.182	−20.283	−17.871	−7.569
1400	−24.463	−27.742	−66.329	−14.609	−16.099	−13.842	−6.270
1600	−19.637	−22.285	−56.055	−11.921	−13.066	−10.830	−5.294
1800	−15.866	−18.030	−48.051	−9.826	−10.657	−8.497	−4.536
2000	−12.840	−14.622	−41.645	−8.145	−8.728	−6.635	−3.931
2200	−10.353	−11.827	−36.391	−6.768	−7.148	−5.120	−3.433
2400	−8.276	−9.497	−32.011	−5.619	−5.832	−3.860	−3.019
2600	−6.517	−7.521	−28.304	−4.648	−4.719	−2.801	−2.671
2800	−5.002	−5.826	−25.117	−3.812	−3.763	−1.894	−2.372
3000	−3.685	−4.357	−22.359	−3.086	−2.937	−1.111	−2.114
3200	−2.534	−3.072	−19.937	−2.451	−2.212	−0.429	−1.888
3400	−1.516	−1.935	−17.800	−1.891	−1.576	0.169	−1.690
3600	−0.609	−0.926	−15.898	−1.392	−1.088	0.701	−1.513
3800	0.202	−0.019	−14.199	−0.945	−0.501	1.176	−1.356
4000	0.934	0.796	−12.660	−0.542	−0.044	1.599	−1.216
4500	2.486	2.513	−9.414	0.312	0.920	2.490	−0.921
5000	3.725	3.895	−6.807	0.996	1.689	3.197	−0.686
5500	4.743	5.023	−4.666	1.560	2.318	3.771	−0.497
6000	5.590	5.963	−2.865	2.032	2.843	4.245	−0.341

Source: Gordon J. Van Wylen and Richard E. Sonntag, *Fundamentals of Classical Thermodynamics,* English/SI Version, 3d ed., Wiley, New York, 1986, p. 723, table A.14. Based on thermodynamic data given in JANAF, *Thermochemical Tables,* Thermal Research Laboratory, The Dow Chemical Company, Midland, Mich.

TABLE A-29

Constants that appear in the Beattie-Bridgeman and the Benedict-Webb-Rubin equations of state

(*a*) The Beattie-Bridgeman equation of state is

$$P = \frac{R_u T}{\bar{v}^2}\left(1 - \frac{c}{\bar{v}T^3}\right)(\bar{v} + B) - \frac{A}{\bar{v}^2} \qquad \text{where} \qquad A = A_0\left(1 - \frac{a}{\bar{v}}\right) \qquad \text{and} \qquad B = B_0\left(1 - \frac{b}{\bar{v}}\right)$$

When P is in kPa, $\bar{v}$ is in m³/kmol, T is in K, and R_u = 8.314 kPa·m³/(kmol·K), the five constants in the Beattie-Bridgeman equation are as follows:

Gas	A_0	a	B_0	b	$c \times 10^{-4}$
Air	131.8441	0.019 31	0.046 11	−0.001 101	4.34
Argon	130.7802	0.023 28	0.039 31	0.0	5.99
Carbon dioxide	507.2836	0.071 32	0.104 76	0.072 35	66.00
Helium	2.1886	0.059 84	0.014 00	0.0	0.0040
Hydrogen	20.0117	−0.005 06	0.020 96	−0.043 59	0.0504
Nitrogen	136.2315	0.026 17	0.050 46	−0.006 91	4.20
Oxygen	151.0857	0.025 62	0.046 24	0.004 208	4.80

Source: Gordon J. Van Wylen and Richard E. Sonntag, *Fundamentals of Classical Thermodynamics,* English/SI Version, 3d ed., Wiley, New York, 1986, p. 46, table 3.3.

(*b*) The Benedict-Webb-Rubin equation of state is

$$P = \frac{R_u T}{\bar{v}} + \left(B_0 R_u T - A_0 - \frac{C_0}{T^2}\right)\frac{1}{\bar{v}^2} + \frac{bR_u T - a}{\bar{v}^3} + \frac{a\alpha}{\bar{v}^6} + \frac{c}{\bar{v}^3 T^2}\left(1 + \frac{\gamma}{\bar{v}^2}\right)e^{-\gamma/\bar{v}^2}$$

When P is in kPa, $\bar{v}$ is in m³/kmol, T is in K, and R_u = 8.314 kPa·m³/(kmol·K), the eight constants in the Benedict-Webb-Rubin equation are as follows:

Gas	a	A_0	b	B_0	c	C_0	α	γ
n-Butane, C_4H_{10}	190.68	1021.6	0.039 998	0.124 36	3.205×10^7	1.006×10^8	1.101×10^{-3}	0.0340
Carbon dioxide, CO_2	13.86	277.30	0.007 210	0.049 91	1.511×10^6	1.404×10^7	8.470×10^{-5}	0.00539
Carbon monoxide, CO	3.71	135.87	0.002 632	0.054 54	1.054×10^5	8.673×10^5	1.350×10^{-4}	0.0060
Methane, CH_4	5.00	187.91	0.003 380	0.042 60	2.578×10^5	2.286×10^6	1.244×10^{-4}	0.0060
Nitrogen, N_2	2.54	106.73	0.002 328	0.040 74	7.379×10^4	8.164×10^5	1.272×10^{-4}	0.0053

Source: Kenneth Wark, *Thermodynamics,* 4th ed., McGraw-Hill, New York, 1983, p. 815, table A-21M. Originally published in H. W. Cooper and J. C. Goldfrank, *Hydrocarbon Processing,* vol. 46, p. 141, 1967.

FIGURE A-30*a*

Generalized compressibility chart—*low pressures*. (Used with permission of Dr. Edward E. Obert, University of Wisconsin.)

(*a*) $0 < P_r < 1.0$

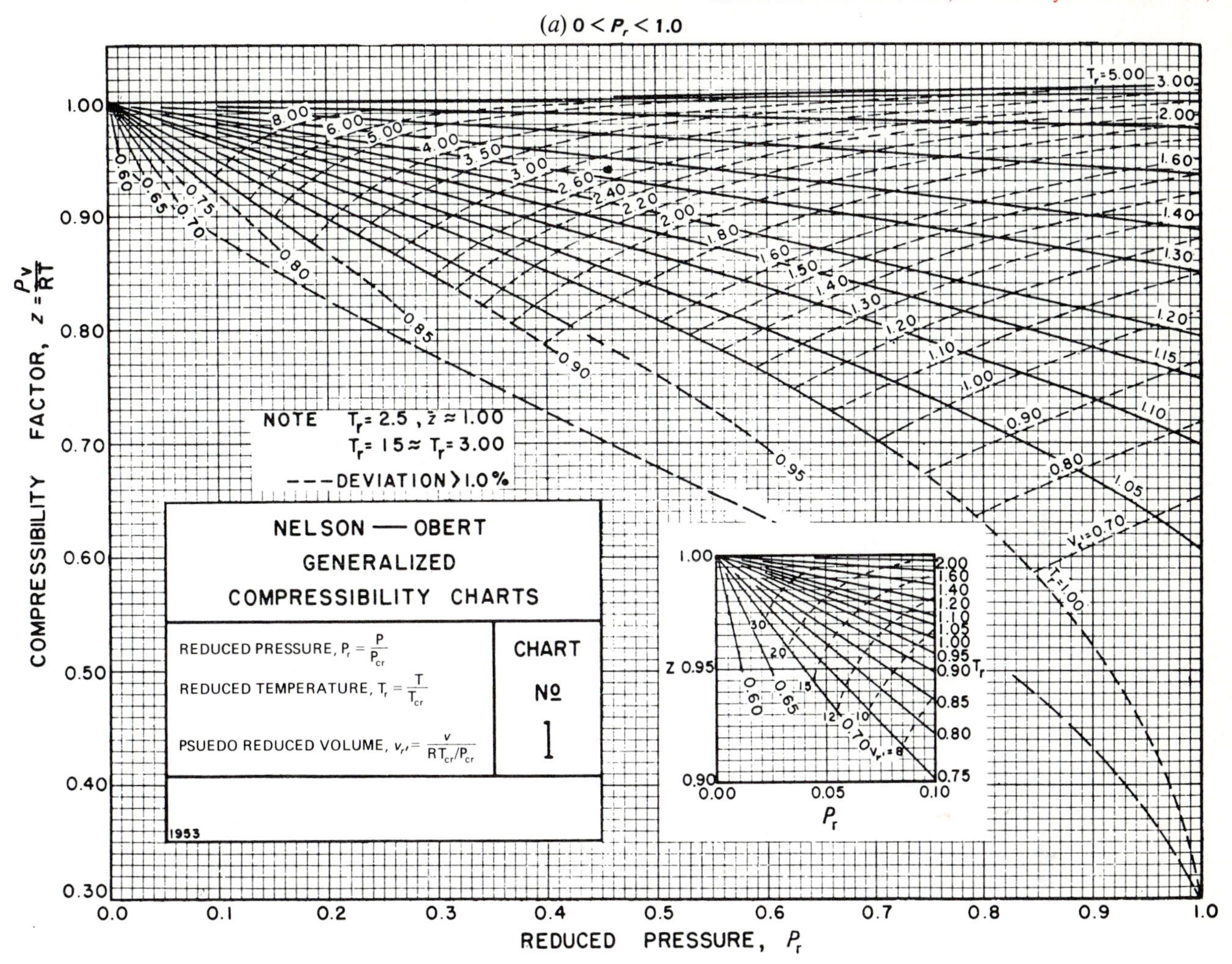

FIGURE A-30b

Generalized compressibility chart—*intermediate pressures*. (Used with permission of Dr. Edward E. Obert, University of Wisconsin.)

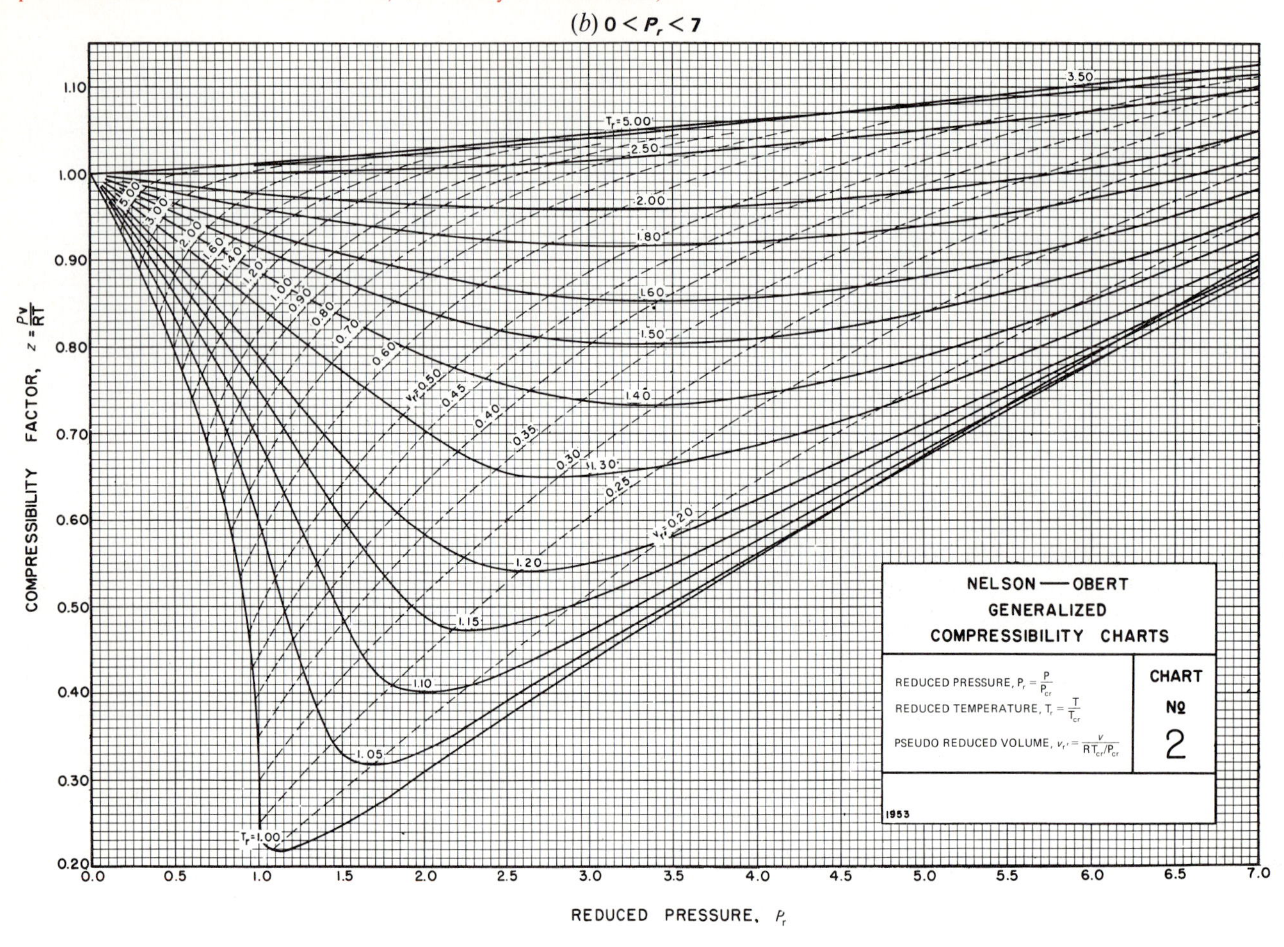

FIGURE A-30c

Generalized compressibility chart—*high pressures*. (Used with permission of Dr. Edward E. Obert, University of Wisconsin.)

(c) $0 < P_r < 40$

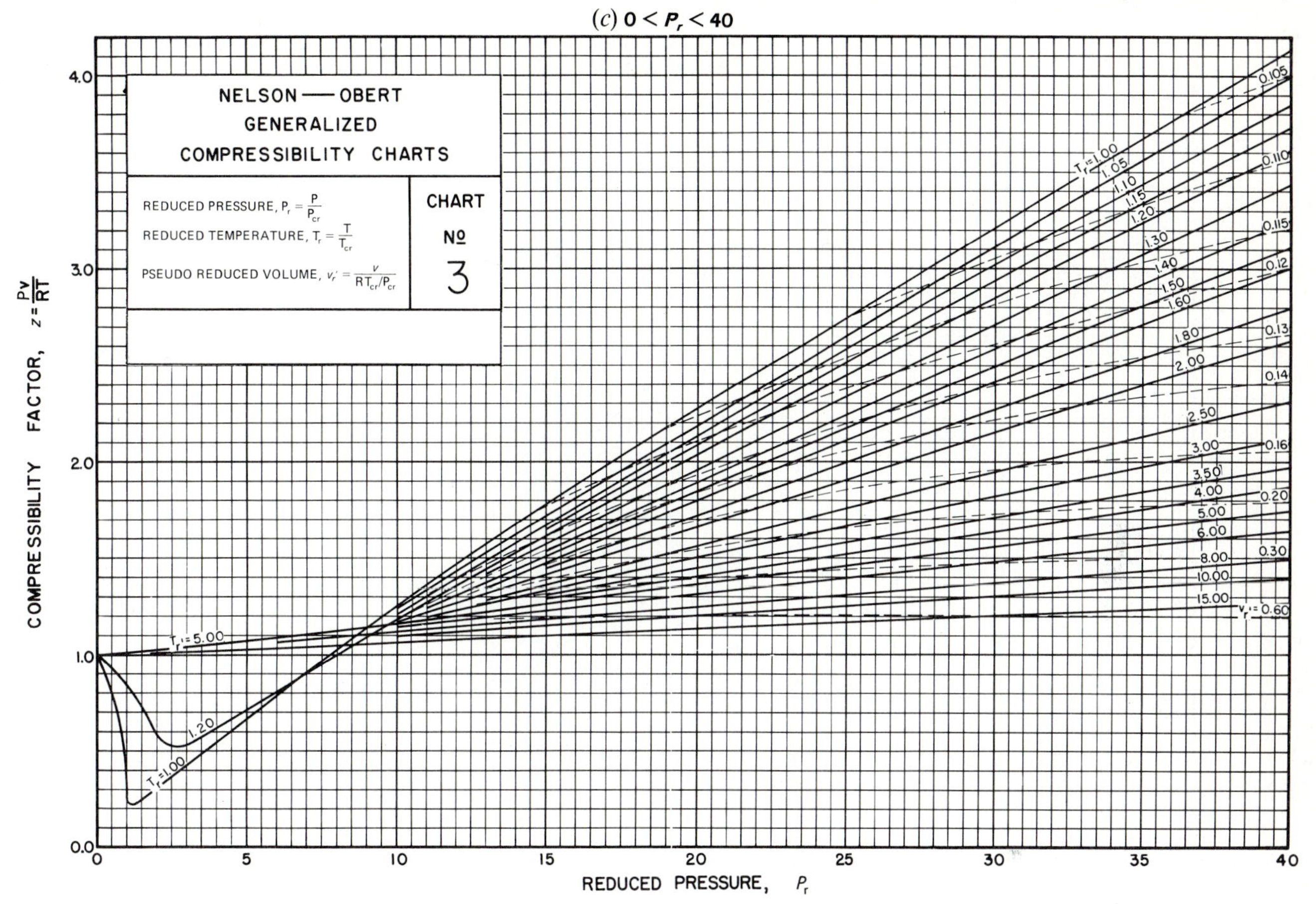

FIGURE A-31

Generalized enthalpy departure chart. (*Source:* John R. Howell and Richard O. Buckius, *Fundamentals of Engineering Thermodynamics,* SI Version, McGraw-Hill, New York, 1987, p. 558, fig. C.2, and p. 561, fig. C.5.)

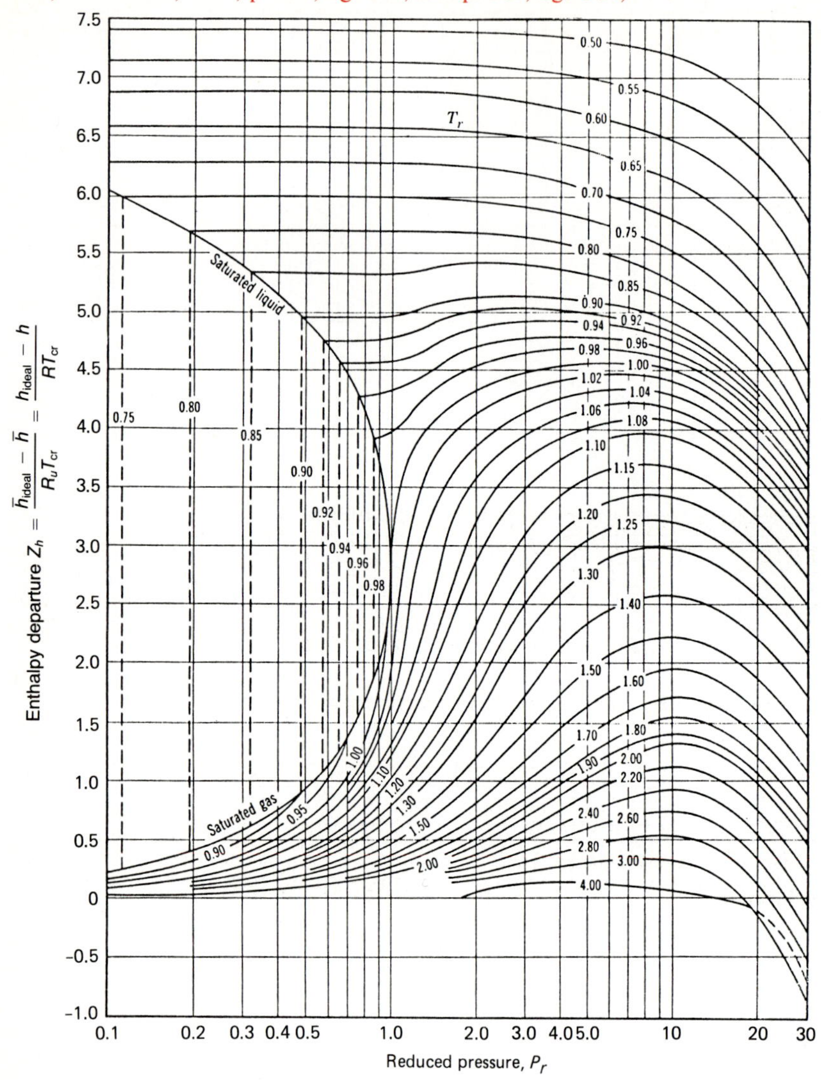

FIGURE A-31

(*Continued*)

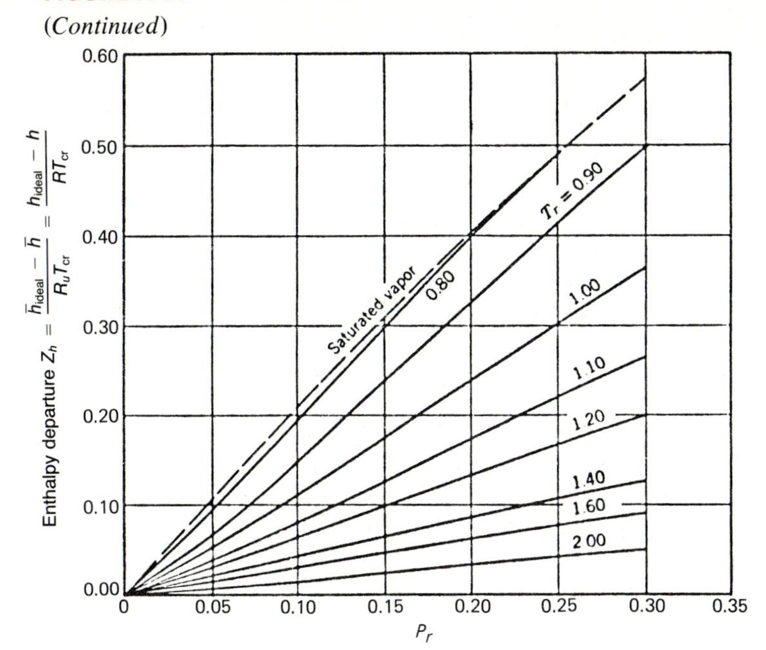

FIGURE A-32

Generalized entropy departure chart. (*Source*: John R. Howell and Richard O. Buckius, *Fundamentals of Engineering Thermodynamics*, SI Version, McGraw-Hill, New York, 1987, p. 559, fig. C.3, and p. 561, fig. C.5.)

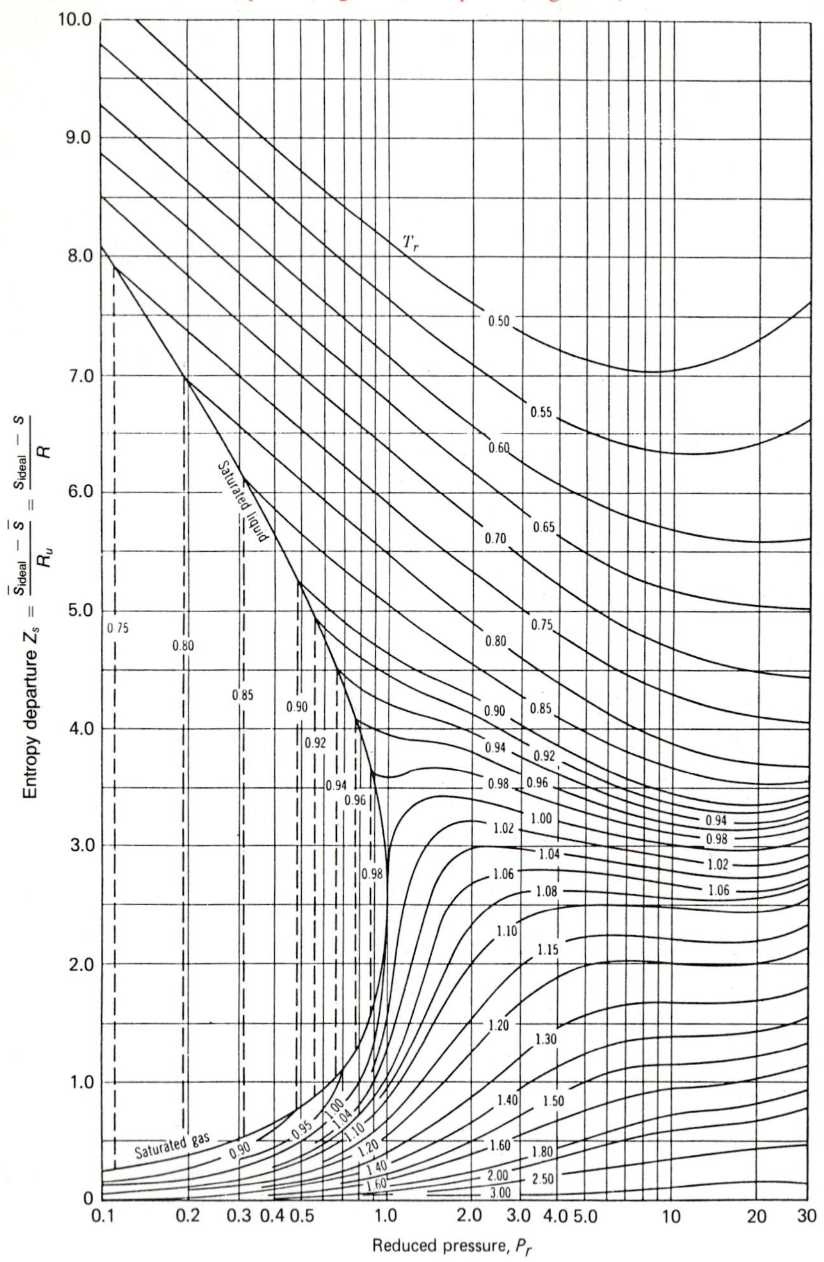

FIGURE A-32
(Continued)

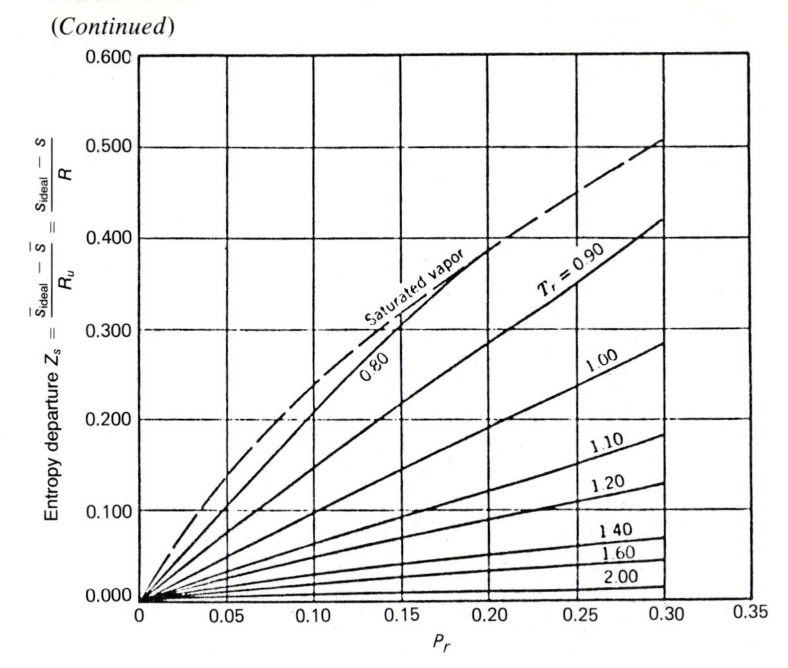

FIGURE A-33

Psychrometric chart at 1-atm total pressure. (Reprinted by permission of the American Society of Heating, Refrigerating and Air-Conditioning Engineers, Inc., Atlanta.)

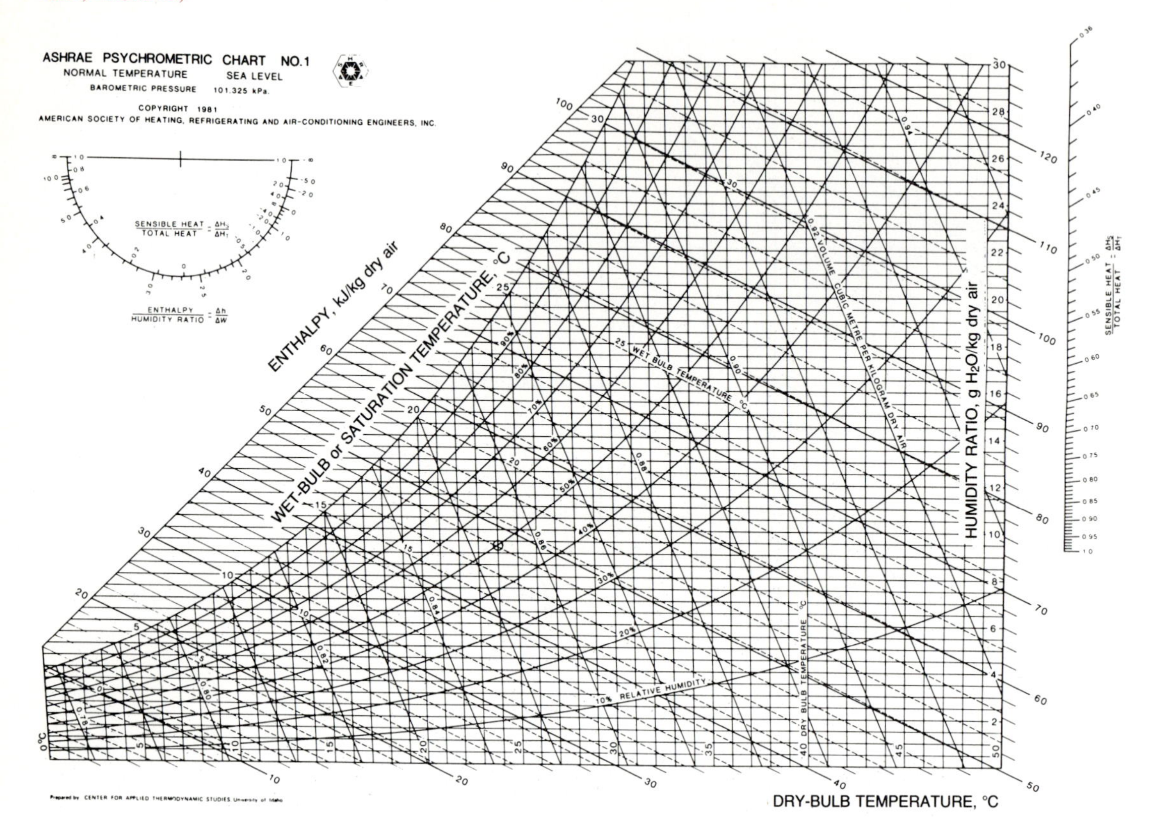

TABLE A-34

One-dimensional isentropic compressible-flow functions for an ideal gas with constant specific heat and molecular weight and k = 1.4†

M	M^*	$\frac{A}{A^*}$	$\frac{P}{P_0}$	$\frac{\rho}{\rho_0}$	$\frac{T}{T_0}$
0	0	∞	1.000 00	1.000 00	1.000 00
0.10	0.109 43	5.8218	0.993 03	0.995 02	0.998 00
0.20	0.218 22	2.9635	0.972 50	0.980 27	0.992 06
0.30	0.325 72	2.0351	0.939 47	0.956 38	0.982 32
0.40	0.431 33	1.5901	0.895 62	0.924 28	0.968 99
0.50	0.534 52	1.3398	0.843 02	0.885 17	0.952 38
0.60	0.634 80	1.1882	0.784 00	0.840 45	0.932 84
0.70	0.731 79	1.094 37	0.720 92	0.791 58	0.910 75
0.80	0.825 14	1.038 23	0.656 02	0.740 00	0.886 52
0.90	0.914 60	1.008 86	0.591 26	0.687 04	0.860 58
1.00	1.000 00	1.000 00	0.528 28	0.633 94	0.833 33
1.10	1.081 24	1.007 93	0.468 35	0.581 69	0.805 15
1.20	1.1583	1.030 44	0.412 38	0.531 14	0.776 40
1.30	1.2311	1.066 31	0.360 92	0.482 91	0.747 38
1.40	1.2999	1.1149	0.314 24	0.437 42	0.718 39
1.50	1.3646	1.1762	0.272 40	0.394 98	0.689 65
1.60	1.4254	1.2502	0.235 27	0.355 73	0.661 38
1.70	1.4825	1.3376	0.202 59	0.319 69	0.633 72
1.80	1.5360	1.4390	0.174 04	0.286 82	0.606 80
1.90	1.5861	1.5552	0.149 24	0.256 99	0.580 72
2.00	1.6330	1.6875	0.127 80	0.230 05	0.555 56
2.10	1.6769	1.8369	0.109 35	0.205 80	0.531 35
2.20	1.7179	2.0050	0.093 52	0.184 05	0.508 13
2.30	1.7563	2.1931	0.079 97	0.164 58	0.485 91
2.40	1.7922	2.4031	0.068 40	0.147 20	0.464 68
2.50	1.8258	2.6367	0.058 53	0.131 69	0.444 44
2.60	1.8572	2.8960	0.050 12	0.117 87	0.425 17
2.70	1.8865	3.1830	0.042 95	0.105 57	0.406 84
2.80	1.9140	3.5001	0.036 85	0.094 62	0.389 41
2.90	1.9398	3.8498	0.031 65	0.084 89	0.372 86
3.00	1.9640	4.2346	0.027 22	0.076 23	0.357 14
3.50	2.0642	6.7896	0.013 11	0.045 23	0.289 86
4.00	2.1381	10.719	0.006 58	0.027 66	0.238 10
4.50	2.1936	16.562	0.003 46	0.017 45	0.198 02
5.00	2.2361	25.000	$189(10)^{-5}$	0.011 34	0.166 67
6.00	2.2953	53.180	$633(10)^{-6}$	0.005 19	0.121 95
7.00	2.3333	104.143	$242(10)^{-6}$	0.002 61	0.092 59
8.00	2.3591	190.109	$102(10)^{-6}$	0.001 41	0.072 46
9.00	2.3772	327.189	$474(10)^{-7}$	0.000 815	0.058 14
10.00	2.3904	535.938	$236(10)^{-7}$	0.000 495	0.047 62
∞	2.4495	∞	0	0	0

†Calculated from the following relations for $k = 1.4$:

$$M^* = M\sqrt{\frac{k+1}{2+(k-1)M^2}} \qquad \frac{A}{A^*} = \frac{1}{M}\left[\left(\frac{2}{k+1}\right)\left(1+\frac{k-1}{2}M^2\right)\right]^{(k+1)/[2(k-1)]}$$

$$\frac{P}{P_0} = \left(1+\frac{k-1}{2}M^2\right)^{-k/(k-1)} \qquad \frac{\rho}{\rho_0} = \left(1+\frac{k-1}{2}M^2\right)^{-1/(k-1)}$$

$$\frac{T}{T_0} = \left(1+\frac{k-1}{2}M^2\right)^{-1}$$

A more extensive table is given in Joseph H. Keenan and Joseph Kaye, *Gas Tables*, Wiley, New York, 1948, table 30.

TABLE A-35

One-dimensional normal-shock functions for an ideal gas with constant specific heat and molecular weight and k = 1.4†

M_x	M_y	$\frac{P_y}{P_x}$	$\frac{\rho_y}{\rho_x}$	$\frac{T_y}{T_x}$	$\frac{P_{0y}}{P_{0x}}$	$\frac{P_{0y}}{P_x}$
1.00	1.000 00	1.0000	1.0000	1.0000	1.000 00	1.8929
1.10	0.911 77	1.2450	1.1691	1.0649	0.998 92	2.1328
1.20	0.842 17	1.5133	1.3416	1.1280	0.992 80	2.4075
1.30	0.785 96	1.8050	1.5157	1.1909	0.979 35	2.7135
1.40	0.739 71	2.1200	1.6896	1.2547	0.958 19	3.0493
1.50	0.701 09	2.4583	1.8621	1.3202	0.929 78	3.4133
1.60	0.668 44	2.8201	2.0317	1.3880	0.895 20	3.8049
1.70	0.640 55	3.2050	2.1977	1.4583	0.855 73	4.2238
1.80	0.616 50	3.6133	2.3592	1.5316	0.812 68	4.6695
1.90	0.595 62	4.0450	2.5157	1.6079	0.767 35	5.1417
2.00	0.577 35	4.5000	2.6666	1.6875	0.720 88	5.6405
2.10	0.561 28	4.9784	2.8119	1.7704	0.674 22	6.1655
2.20	0.547 06	5.4800	2.9512	1.8569	0.628 12	6.7163
2.30	0.534 41	6.0050	3.0846	1.9468	0.583 31	7.2937
2.40	0.523 12	6.5533	3.2119	2.0403	0.540 15	7.8969
2.50	0.512 99	7.1250	3.3333	2.1375	0.499 02	8.5262
2.60	0.503 87	7.7200	3.4489	2.2383	0.460 12	9.1813
2.70	0.495 63	8.3383	3.5590	2.3429	0.423 59	9.8625
2.80	0.488 17	8.9800	3.6635	2.4512	0.389 46	10.569
2.90	0.481 38	9.6450	3.7629	2.5632	0.357 73	11.302
3.00	0.475 19	10.333	3.8571	2.6790	0.328 34	12.061
4.00	0.434 96	18.500	4.5714	4.0469	0.138 76	21.068
5.00	0.415 23	29.000	5.0000	5.8000	0.061 72	32.654
10.00	0.387 57	116.50	5.7143	20.388	0.003 04	129.217
∞	0.377 96	∞	6.000	∞	0	∞

†Calculated from the following relations for k = 1.4:

$$M_y = \sqrt{\frac{M_x^2 + 2/(k-1)}{2M_x^2 k/(k-1) - 1}} \qquad \frac{T_y}{T_x} = \frac{1 + M_x^2 (k-1)/2}{1 + M_y^2 (k-1)/2}$$

$$\frac{P_y}{P_x} = \frac{1 + kM_x^2}{1 + kM_y^2} \qquad \frac{P_{0y}}{P_{0x}} = \frac{M_x}{M_y}\left[\frac{1 + M_y^2 (k-1)/2}{1 + M_x^2 (k-1)/2}\right]^{(k+1)/[2(k-1)]}$$

$$\frac{\rho_y}{\rho_x} = \frac{P_y/P_x}{T_y/T_x} \qquad \frac{P_{0y}}{P_x} = \frac{(1 + kM_x^2)[1 + M_y^2(k-1)/2]^{k/(k-1)}}{1 + kM_y^2}$$

A more extensive table is given in Joseph H. Keenan and Joseph Kaye, *Gas Tables*, Wiley, New York, 1948, table 30.

Property Tables, Figures, and Charts (English Units)

APPENDIX 2

TABLE A-1E

Molar mass, gas constant, and critical-point properties

Substance	Formula	Molar mass lbm/lbmol	Btu/(lbm·R)*	psia·ft³/lbm·R)*	Temp. R	Pressure psia	Volume ft³/lbmol
Ammonia	NH_3	17.03	0.1166	0.6301	729.8	1636	1.16
Argon	Ar	39.948	0.049 71	0.2686	272	705	1.20
Bromine	Br_2	159.808	0.012 43	0.067 14	1052	1500	2.17
Carbon dioxide	CO_2	44.01	0.045 13	0.2438	547.5	1071	1.51
Carbon monoxide	CO	28.011	0.070 90	0.3831	240	507	1.49
Chlorine	Cl_2	70.906	0.028 01	0.1517	751	1120	1.99
Deuterium (normal)	D_2	4.00	0.4965	2.6825	69.1	241	—
Helium	He	4.003	0.4961	2.6805	9.5	33.2	0.926
Hydrogen (normal)	H_2	2.016	0.9851	5.3224	59.9	188.1	1.04
Krypton	Kr	83.80	0.023 70	0.1280	376.9	798	1.48
Neon	Ne	20.183	0.098 40	0.5316	80.1	395	0.668
Nitrogen	N_2	28.013	0.070 90	0.3830	227.1	492	1.44
Nitrous oxide	N_2O	44.013	0.045 12	0.2438	557.4	1054	1.54
Oxygen	O_2	31.999	0.062 06	0.3353	278.6	736	1.25
Sulfur dioxide	SO_2	64.063	0.031 00	1.1675	775.2	1143	1.95
Water	H_2O	18.015	0.1102	0.5956	1165.3	3204	0.90
Xenon	Xe	131.30	0.015 13	0.081 72	521.55	852	1.90
Benzene	C_6H_6	78.115	0.025 42	0.1374	1012	714	4.17
n-Butane	C_4H_{10}	58.124	0.034 17	0.1846	765.2	551	4.08
Carbon tetrachloride	CCl_4	153.82	0.012 91	0.069 76	1001.5	661	4.42
Chloroform	$CHCl_3$	119.38	0.016 64	0.089 88	965.8	794	3.85
Dichlorodifluoromethane	CCl_2F_2 (R-12)	120.91	0.016 43	0.088 74	692.4	582	3.49
Dichlorofluoromethane	$CHCl_2F$	102.92	0.019 30	0.1043	813.0	749	3.16
Ethane	C_2H_6	30.020	0.066 16	0.3574	549.8	708	2.37
Ethyl alcohol	C_2H_5OH	46.07	0.043 11	0.2329	929.0	926	2.68
Ethylene	C_2H_4	28.054	0.070 79	0.3825	508.3	742	1.99
n-Hexane	C_6H_{14}	86.178	0.023 05	0.1245	914.2	439	5.89
Methane	CH_4	16.043	0.1238	0.6688	343.9	673	1.59
Methyl alcohol	CH_3OH	32.042	0.061 98	0.3349	923.7	1154	1.89
Methyl chloride	CH_3Cl	50.488	0.039 34	0.2125	749.3	968	2.29
Propane	C_3H_8	44.097	0.045 04	0.2433	665.9	617	3.20
Propene	C_3H_6	42.081	0.047 19	0.2550	656.9	670	2.90
Propyne	C_3H_4	40.065	0.049 57	0.2678	722	776	—
Trichlorofluoromethane	CCl_3F	137.37	0.014 46	0.078 11	848.1	635	3.97
Air	—	28.97	0.068 55	0.3704	—	—	—

*Calculated from $R = R_u/M$, where R_u = 1.986 Btu/(lbmol·R) = 10.73 psia·ft³/(lbmol·R) and M is the molar mass.

Source: Gordon J. Van Wylen and Richard E. Sonntag, *Fundamentals of Classical Thermodynamics,* English/SI Version, 3d ed., Wiley, New York, 1986, p. 684, table A.6E. Originally published in K. A. Kobe and R. E. Lynn, Jr., *Chemical Review,* vol. 52, pp. 117–236, 1953.

TABLE A-2E

Ideal-gas specific heats of various common gases

(*a*) At 80°F

Gas	Formula	C_{p0} Btu/(lbm·R)	C_{v0} Btu/(lbm·R)	k
Air	—	0.240	0.171	1.400
Argon	Ar	0.1253	0.0756	1.667
Butane	C_4H_{10}	0.415	0.381	1.09
Carbon dioxide	CO_2	0.203	0.158	1.285
Carbon monoxide	CO	0.249	0.178	1.399
Ethane	C_2H_6	0.427	0.361	1.183
Ethylene	C_2H_4	0.411	0.340	1.208
Helium	He	1.25	0.753	1.667
Hydrogen	H_2	3.43	2.44	1.404
Methane	CH_4	0.532	0.403	1.32
Neon	Ne	0.246	0.1477	1.667
Nitrogen	N_2	0.248	0.177	1.400
Octane	C_8H_{18}	0.409	0.392	1.044
Oxygen	O_2	0.219	0.157	1.395
Propane	C_3H_8	0.407	0.362	1.124
Steam	H_2O	0.445	0.335	1.329

Source: Gordon J. Van Wylen and Richard E. Sonntag, *Fundamentals of Classical Thermodynamics,* English/SI Version, 3d ed., Wiley, New York, 1986, p. 687, table A-8E.

TABLE A-2E

(continued)

(*b*) At various temperatures

Temp. °F	c_{p0} Btu/(lbm·R)	c_{v0} Btu/(lbm·R)	k	c_{p0} Btu/(lbm·R)	c_{v0} Btu/(lbm·R)	k	c_{p0} Btu/(lbm·R)	c_{v0} Btu/(lbm·R)	k
		Air			**Carbon dioxide, CO_2**			**Carbon monoxide, CO**	
40	0.240	0.171	1.401	0.195	0.150	1.300	0.248	0.177	1.400
100	0.240	0.172	1.400	0.205	0.160	1.283	0.249	0.178	1.399
200	0.241	0.173	1.397	0.217	0.172	1.262	0.249	0.179	1.397
300	0.243	0.174	1.394	0.229	0.184	1.246	0.251	0.180	1.394
400	0.245	0.176	1.389	0.239	0.193	1.233	0.253	0.182	1.389
500	0.248	0.179	1.383	0.247	0.202	1.223	0.256	0.185	1.384
600	0.250	0.182	1.377	0.255	0.210	1.215	0.259	0.188	1.377
700	0.254	0.185	1.371	0.262	0.217	1.208	0.262	0.191	1.371
800	0.257	0.188	1.365	0.269	0.224	1.202	0.266	0.195	1.364
900	0.259	0.191	1.358	0.275	0.230	1.197	0.269	0.198	1.357
1000	0.263	0.195	1.353	0.280	0.235	1.192	0.273	0.202	1.351
1500	0.276	0.208	1.330	0.298	0.253	1.178	0.287	0.216	1.328
2000	0.286	0.217	1.312	0.312	0.267	1.169	0.297	0.226	1.314
		Hydrogen, H_2			**Nitrogen, N_2**			**Oxygen, O_2**	
40	3.397	2.412	1.409	0.248	0.177	1.400	0.219	0.156	1.397
100	3.426	2.441	1.404	0.248	0.178	1.399	0.220	0.158	1.394
200	3.451	2.466	1.399	0.249	0.178	1.398	0.223	0.161	1.387
300	3.461	2.476	1.398	0.250	0.179	1.396	0.226	0.164	1.378
400	3.466	2.480	1.397	0.251	0.180	1.393	0.230	0.168	1.368
500	3.469	2.484	1.397	0.254	0.183	1.388	0.235	0.173	1.360
600	3.473	2.488	1.396	0.256	0.185	1.383	0.239	0.177	1.352
700	3.477	2.492	1.395	0.260	0.189	1.377	0.242	0.181	1.344
800	3.494	2.509	1.393	0.262	0.191	1.371	0.246	0.184	1.337
900	3.502	2.519	1.392	0.265	0.194	1.364	0.249	0.187	1.331
1000	3.513	2.528	1.390	0.269	0.198	1.359	0.252	0.190	1.326
1500	3.618	2.633	1.374	0.283	0.212	1.334	0.263	0.201	1.309
2000	3.758	2.773	1.355	0.293	0.222	1.319	0.270	0.208	1.298

Source: Kenneth Wark, *Thermodynamics*, 4th ed., McGraw-Hill, New York, 1983, p. 830, table A-4. Originally published in *Tables of Properties of Gases*, NBS *Circular*, 564, 1955.

TABLE A-2E

(continued)

(*c*) As a function of temperature

$\bar{C}_{p0} = a + bT + cT^2 + dT^3$
[T in R, $\bar{C}_{p0}$ in Btu/(lbmol·R)]

Substance	Formula	a	b	c	d	Temperature range R	% error Max.	% error Avg.
Nitrogen	N_2	6.903	-0.02085×10^{-2}	0.05957×10^{-5}	-0.1176×10^{-9}	491–3240	0.59	0.34
Oxygen	O_2	6.085	0.2017×10^{-2}	-0.05275×10^{-5}	0.05372×10^{-9}	491–3240	1.19	0.28
Air	—	6.713	0.02609×10^{-2}	0.03540×10^{-5}	-0.08052×10^{-9}	491–3240	0.72	0.33
Hydrogen	H_2	6.952	-0.02542×10^{-2}	0.02952×10^{-5}	-0.03565×10^{-9}	491–3240	1.01	0.26
Carbon monoxide	CO	6.726	0.02222×10^{-2}	0.03960×10^{-5}	-0.09100×10^{-9}	491–3240	0.89	0.37
Carbon dioxide	CO_2	5.316	0.79361×10^{-2}	-0.2581×10^{-5}	0.3059×10^{-9}	491–3240	0.67	0.22
Water vapor	H_2O	7.700	0.02552×10^{-2}	0.07781×10^{-5}	-0.1472×10^{-9}	491–3240	0.53	0.24
Nitric oxide	NO	7.008	-0.01247×10^{-2}	0.07185×10^{-5}	-0.1715×10^{-9}	491–2700	0.97	0.36
Nitrous oxide	N_2O	5.758	0.7780×10^{-2}	-0.2596×10^{-5}	0.4331×10^{-9}	491–2700	0.59	0.26
Nitrogen dioxide	NO_2	5.48	0.7583×10^{-2}	-0.260×10^{-5}	0.322×10^{-9}	491–2700	0.46	0.18
Ammonia	NH_3	6.5846	0.34028×10^{-2}	0.073034×10^{-5}	-0.27402×10^{-9}	491–2700	0.91	0.36
Sulfur	S_2	6.499	0.2943×10^{-2}	-0.1200×10^{-5}	0.1632×10^{-9}	491–3240	0.99	0.38
Sulfur dioxide	SO_2	6.157	0.7689×10^{-2}	-0.2810×10^{-5}	0.3527×10^{-9}	491–3240	0.45	0.24
Sulfur trioxide	SO_3	3.918	1.935×10^{-2}	-0.8256×10^{-5}	1.328×10^{-9}	491–2340	0.29	0.13
Acetylene	C_2H_2	5.21	1.2227×10^{-2}	-0.4812×10^{-5}	0.7457×10^{-9}	491–2700	1.46	0.59
Benzene	C_6H_6	−8.650	6.4322×10^{-2}	-2.327×10^{-5}	3.179×10^{-9}	491–2700	0.34	0.20
Methanol	CH_4O	4.55	1.214×10^{-2}	-0.0898×10^{-5}	-0.329×10^{-9}	491–1800	0.18	0.08
Ethanol	C_2H_6O	4.75	2.781×10^{-2}	-0.7651×10^{-5}	0.821×10^{-9}	491–2700	0.40	0.22
Hydrogen chloride	HCl	7.244	-0.1011×10^{-2}	0.09783×10^{-5}	-0.1776×10^{-9}	491–2740	0.22	0.08
Methane	CH_4	4.750	0.6666×10^{-2}	0.09352×10^{-5}	-0.4510×10^{-9}	491–2740	1.33	0.57
Ethane	C_2H_6	1.648	2.291×10^{-2}	-0.4722×10^{-5}	0.2984×10^{-9}	491–2740	0.83	0.28
Propane	C_3H_8	−0.966	4.044×10^{-2}	-1.159×10^{-5}	1.300×10^{-9}	491–2740	0.40	0.12
n-Butane	C_4H_{10}	0.945	4.929×10^{-2}	-1.352×10^{-5}	1.433×10^{-9}	491–2740	0.54	0.24
i-Butane	C_4H_{10}	−1.890	5.520×10^{-2}	-1.696×10^{-5}	2.044×10^{-9}	491–2740	0.25	0.13
n-Pentane	C_5H_{12}	1.618	6.028×10^{-2}	-1.656×10^{-5}	1.732×10^{-9}	491–2740	0.56	0.21
n-Hexane	C_6H_{14}	1.657	7.328×10^{-2}	-2.112×10^{-5}	2.363×10^{-9}	491–2740	0.72	0.20
Ethylene	C_2H_4	0.944	2.075×10^{-2}	-0.6151×10^{-5}	0.7326×10^{-9}	491–2740	0.54	0.13
Propylene	C_3H_6	0.753	3.162×10^{-2}	-0.8981×10^{-5}	1.008×10^{-9}	491–2740	0.73	0.17

Source: B. G. Kyle, *Chemical and Process Thermodynamics*, Prentice-Hall, Englewood Cliffs, N.J., 1984. Used with permission.

TABLE A-3E

Specific heats and densities of common solids and liquids

(*a*) At 80°F

Solid	C_p Btu/(lbm·R)	ρ lbm/ft³	Liquid	C_p Btu/(lbm·R)	ρ lbm/ft³
Aluminum	0.215	170	Ammonia	1.146	38
Copper	0.092	555	Ethanol	0.587	49
Granite	0.243	170	Refrigerant-12	0.233	82
Graphite	0.170	155	Mercury	0.033	847
Iron	0.107	490	Methanol	0.609	49
Lead	0.030	705	Oil (light)	0.430	57
Rubber (soft)	0.439	70	Water	1.000	62
Silver	0.056	655			
Tin	0.052	360			
Wood (most)	0.420	22–45			

Source: Gordon J. Van Wylen and Richard E. Sonntag, *Fundamentals of Classical Thermodynamics*, English/SI Version, 3d ed., Wiley, New York, 1986, p. 686, table A-7E.

(*b*) At various temperatures

Solids

Substance	Temp. °F	C_p Btu/(lbm·R)	Substance	Temp. °F	C_p Btu/(lbm·R)
Ice	−100	0.375	Lead	−455	0.0008
	−50	0.424		−435	0.0073
	0	0.471		−150	0.0283
	20	0.491		32	0.0297
	32	0.502		210	0.0320
Aluminum	−150	0.167		570	0.0356
	−100	0.192	Copper	−240	0.0674
	32	0.212		−150	0.0784
	100	0.218		−60	0.0862
	200	0.224		0	0.0893
	300	0.229		100	0.0925
	400	0.235		200	0.0938
	500	0.240		390	0.0963
Iron	68	0.107	Silver	68	0.0558

Liquids

Substance	State	C_p Btu/(lbm·R)	Substance	State	C_p Btu/(lbm·R)
Water	1 atm, 32°F	1.007	Glycerin	1 atm, 50°F	0.554
	1 atm, 77°F	0.998		1 atm, 120°F	0.617
	1 atm, 212°F	1.007	Bismuth	1 atm, 800°F	0.0345
Ammonia	Sat., 0°F	1.08		1 atm, 1400°F	0.0393
	Sat., 120°F	1.22	Mercury	1 atm, 50°F	0.033
Refrigerant-12	Sat., −40°F	0.211		1 atm, 600°F	0.032
	Sat., 0°F	0.217	Sodium	1 atm, 200°F	0.33
	Sat., 120°F	0.244		1 atm, 1000°F	0.30
Benzene	1 atm, 60°F	0.43	Propane	1 atm, 32°F	0.576
	1 atm, 150°F	0.46			

Source: Kenneth Wark, *Thermodynamics*, 4th ed., McGraw-Hill, New York, 1983, p. 862, table A-19.

TABLE A-4E

Saturated water–Temperature table

		Specific volume ft³/lbm		Internal energy Btu/lbm			Enthalpy Btu/lbm			Entropy Btu/(lbm·R)		
Temp. °F. T	Sat. press. psia P_{sat}	Sat. liquid v_f	Sat. vapor v_g	Sat. liquid u_f	Evap. u_{fg}	Sat. vapor u_g	Sat. liquid h_f	Evap. h_{fg}	Sat. vapor h_g	Sat. liquid s_f	Evap. s_{fg}	Sat. vapor s_g
32.018	0.088 66	0.016 022	3302	0.00	1021.2	1021.2	0.01	1075.4	1075.4	0.000 00	2.1869	2.1869
35	0.099 92	0.016 021	2948	2.99	1019.2	1022.2	3.00	1073.7	1076.7	0.006 07	2.1704	2.1764
40	0.121 66	0.016 020	2445	8.02	1015.8	1023.9	8.02	1070.9	1078.9	0.016 17	2.1430	2.1592
45	0.147 48	0.016 021	2037	13.04	1012.5	1025.5	13.04	1068.1	1081.1	0.026 18	2.1162	2.1423
50	0.178 03	0.016 024	1704.2	18.06	1009.1	1027.2	18.06	1065.2	1083.3	0.036 07	2.0899	2.1259
60	0.2563	0.016 035	1206.9	28.08	1002.4	1030.4	28.08	1059.6	1087.7	0.055 55	2.0388	2.0943
70	0.3632	0.016 051	867.7	38.09	995.6	1033.7	38.09	1054.0	1092.0	0.074 63	1.9896	2.0642
80	0.5073	0.016 073	632.8	48.08	988.9	1037.0	48.09	1048.3	1096.4	0.093 32	1.9423	2.0356
90	0.6988	0.016 099	467.7	58.07	982.2	1040.2	58.07	1042.7	1100.7	0.111 65	1.8966	2.0083
100	0.9503	0.016 130	350.0	68.04	975.4	1043.5	68.05	1037.0	1105.0	0.129 63	1.8526	1.9822
110	1.2763	0.016 166	265.1	78.02	968.7	1046.7	78.02	1031.3	1109.3	0.147 30	1.8101	1.9574
120	1.6945	0.016 205	203.0	87.99	961.9	1049.9	88.00	1025.5	1113.5	0.164 65	1.7690	1.9336
130	2.225	0.016 247	157.17	97.97	955.1	1053.0	97.98	1019.8	1117.8	0.181 72	1.7292	1.9109
140	2.892	0.016 293	122.88	107.95	948.2	1056.2	107.96	1014.0	1121.9	0.198 51	1.6907	1.8892
150	3.722	0.016 343	96.99	117.95	941.3	1059.3	117.96	1008.1	1126.1	0.215 03	1.6533	1.8684
160	4.745	0.016 395	77.23	127.94	934.4	1062.3	127.96	1002.2	1130.1	0.231 30	1.6171	1.8484
170	5.996	0.016 450	62.02	137.95	927.4	1065.4	137.97	996.2	1134.2	0.247 32	1.5819	1.8293
180	7.515	0.016 509	50.20	147.97	920.4	1068.3	147.99	990.2	1138.2	0.263 11	1.5478	1.8109
190	9.343	0.016 570	40.95	158.00	913.3	1071.3	158.03	984.1	1142.1	0.278 66	1.5146	1.7932
200	11.529	0.016 634	33.63	168.04	906.2	1074.2	168.07	977.9	1145.9	0.294 00	1.4822	1.7762
210	14.125	0.016 702	27.82	178.10	898.9	1077.0	178.14	971.6	1149.7	0.309 13	1.4508	1.7599
212	14.698	0.016 716	26.80	180.11	897.5	1077.6	180.16	970.3	1150.5	0.312 13	1.4446	1.7567
220	17.188	0.016 772	23.15	188.17	891.7	1079.8	188.22	965.3	1153.5	0.324 06	1.4201	1.7441
230	20.78	0.016 845	19.386	198.26	884.3	1082.6	198.32	958.8	1157.1	0.338 80	1.3901	1.7289
240	24.97	0.016 922	16.327	208.36	876.9	1085.3	208.44	952.3	1160.7	0.353 35	1.3609	1.7143
250	29.82	0.017 001	13.826	218.49	869.4	1087.9	218.59	945.6	1164.2	0.367 72	1.3324	1.7001
260	35.42	0.017 084	11.768	228.64	861.8	1090.5	228.76	938.8	1167.6	0.381 93	1.3044	1.6864
270	41.85	0.017 170	10.066	238.82	854.1	1093.0	238.95	932.0	1170.9	0.395 97	1.2771	1.6731
280	49.18	0.017 259	8.650	249.02	846.3	1095.4	249.18	924.9	1174.1	0.409 86	1.2504	1.6602
290	57.53	0.017 352	7.467	259.25	838.5	1097.7	259.44	917.8	1177.2	0.423 60	1.2241	1.6477
300	66.98	0.017 448	6.472	269.52	830.5	1100.0	269.73	910.4	1180.2	0.437 20	1.1984	1.6356
310	77.64	0.017 548	5.632	279.81	822.3	1102.1	280.06	903.0	1183.0	0.450 67	1.1731	1.6238
320	89.60	0.017 652	4.919	290.14	814.1	1104.2	290.43	895.3	1185.8	0.464 00	1.1483	1.6123
330	103.00	0.017 760	4.312	300.51	805.7	1106.2	300.84	887.5	1188.4	0.477 22	1.1238	1.6010
340	117.93	0.017 872	3.792	310.91	797.1	1108.0	311.30	879.5	1190.8	0.490 31	1.0997	1.5901
350	134.53	0.017 988	3.346	321.35	788.4	1109.8	321.80	871.3	1193.1	0.503 29	1.0760	1.5793
360	152.92	0.018 108	2.961	331.84	779.6	1111.4	332.35	862.9	1195.2	0.516 17	1.0526	1.5688
370	173.23	0.018 233	2.628	342.37	770.6	1112.9	342.96	854.2	1197.2	0.528 94	1.0295	1.5585
380	195.60	0.018 363	2.339	352.95	761.4	1114.3	353.62	845.4	1199.0	0.541 63	1.0067	1.5483
390	220.2	0.018 498	2.087	363.58	752.0	1115.6	364.34	836.2	1200.6	0.554 22	0.9841	1.5383
400	247.1	0.018 638	1.8661	374.27	742.4	1116.6	375.12	826.8	1202.0	0.566 72	0.9617	1.5284
410	276.5	0.018 784	1.6726	385.01	732.6	1117.6	385.97	817.2	1203.1	0.579 16	0.9395	1.5187
420	308.5	0.018 936	1.5024	395.81	722.5	1118.3	396.89	807.2	1204.1	0.591 52	0.9175	1.5091
430	343.3	0.019 094	1.3521	406.68	712.2	1118.9	407.89	796.9	1204.8	0.603 81	0.8957	1.4995
440	381.2	0.019 260	1.2192	417.62	701.7	1119.3	418.98	786.3	1205.3	0.616 05	0.8740	1.4900
450	422.1	0.019 433	1.1011	428.6	690.9	1119.5	430.2	775.4	1205.6	0.6282	0.8523	1.4806
460	466.3	0.019 614	0.9961	439.7	679.8	1119.6	441.4	764.1	1205.5	0.6404	0.8308	1.4712
470	514.1	0.019 803	0.9025	450.9	668.4	1119.4	452.8	752.4	1205.2	0.6525	0.8093	1.4618
480	565.5	0.020 002	0.8187	462.2	656.7	1118.9	464.3	740.3	1204.6	0.6646	0.7878	1.4524
490	620.7	0.020 211	0.7436	473.6	644.7	1118.3	475.9	727.8	1203.7	0.6767	0.7663	1.4430

TABLE A-4E

(*Continued*)

Temp. °F. T	Sat. press. psia P_{sat}	Specific volume ft³/lbm Sat. liquid v_f	Sat. vapor v_g	Internal energy Btu/lbm Sat. liquid u_f	Evap. u_{fg}	Sat. vapor u_g	Enthalpy Btu/lbm Sat. liquid h_f	Evap. h_{fg}	Sat. vapor h_g	Entropy Btu/(lbm·R) Sat. liquid s_f	Evap. s_{fg}	Sat. vapor s_g
500	680.0	0.020 43	0.6761	485.1	632.3	1117.4	487.7	714.8	1202.5	0.6888	0.7448	1.4335
520	811.4	0.020 91	0.5605	508.5	606.2	1114.8	511.7	687.3	1198.9	0.7130	0.7015	1.4145
540	961.5	0.021 45	0.4658	532.6	578.4	1111.0	536.4	657.5	1193.8	0.7374	0.6576	1.3950
560	1131.8	0.022 07	0.3877	557.4	548.4	1105.8	562.0	625.0	1187.0	0.7620	0.6129	1.3749
580	1324.3	0.022 78	0.3225	583.1	515.9	1098.9	588.6	589.3	1178.0	0.7872	0.5668	1.3540
600	1541.0	0.023 63	0.2677	609.9	480.1	1090.0	616.7	549.7	1166.4	0.8130	0.5187	1.3317
620	1784.4	0.024 65	0.2209	638.3	440.2	1078.5	646.4	505.0	1151.4	0.8398	0.4677	1.3075
640	2057.1	0.025 93	0.1805	668.7	394.5	1063.2	678.6	453.4	1131.9	0.8681	0.4122	1.2803
660	2362	0.027 67	0.144 59	702.3	340.0	1042.3	714.4	391.1	1105.5	0.8990	0.3493	1.2483
680	2705	0.030 32	0.111 27	741.7	269.3	1011.0	756.9	309.8	1066.7	0.9350	0.2718	1.2068
700	3090	0.036 66	0.074 38	801.7	145.9	947.7	822.7	167.5	990.2	0.9902	0.1444	1.1346
705.44	3204	0.050 53	0.050 53	872.6	0	872.6	902.5	0	902.5	1.0580	0	1.0580

Source: Gordon J. Van Wylen and Richard E. Sonntag, *Fundamentals of Classical Thermodynamics,* English/SI Version, 3d ed., Wiley, New York, 1986, pp. 619–621, table A.1.1E. Originally published in Joseph H. Keenan, Frederick G. Keyes, Philip G. Hill, and Joan G. Moore, *Steam Tables,* Wiley, New York, 1969.

TABLE A-5E

Saturated water–Pressure table

Press. psia P	Sat. temp. °F T_{sat}	Specific volume ft³/lbm Sat. liquid v_f	Sat. vapor v_g	Internal energy Btu/lbm Sat. liquid u_f	Evap. u_{fg}	Sat. vapor u_g	Enthalpy Btu/lbm Sat. liquid h_f	Evap. h_{fg}	Sat. vapor h_g	Entropy Btu/(lbm·R) Sat. liquid s_f	Evap. s_{fg}	Sat. vapor s_g
1.0	101.70	0.016 136	333.6	69.74	974.3	1044.0	69.74	1036.0	1105.8	0.132 66	1.8453	1.9779
2.0	126.04	0.016 230	173.75	94.02	957.8	1051.8	94.02	1022.1	1116.1	0.174 99	1.7448	1.9198
3.0	141.43	0.016 300	118.72	109.38	947.2	1056.6	109.39	1013.1	1122.5	0.200 89	1.6852	1.8861
4.0	152.93	0.016 358	90.64	120.88	939.3	1060.2	120.89	1006.4	1127.3	0.219 83	1.6426	1.8624
5.0	162.21	0.016 407	73.53	130.15	932.9	1063.0	130.17	1000.9	1131.0	0.234 86	1.6093	1.8441
6.0	170.03	0.016 451	61.98	137.98	927.4	1065.4	138.00	996.2	1134.2	0.247 36	1.5819	1.8292
8.0	182.84	0.016 526	47.35	150.81	918.4	1069.2	150.84	988.4	1139.3	0.267 54	1.5383	1.8058
10	193.19	0.016 590	38.42	161.20	911.0	1072.2	161.23	982.1	1143.3	0.283 58	1.5041	1.7877
14.696	211.99	0.016 715	26.80	180.10	897.5	1077.6	180.15	970.4	1150.5	0.312 12	1.4446	1.7567
15	213.03	0.016 723	26.29	181.14	896.8	1077.9	181.19	969.7	1150.9	0.313 67	1.4414	1.7551
20	227.96	0.016 830	20.09	196.19	885.8	1082.0	196.26	960.1	1156.4	0.335 80	1.3962	1.7320

TABLE A-5E

(Continued)

Press. psia P	Sat. temp. °F T_{sat}	Specific volume ft³/lbm: Sat. liquid v_f	Sat. vapor v_g	Internal energy Btu/lbm: Sat. liquid u_f	Evap. u_{fg}	Sat. vapor u_g	Enthalpy Btu/lbm: Sat. liquid h_f	Evap. h_{fg}	Sat. vapor h_g	Entropy Btu/(lbm·R): Sat. liquid s_f	Evap. s_{fg}	Sat. vapor s_g
25	240.08	0.016 922	16.306	208.44	876.9	1085.3	208.52	952.2	1160.7	0.353 45	1.3607	1.7142
30	250.34	0.017 004	13.748	218.84	869.2	1088.0	218.93	945.4	1164.3	0.368 21	1.3314	1.6996
35	259.30	0.017 073	11.900	227.93	862.4	1090.3	228.04	939.3	1167.4	0.380 93	1.3064	1.6873
40	267.26	0.017 146	10.501	236.03	856.2	1092.3	236.16	933.8	1170.0	0.392 14	1.2845	1.6767
45	274.46	0.017 209	9.403	243.37	850.7	1094.0	243.51	928.8	1172.3	0.402 18	1.2651	1.6673
50	281.03	0.017 269	8.518	250.08	845.5	1095.6	250.24	924.2	1174.4	0.411 29	1.2476	1.6589
55	287.10	0.017 325	7.789	256.28	840.8	1097.0	256.46	919.9	1176.3	0.419 63	1.2317	1.6513
60	292.73	0.017 378	7.177	262.06	836.3	1098.3	262.25	915.8	1178.0	0.427 33	1.2170	1.6444
65	298.00	0.017 429	6.657	267.46	832.1	1099.5	267.67	911.9	1179.6	0.434 50	1.2035	1.6380
70	302.96	0.017 478	6.209	272.56	828.1	1100.6	272.79	908.3	1181.0	0.441 20	1.1909	1.6321
75	307.63	0.017 524	5.818	277.37	824.3	1101.6	277.61	904.8	1182.4	0.447 49	1.1790	1.6265
80	312.07	0.017 570	5.474	281.95	820.6	1102.6	282.21	901.4	1183.6	0.453 44	1.1679	1.6214
85	316.29	0.017 613	5.170	286.30	817.1	1103.5	286.58	898.2	1184.8	0.459 07	1.1574	1.6165
90	320.31	0.017 655	4.898	290.46	813.8	1104.3	290.76	895.1	1185.9	0.464 42	1.1475	1.6119
95	324.16	0.017 696	4.654	294.45	810.6	1105.0	294.76	892.1	1186.9	0.469 52	1.1380	1.6076
100	327.86	0.017 736	4.434	298.28	807.5	1105.8	298.61	889.2	1187.8	0.474 39	1.1290	1.6034
110	334.82	0.017 813	4.051	305.52	801.6	1107.1	305.88	883.7	1189.6	0.483 55	1.1122	1.5957
120	341.30	0.017 886	3.730	312.27	796.0	1108.3	312.67	878.5	1191.1	0.492 01	1.0966	1.5886
130	347.37	0.017 957	3.457	318.61	790.7	1109.4	319.04	873.5	1192.5	0.499 89	1.0822	1.5821
140	353.08	0.018 024	3.221	324.58	785.7	1110.3	325.05	868.7	1193.8	0.507 27	1.0688	1.5761
150	358.48	0.018 089	3.016	330.24	781.0	1111.2	330.75	864.2	1194.9	0.514 22	1.0562	1.5704
160	363.60	0.018 152	2.836	335.63	776.4	1112.0	336.16	859.8	1196.0	0.520 78	1.0443	1.5651
170	368.47	0.018 214	2.676	340.76	772.0	1112.7	341.33	855.6	1196.9	0.527 00	1.0330	1.5600
180	373.13	0.018 273	2.533	345.68	767.7	1113.4	346.29	851.5	1197.8	0.532 92	1.0223	1.5553
190	377.59	0.018 331	2.405	350.39	763.6	1114.0	351.04	847.5	1198.6	0.538 57	1.0122	1.5507
200	381.86	0.018 387	2.289	354.9	759.6	1114.6	355.6	843.7	1199.3	0.5440	1.0025	1.5464
250	401.04	0.018 653	1.8448	375.4	741.4	1116.7	376.2	825.8	1202.1	0.5680	0.9594	1.5274
300	417.43	0.018 896	1.5442	393.0	725.1	1118.2	394.1	809.8	1203.9	0.5883	0.9232	1.5115
350	431.82	0.019 124	1.3267	408.7	710.3	1119.0	409.9	795.0	1204.9	0.6060	0.8917	1.4978
400	444.70	0.019 340	1.1620	422.8	696.7	1119.5	424.2	781.2	1205.5	0.6218	0.8638	1.4856
450	456.39	0.019 547	1.0326	435.7	683.9	1119.6	437.4	768.2	1205.6	0.6360	0.8385	1.4746
500	467.13	0.019 748	0.9283	447.7	671.7	1119.4	449.5	755.8	1205.3	0.6490	0.8154	1.4645
550	477.07	0.019 943	0.8423	458.9	660.2	1119.1	460.9	743.9	1204.8	0.6611	0.7941	1.4551
600	486.33	0.020 13	0.7702	469.4	649.1	1118.6	471.7	732.4	1204.1	0.6723	0.7742	1.4464
700	503.23	0.020 51	0.6558	488.9	628.2	1117.0	491.5	710.5	1202.0	0.6927	0.7378	1.4305
800	518.36	0.020 87	0.5691	506.6	608.4	1115.0	509.7	689.6	1199.3	0.7110	0.7050	1.4160
900	532.12	0.021 23	0.5009	523.0	589.6	1112.6	526.6	669.5	1196.0	0.7277	0.6750	1.4027
1000	544.75	0.021 59	0.4459	538.4	571.5	1109.9	542.4	650.0	1192.4	0.7432	0.6471	1.3903
1200	567.37	0.022 32	0.3623	566.7	536.8	1103.5	571.7	612.3	1183.9	0.7712	0.5961	1.3673
1400	587.25	0.023 07	0.3016	592.7	503.3	1096.0	598.6	575.5	1174.1	0.7964	0.5497	1.3461
1600	605.06	0.023 86	0.2552	616.9	470.5	1087.4	624.0	538.9	1162.9	0.8196	0.5062	1.3258
1800	621.21	0.024 72	0.2183	640.0	437.6	1077.7	648.3	502.1	1150.4	0.8414	0.4645	1.3060
2000	636.00	0.025 65	0.188 13	662.4	404.2	1066.6	671.9	464.4	1136.3	0.8623	0.4238	1.2861
2500	668.31	0.028 60	0.130 59	717.7	313.4	1031.0	730.9	360.5	1091.4	0.9131	0.3196	1.2327
3000	695.52	0.034 31	0.084 04	783.4	185.4	968.8	802.5	213.0	1015.5	0.9732	0.1843	1.1575
3203.6	705.44	0.050 53	0.050 53	872.6	0	872.6	902.5	0	902.5	1.0580	0	1.0580

Source: Gordon J. Van Wylen and Richard E. Sonntag, *Fundamentals of Classical Thermodynamics,* English/SI Version, 3d ed., Wiley, New York, 1986, pp. 622–623, table A.1.2E. Originally published in Joseph H. Keenan, Frederick G. Keyes, Philip G. Hill, and Joan G. Moore, *Steam Tables,* Wiley, New York, 1969.

TABLE A-6E

Superheated water

T °F	*v* ft³/lbm	*u* Btu/lbm	*h* Btu/lbm	*s* Btu/(lbm·R)	*v* ft³/lbm	*u* Btu/lbm	*h* Btu/lbm	*s* Btu/(lbm·R)	*v* ft³/lbm	*u* Btu/lbm	*h* Btu/lbm	*s* Btu/(lbm·R)
	P = **1.0 psia (101.70°F)**				*P* = **5.0 psia (162.21°F)**				*P* = **10.0 psia (193.19°F)**			
Sat.	333.6	1044.0	1105.8	1.9779	73.53	1063.0	1131.0	1.8441	38.42	1072.2	1143.3	1.7877
200	392.5	1077.5	1150.1	2.0508	78.15	1076.3	1148.6	1.8715	38.85	1074.7	1146.6	1.7927
240	416.4	1091.2	1168.3	2.0775	83.00	1090.3	1167.1	1.8987	41.32	1089.0	1165.5	1.8205
280	440.3	1105.0	1186.5	2.1028	87.83	1104.3	1185.5	1.9244	43.77	1103.3	1184.3	1.8467
320	464.2	1118.9	1204.8	2.1269	92.64	1118.3	1204.0	1.9487	46.20	1117.6	1203.1	1.8714
360	488.1	1132.9	1223.2	2.1500	97.45	1132.4	1222.6	1.9719	48.62	1131.8	1221.8	1.8948
400	511.9	1147.0	1241.8	2.1720	102.24	1146.6	1241.2	1.9941	51.03	1146.1	1240.5	1.9171
440	535.8	1161.2	1260.4	2.1932	107.03	1160.9	1259.9	2.0154	53.44	1160.5	1259.3	1.9385
500	571.5	1182.8	1288.5	2.2235	114.20	1182.5	1288.2	2.0458	57.04	1182.2	1287.7	1.9690
600	631.1	1219.3	1336.1	2.2706	126.15	1219.1	1335.8	2.0930	63.03	1218.9	1335.5	2.0164
700	690.7	1256.7	1384.5	2.3142	138.08	1256.5	1384.3	2.1367	69.01	1256.3	1384.0	2.0601
800	750.3	1294.9	1433.7	2.3550	150.01	1294.7	1433.5	2.1775	74.98	1294.6	1433.3	2.1009
1000	869.5	1373.9	1534.8	2.4294	173.86	1373.9	1534.7	2.2520	86.91	1373.8	1534.6	2.1755
1200	988.6	1456.7	1639.6	2.4967	197.70	1456.6	1639.5	2.3192	98.84	1456.5	1639.4	2.2428
1400	1107.7	1543.1	1748.1	2.5584	221.54	1543.1	1748.1	2.3810	110.76	1543.0	1748.0	2.3045
	P = **14.696 psia (211.99°F)**				*P* = **20 psia (227.96°F)**				*P* = **40 psia (267.26°F)**			
Sat.	26.80	1077.6	1150.5	1.7567	20.09	1082.0	1156.4	1.7320	10.501	1092.3	1170.0	1.6767
240	28.00	1087.9	1164.0	1.7764	20.47	1086.5	1162.3	1.7405				
280	29.69	1102.4	1183.1	1.8030	21.73	1101.4	1181.8	1.7676	10.711	1097.3	1176.6	1.6857
320	31.36	1116.8	1202.1	1.8280	22.98	1116.0	1201.0	1.7930	11.360	1112.8	1196.9	1.7124
360	33.02	1131.2	1221.0	1.8516	24.21	1130.6	1220.1	1.8168	11.996	1128.0	1216.8	1.7373
400	34.67	1145.6	1239.9	1.8741	25.43	1145.1	1239.2	1.8395	12.623	1143.0	1236.4	1.7606
440	36.31	1160.1	1258.8	1.8956	26.64	1159.6	1258.2	1.8611	13.243	1157.8	1255.8	1.7828
500	38.77	1181.8	1287.3	1.9263	28.46	1181.5	1286.8	1.8919	14.164	1180.1	1284.9	1.8140
600	42.86	1218.6	1335.2	1.9737	31.47	1218.4	1334.8	1.9395	15.685	1217.3	1333.4	1.8621
700	46.93	1256.1	1383.8	2.0175	34.47	1255.9	1383.5	1.9834	17.196	1255.1	1382.4	1.9063
800	51.00	1294.4	1433.1	2.0584	37.46	1294.3	1432.9	2.0243	18.701	1293.7	1432.1	1.9474
1000	59.13	1373.7	1534.5	2.1330	43.44	1373.5	1534.3	2.0989	21.70	1373.1	1533.8	2.0223
1200	67.25	1456.5	1639.3	2.2003	49.41	1456.4	1639.2	2.1663	24.69	1456.1	1638.9	2.0897
1400	75.36	1543.0	1747.9	2.2621	55.37	1542.9	1747.9	2.2281	27.68	1542.7	1747.6	2.1515
1600	83.47	1633.2	1860.2	2.3194	61.33	1633.2	1860.1	2.2854	30.66	1633.0	1859.9	2.2089
	P = **60 psia (292.73°F)**				*P* = **80 psia (312.07°F)**				*P* = **100 psia (327.86°F)**			
Sat.	7.177	1098.3	1178.0	1.6444	5.474	1102.6	1183.6	1.6214	4.434	1105.8	1187.8	1.6034
320	7.485	1109.5	1192.6	1.6634	5.544	1106.0	1188.0	1.6271				
360	7.924	1125.3	1213.3	1.6893	5.886	1122.5	1209.7	1.6541	4.662	1119.7	1205.9	1.6259
400	8.353	1140.8	1233.5	1.7134	6.217	1138.5	1230.6	1.6790	4.934	1136.2	1227.5	1.6517
440	8.775	1156.0	1253.4	1.7360	6.541	1154.2	1251.0	1.7022	5.199	1152.3	1248.5	1.6755
500	9.399	1178.6	1283.0	1.7678	7.017	1177.2	1281.1	1.7346	5.587	1175.7	1279.1	1.7085
600	10.425	1216.3	1332.1	1.8165	7.794	1215.3	1330.7	1.7838	6.216	1214.2	1329.3	1.7582
700	11.440	1254.4	1381.4	1.8609	8.561	1253.6	1380.3	1.8285	6.834	1252.8	1379.2	1.8033
800	12.448	1293.0	1431.2	1.9022	9.321	1292.4	1430.4	1.8700	7.445	1291.8	1429.6	1.8449
1000	14.454	1372.7	1533.2	1.9773	10.831	1372.3	1532.6	1.9453	8.657	1371.9	1532.1	1.9204
1200	16.452	1455.8	1638.5	2.0448	12.333	1455.5	1638.1	2.0130	9.861	1455.2	1637.7	1.9882
1400	18.445	1542.5	1747.3	2.1067	13.830	1542.3	1747.0	2.0749	11.060	1542.0	1746.7	2.0502
1600	20.44	1632.8	1859.7	2.1641	15.324	1632.6	1859.5	2.1323	12.257	1632.4	1859.3	2.1076
1800	22.43	1726.7	1975.7	2.2179	16.818	1726.5	1975.5	2.1861	13.452	1726.4	1975.3	2.1614
2000	24.41	1824.0	2095.1	2.2685	18.310	1823.9	2094.9	2.2367	14.647	1823.7	2094.8	2.2121

TABLE A-6E

(Continued)

T °F	v ft^3/lbm	u Btu/lbm	h Btu/lbm	s Btu/(lbm·R)	v ft^3/lbm	u Btu/lbm	h Btu/lbm	s Btu/(lbm·R)	v ft^3/lbm	u Btu/lbm	h Btu/lbm	s Btu/(lbm·R)
	P = 120 psia (341.30°F)				**P = 140 psia (353.08°F)**				**P = 160 psia (363.60°F)**			
Sat.	3.730	1108.3	1191.1	1.5886	3.221	1110.3	1193.8	1.5761	2.836	1112.0	1196.0	1.5651
360	3.844	1116.7	1202.0	1.6021	3.259	1113.5	1198.0	1.5812				
400	4.079	1133.8	1224.4	1.6288	3.466	1131.4	1221.2	1.6088	3.007	1128.8	1217.8	1.5911
450	4.360	1154.3	1251.2	1.6590	3.713	1152.4	1248.6	1.6399	3.228	1150.5	1246.1	1.6230
500	4.633	1174.2	1277.1	1.6868	3.952	1172.7	1275.1	1.6682	3.440	1171.2	1273.0	1.6518
550	4.900	1193.8	1302.6	1.7127	4.184	1192.6	1300.9	1.6944	3.646	1191.3	1299.2	1.6784
600	5.164	1213.2	1327.8	1.7371	4.412	1212.1	1326.4	1.7191	3.848	1211.1	1325.0	1.7034
700	5.682	1252.0	1378.2	1.7825	4.860	1251.2	1377.1	1.7648	4.243	1250.4	1376.0	1.7494
800	6.195	1291.2	1428.7	1.8243	5.301	1290.5	1427.9	1.8068	4.631	1289.9	1427.0	1.7916
1000	7.208	1371.5	1531.5	1.9000	6.173	1371.0	1531.0	1.8827	5.397	1370.6	1530.4	1.8677
1200	8.213	1454.9	1637.3	1.9679	7.036	1454.6	1636.9	1.9507	6.154	1454.3	1636.5	1.9358
1400	9.214	1541.8	1746.4	2.0300	7.895	1541.6	1746.1	2.0129	6.906	1541.4	1745.9	1.9980
1600	10.212	1632.3	1859.0	2.0875	8.752	1632.1	1858.8	2.0704	7.656	1631.9	1858.6	2.0556
1800	11.209	1726.2	1975.1	2.1413	9.607	1726.1	1975.0	2.1242	8.405	1725.9	1974.8	2.1094
2000	12.205	1823.6	2094.6	2.1919	10.461	1823.5	2094.5	2.1749	9.153	1823.3	2094.3	2.1601
	P = 180 psia (373.13°F)				**P = 200 psia (381.86°F)**				**P = 225 psia (391.87°F)**			
Sat.	2.533	1113.4	1197.8	1.5553	2.289	1114.6	1199.3	1.5464	2.043	1115.8	1200.8	1.5365
400	2.648	1126.2	1214.4	1.5749	2.361	1123.5	1210.8	1.5600	2.073	1119.9	1206.2	1.5427
450	2.850	1148.5	1243.4	1.6078	2.548	1146.4	1240.7	1.5938	2.245	1143.8	1237.3	1.5779
500	3.042	1169.6	1270.9	1.6372	2.724	1168.0	1268.8	1.6239	2.405	1165.9	1266.1	1.6087
550	3.228	1190.0	1297.5	1.6642	2.893	1188.7	1295.7	1.6512	2.558	1187.0	1293.5	1.6366
600	3.409	1210.0	1323.5	1.6893	3.058	1208.9	1322.1	1.6767	2.707	1207.5	1320.2	1.6624
700	3.763	1249.6	1374.9	1.7357	3.379	1248.8	1373.8	1.7234	2.995	1247.7	1372.4	1.7095
800	4.110	1289.3	1426.2	1.7781	3.693	1288.6	1425.3	1.7660	3.276	1287.8	1424.2	1.7523
900	4.453	1329.4	1477.7	1.8175	4.003	1328.9	1477.1	1.8055	3.553	1328.3	1476.2	1.7920
1000	4.793	1370.2	1529.8	1.8545	4.310	1369.8	1529.3	1.8425	3.827	1369.3	1528.6	1.8292
1200	5.467	1454.0	1636.1	1.9227	4.918	1453.7	1635.7	1.9109	4.369	1453.4	1635.3	1.8977
1400	6.137	1541.2	1745.6	1.9849	5.521	1540.9	1745.3	1.9732	4.906	1540.7	1744.9	1.9600
1600	6.804	1631.7	1858.4	2.0425	6.123	1631.6	1858.2	2.0308	5.441	1631.3	1857.9	2.0177
1800	7.470	1725.8	1974.6	2.0964	6.722	1725.6	1974.4	2.0847	5.975	1725.4	1974.2	2.0716
2000	8.135	1823.2	2094.2	2.1470	7.321	1823.0	2094.0	2.1354	6.507	1822.9	2093.8	2.1223
	P = 250 psia (401.04°F)				**P = 275 psia (409.52°F)**				**P = 300 psia (417.43°F)**			
Sat.	1.8448	1116.7	1202.1	1.5274	1.6813	1117.5	1203.1	1.5192	1.5442	1118.2	1203.9	1.5115
450	2.002	1141.1	1233.7	1.5632	1.8026	1138.3	1230.0	1.5495	1.6361	1135.4	1226.2	1.5365
500	2.150	1163.8	1263.3	1.5948	1.9407	1161.7	1260.4	1.5820	1.7662	1159.5	1257.5	1.5701
550	2.290	1185.3	1291.3	1.6233	2.071	1183.6	1289.0	1.6110	1.8878	1181.9	1286.7	1.5997
600	2.426	1206.1	1318.3	1.6494	2.196	1204.7	1316.4	1.6376	2.004	1203.2	1314.5	1.6266
650	2.558	1226.5	1344.9	1.6739	2.317	1225.3	1343.2	1.6623	2.117	1224.1	1341.6	1.6516
700	2.688	1246.7	1371.1	1.6970	2.436	1245.7	1369.7	1.6856	2.227	1244.6	1368.3	1.6751
800	2.943	1287.0	1423.2	1.7401	2.670	1286.2	1422.1	1.7289	2.442	1285.4	1421.0	1.7187
900	3.193	1327.6	1475.3	1.7799	2.898	1327.0	1474.5	1.7689	2.653	1326.3	1473.6	1.7589
1000	3.440	1368.7	1527.9	1.8172	3.124	1368.2	1527.2	1.8064	2.860	1367.7	1526.5	1.7964
1200	3.929	1453.0	1634.8	1.8858	3.570	1452.6	1634.3	1.8751	3.270	1452.2	1633.8	1.8653
1400	4.414	1540.4	1744.6	1.9483	4.011	1540.1	1744.2	1.9376	3.675	1539.8	1743.8	1.9279
1600	4.896	1631.1	1857.6	2.0060	4.450	1630.9	1857.3	1.9954	4.078	1630.7	1857.0	1.9857
1800	5.376	1725.2	1974.0	2.0599	4.887	1725.0	1973.7	2.0493	4.479	1724.9	1973.5	2.0396
2000	5.856	1822.7	2093.6	2.1106	5.323	1822.5	2093.4	2.1000	4.879	1822.3	2093.2	2.0904

TABLE A-6E

(Continued)

T °F	v ft³/lbm	u Btu/lbm	h Btu/lbm	s Btu/(lbm·R)	v ft³/lbm	u Btu/lbm	h Btu/lbm	s Btu/(lbm·R)	v ft³/lbm	u Btu/lbm	h Btu/lbm	s Btu/(lbm·R)
	P = 350 psia (431.82°F)				**P = 400 psia (444.70°F)**				**P = 450 psia (456.39°F)**			
Sat.	1.3267	1119.0	1204.9	1.4978	1.1620	1119.5	1205.5	1.4856	1.0326	1119.6	1205.6	1.4746
450	1.3733	1129.2	1218.2	1.5125	1.1745	1122.6	1209.6	1.4901				
500	1.4913	1154.9	1251.5	1.5482	1.2843	1150.1	1245.2	1.5282	1.1226	1145.1	1238.5	1.5097
550	1.5998	1178.3	1281.9	1.5790	1.3833	1174.6	1277.0	1.5605	1.2146	1170.7	1271.9	1.5436
600	1.7025	1200.3	1310.6	1.6068	1.4760	1197.3	1306.6	1.5892	1.2996	1194.3	1302.5	1.5732
650	1.8013	1221.6	1338.3	1.6323	1.5645	1219.1	1334.9	1.6153	1.3803	1216.6	1331.5	1.6000
700	1.8975	1242.5	1365.4	1.6562	1.6503	1240.4	1362.5	1.6397	1.4580	1238.2	1359.6	1.6248
800	2.085	1283.8	1418.8	1.7004	1.8163	1282.1	1416.6	1.6844	1.6077	1280.5	1414.4	1.6701
900	2.267	1325.0	1471.8	1.7409	1.9776	1323.7	1470.1	1.7252	1.7524	1322.4	1468.3	1.7113
1000	2.446	1366.6	1525.0	1.7787	2.136	1365.5	1523.6	1.7632	1.8941	1364.4	1522.2	1.7495
1200	2.799	1451.5	1632.8	1.8478	2.446	1450.7	1631.8	1.8327	2.172	1450.0	1630.8	1.8192
1400	3.148	1539.3	1743.1	1.9106	2.752	1538.7	1742.4	1.8956	2.444	1538.1	1741.7	1.8823
1600	3.494	1630.2	1856.5	1.9685	3.055	1629.8	1855.9	1.9535	2.715	1629.3	1855.4	1.9403
1800	3.838	1724.5	1973.1	2.0225	3.357	1724.1	1972.6	2.0076	2.983	1723.7	1972.1	1.9944
2000	4.182	1822.0	2092.8	2.0733	3.658	1821.6	2092.4	2.0584	3.251	1821.3	2092.0	2.0453
	P = 500 psia (467.13°F)				**P = 600 psia (486.33°F)**				**P = 700 psia (503.23°F)**			
Sat.	0.9283	1119.4	1205.3	1.4645	0.7702	1118.6	1204.1	1.4464	0.6558	1117.0	1202.0	1.4305
500	0.9924	1139.7	1231.5	1.4923	0.7947	1128.0	1216.2	1.4592				
550	1.0792	1166.7	1266.6	1.5279	0.8749	1158.2	1255.4	1.4990	0.7275	1149.0	1243.2	1.4723
600	1.1583	1191.1	1298.3	1.5585	0.9456	1184.5	1289.5	1.5320	0.7929	1177.5	1280.2	1.5081
650	1.2327	1214.0	1328.0	1.5860	1.0109	1208.6	1320.9	1.5609	0.8520	1203.1	1313.4	1.5387
700	1.3040	1236.0	1356.7	1.6112	1.0727	1231.5	1350.6	1.5872	0.9073	1226.9	1344.4	1.5661
800	1.4407	1278.8	1412.1	1.6571	1.1900	1275.4	1407.6	1.6343	1.0109	1272.0	1402.9	1.6145
900	1.5723	1321.0	1466.5	1.6987	1.3021	1318.4	1462.9	1.6766	1.1089	1315.6	1459.3	1.6576
1000	1.7008	1363.3	1520.7	1.7371	1.4108	1361.2	1517.8	1.7155	1.2036	1358.9	1514.9	1.6970
1100	1.8271	1406.0	1575.1	1.7731	1.5173	1404.2	1572.7	1.7519	1.2960	1402.4	1570.2	1.7337
1200	1.9518	1449.2	1629.8	1.8072	1.6222	1447.7	1627.8	1.7861	1.3868	1446.2	1625.8	1.7682
1400	2.198	1537.6	1741.0	1.8704	1.8289	1536.5	1739.5	1.8497	1.5652	1535.3	1738.1	1.8321
1600	2.442	1628.9	1854.8	1.9285	2.033	1628.0	1853.7	1.9080	1.7409	1627.1	1852.6	1.8906
1800	2.684	1723.3	1971.7	1.9827	2.236	1722.6	1970.8	1.9622	1.9152	1721.8	1969.9	1.9449
2000	2.926	1820.9	2091.6	2.0335	2.438	1820.2	2090.8	2.0131	2.0887	1819.5	2090.1	1.9958
	P = 800 psia (518.36°F)				**P = 1000 psia (544.75°F)**				**P = 1250 psia (572.56°F)**			
Sat.	0.5691	1115.0	1199.3	1.4160	0.4459	1109.9	1192.4	1.3903	0.3454	1101.7	1181.6	1.3619
550	0.6154	1138.8	1229.9	1.4469	0.4534	1114.8	1198.7	1.3966				
600	0.6776	1170.1	1270.4	1.4861	0.5140	1153.7	1248.8	1.4450	0.3786	1129.0	1216.6	1.3954
650	0.7324	1197.2	1305.6	1.5186	0.5637	1184.7	1289.1	1.4822	0.4267	1167.2	1266.0	1.4410
700	0.7829	1222.1	1338.0	1.5471	0.6080	1212.0	1324.6	1.5135	0.4670	1198.4	1306.4	1.4767
750	0.8306	1245.7	1368.6	1.5730	0.6490	1237.2	1357.3	1.5412	0.5030	1226.1	1342.4	1.5070
800	0.8764	1268.5	1398.2	1.5969	0.6878	1261.2	1388.5	1.5664	0.5364	1251.8	1375.8	1.5341
900	0.9640	1312.9	1455.6	1.6408	0.7610	1307.3	1448.1	1.6120	0.5984	1300.0	1438.4	1.5820
1000	1.0482	1356.7	1511.9	1.6807	0.8305	1352.2	1505.9	1.6530	0.6563	1346.4	1498.2	1.6244
1100	1.1300	1400.5	1567.8	1.7178	0.8976	1396.8	1562.9	1.6908	0.7116	1392.0	1556.6	1.6631
1200	1.2102	1444.6	1623.8	1.7526	0.9630	1441.5	1619.7	1.7261	0.7652	1437.5	1614.5	1.6991
1400	1.3674	1534.2	1736.6	1.8167	1.0905	1531.9	1733.7	1.7909	0.8689	1529.0	1730.0	1.7648
1600	1.5218	1626.2	1851.5	1.8754	1.2152	1624.4	1849.3	1.8499	0.9699	1622.2	1846.5	1.8243
1800	1.6749	1721.0	1969.0	1.9298	1.3384	1719.5	1967.2	1.9046	1.0693	1717.6	1965.0	1.8791
2000	1.8271	1818.8	2089.3	1.9808	1.4608	1817.4	2087.7	1.9557	1.1678	1815.7	2085.8	1.9304

TABLE A-6E

(Continued)

T °F	*v* ft^3/lbm	*u* Btu/lbm	*h* Btu/lbm	*s* Btu/(lbm·R)	*v* ft^3/lbm	*u* Btu/lbm	*h* Btu/lbm	*s* Btu/(lbm·R)	*v* ft^3/lbm	*u* Btu/lbm	*h* Btu/lbm	*s* Btu/(lbm·R)
	***P* = 1500 psia (596.39°F)**				***P* = 1750 psia (617.31°F)**				***P* = 2000 psia (636.00°F)**			
Sat.	0.2769	1091.8	1168.7	1.3359	0.2268	1080.2	1153.7	1.3109	0.18813	1066.6	1136.3	1.2861
600	0.2816	1096.6	1174.8	1.3416								
650	0.3329	1147.0	1239.4	1.4012	0.2627	1122.5	1207.6	1.3603	0.2057	1091.1	1167.2	1.3141
700	0.3716	1183.4	1286.6	1.4429	0.3022	1166.7	1264.6	1.4106	0.2487	1147.7	1239.8	1.3782
750	0.4049	1214.1	1326.5	1.4767	0.3341	1201.3	1309.5	1.4485	0.2803	1187.3	1291.1	1.4216
800	0.4350	1241.8	1362.5	1.5058	0.3622	1231.3	1348.6	1.4802	0.3071	1220.1	1333.8	1.4562
850	0.4631	1267.7	1396.2	1.5320	0.3878	1258.8	1384.4	1.5081	0.3312	1249.5	1372.0	1.4860
900	0.4897	1292.5	1428.5	1.5562	0.4119	1284.8	1418.2	1.5334	0.3534	1276.8	1407.6	1.5126
1000	0.5400	1340.4	1490.3	1.6001	0.4569	1334.3	1482.3	1.5789	0.3945	1328.1	1474.1	1.5598
1100	0.5876	1387.2	1550.3	1.6399	0.4990	1382.2	1543.8	1.6197	0.4325	1377.2	1537.2	1.6017
1200	0.6334	1433.5	1609.3	1.6765	0.5392	1429.4	1604.0	1.6571	0.4685	1425.2	1598.6	1.6398
1400	0.7213	1526.1	1726.3	1.7431	0.6158	1523.1	1722.6	1.7245	0.5368	1520.2	1718.8	1.7082
1600	0.8064	1619.9	1843.7	1.8031	0.6896	1617.6	1841.0	1.7850	0.6020	1615.4	1838.2	1.7692
1800	0.8899	1715.7	1962.7	1.8582	0.7617	1713.9	1960.5	1.8404	0.6656	1712.0	1958.3	1.8249
2000	0.9725	1814.0	2083.9	1.9096	0.8330	1812.3	2082.0	1.8919	0.7284	1810.6	2080.2	1.8765
	***P* = 2500 psia (668.31°F)**				***P* = 3000 psia (695.52°F)**				***P* = 3500 psia**			
Sat.	0.130 59	1031.0	1091.4	1.2327	0.084 04	968.8	1015.5	1.1575				
650									0.024 91	663.5	679.7	0.8630
700	0.168 39	1098.7	1176.6	1.3073	0.097 71	1003.9	1058.1	1.1944	0.030 58	759.5	779.3	0.9506
750	0.2030	1155.2	1249.1	1.3686	0.148 31	1114.7	1197.1	1.3122	0.104 60	1058.4	1126.1	1.2440
800	0.2291	1195.7	1301.7	1.4112	0.175 72	1167.6	1265.2	1.3675	0.136 26	1134.7	1223.0	1.3226
850	0.2513	1229.5	1345.8	1.4456	0.197 31	1207.7	1317.2	1.4080	0.158 18	1183.4	1285.9	1.3716
900	0.2712	1259.9	1385.4	1.4752	0.2160	1241.8	1361.7	1.4414	0.176 25	1222.4	1336.5	1.4096
950	0.2896	1288.2	1422.2	1.5018	0.2328	1272.7	1402.0	1.4705	0.192 14	1256.4	1380.8	1.4416
1000	0.3069	1315.2	1457.2	1.5262	0.2485	1301.7	1439.6	1.4967	0.2066	1287.6	1421.4	1.4699
1100	0.3393	1366.8	1523.8	1.5704	0.2772	1356.2	1510.1	1.5434	0.2328	1345.2	1496.0	1.5193
1200	0.3696	1416.7	1587.7	1.6101	0.3036	1408.0	1576.6	1.5848	0.2566	1399.2	1565.3	1.5624
1400	0.4261	1514.2	1711.3	1.6804	0.3524	1508.1	1703.7	1.6571	0.2997	1501.9	1696.1	1.6368
1600	0.4795	1610.2	1832.6	1.7424	0.3978	1606.3	1827.1	1.7201	0.3395	1601.7	1821.6	1.7010
1800	0.5312	1708.2	1954.0	1.7986	0.4416	1704.5	1949.6	1.7769	0.3776	1700.8	1945.4	1.7583
2000	0.5820	1807.2	2076.4	1.8506	0.4844	1803.9	2072.8	1.8291	0.4147	1800.6	2069.2	1.8108
	***P* = 4000 psia**				***P* = 5000 psia**				***P* = 6000 psia**			
650	0.024 47	657.7	675.8	0.8574	0.023 77	648.0	670.0	0.8482	0.023 22	640.0	665.8	0.8405
700	0.028 67	742.1	763.4	0.9345	0.026 76	721.8	746.6	0.9156	0.025 63	708.1	736.5	0.9028
750	0.063 31	960.7	1007.5	1.1395	0.033 64	821.4	852.6	1.0049	0.029 78	788.6	821.7	0.9746
800	0.105 22	1095.0	1172.9	1.2740	0.059 32	987.2	1042.1	1.1583	0.039 42	896.9	940.7	1.0708
850	0.128 33	1156.5	1251.5	1.3352	0.085 56	1092.7	1171.9	1.2596	0.058 18	1018.8	1083.4	1.1820
900	0.146 22	1201.5	1309.7	1.3789	0.103 85	1155.1	1251.1	1.3190	0.075 88	1102.9	1187.2	1.2599
950	0.161 51	1239.2	1358.8	1.4144	0.118 53	1202.2	1311.9	1.3629	0.090 08	1162.0	1262.0	1.3140
1000	0.175 20	1272.9	1402.6	1.4449	0.131 20	1242.0	1363.4	1.3988	0.102 07	1209.1	1322.4	1.3561
1100	0.199 54	1333.9	1481.6	1.4973	0.153 02	1310.6	1452.2	1.4577	0.122 18	1286.4	1422.1	1.4222
1200	0.2213	1390.1	1553.9	1.5423	0.171 99	1371.6	1530.8	1.5066	0.139 27	1352.7	1507.3	1.4752
1300	0.2414	1443.7	1622.4	1.5823	0.189 18	1428.6	1603.7	1.5493	0.154 53	1413.3	1584.9	1.5206
1400	0.2603	1495.7	1688.4	1.6188	0.205 17	1483.2	1673.0	1.5876	0.168 54	1470.5	1657.6	1.5608
1600	0.2959	1597.1	1816.1	1.6841	0.2348	1587.9	1805.2	1.6551	0.194 20	1578.7	1794.3	1.6307
1800	0.3296	1697.1	1941.1	1.7420	0.2626	1689.8	1932.7	1.7142	0.218 01	1682.4	1924.5	1.6910
2000	0.3625	1797.3	2065.6	1.7948	0.2895	1790.8	2058.6	1.7676	0.240 87	1784.3	2051.7	1.7450

Source: Gordon J. Van Wylen and Richard E. Sonntag, *Fundamentals of Classical Thermodynamics,* English/SI Version, 3d ed., Wiley, New York, 1986, pp. 624–631, table A.1.3E. Originally published in Joseph H. Keenan, Frederick G. Keyes, Philip G. Hill, and Joan G. Moore, *Steam Tables,* Wiley, New York, 1969.

TABLE A-7E

Compressed liquid water

T °F	v ft³/lbm	u Btu/lbm	h Btu/lbm	s Btu/(lbm·R)	v ft³/lbm	u Btu/lbm	h Btu/lbm	s Btu/(lbm·R)	v ft³/lbm	u Btu/lbm	h Btu/lbm	s Btu/(lbm·R)
	P = 500 psia (467.13°F)				P = 1000 psia (544.75°F)				P = 1500 psia (596.39°F)			
Sat.	0.019 748	447.70	449.53	0.649 04	0.021 591	538.39	542.38	0.743 20	0.023 461	604.97	611.48	0.808 24
32	0.015 994	0.00	1.49	0.000 00	0.015 967	0.03	2.99	0.000 05	0.015 939	0.05	4.47	0.000 07
50	0.015 998	18.02	19.50	0.035 99	0.015 972	17.99	20.94	0.035 92	0.015 946	17.95	22.38	0.035 84
100	0.016 106	67.87	69.36	0.129 32	0.016 082	67.70	70.68	0.129 01	0.016 058	67.53	71.99	0.128 70
150	0.016 318	117.66	119.17	0.214 57	0.016 293	117.38	120.40	0.214 10	0.016 268	117.10	121.62	0.213 64
200	0.016 608	167.65	169.19	0.293 41	0.016 580	167.26	170.32	0.292 81	0.016 554	166.87	171.46	0.292 21
250	0.016 972	217.99	219.56	0.367 02	0.016 941	217.47	220.61	0.366 28	0.016 910	216.96	221.65	0.365 54
300	0.017 416	268.92	270.53	0.436 41	0.017 379	268.24	271.46	0.435 52	0.017 343	267.58	272.39	0.434 63
350	0.017 954	320.71	322.37	0.502 49	0.017 909	319.83	323.15	0.501 40	0.017 865	318.98	323.94	0.500 34
400	0.018 608	373.68	375.40	0.566 04	0.018 550	372.55	375.98	0.564 72	0.018 493	371.45	376.59	0.563 43
450	0.019 420	428.40	430.19	0.627 98	0.019 340	426.89	430.47	0.626 32	0.019 264	425.44	430.79	0.624 70
500					0.020 36	483.8	487.5	0.6874	0.020 24	481.8	487.4	0.6853
550									0.021 58	542.1	548.1	0.7469
	P = 2000 psia (636.00°F)				P = 3000 psia (695.52°F)				P = 5000 psia			
Sat.	0.025 649	662.40	671.89	0.862 27	0.034 310	783.45	802.50	0.973 20				
32	0.015 912	0.06	5.95	0.000 08	0.015 859	0.09	8.90	0.000 09	0.015 755	0.11	14.70	−0.000 01
50	0.015 920	17.91	23.81	0.035 75	0.015 870	17.84	26.65	0.035 55	0.015 773	17.67	32.26	0.035 08
100	0.016 034	67.37	73.30	0.128 39	0.015 987	67.04	75.91	0.127 77	0.015 897	66.40	81.11	0.126 51
200	0.016 527	166.49	172.60	0.291 62	0.016 476	165.74	174.89	0.290 46	0.016 376	164.32	179.47	0.288 18
300	0.017 308	266.93	273.33	0.433 76	0.017 240	265.66	275.23	0.432 05	0.017 110	263.25	279.08	0.428 75
400	0.018 439	370.38	377.21	0.562 16	0.018 334	368.32	378.50	0.559 70	0.018 141	364.47	381.25	0.555 06
450	0.019 191	424.04	431.14	0.623 13	0.019 053	421.36	431.93	0.620 11	0.018 803	416.44	433.84	0.614 51
500	0.020 14	479.8	487.3	0.6832	0.019 944	476.2	487.3	0.679 4	0.019 603	469.8	487.9	0.6724
560	0.021 72	551.8	559.8	0.7565	0.021 382	546.2	558.0	0.750 8	0.020 835	536.7	556.0	0.7411
600	0.023 30	605.4	614.0	0.8086	0.022 74	597.0	609.6	0.8004	0.021 91	584.0	604.2	0.7876
640					0.024 75	654.3	668.0	0.8545	0.023 34	634.6	656.2	0.8357
680					0.028 79	728.4	744.3	0.9226	0.025 35	690.6	714.1	0.8873
700									0.026 76	721.8	746.6	0.9156

Source: Gordon J. Van Wylen and Richard E. Sonntag, *Fundamentals of Classical Thermodynamics,* English/SI Version, 3d ed., Wiley, New York, 1986, pp. 632–633, table A.1.4E. Originally published in Joseph H. Keenan, Frederick G. Keyes, Philp G. Hill, and Joan G. Moore, *Steam Tables,* Wiley, New York, 1969.

TABLE A-8E

Saturated ice-water vapor

Temp. °F T	Sat. Press. psia P_{sat}	Specific volume ft³/lbm Sat. ice v_i	Sat. vapor $v_g \times 10^{-3}$	Internal energy Btu/lbm Sat. ice u_i	Subl. u_{ig}	Sat. vapor u_g	Enthalpy Btu/lbm Sat. ice h_i	Subl. h_{ig}	Sat. vapor h_g	Entropy Btu/(lbm·R) Sat. ice s_i	Subl. s_{ig}	Sat. vapor s_g
32.018	0.0887	0.017 47	3.302	−143.34	1164.6	1021.2	−143.34	1218.7	1075.4	−0.292	2.479	2.187
32	0.0886	0.017 47	3.305	−143.35	1164.6	1021.2	−143.35	1218.7	1075.4	−0.292	2.479	2.187
30	0.0808	0.017 47	3.607	−144.35	1164.9	1020.5	−144.35	1218.9	1074.5	−0.294	2.489	2.195
25	0.0641	0.017 46	4.506	−146.84	1165.7	1018.9	−146.84	1219.1	1072.3	−0.299	2.515	2.216
20	0.0505	0.017 45	5.655	−149.31	1166.5	1017.2	−149.31	1219.4	1070.1	−0.304	2.542	2.238
15	0.0396	0.017 45	7.13	−151.75	1167.3	1015.5	−151.75	1219.7	1067.9	−0.309	2.569	2.260
10	0.0309	0.017 44	9.04	−154.17	1168.1	1013.9	−154.17	1219.9	1065.7	−0.314	2.597	2.283
5	0.0240	0.017 43	11.52	−156.56	1168.8	1012.2	−156.56	1220.1	1063.5	−0.320	2.626	2.306
0	0.0185	0.017 43	14.77	−158.93	1169.5	1010.6	−158.93	1220.2	1061.2	−0.325	2.655	2.330
−5	0.0142	0.017 42	19.03	−161.27	1170.2	1008.9	−161.27	1220.3	1059.0	−0.330	2.684	2.354
−10	0.0109	0.017 41	24.66	−163.59	1170.9	1007.3	−163.59	1220.4	1056.8	−0.335	2.714	2.379
−15	0.0082	0.017 40	32.2	−165.89	1171.5	1005.6	−165.89	1220.5	1054.6	−0.340	2.745	2.405
−20	0.0062	0.017 40	42.2	−168.16	1172.1	1003.9	−168.16	1220.6	1052.4	−0.345	2.776	2.431
−25	0.0046	0.017 39	55.7	−170.40	1172.7	1002.3	−170.40	1220.6	1050.2	−0.351	2.808	2.457
−30	0.0035	0.017 38	74.1	−172.63	1173.2	1000.6	−172.63	1220.6	1048.0	−0.356	2.841	2.485
−35	0.0026	0.017 37	99.2	−174.82	1173.8	988.9	−174.82	1220.6	1045.8	−0.361	2.874	2.513
−40	0.0019	0.017 37	133.8	−177.00	1174.3	997.3	−177.00	1220.6	1043.6	−0.366	2.908	2.542

Source: Gordon J. Van Wylen and Richard E. Sonntag, *Fundamentals of Classical Thermodynamics,* English/SI Version, 3d ed., Wiley, New York, 1986, p. 633, table A.1.5E. Originally published in Joseph H. Keenan, Frederick G. Keyes, Philip G. Hill, and Joan G. Moore, *Steam Tables,* Wiley, New York, 1969.

FIGURE A-9E

T-s diagram for water. (*Source*: Joseph H. Keenan, Frederick G. Keyes, Philip G. Hill, and Joan G. Moore, *Steam Tables,* Wiley, New York, 1969).

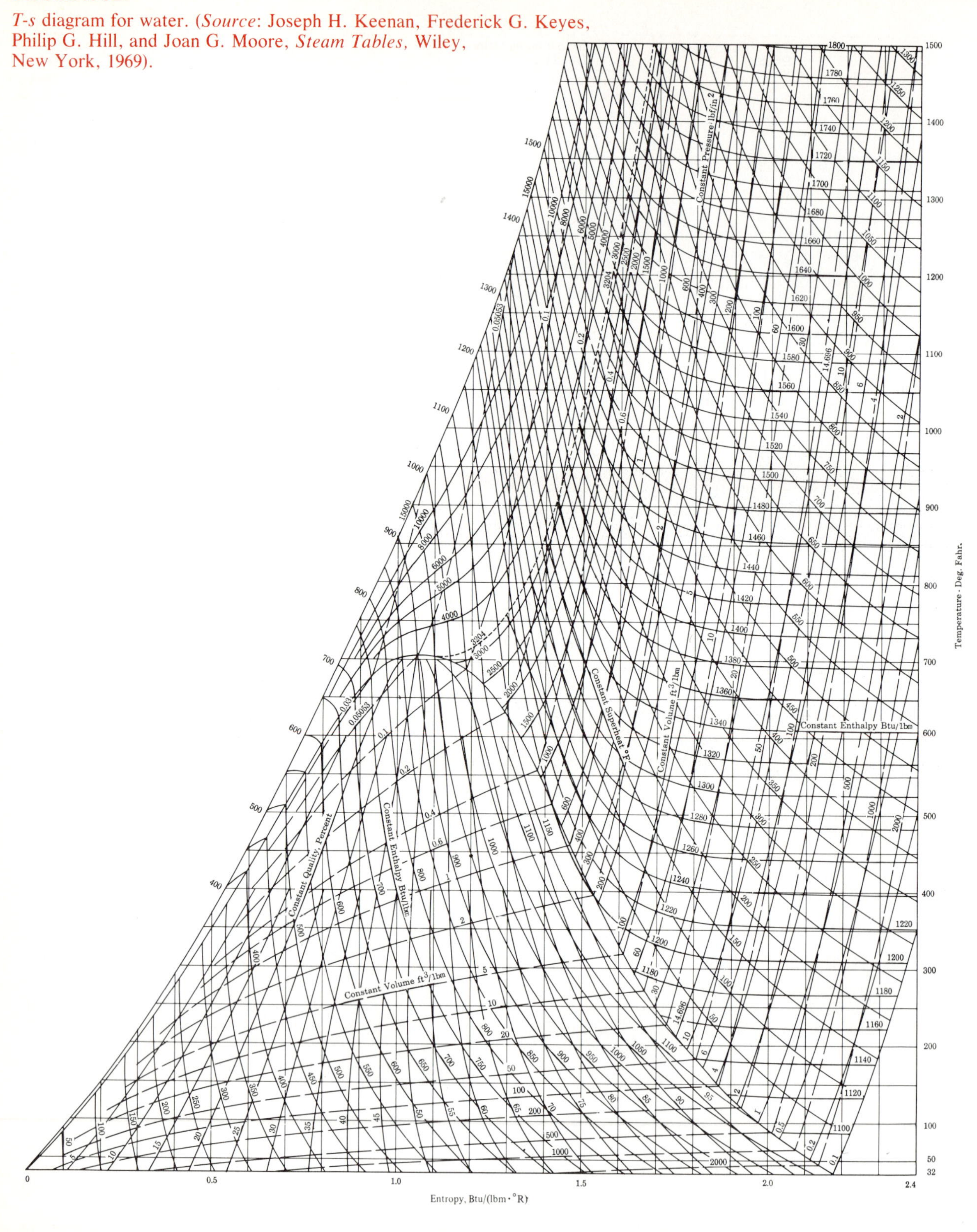

FIGURE A-10E

Mollier diagram for water. (*Source*: Joseph H. Keenan, Frederick G. Keyes, Philip G. Hill, and Joan G. Moore, *Steam Tables*, Wiley, New York, 1969.)

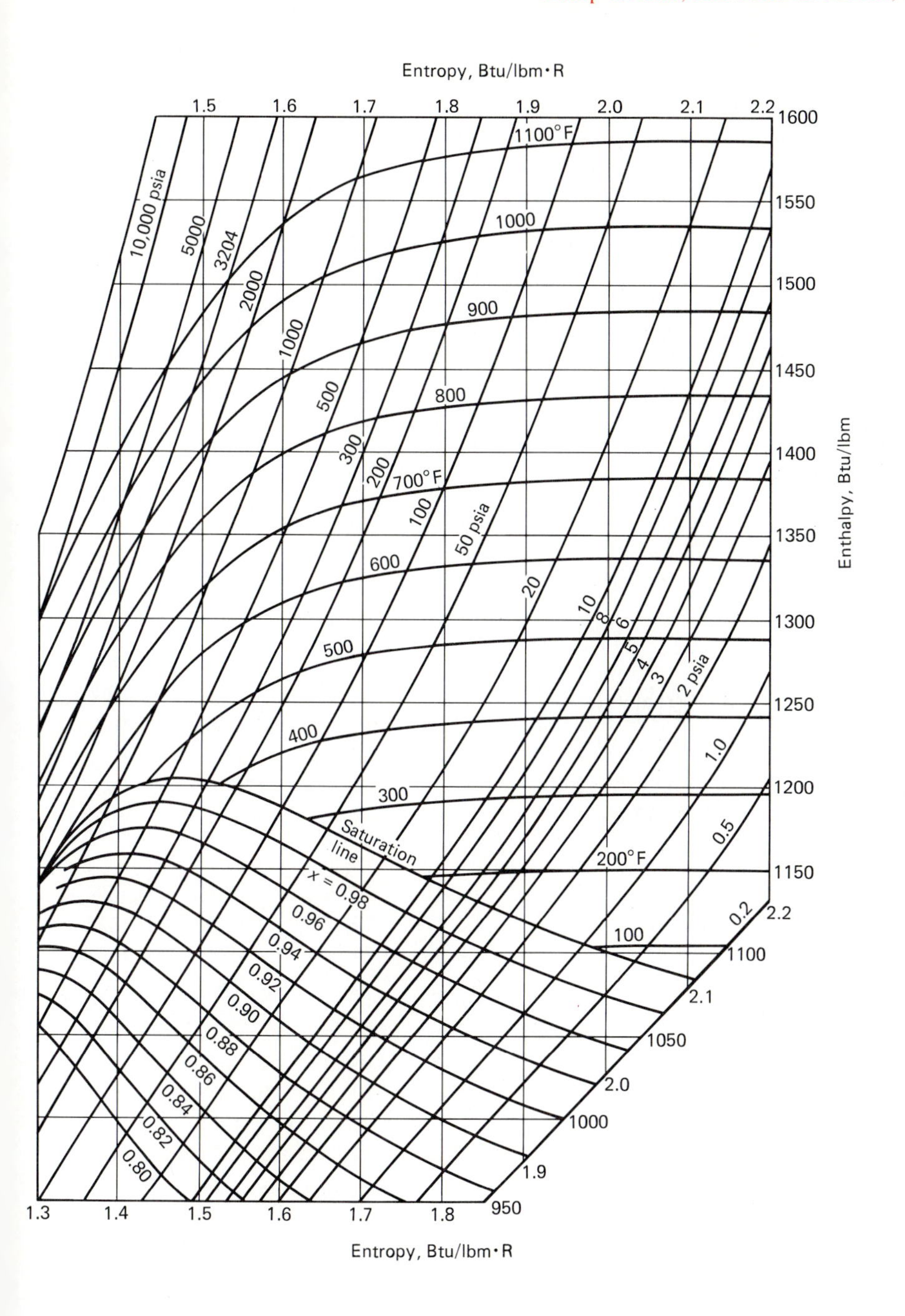

TABLE A-11E

Saturated refrigerant-12–Temperature table

Temp. °F T	Sat. press. psia P_{sat}	Specific volume ft³/lbm Sat. liquid v_f	Sat. vapor v_g	Internal energy Btu/lbm Sat. liquid u_f	Sat. vapor u_g	Enthalpy Btu/lbm Sat. liquid h_f	Evap. h_{fg}	Sat. vapor h_g	Entropy Btu/(lbm·R) Sat. liquid s_f	Sat. vapor s_g
−40	9.308	0.010 56	3.8750	−0.02	66.24	0	72.91	72.91	0	0.1737
−30	11.999	0.010 67	3.0585	1.93	67.22	2.11	71.90	74.01	0.0050	0.1723
−20	15.267	0.010 79	2.4429	4.21	68.21	4.24	70.87	75.11	0.0098	0.1710
−15	17.141	0.010 85	2.1924	5.27	68.69	5.30	70.35	75.65	0.0122	0.1704
−10	19.189	0.010 91	1.9727	6.33	69.19	6.37	69.82	76.19	0.0146	0.1699
−5	21.422	0.010 97	1.7794	7.40	69.67	7.44	69.29	76.73	0.0170	0.1694
0	23.849	0.011 03	1.6089	8.47	70.17	8.52	68.75	77.27	0.0193	0.1689
5	26.483	0.011 09	1.4580	9.55	70.65	9.60	68.20	77.80	0.0216	0.1684
10	29.335	0.011 16	1.3241	10.62	71.15	10.68	67.65	78.33	0.0240	0.1680
20	35.736	0.011 30	1.0988	12.79	72.12	12.86	66.52	79.38	0.0285	0.1672
25	39.310	0.011 37	1.0039	13.88	72.61	13.96	65.95	79.91	0.0308	0.1668
30	43.148	0.011 44	0.9188	14.97	73.08	15.06	65.36	80.42	0.0330	0.1665
40	51.667	0.011 59	0.7736	17.16	74.04	17.27	64.16	81.43	0.0375	0.1659
50	61.394	0.011 75	0.6554	19.38	74.99	19.51	62.93	82.44	0.0418	0.1653
60	72.433	0.011 91	0.5584	21.61	75.92	21.77	61.64	83.41	0.0462	0.1648
70	84.89	0.012 09	0.4782	23.86	76.85	24.05	60.31	84.36	0.0505	0.1643
80	98.87	0.012 28	0.4114	26.14	77.76	26.37	58.92	85.29	0.0548	0.1639
85	106.47	0.012 38	0.3821	27.29	78.20	27.53	58.20	85.73	0.0569	0.1637
90	114.49	0.012 48	0.3553	28.45	78.65	28.71	57.46	86.17	0.0590	0.1635
95	122.95	0.012 58	0.3306	29.61	79.09	29.90	56.71	86.61	0.0611	0.1633
100	131.86	0.012 69	0.3079	30.79	79.51	31.10	55.93	87.03	0.0632	0.1632
105	141.25	0.012 81	0.2870	31.98	79.94	32.31	55.13	87.44	0.0653	0.1630
110	151.11	0.012 92	0.2677	33.16	80.36	33.53	54.31	87.84	0.0675	0.1628
115	161.47	0.013 05	0.2498	34.37	80.76	34.76	53.47	88.23	0.0696	0.1626
120	172.35	0.013 17	0.2333	35.59	81.17	36.01	52.60	88.61	0.0717	0.1624
140	221.32	0.013 75	0.1780	40.60	82.68	41.16	48.81	89.97	0.0802	0.1616
160	279.82	0.014 45	0.1360	45.88	83.96	46.63	44.37	91.00	0.0889	0.1605
180	349.00	0.015 36	0.1033	51.57	84.89	52.56	39.00	91.56	0.0980	0.1590
200	430.09	0.016 66	0.0767	57.87	85.17	59.20	32.08	91.28	0.1079	0.1565
233.6	596.9	0.0287	0.0287	75.69	75.69	78.86	0	78.86	0.1359	0.1359

Source: Kenneth Wark, *Thermodynamics,* 4th ed., McGraw-Hill, New York, 1983, p. 857, table A-16. Originally published by Freon Products Division, E. I. du Pont de Nemours & Company, 1956.

TABLE A-12E

Saturated refrigerant-12–Pressure table

		Specific volume ft³/lbm		Internal energy Btu/lbm		Enthalpy Btu/lbm			Entropy Btu/(lbm·R)	
Press. psia P	Sat. Temp. °F T_{sat}	Sat. liquid v_f	Sat. vapor v_g	Sat. liquid u_f	Sat. vapor u_g	Sat. liquid h_f	Evap. h_{fg}	Sat. vapor h_g	Sat. liquid s_f	Sat. vapor s_g
5	−62.35	0.0103	6.9069	−4.69	64.04	−4.68	75.11	70.43	−0.0114	0.1776
10	−37.23	0.0106	3.6246	0.56	66.51	0.58	72.64	73.22	0.0014	0.1733
15	−20.75	0.0108	2.4835	4.05	68.13	4.08	70.95	75.03	0.0095	0.1711
20	−8.13	0.0109	1.8977	6.73	69.37	6.77	69.63	76.40	0.0155	0.1697
30	11.11	0.0112	1.2964	10.86	71.25	10.93	67.53	78.45	0.0245	0.1679
40	25.93	0.0114	0.9874	14.08	72.69	14.16	65.84	80.00	0.0312	0.1668
50	38.15	0.0116	0.7982	16.75	73.86	16.86	64.39	81.25	0.0366	0.1660
60	48.64	0.0117	0.6701	19.07	74.86	19.20	63.10	82.30	0.0413	0.1654
70	57.90	0.0119	0.5772	21.13	75.73	21.29	61.92	83.21	0.0453	0.1649
80	66.21	0.0120	0.5068	23.00	76.50	23.18	60.82	84.00	0.0489	0.1645
90	73.79	0.0122	0.4514	24.72	77.20	24.92	59.79	84.71	0.0521	0.1642
100	80.76	0.0123	0.4067	26.31	77.82	26.54	58.81	85.35	0.0551	0.1639
120	93.29	0.0126	0.3389	29.21	78.93	29.49	56.97	86.46	0.0604	0.1634
140	104.35	0.0128	0.2896	31.82	79.89	32.15	55.24	87.39	0.0651	0.1630
160	114.30	0.0130	0.2522	34.21	80.71	34.59	53.59	88.18	0.0693	0.1626
180	123.38	0.0133	0.2228	36.42	81.44	36.86	52.00	88.86	0.0731	0.1623
200	131.74	0.0135	0.1989	38.50	82.08	39.00	50.44	89.44	0.0767	0.1620
220	139.51	0.0137	0.1792	40.48	82.08	41.03	48.90	89.94	0.0816	0.1616
240	146.77	0.0140	0.1625	42.35	83.14	42.97	47.39	90.36	0.0831	0.1613
260	153.60	0.0142	0.1483	44.16	83.58	44.84	45.88	90.72	0.0861	0.1609
280	160.06	0.0145	0.1359	45.90	83.97	46.65	44.36	91.01	0.0890	0.1605
300	166.18	0.0147	0.1251	47.59	84.30	48.41	42.83	91.24	0.0917	0.1601

Source: Kenneth Wark, *Thermodynamics,* 4th ed., McGraw-Hill, New York, 1983, p. 858, table A-16. Originally published by Freon Products Division, E. I. du Pont de Nemours & Company, 1956.

TABLE A-13E

Superheated refrigerant-12

Temp. °F	v ft³/lbm	u Btu/lbm	h Btu/lbm	s Btu/(lbm·R)	v ft³/lbm	u Btu/lbm	h Btu/lbm	s Btu/(lbm·R)
	10 psia (T_{sat} = −37.23°F)				**15 psia (T_{sat} = −20.75°F)**			
Sat.	3.6246	66.512	73.219	0.1733	2.4835	68.134	75.028	0.1711
0	3.9809	70.879	78.246	0.1847	2.6201	70.629	77.902	0.1775
20	4.1691	73.299	81.014	0.1906	2.7494	73.080	80.712	0.1835
40	4.3556	75.768	83.828	0.1964	2.8770	75.575	83.561	0.1893
60	4.5408	78.286	86.689	0.2020	3.0031	78.115	86.451	0.1950
80	4.7248	80.853	89.596	0.2075	3.1281	80.700	89.383	0.2005
100	4.9079	83.466	92.548	0.2128	3.2521	83.330	92.357	0.2059
120	5.0903	86.126	95.546	0.2181	3.3754	86.004	95.373	0.2112
140	5.2720	88.830	98.586	0.2233	3.4981	88.719	98.429	0.2164
160	5.4533	91.578	101.669	0.2283	3.6202	91.476	101.525	0.2215
180	5.6341	94.367	104.793	0.2333	3.7419	94.274	104.661	0.2265
200	5.8145	97.197	107.957	0.2381	3.8632	97.112	107.835	0.2314
	20 psia (T_{sat} = −8.13°F)				**30 psia (T_{sat} = 11.11°F)**			
Sat.	1.8977	69.374	76.397	0.1697	1.2964	71.255	78.452	0.1679
20	2.0391	72.856	80.403	0.1783	1.3278	72.394	79.765	0.1707
40	2.1373	75.379	83.289	0.1842	1.3969	74.975	82.730	0.1767
60	2.2340	77.942	86.210	0.1899	1.4644	77.586	85.716	0.1826
80	2.3295	80.546	89.168	0.1955	1.5306	80.232	88.729	0.1883
100	2.4241	83.192	92.164	0.2010	1.5957	82.911	91.770	0.1938
120	2.5179	85.879	95.198	0.2063	1.6600	85.627	94.843	0.1992
140	2.6110	88.607	98.270	0.2115	1.7237	88.379	97.948	0.2045
160	2.7036	91.374	101.380	0.2166	1.7868	91.166	101.086	0.2096
180	2.7957	94.181	104.528	0.2216	1.8494	93.991	104.258	0.2146
200	2.8874	97.026	107.712	0.2265	1.9116	96.852	107.464	0.2196
220	2.9789	99.907	110.932	0.2313	1.9735	99.746	110.702	0.2244
	40 psia (T_{sat} = 25.93°F)				**50 psia (T_{sat} = 38.15°F)**			
Sat.	0.9874	72.691	80.000	0.1668	0.7982	73.863	81.249	0.1660
40	1.0258	74.555	82.148	0.1711	0.8025	74.115	81.540	0.1666
60	1.0789	77.220	85.206	0.1771	0.8471	76.838	84.676	0.1727
80	1.1306	79.908	88.277	0.1829	0.8903	79.574	87.811	0.1786
100	1.1812	82.624	91.367	0.1885	0.9322	82.328	90.953	0.1843
120	1.2309	85.369	94.480	0.1940	0.9731	85.106	94.110	0.1899
140	1.2798	88.147	97.620	0.1993	1.0133	87.910	97.286	0.1953
160	1.3282	90.957	100.788	0.2045	1.0529	90.743	100.485	0.2005
180	1.3761	93.800	103.985	0.2096	1.0920	93.604	103.708	0.2056
200	1.4236	96.674	107.212	0.2146	1.1307	96.496	106.958	0.2106
220	1.4707	99.583	110.469	0.2194	1.1690	99.419	110.235	0.2155
240	1.5176	102.524	113.757	0.2242	1.2070	102.371	113.539	0.2203

TABLE A-13E

(Continued)

Temp. °F	v ft³/lbm	u Btu/lbm	h Btu/lbm	s Btu/(lbm·R)	v ft³/lbm	u Btu/lbm	h Btu/lbm	s Btu/(lbm·R)
	60 psia (T_{sat} = 48.64°F)				**70 psia (T_{sat} = 57.90°F)**			
Sat.	0.6701	74.859	82.299	0.1654	0.5772	75.729	83.206	0.1649
60	0.6921	76.442	84.126	0.1689	0.5809	76.027	83.552	0.1656
80	0.7296	79.229	87.330	0.1750	0.6146	78.871	86.832	0.1718
100	0.7659	82.024	90.528	0.1808	0.6469	81.712	90.091	0.1777
120	0.8011	84.836	93.731	0.1864	0.6780	84.560	93.343	0.1834
140	0.8335	87.668	96.945	0.1919	0.7084	87.421	96.597	0.1889
160	0.8693	90.524	100.776	0.1972	0.7380	90.302	99.862	0.1943
180	0.9025	93.406	103.427	0.2023	0.7671	93.205	103.141	0.1995
200	0.9353	96.315	106.700	0.2074	0.7957	96.132	106.439	0.2046
220	0.9678	99.252	109.997	0.2123	0.8240	99.083	109.756	0.2095
240	0.9998	102.217	113.319	0.2171	0.8519	102.061	113.096	0.2144
260	1.0318	105.210	116.666	0.2218	0.8796	105.065	116.459	0.2191
	80 psia (T_{sat} = 66.21°F)				**90 psia (T_{sat} = 73.79°F)**			
Sat.	0.5068	76.500	84.003	0.1645	0.4514	77.194	84.713	0.1642
80	0.5280	78.500	86.316	0.1689	0.4602	78.115	85.779	0.1662
100	0.5573	81.389	89.640	0.1749	0.4875	81.056	89.175	0.1723
120	0.5856	84.276	92.945	0.1807	0.5135	83.984	92.536	0.1782
140	0.6129	87.169	96.242	0.1863	0.5385	86.911	95.879	0.1839
160	0.6394	90.076	99.542	0.1917	0.5627	89.845	99.216	0.1894
180	0.6654	93.000	102.851	0.1970	0.5863	92.793	102.557	0.1947
200	0.6910	95.945	106.174	0.2021	0.6094	95.755	105.905	0.1998
220	0.7161	98.912	109.513	0.2071	0.6321	98.739	109.267	0.2049
240	0.7409	101.904	112.872	0.2119	0.6545	101.743	112.644	0.2098
260	0.7654	104.919	116.251	0.2167	0.6766	104.771	116.040	0.2146
280	0.7898	107.960	119.652	0.2214	0.6985	107.823	119.456	0.2192
	100 psia (T_{sat} = 80.76°F)				**120 psia (T_{sat} = 93.29°F)**			
Sat.	0.4067	77.824	85.351	0.1639	0.3389	78.933	86.459	0.1634
100	0.4314	80.711	88.694	0.1700	0.3466	79.978	87.675	0.1656
120	0.4556	83.685	92.116	0.1760	0.3684	83.056	91.237	0.1718
140	0.4788	86.647	95.507	0.1817	0.3890	86.098	94.736	0.1778
160	0.5012	89.610	98.884	0.1873	0.4087	89.123	98.199	0.1835
180	0.5229	92.580	102.257	0.1926	0.4277	92.144	101.642	0.1889
200	0.5441	95.564	105.633	0.1978	0.4461	95.170	105.076	0.1942
220	0.5649	98.564	109.018	0.2029	0.4640	98.205	108.509	0.1993
240	0.5854	101.582	112.415	0.2078	0.4816	101.253	111.948	0.2043
260	0.6055	104.622	115.828	0.2126	0.4989	104.317	115.396	0.2092
280	0.6255	107.684	119.258	0.2173	0.5159	107.401	118.857	0.2139
300	0.6452	110.768	122.707	0.2219	0.5327	110.504	122.333	0.2186

TABLE A-13E
(Continued)

Temp. °F	v ft³/lbm	u Btu/lbm	h Btu/lbm	s Btu/(lbm·R)	v ft³/lbm	u Btu/lbm	h Btu/lbm	s Btu/(lbm·R)
	140 psia (T_{sat} = 104.35°F)				**160 psia (T_{sat} = 114.30°F)**			
Sat.	0.2896	79.886	87.389	0.1630	0.2522	80.713	88.180	0.1626
120	0.3055	82.382	90.297	0.1681	0.2576	81.656	89.283	0.1645
140	0.3245	85.516	93.923	0.1742	0.2756	84.899	93.059	0.1709
160	0.3423	88.615	97.483	0.1801	0.2922	88.080	96.732	0.1770
180	0.3594	91.692	101.003	0.1857	0.3080	91.221	100.340	0.1827
200	0.3758	94.765	104.501	0.1910	0.3230	94.344	103.907	0.1882
220	0.3918	97.837	107.987	0.1963	0.3375	97.457	107.450	0.1935
240	0.4073	100.918	111.470	0.2013	0.3516	100.570	110.980	0.1986
260	0.4226	104.008	114.956	0.2062	0.3653	103.690	114.506	0.2036
280	0.4375	107.115	118.449	0.2110	0.3787	106.820	118.033	0.2084
300	0.4523	110.235	121.953	0.2157	0.3919	109.964	121.567	0.2131
320	0.4668	113.376	125.470	0.2202	0.4049	113.121	125.109	0.2177
	180 psia (T_{sat} = 123.38°F)				**200 psia (T_{sat} = 131.74°F)**			
Sat.	0.2228	81.436	88.857	0.1623	0.1989	82.077	89.439	0.1620
140	0.2371	84.238	92.136	0.1678	0.2058	83.521	91.137	0.1648
160	0.2530	87.513	95.940	0.1741	0.2212	86.913	95.100	0.1713
180	0.2678	90.727	99.647	0.1800	0.2354	90.211	98.921	0.1774
200	0.2818	93.904	103.291	0.1856	0.2486	93.451	102.652	0.1831
220	0.2952	97.063	106.896	0.1910	0.2612	96.659	106.325	0.1886
240	0.3081	100.215	110.478	0.1961	0.2732	99.850	109.962	0.1939
260	0.3207	103.364	114.046	0.2012	0.2849	103.032	113.576	0.1990
280	0.3329	106.521	117.610	0.2061	0.2962	106.214	117.178	0.2039
300	0.3449	109.686	121.174	0.2108	0.3073	109.402	120.775	0.2087
320	0.3567	112.863	124.744	0.2155	0.3182	112.598	124.373	0.2134
340	0.3683	116.053	128.321	0.2200	0.3288	115.805	127.974	0.2179
	300 psia (T_{sat} = 166.18°F)				**400 psia (T_{sat} = 192.93°F)**			
Sat.	0.1251	84.295	91.240	0.1601	0.0856	85.178	91.513	0.1576
180	0.1348	87.071	94.556	0.1654				
200	0.1470	90.816	98.975	0.1722	0.0910	86.982	93.718	0.1609
220	0.1577	94.379	103.136	0.1784	0.1032	91.410	99.046	0.1689
240	0.1676	97.835	107.140	0.1842	0.1130	95.371	103.735	0.1757
260	0.1769	101.225	111.043	0.1897	0.1216	99.102	108.105	0.1818
280	0.1856	104.574	114.879	0.1950	0.1295	102.701	112.286	0.1876
300	0.1940	107.899	118.670	0.2000	0.1368	106.217	116.343	0.1930
320	0.2021	111.208	122.430	0.2049	0.1437	109.680	120.318	0.1981
340	0.2100	114.512	126.171	0.2096	0.1503	113.108	124.235	0.2031
360	0.2177	117.814	129.900	0.2142	0.1567	116.514	128.112	0.2079

Source: Kenneth Wark, *Thermodynamics,* 4th ed., McGraw-Hill, New York, 1983, pp. 859–861, table A-18. Originally published by Freon Products Division, E. I. du Pont de Nemours & Company, 1956.

FIGURE A-14E

P-h diagram for refrigerant-12. (Freon 12 is the DuPont trademark for refrigerant-12. Copyright E. I. du Pont de Nemours & Company; used with permission.)

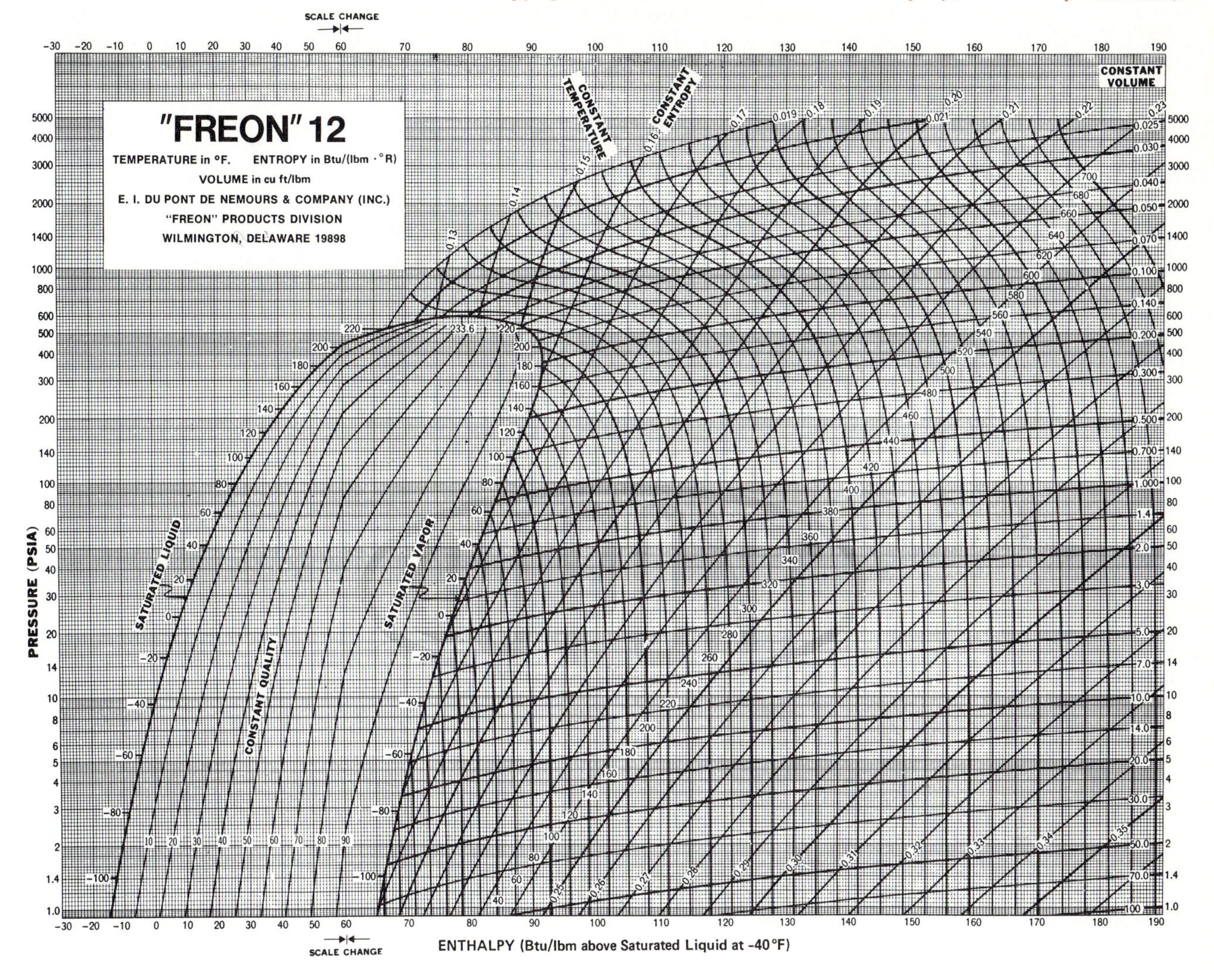

TABLE A-15E

Saturated nitrogen–Temperature table

Temp. R T	Sat. press. psia P_{sat}	Specific volume ft³/lbm: Sat. liquid v_f	Sat. vapor v_g	Internal energy Btu/lbm: Sat. liquid u_f	Sat. vapor u_g	Enthalpy Btu/lbm: Sat. liquid h_f	Evap. h_{fg}	Sat. vapor h_g	Entropy Btu/(lbm·R): Sat. liquid s_f	Sat. vapor s_g
113.7*	1.82	0.0185	23.73	35.32	119.91	35.33	92.59	127.90	0.5802	1.395
120.0	3.34	0.0188	13.56	38.33	120.93	38.34	90.98	129.32	0.6060	1.365
130.0	7.67	0.0193	6.321	43.19	122.47	43.22	88.22	131.44	0.6449	1.324
139.2	14.70	0.0198	3.473	47.71	123.77	47.77	85.45	133.22	0.6785	1.293
140.0	15.46	0.0199	3.315	48.09	123.87	48.15	85.21	133.36	0.6812	1.290
150.0	28.19	0.0205	1.899	53.02	125.11	53.13	81.89	135.02	0.7153	1.262
160.0	47.52	0.0213	1.164	58.00	126.14	58.19	78.19	136.38	0.7474	1.236
170.0	75.18	0.0222	0.750	63.08	126.91	63.39	73.96	137.35	0.7782	1.214
180.0	113.0	0.0233	0.502	68.30	127.36	68.79	69.07	137.86	0.8082	1.192
190.0	162.8	0.0246	0.344	73.75	127.40	74.49	63.28	137.77	0.8378	1.171
200.0	226.9	0.0262	0.239	79.52	126.84	80.62	56.24	136.86	0.8677	1.149
210.0	307.3	0.0285	0.164	85.86	125.33	87.48	47.20	134.68	0.8992	1.124
220.0	406.9	0.0325	0.107	93.57	121.80	96.01	33.85	129.86	0.9363	1.090
226.0	477.9	0.0394	0.071	101.46	115.66	104.95	16.97	121.92	0.9742	1.049
227.2*	493.1	0.051	0.051	108.59	108.59	113.25	0	113.25	1.010	1.010

*113.7 R is the triple state, and 227.2 R is the critical state.

Source: Kenneth Wark, *Thermodynamics,* 4th ed., McGraw-Hill, New York, 1983, p. 869, table A-26. Originally published in NBS Technical Note 648, 1973.

TABLE A-16E

Superheated nitrogen

Temp. R	v ft³/lbm	u Btu/lbm	h Btu/lbm	s Btu/(lbm·R)	v ft³/lbm	u Btu/lbm	h Btu/lbm	s Btu/(lbm·R)
	20 psia (T_{sat} = 144.1 R)				**50 psia (T_{sat} = 161.0 R)**			
200	3.755	134.83	148.73	1.364	1.454	133.97	147.44	1.295
250	4.740	143.87	161.42	1.421	1.866	143.30	160.58	1.354
300	5.714	152.82	173.98	1.467	2.266	152.40	173.38	1.400
350	6.682	161.74	186.49	1.505	2.659	161.41	186.03	1.439
400	7.647	170.65	198.97	1.539	3.050	170.37	198.61	1.473
450	8.610	179.54	211.4	1.568	3.438	179.30	211.1	1.502
500	9.572	188.42	223.9	1.594	3.826	188.22	223.6	1.529
550	10.53	197.31	236.3	1.618	4.212	197.13	236.1	1.553
	100 psia (T_{sat} = 176.9 R)				**200 psia**			
200	0.684	132.40	145.07	1.238	0.289	128.33	139.04	1.166
250	0.908	142.31	159.13	1.300	0.427	140.18	156.01	1.243
300	1.116	151.68	172.36	1.349	0.542	150.19	170.25	1.295
350	1.319	160.84	185.26	1.389	0.649	159.69	183.71	1.336
400	1.518	169.90	198.01	1.423	0.752	168.96	196.81	1.371
450	1.715	178.91	210.7	1.452	0.853	178.12	209.7	1.401
500	1.910	187.88	223.3	1.479	0.953	187.20	222.5	1.428
550	2.105	196.83	235.8	1.503	1.052	196.23	235.2	1.453
	500 psia				**1000 psia**			
250	0.132	131.52	143.78	1.141	0.038	106.30	113.40	0.994
300	0.197	145.25	163.45	1.213	0.083	135.26	150.59	1.131
350	0.247	156.09	178.96	1.261	0.115	149.67	170.95	1.194
400	0.293	166.10	193.23	1.299	0.142	161.23	187.45	1.238
450	0.337	175.73	206.9	1.331	0.166	171.77	202.5	1.274
500	0.379	185.16	220.3	1.359	0.189	181.81	216.8	1.304
550	0.421	194.46	233.4	1.384	0.211	191.56	230.6	1.330
	2000 psia				**3000 psia**			
250	0.029	95.00	115.61	0.939	0.026	90.47	114.97	0.917
300	0.040	118.41	133.17	1.040	0.032	110.73	128.60	1.003
350	0.055	137.38	157.78	1.116	0.040	128.97	151.34	1.073
400	0.070	151.93	177.78	1.169	0.049	144.47	171.81	1.128
450	0.083	164.27	195.11	1.210	0.058	157.89	190.16	1.171
500	0.096	175.50	211.0	1.244	0.066	169.99	207.0	1.206
550	0.108	186.11	226.0	1.272	0.075	181.28	222.8	1.237

Source: Kenneth Wark, *Thermodynamics*, 4th ed., McGraw-Hill, New York, 1983, p. 870, table A-27. Originally published in NBS Technical Note 648, 1973.

TABLE A-17E

Ideal-gas properties of air

T R	h Btu/lbm	P_r	u Btu/lbm	v_r	$s°$ Btu/(lbm·R)
360	85.97	0.3363	61.29	396.6	0.503 69
380	90.75	0.4061	64.70	346.6	0.516 63
400	95.53	0.4858	68.11	305.0	0.528 90
420	100.32	0.5760	71.52	270.1	0.540 58
440	105.11	0.6776	74.93	240.6	0.551 72
460	109.90	0.7913	78.36	215.33	0.562 35
480	114.69	0.9182	81.77	193.65	0.572 55
500	119.48	1.0590	85.20	174.90	0.582 33
520	124.27	1.2147	88.62	158.58	0.591 73
537	128.10	1.3593	91.53	146.34	0.599 45
540	129.06	1.3860	92.04	144.32	0.600 78
560	133.86	1.5742	95.47	131.78	0.609 50
580	138.66	1.7800	98.90	120.70	0.617 93
600	143.47	2.005	102.34	110.88	0.626 07
620	148.28	2.249	105.78	102.12	0.633 95
640	153.09	2.514	109.21	94.30	0.641 59
660	157.92	2.801	112.67	87.27	0.649 02
680	162.73	3.111	116.12	80.96	0.656 21
700	167.56	3.446	119.58	75.25	0.663 21
720	172.39	3.806	123.04	70.07	0.670 02
740	177.23	4.193	126.51	65.38	0.676 65
760	182.08	4.607	129.99	61.10	0.683 12
780	186.94	5.051	133.47	57.20	0.689 42
800	191.81	5.526	136.97	53.63	0.695 58
820	196.69	6.033	140.47	50.35	0.701 60
840	201.56	6.573	143.98	47.34	0.707 47
860	206.46	7.149	147.50	44.57	0.713 23
880	211.35	7.761	151.02	42.01	0.718 86
900	216.26	8.411	154.57	39.64	0.724 38
920	221.18	9.102	158.12	37.44	0.729 79
940	226.11	9.834	161.68	35.41	0.735 09
960	231.06	10.61	165.26	33.52	0.740 30
980	236.02	11.43	168.83	31.76	0.745 40
1000	240.98	12.30	172.43	30.12	0.750 42
1040	250.95	14.18	179.66	27.17	0.760 19
1080	260.97	16.28	186.93	24.58	0.769 64
1120	271.03	18.60	194.25	22.30	0.778 80
1160	281.14	21.18	201.63	20.29	0.787 67
1200	291.30	24.01	209.05	18.51	0.796 28
1240	301.52	27.13	216.53	16.93	0.804 66
1280	311.79	30.55	224.05	15.52	0.812 80
1320	322.11	34.31	231.63	14.25	0.820 75
1360	332.48	38.41	239.25	13.12	0.828 48
1400	342.90	42.88	246.93	12.10	0.836 04
1440	353.37	47.75	254.66	11.17	0.843 41
1480	363.89	53.04	262.44	10.34	0.850 62
1520	374.47	58.78	270.26	9.578	0.857 67
1560	385.08	65.00	278.13	8.890	0.864 56

T R	h Btu/lbm	P_r	u Btu/lbm	v_r	$s°$ Btu/(lbm·R)
1600	395.74	71.73	286.06	8.263	0.871 30
1650	409.13	80.89	296.03	7.556	0.879 54
1700	422.59	90.95	306.06	6.924	0.887 58
1750	436.12	101.98	316.16	6.357	0.895 42
1800	449.71	114.0	326.32	5.847	0.903 08
1850	463.37	127.2	336.55	5.847	0.910 56
1900	477.09	141.5	346.85	4.974	0.917 88
1950	490.88	157.1	357.20	4.598	0.925 04
2000	504.71	174.0	367.61	4.258	0.932 05
2050	518.71	192.3	378.08	3.949	0.938 91
2100	532.55	212.1	388.60	3.667	0.945 64
2150	546.54	223.5	399.17	3.410	0.952 22
2200	560.59	256.6	409.78	3.176	0.959 19
2250	574.69	281.4	420.46	2.961	0.965 01
2300	588.82	308.1	431.16	2.765	0.971 23
2350	603.00	336.8	441.91	2.585	0.977 32
2400	617.22	367.6	452.70	2.419	0.983 31
2450	631.48	400.5	463.54	2.266	0.989 19
2500	645.78	435.7	474.40	2.125	0.994 97
2550	660.12	473.3	485.31	1.996	1.000 64
2600	674.49	513.5	496.26	1.876	1.00623
2650	688.90	556.3	507.25	1.765	1.011 72
2700	703.35	601.9	518.26	1.662	1.017 12
2750	717.83	650.4	529.31	1.566	1.022 44
2800	732.33	702.0	540.40	1.478	1.027 67
2850	746.88	756.7	551.52	1.478	1.027 67
2900	761.45	814.8	562.66	1.318	1.037 88
2950	776.05	876.4	573.84	1.247	1.042 88
3000	790.68	941.4	585.04	1.180	1.047 79
3050	805.34	1011	596.28	1.118	1.052 64
3100	820.03	1083	607.53	1.060	1.057 41
3150	834.75	1161	618.82	1.006	1.062 12
3200	849.48	1242	630.12	0.955	1.066 76
3250	864.24	1328	641.46	0.907	1.071 34
3300	879.02	1418	652.81	0.8621	1.075 85
3350	893.83	1513	664.20	0.8202	1.080 31
3400	908.66	1613	675.60	0.7807	1.084 70
3450	923.52	1719	687.04	0.7436	1.089 04
3500	938.40	1829	698.48	0.7087	1.093 32
3550	953.30	1946	709.95	0.6759	1.097 55
3600	968.21	2068	721.44	0.6449	1.101 72
3650	983.15	2196	732.95	0.6157	1.105 84
3700	998.11	2330	744.48	0.5882	1.109 91
3750	1013.1	2471	756.04	0.5621	1.113 93
3800	1028.1	2618	767.60	0.5376	1.117 91
3850	1043.1	2773	779.19	0.5143	1.121 83
3900	1058.1	2934	790.80	0.4923	1.125 71
3950	1073.2	3103	802.43	0.4715	1.129 55

TABLE A-17E

(Continued)

T R	h Btu/lbm	P_r	u Btu/lbm	v_r	s° Btu/(lbm·R)
4000	1088.3	3280	814.06	0.4518	1.133 34
4050	1103.4	3464	825.72	0.4331	1.137 09
4100	1118.5	3656	837.40	0.4154	1.140 79
4150	1133.6	3858	849.09	0.3985	1.144 46
4200	1148.7	4067	860.81	0.3826	1.148 09
4300	1179.0	4513	884.28	0.3529	1.155 22
4400	1209.4	4997	907.81	0.3262	1.162 21
4500	1239.9	5521	931.39	0.3019	1.169 05
4600	1270.4	6089	955.04	0.2799	1.175 75
4700	1300.9	6701	978.73	0.2598	1.182 32
4800	1331.5	7362	1002.5	0.2415	1.188 76
4900	1362.2	8073	1026.3	0.2248	1.195 08
5000	1392.9	8837	1050.1	0.2096	1.201 29
5100	1423.6	9658	1074.0	0.1956	1.207 38
5200	1454.4	10,539	1098.0	0.1828	1.213 36
5300	1485.3	11,481	1122.0	0.1710	1.219 23

Source: Kenneth Wark, *Thermodynamics,* 4th ed., McGraw-Hill, New York, 1983, pp. 832–833, table A-5. Originally published in J. H. Keenan and J. Kaye, *Gas Tables,* Wiley, New York, 1945.

TABLE A-18E

Ideal-gas properties for nitrogen, N_2

T R	$\overline{h}$ Btu/lbmol	$\overline{u}$ Btu/lbmol	$\overline{s}^{\circ}$ Btu/(lbmol·R)
300	2,082.0	1,486.2	41.695
320	2,221.0	1,585.5	42.143
340	2,360.0	1,684.4	42.564
360	2,498.9	1,784.0	42.962
380	2,638.0	1,883.4	43.337
400	2,777.0	1,982.6	43.694
420	2,916.1	2,082.0	44.034
440	3,055.1	2,181.3	44.357
460	3,194.1	2,280.6	44.665
480	3,333.1	2,379.9	44.962
500	3,472.2	2,479.3	45.246
520	3,611.3	2,578.6	45.519
537	3,729.5	2,663.1	45.743
540	3,750.3	2,678.0	45.781
560	3,889.5	2,777.4	46.034
580	4,028.7	2,876.9	46.278
600	4,167.9	2,976.4	46.514
620	4,307.1	3,075.9	46.742
640	4,446.4	3,175.5	46.964
660	4,585.8	3,275.2	47.178
680	4,725.3	3,374.9	47.386
700	4,864.9	3,474.8	47.588
720	5,004.5	3,574.7	47.785
740	5,144.3	3,674.7	47.977
760	5,284.1	3,774.9	48.164
780	5,424.2	3,875.2	48.345
800	5,564.4	3,975.7	48.522
820	5,704.7	4,076.3	48.696
840	5,845.3	4,177.1	48.865
860	5,985.9	4,278.1	49.031
880	6,126.9	4,379.4	49.193
900	6,268.1	4,480.8	49.352
920	6,409.6	4,582.6	49.507
940	6,551.2	4,684.5	49.659
960	6,693.1	4,786.7	49.808
980	6,835.4	4,889.3	49.955
1000	6,977.9	4,992.0	50.099
1020	7,120.7	5,095.1	50.241
1040	7,263.8	5,198.5	50.380
1060	7,407.2	5,302.2	50.516
1080	7,551.0	5,406.2	50.651
1100	7,695.0	5,510.5	50.783
1120	7,839.3	5,615.2	50.912
1140	7,984.0	5,720.1	51.040
1160	8,129.0	5,825.4	51.167
1180	8,274.4	5,931.0	51.291
1200	8,420.0	6,037.0	51.413
1220	8,566.1	6,143.4	51.534
1240	8,712.6	6,250.1	51.653
1260	8,859.3	6,357.2	51.771
1280	9,006.4	6,464.5	51.887
1300	9,153.9	6,572.3	51.001
1320	9,301.8	6,680.4	52.114
1340	9,450.0	6,788.9	52.225
1360	9,598.6	6,897.8	52.335
1380	9,747.5	7,007.0	52.444
1400	9,896.9	7,116.7	52.551
1420	10,046.6	7,226.7	52.658
1440	10,196.6	7,337.0	52.763
1460	10,347.0	7,447.6	52.867
1480	10,497.8	7,558.7	52.969
1500	10,648.0	7,670.1	53.071
1520	10,800.4	7,781.9	53.171
1540	10,952.2	7,893.9	53.271
1560	11,104.3	8,006.4	53.369
1580	11,256.9	8,119.2	53.465
1600	11,409.7	8,232.3	53.561
1620	11,562.8	8,345.7	53.656
1640	11,716.4	8,459.6	53.751
1660	11,870.2	8,573.6	53.844
1680	12,024.3	8,688.1	53.936
1700	12,178.9	8,802.9	54.028
1720	12,333.7	8,918.0	54.118
1740	12,488.8	9,033.4	54.208
1760	12,644.3	9,149.2	54.297
1780	12,800.2	9,265.3	54.385
1800	12,956.3	9,381.7	54.472
1820	13,112.7	9,498.4	54.559
1840	13,269.5	9,615.5	54.645
1860	13,426.5	9,732.8	54.729

TABLE A-18E

(Continued)

T R	$\bar{h}$ Btu/lbmol	$\bar{u}$ Btu/lbmol	$\bar{s}^\circ$ Btu/(lbmol·R)
1900	13,742	9,968	54.896
1940	14,058	10,205	55.061
1980	14,375	10,443	55.223
2020	14,694	10,682	55.383
2060	15,013	10,923	55.540
2100	15,334	11,164	55.694
2140	15,656	11,406	55.846
2180	15,978	11,649	55.995
2220	16,302	11,893	56.141
2260	16,626	12,138	56.286
2300	16,951	12,384	56.429
2340	17,277	12,630	56.570
2380	17,604	12,878	56.708
2420	17,932	13,126	56.845
2460	18,260	13,375	56.980
2500	18,590	13,625	57.112
2540	18,919	13,875	57.243
2580	19,250	14,127	57.372
2620	19,582	14,379	57.499
2660	19,914	14,631	57.625
2700	20,246	14,885	57.750
2740	20,580	15,139	57.872
2780	20,914	15,393	57.993
2820	21,248	15,648	58.113
2860	21,584	15,905	58.231
2900	21,920	16,161	58.348
2940	22,256	16,417	58.463
2980	22,593	16,675	58.576
3020	22,930	16,933	58.688
3060	23,268	17,192	58.800
3100	23,607	17,451	58.910
3140	23,946	17,710	59.019
3180	24,285	17,970	59.126
3220	24,625	18,231	59.232
3260	24,965	18,491	59.338
3300	25,306	18,753	59.442
3340	25,647	19,014	59.544
3380	25,989	19,277	59.646
3420	26,331	19,539	59.747
3460	26,673	19,802	59.846
3500	27,016	20,065	59.944
3540	27,359	20,329	60.041
3580	27,703	20,593	60.138
3620	28,046	20,858	60.234
3660	28,391	21,122	60.328
3700	28,735	21,387	60.422
3740	29,080	21,653	60.515
3780	29,425	21,919	60.607
3820	29,771	22,185	60.698
3860	30,117	22,451	60.788
3900	30,463	22,718	60.966
3940	30,809	22,985	60.877
3980	31,156	23,252	61.053
4020	31,503	23,520	61.139
4060	31,850	23,788	61.225
4100	32,198	24,056	61.310
4140	32,546	24,324	61.395
4180	32,894	24,593	61.479
4220	33,242	24,862	61.562
4260	33,591	25,131	61.644
4300	33,940	25,401	61.726
4340	34,289	25,670	61.806
4380	34,638	25,940	61.887
4420	34,988	26,210	61.966
4460	35,338	26,481	62.045
4500	35,688	26,751	62.123
4540	36,038	27,022	62.201
4580	36,389	27,293	62.278
4620	36,739	27,565	62.354
4660	37,090	27,836	62.429
4700	37,441	28,108	62.504
4740	37,792	28,379	62.578
4780	38,144	28,651	62.652
4820	38,495	28,924	62.725
4860	38,847	29,196	62.798
4900	39,199	29,468	62.870
5000	40,080	30,151	63.049
5100	40,962	30,834	63.223
5200	41,844	31,518	63.395
5300	42,728	32,203	63.563

Source: Kenneth Wark, *Thermodynamics,* 4th ed., McGraw-Hill, New York, 1983, pp. 834–835, table A-6. Originally published in J. H. Keenan and J. Kaye, *Gas Tables,* Wiley, New York, 1945.

TABLE A-19E

Ideal-gas properties of oxygen, O_2

T R	$\bar{h}$ Btu/lbmol	$\bar{u}$ Btu/lbmol	$\bar{s}^\circ$ Btu/(lbmol·R)	T R	$\bar{h}$ Btu/lbmol	$\bar{u}$ Btu/lbmol	$\bar{s}^\circ$ Btu/(lbmol·R)
300	2,073.5	1,477.8	44.927	1080	7,696.8	5,552.1	54.064
320	2,212.6	1,577.1	45.375	1100	7,850.4	5,665.9	54.204
340	2,351.7	1,676.5	45.797	1120	8,004.5	5,780.3	54.343
360	2,490.8	1,775.9	46.195	1140	8,159.1	5,895.2	54.480
380	2,630.0	1,875.3	46.571	1160	8,314.2	6,010.6	54.614
400	2,769.1	1,974.8	46.927	1180	8,469.8	6,126.5	54.748
420	2,908.3	2,074.3	47.267	1200	8,625.8	6,242.8	54.879
440	3,047.5	2,173.8	47.591	1220	8,782.4	6,359.6	55.008
460	3,186.9	2,273.4	47.900	1240	8,939.4	6,476.9	55.136
480	3,326.5	2,373.3	48.198	1260	9,096.7	6,594.5	55.262
500	3,466.2	2,473.2	48.483	1280	9,254.6	6,712.7	55.386
520	3,606.1	2,573.4	48.757	1300	9,412.9	6,831.3	55.508
537	3,725.1	2,658.7	48.982	1320	9,571.6	6,950.2	55.630
540	3,746.2	2,673.8	49.021	1340	9,730.7	7,069.6	55.750
560	3,886.6	2,774.5	49.276	1360	9,890.2	7,189.4	55.867
580	4,027.3	2,875.5	49.522	1380	10,050.1	7,309.6	55.984
600	4,168.3	2,976.8	49.762	1400	10,210.4	7,430.1	56.099
620	4,309.7	3,078.4	49.993	1420	10,371.0	7,551.1	56.213
640	4,451.4	3,180.4	50.218	1440	10,532.0	7,672.4	56.326
660	4,593.5	3,282.9	50.437	1460	10,693.3	7,793.9	56.437
680	4,736.2	3,385.8	50.650	1480	10,855.1	7,916.0	56.547
700	4,879.3	3,489.2	50.858	1500	11,017.1	8,038.3	56.656
720	5,022.9	3,593.1	51.059	1520	11,179.6	8,161.1	56.763
740	5,167.0	3,697.4	51.257	1540	11,342.4	8,284.2	56.869
760	5,311.4	3,802.2	51.450	1560	11,505.4	8,407.4	56.975
780	5,456.4	3,907.5	51.638	1580	11,668.8	8,531.1	57.079
800	5,602.0	4,013.3	51.821	1600	11,832.5	8,655.1	57.182
820	5,748.1	4,119.7	52.002	1620	11,996.6	8,779.5	57.284
840	5,894.8	4,226.6	52.179	1640	12,160.9	8,904.1	57.385
860	6,041.9	4,334.1	52.352	1660	12,325.5	9,029.0	57.484
880	6,189.6	4,442.0	52.522	1680	12,490.4	9,154.1	57.582
900	6,337.9	4,550.6	52.688	1700	12,655.6	9,279.6	57.680
920	6,486.7	4,659.7	52.852	1720	12,821.1	9,405.4	57.777
940	6,636.1	4,769.4	53.012	1740	12,986.9	9,531.5	57.873
960	6,786.0	4,879.5	53.170	1760	13,153.0	9,657.9	57.968
980	6,936.4	4,990.3	53.326	1780	13,319.2	9,784.4	58.062
1000	7,087.5	5,101.6	53.477	1800	13,485.8	9,911.2	58.155
1020	7,238.9	5,213.3	53.628	1820	13,652.5	10,038.2	58.247
1040	7,391.0	5,325.7	53.775	1840	13,819.6	10,165.6	58.339
1060	7,543.6	5,438.6	53.921	1860	13,986.8	10,293.1	58.428

TABLE A-19E
(Continued)

T R	$\bar{h}$ Btu/lbmol	$\bar{u}$ Btu/lbmol	$\bar{s}^\circ$ Btu/(lbmol·R)	T R	$\bar{h}$ Btu/lbmol	$\bar{u}$ Btu/lbmol	$\bar{s}^\circ$ Btu/(lbmol·R)
1900	14,322	10,549	58.607	3500	28,273	21,323	63.914
1940	14,658	10,806	58.782	3540	28,633	21,603	64.016
1980	14,995	11,063	58.954	3580	28,994	21,884	64.114
2020	15,333	11,321	59.123	3620	29,354	22,165	64.217
2060	15,672	11,581	59.289	3660	29,716	22,447	64.316
2100	16,011	11,841	59.451	3700	30,078	22,730	64.415
2140	16,351	12,101	59.612	3740	30,440	23,013	64.512
2180	16,692	12,363	59.770	3780	30,803	23,296	64.609
2220	17,036	12,625	59.926	3820	31,166	23,580	64.704
2260	17,376	12,888	60.077	3860	31,529	23,864	64.800
2300	17,719	13,151	60.228	3900	31,894	24,149	64.893
2340	18,062	13,416	60.376	3940	32,258	24,434	64.986
2380	18,407	13,680	60.522	3980	32,623	24,720	65.078
2420	18,572	13,946	60.666	4020	32,989	25,006	65.169
2460	19,097	14,212	60.808	4060	33,355	25,292	65.260
2500	19,443	14,479	60.946	4100	33,722	25,580	65.350
2540	19,790	14,746	61.084	4140	34,089	25,867	64.439
2580	20,138	15,014	61.220	4180	34,456	26,155	65.527
2620	20,485	15,282	61.354	4220	34,824	26,444	65.615
2660	20,834	15,551	61.486	4260	35,192	26,733	65.702
2700	21,183	15,821	61.616	4300	35,561	27,022	65.788
2740	21,533	16,091	61.744	4340	35,930	27,312	65.873
2780	21,883	16,362	61.871	4380	36,300	27,602	65.958
2820	22,232	16,633	61.996	4420	36,670	27,823	66.042
2860	22,584	16,905	62.120	4460	37,041	28,184	66.125
2900	22,936	17,177	62.242	4500	37,412	28,475	66.208
2940	23,288	17,450	62.363	4540	37,783	28,768	66.290
2980	23,641	17,723	62.483	4580	38,155	29,060	66.372
3020	23,994	17,997	62.599	4620	38,528	29,353	66.453
3060	24,348	18,271	62.716	4660	38,900	29,646	66.533
3100	24,703	18,546	62.831	4700	39,274	29,940	66.613
3140	25,057	18,822	62.945	4740	39,647	30,234	66.691
3180	25,413	19,098	63.057	4780	40,021	30,529	66.770
3220	25,769	19,374	63.169	4820	40,396	30,824	66.848
3260	26,175	19,651	63.279	4860	40,771	31,120	66.925
3300	26,412	19,928	63.386	4900	41,146	31,415	67.003
3340	26,839	20,206	63.494	5000	42,086	32,157	67.193
3380	27,197	20,485	63.601	5100	43,021	32,901	67.380
3420	27,555	20,763	63.706	5200	43,974	33,648	67.562
3460	27,914	21,043	63.811	5300	44,922	34,397	67.743

Source: Kenneth Wark, *Thermodynamics,* 4th ed., McGraw-Hill, New York, 1983, pp. 836–837, table A-7. Originally published in J. H. Keenan and J. Kaye, *Gas Tables,* Wiley, New York, 1945.

TABLE A-20E

Ideal-gas properties of carbon dioxide, CO_2

T R	$\bar{h}$ Btu/lbmol	$\bar{u}$ Btu/lbmol	$\bar{s}^\circ$ Btu/(lbmol·R)	T R	$\bar{h}$ Btu/lbmol	$\bar{u}$ Btu/lbmol	$\bar{s}^\circ$ Btu/(lbmol·R)
300	2,108.2	1,512.4	46.353	1080	9,575.8	7,431.1	58.072
320	2,256.6	1,621.1	46.832	1100	9,802.6	7,618.1	58.281
340	2,407.3	1,732.1	47.289	1120	10,030.6	7,806.4	58.485
360	2,560.5	1,845.6	47.728	1140	10,260.1	7,996.2	58.689
380	2,716.4	1,961.8	48.148	1160	10,490.6	8,187.0	58.889
400	2,874.7	2,080.4	48.555	1180	10,722.3	8,379.0	59.088
420	3,035.7	2,201.7	48.947	1200	10,955.3	8,572.3	59.283
440	3,199.4	2,325.6	49.329	1220	11,189.4	8,766.6	59.477
460	3,365.7	2,452.2	49.698	1240	11,424.6	8,962.1	59.668
480	3,534.7	2,581.5	50.058	1260	11,661.0	9,158.8	59.858
500	3,706.2	2,713.3	50.408	1280	11,898.4	9,356.5	60.044
520	3,880.3	2,847.7	50.750	1300	12,136.9	9,555.3	60.229
537	4,027.5	2,963.8	51.032	1320	12,376.4	9,755.0	60.412
540	4,056.8	2,984.4	51.082	1340	12,617.0	9,955.9	60.593
560	4,235.8	3,123.7	51.408	1360	12,858.5	10,157.7	60.772
580	4,417.2	3,265.4	51.726	1380	13,101.0	10,360.5	60.949
600	4,600.9	3,409.4	52.038	1400	13,344.7	10,564.5	61.124
620	4,786.6	3,555.6	52.343	1420	13,589.1	10,769.2	61.298
640	4,974.9	3,704.0	52.641	1440	13,834.5	10,974.8	61.469
660	5,165.2	3,854.6	52.934	1460	14,080.8	11,181.4	61.639
680	5,357.6	4,007.2	53.225	1480	14,328.0	11,388.9	61.800
700	5,552.0	4,161.9	53.503	1500	14,576.0	11,597.2	61.974
720	5,748.4	4,318.6	53.780	1520	14,824.9	11,806.4	62.138
740	5,946.8	4,477.3	54.051	1540	15,074.7	12,016.5	62.302
760	6,147.0	4,637.9	54.319	1560	15,325.3	12,227.3	62.464
780	6,349.1	4,800.1	54.582	1580	15,576.7	12,439.0	62.624
800	6,552.9	4,964.2	54.839	1600	15,829.0	12,651.6	62.783
820	6,758.3	5,129.9	55.093	1620	16,081.9	12,864.8	62.939
840	6,965.7	5,297.6	55.343	1640	16,335.7	13,078.9	63.095
860	7,174.7	5,466.9	55.589	1660	16,590.2	13,293.7	63.250
880	7,385.3	5,637.7	55.831	1680	16,845.5	13,509.2	63.403
900	7,597.6	5,810.3	56.070	1700	17,101.4	13,725.4	63.555
920	7,811.4	5,984.4	56.305	1720	17,358.1	13,942.4	63.704
940	8,026.8	6,160.1	56.536	1740	17,615.5	14,160.1	63.853
960	8,243.8	6,337.4	56.765	1760	17,873.5	14,378.4	64.001
980	8,462.2	6,516.1	56.990	1780	18,132.2	14,597.4	64.147
1000	8,682.1	6,696.2	57.212	1800	18,391.5	14,816.9	64.292
1020	8,903.4	6,877.8	57.432	1820	18,651.5	15,037.2	64.435
1040	9,126.2	7,060.9	57.647	1840	18,912.2	15,258.2	64.578
1060	9,350.3	7,245.3	57.861	1860	19,173.4	15,479.7	64.719

TABLE A-20E
(Continued)

T R	$\bar{h}$ Btu/lbmol	$\bar{u}$ Btu/lbmol	$\bar{s}°$ Btu/(lbmol·R)	T R	$\bar{h}$ Btu/lbmol	$\bar{u}$ Btu/lbmol	$\bar{s}°$ Btu/(lbmol·R)
1900	19,698	15,925	64.999	3500	41,965	35,015	73.462
1940	20,224	16,372	65.272	3540	42,543	35,513	73.627
1980	20,753	16,821	65.543	3580	43,121	36,012	73.789
2020	21,284	17,273	65.809	3620	43,701	36,512	73.951
2060	21,818	17,727	66.069	3660	44,280	37,012	74.110
2100	22,353	18,182	66.327	3700	44,861	37,513	74.267
2140	22,890	18,640	66.581	3740	45,442	38,014	74.423
2180	23,429	19,101	66.830	3780	46,023	38,517	74.578
2220	23,970	19,561	67.076	3820	46,605	39,019	74.732
2260	24,512	20,024	67.319	3860	47,188	39,522	74.884
2300	25,056	20,489	67.557	3900	47,771	40,026	75.033
2340	25,602	20,955	67.792	3940	48,355	40,531	75.182
2380	26,150	21,423	68.025	3980	48,939	41,035	75.330
2420	26,699	21,893	68.253	4020	49,524	41,541	75.477
2460	27,249	22,364	68.479	4060	50,109	42,047	75.622
2500	27,801	22,837	68.702	4100	50,695	42,553	75.765
2540	28,355	23,310	68.921	4140	51,282	43,060	75.907
2580	28,910	23,786	69.138	4180	51,868	43,568	76.048
2620	29,465	24,262	69.352	4220	52,456	44,075	76.188
2660	30,023	24,740	69.563	4260	53,044	44,584	76.327
2700	30,581	25,220	69.771	4300	53,632	45,093	76.464
2740	31,141	25,701	69.977	4340	54,221	45,602	76.601
2780	31,702	26,181	70.181	4380	54,810	46,112	76.736
2820	32,264	26,664	70.382	4420	55,400	46,622	76.870
2860	32,827	27,148	70.580	4460	55,990	47,133	77.003
2900	33,392	27,633	70.776	4500	56,581	47,645	77.135
2940	33,957	28,118	70.970	4540	57,172	48,156	77.266
2980	34,523	28,605	71.160	4580	57,764	48,668	77.395
3020	35,090	29,093	71.350	4620	58,356	49,181	77.581
3060	35,659	29,582	71.537	4660	58,948	49,694	77.652
3100	36,228	30,072	71.722	4700	59,541	50,208	77.779
3140	36,798	30,562	71.904	4740	60,134	50,721	77.905
3180	37,369	31,054	72.085	4780	60,728	51,236	78.029
3220	37,941	31,546	72.264	4820	61,322	51,750	78.153
3260	38,513	32,039	72.441	4860	61,916	52,265	78.276
3300	39,087	32,533	72.616	4900	62,511	52,781	78.398
3340	39,661	33,028	72.788	5000	64,000	54,071	78.698
3380	40,236	33,524	72.960	5100	65,491	55,363	78.994
3420	40,812	34,020	73.129	5200	66,984	56,658	79.284
3460	41,388	34,517	73.297	5300	68,471	57,954	79.569

Source: Kenneth Wark, *Thermodynamics,* 4th ed., McGraw-Hill, New York, 1983, pp. 840–841, table A-9. Originally published in J. H. Keenan and J. Kaye, *Gas Tables,* Wiley, New York, 1945.

TABLE A-21E

Ideal-gas properties of carbon monoxide, CO

T R	$\bar{h}$ Btu/lbmol	$\bar{u}$ Btu/lbmol	$\bar{s}^\circ$ Btu/(lbmol·R)	T R	$\bar{h}$ Btu/lbmol	$\bar{u}$ Btu/lbmol	$\bar{s}^\circ$ Btu/(lbmol·R)
300	2,081.9	1,486.1	43.223	1080	7,571.1	5,426.4	52.203
320	2,220.9	1,585.4	43.672	1100	7,716.8	5,532.3	52.337
340	2,359.9	1,684.7	44.093	1120	7,862.9	5,638.7	52.468
360	2,498.8	1,783.9	44.490	1140	8,009.2	5,745.4	52.598
380	2,637.9	1,883.3	44.866	1160	8,156.1	5,851.5	52.726
400	2,776.9	1,982.6	45.223	1180	8,303.3	5,960.0	52.852
420	2,916.0	2,081.9	45.563	1200	8,450.8	6,067.8	52.976
440	3,055.0	2,181.2	45.886	1220	8,598.8	6,176.0	53.098
460	3,194.0	2,280.5	46.194	1240	8,747.2	6,284.7	53.218
480	3,333.0	2,379.8	46.491	1260	8,896.0	6,393.8	53.337
500	3,472.1	2,479.2	46.775	1280	9,045.0	6,503.1	53.455
520	3,611.2	2,578.6	47.048	1300	9,194.6	6,613.0	53.571
537	3,725.1	2,663.1	47.272	1320	9,344.6	6,723.2	53.685
540	3,750.3	2,677.9	47.310	1340	9,494.8	6,833.7	53.799
560	3,889.5	2,777.4	47.563	1360	9,645.5	6,944.7	53.910
580	4,028.7	2,876.9	47.807	1380	9,796.6	7,056.1	54.021
600	4,168.0	2,976.5	48.044	1400	9,948.1	7,167.9	54.129
620	4,307.4	3,076.2	48.272	1420	10,100.0	7,280.1	54.237
640	4,446.9	3,175.9	48.494	1440	10,252.2	7,392.6	54.344
660	4,586.6	3,275.8	48.709	1460	10,404.8	7,505.4	54.448
680	4,726.2	3,375.8	48.917	1480	10,557.8	7,618.7	54.522
700	4,866.0	3,475.9	49.120	1500	10,711.1	7,732.3	54.665
720	5,006.1	3,576.3	49.317	1520	10,864.9	7,846.4	54.757
740	5,146.4	3,676.9	49.509	1540	11,019.0	7,960.8	54.858
760	5,286.8	3,777.5	49.697	1560	11,173.4	8,075.4	54.958
780	5,427.4	3,878.4	49.880	1580	11,328.2	8,190.5	55.056
800	5,568.2	3,979.5	50.058	1600	11,483.4	8,306.0	55.154
820	5,709.4	4,081.0	50.232	1620	11,638.9	8,421.8	55.251
840	5,850.7	4,182.6	50.402	1640	11,794.7	8,537.9	55.347
860	5,992.3	4,284.5	50.569	1660	11,950.9	8,654.4	55.411
880	6,134.2	4,386.6	50.732	1680	12,107.5	8,771.2	55.535
900	6,276.4	4,489.1	50.892	1700	12,264.3	8,888.3	55.628
920	6,419.0	4,592.0	51.048	1720	12,421.4	9,005.7	55.720
940	6,561.7	4,695.0	51.202	1740	12,579.0	9,123.6	55.811
960	6,704.9	4,798.5	51.353	1760	12,736.7	9,241.6	55.900
980	6,848.4	4,902.3	51.501	1780	12,894.9	9,360.0	55.990
1000	6,992.2	5,006.3	51.646	1800	13,053.2	9,478.6	56.078
1020	7,136.4	5,110.8	51.788	1820	13,212.0	9,597.7	56.166
1040	7,281.0	5,215.7	51.929	1840	13,371.0	9,717.0	56.253
1060	7,425.9	5,320.9	52.067	1860	13,530.2	9,836.5	56.339

TABLE A-21E

(Continued)

T R	$\bar{h}$ Btu/lbmol	$\bar{u}$ Btu/lbmol	$\bar{s}^\circ$ Btu/(lbmol·R)	T R	$\bar{h}$ Btu/lbmol	$\bar{u}$ Btu/lbmol	$\bar{s}^\circ$ Btu/(lbmol·R)
1900	13,850	10,077	56.509	3500	27,262	20,311	61.612
1940	14,170	10,318	56.677	3540	27,608	20,576	61.710
1980	14,492	10,560	56.841	3580	27,954	20,844	61.807
2020	14,815	10,803	57.007	3620	28,300	21,111	61.903
2060	15,139	11,048	57.161	3660	28,647	21,378	61.998
2100	15,463	11,293	57.317	3700	28,994	21,646	62.093
2140	15,789	11,539	57.470	3740	29,341	21,914	62.186
2180	16,116	11,787	57.621	3780	29,688	22,182	62.279
2220	16,443	12,035	57.770	3820	30,036	22,450	62.370
2260	16,722	12,284	57.917	3860	30,384	22,719	62.461
2300	17,101	12,534	58.062	3900	30,733	22,988	62.511
2340	17,431	12,784	58.204	3940	31,082	23,257	62.640
2380	17,762	13,035	58.344	3980	31,431	23,527	62.728
2420	18,093	13,287	58.482	4020	31,780	23,797	62.816
2460	18,426	13,541	58.619	4060	32,129	24,067	62.902
2500	18,759	13,794	58.754	4100	32,479	24,337	62.988
2540	19,093	14,048	58.885	4140	32,829	24,608	63.072
2580	19,427	14,303	59.016	4180	33,179	24,878	63.156
2620	19,762	14,559	59.145	4220	33,530	25,149	63.240
2660	20,098	14,815	59.272	4260	33,880	25,421	63.323
2700	20,434	15,072	59.398	4300	34,231	25,692	63.405
2740	20,771	15,330	59.521	4340	34,582	25,934	63.486
2780	21,108	15,588	59.644	4380	34,934	26,235	63.567
2820	21,446	15,846	59.765	4420	35,285	26,508	63.647
2860	21,785	16,105	59.884	4460	35,637	26,780	63.726
2900	22,124	16,365	60.002	4500	35,989	27,052	63.805
2940	22,463	16,225	60.118	4540	36,341	27,325	63.883
2980	22,803	16,885	60.232	4580	36,693	27,598	63.960
3020	23,144	17,146	60.346	4620	37,046	27,871	64.036
3060	23,485	17,408	60.458	4660	37,398	28,144	64.113
3100	23,826	17,670	60.569	4700	37,751	28,417	64.188
3140	24,168	17,932	60.679	4740	38,104	28,691	64.263
3180	24,510	18,195	60.787	4780	38,457	28,965	64.337
3220	24,853	18,458	60.894	4820	38,811	29,239	64.411
3260	25,196	18,722	61.000	4860	39,164	29,513	64.484
3300	25,539	18,986	61.105	4900	39,518	29,787	64.556
3340	25,883	19,250	61.209	5000	40,403	30,473	64.735
3380	26,227	19,515	61.311	5100	41,289	31,161	64.910
3420	26,572	19,780	61.412	5200	42,176	31,849	65.082
3460	26,917	20,045	61.513	5300	43,063	32,538	65.252

Source: Kenneth Wark, *Thermodynamics,* 4th ed., McGraw-Hill, New York, 1983, pp. 838–839, table A-8. Originally published in J. H. Keenan and J. Kaye, *Gas Tables,* Wiley, New York, 1945.

TABLE A-22E

Ideal-gas properties of hydrogen, H_2

T R	$\bar{h}$ Btu/lbmol	$\bar{u}$ Btu/lbmol	$\bar{s}°$ Btu/(lbmol·R)	T R	$\bar{h}$ Btu/lbmol	$\bar{u}$ Btu/lbmol	$\bar{s}°$ Btu/(lbmol·R)
300	2,063.5	1,467.7	27.337	1400	9,673.8	6,893.6	37.883
320	2,189.4	1,553.9	27.742	1500	10,381.5	7,402.7	38.372
340	2,317.2	1,642.0	28.130	1600	11,092.5	7,915.1	38.830
360	2,446.8	1,731.9	28.501	1700	11,807.4	8,431.4	39.264
380	2,577.8	1,823.2	28.856	1800	12,526.8	8,952.2	39.675
400	2,710.2	1,915.8	29.195	1900	13,250.9	9,477.8	40.067
420	2,843.7	2,009.6	29.520	2000	13,980.1	10,008.4	40.441
440	2,978.1	2,104.3	29.833	2100	14,714.5	10,544.2	40.799
460	3,113.5	2,200.0	30.133	2200	15,454.4	11,085.5	41.143
480	3,249.4	2,296.2	30.424	2300	16,199.8	11,632.3	41.475
500	3,386.1	2,393.2	30.703	2400	16,950.6	12,184.5	41.794
520	3,523.2	2,490.6	30.972	2500	17,707.3	12,742.6	42.104
537	3,640.3	2,573.9	31.194	2600	18,469.7	13,306.4	42.403
540	3,660.9	2,588.5	31.232	2700	19,237.8	13,876.0	42.692
560	3,798.8	2,686.7	31.482	2800	20,011.8	14,451.4	42.973
580	3,937.1	2,785.3	31.724	2900	20,791.5	15,032.5	43.247
600	4,075.6	2,884.1	31.959	3000	21,576.9	15,619.3	43.514
620	4,214.3	2,983.1	32.187	3100	22,367.7	16,211.5	43.773
640	4,353.1	3,082.1	32.407	3200	23,164.1	16,809.3	44.026
660	4,492.1	3,181.4	32.621	3300	23,965.5	17,412.1	44.273
680	4,631.1	3,280.7	32.829	3400	24,771.9	18,019.9	44.513
700	4,770.2	3,380.1	33.031	3500	25,582.9	18,632.4	44.748
720	4,909.5	3,479.6	33.226	3600	26,398.5	19,249.4	44.978
740	5,048.8	3,579.2	33.417	3700	27,218.5	19,870.8	45.203
760	5,188.1	3,678.8	33.603	3800	28,042.8	20,496.5	45.423
780	5,327.6	3,778.6	33.784	3900	28,871.1	21,126.2	45.638
800	5,467.1	3,878.4	33.961	4000	29,703.5	21,760.0	45.849
820	5,606.7	3,978.3	34.134	4100	30,539.8	22,397.7	46.056
840	5,746.3	4,078.2	34.302	4200	31,379.8	23,039.2	46.257
860	5,885.9	4,178.0	34.466	4300	32,223.5	23,684.3	46.456
880	6,025.6	4,278.0	34.627	4400	33,070.9	24,333.1	46.651
900	6,165.3	4,378.0	34.784	4500	33,921.6	24,985.2	46.842
920	6,305.1	4,478.1	34.938	4600	34,775.7	25,640.7	47.030
940	6,444.9	4,578.1	35.087	4700	35,633.0	26,299.4	47.215
960	6,584.7	4,678.3	35.235	4800	36,493.4	26,961.2	47.396
980	6,724.6	4,778.4	35.379	4900	35,356.9	27,626.1	47.574
1000	6,864.5	4,878.6	35.520	5000	38,223.3	28,294.0	47.749
1100	7,564.6	5,380.1	36.188	5100	39,092.8	28,964.9	47.921
1200	8,265.8	5,882.8	36.798	5200	39,965.1	29,638.6	48.090
1300	8,968.7	6,387.1	37.360	5300	40,840.2	30,315.1	48.257

Source: Kenneth Wark, *Thermodynamics,* 4th ed., McGraw-Hill, New York, 1983, p. 844, table A-11. Originally published in J. H. Keenan and J. Kaye, *Gas Tables,* Wiley, New York, 1945.

TABLE A-23E

Ideal-gas properties of water vapor, H_2O

T R	$\bar{h}$ Btu/lbmol	$\bar{u}$ Btu/lbmol	$\bar{s}^\circ$ Btu/(lbmol·R)	T R	$\bar{h}$ Btu/lbmol	$\bar{u}$ Btu/lbmol	$\bar{s}^\circ$ Btu/(lbmol·R)
300	2,367.6	1,771.8	40.439	1080	8,768.2	6,623.5	50.854
320	2,526.8	1,891.3	40.952	1100	8,942.0	6,757.5	51.013
340	2,686.0	2,010.8	41.435	1120	9,116.4	6,892.2	51.171
360	2,845.1	2,130.2	41.889	1140	9,291.4	7,027.5	51.325
380	3,004.4	2,249.8	42.320	1160	9,467.1	7,163.5	51.478
400	3,163.8	2,369.4	42.728	1180	9,643.4	7,300.1	51.630
420	3,323.2	2,489.1	43.117	1200	9,820.4	7,437.4	51.777
440	3,482.7	2,608.9	43.487	1220	9,998.0	7,575.2	51.925
460	3,642.3	2,728.8	43.841	1240	10,176.1	7,713.6	52.070
480	3,802.0	2,848.8	44.182	1260	10,354.9	7,852.7	52.212
500	3,962.0	2,969.1	44.508	1280	10,534.4	7,992.5	52.354
520	4,122.0	3,089.4	44.821	1300	10,714.5	8,132.9	52.494
537	4,258.0	3,191.9	45.079	1320	10,895.3	8,274.0	52.631
540	4,282.4	3,210.0	45.124	1340	11,076.6	8,415.5	52.768
560	4,442.8	3,330.7	45.415	1360	11,258.7	8,557.9	52.903
580	4,603.7	3,451.9	45.696	1380	11,441.4	8,700.9	53.037
600	4,764.7	3,573.2	45.970	1400	11,624.8	8,844.6	53.168
620	4,926.1	3,694.9	46.235	1420	11,808.8	8,988.9	53.299
640	5,087.8	3,816.8	46.492	1440	11,993.4	9,133.8	53.428
660	5,250.0	3,939.3	46.741	1460	12,178.8	9,279.4	53.556
680	5,412.5	4,062.1	46.984	1480	12,364.8	9,425.7	53.682
700	5,575.4	4,185.3	47.219	1500	12,551.4	9,572.7	53.808
720	5,738.8	4,309.0	47.450	1520	12,738.8	9,720.3	53.932
740	5,902.6	4,433.1	47.673	1540	12,926.8	9,868.6	54.055
760	6,066.9	4,557.6	47.893	1560	13,115.6	10,017.6	54.117
780	6,231.7	4,682.7	48.106	1580	13,305.0	10,167.3	54.298
800	6,396.9	4,808.2	48.316	1600	13,494.4	10,317.6	54.418
820	6,562.6	4,934.2	48.520	1620	13,685.7	10,468.6	54.535
840	6,728.9	5,060.8	48.721	1640	13,877.0	10,620.2	54.653
860	6,895.6	5,187.8	48.916	1660	14,069.2	10,772.7	54.770
880	7,062.9	5,315.3	49.109	1680	14,261.9	10,925.6	54.886
900	7,230.9	5,443.6	49.298	1700	14,455.4	11,079.4	54.999
920	7,399.4	5,572.4	49.483	1720	14,649.5	11,233.8	55.113
940	7,568.4	5,701.7	49.665	1740	14,844.3	11,388.9	55.226
960	7,738.0	5,831.6	49.843	1760	15,039.8	11,544.7	55.339
980	7,908.2	5,962.0	50.019	1780	15,236.1	11,701.2	55.449
1000	8,078.9	6,093.0	50.191	1800	15,433.0	11,858.4	55.559
1020	8,250.4	6,224.8	50.360	1820	15,630.6	12,016.3	55.668
1040	8,422.4	6,357.1	50.528	1840	15,828.7	12,174.7	55.777
1060	8,595.0	6,490.0	50.693	1860	16,027.6	12,333.9	55.884

TABLE A-23E
(Continued)

T R	$\overline{h}$ Btu/lbmol	$\overline{u}$ Btu/lbmol	$\overline{s}^\circ$ Btu/(lbmol·R)	T R	$\overline{h}$ Btu/lbmol	$\overline{u}$ Btu/lbmol	$\overline{s}^\circ$ Btu/(lbmol·R)
1900	16,428	12,654	56.097	3500	34,324	27,373	62.876
1940	16,830	12,977	56.307	3540	34,809	27,779	63.015
1980	17,235	13,303	56.514	3580	35,296	28,187	63.153
2020	17,643	13,632	56.719	3620	35,785	28,596	63.288
2060	18,054	13,963	56.920	3660	36,274	29,006	63.423
2100	18,467	14,297	57.119	3700	36,765	29,418	63.557
2140	18,883	14,633	57.315	3740	37,258	29,831	63.690
2180	19,301	14,972	57.509	3780	37,752	30,245	63.821
2220	19,722	15,313	57.701	3820	38,247	30,661	63.952
2260	20,145	15,657	57.889	3860	38,743	31,077	64.082
2300	20,571	16,003	58.077	3900	39,240	31,495	64.210
2340	20,999	16,352	58.261	3940	39,739	31,915	64.338
2380	21,429	16,703	58.445	3980	40,239	32,335	64.465
2420	21,862	17,057	58.625	4020	40,740	32,757	64.591
2460	22,298	17,413	58.803	4060	41,242	33,179	64.715
2500	22,735	17,771	58.980	4100	41,745	33,603	64.839
2540	23,175	18,131	59.155	4140	42,250	34,028	64.962
2580	23,618	18,494	59.328	4180	42,755	34,454	65.084
2620	24,062	18,859	59.500	4220	43,267	34,881	65.204
2660	24,508	19,226	59.669	4260	43,769	35,310	65.325
2700	24,957	19,595	59.837	4300	44,278	35,739	65.444
2740	25,408	19,967	60.003	4340	44,788	36,169	65.563
2780	25,861	20,340	60.167	4380	45,298	36,600	65.680
2820	26,316	20,715	60.330	4420	45,810	37,032	65.797
2860	26,773	21,093	60.490	4460	46,322	37,465	65.913
2900	27,231	21,472	60.650	4500	46,836	37,900	66.028
2940	27,692	21,853	60.809	4540	47,350	38,334	66.142
2980	28,154	22,237	60.965	4580	47,866	38,770	66.255
3020	28,619	22,621	61.120	4620	48,382	39,207	66.368
3060	29,085	23,085	61.274	4660	48,899	39,645	66.480
3100	29,553	23,397	61.426	4700	49,417	40,083	66.591
3140	30,023	23,787	61.577	4740	49,936	40,523	66.701
3180	30,494	24,179	61.727	4780	50,455	40,963	66.811
3220	30,967	24,572	61.874	4820	50,976	41,404	66.920
3260	31,442	24,968	62.022	4860	51,497	41,856	67.028
3300	31,918	25,365	62.167	4900	52,019	42,288	67.135
3340	32,396	25,763	62.312	5000	53,327	43,398	67.401
3380	32,876	26,164	62.454	5100	54,640	44,512	67.662
3420	33,357	26,565	62.597	5200	55,957	45,631	67.918
3460	33,839	26,968	62.738	5300	57,279	46,754	68.172

Source: Kenneth Wark, *Thermodynamics*, 4th ed., McGraw-Hill, New York, 1983, pp. 842–843, table A-10. Originally published in J. H. Keenan and J. Kaye, *Gas Tables*, Wiley, New York, 1945.

TABLE A-26E

Enthalpy of formation, Gibbs function of formation, and absolute entropy at 77°F, 1 atm

Substance	Formula	$\overline{h}_f^\circ$ Btu/lbmol	$\overline{g}_f^\circ$ Btu/lbmol	$\overline{s}^\circ$ Btu/(lbmol·R)
Carbon	$C(s)$	0	0	1.36
Hydrogen	$H_2(g)$	0	0	31.21
Nitrogen	$N_2(g)$	0	0	45.77
Oxygen	$O_2(g)$	0	0	49.00
Carbon monoxide	$CO(g)$	−47,540	−59,010	47.21
Carbon dioxide	$CO_2(g)$	−169,300	−169,680	51.07
Water	$H_2O(g)$	−104,040	−98,350	45.11
Water	$H_2O(l)$	−122,970	−102,040	16.71
Hydrogen peroxide	$H_2O_2(g)$	−58,640	−45,430	55.60
Ammonia	$NH_3(g)$	−19,750	−7,140	45.97
Methane	$CH_4(g)$	−32,210	−21,860	44.49
Acetylene	$C_2H_2(g)$	+97,540	+87,990	48.00
Ethylene	$C_2H_4(g)$	+22,490	+29,306	52.54
Ethane	$C_2H_6(g)$	−36,420	−14,150	54.85
Propylene	$C_3H_6(g)$	+8,790	+26,980	63.80
Propane	$C_3H_8(g)$	−44,680	−10,105	64.51
n-Butane	$C_4H_{10}(g)$	−54,270	−6,760	74.11
n-Octane	$C_8H_{18}(g)$	−89,680	+7,110	111.55
n-Octane	$C_8H_{18}(l)$	−107,530	+2,840	86.23
n-Dodecane	$C_{12}H_{26}(g)$	−125,190	+21,570	148.86
Benzene	$C_6H_6(g)$	+35,680	+55,780	64.34
Methyl alcohol	$CH_3OH(g)$	−86,540	−69,700	57.29
Methyl alcohol	$CH_3OH(l)$	−102,670	−71,570	30.30
Ethyl alcohol	$C_2H_5OH(g)$	−101,230	−72,520	67.54
Ethyl alcohol	$C_2H_5OH(l)$	−119,470	−75,240	38.40
Oxygen	$O(g)$	+107,210	+99,710	38.47
Hydrogen	$H(g)$	+93,780	+87,460	27.39
Nitrogen	$N(g)$	+203,340	+195,970	36.61
Hydroxyl	$OH(g)$	+16,790	+14,750	43.92

Source: From the JANAF, *Thermochemical Tables,* Dow Chemical Co., 1971; *Selected Values of Chemical Thermodynamic Properties,* NBS Technical Note 270-3, 1968; and *API Research Project 44,* Carnegie Press, 1953.

TABLE A-27E

Enthalpy of combustion and enthalpy of vaporization at 77°F, 1 atm
(Water appears as a liquid in the products of combustion)

Substance	Formula	$\Delta\bar{h}_c^\circ = -$HHV Btu/lbmol	$\bar{h}_{fg}$ Btu/lbmol
Hydrogen	$H_2(g)$	−122,970	
Carbon	$C(s)$	−169,290	
Carbon monoxide	$CO(g)$	−121,750	
Methane	$CH_4(g)$	−383,040	
Acetylene	$C_2H_2(g)$	−559,120	
Ethylene	$C_2H_4(g)$	−607,010	
Ethane	$C_2H_6(g)$	−671,080	
Propylene	$C_3H_6(g)$	−885,580	
Propane	$C_3H_8(g)$	−955,070	6,480
n-Butane	$C_4H_{10}(g)$	−1,237,800	9,060
n-Pentane	$C_5H_{12}(g)$	−1,521,300	11,360
n-Hexane	$C_6H_{14}(g)$	−1,804,600	13,563
n-Heptane	$C_7H_{16}(g)$	−2,088,000	15,713
n-Octane	$C_8H_{18}(g)$	−2,371,400	17,835
Benzene	$C_6H_6(g)$	−1,420,300	14,552
Toluene	$C_7H_8(g)$	−1,698,400	17,176
Methyl alcohol	$CH_3OH(g)$	−328,700	16,092
Ethyl alcohol	$C_2H_5OH(g)$	−606,280	18,216

Source: Kenneth Wark, *Thermodynamics*, 3d ed., McGraw-Hill, New York, 1977, p. 879, table A-23.

Constants that appear in the Beattie-Bridgeman and the Benedict-Webb-Rubin equations of state

(*a*) The Beattie-Bridgeman equation of state is

$$P = \frac{R_u T}{\bar{v}^2}\left(1 - \frac{c}{\bar{v}T^3}\right)(\bar{v} + B) - \frac{A}{\bar{v}^2}$$

where

$$A = A_0\left(1 - \frac{a}{\bar{v}}\right) \quad \text{and} \quad B = B_0\left(1 - \frac{b}{\bar{v}}\right)$$

When P is in psia, $\bar{v}$ is in ft³/lbmol, T is in R, and R_u = 10.73 psia·ft³/(lbmol·R), the five constants in the Beattie-Bridgeman equation are as follows:

Gas	A_0	a	B_0	b	c
Air	4,905.096	0.3093	0.7386	−0.017 64	4.054×10^6
Argon	4,865.515	0.3729	0.6297	0.0	5.596×10^6
Carbon dioxide	18,872.857	1.142	1.678	1.159	6.166×10^7
Helium	81.424	0.9587	0.2243	0.0	3.737×10^3
Hydrogen	744.510	−0.081 05	0.3357	−0.6982	4.708×10^4
Nitrogen	5,068.324	0.4192	0.8083	−0.1107	3.924×10^6
Oxygen	5,620.956	0.4104	0.7407	0.067 41	4.484×10^6

Source: Computed from Table A-29a by using the proper conversion factors.

(*b*) The Benedict-Webb-Rubin equation of state is

$$P = \frac{R_u T}{\bar{v}} + \left(B_0 R_u T - A_0 - \frac{C_0}{T^2}\right)\frac{1}{\bar{v}^2} + \frac{bR_u T - a}{\bar{v}^3} + \frac{a\alpha}{\bar{v}^6} + \frac{c}{\bar{v}^3 T^2}\left(1 + \frac{\gamma}{\bar{v}^2}\right)e^{-\gamma/\bar{v}^2}$$

When P is in atm, $\bar{v}$ is in ft³/lbmol, T is in R, and R_u = 0.730 atm·ft³/(lbmol·R), the eight constants in the Benedict-Webb-Rubin equation are as follows:

Gas	a	A_0	b	B_0	c	C_0	α	γ
n-Butane, C_4H_{10}	7747	2590	10.27	1.993	4.219×10^9	8.263×10^8	4.531	8.732
Carbon dioxide, CO_2	563.1	703.0	1.852	0.7998	1.989×10^8	1.153×10^8	0.3486	1.384
Carbon monoxide, CO	150.7	344.5	0.676	0.8740	1.387×10^7	7.124×10^6	0.5556	1.541
Methane, CH_4	203.1	476.4	0.868	0.6827	3.393×10^7	1.878×10^7	0.5120	1.541
Nitrogen, N_2	103.2	270.6	0.598	0.6529	9.713×10^6	6.706×10^6	0.5235	1.361

Source: Kenneth Wark, *Thermodynamics,* 4th ed., McGraw-Hill, New York, 1983, p. 864, table A-21. Originally published in H. W. Cooper and J. C. Goldfrank, *Hydrocarbon Processing,* vol. 46, no. 12, p. 141, 1967.

FIGURE A-33E

Psychrometric chart at 1-atm total pressure. (From the American Society of Heating, Refrigerating and Air-Conditioning Engineers; used with permission.)

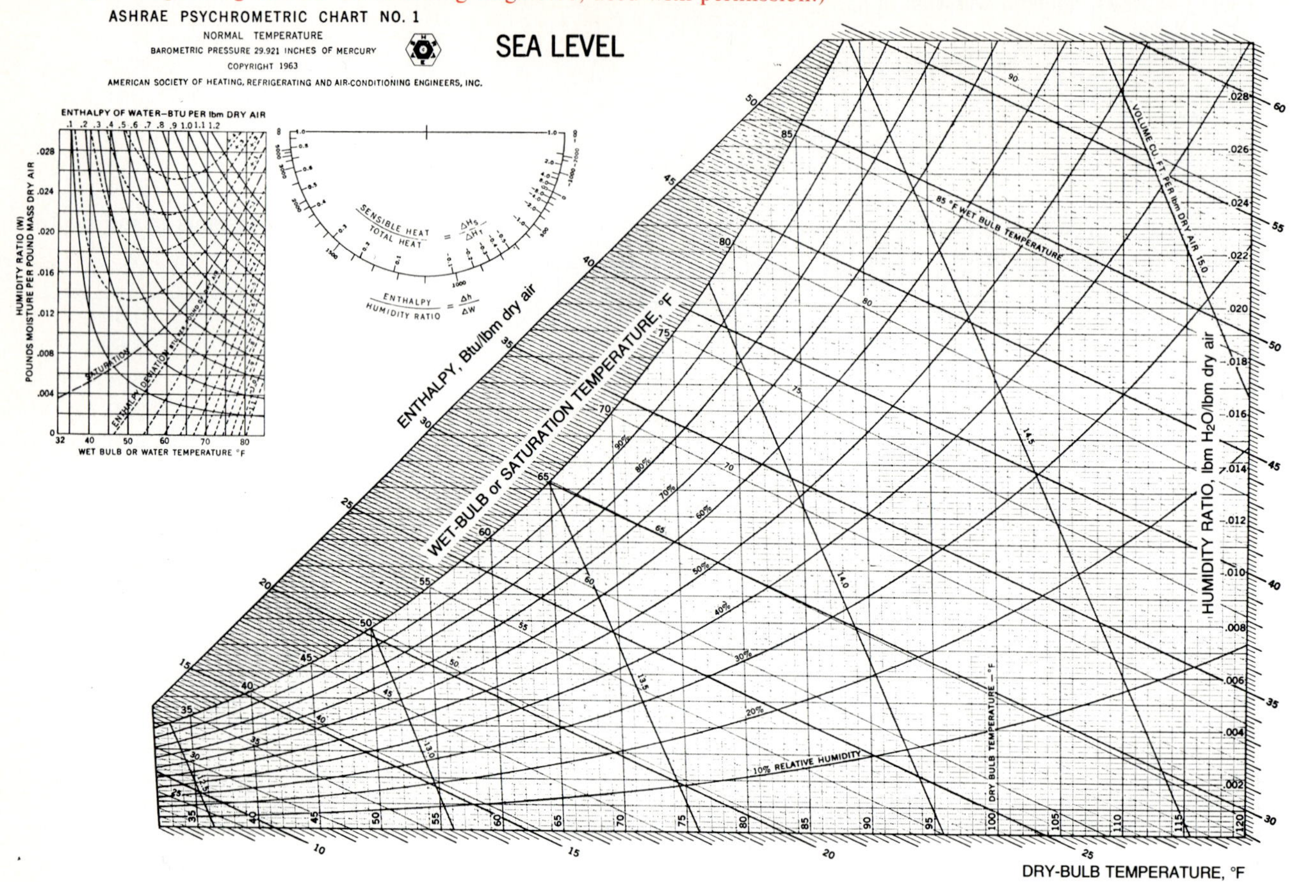

Index